THIS BELONGS TO
JACKIE

D1354307

THE MOSS FLORA OF BRITAIN AND IRELAND

THE MOSS FLORA OF BRITAIN AND IRELAND

A. J. E. SMITH

Lecturer, School of Plant Biology
University College of North Wales, Bangor

WITH ILLUSTRATIONS BY

RUTH SMITH

CAMBRIDGE UNIVERSITY PRESS
CAMBRIDGE
LONDON NEW YORK NEW ROCHELLE
MELBOURNE SYDNEY

Published by the Press Syndicate of the University of Cambridge
The Pitt Building, Trumpington Street, Cambridge CB2 1RP
32 East 57th Street, New York, NY 10022, USA
296 Beaconsfield Parade, Middle Park, Melbourne 3206, Australia

First published 1978
First paperback edition 1980

Printed in Great Britain
by W & J Mackay Ltd, Chatham

British Library Cataloguing in Publication Data

Smith, Anthony John Edwin
The moss flora of Britain and Ireland.

1. Mosses – Great Britain – Identification
I. Title II. Smith, Ruth
588'.2'0941 QK543 77–71428
ISBN 0 521 21648 6 hard covers
ISBN 0 521 29973 X paperback (not for sale in USA or Canada)

CONTENTS

PREFACE

It is now more than fifty years ago that the third edition of *The Student's Handbook of British Mosses* by H. N. Dixon & H. G. Jameson was published in 1924. Since that time the very considerable taxonomic and nomenclatural changes have been such that it is often difficult for the non-expert to equate the 1924 taxa with those of today. Taxonomic revisions have resulted in the splitting of genera (and Dixon had an extraordinarily broad concept of the genus in *The Handbook*) and the recognition of numerous additional species especially in the *Bryum capillare*, *B. erythrocarpum* and *Plagiothecium denticulatum-sylvaticum* complexes. As a consequence of the marked increase in interest in field bryology since the end of the Second World War, numerous new species have been added to the British and Irish lists. Dixon recognised 115 genera and 625 species (in present-day terms) compared with the 175 genera and 692 species described here. There is also an increasing awareness of the extent of morphological variation and the existence of taxonomically difficult groups requiring further study.

It was felt that a new moss flora was long overdue, hence the production of this book. It must be stressed that this is not a revision of Dixon & Jameson but is a completely new moss flora embodying recent ideas and views, some doubtless controversial, on the taxonomy of mosses. Although I have sought opinions and guidance from many sources, except for the three genera contributed by other authors (*Sphagnum*, *Campylopus* and *Pottia*) the views expressed here are entirely my own.

A. J. E. SMITH

Bangor, August 1976

ACKNOWLEDGEMENTS

I am greatly indebted to Mr A. C. Crundwell for the great assistance afforded me during the preparation of this flora. He has been unstinting in his help over nomenclature, in taxonomic discussion and in the loan of specimens. I have made free use of his manuscripts on certain genera of the Dicranaceae, *Mnium* and accounts of species not previously described from Britain. Without his assistance the preparation of this book would have been a much more onerous task.

I have received much comment and helpful criticism from Dr E. V. Watson; Professor P. W. Richards allowed me free access to his herbarium and library; to both these bryologists I am most grateful. I wish also to tender my thanks to the following individuals for the loan of specimens or provision of information: Mrs J. Appleyard, Mr M. F. V. Corley, Mr L. Derrick, Dr U. K. Duncan, Dr J. G. Duckett, Mr Alan Eddy, Mr M. O. Hill, Mrs J. A. Paton, Mr A. R. Perry, Dr F. Rose, Mr E. C. Wallace and Dr H. L. K. Whitehouse.

The Directors or Curators of the following institutions have also been most helpful in lending material: British Museum (Natural History); Royal Botanic Gardens, Edinburgh; National Museum of Wales, Cardiff; National Botanic Gardens, Dublin; the Manchester Museum, University of Manchester; Department of Botany, University of Oxford; Botanical Museum, University of Helsinki; Museum of Natural History, Stockholm; and the New York Botanical Gardens.

I am indebted to Professor J. L. Harper of the School of Plant Biology, University College of North Wales for providing facilities and for his forbearance during the progress of this work.

Finally I would like to thank my wife for all the time and trouble spent in preparing illustrations and my mother who typed much of the manuscript, often from almost illegible handwriting.

INTRODUCTION

As there are a number of text-books dealing in whole or in part with the various aspects of Bryophyta it is superfluous here to detail the structure and life-cycle of mosses. An excellent general account is available in *Structure and Life of Bryophytes* by E. V. Watson (Hutchinson University Library, London, 1972).

Examination of material

Fresh or dried specimens may be examined and the only drawback of the latter is the tendency to fade. Dried plants unless heavily pressed (a process not recommended) recover their form on moistening. Where species are difficult to wet the process may be expedited by soaking in a 20 % solution of household liquid detergent; detergent will also help to remove air bubbles from cells.

For study of leaf shape, cells, etc., leaves are best dissected from the stem, but it should be remembered that auricles, if present, may remain attached to the stem and in genera such as *Plagiomnium* and *Plagiothecium* are best studied *in situ*.

Sections of stem and leaves are best obtained by sandwiching a small bunch of shoots between two halves of a length of elder pith (or carrot). By resting a razor blade flat on one half of the pith and cutting successive slices from the specimen by moving the blade in a rotary fashion and very gradually increasing pressure on the blade, at least some very thin sections will be obtained. A double-sided safety-razor blade, though of short life, will produce much better results than a single-sided blade or old fashioned cut-throat razor.

Staining is unnecessary except for *Sphagnum*. A solution of crystal violet or gentian violet stains *Sphagnum* extremely well, as also does the solution made from scrapings of the lead of an indelible pencil. Where cells are very obscure because of papillae or cell contents, gum chloral* acts as a very effective clearing agent as well as being useful for semi-permanent preparations.

Nomenclature

I have followed for the most part the nomenclature of *Index Muscorum* (Wijk, Margadant & Florschütz, 1959–69) only deviating where that work is incorrect or I differ taxonomically. Synonyms are cited where names differ from those in Dixon & Jameson (1924), Richards & Wallace (an annotated list of British mosses, *Trans. Br. bryol. Soc.* **1** (4), i–xxxi, 1950), Warburg (1963) and other recent works including *Index Muscorum*.

The author citations of some taxa are long and cumbersome but may be abbreviated for convenience. Where 'ex' is used the name preceding 'ex' may be omitted (e.g. *Bryum salinum* Hagen ex Limpr. may be shortened to *B. salinum* Limpr.). Where 'in' is used the authority following 'in' may be left out (e.g. *Sphagnum subsecundum* Nees in Sturm may be written *S. subsecundum* Nees). If there are two sets of authori-

* Gum chloral (Hoyer's solution): distilled water, 50 ml; gum arabic (preferably U.S.P. flake), 30 g; chloral hydrate, 200 g; glycerine, 20 ml. Mix ingredients in above order (U.S.P. flake gum arabic goes into solution more easily than powder or crystals). Use an electric rotary magnetic mixer and do not heat or filter. This recipe and technique has been found to give clearer gum chloral than other recipes tried.

ties then those outside brackets may be omitted, e.g. *Bryum warneum* (Röhl.) Bland. ex Brid. can be written *B. warneum* (Röhl.). Thus such cumbersome citations as *Mielichhoferia elongata* (Hoppe & Hornsch. ex Hook.) Hornsch., or *Buxbaumia viridis* (Moug. ex DC.) Brid. ex Moug. & Nestl., may be reduced to *M. elongata* (Hook.) or *B. viridis* (DC.). Where accuracy is required, however, the full citation should be given.

Taxonomic categories

The nature and status of intraspecific variants in many moss species are very debatable, partly because it is unknown whether variation is genotypic or phenotypic and partly because nothing is known of the breeding systems involved. It is with some trepidation that I have used two intraspecific categories, variety and subspecies, as many authorities question the value of differentiating between the two; also, in a number of instances, further investigation may show that the various taxa so treated may merit different status.

I have used the category 'variety' where there are two or more reasonably well marked peaks in the range of morphological variation within a species and where these appear to have a sound genetical basis. Varieties may or may not show ecological or geographical differentiation. 'Subspecies' is intended to categorise two types of variation: first for intraspecific taxa which are usually distinct morphologically and either ecologically or geographically but which intergrade in one part of their range (e.g. the subspecies of *Thuidium abietinum*), and secondly to indicate the presence of two or more sets of morphological variants within a species (as in *Pohlia elongata*).

I have used the term 'form' loosely, not as a taxonomic category, but as a term to indicate a variant which has no, or very slight, genetical basis.

Classification

I have adopted, in certain instances, a rather broader concept of the genus than some recent authors. The reasons for this are given where relevant but basically it is because no useful purpose is served by splitting taxa unless this leads to easier identification or to a better understanding of the group concerned.

The taxonomic arrangement adopted is indicated on pp. 4–7. I have as far as possible taken into account both gametophyte and sporophyte characters. In the past great importance has been attached to the sporophyte and this has resulted in taxonomic peculiarities. In the production of a natural classification cognizance must be taken of as many characters as possible and undue weight should not be placed on any one of these. Because of the paucity of characters in some taxa the use of only a few or even a single character has to be accepted and this sometimes leads to anomalies to which there is, as yet, no answer. I have, where appropriate, made use of cytological evidence, accepting the argument that chromosome morphology and behaviour are of greater significance than other individual characters.

Frequency and distribution

Only six estimates of frequency have been used: very rare, rare, occasional, frequent, common, very common. Into which of these categories any particular species is placed is to some extent a matter of personal opinion and sometimes the selection of which of two categories is arbitrary; also frequency is not constant in different parts of the country. It was felt, however, that this system is more satisfactory than the use of a multiplicity of phrases such as 'not infrequent', 'not common', etc., which are met with elsewhere.

The figures given at the end of habitat details are the number of vice-counties from which a taxon has been recorded. Thus '17, H3, C' means a plant has been recorded from 17 British, 3 Irish vice-counties and the Channel Islands. The data have been obtained from Warburg (1963) and publications of the British Bryological Society and are complete to the end of 1975.

World distributions are only approximate and are compiled from Podpera (1954), published lists and other floras and no detailed search of the literature has been made. It should be noted that some European species are not necessarily conspecific with plants to which the same name has been applied outside Europe. 'North America' refers to the United States and Canada; 'cosmopolitan' refers to records from all the continents; and 'world-wide' means most but not all major land masses.

ILLUSTRATIONS

With the exception of a small number of species (*Barbula asperifolia*, *Bryum purpurascens*, *Buxbaumia viridis*, *Cinclidotus riparius*, *Cynodontium gracilescens*, *Dicranum elongatum*, *Eurhynchium pulchellum* var. *pulchellum* and var. *diversifolium*, *Fontinalis dalecarlica*, *Grimmia crinita*, *G. elatior*, *Lescuraea saxicola*, *Meesia triquetra*, *Mnium medium*, *Neckera pennata*, *Orthotrichum gymnostomum*, *Paraleucobryum longifolium*, *Tetradontium repandum*, *Tortula norvegica*, *Trematodon ambiguus*, *Weissia wimmerana*) all figures have been prepared from British or Irish specimens, although some have been augmented from other sources. Material for this purpose and for the species listed above came from Scandinavia or Central Europe. The illustrations, except of *Pottia* which were drawn by Miss Gillian A. Meadows of the Royal Botanic Garden, Edinburgh, were prepared by Mrs Ruth Smith using a Leitz Laborlux microscope with either a Leitz camera lucida or a Leitz drawing apparatus. For reasons of economy only the minimum of illustrations has been used and habit drawings have mostly been omitted as, to be of use, they must be of such a size that they take up an amount of space far in excess of their value.

CONSPECTUS OF CLASSIFICATION

The classification used in this work follows basically that of Fleischer (1915–22), *Musci der Flora von Buitenzorg*, I–IV, although there are modifications both of mine and of other authors which might be considered to produce a more natural classification. I have treated the Sphagnopsida, Andreaeopsida and Bryopsida as separate classes, as recent evidence tends to indicate that there is little relationship between them other than the small size of the plants and the dominance of the gametophyte generation. It must be stressed that the classification of taxa is strictly natural and no phylogenetic trends should be construed therefrom; in the almost complete absence of genetical and fossil evidence in the mosses any classification must perforce be natural and is likely to remain so unless some completely new means of tracing phylogeny is devised.

DIVISION BRYOPHYTA

CLASS I SPHAGNOPSIDA

ORDER 1. SPHAGNALES

FAMILY 1. SPHAGNACEAE

1. SPHAGNUM

CLASS II ANDREAEOPSIDA

ORDER 2. ANDREAEALES

FAMILY 2. ANDREAEACEAE

2. ANDREAEA

CLASS III BRYOPSIDA

Subclass 1 Polytrichideae

ORDER 3. TETRAPHIDALES

FAMILY 3. TETRAPHIDACEAE

3. TETRAPHIS
4. TETRODONTIUM

ORDER 4. POLYTRICHALES

FAMILY 4. POLYTRICHACEAE

5. POLYTRICHUM
6. POGONATUM
7. OLIGOTRICHUM
8. ATRICHUM

Subclass 2 Buxaumiideae

ORDER 5. BUXBAUMIALES

FAMILY 5. BUXBAUMIACEAE

9. DIPHYSCIUM
10. BUXBAUMIA

Subclass 3 Eubryideae

ORDER 6. ARCHIDIALES

FAMILY 6. ARCHIDIACEAE

11. Archidium

ORDER 7. DICRANALES

FAMILY 7. DITRICHACEAE

12. Pleuridium
13. Pseudephemerum
14. Ditrichum
15. Distichium
16. Trematodon

FAMILY 8. SELIGERIACEAE

17. Brachydontium
18. Seligeria
19. Blindia

FAMILY 9. DICRANACEAE

20. Saelania
21. Ceratodon
22. Cheilothela
23. Rhabdoweisia
24. Cynodontium
25. Oncophorus
26. Dichodontium
27. Aongstroemia
28. Dicranella
29. Dicranoweisia
30. Arctoa
31. Kiaeria
32. Dicranum
33. Dicranodontium
34. Campylopus
35. Paraleucobryum

FAMILY 10. LEUCOBRYACEAE

36. Leucobryum

ORDER 8. FISSIDENTALES

FAMILY 11. FISSIDENTACEAE

37. Fissidens
38. Octodiceras

ORDER 9. ENCALYPTALES

FAMILY 12. ENCALYPTACEAE

39. Encalypta

ORDER 10. POTTIALES

FAMILY 13. POTTIACEAE

40. Tortula
41. Aloina
42. Desmatodon
43. Pterygoneurum
44. Stegonia
45. Pottia
46. Phascum
47. Acaulon
48. Hyophila
49. Barbula
50. Gymnostomum
51. Gyroweisia
52. Anoectangium
53. Eucladium
54. Weissia
55. Oxystegus
56. Trichostomum
57. Tortella
58. Pleurochaete
59. Trichostomopsis
60. Leptodontium
61. Cinclidotus

ORDER 11. GRIMMIALES

FAMILY 14. GRIMMIACEAE

62. Coscinodon
63. Schistidium
64. Grimmia
65. Dryptodon
66. Racomitrium

FAMILY 15. PTYCHOMITRIACEAE

67. Ptychomitrium
68. Glyphomitrium
69. Campylostelium

ORDER 12. FUNARIALES

FAMILY 16. DISCELIACEAE

70. Discelium

FAMILY 17. FUNARIACEAE

71. Funaria
72. Physcomitrium
73. Physcomitrella

FAMILY 18. EPHEMERACEAE

74. Micromitrium
75. Ephemerum

FAMILY 19. OEDIPODIACEAE

76. Oedipodium

FAMILY 20. SPLACHNACEAE

77. Tayloria
78. Tetraplodon
79. Aplodon
80. Splachnum

ORDER 13. SCHISTOSTEGALES

FAMILY 21. SCHISTOSTEGACEAE

81. Schistostega

ORDER 14. BRYALES

FAMILY 22. BRYACEAE

82. Mielichhoferia
83. Orthodontium
84. Leptobryum
85. Pohlia
86. Epipterygium
87. Plagiobryum
88. Anomobryum
89. Bryum
90. Rhodobryum

FAMILY 23. MNIACEAE

91. Mnium
92. Cinclidium
93. Rhizomnium
94. Plagiomnium
95. Pseudobryum

FAMILY 24. AULACOMNIACEAE

96. Aulacomnium

FAMILY 25. MEESIACEAE

97. Paludella
98. Meesia
99. Amblyodon

FAMILY 26. CATOSCOPIACEAE

100. Catoscopium

FAMILY 27. BARTRAMIACEAE

101. Plagiopus
102. Bartramia
103. Conostomum
104. Bartramidula
105. Philonotis
106. Breutelia

FAMILY 28. TIMMIACEAE

107. Timmia

ORDER 15. ORTHOTRICHALES

FAMILY 29. ORTHOTRICHACEAE

108. Amphidium
109. Zygodon
110. Orthotrichum
111. Ulota

FAMILY 30. HEDWIGIACEAE

112. Hedwigia

ORDER 16. ISOBRYALES

FAMILY 31. FONTINALACEAE

113. Fontinalis

FAMILY 32. CLIMACIACEAE

114. Climacium

FAMILY 33. CRYPHAEACEAE

115. Cryphaea

FAMILY 34. LEUCODONTACEAE

116. Leucodon
117. Antitrichia
118. Pterogonium

FAMILY 35. MYURIACEAE

119. Myurium

FAMILY 36. NECKERACEAE

120. Leptodon
121. Neckera
122. Homalia

FAMILY 37. THAMNIACEAE

123. Thamnobryum

ORDER 17. HOOKERIALES

FAMILY 38. HOOKERIACEAE

124. Hookeria
125. Eriopus
126. Cyclodictyon

FAMILY 39. DALTONIACEAE

127. Daltonia

ORDER 18. THUIDIALES

FAMILY 40. THELIACEAE

128. Myurella

FAMILY 41. FABRONIACEAE

129. Myrinia
130. Habrodon

FAMILY 42. LESKEACEAE

131. Pseudoleskeella
132. Leskea
133. Lescuraea
134. Pterigynandrum

FAMILY 43. THUIDIACEAE

135. Heterocladium
136. Anomodon
137. Thuidium
138. Helodium

ORDER 19. HYPNOBRYALES

FAMILY 44. AMBLYSTEGIACEAE

139. Cratoneuron
140. Campylium
141. Amblystegium
142. Platydictya
143. Drepanocladus
144. Hygrohypnum
145. Scorpidium
146. Calliergon

FAMILY 45. BRACHYTHECIACEAE

147. Isothecium
148. Scorpiurium
149. Homalothecium
150. Brachythecium
151. Pseudoscleropodium
152. Scleropodium
153. Cirriphyllum
154. Rhynchostegium
155. Eurhynchium
156. Rhynchostegiella

FAMILY 46. ENTODONTACEAE

157. Orthothecium
158. Entodon

FAMILY 47. PLAGIOTHECIACEAE

159. Plagiothecium
160. Herzogiella
161. Isopterygium
162. Isopterygiopsis
163. Taxiphyllum

FAMILY 48. SEMATOPHYLLACEAE

164. Sematophyllum

FAMILY 49. HYPNACEAE

165. Pylaisia
166. Platygyrium
167. Homomallium
168. Hypnum
169. Ptilium
170. Ctenidium
171. Hyocomium
172. Rhytidium
173. Rhytidiadelphus
174. Pleurozium
175. Hylocomium

ABBREVIATIONS

ca	approximately
cm	centimetre
fr	fruit or, more correctly, sporophyte
lf, lvs	leaf, leaves
m	metre
mm	millimetre
n	functional haploid chromosome number
sp.	species (plural spp.)
ssp.	subspecies
var.	variety
x	basic haploid chromosome number
μm	micrometre; colloquially micron or mu; 1/1000 mm
$\pm$	more or less
*	following a chromosome number indicates that count is based on British or Irish material

ARTIFICIAL KEY TO GENERA

It must be stressed that this key is merely a guide to the allocation of specimens to genera and is not infallible. Whilst every effort has been made to avoid the use of vague or relative characters, because of the poor distinctions between some moss genera, precise definition is sometimes impossible. Where the generic characters are based on the sporophyte, as between *Brachythecium* and *Eurhynchium* and in the Funariaceae, identification of sterile material is carried as far as possible.

Where a specimen exhibits both states of a section in the key, for example nerve excurrent/nerve not excurrent or margin recurved/margin plane, the first alternative should be taken. Where one individual of a species may possess one state of a character and another individual the alternative state this is allowed for.

In acrocarpous mosses leaf characters are based on leaves from the upper part of a shoot or on comal leaves; it should be borne in mind that perichaetial or perigonial leaves may be very different from vegetative or stem leaves. In pleurocarpous mosses leaf characters refer to stem leaves, particularly in the Amblystegiaceae and Brachytheciaceae.

1 Lvs nerveless, composed of network of green cells surrounding hyaline cells with spiral or annular thickenings **1. Sphagnum (p. 30)**
Lvs with or without nerve, green cells not interspersed with large hyaline cells 2

2 Capsules dehiscing by 4(–8) slits, plants saxicolous, deep red to blackish, fragile when dry, cells very incrassate with reddish to brownish walls **2. Andreaea (p. 79)**
Capsules with lid or cleistocarpous, plants not as above 3

3 Lvs distichous or strongly complanate **42 (p. 11)**
Lvs arranged in 3 or more ranks or if complanate then not strongly so

4 Lvs with filaments or lamellae on upper (ventral) surface at least towards apex **56 (p. 12)**
Lvs without filaments or lamellae on upper surface 5

5 Plants acrocarpous, lvs lingulate to spathulate, apex rounded to acute, basal cells rectangular, thin-walled, hyaline or brownish, cells above strongly papillose, obscure, calyptra enclosing cylindrical capsule **39. Encalypta (p. 206)**
Plants lacking above combination of characters 6

6 Basal cells of lvs narrow, sinuose-nodulose, cells above strongly sinuose **66. Racomitrium (p. 328)**
Areolation not as above 7

7 Lf apex hyaline, whitish when dry or nerve excurrent in hair-point **62 (p. 12)**
Lf apex not hyaline, nerve if excurrent not forming hair-point 8

8 Lvs bordered with long narrow cells or margin several-stratose at least above **72 (p. 13)**
Lvs unbordered, margin 1–2-stratose 9

9 Plants acrocarpous or cladocarpous 10
Plants pleurocarpous 34

10 Lvs squarrose-recurved or squarrose-flexuose when moist **84 (p. 14)**
Lvs imbricate to reflexed when moist but not squarrose 11

11 Gemmae present on stem or lf tips or in lf axils or gemmae cups **88 (p. 14)**
Plants without such gemmae but rhizoidal gemmae or gemmae scattered on stem or lvs sometimes present 12

12 Nerve $\frac{1}{3}$ or more width of lf near lf base **93 (p. 14)**
Nerve less than $\frac{1}{3}$ width of lf near base 13

13 Angular cells differentiated from other basal cells of lf, usually inflated or coloured **98 (p. 15)**
Angular cells not differentiated 14

14 Capsules present 15
Capsules lacking or imperfect 23

15 Capsules cleistocarpous **104 (p. 15)**
Capsules dehiscent 16

16 Capsules immersed in perichaetial lvs **113 (p. 16)**
Capsules emergent to longly exserted 17

17 Capsule ± globose or sub-globose when moist **117 (p. 16)**
Capsule ovoid to cylindrical or pyriform 18

18 Seta arcuate or cygneous **124 (p. 16)**
Seta straight or flexuose 19

19 Capsule inclined to pendulous **128 (p. 16)**
Capsule ± erect 20

20 Capsule longitudinally striate, ribbed or furrowed at least when dry **143 (p. 17)**
Capsule smooth or rugose when dry but not as above 21

21 Capsule gymnostomous **151 (p. 18)**
Capsule with at least a rudimentary peristome 22

22 Peristome teeth single or in pairs, 4 or 16, entire or divided at tips only, inner peristome present or not **159 (p. 18)**
Peristome teeth 16, divided to halfway or more, sometimes into filiform segments, inner peristome lacking **175 (p. 19)**

23 Lvs nerveless **185 (p. 20)**
Lvs with distinct nerve 24

24 Mid-lf cells ± isodiametric 25
Mid-lf cells longer than wide 30

25 Nerve excurrent 26
Nerve ending in or below apex 27

26 Margin recurved at least below **187 (p. 20)**
Margin plane or incurved **194 (p. 20)**

27 Margin denticulate or dentate from middle or below **204 (p. 21)**
Margin entire or denticulate near apex only 28

28 Apex obtuse or rounded, apiculate or not **209 (p. 21)**
Apex sub-acute to longly acuminate 29

29 Margin recurved at least below **215 (p. 22)**
Margin plane or incurved **228 (p. 22)**

30 Lf apex obtuse or rounded, with or without mucro or apiculus **243 (p. 23)**
Lf apex sub-acute to acuminate 31

31 Cells distinctly papillose **248 (p. 24)**
Cells ± smooth 32

32 Cells lax, thin-walled **251 (p. 24)**
Cells firm, thin-walled or not 33

33 Lf apex longly acuminate or subulate, consisting largely or entirely of nerve **254 (p. 24)**
Lf apex wider, not consisting largely or entirely of nerve **258 (p. 24)**

34 Cells less than twice as long as wide at least towards margin at widest part of lf 35
Cells more than twice as long as wide 36

35 Nerve single, extending at least ½ way up lf **270 (p. 25)**
Nerve short and ceasing less than ½ way up lf or double or lacking **277 (p. 25)**

36 Nerve single, extending at least ½ way up lf, sometimes forked above 37
Nerve short and ceasing less than ½ way up lf or double or absent 40

37 Lvs falcato-secund at least at stem and branch tips **283 (p. 26)**
Lvs various but not falcato-secund 38

38 Lvs longitudinally plicate **286 (p. 26)**
Lvs not or hardly plicate 39

39 Stem lvs shortly pointed, apex acute to rounded or abruptly narrowed to long or short apiculus **293 (p. 26)**
Stem lvs gradually tapering to acute to acuminate apex or if apex obtuse then nerve very stout **304 (p. 27)**

40 Lf apex obtuse or rounded, apiculate or not **310 (p. 28)**
Lvs tapering to apex 41

41 Lvs falcato-secund to circinate-secund **315 (p. 28)**
Lvs various but not falcato-secund or circinate-secund **319 (p. 28)**

. . . .

42 Lower part of lf with conduplicate portion (sheathing laminae) 43
Lvs without sheathing laminae 44

43 Sheathing laminae about ½ total lf length, seta long **37. Fissidens (p. 189)**
Sheathing laminae about ⅓ total lf length, seta short, capsule barely emerging from perichaetial lvs **38. Octodiceras (p. 204)**

44 Lvs abruptly narrowed from broad sheathing basal part to long fine apex consisting largely of nerve **15. Distichium (p. 116)**
Lvs not as above 45

45 Lvs with border of narrow elongated cells 46
Lvs unbordered 48

46 Nerve single, extending almost to apex or excurrent **94. Plagiomnium (p. 439)**
Nerve single, short and faint or double 47

47 Nerve very short, cells in mid-lf 25–40 μm wide **125. Eriopus (p. 512)**
Nerve extending about ¾ way up lf, cells 15–30 μm wide **126. Cyclodictyon (p. 514)**

48 Cells in mid-lf 15–100 μm wide, to twice as long as wide 49
Cells in mid-lf 6–22 μm wide, more than twice as long as wide 50

49 Lvs distichous, cells 15–30 μm wide in mid-lf **81. Schistostega (p. 358)**
Lvs strongly complanate, cells 60–100 μm wide in mid-lf **124. Hookeria (p. 512)**

50 Nerve single, extending to ½–⅔ way up lf **122. Homalia (p. 508)**
Nerve double and usually short or ± absent 51

51 Lvs often sub-falcate, often transversely undulate but not whitish-green, base not decurrent **121. Neckera (p. 506)**
Lvs not sub-falcate, if transversely undulate then plants whitish-green, lf base decurrent or not 52

52 Lf base decurrent, angular cells hyaline or greenish 53
Lf base not decurrent, angular cells greenish 54

53 Lf margin entire or denticulate towards apex only **159. Plagiothecium (p. 622)**
Lf margin denticulate ± throughout **160. Herzogiella (p. 632)**

54 Lf apex acute, mid-lf cells 6–10 μm wide, deciduous filiform shoots absent **163. Taxiphyllum (p. 638)**
Lf apex filiform, mid-lf cells 4–7 μm wide, deciduous filiform shoots present or not 55

55 Epidermal cells of stem small, thick-walled, 10–12 μm wide, filiform deciduous shoots often present **161. Isopterygium (p. 634)**
Epidermal cells large, thin-walled, 16–30 μm wide, filiform deciduous shoots lacking **162. Isopterygiopsis (p. 636)**

. . . .

56 Lvs with branched filaments on upper surface of nerve at least in upper part 57
Lvs with longitudinal lamellae on upper surface of nerve at least in upper part 58

57 Margin strongly inflexed above, upper surface of lf densely covered with filaments except near base **41. Aloina (p. 225)**
Lf margin recurved, filaments on upper part of lf only **42. Desmatodon (p. 230)**

58 Nerve narrow with 2–4 lamellae on upper part of lf, lvs unbordered, capsule without epiphragm **43. Pterygoneurum (p. 232)**
Nerve occupying $\frac{1}{2}$ or more of lf width or lf bordered, capsule with epiphragm 59

59 Lvs with border of long narrow cells, nerve narrow **8. Atrichum (p. 99)**
Lvs unbordered, nerve wide 60

60 Lamellae sinuose **7. Oligotrichum (p. 99)**
Lamellae ± straight 61

61 Capsule usually at least obscurely angled, apophysis present **5. Polytrichum* (p. 89)**
Capsule not angled in section, apophysis absent **6. Pogonatum* (p. 97)**

. . . .

62 Lvs nerveless **112. Hedwigia (p. 492)**
Lvs nerved 63

63 Nerve occupying more than $\frac{1}{4}$ width of lf base **34. Campylopus (p. 166)**
Nerve narrower 64

64 Lf cells with pointed ends, rhomboidal or narrowly hexagonal, lf margin denticulate or not above, capsule cernuous or pendulous **89. Bryum (p. 383)**
Cells ± isodiametric, ends not pointed, if longer than wide then lf margin entire, capsule erect or inclined 65

65 Nerve excurrent in hair-point 66
Lf apex hyaline, whitish when dry, nerve ending in or below apex 69

66 Cells not or only faintly papillose, pellucid 67
Cells strongly papillose, ± obscure 68

67 Capsule exserted, dehiscent **45. Pottia (p. 234)**
Capsule immersed, cleistocarpous **46. Phascum (p. 242)**

68 Lf apex rounded or if acute then hair-point denticulate to spinulose **40. Tortula (p. 211)**
Lf apex acute, hair-point ± smooth **42. Desmatodon (p. 230)**

* For joint key to the species of *Polytrichum* and *Pogonatum* see p. 90.

69 Cells in mid-lf 10–16 μm wide, hexagonal, neither incrassate nor sinuose, capsule immersed, ribbed, peristome double **110. Orthotrichum (p. 473)**
Cells in mid-lf 6–12μm wide, rounded or quadrate, usually incrassate and sinuose, capsule if immersed smooth with single peristome 70

70 Lvs with thickened plicae on either side of nerve, margin plane, capsule immersed, plants dioecious **62. Coscinodon (p. 304)**
Lvs without plicae, margin plane, incurved or recurved on one or both sides, capsule immersed or exserted, if immersed then plants autoecious 71

71 Capsule immersed, erect, plants autoecious **63. Schistidium (p. 306)**
Capsule exserted or if immersed then inclined, or fruit lacking, plants autoecious or dioecious **64. Grimmia (p. 312)**

....

72 Marginal cells 3–5 stratose at least in upper part of lf, not elongated **61. Cinclidotus (p. 302)**
Marginal cells elongated, forming distinct border 73

73 Lf cells not more than 20 μm wide, ± isodiametric, if lf margin strongly toothed then cells papillose 74
Cells usually more than 20 μm wide, hexagonal to linear, smooth, or if less than 20 μm then margin strongly toothed 77

74 Lf margin irregularly toothed towards apex, rhizoidal gemmae present, capsule gymnostomous **48. Hyophila (p. 247)**
Lf margin entire or denticulate above rhizoidal gemmae lacking, peristome present or not 75

75 Capsule inclined, border bistratose below **42. Desmatodon (p. 230)**
Capsule erect, border unistratose below 76

76 Capsule cylindrical, peristome teeth filiform, lf margin plane or recurved **40. Tortula (p. 211)**
Capsule ± obloid, gymnostomous, margin plane **45. Pottia (p. 234)**

77 Plants pleurocarpous, lvs linear-lanceolate, entire, seta papillose **127. Daltonia (p. 514)**
Plants acrocarpous, lvs narrowly lanceolate to orbicular, entire or toothed, seta smooth 78

78 Capsule erect, peristome rudimentary or absent, lf cells lax, thin-walled, rectangular to rhomboidal **71. Funaria (p. 338)**
Capsule inclined to pendulous, peristome well developed, cells firm, ± hexagonal to elongate-hexagonal 79

79 Nerve ending well below lf apex, border weak, unistratose, margin entire, cells 20–40 μm wide in mid-lf **86. Epipterygium (p. 380)**
Lvs not as above 80

80 Lf margin toothed, sometimes spinosely so or if entire then sterile shoots arcuate to procumbent 81
Lf margin entire or denticulate, sterile shoots erect 82

81 Sterile shoots erect, lf cells ± isodiametric, border 2–3-stratose, nerve often toothed at back above **91. Mnium (p. 430)**
Sterile shoots often procumbent or arcuate, cells 1–3 times as long as wide in mid-lf, border unistratose, nerve not toothed at back **94. Plagiomnium (p. 439)**

82 Lvs ovate to ovate-lanceolate, acute to acuminate, margin ± denticulate above **89. Bryum (p. 383)**
Lvs elliptical to orbicular or obovate, apex rounded, margin entire 83

83 Cells 15–35 μm wide in mid-lf, teeth of inner peristome joined to form conical structure **92. Cinclidium (p. 437)**
Cells 35–50 μm wide, teeth of inner peristome not joined **93. Rhizomnium (p. 437)**

....

84 Lf margin plane, cells smooth 85
Margin plane or recurved, cells papillose 86
85 Lf acumen long, flexuose, denticulate all round, capsule ± erect, cells in upper part of lf ± rectangular **14. Ditrichum (p. 109)**
Lf acumen if long only denticulate at margin, capsule inclined, or if erect then cells in upper part of lf linear **28. Dicranella (p. 139)**
86 Lf margin plane, hyaline basal cells ascending up margin of lf **58. Pleurochaete (p. 298)**
Margin recurved at least below, hyaline basal cells not ascending up margin 87
87 Lvs entire **49. Barbula (p. 248)**
Lvs toothed above **97. Paludella (p. 451)**

....

88 Gemmae in cups at ends of stems **3. Tetraphis (p. 87)**
Gemmae not in cups 89
89 Gemmae in clusters on ends of pseudopodia **96. Aulacomnium (p. 447)**
Gemmae on lf tips or in lf axils 90
90 Gemmae in clusters at tips of upper lvs 91
Gemmae axillary 92
91 Lvs toothed **60. Leptodontium (p. 299)**
Lvs entire **111. Ulota (p. 486)**
92 Lf margin toothed above, unbordered, cells narrowly hexagonal to linear **85. Pohlia (p. 362)**
Lf margin ± entire or bordered, cells narrowly hexagonal **89. Bryum (p. 383)**

....

93 Plants whitish when dry, lf consisting almost entirely of nerve, margin inflexed above **36. Leucobryum (p. 186)**
Plants not as above 94
94 Lvs ligulate or ligulate-lanceolate, apex rounded **98. Meesia (p. 451)**
Lvs with long fine acumen 95
95 Seta straight, capsule pendulous, smooth, lf cells ± linear-rhomboidal, rhizoidal gemmae present **84. Leptobryum (p. 361)**
Seta straight or arcuate, capsule inclined, striate or furrowed, basal cells of lf shorter and wider than upper cells 96
96 Lamina cells extending only short distance up lf, nerve in section of 3 layers of hyaline cells interspersed with small green cells, stereids lacking **35. Paraleucobryum (p. 186)**
Lamina cells extending part or all way up lf, nerve in section without green cells, with or without stereids 97
97 Seta straight, angular cells not inflated or coloured **28. Dicranella (p. 139)**
Seta arcuate or straight, angular cells coloured or inflated **33. Dicranodontium** and **34. Campylopus***

....

*For key to the species of these 2 genera see p. 166

98 Nerve more than ¼ width of lf base **33. Dicranodontium** and **34. Campylopus***
Nerve less than ¼ width of lf base 99

99 Angular cells enlarged, hyaline **25. Oncophorus (p. 135)**
Angular cells orange to brown at least in older lvs 100

100 Cells in upper part of lf ± linear, nerve excurrent in subulate point, capsule erect **19. Blindia (p. 125)**
Cells in upper part of lf quadrate to rectangular, nerve if excurrent not forming subulate point, capsule erect to horizontal 101

101 Lvs crisped when dry, margin entire, bistratose above **29. Dicranoweisia (p. 147)**
Lvs not or only slightly crisped or if strongly crisped then margin denticulate, margin unistratose 102

102 Seta short, stout, capsule ovoid, when dry wide-mouthed, lvs rapidly narrowed above basal part, nerve longly excurrent **30. Arctoa (p. 149)**
Sporophyte not as above, lvs more gradually tapering, nerve if excurrent not longly so 103

103 Autoecious, capsule strumose, lf cells not or only slightly porose, nerve without stereids in section and lvs not brittle, alpine plants **31. Kiaeria (p. 150)**
Dioecious or rarely autoecious, capsule not strumose, cells strongly porose or not, nerve in section with stereids or very brittle, lowland or alpine plants **32. Dicranum (p. 152)**

....

104 Capsule ± without apiculus 105
Capsule with distinct apiculus or beak 107

105 Lvs nerveless, spores 20–30 μm **74. Micromitrium (p. 346)**
Lvs nerved, spores 30 μm or more 106

106 Spores *ca* 16 in number, 100–120 μm diameter, perichaetial lvs linear-lanceolate **11. Archidium (p. 106)**
Spores numerous, 30–40 μm, perichaetial lvs wide **47. Acaulon (p. 245)**

107 Lvs ovate to ovate-lanceolate, with or without nerve 108
At least upper lvs narrowly lanceolate or linear-lanceolate, nerve present 111

108 Plants very minute, arising from persistent protonema, capsule, globose, not concealed by perichaetial lvs **75. Ephemerum (p. 346)**
Plants minute or small, protonema not persistent, capsule ovoid or ellipsoid, or if globose then completely surrounded by perichaetial lvs 109

109 Lvs toothed from about middle, nerve ending well below apex **73. Physcomitrella (p. 346)**
Margin entire or toothed towards apex only, nerve ending in apex or excurrent at least in upper lvs 110

110 Capsule immersed or seta curved **46. Phascum (p. 242)**
Capsule exserted, seta straight **45. Pottia (p. 234)**

111 Lf cells papillose, ± isodiametric **54. Weissia (p. 274)**
Cells smooth, long and narrow in upper part of lf 112

112 Perichaetial lvs longer than stem lvs, nerve ending in apex or excurrent **12. Pleuridium (p. 108)**
Perichaetial lvs about same length as stem lvs, nerve ending below apex **13. Pseudephemerum (p. 109)**

....

* For key to the species of these 2 genera see p. 166.

113 Lvs nerveless **112. Hedwigia (p. 492)**
Lvs nerved 114

114 Inner perichaetial lvs with cilia, capsule oblique **9. Diphyscium (p. 104)**
Inner perichaetial lvs not ciliate, capsule erect 115

115 Lvs strongly incurved when dry, narrowly lanceolate, acuminate **54. Weissia (p. 274)**
Lvs not as above 116

116 Capsule smooth, peristome single **63. Schistidium (p. 306)**
Capsule striate or ribbed, peristome double **110. Orthotrichum (p. 473)**

. . . .

117 Plants bud-like, lvs nerveless **70. Discelium (p. 338)**
Plants not bud-like, lvs nerved 118

118 Seta arcuate or cygneous when moist 119
Seta straight 120

119 Plants very small, to 5 mm tall, lvs not plicate, cells smooth, peristome lacking **104. Bartramidula (p. 458)**
Plants larger, lvs plicate, cells papillose, peristome present **106. Breutelia (p. 465)**

120 Capsule minute, scarcely 1 mm long, blackish, horizontal, smooth **100. Catoscopium (p. 453)**
Capsule more than 1 mm in diameter, green to brown, erect to horizontal, smooth or sulcate when dry 121

121 Lvs erect when moist, 5-ranked, peristome teeth joined at tips **103. Conostomum (p. 458)**
Lvs erecto-patent to patent or secund when moist, not obviously 5-ranked peristome teeth free 122

122 Lvs lanceolate to ovate, if narrower then plants very small, to 1.5 cm **105. Philonotis (p. 459)**
Lvs linear-lanceolate, plants medium-sized to large, more than 1.5 cm 123

123 Lf cells smooth **101. Plagiopus (p. 454)**
Lf cells papillose **102. Bartramia (p. 454)**

. . . .

124 Capsule asymmetrical, lvs obovate to lanceolate-spathulate, **71. Funaria (p. 338)**
Capsule symmetrical, lvs ovate to linear-lanceolate 125

125 Lvs concave, imbricate or erect when moist, ovate or lanceolate, shortly pointed **87. Plagiobryum (p. 381)**
Lvs not concave or imbricate when moist, lanceolate to linear, gradually tapering to acute, to acuminate apex 126

126 Plants more than 3 mm tall, nerve strongly 2-winged at back above **65. Dryptodon (p. 326)**
Plants minute, not more than 3 mm tall, nerve not winged at back above 127

127 Nerve ending in apex or excurrent, cells in upper part of lf quadrate-rectangular, capsule ovoid **18. Seligeria (p. 118)**
Nerve ending below apex, cells in upper part of lf rounded-quadrate, capsule narrowly ellipsoid **69. Campylostelium (p. 336)**

. . . .

128 Shoots julaceous with concave imbricate lvs when moist 129
Shoots not julaceous 130

129 Cells lax, elongate-hexagonal, plants reddish or reddish-brown below **87. Plagiobryum (p. 381)**
Cells firm, linear-vermicular, plants pale brown below **88. Anomobryum (p. 383)**

130 Seta papillose, lvs minute, ± decayed by time of maturity of obliquely ovoid capsule **10. Buxbaumia (p. 105)**
Seta smooth, lvs well developed, not decayed when capsule mature 131

131 Capsule smooth when moist 132
Capsule longitudinally striate or furrowed when moist 140

132 Nerve excurrent, thickened in upper part **42. Desmatodon (p. 230)**
Nerve excurrent or not, not thickened in upper part 133

133 Lf cells ± linear or capsule cernuous to pendulous 134
Lf cells shorter, capsule inclined to horizontal 136

134 Lvs linear-lanceolate to linear-setaceous **83. Orthodontium (p. 361)**
Lvs wider 135

135 Cilia simple or rudimentary, stem lvs, acute with nerve usually ending below apex **85. Pohlia (p. 362)**
Cilia appendiculate or lvs obtuse or nerve longly excurrent **89. Bryum (p. 383)**

136 Neck of capsule as long as theca **16. Trematodon (p. 118)**
Neck of capsule shorter than theca 137

137 Capsule strumose **25. Oncophorus (p. 135)**
Capsule not strumose 138

138 Peristome single, lf cells papillose **26. Dichodontium (p. 137)**
Peristome double, cells smooth 139

139 Cells in upper part of lf 9–14 μm wide **98. Meesia (p. 451)**
Cells in upper part of lf 15–30 μm wide **99. Amblyodon (p. 453)**

140 Capsule strumose **21. Ceratodon (p. 126)**
Capsule not strumose 141

141 Lid of capsule conical, peristome double, lf cells papillose **96. Aulacomnium (p. 447)**
Lid with oblique beak, peristome single, cells smooth or mamillose 142

142 Cells in upper part of lf ± isodiametric, smooth or mamillose **24. Cynodontium (p. 130)**
Cells in upper part of lf rectangular to linear, smooth **28. Dicranella (p. 139)**

. . . .

143 Plants minute, to 3 mm tall **17. Brachydontium (p. 118)**
Plants larger 144

144 Capsule cylindrical or shortly cylindrical, caplytra naked 145
Capsule ovoid, ellipsoid or pyriform, calyptra naked or hairy 146

145 Plants glaucous green, upper lvs longer than lower, cells smooth **20. Saelania (p. 126)**
Plants dull or bright green, upper lvs not much larger than lower, cells smooth or mamillose **24. Cynodontium (p. 130)**

146 Cells in mid-lf linear **83. Orthodontium (p. 361)**
Cells in mid-lf quadrate or rounded 147

147 Cells in upper part of lf ± quadrate or quadrate-hexagonal, margin crenulate to dentate **23. Rhabdoweisia (p. 128)**
Cells in upper part of lf rounded or if quadrate then margin entire 148

148 Capsule emergent or if exserted then capsule and seta deep red, lvs usually ± straight, appressed when dry **110. Orthotrichum (p. 473)**
Capsule exserted, not red, lvs curved to crisped when dry 149
149 Calyptra hairy, capsule with long tapering neck, peristome present **111. Ulota (p. 486)**
Calyptra naked, neck not long and tapering, peristome present or not 150
150 Seta short, to 3 mm long, capsule barely emergent from upper lvs, spores 8–12 μm **108. Amphidium (p. 468)**
Seta 3–8 mm long, capsules exserted, spores (10–)14–20 μm **109. Zygodon (p. 470)**

. . . .

151 Plants minute, to 3 mm tall 152
Plants larger 153
152 Cells smooth, seta 1–2 mm long **18. Seligeria (p. 118)**
Cells papillose, seta to 7 mm long **51. Gyroweisia (p. 270)**
153 Lvs obovate-spathulate, apophysis longly tapering into seta, seta succulent **76. Oedipodium (p. 350)**
Lvs ovate to linear, neck or apophysis if present not longly tapering, seta not succulent 154
154 Cells lax, rectangular or irregularly hexagonal, (12–)20–50 μm wide in mid-lf 155
Cells firm, rounded or quadrate, mostly less than 20 μm wide in mid-lf 156
155 Lid of capsule convex, calyptra cucullate, oblique **71. Funaria (p. 338)**
Lid apiculate or rostellate, calyptra mitriform, symmetrical **72. Physcomitrium (p. 344)**
156 Sporophyte axillary **52. Anoectangium (p. 271)**
Sporophyte terminal 157
157 Lf cells pellucid, base not sheathing **45. Pottia (p. 234)**
Cells obscure with papillae or if pellucid then lf base sheathing 158
158 Nerve ending below lf apex **50. Gymnostomum (p. 268)**
Nerve excurrent **54. Weissia (p. 274)**

. . . .

159 Peristome teeth 4, solid 160
Outer peristome teeth 16, articulated 161
160 Stem well developed, nerve strong, protonemal lvs lacking **3. Tetraphis (p. 87)**
Stem almost lacking, nerve short or absent, protonemal lvs present **4. Tetrodontium (p. 87)**
161 Capsule with conspicuous apophysis as wide as or wider than theca 162
Capsule without apophysis or apophysis narrower than theca 164
162 Apophysis purplish, wider than theca and plants autoecious or if only as wide as theca then plants dioecious **80. Splachnum (p. 354)**
Apophysis and theca of similar colour and width, plants autoecious 163
163 Lvs ovate-lanceolate to lanceolate or oblanceolate, seta succulent, yellowish to red, male inflorescence terminal **78. Tetraplodon (p. 352)**
Lvs usually ovate to obovate, seta thin, pale, male inflorescences on slender axillary branches **79. Aplodon (p. 354)**
164 Cells lax, rectangular to elongate-hexagonal smooth, 165
Cells firm, rounded or quadrate, smooth or papillose 166

165 Lvs acute, margin bluntly toothed, capsule without apophysis **71. Funaria (p. 338)**
Lvs obtuse or margin sharply toothed, capsule with long apophysis **77. Tayloria (p. 352)**

166 Stem lvs concave, obtuse, appressed when moist **27. Aongstroemia (p. 139)**
Lvs acuminate to obtuse, erecto-patent to spreading when moist 167

167 Lvs tapering to subulate apex **18. Seligeria (p. 118)**
Lf apex acute to acuminate but not subulate 168

168 Lf apex rounded or obtuse, peristome teeth not united in pairs 169
Lf apex acute to acuminate or if obtuse then peristome teeth united in pairs 170

169 Lvs obovate to broadly obovate-spathulate **44. Stegonia (p. 234)**
Lvs lingulate or narrowly lanceolate **64. Grimmia (p. 312)**

170 Peristome teeth united in pairs, reflexed when dry **68. Glyphomitrium (p. 336)**
Peristome teeth not united in pairs, ± erect when dry 171

171 Lvs dentate above **60. Leptodontium (p. 299)**
Lvs entire or only slightly denticulate 172

172 Cells in upper part of lf ± hexagonal, 10–18 μm wide **45. Pottia (p. 234)**
Cells in upper part of lf quadrate or rounded, 8–10 μm wide 173

173 Lf margin recurved **29. Dicranoweisia (p. 147)**
Lf margin plane or incurved 174

174 Margin of stem lvs toothed near base, plane above **53. Eucladium (p. 274)**
Lf margin entire near base, plane or involute above **54. Weissia (p. 274)**

. . . .

175 Segments of peristome teeth spirally curved or if straight then hyaline basal cells of lf extending up margin 176
Peristome teeth or segments straight, hyaline basal cells not extending up margin of lf 178

176 Hyaline basal cells of lf extending up margin, transition between basal and upper cells abrupt **57. Tortella (p. 291)**
Hyaline basal cells not extending up margin, transition between basal and upper cells gradual 177

177 Lvs broadly pointed, lingulate, spathulate or obovate, nerve not excurrent **40. Tortula (p. 211)**
Lvs gradually tapering to apex or nerve excurrent **49. Barbula (p. 248)**

178 Upper cells longer than wide 179
Upper cells quadrate or rounded 180

179 Peristome teeth divided ± to base into 2 filiform segments **14. Ditrichum (p. 109)**
Peristome teeth divided to about halfway **28. Dicranella (p. 139)**

180 Lvs toothed above, plants not rusty-red below 181
Lvs entire or if toothed then plants rusty-red below 183

181 Lower part of lf plicate, longitudinal walls of basal cells more heavily thickened than transverse walls **67. Ptychomitrium (p. 336)**
Lvs not plicate, basal cells with ± uniformly thickened walls 182

182 Lvs coarsely toothed from middle or below **26. Dichodontium (p. 137)**
Lvs denticulate above **24. Cynodontium (p. 130)**

183 Lf margin recurved or revolute **49. Barbula (p. 248)**
Lf margin plane 184

184 Lf apex flat, nerve if excurrent not forming a mucro, margin irregularly notched above or undulate **55. Oxystegus (p. 285)**
Apex cucullate or nerve excurrent in mucro, margin neither notched nor undulate **56. Trichostomum (p. 289)**

....

185 Lf margin recurved, cells papillose **112. Hedwigia (p. 492)**
Lf margin plane, cells smooth 186

186 Plants not bud-like, protonemal lvs present **4. Tetrodontium (p.87)**
Plants bud-like, protonemal lvs lacking **70. Discelium (p. 338)**

....

187 Lvs ± ovate, lingulate or spathulate, ± shortly pointed 188
Lvs lanceolate or linear-lanceolate or if wider then gradually tapering to apex 191

188 Nerve widened in upper part of lf **42. Desmatodon (p. 230)**
Nerve not widened above 189

189 Lvs lingulate, narrowly lingulate or lingulate-lanceolate, cells 10–14 μm wide in mid-lf **49. Barbula (p. 248)**
Lf shape not as above or if so then lf cells 16–20 μm wide 190

190 Lf apex rounded **40. Tortula (p. 211)**
Apex acute **45. Pottia (p. 234)**

191 Margin denticulate or dentate ± throughout **102. Bartramia (p. 454)**
Margin entire or denticulate towards apex only 192

192 Lf cells bistratose above, strongly mamillose, obscure **22. Cheilothela (p. 128)**
Cells unistratose or if bistratose above then not strongly mamillose 193

193 Lf cells ± sinuose, opaque and bistratose above but not markedly papillose, capsule immersed **63. Schistidium (p. 306)**
Lf cells not sinuose; unistratose or if bistratose then only so at margin, pellucid or not, capsule exserted 227

....

194 Lvs broad above, shortly pointed or apex obtuse or rounded 195
Lvs gradually tapering to acute to acuminate apex 196

195 Lf apex rounded or if acute then plant dioecious and with rhizoidal gemmae **40. Tortula (p. 211)**
Lf apex acute, plants synoecious or autoecious, rhizoidal gemmae lacking **45. Pottia (p. 234)**

196 Lf margin toothed from near base **60. Leptodontium (p. 299)**
Margin entire, or denticulate near base or towards apex only 197

197 Plants minute, to 2 mm tall (*Brachydontium* or *Seligeria* – indeterminable without fruit) or lvs trifarious or if plants more than 2 mm tall then lf apex subulate **18. Seligeria (p. 118)**
Plants more than 2 mm, lvs neither trifarious nor with subulate apex 198

198 Hyaline basal cells ascending up margin, transition from basal to upper cells abrupt **57. Tortella (p. 291)**
Basal cells not ascending up margin, transition from basal to upper cells gradual 199

199 Lvs denticulate above, plants glaucous, bluish-green **20. Saelania (p. 126)**
Lvs entire above or if denticulate then plants not glaucous and bluish-green 200

200 Lvs denticulate near base, linear-lanceolate, margin plane **53. Eucladium (p. 274)**
Lvs entire near base or if denticulate then shorter and wider or margin inrolled above 201

201 Lf cells with rounded lumens except near base **109. Zygodon (p. 470)**
Cell lumens not rounded 202

202 Lvs crisped or curled when dry, margin plane above, often irregularly notched or toothed **55. Oxystegus (p. 285)**
Lvs ± tightly incurved when dry, margin inrolled above or not, entire or papillose-crenulate 203

203 Lf margin plane or inrolled above, apex not cucullate, plants autoecious **54. Weissia (p. 274)**
Lf margin plane and apex flat or margin inflexed and apex cucullate, plants dioecious **56. Trichostomum (p. 289)**

. . . .

204 Lf cells obscure with papillae 205
Cells pellucid, papillose or not 206

205 Basal cells narrowly rectangular, upper cells shortly rectangular to quadrate, marginal cells not pellucid **26. Dichodontium (p. 137)**
Basal cells rectangular, upper cells hexagonal, marginal cells pellucid **60. Leptodontium (p. 299)**

206 Lvs shortly pointed or obtuse **23. Rhabdoweisia (p. 128)**
Lvs tapering to narrow apex 207

207 Lvs plicate below, longitudinal walls of basal cells more heavily thickened than transverse walls **67. Ptychomitrium (p. 336)**
Lvs not plicate, walls of basal cells uniformly thickened 208

208 Lvs curled or curved when dry but not crisped, nerve with teeth or papillae at back above **107. Timmia (p. 467)**
Lvs crisped when dry, nerve smooth at back 226

. . . .

209 Lf apex flat, nerve poorly defined, cells 2–3-stratose above, perichaetial lvs ciliate, capsule immersed **9. Diphyscium (p. 104)**
Lvs differing from above or if similar then apex cucullate and nerve well defined, perichaetial lvs not ciliate, capsule exserted 210

210 Margin bistratose at middle of lf, basal marginal cells narrow, upper cells obscure with papillae **59. Trichostomopsis (p. 298)**
Lvs not as above 211

211 Lvs obovate-spathulate to lingulate-spathulate, mostly widest above middle, cell lumens not stellate **40. Tortula (p. 211)**
Lf shape various but widest below middle or cell lumens ± stellate 212

212 Cells in upper part of lf not sinuose 212*a*
Cells in upper part of lf ± sinuose or walls unevenly thickened 213

212*a* Lf margin plane **52. Anoectangium (p. 271)**
Lf margin recurved ± from base to near apex **110. Orthotrichum (p. 473)**

213 Cells at extreme base of lf enlarged, 2–3 stratose, cells above with conical papillae or stellate lumens **96. Aulacomnium (p. 447)**
Extreme basal cells not enlarged, unistratose, cells above without conical papillae or stellate lumens 214

214 Lvs lingulate-lanceolate to lanceolate or ovate, capsule immersed **63. Schistidium (p. 306)**
Lvs lingulate or narrowly lanceolate, capsule exserted **64. Grimmia (p. 312)**

. . . .

215 Lf margin crenulate or papillose-crenulate above 216
Margin entire or denticulate above but not crenulate or papillose-crenulate 219

216 Lvs ± broadly pointed **49. Barbula (p. 248)**
Lvs gradually tapering to apex 217

217 Stems with dense reddish-brown tomentum **96. Aulacomnium (p. 447)**
Stems without tomentum 218

218 Lf margin notched and with irregular teeth above **55. Oxystegus (p. 285)**
Margin not notched or toothed **50. Gymnostomum (p. 268)**

219 Nerve 2-winged at back above, basal cells linear, ± sinuose **65. Dryptodon (p. 326)**
Nerve not 2-winged at back, basal cells shorter, not sinuose 220

220 Upper cells with ± rounded lumens, not sinuose 221
Cells with quadrate or hexagonal lumens, sinuose or not 224

221 Lvs linear-lanceolate or narrowly lanceolate, not widened above base, capsule exserted 222
Lvs lanceolate, widened above base or not, capsule exserted or not 223

222 Basal cells thin-walled, calyptra concealing capsule **68. Glyphomitrium (p. 336)**
Basal cells incrassate, calyptra small **108. Amphidium (p. 468)**

223 Capsule immersed, emergent or shortly exserted, calyptra naked or sparsely hairy, basal part of lf not widened, nor concave or plicate, basal cells near margin not differentiated **110. Orthotrichum (p. 473)**
Capsule exserted, calyptra hairy, basal part of lf widened, concave, sometimes plicate marginal cells near base widened, hyaline **111. Ulota (p. 486)**

224 Upper lvs crisped when dry, cells pellucid 225
Lvs not crisped when dry, or if crisped then upper cells opaque 227

225 Lf margin entire, capsule erect **29. Dicranoweisia (p. 147)**
Margin denticulate towards apex, capsule erect or inclined 226

226 Lf base not sheathing, upper cells mamillose or not **24. Cynodontium (p. 130)**
Lf base ± sheathing, cells smooth above **25. Oncophorus (p. 135)**

227 Lvs toothed near apex, cells smooth, shoots not rusty-red below **21. Ceratodon (p. 126)**
Lvs entire or if denticulate near apex then shoots rusty-red below **49. Barbula (p. 249)**

. . . .

228 Lf margin crenulate or papillose-crenulate 229
Lf margin entire, denticulate or toothed above but not crenulate or papillose-crenulate 237

229 Nerve poorly defined, cells 2–3-stratose, perichaetial lvs ciliate, capsule immersed **9. Diphyscium (p. 104)**
Nerve well defined, cells unistratose except sometimes at margin, perichaetial lvs not ciliate, capsule exserted 230

230 Hyaline basal cells extending up margin, transition from basal to upper cells abrupt **57. Tortella (p. 291)**
Hyaline basal cells not extending up margin, transition to upper cells gradual 231

231 Stems matted below with reddish-brown tomentum, sporophyte axillary, cells very obscure with papillae **52. Anoectangium (p. 271)**
Stems not matted with reddish-brown tomentum, sporophyte terminal, cells obscure or not 232

232 Lf margin papillose-crenulate, cells conspicuously papillose 233
Lf margin crenulate, cells not or only faintly papillose 236

233 Plants gregarious, very small, to 2.5 mm tall, basal cells of lvs narrowly rectangular, upper cells ± pellucid **51. Gyroweisia (p. 270)**
Plants gregarious or tufted, larger or if as small then basal cells rectangular and upper cells obscure 234

234 Lf cells with ± rounded lumens **108. Amphidium (p. 468)**
Cells with ± quadrate lumens 235

235. Lvs not more than 2 mm long, not undulate when moist **50. Gymnostomum (p. 268)**
Lvs 3–7 mm long, often undulate when moist **55. Oxystegus (p. 285)**

236 Marginal cells near lf base not narrowed, marginal cells above unistratose **23. Rhabdoweisia (p. 128)**
Basal marginal cells narrowed, marginal cells at middle of lf bistratose **59. Trichostomopsis (p. 298)**

237 Lvs broadly ovate to ovate-lanceolate, irregularly toothed, cells 20–34 μm wide in mid-lf **91. Mnium (p. 430)**
Lvs not as above 238

238 Lvs ovate to lanceolate, not longly tapering **109. Zygodon (p. 470)**
Lvs ligulate to linear-lanceolate, ± longly tapering 239

239 Lvs denticulate near base **53. Eucladium (p. 274)**
Lvs entire throughout or denticulate only above 240

240. Plants glaucous, bluish-green, lvs slightly twisted, flexuose when dry **20. Saelania (p. 126)**
Plants not glaucous or bluish-green, lvs crisped when dry 241

241 Basal cells incrassate **108. Amphidium (p. 468)**
Basal cells thin-walled 242

242 Upper cells with ± quadrate lumens, plants 3–9 cm tall **25. Oncophorus (p. 135)**
Upper cells with ± rounded lumens, plants to 1 cm tall **68. Glyphomitrium (p. 336)**

. . . .

243 Shoots julaceous with concave imbricate lvs 129
Shoots not julaceous 244

244 Lvs lingulate **77. Tayloria (p. 352)**
Lvs ovate to orbicular 245

245 Base of lf very narrow, ciliate, cells 40–100 μm wide in mid-lf **76. Oedipodium (p. 350)**
Lf base not ciliate, cells smaller 246

246 Lvs imbricate when moist, cells ± rectangular, 10–15 μm wide in mid-lf **27. Aongstroemia (p. 139)**
Lvs not imbricate, cells with pointed ends or 20–30 μm wide in mid-lf 247

247 Cells not in radiating rows **89. Bryum (p. 383)**
Cells in rows radiating from nerve **95. Pseudobryum (p. 447)**

. . . .

248 Cells and nerve scabrous with papillae in upper part of lf only **14. Ditrichum (p. 109)**
Cells mamillose except near lf base 249
249 Lvs longitudinally plicate **106. Breutelia (p. 465)**
Lvs smooth or plicate near base only 250
250 Lvs narrowly to linear-lanceolate, plants to 10(–15) cm tall **102. Bartramia (p. 454)**
Lanceolate lvs ovate to ovate-lanceolate or if narrower then plants not more than 1.5 cm tall **105. Philonotis (p. 459)**

. . . .

251 Plants occurring on dung or decaying animal remains 252
Plants of soil or humus (*Funaria, Physcomitrium* and *Tayloria*, not determinable in the absence of fruit)
252 Apical cell of lf 1–2(–3) times as long as wide, antheridia on slender lateral branches **79. Aplodon (p. 354)**
Apical cell (3–)5–10 times as long as wide, antheridia terminal (but sometimes apparently lateral by overgrowth of branches) 253
253 Lvs narrowing to long, fine acumen, plants of dung or decaying animal remains **78. Tetraplodon (p. 352)**
Acumen shorter and wider, plants of dung **80. Splachnum (p. 354)**

. . . .

254 Nerve ¼ or more width of lf base **84. Leptobryum (p. 361)**
Nerve narrower 255
255 Plants monoecious, nerve ending below apex, cells in mid-lf linear **83. Orthodontium (p. 361)**
Nerve ending in apex or excurrent or if ending below apex then cells rectangular, plants dioecious or if paroecious then nerve excurrent 256
256 Plant not more than 2 mm tall or if more then nerve excurrent in blunt subula **18. Seligeria (p. 118)**
Plants more than 2 mm tall, nerve if excurrent not forming blunt subula 257
257 Mid-lf cells ± rectangular or if longer then margin entire, nerve excurrent or ending in apex **14. Ditrichum (p. 109)**
Mid-lf cells narrowly rectangular to linear, margin ± denticulate, if cells shorter then nerve ending below apex and abundant rhizoidal gemmae present **28. Dicranella (p. 139)**

. . . .

258 Plants not more than 2 mm tall, nerve ending well below apex, protonemal lvs present **4. Tetradontium (p. 87)**
Plants larger or if small then nerve reaching ± to apex and protonemal lvs absent 259
259 Cells with pointed ends 260
Cell ends not pointed 265
260 Plants with underground rhizomatous stems, erect stems with scale-like lower lvs and large crowded upper lvs **90. Rhodobryum (p. 430)**
Plants not as above 261
261 Vegetative shoots slender with small distant lvs, fertile shoots with longer crowded lvs **11. Archidium (p. 106)**
Plants not as above 262

262 Plants with very crowded very slender shoots forming dense tufts, capsules borne on lateral shoots **82. Mielichhoferia (p. 359)**
Plants not forming dense tufts, capsules terminal 263

263 Mid-lf cells narrowly hexagonal, 15–30 μm wide, lvs not concave, base not decurrent **99. Amblyodon (p. 453)**
Mid-lf cells narrower or if as wide then cells linear or lf base longly decurrent or lvs concave 264

264 Lf margin denticulate ± from middle **85. Pohlia (p. 362)**
Margin entire or denticulate only near apex **89. Bryum (p. 383)**

265 Lvs quinquefarious, imbricate when moist **103. Conostomum (p. 458)**
Lvs not quinquefarious, imbricate or not when moist 266

266 Lvs 3–5 mm long **101. Plagiopus (p. 454)**
Lvs less than 3 mm long 267

267 Margin entire or with a few obscure denticulations near apex 268
Margin denticulate or dentate 269

268 Plants 2–15 mm **14. Ditrichum (p. 109)**
Plants taller **100. Catoscopium (p. 453)**

269 Synoecious, plants 2–5 mm tall **104. Bartramidula (p. 458)**
Dioecious, plants 5 mm or more **105. Philonotis (p. 459)**

. . . .

270 Stems ± regularly pinnately or bipinnately branched, with abundant paraphyllia **137. Thuidium (p. 531)**
Stems pinnately branched or not, never regularly bipinnate, paraphyllia few or absent 271

271 Lf margin distinctly toothed, secondary stems erect, often dendroid with lower lvs scale-like and scarious **123. Thamnobryum (p. 510)**
Lf margin entire or crenulate or if denticulate then only near apex, secondary stems not dendroid nor with scarious scale lvs 272

272 Lvs longitudinally plicate **133. Lescuraea (p. 522)**
Lvs not plicate 273

273 Lf apex rounded **120. Leptodon (p. 504)**
Apex acute to obtuse 274

274 Cells smooth or if papillose only slightly so with papillae on end walls of cells at back of lf 275
Cells papillose, papillae central on both sides of cells 276

275 Plants medium-sized, lvs mostly more than 1 mm long, capsule immersed **115. Cryphaea (p. 500)**
Plants slender, lvs less than 1 mm long, capsule exserted **131. Pseudoleskeella (p. 519)**

276 Primary stems stoloniform, lf margin plane or if recurved then plants robust and cells with 2–3 papillae **136. Anomodon (p. 529)**
Primary stems not stoloniform, lf margin recurved, plants slender, cells unipapillose **132. Leskea (p. 520)**

. . . .

277 Lvs longitudinally plicate **116. Leucodon (p. 501)**
Lvs not plicate 278

278 Cells papillose 279
Cells smooth 282

279 Lvs more than 1 mm long — 280
Lvs less than 1 mm long — 281

280 Lf margin strongly recurved ± from base to apex, entire — **112. Hedwigia (p. 492)**
Margin plane, toothed above — **118. Pterogonium (p. 503)**

281 Lvs very concave, cochleariform, apex obtuse or rounded, with or without apiculus — **128. Myurella (p. 516)**
Lvs not or only slightly concave, apex acute — **135. Heterocladium (p. 526)**

282 Lf apex acute, mid-lf cells 12–16 μm wide — **129. Myrinia (p. 517)**
Apex acuminate, cells *ca* 10 μm wide — **130. Habrodon (p. 519)**

. . . .

283 Lvs rugose when moist, cells papillose at back — **172. Rhytidium (p. 662)**
Lvs smooth or plicate when moist, cells smooth — 284

284 Lvs shortly pointed, concave — **144. Hygrohypnum (p. 569)**
Lvs tapering from below middle, concave or not — 285

285 Lvs with cordate-auriculate base and very stout nerve, stems with or without paraphyllia — **139. Cratoneuron (p. 539)**
Lf base not cordate, auricles present or not, paraphyllia lacking — **143. Drepanocladus (p. 561)**

. . . .

286 Secondary stems erect, dendroid, arising from underground rhizomatous stem — **114. Climacium (p. 499)**
Secondary stems not obviously dendroid, primary stems not rhizomatous — 287

287 Lvs with reflexed teeth at apex — **117. Antitrichia (p. 503)**
Lvs entire or teeth not reflexed — 288

288 Stems with abundant paraphyllia — 289
Paraphyllia lacking — 290

289 Lf cells papillose — **138. Helodium (p. 536)**
Cells smooth — **175. Hylocomium (p. 669)**

290 Basal and angular cells small, opaque, forming distinctive group — **147. Isothecium (p. 582)**
Basal and angular cells not as above — 291

291 Cells ± uniform throughout lf — **149. Homalothecium (p. 587)**
Cells near lf base shorter and wider than cells above — 292

292 Stem lvs cordate-triangular, shortly pointed, nerve of branch lvs ending in small projection at back of lf, lid longly rostrate — **155. Eurhynchium (p. 607)**
Stem lvs with long acumen, nerve of branch lvs not ending in projection, lid conical-rostellate — **150. Brachythecium (p. 589)**

. . . .

293 Lf apex rounded — 294
Lf apex acute to obtuse, with or without short or long apiculus — 295

294 Angular cells inflated, hyaline, forming distinct auricles — **146. Calliergon (p. 577)**
Angular cells not as above — **144. Hygrohypnum (p. 569)**

295 Basal and angular cells small and opaque, forming distinct group — **147. Isothecium (p. 582)**
Basal and angular cells not as above — 296

296 Lvs very concave 297
Lvs not or only slightly concave 301

297 Lf apex ± abruptly narrowed to long or short apiculus 298
Lf apex not apiculate 299

298 Shoots julaceous with imbricate lvs, apiculus short, reflexed **151. Pseudoscleropodium (p. 598)**
Shoots julaceous and lvs with long piliferous acumen or shoots not julaceous and apiculus not reflexed **153. Cirriphyllum (p. 601)**

299 Stems with numerous short branches, seta papillose **152. Scleropodium (p. 599)**
Stems irregularly branched, branches of various lengths, seta smooth 300

300 Lf margin entire, angular cells incrassate, opaque or forming small distinct auricles, plants of aquatic habitats **144. Hygrohypnum (p. 569)**
Lf margin denticulate or if entire then angular cells not as above, plants terrestrial **154. Rhynchostegium (p. 603)**

301 Stem and branch lvs differing in shape 302
Stem and branch lvs hardly differing in shape though sometimes differing in size 303

302 Branches numerous and crowded, curved, stem lvs smaller than branch lvs **148. Scorpiurium (p. 585)**
Branches few or numerous but not crowded, not curved, stem lvs of similar size to or larger than branch lvs **155. Eurhynchium (p. 607)**

303 Seta rough, lid conical-rostellate, lf base decurrent **150. Brachythecium (p. 589)**
Seta smooth, lid rostrate, lf base not decurrent **154. Rhynchostegium (p. 603)**

. . . .

304 Basal and angular cells small, opaque, forming distinct group, seta smooth **147. Isothecium (p. 582)**
Basal and angular cells not as above, seta smooth or papillose 305

305 Lvs with fine channelled acumen and angular cells forming distinct auricles **140. Campylium (p. 545)**
Acumen if present not channelled or angular cells not forming distinct auricles 306

306 Mid-lf cells short, 2–6 times as long as wide, lvs tapering to acuminate apex or nerve very stout, if cells longer than lvs spreading and sub-complanate **141. Amblystegium (p. 551)**
Mid-lf cells longer or if shorter then lvs shortly pointed, lvs not spreading or sub-complanate 307

307 Nerve ending in small projection at back of branch lvs, lid rostrate **155. Eurhynchium (p. 607)**
Nerve not ending in projection at back of branch lvs, lid conical to rostrate 308

308 Seta smooth, lid rostrate, lf margin denticulate ± from base to apex or cells only 3–5 times as long as wide, lvs ovate to ovate-lanceolate **154. Rhynchostegium (p. 603)**
Seta papillose or if smooth then lvs linear-lanceolate, lid rostrate or conical-rostellate, lvs denticulate in upper part only or ± entire, cells more than 5 times as long as wide 309

309 Seta papillose, lid conical-rostellate, plants slender to robust, when slender then lvs lanceolate or wider with long acumen **150. Brachythecium (p. 589)**
Plants very slender, seta smooth or papillose, lid rostrate, lvs linear-lanceolate or else lanceolate or oblong-lanceolate and shortly pointed **156. Rhynchostegiella (p. 616)**

. . . .

310 Lvs hardly concave, angular cells forming narrow longly decurrent auricles **159. Plagiothecium (p. 622)**
Lvs very concave, angular cells not as above 311

311 Angular cells inflated, hyaline, forming distinct auricles, lvs not falcate **146. Calliergon (p. 577)**
Angular cells not as above or if inflated and hyaline then lvs falcate 312

312 Stems reddish **174. Pleurozium (p. 669)**
Stems yellow to green or brown but not reddish 313

313 Stems pinnately branched, branches numerous, lf margin recurved near base, plants of dry habitats **158. Entodon (p. 621)**
Stems sparsely or irregularly branched, lf margin plane, plants of wet or aquatic habitats 314

314 Lvs neither falcate nor apiculate **144. Hygrohypnum (p. 569)**
Lvs falcate or apiculate **145. Scorpidium (p. 575)**

. . . .

315 Lvs shortly pointed, plants of streams and rivers **144. Hygrohypnum (p. 569)**
Lvs ± longly tapering to apex, plants not aquatic 316

316 Lvs not plicate nor lf base cordate **168. Hypnum (p. 643)**
Lvs plicate or if not then lf base cordate 317

317 Shoots arcuate, stems reddish, stem and branch lvs ± similar in shape **173. Rhytidiadelphus (p. 664)**
Shoots not arcuate, stems not reddish, branch lvs much narrower than stem lvs 318

318 Stems closely pinnately branched, branches complanate, of ± uniform length, lvs strongly plicate, paraphyllia abundant **169. Ptilium (p. 657)**
Branching various, lvs not or only slightly plicate, paraphyllia lacking **170. Ctenidium (p. 657)**

. . . .

319 Lvs very concave with long apiculus **119. Myurium (p. 504)**
Lvs not as above 320

320 Cells papillose at back above, plants slender **134. Pterigynandrum (p. 524)**
Cells smooth or if papillose then plants robust 321

321 Plants very slender, lvs rarely more than 0.3 mm long, cells mostly 2–4 times as long as wide **142. Platydictya (p. 560)**
Plants slender or not, lvs more than 0.3 mm long, cells more than 4 times as long as wide 322

322 Stems red (but whole plant greenish) 323
Stems yellowish to green or brown or whole plant red 324

323 Stems without paraphyllia **173. Rhytidiadelphus (p. 664)**
Stems with abundant paraphyllia **175. Hylocomium (p. 669)**

324 Lvs spreading to squarrose when moist **140. Campylium (p. 545)**
Lvs erect to patent when moist 325

325 Stem lvs cordate-triangular, rapidly narrowed to apex, margin sharply toothed, seta papillose **171. Hyocomium (p. 661)**
Stem lvs narrower, more gradually tapering, margin entire or denticulate, seta smooth 326

326 Lf margin denticulate ± throughout **160. Herzogiella (p. 632)**
Margin entire or denticulate towards apex only 327

327 Capsule immersed, plants aquatic, robust, lvs keeled or not **113. Fontinalis (p. 495)**
Capsule longly exserted, plants terrestrial, size various, lvs not keeled 328

328 Lf base longly and narrowly decurrent **159. Plagiothecium (p. 622)**
Lf base not decurrent 329

329 Angular cells differentiated and forming a distinct group of cells 330
Angular cells not or hardly differentiated 334

330 A few angular cells inflated, hyaline, forming small auricles **164. Sematophyllum (p. 639)**
Numerous angular cells differentiated but not inflated and hyaline 331

331 Some basal cells between nerve and angular cells with flat end walls, plants autoecious, fertile branches numerous and crowded, lid conical **165. Pylaisia (p. 641)**
Basal cells with pointed ends, plants autoecious or dioecious, fertile branches not crowded, lid conical or rostrate. 332

332 Lf margin recurved, gemmae often present at stem and branch tips **166. Platygyrium (p. 641)**
Margin not or hardly recurved, gemmae lacking 333

333 Autoecious, plants slender, capsule strongly inclined, curved **167. Homomallium (p. 643)**
Dioecious, plants slender or robust, capsules not as above **168. Hypnum (p. 643)**

334 Plants reddish **157. Orthothecium (p. 619)**
Plants green **161. Isopterygium (p. 634)**

. . . .

1. SPHAGNOPSIDA

By M. O. Hill

Protonema thalloid or filamentous, depending on environment. Rhizoids resembling those of the Bryopsida, consisting of branched or unbranched rows of cells with oblique dividing walls. Antheridia axillary on ± unspecialised branches, globose, with a narrow stalk of 2–4 rows of cells. Antherozoids biflagellate, with elongated body coiled into 2 turns of a sinistrose spiral. Archegonia in groups of 1–5 at apex of a short specialised branch. Seta absent, capsule joined directly to foot and exserted from perichaetium by elongation of pseudopodium (branch subtending archegonia). Capsule globose (urn-shaped when empty), with a convex lid, lacking a peristome; exothecial cells brown at maturity, with scattered non-functional stomata; spore sac amphithecial in origin, overarching columella; calyptra a thin hyaline membrane, irregularly ruptured at maturity by growth of capsule. Spores arising in tetrads, each with a conspicuous triradiate scar on inner face.

1. SPHAGNALES

With the characters of the class.

1. SPHAGNACEAE

A family of about 150 species with the characters of the class. It may be divided into about 11 well defined groups, of which 6 are represented in Britain. In the present account these groups are treated as sections of a single diverse genus, but the structural differences between them are as great, if not greater, than those used to define genera in the Hypnobryales (in the sense used here).

1. Sphagnum L., Sp. Pl., 1753

Autoecious or dioecious. Rhizoids lacking except on protonema (when thalloid) and at earliest stages of leafy shoot development. Stem differentiated into a cortex of 1–4 layers of often hyaline parenchymatous cells surrounding a central cylinder; cylinder with thin-walled parenchymatous cells at centre, merging gradually outwards into thick-walled, smaller and often pigmented prosenchymatous cells inside cortex. Branches arising in fascicles of (1–)2–8, differentiated in most species into spreading branches which diverge from stem and bear the main photosynthetic leaves, and pendent branches which are appressed to stem and may assist in water conduction. Branch cellular anatomy similar to that of stems, or cells of branch cortex dimorphic, consisting of large, often protuberant retort cells, each with a pore at distal end and smaller non-protuberant cells usually without pores (branch cortex is unistratose). The stems do not fork, an enlarged fascicle branch developing into a new stem. Branch and stem lvs spirally arranged with a 2/5 phyllotaxy; fascicles of branches arise in a position adjacent to every fourth lf in stem lf spiral, and hence themselves arise in a reverse 2/5 'phyllotaxy'. Leaves nerveless, unistratose, with three main types of cells. Elongated cells, dead at maturity, constitute a border 1–9 cells wide (sometimes $\pm$ completely destroyed by resorption during development). The main part of the lamina (except sometimes in perichaetial lvs) is composed of a mesh of two types of cell, narrow living green cells, and inflated dead hyaline cells which provide mechanical support for the green cells. The walls of the hyaline cells may be ornamented. These ornamentations are fibrils (spiral or annular thickenings which give the cell its mechanical strength) and pores (round perforations or membrane thinnings in which the cell wall is resorbed). In some species fibrils and pores are present also in stem and branch cortex. After fertilisation the perichaetial lvs grow rapidly, but the pseudopodium elongates only at maturity. Spore discharge is by an 'air gun' mechanism, the ripe capsule shrinking in dry weather to build up an internal pressure, reputedly 4–6 atmospheres, blowing off lid and ejecting the spores several centimetres into the air. The mechanism frequently does not work, and then the lid merely falls off, or the capsule disintegrates. *Sphagnum* species occur as variously coloured tussocks and 'lawns' in bogs, marshes, pools, wet woodland, moors and damp grassland, rarely if ever in localities with a pH exceeding 6.0. Distribution world-wide, but rare in tropical Africa.

Notes on the examination of Sphagna

Most species can be determined by examination of the stem lvs and branch lvs. Transverse sections of the stem are often useful, particularly in Section *Subsecunda*. Lf sections are seldom necessary, as the exposure of the green cells on the dorsal and ventral surface can be assessed by focussing up and down on 2 lvs, one placed with the dorsal (convex) surface facing upwards, the other with the ventral (concave) surface facing upwards. Details of pore structure are not easy to see without stain unless the pores are surrounded by a fibril ring. Dissection is often necessary to observe pores in the stem cortex, as the cylinder is opaque. Although most species can normally be determined without stain, staining very much simplifies identification. In the laboratory an aqueous solution of crystal violet or gentian violet is suitable. For home or excursion use, the purple dye emitted by a wetted 'indelible' pencil is quite adequate.

Branch lf dimensions refer to the larger lvs of the spreading branches. Pore measurements are taken longitudinally (relative to lf shape). In the branch lvs, pore dimensions refer to those from the central part of the lf unless otherwise stated. Pores are often absent in stem lvs except near apex. In the stem lvs, therefore, pore dimensions refer to the part of the lf about $\frac{2}{3}$–$\frac{3}{4}$ of the distance from lf base, or higher in lf if necessary.

Keys to sections of *Sphagnum*

Key 1 uses reliable structural characters but may be found difficult by beginners. Key 2 is more artificial, but uses characters which may be found easier. Key 2 should be reliable with well grown plants, but may occasionally give the wrong answer when applied to abnormal growth forms. When in doubt use Key 1.

Key 1

1 Cortical cells of stems and branches with spiral fibrils (always easy to see in pendent branches) — Sect. **Sphagnum** **(p. 32)**
  Cortical cells without fibrils — 2

2 Green cells of branch lvs triangular or trapezoid in section, broadly exposed on ventral surface, less exposed on dorsal surface — Sect. **Acutifolia** **(p. 42)**

Green cells not, or slightly exposed on ventral surface, or if with a moderate ventral exposure then with a greater dorsal exposure 3

3 Branch lvs denticulate at margin owing to resportion of cells in border; cells of branch cortex not dimorphic, all with a pore at distal end Sect. **Rigida (p. 55)**

Branch lvs with intact border; cells of branch cortex dimorphic, consisting of large retort cells with a pore at distal end, and smaller cells which usually lack pores 4

4 Pores in middle of branch lvs large, numerous and conspicuous, 12–40 μm Sect. **Squarrosa (p. 39)**

Pores small, less than 12 μm, or if larger then not more than 1 per cell 5

5 Green cells of branch lvs triangular or trapezoid in section, broadly exposed on dorsal surface; ventral surface of stem lvs with extensive resorption near apex, or else with pores mostly greater than 15μm (use stain); stem cortex 2–3 layers, sometimes so indistinctly differentiated from cortex as to appear absent Sect. **Cuspidata (p. 66)**

Green cells of branch lvs ± barrel-shaped in section, with a narrow exposure on dorsal surface; ventral surface of stem lvs intact or with pores mostly less than 12 μm; stem cortex very distinct, 1–3 layers (often only 1) Sect. **Subsecunda (p. 58)**

Key 2

1 Branch lvs broad, cucullate, rarely less than 1 mm wide; stem cortex occupying $\frac{1}{3}$–$\frac{1}{2}$ diameter of stem Sect. **Sphagnum (p. 32)**

Branch lvs narrow, or if broader than 1 mm then stem cortex less than $\frac{1}{4}$ diameter of stem 2

2 Lvs without reddish pigments 3

Lvs with rose-pink, crimson or red pigments 8

3 Stem lvs with patches of narrow cells at basal angles, these being roughly the same width as cells of border, so that border appears to merge with them (border may be ± absent in *S. fimbriatum*) 4

Border distinct to base, not appearing to merge with patches of narrow cells 5

4 Branch lvs with large pores, 8–30 μm, normally 3 or more per cell; stem lvs erect and appressed to stem (except compact forms) Sect. **Acutifolia (p. 42)**

Branch lvs with pores 9 μm or less (sometimes absent), or if pores larger then not more than 1 per cell; stem lvs spreading or hanging Sect. **Cuspidata (p. 66)**

5 Stem lvs large, lingulate, 1.2–2.0 mm long, lacking fibrils Sect. **Squarrosa (p. 39)**

Stem lvs various, if large and lingulate then with conspicuous fibrils 6

6 Stem lvs spreading or hanging 7

Stem lvs erect and appressed to stem 8

7 Branch lvs with a denticulate margin; pores medium-sized or large, 5–25 μm Sect. **Rigida (p. 55)**

Branch lvs with an intact border; pores small, 2–6 μm, or sometimes absent Sect. **Subsecunda (p. 58)**

8 Pores small, less than 6 μm; stem cortex with a single layer of hyaline cells Sect. **Subsecunda (p. 58)**

Pores 8–30 μm; stem cortex 2–4 layers Sect. **Acutifolia (p. 42)**

Section *Sphagnum*

Plants medium-sized or robust. Stem typically 0.6–1.2 mm diameter; cortex 3–4 layers, hyaline, (60–)100–300 μm wide, occupying about $\frac{1}{2}$ diameter of stem; cortical cell walls with sinistrally spiralling fibrils (sometimes absent in surface layer but always present in inner cells); outer surface with 1–8 pores per cell; diameter of cortical cells measured at right-angles to radius 25–90(–120) μm, numerous cells always exceeding 50 μm; cylinder strongly differentiated from cortex. Branches in fascicles of 3–7, (1–)2(–3) spreading, 1–4 pendent; branch cortex with 0–1 pores per cell, the cells not dimorphic, their walls with sinistrally spiralling fibrils (not always obvious in spreading branches, easily seen in pendent branches). Stem lvs erect or hanging, typically 1.1–2.0 mm × 0.7–1.4 mm, ± rectangular; border 2–5 cells wide, rather ill-defined, denticulate along sides of lf, becoming irregular and almost unrecognisable near apex; cells of border 8–14 μm wide, those at margin extensively resorbed on outer surface and usually represented only by a resorption furrow, or sometimes ± intact near base; hyaline cells septate or not; dorsal surface almost completely resorbed and lacking through much of lf; ventral surface intact, with or without fibrils in upper lf. Branch lvs cucullate, elliptic or broadly ovate, typically 1.1–2.8 mm × 0.9–1.9 mm; margin denticulate, the border 1 cell wide, completely resorbed on outer surface and represented only by a resorption furrow (margin appears denticulate owing to projection of transverse cell walls which persist after destruction of border); green cells triangular, barrel-shaped or lens-shaped in section, varying from completely exposed on both surfaces to broadly enclosed on ventral surface with a slight exposure on dorsal surface; hyaline cells in cucullate region near lf apex broader than long, their dorsal surface with a large membrane gap in distal half of cell, so that lf apex appears scabrous because of projecting cell walls; dorsal pores in mid-lf 8–25 μm, (0–)3–6(–16) per cell, in mid-lf normally occurring in triplets, 1 at each corner of three adjacent hyaline cells, towards margin of lf (and rarely in centre) more numerous and not confined to cell angles; ventral pores few or absent in mid-lf, more numerous towards margins, positioned along commissures. Antheridia on spreading branches. Perichaetial lvs oblong, rounded and tattered at apex; border 2–3 cells wide, ± intact in lower half of lf, strongly resorbed above and absent at apex; hyaline cells fibrillose and porose near apex, not resorbed on either surface below, above ± lacunose on dorsal surface.

Key to *Sphagnum* Section *Sphagnum*

1 Inside walls of hyaline cells of branch lvs ornamented with papillae or lamellae where they abut on green cells, so that in surface view of lf the green cells appear to be papillose or lamellate; cells near centre of stem lvs mostly septate; plant green, yellowish, ochre or brown with no red tinge — 2

Surface of green cells appearing smooth; cells near centre of stem lvs not, or hardly septate; plant green, yellowish, pinkish-orange or dull crimson — 3

2 Surface of green cells appearing papillose; green cells in cross-section barrel-shaped with strongly bulging sides — **2. S. papillosum**

Surface of green cells with transverse lamellae, which project into hyaline cells like the teeth of a comb; green cells in cross-section triangular, with straight sides — **1. S. imbricatum**

3 Green cells of branch lvs completely enclosed on both surfaces of lf; plant dull crimson, changing to mud brown in weak alkali (e.g. household bicarbonate). — **4. S. magellanicum**

Green cells of branch lvs reaching surface on both sides of lf, often fully exposed on ventral face; colour various, sometimes pinkish-orange, but not dull crimson, and not altering in weak alkali — **3. S. palustre**

1. S. imbricatum Hornsch. ex Russ., Beitr., 1865
S. austini Sull. ex Aust.

Dioecious. Shoots to 15 cm, green, ochre or fulvous brown. Stem (0.3–)0.5–0.9 mm diameter; cortex 3–4 layers; pores on outer surface 1–6 per cell; cylinder dark brown. Branches 3–4 per fascicle, 2 spreading, 1–2 pendent; fibrils of cortex 2–3 times more numerous on inner surface adjacent to cylinder than they are on outer surface (in other British members of Sect. *Sphagnum* they are equally numerous on both surfaces). Stem lvs 1.1–1.8 mm × 0.8–1.0 mm, lingulate or rectangular; cells near middle of lf 50–100 % septate. Branch lvs 1.1–2.2 mm × 0.9–1.5 mm, concave, broadly ovate, the apex broad and rounded, cucullate; thickness of lamina in mid-lf 25–42 μm in middle of hyaline cells, 12–16 μm where these abut on green cells; internal walls of

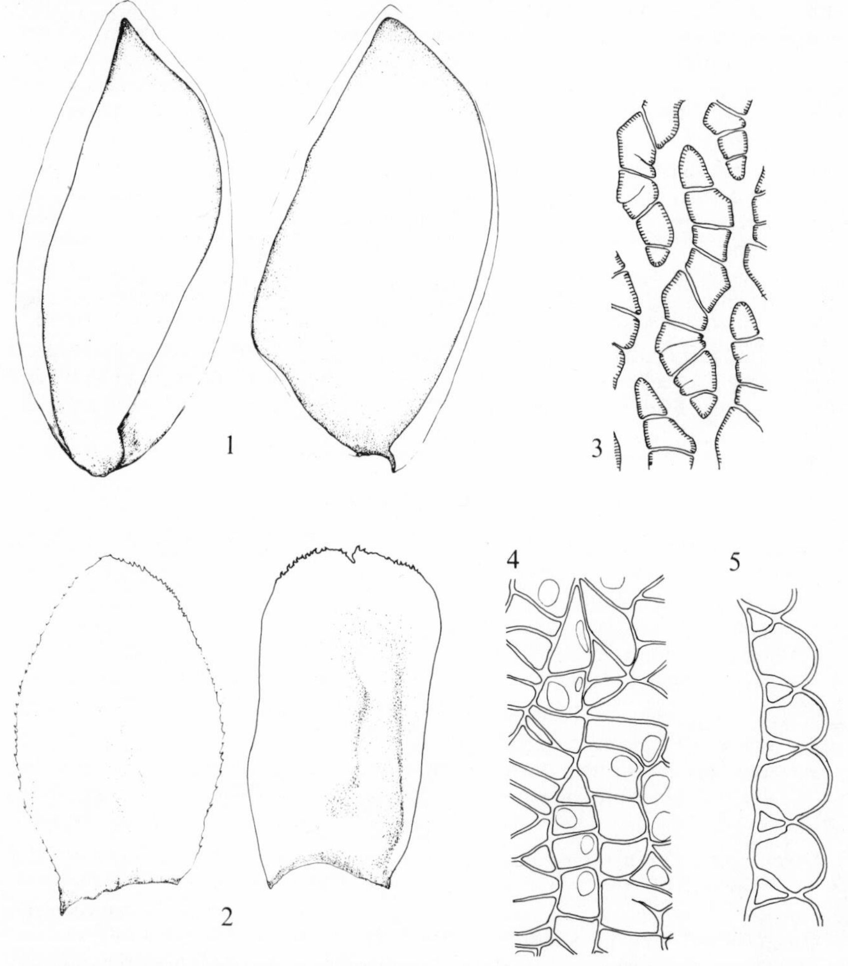

Fig. **1**. *Sphagnum imbricatum*: 1, branch leaves; 2, stem leaves; 3 and 4, cells at middle of branch leaf on ventral and dorsal surfaces; 5, section of branch leaf. Leaves × 27, cells × 280.

hyaline cells where these abut on green cells with conspicuous lamellae (morphologically fibrils), which mostly run perpendicular to the green cells, projecting to give a comb-like appearance to the green cells in surface view of lf; green cells in section equilaterally triangular, straight-sided, reaching surface on both sides of lf, broadly exposed on ventral face; dorsal pores in mid lf 11–18 μm, (0–)3–8 per cell, ventral pores usually absent in mid-lf, occasionally to 5 per cell. Antheridial lvs densely imbricated, ochre. Spores 24–28 μm. Fr rare, summer. Greenish or ochre tussocks in marshes and by streams; occasionally also in large hummocks on well developed bogs. Devon to Shetland, confined to the north and west; widespread in Ireland; local and seldom abundant. Sub-fossil *S. imbricatum* is a major component of the peat of raised bogs in many parts of the British Isles. 39, H20. Circumpolar main distribution sub-Oceanic boreal; Himalayas, West Indies, Chile.

Resembling small forms of *S. papillosum* but usually distinct in its more tapering branches and stronger colour. When growing on bogs it sometimes forms very large and conspicuous hummocks. Green forms in marshes resemble *S. palustre*, but are smaller, with more concave branch lvs and fewer branches per fascicle. The lamellae in the branch lvs are sometimes ± obsolete except near margins at base. The characteristic ornamentation of the inner surface of the branch cortex is, however, constant.

2. S. papillosum Lindb., Acta Soc. Sc. Fenn., 1872

Dioecious. Shoots to 20 cm, green, yellowish or ochre. Stem 0.7–1.0 mm diameter; cortex 3–4 layers; pores on outer surface 1–5(–6) per cell; cylinder green or brown. Branches 3–4 per fascicle, (1–)2 spreading, 1–2 pendent. Stem lvs 1.1–1.8 mm × 0.8–1.4 mm, obovate, lingulate or ± rectangular; cells near middle of lf 50–100 % septate, or rarely not septate. Branch lvs 1.5–2.5 × 1.3–1.9 mm, concave, broadly ovate, the apex cucullate, broad and rounded; lamina in mid-lf 27–48 μm thick in middle of hyaline cells, (16–)20–32 μm thick where these abut on green cells; internal walls of hyaline cells where these abut on green cells papillose; green cells in section reaching surface on both sides of lf, either barrel-shaped and moderately exposed on ventral face or lens-shaped with a negligible exposure on ventral face; dorsal pores in mid-lf 13–22 μm, 3–10 per cell; ventral pores absent in mid-lf. Antheridial lvs densely imbricated, ochre. Spores 27–30 μm. Fr occasional, summer. $n=38+4$*. Green, yellowish or ochre tussocks on moors and bogs, often hummock-forming; confined to strongly acid localities. Abundant in the north and west, local in the south-east. 99, H40. Circumpolar, mainly boreal but extending south to India.

Generally easy to recognise, with a characteristic ochre colour and stubby branches. Partially shaded forms strongly resemble *S. palustre*, and can be identified in the field only after considerable experience. Well grown plants of *S. palustre* differ from *S. papillosum* in having 3–4 pendent branches per fascicle; but the compact forms which are most likely to be confused with *S. papillosum* have only 1–2 pendent branches, so the difference is not much use in practice. Compact forms of both species superficially resemble *S. compactum*, but differ in the more cucullate branch lvs and large stem lvs. For differences from *S. imbricatum* see notes under that species. Non-papillose states of *S. papillosum* are reported from elsewhere in its range, but are not known in Britain. Weakly papillose forms do, however, occur rarely. In these papillae may be ± obsolete except near margins in lower lf.

3. S. palustre L., Sp. Pl., 1753
S. cymbifolium Hedw., *S. latifolium* Hedw., *S. obtusifolium* Ehrh., *S. centrale* C. Jen., *S. subbicolor* auct.

Dioecious. Shoots to 25 cm, variously green, pinkish-orange or slightly yellowish. Stem 0.6–1.2 mm diameter; cortex 3–4 layers; pores on outer surface 1–8(–10) per

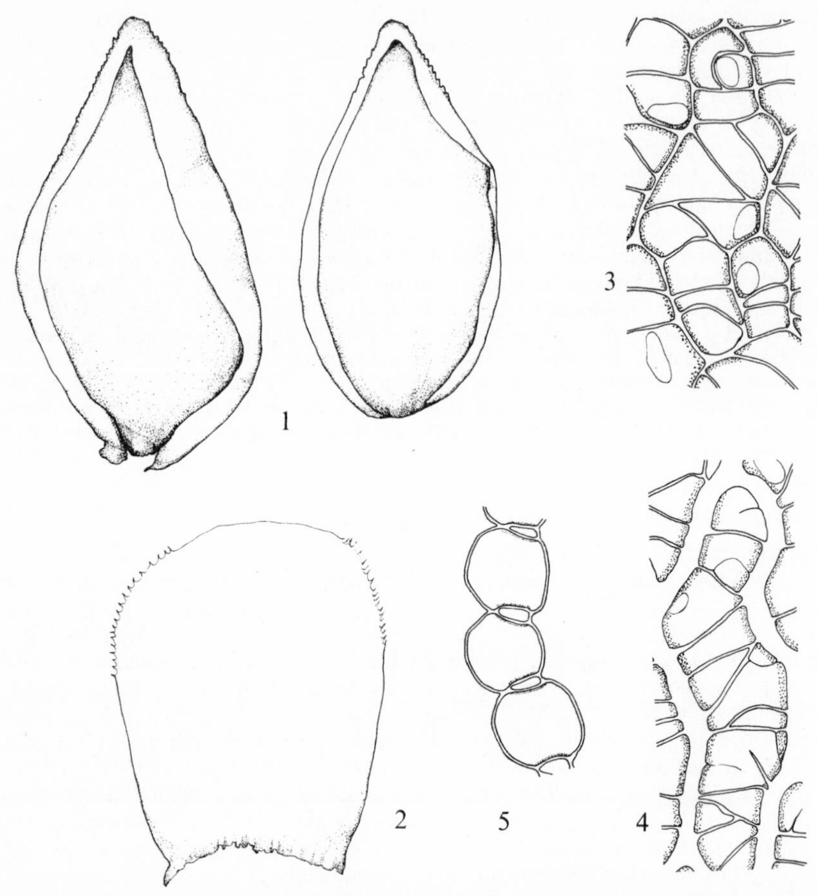

Fig. 2. *Sphagnum papillosum*: 1, branch leaves; 2, stem leaf; 3 and 4, cells at middle of branch leaf on dorsal and ventral surfaces; 5, section of branch leaf. Leaves ×27, cells ×280.

cell; cylinder green or brown. Branches 3–6(–7) per fascicle, 2(–3) spreading, 1–4 pendent. Stem lvs 1.2–2.0 mm×0.9–1.4 mm, obovate, lingulate or ± rectangular; cells not septate, or rarely a few septate near middle of lf. Branch lvs 1.7–2.8 mm× 1.1–1.8 mm, concave, ovate or broadly ovate, the apex cucullate, normally broad and rounded, but in shade forms narrower, ± acute and somewhat squarrose (though the extreme tip is always cucullate); lamina in mid-lf 30–50 μm thick in middle of hyaline cells, 19–30 μm thick where these abut on green cells; internal walls of hyaline cells lacking papillae or lamellae; green cells reaching surface on both sides of lf, in section very variable even within a single gathering, ranging in shape from narrowly triangular and fully exposed on ventral face, through barrel-shaped and broadly exposed on ventral face, to lens-shaped with a negligible ventral exposure; dorsal pores in mid-lf variable, 8–25 μm, 3–16 per cell; ventral pores in mid-lf absent or very few. Antheridial lvs densely imbricate, bright pinkish-orange in autumn and winter, changing to yellowish in summer. Spores 26–32 μm. Fr occasional, summer. $n=38+4$*. Whitish or green tussocks in mesotrophic marshes, streamsides and wet

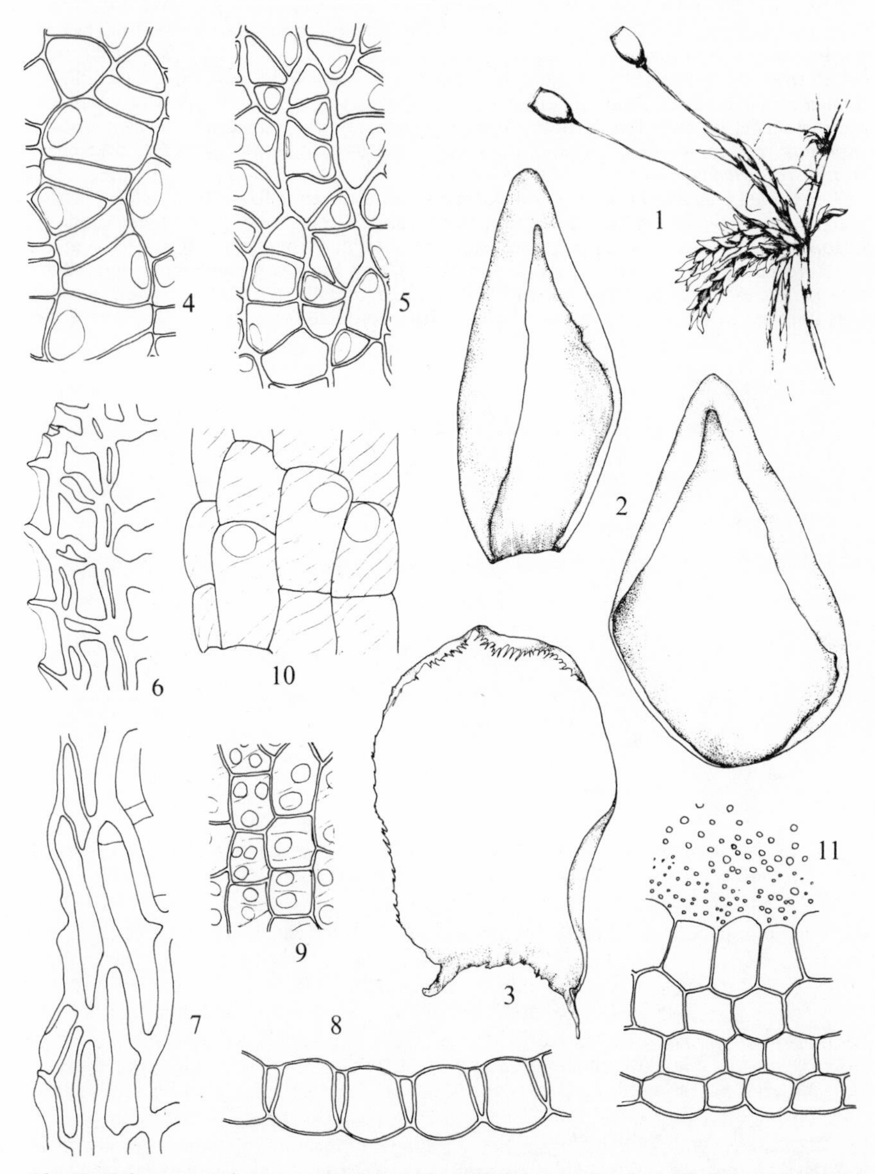

Fig. 3. *Sphagnum palustre*: 1, fascicle with dehisced capsules (×1.5); 2, branch leaves; 3, stem leaf; 4 and 5, cells at middle of branch leaf on dorsal and ventral surfaces; 6, margin of branch leaf near apex showing resorption of cells; 7, marginal cells at widest part of stem leaf; 8, section of branch leaf; 9, stem cortex in surface view; 10, branch cortex in surface view; 11, section of stem. Leaves ×27, leaf cells ×280, stem and branch cells ×110.

woodland. Abundant in the north and west, frequent in the south-east. 110, H39, C. Northern hemisphere south to Mexico and Taiwan.

In the field it can be confused with *S. auriculatum*, *S. compactum* and *S. squarrosum*; but the members of Section *Sphagnum* are distinct not only in the cucullate apex of the branch lvs, but also in the broad hyaline cortex, occupying $\frac{1}{3}$–$\frac{1}{2}$ the diameter of the stem (pull stem in half to see this). The field distinction between *S. papillosum* and *S. palustre* can be difficult, though with experience almost all plants can be named correctly. The branches of *S. palustre* are more tapering, its colour is more commonly green, and the branches at the centre of the capitulum are commonly pinkish-orange, not ochre as in *S. papillosum*. For differences from *S. magellanicum* see under that species.

S. centrale (*S. subbicolor* auct.) is the name commonly given to forms of *S. palustre* whose green cells are ± lens-shaped in section. Such forms are frequent in the boreal zone of Europe and N. America. In other respects they do not differ from typical *S. palustre* and are therefore not given taxonomic recognition here. Many European plants are intermediate between the extremes of *S. palustre* and *S. centrale*, and really well marked *S. centrale* such as is frequent in N. America has not hitherto been recorded from Britain. However, some

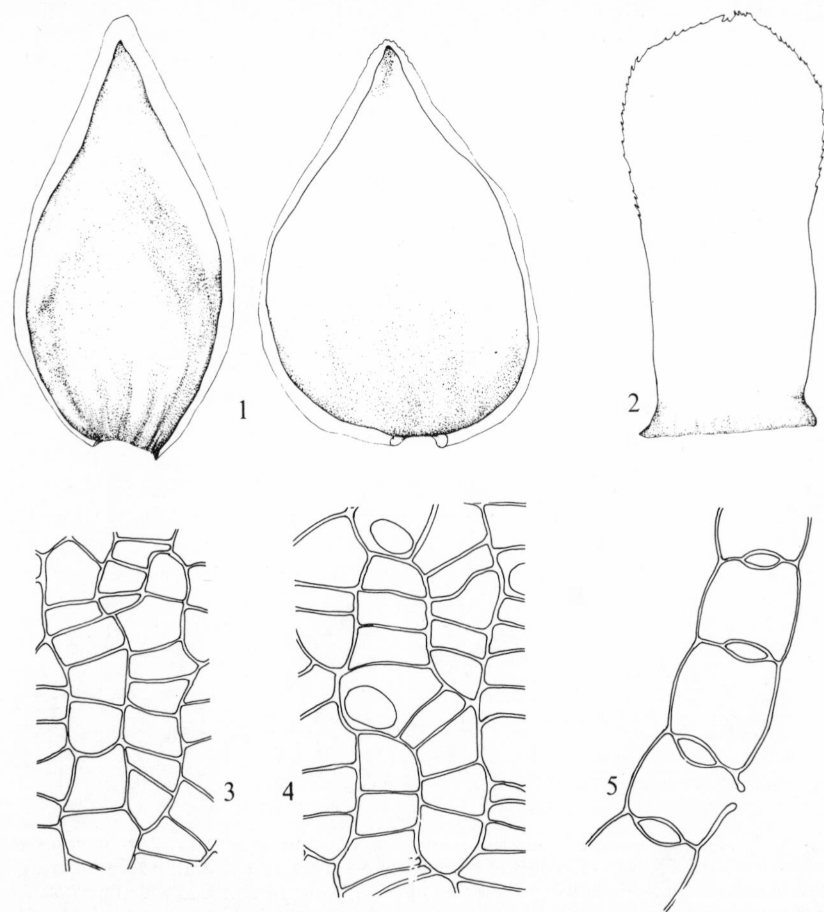

Fig. 4. *Sphagnum magellanicum*: 1, branch leaves; 2, stem leaf; 3 and 4, cells in middle of branch leaf on ventral and dorsal surfaces; 5, section of branch leaf. Leaves ×27, cells ×280.

British gatherings are indistinguishable from plants that have been called *S. centrale* in other parts of Europe.

4. S. magellanicum Brid., Musc. Rec. II, 1798
S. medium Limpr.

Dioecious. Shoots to 20 cm, dull crimson or occasionally green. Stem 0.7–1.1 mm diameter, cortex 3–4 layers; pores on outer surface (0–)1(–3) per cell; cylinder blackish-red. Branches 3–5 per fascicle, 2 spreading, 1–3 pendent. Stem lvs 1.4–1.8 mm × 0.7–1.2 mm, ± rectangular; cells near middle of lf not septate, or rarely a few septate. Branch lvs 1.7–2.7 mm × 1.5–2.0 mm, concave, broadly ovate, the apex broad and rounded, cucullate; thickness of lamina in mid-lf 30–45 μm in middle of hyaline cells, 27–40 μm at cell junctions; internal walls of hyaline cells without papillae or lamellae; green cells in section lens-shaped completely enclosed by the hyaline cells and not reaching either face of lf; dorsal pores in mid-lf 11–19 μm, (0–)3–6 per cell; ventral pores in mid-lf absent or very few. Antheridial lvs densely imbricated, crimson. Spores 26–30 μm. Fr rare, summer. $n=17$–19. Reddish tussocks and hummocks, usually mixed with *S. papillosum*; often abundant on well developed convex bogs, local on wet open moors and in valley bogs. Throughout the British Isles, but rare in south-east England. 68, H33. Circumpolar, mainly boreal, also C. and S. America to Tierra del Fuego, Himalayas, Malagasay Republic.

The branches are stubby, so that the habit resembles *S. papillosum*. The dull crimson colour, once known, is unmistakable, but can easily be confused by the inexperienced with the pinkish-orange (occasionally brick-red) of *S. palustre*. *S. magellanicum* changes colour in weak alkali, turning a mud brown; the colour of *S. palustre* is hardly altered.

Section *Squarrosa* (Russ.) Schimp.

Plants slender to robust. Stem typically 0.5–1.2 mm diameter; cortex 2–4 layers, hyaline, 30–70 μm wide, cell walls without fibrils; outer surface mostly without pores, or with scattered patches of cells having a single indistinct pore at upper end; diameter of cortical cells measured at right-angles to radius, 15–55 μm; cylinder strongly differentiated from cortex. Branches in fascicles of 4–6(–7), 2–3(–4) spreading, 2–3(–4) pendent; branch cortex with 0–1 pores per cell, the cells dimorphic (dimorphism often weak in spreading branches), cell walls without spiral fibrils; retort cells flat or with an indistinct neck, in groups of 1–4(–6). Stem lvs hanging, spreading or erect, typically 1.2–2.0 mm × 0.8–1.3 mm, lingulate; border 2–6 cells wide, ceasing below rounded region of apex where marginal cells are resorbed and tattered; patches of narrow cells at basal angles absent; hyaline cells hardly septate except near basal angles, lacking fibrils, or rarely with weak fibrils near basal angles; dorsal surface almost completely resorbed, lacking or (less often) very thin; ventral surface ± intact. Branch lvs ovate or broadly ovate, not cucullate, typically 1.0–3.1 mm × 0.5–1.8 mm, with an intact border of 1–3 narrow cells; green cells ± trapezoid in section, more exposed on dorsal than ventral surface; hyaline cells near apex longer than broad, their dorsal surface often with a pore at distal angle but not lacunose; dorsal pores in mid-lf 12–40 μm, (0–)1–12(–16) per cell, not normally in triplets at cell angles, sometimes centrally placed and coalescing into large membrane gaps; ventral pores mostly along commissures, 0–11(–15) per cell in mid-lf, more numerous towards margins, near apex commonly in triplets, 1 at each corner of 3 adjacent hyaline cells. Antheridia on spreading branches. Perichaetial lvs oblong or cuneate; border 1–2 cells wide, intact in lower half of lf, ± resorbed in upper half and absent at apex. Hyaline cells lacking fibrils and pores; dorsal surface extensively resorbed, thin or lacking throughout lf; ventral surface ± intact.

Key to *Sphagnum* Section *Squarrosa*

1 Robust green or slightly brownish plant; stem 0.7–1.2 mm diameter; branch lvs usually squarrose, 1.7–3.1 mm × 1.0–1.8 mm **5. S. squarrosum**
Plant slender or medium-sized, greenish-yellow or brown, rarely completely green; stem 0.5–0.7 mm diameter; branch lvs ± appressed, rarely squarrose, 1.0–2.3 mm × 0.5–1.2 mm **6. S. teres**

5. S. squarrosum Crome, Samml. Deutsch. Laubm., 1803

Autoecious. Shoots to 20 cm, green or somewhat brownish, with a medium-sized terminal bud. Stem 0.7–1.2 mm diameter; cortex 2–4 layers; outer surface without pores, or with pores in a few scattered patches of cells; cylinder green or brown. Branches 4–6(–7) per fascicle, 2–3(–4) spreading, 2–3(–4) pendent; retort cells in groups of 1–6, often weakly differentiated from others on spreading branches. Stem

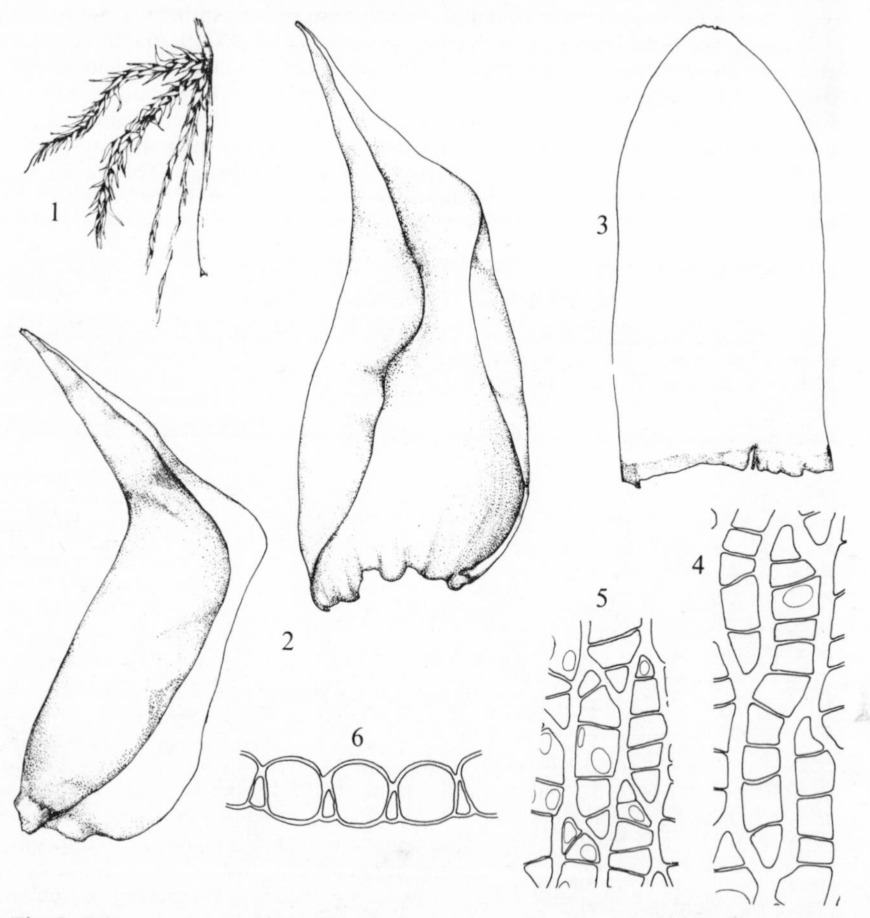

Fig. 5. *Sphagnum squarrosum*: 1, fascicle (×1.5); 2, branch leaves; 3, stem leaf; 4, cells at middle of branch leaf on dorsal surface; 5, cells on ventral surface near apex of branch leaf; 6, section of branch leaf. Leaves ×27, cells ×280.

lvs erect, spreading or hanging, 1.7–2.0 mm×0.8–1.3 mm, lingulate; border 2–6 cells wide, ceasing below rounded region of apex, patches of narrow cells at basal angles absent or rarely to about 8 cells wide; hyaline cells near basal angles often septate, above not, or hardly septate; dorsal surface lacking or very thin; ventral surface intact, without fibrils or pores. Branch lvs 1.7–3.1 mm×1.0–1.8 mm, ovate or broadly ovate, the apex normally bent back sharply in mid-lf so that lf is strongly squarrose; border 1–3 cells wide; green cells in section ± trapezoid with bulging sides, the greater exposure being on the dorsal surface; interior walls of hyaline cells where they abut on green cells obscurely papillose; dorsal pores in mid-lf variable, 13–50 μm, (0–)1–12(–16) per cell, perforate or imperforate, centrally placed or along commissures; ventral pores in mid-lf imperforate, 12–25 μm, 0–11(–15) per cell. Antheridial lvs not markedly squarrose, green or slightly yellowish. Perichaetial lvs with the characters of the Section, cuneate, with a truncate or retuse apex. Spores 27–30 μm. Fr common, summer. $n=19+2$, $19+4$, $38+4$*. Green 'lawns' among rushes and other large higher plants, sometimes abundant in swampy woodland or on swampy ground by lakes, or in smaller quantity in upland flushes, confined to markedly eutrophic situations. Frequent to common throughout the British Isles. 103, H34. Circumpolar, arctic and north temperate, New Zealand.

With its robust habit, green colour and strongly squarrose branch lvs it is usually easy to recognise in the field. It can resemble shade forms of *S. palustre* but differs in the relatively narrow stem cortex (occupying less than ¼ of the diameter of the stem, as opposed to nearly half in *S. palustre*). *S. compactum* occasionally has squarrose branch lvs but differs in its small stem lvs. For distinction from *S. teres* see under that species.

6. S. teres (Schimp.) Ångstr. in Hartm., Skand. Fl., 1861
S. squarrosum var. *teres* Schimp.

Dioecious. Shoots to 20 cm, green, yellowish or brown, with a large terminal bud. Stem 0.5–0.7 mm diameter; cortex 2–4 layers; outer surface with indistinct pores in scattered patches of cells; cylinder brown. Branches 4–5(–7) per fascicle, usually 3 spreading, 2 pendent; retort cells in groups of 1–2(–5). Stem lvs erect, spreading or hanging, 1.2–2.0 mm×0.8–1.1 mm, lingulate; border 2–5 cells wide, ceasing below rounded region of apex; patches of narrow cells at basal angles absent; hyaline cells near basal angles often septate, above not or hardly septate; dorsal surface lacking or very thin; ventral surface intact, without fibrils or pores. Branch lvs 1.0–2.3 mm× 0.5–1.2 mm, ovate, the apex bent back only slightly in mid-lf, or in shade forms sometimes strongly squarrose; border 1–3 cells wide; green cells in section trapezoid with the greater exposure on the dorsal surface; interior walls of hyaline cells where they abut on green cells sometimes finely papillose; dorsal pores in mid-lf 12–40 μm, 3–8 per cell, often centrally placed and coalescing into ± extensive membrane gaps; ventral pores in mid-lf imperforate, 12–25 μm, 1–7 per cell, or sometimes absent (though numerous near margins). Antheridial lvs greenish. Perichaetial lvs with the characters of the Section, oblong with a retuse apex. Spores 24–26 μm. Fr rare, summer. $n=19+2$. Green or brownish 'lawns' among rushes and small sedges, indicative of base-rich flushing. Throughout the British Isles; rare in the lowlands, frequent to common in the uplands. 66, H7. Circumpolar, mainly arctic and boreal, extending south to Italy and California.

Can normally be recognised in the field by its rather rigid branches, with lvs slightly bent backwards in mid-lf, the yellowish or brown colour, and large terminal bud. It can have a strong superficial resemblance to *S. girgensohnii*, but differs when well illuminated in its dark brown stem and in normally having 3 + 2 branches per fascicle. Robust greenish forms can be exceedingly hard to separate from *S. squarrosum*. The best policy is to make a careful search in the field to find typical *S. teres* or *S. squarrosum* in the immediate vicinity. There is

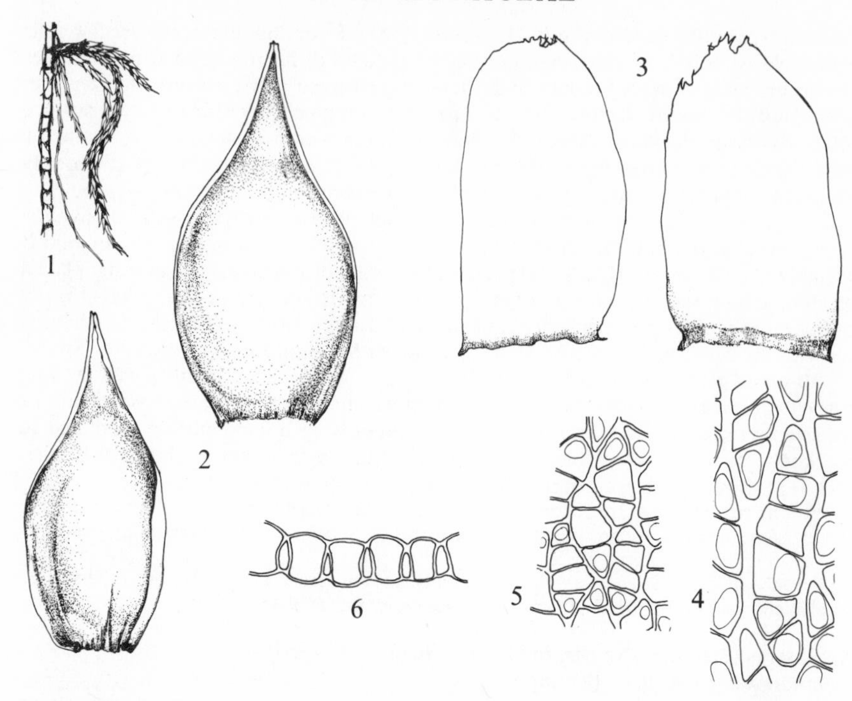

Fig. 6. *Sphagnum teres*: 1, fascicle (×1.5); 2, branch leaves; 3, stem leaves; 4, cells at middle of branch leaf on dorsal surface; 5, cells on ventral surface near apex of branch leaf; 6, section of branch leaf. Leaves ×27, cells ×280.

no reliable microscopic distinction, and differences given by various authors break down when applied to the abnormal forms which they might be useful to distinguish. It should be appreciated, however, that these forms are unusual, and that *S. teres* normally looks so different from *S. squarrosum* that their close relationship is not apparent without microscopic examination.

Section *Acutifolia* Wils.

Plants slender to medium-sized, with red or brown pigments. Stem typically 0.3–0.9 mm diameter; cortex 2–4 layers, hyaline, 40–200 μm wide, cell walls without fibrils; outer surface without pores or with 1(–2) pores per cell, diameter of cortical cells measured at right-angles to radius, 20–85 μm; cylinder strongly differentiated from cortex. Branches in fascicles of 3–5(–6), 2–3 spreading, (0–)1–2(–3) pendent; branch cortex with clearly dimorphic cells, cell walls without spiral fibrils; retort cells with a moderate to strong neck, in groups of 1(–2), except *S. molle*. Stem lvs erect, or in compact forms spreading, typically 0.8–1.7 mm×0.5–1.2 mm (larger in the nearly isophyllous *S. molle*), spathulate, triangular or lingulate; border 2–9 cells wide, entire or denticulate (± lacking in *S. fimbriatum*); patches of narrow cells at basal angles occupying (0–)20–80% of lf base; hyaline cells in upper lf (0–)20–100% septate, with or without fibrils, extensively resorbed on ventral surface, dorsal surface in most species ± intact. Branch lvs ovate or narrowly ovate, not cucullate, typically 0.8–2.7 mm×0.4–1.8 mm; border 1–3(–4) cells wide, intact (except *S. molle*); green cells in section triangular, broadly exposed on ventral surface; hyaline cells near apex longer than broad, not lacunose; dorsal pores in mid-lf 8–30(–55) μm, 1–16 per cell, towards lf apex often in triplets, 1 at each corner of 3 adjacent hyaline cells; ventral pores in mid-lf often absent, when present centrally placed, 10–20 μm, to 5 per cell, more numerous towards margins. Antheridia on spreading branches. Perichaetial lvs ovate or oblong; border intact to apex; hyaline cells intact on both surfaces, lacking fibrils and pores (except sometimes *S. molle*).

Key to *Sphagnum* Section *Acutifolia*

1 Branch lvs denticulate at margin in upper ½ owing to resorption of cell walls in border; stem lvs strongly fibrillose, 1.5–2.8 mm **15. S. molle**
Branch lvs with intact border; stem lvs to 1.4 mm, or if larger then lacking fibrils 2

2 Stem lvs triangular or triangular-lingulate, margin plane or inrolled at apex 3
Stem lvs lingulate or spathulate, margin ± plane at apex 5

3 Stem cylinder predominantly green with at most a few red flecks, invariably green in plants whose lvs are green; stem cortex with scattered pores (use stain); branches (except in obviously depauperate forms) arising mainly in fascicles of 4–5, of which 2–3 spreading **10. S. quinquefarium**
Stem cylinder usually with red or brown pigment, even in plants whose lvs are green; stem cortex without pores; branches mainly in fascicles of 3–4, rarely with more than 2 spreading branches 4

4 Stem lvs without fibrils or fibrils very weak, apex acute with inrolled lf margins; branch lvs (0.9–)1.2–2.7 mm **14. S. subnitens**
Stem lvs with conspicuous fibrils, or if not then apex ± plane; branch lvs (0.6–) 0.9–1.4 mm 9

5 Brown plant with dark brown stem **13. S. fuscum**
Stem red or pale 6

6 Green plant with no trace of red pigment; stem lvs lacking fibrils, often extensively tattered near apex, equally resorbed on both surfaces so that when stained extensive gaps are visible both near apex and at base 7
Red pigment often present; stem lvs with or without fibrils, not extensively tattered near apex, with extensive resorption only on ventral surface and hence not showing gaps when stained 8

7 Stem lvs spathulate, widest above base, border ceasing at or below mid-lf **7. S. fimbriatum**
Stem lvs lingulate, widest at base, border continued to near apex **8. S. girgensohnii**

8 Stem lvs lingulate with a broad rounded apex; many cells in middle part of branch lvs having more than 8 dorsal pores; surface of stem cortex with scattered pores (use stain) **9. S. russowii**
Stem lvs various, sometimes lingulate with a broad rounded apex; cells with more than 8 pores few or absent; stem cortex without pores 9

9 Pores on dorsal side of branch lvs near apex small, 2–6(–8) μm, each bordered by a thick ring and appearing round in surface view of cell; stem lvs without fibrils or fibrils weak **11. S. warnstorfii**
Pores larger, (5–)8–13 μm, not normally with thick rings, often flattened against margin of cell and appearing semicircular in surface view; fibrils of stem lvs usually conspicuous, occasionally weak or absent **12. S. capillifolium**

7. S. fimbriatum Wils. in J. D. Hook. & Wils., Fl. Antarct., 1846

Autoecious. Shoots to 20 cm, green, with a large terminal bud. Stem 0.4–0.9 mm diameter; cortex 2–3 layers, pores present in 40–100 % of cells on outer surface, 1(–2) per cell, round; cylinder green. Branches 3–4(–5) per fascicle, 2(–3) spreading, 1–2 pendent. Stem lvs 0.8–1.3 mm × 0.8–1.2 mm, spathulate, erect and appressed to stem; margin plane, tattered round most of upper lf; border absent, or else 2–4 cells wide, ceasing below mid-lf; patches of narrow cells at basal angles not pigmented, occupying 30–60 % of lf base; hyaline cells lacking fibrils and pores, extensively resorbed on both surfaces; upper cells 20–90 % septate, grossly distorted in consequence of lf margin having grown more than middle. Branch lvs 0.8–2.0 mm × 0.4–1.2 mm, ovate, not 5-ranked; border 1–3 cells wide; green cells in section narrowly trapezoid, reaching both surfaces, with the greater exposure on the ventral face; dorsal pores 10–20 μm, 5–13 per cell, near lf base larger, often coalescing into extensive membrane gaps; ventral pores 11–16 μm, 2–5 per cell, or sometimes absent in mid-lf though always present near apex. Antheridial lvs green or yellowish-brown. Perichaetial lvs ± rectangular; border intact to apex; hyaline cells intact on both surfaces, lacking fibrils and pores. Spores 25–28 μm. Fr common, summer. $n = 17$–19, 19 + 2, 38 + 4*. Extensive green 'lawns' among *Betula* and *Molinia* in damp woodland, or in smaller quantity on moors, banks, streamsides and degenerating bogs; frequent

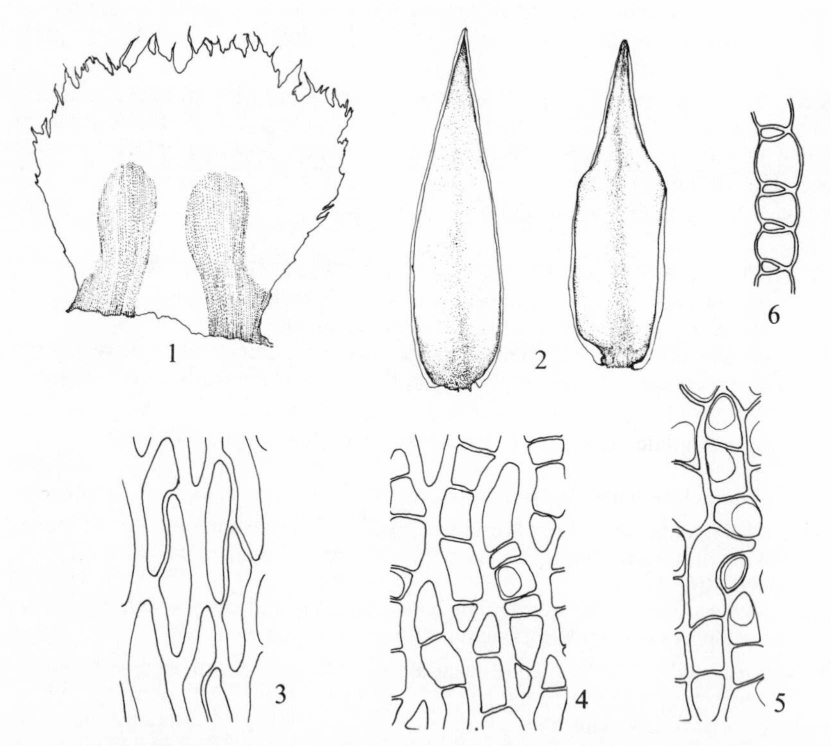

Fig. 7. *Sphagnum fimbriatum*: 1, stem leaf; 2, branch leaves; 3, stem leaf cells, $\frac{2}{3}$ from base, ventral surface; 4 and 5, branch leaf cells from middle of leaf on ventral and dorsal surfaces; 6, section of branch leaf. Leaves × 27, cells × 280.

to common throughout the British Isles. 103, H24. Arctic and temperate regions of the world in both hemispheres.

Readily identified in the field by pulling off the top of the capitulum. The spathulate stem lvs appear as a conspicuous 'ruff' at the top of the decapitated stem. The large terminal bud is also a useful field character. In most species of *Sphagnum* a tattered margin is a sign of resorption – i.e. that some of the cells have been destroyed in the course of development. In *S. fimbriatum*, however, the margin of the stem lvs is disrupted merely by unequal growth of the green cells.

8. S. girgensohnii Russ., Beitr., 1865
S. strictum Lindb., non Sull.

Dioecious in Europe. Shoots to 20 cm, green or faintly brownish, with a medium-sized terminal bud. Stem 0.5–0.8 mm diameter; cortex 2–3(–4) layers, pores present in 60–100 % of cells on outer surface, 1(–2) per cell, round or transversely elliptical; cylinder green or pale brown. Branches 3–4(–5) per fascicle, 2(–3) spreading, 1–2 pendent. Stem lvs 0.8–1.3 mm × 0.6–0.9 mm, lingulate, erect and appressed to stem;

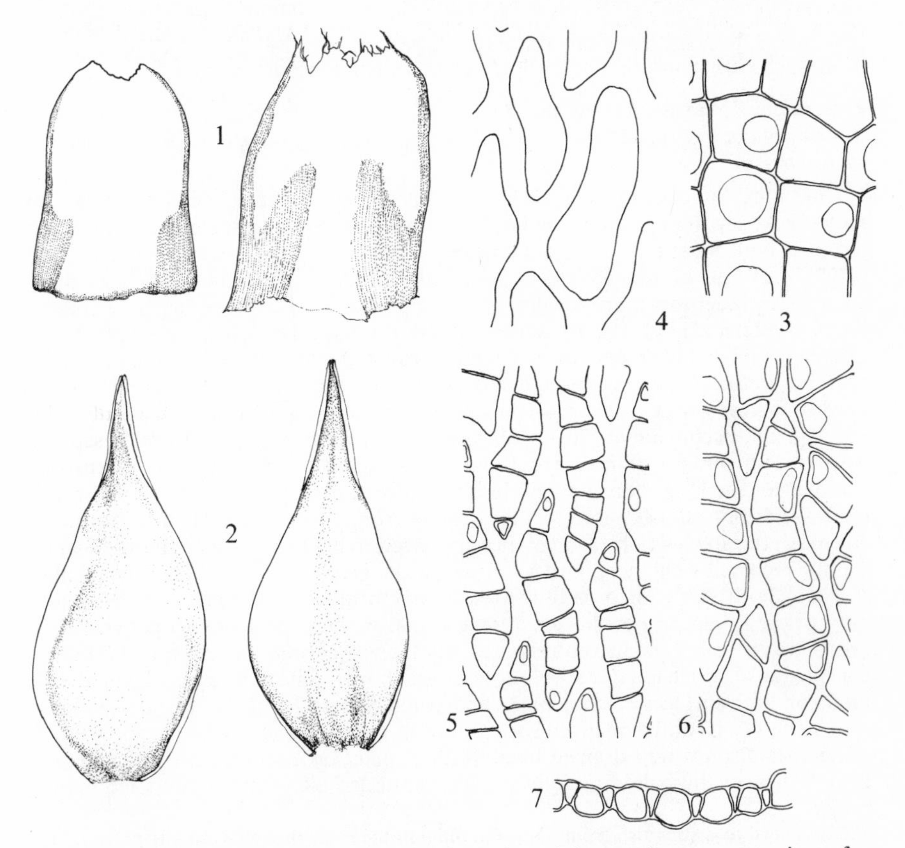

Fig. **8.** *Sphagnum girgensohnii*: 1, stem leaves; 2, branch leaves; 3, stem cortex in surface view; 4, stem leaf cells $\frac{2}{3}$ from base, ventral surface; 5 and 6, branch leaf cells from middle of leaf on ventral and dorsal surfaces; 7, section of stem leaf. Leaves × 27, cells × 280.

apex plane, truncate and tattered or sometimes entire and rounded; border 2–5 cells wide, intact to near apex; patches of narrow cells at basal angles usually with brown pigment, occupying 40–70 % of lf base; hyaline cells lacking fibrils and pores, equally and extensively resorbed on both surfaces; upper cells not, or hardly septate, some usually distorted by margin of lf having grown more than middle. Branch lvs 1.1–1.6 mm × 0.5–0.8 mm, ovate, not 5-ranked; margins strongly inrolled above so that the apex forms a snout which is slightly but distinctly bent backwards; border 1–2 cells wide; green cells in section narrowly trapezoid, reaching both surfaces, with the greater exposure on ventral face; dorsal pores 8–16 μm, 4–15 per cell; ventral pores 10–15 μm, 3–5 per cell, or sometimes absent in mid-lf though always present near apex. Antheridial lvs yellowish-brown. Perichaetial lvs oblong; border intact; hyaline cells intact on both surfaces, lacking fibrils and pores. Spores 22–25 μm. Fr rare, summer. $n = 19 + 2$. Green tussocks and 'lawns' on ditch sides, damp grassy banks and in mesotrophic marshes; rare in the south, frequent to common in Wales, N. England and Scotland. 85, H9. Circumpolar, mainly boreal.

Distinguished from *S. fimbriatum* in the field by the stem lvs widest at base and tattered only at apex, and by the often conspicuous brown pigmentation at their basal angles. The branches are stiffer and more rigid, and the snout-like apex of the branch lvs is usually more pronounced. *S. girgensohnii* can also resemble both *S. teres* and *S. russowii* (q.v. for differences), but the stem lvs are normally more torn than in either of these species.

9. S. russowii Warnst., Hedwigia, 1886

S. acutifolium var. *robustum* Russ., *S. girgensohnii* var. *robustum* (Russ.) Dix., *S. robustum* Röll

Dioecious. Shoots to 20 cm, variously crimson, variegated crimson and green, or green flecked with pink, rarely without any trace of red pigment, lacking an enlarged terminal bud. Stem 0.5–0.8 mm diameter; cortex 2–4 layers, pores present in 5–50 (–100) % of cells on outer surface, 1 per cell, round, transversely elliptical or semi-circular; cylinder green or red. Branches 3–4 per fascicle, 2 spreading, 1–2 pendent. Stem lvs erect and appressed to stem, 1.1–1.4 mm × 0.6–0.9 mm, lingulate; apex plane, usually rounded and ± entire, occasionally truncate and tattered because of resorption (but unlike *S. girgensohnii* not torn by unequal growth of different parts of lf); border 3–5 cells wide, intact; patches of narrow cells at basal angles usually with red pigment, occupying 40–70 % of lf base; hyaline cells in upper lf 0–50 % septate, usually with obscure fibrils; dorsal surface ± intact; ventral surface extensively resorbed and lacking; dorsal pores absent, or sometimes 1 per cell, 10–15 μm. Branch lvs 1.1–1.8 mm × 0.5–0.9 mm, ovate, varying from strongly 5-ranked to not at all 5-ranked, the apex variable, sometimes appressed to branch, at others snout-like and slightly recurved when dry; border 1–3 cells wide; green cells in section ± triangular, reaching both surfaces, broadly exposed on ventral face; dorsal pores 9–18 μm, 7–13 per cell; ventral pores often absent in mid-lf, when present to 4 per cell, 8–16 μm (there are usually some ventral pores near lf apex). Antheridial lvs bright crimson, conspicuous. Perichaetial lvs oblong, obtuse; border intact to apex; hyaline cells intact on both surfaces, lacking fibrils and pores. Spores 24–27 μm. Fr rare, summer. $n = 38 + 8$, 42. Reddish or greenish tussocks in marshes and on wet rocky ground, sometimes also on well drained banks with *S. quinquefarium*; frequent to common from Wales northwards, especially at higher altitudes. 52, H13. Circumpolar, mainly boreal.

Often hard to distinguish from *S. capillifolium* in the field, though normally more robust and with relatively broader branch lvs. The pores in the stem cortex and the more numerous pores in the branch lvs distinguish it microscopically. Green forms are occasionally hard to separate from *S. girgensohnii*. Usually they are distinct in the weakly fibrillose stem lvs.

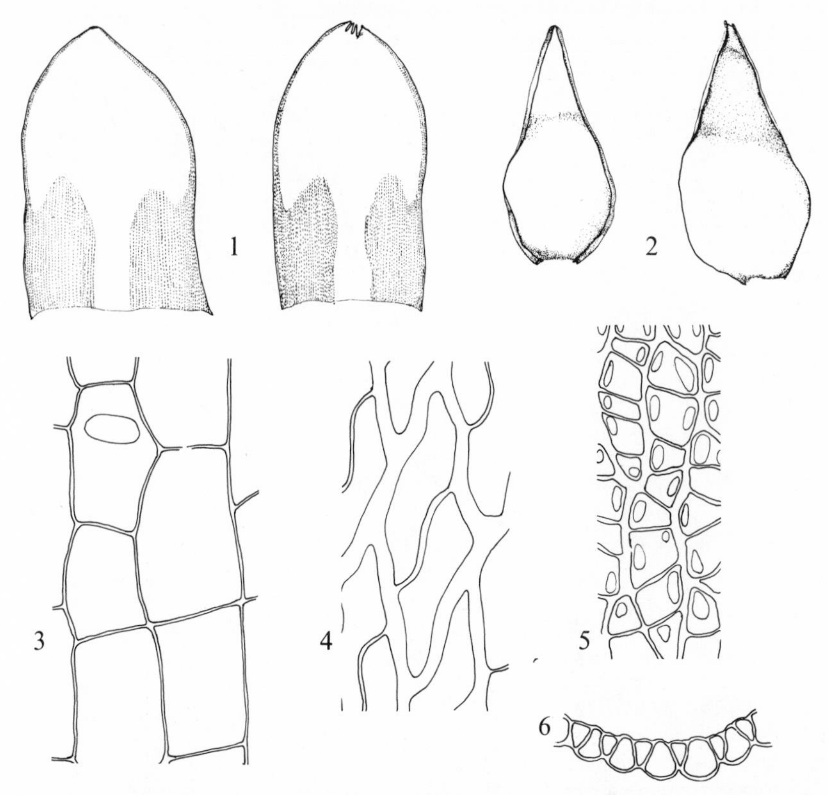

Fig. 9. *Sphagnum russowii*: 1, stem leaves; 2, branch leaves; 3, stem cortex in surface view; 4, stem leaf cells $\frac{2}{3}$ from base, ventral surface; 5, branch leaf cells in middle of leaf, dorsal surface; 6, section of branch leaf. Leaves ×27, cells ×280.

When stained, the different pattern of dorsal and ventral resorption provides a definite structural difference.

10. S. quinquefarium (Braithw.) Warnst., Hedwigia 1886
S. acutifolium var. *quinquefarium* Lindb. ex Braithw.

Autoecious. Shoots to 20 cm, green or pink. Stem 0.4–0.8 mm diameter; cortex 2–3 layers; pores present on outer surface, 1 per cell in 10–30 % of cells, semicircular or occasionally round; cylinder green, or in plants whose lvs are pink sometimes with a few red flecks. Branches (3–)4–5(–6) per fascicle, 2–3 spreading, 1–2(–3) pendent. Stem lvs erect and appressed to stem, 0.9–1.4 mm × 0.6–0.9 mm, triangular or triangular-lingulate; apex rounded or acute, plane or somewhat inrolled; border 3–7 cells wide, entire; patches of narrow cells at basal angles occupying 20–70 % of lf base; hyaline cells in upper lf 30–100 % septate, weakly fibrillose or with fibrils absent; dorsal surface intact; ventral surface extensively resorbed and lacking; dorsal pores normally absent, occasionally to 2 per cell, 15–25 μm. Branch lvs 0.8–1.6 mm × 0.4–0.6 mm, ovate or narrowly ovate, often strongly 5-ranked; green cells in section triangular, broadly exposed on ventral face; dorsal pores 11–22 μm, 4–7 per cell;

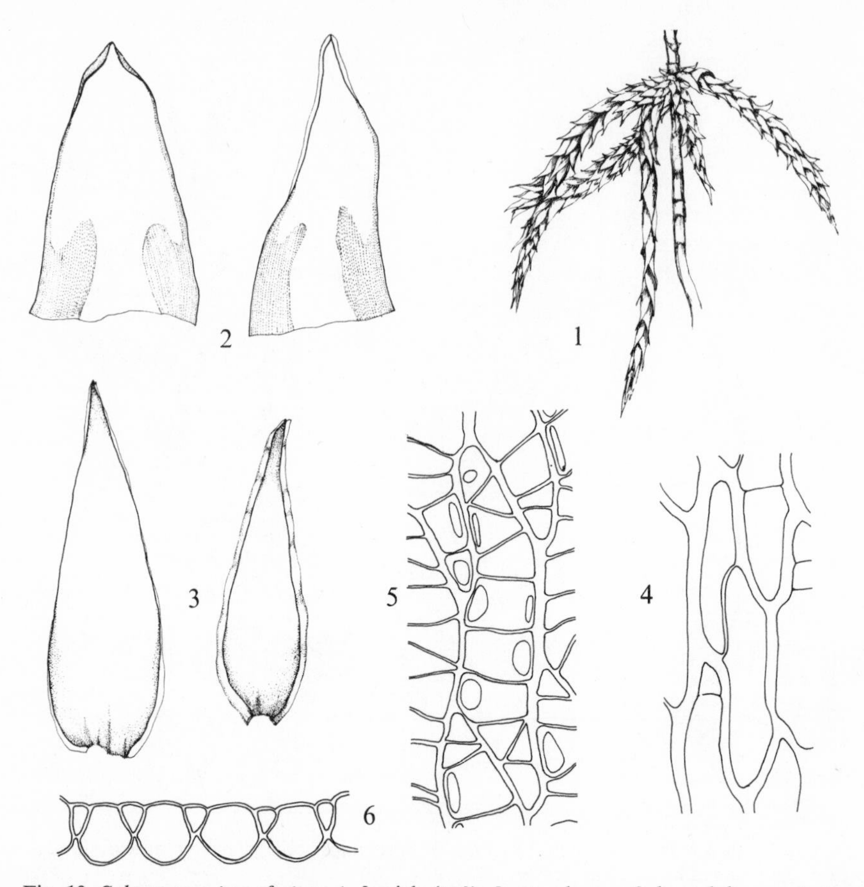

Fig. **10**. *Sphagnum quinquefarium*: 1, fascicle (×3); 2, stem leaves; 3, branch leaves; 4, stem leaf cells $\frac{2}{3}$ from base, ventral surface; 5, branch leaf cells at middle of leaf, dorsal surface; 6, section of branch leaf. Leaves ×27, cells ×280.

ventral pores absent. Antheridial lvs conspicuous, crimson or bright pink. Perichaetial lvs ovate; border intact; hyaline cells intact on both surfaces, lacking fibrils and pores. Spores 22–25 μm. Fr occasional, summer or autumn. $n=19+2$. Green or pinkish tussocks on well drained sheltered banks and in steeply sloping acid woodland, avoiding waterlogged ground; frequent to common in the north and west, absent from the Midlands and south-east. 58, H19. Circumpolar, boreal and north temperate.

Can be confused with *S. capillifolium* and *S. subnitens*. For differences see key. In the field, green forms can be confused with *S. recurvum*. The erect stem lvs are a good macroscopic character separating it from members of Section *Cuspidata*. All fruiting plants examined were autoecious, but non-fruiting material is often apparently male.

11. S. warnstorfii Russ., Sitz.-ber. Nat.-Ges., 1887
S. acutifolium var. *gracile* Russ., *S. warnstorfianum* Du Rietz

Dioecious. Shoots to 15 cm, variously crimson, rose-pink or green flecked with

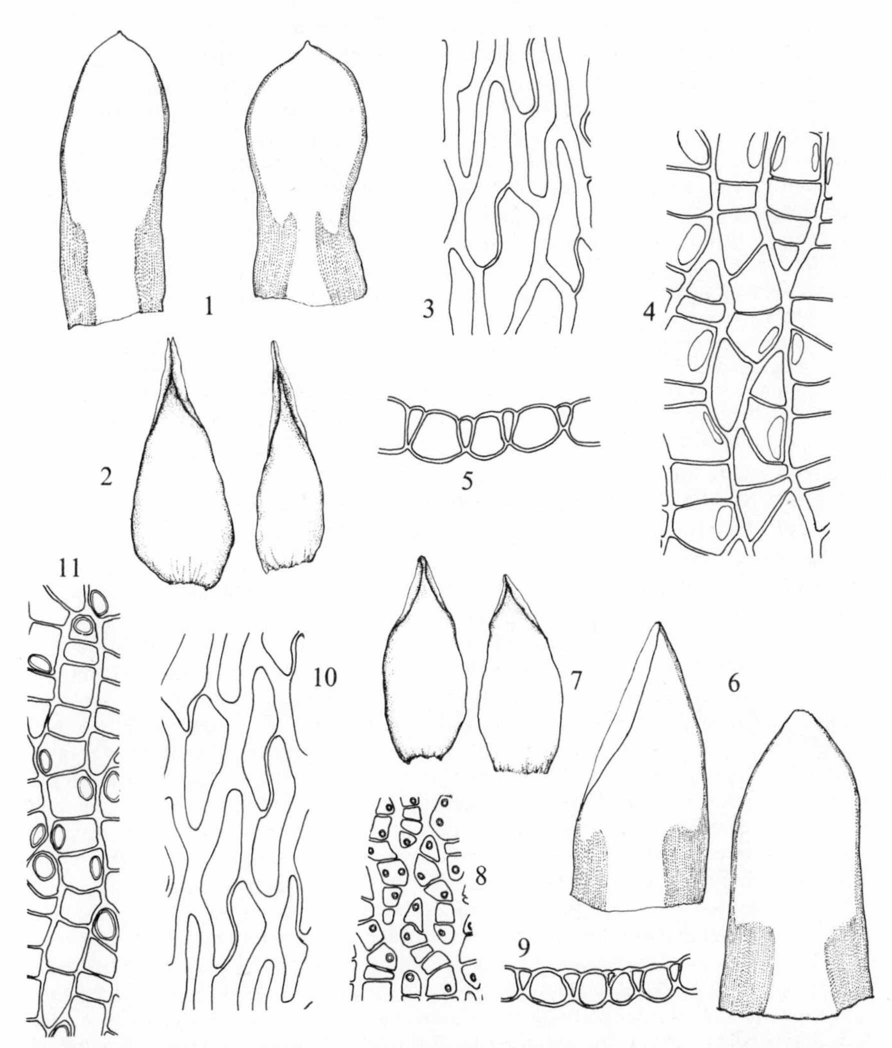

Fig. **11**. 1–5, *Sphagnum fuscum*: 1, stem leaves; 2, branch leaves; 3, stem leaf cells $\frac{2}{3}$ from base, ventral surface; 4, branch leaf cells in middle of leaf, dorsal surface; 5, section of branch leaf. 6–11, *S. warnstorfii*: 6, stem leaves; 7, branch leaves; 8, branch leaf cells near apex on dorsal surface; 9, section of branch leaf; 10, stem leaf cells $\frac{2}{3}$ from base, ventral surface; 11, branch leaf cells at middle of leaf, dorsal surface. Leaves ×27, cells ×280.

pink. Stem 0.4–0.5 mm diameter; cortex 2–3 layers, lacking pores on outer surface; cylinder red or green. Branches 3–4 per fascicle, 2 spreading, 1–2 pendent. Stem lvs erect and appressed to stem, 0.9–1.3 mm×0.5–0.8 mm, lingulate or triangular-lingulate; apex plane and rounded; border 3–7 cells wide, entire; patches of narrow cells at basal angles occupying 50–70 % of lf base; hyaline cells in upper lf 30–100 % septate, without fibrils or sometimes weakly fibrillose; dorsal surface intact; ventral surface extensively resorbed and lacking; pores absent. Branch lvs 0.6–1.4 mm×0.3–0.7 mm, ovate or narrowly ovate, usually strongly 5-ranked; border 1–4 cells wide; green cells in section triangular, reaching both surfaces, broadly exposed on ventral face; dorsal pores in mid-lf 8–16 μm, 3–6 per cell, near lf apex 2–6(–8) μm, (1–)2–6 per cell, thick-ringed, round, sometimes dimorphic with the pore at distal angle of cell then 6–8 μm when the lateral ones are 2–5 μm; ventral pores absent in mid-lf, above variably present or absent, 2–5 μm, to 4 per cell. Antheridial lvs conspicuous, crimson or bright pink. Perichaetial lvs as in *S. capillifolium*. Fr unknown in Britain or Ireland. Red or pink carpets and tussocks in base-rich flushes. N. Wales, N. England and Scotland, mainly above 400 m, but descending to sea level in the west; frequent in upland areas with sufficiently base-rich water; rare in Ireland. 31, H3. Circumpolar, arctic and boreal.

Closely resembling *S. capillifolium*, both anatomically and in the field. It often grows with *S. contortum* and *S. teres*, and may be easier to recognise from these associates than by its appearance. The branch lvs of *S. warnstorfii* are usually more markedly 5-ranked than those of *S. capillifolium*, but this is not constant, and microscopic examination is necessary until much field experience has been acquired.

12. S. capillifolium (Ehrh.) Hedw., Fund. Musc., 1782

S. acutifolium Ehrh. ex Schrad., *S. capillaceum* (Weiss) Schrank, *S. nemoreum* auct., *S. rubellum* Wils., *S. capillifolium* var. *tenellum* (Schimp.) Crum

Normally dioecious in Britain. Shoots to 15 cm, variously crimson, rose pink or green flecked with pink. Stem 0.4–0.6 mm diameter; cortex 3–4 layers, lacking pores on outer surface; cylinder red or green. Branches 3–4(–5) per fascicle, 2(–3) spreading, 1–2 pendent. Stem lvs erect and appressed to stem, or in compact forms sometimes spreading, 0.9–1.4 mm×0.5–0.9 mm, widest at base, lingulate or triangular-lingulate, rarely triangular; apex varying from plane and rounded to acute and somewhat inrolled; border 3–8 cells wide, entire; patches of narrow cells at basal angles occuping 20–70 % of lf base; hyaline cells in upper lf 30–100 % septate, a few often divided more than once, weakly or strongly fibrillose; dorsal surface intact; ventral surface extensively resorbed and lacking; pores absent or sometimes a few ill-defined dorsal pores present. Branch lvs 0.8–1.4 mm×0.4–0.6 mm, ovate or narrowly ovate, 5-ranked or not; green cells in section triangular, broadly exposed on ventral face; dorsal pores in mid-lf 8–25 μm, 4–7(–9) per cell, near lf apex 5–13(–16) μm, 3–8 per cell, not normally thick-ringed, mostly flattened against commissures except at upper cell angles; ventral pores absent or very few. Antheridial lvs bright crimson, conspicuous. Perichaetial lvs ovate; border intact; hyaline cells intact on both surfaces, lacking fibrils and pores. Spores 24–27 μm. Fr occasional, summer. $n=19+2$*. Often in compact tussocks but can form carpets under heather. Bogs, marshes and moors, avoiding localities which are either very wet or heavily shaded; abundant in the north and west; frequent on bogs in the south and east. 104, H38. Circumpolar, extending south to Mexico.

Dillenius's type specimen, presumably British, is autoecious, but the great majority of British populations are dioecious. *S. capillifolium* varies considerably in size and stem leaf morphology. Smaller forms are sometimes referred to *S. rubellum* Wils. (var. *tenellum* (Schimp.) Crum, 1975). Typically these plants have lingulate stem lvs, 0.9–1.2 mm, with a broadly rounded apex, weak fibrils and large patches of narrow cells at the basal angles. Larger plants, with longer, more strongly fibrillose stem lvs are referred to *S. capillifolium sensu stricto*. But in Britain intermediates are too common for the distinction to be workable. The pattern of variation is, however, useful in separating *S. subnitens*, as the forms of *S. capillifolium* which resemble it in their larger size and ± triangular stem lvs differ in the conspicuous presence of fibrils. For the differences from *S. quinquefarium*, *S. russowii* and *S. warnstorfii* see key, and notes under these species.

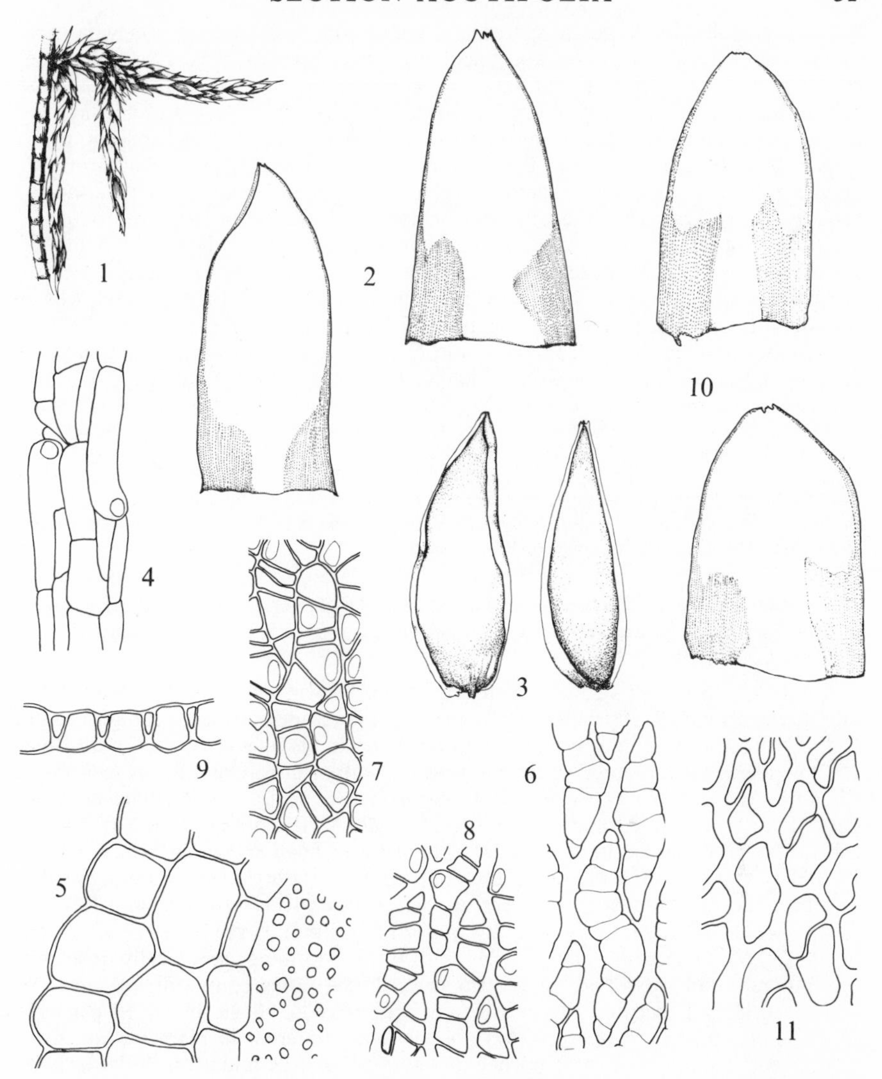

Fig. 12. *Sphagnum capillifolium*: 1, fascicle (×3); 2 and 10, stem leaves; 3, branch leaves; 4, branch cortex in surface view (×110); 5, section of stem; 6 and 11, stem leaf cells $\frac{2}{3}$ from base, ventral side; 7, branch leaf cells in middle of leaf, dorsal side; 8, branch leaf cells near apex, dorsal side; 9, section of branch leaf. Leaves ×27, cells ×280.

13. S. fuscum (Schimp.) Klinggr., Phys.-ök. Ges. Königsb., 1872
S. acutifolium var. *fuscum* Schimp.

Dioecious. Shoots to 15 cm, brown. Stem 0.3–0.5 mm diameter; cortex 3–4 layers, lacking pores on outer surface; cylinder dark brown. Branches 3–4 per fascicle, 2 spreading, 1–2 pendent. Stem lvs erect and appressed to stem, 0.8–1.3 mm×0.4–

0.7 mm, lingulate or lingulate-spathulate, either widest at base or somewhat narrowed in mid-lf and widest near apex; apex plane and rounded; border 5–9 cells wide, entire; patches of narrow cells at basal angles occupying 40–80 % of lf base; hyaline cells in upper lf (40–)80–100 % septate, sometimes divided more than once, lacking fibrils; dorsal surface intact; ventral surface extensively resorbed and lacking; pores absent. Branch lvs 0.9–1.3 mm×0.3–0.5 mm, ovate or narrowly ovate, not 5-ranked; border 1–2 cells wide; green cells in section triangular, reaching both surfaces, broadly exposed on ventral face; dorsal pores 12–30 μm, 3–7 per cell; ventral pores 10–20 μm, 0(–1) per cell. Antheridial lvs brown, somewhat darker than others. Perichaetial lvs ovate; border intact; hyaline cells intact on both surfaces, lacking fibrils and pores. Spores 24–27 μm. Fr rare, summer. $n=19+2$. Brown hummocks or occasionally carpets. In most of Britain and Ireland a rather rare plant of raised bogs; more frequent in N.E. England and E. Scotland, occurring in flushes and on blanket bogs, mainly above 400 m. 36, H18. Circumpolar, mainly boreal.

Easily recognised in the field, resembling *S. capillifolium* in habit, but brown. It is sometimes confused with brown forms of *S. subnitens*, but is at once distinct in the shape of the stem lvs. In most of the characters given above it comes close to *S. capillifolium*, but *S. fuscum* has markedly bigger pores near the base of the branch lvs (15–40 μm as opposed to 8–20 (–25) μm), and there is a greater contrast between the size of the cells in the lf base and those at the apex than in *S. capillifolium*. In *S. fuscum*, the dorsal pores near the lf apex are sometimes small, resembling those of *S. warnstorfii*.

14. S. subnitens Russ. & Warnst. in Warnst., Abh. Bot. Ver. Prov. Brandenb., 1888
S. acutifolium var. *luridum* Hüb., *S. acutifolium* var. *subnitens* (Russ. & Warnst.) Dix., *S. plumulosum* Röll

Autoecious. Shoots to 20 cm, variously green, brown, red or pink. Stem 0.4–0.7 mm diameter; cortex (2–)3–4 layers, lacking pores on outer surface; cylinder usually coppery-brown, sometimes red or green. Branches in fascicles of 3(–4), 2 spreading, 1(–2) pendent. Stem lvs erect and appressed to stem, or in compact forms sometimes spreading, 1.1–1.7 mm×0.6–1.0 mm, triangular; apex acute, often somewhat cuspidate because of inrolled margins; border 3–7 cells wide, entire; patches of narrow cells at basal angles occupying 20–60 % of lf base; hyaline cells in upper lf (50–) 80–100 % septate, often divided more than once, without fibrils or rarely weakly fibrillose; dorsal surface intact; ventral surface extensively resorbed and lacking; pores absent. Branch lvs (0.9–)1.2–2.7 mm×0.5–1.3 mm, ovate or narrowly ovate, not 5-ranked, or in small plants rarely with weak 5-ranking; green cells in section triangular, broadly exposed on ventral face; dorsal pores commonly ± invisible without folding lf because of protuberant hyaline cells, 16–30 μm, 4–10 per cell; ventral pores absent. Antheridial lvs not markedly different in pigmentation from others. Perichaetial lvs ovate; border intact; hyaline cells intact on both surfaces, lacking fibrils and pores. Spores 26–30 μm. Fr common, summer. Varying in habit from compact flesh-pink or brownish tussocks, through lax pink cushions to greenish 'lawns', on moors, heaths, rocky banks and in woods, avoiding flat acid bogs; abundant in the north and west, common in woods and on heaths in the Midlands and south. 108, H39, C. Circumpolar, extending south to the Azores, New Zealand.

The name *subnitens* refers to the pronounced lustre which the plant often has when dry. This can be a useful diagnostic character in the herbarium. Green forms of ditches and damp woods can resemble *S. auriculatum* or *S. recurvum*. *S. auriculatum* has rounded stem lvs, and in *S. recurvum* the stem lvs are hanging. For differences from *S. quinquefarium* see key. As well as the usually smaller size, the pigmentation is often useful in distinguishing *S. capillifolium*. Thus *S. capillifolium* generally has a pronounced concentration of red pigment in the central part of the capitulum. In *S. subnitens* the centre of the capitulum is seldom redder

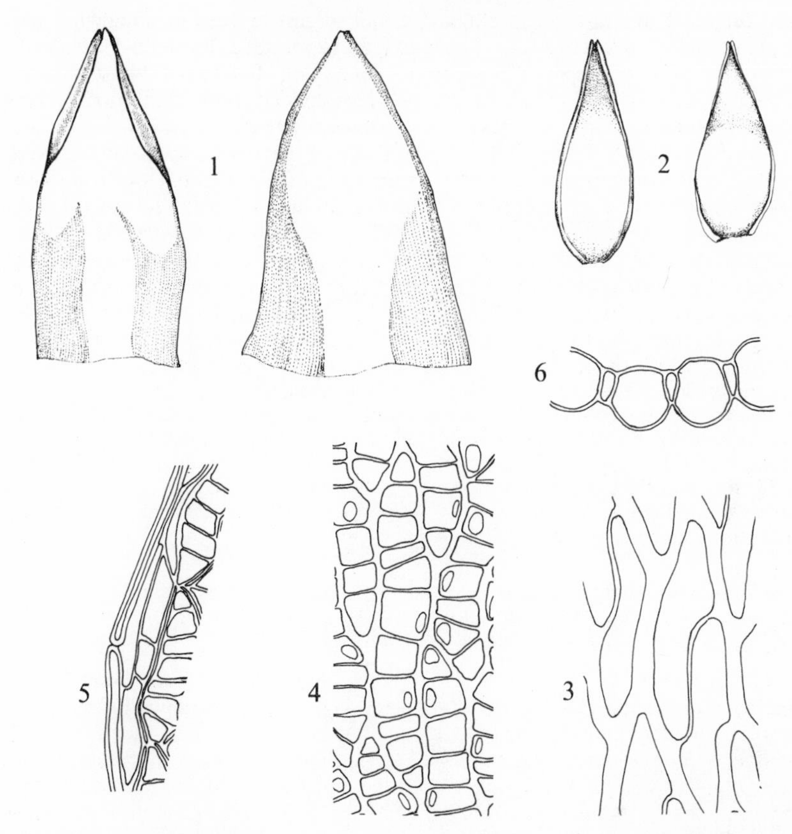

Fig. 13. *Sphagnum subnitens*: 1, stem leaves; 2, branch leaves; 3, stem leaf cells $\frac{2}{3}$ from base, ventral surface; 4, branch leaf cells in middle of leaf, dorsal surface; 5, margin of branch leaf near apex, dorsal surface; 6, section of branch leaf. Leaves ×27, cells ×280.

than the surrounding branches, and is often paler. Moreover, *S. subnitens* often contains some brown pigment, making the red rather dingy. Occasionally brown is the dominant colour but there are always undertones of red.

S. subfulvum Sjörs, a closely allied boreal and arctic species, lacks all trace of red pigment, and has been reported from Llanberis, N. Wales on the basis of a nineteenth-century herbarium specimen. On phytogeographical grounds the locality is improbable and the record must be regarded as doubtful. If *S. subfulvum* occurs in Britain at all, it would be expected in the eastern Highlands of Scotland.

15. S. molle Sull., Musc., Allegh., 1845

Autoecious. Shoots to 10 cm, whitish or pink. Stem 0.2–0.5 mm diameter; cortex (1–)2–3 layers, with pores in 10–50 % of cells on outer surface; cylinder green or pale brown. Branches 2–3(–4) per fascicle, mostly squeezed together in the tufts and pointing upwards, sometimes 1(–2) pendent; retort cells in groups of (1–)2–5 (mostly solitary in other Section *Acutifolia*). Stem lvs spreading or erect, 1.5–2.8 mm × 0.4–1.5 mm, rhomboid, oblong or lingulate; apex acute, plane or with ± inrolled margins; border (1–)2–3 cells wide, denticulate above because of resorption of cell

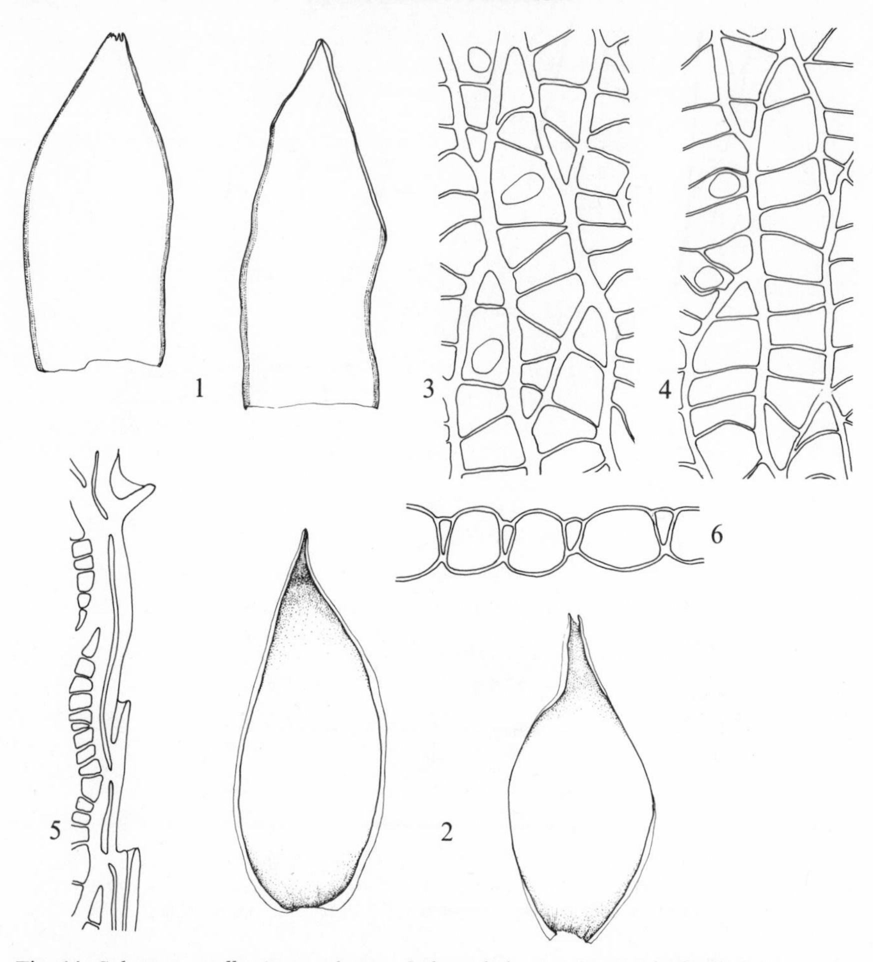

Fig. **14**. *Sphagnum molle*: 1, stem leaves; 2, branch leaves; 3, stem leaf cells $\frac{2}{3}$ from base, ventral surface; 4, branch leaf cells at middle of leaf, dorsal surface; 5, margin of branch leaf near apex, dorsal surface; 6, section of branch leaf. Leaves ×27, cells ×280.

walls along margin; patches of narrow cells at basal angles absent; hyaline cells fibrillose in a zone 40–100 % of lf, 0–40 % septate in mid-lf; dorsal pores ringed, 13–22 μm, 0–8 per cell; ventral pores large, 15–30 μm, unringed, 0–3 per cell, or sometimes replaced by larger membrane gaps. Branch lvs 1.4–2.3 mm × 0.5–1.2 mm, ovate, not 5-ranked; border 1–2 cells wide, denticulate at least in upper $\frac{1}{2}$ because of resorption of outer surface of cells along margin (border is intact in other British Section *Acutifolia*); green cells in section triangular, broadly exposed on ventral surface, not, or hardly, exposed on dorsal surface; hyaline cells strongly protuberant on dorsal surface; dorsal pores ± invisible without folding lf because of highly protuberant hyaline cells, 14–22 μm, 6–12 per cell; ventral pores absent. Antheridial lvs greenish, not different in pigmentation from others. Perichaetial lvs ovate, acuminate; border 3–4 cells wide; hyaline cells lacking fibrils and pores, or a few cells with fibrils and dorsal pores. Spores 28–32 μm. Fr common, summer. Compact, whitish or slightly pink tussocks on damp ground by streams and on heaths. Throughout the British Isles but rare in most districts except western Scotland. 59, H17. W. and C. Europe; eastern N. America.

In habit resembling *S. compactum*, but immediately distinct in the large stem lvs, pale stem and traces of pink colouring. Lax forms may resemble *S. subnitens*, but the large stem lvs and lack of well defined pendent branches will usually separate *S. molle*. The acuminate perichaetial lvs are also a distinctive feature. In *S. molle* the tufts are usually interwoven with ± unbranched julaceous stems, whose cortex has well marked retort cells. This is one aspect of a general lack of differentiation between stems and branches which is manifested also in the lvs.

Section *Rigida* (Lindb.) Schlieph.

Plants medium-sized. Stem typically 0.4–0.8 mm diameter; cortex 2–3 layers, hyaline, 30–100 μm wide, cell walls without spiral fibrils, those on outer surface each with one ± semicircular pore at upper end; diameter of cortical cells measured at right-angles to radius, 20–70 μm; cylinder strongly differentiated from cortex. Branches in fascicles of 2–7, of which 1–3 spreading; cells of branch cortex each with an apical pore, not dimorphic, lacking spiral fibrils. Stem lvs hanging, typically 0.3–0.7 mm × 0.4–0.7 mm, triangular or trapezoid, with a border of 3–8 narrow cells; border intact below, above varying from ± entire to extensively resorbed and tattered; patches of narrow cells at basal angles absent; hyaline cells not, or a few, septate; dorsal surface with or without fibrils; ventral surface almost completely resorbed, and either lacking or very thin with large pores. Branch lvs varying from oblong and ± cucullate to ovate-acuminate with a squarrose acumen, typically 1.7–3.0 mm × 0.8–1.8 mm; apex rounded or broadly truncate; border 1(–3) cells wide, usually completely resorbed and represented only by a resorption furrow (margin appears denticulate owing to projection of transverse cell walls which persist after resorption of border); green cells lens-shaped or elliptical in section, either enclosed by hyaline cells or slightly exposed on dorsal surface; hyaline cells near lf apex longer than broad (a few at extreme apex ± as broad as long), their dorsal surface sometimes with a pore at distal angle but not lacunose; dorsal pores in mid-lf 5–25 μm, 0–4(–8) per cell, positioned at cell angles or more generally along commissures (a few large pores are sometimes centrally placed, but these are never the majority); ventral pores absent in mid-lf, near apex present or absent, either in triplets at corners of 3 adjacent hyaline cells, or distributed more generally along commissures. (In mid-lf the bounding fibrils of ventral pores may be present but without any marked membrane thinning within them.) Antheridia on pendent branches. Perichaetial lvs resembling branch lvs, ovate, acute; border intact, 2–6 cells wide, distinct to apex; hyaline cells not extensively resorbed on either surface, fibrillose and porose in at least the upper half of lf, often throughout; pores various, commonly present on both surfaces, positioned at cell angles or distributed more generally along commissures.

Many authors place the Section *Rigida* close to Section *Sphagnum* because of the resorbed border of the branch lvs and the lack of dimorphism in the cells of the branch cortex. But there the similarity ends. Other structural features are closer to Section *Subsecunda*. In particular the pendent branch lvs have abundant ringed commisural pores similar to those found in Section *Subsecunda*.

Key to *Sphagnum* Section *Rigida*

1 Stem greenish or pale brown; branch lvs squarrose, the green cells just reaching dorsal surface; dorsal pores in upper $\frac{1}{2}$ of lf 1(–6) per cell, 10–25 μm, mostly positioned at distal angles of hyaline cells **16. S. strictum**

Stem dark brown or black, rarely pale; branch lvs usually ± appressed, occasionally squarrose, green cells completely enclosed on both surfaces; dorsal pores in upper $\frac{1}{2}$ of lf few to numerous, (1–)3–10(–12) per cell, 5–12(–16) μm, distributed along commissures and not concentrated on distal angles of hyaline cells **17. S. compactum**

16. S. strictum Sull., Musc. Allegh., 1845

Autoecious. Shoots to 15 cm, green or brownish. Stem 0.5–0.8 mm diameter; cortex 2–3 layers; pores on outer surface 1 per cell; cylinder green or pale brown. Branches 3–6 per fascicle, of which 2–3 spreading, 1–3 pendent. Stems lvs hanging, 0.5–0.7 mm × 0.5–0.6 mm, triangular, the apex rounded and ± cucullate; border 4–8 cells wide, near apex denticulate but not tattered; patches of narrow cells at basal angles absent; hyaline cells not, or a few, septate, lacking fibrils; dorsal surface intact; ventral surface ± completely resorbed and lacking. Branch lvs 1.7–2.7 mm × 0.8–1.5 mm, ovate, with a squarrose acumen whose extreme apex is abruptly truncate; border 1 cell wide, denticulate because of resorption; green cells just reaching dorsal surface of lf, enclosed on ventral surface, in section lens-shaped, measuring (15–)17–27 μm along long axis; interior walls of hyaline cells where they abut on green cells obscurely papillose; dorsal pores in mid-lf often absent, when present 10–25 μm, mostly 1 per cell at distal angle, sometimes more numerous,

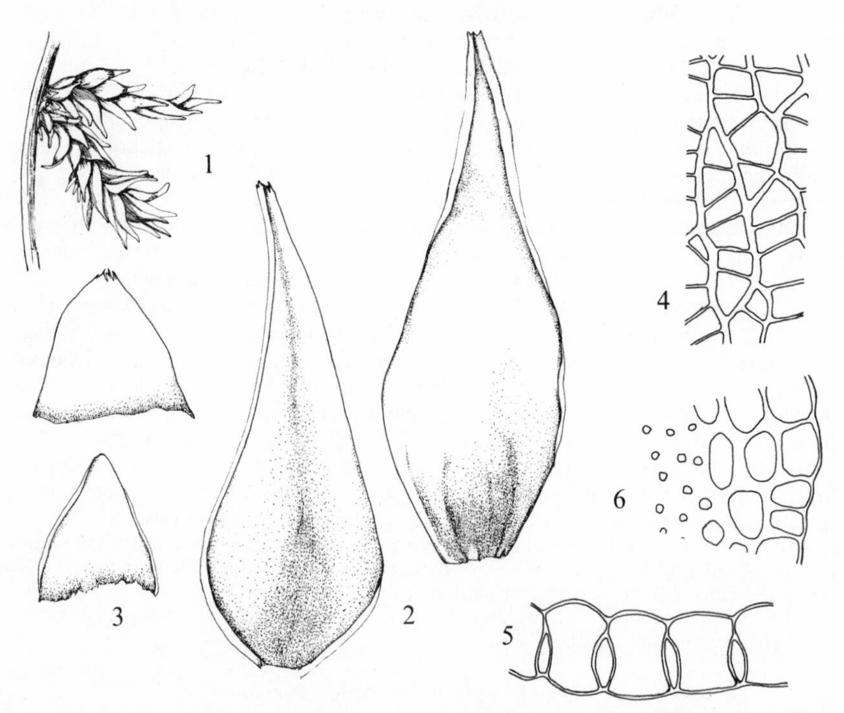

Fig. **15**. *Sphagnum strictum*: 1, fascicle (×3); 2, branch leaves; 3, stem leaves; 4, cells at middle of branch leaf on dorsal surface; 5, section of branch leaf; 6, section of stem. Leaves ×27, cells ×280.

especially near lf apex; ventral pores normally absent in mid-lf, near apex occurring in triplets, 1 at the corner of each of 3 adjacent hyaline cells. Antheridial lvs whitish, not contrasting with those on sterile pendent branches. Perichaetial lvs ovate, acute, fibrillose; border 3–5 cells wide; dorsal pores large, mostly 1 per cell at distal angle, or sometimes absent; ventral pores also large, near lf apex forming well defined triplets 1 at each corner of 3 adjacent hyaline cells. Spores 28–30 μm. Fr frequent, summer. Tussocks on blanket bog and wet moorland, especially among *Molinia*. N. Wales, Lake District, N. and W. Scotland, N. and W. Ireland; rare except in the hyper-Oceanic districts of Scotland and Ireland. 20, H9. Atlantic and sub-Atlantic regions of Europe and N. America; C. and S. America.

With its squarrose branch lvs it has some resemblance to *S. squarrosum*, but differs in its small stem leaves. In the field, it can be distinguished from squarrose-leaved forms of *S. compactum* by its pale stem and usually laxer habit.

17. S. compactum DC. in Lam. & DC., Flore Française, 1805
S. rigidum (Nees & Hornsch.) Schimp.

Autoecious. Shoots to 10 cm, variously whitish-green, yellowish or bright ochre. Stem 0.4–0.7 mm diameter; cortex 2–3 layers; pores on outer surface 1 per cell; cylinder dark brown or black, rarely pale. Branches 2–7 per fascicle, of which 1–2 spreading, 1–5 pendent. Stem lvs hanging, 0.3–0.7 mm × 0.4–0.7 mm, triangular or trapezoid, the apex rounded, truncate or retuse; border 3–8 cells wide, intact in

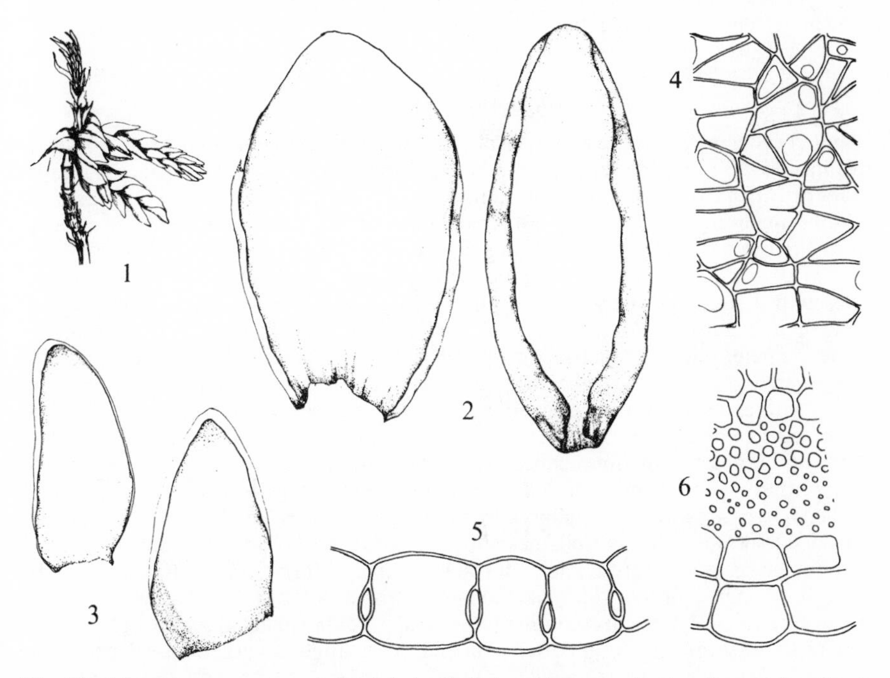

Fig. 16. *Sphagnum compactum*: 1, fascicle (× 3); 2, branch leaves; 3, stem leaves; 4, cells at middle of branch leaf on dorsal surface; 5, section of branch leaf; 6, section of stem. Leaves × 27, cells × 280.

lower part of lf, near apex variable, usually strongly resorbed and tattered; patches of narrow cells at basal angles absent; hyaline cells not, or slightly, septate, with or without fibrils; dorsal surface intact; ventral surface ± completely resorbed and lacking. Branch lvs 1.8–3.0 mm × 0.9–1.8 mm, varying from oblong and ± cucullate to ovate with a squarrose acumen, extreme apex rounded or truncate; border 1(–3) cells wide, denticulate because of resorption of outer surface of cells along margin; green cells completely enclosed by hyaline cells, in section elliptical, measuring 11–16 μm along long axis; interior walls of hyaline cells smooth; dorsal pores in mid-lf very variable, perforate or imperforate, 5–12(–16) μm, 0–4(–8) per cell, distributed along commissures, near lf apex more numerous, to 12 per cell; ventral pores absent in mid-lf. (Numerous bounding fibrils are often present along commissures on both surfaces, but the membrane within them is mostly not thinned.) Antheridial lvs white, not contrasting with those on sterile pendent branches. Perichaetial lvs ovate, acute; border 2–6 cells wide; hyaline cells fibrillose, with abundant, mostly imperforate, small, ringed commissural pores on both surfaces, or those on ventral surface sometimes larger and less abundant. Spores 30–35 μm. Fr frequent, summer. $n = 19 + 2$. Compact tussocks on wet heaths, blanket bogs and among rocks, favouring moist but not wet ground, especially where there has been some disturbance such as burning. Frequent to common throughout the British Isles. 93, H34. Circumpolar, extending south to Madeira and the Gulf of Mexico.

Usually easy to recognise in the field, the shoots being so densely crowded that the upper branches point upwards and the individual capitula are hard to discern. The habit resembles *Leucobryum glaucum*. *S. compactum* can resemble compact forms of *S. molle*, *S. palustre*, *S. papillosum* or *S. squarrosum*, but differs in its minute hanging stem leaves. For field differences from *S. strictum* see under that species.

Section *Subsecunda* (Lindb.) Schlieph.

Plants slender to robust. Stem typically 0.4–0.8 mm diameter; cortex 1–3 layers, hyaline, 15–60 μm wide, cell walls without fibrils, those on outer surface without pores or with a single indistinct pore at upper end; diameter of cortical cells measured at right-angles to radius, 15–50(–60) μm; cylinder strongly differentiated from cortex. Branches often curved, in fascicles of 1–6(–7), of which 1–3 spreading, 0–4 pendent; branch cortex with clearly dimorphic cells, cell walls without spiral fibrils; retort cells with a moderate neck, in groups of (1–)2(–4). Stem lvs erect, spreading or hanging, typically 0.4–2.0 mm × 0.4–1.1 mm; border 3–6 cells wide, near apex entire or denticulate; patches of narrow cells at basal angles absent; hyaline cells septate or not, strongly or weakly fibrillose, not extensively resorbed on either surface, or rarely with extensive ventral resorption; dorsal pores ringed, 2–8 μm, absent or to 40 per cell; ventral pores ringed or not, 3–15 μm, absent or to 20 per cell. Branch lvs ovate or elliptical, sometimes concave and cucullate, typically 0.7–2.5 mm × 0.4–1.5 mm, with an intact border of 2–4 narrow cells; green cells lens-shaped or barrel-shaped in cross-section, ± equally exposed on both surfaces, or slightly more exposed on dorsal surface; hyaline cells near apex longer than broad, their dorsal surface without a large pore at distal angle; dorsal pores in mid-lf (1–)2–6 μm, (0–)12–40(–50) per cell, strongly ringed, positioned along commissures, typically with 2 corresponding to each fibril, 1 on each commissure; ventral pores normally absent, when present similar to dorsal pores though often confined to cell angles. Antheridia on spreading branches. Perichaetial lvs triangular or oblong; border 2–3 cells wide, distinct almost to apex; hyaline cells fibrillose near lf apex, not extensively resorbed on either surface; dorsal pores numerous, resembling those of branch lvs; ventral pores few, similar.

Key to *Sphagnum* Section *Subsecunda*

1 Stem lvs triangular or lingulate, 0.4–1.3(–1.5) mm, fibrillose for 40 % or less of length of lf, ventral pores equally or more developed than dorsal pores (dorsal pores may be absent); pendent branches strongly or weakly differentiated from spreading branches; colour green or yellowish, not coppery-red 2
Stem lvs lingulate, oblong or elliptical, (1.0–)1.3–2.3 mm, fibrillose in upper 40 % or more, dorsal pores numerous, ventral pores usually much fewer, occasionally equally numerous; pendent branches weakly or not at all differentiated from spreading branches; colour green, yellowish or coppery-red 4

2 Stem cortex consisting of 2–3 layers of hyaline cells; cylinder green or pale brown; ventral pores of stem lvs 3–6 μm, ringed, lying along commissures **21. S. contortum**
Stem cortex 1 layer; cylinder green, pale brown, or in well illuminated plants usually dark drown; ventral pores of stem lvs 5–15 μm, some usually unringed and centrally placed 3

3 Slender plant (size similar to *S. capillifolium*); stem lvs 0.4–0.9 mm × 0.4–0.6 mm; upper cells fibrillose in a zone 0–25 % of length of lf; branch lvs 0.7–1.5 mm × 0.4–0.8 mm **20. S. subsecundum**
Medium-sized; stem lvs 0.9–1.3(–1.5) mm × 0.6–0.9 mm; upper cells fibrillose in a zone 20–40 % of lf length; branch lvs (1.0–)1.3–2.2 mm × 0.6–1.5 mm **19. S. auriculatum** var. **inundatum**

4 Stem cortex uniformly 1 cell thick; stem lvs lingulate or oblong, often ± plane, 40–100 % fibrillose; branches tumid or not, arising in fascicles of 3–5 **19. S. auriculatum** var. **auriculatum**
Stem cortex irregularly 1–2(–3) cells thick; stem lvs elliptical, concave, 80–100 % fibrillose; branches tumid, in fascicles of 1–3(–4) **18. S. platyphyllum**

18. S. platyphyllum (Braithw.) Warnst., Flora, 1884
S. contortum var. *platyphyllum* (Lindb. ex Braithw.) Åberg, *S. laricinum* var. *platyphyllum* Lindb. ex Braithw.

Dioecious. Shoots to 15 cm, green or yellowish. Stem 0.3–0.7 mm diameter; cortex irregularly 1–2(–3) layers, cells of outer surface mostly with a pore at upper end; cylinder green or brown. Branches 1–3(–4) per fascicle, to 13 mm, not differentiated into pendent and spreading. Stem lvs erect or spreading, 1.2–2.3 mm × 0.8–2.0 mm, concave, elliptical, resembling branch lvs; border 2–4(–6) cells wide; patches of narrow cells at basal angles absent; hyaline cells not septate, fibrillose in a zone 80–100 % of lf; dorsal pores 2–5 μm, to 25 per cell, or else ill-defined and ± obsolete; ventral pores absent, or else 2–4 μm, 0–6 per cell, ringed and confined to cell angles. Branch lvs 0.9–2.1 mm × 0.7–2.0 mm, concave, elliptical; border 1–3 cells wide; green cells in section barrel-shaped or lens-shaped, exposed on both surfaces, or just enclosed on ventral surface; dorsal pores in mid-lf 2–5 μm, to 40 per cell, or else ill-defined and ± obsolete; ventral pores absent. Perichaetial lvs broadly elliptical-oblong, blunt, fibrillose and porose above. Spores 32–35 μm. Fr unknown in Britain. Untidy green or yellowish carpets in ± base-rich, seasonally wet, peaty runnels and swamps, often with *Drepanocladus revolvens*. N. Wales to W. Scotland, Fermanagh, rare but probably overlooked. 9, H1. Circumpolar, boreal and north temperate, with a somewhat continental distribution on the mainland of Europe.

Resembling semi-submerged forms of *S. auriculatum*, but with relatively broader and more concave stem lvs, and with branches often in fascicles of 1–2. The stem cylinder is usually pale, but may be blackened by swampy habitat. More than 1 section may be necessary to

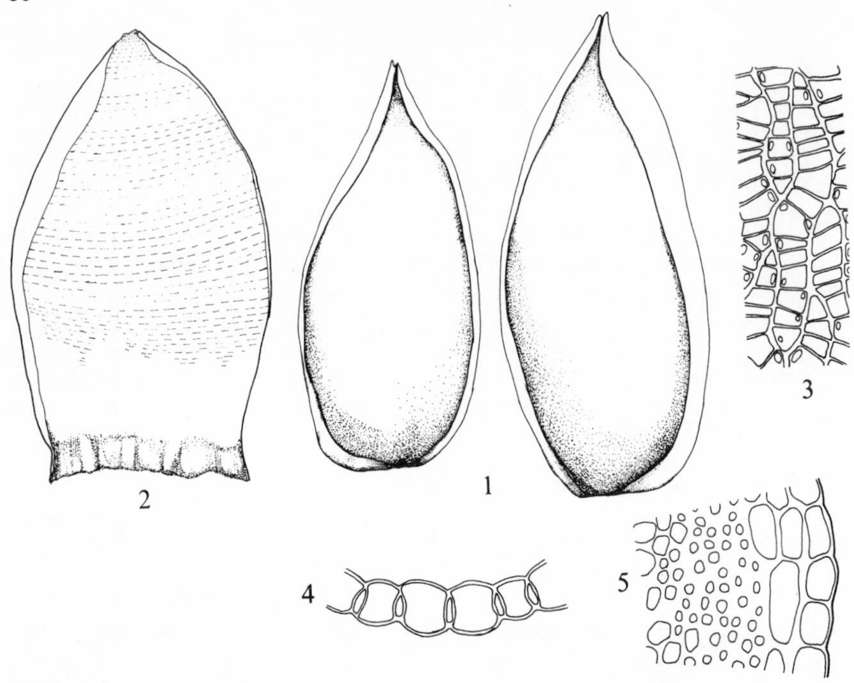

Fig. 17. *Sphagnum platyphyllum*: 1, branch leaves; 2, stem leaf; 3, cells from middle of branch leaf on dorsal surface; 4, section of branch leaf; 5, section of stem. Leaves ×27, cells ×280.

demonstrate the 2-layered stem cortex. Often on a single stem there are parts which have a 2-layered cortex and others which have a 1-layered cortex.

19. S. auriculatum Schimp., Mém. Prés. Div. Sav. Acad. Sci. Inst. Imp. Fr., 1857
S. contortum auct. non Schultz, *S. subsecundum* auct. Brit., p.p., *S. lescurii* Sull.

Dioecious. Shoots to 20 cm, green, yellow or coppery red. Stem 0.4–0.8 mm diameter; cortex 1 layer, cells of outer surface without pores, or occasionally with a few indistinct pores at upper end; cylinder dark brown or black, or in shade and submergence forms green. Branches 3–6(–7) per fascicle, 2–3 spreading, 0–4 pendent; spreading branches often markedly curved, pendent branches variously strongly, weakly or not at all differentiated from spreading. Stem lvs erect, spreading or hanging, 0.9–2.0(–2.7) mm×0.6–1.0(–1.5) mm, triangular-lingulate, lingulate or oblong; border 3–5 cells wide, ±.fringed or toothed near apex; patches of narrow cells near basal angles absent; hyaline cells 0–100 % septate, fibrillose in a zone 20–100 % of lf; dorsal pores 3–8(–11) μm, 0–30 per cell; ventral pores 3–12(–15) μm, 0–25 per cell. Branch lvs (1.0–)1.3–2.5(–3.5) mm×0.6–1.5(–2.0) mm, ovate or elliptical, imbricate, patent or secund; border 2–3 cells wide; green cells in section barrel-shaped or lens-shaped, exposed on both surfaces or slightly enclosed on ventral surface; dorsal pores 2–6 μm, (0–)20–40(–50) per cell, ringed, lying along commissures; ventral pores absent or rarely 2–6 μm, to 6 per cell. Antheridial lvs yellow (but not observed in coppery forms). Perichaetial lvs ovate or triangular, acute, fibrillose and porose above. Spores 28–33 μm. Fr occasional, summer.

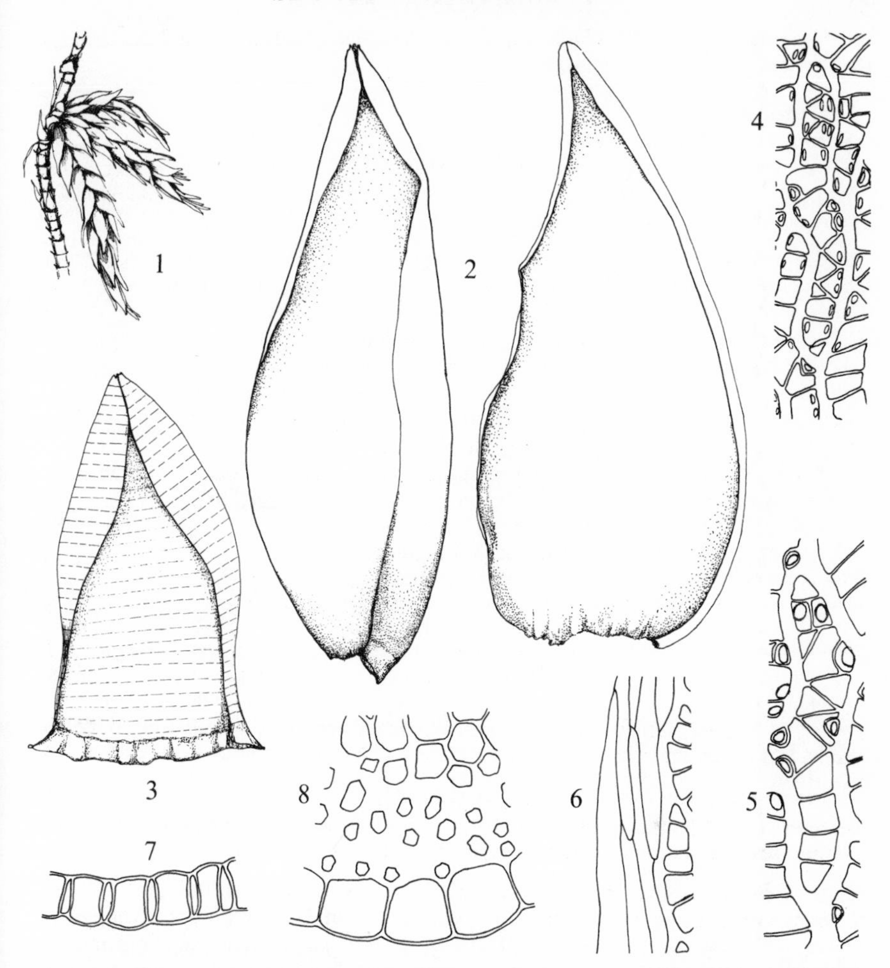

Fig. **18**. *Sphagnum auriculatum* var. *auriculatum*: 1, fascicle (×3); 2, branch leaves; 3, stem leaves (hatching indicates distribution of fibrils); 4, cells from middle of branch leaf, dorsal side; 5, stem leaf 80 % from base, ventral surface; 6, cells from stem leaf near apex, dorsal surface; 7, section of branch leaf; 8, section of stem. Leaves ×27, cells ×280.

Exceptionally variable. There is a well marked morphological cline from relatively slender yellowish plants with well differentiated pendent branches and small stem lvs, to gross coppery-red plants with no pendant branches and large stem lvs resembling branch lvs. But the characters are imperfectly correlated, and the line of demarcation between the varieties is perforce arbitrary. Examination of a range of material leaves an impression that in spite of gross differences in habit, genetic differences may be small. In particular, the degree of differentiation between stem lvs and branch lvs (which is here used as the main criterion to distinguish the varieties) is probably controlled by only a few genes. Differentiation of stem lvs appears to be linked to differentiation of pendent branches. In *S. molle* and *S. platyphyllum* there is an analogous tendency to isophylly, and the pendent branches are also poorly developed. In *S. pylaesii* Brid., an aberrant member of the *Subsecunda* known from Brittany, there is a general loss of differentiation, the branches being rudimentary and the stem lvs resembling branch lvs of a normal *Sphagnum*.

Var. **auriculatum**

S. contortum var. *obesum* Wils., *S. contortum* var. *rufescens* Nees & Hornsch., *S. crassicladum* Warnst., *S. obesum* (Wils.) Warnst., *S. rufescens* (Nees & Hornsch.) Warnst., *S. subsecundum* var. *auriculatum* (Schimp.) Lindb., *S. subsecundum* var. *rufescens* (Nees & Hornsch.) Åberg.

Medium-sized or robust; colour green, yellowish or coppery-red. Branches 3–5 per fascicle, either not differentiated, or weakly differentiated into 2–3 spreading, 1–2(–3) pendent. (When there are 5 branches in a fascicle, 1 or 2 of the 5 are often small and poorly developed.) Stem lvs erect, spreading or hanging, (1.0–)1.3–2.0(–2.7) mm × 0.6–1.0(–1.5) mm, lingulate or oblong, apex rounded; hyaline cells 0–100 % septate, fibrillose in a zone 40–100 % of length of lf; dorsal pores near apex numerous, 3–8 μm, to 30 per cell; ventral pores variously absent or to *ca* 20 per cell, 3–8(–12) μm, when present mostly ringed and lying along commissures. Branch lvs 1.3–2.5(–3.5) mm × 0.8–1.5(–2.0) mm, imbricated or occasionally patent. Green, yellowish or coppery 'lawns' and tussocks, in woods, on moors and submerged in pools. Throughout the British Isles; very common in the north and west; frequent to common in the south and east. 106, H38. N.W. Europe and eastern N. America.

The coppery forms of flushed base-poor upland banks normally have strongly curled branches, and can be recognised at a glance. Turgid forms submerged in pools have a superficial resemblance to *S. palustre*, but differ in the narrow stem without a wide layer of cortex. Green or yellowish forms of woods and marshes intergrade with var. *inundatum* and may require microscopic examination. The best single character is the disposition of the pores in the stem lvs – mainly dorsal in var. *auriculatum*, mainly ventral in var. *inundatum*. Occasional plants have numerous pores on both surfaces, and must be determined by the balance of the other characters.

Var. **inundatum** (Russ.) M. O. Hill, J. Bryol., 1975

S. bavaricum Warnst., *S. inundatum* Russ., *S. subsecundum* var. *bavaricum* (Warnst.) Åberg, *S. subsecundum* var. *inundatum* (Russ.) C. Jens.

Medium-sized; colour green or yellow. Branches 4–6(–7) per fascicle, usually differentiated into 2–3 spreading, 2–4 pendent. Stem lvs spreading or hanging, 0.9–1.3(–1.5) mm × 0.6–0.9 mm lingulate or triangular-lingulate, apex rounded; hyaline cells 0–50 % septate, fibrillose in a zone 20–40 % of length of lf; dorsal pores near apex variously absent to numerous, 3–8(–11) μm; ventral pores well developed, 5–12(–15) μm, centrally placed or positioned along commissures, usually at least a few centrally placed, unringed and relatively large (8–12 μm). Branch lvs (1.0–)1.3–2.2 mm × 0.6–1.5 mm, patent or imbricated. $n=38+4^*$, *ca* 42. Green or yellowish 'lawns' in marshes, heathy depressions, fens and flushes. Frequent to common throughout the British Isles. 98, H30. N.W. Europe and eastern N. America.

The name is misleading; submerged *S. auriculatum* is almost always var. *auriculatum*. When well illuminated, the dark stem is a useful field character separating it from unrelated species. Well marked var. *inundatum* can be separated from var. *auriculatum* by its more slender habit and well developed pendent branches. But because of intergradation it must often be determined by the microscopic characters of stem lf anatomy. For differences from *S. subsecundum* see under that species.

The relationships of var. *inundatum* are controversial. Hill (*J. Bryol.* **8**, 435–41, 1975) presents evidence for its relationship with *S. auriculatum*. Rahman (*J. Bryol.* **7**, 169–79, 1972) and Eddy (*J. Bryol.* **9**, 309–19, 1977) suggest it is related to *S. subsecundum*. Ecological and phytogeographical data support the former view.

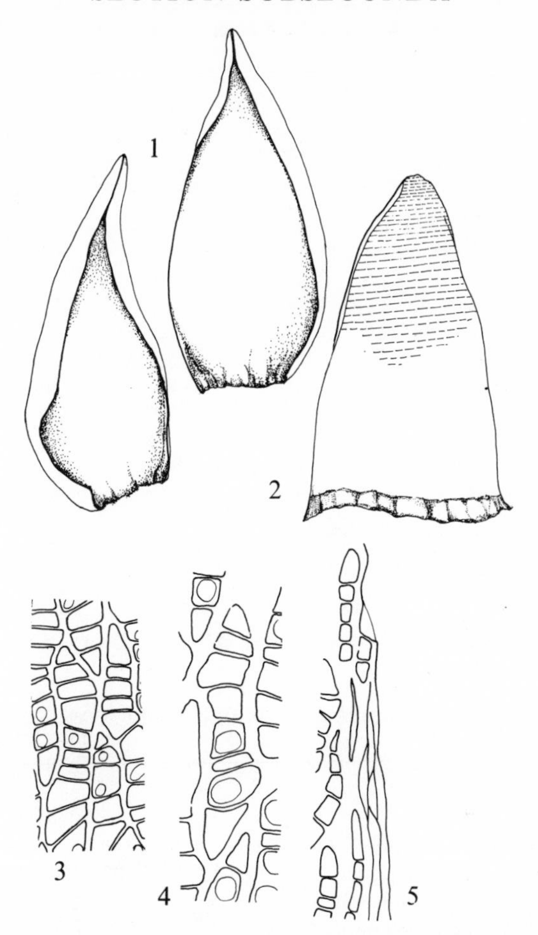

Fig. **19**. *Sphagnum auriculatum* var. *inundatum*: 1, branch leaves; 2, stem leaf; 3, cells from middle of branch leaf, dorsal surface; 4, stem leaf cells 80 % from base, ventral surface; 5, cells at margin of stem leaf near apex. Leaves ×27, cells ×280.

20. S. subsecundum Nees in Sturm, Deutschl. Fl., 1819

Dioecious. Shoots to 15 cm, green, yellow or ochre. Stem 0.4–0.7 mm diameter; cortex 1 layer, cells of outer surface with indistinct pores at upper end or pores absent; cylinder dark brown, or in shade forms green. Branches 4–6(–7) per fascicle, 2–3 spreading, 2–4 pendent; spreading branches clearly differentiated from pendent branches, often markedly curved. Stem lvs hanging, or a few sometimes spreading, 0.4–0.9 mm × 0.4–0.6 mm, triangular or triangular-lingulate, apex plane or cucullate, rounded; border 3–6 cells wide, ± strongly fringed round apex; patches of narrow cells at basal angles absent; hyaline cells not septate; upper cells fibrillose in a zone 0–25 % of length of lf; dorsal pores absent; ventral pores 5–14 μm, 0–10 per cell, unringed, centrally placed or lying along commissures. Branch lvs 0.7–1.5 mm × 0.4–0.8 mm, ovate, secund and pointing towards centre of curve of branch; border 2–3

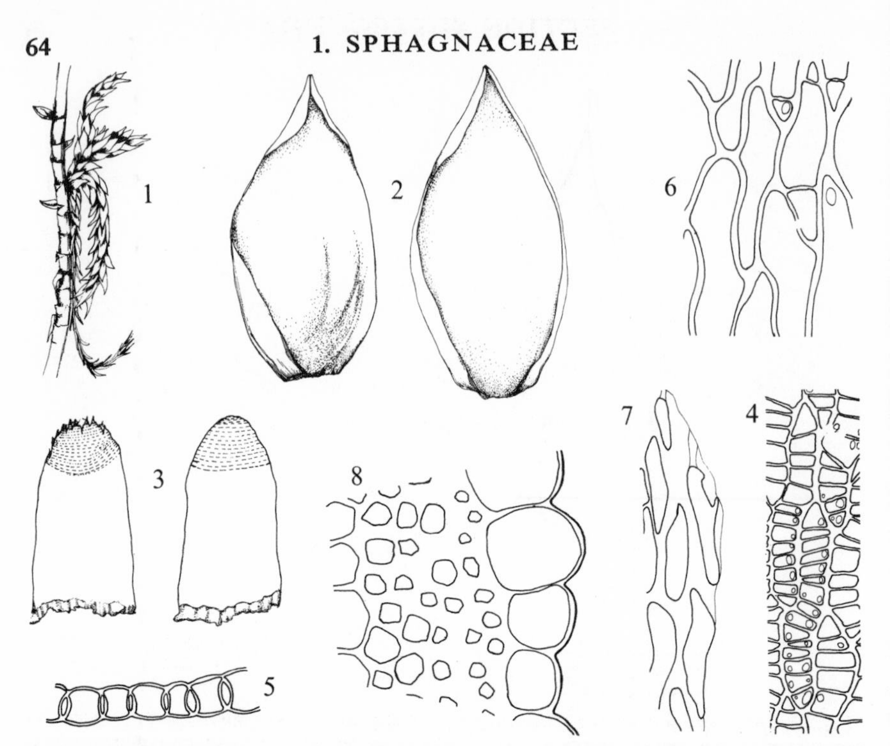

Fig. **20**. *Sphagnum subsecundum*: 1, fascicle (×3); 2, branch leaves; 3, stem leaves; 4, cells from middle of branch leaf on dorsal surface; 5, section of branch leaf; 6, stem leaf cells 80% from base; 7, marginal cells of stem leaf near apex; 8, section of stem. Leaves ×27, cells ×280.

cells wide; green cells in section lens-shaped, reaching both surfaces or just enclosed on ventral surface; dorsal pores 2–5 μm, (0–)20–40 per cell, ringed, lying along commissures, ventral pores absent or rarely 4–6 μm, to 6 per cell. Antheridial lvs bright yellow. Perichaetial lvs narrowly ovate, fibrillose and porose near apex. Spores 27–31 μm. Fr unknown in Britain. Yellow or ochre 'lawns', or sometimes scattered stems (green only in shade forms), in base-rich flushes and fens, favouring slightly more acid conditions than *S. contortum*. Throughout Britain, frequent in some upland districts but mostly rather rare. 23. Circumpolar, mainly in the north temperate zone.

Similar to *S. auriculatum* var. *inundatum*, but normally about half the size, and differing in stem lf anatomy and chromosome number. Depauperate *S. auriculatum* var. *inundatum* is often mistaken for it, and sometimes produces stem lvs which are similar. Branch lf size will usually distinguish these forms. Moreover, *S. subsecundum* is quite stable in its small size, but depauperate *S. auriculatum* var. *inundatum* almost always betrays its true identity by producing some normal-sized stem and branch lvs in parts of the plant. In the field, *S. subsecundum* is usually easy to recognise, with its bright yellow or ochre colour, dark brown stem, curved branches, and slender habit resembling *S. capillifolium*.

21. S. contortum Schultz, Prodr. Fl. Star. Suppl. I, 1819
S. laricinum (Wils.) Spruce

Dioecious. Shoots to 15 cm, green, yellow or brownish. Stem 0.4–0.7 mm diameter; cortex 2–3 layers, cells of outer surface mostly with a rather faint pore at upper end;

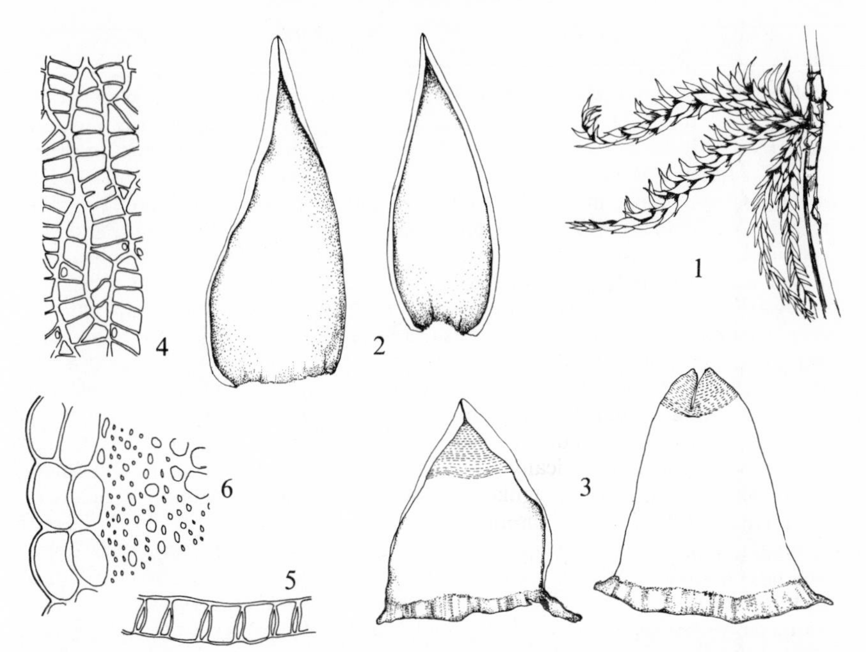

Fig. **21**. *Sphagnum contortum*: 1, fascicle (×3); 2, branch leaves; 3, stem leaves; 4, cells from middle of branch leaf on dorsal surface; 5, section of branch leaf; 6, section of stem. Leaves ×27, cells ×280.

cylinder green or pale brown. Branches 4–6(–7) per fascicle, 2–3 spreading, (1–)2–4 pendent; spreading branches clearly differentiated from pendent branches, often markedly curved. Stem lvs spreading or hanging, 0.7–1.3 mm×0.5–1.0 mm, triangular or triangular-lingulate, apex rounded or retuse; border 3–5 cells wide, partially resorbed and denticulate or fringed round apex; patches of narrow cells at basal angles absent; hyaline cells not septate; upper cells fibrillose in a zone 10–35 % of length of lf; dorsal pores near lf apex 2–4 μm, 0–4(–6) per cell; ventral pores ringed, 3–6 μm, 3–12 per cell, lying along commissures. Branch lvs 1.0–1.9 mm×0.5–0.9 mm, ovate or narrowly ovate, often distinctly secund and pointing towards centre of curve of branch; border 2–4 cells wide; green cells in section lens-shaped, reaching dorsal surface, just enclosed on ventral surface; dorsal pores in mid-lf very small, 1–3 μm, sometimes to 30 per cell, often ± indistinct or obsolete; ventral pores absent. Antheridial lvs ochre with a slight pink tinge. Perichaetial lvs ovate, fibrillose and porose above. Spores 24–27 μm. Fr unknown in Britain. Greenish to ochre 'lawns', or sometimes scattered stems, confined to base-rich habitats, mainly upland flushes, less often lowland fens. Throughout the British Isles, frequent in the north and west, rare in the south and east. 57, H23. Circumpolar, mainly in the boreal zone.

In the field, the curved branches and normally ochre colour will suggest this species, *S. subsecundum* or *S. auriculatum* var. *inundatum*. Well illuminated *S. contortum* can be distinguished by its pale stem, but greenish shade forms are hard to recognise in the field. The base-rich habitat is characteristic; *S. contortum* is usually associated with such indicators as *Carex pulicaris*, *Drepanocladus revolvens*, *Scorpidium scorpioides* and *Sphagnum teres*.

Section *Cuspidata* (Lindb.) Schlieph.

Plants slender to moderately robust. Stem typically 0.3–1.0 mm diameter; cortex 2–3(–4) layers, 25–80 μm wide, hyaline or not, cell walls without fibrils, outer surface without pores; diameter of cortical cells measured at right-angles to radius, 15–40(–60) μm; cylinder in some species strongly differentiated from cortex, in others weakly so. Branches in fascicles of (2–)3–5(–6), in some species not differentiated into pendent and spreading, in those which have differentiation 2(–3) spreading, 1–3 pendent; branch cortex with clearly dimorphic cells (except *S. lindbergii*, and submerged forms), cell walls without spiral fibrils; retort cells various, with almost no neck or a moderate or large one, in groups of 1–2. Stem lvs hanging or spreading, typically 0.8–1.6 mm×0.5–1.3 mm; border 4–10 cells wide, near apex entire or resorbed; patches of narrow cells at basal angles large, occupying 50–100% of lf base; hyaline cells not, or hardly septate, variously fibrillose or not; dorsal surface ± intact (except *S. lindbergii*); ventral surface extensively resorbed near lf apex, or more rarely with large pores. Branch lvs ovate or lanceolate, concave or straight, in several species markedly recurved when dry (a character confined to this Section among British *Sphagna*), typically 0.8–3.5 mm×0.4–0.9(–1.2) mm; border intact, 2–6 cells wide; green cells triangular or trapezoid in section, broadly exposed on dorsal surface, with a lesser or absent exposure on ventral surface; hyaline cells near apex longer than broad (sometimes ± isodiametric in *S. tenellum*), their dorsal surface near lf apex sometimes with a large pore at distal angle (usually not); dorsal pores in mid-lf 0–1(–3) per cell, 2–9 μm (larger in *S. tenellum*), confined to ends of cell (especially distal end), or occasionally a few at other angles (in some species there are more numerous dorsal pores, but these are then 2–8 μm, unringed, not appressed to commissures – a feature confined to this Section of British *Sphagna*); ventral pores (2–)4–9(–18) μm, 0–10(–12) per cell, faint, unringed, often visible only with intense staining, appressed to commissures (except *S. obtusum*). Antheridia on spreading branches. Perichaetial lvs with border intact to apex (except *S. lindbergii*), hyaline cells fibrillose or not, often resorbed on ventral surface.

Key to *Sphagnum* Section *Cuspidata*

1 Stem lvs with fibrils near apex — 2
  Stem lvs without fibrils near apex — 6

2 Small greenish plant with branches 4–8(–10) mm; branch lvs concave, ovate or broadly ovate, to 1.5 mm; hyaline cells near apex of branch lvs 20–40 μm wide, 1–4 times as long as wide — **24. S. tenellum**
  Habit and colour various, branches often exceeding 10 mm; branch lvs concave or not, ovate or lanceolate, often exceeding 1.5 mm; hyaline cells near apex of branch lvs less than 20 μm wide, 3–9 times as long as wide — 3

3 Colour orange, yellow or golden-brown — 4
  Colour green or dingy olive — 8

4 Stems darker than lvs, brown or dark brown (occasionally this pigmentation is manifested only in patches); relatively robust plants with branches in dry state 1–2 mm diameter — **26. S. pulchrum**
  Stem pale, plant slender or medium-sized, the branches in dry specimens rarely exceeding 1.3 mm diameter — 5

5 Stem cortex strongly differentiated; stem lvs ± spreading; branches in fascicles of 3(–4) — **25. S. balticum**
  Stem cortex weakly differentiated; stem lvs hanging; branches in fascicles of 4–5 — 8

6 Brown plant with dark brown stem; stem lvs widest near the fringed retuse apex **30. S. lindbergii**
Colour various; stem lvs widest near base 7
7 Stem lvs conspicuously cleft at apex **29. S. riparium**
Stem lvs entire or somewhat fringed at apex, not cleft 8
8 Pendent branches weakly or not at all differentiated, not appressed to stem, their lvs resembling those of spreading branches; plant often ± submerged; when growing in drier places the lvs at ends of branches are rolled into a sharp cusp; stem lvs usually with fibrils; stem cortex ± clearly differentiated 9
Pendent branches with whitish lvs, clearly differentiated from spreading branches, appressed to stem and ± concealing it; plant rarely submerged; lvs at branch apex not rolled into a cusp; stem lvs usually without fibrils; stem cortex usually very indistinct, occasionally well differentiated 10
9 Colour dingy olive; hyaline cells of branch lvs with abundant (usually *ca* 10 per cell) unringed dorsal pores (use stain) **23. S. majus**
Colour green, rarely dingy olive, hyaline cells of branch lvs with up to 3 ringed dorsal pores per cell, unringed dorsal pores absent **22. S. cuspidatum**
10 Stem lvs acute, with margins inrolled at apex to form an apiculus **27. recurvum** var. **mucronatum**
Stem lvs obtuse, apex cucullate or plane and rounded 11
11 Dorsal pores of lvs on pendent branches 3–10(–14) μm, usually none exceeding 12 μm (use stain) 12
Dorsal pores of pendent branch lvs 8–20(–32) μm, at least some exceeding 12 μm 13
12 Stem lvs rounded, ± fringed at apex, ventral pores on lvs of spreading branches very indistinct even with heavy staining, 2–5 μm **28. S. obtusum**
Stem lvs sub-acute; ventral pores of spreading branch lvs 5–9 μm, clearly visible with heavy staining, at least near apex of lf **27. recurvum** var. **mucronatum**
13 Branch bases red or pink, stems also often pink; stem lvs ± cucullate, not or scarcely fringed at apex; slender plants **27. S. recurvum** var. **tenue**
Red pigment absent; stem lvs mostly plane and ± strongly fringed at apex; plants relatively robust **27. recurvum** var. **amblyphyllum**

22. S. cuspidatum Hoffm., Bot. Taschenb., 1795

Dioecious. Shoots to *ca* 15 cm in terrestrial forms, often much longer in aquatic forms, green. Stem 0.4–0.8 mm diameter; cortex 2–3 layers, ± clearly differentiated from cylinder, lacking pores on outer surface; cylinder green. Branches 3–5(–6) per fascicle, straight or sometimes strongly curved, not or hardly differentiated into pendent and spreading. Stem lvs 0.9–1.3 mm × 0.6–1.0 mm, rounded-triangular, entire, variously hanging or spreading; border 4–9 cells wide, merging at basal angles with patches of narrow cells which take up 60–100 % of lf base; hyaline cells fibrillose in upper (0–)20–40 % of lf; dorsal surface intact; ventral surface near lf apex either with 3–5 large (8–25 μm) pores per cell, or ± extensively lacking. Branch lvs 1.5–3.5 mm × 0.3–0.7 mm, lanceolate or ovate, sometimes little modified when dry, or else strongly recurved except at the cuspidate branch tips; border 3–6 cells wide; green cells in section triangular or trapezoid, broadly exposed on dorsal surface, reaching ventral surface and often moderately exposed on it; dorsal pores in mid-lf 2–6 μm, 0–1(–3) per cell, ringed, usually at distal angle of cell, less often at proximal end or other angles; ventral pores faint, unringed, 4–8 μm, 4–10 per cell, confined to cell angles or distributed along commissures. Antheridial lvs brown, contrasting with

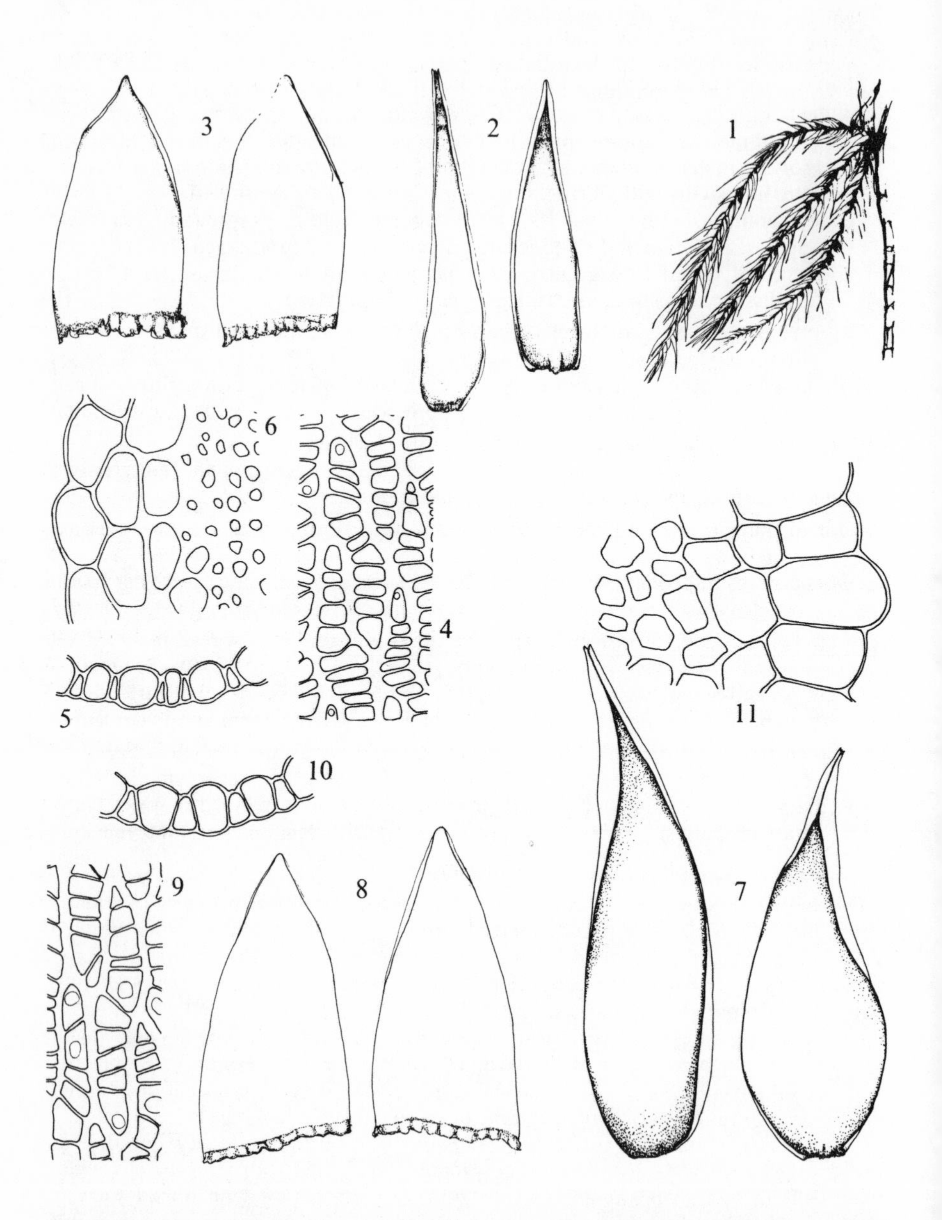

Fig. 22. 1–6, *Sphagnum cuspidatum*: 1, fascicle (×3); 2, branch leaves; 3, stem leaves; 4, cells from middle of branch leaf on dorsal surface, 5, section of branch leaf; 6, section of stem. 7–11, *S. majus*: 7, branch leaves; 8, stem leaves; 9, cells from middle of branch leaf on dorsal surface; 10, section of branch leaf; 11, section of stem. Leaves ×27, cells ×280.

green ordinary lvs. Perichaetial lvs oblong, obtuse, fibrillose above; dorsal surface intact; ventral surface with pores or more extensive resorption near lf apex. Spores 32–37 μm. $n=19+2^*$. Fr occasional, summer. Flaccid green wefts on wet peaty ground, typically in pools and runnels on acid bogs and moorland; also among Cyperaceae by oligotrophic lakes and pools, or free-floating as an aquatic. Throughout the British Isles; common in the north and west, local in S.E. England. 105, H40. Nearly circumpolar, absent from western N. America.

Submerged forms are sometimes highly modified, and may have ± denticulate lf margins. Ordinarily *S. cuspidatum* is an easy plant to recognise, with narrowly lanceolate lvs and a flaccid habit which has aptly been likened to a drowned kitten. In the field, the best marks of distinction from *S. recurvum* are the absence of well defined pendent branches, and, in terrestrial forms, the tightly rolled lvs of the cuspidate branch apices. For distinction from *S. majus* see under that species.

23. S. majus (Russ.) C. Jens., Bot. For. Festskr., 1890
S. cuspidatum var. *majus* Russ., *S. dusenii* Warnst.

Dioecious. Shoots to 25 cm, dull brownish-green. Stem 0.6–0.8 mm diameter; cortex 2–3 layers, ± clearly differentiated from cylinder, lacking pores on outer surface; cylinder green. Branches 4–5 per fascicle, not or hardly differentiated into pendent and spreading. Stem lvs 0.9–1.4 mm×0.8–1.0 mm, rounded-triangular, entire, variously hanging or spreading; border 4–8 cells wide, merging at basal angles with patches of narrow cells which take up 60–100 % of lf base; hyaline cells fibrillose in upper 20–40 % of lf; dorsal surface intact; ventral surface near lf apex extensively resorbed, often ± completely lacking. Branch lvs 1.7–3.0 mm×0.6–0.8 mm, ovate or lanceolate, little modified when dry; border 3–5 cells wide; green cells in section triangular or trapezoid, broadly exposed on dorsal surface, reaching ventral surface and often moderately exposed on it; dorsal pores in mid-lf 5–8 μm, (2–)8–18 per cell, unringed, many of them in middle of cell surface, not appressed to commissures; ventral pores faint, unringed, 5–7 μm, 0–6 per cell, confined to cell angles. Antheridial lvs rusty-brown, contrasting with ordinary lvs. Perichaetial lvs oblong, obtuse, fibrillose near apex; dorsal surface intact; ventral surface of hyaline cells resorbed near lf apex. Spores 27–38 μm. Fr not known in Birtain. Greenish-brown wefts semi-submerged on wet peaty ground, a well known locality being the lagg zone of a Northumberland raised bog. Northumberland and Angus, very rare. 2. Circumpolar in the boreal zone; distribution in Europe mainly continental.

In the field resembling *S. cuspidatum*, to which it is closely allied. The main distinguishing feature is the dingy brownish-green colour. *S. cuspidatum* is normally green except for the antheridial lvs. Microscopically the numerous dorsal pores are an easy mark of distinction, but because they lack rings they are hard to see without stain. For the occurrence of *S. majus* in Britain see Maass, *Bryologist* **68**, 211–17, 1965.

24. S. tenellum (Brid.) Brid., Musc. Rec. Suppl. IV, 1819
S. cymbifolium var. *tenellum* Brid., *S. molluscum* Bruch

Dioecious. Shoots to 10 cm, green or faintly brownish. Stem 0.3–0.5 mm diameter; cortex (1–)2–3 layers, clearly differentiated from cylinder, lacking pores on outer surface; cylinder green. Branches 4–8(–10) mm, (2–)3(–4) per fascicle, of which usually 2 spreading, 1 pendent, the pendent branch small, not appressed to stem, weakly differentiated from spreading branches; retort cells in pairs, with strongly protuberant necks. Stem lvs large for size of plant, 0.9–1.4 mm×0.5–0.7 mm, lingulate, entire, spreading, the apex rounded, often ± cucullate; border 5–10 cells wide, at basal angles merging with patches of narrow cells which take up 50–100 % of lf base; hyaline cells fibrillose in upper 40–100 % of lf; dorsal surface intact, or with a

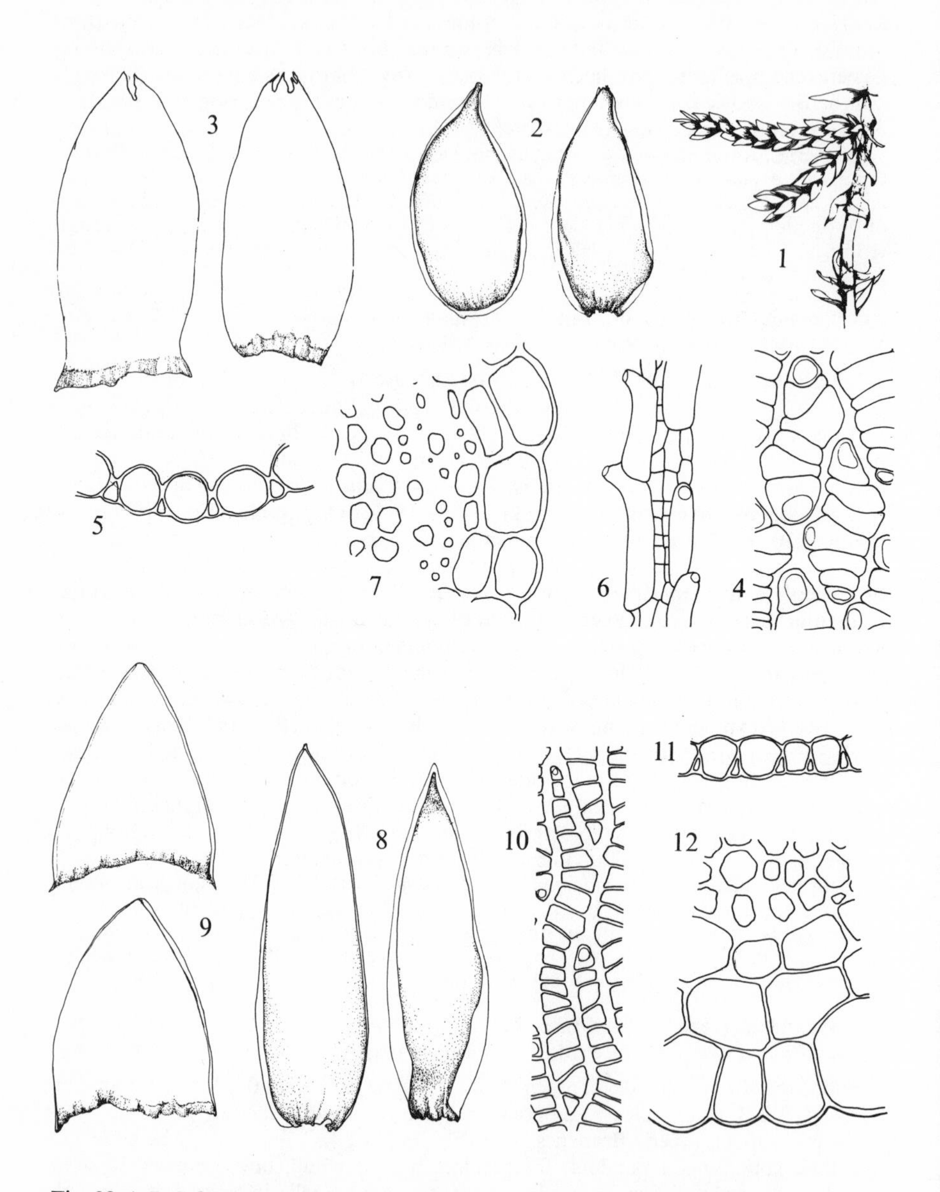

Fig. 23. 1–7, *Sphagnum tenellum*: 1, fascicle (×3); 2, branch leaves; 3, stem leaves; 4, cells from middle of branch leaf on dorsal surface; 5, section of branch leaf; 6, surface view of branch cortex (×70); 7, stem section. 8–11, *S. balticum*: 8, branch leaves; 9, stem leaves; 10, cells from middle of branch leaf on dorsal surface; 11, section of branch leaf; 12 stem section. Leaves ×27, cells ×280.

few large pores at distal angles of hyaline cells; ventral surface near apex usually with a single large pore 13–30 μm at distal angle of each cell, occasionally more extensively resorbed. Branch lvs 0.8–1.5 mm×0.4–0.7 mm, ovate, concave, little modified when dry, apex acute or obtuse; border 2–3(–5) cells wide; green cells in section triangular or trapezoid, broadly exposed on dorsal surface, reaching ventral surface and usually somewhat exposed on it; hyaline cells near lf apex 1–4 times as long as broad; dorsal pores in mid-lf 4–12(–25) μm, 0–1(–2) per cell, when present usually confined to distal angle of cell; ventral pores faint, unringed, 4–18 μm, 0–2 per cell, confined to ends of cell. Antheridial lvs faintly yellowish. Perichaetial lvs ovate, acute, composed either of uniform vermicular cells, or if cells dimorphic then hyaline cells fibrillose above, intact on dorsal surface, on ventral surface with large pores 1 per cell at distal angle. Spores 35–40 μm. Fr frequent, summer. $n=19+2$*. Scattered stems among other sphagna, particularly *S. capillifolium* and *S. papillosum*, on wet heaths and bogs; in high rainfall areas also forming ± pure tussocks on sheltered rocky banks. Common in the north and west, local in S.E. England. 92, H39. Circumpolar, sub-Oceanic; S. America.

Usually easy to recognise in the field. Good field characters are the short branches, the ovate concave branch leaves, the weak differentiation between pendent and spreading branches, and the relatively large spreading stem lvs. Microscopically, the necks of the retort cells are more protuberant than in any other British *Sphagnum*.

25. S. balticum (Russ.) C. Jens., Bot. For. Festskr., 1890
S. recurvum ssp. *balticum* Russ.

Dioecious. Shoots to 15 cm, fulvous-brown. Stem 0.3–0.7 mm diameter; cortex 2–3 layers, clearly differentiated from cylinder, lacking pores on outer surface; cylinder yellowish, pale. Branches 3(–4) per fascicle, of which 2 spreading, 1(–2) pendent; pendent branch sometimes weakly differentiated from spreading, more often whitish, appressed to stem. Stem lvs 0.9–1.2 mm×0.5–0.8 mm, lingulate, entire, ± spreading, the margins at apex inrolled so that until flattened under a coverslip the lf appears triangular; border 4–7(–9) cells wide, merging at basal angles with patches of narrow cells which take up to 60–80% of lf base; hyaline cells fibrillose in upper 20–50% of lf; dorsal surface intact; ventral surface near lf apex resorbed and usually ± completely lacking. Branch lvs 0.9–1.7(–2.4) mm×0.4–0.7 mm, ovate, little modified when dry; border 2–3 cells wide; green cells in section ± triangular, broadly exposed on dorsal surface, reaching ventral surface and sometimes slightly exposed on it; dorsal pores in mid-lf 5–9 μm, 0–1 per cell, when present usually confined to distal angle of cell (but in a Northumberland population forms occur with up to 9 unringed dorsal pores resembling those of *S. majus*); ventral pores faint, unringed, 5–8 μm, 5–12 per cell, distributed along commissures. Antheridial lvs brown, contrasting with ordinary lvs. Perichaetial lvs oblong, weakly fibrillose above, hyaline cells ± extensively resorbed on ventral surface near lf apex. Spores 26–28 μm. Fr not known in Britain. Ditches and depressions in ombrotrophic bogs. Wales, N. England and Scotland, showing a distinct eastern tendency, very rare. 6. Circumpolar, slightly continental.

Superficially like a strongly coloured, slender form of *S. recurvum* var. *mucronatum*. Useful field characters are the ± spreading stem lvs and the smaller number of branches per fascicle. British collections examined all had the branches mainly in fascicles of 3, but one Scandinavian collection had branches regularly in fascicles of 4.

26. S. pulchrum (Braithw.) Warnst., Bot. Centralb., 1900
S. intermedium var. *pulchrum* Lindb. ex Braithw.

Dioecious. Shoots to 15 cm, orange-brown, orange or yellow. Stem 0.5–0.8 mm

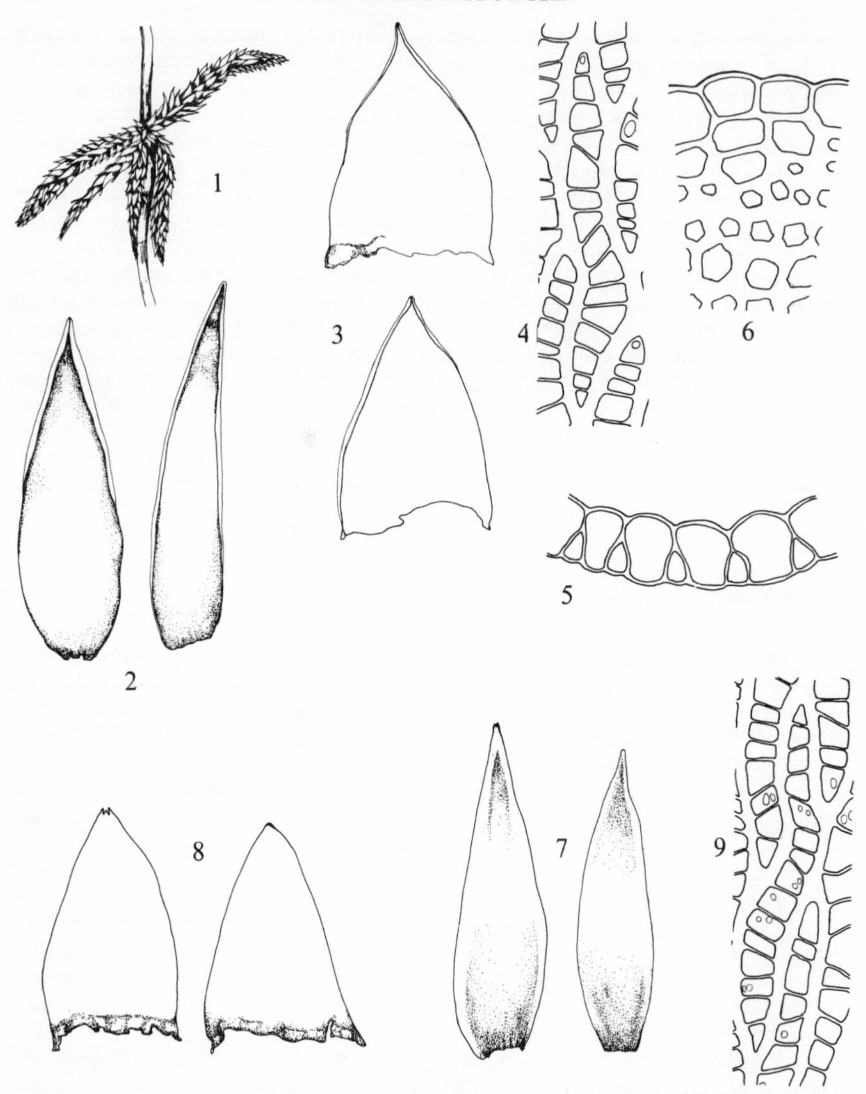

Fig. **24**. 1–6, *Sphagnum pulchrum*: 1, fascicle (×3); 2, branch leaves; 3, stem leaves; 4, cells from middle of branch leaf on dorsal surface; 5, section of branch leaf; 6, stem section. 7–9, *S. obtusum*: 7, branch leaves; 8, stem leaves; 9, cells from middle of branch leaf on dorsal surface. Leaves ×27, cells ×280.

diameter; cortex 2–3 layers, ± clearly differentiated from cylinder but not hyaline, lacking pores on outer surface; cylinder brown, darker than lvs, or sometimes partly green. Branches regularly 4 per fascicle, 2 spreading, 2 pendent; pendent branches short, not exceeding 13 mm; spreading branches also rather short, to *ca* 17 mm, in dried specimens generally 1.5–2.0 mm wide (including lvs) and appearing rather stubby. Stem lvs 0.9–1.2 mm×0.7–0.9 mm, triangular, hanging or spreading, the margins inrolled at apex to form a pronounced cusp; border 5–9 cells wide, merging at basal angles with patches of narrow cells which take up 60–80 % of lf base; hyaline cells fibrillose in upper 20–50 % of lf; dorsal surface intact; ventral surface near lf apex resorbed and lacking. Branch lvs 1.2–1.8 mm × 0.5–0.9 mm, ovate, appressed in 5 distinct rows when wet, much modified and flexuose at margin when dry, bending back so that branch appears stubby; border 2–4 cells wide; green cells in section triangular, broadly exposed on dorsal surface, completely enclosed on ventral surface; dorsal pores in mid-lf 3–8 μm, 0–1(–2) per cell, when present usually confined to distal angle of cell; ventral pores faint, unringed, 4–8 μm, 3–9 per cell, confined to cell angles or distributed along commissures. Antheridial lvs bright orange-brown, contrasting markedly with other lvs. Perichaetial lvs ovate-oblong, intact on both surfaces, lacking fibrils or pores. Spores 27–30 μm. Fr unknown in Britain or Ireland. Bright orange carpets in wetter parts of raised bogs in western Britain and Ireland; also on valley bogs in S.E. Dorset. Rare but locally abundant. 14, H6. Atlantic and sub-Atlantic regions of Europe and eastern N. America, Japan.

A distinctive species, the bright colour rapidly attracting attention. In the field, the 5-ranked branch lvs distinguish it from *S. auriculatum*; the dark stem distinguishes it from *S. balticum* and *S. recurvum*; the triangular stem lvs distinguish it from *S. lindbergii*. Microscopically it is close to *S. recurvum* var. *mucronatum*, but can be distinguished in almost all cases by the combination of fibrillose stem lvs and ventrally enclosed green cells of the branch lvs. The best single character, however, is the pigmentation of the stem.

27. S. recurvum P. Beauv., Prodr., 1805
S. flexuosum Dozy & Molk., *S. intermedium* auct. non Hoffm.

Dioecious. Shoots to 20 cm, green or orange. Stem 0.4–0.9 mm diameter; cortex 2–3 layers, often so indistinctly differentiated from cylinder as to appear absent, lacking pores on outer surface; cylinder green, pale orange or pinkish. Branches (4–)5(–6) per fascicle, straight or hanging, not strongly curved, clearly differentiated into 2(–3) spreading and (2–)3(–4) pendent, pendent branches appressed to stem and

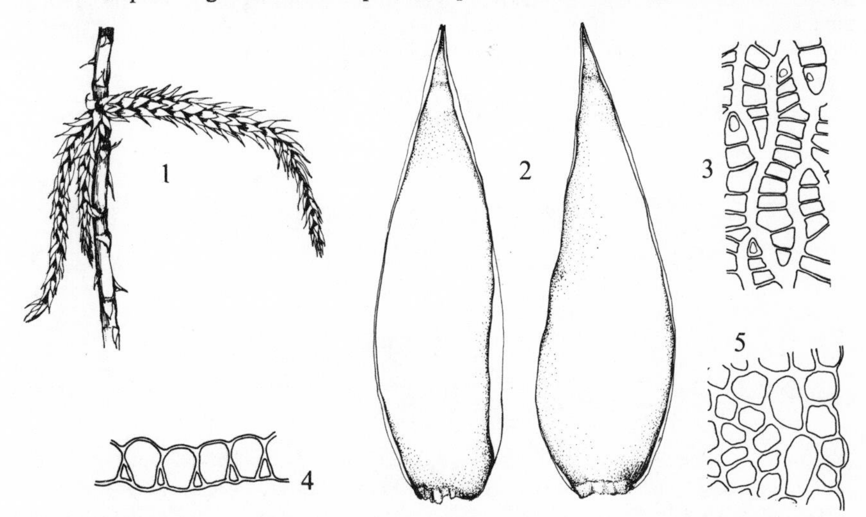

Fig. 25. *Sphagnum recurvum* var. *amblyphyllum*: 1, fascicle (×3); 2, branch leaves (×27); 3, cells from middle of spreading branch leaves, dorsal surface; 4, section of branch leaf; 5, stem section. Cells ×280.

± whitish. Stem lvs 0.7–1.2 mm×0.6–1.0 mm, triangular or triangular-lingulate, mostly hanging and appressed to stem; apex acute or rounded; border 4–8 cells wide, merging at basal angles with patches of narrow cells which take up 60–80 % of base; hyaline cells without fibrils or occasionally with fibrils in upper 20–40 % of lf; dorsal surface intact; ventral surface near lf apex resorbed and lacking. Branch lvs 1.2–2.1 mm×0.3–0.8 mm, ovate or narrowly ovate, 5-ranked or not, usually markedly recurved when dry; border 2–4 cells wide; green cells in section triangular or trapezoid, broadly exposed on dorsal surface, reaching ventral surface or not, variously not exposed or moderately exposed on it; dorsal pores in mid-lf 4–9 μm, 0–1(–2) per cell, ringed, when present usually confined to distal angle of cell; ventral pores faint, unringed, 5–9 μm, 3–7 per cell, ± confined to cell angles, often larger, ringed and perforate near lf apex. Antheridial lvs bright orange, contrasting markedly with ordinary lvs. Perichaetial lvs ovate or oblong, ± obtuse with a small apiculus, intact on both surfaces, lacking fibrils and pores. Spores 26–31 μm. Fr occasional, August.

S. recurvum is one of our commonest species and can usually be recognised by the pale stem and widely spaced, triangular, hanging stem lvs. It is most likely to be confused with *S. cuspidatum* but can almost always be distinguished by its well developed pendent branches. Rarely, when growing submerged, the pendent branches may lose their distinctness and the stem lvs may develop strong fibrils. The stem section and shape of the branch lvs will usually distinguish such plants.

Variation in British *S. recurvum* has been considered by S. Agnew (A study in the experimental taxonomy of some British Sphagna (Section Cuspidata) with observations on their ecology, Ph.D. thesis, University of Wales, 1958). She demonstrated that submerged forms with fibrillose stem lvs can develop non-fibrillose stem lvs when cultivated under drier conditions and that the distinction between rounded and acute stem lvs is genetically determined. Recent authors, both in continental Europe and N. America, have mostly recognised the 3 taxa considered here. The Europeans generally consider them good species, and certainly when var. *mucronatum* and var. *tenue* grow together the differences are striking. But the characters are not perfectly correlated in Britain; e.g. forms of var. *mucronatum* may sometimes develop a red stem. Accordingly a conservative treatment is adopted here. *S. recurvum* var. *recurvum* is not known to occur in Europe.

Var. **amblyphyllum** (Russ.) Warnst., Bot. Gaz., 1890
S. flexuosum Dozy & Molk. var. *flexuosum*, *S. amblyphyllum* (Russ.) Zick.

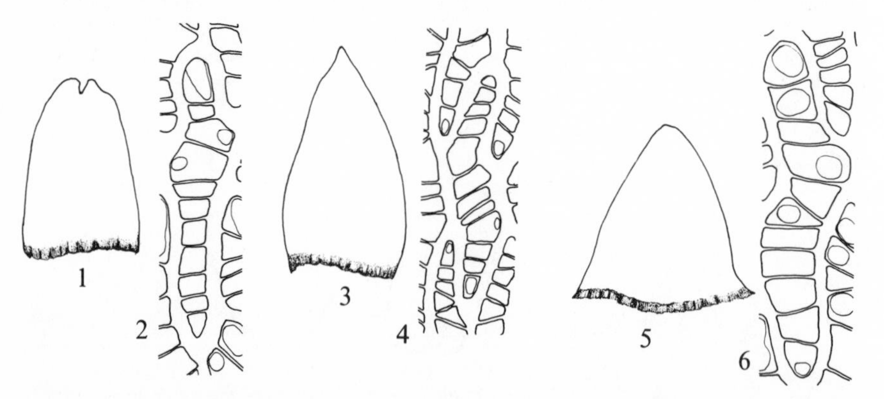

Fig. 26. Stem leaves and cells from middle of pendent branch leaves, dorsal surface: 1 and 2, *Sphagnum recurvum* var. *amblyphyllum*; 3 and 4, *S. recurvum* var. *mucronatum*; 5 and 6, *S. recurvum* var. *tenue*. Leaves ×27, cells ×280.

Shoots normally greenish; stems and branch bases lacking red pigment. Stem lvs plane or slightly cucullate; apex rounded, ± extensively fringed because of resorption. Dorsal pores of pendent branch lvs 5–20(–30) μm, sometimes with many exceeding 12 μm. Greenish 'lawns' in marshes and damp woods. Probably throughout Britain and Ireland but distribution not well known; much less common than var. *mucronatum* in the north and west. Circumpolar.

Var. **tenue** Klinggr., Schrift. Phys. Ök. Ges. Koningsberg, 1872
S. flexuosum var. *tenue* M. O. Hill ex A. J. E. Smith, *S. angustifolium* (Russ.) C. Jens., *S. parvifolium* (Warnst.) Warnst.

Slightly more slender than var. *mucronatum* and var. *amblyphyllum*. Shoots greenish or orange; stems and branches with red pigment, at least when well illuminated. Stem lvs rounded, slightly cucullate, not or scarcely fringed. Dorsal pores of pendent branch lvs 7–20(–30) μm, at least some exceeding 12 μm. Green or orange 'lawns' in soligenous marshes, often mixed with var. *mucronatum*, but not occurring in the more acid habitats of the latter. Wales to W. Scotland; British distribution imperfectly known but probably frequent in upland areas. Circumpolar.

Var. **mucronatum** (Russ.) Warnst., Bot. Gaz., 1890
S. flexuosum var. *fallax* (Klinggr.) M. O. Hill ex A. J. E. Smith, *S. apiculatum* H. Lindb., *S. fallax* (Klinggr.) Klinggr., *S. recurvum* var. *recurvum* plur. auct.

Shoots green or orange; stem lacking red pigment, branch bases with or without weak red pigment. Stem lvs acute; margins usually inrolled above to form a distinct cusp; apex not fringed. Dorsal pores of pendent branch lvs 5–10(–14) μm, none usually exceeding 12 μm. $n=19+2$*. Green or orange 'lawns' on moors, in wet woods, marshes and bogs. Throughout Britain and Ireland; very common and often abundant in the north and west, local in the south-east. Circumpolar.

28. S. obtusum Warnst., Bot. Zeitschr., 1877

Closely allied to *S. recurvum* var. *amblyphyllum*, differing in pore structure and slightly more robust habit. Stem lvs rounded-triangular, obtuse, ± fringed at apex because of resorption. Lvs of spreading branches with very small unringed dorsal pores, 2–5 μm, 0–14 per cell, disposed in 1 or 2 rows along middle of cell surface, not appressed against commissures; ventral pores also very small, unringed, 2–5 μm, 0–8 per cell, appressed against commissures or not, often very indistinct. Dorsal pores of lvs of pendent branches 3–5(–8) μm. Formerly in W. and S. Lancs but thought to be extinct through drainage. 2. Circumpolar; distribution in Europe continental.

The characteristic unringed dorsal pores are usually easiest to see near margin in lower lf but can be present in middle of lf. Intense staining is required; the lf should be almost black. When a lf of *S. recurvum* is stained in this way the relatively large ventral pores are easily visible, at least near lf apex.

29. S. riparium Ångstr., Öfv. K. V. A. Handl., 1864
S. intermedium var. *riparium* (Ångstr.) Lindb.

Dioecious. Shoots to 25 cm, green or faintly brownish. Stem 0.5–1.0 mm diameter; cortex 2–3 layers, weakly differentiated from cylinder, lacking pores on outer surface; cylinder green. Branches in fascicles of 4–5, 2(–3) spreading, 2–3 pendent. Stem lvs mostly hanging, 1.3–1.6 mm × 0.9–1.3 mm, triangular or triangular-lingulate, conspicuously cleft at apex; border 3–8 cells wide, merging at basal angles with patches of narrow cells which take up 60–80 % of lf base; hyaline cells lacking fibrils;

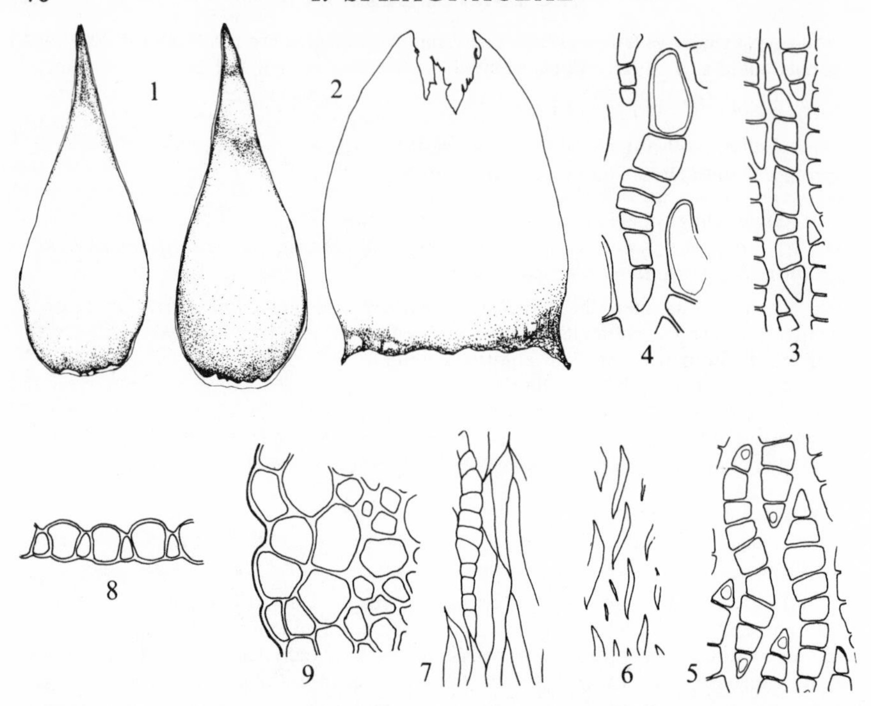

Fig. 27. *Sphagnum riparium*: 1, branch leaves; 2, stem leaf; 3, cells from middle of spreading branch leaf, dorsal side; 4, cells from middle of pendent branch leaf, dorsal side; 5 and 6, cells from near apex of spreading branch leaf on ventral and dorsal surfaces; 7, cells near apex on dorsal surface of pendent branch leaf; 8, section of branch leaf; 9, stem section. Leaves ×27, cells ×280.

ventral surface of hyaline cells extensively resorbed and lacking in upper half of lf, both surfaces being resorbed near apical cleft. Branch lvs 1.3–3.0 mm×0.5–0.9 mm, ovate, somewhat recurved when dry; border 2–4(–6) cells wide; green cells in section triangular or trapezoid, broadly exposed on dorsal surface, reaching ventral surface and often moderately exposed on it; near apex of lf there are no hyaline cells, undifferentiated green cells combining to form a snout 100–500 μm long; dorsal pores dimorphic, in upper and mid-lf 4–6 μm, 0–1 per cell, confined to distal angles of cells, in middle and lower lf near margin often consisting of membrane gaps 14–28 μm, also confined to distal angles, these being repeated on ventral surface to give a colander-like appearance; in the pendent branch lvs, the colander-like membrane gaps occur throughout upper part of lvs, giving them a striking appearance when stained; ventral pores 3–15 μm (except when opposite a dorsal membrane gap), faint, unringed, 0–14 per cell, distributed along commissures. Antheridial lvs brownish. Perichaetial lvs oblong, entire; apex rounded with small apiculus; hyaline cells intact on both surfaces, lacking fibrils and pores. Spores 25–27 μm. Fr very rare, summer. Green carpets in marshes and by streams, often among rushes on moderately nutrient-rich ground. Berks to Shetland, markedly eastern, and showing a preference for higher altitudes; rare. 16. Circumpolar, common in the arctic, more scattered further south.

In the field the stem lvs are unmistakable, but the plant is not conspicuous and could easily be overlooked as *S. girgensohnii* or a robust *S. recurvum*.

30. S. lindbergii Schimp. ex Lindb., Öfv. K.V.A. Förh., 1857

Autoecious. Shoots to 20 cm, fulvous brown. Stem 0.5–0.9 mm diameter, dark brown or almost black; cortex 3–4 layers, strongly differentiated from cylinder, lacking pores on outer surface; cylinder dark brown. Branches in fascicles of 4–5(–6), not clearly differentiated into pendent and spreading, or occasionally well differentiated; cells of branch cortex not clearly differentiated into retort and non-retort cells, almost all cells having a pore at distal end. Stem lvs hanging, 1.3–1.6 mm × 0.9–1.5 mm, cuneate or rectangular, with a retuse, conspicuously tattered apex; border 3–7 cells wide, continued up sides of lf but not extending across apex, at basal angles merging with patches of narrow cells which take up 80–100 % of lf base; hyaline cells lacking fibrils, extensively and equally resorbed, both surfaces being largely absent in upper lf. Branch lvs 1.3–2.5 mm × 0.5–0.8 mm, narrowly ovate, not recurved when dry; border 3–5 cells wide; green cells in section triangular or somewhat barrel-shaped, with a broad dorsal exposure, not exposed on ventral surface; dorsal pores 3–8 μm, 0–1(–2) per cell, when present normally confined to distal angle

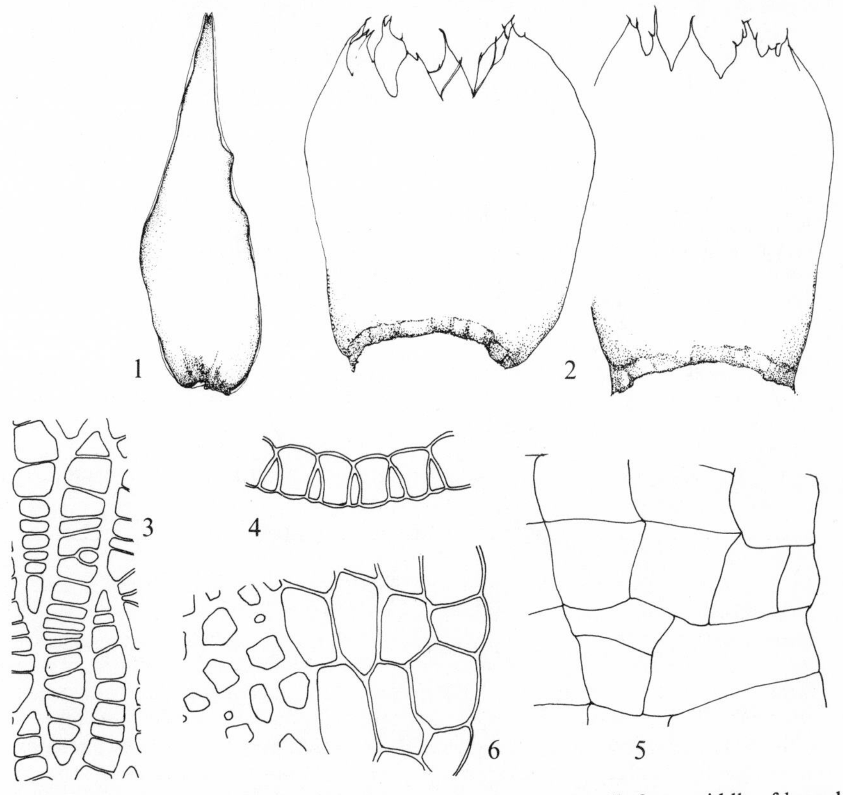

Fig. 28. *Sphagnum lindbergii*: 1, branch leaf; 2, stem leaves; 3, cells from middle of branch leaf on dorsal surface; 4, section of branch leaf; 5, stem cortex in surface view; 6, stem section. Leaves × 27, cells × 280.

of cell; ventral pores 3–8 μm, faint, unringed, (0–)4–10 per cell, in cell angles or distributed along commissures. Antheridial lvs slightly darker than others, not conspicuous. Perichaetial lvs oblong, either obtuse and entire or retuse and ± fringed at apex, lacking fibrils, extensively resorbed on both surfaces near apex. Spores 29–32 μm. Fr very rare, summer. Brown carpets in montane flushes, rarely below 600 m. Highlands of Scotland; Shetland; rare. 8. Circumpolar, common in the arctic, scattered further south.

The bright orange-brown colour makes the plant conspicuous. The dark brown stem and usually cuneate, fringed stem lvs make it unmistakable on close examination.

2. ANDREAEOPSIDA

Protonema thalloid. Plants acrocarpous. Seta absent, capsule joined directly to foot and exserted from perichaetium by elongation of pseudopodium (stalk of archegonium); capsule dehiscing by 4(–8) slits, spore sac endothecial in origin and overarching columella, no air cavities between spore sac and wall.

2. ANDREAEACEAE

A small family with the characters of the class. Two genera only.

2. Andreaea Hedw., Sp. Musc., 1801

Autoecious or dioecious. Tuft- or cushion-forming plants, stems usually slender fragile when dry, without central strand. Lvs spreading to secund or falcato-secund when moist, symmetrical or curved to one side, ovate to narrowly lanceolate or obovate-spathulate or abruptly narrowed above basal part, apex acuminate to rounded, margin entire or denticulate; nerve present, single, excurrent or not, or lacking; basal cells usually elongated, sinuose, cells above quadrate or rounded-quadrate, incrassate, smooth or papillose. Reddish-purple to brown or blackish, tuft-forming, saxicolous, montane or arctic mosses. About 115 species with a world-wide distribution.

Some of the species of *Andreaea* (*A. obovata*, *A. rupestris*, *A. rothii*, *A. crassinervia* and their varieties) do not seem to have been properly understood in Britain and the distributions of *A. crassinervia* and of the varieties of the other species require revision as many gatherings have been incorrectly named. In the treatment of ecostate species I have followed Nyholm (1954–69) and of the costate species, Schultze-Motel (*Willdenowia* **6**, 25–110, 1970).

1 Lvs nerveless — 1
  Lvs with nerve — 5
2 Lvs widest at or above the middle, margin denticulate near base, cells and nerve smooth at back — **1. A. alpina**
  Lvs widest near base, margin entire below, nerve and cells ± smooth to strongly papillose at back — 3
3 Lvs distant, spreading, symmetrical, smooth or obscurely papillose at back — **2. A. obovata** var. **sparsifolia**
  Lvs crowded, regularly spreading to falcato-secund, conspicuously papillose at back (papillae best seen in side view) — 4
4 Lvs regularly spreading to falcato-secund, at least some lvs curved, lf apex obtuse or rounded, rarely acute, stem and perichaetial lvs differing in shape — **3. A. rupestris**
  Shoots very slender, lvs regularly spreading, symmetrical, lf apex acute to acuminate, stem and perichaetial lvs ± similar — **2. A. obovata** var. **papillosa**
5 Stem lvs gradually tapering from near base to apex, cells coarsely papillose on both sides, nerve coarsely papillose at back — **6. A. nivalis**
  Stem lvs abruptly narrowed above basal part, cells and nerve smooth — 6
6 Nerve ending in apex — **4. A. rothii**
  Nerve excurrent, often markedly so — **5. A. crassinervia**

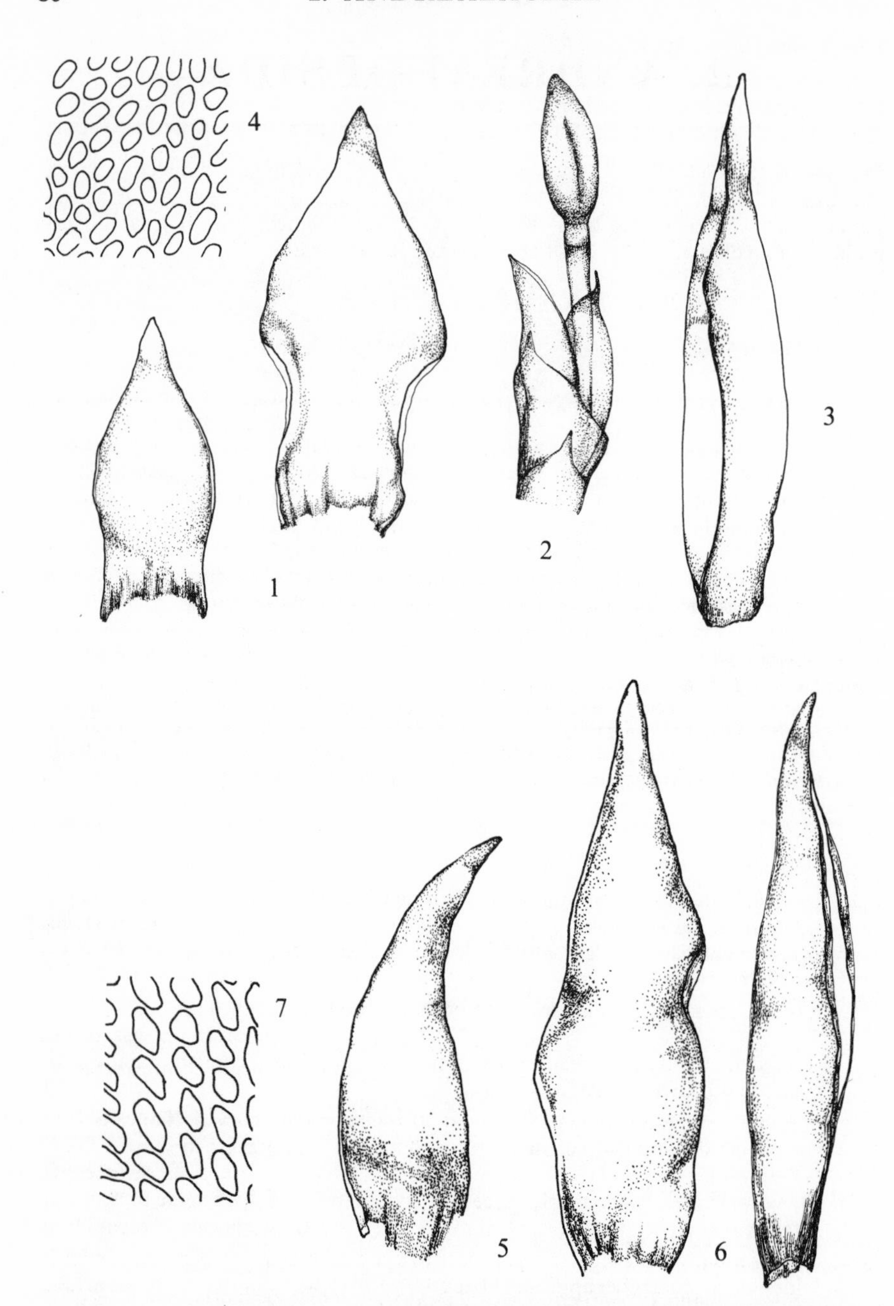

Fig. 29. 1–4, *Andreaea alpina*: 1, leaves (×40); 2, capsule (×15); 3, perichaetial leaf (×40); 4, cells from widest part of leaf. 5–7, *A. obovata* var. *papillosa*: 5, leaves (×62); 6, perichaetial leaf (×40); 7, cells from widest part of leaf. Cells ×415.

1. A. alpina Hedw., Sp. Musc., 1801

Autoecious. Plants 1–8 cm. Lvs imbricate when dry, patent to spreading when moist, larger and less crowded on capsule-bearing branches than elsewhere, concave, obovate-spathulate or panduriform, widest at or above middle, narrowed to a short or long, acute to obtuse apex, margin denticulate near base, entire above; nerve lacking; cells incrassate, basal elongate, sinuose, porose, above shorter, elliptical to ovate, in ± radiating rows, not papillose. Perichaetial lvs larger, convolute, not papillose at back. Spores *ca* 28 μm. Fr occasional, summer. Dark, purplish-red tufts or patches on exposed, damp, often basic, alpine rocks. Occasional to frequent in the mountains from Brecon, Carms and Yorks northwards. 42, H14. Norway, Faroes, Patagonia.

2. A. obovata Thed., Nya Bot. Not. 1849

Autoecious. Lvs patent to spreading when moist, not secund, ± symmetrical, not curved to one side, margin entire; nerveless; cells 12–15 μm wide in upper part of lf. Perichaetial lvs larger than but similar in shape to stem lvs. *A. obovata* var. *obovata* is not known in Britain but occurs in N. Europe, Siberia, N. America, Greenland.

Var. **papillosa** (Lindb.) Nyholm, Moss Fl. Fenn., 1969
A. rupestris var. *acuminata* (Br. Eur.) Sharp, *A. petrophila* var. *acuminata* Br. Eur.
Plants slender, to 3 cm. Lvs ovate-lanceolate, apex acute to acuminate; cells with conspicuous conical papillae at back. Perichaetial lvs larger than but ± similar in shape to stem lvs, apex acute to acuminate. Acidic rocks at high altitudes, rare. N.W. Wales and Yorks northwards, Louth, Down. 16, H2. Europe, N. Asia, N. America.

Var. **sparsifolia** (Zett.) Nyholm, Moss Fl. Fenn., 1969
A. rupestris var. *sparsifolia* (Zett.) Sharp, *A. petrophila* var. *sparsifolia* (Zett.) Lindb.
Plants slender, stems flexuose, to 3 cm. Lvs distant, spreading when moist, ovate, acuminate; cells not or only slightly papillose at back. Acidic alpine rocks at the snow line, very rare, recorded from Perth in 1875. 1. Europe, N. America.

A. obovata and its varieties are more erect and more slender plants than *A. rupestris*. It differs in the symmetrical lvs and except in var. *papillosa* the cells only obscurely papillose. Small forms of *A. rupestris* may have regularly spreading non-secund lvs, but at least some of the younger lvs are asymmetrical, the apices being curved to one side. In var. *papillosa* the slender habit and symmetrical lvs will distinguish it from forms of *A. rupestris* which are usually coarser. *A. rupestris* var. *alpestris* may be mistaken for a depauperate form of *A. obovata* but the cells are narrower and perichaetial lvs, if present, are distinctive. *A. alpina* is a coarser plant with lvs of a different shape and denticulate below.

The only British record of *A. obovata* var. *sparsifolia* is based upon a very depauperate specimen gathered in 1875. It is much smaller than Scandinavian material with stems only 4–5 mm long but nevertheless appears correctly named.

A. obovata var. *hartmanii* (Thed.) Nyholm, as *A. hartmanii* Thed., has been reported from the Isle of Rhum (Heslop-Harrison & Cooke, *J. Bot., Lond.* **80**, 35–8, 1942) but the record has not been substantiated. The lvs are distant, ovate, acute, with entire margins and the plant forms lax, brownish-green tufts on wet montane rocks (see fig. 30, 1–2).

3. A. rupestris Hedw., Sp. Musc., 1801
A. petrophila Ehrh.

Autoecious. Plants 0.5–3.0 cm. Lvs imbricate when dry, regularly patent or often secund or falcato-secund at least in larger forms when moist, widest below middle, variable in shape, narrowly lanceolate to ovate-lanceolate, at least some with apex turned to one side, apex obtuse to rounded, rarely acute, margin plane or incurved,

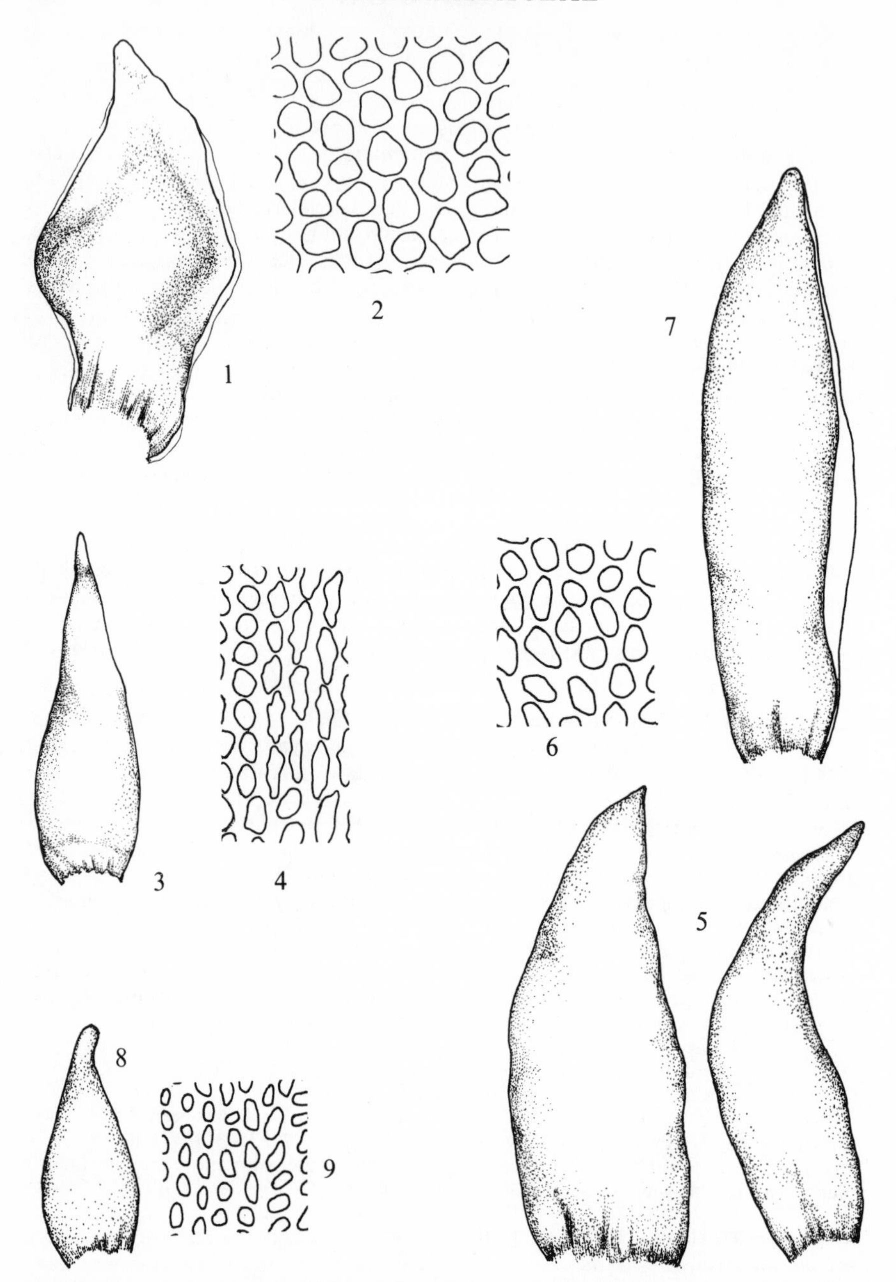

Fig. **30**. 1–2, *Andreaea obovata* var. *hartmanii*: 1, leaf; 2, cells from widest part of leaf. 3–4, *A. obovata* var. *sparsifolia*: 3, leaf; 4, cells from widest part of leaf. 5–7, *A. rupestris* var. *rupestris*: 5, leaves; 6, cells from widest part of leaf; 7, perichaetial leaf (×62). 8–9, *A. rupestris* var. *alpestris*: 8, leaf; 9, cells from widest part of leaf. Leaves ×40, cells ×415.

entire; nerve lacking; basal cells linear to narrowly rectangular, ± sinuose, above ± hexagonal, incrassate, strongly papillose at back. Perichaetial lvs convolute, larger and wider than stem lvs, shortly pointed with ± obtuse apex. Spores 26–40 μm. Fr frequent, summer.

Var. **rupestris**

Plants 0.5–3.0 cm. Lvs regularly patent or often secund or falcato-secund at least in larger forms when moist, narrowly lanceolate to ovate-lanceolate, at least some with apex turned to one side, apex obtuse to rounded, rarely acute; cells strongly papillose at back, 10–14 μm wide in upper part of lf. $n=10$. Brownish-green to dark brown, often coalescing tufts on acidic or slightly basic dry or periodically moist montane rocks and boulders, on scree, cliffs and walls, frequent or common in montane areas. Devon, Wales, Shrops and Derby northwards. 63, H22. Europe, Faroes, Caucasus, N. Asia, Japan, Madeira, Azores, southern Africa, America, Tasmania, New Zealand, Antarctica.

Var. **alpestris** (Thed.) Sharp in Grout, Moss Fl. N. Am., 1936

A. petrophila var. *alpestris* Thed.

Plants scarcely 1 cm, stems very slender. Lvs erecto-patent to spreading when moist, lanceolate to ovate-lanceolate, symmetrical or slightly curved to one side, acute to obtuse; cells in upper part less papillose, smaller, 8–10 μm wide. Small, blackish, glossy tufts on acidic rocks at high altitudes, rare. Caerns, Cumberland and Scottish Highlands. 12. N. Europe, Madeira, N. America, Greenland.

4. A. rothii Web. & Mohr, Bot. Taschenb., 1807

Autoecious. Plants 0.5–2.5 cm. Lvs imbricate when dry, patent to spreading or falcato-secund when moist, from ± ovate base abruptly narrowed to lanceolate to linear-lanceolate upper part or ovate-lanceolate and gradually tapering to apex, margin plane, entire or obscurely denticulate above; nerve ending in apex and occupying most of the upper part of the lf; cells incrassate, smooth throughout, basal near nerve rectangular, elsewhere quadrate or rounded-quadrate, smooth, 8–10 μm above basal part, cells in upper part extending to apex in (1–)2–4 rows but sometimes eroded so that nerve appears excurrent. Perichaetial lvs larger, convolute, smooth at back. Fr occasional, spring, summer.

Ssp. **rothii**

Lvs from ± ovate base abruptly narrowed to lanceolate to linear-lanceolate upper part. Spores 40–42 μm. Fr occasional, spring, summer. $n=10$, 11*. Dark brown to blackish, often coalescing tufts on acidic or mildly basic montane rocks and boulders, on scree, cliffs and walls, frequent or common in the mountains, rare elsewhere. From Cornwall, Glos, Worcs and Shrops northwards. 71, H29. N. and C. Europe, Yugoslavia, Faroes, Iceland, N. America.

Ssp. **frigida** (Hüb.) Schultze-Motel, Herzogia, 1968

A. rothii var. *frigida* (Hüb.) Lindb. ex Braithw., *A. rothii* var. *grimsulana* (Bruch) C. Müll., *A. frigida* Hüb.

Lvs ovate-lanceolate, gradually tapering to apex or if narrowed above basal part then not markedly so. Spores 26–28 μm. Montane rocks, very rare. Cumberland, S. Aberdeen, Inverness. 4. Norway, C. Europe, S. France, Pyrenees, C. Spain, Corsica, Transsylvania.

A. rothii and *A. crassinervia* are very similar in appearance but in the former the cells of the lamina extend to the apex of the lf whereas in the latter the nerve is distinctly excurrent.

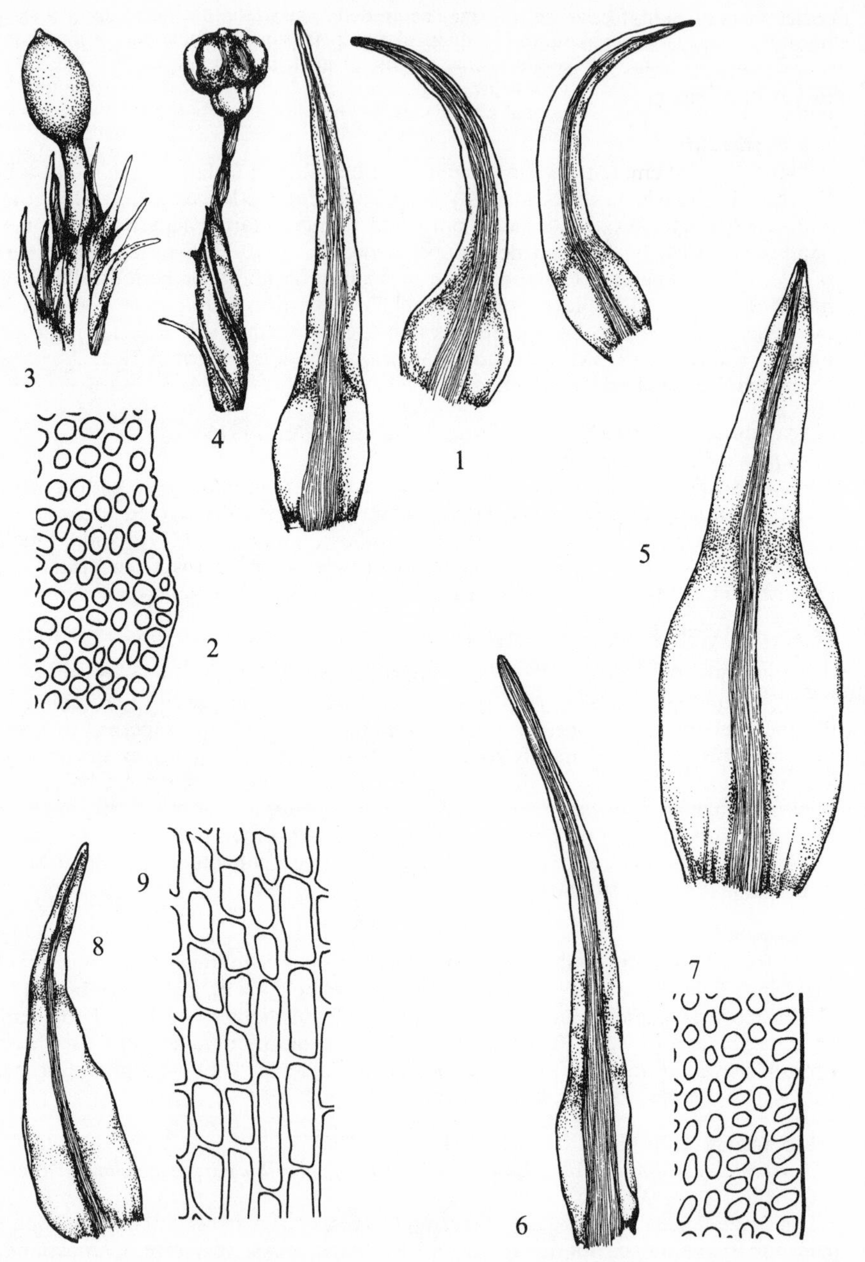

Fig. **31**. 1–4, *Andreaea rothii* ssp. *rothii*: 1, leaves; 2, cells from widest part of leaf; 3 and 4, mature and dehisced capsules (×15). 5, *A. rothii* ssp. *frigida*: leaf. 6–7, *A. crassinervia*: 6, leaves; 7, cells from widest part of leaf. 8–9, *A. blyttii*: 8, leaf; 9, cells from widest part of leaf. Leaves ×40, cells ×415.

In older lvs of *A. rothii* the lamina cells may be eroded away so that the nerve appears excurrent, but the 2 species can be readily distinguished if lvs from the upper part of the stem are examined. Nyholm (1969) reports that spore size in *A. crassinervia* is *ca* 30 μm and Schultze-Motel *op. cit.* that it is 30–40 μm. In all British material of *A. crassinervia* examined spore size was 40–46 μm. Occasional plants with lvs intermediate between the two species and abnormally large and variable sized spores may be encountered (e.g. Breidden Hill, Montgomery). Similar plants have been reported from Sweden (Nyholm, 1969) and their nature is obscure.

A. blyttii has been reported from Harris (Heslop-Harrison & Cooke, *J. Bot. Lond.* **80**, 35–8, 1942) but this has not been substantiated. The plant somewhat resembles *A. rothii* but is dioecious and differs in areolation, the cells being rectangular or elongated throughout the basal part of the lf and ± quadrate and 10–12 μm wide above the basal portion (see fig. 31, 8–9).

5. A. crassinervia Bruch, Abh. Math.-Phys, Classe Kön. Bayer. Akad. Wiss., 1832

Plants 1–6 cm. Lvs similar to *A. rothii* in shape and areolation but upper part of lf composed entirely of nerve.

Ssp. **crassinervia**

A. rothii var. *crassinervia* (Bruch) Mönk.

Plants 1–2 cm. Inner perichaetial lvs smooth at back. Dark brown to blackish tufts on acidic montane rocks, cliffs, etc., occasional. Devon, Pembroke, N. Wales and Derby northwards. 28, H6. N. and C. Europe, eastern N. America, Greenland.

Ssp. **huntii** (Limpr.) Amann, Fl. Mouss. Suisse, 1919

A. rothii var. *huntii* (Limpr.) Dix.

Plants to 6 cm. Inner perichaetial lvs with conspicuous conical papillae at back. Acidic montane rocks, very rare. Merioneth, Westmorland, Cumberland, S. Aberdeen, Down. 4, H1. W. and C. Europe.

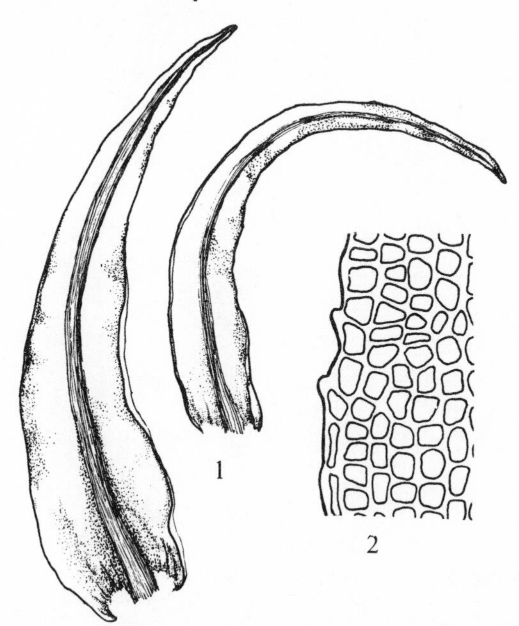

Fig. **32**. *Andreaea nivalis*: 1, leaves (× 40); 2, cells from widest part of leaf (× 415).

6. A. nivalis Hook., Trans. Linn. Soc., 1811

Dioecious. Plants 1–6 cm, shoots slender or stout. Lvs imbricate when dry, patent to spreading or falcato-secund when moist, narrowly lanceolate to ovate-lanceolate, gradually tapering to blunt to acuminate apex, margin plane or recurved on one or both sides, irregularly denticulate; nerve reaching to apex or excurrent, strongly papillose at back; basal cells rectangular or shortly rectangular, above irregularly quadrate or rounded-quadrate, incrassate or not, coarsely papillose on both sides. Perichaetial lvs similar to but a little larger than stem lvs. Fr rare, summer. Dark, reddish-brown tufts or patches on acidic rocks at high altitudes particularly in areas of late snow-lie on the highest Scottish mountains, rare. Angus, S. Aberdeen, Banff, Inverness, Argyll, Ross. 8. Montane parts of Europe, Nova Zemlya, Japan, Kamchatka, Aleutian Is., western N. America, Greenland.

3. BRYOPSIDA

Stems arising from filamentous or rarely thalloid protonema. Spores and columella derived from endothecium; archesporium cylindrical, not overtopping columella; capsule opening by lid, with or without peristome teeth or indehiscent.

Subclass 1. Polytrichideae

Acrocarpous. Usually robust plants. Seta elongate; peristome teeth 4, 16, 32 or 64, not transversely barred, derived from several concentric layers of cells.

3. TETRAPHIDALES

Plants small. Lvs ovate or lanceolate; nerve when present without lamellae on surface; cells rounded-hexagonal, smooth, unistratose. Capsule erect, symmetrical, smooth; annulus absent; calyptra mitriform; peristome teeth 4, basal membrane lacking. Protonema producing frondiform entire or forked outgrowths (protonemal lvs).

3. TETRAPHIDACEAE

With the characters of the order.

3. Tetraphis Hedw., Sp. Musc., 1801

Autoecious. Stems with central strand. Stem lvs numerous with distinct nerve. Capsule without stomata; calyptra plicate. Protonemal lvs not persistent. Two north temperate species.

1. T. pellucida Hedw., Sp. Musc., 1801

Plants to 1.5(–3.5) cm. Lvs loosely appressed when dry, patent when moist, lower orbicular to ovate, upper ovate to lanceolate, acute, margin entire; nerve ending below apex; cells irregularly hexagonal, incrassate, 12–20 μm wide in mid-lf. Distal lvs of sterile stems frequently crowded, orbicular, forming gemma cups containing discoid gemmae *ca* 40 μm diameter. Perichaetial lvs narrowly lanceolate, acuminate. Seta straight, 4–15 mm; capsule cylindrical; lid conical; peristome teeth 4; spores 10–12 μm. Fr summer, rare in drier parts of Britain, occasional elsewhere. $n=6$, 7, 8*. Patches or tufts, bright green above, reddish-brown below, on rotting wood, peat and sandstone rock. Rare in S.E. England and the Midlands where calcareous, rare in N. Scotland, frequent elsewhere. 108, H31. Europe, Caucasus, C. Asia, China, Taiwan, Japan, Korea, N. America.

4. Tetrodontium Schwaegr., Suppl. II, 1824

Autoecious. Stems very short, without central strand. Stem lvs few, nerve weak or

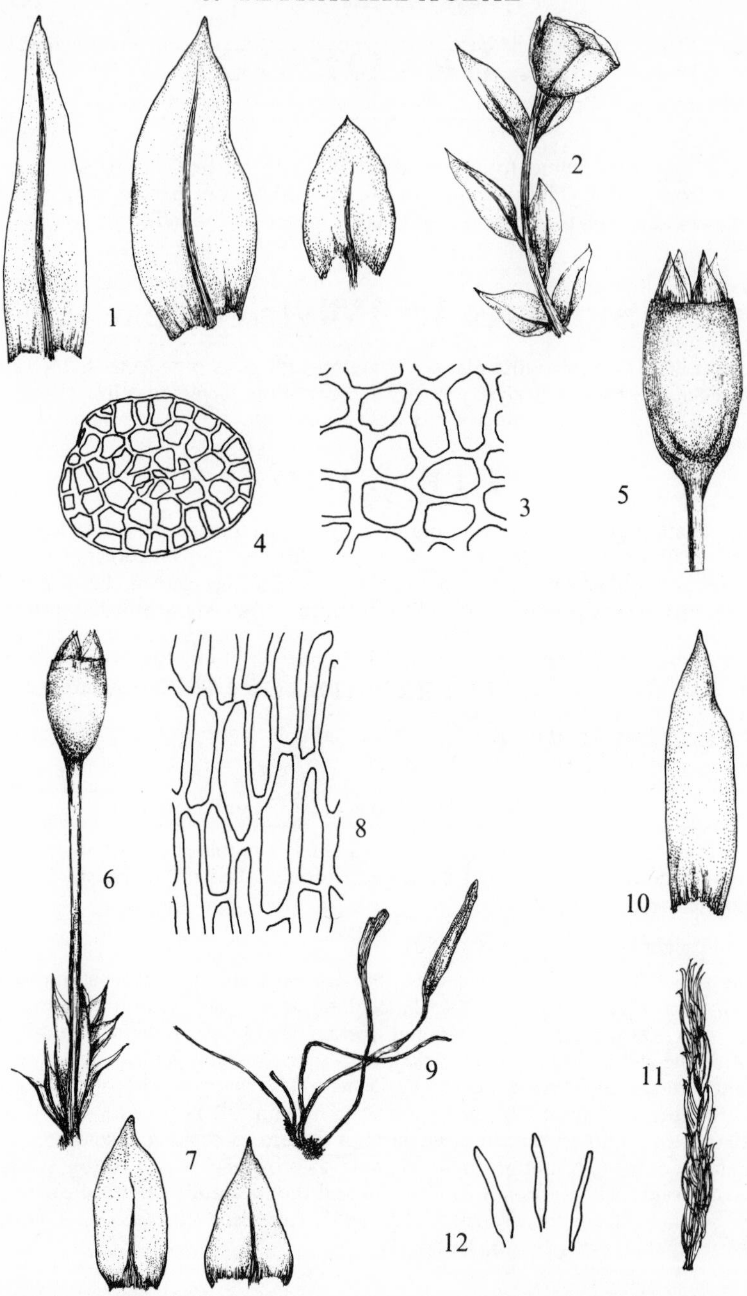

Fig. 33. 1–5, *Tetraphis pellucida*: 1, leaves; 2, shoot with gemma cup; 3, mid-leaf cells; 4, gemma (×80); 5, capsule. 6–9, *Tetrodontium brownianum*. 6, plant (×12.5); 7, leaves; 8, mid-leaf cells; 9, protonemal leaves. 10–12, *T. repandum*: 10, leaf; 11, propaguliferous shoot; 12, protonemal leaves (×12.5). Leaves and capsule ×22.5, cells ×450.

absent. Stomata present at base of capsule; calyptra smooth. Protonemal lvs persistent. Two or 3 species.

Protonemal lvs to 2.5 mm long, plants without flagelliform branches
1. T. brownianum

Protonemal lvs to 0.5 mm long, flagelliform branches often present
2. T. repandum

1. T. brownianum (Dicks.) Schwaegr., Suppl. II., 1824
Tetraphis browniana (Dicks.) Grev.

Plants minute, gregarious, to 2 mm, ± stemless. Protonemal lvs to 2.5 mm long, linear to narrowly lingulate or narrowly spathulate, sometimes forked, entire; cells 2–3-stratose. Stem and perichaetial lvs imbricate, ovate to lanceolate, acuminate; nerve lacking in lower lvs, weak in upper; cells irregularly rectangular, very incrassate. Seta to 4 mm; capsule ± ellipsoid; peristome teeth 4; spores 14–16 μm. Fr common, summer. $n=7$. In crevices and on the surface (often the under-surface) of acidic rocks in heavily shaded sites by streams and rivers. Devon, Sussex, Staffs, Hereford and occasional from Wales, Derby and Yorks northwards. 51, H8. N., W. and C. Europe, Asia Minor, Japan, N. America, New Zealand.

2. T. repandum (Funck) Schwaegr., Suppl. II., 1824
Tetraphis browniana var. *repanda* (Funck) Hampe

Close to *T. brownianum*. Protonemal lvs to 0.5 mm long, entire to coarsely and irregularly toothed; cells unistratose. Perichaetial lvs nerveless or with nerve very poorly developed. Tristichous, flagelliform, propaguliferous shoots often present at base of stem. Similar habitats to *T. brownianum*, very rare. N.E. Yorks. 1. N., W. and C. Europe, Caucasus, Japan, N. America.

For the occurrence of this plant in Britain see Appleyard, *Trans. Br. bryol. Soc.* **3**, 64–5, 1956.

4. POLYTRICHALES

Plants frequently large, stems simple or branched, tough, with internal anatomical differentiation. Lvs with or without sheathing membranous base; nerve with longitudinal lamellae on ventral surface and often constituting most of lf blade. Capsule on long seta, spherical to cylindrical, round or 2–6-angled in section; columella expanded at top into membranous epiphragm joined to tips of the 32 or 64 peristome teeth; calyptra glabrous or hairy.

4. POLYTRICHACEAE

With the characters of the order (if *Dawsonia* is included in this family). For a monograph of the family see G. L. Smith, *Mem. New York Bot. Gard.* **21** (3), 1–82, 1971.

5. Polytrichum Hedw., Sp. Musc., 1801

Dioecious. Shoots arising from underground rhizome-like stem. Lvs with broad sheathing basal part abruptly narrowed into erecto-patent to spreading, lanceolate to linear-lanceolate blade composed mainly of nerve with a very narrow lamina; nerve with lamellae on ventral (upper) surface, lamellae straight, chlorophyllose.

Capsule erect to horizontal, long or short, usually obscurely to sharply 4-angled, apophysis and stomata present at base of capsule; calyptra densely hairy, covering capsule; peristome teeth 64. Protonema ephemeral. A cosmopolitan genus of some 120 mainly calcifuge species.

Key to species of *Polytrichum* and *Pogonatum*

1 Lf margin entire, inflexed over blade, capsule 4-6-angled 2
Lf margin toothed, never inflexed over blade, capsule angled or smooth 5
2 Lf apex cucullate, nerve ending in apex or shortly excurrent, capsule obscurely angled **5. Polytrichum sexangulare**
Lf apex not cucullate, nerve excurrent in arista, capsule sharply angled 3
3 Nerve excurrent in long hyaline arista, not toothed at back above **6. Polytrichum piliferum**
Nerve excurrent in brownish arista, toothed at back above 4
4 Plants mostly 1–7 cm, tomentum if present brownish **7. Polytrichum juniperinum**
Plants mostly 6–20 cm, stems with dense off-white tomentum below **8. Polytrichum alpestre**
5 Plants to 2 cm high, unbranched, shoots arising from persistent protonema, capsule smooth 6
Plants (1–)2–40 cm, stems often branched and arising from underground rhizomes which decay leaving curved subterreanean ends attached to stem bases, capsule smooth or angled 7
6 Capsule ± spherical, exothecial cells finely papillose **1. Pogonatum nanum**
Capsule shortly cylindrical, rarely spherical, exothecial cells coarsely papillose **2. Pogonatum aloides**
7 Apical cells of lamellae grooved or flat-topped in section **4. Polytrichum commune**
Apical cells of lamellae rounded in section 8
8 Apical cells of lamellae papillose, capsule smooth 9
Apical cells of lamellae smooth, capsule angled 10
9 Lvs dull green, acuminate, lamellae 6–8(–9) cells high with apical cells strawberry-shaped in section **1. Polytrichum alpinum**
Lvs glaucous, acute, lamellae 5(–6) cells high with apical cell rounded or elliptical in section **3. Pogonatum urnigerum**
10 Capsule obscurely angled, lamina in mid-blade 6 or more cells wide, lamina cells 14–25 μm wide **2. Polytrichum longisetum**
Capsule sharply 4- to 6-angled, lamina in mid-blade 1–5 cells wide, lamina cells 10–16 μm wide **3. Polytrichum formosum**

1. P. alpinum Hedw., Sp. Musc., 1801
Pogonatum alpinum (Hedw.) Rohl., *Polytrichastrum alpinum* (Hedw.) G. L. Smith

Plants to *ca* 10 cm. Lvs erecto-patent to recurved when moist, 5–10 mm long (including sheathing base), blade linear-lanceolate, acuminate, margin toothed; lamellae 6–9 cells high, apical cells papillose, incrassate, enlarged, strawberry-shaped in section; nerve excurrent. Seta flexuose; calyptra not enclosing capsule; capsule sub-erect to inclined, asymmetrical, globose to sub-cylindrical, smooth when mature, rugose when old, not angled, exothecial cells smooth, a few stomata at base, apophysis hardly distinct; lid with long curved beak; peristome teeth irregular, short; spores 18–20 μm. Fr common, autumn.

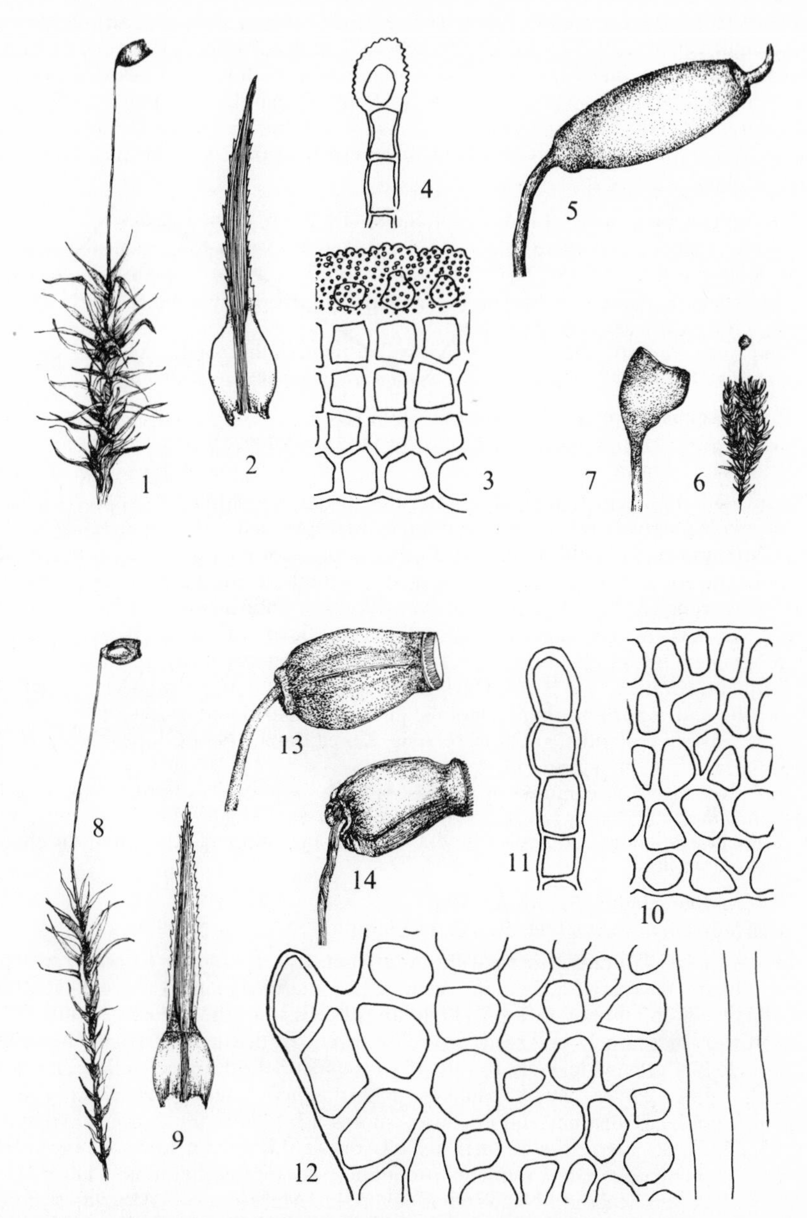

Fig. **34.** 1–5, *Polytrichum alpinum* var. *alpinum*: 1, shoot; 2, leaf; 3, part of lamella in side view; 4, section of lamella; 5, capsule. 6–7, *P. alpinum* var. *septentrionale*: 6, shoot; 7, capsule. 8–14, *P. longisetum*: 8, shoot; 9, leaf; 10, part of lamella in side view; 11, section of lamella; 12, marginal cells of leaf blade; 13 and 14, moist and dry capsules. Shoots $\times 1$, leaves and capsules $\times 5$, cells $\times 450$.

Var. **alpinum**

Plants to 10 cm. Lvs (5–)6–10 mm long. Seta 2–3 cm; capsule inclined, globose to sub-cylindrical, $n=7^*$, 14. Lax, dull green tufts or patches in turf on stony banks, amongst rocks, on cliff ledges and on moorland peat in montane areas. Absent from S.E. England and the Midlands, rare in S.W. England, occasional to frequent elsewhere. 72, H21. Europe, Faroes, Iceland, C. and E. Asia, Celebes, Congo, Kerguelen Is., N. America, Greenland, southern S. America, Australia, Tasmania, New Zealand, Antarctica.

Var. **septentrionale** (Sw.) Lindb., Not. Sällsk. Fl. Fenn. Förh., 1886

Smaller than var. *alpinum*, plants to 2 cm. Lvs to 5 mm long. Seta 0.5–2.0 cm; capsule suberect, ± globose. $n=7$. Montane habitats, very rare and not seen recently. W. Inverness, S. Kerry. E. Mayo, W. Donegal. 1, H3. N. Europe, Faroes, Iceland, Spain, Urals, Caucasus, N. America.

The papillose apical cells of the lamellae will distinguish this plant from other species of *Polytrichum*. For the differences from *Pogonatum urnigerum* see under that plant.

2. P. longisetum Sw. ex Brid., J. Bot. Schrad., 1801

P. aurantiacum Hoppe ex Brid., *P. gracile* Dicks., *Polytrichastrum longisetum* (Sw. ex Brid.) G. L. Smith

Plants 1.5–10.0 cm. Lvs erect, ± flexuose when dry, spreading, recurved when moist, blade narrowly lanceolate, acuminate, margin erect, toothed; lamellae usually 25–35 in number, 5–7 cells high, apical cell smooth, not enlarged, hardly incrassate; nerve excurrent in brown, denticulate arista; lamina at middle of blade 6 or more cells wide, cells 16–20(–25) μm wide. Seta flexuose, reddish-brown below, yellowish above, 1.5–6.0 cm; capsule erect at maturity, inclined with age, obloid, obscurely 5- to 6-angled, apophysis well defined; lid with long curved beak; spores 20–26 μm. Fr frequent, summer. $n=7^*$, 14. Dark green tufts or turfs on acidic, well drained soil on heaths and moorland and in woods, generally distributed, occasional. 94, H30. Europe, Faroes, Iceland, Caucasus, N. Asia, Korea, Japan, New Guinea, N. America, Greenland, Chile, New Zealand.

This plant has been confused in the past with *P. formosum*, as the key character for separating the two given by Dixon (1924) is unreliable. Sterile material of *P. longisetum* may be distinguished from *P. formosum* by the wider lamina with larger cells and the smaller number of lamellae.

3. P. formosum Hedw., Sp. Musc., 1801

Polytrichastrum formosum (Hedw.) G. L. Smith

Plants to 10(–20) cm. Lvs erect-flexuose when dry, spreading to recurved when moist, blade narrowly lanceolate, acuminate, margin plane or erect, toothed; lamellae to 70 in number, 5–7 cells high, apical cells smooth, incrassate, rounded in section, not or scarcely enlarged; nerve excurrent in denticulate arista; lamina in mid-blade 2–5 cells wide, cells 10–14(–16) μm wide. Seta flexuose, yellowish above, reddish below, 2.5–6.0 cm; capsule erect or inclined, rectangular, sharply 4(–6)-angled, apophysis distinct; lid rostrate; spores 12–16 μm. Fr summer, frequent. $n=7^*$, 14. Dark green tufts or turfs, usually on well drained acidic soils on heaths, moorland, in woods, etc. Frequent to common in suitable habitats. 111, H38, C. Europe, Faroes, Iceland, N., W. and C. Asia, Macaronesia. Algeria, southern Africa, N. America, Greenland, New Zealand.

4. P. commune Hedw., Sp. Musc., 1801

Plants 2–40 cm. Lvs flexuose when dry, spreading to squarrose when moist, blade narrowly lanceolate, acuminate, margin toothed; lamellae to *ca* 70 in number, 5–9

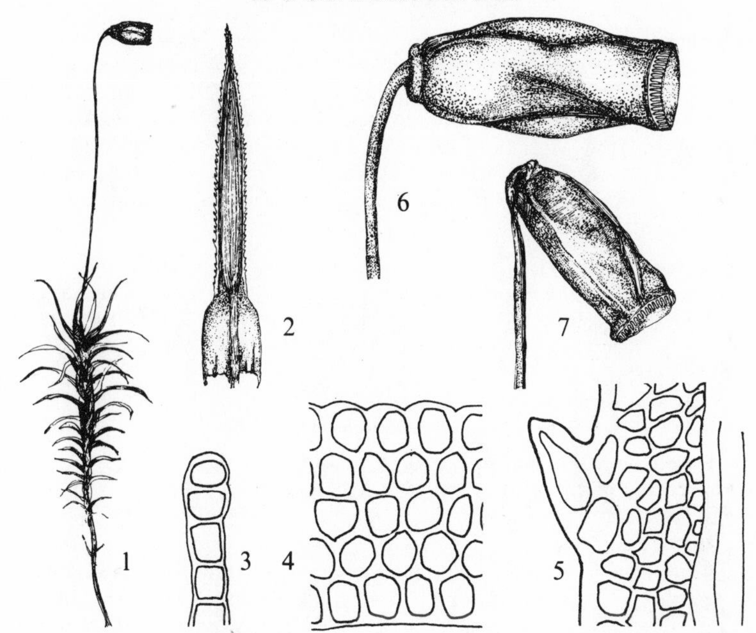

Fig. **35**. *Polytrichum formosum*: 1, shoot (×1); 2, leaf (×5); 3, transverse section of lamella (×450); 4, part of lamella in side view (×450); 5, marginal cells of leaf blade (×450); 6, moist capsule (×5); 7, dry capsule (×5).

cells high, apical cell grooved to flat-topped in section, larger than other cells, smooth; nerve excurrent in denticulate point; lamina in mid-blade 1–3 cells wide, cells 10–16 μm wide. Perichaetial lvs with long sheathing base gradually or abruptly tapering into acuminate apex, Seta flexuose, reddish, 3–9 cm; capsule erect at maturity, inclined with age, cubic to shortly rectangular, sharply 4-angled, slightly trapezoid in section, apophysis distinct; spores 8–12 μm. Fr frequent, summer.

Key to intraspecific taxa of *P. commune*

1 Plants (2–)5–40 cm, apical cells of lamellae grooved in section, 16–20 μm wide, inner perichaetial lvs toothed above — var. **commune**
Plants to 6 cm, apical cells of lamellae hardly grooved, or flat-topped, 10–14 μm wide, inner perichaetial lvs entire or only slightly toothed above — 2

2 Inner perichaetial lvs gradually tapering into long acuminate apex — var. **perigoniale**
Sheathing base of inner perichaetial lvs abruptly narrowed into short acuminate apex — var. **humile**

Var. **commune**

Plants (2–)5–40 cm. Lf margin sharply toothed; apical cells of lamellae grooved in section but those towards base of blade less deeply grooved or flat-topped, 16–20 μm wide. Inner perichaetial lvs denticulate towards apex. Capsule cubic to shortly rectangular. $n=7^*$, 14. Dark green tufts or turfs, sometimes of considerable size in bogs, by streams, on wet heath, moorland, etc., frequent to abundant in suitable

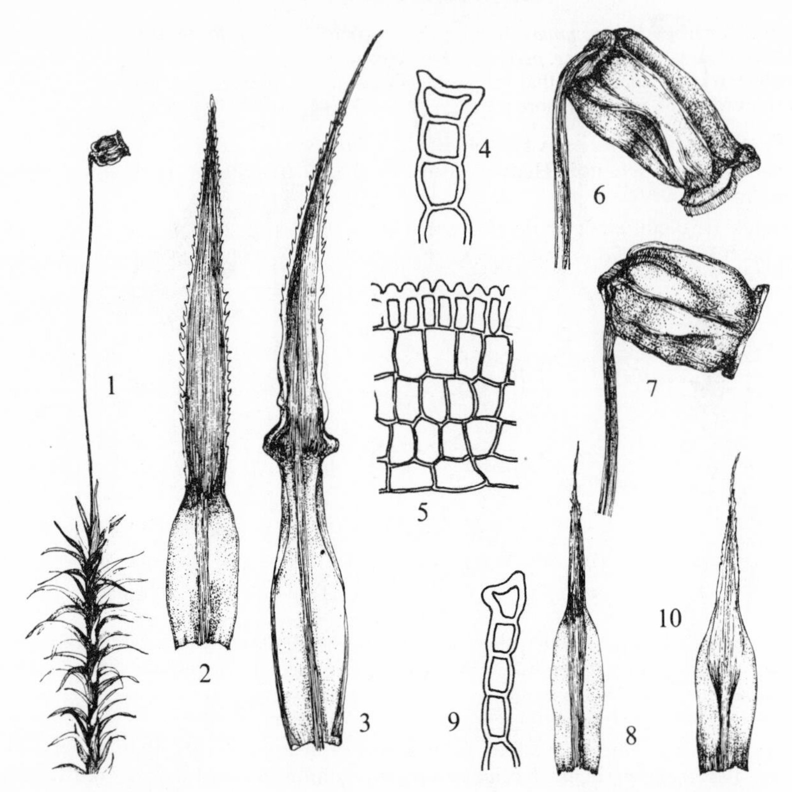

Fig. **36.** 1–7, *Polytrichum commune* var. *commune*: 1, shoot (× 1); 2, leaf; 3, perichaetial leaf; 4, section of lamella; 5, part of lamella in side view; 6 and 7, dry and moist capsules. 8–9, *P. commune* var. *perigoniale*: 8, perichaetial leaf; 9, section of lamella. 10, *P. commune* var. *humile*: perichaetial leaf. Leaves and capsules × 5, cells × 450.

habitats. 112, H40, C. Europe, Faroes, Iceland, Caucasus, N. and E. Asia. Macaronesia, E., W. and southern Africa, N. America, Greenland, Peru, Brazil, Australia, Tasmania, New Zealand, Chatham Is.

Var. **perigoniale** (Michx.) Hampe, Linnaea, 1839

Plants to *ca* 6 cm. Lvs less sharply toothed than in var. *commune*; apical cells of lamellae less deeply grooved to flat-topped, 10–14 μm wide in section. Perichaetial lvs gradually tapering to entire or slightly toothed longly acuminate apex. Capsule cubic. On soil and rock ledges in drier habitats than var. *commune*, rare. Scattered localities from S. Devon and Surrey north to Ross and Orkney, unknown in Ireland. 24. Europe, N. Africa, N. America, Australia.

Var. **humile** Sw., Adnat. Bot., 1829

P. commune var. *minus* De Not.

Similar to var. *perigoniale* but perichaetial lvs abruptly narrowed from sheathing base into short acuminate apex. Similar habitats to var. *perigoniale*, rare. Extending from W. Cornwall and E. Sussex north to Angus and E. Inverness, Down. 25, H1. Europe.

Some forms of var. *commune* have entire inner perichaetial lvs and the only difference between such plants and var. *perigoniale* is their larger size and the wider apical cells of the lamellae. It is very probable that var. *perigoniale* is only a drier habitat form of *P. commune*. The nature of var. *humile* is more problematical.

5. P. sexangulare (Flörke ex Hoppe) Brid., J. Bot. Schrad. 1801

P. norvegicum auct. non Hedw., *Polytrichastrum sexangulare* (Flörke ex Hoppe) G. L. Smith

Plants 1–10 cm, erect or decumbent. Lvs rigid, imbricate, incurved at tips when dry, patent when moist, narrowly to broadly ligulate, apex obtuse to acute, cucullate,

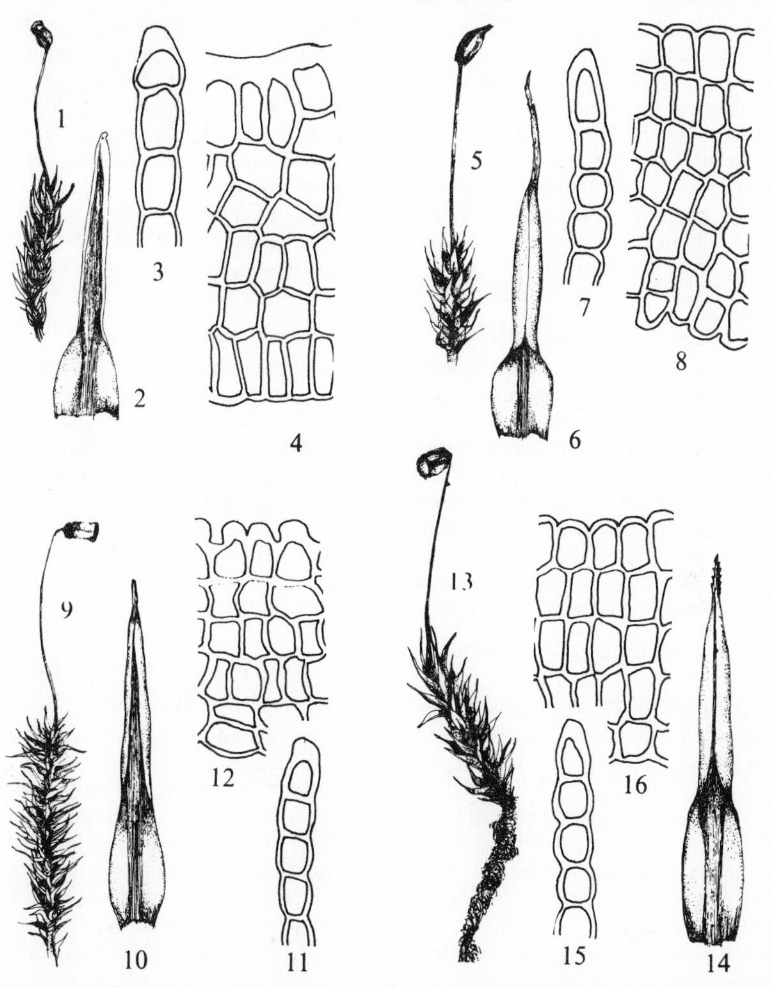

Fig. 37. 1–4, *Polytrichum sexangulare*: 1, shoot; 2, leaf; 3, section of lamella; 4, part of lamella in side view. 5–8, *P. piliferum*: 5, shoot; 6, leaf; 7, section of lamella; 8, part of lamella in side view. 9–12, *P. juniperinum*: 9, shoot; 10, leaf; 11, section of lamella; 12, lamella in side view. 13–16, *P. alpestre*: 13, shoot; 14, leaf; 15, section of lamella; 16, part of lamella in side view. Shoots × 1, leaves × 7.5, cells × 450.

margin entire, erect below, inflexed above; lamellae 30–35 in number, 5–7 cells high, apical cells smooth, incrassate, not or scarcely enlarged, rounded to conical in section; nerve ending in apex to shortly excurrent. Seta 2–3 cm, stout; capsule erect or inclined, obloid, bluntly 6-angled, apophysis scarcely distinct; lid with long beak; spores 18–20 μm. Fr rare, summer. $n=14$. Dark green tufts or patches on soil at altitudes of 900 m or more, often in areas of late snow-lie, in the Scottish Highlands, rare. Perth, Angus, S. Aberdeen, Banff, Inverness, Ross. 9. Europe, Faroes, Iceland, Pyrenees, N. Asia, N. America, Greenland.

6. P. piliferum Hedw., Sp. Musc., 1801

Plants (0.5–)1.0–6.0 cm. Lvs crowded towards stem apex, appressed, straight when dry, patent when moist, blade narrowly lanceolate, acute to acuminate, margins entire, ± meeting over middle of blade; lamellae *ca* 40, 4–8 cells high, crenulate, apical cells incrassate, conical, smooth, rounded in section; nerve not toothed at back above, excurrent in denticulate, hyaline hair-point to as long as blade. Seta deep red, 1–3 cm; capsule erect to inclined, obloid, 4(–6)-angled apophysis well defined; lid rostrate; spores 12–15 μm. Fr common, summer. $n=7$*. Glaucous green tufts or patches on exposed, well drained, acidic, mineral soil on heaths, moorland, etc., frequent to common. 112, H39, C. Europe, Faroes, Iceland, Caucasus, N., C. and E. Asia, Macaronesia, Congo, N. America, Greenland, Patagonia, Falkland and Juan Fernandez Is., Australia, Hawaii, Antarctica.

7. P. juniperinum Hedw., Sp. Musc., 1801

Plants 1–7(–10) cm; stems not or scarcely tomentose below, tomentum when present brownish. Lvs appressed, straight or with apices slightly flexuose when dry, patent when moist, blade narrowly lanceolate to linear-lanceolate, apex acute to acuminate, occasionally obtuse, margin entire, inflexed but not meeting over middle of blade except near apex; lamellae *ca* 70 in number, 5–8 cells high, entire or crenulate, apical cells incrassate, smooth; nerve toothed at back above, excurrent in brownish denticulate arista. Seta stout, red, 3–6 cm; capsule erect or sub-erect, sharply 4-angled, apophysis well defined; spores (8–)10–12 μm. Fr frequent, summer. $n=7$*. Greyish-green patches on usually well drained acidic soil on heaths, moorland, rocks, walls, etc., common. 112, H38, C. Cosmopolitan.

8. P. alpestre Hoppe, Bot. Taschenb., 1801

P. strictum Menz. ex Brid., *P. juniperinum* var. *affine* (Funck.) Brid., *P. juniperinum* var. *gracilius* Wahlenb., *P. juniperinum* ssp. *strictum* (Brid.) Nyl. & Säl.

Plants (3–)6–20 cm, stems slender, densely tomentose below with off-white tomentum. Lvs imbricate when dry, patent when moist, blade narrowly lanceolate, acuminate, margin entire, inflexed but not meeting over middle of blade except near apex; lamellae to *ca* 40 in number, 5–6 cells high, crenulate to serrulate, apical cells incrassate, smooth; nerve toothed at back above, excurrent in brownish arista. Seta slender, 3–4 cm; capsule erect or sub-erect, sharply 4–5-angled; spores (8–)10–12 μm. Fr occasional, summer. $n=7$*. Dense, matted tufts in bogs and on blanket peat, often amongst *Sphagnum*, frequent or common in suitable habitats. 81, H26. Europe, Faroes, Iceland, Caucasus, N. and E. Asia, N. America, Greenland, Patagonia, Antarctica.

The specific status of this plant is open to question. Most non-British authorities treat it as a variety or subspecies of *P. juniperinum* and occasional plants may be encountered that are intermediate in form and cannot be named. It is distinct from other large *Polytrichum* species in the tomentum of off-white or dirty white rhizoids which may clothe much of the stem.

6. Pogonatum P. Beauv., Mag. Enc., 1804

Similar to *Polytrichum* but capsule smooth, without apophysis or stomata; peristome teeth 32. Plants with persistent protonema or underground rhizome-like stem. A cosmopolitan genus for which some 210 species have been described.

For key to British and Irish species of *Pogonatum* see under *Polytrichum* (p. 90).

1. P. nanum (Hedw.) P. Beauv., Prodr., 1805
Polytrichum nanum Hedw.

Plants to 5(–10) mm. Lvs erecto-patent when moist, blade ± shortly oblong, bluntly pointed, margin plane or erect above, bluntly toothed from about mid-blade, teeth mostly composed of only 1 cell; lamellae 4–6(–10) cells high, apical cells smooth, rounded in section, not enlarged; nerve percurrent. Seta deep red, flexuose, 5–30 mm; capsule erect or inclined, spherical to ovoid, urceolate or turbinate when empty; exothecial cells finely papillose, lid rostellate; columella cylindrical; spores 24–27 μm. Fr common, winter. $n=7$. Dark green patches or scattered plants arising from dark green, persistent protonema, on acidic soil on banks, heaths and road-sides, occasional throughout Britain and Ireland. 99, H21, C. Europe, Iceland, Faroes, N. Asia, Canaries, Madeira, Algeria.

Sterile plants of *P. nanum* cannot be distinguished from depauperate specimens of *P. aloides*. *P. aloides* var. *minimum* may be mistaken for *P. nanum* but differs in the coarsely papillose exothecium, the 4-winged columella and spore size. *P. nanum* and *P. aloides* are distinct from *P. urnigerum* and the species of *Polytrichum* in that the shoots arise from a persistent protonema and not from an underground rhizome-like stem.

2. P. aloides (Hedw.) P. Beauv., Prodr., 1805
Polytrichum aloides Hedw.

Plants to *ca* 2 cm, arising from persistent protonema. Lvs erecto-patent to spreading when moist, blade lingulate to oblong-lanceolate, obtuse to acute, margin plane or erect, ± toothed from base of blade to apex, teeth usually of 2 to several cells; lamellae 5–6 cells high, marginal cells smooth, rounded in section, not enlarged; nerve ending in apex to slightly excurrent. Seta deep red; capsule erect or slightly inclined, ovoid to shortly cylindrical, frequently slightly asymmetrical or urceolate when empty; exothecial cells coarsely papillose; columella 4-winged; spores 8–12 μm. Fr common, autumn, winter.

Var. **aloides**

Lvs sharply toothed. Seta 1.5–4.0 cm; capsule obloid to shortly cylindrical. $n=7$*. Dark green patches or scattered shoots arising from dark green, felty, persistent protonema on acidic soil on banks, sides of ditches, often on a vertical or sloping substrate, frequent to common except in basic habitats. 111, H36, C. Europe, Faroes, Asia, Algeria, Congo, Uganda, Macaronesia, New Zealand.

Var. **minimum** (Crome) Mol., Jahresb. Naturw. Ver. Passau, 1875
Polytrichum aloides var. *minimum* (Crome) Rich. & Wall., *Polytrichum aloides* var. *dicksonii* Wallm.

Lvs less sharply toothed than in var. *aloides*. Seta very short, *ca* 1 cm; capsule obovoid. In similar habitats to var. *aloides*, rare. Scattered localities from W. Cornwall and Bucks north to Ross and Orkney, E. Cork, Down, Antrim. 31, H3. Europe, N. America.

3. P. urnigerum (Hedw.) P. Beauv., Prodr., 1805
Polytrichum urnigerum Hedw.

Plants to 7(–10) cm, shoots arising from underground rhizome-like stem, protonema not persisting. Lvs glaucous green, patent to spreading when moist, blade

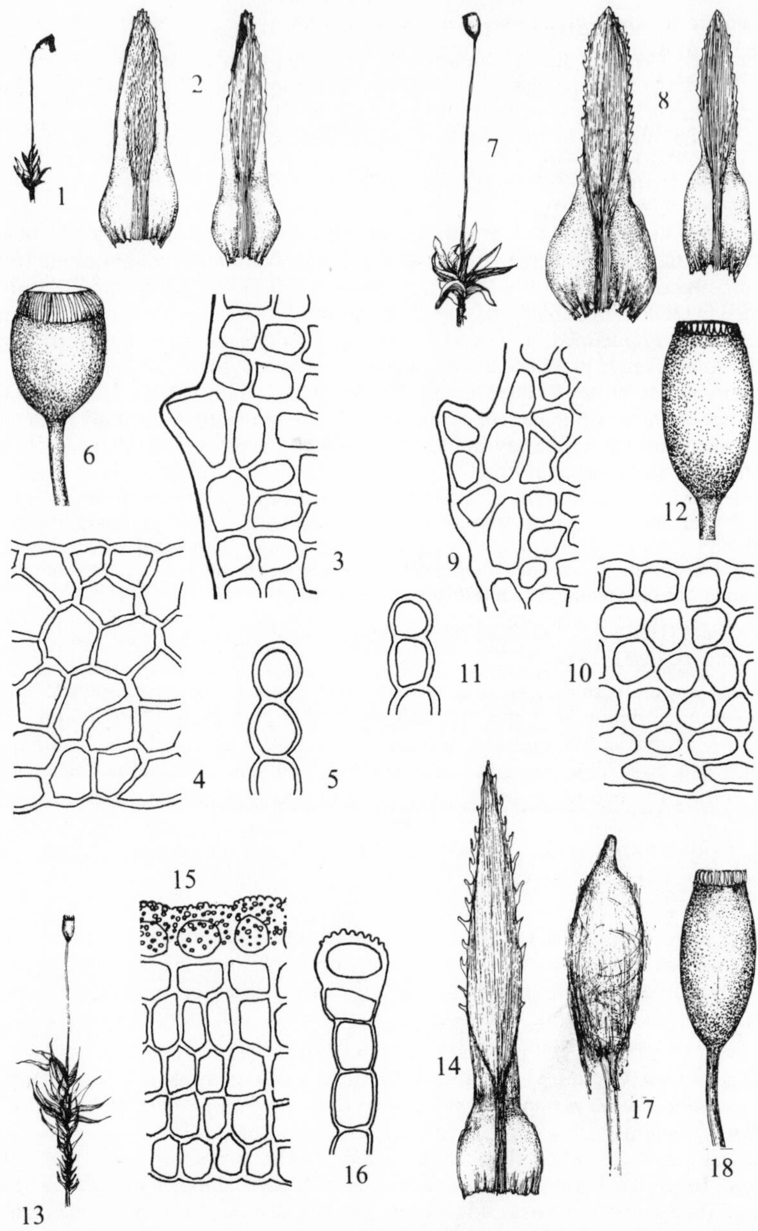

Fig. **38**. 1–6, *Pogonatum nanum*: 1, plant; 2, leaves; 3, leaf margin towards apex; 4, part of lamella in side view; 5, section of lamella; 6, capsule (×10). 7–12, *P. aloides*: 7, shoot; 8, leaves; 9, leaf margin towards apex; 10, part of lamella in side view; 11, section of lamella; 12, capsule (×7.5). 13–18, *P. urnigerum*: 13, shoot; 14, leaf; 15, part of lamella in side view; 16, section of lamella; 17, young sporophyte; 18, capsule (×7.5). Plants ×1, leaves ×7.5, cells ×450.

lanceolate to narrowly lanceolate, acute, margin plane with coarse spinose teeth; lamellae mostly 5–6 cells high, apical cells papillose, incrassate, enlarged; nerve excurrent. Seta long, 1–3(–5) cm; capsule shortly cylindrical, ± erect to inclined; exothecial cells coarsely papillose; spores 10–14 μm. Fr common, autumn, winter. $n=7$*. Usually glaucous green patches or scattered plants on well drained, acidic soil on banks, crevices of walls, roadsides, etc., occasional in S.E. England and the Midlands, frequent to common elsewhere. 102, H32, C. Europe, Iceland, Faroes, W., C. and E. Asia, Azores, Canaries, N. America, Greenland, Caribbean.

Although usually distinct from *Polytrichum alpinum* in its glaucous green colour, taller forms on damp ground may be of a duller green but differ from *P. alpinum* in the relatively shorter, broader lf blade, the shorter lamellae with apical cells rounded or elliptical in section and the ± erect capsule with papillose exothecial cells.

7. Oligotrichum Lam. & Cand., Fl. Fr., 1805

Dioecious. Lvs broadly lanceolate to lingulate, gradually tapering from broad base, concave, margin not bordered; nerve broad with tall, sinuose, longitudinal lamellae on ventral surface and a few longitudinal lamellae on dorsal surface; lamina 1- or 2-stratose. Calyptra with sparse hairs; capsule ovoid to sub-cylindrical, lacking apophysis, stomata present at base; peristome teeth 32. A ± world-wide genus of about 25 species.

1. O. hercynicum (Hedw.) Lam. & Cand., Fl. Fr., 1805

Plants 0.5–4.0(–9.0) cm, stems rigid. Lvs incurved to crisped when dry, erecto-patent, ± incurved when moist, narrowly triangular to lanceolate from broad base, apex cucullate, obtuse to acute, margin plane and obscurely toothed below, erect and bluntly toothed above; lamellae on dorsal and ventral surface of nerve in upper $\frac{2}{3}$ of lf, ventral lamellae to *ca* 12 in number, to 12 cells high, strongly sinuose, margins notched and crenulate, dorsal lamellae 2–4, widely spaced, 1–3 cells high; nerve stout, ending in apex. Seta yellow, 1.5–3.5 cm; capsule erect or slightly inclined, shortly cylindrical; lid rostrate; spores 12–15 μm. Fr occasional, spring. $n=7$*. Scattered plants or lax patches, yellowish-green to dark green or reddish-brown on damp, acidic soil, frequent in montane areas but sometimes descending to sea level. Cornwall, S. Devon, Wales, Stafford, Derby, Lancs and Yorks northwards. 66, H19. Europe, Faroes, Iceland, N. America, Greenland.

8. Atrichum P. Beauv., Prodr., 1805

Autoecious or dioecious. Lvs narrowly lingulate to ovate, crisped when dry, often undulate when moist, lamina often toothed at back above, margin with 1–3 stratose border of narrow cells, toothed, teeth single or double; nerve with 2–9 lamellae, 1–9 cells high on ventral surface, nerve toothed at back above. Calyptra glabrous, papillose at apex; capsule ovoid to cylindrical, straight to arcuate, erect or inclined; lid longly rostrate; peristome with 32 teeth. About 15 mainly temperate species. For a monograph of the genus see Nyholm, *Lindbergia* **1**, 1–33, 1971.

1 Cells in mid-lf 12–18(–20) μm wide, lamellae 4–7 in number, spores 12–14 μm **4. A. angustatum**
  Cells 20–50 μm wide, lamellae 1–6, spores 16–20 μm 2
2 Lvs strongly undulate, lingulate to narrowly lanceolate, not or scarcely narrowed towards insertion **3. A. undulatum**
  Lvs not or only slightly undulate, ± ovate to lanceolate, narrowed towards insertion 3

3 Lf cells 24–50 μm wide, lamellae 1–3(–4), 1–3 cells high, obscure **1. A. crispum**
Lf cells 20–30(–40) μm, lamellae 2–4(–5), (3–)6–9 cells high, obvious **2. A. tenellum**

1. A. crispum (James) Sull. & Lesq. in Grey, Man., 1856
Catharinea crispa James

Dioecious. Plants (0.5–)1.0–7.0 cm. Lvs crisped when dry, soft, patent, not or scarcely undulate when moist, ovate or oblong to oblong-lanceolate, sometimes obovate or oblanceolate, acute, narrowed at base, lamina without teeth at back above, margin toothed almost from base, teeth single or occasionally double; lamellae 1–3(–4) in number in upper half of lf only, 1–3 cells high, often obscure; nerve ending below apex; cells irregularly quadrate to ± hexagonal, obscurely papillose, longer near base, 24–50 μm wide in mid-lf, smaller towards margin. Female plants unknown in Britain or Ireland. $n=7$. Yellowish-green to green patches on gravelly or sandy soil by water, less commonly on peaty soil. Frequent and sometimes locally abundant in N.W. Wales and parts of N.W. England, rare or occasional in W. Cornwall, Devon, Staffs, I. of Man, Wicklow. 20, H1. N. America.

Differs from other British species of *Atrichum* in the lvs being crowded towards the stem apex and the obscure lamellae. Likely to be mistaken for a species of *Mnium* but differs in areolation, the nerve ending below the apex and the presence of lamellae. It seems probable in view of the absence of the species from other parts of Europe that it is an introduction from N. America. Although *A. crispum* spreads very rapidly in a suitable habitat, its mode of dispersal in Britain is unknown.

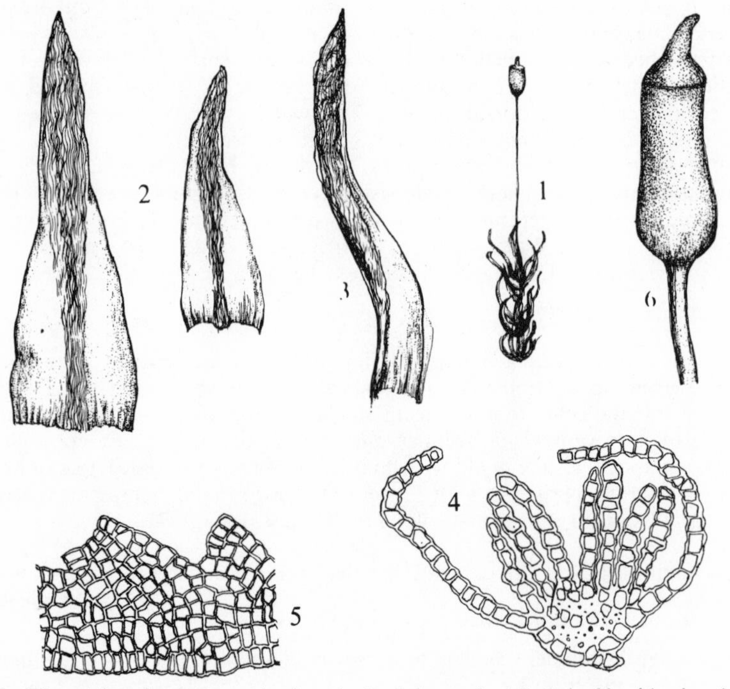

Fig. 39. *Oligotrichum hercynicum*: 1, plant (×1); 2, leaves (×15); 3, leaf in side view (×15); 4, transverse section of nerve (×180); 5, part of lamella in side view (×360); 6, capsule (×7.5).

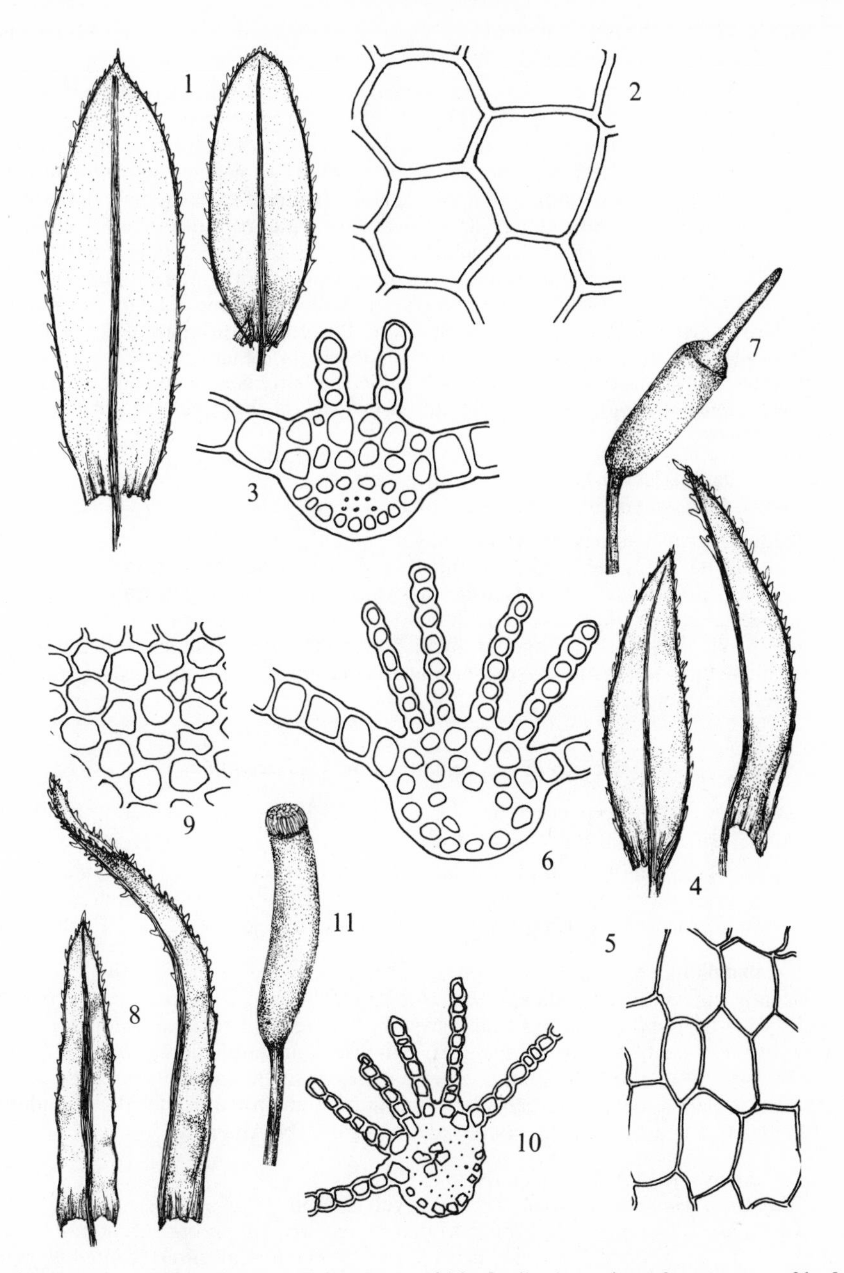

Fig. **40**. 1–3, *Atrichum crispum*: 1, leaves; 2, mid-leaf cells; 3, section of upper part of leaf. 4–7, *A. tenellum*: 4, leaves; 5, mid-leaf cells; 6, section of upper part of leaf; 7, capsule. 8–11, *A. angustatum*: 8, leaves; 9, mid-leaf cells; 10, section of upper part of leaf; 11, capsule. Leaves ×10, sections ×250, cells ×450, capsules ×7.5.

2. A. tenellum (Röhl.) Br. Eur., 1844
Catherinea tenella Röhl.

Dioecious. Plants to *ca* 1.5 cm. Lvs crisped when dry, soft, erecto-patent, not or scarcely undulate when moist, ovate to lanceolate, narrowed towards insertion, apex acute to acuminate, obtuse in the lower lvs, lamina with a few teeth on back above, margin with single or double spinose teeth from mid-lf; lamellae 2–4(–5) in number, (3–)6–9 cells high; nerve ending in apex; cells ± quadrate to irregularly hexagonal, larger, rectangular towards base, 20–30 (–40) μm wide in mid-lf, smaller towards margin. Seta yellowish; capsule inclined, shortly cylindrical, straight or curved; spores 20–25 μm. Fr rare, late summer to winter. $n=14$. Dull green patches on acidic soil in open places in woodland and by tracks. A few localities in Kent and Sussex, very rare elsewhere. W. Cornwall, Wilts, Ayr, Peebles, Midlothian, Fife, E. Perth, W. Inverness, E. Ross. 14. N. and C. Europe, Siberia, N. America.

May be confused with small forms of *A. undulatum* but differs in the relatively wider, scarcely undulate lvs with only a few teeth on the back of the lamina and in the fewer, taller lamellae.

3. A. undulatum (Hedw.) P. Beauv., Prodr., 1805
Catharinea undulata (Hedw.) Web. & Mohr

Plants to 7 cm. Lvs crisped when dry, soft, patent, strongly undulate when moist, lingulate to narrowly lanceolate, acuminate, not narrowed at insertion, lamina with rows of teeth on dorsal (under) surface above, margin spinosely dentate with single or double teeth from near base; lamellae 3–6 in number, 3–7 cells high; nerve ending in apex; cells irregularly hexagonal, incrassate, rectangular, thinner-walled towards base, 20–40 μm wide in mid-lf, smaller towards margin. Capsule inclined, cylindrical, arcuate. Fr common, winter.

Key to intraspecific taxa of *A. undulatum*

1 Seta yellowish, plants paroecious — var. **gracilisetum**
Seta reddish, plants usually autoecious — 2

2 Plants to 7 cm, seta 1.5–3.0 cm, capsule cylindrical, spores normal — var. **undulatum**
Plants rarely more than 2 cm high, seta to 1 cm, capsule shortly cylindrical, often malformed, spores irregular in size and often shrunken — var. **minus**

Var. **undulatum**

Usually autoecious. Plants to 7 cm. Antheridia terminal on first year's growth, archegonia terminal on second year's growth. Seta red, 1.5–3.0 cm, frequently 2 or more per perichaetium; capsule inclined, cylindrical, arcuate; spores 16–20(–28) μm. $n=7^*, 14^*, 21^*$. Yellowish-green to green patches or scattered plants on loose soil in woods, on banks, heaths, in fields, etc., common and sometimes locally abundant. 112, H40, C. Europe, Asia, Algeria, Madeira, Azores, N. America.

Var. **minus** (Hedw.) Paris, Ind. Bryol., 1903
Catherinea undulata var. *minor* (Hedw.) Web & Mohr

Autoecious. Gametophyte similar to that of var. *undulatum* but rarely more than 2 cm high. Seta reddish, to 1 cm; capsule shortly cylindrical, often stunted or malformed in appearance; spores variable in size, 20–40 μm, many apparently aborted. Usually in very small quantity in similar habitats to and growing with var. *undulatum*, occasional, generally distributed in Britain, unknown in Ireland. 49. Europe, N. Asia, N. America.

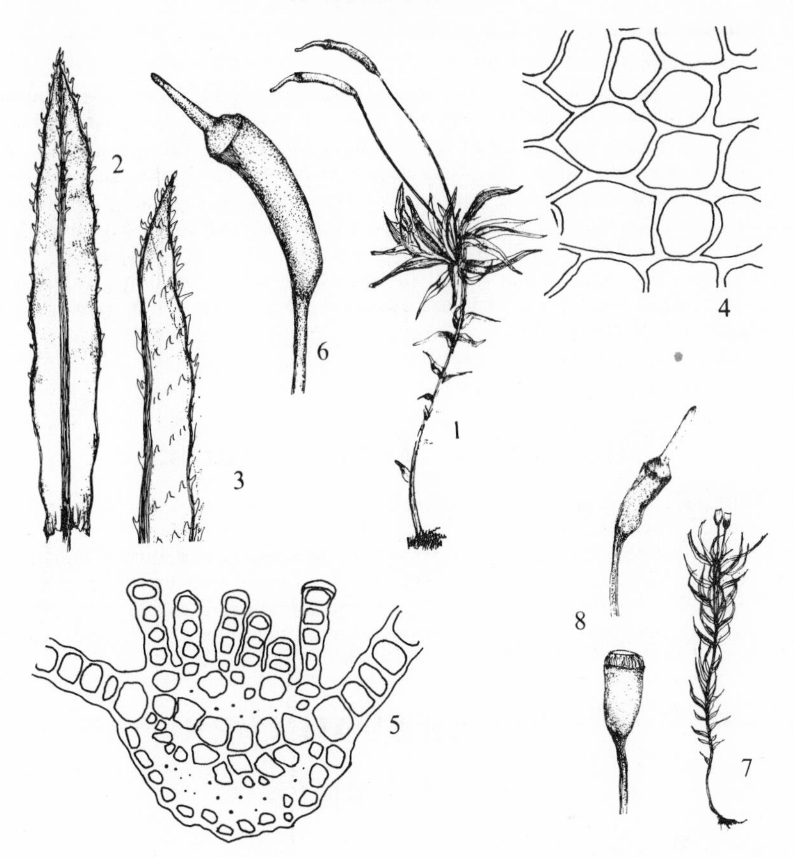

Fig. **41**. 1–6, *Atrichum undulatum* var. *undulatum*: 1, shoot (× 1); 2, leaf (× 7.5); 3, upper part of leaf in side view (× 15); 4, mid-leaf cells (× 450); 5, section of nerve (× 180); 6, capsule (× 5). 7–8, *A. undulatum* var. *minus*: 7, shoot; 8, capsules from different plants (× 5).

Var. **gracilisetum** Besch., Ann. Sc. Nat. Bot., 1893
Var. *haussknechtii* (Jur. & Milde) Frye

Paroecious. Antheridia terminal, archegonia outside perigonial lvs; stem continuing growth for 2 or more years so that persistent setae appear lateral. Seta yellow, to 2 cm, often several per perichaetium; spores 16–20(–28) μm. Similar habitats to the type, very rare and not seen recently. Surrey, Herts, Warwick, Worcs, Shrops, Notts, S. Lancs, Cumberland. 8. Europe, Caucasus, Siberia, Japan, Philippines, N. America.

The nature and relationships of the various chromosome races of var. *undulatum* require further investigation, although they are apparently morphologically indistinguishable. The frequently malformed capsule, variation in spore size and the presence of aborted spores in var. *minus* suggests that the sporophyte is genetically abnormal in some way. It is possible that it is the result of hybridisation between different cytotypes of var. *undulatum*. The sex of the plants is physiologically controlled and paroecious plants of var. *undulatum* may be encountered, but var. *gracilisetum* differs from these in its yellow setae.

4. A. angustatum (Brid.) Br. Eur., 1844
Catharinea angustata (Brid.) Brid.

Dioecious. Plants to 3 cm. Lvs crisped when dry, stiff, patent, ± smooth to strongly undulate when moist, narrowly lingulate to lingulate-lanceolate, obtuse to acute or acuminate, not narrowed towards insertion, lamina toothed at back above, margin dentate, sometimes spinosely so, from below mid-lf; lamellae (3–)4–7 in number, 5–9 cells high; nerve ending in apex; cells ± hexagonal, 12–18(–20) μm wide in mid-lf, smaller towards margin. Seta yellowish to purple; capsule narrowly cylindrical, erect and straight to inclined and curved; spores 12–14 μm. Fr rare, winter. $n=7$. Patches on acidic soil in woods, especially on damp rides, locally frequent in Kent and Sussex, very rare elsewhere. Dorset, Surrey, N. Essex, E. Glos, Worcs, Carms, W. Perth, Tyrone, Down. 11, H2. Europe, Caucasus, Japan, Azores, Madeira, N. and C. America, Caribbean.

Subclass 2. Buxbaumiideae

Acrocarpous. Plants small or minute. Lvs well developed or not; cells 1–3-stratose; perichaetial lvs ciliate. Seta short or long, straight; capsule large, oblique, immersed or exserted; peristome single or double, outer peristome of 1–4 rows of filiform teeth derived from several concentric layers of cells, faintly transversely barred, inner peristome membranous, conical, truncate, longitudinally plicate; calyptra very small.

A small group of reduced mosses placed between the *Polytrichideae* and *Eubryideae* because of the intermediate nature of the peristome.

5. BUXBAUMIALES

5. BUXBAUMIACEAE

The order and family with the characters of the subclass.

9. Diphyscium Mohr, Obs. Bot., 1803

Autoecious or dioecious. Plants gregarious, stems very short. Lvs lingulate to lanceolate, nerved, cells 2–3-stratose. Perichaetial lvs larger, ciliate at apex, with longly excurrent nerve. Antheridia several together. Capsule ± sessile, ovoid, oblique, asymmetrical; outer peristome absent or rudimentary, inner membranous with 16 longitudinal plicae. About 20 mainly northern hemisphere species.

1. D. foliosum (Hedw.) Mohr, Obs. Bot., 1803

Dioecious. Stems very short. Lvs crisped when dry, spreading, soft when moist, fragile, longly lingulate from a sometimes slightly sheathing base, margin finely crenulate, apex obtuse to acute; nerve ending below apex; cells incrassate, rounded-hexagonal, papillose, 2–3-stratose. 10–12 μm wide, basal cells thinner-walled, smooth, marginal band hyaline, rectangular. Perichaetial lvs ± lanceolate, acute to obtuse, ciliate; nerve excurrent in long, smooth or slightly denticulate setaceous point. Capsule immersed or slightly exserted, ovoid, asymmetrical; peristome white, becoming brown with age; spores 8–13 μm. Fr frequent, spring to autumn. $n=9$*. Patches or scattered plants on usually peaty soil, on banks, in rock crevices and by tracks, frequent in montane areas, very rare elsewhere. S.W. England, E. Sussex,

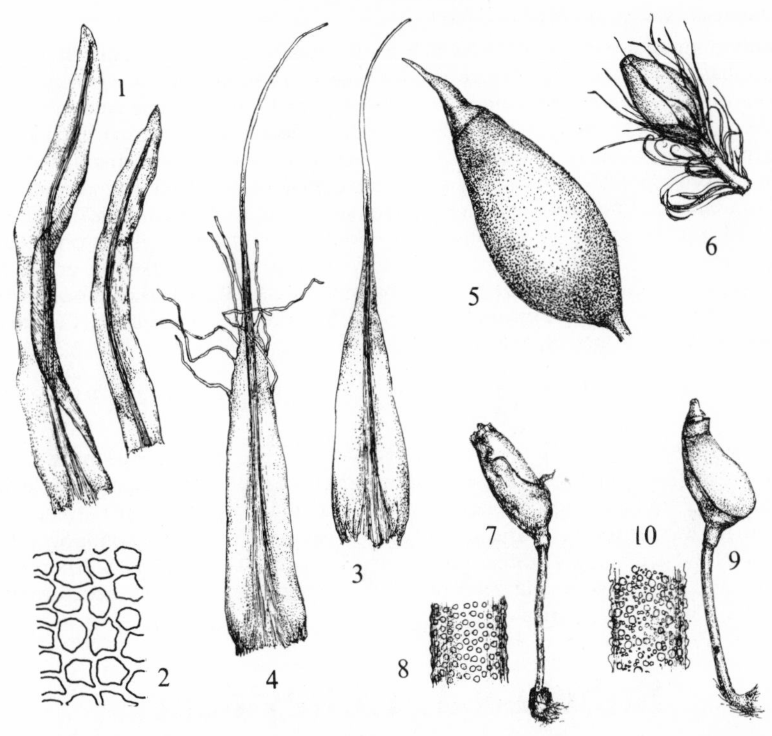

Fig. **42**. 1–6, *Diphyscium foliosum*: 1, leaves (×10); 2, leaf cells; 3 and 4, outer and inner perichaetial leaves (×7.5); 5, capsule (×7.5); 6, stem with capsule (×5). 7–8, *Buxbaumia viridis*: 7, sporophyte (×3); 8, part of seta (×25). 9–10, *B. aphylla*: 9, sporophyte (×3); 10, part of seta (×25).

Oxford, Wales, Staffs and Shrops northwards. 68, H14. N., W. and C. Europe, Faroes, Iceland, Caucasus, Japan, Madeira, Azores, N. and C. America.

When fertile readily recognised by the oblique sessile capsule but sterile plants may be mistaken for *Tortella* species. It differs from these in the 2–3 stratose lvs with poorly defined nerve. The var. *acutifolium* (Lindb. & Braithw.) Dix. intergrades with the type to such an extent that it is not worth maintaining.

10. Buxbaumia Hedw., Sp. Musc., 1801

Dioecious. Plants arising from brownish protonema, stems minute. Lvs and perichaetial lvs minute, ciliate, nerveless. Male plant consisting of single lf enclosing 1 spherical antheridium. Seta long, stout, papillose; capsule large, inclined, obliquely ovoid, asymmetrical; outer peristome of 1–4 rows irregular filiform teeth, inner tubular, conical, truncate, with 32 longitudinal plicae. Ten species distributed through Europe, E. Asia, N. Africa, N. America. Australia and New Zealand.

Capsule glossy, upper surface flattened, cuticle not peeling from back of capsule **1. B. aphylla**

Capsule dull, upper surface scarcely flattened, cuticle splitting longitudinally and peeling from back of ripe capsule **2. B. viridis**

1. B. aphylla Hedw., Sp. Musc., 1801

Plants minute, arising from brownish protonemal mat. Perichaetial lvs minute, ovate, ciliate, cilia becoming filamentous with age and forming protonemal-like mass at base of seta. Male plants on side of archegonial shoots. Seta erect, coarsely papillose, (3–)5–10(–20) mm; capsule with short neck, glossy brown, inclined to horizontal, ± obliquely ovoid, upper surface flattened with ± angular edges, exothecial cells sparsely and bluntly papillose, cuticle peeling back from mouth but not from back of ripe capsule; outer peristome teeth in single row, filiform; spores (8–)10–13 μm. Fr spring to autumn. $n=8$. Solitary or scattered plants on usually humus-rich, acidic, sandy soil or occasionally on rotting wood in open or shaded places, especially coniferous woods, also coal-mine waste. Rare and of sporadic and often ephemeral occurrence throughout Britain, S. Kerry. 40, H1. Europe, Caucasus, N. Asia, Japan, N. America, New Zealand.

2. B. viridis (Moug. ex DC) Brid. ex Moug. & Nestl., Stirp. Crypt. Vog. Rhen., 1823
B. indusiata Brid., *B. foliosa* auct.

Similar to *B. aphylla* but seta less coarsely papillose, capsule pale brown, not glossy, ± obliquely ellipsoid, scarcely flattened or angled on back, cuticle splitting longitudinally and peeling off back of capsule at maturity; outer peristome of 4 concentric rows of teeth; spores *ca* 10 μm. Fr summer. $n=8$. Decaying wood in coniferous and deciduous woodland. Very rare. Recorded from Angus, S. Aberdeen, E. Ross and E. Inverness but seen recently only from the last. Europe, Corsica, Caucasus, China, N. America

Subclass 3. Eubryideae

6. ARCHIDIALES

Plants small, stems innovating from below perichaetium. Stem lvs small, lanceolate, perichaetial lvs larger, base sheathing. Seta very short; capsule cleistocarpous, globose, wall of 1 layer of translucent cells, columella lacking, spores 16–20 in number, arising from single basal cell.

6. ARCHIDIACEAE

The only family.

11. ARCHIDIUM BRID., BR. UNIV., 1826

The only genus; about 35 species distributed ± throughout the world.

1. A. alternifolium (Hedw.) Schimp., Syn., ed. 2, 1876

Plants to 2 cm, stems branched or simple, often innovating from below perichaetia. Lvs distant, erecto-patent when moist, lanceolate to lanceolate-subulate, acute to acuminate, margin plane, denticulate; nerve ± ending in apex; cells ± incrassate, basal rectangular, above rhomboidal or narrowly hexagonal. Perichaetial lvs larger with broad sheating base. Capsule immersed, cleistocarpous, globose, pellucid; spores angular, 100–200 μm. Fr frequent, spring. Dense tufts or patches on damp, often sandy soil in fields, on moorland, wood rides and by rivers, frequent. 92, H22, C. Europe, Faroes, Iceland, Middle East, Azores, southern Africa, N. America.

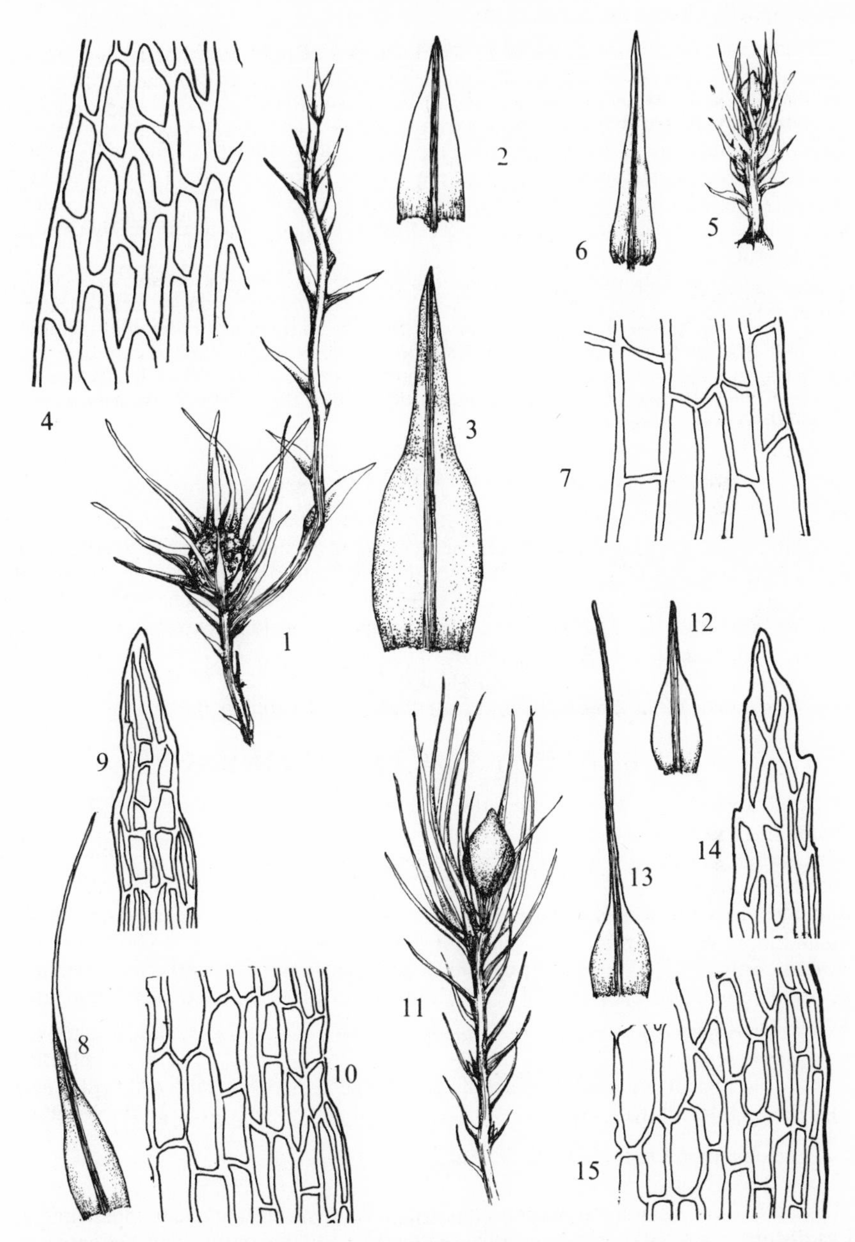

Fig. **43**. 1–4, *Archidium alternifolium*: 1, plant (×15); 2, leaf (×50); 3, perichaetial leaf (×50); 4, leaf cells. 5–7, *Pseudephemerum nitidum*: 5, plant (×10); 6, leaf; 7, cells near leaf base. 8–10, *Pleuridium acuminatum*: 8, perichaetial leaf; 9, leaf apex; 10, cells near leaf base. 11–15, *P. subulatum*: 11, plant (×10); 12, stem leaf; 13, perichaetial leaf; 14, leaf apex; 15, cells near leaf base. Leaves ×20, cells ×450.

7. DICRANALES

Plants large or small, apical cell of stem with 3 cutting faces. Lvs subulate or linear to broadly ovate; nerve present; upper cells quadrate to rectangular, usually smooth, basal cells quadrate to linear, sometimes porose, differentiated angular cells sometimes present. Capsule on long seta, rarely sessile, ovoid to cylindrical, frequently curved, columella present; peristome teeth 16, usually bifid, transversely articulated with fine vertical striae between the articulations (i.e. dicranoid teeth), rarely capsule cleistocarpous.

The disposition of the genera of the order into families is somewhat problematical as views on the delimitation of the families vary considerably. Attempts at producing a classification using numerical techniques have been unsuccessful. The conventional division of the order into Ditrichaceae, Seligeriaceae, Dicranaceae and Leucobryaceae is followed here, although the Leucobryaceae would probably be better regarded as a subfamily of the Dicranaceae, the difference between the 2 families being mainly the form of the nerve with *Paralencobryum* being intermediate between them.

7. DITRICHACEAE

Plants usually small, stem with central strand. Lvs usually lanceolate, acuminate or subulate, upper lvs longer than lower, nerve percurrent or excurrent, in section with guide cells and 2 stereid bands; cells smooth, quadrate to elongate in upper part of lf, larger, longer towards base, angular cells not differentiated. Seta short to long; capsule immersed or emergent and cleistocarpous, or long exserted and stegocarpous, globose to cylindrical, straight or slightly curved; lid conical; annulus of large cells or absent; peristome teeth 16, usually perforated or bifid nearly to base, papillose above, striate below; calyptra cucullate or occasionally mitriform.

SUBFAMILY 1. DITRICHOIDEAE

Neck of capsule short with few stomata; annulus present in stegocarpous genera.

12. Pleuridium Brid., Mant. Musc., 1819

Small plants with usually unbranched stems. Perichaetial lvs with subulate points, much longer than stem lvs. Seta very short; capsule erect, cleistocarpous, immersed, stomata only at base of capsule; calyptra cucullate. A mainly northern hemisphere genus of about 30 species, a few of which, however, occur in S. America and Australia.

Antheridia naked amongst perichaetial lvs, perichaetial lvs gradually narrowed into subula **1. P. acuminatum**
Antheridia in persistent axillary dwarf branches, perichaetial lvs abruptly narrowed into subula **2. P. subulatum**

1. P. acuminatum Lindb., Öfv. K. V. A. Förh., 1863
P. subulatum (Huds.) Rabenh.

Paroecious. Lower lvs spreading, lanceolate or ovate, acuminate, upper longer, subulate, ± erect or slightly secund, merging into perichaetial lvs, margin entire or slightly denticulate, nerve broad, obscure below, excurrent in upper lvs; basal cells rectangular, cells in subula ± linear. Perichaetial lvs gradually narrowing from ovate basal part into long subula. Antheridia naked, amongst perichaetial lvs. Capsule immersed, cleistocarpous, ovoid with blunt apiculus; spores 22–30 μm. Fr common, spring, summer. $n=13$*, 26. Green or yellowish-green tufts or patches on soil in arable fields, waste places, heaths, moorland, open places in woodland, etc. Common, sometimes locally abundant. 106, H22, C. Europe, China, Macaronesia, N. America.

This and the next species have sometimes been confused, as intermediate forms occur and the situation has not been helped by nomenclatural confusion. The 2 species sometimes grow together and it is possible that the intermediate forms are hybrids. Plants of *P. acuminatum* with capsules are usually devoid of antheridia but the dwarf antheridial shoots of *P. subulatum* persist.

2. P. subulatum (Hedw.) Lindb., Öfv. K. V. A. Förh., 1863
P. alternifolium of Dix., Handb., not Rabenh.

Very close to *P. acuminatum*. Autoecious. Perichaetial lvs abruptly narrowed from ovate basal part into long, denticulate setaceous subula; cells of subula rectangular to narrowly rectangular. Antheridia in persistent, axillary, dwarf shoots. Fr common, spring, summer. $n=13$*. In green to yellowish-green tufts or patches on damp soil in fields, waste places, woods, etc. Occasional to frequent, less common than *P. acuminatum*. 80, H8, C. Europe, Caucasus, Middle East, N. Africa, Macaronesia, N. America, New Zealand.

13. Pseudephemerum Lindb., Hag., K. Norsk. Vid. Selsk. Skrift., 1910

Synoecious, antheridia naked among perichaetial lvs. Upper and perichaetial lvs similar, lanceolate, acuminate; nerve ending below apex. Capsule cleistocarpous, immersed.

Although usually placed in the Dicranaceae the genus is here placed in the Ditrichaceae because of its very close affinity to *Pleuridium* on cytological grounds.

1. P. nitidum (Hedw.) Reim., Verh. Bot. Ver. Brandenburg, 1933
Pleuridium axillare (Sm.) Lindb.

Plants variable in size, to 5 mm. Lvs increasing in size up stem, upper and perichaetial patent when moist, lanceolate, longly acuminate, margin entire below, denticulate above; nerve ending below apex; cells thin-walled, pellucid, rectangular to narrowly rectangular throughout. Seta very short, *ca* 0.5 mm; capsule immersed, cleistocarpous, ovoid with conical beak, thin-walled, orange-brown when ripe; spores 20–32 μm. Fr common, throughout year but especially late summer and autumn. $n=13$*. Soft, pale green patches or scattered plants on soil on banks, wood rides, damp cultivated ground, etc., common at low altitudes. 100, H22, C. Europe, Azores, Algeria, Morocco, C. Africa, N. America.

Distinguished from *Pleuridium* species by the upper lvs being of similar length to the perichaetial lvs and, because of the shorter perichaetial lvs, the more conspicuous capsules.

14. Ditrichum Hampe, Flora, 1876

British species dioecious except *D. subulatum*. Lvs from broad base gradually or abruptly narrowed into long or short, frequently channelled subula; nerve often broad, ending just below apex to excurrent; cells in upper part of lf quadrate or rectangular, smooth (except *D. zonatum* var. *scabrifolium*), below rectangular to ± linear. Seta long; capsule erect or inclined; peristome teeth 16, divided into 2 filiform, papillose segments. A cosmopolitan genus of about 90 species.

1 Upper lvs squarrose, abruptly narrowed into long subula **1. D. cylindricum**
Upper lvs not squarrose, variously narrowed into subula 2

2 Plants usually 2–6(–10) cm tall, lvs with long fine subula, cells in upper part of lf oval, unistratose **2. D. flexicaule**
Plants to 1.5 cm or if more then lvs without long fine subula and lamina cells bistratose above, upper cells quadrate to rectangular 3
3 Paroecious, lvs abruptly narrowed into long subula **3. D. subulatum**
Dioecious, lvs gradually tapering, upper part subulate or not 4
4 Lf margin recurved, denticulate above **6. D. pusillum**
Lf margin plane or narrowly recurved, entire or with a few denticulations near apex only 5
5 Lvs tapering to long acumen composed largely of percurrent to excurrent nerve, or if shortly pointed then lamina cells in upper part of lf bistratose 6
Lvs ± shortly pointed, nerve ending below apex to percurrent and upper part of lf not composed largely of nerve, lamina cells unistratose 7
6 Nerve occupying most of upper part of lf, cells in lower part of lf narrowly rectangular to linear, plants to *ca* 1 cm **4. D. heteromallum**
Cells of lamina extending ± to lf apex, cells in lower part of lf quadrate to rectangular, plants 0.5–5.0 cm **5. D. zonatum**
7 Lvs erecto-patent to patent when moist, cells in lower part of lf 7–12 μm×15–25 μm, rhizoidal gemmae present **9. D. cornubicum**
Lvs appressed when moist, cells in lower part of lf 5–9 μm×15–42 μm, rhizoidal gemmae absent 8
8 Upper lvs 3–4 times as long as wide, margin recurved on one or both sides **7. D. lineare**
Upper lvs 2–3 times as long as wide, margin plane **8. D. plumbicola**

1. D. cylindricum (Hedw.) Grout, Moss. Fl. N. Am. 1936
D. tenuifolium Lindb., *Trichodon cylindricus* (Hedw.) Schimp.

Plants to 5 mm, stems usually simple. Upper lvs squarrose, flexuose, from oblong sheathing basal portion abruptly narrowed to long subula composed mainly of nerve, denticulate all round, not just at margin, margin plane; cells in sheathing base rhomboidal, 2–3 marginal rows narrowly rectangular, upper cells rectangular. Brownish, ovoid rhizoidal gemmae, to *ca* 120 μm long, often present. Seta pale red; capsule cylindrical, straight or slightly curved. Fr very rare, summer. $n=12$, 13, 24. Dull, yellowish-green tufts, patches or scattered plants on damp soil in fields, roadsides, banks, open spaces in woods, etc., common in lowland areas, rare elsewhere. 100, H20, C. Europe, Faroes, Siberia, Japan, N. America, New Zealand.

The squarrose, longly subulate lvs will easily distinguish *D. cylindricum* from most other arable mosses except *Dicranella schreberana* (q.v.). *Pleuridium* species have immersed capsules and are autoecious or paroecious; *Dicranella staphylina* has much shorter, non-squarrose lvs.

2. D. flexicaule (Schimp.) Hampe, Flora, 1867

Plants to 6(–10) cm, stems fragile, tomentose below. Lvs erect to secund or falcato-secund, 1–4 mm long, gradually tapering from sheathing base to very long, channelled, entire or scabrous subula, margin erect or inflexed; nerve broad, indistinct below, longly excurrent; basal cells rectangular, becoming rhomboidal or oval above, basal marginal cells sometimes very narrow. Capsule ellipsoid, slightly curved. Fr very rare. Yellowish-green to dark green, glossy, silky, loose or dense tufts or patches on basic soil in grassland, on cliffs, sand-dunes, etc., rarely on acidic soil.

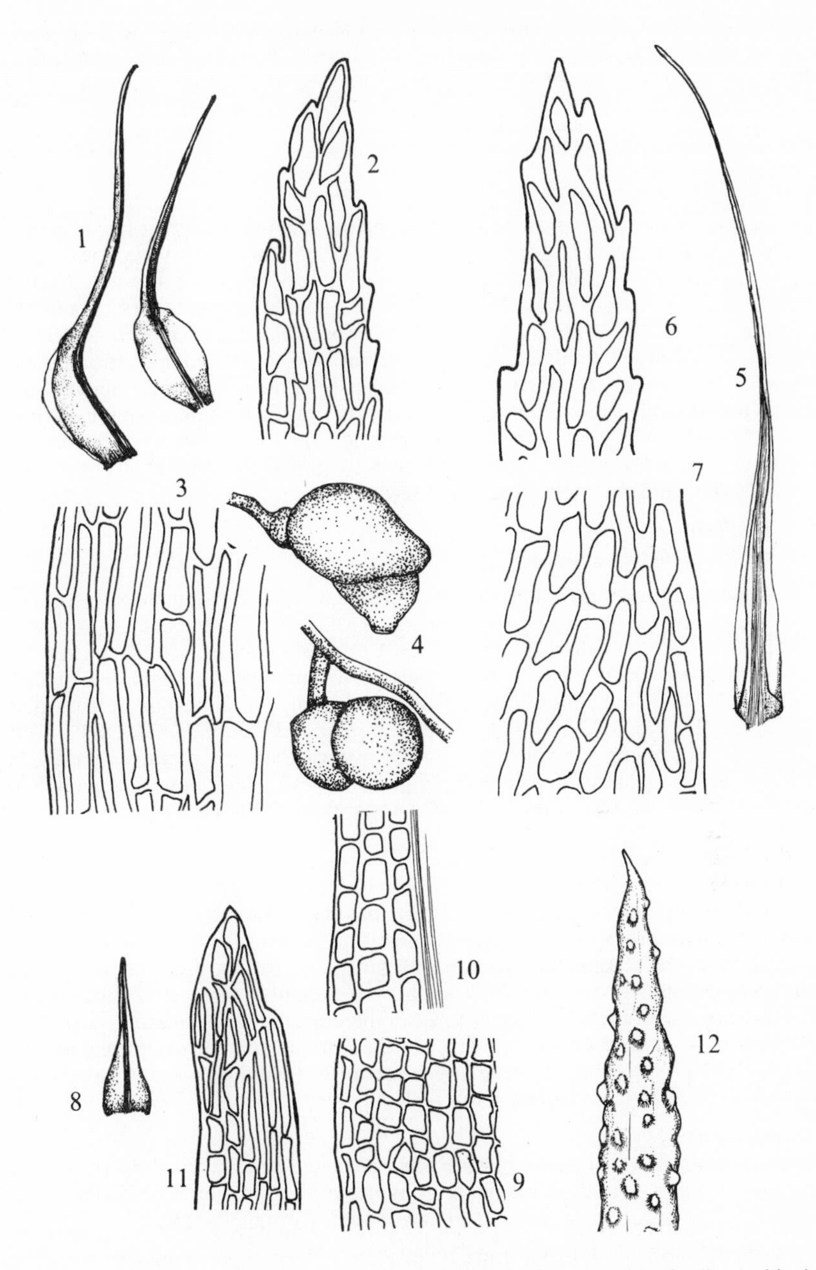

Fig. 44. 1–4, *Ditrichum cylindricum*: 1, leaves (×20); 2, leaf apex; 3, basal cells; 4, rhizoidal gemmae (×250). 5–7, *D. flexicaule*: 5, leaf (×15); 6, leaf apex; 7, basal cells. 8–11, *D. zonatum* var. *zonatum*: 8, leaf (×20); 9, basal cells; 10, cells at middle of leaf; 11, leaf apex. 12, *D. zonatum* var. *scabrifolium*: leaf apex (×250). Cells ×450.

Common in basic areas, rare elsewhere, 102, H38, C. Europe, Faroes, Caucasus, Siberia, China, Algeria, Canaries, Madeira, N. America, Greenland, New Zealand.

May be confused with *Distichium inclinatum*, *Dicranodontium asperulum* or *D. uncinatum*. The first differs in the distichous lvs and cells rectangular in upper part of lf. The *Dicranodontium* species have the lf margin denticulate from about halfway and the angular cells inflated, hyaline and extending to nerve.

3. D. subulatum Hampe, Flora, 1867

Paroecious. Plants to 8 mm. Lvs distant below, crowded above, lower small, narrowly lanceolate, upper larger, from ovate or oblong base abruptly narrowed into long, channelled subula composed largely of nerve, uppermost lvs and perichaetial lvs falcato-secund, margin plane below, erect or inflexed above, entire or slightly denticulate above; nerve longly excurrent in upper lvs; basal cells in lower lvs quadrate to quadrate-rectangular, in upper lvs variable, rectangular or rhomboidal to narrowly rectangular, 8–14 μm wide, narrower towards margin and above. Seta yellowish-red; capsule ovate-ellipsoid, spores 14–18 μm. Fr common, winter, spring. $n = 13, 14^*$. Yellowish-green, silky tufts or patches on shaley soil, in rock crevices and earthy or rocky roadside banks in Cornwall, S. Devon and E. Kent. 4. W. Europe, Mediterranean region, Madeira, Tenerife.

4. D. heteromallum (Hedw.) Britt., N. Am. Fl., 1913
D. homomallum (Hedw.) Hampe

Plants seldom more than 1 cm, loosely tufted. Lvs erecto-patent to sub-secund, from ovate to lanceolate basal part tapering to channelled subula, margin plane, entire; nerve ill-defined below, occupying most of upper part of lf and excurrent in slightly denticulate point; cells in lower part of lf narrowly rectangular to linear, 6–10 μm × (24–)40–80 μm, above linear by nerve, rectangular at margin. Seta purple; capsule narrowly ellipsoid, ± straight. Fr common. $n = 13^*$. Slightly silky, yellowish-green tufts or patches on sandy or gravelly soil by paths, among rocks, on banks, in woods, etc. Rare to occasional in S.E. England and the Midlands, occasional to frequent elsewhere. 93, H27. Europe, Faroes, Iceland, N. America.

5. D. zonatum (Brid.) Limpr., Laubm. I., 1887
D. heteromallum var. *zonatum* (Brid.) Lindb.

Dioecious. Plants fragile, shoots 0.5–5.0 cm, forming dense tufts, often with annual growth zones. Lvs erect to erecto-patent, tapering from ovate or ovate-lanceolate basal part to short, channelled acumen, margin plane below, erect or sometimes slightly incurved above, apex entire or slightly denticulate; nerve stout but lamina cells extending ± to apex, ending in apex or percurrent; cells in lower part of lf rectangular, (5–)6–8 μm × 10–24(–32) μm, towards margin and above shorter, shortly rectangular or quadrate, bistratose in upper part of lf. Fr unknown in Britain. Tufts glossy green above, reddish-brown below.

Var. **zonatum**

Upper cells and back of nerve smooth. On rocks and in rock crevices, usually at high altitudes, rare. Scattered localities in Wales and from Northumberland and Westmorland northwards, Antrim. 20, H1. W. and C. Europe, Pyrenees.

Var. **scabrifolium** Dix., J. Bot. Lond., 1902

Cells in upper part of lf and back of nerve scabrous with large conical papillae. Damp rock crevices, especially where basic, at high altitudes, rare. Merioneth, Caerns, Kirkcudbright, Scottish Highlands, Outer Hebrides, W. Galway, W. Donegal. 10, H2. Endemic.

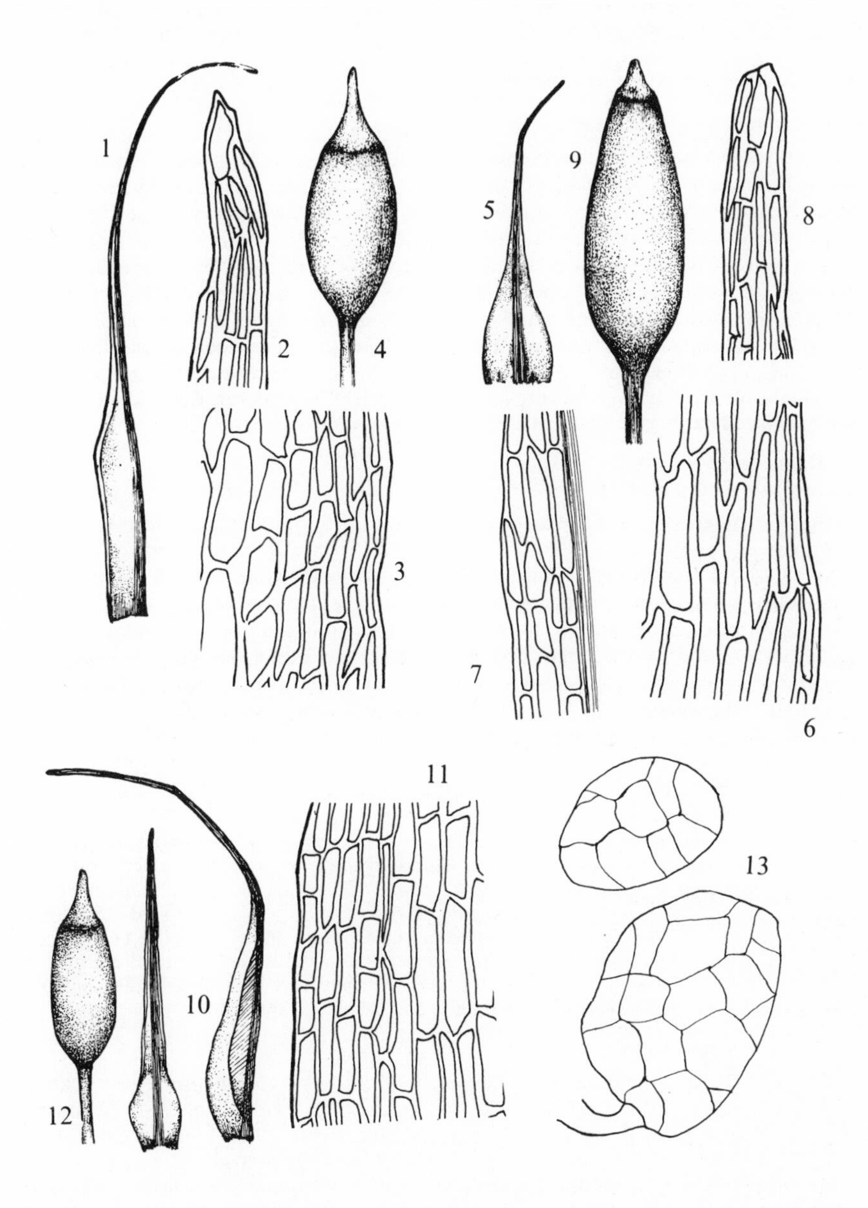

Fig. **45**. 1–4, *Ditrichum subulatum*: 1, leaf; 2, leaf apex; 3, basal cells; 4, capsule. 5–9, *D. heteromallum*: 5, leaf; 6, basal cells; 7, cells at middle of leaf; 8, leaf apex; 9, capsule. 10–13, *D. pusillum*: 10, leaves; 11, basal cells; 12, capsule; 13, rhizoidal gemmae (×250). Leaves ×20, cells ×450, capsules ×20.

D. zonatum is distinct from *D. heteromallum* in the shorter cells and the lamina extending ± to lf apex. Var. *scabrifolium* is very distinctive although the papillae may be lost in old specimens; as there may be some ecological differences between it and var. *zonatum* its taxonomic status requires further investigation.

6. D. pusillum (Hedw.) Hampe, Flora, 1867
D. tortile (Schrad.) Brockm.

Plants small, to 10 mm, stems usually unbranched. Lvs erecto-patent, slightly secund, upper lanceolate, narrowing to channelled subula, lower smaller, ovate-lanceolate, acute, margin denticulate, recurved at least in middle of lf; nerve stout, percurrent or excurrent; cells in lower part of lf narrowly rectangular, 10–12 μm wide, narrower and shorter towards margin, cells above rectangular. Yellow or yellowish-brown, ± pyriform rhizoidal gemmae, 100–150 μm × 75–100 μm, abundant on sterile plants, occasional on fertile plants. Seta red; capsule ovoid to cylindrical. Fr common, winter. $n=13$*. Dull green tufts, rare on sandy soil on banks and in quarries, occasional and possibly overlooked in non-calcareous arable fields. 20, H6. Europe, Iceland, Caucasus, Siberia, Amur, India, Algeria, N. America.

Plants in arable fields may be very depauperate, resembling juvenile *Ceratodon purpureus*, but are readily recognised by the rhizoidal gemmae (see Whitehouse, *J. Bryol.* **9**, 7–11, 1976).

7. D. lineare (Sw.) Lindb., Act. Soc. Sc. Fenn., 1871
D. vaginans (Sull.) Hampe

Plants slender, shoots innovating from below, 5–15 mm. Lvs rigid, loosely appressed, scarcely altered when dry, lower lvs small, ovate to ovate-lanceolate, upper larger, 0.6–1.4 mm long, 3–4 times as long as wide, widest about ¼ from base, channelled above, lanceolate, tapering to bluntly pointed apex, margin frequently narrowly recurved on one or both sides, entire or obscurely denticulate towards apex; nerve broad, percurrent or shortly excurrent; cells thick-walled, in lower part of lf narrowly rectangular, becoming narrower or not towards margin, cells above rectangular to quadrate, 5–8 μm × 15–40 μm at widest part of lf. Seta yellowish; capsule narrowly ellipsoid. Only female plants known in Britain and Ireland. $n=13+1$. Dense, green or yellowish-green tufts, patches or scattered shoots on acidic soils. A few localities in Wales, N. England and Scotland, Sligo, Antrim. 11, H2. Europe, N. America.

This and the next 2 species are close and easily confused. *D. plumbicola* has more shortly pointed lvs with plane margins, the stem is less triangular in section and the nerve is thinner and less prominent at the back of the lf. *D. plumbicola* occurs on lead-containing soils, a habitat from which *D. lineare* is not known. *D. cornubicum* has lvs intermediate in shape between those of *D. lineare* and *D. plumbicola* but the plant differs in the lvs being erecto-patent to patent when moist, the shorter, wider cells in the lower part of the lf and the presence of rhizoidal gemmae. *D. lineare* and *D. plumbicola* are superficially similar in appearance to *Aongstroemia longipes* but differ in lf shape and areolation.

8. D. plumbicola Crundw., J. Bryol., 1976

Close to *D. lineare*. Shoots very slender, 5–15 mm. Lvs closely appressed, scarcely altered when dry, 0.4–0.7 mm long, 2–3 times as long as wide, widest ⅓–½ from base, of ± similar size except at base of stem, concave, ovate to ovate-lanceolate, not or hardly channelled above, shortly tapering to blunt apex, margin plane or incurved above in young lvs, entire or with a few obscure denticulations towards apex; nerve wide, percurrent; cells similar to those of *D. lineare* ,5–9 μm × 15–42 μm at widest part of lf. Gametangia and sporophyte unknown. Dense, pale, yellowish-green, ± glossy patches on lead-mine spoil, very rare. Caerns, S. Northumberland, I. of Man. 3. Endemic.

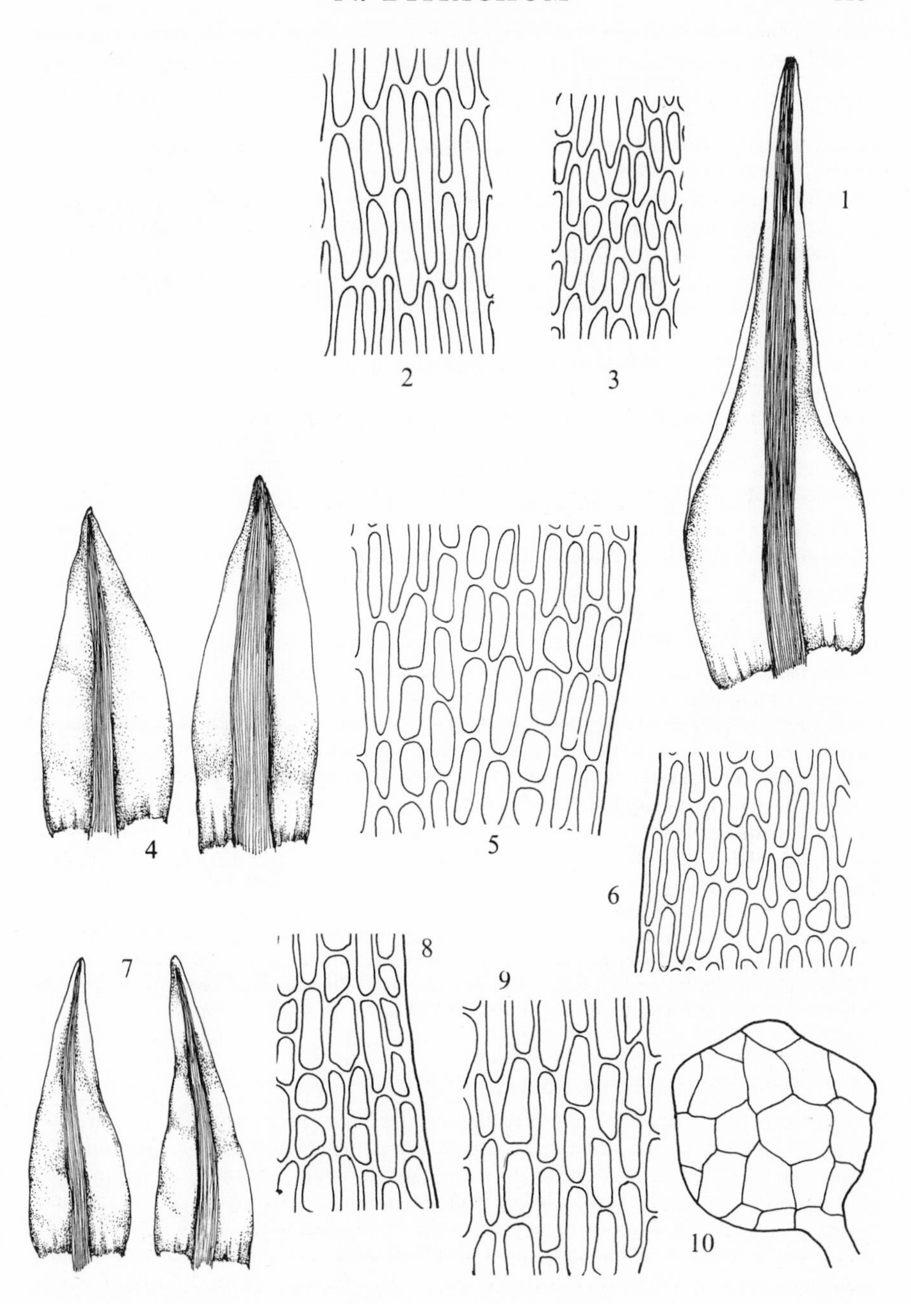

Fig. 46. 1–3, *Ditrichum lineare*: 1, leaf; 2, basal cells; 3, mid-leaf cells. 4–6, *D. plumbicola*: 4, leaves; 5, basal cells; 6, mid-leaf cells. 7–10, *D. cornubicum*: 7, leaves; 8, basal cells; 9, mid-leaf cells; 10, gemmae (×250). Leaves ×65, cells ×415.

9. D. cornubicum Paton, J. Bryol., 1976

Plants very small, 1–5 mm. Lower lvs distant, upper more crowded, appressed, uppermost on well grown stems slightly secund when dry, erecto-patent to patent when moist, 0.4–0.9 mm long, about 3 times as long as wide, widest about ¼ from base, concave, lanceolate, tapering to obtuse to sub-acute apex, margin plane, with a few obscure denticulations; nerve broad, ending just below apex; cells in lower part of lf irregularly rectangular, 1–2 marginal rows narrower, cells above rectangular, narrowly rectangular by nerve, 7–10(–12) μm × 15–25 μm at widest part. Brownish multicellular rhizoidal gemmae, spherical to ovoid or irregular in shape, 80–160 μm long, present. Only male plants known. Dull green patches or scattered shoots on peaty soil. Known only from 2 localities in Cornwall. 2. Endemic.

15. Distichium Br. Eur., 1846

Lvs distichous, subulate from broad, whitish sheathing basal part; cells of subula quadrate to rectangular, papillose. Capsule ovoid to cylindrical; peristome teeth strongly perforated or irregularly divided. A cosmopolitan genus of *ca* 18 species.

Plants usually glossy, bright green above, reddish-brown below, capsule erect or slightly inclined, spores 17–22 μm, plants paroecious **1. D. capillaceum**
Plants dull green, capsule inclined, spores 30–40 μm, plants autoecious **2. D. inclinatum**

1. D. capillaceum (Hedw.) Br. Eur., 1846
Swartzia montana Lindb.

Paroecious. Plants to 10(–15) cm, stems radiculose. Lvs markedly distichous, erect to spreading, flexuose when dry, reflexed, often sharply so when moist, basal part sheathing, whitish, abruptly narrowed into sharply reflexed setaceous subula composed mainly of nerve, margin of sheathing base entire, of subula denticulate; nerve excurrent; basal cells of sheath narrowly rectangular, above rectangular, cells in subula shortly rectangular. Antheridia naked in axils of upper lvs. Seta to 15 mm; capsule ovoid to cylindrical, erect or slightly inclined; peristome teeth 30 μm wide at base, irregularly split into filiform segments; spores 17–22 μm. Fr frequent, summer. $n=14$*, 28. Silky, green or bright green, loose or compact tufts, reddish-brown below, on basic soil, rock ledges and boulders, mainly in montane habitats. N. Somerset, Surrey, Wales, Derby and Yorks northwards. 56, H14. Cosmopolitan.

The posture of the lvs in this and the next species is variable and in small compact forms of *D. capillaceum* the subula is scarcely reflexed from the sheathing base, but colour and the form of the capsule will distinguish the 2 species.

2. D. inclinatum (Hedw.) Br. Eur., 1846
Swartzia inclinata (Hedw.) P. Beauv.

Autoecious. Plants to 5 cm. Lvs distichous, leaf base sheathing, abruptly narrowed into ± reflexed subula; nerve and areolation as in *D. capillaceum*. Antheridia in dwarf axillary branches. Seta to 10 mm; capsule ovoid, inclined; peristome teeth 40–50 μm wide at base, perforated or irregularly split; spores 30–40 μm. Fr frequent, summer. $n=13$, 14. Dull or dark green tufts or patches in turf on basic soil and in dune slacks, occasional but sometimes locally abundant. Scattered localities in Wales and from Derby and Yorks northwards. 27, H7. Europe, N. and C. Asia, Japan, N. America.

Superficially similar to *Ditrichum flexicaule* but distinct in the shorter, distichous lvs and in the areolation.

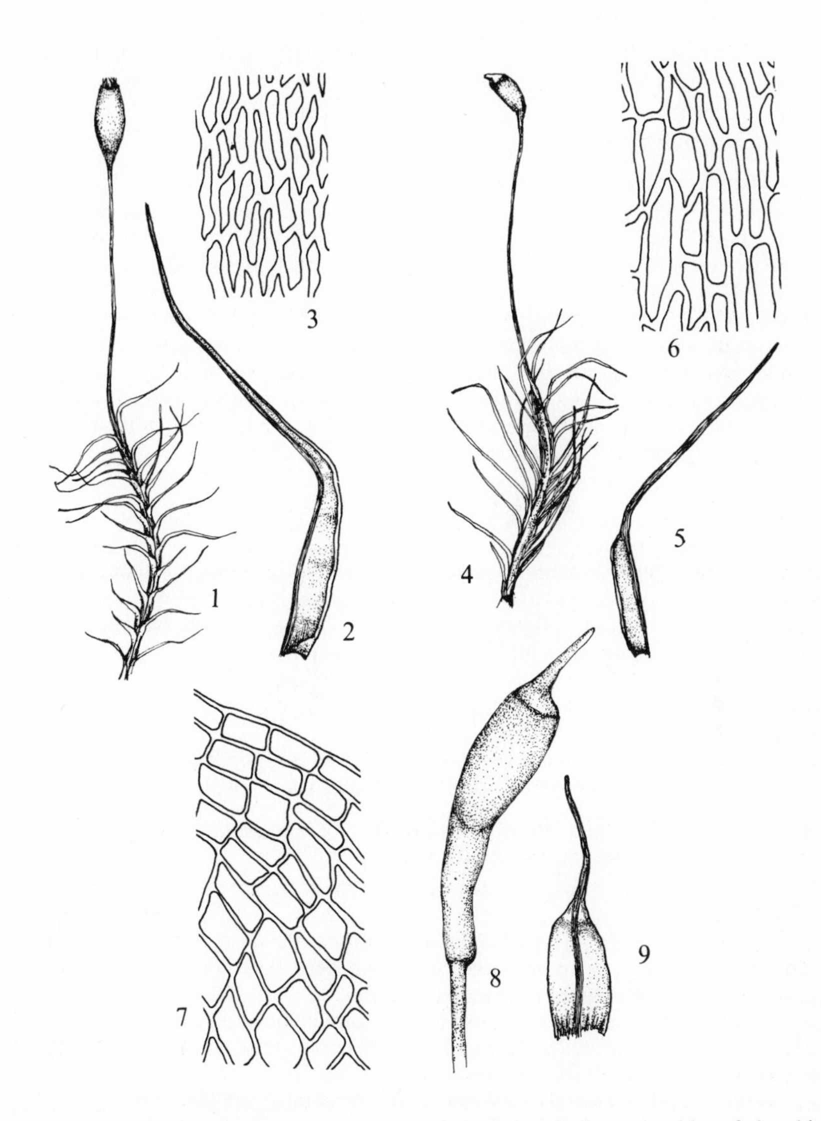

Fig. **47**. 1–3, *Distichium capillaceum*: 1, shoot; 2, leaf; 3, cells from shoulder of sheathing base. 4–6, *D. inclinatum*: 4, shoot; 5, leaf; 6, cells from shoulder of sheathing base. 7–9, *Trematodon ambiguus*: 7, cells near leaf shoulder; 8, capsule; 9, leaf. Shoots $\times 5$, leaves $\times 15$, cells $\times 450$.

SUBFAMILY 2. TREMATODONTOIDEAE

Neck of capsule ½ or more length of capsule with numerous stomata, annulus absent in *Trematodon* (capsule cleistocarpous in *Bruchia*).

16. TREMATODON MICH., FL. BOR. AMER., 1803

Autoecious. Lvs ± abruptly narrowed from sheathing base into subula; nerve excurrent. Capsule sub-clavate, slightly curved, neck ½–1 length of capsule with numerous stomata; peristome teeth ± entire to deeply bifid. A ± cosmopolitan genus of nearly 100 species.

1. T. ambiguus (Hedw.) Hornsch., Flora, 1819

Shoots to 1 cm. Lvs erecto-patent when moist, from oblong to oblong-lanceolate, sheathing base rapidly narrowed to obtuse subula composed mainly of nerve, subula about ½ length of lamina in lvs from middle part of stem, shorter in upper and perichaetial lvs, margin entire; nerve stout, excurrent, obscurely toothed towards apex; basal cells of sheathing portion narrowly rectangular, narrower towards margin, above shorter, rectangular at shoulder, cells in subula shortly rectangular. Seta yellow, 1–3 cm; capsule slightly curved, neck about same length as capsule, slightly strumose; spores 30–36 μm. Fr common, summer. Green or brownish tufts on moist peaty soil in montane areas. Very rare. Mid Perth, 1883, not seen since. Europe, Iceland, Japan, N. America.

8. SELIGERIACEAE

Plants minute and gregarious or larger and tufted, saxicolous. Lvs lanceolate, subulate; nerve without stereids, ending below apex to excurrent, occupying most of subula; cells smooth, usually not differentiated at basal angles. Seta usually long, straight or curved; capsule ovoid, ellipsoid or pyriform, smooth; annulus usually lacking; peristome teeth entire, smooth, or absent; calyptra usually cucullate.

17. BRACHYDONTIUM FÜRNR., FLORA, 1827

Autoecious. Similar to *Seligeria* but capsule striate; calyptra mitriform, 3–4 lobed at base; peristome teeth very short, truncate, papillose. Europe, E. Asia, N. America.

1. B. trichodes (Web.) Milde, Bryol. Siles., 1869
Brachyodus trichodes (Web.) Nees & Hornsch.

Plants minute. Lvs from ovate or lanceolate basal part, narrowed into long, acute, entire subula, margin entire; nerve excurrent; cells in basal part rectangular, in subula quadrate, opaque. Seta straight when moist, flexuose when dry, 2–3 mm; capsule ovoid to obloid, striate; peristome teeth very short, not projecting above annulus; spores 10–12 μm. Fr common, autumn. $n = 11$*. Dull green patches on damp, shaded, basic or acidic rocks and stones. Rare in most of England and Ireland, occasional to frequent in Wales, N. England and Scotland. 50, H5. Europe, Caucasus, N. America.

Resembling some *Seligeria* spp. but distinguished by the finely striate capsule.

18. SELIGERIA BR. EUR., 1846

Autoecious or paroecious. Usually minute, gregarious plants. Upper lvs much larger

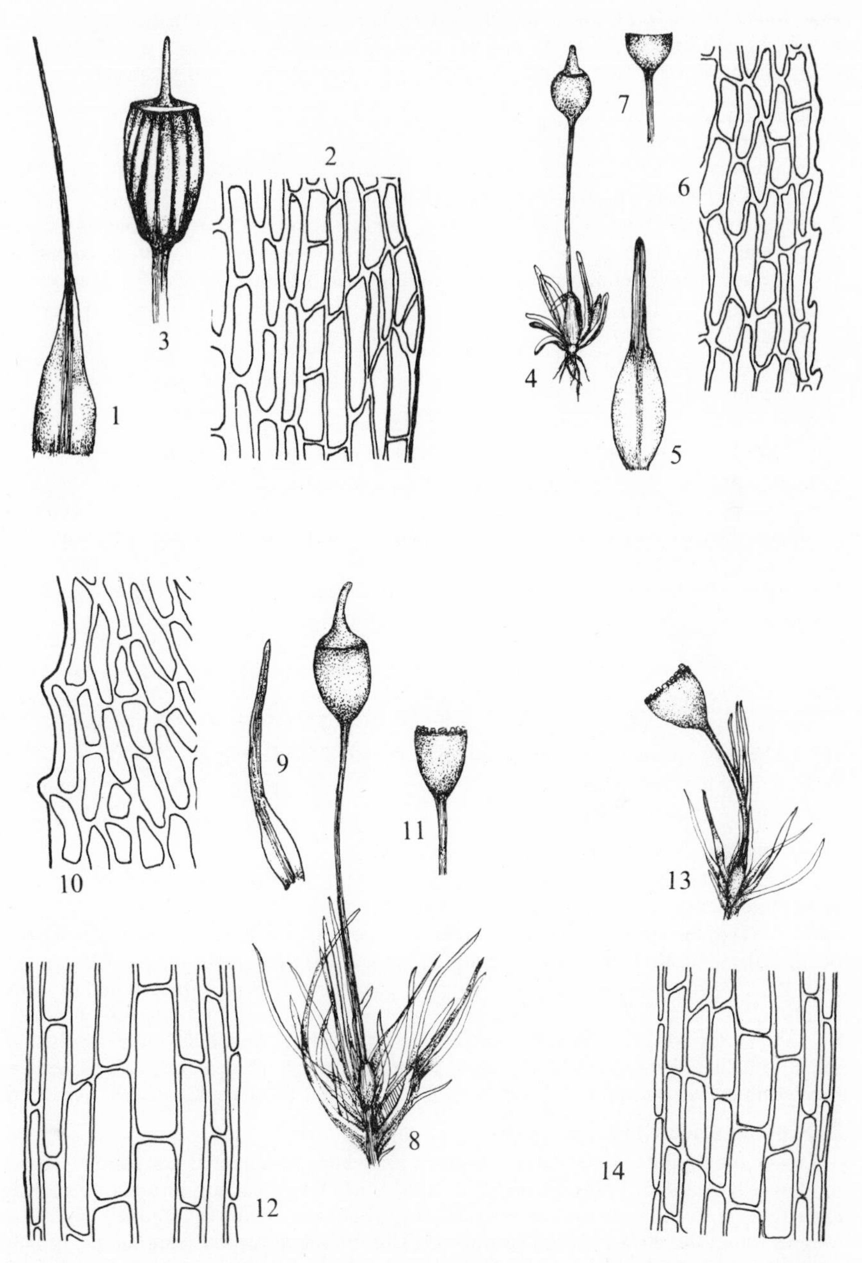

Fig. **48**. 1–3, *Brachydontium trichodes*: 1, leaf; 2, leaf cells; 3, capsule (×25). 4–7, *Seligeria donniana*: 4, plant; 5, leaf; 6, leaf cells; 7, old capsule (×15). 8–12, *S. pusilla*: 8, plant; 9, leaf; 10, leaf cells; 11, old capsule (×15); 12, cells of seta. 13–14, *S. acutifolia* 13, plant; 14, cells of seta. Plants ×15, leaves ×30, cells ×450.

than lower, lanceolate, subulate, without or with very slightly differentiated angular cells. Capsule smooth, with or without stomata; calyptra cucullate; peristome teeth smooth, entire or lacking. A mainly northern hemisphere genus of 18 species.

1 Seta very stout relative to size of capsule, capsule ± globose, columella falling with lid, spores 20–30 μm — 2
Seta relatively slender, capsule ovoid to ellipsoid or pyriform, columella persistent, spores 10–18 μm or if more than lvs trifarious — 3
2 Upper and perichaetial lvs with secund, setaceous subula several times as long as sheathing base — **10. S. carniolica**
Upper and perichaetial lvs not secund and with short, stout subula — **9. S. oelandica**
3 Seta arcuate when moist — **8. S. recurvata**
Seta straight or flexuose when moist — 4
4 Lf margin entire or nearly so — 5
Lf margin denticulate below — 8
5 Lvs trifarious, nerve excurrent, spores 16–28 μm — **4. S. trifaria**
Lvs not trifarious, nerve usually ending below apex, spores 9–18 μm — 6
6 Upper part of upper lvs ligulate, obtuse, capsule wide-mouthed when dry and empty, spores 14–18 μm — **6. S. calcarea**
Upper part of lvs subulate, ± acute, capsule not wide-mouthed when dry and empty, spores 9–14 μm — 7
7 Capsule with long neck gradually tapering into seta — **5. S. paucifolia**
Capsule with short neck abruptly narrowed into seta — **7. S. diversifolia**
8 Nerve of upper lvs longly excurrent, seta 1.0–1.5 mm, perichaetial lvs almost reaching base of capsule, seta cells quadrate-rectangular, mostly less than 25 μm long — **3. S. acutifolia**
Nerve of upper lvs percurrent or shortly excurrent, seta 2–3 mm, lvs ending well below capsule, seta cells rectangular, mostly 25–45 μm long — 9
9 Peristome lacking, lid conical — **1. S. donniana**
Peristome present, lid rostrate — **2. S. pusilla**

1. S. donniana (Sm.) C. Müll., Syn. Musc., 1848

Lower lvs lanceolate, obtuse, upper and perichaetial lvs tapering from ovate or lanceolate denticulate basal part to obtuse to acute subula composed mainly of nerve; nerve percurrent or slightly excurrent; cells in basal part of leaf rectangular or rhomboidal, shorter towards margin, cells in subula quadrate to rectangular. Seta to 2(–3) mm; capsule shortly ovoid to pyriform, wide-mouthed when dry and empty; lid conical; peristome lacking; spores 10–14 μm. Fr common, summer. $n=13$*. Yellowish-green or green patches on chalk, limestone, calcareous sandstone, schist and slate in moist shaded places. Rare in S. England and Ireland, frequent in the Pennines, occasional in Scotland. 42, H4. Europe, Caucasus, C. Asia, N. America.

2. S. pusilla (Hedw.) Br. Eur., 1846

Lower lvs ligulate to ± linear, acute, upper and perichaetial lvs subulate from narrowly lanceolate, finely denticulate basal part; nerve ending in apex or slightly excurrent; cells in basal part rectangular, in subula narrowly rectangular, 16–24 μm long in upper lvs. Seta 2–3 mm, surface cells in middle narrowly rectangular, mostly more than 25 μm long; capsule ovoid, wide-mouthed and turbinate when empty; spores 10–12 μm. Fr common, summer. Dull green patches on basic rock in moist shaded places. Occasional to frequent in suitable habitats in Britain, rare in Ireland. 49, H4. Europe, Caucasus, Jenisei, N. America.

3. S. acutifolia Lindb. in Hartm., Skand. Fl., 1864

Close to *S. pusilla.* Nerve in upper lvs longly excurrent; lamina cells in subula 16–24 μm long. Perichaetial lvs sometimes reaching capsule. Seta variable in length, to 1.5 mm long, surface cells rectangular, rarely more than 25 μm long. Patches on shaded basic rocks. A few localities in W. and N. England, Brecon, N. Ebudes, Antrim. 16, H1. Sweden, Norway, Italy, N. Asia, N. Africa.

British plants have in the past been referred to var. *longiseta* Lindb., but even in Britain the length of the seta is very variable and the variety is clearly worthless. The plant is most easily distinguished from *S. pusilla* and *S. diversifolia* by the shorter seta with shorter surface cells.

4. S. trifaria (Brid.) Lindb., Öfv. K. V. A. Förh., 1863
S. tristicha (Brid.) Br. Eur.

Plants to 4 mm, stems brittle. Lvs imbricate, trifarious, ovate-lanceolate to lanceolate, tapering to short, acute to obtuse subula, margin entire; nerve excurrent; cells in basal part narrowly rectangular, shorter above. Seta to 2 mm; capsule ovoid, long-necked, wide-mouthed and turbinate when empty; spores 16–28 μm. Fr occasional, summer. In brownish-green patches on shaded, flushed limestone. Occasional in Yorks, very rare elsewhere, Brecon, Derby, Westmorland, E. Perth, W. Sutherland, Sligo, Leitrim, Fermanagh. 8, H3. Europe, Caucasus, Siberia, Jenisei.

Although *S. trifaria* fruits less commonly than other *Seligeria* species it may readily be recognised when sterile by its small size and imbricate trifarious lvs.

5. S. paucifolia (Dicks.) Carruthers, Lond. J. Bot., 1866

Lvs ovate to narrowly lanceolate, acuminate to longly subulate, acute, margin entire; nerve usually ending below apex; cells in lower part of lf rectangular to linear, shorter towards margin and above. Seta to 2 mm; capsule ovoid to ellipsoid or pyriform, with long neck, narrowed at mouth when empty; spores 9–12 μm. Fr common, summer. $n = 12$*. Patches on shaded chalk. Frequent and sometimes locally common in S.E. England, rare elsewhere, extending west to N. Somerset and north to N. Lincs and S.E. Yorks, Londonderry, 22, H1. France, Italy.

6. S. calcarea (Hedw.) Br. Eur., 1846

Lvs ovate to ovate-lanceolate, narrowing to ligulate, obtuse apex, margin entire; nerve broad, faint, ending below apex; cells in lower part of lf rectangular, above quadrate. Seta 1–2 mm; capsule ellipsoid or pyriform with long neck, wide-mouthed when empty; spores 14–18 μm. Fr common, summer. $n = 12 + 1$. On shaded chalk or limestone, frequent in S.E. England, rare elsewhere, extending north to S.E. Yorks and Derby, Antrim, Londonderry. 31, H2. Sweden, Denmark, France, Germany, W. Russia, N. America.

7. S. diversifolia Lindb., Öfv. K. V. A. Förh., 1861

Lower lvs narrowly lanceolate, obtuse, upper lvs longer, *ca* 1 mm long in fertile stems, 2.5 mm in sterile stems, basal part narrowly lanceolate, entire, tapering into long, channelled, ± acute subula; nerve stout, ending in apex; cells in basal part rectangular, in subula shortly rectangular, 7–14 μm long. Seta flexuose, 2.4–3.0 mm, surface cells in middle of seta narrowly rectangular, 25–45 μm long; capsule ovoid, scarcely altered when dry and empty; peristome present; spores 10–14 μm. Fr common, spring. Yellowish-green patches on limestone cliffs, very rare. Yorks. 1. Norway, Sweden, Finland, Switzerland, northern Russia, Siberia, Caucasus.

Although most closely related to *S. recurvata* this species is most likely to be confused with *S. pusilla* or *S. acutifolia.* It differs from both these in the ± entire and more bluntly pointed lvs and the cells in the subula being shorter. *S. acutifolia* also has a shorter seta with shorter surface cells. For the occurrence of this plant in Britain see Crundwell, *J. Bryol.* **7**, 261–3, 1973.

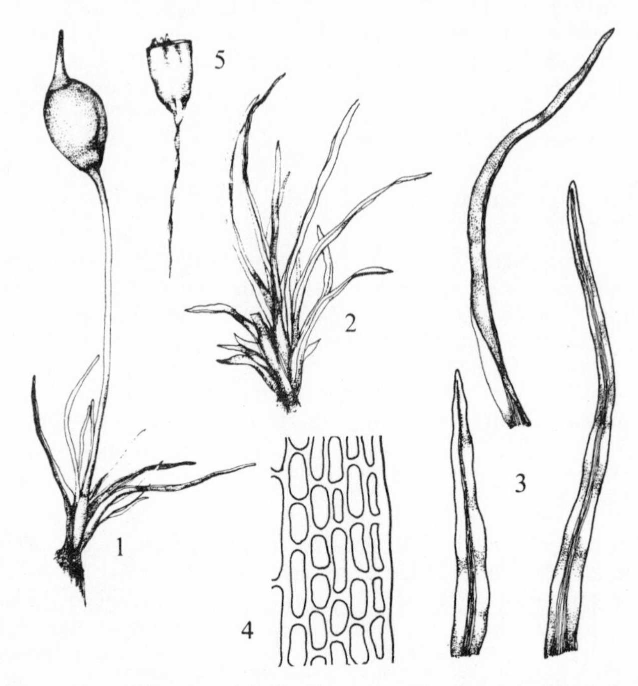

Fig. **49**. *Seligeria diversifolia*: 1 and 2, fertile and sterile plants (×15); 3, leaves (×30); 4, leaf cells (×415); 5, old capsule (×15).

8. S. recurvata (Hedw.) Br. Eur., 1846

Upper and perichaetial lvs flexuose, ovate-lanceolate to narrowly lanceolate, tapering to acute, ± setaceous subula, margin entire; nerve excurrent; cells in basal part rectangular, shorter above. Seta arcuate when moist, flexuose when dry, to 5 mm; capsule ovoid, wide-mouthed when empty; spores *ca* 10 μm. Fr common, spring. $n = 13, 14^*$. Patches on shaded basic or acidic rock, especially sandstone. Rare in S.E. England and the Midlands, occasional to frequent elsewhere. 61, H11. Europe, Caucasus, E. and N. Asia, N. America.

A form has been recorded from Bucks (see Warburg, *Trans. Br. bryol. Soc.* **1**, 14–15, 1947) in which the plants are bright green, the lamina remains distinct to the lf apex which is obtuse and the margin near the apex is sometimes obscurely toothed.

9. S. oelandica Jens. & Medel., Bot. Not., 1929
S. lapponica Nyman & Uggla

Plants 5–15 mm. Lower lvs ± ovate, obtuse, upper and perichaetial lvs from ovate or lanceolate base, narrowing to stout, obtuse subula, margin entire; nerve excurrent; cells in basal part rectangular, shorter above. Seta stout, *ca* 2.5 mm; capsule ± spherical, wide-mouthed and turbinate when empty, without stomata; columella attached to and falling with lid; peristome teeth truncate; spores 22–30 μm. Fr

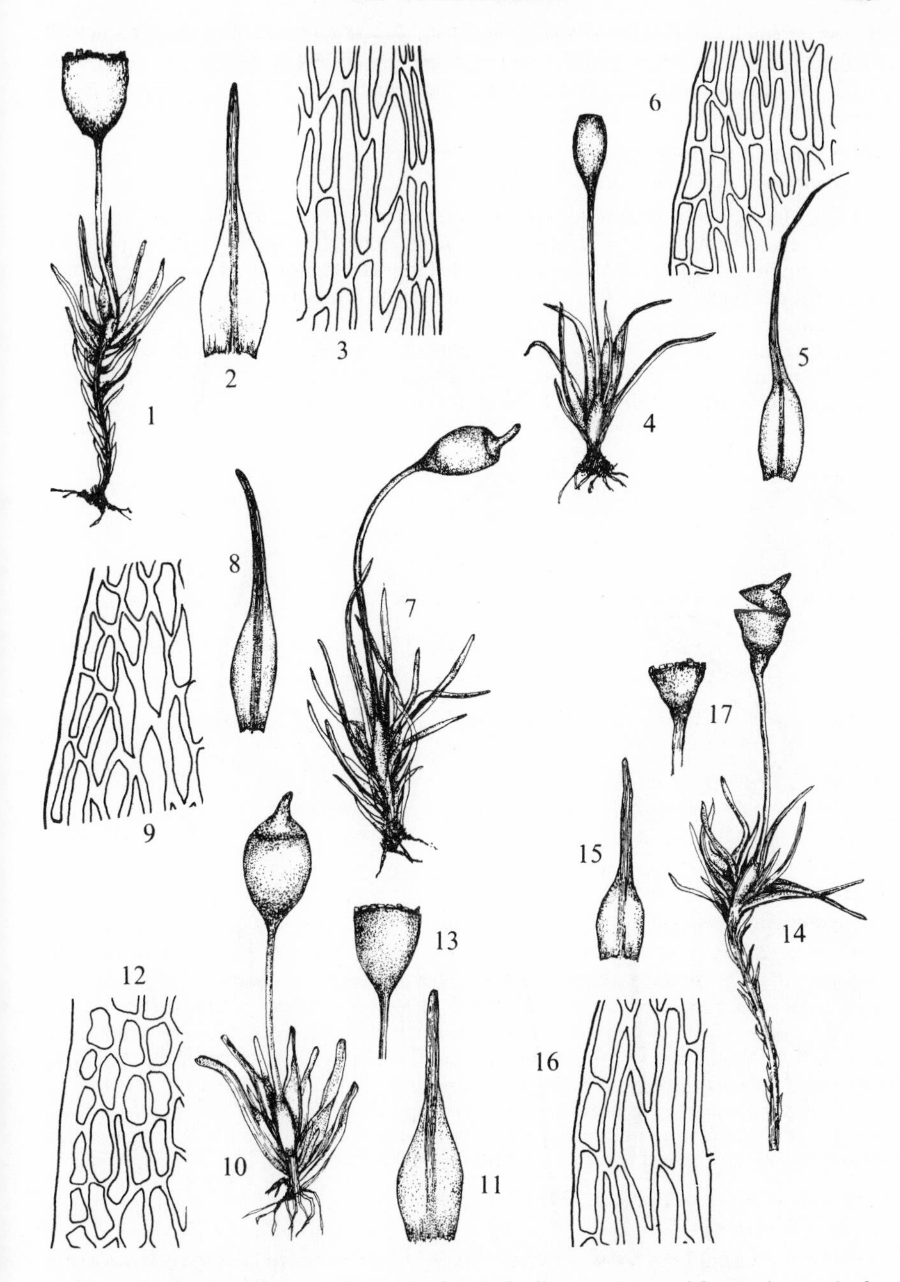

Fig. 50. 1–3, *Seligeria trifaria*: 1, plant; 2, leaf; 3, leaf cells. 4–6, *S. paucifolia*: 4, plant; 5, leaf; 6, leaf cells. 7–9, *S. recurvata*: 7, plant; 8, leaf; 9, leaf cells. 10–13, *S. calcarea*: 10, plant; 11, leaf; 12, leaf cells; 13, old capsule (×15). Plants ×15, leaves ×30. 14–17, *S. oelandica*: 14, plant (×10); 15, leaf (×20); 16, leaf cells; 17, old capsule (×10). Cells ×450.

frequent, summer. Loose or dense, blackish-green patches on flushed, montane limestone, very rare. Sligo, Leitrim. H2. Swedish Lapland, S.E. Sweden, N. America.

For the occurrence of this plant in Britain see Crundwell & Warburg, *Trans. Br. bryol. Soc.* **4**, 426–8, 1963.

10. S. carniolica (Breidl. & Beck) Nyholm, Moss Fl. Fenn., 1954
Trochobryum carniolicum Breidl. & Beck

Lvs of fertile stems secund, ovate, abruptly narrowed to acute, smooth, setaceous subula several times length of basal portion in upper lvs; nerve faint below, stout above, longly excurrent; cells in basal part rectangular or rhomboidal. Seta to 2.5 mm, straight or curved, very stout; capsule ± spherical, wide-mouthed, wider than deep when empty, without stomata; columella attached to and falling with lid; peristome teeth truncate; spores 20–27 μm. Fr frequent, summer. Dark green to blackish patches on damp, shaded, calcareous sandstone by streams (in Britain). Very rare. Northumberland, Roxburgh. 2. Gotland, C. Europe.

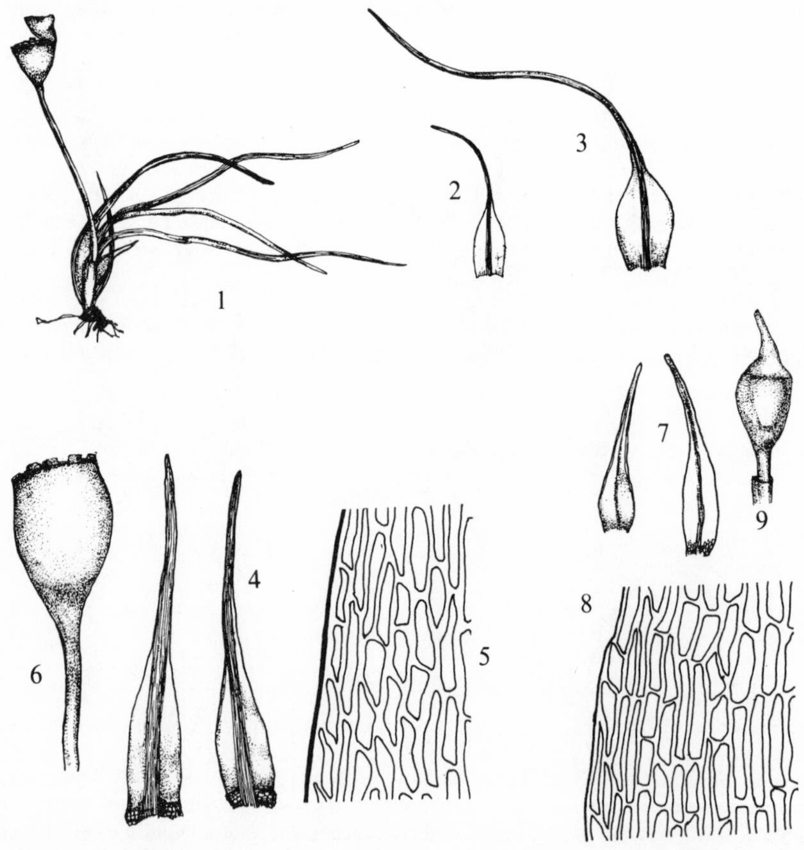

Fig. **51**. 1–3, *Seligeria carniolica*: 1, plant (×10); 2, leaf (×20); 3, perichaetial leaf (×20). 4–6, *Blindia acuta*: 4, leaves (×15); 5, leaf cells; 6, capsule. 7–9, *B. caespiticia*: 7, leaves (×15); 8, leaf cells; 9, capsule. Cells ×450, capsules ×15.

Although the sporophyte of *S. carniolica* and *S. oelandica* differs somewhat from that of other British species of *Seligeria*, there is a Scandinavian species, S. *tristichoides* Kindb., which is intermediate. For the occurrence of *S. carniolica* in Britain see Warburg, *Trans. Br. bryol. Soc.* **1**, 199–201, 1949.

19. Blindia Br. Eur., 1846

Lvs lanceolate, subulate; angular cells inflated, orange-brown. Capsule usually exserted, ovoid or pyriform, stomata present; peristome teeth entire, smooth or absent. About 35 species occurring in Eurasia, America and Australia.

Plants (0.5–)1.5–10.0 cm, dioecious, capsule exserted, peristome present **1. B. acuta**

Plants 0.5–1.5 mm, autoecious, capsule immersed, peristome lacking **2. B. caespiticia**

1. B. acuta (Hedw.) Br. Eur., 1846

Dioecious. Plants (0.5–)1.5–10.0 cm. Lvs erecto-patent, sometimes secund when moist, hardly altered when dry, lower lanceolate, subulate, upper ovate or ovate-lanceolate, narrowed into entire or faintly denticulate subula composed mainly of nerve and up to 3 times length of basal portion, acute to obtuse; angular cells inflated, orange-brown, basal cells narrowly rectangular, becoming elliptical above, incrassate, 6–8 μm wide in mid-lf, narrower at margin. Perichaetial lvs with broader base more sharply narrowed into subula. Seta erect, flexuose, 4–5 mm; capsule ovoid to pyriform; stomata present; lid oblique, longly rostrate; peristome present; spores 18–20 μm. Fr frequent, summer. Dark green to yellowish-green tufts or patches, blackish below, on damp or wet rocks, frequent in montane areas. S.W. England, Wales, Derby and Yorks northwards. 76, H23. Europe, Caucasus, N. Asia, Madeira, Azores, N. America, Greenland, Tasmania.

The degree of excurrence of the nerve is very variable. Forms with a very long point have been recognised as var. *arenacea* Mol. (var. *trichodes* (Wils.) Braithw.), but the taxon is not recognised here as the plants are merely the extreme of a range of forms.

2. B. caespiticia (Web. & Mohr) C. Müll., Deutschl. Moose, 1853
Stylostegium caespiticium (Web. & Mohr.) Br. Eur.

Autoecious. Plants 5–15 mm. Lvs secund, appressed when dry, erecto-patent or secund when moist, similar to *B. acuta* but subula entire. Seta very short, *ca* 0.5 mm; capsule ± immersed, ovoid, turbinate when dry and empty; lid oblique, longly rostrate; columella falling with lid; peristome and stomata lacking; spores 10–12 μm. Fr frequent, summer. Dull, yellowish-green tufts on high alpine rocks, very rare. Mid Perth, W. Inverness. 2. N., W. and C. Europe, C. Africa.

Hardly to be distinguished from the former species when sterile. *B. acuta* is, however, usually a larger plant with the lf subula often faintly denticulate.

9. DICRANACEAE

Autoecious or dioecious. Plants small or large, stems with central strand, often tomentose. Lvs of ± uniform size or, rarely, upper larger, straight to falcate, narrowly to broadly lanceolate; nerve usually ± excurrent, with stereids; upper cells quadrate to elongate, smooth or occasionally papillose or mamillose, basal cells elongate, often porose, angular cells differentiated or not. Seta usually long, straight or arcuate; capsule stegocarpous, ovoid to cylindrical, straight or curved, strumose or not, smooth or striate; annulus and stomata present or not; peristome dicranoid (see p. 108).

A somewhat heterogeneous family the exact constitution of which depends to some extent upon the characters used to define it. *Saelania, Ceratodon* and *Cheilothela* are sometimes placed in the Ditrichaceae but on morphological grounds seem better placed here. The affinities of *Rhabdoweisia* are uncertain. *Amphidium* is sometimes included in the Dicranaceae; on cytological grounds it could equally well be placed in the Grimmiales. The differences between the subfamilies listed below approach those separating the 3 British families of the Dicranales in magnitude and they could equally well be treated as separate families.

SUBFAMILY 1. CYNODONTOIDEAE

Lvs ovate to lanceolate, acute to acuminate; without distinct angular cells, cells mamillose or papillose. Capsule ovoid to ellipsoid, straight or more usually curved, sometimes asymmetrical and strumose, often striate or sulcate.

20. Saelania Lindb., Utkast, 1878

A monotypic genus with the characters of the species.

1. S. glaucescens (Hedw.) Broth., Nat. Pfl. Fam., 1924
S. caesia (P. Beauv.) Lindb.

Autoecious. Stems to 3.5 cm. Lvs appressed, upper flexuose when dry, erecto-patent when moist, lower tapering ± from base to apex, upper longer, gradually tapering from narrowly lanceolate basal part to narrow, acuminate apex, margin serrate throughout, sparsely so below; nerve ending in or below apex in lower lvs, percurrent or excurrent in upper; basal cells in lower lvs rectangular, becoming ± quadrate above, in upper lvs basal cells narrowly rectangular to rectangular, becoming rectangular to quadrate-rectangular in mid-lf, longer above, *ca* 10 μm wide in mid-lf. Seta to 1 cm; capsule erect, narrowly ellipsoid, not strumose; spores 14–16 μm. Fr occasional, summer. $n=13$. Glaucous, bluish-green tufts on usually damp, shaded soil over basic, montane rock, very rare. A few localities in Angus and E. Inverness. 2. Europe, Iceland, Asia, southern Africa, N. America, Greenland, New Zealand, Hawaii.

21. Ceratodon Brid., Br. Univ., 1826

Dioecious. Lvs ovate to narrowly lanceolate, margin recurved; nerve percurrent or excurrent; cells uniform throughout or longer towards base, smooth or slightly papillose, unistratose. Seta purple or yellow; capsule inclined, straight or curved, strumose, sulcate when dry; peristome teeth bifid into papillose, filiform segments. Cosmopolitan, about 22 species.

1. C. purpureus (Hedw.) Brid., Br. Univ., 1826

Lower lvs loosely appressed, ± straight, upper slightly twisted, flexuose when dry, erecto-patent when moist, ovate to narrowly lanceolate, acute, concave, margin recurved nearly to apex; nerve stout, percurrent or excurrent, sometimes longly so; basal cells rectangular, above becoming quadrate to irregularly hexagonal, smooth. Perichaetial lvs longer with expanded base. Seta purple; capsule reddish-brown, obloid, straight or curved, when dry inclined to horizontal, sulcate; lid conical; spores 10–14 μm. Fr common, spring.

Ssp. **purpureus**

Plants to 3(–5) cm. Lvs ovate to narrowly lanceolate, margin often bluntly toothed towards apex; nerve ending in apex to excurrent; cells 9–12 μm wide in mid-lf. Capsule inclined, curved or straight, strumose, sulcate when dry; peristome segments with pale margins to about middle. $n=13$*. Tufts or patches varying in colour from yellowish-green to vinous or brownish-red, usually calcifuge, on various substrata, especially soil, throughout Britain and Ireland, probably the commonest native acrocarpous moss, one of the few species occurring in urban and industrial areas. 112, H40, C. Cosmopolitan.

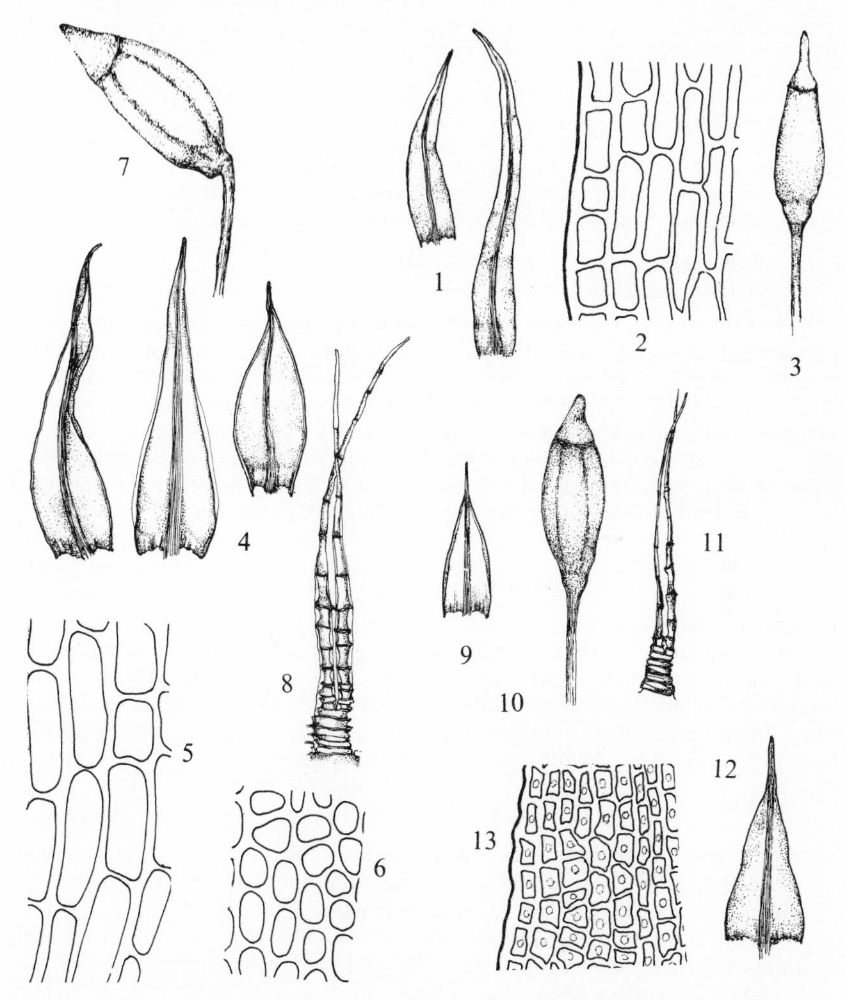

Fig. **52**. 1–3, *Saelania glaucescens*: 1, leaves; 2, mid-leaf cells; 3, capsule (×10). 4–8, *Ceratodon purpureus* ssp. *purpureus*: 4, leaves, 5, basal cells; 6, mid-leaf cells; 7, capsule (×15); 8, peristome tooth. 9–11, *C. purpureus* ssp. *conicus*: 9, leaf; 10, capsule (×15); 11, peristome tooth. 12–13, *Cheilothela chloropus*: 12, leaf; 13, mid-leaf cells. Leaves ×15, cells ×415.

Ssp. **conicus** (Hampe) Dix., Stud. Handb. Br. Moss., 1896
C. conicus (Hampe) Lindb., *C. purpureus* var. *conicus* (Hampe) Husn.

Plants to 1 cm. Lvs ovate to lanceolate, usually entire; nerve usually longly excurrent in upper lvs; cells 5–12 μm wide in mid-lf. Capsule ± erect, straight, not or only slightly strumose, slightly sulcate when dry; segments of peristome teeth without pale border. Fr occasional, spring. Bare soil, especially on mud-capped walls, rare and apparently decreasing, calcicole. Scattered localities from Oxford, Berks and East Anglia to Westmorland and Yorks, E. Perth, E. Inverness, S. Ebudes, Dublin, W. Mayo. 22, H2, C. N. and W. Europe, Caucasus, Madeira, N. America.

A very variable species for which some 39 intraspecific taxa have been described. The relationship between the 2 subspecies requires further investigation as, although they intergrade morphologically, ssp. *purpureus* is calcifuge and ssp. *conicus* is calcicole.

22. Cheilothela (Lindb.) Broth., Nat. Pfl., 1901

Dioecious. Lvs rigid when moist; nerve excurrent; cells smooth in the base of lf, mamillose and bistratose above. Capsule similar to *Ceratodon* but not strumose. Europe, Far East, Africa, S. America, New Zealand. Six species.

1. C. chloropus (Brid.) Broth., Nat. Pfl., 1901
Ceratodon chloropus (Brid.) Brid.

Stems to 1 cm. Lvs appressed, straight when dry, erecto-patent, rigid when moist, ± narrowly triangular to ovate, with long acuminate apex, margin plane, finely crenulate below, papillose-crenulate above; nerve broad, excurrent in stout point; basal cells shortly rectangular, smooth, hyaline, becoming gradually smaller, quadrate, mamillose, bistratose, very obscure above, 6–8 μm wide in mid-lf. Seta pale yellow; capsule ellipsoid, not strumose. Fr unknown in Britain. Dull, yellowish-green patches on thin soil overlying basic rock, very rare. S. Devon, N. Somerset. 2. S.W. and Mediterranean Europe, Algeria, southern Africa.

23. Rhabdoweisia Br. Eur., 1846

Autoecious. Stem with central strand. Lvs spreading when moist, crisped when dry, narrowly linear-lanceolate to lingulate, acute to obtuse, margin plane or recurved below, entire or toothed above; nerve ending below apex; basal cells rectangular, angular not differentiated, cells above ± quadrate, slightly mamillose or not. Seta short; capsule erect, ovoid, striate; lid with oblique beak; annulus absent; peristome teeth 16, not divided; calyptra cucullate. A small genus of 5 species occurring in Europe, E. and C. Asia, Congo, Cameroon, Macaronesia and the Americas.

1 Cells in upper part of lf 12–20 μm wide, margin toothed **3. R. crenulata**
Cells in upper part of lf 8–12(–14) μm wide, margin entire or denticulate above 2
2 Lamina 3–4(–5) cells wide on either side of nerve 220 μm from apex, peristome teeth abruptly narrowed above base **1. R. fugax**
Lamina 5–7 cells wide on either side of nerve 220 μm from apex, peristome teeth gradually tapering from base to apex **2. R. crispata**

1. R. fugax (Hedw.) Br. Eur., 1846

Plants 0.5–1.0 cm. Lvs crisped when dry, flexuose-spreading when moist, ligulate, gradually tapering to acute to acuminate apex, lamina 6–9 cells wide on each side of

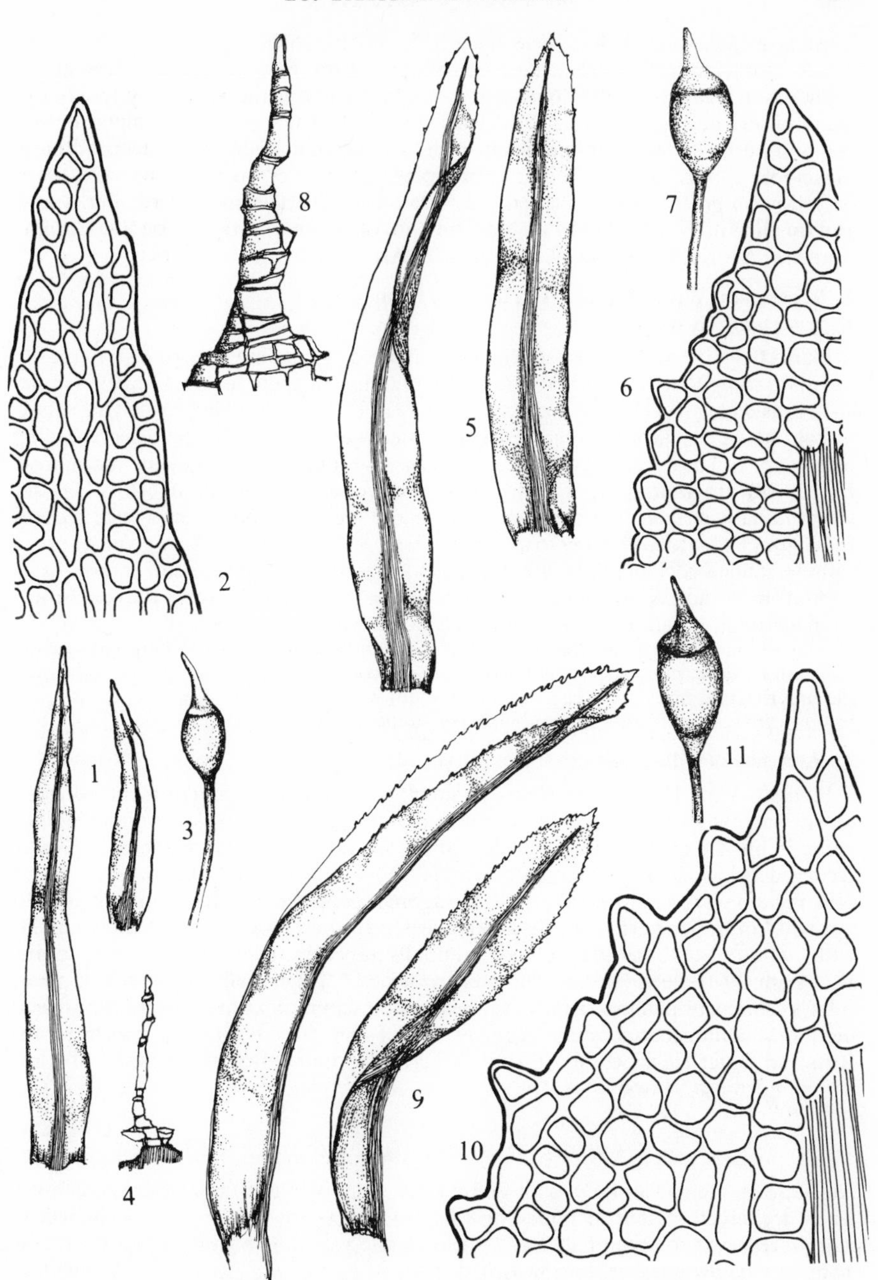

Fig. **53**. 1–4, *Rhabdoweisia fugax*: 1, leaves; 2, cells at leaf apex; 3, capsule; 4, peristome tooth. 5–8, *R. crispata*: 5, leaves; 6, cells at leaf apex; 7, capsule; 8, peristome tooth. 8–11, *R. crenulata*: 9, leaves; 10, cells at leaf apex; 11, capsule. Cells ×415, capsules ×15, teeth ×175, leaves ×25.

nerve at middle of lf, 3–4(–5) cells wide 220 μm from apex, margin plane, entire or denticulate near apex; nerve ending below apex; basal cells rectangular, cells above rounded-quadrate or wider than long, mamillose, 8–12 μm wide in upper part of lf. Seta yellowish; capsule ovoid, when dry longitudinally striate, contracted or not below mouth; lid with oblique beak; peristome teeth subulate from a broad base, fugacious; spores 14–20(–22) μm. Fr frequent, late winter, spring. $n = ca$ 12. Bright or yellowish-green tufts in shaded, non-basic rock crevices in ravines and woods and on cliffs in W. and N. Britain and Ireland, occasional or rare. 49, H11. Europe, Madeira, Azores, S. America, Guatemala, S. Africa (?).

2. R. crispata (With.) Lindb., Act. Soc. Sc. Fenn., 1871
R. denticulata (Brid.) Br. Eur.

Plants 0.5–1.5 cm. Lvs crisped when dry, flexuose-spreading when moist, narrowly lingulate to ligulate, abruptly to gradually tapering to acute to obtuse apex, lamina 8–12 cells wide on each side of nerve at middle of lf, 5–7 cells wide 220 μm from apex, margin plane or narrowly recurved below, denticulate towards apex; nerve ending below apex; basal cells rectangular, cells above quadrate-hexagonal, mamillose, 10–12(–14) μm wide in upper part of lf. Seta yellowish; capsule ovoid, when dry longitudinally striate; peristome teeth gradually tapering from base to apex; spores 18–20 μm. Fr frequent, spring to autumn. $n = 13 + 1$*. Bright yellowish-green or green tufts in shaded rock crevices, among rocks in ravines, rarely on peat, in non-basic habitats in W. and N. Britain and Ireland, frequent. 52, H18. Europe, Faroes, Amur, China, Korea, Japan, Java, N. America, Bolivia, Hawaii.

Intermediate between the other 2 species of *Rhabdoweisia. R. crenulata* differs in its usually larger size, wider, more shortly pointed and more coarsely toothed lvs, though sometimes the only reliable character is cell size and lf width near the apex. *R. fugax* has the lvs narrower near the apex and abruptly subulate peristome teeth.

3. R. crenulata (Mitt.) Jameson, Rev. Bryol., 1890

Plants 1–3 cm. Lvs crisped when dry, erecto-patent to spreading or recurved when moist, narrowly lingulate, acute to obtuse, lamina 9–14 cells wide on either side of nerve in middle of lf, 6–10 cells wide 220 μm from apex, margin plane or narrowly recurved below, strongly crenulate-dentate above; nerve ending below apex; basal cells rectangular, above ± quadrate-hexagonal, sometimes wider than long, 12–20 μm wide in upper part of lf. Seta yellowish; capsule ovoid or ovate ellipsoid; lid with oblique beak; peristome teeth gradually tapering from base to apex; spores 18–24 μm. Fr frequent, spring, summer. $n = 13 + 1$*. Bright yellowish-green or green tufts, cushions or patches in rock crevices and on damp rocks in sheltered, non-basic montane habitats, occasional. From N. Wales and the Lake District north to W. Ross and Outer Hebrides, W. Ireland. 30, H11. Germany, Belgium, France, Sikkim, China, Formosa, Hawaii.

24. Cynodontium Schimp., Coroll., 1855

Autoecious. Plants forming lax or dense tufts; stem with central strand. Lvs crisped when dry, erecto-patent to spreading when moist, narrowly lanceolate, channelled, margin recurved to about half way, ± toothed above, often bistratose; nerve ending at or just below apex or, in non-British species, excurrent; basal cells rectangular, shorter at margin, angular cells not or scarcely differentiated, cells above quadrate, thick-walled, slightly collenchymatous, usually mamillose. Seta long; capsule ± erect or inclined, ± straight or curved, usually striate, sometimes strumose; lid with long oblique beak; peristome teeth 16, divided to half-way or below, sometimes irregular and short, longitudinally striate below, papillose above. About 12 species occurring mainly in Europe, Kilimanjaro.

This genus has been misunderstood in Britain. For a brief account of *Cynodontium* species see A. C. Crundwell, *Trans. Br. bryol. Soc.* **3**, 706–12, 1960. The present account of the genus is based upon an unpublished MS by Mr A. C. Crundwell.

1 Capsule smooth when moist, faintly plicate when dry, peristome teeth irregular, to 160 μm long, lvs curled inwards when dry — **7. C. bruntonii**
Capsule distinctly striate when moist, peristome teeth regular, more than 200 μm long, lvs crisped but hardly incurved when dry — 2

2 Lf cells 11–22 μm wide, smooth — **3. C. jenneri**
Lf cells 7–14 μm wide, ± mamillose — 3

3 Lf margin unistratose throughout — 4
Lf margin bistratose above — 5

4 Lvs broadly pointed, cells 7–11 μm, coarsely mamillose on both surfaces, seta arcuate when moist, peristome teeth entire below — **5. C. gracilescens**
Lvs narrowly pointed, cells 9–14 μm, almost smooth on lower surface, seta straight, peristome teeth split and perforated — **6. C. fallax**

5 Capsule inclined, curved, strumose — **1. C. strumiferum**
Capsule erect, ± straight and symmetrical, struma absent or rudimentary — 6

6 Lvs distinctly mamillose especially on upper surface, perigonial lvs ± acute, annulus separating, lid crenulate at base — **2. C. polycarpon**
Lvs smooth or slightly mamillose, mamillae mainly on lower surface, perigonial lvs obtuse, annulus persistent, lid entire at base — **4. C. tenellum**

1. C. strumiferum (Hedw.) De Not., Epil., 1869

Plants 1–4 cm. Lvs somewhat crisped when dry, spreading, sub-secund when moist, narrowly lanceolate, gradually tapering to acute apex, margin strongly recurved to middle of lf or above, bistratose and irregularly crenulate-dentate above; upper cells 9–12 μm, usually strongly mamillose on both surfaces. Perigonial lvs sub-acute. Seta yellowish; capsule oblong-ellipsoid, inclined, curved, striate, conspicuously strumose; lid crenulate at base; annulus of 3 rows large cells, separating; peristome teeth all of ± similar length, 300–400 μm long, divided to halfway or below; spores 18–24 μm. Fr common, early summer. $n = 14$, 15. Dense, green tufts on non-basic rocks and screes, rare, western Scottish Highlands. E. Perth, Angus, Aberdeen, E. Inverness, Skye. 6. N., C. and E. Europe, Caucasus, Siberia, N. America, Greenland.

Sterile plants cannot be distinguished from *C. polycarpon* but both species fruit freely and differ markedly in capsule form. *C. strumiferum* is the only British species with peristome teeth of equal length.

2. C. polycarpon (Hedw.) Schimp., Coroll., 1856

Plants 2–4 cm. Lvs crisped when dry, erect-spreading when moist, narrowly lanceolate, gradually tapering to acute apex, margin recurved to above halfway, bistratose and crenulate-dentate above; nerve slightly papillose at back above; upper cells 8–14 μm, distinctly papillose especially on upper surface. Perigonial lvs ± acute. Seta yellowish; capsule ± erect, ellipsoid with short neck, symmetrical, striate, not or scarcely strumose; lid crenulate at base; annulus of 3 rows of separating cells; peristome teeth shorter on one side of capsule than other, 300–400 μm, bifid to below middle; spores 20–25 μm. Fr common, early summer. Dense or lax green tufts on acidic rocks, very rare. Merioneth, N. Northumberland, Cumberland, Angus, E. Inverness. 5. Europe, Iceland, Caucasus, C. Asia, Japan, N. America, Greenland.

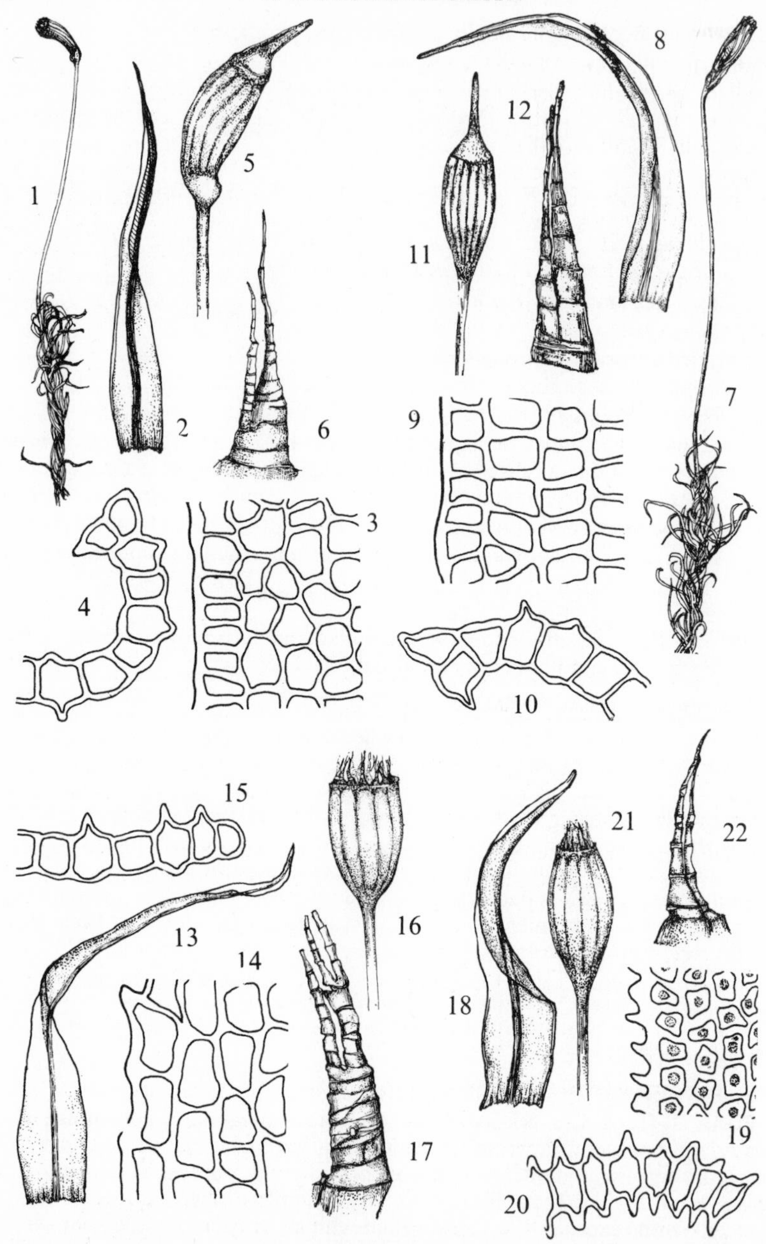

Fig. 54. 1–6, *Cynodontium strumiferum*: 1, dry shoot; 2, leaf (×20); 3, leaf cells; 4, section of leaf margin; 5, capsule; 6, peristome tooth. 7–12, *C. polycarpon*: 7, dry shoot; 8, leaf (×15); 9, leaf cells; 10, section of leaf margin; 11, capsule; 12, peristome tooth. 13–17, *C. fallax*: 13, leaf (×15); 14, leaf cells; 15, section of leaf margin; 16, capsule; 17, peristome tooth. 18–22, *C. gracilescens*: 18, leaf (×15); 19, leaf cells; 20, section of leaf margin; 21, capsule; 22, peristome tooth. Cells ×450, capsules ×10, teeth ×100.

3. C. jenneri (Pöch) Limpr., Laubm., 1886

Plants 1–4 cm. Lvs crisped when dry, erect-spreading when moist, narrowly lanceolate, gradually tapering to acute apex, margin narrowly recurved to *ca* halfway, rarely plane, unistratose throughout, coarsely dentate above to almost entire; nerve usually slightly toothed at back above; upper cells 11–22 μm, not mamillose. Perigonial lvs ± acute. Seta yellowish; capsule ± erect and symmetrical, ellipsoid with long neck, striate, not or only very slightly strumose; lid crenulate at base; annulus of 2 rows large cells, separating; peristome longer on one side of capsule than other, 300–400 μm long; spores 20–25 μm. Fr common, summer. Dense or lax, bright green tufts on non-basic, sheltered rocks or rock ledges, occasionally on soil in woodland. N. Wales, Durham and Westmorland north to Orkney, occasional. 31. Scandinavia, Westphalia.

Readily distinguished by the unistratose margin and large cells.

4. C. tenellum (Br. Eur.) Limpr., Krypt. Fl. Schlesien, 1877

Plants 1–3 cm. Lvs crisped when dry, erect-spreading when moist, narrowly lanceolate, gradually tapering to acute apex, margin recurved to more than halfway, bistratose and irregularly crenulate-dentate or rarely entire above; nerve smooth or slightly roughened at back above; upper cells 7–12 μm, ± smooth on upper surface, smooth to slightly mamillose on lower. Perigonial lvs obtuse. Seta yellowish; capsule ± erect and symmetrical, ellipsoid with short neck, striate, not strumose; lid entire at base; annulus of 2 rows small cells, persistent; peristome teeth longer on one side of capsule than other, 220–280 μm long, divided nearly to base; spores 16–20 μm. Fr common, early summer. Dense green tufts on sheltered, non-basic rocks and screes, rare. Berwick and eastern Scottish Highlands. 9. C. and N. Europe, Caucasus, N. America.

5. C. gracilescens (Web. & Mohr) Schimp., Coroll, 1856

Plants 1–3 cm. Lvs crisped when dry, spreading when moist, lanceolate, tapering to acute or sub-acute apex, margin recurved to above halfway, unistratose throughout, irregularly crenulate-dentate above; nerve strongly mamillose at back; upper cells 7–11 μm, obscure, coarsely mamillose on both surfaces. Perigonial lvs acuminate. Seta yellowish, arcuate when moist; capsule as in *C. tenellum*; spores 17–20 μm. Fr common, summer. On non-basic rocks, Wales. Norway, Europe, Portugal, Caucasus, Sikkim, Amur, Yunnan, N. America.

The record for this plant is based upon a single unlocalised gathering from Wales by Evans in 1821. Its status as a British plant is questionable.

6. C. fallax Limpr., Laubm., 1886

Plants 2–6 cm. Lvs crisped when dry, erect-spreading when moist, narrowly lanceolate to linear-lanceolate, gradually tapering to acute apex, margin recurved to *ca* halfway, unistratose throughout, crenulate-dentate above; nerve strongly mamillose at back above; upper cells 9–14 μm with numerous pointed mamillae on upper surface, smooth to slightly mamillose on lower. Perichaetial lvs acuminate. Seta yellowish, straight; capsule ± erect, almost symmetrical, ovate-ellipsoid with short neck, striate, not strumose; lid entire at base; annulus a single row of small cells, persistent; peristome teeth shorter on one side of capsule than other, 250–350(–450) μm long, bifid to below halfway and usually with slits and perforations below; spores 18–23 μm. Fr common, summer. Dense or lax green tufts on non-basic rocks, Angus (1868). Norway, Sweden, C. Europe, Altai.

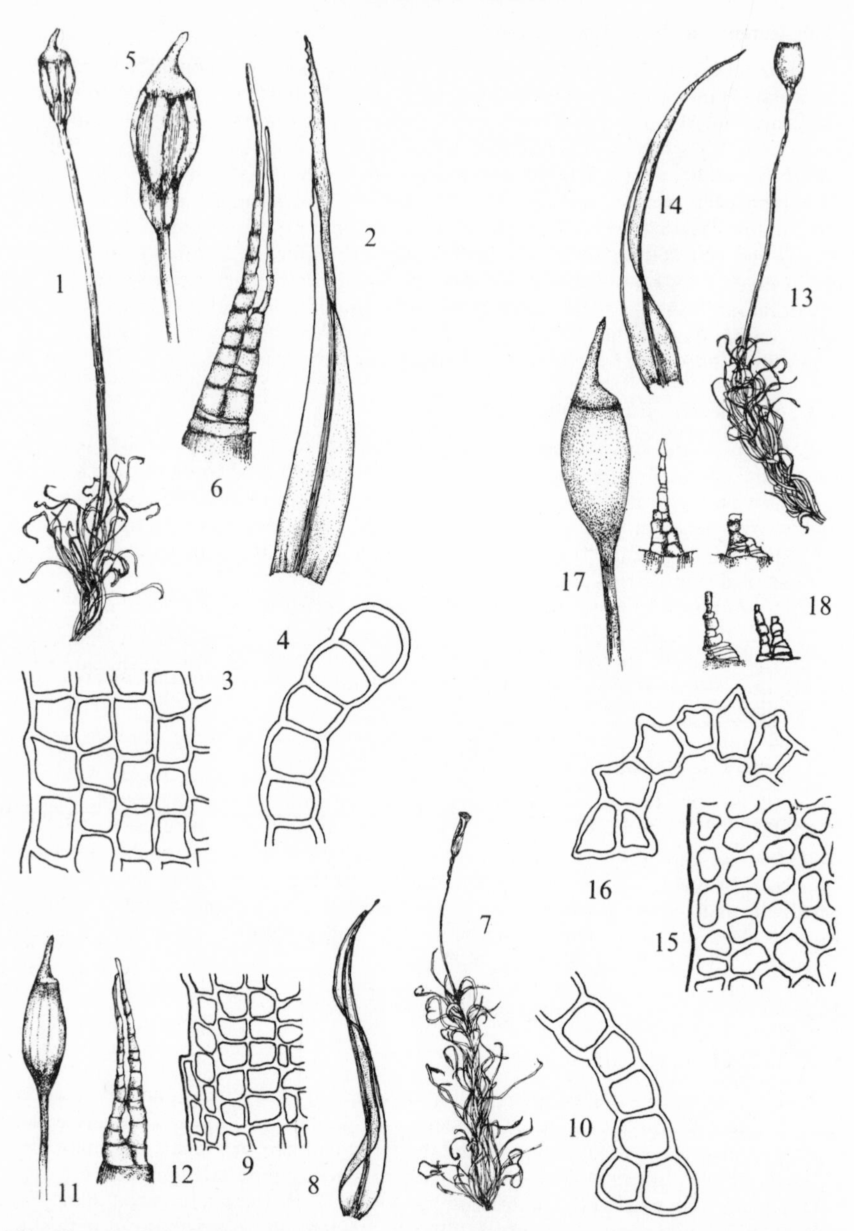

Fig. **55**. 1–6, *Cynodontium jenneri*: 1, dry shoot; 2, leaf; 3, mid-leaf cells; 4, section of margin of upper part of leaf; 5, capsule; 6, peristome tooth. 7–12, *Cynodontium tenellum*: 7, dry shoot; 8, leaf; 9, mid-leaf cells; 10, section of leaf margin; 11, capsule; 12, peristome tooth. 13–18, *C. bruntonii*: 13, dry shoot; 14, leaf; 15, mid-leaf cells; 16, section of leaf margin; 17, capsule; 18, peristome teeth. Leaves × 15, cells × 450, capsules × 10, teeth × 115, shoots × 5.

7. C. bruntonii (Sm.) Br. Eur., 1846

Plants 1–3 cm. Lvs strongly curled inwards when dry, erect-spreading with tips often recurved when moist, narrowly lanceolate, gradually tapering to acute apex, margin recurved to *ca* halfway, bistratose and remotely crenulate-dentate above; nerve slightly rough at back above; upper cells 7–14 μm wide, obscure, mamillose on both surfaces. Perigonial lvs sub-acute. Seta yellowish; capsule pale, erect, ovoid, symmetrical or slightly one-sided, slightly contracted at mouth, smooth when moist, often slightly plicate when dry; annulus persistent; lid entire at base; peristome teeth irregular, to 160 μm long, variously divided; spores 15–19 μm. Fr frequent, late spring, early summer. Dense, dark green cushions in clefts and crevices of sheltered non-basic rocks and walls. Occasional in W. and N. Britain, very rare in Ireland, rare elsewhere, E. Kent, Jersey. 65, H11, C. Europe, Canaries.

Distinct from other *Cynodontium* species in capsule characters and most likely to be confused with *Dicranoweisia cirrata* which, however, grows on rock surfaces and bark rather than in sheltered crevices. When sterile it may usually be distinguished from *C. strumiferum* and *C. polycarpon* by the more densely tufted habit, the lvs incurved when dry and the lower lvs much shorter than the upper.

25. Oncophorus (Brid.) Brid., Br. Univ., 1826

Autoecious. Lvs crisped when dry, spreading when moist, from ± sheathing base narrowed to lanceolate or subulate limb; nerve reaching apex or excurrent; basal cells rectangular, angular cells differentiated, cells above rounded-quadrate to rectangular, ± smooth. Seta straight; capsule inclined, arcuate, gibbous, strumose, not striate but often sulcate when dry; peristome teeth divided almost to base. A small genus of about 13 species occurring in north temperate and arctic regions, Sri Lanka, southern S. America.

Lf margin recurved, angular cells enlarged, often forming auricles, capsules mostly 2–3 times as long as wide **1. O. virens**
Lf margin plane, angular cells not enlarged, capsules 1.5–1.7 times as long as wide **2. O. wahlenbergii**

1. O. virens (Hedw.) Brid., Br. Univ., 1826
Cynodontium virens (Hedw.) Schimp.

Plants 3–9 cm. Lvs flexuose to crisped when dry, spreading when moist, upper longer than lower, ± reflexed and gradually tapering from ovate or oblong basal part to narrowly lanceolate limb, margin recurved below, entire or toothed above; nerve percurrent or shortly excurrent; basal cells rectangular or narrowly rectangular, angular cells shorter, wider, 2-3-stratose, sometimes forming weak auricles; cells above rounded-quadrate, bistratose at margin, not papillose, 8–12 μm wide in mid-lf. Seta reddish; capsule inclined, arcuate, ellipsoid, smooth, strumose, 1.5–2.0 mm long, (1.5–)2.0–3.0 times as long as wide; lid with oblique beak; peristome teeth bifid; spores 20–28 μm. Fr frequent, summer. $n=14$. Dense tufts or patches, bright or yellowish-green above, brownish below, on damp rocks and on wet ground by streams and in flushes in basic places above 700 m in the Scottish Highlands, frequent in Perth, rare elsewhere, Westmorland, Cumberland. 13. Europe, Caucasus, Himalayas, arctic Asia, N. America, Greenland.

2. O. wahlenbergii Brid., Br. Univ., 1826
Cynodontium wahlenbergii (Brid.) Hartm.

Plants to 3 cm. Lvs crisped when dry, patent when moist, upper scarcely longer than lower, abruptly or gradually tapering from short basal part to lanceolate or

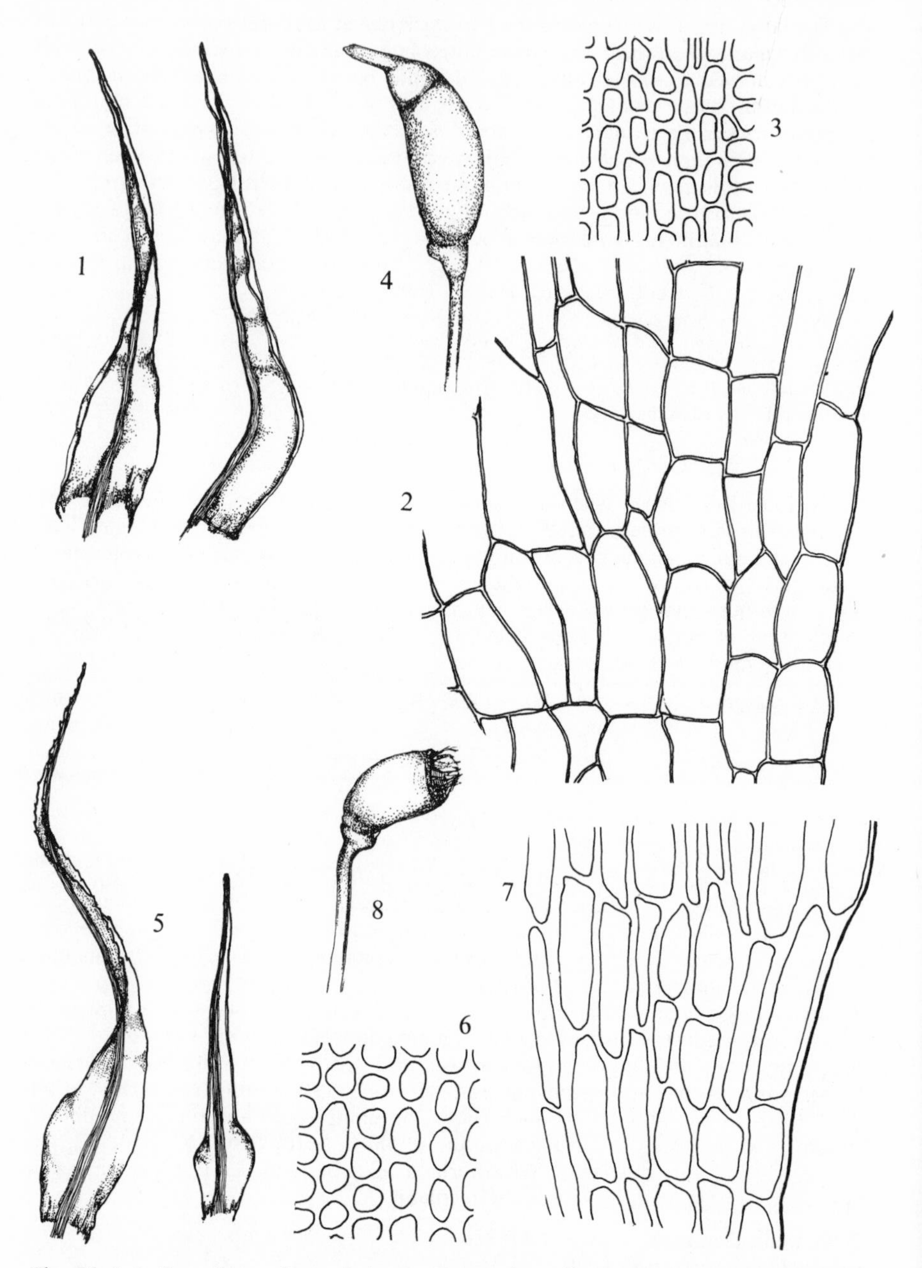

Fig. **56**. 1–4, *Oncophorus virens*: 1, leaves; 2, angular cells; 3, mid-leaf cells; 4, capsule. 5–8, *O. wahlenbergii*: 5, leaves; 6, angular cells; 7, mid-leaf cells; 8, capsule. Leaves ×15, cells ×415, capsules ×10.

linear-lanceolate limb, margin toothed just above base, entire or irregularly toothed above, plane; nerve ending below apex to excurrent; basal cells rectangular, shorter but not enlarged, unistratose towards basal angles, above irregularly rounded-quadrate, incrassate, not papillose, bistratose at margin, 8–12 μm wide in mid-lf. Seta reddish; capsule inclined, ovoid, curved, smooth, strumose, 1.2–1.7 mm long, 1.5–1.7 times as long as wide; lid with oblique beak; spores 20–28 μm. Fr frequent, summer. *n*=14. Green or yellowish tufts or patches on damp rocks, in rock crevices and on peaty soil usually above 700 m, rare. Perth, Angus, S. Aberdeen, E. Inverness, Argyll, Dunbarton, Ross. 9. Montane and arctic Europe, Caucasus, Himalayas, arctic Asia and N. America, Greenland.

26. DICHODONTIUM SCHIMP., COROLL., 1856

Dioecious. Plants forming tufts or patches, stem with central strand. Lvs ovate to narrowly lanceolate, acute to obtuse, margin usually toothed; cells opaque, mamillose, in upper part of lf rounded-quadrate. Capsule ovoid to cylindrical, straight or curved, smooth; peristome dicranoid; calyptra cucullate. Six species occurring in Europe, Asia, America, New Zealand.

Capsule asymmetrical, ovoid to ellipsoid, 1–2 times as long as wide, lvs narrowly lanceolate and acute to ovate and obtuse **1. D. pellucidum**
Capsule symmetrical or nearly so, narrowly ellipsoid to cylindrical, 1½–4 times as long as wide, lvs narrowly lanceolate, acute **2. D. flavescens**

1. D. pellucidum (Hedw.) Schimp., Coroll., 1856

Plants to 7 cm. Lvs incurved when dry, spreading to sharply reflexed when moist, narrowly lanceolate to ovate from broad base, apex acute to obtuse, margin recurved below, ± entire to coarsely toothed, sometimes undulate above; nerve stout, ending below apex; cells incrassate, lower narrowly rectangular, becoming smaller and coarsely mamillose above, towards apex ± quadrate in narrow-leaved forms, ± quadrate from middle in wide-leaved forms, 8–10 μm wide in mid-lf. Seta straight, thick, 5–10 mm long; capsule usually curved and asymmetrical, sometimes strongly so, ovoid to ellipsoid, 1–2 times as long as wide (not including lid); spores 10–22 μm. Fr occasional, autumn to spring. *n*=14*. Dull or bright green, lax tufts or patches on damp soil and rocks by streams and rivers, occasional in southern England and the Midlands, frequent to common elsewhere. 95, H33. Europe, Faroes, Iceland, Caucasus, Siberia, China, Japan, Azores, N. America.

An exceedingly variable species. Plants with narrowly lanceolate, acute lvs can only be distinguished from *D. flavescens* when in fruit; sterile plants with wider, more shortly pointed lvs belong to *D. pellucidum*. *D. flavescens* is regarded by some authorities as a variety of, or even as synonymous with, *D. pellucidum*, but with the consistent difference in capsule shape it appears to merit specific status. The similarity of the gametophytes has led to the other view and there has been a tendency in Britain to determine narrow-leaved forms as *D. flavescens*, but as these can only be determined when in fruit it is necessary that the distribution of that plant in Britain and Ireland be revised. Stunted forms of *D. pellucidum* may be mistaken for *Leptodontium flexifolium* which differs in the shorter basal cells, rounded-hexagonal upper cells and row of pale marginal cells near the base. Plants forming bright green tufts and with small, sharply reflexed, ovate lvs have been named var. *fagimontanum* (Brid.) Schimp., but because of the large number of intermediates the variety is not recognised here.

2. D. flavescens (With.) Lindb., Bot. Not., 1878
D. pellucidum var. *flavescens* (With.) Husn.

Close to *D. pellucidum*. Lvs from wide, erect basal part narrowly lanceolate, acute,

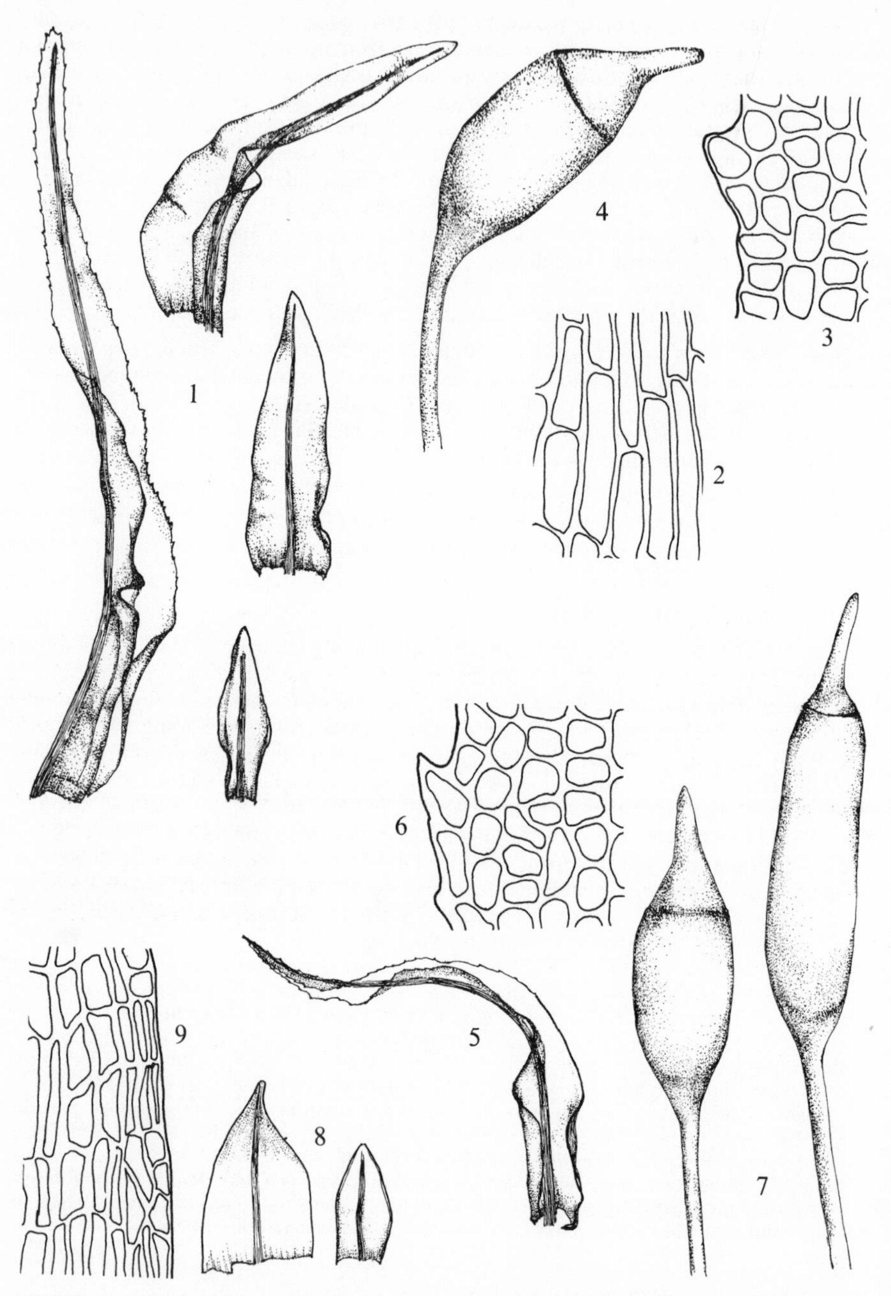

Fig. 57. 1–4, *Dichodontium pellucidum*: 1, leaves (×15); 2, basal cells; 3, mid-leaf cells; 4, capsule. 5–7, *D. flavescens*: 5, leaf (×15): 6, mid-leaf cells; 7, capsules. 8–9, *Aongstroemia longipes*: 8, leaves (×25); 9, mid-leaf cells. Cells ×415, capsules ×15.

margin entire to coarsely toothed. Capsule erect and symmetrical or nearly so, narrowly ellipsoid to cylindrical, 1.5–4.0 times as long as wide (not including lid); spores 10–24 μm. Fr occasional, autumn to spring. $n=7$*, 14. Yellowish to dark green tufts on wet soil and rocks by streams and waterfalls. Rare in S.E. England and the Midlands, occasional elsewhere. 68, H27. N., W. and C. Europe, Faroes, N. America.

SUBFAMILY 2. DICRANELLOIDEAE

Plants usually small. Lvs lanceolate, subulate, more rarely ovate, acuminate to obtuse; without differentiated angular cells, cells smooth. Capsule erect or inclined, straight or curved, sometimes strumose, smooth or striate.

27. Aongstroemia Br. Eur., 1846

Dioecious. Plants small, slender. Lvs small, imbricate, ± ovate with wide base, obtuse to acute; nerve thin; cells lax, thin-walled. Perichaetia and perigonia bud-like with lvs larger and more acute with stronger nerve than stem lvs. Capsule erect, ovoid; lid obliquely rostrate; peristome teeth entire or irregularly cleft to halfway. A mainly tropical genus with about 37 species.

1. A. longipes (Somm.) Br. Eur., 1846

Plants to 10(–15) mm, stems innovating from below inflorescences. Lvs imbricate both when wet and dry, difficult to detach from fragile stems, concave, ovate-oblong, obtuse, margin plane, entire; nerve ending below apex; cells rectangular, thick-walled, decreasing in size from nerve to margin, shorter and smaller towards apex, in mid-lf 10–15 μm wide. Perichaetia and perigonia swollen, bud-like, lvs larger than stem lvs, tapering to blunt to acute apex, nerve stouter, ending in apex. Seta red; capsule ± erect, ovoid; spores 15–20 μm. Fr rare, summer. Dense, bright green tufts, patches or scattered shoots on damp, sandy soil, very rare. Perth, Elgin, E. Inverness, Argyll. 5. N. and C. Europe, Iceland, Siberia, N. America, Greenland.

Somewhat resembling *Pohlia filum* in general appearance but differing in the obtuse lvs, areolation and lack of axillary bulbils. For the occurrence of this plant in Britain see Crundwell, *Trans. Br. bryol. Soc.* **4**, 767–74, 1965.

28. Dicranella (C. Müll.) Schimp., Coroll., 1856

Dioecious or autoecious. Lvs narrow, differentiated into sheathing base and subulate limb or gradually tapering from insertion, ± secund or squarrose; cells rectangular, smooth, not differentiated at basal angles. Seta straight; capsule erect or inclined, symmetrical or gibbous, sometimes strumose; peristome teeth 16, reddish, bi- or trifid to about middle. About 100 mainly terrestrial species, cosmopolitan.

Many authorities place the species in 2 genera, *Dicranella* and *Anisothecium* Mitt., but the latter genus is not recognised here as it separates the species in an unnatural fashion and is not supported by cytological data.

1 Seta yellow at maturity (turning brown with age), nerve occupying ⅓ or more of lf base — 2
Seta red to purple, nerve occupying up to ⅕ of lf base — 3

2 Capsule not strumose, lvs dentate, basal cells 30–50 μm long **10. D. heteromalla**
Capsule strumose, lvs entire except near apex, basal cells 70–115 μm long **9. D. cerviculata**
3 Lvs squarrose or reflexed from erect sheathing base 4
Lvs ± secund 7
4 Lvs broad, obtuse, not abruptly narrowed above sheathing base **1. D. palustris**
Lvs narrower, abruptly narrowed above sheathing base or gradually tapering to acute or acuminate apex 5
5 Capsule erect, symmetrical, cells in mid-lf 4–6 μm wide **4. D. crispa**
Capsule arcuate, gibbous, mid-lf cells 6–14 μm wide 6
6 Lvs usually denticulate, cells 8–14 μm wide in mid-lf, capsule smooth **2. D. schreberana**
Lvs usually entire, cells *ca* 6 μm wide in mid-lf, capsule striate **3. D. grevilleana**
7 Lvs abruptly narrowed to gradually tapering from erect sheathing base, capsule striate **5. D. subulata**
Lvs ± gradually tapering from insertion, base not sheathing, capsule smooth 8
8 Lvs usually secund, reddish-tinged, margin plane, capsule ± erect **6. D. rufescens**
Lvs erect-spreading to slightly secund, margin narrowly recurved below 9
9 Upper lvs linear-lanceolate, cells 4–9 μm wide in mid-lf, perichaetial lvs similar in shape to stem lvs, capsule inclined **7. D. varia**
Upper lvs lanceolate, cells 10–14 μm wide in mid-lf, perichaetial lvs differing in shape from stem lvs **8. D. staphylina**

1. D. palustris (Dicks.) Crundw. ex Warb., Trans. Br. bryol. Soc., 1962
D. squarrosa (Starke) Schimp., *Anisothecium palustris* (Dicks.) Hagen

Dioecious. Plants 1–10(–15) cm. Lvs squarrose when moist, base erect, sheathing, decurrent, limb oblong-lanceolate, apex obtuse, margin plane, crenulate or rarely entire above; nerve thin, ending well below apex; cells in sheathing base narrowly rectangular or narrowly rhomboidal, ± hyaline, gradually becoming smaller in limb, irregular in shape and more incrassate, 8–12 μm wide in mid-lf. Spherical, reddish-brown to brown rhizoidal gemmae, 200–250 μm diameter, often present. Seta stout, purple; capsule ovoid, inclined, gibbous, smooth; lid with oblique beak; spores 16–22 μm. Fr rare, autumn to spring. $n=15$*. Yellowish-green to bright green tufts or patches in marshes, springs, and at edges of small streams. Rare in S.E. England and the Midlands, frequent or common elsewhere. 78, H31. Europe, Caucasus, Japan, N. America, Greenland.

Stunted forms may be confused with *Dichodontium pellucidum* which, however, differs in its toothed lvs and mamillose cells.

2. D. schreberana (Hedw.) Dix., Rev. Bryol., 1933
D. schreberi (Hedw.) Schimp., *Anisothecium schreberianum* (Hedw.) Dix.

Dioecious. Plants 3–10(–30) mm. Lvs loosely crisped when dry, squarrose-flexuose when moist, from oblong, sheathing base abruptly narrowed to lanceolate to subulate limb, shorter, wider, more strongly toothed and less sharply reflexed in tall forms, margin plane, denticulate at least towards apex, rarely entire; nerve thin, ending in apex to excurrent; cells in basal part rectangular to narrowly rectangular, in limb shorter, smaller, ± rectangular, 8–14 μm wide in mid-lf. Brown to reddish-brown, ± spherical gemmae, 90–140 μm diameter, present on rhizoids and sometimes in lf axils. Seta purple; capsule inclined, ovoid, gibbous, not strumose, smooth; lid

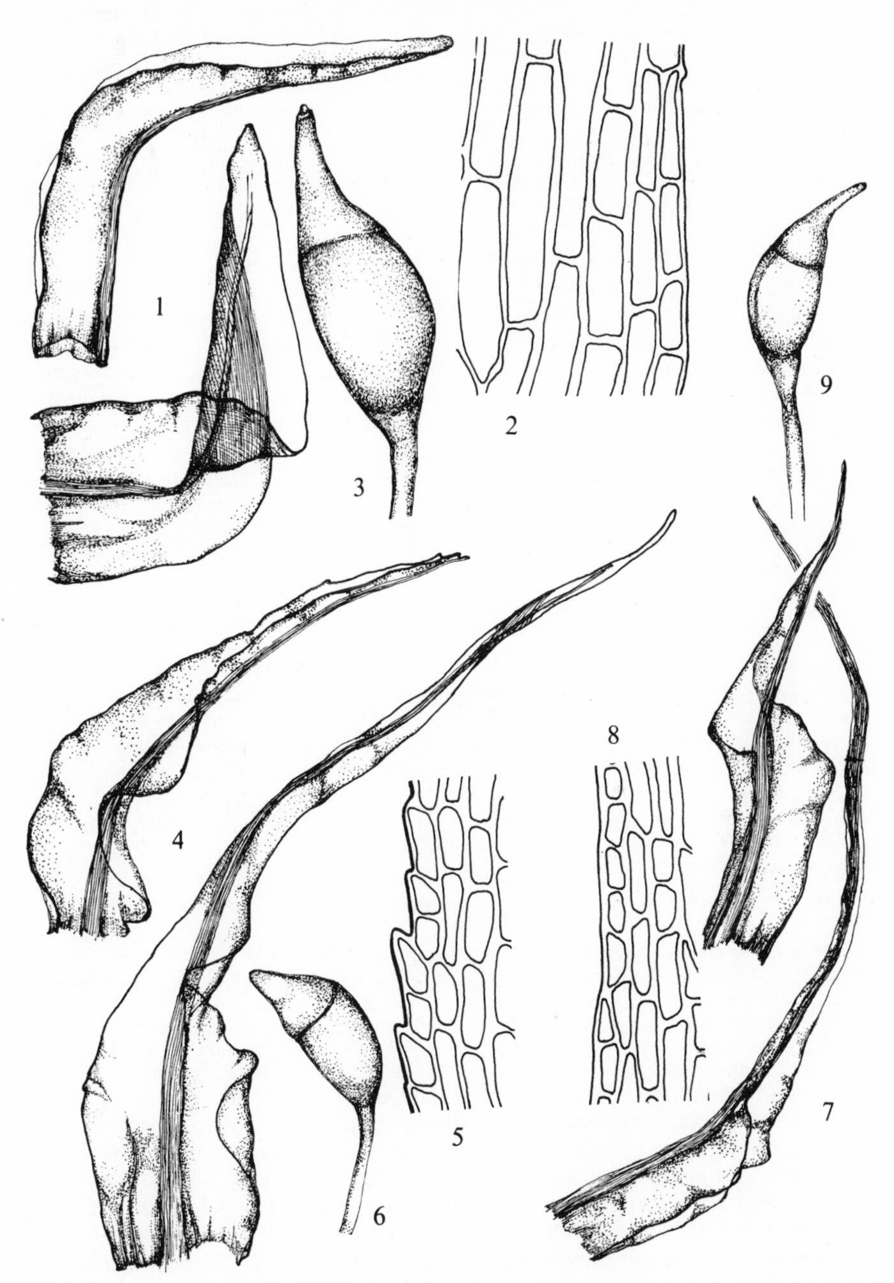

Fig. **58**. 1–3, *Dicranella palustris*: 1, leaves (×12.5); 2, mid-leaf cells; 3, capsule. 4–6, *D. schreberana*: 4, leaves (×40); 5, mid-leaf cells; 6, capsule. 7–9, *D. grevilleana*: 7, leaves (×40); 8, mid-leaf cells; 9, capsule. Cells ×415, capsules ×15.

with oblique conical to subulate beak; longitudinal walls of exothecial cells more heavily thickened than transverse walls; spores 12–16 μm. Fr frequent, autumn, winter. $n=14$. Yellowish-green patches or scattered plants on damp, heavy, often calcareous soil in fields, gardens, wood rides, banks of ditches, etc., occasional to frequent in lowland areas, rare elsewhere. 100, H30. C. and N. Europe, Caucasus, Siberia, N. America.

D. grevilleana differs in its montane habitat, smaller lf cells and evenly thickened exothecial cells. *Ditrichum cylindricum* has the lf subula denticulate all round with projecting cell walls and not just at margin.

3. D. grevilleana (Brid.) Schimp., Coroll., 1856
Anisothecium grevilleanum (Brid.) Broth.

Dioecious or autoecious. Plants 3–8 mm. Lvs squarrose-flexuose, from oblong, sheathing base abruptly narrowed to channelled subula, margin entire or obscurely denticulate near apex; nerve thin, percurrent; cells *ca* 6 μm wide, 15–25 μm long in mid-lf. Gemmae similar to those of *D. schreberana* sometimes present. Seta purple; capsule inclined, ovoid, slightly curved, very slightly strumose, faintly striate when dry; exothecial cells with uniformly thickened walls; lid with long beak; spores 12–22 μm. Fr frequent, late summer, early autumn. $n=15$. Yellowish-green tufts or scattered plants on disturbed, damp, basic soil, usually at high altitudes, very rare. Perth, Argyll, W. Ross, Dublin, Sligo, Leitrim. 4, H3. Pyrenees, C. and N. Europe, Iceland, Caucasus, Siberia, N. America.

4. D. crispa (Hedw.) Schimp., Coroll., 1856
Anisothecium vaginale (With.) Loeske

Dioecious or autoecious. Plants 5–10 mm. Lvs loosely crisped when dry, squarrose-flexuose when moist, base oblong, sheathing, abruptly narrowed to long, channelled, linear-subulate limb, margin plane, entire below, erect or inflexed above, entire or denticulate towards apex; nerve thin, ending in apex to excurrent, occupying most of subula; cells in basal part linear to rectangular or rhomboidal, above very narrow, linear, 4–6 μm wide in mid-lf. Seta reddish-brown; capsule erect, ellipsoid or obovoid, ± symmetrical, striate when dry; lid with long oblique beak; spores 16–20(–26) μm. Fr common, autumn, winter. $n=14, 15$. Yellowish-green to green tufts or patches on soil on banks of streams and ditches, and in dune slacks, generally distributed but occasional. 47, H9. Europe, Siberia, N. America, Greenland.

5. D. subulata (Hedw.) Schimp., Coroll., 1856
D. curvata (Hedw.) Schimp., *D. secunda* Lindb., *D. subulata* var. *curvata* (Hedw.) Rabenh.

Dioecious. Plants 5–20 mm. Lvs erect to curved or secund when dry, erecto-patent or more usually secund when moist, basal part sheathing, oblong, abruptly narrowed to gradually tapering to long channelled subula, margin plane below, erect or incurved above, towards apex entire or denticulate; nerve thin, excurrent; cells in basal part narrowly rectangular, above smaller, linear, 4–6 μm wide in mid-lf. Perichaetial lvs sheathing or not. Dark brown rhizoidal gemmae, 110–170 μm × 95–150 μm, often present. Seta reddish-purple to brownish-purple; capsule ovoid, erect and symmetrical to inclined and gibbous, striate or sulcate when dry; lid with long subulate beak; spores 16–20 μm. Fr common, autumn, winter. $n=13^*, 14+1$. Yellowish-green to dark green tufts on damp, sandy or stony ground in turf on banks and by streams, occasional in montane areas, rare elsewhere. 58, H13. Europe, Turkey, Siberia, China, N. America, Greenland.

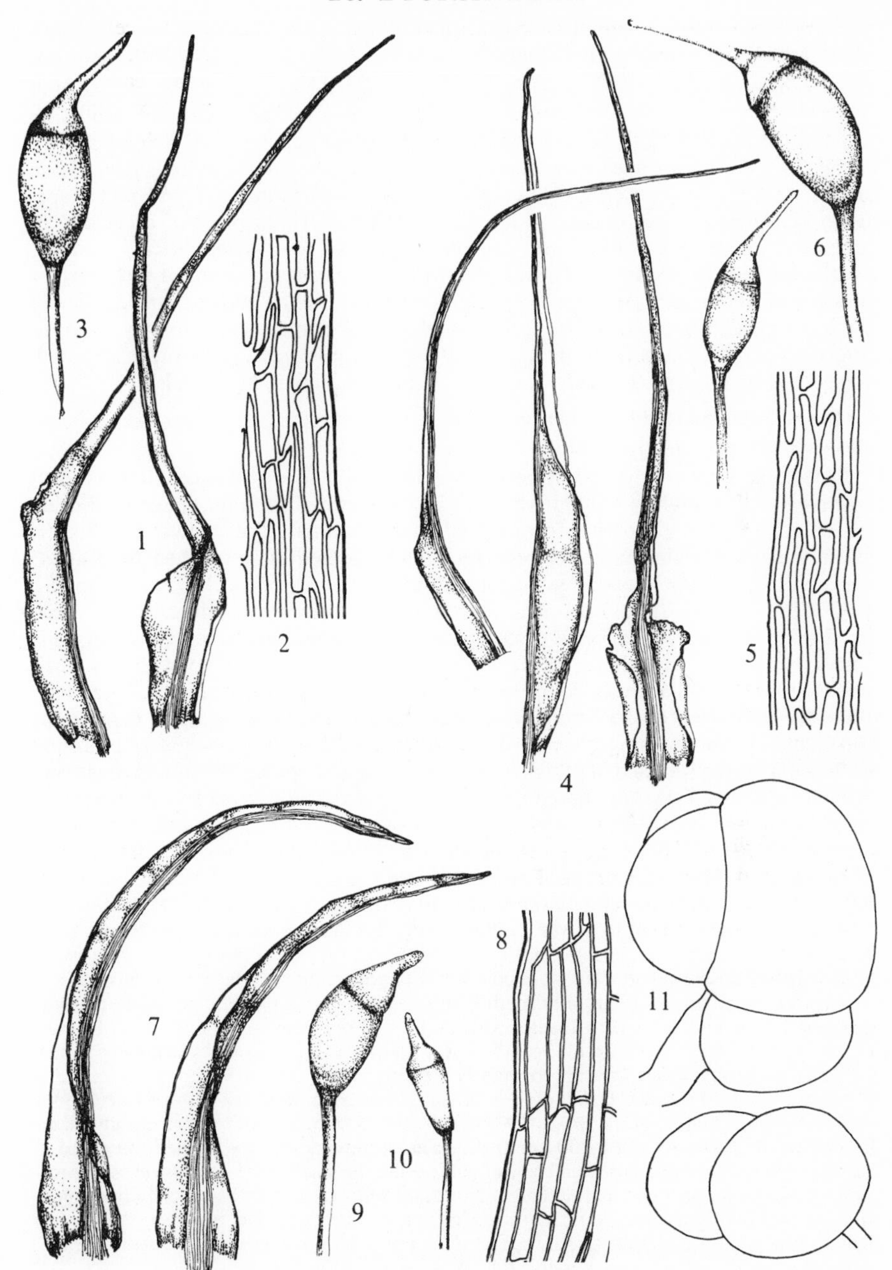

Fig. 59. 1–3, *Dicranella crispa*: 1, leaves (×25); 2, mid-leaf cells; 3, capsule. 4–6, *D. subulata*: 4, leaves (×25); 5, mid-leaf cells; 6, capsules. 7–11, *D. varia*; 7, leaves (×40); 8, mid-leaf cells; 9, normal capsule; 10, 'apogamous' capsule; 11, gemmae (×250). Cells × 415; capsules ×15.

6. D. rufescens (With.) Schimp., Coroll., 1856
Anisothecium rufescens (With.) Lindb.

Dioecious. Plants to *ca* 1 cm, often reddish-tinged. Lvs erect to flexuose-secund when dry, secund when moist, narrowly lanceolate, base not sheathing, gradually tapering to acuminate to subulate apex, margin plane, denticulate, at least above; nerve thin, ending in apex to excurrent; basal cells irregularly rectangular, above smaller, 8–14 μm wide in mid-lf. Perichaetial lvs similar to upper stem lvs in shape. Rhizoidal gemmae, composed of 1–3 large cells sometimes present, especially in senescent plants. Seta deep red; capsule erect, ellipsoid, symmetrical, smooth; exothecial cells uniformly thickened; lid with oblique beak; spores 12–16 μm. Fr occasional to frequent, autumn to spring. $n=10$, 14*, 15*. Yellowish-green, usually reddish-tinged tufts, patches or scattered plants on damp, acidic soil by streams, pools, on paths, beds of reservoirs, occasionally on wood rides, generally distributed, occasional to frequent. 96, H25, C. Europe Caucasus, Japan, N. America.

7. D. varia (Hedw.) Schimp., Coroll., 1856
Anisothecium varium (Hedw.) Mitt.

Dioecious. Plants 2–10(–30) mm. Lvs rigid, straight to slightly secund when dry, erecto-patent to slightly secund when moist, base not sheathing, upper lvs linear-lanceolate, gradually tapering from insertion to acuminate apex, lower lvs shorter, wider, margin usually narrowly recurved, entire or minutely toothed near apex; nerve ending in apex to excurrent, about 55–85(–100) μm wide at lf base, occupying ⅕ of width of base, sharply defined, in section with 2 rows of guide cells with scattered stereids above; basal cells ± rectangular, incrassate, smaller near margin, cells above narrowly rectangular to linear, unistratose throughout, 4–9 μm wide in mid-lf. Perichaetial lvs similar in shape to stem lvs. Irregular, pale brown rhizoidal gemmae, 100–140(–250) μm × 60–95 μm often present. Seta deep reddish-brown; capsule inclined, ovoid, gibbous, smooth when dry; exothecial cells with ± straight longitudinal walls, which are more heavily thickened than transverse walls; lid conical or shortly beaked; spores 14–18 μm. Fr common, autumn to spring. $n=14$*, 15. Bright to yellowish-green, lax to dense tufts or patches or scattered plants on usually damp, often basic soil on banks, by roadsides, ditches, in fields, quarries, etc., frequent to common. 109, H40, C. Europe, Faroes, Middle East, Kashmir, China, Siberia, N. Africa, Macaronesia, N. America south to Guatemala, Cuba, Jamaica, Haiti.

A curious form referred to as var. *callistoma* (With.) Schimp., is occasionally met with. The seta is very short, the capsule erect, symmetrical and very small but with a normal-sized lid; the spores are shrunken. This plant possibly has apogamous or hybrid sporophytes.

D. rufescens differs in its usually reddish tinge, the lvs with plane, denticulate margins and larger cells and the erect capsules with uniformly thickened exothecial cells. In both *D. varia* and *D. rufescens* rhizoidal gemmae seem only to be produced in senescing plants, unlike *D. staphylina* where they are constantly present.

A closely related species, *D. howei* Ren. & Card., occurs in southern Europe (see Crundwell & Nyholm, *Lindbergia* 4, 34–8, 1977). This differs from *D. varia* in the usually plane lf margin, the nerve about 85–100 μm at the base, occupying about ⅓ of the lf base and in section with only a single superficial row of guide cells; the lf cells are bistratose except sometimes in the lower part of the leaf. The longitudinal walls of the exothecial cells are slightly sinuose and hardly thicker than the transverse walls. This plant has not been detected in Britain but plants intermediate between *D. howei* and *D. varia*, possibly of hybrid origin, have been found, and it is very possible that *D. howei* occurs in Britain.

8. D. staphylina Whitehouse, Trans. Br. bryol. Soc. 1969
Anisothecium staphylinum (Whitehouse) Sipman, Rubers & Riemann

Apparently dioecious. Plants to 5 mm. Lvs erect when dry, erecto-patent to

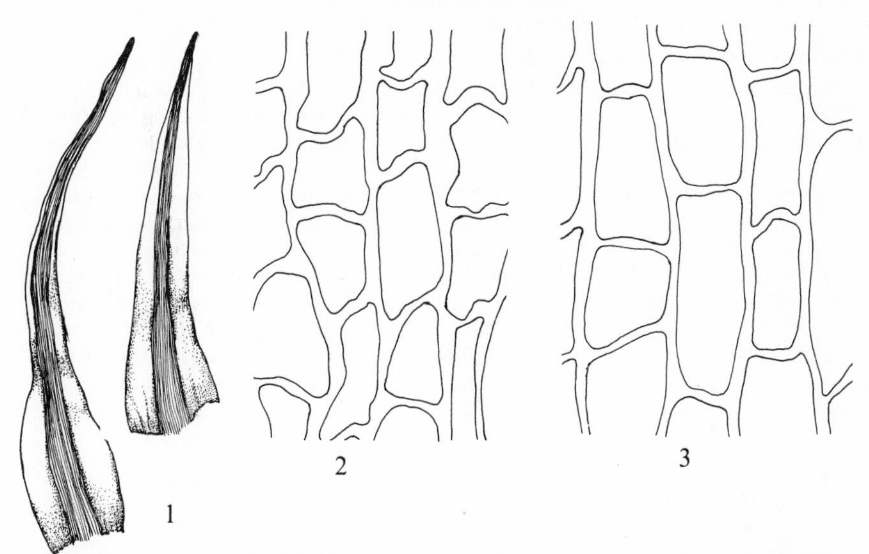

Fig. **60**. 1–2, plant intermediate between *D. varia* and *D. howei*: 1, leaves (×40); 2, exothecial cells. 3, *D. varia*: exothecial cells. Cells ×415.

spreading, rarely secund when moist, lanceolate, tapering to acute apex, margin plane or recurved below, with a few obscure teeth towards apex; nerve thin, ending below apex; cells ± rectangular, 10–14 μm wide in mid-lf. Perichaetial lvs with sheathing base, ± abruptly narrowed to long flexuose or squarrose limb. Brownish rhizoidal gemmae, 80–100 μm×50–80 μm, always present. Sporophyte unknown. Green, dense tufts or scattered plants on disturbed soil in arable fields, gardens, gravel pits, bare patches in grassland, by paths and roads, on banks and wood rides, probably common. 56, H3, C. Belgium, Denmark, Germany, Holland, N. America.

Similar to *D. varia* and *D. rufescens*. The former differs in the narrower, more secund lvs, stronger nerve, narrower cells and perichaetial lvs similar to stem lvs. *D. rufescens* has longer, narrower, secund lvs with more obviously toothed, plane margins and perichaetial lvs similar to the stem lvs and most particularly the reddish coloration of older parts of the plant. For an account of the plant see Whitehouse, *Trans. Br. bryol. Soc.* **5**, 757–65, 1969.

9. D. cerviculata (Hedw.) Schimp., Coroll., 1856

Dioecious. Plants 5–20 mm. Lvs flexuose when dry, erecto-patent to secund when moist, base ovate or oblong, half-sheathing, gradually or rapidly tapering to long channelled subula, lower lvs smaller, more gradually tapering, margin plane below, erect or inflexed above, entire or finely denticulate towards apex; nerve stout, occupying $\frac{1}{3}$–$\frac{1}{2}$ lf base, excurrent in entire or denticulate point in upper lvs, ending below apex in lower; cells in basal part narrowly rectangular, 70–115 μm long, above smaller, bistratose, narrowly rectangular, 4–8 μm wide in mid-lf. Seta yellowish, turning brown with age; capsule inclined, ovoid, gibbous, strumose, sulcate when dry; lid with curved, subulate beak; spores 16–22 μm. Fr common, summer to winter. $n=13$*, 15. Yellowish-green or green patches on partially dried peat especially in peat cuttings and ditch banks and on damp, acidic sandy or gravelly soil, generally frequent in suitable habitats. 93, H32. C. and N. Europe, Siberia, N. America, Greenland.

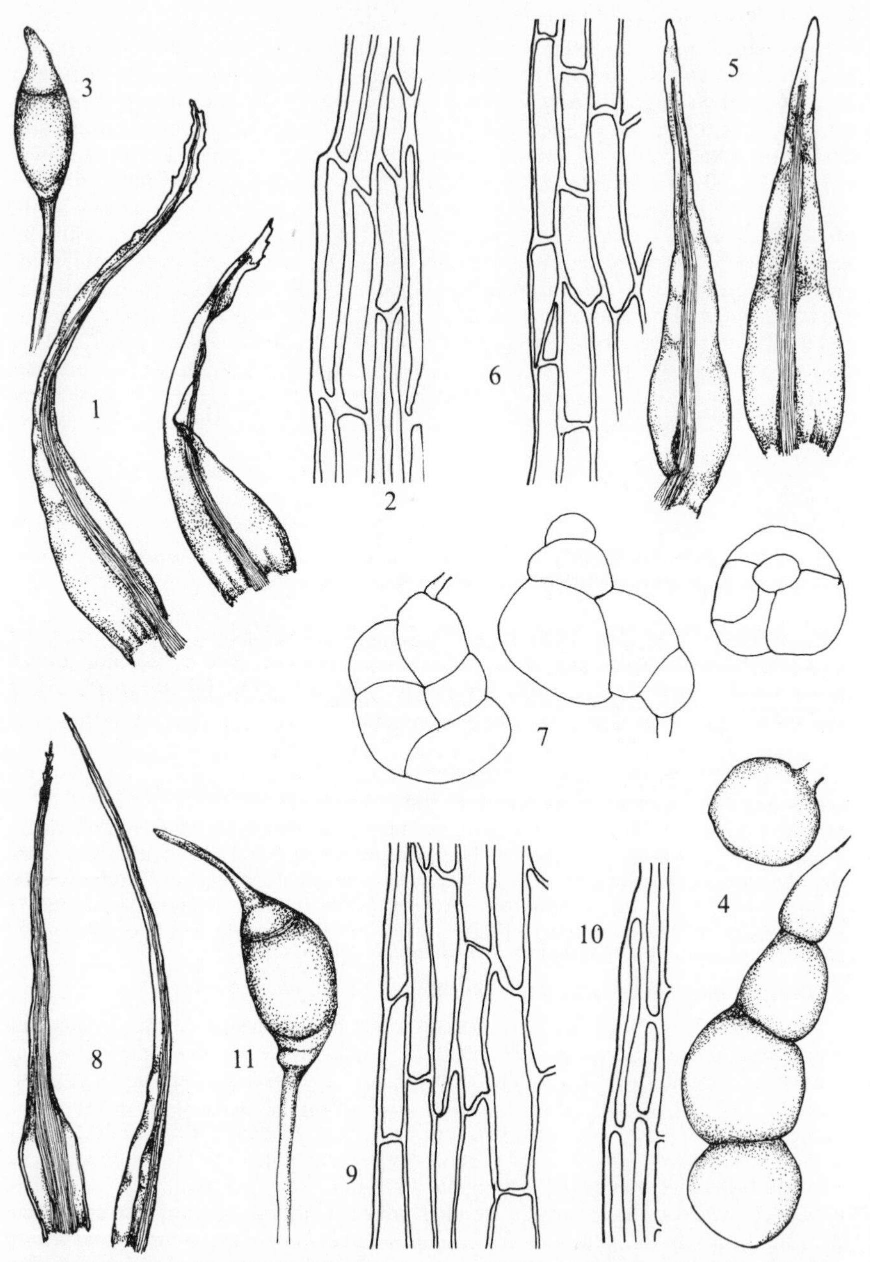

Fig. **61**. 1–4, *Dicranella rufescens*: 1, leaves (×40); 2, mid-leaf cells; 3, capsule; 4, rhizoidal gemmae (×250). 5–7, *D. staphylina*: 5, leaves (×63); 6, mid-leaf cells; 7, rhizoidal gemmae (×250). 8–11, *D. cerviculata*: 8, leaves (×25); 9, cells from near leaf base; 10, cells from upper part of leaf; 11, capsule. Cells ×415, capsules ×15.

10. D. heteromalla (Hedw.) Schimp., Coroll., 1856

Dioecious. Plants 1–3(–6) cm. Lvs falcato-secund, rarely erecto-patent when moist, scarcely altered when dry, from ovate or lanceolate basal part gradually tapering to long channelled subula, margin plane, entire below, dentate above, especially towards apex; nerve strong, occupying *ca* $\frac{1}{3}$ of lf base, percurrent or excurrent and occupying greater part of subula; cells in basal part rhomboidal to narrowly rectangular, 30–50 μm long, smaller and narrower towards margin, above smaller, rectangular to linear and extending in narrow band towards apex, 4–6 μm wide in mid-lf. Perichaetial lvs larger abruptly narrowed from sheathing base to long subula. Seta yellowish, becoming brown with age; capsule inclined to ± horizontal, ellipsoid, gibbous, not strumose, sulcate when dry and empty; lid with long curved subula; spores 12–17 μm. Fr common, winter, spring. $n = 13^*$, $13 + 2$–3, 14. Dull or yellowish-green tufts or patches, sometimes extensive and locally dominant on acidic soil on banks, in hedgerows, woods, ditches, among rocks, on tree stumps, etc., common and sometimes abundant except on calcareous substrata. 112, H39, C. Europe, Asia from Lebanon, the Caucasus and Himalayas northwards, Madeira, Kenya, N. America, Bolivia.

Readily recognised by the silky, falcato-secund lvs and the yellow setae. Plants on sandstone may have narrower and erect-spreading or sub-secund lvs and have been called var. *sericea* (Schimp.) Schimp., but such plants are merely an extreme form of a continuous series.

SUBFAMILY 3. DICRANOIDEAE

Plants usually large. Lvs straight to falcate, narrowly lanceolate, acuminate or subulate; basal cells often porose, angular cells enlarged, hyaline or coloured, upper cells smooth or papillose. Capsule ovoid to cylindrical, straight or curved, rarely strumose, smooth or striate.

29. Dicranoweisia Lindb. ex Milde., Bryol. Siles., 1869

Autoecious. Plants forming dense tufts or cushions, stem with central strand. Lvs crisped when dry, patent to spreading when moist, from ovate or lanceolate basal part abruptly or gradually tapering to apex, margin plane or recurved, entire; nerve ending below apex; cells at basal angles enlarged or not, cells in limb quadrate or shortly rectangular, mamillose or not. Seta straight; capsule erect, symmetrical, smooth; peristome teeth deeply inserted, entire or bifid at tip. Twenty-four species occurring in temperate, arctic, antarctic and montane tropical regions.

Lvs with recurved, unistratose margins, without auricles, cells in mid-lf 10–14 μm wide, capsule narrowly ellipsoid or sub-cylindrical **1. D. cirrata**
Lf margin plane, bistratose above, at least some lvs with distinct brownish auricles, cells in mid-lf 6–8(–10) μm wide, capsule ovate-ellipsoid **2. D. crispula**

1. D. cirrata (Hedw.) Lindb. ex Milde, Bryol. Siles., 1869

Plants 0.5–2.0 cm. Lvs crisped and incurved when dry, erecto-patent to spreading when moist, lower lanceolate, upper narrowly linear-lanceolate, gradually tapering to acute apex, margin entire, recurved at middle of lf; nerve ending below apex; basal cells rectangular, (10–)12–18 μm wide, at basal angles shorter and wider but not forming auricles, cells above quadrate or rounded-quadrate, ± incrassate, smooth, 10–14 μm wide in mid-lf. Gemmae often present and sometimes abundant

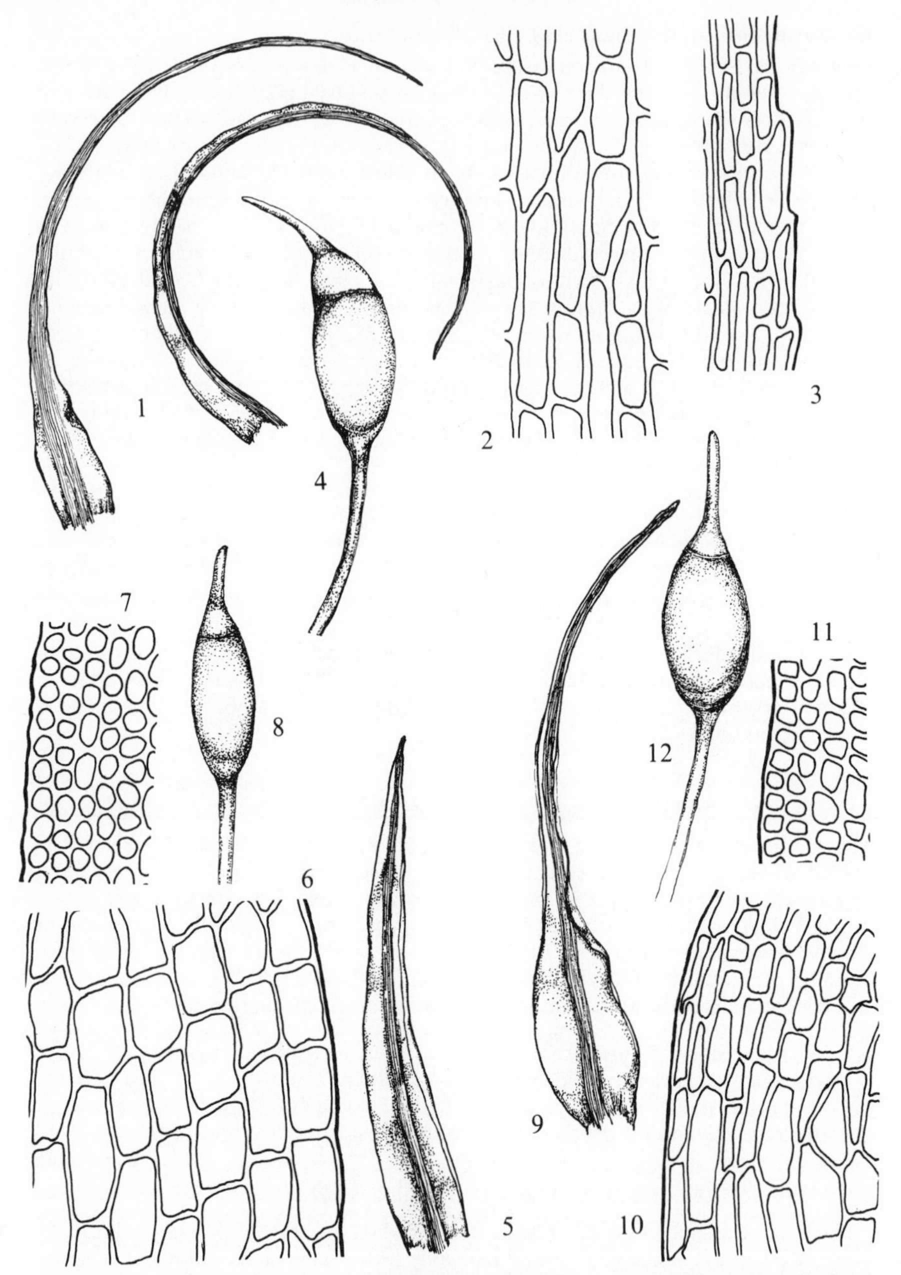

Fig. **62**. 1–4, *Dicranella heteromalla*: 1, leaves; 2, cells from near leaf base; 3, cells from upper part of leaf; 4, capsule. 5–8, *Dicranoweisia cirrata*: 5, leaf; 6, cells near leaf base; 7, mid-leaf cells; 8, capsule. 9–12, *D. crispula*: 9, leaf; 10, cells near leaf base; 11, mid-leaf cells, 12, capsule. Cells ×415, capsules ×15.

on dorsal surface of lamina near lf base, ellipsoid to cylindrical, to 150 μm long. Seta yellow; capsule narrowly ellipsoid to sub-cylindrical; lid obliquely rostrate; spores 14–20 μm. Fr frequent or common, autumn to spring. $n=11$, $11+1$, 13*. Yellowish-green or green tufts or cushions on trees, wood, thatch, rocks and walls, common at low altitudes, rare in N. and N.E. Scotland and Ireland. 109, H17, C. Europe, Caucasus, Himalayas, Mongolia, Algeria, Madeira, Canaries, N. America, Hawaii, Tasmania.

Only likely to be confused with *Cynodontium bruntonii* and *Dicranum montanum*. *C. bruntonii* has toothed lvs with smaller, more incrassate cells, a shorter, slightly asymmetrical capsule and occurs in rock crevices and cavities in walls rather than on rock surfaces or trees. *D. montanum* also has toothed lvs and the nerve is denticulate at the back above.

2. D. crispula (Hedw.) Milde, Bryol. Siles., 1869

Plants 1–3 cm. Lvs crisped, incurved when dry, flexuose-spreading, sometimes secund when moist, from ovate or lanceolate base abruptly narrowed or gradually tapering to linear-subulate limb, margin plane, entire; nerve shortly excurrent; basal cells linear, 6–8 μm wide, at basal angles enlarged or inflated and forming brownish auricles in most lvs, upper cells irregularly quadrate, bistratose at margin, 6–8(–10) μm wide in mid-lf. Seta yellow; capsule ovate-ellipsoid; spores 12–20 μm. Fr common, spring, summer. $n=11$, 14. Yellowish-green to green, sometimes blackish tufts or cushions on calcareous and siliceous rocks and boulders, usually above 500 m, frequent in montane areas. Shrops, N. Wales, Lake District, and Scottish Highlands. 28. Mountains of C. and N. Europe, Caucasus, Himalayas, Siberia, Japan, N. America, Greenland.

30. ARCTOA Br. Eur., 1846

Close to *Kiaeria*. Autoecious. Lvs as in *Dicranum* but nerve longly excurrent, without stereids; cells not porose, angular differentiated, brownish. Seta short, thick; capsule erect, slightly asymmetrical, furrowed when dry. Two or 3 north temperate or arctic species.

1. A. fulvella (Dicks.) Br. Eur., 1846
Dicranum fulvellum (Dicks.) Sm.

Plants 0.5–3.0 cm. Lvs slightly crisped to secund when dry, erect-flexuose to falcato-secund when moist, from ovate or lanceolate basal part narrowed to long entire or faintly denticulate subula composed mainly of longly excurrent nerve; basal cells elongate rectangular, angular cells quadrate, brownish, cells above narrowly rectangular, in upper part irregularly quadrate at margin, narrowly rectangular by nerve, 6–8 μm wide in mid-lf. Seta yellowish, thick; capsule erect, ovoid, slightly asymmetrical, contracted below mouth and wide-mouthed with spreading peristome teeth, striate when dry and empty; lid with oblique beak; spores *ca* 20 μm. Fr common, summer. Yellowish-green to green, dense tufts, brownish or blackish below on sheltered rocks and in rock crevices, mainly in basic habitats, rarely below 450 m, occasional. N. Wales, Yorks, Lake District, S.W. Scotland, Scottish Highlands, Kerry. 23, H1. Europe, Iceland, Japan, N. America.

When fertile readily distinguished from species of *Kiaeria* by the erect, ± symmetrical, non-strumose capsules. When sterile it may be separated from *Kiaeria* species by the lvs more abruptly contracted above the basal part and the very long excurrent nerve. *Blindia acuta* is a more rigid plant of damper habitats and the lvs have orange auricles.

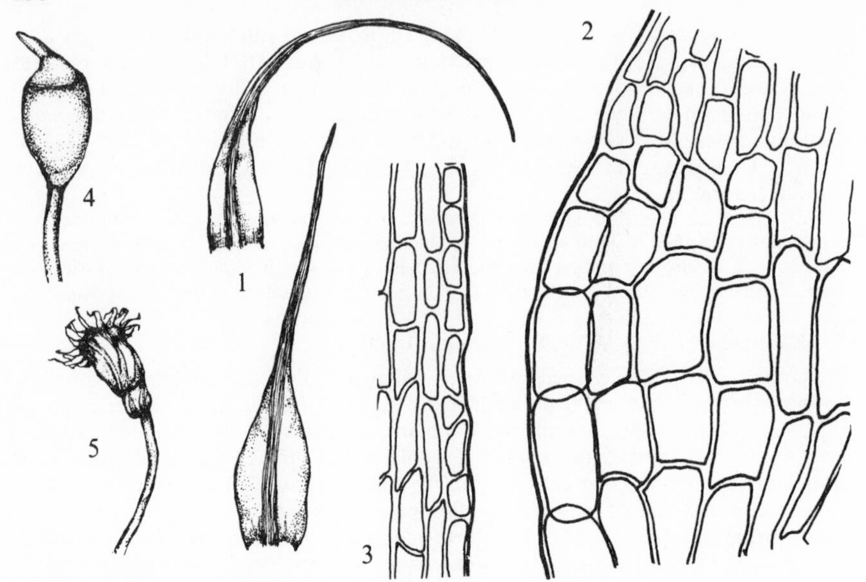

Fig. **63**. *Arctoa fulvella*: 1, leaves (×15), 2, angular cells, 3, cells *ca* ⅓ way up leaf, 4 and 5, mature and dehisced capsules (×15). Cells ×415.

31. Kiaeria Hagen, K. N. V. Selsk. Skr., 1914

Autoecious. Lvs similar to *Dicranum* but nerve without stereids; angular cells brownish, distinct or not, basal cells porose or not. Capsule ± curved, strumose. A northern hemisphere, arctic–alpine genus of about 4 species.

Angular cells sharply distinct, capsule ellipsoid to sub-cylindrical, striate when dry **3. K. starkei**
Angular cells intergrading with basal cells, capsule ovoid to ovate-ellipsoid, sometimes irregularly furrowed but not striate when dry 2
2 Lvs scarcely altered when dry, usually yellowish-green, exothecial cells incrassate, annulus persistent **1. K. falcata**
Lvs ± crisped when dry, dull or dark green, exothecial cells thin-walled, annulus fugacious **2. K. blyttii**

1. K. falcata (Hedw.) Hagen, K. N. V. Selsk. Skr., 1914
Dicranum falcatum Hedw.

Androecium immediately beneath perichaetium. Plants to 3 cm, yellowish-green to green above. Lvs falcato-secund, hardly altered when dry, lanceolate, tapering to channelled subula, margin entire or slightly denticulate above; nerve excurrent; cells not porose, basal rectangular to narrowly rectangular, angular cells brownish, quadrate, intergrading with other cells and not sharply distinct, upper cells quadrate to quadrate-rectangular, or rhomboidal, slightly papillose. Seta to 10 mm; capsule inclined, ovoid, curved, strumose, sometimes furrowed but not striate when dry; exothecial cells mostly less than twice as long as wide, heavily incrassate; annulus of 2–3 rows small persistent cells; spores 14–16 μm. Fr common, summer. $n=7$.

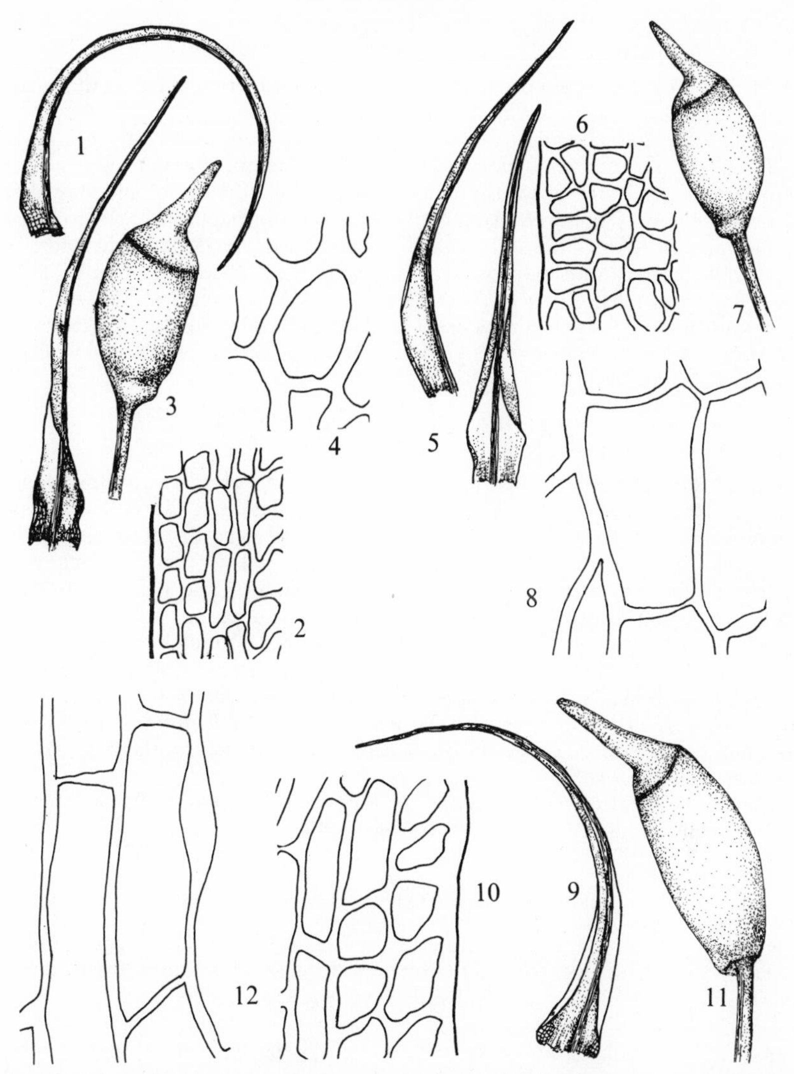

Fig. **64**. 1–4, *Kiaeria falcata*: 1, leaves; 2, mid-leaf cells; 3, capsule; 4, exothecial cells. 5–8, *K. blyttii*: 5, leaves; 6, mid-leaf cells; 7, capsule; 8, exothecial cells. 9–12, *K. starkei*: 9, leaf; 10, mid-leaf cells; 11, capsule; 12, exothecial cells. Leaves ×15, cells ×450, capsules ×15.

Yellowish or yellowish-green to green tufts or mats on rocks and soil at high altitudes. Frequent on the higher Scottish mountains, Snowdon. 21. N., W. and C. Europe, N. America.

Differs from *K. blyttii* and *K. starkei* in the persistent annulus and thick-walled exothecial cells. When sterile may be distinguished from *K. blyttii* by the lvs scarcely altered when dry and the colour of the plants and from *K. starkei* by its smaller size and indistinct angular cells.

2. K. blyttii (Schimp.) Broth., Laubm. Fennosk., 1923
Dicranum blyttii Schimp., *D. schistii* Lindb.

Androecium distant from perichaetium or on separate branch. Plants to 3 cm, dull to blackish-green. Lvs ± erect to falcato-secund when moist, crisped when dry, lanceolate, tapering to acute to obtuse subula, margin ± entire; nerve excurrent; basal cells rectangular to quadrate-rectangular, occasionally porose near nerve, angular cells quadrate, brownish, intergrading with other cells and not sharply distinct, upper cells quadrate-rectangular to quadrate, papillose. Seta to 15 mm; capsule inclined, ovate-ellipsoid, curved, ± strumose, not striate when dry; exothecial cells mostly more than twice as long as wide, thin-walled; annulus a single row of fugacious cells; spores 16–18 μm. Fr occasional, summer. $n = 14$. Dull to dark green patches on rocks and in rock crevices at high altitudes. Frequent in the Scottish Highlands, rare in N. Wales, N. England and Ireland. 29, H4. N., W. and C. Europe, Himalayas, N. America.

3. K. starkei (Web. & Mohr) Hagen, K. N. V. Selsk. Skr., 1914
Dicranum starkei Web. & Mohr

Androecium immediately below perichaetium. Plants to 10 cm, yellowish-green Lvs erect to falcato-secund when moist, flexuose when dry, lanceolate, subulate, margin entire or denticulate; nerve excurrent; basal cells quadrate to quadrate-rectangular, occasionally porose near nerve, angular cells forming distinct auricles, upper cells quadrate to rectangular, not porose or papillose. Seta to 16 mm; capsule inclined, ellipsoid to sub-cylindrical, curved, ± strumose, striate when dry; exothecial cells mostly twice as long as wide, thin-walled; annulus of 1 row of large fugacious cells; spores 14–16 μm. Fr common. $n = 7$. Yellowish-green to green tufts or patches on rocks, soil and amongst boulders. Frequent in the higher Scottish mountains, very rare elsewhere, Caerns, Westmorland. 13. Europe, Faroes, Iceland, Caucasus, Asia Minor, C. Asia, N. America, Greenland, southern S. America, Australia.

Distinct from *K. blyttii* in capsule shape and capsule striate when dry. When sterile it may be distinguished by the colour and distinct angular cells.

32. Dicranum Hedw., Sp. Musc., 1801

Dioecious or rarely autoecious. Lvs erect to falcato-secund, lanceolate, sometimes subulate; nerve sometimes with teeth or lamellae at back above, usually with stereids; angular cells brownish or rarely hyaline, sometimes inflated, usually 2 or more stratose, other cells usually incrassate, porose or not, smooth or mamillose. Male plants minute to as large as female. Seta straight; capsule erect or inclined, straight or curved, not strumose; peristome teeth divided to about halfway, point-striated below, papillose above; calyptra cucullate, entire. A cosmopolitan genus of about 150 species.

D. montanum, *D. flagellare* and *D. tauricum* and sometimes *D. scottianum* are often placed in the genus *Orthodicranum* (Br. Eur.) Loeske, but the characters separating this from *Dicranum* are poorly defined and the genus is not recognised here.

1	Lvs rugose or transversely undulate when moist, at least towards apex	2
	Lvs not rugose or undulate	6
2	Cells porose throughout lf	3
	Upper cells not porose	5

3 Lf margin recurved below, spinosely toothed above **2. D. polysetum**
Lf margin plane below, denticulate to serrate above 4
4 Nerve not or scarcely toothed at back above, not more than $\frac{1}{15}$ width of lf at widest part of lf **3. D. bonjeanii**
Nerve toothed at back above, $\frac{1}{8}$–$\frac{1}{12}$ width of lf at widest part of lf **5. D. scoparium**
5 Lvs incurved, crisped when dry, cells mamillose at back towards apex **7. D. spurium**
Lvs ± erect when dry, cells not mamillose **8. D. undulatum**
6 At least basal cells of lf porose 7
Lf cells not porose 12
7 Cells porose throughout lf 8
Upper cells not porose 11
8 Lvs falcato-secund, 9–15 mm long, seta yellowish-green **6. D. majus**
Lvs erect to secund, to 10 mm long, seta reddish at least below 9
9 Lvs serrate and with low, toothed lamellae at back above **5. D. scoparium**
Lvs without lamellae or teeth at back, margin entire or denticulate above 10
10 Autoecious, plants without flagelliform branches, lvs channelled above **1. D. glaciale**
Dioecious, plants often with flagelliform branches, lvs deeply channelled to ± tubular above **4. D. leioneuron**
11 Lvs ± flexuose or crisped when dry, 4–8 mm long **9. D. fuscescens**
Lvs straight, ± appressed when dry, to 3 mm long **10. D. elongatum**
12 Lvs straight when dry, with fragile apices usually broken off **14. D. tauricum**
Lvs flexuose, curved or crisped when dry, not fragile 13
13 Nerve percurrent or excurrent, cells immediately above angular cells longer than angular cells 14
Nerve ending below apex, angular and basal cells of ± similar shape 15
14 Capsule inclined, lf margin usually denticulate above, basal cells 3–8 times as long as wide **9. D. fuscescens**
Capsule erect, lf margin entire above, basal cells 2–4 times as long as wide **11. D. scottianum**
15 Cells in upper part of lf mamillose, margin and back of nerve toothed above **12. D. montanum**
Cells smooth, nerve smooth, margin denticulate near apex only **13. D. flagellare**

1. D. glaciale Berggr., Lunds Niv. Ars-Skr. Afd. Math. Nat,. 1866
D. molle (Wils.) Lindb., *Kiaeria glaciale* (Berggr.) Hagen

Autoecious, androecium immediately beneath perichaetium. Plants to 12 cm. Lvs flexuose, erect to secund when dry, erecto-patent to secund when moist, 5–7 mm long, lanceolate to ovate-lanceolate, tapering to entire to obscurely denticulate apex, channelled above; nerve ending below apex, without lamellae or teeth at back above, with stereids in section; cells ± porose throughout, basal narrowly rectangular, angular cells brownish, distinct, upper cells narrowly rectangular to rectangular, not papillose, 6–13 μm wide in mid-lf. Seta to 15 mm; capsule inclined, to sub-cylindrical, curved, not or slightly strumose, exothecial cells more than twice as long as wide, incrassate; spores *ca* 16 μm. Fr occasional, autumn. Loose tufts on soil or rocks at high altitudes, rare. Higher mountains in Perth, Angus, S. Aberdeen, Inverness, Argyll, E. Ross. 9. Europe, Faroes, Iceland, N. Asia, N. America, Greenland.

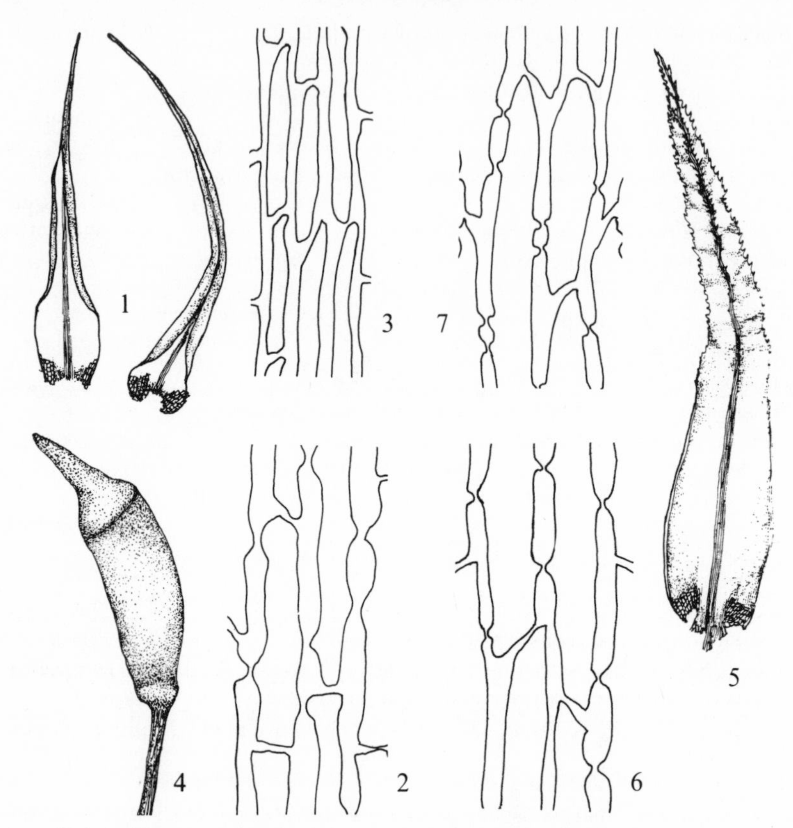

Fig. .65 1–4, *Dicranum glaciale*: 1, leaves; 2, basal cells; 3, upper cells; 4, capsule (×10). 5–7, *D. polysetum*: 5, leaf; 6, basal cells; 7, upper cells. Leaves ×10, cells ×450.

Differs from the species of *Kiaeria* in the markedly porose cells, the nerve with stereids, the exothecial cells more than twice as long as wide and incrassate. For the differences from *D. scoparium* see under that species.

2. D. polysetum Sw., Month. Rev., 1801
D. undulatum Erhr. ex Web. & Mohr, *D. rugosum* (Funck.) Hoffm. ex Brid.

Plants to 15 cm, stems tomentose. Lvs spreading, occasionally sub-secund, strongly transversely undulate, scarcely altered when dry, to 12 mm long, lanceolate, acuminate, margin recurved below, plane, spinosely toothed above; nerve with 2 strongly toothed lamellae at back above; angular cells brownish, distinct, other cells ± uniformly narrowly elliptical, porose throughout not papillose. Setae 1–5 per perichaetium. Fr unknown in Britain. $n=11$, 12, $12+1$. Lax, yellowish-green tufts or patches in conifer and birch woods, on heaths and raised bogs, rare but apparently increasing. Surrey, E. Anglia and W. Glos north to Elgin. 27. Europe, Siberia, N. and S. America.

D. spurium has the lvs incurved and crisped when dry and, as in *D. undulatum*, the upper cells are not porose.

3. D. bonjeanii De Not. in Lisa, Elenco Muschi Torino, 1837

Plants to 13 cm, rarely more, stems tomentose. Lvs erecto-patent to sub-secund, transversely undulate above, scarcely altered when dry, 4–9 mm long, lanceolate tapering to acute to obtuse apex, margin plane below, erect and dentate above; nerve narrow, $\frac{1}{15}$–$\frac{1}{23}$ lf width at widest part of lf, with 2(–4) low, scarcely toothed lamellae at back above, in section with 1 row of large empty cells; cells porose throughout, angular cells brownish, distinct, basal cells linear to narrowly rectangular, similar to elliptic-hexagonal above, not papillose. Seta reddish below, yellowish above, 1–2 per perichaetium; capsule inclined, cylindrical, curved; spores 18–20 μm. Fr very rare, autumn. Usually bright green tufts or turfs on wet or marshy ground, wet moorland, in damp woods and well drained basic grassland, rarely on acidic soils. Widespread, occasional to frequent. 109, H37, C. Europe, Faroes, Iceland, Caucasus, N. America, Azores, N. and C. Africa, Greenland.

Forms of *D. scoparium* with transversely undulate lvs may be confused with *D. bonjeanii* but differ in the narrower lf apex, wider nerve with more marked lamellae and teeth and the shorter cells in the upper part of the lf.

4. D. leioneuron Kindb., Bull. Torr. Bot. Club, 1889

Close to *D. scoparium* and *D. bonjeanii.* Plants to 8 cm or more, stems sparsely tomentose, frequently with flagelliform branches. Lvs ± erect, narrowly lanceolate, subulate, deeply channelled to ± tubular above, not undulate, margin entire or denticulate above; nerve without lamellae, not toothed at back above. Lvs of flagelliform branches ovate to lanceolate, concave, appressed, smaller than ordinary lvs. Fr unknown in Britain. Yellowish-green tufts in *Sphagnum* hummocks on blanket or raised bog, very rare. Cardigan, S. Northumberland, Westmorland, Mid Perth, S. Aberdeen. 5. N.W. Europe, north-east N. America.

For the occurrence of this plant in Britain see Ahti, *Bryologist*, **68**, 197–201, 1965. Distinct from *D. bonjeanii* in the lvs not undulate and from *D. scoparium* (paludal forms of which often have undulate lvs) in the nerve without teeth or lamellae. When present the flagelliform shoots give the plant a distinctive appearance.

5. D. scoparium Hedw., Sp. Musc., 1801

Plants to *ca* 10 cm, stems tomentose, Lvs crowded, erect to secund, flexuose, scarcely altered when dry, sometimes undulate above, 5–10 mm long, lanceolate, tapering to channelled subula, margin entire to serrate above; nerve usually $\frac{1}{8}$–$\frac{1}{12}$ width of lamina at widest part of lf, with low, toothed lamellae at back above, in section with 1 row of large cells; cells porose, incrassate throughout, angular cells brown, distinct, basal narrowly rectangular, becoming shorter above, upper cells rectangular to irregularly quadrate, rhomboidal at margin, mid-lf cells 10–16 μm wide, cells in lower $\frac{1}{2}$ of lf ± quadrate in section. Seta yellowish above, reddish below, usually 1 per perichaetium; capsule sub-erect, cylindrical, curved, 3–4 mm long (not including lid); spores 12–22 μm. Fr frequent in Ireland and wetter parts of Britain, rare elsewhere, usually in woods, summer to winter. $n = 11, 12^*, 12+1^*, 13, 14$. Yellowish-green to dark green patches on acidic or leached soil, rocks, walls, trees, logs, in woodland, on heaths, moorland, grassland, sand-dunes, bogs and marshes, frequent to common. 112, H40, C. Europe, Faroes, Iceland, Macaronesia, America, New Zealand.

A very variable plant for which several varieties have been described, but because of phenotypic plasticity these cannot be satisfactorily discriminated (see Briggs, *New Phytol.* **64**, 366–86, 1965). Luxuriant plants may resemble small forms of *D. majus* but differ in the presence of lamellae on the back of the nerve, the single row of large cells in the nerve section and the lamina cells being ± quadrate in section. Montane forms with entire lvs and the

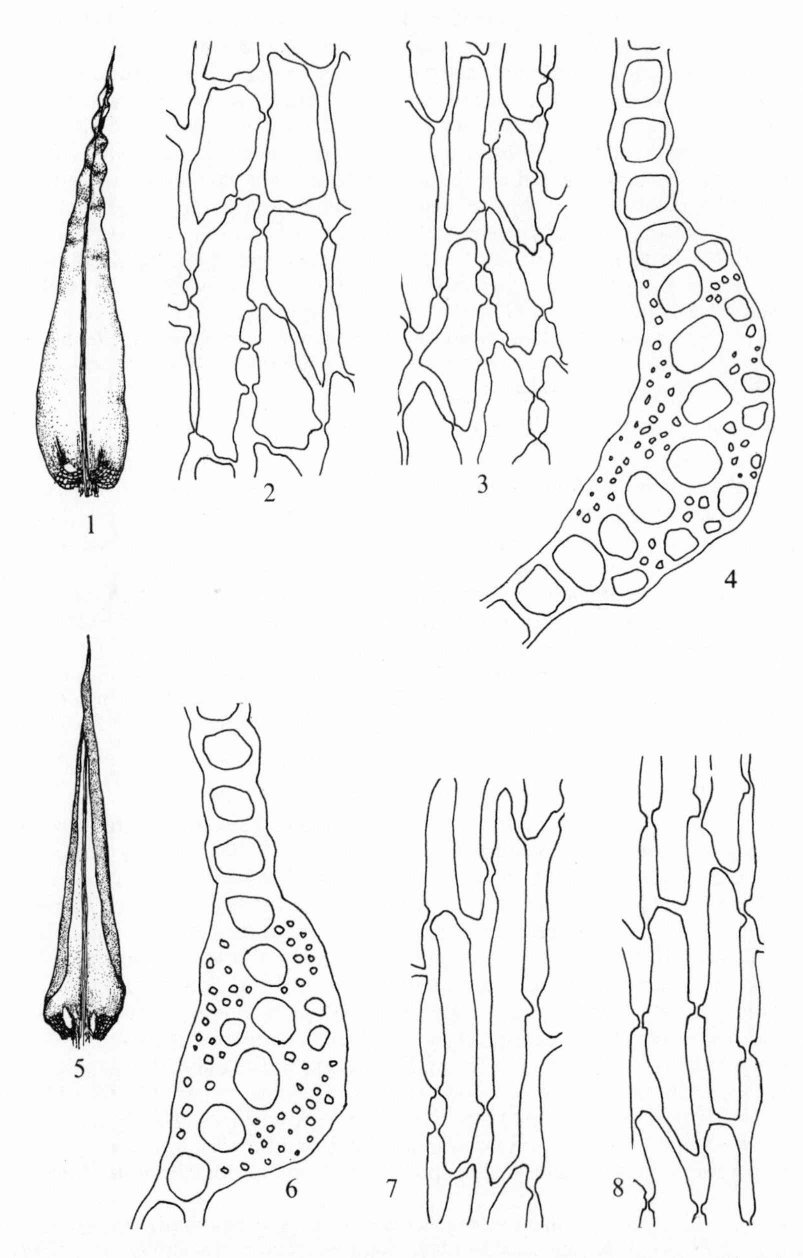

Fig. 66. 1–4, *Dicranum bonjeanii*: 1, leaf; 2, basal cells; 3, mid-leaf cells; 4, section through upper part of nerve. 5–8, *D. leioneuron*: 5, leaf; 6, section through upper part of nerve; 7, basal cells; 8, mid-leaf cells. Leaves ×7.5, cells ×450.

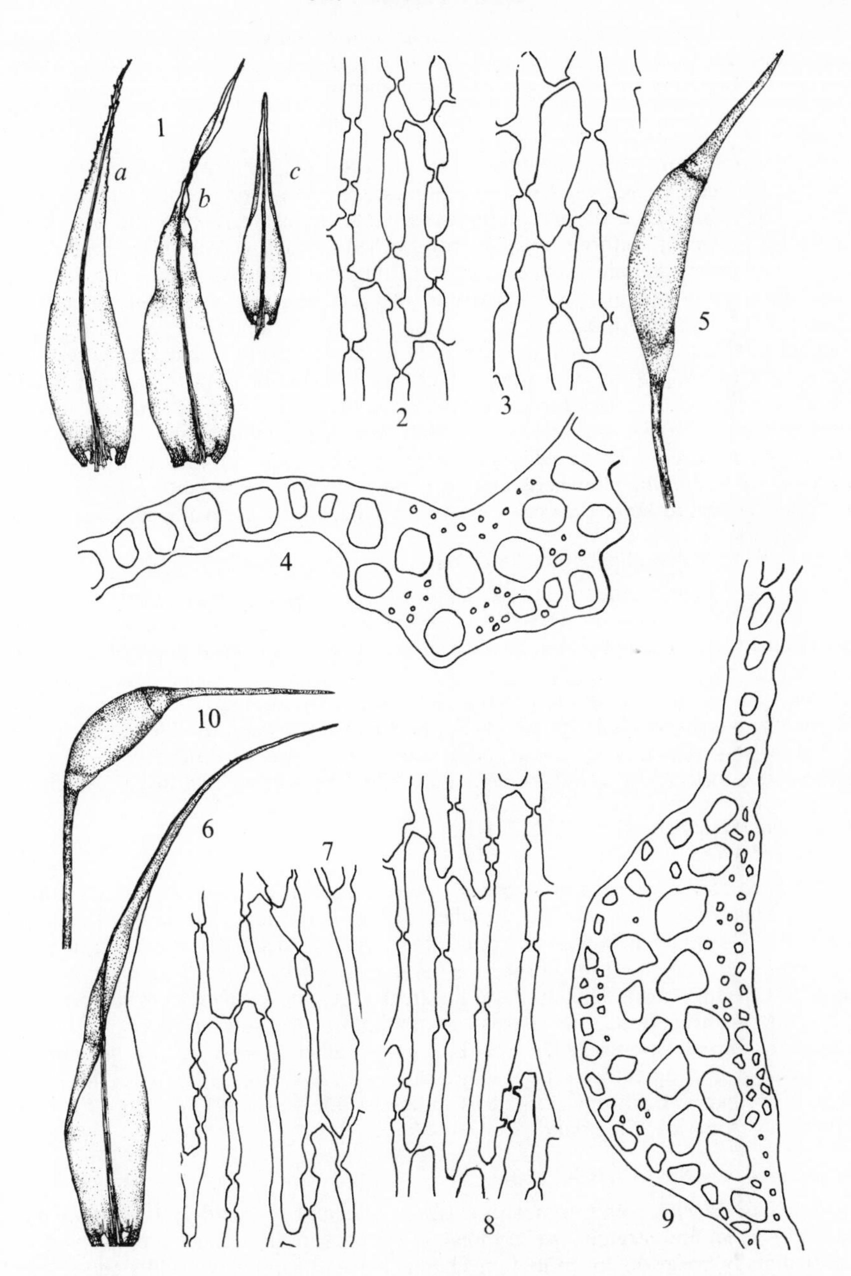

Fig. 67. 1–5, *Dicranum scoparium*: 1*a*, leaf of woodland plant; 1*b*, leaf of marsh plant; 1*c*, leaf of exposed, dry ground plant; 2, basal cells; 3, mid-leaf cells; 4, section through upper part of nerve; 5, capsule. 6–10, *D. majus*: 6, leaf; 7, basal cells; 8, mid-leaf cells; 9, section through upper part of nerve; 10, capsule. Leaves and capsules ×7.5, cells ×450.

nerve without teeth or lamellae may be confused with *D. glaciale*. That plant differs in the autoecious inflorescence and the narrower mid-lf cells. *D. fuscescens* and *D. scottianum* have narrower, non-porose, mid-lf cells and lvs crisped when dry.

6. D. majus Sm., Fl. Brit., 1804

Plants to 15 cm, stems tomentose. Lvs not crowded, uniformly falcato-secund, flexuose, scarcely altered when dry, not undulate, 9–15 mm long, lanceolate, narrowing to long, channelled subula, margin serrate above; nerve $\frac{1}{11}$–$\frac{1}{14}$ width of lamina at widest part of lf, without lamellae but toothed at back above, in section with 2 rows large cells; cells incrassate, porose throughout, angular brownish, distinct, basal narrowly rectangular, cells above narrowly rectangular to rectangular, marginal rhomboidal, cells in lower half of lf rectangular in section. Seta yellowish-green, 1–5 per perichaetium; capsule inclined to horizontal, shortly cylindrical, curved, to 3 mm long (not including lid); spores 14–26 μm. Fr occasional in Ireland and wetter parts of Britain, rare elsewhere, summer, autumn. $n=11$, 12*, 12+1*, 13. Loose tufts or turfs on rocks, soil or logs in woods, amongst boulders, on cliff ledges and banks in sheltered places, calcifuge. Rare in S.E. England and the Midlands, frequent to common elsewhere and sometimes locally abundant. 108, H31, C. Europe, Faroes, Iceland, Caucasus, C. Asia, China, Japan, N. America, Greenland.

7. D. spurium Hedw., Sp. Musc., 1801

Plants to 10 cm, stems tomentose. Lvs incurved, crisped, rugose when dry, erect on new shoots, otherwise spreading, rugose when moist, 5–7 mm long, ovate, acuminate, apex often twisted, margin entire to serrate; nerve papillose but not toothed at back above; angular cells brownish, distinct, cells near base ± linear, porose, cells above quadrate to rhomboidal, not porose, coarsely mamillose at back. Capsule sub-erect, curved; spores 14–24 μm. Fr rare, summer. $n=12$. Tumid, green or yellowish-green tufts on heaths, occasional in scattered localities from S. Devon and Sussex north to Elgin. 30. N., W. and C. Europe, Siberia, Sikkim, N. America.

8. D. undulatum Schrad. ex Brid., J. f. Bot., 1801
D. bergeri Bland.

Plants to 15 cm, stems tomentose. Lvs erecto-patent, transversely undulate and rugose when moist, scarcely altered when dry, 4–7 mm long, lanceolate to ovate-lanceolate, tapering to acute to obtuse apex, margin denticulate or occasionally entire above; nerve not papillose, not or only slightly toothed at back above; angular cells brownish, distinct, basal linear, porose, above quadrate-rectangular to elliptical or rhomboidal, not porose or mamillose. Capsule sub-erect, narrowly ellipsoid, curved; spores 22–24 μm. Fr rare, summer. $n=11$, 12. Large, dull to yellowish-green, tomentose tufts in bogs, rare and decreasing, in scattered localities from Cardigan and Shrops north to Caithness, Offaly. 14, H1. Europe, C. Asia, Siberia, N. America, Greenland.

9. D. fuscescens Sm., Fl. Brit., 1804

Plants to 12 cm; stems tomentose. Lvs sub-secund to strongly falcato-secund, rarely ± erect and straight, when moist, crisped when dry, lanceolate, tapering to denticulate or occasionally entire long channelled acumen; nerve usually denticulate at back above; angular cells brownish, distinct, cells near base porose or not, narrowly rectangular, 3–8 times as long as wide, upper cells variable, ± quadrate or elliptical or rhomboidal, smooth to strongly mamillose, not porose. Capsule inclined, sub-cylindrical, curved; spores 18–24 μm. Fr occasional, summer.

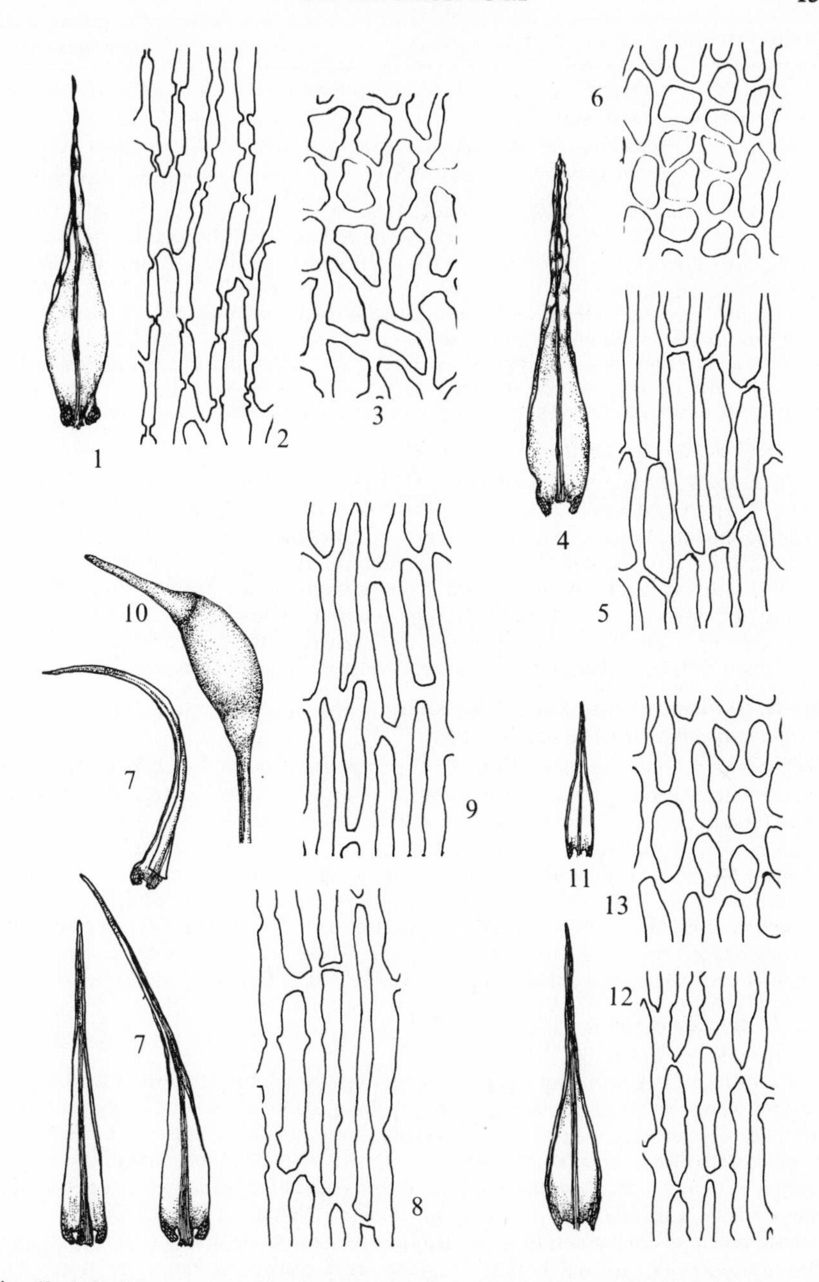

Fig. **68**. 1–3, *Dicranum spurium*: 1, leaf (×7.5); 2, basal cells; 3, mid-leaf cells. 4–6, *D. undulatum*: 4, leaf (×7.5); 5, basal cells; 6, mid-leaf cells. 7–10, *D. fuscescens*: 7, leaves (×7.5); 8, basal cells; 9, mid-leaf cells; 10, capsule (×7.5). 11–13, *D. elongatum*: 11, leaves (×15); 12, basal cells; 13, mid-leaf cells. Cells ×450.

Var. **fuscescens**

Lvs 4–8 mm long; mid-lf cells 8–12(–16) μm wide. *n*=9*, 10*, 10+1*, 12, 24. Dull or yellowish-green tufts, patches or turfs on soil, rocks, logs and trees in woodland and, in northern and western Britain, in open habitats. Rare in southern England and the Midlands, occasional to frequent elsewhere, sometimes locally abundant. 79, H23. Europe, Faroes, Iceland, Caucasus, N. and C. Asia, N. America, Greenland.

Var. **congestum** (Brid.) Husn., Fl. Gall., 1884

Lvs shorter, wider; nerve narrower. Montane habitats. Merioneth, Durham and from Perth northwards, W. Mayo. 11, H1. Europe, Iceland, N. and C. Asia, N. America, Greenland.

A variable species for which a number of varieties have been described, but most of these appear to be environmental variants, except var. *congestum* which might have a genetical basis. Straight, entire-leaved forms may be difficult to distinguish from *D. scottianum* when sterile but may be recognised by the nerve denticulate or papillose at back above. Usually, however, *D. fuscescens* has a distinctly denticulate lf margin, at least towards the apex.

10. D. elongatum Schleich. ex Schwaegr., Suppl. I., 1811

Close to *D. fuscescens.* Plants to 6(–12) cm, stems densely tomentose. Lvs ± straight, appressed when dry, erecto-patent, sometimes sub-secund when moist, to 3 mm long, narrowly lanceolate to lanceolate, tapering to short, entire or denticulate subula; cells near base narrowly rectangular to rectangular, porose, upper cells 6–10(–12) μm wide, rhomboidal, not porose or mamillose, Capsule sub-erect, ovoid, curved. Fr not known in Britain, Densely matted, yellowish-green tufts on damp, montane rocks and soil, very rare and not seen recently. S. Aberdeen, Inverness, W. Ross, Caithness, Shetland. 6. Europe, W. and C. Asia, N. America.

11. D. scottianum Turn., Musc. Hib., 1804
Orthodicranum scottianum (Turn.) Roth

Plants to 7.5 cm, stems tomentose. Lvs appressed, twisted to crisped when dry, erecto-patent, sometimes slightly secund when moist, narrowly lanceolate to lanceolate, tapering to entire or rarely denticulate, channelled subula; nerve smooth at back above; angular cells brownish, distinct, cells near base rectangular, 2–6 times as long as wide, very rarely porose, above ± uniformly quadrate, not porose or mamillose, 8–13 μm wide in mid-lf. Capsule erect, cylindrical, straight or slightly curved; spores 24–30 μm. Fr occasional, summer. *n*=22. Dull green tufts or cushions on dry shaded rocks, or occasionally on soil. Occasional in W. and N. Britain and Ireland, very rare elsewhere. 47, H22, C. N., W. and C. Europe, Macaronesia.

12. D. montanum Hedw., Sp. Musc., 1801
Orthodicranum montanum (Hedw.) Loeske

Plants to 3 cm. Lvs strongly crisped when dry, spreading, flexuose, sometimes sub-secund when moist, lanceolate, tapering to long, denticulate subula; nerve toothed at back above; cells near base rectangular, not porose, angular cells hyaline or brown, unistratose, cells above shortly rectangular to quadrate, mamillose at back, not porose. Small propaguliferous shoots with deciduous lvs sometimes produced at stem apices. Capsule erect, cylindrical, straight. Fr not known in Britain. *n*=14. Small, green or dark green tufts or cushions on exposed roots and on tree trunks at low altitudes. Occasional in S.E. England, rare elsewhere, extending north to E. Inverness, Elgin, Shetland. 45. Europe, Siberia, Kashmir, Himalayas, China, N. America.

Most likely to be mistaken for *Dicranoweisia cirrata* but differs in the lvs denticulate a bove.

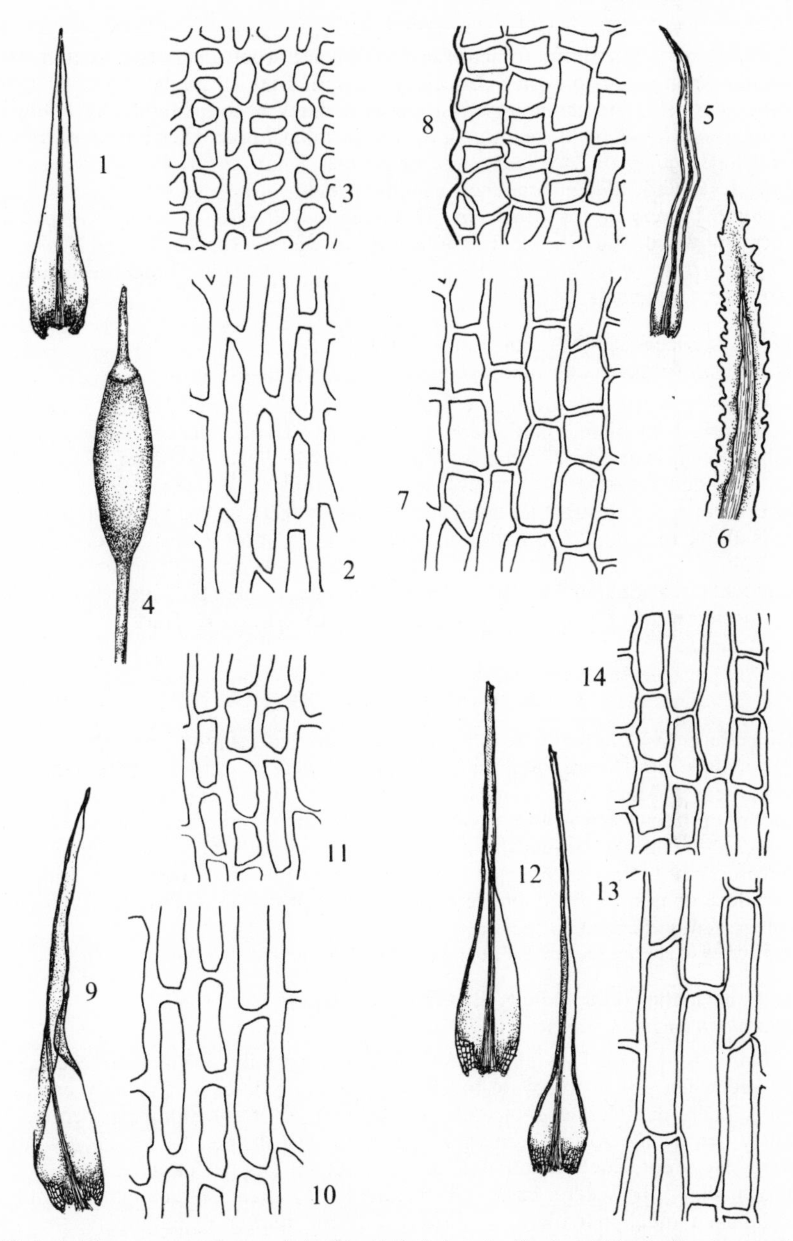

Fig. **69**. 1–4, *Dicranum scottianum*: 1, leaf (×7.5); 2, basal cells; 3, mid-leaf cells; 4, capsule (×7.5). 5–8, *D. montanum*: 5, leaf (×15); 6, leaf apex (×180); 7, basal cells; 8, mid-leaf cells. 9–11, *D. flagellare*: 9, leaf (×15); 10, basal cells; 11, mid-leaf cells. 12–14, *D. tauricum*: 12, leaves (×15); 13, basal cells; 14, mid-leaf cells. Cells ×450.

13. D. flagellare Hedw., Sp. Musc., 1801
Orthodicranum flagellare (Hedw.) Loeske

Plants to 6 cm, stems tomentose. Lvs crisped when dry, erecto-patent to subsecund when moist, narrowly lanceolate, tapering to ± tubular, smooth or faintly denticulate subula; nerve slightly toothed at back above; angular cells brownish, unistratose, basal cells rectangular, not porose, upper variable, rhomboidal to irregularly quadrate, not mamillose or porose. Small, erect, straight-leaved propaguliferous flagelliform branches usually present. Capsule erect, sub-cylindrical, straight. Fr very rare, summer. $n=23$. Green or yellowish-green tufts or patches on decaying wood. Rare in S.E. England, very rare elsewhere. S. Devon, N. Hants and Kent north to Mid Perth, N. Kerry. 17, H1. Europe, Asia, Madeira, Canaries, N. America, Guatemala, Mexico.

14. D. tauricum Sapehin, Bot. Jahrb., 1911
D. strictum Schleich., *Orthodicranum strictum* (Schleich.) Culm.

Plants to 4 cm. Lvs straight when dry, erecto-patent when moist, fragile, apices frequently broken off, lanceolate, tapering to narrow, channelled, fragile, rough subula composed mainly of nerve, margin entire or finely denticulate; nerve smooth or finely denticulate at back above, without stereids in section; angular cells hyaline or brownish, ± unistratose, basal cells narrowly rectangular, not or slightly porose, cells above rectangular to quadrate, smooth, not porose. Capsule erect, cylindrical, straight. Fr very rare, spring. $n=12$*. Light green tufts on wood or tree trunks, rarely on rock. Occasional in S.E. and C. England, rare elsewhere but apparently increasing. N. Somerset and Kent north to Elgin and E. Inverness. 42. Europe, Kerguelen Is., N. America.

33. Dicranodontium Br. Eur., 1847

Dioecious. Lvs gradually or abruptly narrowed to long, channelled, ± entire to denticulate subula composed mainly of nerve; nerve in section with a median row of guide cells and dorsal and ventral layer of stereids; basal cells near nerve rectangular, becoming narrower towards margin, angular cells ± inflated, sometimes extending to nerve, hyaline or brownish, often not persisting, cells above becoming narrower, porose or not. Seta straight or cygneous; peristome teeth vertically striate below, papillose or not above; calyptra entire or fringed below. A mainly northern hemisphere genus of about 38 species.

For key to species see under *Campylopus* (p. 166).

1. D. uncinatum (Harv.) Jaeg., Ber. S. Gall. Naturw. Ges., 1880
Dicranum uncinatum (Harv.) C. Müll.

Plants to 12 cm. Lvs falcato-secund, less commonly ± erect or secund, hardly altered when dry, from ovate to oblong sheathing basal part rapidly narrowed to long, finely denticulate, setaceous subula, margin of sheathing part entire, above finely denticulate from about midway; nerve well defined below, *ca* $\frac{1}{4}$ width of lf base, excurrent, finely denticulate at back above; angular cells inflated, hyaline to brownish, not persistent, basal cells rectangular to quadrate-rectangular, not porose, at margin much narrower, ± linear, forming border extending to shoulder of sheathing portion, cells above narrowly rectangular to linear, single marginal row shorter. Seta arcuate or straight, capsule erect; calyptra fringed below. Fr unknown in Britain. Silky tufts, golden-yellow above, brownish below, occasional on damp, shaded, montane rocks. Lake District, Kirkcudbright, Stirling and Perth northwards,

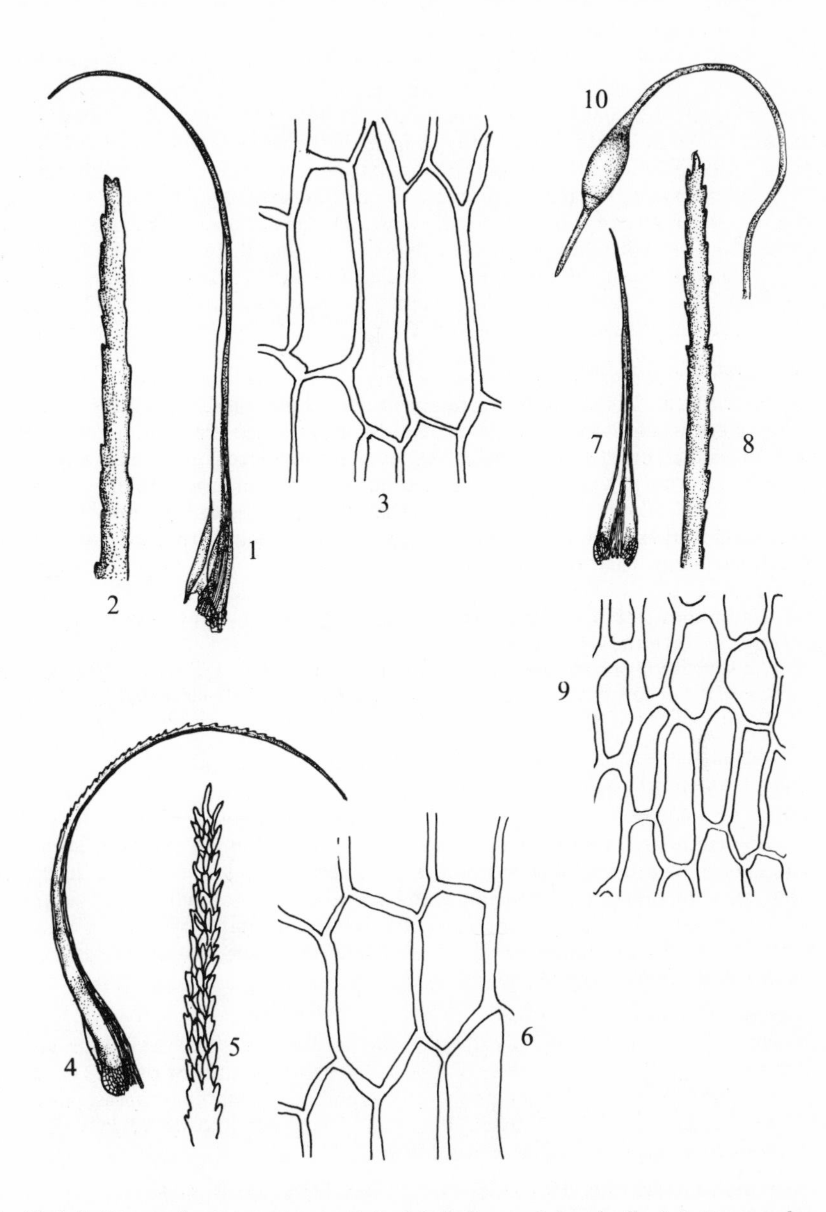

Fig. **70**. 1–3, *Dicranodontium uncinatum*: 1, leaf; 2, leaf apex; 3, basal cells. 4–6, *D. asperulum*: 4, leaf; 5, leaf apex; 6, basal cells. 7–10, *D. denudatum*: 7, leaf; 8, leaf apex; 9, basal cells; 10, capsule (×5). Leaves ×7.5, apices ×180, cells ×450.

N. Kerry, W. Galway, W. Mayo. 21, H3. N., W. and C. Europe, Himalayas, Nepal, Assam, Sri Lanka, Java, Moluccas.

2. D. asperulum (Mitt.) Broth., Nat. Pfl., 1901
Dicranum asperulum Mitt.

Plants to 7.5 cm. Lvs erecto-patent to secund, occasionally falcato-secund, when dry flexuose, often deciduous, from sheathing basal part ± abruptly narrowed to long, spinosely denticulate setaceous subula, margin denticulate from shoulder of sheathing portion or sometimes almost from base; nerve well defined below, $\frac{1}{5}$–$\frac{1}{4}$ width of lf base, excurrent, spinosely denticulate at back above; angular cells inflated, hyaline to brownish, basal cells quadrate to quadrate-rectangular, not porose, much narrower at margin, cells above narrowly rectangular to linear, single marginal row shorter. Seta straight; capsule erect; calyptra not fringed. Fr unknown in Britain. Silky, yellowish-green tufts on damp, shaded, montane rocks and soil, rare. Roxburgh, Stirling and Perth northwards, Leitrim, Cavan, Fermanagh. 16, H3. Norway, France, Germany, Austria, Switzerland, Carpathians, Sikkim, Yunnan, N. America.

3. D. subporodictyon Broth., Symb. Sin., 1929

Plants *ca* 5 cm. Lvs appressed, straight or with slightly flexuose apices when dry, erecto-patent when moist, from lanceolate basal part tapering to long, channelled, subulate acumen consisting mainly of nerve, margin inflexed, entire below, denticulate near apex; nerve well defined below, about $\frac{3}{4}$ width of lf base, ending in apex, not denticulate at back above; basal cells rectangular, incrassate, deep brown, angular cells longer and narrower, not inflated, cells above elongate-rhomboidal, sinuose, porose, becoming narrower in upper part of lf, marginal cells below very narrow, cells in mid-lf 10–12 μm wide. Scattered dark brown rhizoids on back of lf near lf base. Fr unknown. Glossy, yellowish-green tufts, reddish-brown below, on rocks, very rare. W. Inverness. 1. Yunnan.

Very distinctive in the conspicuous reddish-brown basal cells and the scattered rhizoids near base of lf. For an account of this plant in Britain see Corley & Wallace, *J. Bryol.* **8**, 185–9, 1974.

4. D. denudatum (Brid.) Broth., Nat. Pfl., 1901
D. longirostre (Web. & Mohr) Br. Eur.

Stems with reddish-brown tomentum. Lvs often deciduous, ± erect to secund, flexuose, gradually tapering from sheathing base to long ± smooth to faintly denticulate subula, margin incurved, ± entire; nerve $\frac{1}{3}$–$\frac{1}{2}$ width of lf base, ill-defined below, smooth or faintly denticulate at back above; basal cells rectangular, not porose, narrower at margin, angular cells inflated, forming hyaline or brownish auricles, cells above rectangular to rhomboidal. Seta straight or cygneous; capsule cylindrical, straight; calyptra entire. Fr very rare.

Var. **denudatum**

Plants to 5(–8) cm, forming lax tufts or patches. Lvs soft, often deciduous, spreading to secund. n=11–12, 13. Silky, dull or yellowish-green tufts or patches on damp, shaded, acidic rocks, wood or peaty soil, common in montane areas, very rare elsewhere. S. Devon, Dorset, Sussex, Wales and Derby northwards. 50, H19. Europe, Caucasus, Siberia, Japan, N. America, Mexico.

Var. **alpinum** (Schimp.) Hagen, K. Norsk. Ved. Selsk. Skrift., 1915

Plants to 10 cm forming dense tufts. Lvs rigid, not deciduous, ± erect. Damp peaty montane soil, occasional. N. Wales, Derby and Yorks northwards. 22, H9. Norway, Germany, Switzerland, Italy.

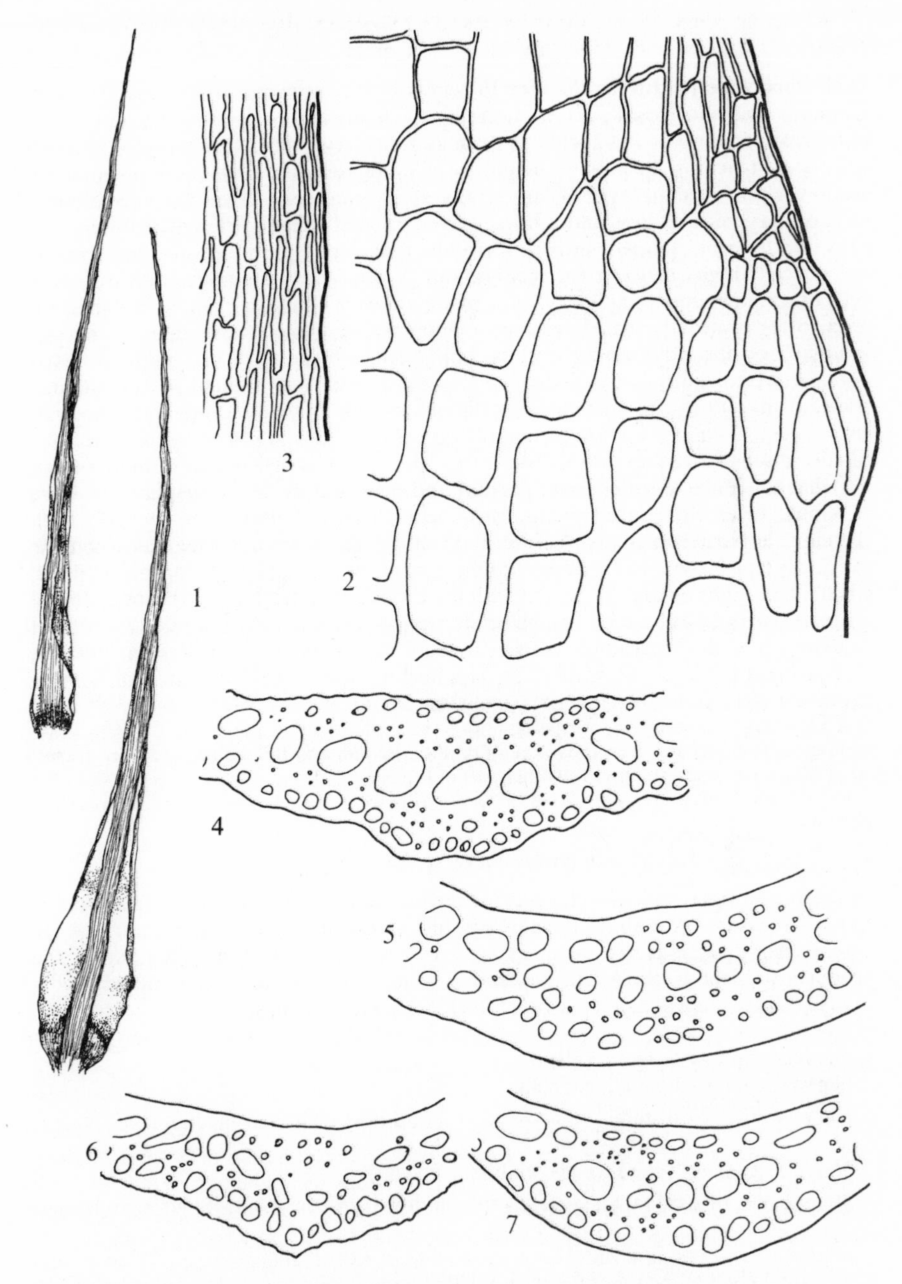

Fig. **71**. 1–3, *Dicranodontium subporodictyon*: 1, leaves (×7.5); 2, angular cells; 3, cells about ¼ way up leaf. 4–7, transverse sections of nerve about ¼ way up leaf: 4, *D. subporodictyon*; 5, *D. asperulum*; 6, *D. uncinatum*; 7, *D. denudatum*. Cells ×415.

34. Campylopus Brid., Mant. Musc., 1819
By M. F. V. Corley

Dioecious or sterile. Moderate-sized to very robust plants; stem with central strand. Lvs long with wide basal portion, tapering to channelled or tubular subulate upper part; margin more or less incurved throughout. Nerve wide, $\frac{1}{3}$ or more of lf width near base, reaching apex or excurrent, sometimes forming a hyaline hair-point; in section with a median layer of large transversely oval cells, a ventral layer of clear cells of very variable size and a dorsal layer of small clear cells of which alternate cells usually project downwards to form low ribs, rarely produced to form lamellae several cells high; between the median and dorsal layers is either a band of clear cells or an interrupted band of groups of stereid cells. Angular cells often differentiated to form auricles, basal cells rectangular, wider beside nerve than at margin, gradually or abruptly passing into the upper cells which are of very variable shape and size, vermicular, or more frequently consisting of longitudinal rows of mainly trapeziform and shortly rectangular cells mixed with a few quadrate and triangular cells. Vegetative reproduction commonly occurs by means of 1 or more of the following deciduous structures: whole lvs, lf tips, shoot tips or specialised diminutive lvs. Perichaetial lvs have wider basal portion and narrower nerve. Male plants as large as female, or rarely reduced, gemmiform. Seta yellow to light brown, cygneous with the capsule buried in the lvs before maturation, but erect and flexuous when the capsule is ripe (except in *C. subulatus* where it is erect throughout); capsule inclined, ellipsoid, straight or slightly curved with the mouth slightly oblique, striate; calyptra cucullate, fringed at base but fringe often caducous; lid conical-rostrate; annulus of 2–3 rows of cells, separating; peristome teeth 16, divided to a little over halfway, orange-brown below with transverse bars and minute vertical striations, becoming hyaline and papillose above; spores finely verrucose.

A very large genus of *ca* 750 species, principally occurring in the tropics; in temperate regions most numerous in regions of high rainfall. Britain and Ireland are richer in species, all of which are calcifuge, than any other part of Europe.

Hybrid sporophytes (*C. paradoxus* × *C. fragilis*) have been found once; the spores are abortive.

Key to *Campylopus* and *Dicranodontium* species

In the following keys and descriptions, measurements are made as follows: nerve width is measured just above the auricles, or in an equivalent position if auricles are absent; nerve sections are taken at *ca* $\frac{1}{4}$ way up lf, unless otherwise stated; the proportion of the thickness of the nerve (in the middle) in section occupied by the ventral cells is given as a percentage of the total nerve thickness.

1 Nerve section with stereid cells — 2
Nerve section without stereid cells — 15

2 Nerve section with stereid cells on ventral and dorsal sides of median layer of cells — 3
Nerve section with stereid cells on dorsal side only — 8

3 Stereid cells rather few on ventral side; on dorsal side in groups of a few cells each — 4
Stereid cells forming a band on ventral side of nerve; more numerous on dorsal side, not grouped, but forming a wide band — **(Dicranodontium)** 5

4 Lvs often with short hair-points; margin recurved in some lvs; cells shortly vermicular — **12. C. brevipilus**

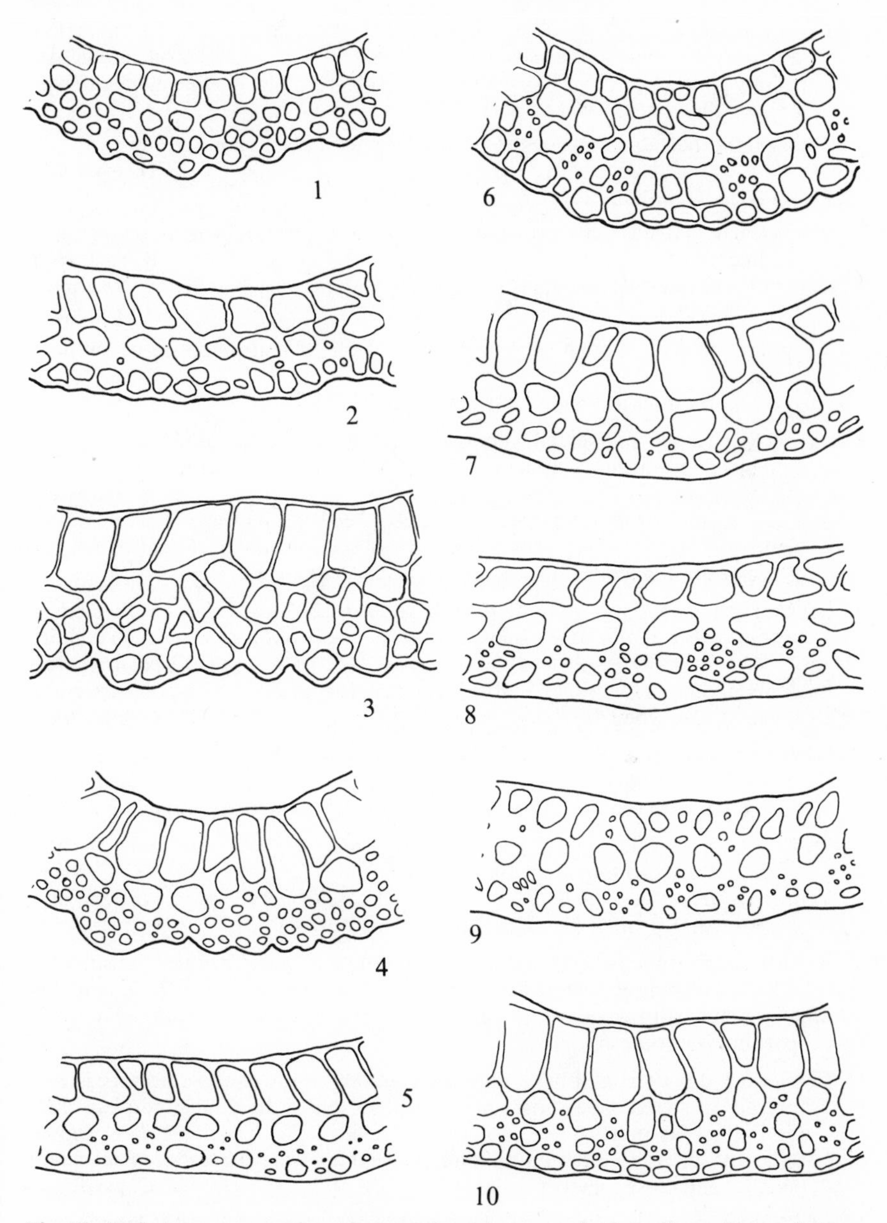

Fig. **72**. 1–10, transverse section of nerve about $\frac{1}{4}$ way up leaf: 1, *Campylopus subulatus*; 2, *C. schimperi*; 3, *C. schwarzii*; 4, *C. fragilis*; 5, *C. setifolius*; 6, *C. pyriformis*, from typical plant; 7. *C. pyriformis*, from slender plant; 8, *C. paradoxus*, from typical plant; 9, *C. paradoxus*, from xeromorphic form; 10, *C. shawii*. All ×415.

Lvs without hair-point; margin never recurved; cells not vermicular **6. C. paradoxus**

5 Lf from lanceolate base gradually tapering to subula; cells elongate, incrassate and porose throughout almost whole lamina **3. D. subporodictyon**

Lf abruptly contracted to very slender subula; cells variable, if some incrassate and porose, then only in a small part of lamina 6

6 Lf margin toothed almost throughout, usually right down to laminal wing **2. D. asperulum**

Lf toothed in upper half only, laminal wing entire 7

7 Nerve clearly defined; cells above auricles beside the nerve broadly rectangular, hyaline **1. D. uncinatum**

Nerve not very clearly defined; cells beside nerve above auricles narrowly rectangular, not hyaline **4. D. denudatum**

8 Hair-points present, though sometimes broken off; lf margin entire below hair-point 9

Hair-points absent, lf toothed at least near apex 11

9 Basal cells strongly pigmented, incrassate and porose, especially towards nerve and just above auricles, upper cells long, usually linear or vermicular **9. C. atrovirens**

Basal cells hyaline or lightly pigmented, thin-walled, usually sharply differentiated from pigmented upper cells along an oblique outwardly directed line (cf. *Tortella* spp.), upper cells rather short, trapeziform or rectangular 10

10 Back of nerve in upper part of lf with lamellae 2–4 cells high (not visible in sections from near lf base), stems not nodose, hair-point erect when dry, short **10. C. polytrichoides**

Back of nerve with low ribs, 1 cell high, stems often nodose, hair-points of nodal lvs squarrose when dry, long **11. C. introflexus**

11 Nerve section with ventral row of cells more numerous than median cells **6. C. paradoxus**

Nerve section with ventral cells approximately equal to median cells in number 12

12 Ventral layer of cells very large, occupying over half total thickness of nerve section, lvs over 7 mm long, auricles present **8. C. shawii**

Ventral layer of cells smaller, half total thickness of nerve section or less, or if larger, then lvs under 6 mm long and lacking auricles 13

13 Robust plant (mostly 3–13 cm tall) with angular cells forming conspicuous auricles, lf spinose-dentate in upper part **7. C. setifolius**

Smaller plants with angular cells very rarely forming auricles, lf toothed in upper part but not spinose 14

14 Lf widest $\frac{1}{8}$–$\frac{1}{4}$ way from lf base, tapering gradually towards base, lvs very closely placed on stems, spreading when moist, so that individual shoots are thick and rather far apart **4. C. fragilis**

Lf widest at, or very shortly above base, lvs patent when moist; individual shoots slender and close together **5. C. pyriformis**

15 Lf toothed in whole upper part, basal wing short, less than $\frac{1}{4}$ of lf length, cells wide (3–24 μm) **5. C. pyriformis**

Lf entire or minutely toothed close to apex only, basal wing at least $\frac{1}{3}$ of lf length, cells rather narrow (3–16 μm) 16

16 Auricles present; nerve generally $\frac{2}{3}$ or more of lf width, excurrent from about mid-lf, tomentum absent **3. C. schwarzii**
Auricles usually absent; nerve generally under $\frac{2}{3}$ of lf width, only excurrent near apex, tomentum present or absent 17

17 Slender, usually yellowish non-tomentose plants, to 3 cm tall, not forming dense tufts, at low altitudes, lvs 2–4 mm long **1. C. subulatus**
Small to robust light or dark green plants, 2.5–8.0 cm tall, forming dense, often abundantly tomentose tufts, generally at high altitudes but descending to sea level in Shetland, lvs 2.5–7.5 mm long **2. C. schimperi**

SUBGENUS PSEUDOCAMPYLOPUS LIMPR.

Nerve section without stereids.

1. C. subulatus Schimp. in Rabenh., Bryoth. Eur., 1861

Plants 0.4–3.0 cm, rarely to 5 cm, tomentum lacking, or rarely present at base in taller plants. Lvs 2–4 mm long, erect, straight when moist, more appressed when dry, parallel-sided at base for ($\frac{1}{10}$–)$\frac{1}{5}$–$\frac{1}{3}$ of length, tapering above, at first strongly then gradually to apex, subula rather short, margin entire below, faintly denticulate near apex which ends in 1–2 larger teeth; lamina cells in 2–9 rows at mid-lf. Nerve shortly excurrent, $\frac{1}{2}$–$\frac{2}{3}$ lf width; section with ventral cells larger than median, occupying 32–47 % of nerve thickness; stereid cells absent, back of nerve rough with low ribs. Basal cells thin-walled, hyaline, linear to rectangular, with hyaline cells continuing up margin in a tapering band; angular cells slightly wider and more incrassate than basal cells but generally little differentiated, rarely forming more or less distinct hyaline auricles; upper cells chlorophyllose, sub-quadrate, shortly rectangular and trapeziform, 4–9 μm wide. Vegetative reproduction by means of deciduous shoot tips. Seta erect throughout, 1 cm long; capsule erect, ellipsoid, symmetrical or slightly asymmetrical. Fr very rare, found only once in Britain. Slender, golden-yellow or yellowish-green, loose turfs or low tufts or as scattered plants in sandy or gravelly places on roadsides or beside streams, rarely on damp rocks, generally at low altitudes, occasional. Cornwall and Sussex north to Orkney but very rare in S. England and absent from E. England and the Midlands; N. and W. Ireland, Channel Is. 35, H10, C. N., W. and C. Europe, Yunnan.

2. C. schimperi Milde, Bot. Zeit., 1864
C. subulatus var. *schimperi* (Milde) Husn.

Plant 2.5–8.0 cm, matted with reddish tomentum in lower part or rarely almost without tomentum. Lvs 2.5–7.5 mm long, erect, often slightly secund when moist, becoming more appressed when dry, parallel-sided at base for up to $\frac{1}{3}$ of lf length then tapering gradually to apex which is entire or minutely denticulate, ending in 1–3 teeth; upper part of lf sometimes finely papillose at back; lamina cells in 1–3(–8) rows at mid-lf. Nerve shortly excurrent, $\frac{1}{2}$–$\frac{2}{3}$ width of lf; section with ventral cells much larger than median, occupying 46–55 % of nerve thickness; stereids absent, back of nerve with low ribs which are occasionally obsolete. Basal cells hyaline or slightly chlorophyllose, thin-walled, linear or rectangular, often not clearly delimited from upper cells; angular cells fragile, slightly wider than basal cells, hyaline or brownish, sometimes forming distinct hyaline auricles; upper cells shortly rectangular or trapeziform, 3–10 μm wide. Vegetative reproduction by means of fragile lf tips and occasionally deciduous lvs or shoot tips. Fr not known in Britain or Ireland. Densely matted, green to yellowish-green tufts on stones and gravelly soil in montane

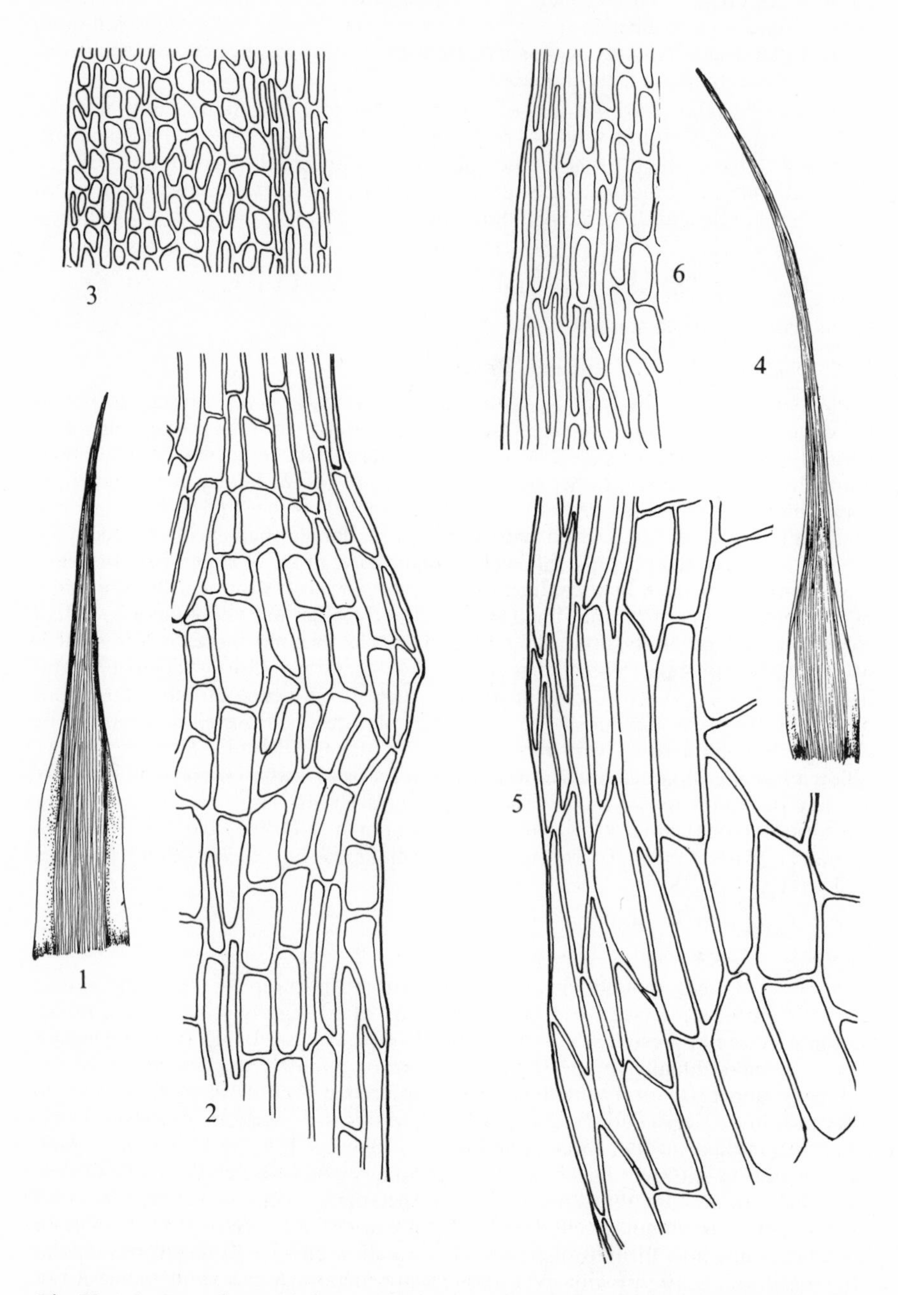

Fig. 73. 1–3, *Campylopus subulatus*: 1, leaf; 2, angular cells; 3, cells *ca* $\frac{1}{4}$ way up leaf. 4–6, *C. schimperi*: 4, leaf; 5, angular cells; 6, cells *ca* $\frac{1}{4}$ way up leaf. Leaves ×25, cells ×415.

habitats, rarely below 600 m, very rare, occasionally locally abundant. From Stirling and Arran north to Shetland; Kerry, Wicklow and Donegal. 17, H4. W. Europe from Iceland and Norway to the Pyrenees, C. Europe, Caucasus, China, Japan, Alaska, Greenland.

This species varies considerably in size. Plants from lower altitudes in W. Scotland and Ireland being taller and longer leaved, while those from C. and E. Scotland and high altitudes in the west are more stunted with short lvs and more abundant tomentum. *C. schimperi* is very close on one side to *C. subulatus* and on the other to *C. schwarzii*. Some authors have relegated it to varietal status under *C. subulatus*, but it differs markedly in size, lf length, distinctness of basal areolation, habit and habitat from that species and is not very likely to be confused with it so that it seems best to keep them apart. Confusion is more likely with *C. schwarzii* but this has the nerve wider and excurrent for about half the lf length, besides having auricles constantly present and generally lacking tomentum. The lf base of *C. schimperi* is longer and more abruptly contracted into the subula than that of *C. schwarzii*.

3. C. schwarzii Schimp., Musci Eur. Nov. Bryol. Eur. Suppl., 1864

Sterile. Plant 1–8 cm, tomentum absent or very scanty. Lvs 2.5–8.0 mm long, erecto-patent, straight or slightly secund, rarely strongly falcate when moist, becoming more erect and appressed when dry; tapering gradually from just above auricles to entire or faintly toothed apex, ending in 1–3 teeth; lamina cells in 0–2 rows at mid-lf. Nerve excurrent from about mid-lf, very wide, $\frac{3}{5}$–$\frac{4}{5}$ lf width; section with ventral cells large, occupying 40–52 % of nerve thickness; stereids absent; back of nerve with low ribs. Basal cells hyaline or slightly chlorophyllose, thin-walled, linear or rectangular, not clearly differentiated from upper cells; angular cells forming conspicuous hyaline or mauve-red auricles, upper cells thick-walled, rectangular or trapeziform, linear at margin, 3–12 μm wide. Vegetative reproduction by means of fragile lf tips and shoot tips. Fr unknown. Robust, dense, glossy yellow-green or golden tufts or turfs in shallow bogs and flushes, on soil among rocks on steep slopes and on wet rocks. Occasional to locally frequent on upland moorland and on mountains, chiefly in the west. From Cardigan north to Sutherland, Ireland from Cork and Tipperary to Donegal and Tyrone, Wicklow. 24, H14. Faroes, Norway, C. Europe, Himalayas, Formosa, Japan, Korea, Kamchatka, British Columbia.

Var. *huntii* (Stirt.) Dix., is a highly aberrant form, possibly belonging to this species.

Subgenus Campylopus

Nerve section with stereids on dorsal side only.

4. C. fragilis (Brid.) Br. Eur., 1847

Plant 0.5–8.0 cm, with abundant red or red-brown tomentum in large forms. Lvs 3–6 mm long, erecto-patent, straight, rather rigid when moist, more erect and appressed, slightly flexuose when dry, often strongly imbricated at apex of shoot, base obcuneate, widest at $\frac{1}{8}$–$\frac{1}{4}$ of lf length from base, then rapidly contracted to a tapering subula, finely toothed in upper part with a few coarser teeth at apex; lamina cells in 1–6 rows at mid-lf. Nerve $\frac{1}{3}$–$\frac{2}{3}$ lf width, ending in apex; section with ventral cells larger than median, occupying 35–66 % of nerve thickness, about equal in number to median cells; back of nerve with low ribs. Basal cells rectangular, hyaline, in dried material tending to remain full of air when wetted, angular cells not differentiated, upper cells irregular, mainly trapeziform and rectangular, 3–19 μm wide. Abundant minute deciduous lvs often present at shoot tips, normal lvs and occasionally lf tips may also break off. Capsule straight or slightly curved; peristome

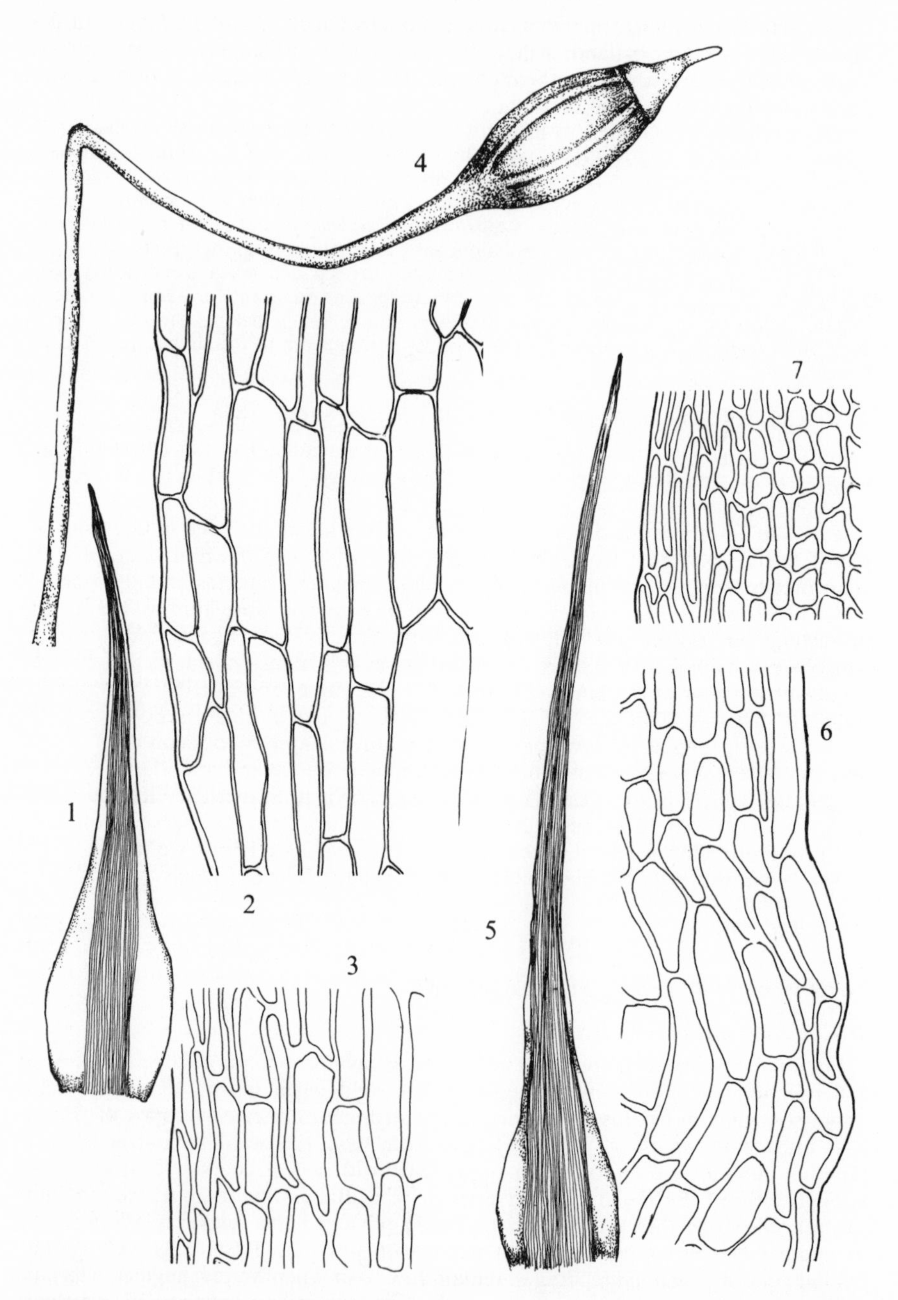

Fig. 74. 1–4, *Campylopus fragilis*: 1, leaf (×15); 2, angular cells; 3, cells from shoulder of leaf; 4, capsule (×15). 5–7, *C. schwarzii*: 5, leaf (×20); 6, angular cells; 7, cells about $\frac{1}{4}$ way up leaf. Cells ×415.

teeth 300–470 μm long; spores 9–20 μm. Fr occasional, spring. Lax green turfs or dense tufts. More base-tolerant than the other species of the genus, most commonly on soil in rock crevices, often near the sea where it may grow on gravelly or stony soil; also on peat and rotten wood. Frequent in S.W., W. and N. of Britain but rare or absent from much of S.E. England, the Midlands and the north-east. 84, H26, C. Europe, Macaronesia, Japan, Florida, N.W. United States and British Columbia.

Often confused with *C. pyriformis* which has similar areolation and also lacks auricles, but differing markedly in lf shape and habit. *C. fragilis* has the lf widest at some distance from the base, tapering gradually towards the base and rapidly above the widest point to the subula. The wide lf bases are conspicuous and usually whitish in the moist plant. The shoots are thick and pencil-like (if of any length), with the lvs very closely placed on them and spreading widely when moist. The thickness of the shoots causes them to be rather few in number on a given area of ground. *C. pyriformis* has the lf widest at or very shortly above the base, soon contracted to the slender subula; the lf may or may not be contracted at the extreme base. The short basal lamina sometimes appears whitish as in *C. fragilis* but is never so conspicuous. The shoots are slender with the lvs rather less densely placed and more erect, allowing the shoots to grow closer together. The whitish lf bases and the habit should distinguish *C. fragilis* from all the other non-piliferous *Campylopus* species.

5. C. pyriformis (Schultz) Brid., Br. Univ., 1826
C. fragilis var. *pyriformis* (Schultz) Agst.

Plant 0.2–3.5 cm; tomentum reddish, present only in large forms. Lvs 2.5–7.5 mm long, (perichaetial lvs to 9.5 mm), erecto-patent when moist, ± straight, the lower more appressed, the upper somewhat flexuose when dry, basal part ovate to ovate-lanceolate, abruptly contracted to long slender subula, widest at or very near the insertion, margin toothed in upper half of lf, or only towards apex; laminal cells in 1–3(–4) rows at mid-lf. Nerve reaching apex or shortly excurrent, $\frac{1}{3}$–$\frac{3}{5}$ lf width; section with ventral cells as large as or larger than median, occupying 32–54 % of nerve thickness, the 2 rows of cells about equal in number; stereids present on dorsal side of median cells in well grown plants but these cells are often thin-walled; back of nerve with low ribs. Basal cells rectangular, hyaline, sometimes remaining air-filled when wetted, angular cells variable, usually undifferentiated, hyaline or pale red, rarely forming distinct auricles; upper cells irregular, mostly trapeziform or rectangular, 3–23 μm wide. Vegetative reproduction by means of deciduous lvs or less frequently by diminutive lvs, lf tips or shoot tips. Male plants normal or sometimes gemmiform. Capsule straight or somewhat asymmetrical, mouth slightly oblique; peristome teeth 260–400 μm long; spores 8–17 μm. Fr occasional in well developed plants, spring. $n=10$*, 12. Forming low, bright or yellowish-green mats, often only a few millimetres high on light acid soil, especially in open woods, on bare peat and on tussocks in bogs, rarely on rotten wood. Rather common throughout the British Isles in suitable habitats, the commonest species of *Campylopus* in much of the south and east.

Key to varieties of *C. pyriformis*

1 Lf with well defined hyaline or coloured auricles — var. **fallaciosus**
Lf without auricles, though the undifferentiated basal cells may sometimes be faintly coloured — 2

2 Lf apex very slender, 11–32 μm wide at 200 μm from the tip, nerve usually less than $\frac{2}{5}$ of lf width, in section with ventral cells under 14 μm high — var. **azoricus**
Lf apex 20–42 μm wide at 200 μm from tip, nerve usually more than $\frac{2}{5}$ of lf width, in section with ventral cells over 14 μm high — var. **pyriformis**

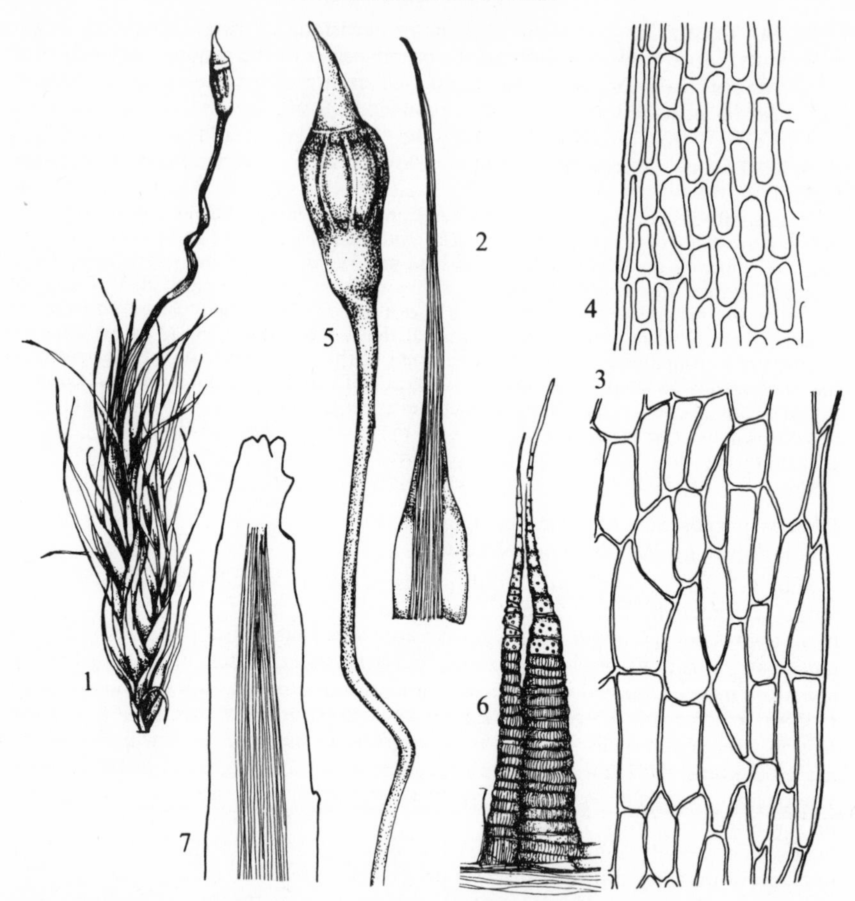

Fig. 75. *Campylopus pyriformis* var. *pyriformis*: 1, shoot (×10); 2, leaf (×15); 3, angular cells; 4, cells at shoulder of leaf; 5, capsule (×15); 6, peristome tooth (×250); leaf apex (×165). Cells ×415.

Var. **pyriformis**

Short, rarely over 1.5 cm high; tomentum sparse or absent. Lf tip (16–)20–42 μm wide at 200 μm from extreme apex, toothed in upper part of lf with teeth rather close together. Nerve $\frac{2}{5}$–$\frac{1}{2}$ lf width; section with ventral cells 14–19 μm high. Angular cells colourless, undifferentiated. Vegetative reproduction by means of deciduous lvs, sometimes of reduced size, rarely by shoot tips or fragile lf tips. Pale or yellowish-green, rarely dark green turfs on light acid soils, especially in open woods and on peat throughout Britain and Ireland. 112, H39, C. Europe to C. Asia.

Var. **azoricus** (Mitt.) M. Corley, J. Bryol., 1976

C. azoricus Mitt. in Godm., Nat. Hist. Azores, 1870

Often taller, 0.5–3.5 cm; tomentum sparse to abundant. Lvs long and slender, lf tip narrow on all lvs but most markedly so in perichaetial lvs, 11–32 μm wide at 200 μm from extreme apex, toothing very variable, sometimes from mid-lf but most often only in upper quarter of lf, rarely entire, teeth rather distant. Nerve more excurrent, ⅓–⅔ of lf width; section with ventral cells 9–14 μm high. Angular cells rarely different in shape from cells in rest of lf base, frequently pale red or brown. Vegetative reproduction by means of fragile lf tips, occasionally also by deciduous lvs but these never of reduced size. Yellowish-green, loose turfs in bogs, on tussocks, particularly of *Molinia* and on damp peat, rarely on rotten wood, widely distributed throughout the British Isles but detailed distribution very incompletely known. Cornwall, Kent north to Caithness and Sutherland, Wicklow, W. Mayo and Donegal. 25, H3. France (Normandy), Germany (Bavaria), Denmark, Azores.

Var. **fallaciosus** (Thér,) M. Corley, J. Bryol., 1976
C. turfaceus var. *fallaciosus* Thér., *C. fallaciosus* (Thér.) Podp.

Plant 0.5–1.8 cm; taller stems with reddish tomentum and often with lvs small and distant, except in comal tuft. Lf tip 22–38 μm wide at 200 μm from extreme apex; toothing variable from almost absent to very strong; lamina cells in 1–4 rows at mid-lf. Nerve as in type but ventral cells 11–21 μm high. Angular cells forming clearly defined reddish-brown or hayline auricles. Vegetative reproduction by deciduous lvs, rarely lf tips. Yellowish to dark green plants on peat, rotten wood and tussocks in bogs, rare, distribution very imperfectly known. N. Hants, Surrey, Berks. 3. France (Sarthe).

Some shoots of some plants superficially resemble *Dicranodontium denudatum* but the stems are not in fact bare of lvs. *C. paradoxus* has similar auricles but normally has more rows of laminal cells at mid-lf and different nerve section.

C. pyriformis is a distinct species rarely confused in its typical form with any other species. The numerous deciduous lvs scattered on top of the plants are characteristic but not always present. *Dicranella heteromalla* has all the cells rectangular and strongly chlorophyllose and the lvs usually secund. In some plants there are no stereid cells in the nerve section, the relevant cells being thin-walled; such plants could be mistaken for *C. subulatus* but this has these cells smaller and narrower, the upper part of the lf almost entire and the whole plant is of a golden-yellow colour and often taller than *C. pyriformis*. For the differences from *C. fragilis* and *C. paradoxus* see notes under these species. Most of the confusion that has arisen in the past between these two species and *C. pyriformis* has been due to var. *azoricus* not having been recognised as a distinct taxon. This variety does not resemble *C. fragilis* any more than does the type, except that the lf bases are sometimes more conspicuous and whitish, especially in the perichaetial lvs. Confusion with *C. paradoxus* is due to the existence of forms of this variety with coloured angular cells and the absence of deciduous lvs. Otherwise the variety differs from *C. paradoxus* in the same ways as does the type of *C. pyriformis*. Var. *azoricus* intergrades with var. *pyriformis* and not all plants can be assigned with certainty to one variety or other.

6. C. paradoxus Wils. in Hardy, Berwickshire Nat. Hist. Cl., 1868
C. flexuosus auct. non *Dicranum flexuosum* Hedw.

Plant 0.5–9.0 cm; tomentum present but very variable in abundance, reddish, occasionally zoned in different colours. Lvs 2–7 mm long, erecto-patent, straight or falcato-secund when moist, more appressed below, somewhat flexuose above when dry, parallel-sided or slightly tapering in basal ⅛–¼, then gradually tapering to apex; upper part of lf toothed, often only close to apex; lamina cells in (3–)4–19 rows at mid-lf. Nerve reaching apex but not excurrent, ⅓–⅔ lf width; section with ventral cells more numerous than median cells (1.3:1 to 1.9:1) and smaller than them (16–36 % of nerve thickness); lvs of large plants occasionally have a few of the ventral cells divided transversely, giving the impression of stereid cells on ventral side of nerve; back of nerve with or without low ribs. Basal cells rectangular, usually thick-walled; angular cells differentiated to form hyaline or reddish-brown auricles,

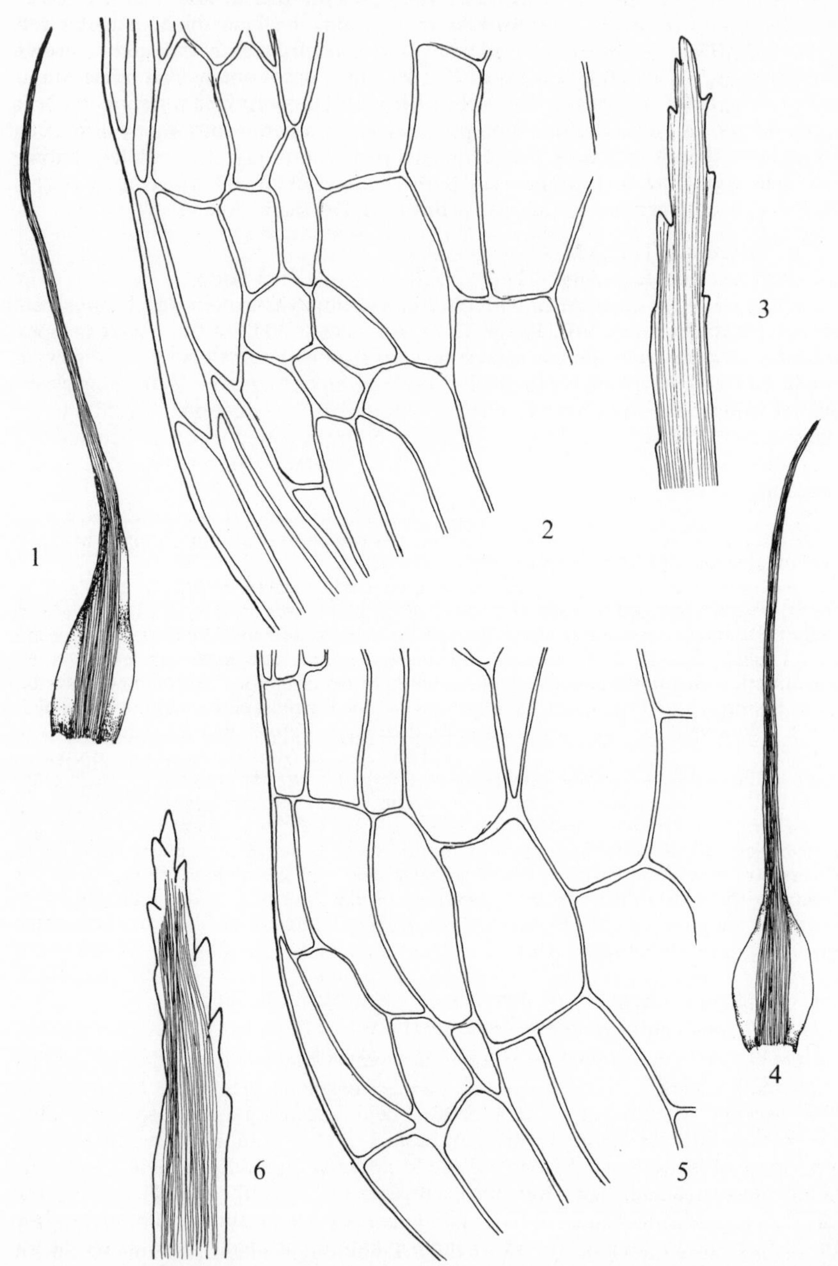

Fig. 76. 1–3, *Campylopus pyriformis* var. *azoricus*: 1, leaf; 2, angular cells; 3, leaf apex. 4–6, *C. pyriformis* var. *fallaciosus*: 4, leaf; 5, angular cells; 6, leaf apex. Leaves ×15, apices ×165, cells ×415.

upper cells irregular, mainly trapeziform and shortly rectangular, 3–19 μm wide. Vegetative reproduction by deciduous lvs or shoot tips, rarely lf tips. Capsule curved and asymmetrical with mouth ± oblique; peristome teeth 290–580 μm long, spores 10–17(–24) μm. Fr occasional, spring. $n=10^*$, 11^*, 12^*. Forming dark green tufts, sometimes with the stems very uneven in height, on bare peat and rotten wood, less frequently on rocks on moorland and on shaded vertical rock faces. Frequent throughout the British Isles but commonest in the west. 112, H39, C. W. and C. Europe, Macaronesia, East African Is., British Columbia, Mexico.

Differs from all the other British species of the subgenus in the nerve section, with the ventral cells smaller and more numerous than the cells of the median layer. Confusion can only arise in the case of abnormal plants. Some robust forms have a few stereid-like cells on the ventral side of nerve, which might lead to their being taken for a muticous form of *C. brevipilus* which, however, has quite different areolation. Stunted forms of *C. paradoxus* have poorly developed auricles and the ventral cells of the nerve section about as large as the median cells. Such plants might be named as *C. pyriformis* but differ markedly in lf shape, having a gradually tapering lf with many rows of cells at mid-lf, whereas *C. pyriformis* has a rapidly contracted lf with slender subula and only 1–3(–4) rows of lamina cells at mid-lf.

The type specimen of *Dicranum flexuosus* Hedw. is *C. pyriformis* and the earliest available name for the plant which has been known as *C. flexuosus* is *C. paradoxus* Wils.

7. C. setifolius Wils., Br. Brit., 1855

Plant very robust, (1–)3–13(–25) cm; tomentum absent or a few pale rhizoids present. Lvs 3.0–8.5 mm long, when moist straight, never falcate, erecto-patent, often very laxly placed on stem, more flexuose when dry; parallel-sided or slightly narrowed towards base, gradually tapering to long subula from $\frac{1}{12}$–$\frac{1}{8}$ from base, upper quarter of lf (rarely less), with numerous spinous teeth, apex crowned with 1–4 large teeth; lamina cells in 2–9 rows at mid-lf. Nerve not excurrent, $\frac{2}{5}$–$\frac{3}{5}$ width of lf; section with ventral cells about equal in number and of similar size to median cells, occupying 30–52% of nerve thickness; back of nerve smooth or with low toothed ribs near apex. Basal cells rectangular, usually thick-walled with few to many cells porose, rarely hyaline, becoming thick-walled above; angular cells forming large orange-red auricles, upper cells thick-walled, irregular, mostly trapeziform or rectangular, occasionally shortly vermicular, 4–23 μm wide. Vegetative reproduction by broken-off lf tips or whole lvs, very rarely by shoot tips. Male plants much rarer than female. Fr unknown. Forming deep, loose, dark green or olive-green tufts, often tinged dark brown, in very humid places, on rock ledges and in block screes, often under tall heather, in spray zones of waterfalls and elsewhere where water drips from above, and in shallow bogs, rare but very locally frequent. W. Britain from Merioneth to the Outer Hebrides and W. Sutherland, montane areas of Ireland. 13, H11. Spain (Asturias).

C. setifolius can only be confused with robust forms of *C. paradoxus*, *C. shawii* or possibly *C. atrovirens*. Muticous forms of the last species are entire-leaved, and *C. shawii* is almost so. The latter also has the ventral cells of the nerve section greatly enlarged and is normally tomentose. *C. paradoxus* is also tomentose, and has the ventral cells of the nerve section more numerous and smaller, and robust forms generally have falcate lvs.

8. C. shawii Wils. ex Hunt, Mem. Lit. Phil. Soc. Manch., 1868
C. setifolius var. *shawii* (Hunt) Mönk.

Sterile. Plant very robust, 3–11 cm; tomentum abundant, reddish-brown, ± conspicuous, very rarely absent. Lvs 5.0–11.5 mm long, widely spreading to almost erect, straight or falcate, when moist becoming slightly appressed and flexuose when dry; parallel-sided at base for $\frac{1}{9}$–$\frac{1}{5}$ of lf length or slightly narrowed towards base, gradually tapering above to long slender subula, finely toothed at extreme apex only,

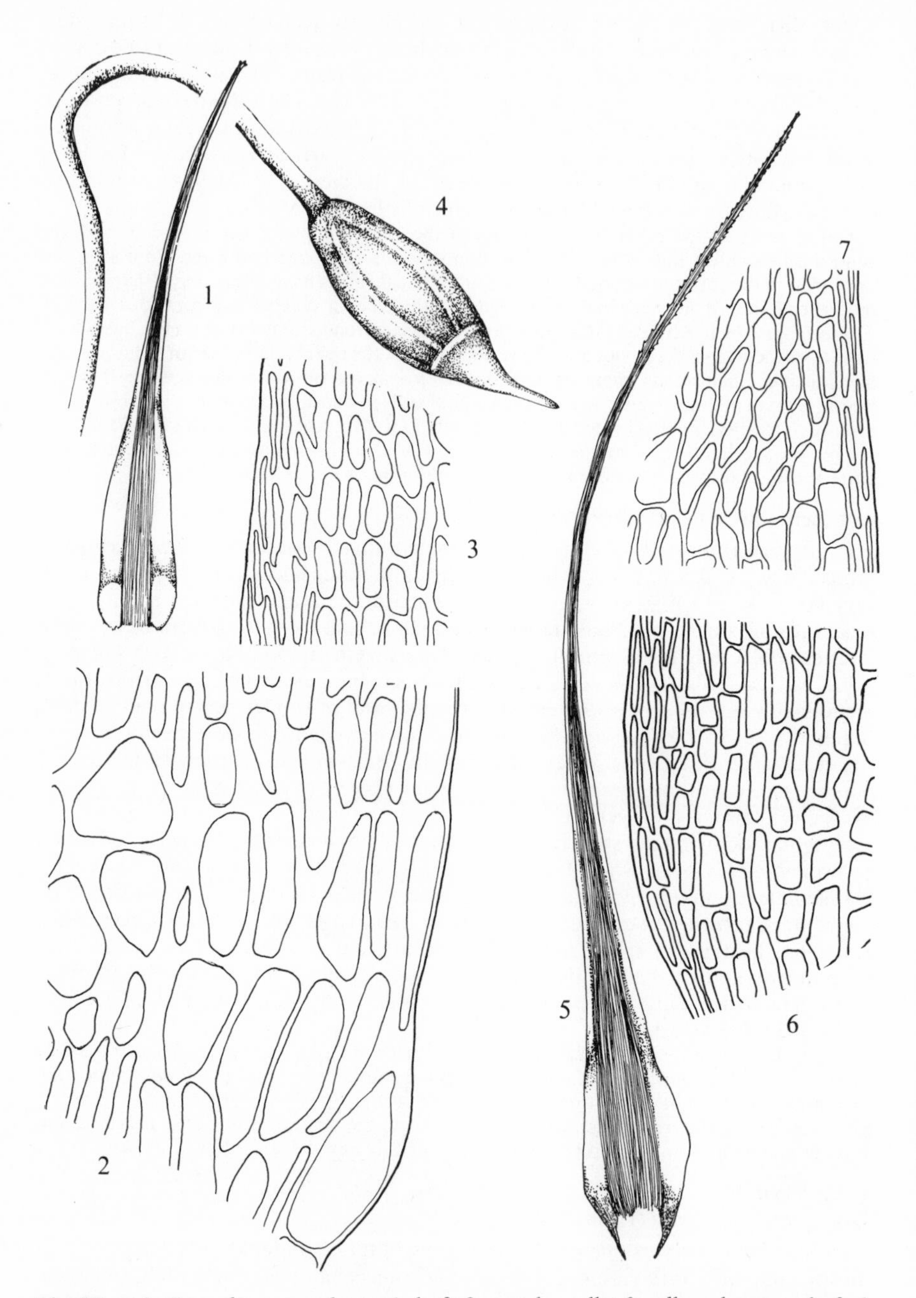

Fig. 77. 1–4, *Campylopus paradoxus*: 1, leaf; 2, angular cells; 3, cells *ca* ⅓ way up leaf; 4, capsule (×15). 5–7, *C. setifolius*: 5, leaf; 6, angular cells; 7, cells *ca* ⅓ way up leaf. Leaves ×15, cells ×415.

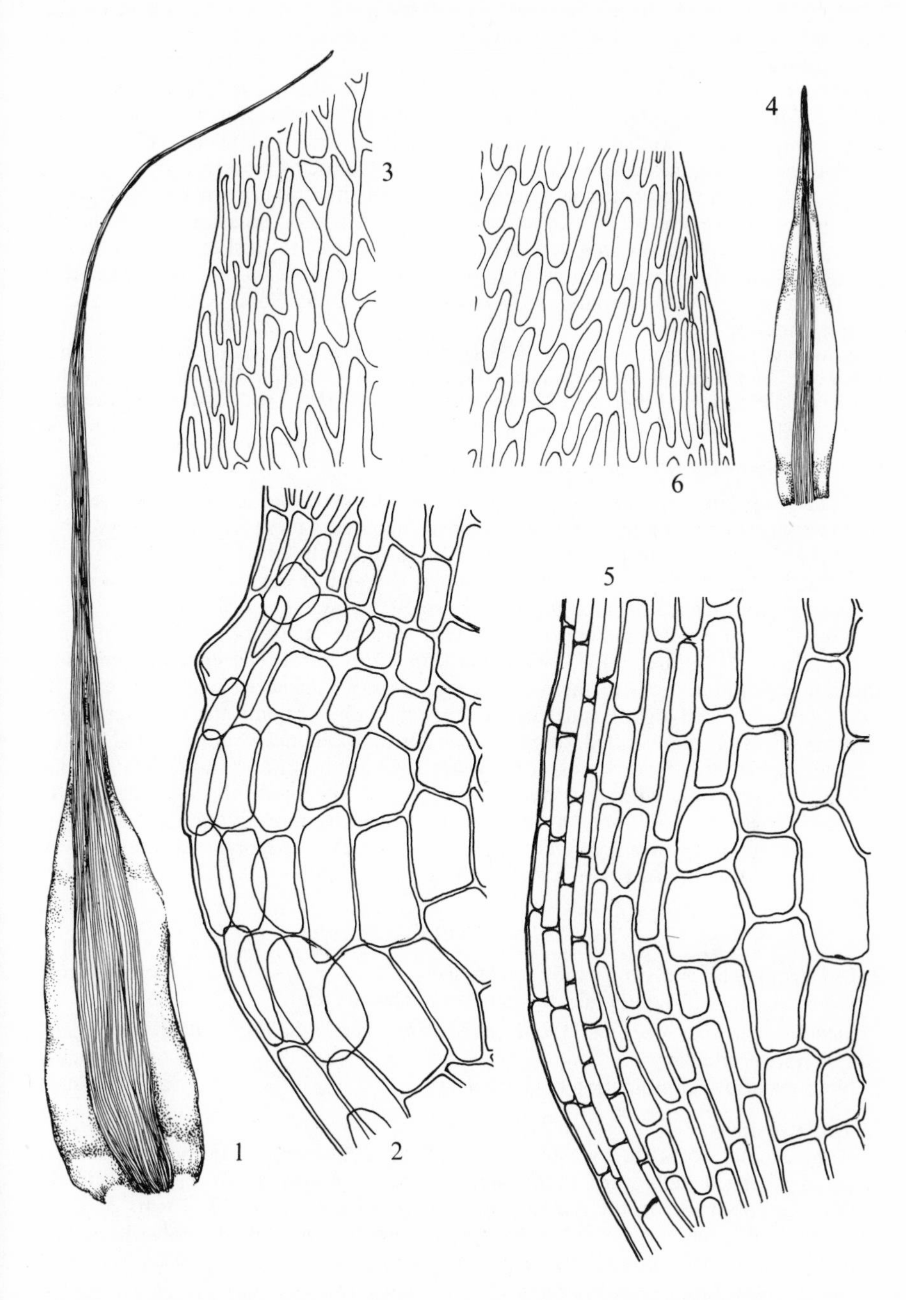

Fig. 78. 1–3, *Campylopus shawii*: 1, leaf; 2, angular cells; 3, cells near shoulder of leaf. 4–6, *C. brevipilus*: 4, leaf; 5, angular cells; 6, cells $\frac{1}{2}$ way up leaf. Leaves ×15, cells ×415.

rarely quite entire or toothed in upper 1/8 of lf, ending in 1–2(–3) small teeth; lamina cells in 1–4(–6) rows at mid-lf. Nerve not or shortly excurrent, 2/5–2/3 of lf width; section with ventral cells equal in number to median cells, but very much larger, occupying 48–70% of nerve thickness; back of nerve smooth or slightly roughened with very low ribs. Basal cells rectangular, thick- or thin-walled, if thin-walled then these cells giving way to thick-walled cells further up the lf, lowest thick-walled cells with or without pores, angular cells forming conspicuous hyaline or reddish-brown auricles, upper cells rather thin-walled, sometimes porose, mostly trapeziform or rectangular, 3–22 μm wide. Vegetative reproduction by fragile lf tips or occasionally whole lvs. Forming extensive yellowish-green turfs in shallow bogs and flushes at low altitudes in high-rainfall districts, very local but abundant in some areas. I. of Man, Mull and W. Inverness to W. Sutherland and Outer Hebrides, W. Cork, Kerry, W. Mayo. 7, H4. Endemic.

A striking plant, more resembling the larger *Dicranum* species than a *Campylopus*, but yellower than is usual in *Dicranum* and with wide nerve. *C. schwarzii* is somewhat similar but has wider nerve and shorter lf. The nerve section of *C. shawii* with its very large ventral cells will distinguish it at once from all the other auriculate species.

9. C. atrovirens De Not., Syll. Musc. Ital., 1838

Plant 1–13(–20) cm; scarcely tomentose. Lvs 3.5–9.0 mm, the terminal piece of stem sometimes becoming detached and when this occurs these lvs may be very long, up to 18 mm. Lvs erecto-patent to erect, straight or rarely falcate when moist, more appressed, slightly flexuous or straight when dry, basal part lanceolate, tapering above to a long entire subula; lamina cells in 5–16 rows at mid-lf. Nerve excurrent in a hyaline spinose-dentate hair-point of very variable length, often lacking in some lvs, rarely on all lvs; nerve 2/5–3/5 lf width; section with ventral cells slightly larger than median and about equal to them in number, 24–40% of nerve thickness; back of nerve with ribs 1 cell high. Basal cells shortly rectangular or quadrate, very incrassate, ± porose, rarely thin-walled, angular cells forming distinct hyaline or red-brown auricles, upper cells from very shortly above the base becoming trapeziform, linear or vermicular, rarely without some vermicular cells, incrassate throughout, usually porose in lower part of lf only, 4–18 μm wide. Vegetative reproduction by deciduous shoot tips and lf tips, which often break off at the base of the hair-point, occasionally whole lvs are deciduous. Fr not found in Britain or Ireland.

Key to varieties of *C. atrovirens*

1	Lvs strongly falcate, not flexuous when dry	var. **falcatus**
	Lvs straight, or if slightly falcate, then flexuous when dry	2
2	Slender yellowish-green plant with narrow lvs; basal cells thin-walled, upper cells not vermicular	var. **gracilis**
	Not with above combination of characters	var. **atrovirens**

Var. **atrovirens**

Plant usually blackish. Lvs straight, erect, or if falcate then becoming flexuous when dry. Basal cells very incrassate, porose, thin-walled; upper cells mostly vermicular or linear, rarely rectangular or trapeziform. $n=11$. Tall tufts or deep turfs, brownish-black or green at the tips, yellowish-brown or blackish within, the whole plant commonly having a deep brown appearance, in bogs and on wet rocks on moors and mountains. Frequent in the north and west from Cornwall, Devon, Wales, Cheshire and Derby northwards. 61, H30. Norway, Alps, Pyrenees, N. Turkey, China, Japan, Pacific Coast of N. America from Washington to Alaska, N. Carolina.

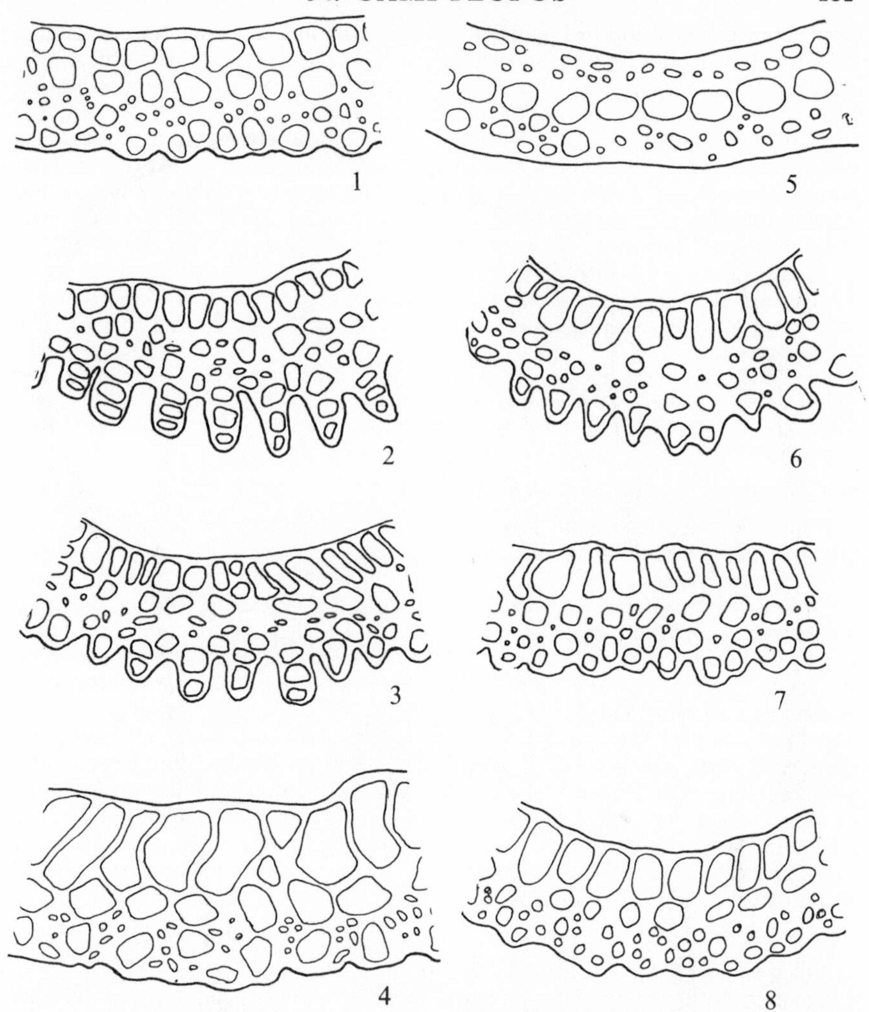

Fig. **79**. 1–8, transverse sections of nerve: 1, *Campylopus atrovirens*, $\frac{1}{4}$ way up leaf; 2–4, *C. polytrichoides*: 2, near leaf tip; 3, $\frac{1}{2}$ way up leaf; 4, near leaf base; 5, *C. brevipilus*, $\frac{1}{4}$ way up leaf; 6–8, *C. introflexus*: 6, near leaf tip; 7, $\frac{1}{2}$ way up leaf; 8, near leaf base. All $\times$415.

Var. **falcatus** Braithw., Brit. Moss. Fl., 1882

As type but lvs strongly falcato-secund, with different stems facing in different directions, unaltered when dry. In bogs and on wet rocks, rare, in areas of high rainfall near the west coast from Westmorland to Sutherland, Outer Hebrides, Shetland, Ireland, from Cork to Donegal. 9, H4. Norway.

Var. **gracilis** Dix., J. Bot., 1902

Plants more slender than the type; yellowish-green, brown below, never blackish. Lvs sometimes slightly flexuous when dry, very slender, narrower than in type. Basal cells thin-walled, scarcely porose; upper cells rectangular or trapeziform, not

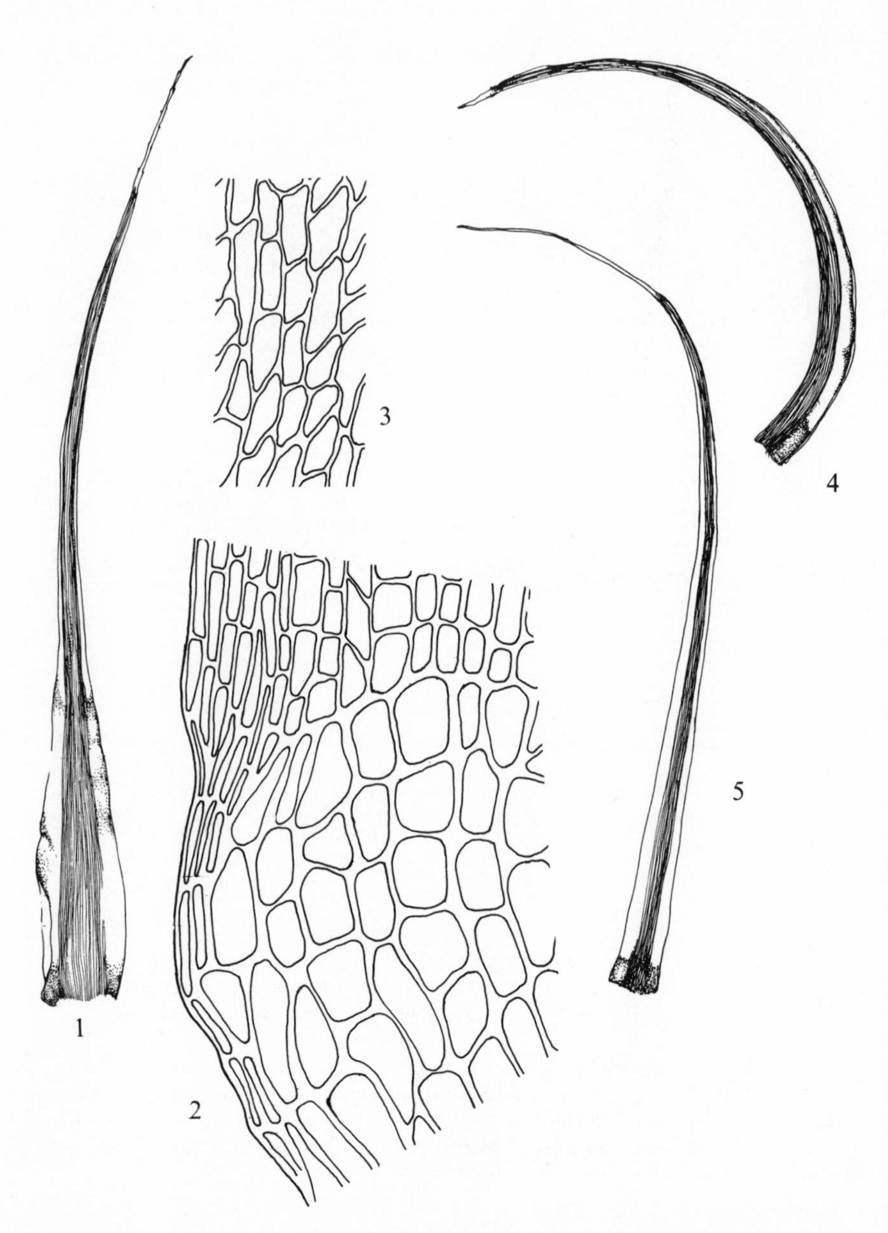

Fig. **80**. 1–3, *Campylopus atrovirens* var. *atrovirens*: 1, leaf; 2, angular cells; 3, cells *ca* $\frac{1}{3}$ way up leaf. 4, *C. atrovirens* var. *falcatus*: leaf. 5, *C. atrovirens* var. *gracilis*: leaf. Leaves ×15, cells ×415.

vermicular. Very rare; in rocky and stony places from Merioneth to Angus and Skye, W. Cork. 9, H1.

C. setifolius and *C. brevipilus* are often similarly coloured. The first has the lf margin strongly toothed, whilst it is entire in *C. atrovirens*, teeth being confined to the hair-point. *C. brevipilus* has the basal cells thin-walled and hyaline, these cells ascending higher at the margin than next to the nerve, but the nerve section is the surest way of distinguishing them.

10. C. polytrichoides De Not., Syll. Musc. Ital., 1838
C. introflexus var. *polytrichoides* (De Not.) Giac.

Plant 0.5–5.0(–9.0) cm; tomentum reddish, variable in abundance. Lvs 3.0–5.5 mm long, straight, erecto-patent when moist, more tightly appressed when dry; parallel-sided at base for $\frac{1}{4}$–$\frac{2}{3}$ of lf length, with a short evenly tapering entire subula; lamina cells in 5–16 rows at mid-lf. Nerve excurrent in a short straight spinose-toothed hyaline hair-point, nerve $\frac{1}{2}$–$\frac{3}{4}$ lf width; section with ventral cells large, occupying 30–50 % of nerve thickness, approximately equal to median cells in number; numerous stereid cells in each group, with rather thinner walls than is usual; dorsal rib cells large in lower part of lf, smaller and thicker in upper part, and forming lamellae 2–3(–4) cells high. Basal cells rectangular, thin-walled, hyaline or slightly chlorophyllose, extending higher at margin than by nerve, usually sharply demarcated from upper cells along a straight oblique line (as in *Tortella* spp.); angular cells thin-walled, hyaline, or pale reddish-brown, forming auricles of very variable size; upper cells thicker-walled, trapeziform or rectangular, rather short, 4–15 μm wide. Vegetative reproduction by deciduous lvs and shoot tips. Seta 4 mm long; capsule small; spores *ca* 17 μm. Fr found only once, in Ireland. Loose, golden-green tufts or patches, brown or blackish below, with the individual stems densely leaved and rather thick as in *C. fragilis*. On acid rocks or stony ground, usually near the sea, in warm situations, very rare on south-western coasts from Cornwall and S. Devon to N. Wales, Lismore and Skye, W. Cork, Kerry, W. Galway, W. Mayo. 8, H5, C. Belgium, France, Italy and Spain, Macaronesia, southern Africa, Malagasay Republic, India, Sri Lanka, Sumatra and Java, Mexico to Venezuela, Galapagos Is.

Similar in appearance to some forms of *C. brevipilus* but differs in the nerve section with no ventral stereids and in the areolation which is of short trapeziform cells. Muticous forms are known from the Channel Is. and the Azores. Some plants are blackish below, and might be mistaken for *C. atrovirens*, but this differs in the areolation, having the basal cells incrassate, and the upper cells generally vermicular; it is also normally taller with longer lvs.

11. C. introflexus (Hedw.) Brid., Mant. Musc., 1819

Plant 0.5–5.0 cm; sexual plants with swollen perichaetial or perigonial nodes and slender internodes; tomentum reddish-brown, often rather scanty. Lvs 2.5–6.5 mm long, straight, erecto-patent when moist, more tightly appressed when dry with the hair-points strongly squarrose, particularly on nodal lvs, lanceolate, widest at $\frac{1}{6}$–$\frac{1}{3}$ from base, or parallel-sided for a similar length with the upper part of the lf tapering, subulate or acuminate, entire; lamina cells in 5–13 rows at mid-lf. Nerve excurrent in a strongly toothed hyaline hair-point, $\frac{3}{4}$–$\frac{1}{2}$ length of rest of lf but sometimes very short or lacking in some lvs, especially in the internodes; very variable in width, $\frac{1}{3}$–$\frac{3}{4}$ lf width; section with ventral cells large, occupying 23–50 % of nerve thickness, approximately equal to median cells in number or slightly more numerous; stereid cells few in each group and with large lumens; upper part of nerve with ribs 1 cell high. Basal cells rectangular, becoming shorter and more trapeziform above, thin-walled, hyaline, rarely slightly chlorophyllose, extending higher at margin than by nerve, demarcated from upper cells along a curved or irregular line; angular cells thin-walled, hyaline to dark red-brown, forming auricles of variable size or auricles

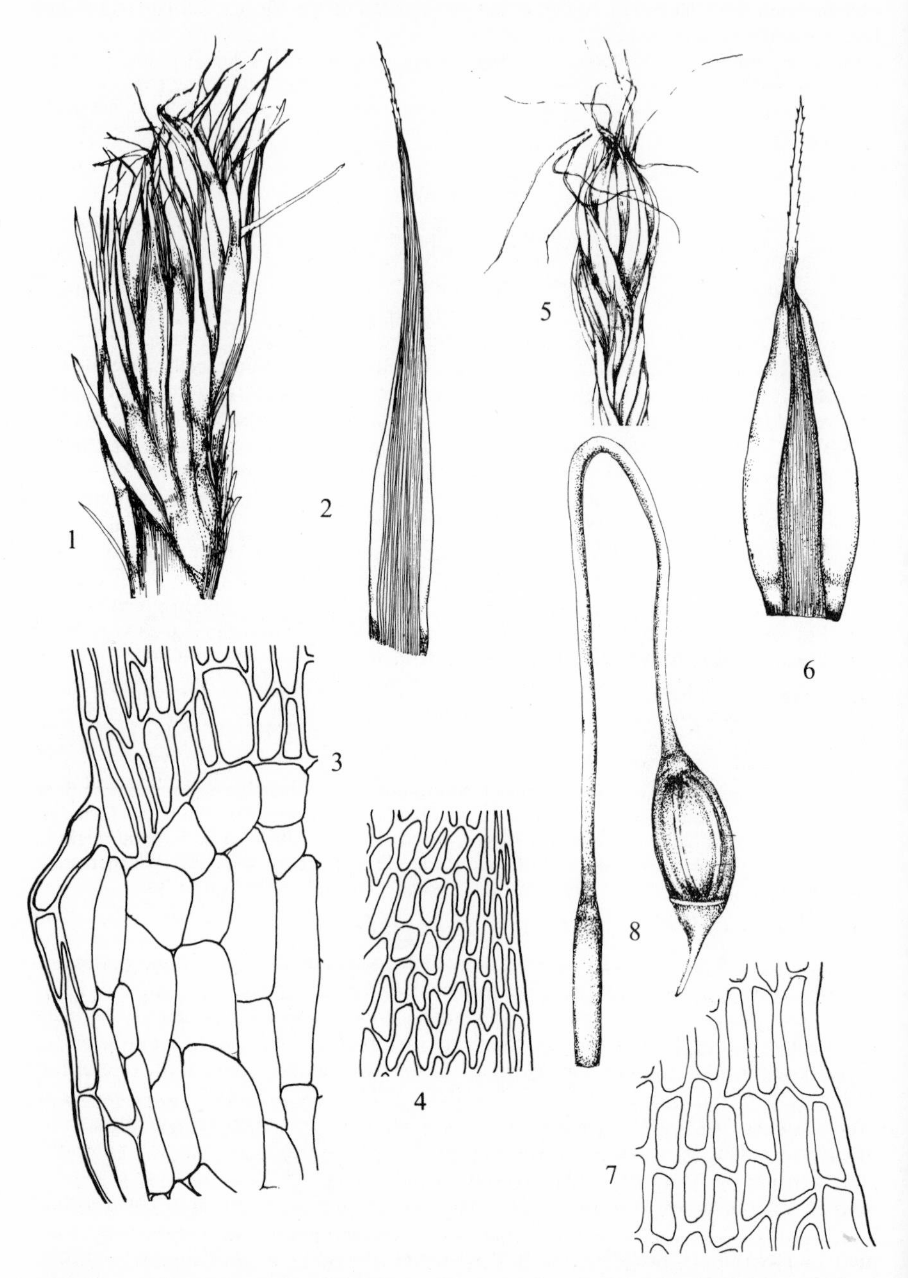

Fig. **81**. 1–4, *Campylopus polytrichoides*: 1, dry shoot tip; 2, leaf; 3, angular cells; 4, cells *ca* ⅓ way up leaf. 5–8, *C. introflexus*: 5, dry shoot tip; 6, leaf; 7, cells *ca* ⅓ way up leaf; 8, capsule (×15). Leaves ×15, shoot tips ×5, cells ×415.

absent when the lf base may be decurrent, upper cells very irregular, shortly rectangular, trapeziform or triangular, strongly incrassate, 8–15 μm wide. Vegetative reproduction by deciduous lvs. Seta 7–12 mm, shoots sometimes appearing polysetous, (1 seta from each of several perichaetia at 1 node); peristome teeth 280–420 μm long; spores 10–14 μm. Fr occasional, spring. $n=12$*. Olive-green or rarely green patches or loose tufts, most commonly on bare peat, especially in turbaries and on burnt moorland, occasionally on acid sandy or clay soil, rotten wood or tree boles, frequent except in the Midlands and parts of eastern Britain, common and often very abundant in Ireland. 91, H39, C. France, Belgium, Holland, Germany, Denmark and Faroes, N. America south of Arizona and Tennessee, C. and S. America, Galapagos and Falkland Is., Kerguelen Is., Australia, New Zealand.

Introduced, first found in Sussex in 1941 and in Co. Dublin in 1942, since when it has spread very rapidly. Hardly likely to be confused with any other species except *C. polytrichoides*, from which it differs in the stems commonly having slender internodes and swollen nodes, whereas in *C. polytrichoides* the stems are uniformly thick. Some hair-points are usually strongly reflexed when dry, often to an angle of 90° or more; in *C. polytrichoides* they are straight or sometimes slightly reflexed. The 2 species also differ in colour and in habitat but the most useful distinction lies in the nerve section, taken in the upper part of the lf, where *C. polytrichoides* has lamellae 2–3 cells high while *C. introflexus* has low ribs 1 cell high. The seta and capsule are much larger in *C. introflexus* and fruit is so rare in *C. polytrichoides* that fruiting plants are almost certain to be *C. introflexus*. A muticous form is known from Donegal. For the occurrence and spread of this plant in Britain and Ireland see Richards, *Trans. Br. bryol. Soc.* **4**, 404–17, 1963, and Richards & Smith, *J. Bryol*, **8**, 293–8, 1975.

Subgenus Palinocraspis Limpr.

Nerve section with stereids on ventral and dorsal sides.

12. C. brevipilus Br. Eur., 1847

Plant 0.5–5.0 cm; tomentum fairly plentiful, but inconspicuous, pale brown. Lvs 2.5–6.0 mm long, straight, erect, rarely falcate when moist, more tightly appressed when dry but with the lf tips free, basal part of lf oblong, elliptical or lanceolate, gradually tapering to entire subulate apex, widest at $\frac{1}{5}$–$\frac{1}{3}$ lf length from base, very variable in width relative to length, margin often narrowly recurved about mid-lf, especially in perichaetial lvs; lamina cells in 6–25 rows at mid-lf. Nerve excurrent, sometimes entire, but generally more or less toothed, often forming a hyaline hair-point of variable length but never very long; nerve $\frac{1}{4}$–$\frac{1}{2}$ lf width, down to $\frac{1}{7}$ of lf width in some perichaetial lvs; section occasionally biconvex, ventral layer of cells smaller than median, with a band of stereid cells between ventral and median cells, often reduced to a few cells only; dorsal layer with more numerous stereids, arranged in groups alternating with dorsal cells; back of nerve slightly rough or smooth. Basal cells thin-walled, hyaline, rectangular, extending further up margin than near nerve; angular cells forming small, rather fragile auricles, hyaline or brownish, often not clearly differentiated from basal cells; upper cells from shortly above the base becoming trapeziform, linear or shortly vermicular, 4–12 μm wide, usually strongly incrassate and often porose. Vegetative reproduction by means of fragile lf tips and occasionally whole lvs. Perichaetial lvs from wider base more suddenly contracted, with hair-points longer than in vegetative lvs. Peristome teeth 350–440 μm long; spores 11–13 μm. Fr very rare. Yellowish-green, olive or blackish brown, close turfs or looser patches on heaths and moors, rarely on acid sand, generally distributed except in the Midlands, always local, very common in some areas but unaccountably absent or very rare in other apparently equally suitable districts. 65, H25, C. W. Europe from Denmark to Spain, Switzerland, Azores, Madeira, Algeria.

Perhaps the most variable species of the genus, showing great diversity in almost every character, particularly lf width relative to length; presence or absence of hair-point and length of hair-point; cell length and hence cell shape. The hair-point may vary not only from plant to plant, but also up the stem: whole plants may lack hair-points altogether, such muticous forms are very much more frequent than equivalent forms in other species, and in some districts may form whole populations. The areolation in the best-developed forms closely resembles that of *C. atrovirens* with its incrassate, porose, vermicular cells, but the basal cells of *C. brevipilus* are thin-walled and hyaline. The narrow nerve, areolation and especially the nerve section should distinguish *C. brevipilus* at all times.

35. Paraleucobryum (Limpr.) Loeske, Hedwigia, 1908

Dioecious. Lvs lanceolate-subulate; nerve very wide, in section of 2–4 rows large, hyaline cells and a middle layer of chlorophyllose cells, stereids lacking; angular cells inflated, hyaline or brownish. Capsule erect, ± cylindrical. Intermediate between *Dicranum* and *Leucobryum*. A mainly northern hemisphere genus of 7 species.

1. P. longifolium (Hedw.) Loeske, Hedwigia, 1908
Dicranum longifolium Hedw.

Plants to 4(–6) cm. Lvs secund to falcato-secund when moist, flexuose when dry, from lanceolate basal part gradually tapering to channelled or ± tubular, denticulate subula composed entirely of nerve, margin denticulate above; nerve $\frac{1}{3}$–$\frac{1}{2}$ width of base of lf, denticulate at back above, spinosely so near apex, in section with 3 rows of hyaline cells and a layer of small chlorophyllose cells; angular cells inflated, hyaline or brownish, basal cells rectangular, narrower towards margin, above lamina composed of only 2–3 rows shortly rectangular cells, extending short distance up lf. Capsule erect, cylindrical, straight or curved. Fr unknown in Britain. $n=12$. Silky tufts, pale green when moist, whitish when dry, on acidic montane rocks, very rare and not seen recently. Dumfries, Mid Perth, Angus. 3. Europe, Caucasus, Siberia, Japan, N. America, Greenland.

10. LEUCOBRYACEAE

Plants large, albescent. Lvs consisting mainly of nerve, nerve composed of 2 or more layers of large, hyaline cells with large circular pores on inner faces and a central layer of chlorophyllose cells. Capsule erect or inclined, straight or curved, without stomata; peristome dicranoid.

36. Leucobryum Brid., Br. Univ., 1826

Plants robust, forming rounded, compact, glaucous-green or albescent cushions. Lvs several-stratose composed largely of nerve with a few rows of narrow, hyaline cells towards base representing lamina; nerve in section with 2 to several layers of hyaline cells with large circular pores on the inner surfaces, with small central layer of chlorophyllose cells. Sporophyte relatively small, as in *Dicranum*. A mainly tropical genus of *ca* 150 species.

Capsule markedly strumose, arcuate, plants to 15 cm, basal part of lf usually longer than narrower upper part, hyaline cells in middle of lf 24–48 μm wide **1. L. glaucum**

Capsule slightly strumose, slightly curved, plants to 6 cm, basal part of lf shorter than upper part, hyaline cells in middle of lf (15–)22–33 μm wide **2. L. juniperoideum**

1. L. glaucum (Hedw.) Ångstr. in Fries, Summ. Veg. Scand., 1846

Dioecious. Plants to 15 cm, occasionally more. Lvs ± imbricate when dry, erecto-patent or sometimes sub-secund in large forms when moist, mostly 5.0–9.0 mm × 1.2–2.1 mm, composed largely of nerve, basal part ovate to lanceolate, usually longer than the narrowly triangular, ± tubular upper part, entire; nerve in section in middle of basal part of lf of 4–6 layers of hyaline cells, in middle of upper part of lf hyaline cells 24–48 μm wide. Seta 10–18 mm; capsule arcuate, asymmetrical, markedly strumose, sulcate when dry and empty, 1.5–2.1 mm long; spores 16–20 μm. Fr rare, autumn. *n* = 11.* Glaucous-green to albescent, rounded cushions, sometimes of considerable size on soil, rocks, tree stumps in woods, on heaths, in bogs, in wet or dry places, calcifuge. Frequent to common except in basic areas, especially in western and northern Britain and Ireland. 110, H39, C. Europe, Caucasus, Hong Kong, Japan, Macaronesia, N. America, Jamaica, Andes, Hawaii.

Very close to the next species and as it is usually sterile can be differentiated by the combination of lf characters. *L. glaucum* does, however, form cushions of a more compact and neater appearance and when the 2 species occur together the difference in general appearance is very obvious. Not likely to be confused with any other moss genus other than *Sphagnum* which is quite different in the unistratose, nerveless lvs and areolation.

2. L. juniperoideum (Brid.) C. Müll., Linnaea, 1845

Dioecious. Plants to 6 cm. Lvs often slightly crisped when dry, often secund or sub-secund when moist, mostly 5.0–7.0 mm × 1.0–1.6 mm, basal part ovate-lanceolate, usually shorter than the narrow, ± parallel-sided upper part, entire; nerve in middle of basal part of 2 layers of hyaline cells, hyaline cells in middle of upper part of lf (15–)22–35(–44) μm wide. Seta 8–12(–15) mm; capsule slightly curved and inclined, weakly strumose, 2.0–2.6 mm long; spores 18–20 μm. Fr frequent, spring. Dull, glaucous-green to albescent, rounded cushions on base-poor, sandy soil in old woodland at low altitudes, often with *L. glaucum*, rare. S. England, W. Glos, Radnor, Merioneth, Caerns, Westmorland, Stirling, W. Inverness. 15. Europe, N.E. Turkey, Macaronesia, Mauritius, Réunion, Malagasay Republic.

This is the plant referred to as *L. albidum* (P. Beauv.) Lindb. and *L. minus* Hampe by Dixon (1924) and Braithwaite (1887). For a detailed account of the differences between *L. glaucum* and *L. juniperoideum* see Crundwell, *J. Bryol.* 7, 1–5, 1972.

8. FISSIDENTALES

Apical cell of stem bilateral. Lvs alternate, distichous in 1 plane. Each lf consists of 3 parts; a conduplicate part known as the *sheathing laminae* (or true lamina); the *apical lamina* (or superior lamina) continuing beyond the sheathing laminae; and the *dorsal lamina* (or inferior lamina) forming the whole of the dorsal part of the lf. Nerve single; cells in upper half of lf ± hexagonal, unistratose or bistratose in patches, smooth, papillose or mamillose, 1–3-stratose border of narrow cells sometimes present. Perichaetial lvs similar in form to stem lvs. Capsule terminal or lateral on long or short seta, erect or inclined, symmetrical or not; calyptra cucullate or mitriform, entire at base; peristome dicranoid, of 16 teeth, rarely rudimentary.

A natural order of only 1 family, with affinities with the Dicranales but with greater specialisation of the gametophyte.

11. FISSIDENTACEAE

The only family, containing 5 genera and *ca* 1100 species, the great majority of which belong to *Fissidens*.

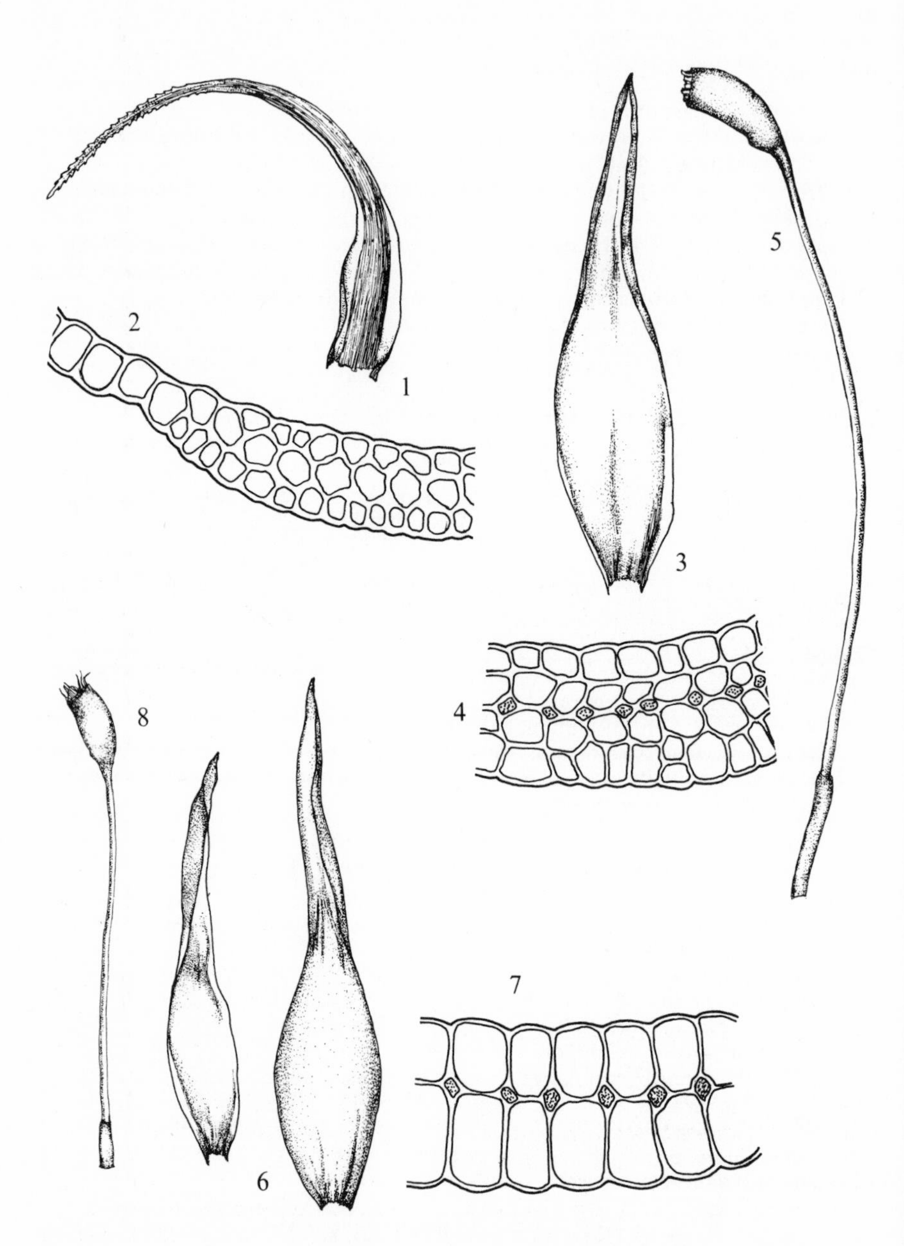

Fig. **82**. 1–2, *Paraleucobryum longifolium*: 1, leaf; 2, leaf section (×415). 3–5, *Leucobryum glaucum*: 3, leaf; 4, section of basal part of leaf (×250); 5, capsule. 6–8, *L. juniperoideum*: 6, leaves; 7, section of basal part of leaf (×250); 8, capsule. Leaves ×15, capsules ×10.

37. FISSIDENS HEDW., SP. MUSC., 1801

Stem with central strand; sheathing laminae about ½ total length of lf. Seta long; capsule longly exserted, with stomata. A taxonomically difficult genus, the subdivisions of which are somewhat artificial.

1 Lvs with border of narrow, elongated cells at least on sheathing laminae of perichaetial lvs, margin entire or obscurely denticulate above 2
Lvs unbordered, margin crenulate, often denticulate or irregularly dentate towards apex 11

2 Lvs ± tapering from below middle to acuminate apex, cells about twice as long as wide **10. F. algarvicus**
Stem lvs ± abruptly narrowed to acute to obtuse or obtuse and apiculate apex, cells about as long as wide 3

3 Border present only on sheathing lamina of perichaetial lvs, plants 1.5–2.5 mm tall **1. F. exiguus**
Border present on all laminae at least of perichaetial lvs, plants 2–25 mm 4

4 Capsule ± horizontal, antheridia in small branch at base of female shoot **3. F. incurvus**
Capsule erect or inclined but not horizontal, antheridia usually in small basal shoot, in bud-like axillary shoots or plants synoecious or dioecious 5

5 Fertile plants small, 2–6 mm, border of stem lvs confluent or not with nerve at lf apex, antheridia never in bud-like axillary shoots **2. F. viridulus**
Plants 3–20 mm, if plants less than 5 mm then border of stem lvs confluent with nerve and antheridia in bud-like axillary branches 6

6 Border of stem lvs confluent with nerve at lf apex, antheridia usually in bud-like axillary branches 7
Border not confluent with nerve, antheridia never in axillary branches 10

7 Older parts of stems matted with deep-red rhizoids **5. F. curnovii**
Older parts of stems not matted with rhizoids, rhizoids reddish-brown or brown 8

8 Border thin, unistratose, usually colourless, spores 10–14 μm **4. F. bryoides**
Border 2 or more stratose, stout, usually yellowish, spores 14–20 μm 9

9 Perichaetial lvs ± similar to upper stem lvs, cells 8–10 μm wide, capsule ± erect **6. F. rivularis**
Perichaetial lvs markedly narrower than upper stem lvs, cells mostly 10–14 μm wide, capsule inclined **7. F. monguillonii**

10 Lf cells mostly 10–18 μm, a row of chlorophyllose cells outside border of sheathing laminae **8. F. crassipes**
Cells mostly 6–10 μm wide, no chlorophyllose cells outside border of sheathing laminae **9. F. rufulus**

11 Plants not exceeding 5 mm, nerve ending below apex to percurrent 12
Plants larger, branches mostly more than 5 mm, nerve ending below apex to excurrent 13

12 Stems with 4–8 lvs, nerve straight, cells 8–12 μm wide **11. F. exilis**
Stems with up to 36 lvs, nerve with distinct bend midway, cells 12–20 μm wide **12. F. celticus**

13 Lvs irregularly dentate towards apex, 3–4 rows marginal cells forming pale band, especially in older lvs 14

Lvs crenulate or regularly serrate towards apex, marginal cells not differentiated 16

14 Lf cells conically mamillose **17. F. serrulatus**
Lf cells not mamillose 15

15 Cells in upper part of lf 6–12 μm wide, partially bistratose, spores 10–16 μm **15. F. cristatus**
Cells 12–20 μm, unistratose throughout, spores 18–24 μm **16. F. adianthoides**

16 Nerve excurrent, cells 6–10 μm, branches to 5 cm long, seta red **14. F. taxifolius**
Nerve ending in or below apex, cells 8–22 μm, branches 2–10 cm long, seta purple or yellow 17

17 Lvs lingulate, cells 12–22 μm wide, seta purple **13. F. osmundoides**
Lvs narrowly lingulate or narrowly lanceolate, cells 8–14 μm, seta yellow **18. F. polyphyllus**

Section I

At least sheathing laminae of perichaetial lvs bordered, colourless to coloured, 1–3-stratose; cells in upper part of lf ± hexagonal, smooth. Sporophyte terminal; peristome teeth spirally thickened and papillose. This section includes sect. *Pachylomidium* and sect. *Bryoideum* of other authors, but these are ill-defined and arbitrary and are not recognised here.

1. F. exiguus Sull., Musc. Allegh., 1846

Dioecious. Fertile plants 1.5–2.5 mm with 4–8 lvs. Lvs ovate to lanceolate, acute, margin entire, unbordered except for sheathing laminae of perichaetial lvs; nerve ending below apex; cells 8–10 μm wide; perichaetial lvs longer and narrower than stem lvs. Seta terminal, red; capsule erect or inclined, ellipsoid; spores 12–16 μm. Fr common, spring. Patches on shaded, wet or submerged rocks in streams and rivers, rare. N. Somerset, E. Sussex, Kent, Warwick, Brecon, Cumberland. 7. W. Europe, N., C. and S. America.

Only likely to be mistaken for weakly bordered forms of *F. viridulus*, but aquatic or wet rock forms of that species have a well developed border on all laminae. *F. exilis* has no trace of border, the margin is crenulate and the plants grow on soil. This is not the plant that has been called *F. viridulus* var. *lylei* which is merely a small form of *F. viridulus*.

2. F. viridulus (Sw.) Wahlenb., Fl. Lapp., 1812
F. bryoides ssp. *viridulus* (Sw.) Kindb.

Fertile shoots 1.5–6.0 mm with 5–10(–16) lvs, sterile shoots sometimes longer with more lvs. Perichaetial lvs usually longer and narrower than upper stem lvs, margin entire or obscurely denticulate towards apex, border colourless or occasionally yellowish or reddish, 1(–3)-stratose; nerve ending below apex in stem lvs, below apex to excurrent in perichaetial lvs; cells hexagonal, thin-walled to incrassate. Seta terminal, red; capsule erect or slightly inclined, ellipsoid; spores 8–20 μm, many often aborted.

Key to varieties of *F. viridulus*

1 Perichaetial lvs linear-lanceolate, 7–9 times as long as wide, longly acuminate, nerve percurrent to excurrent var. **tenuifolius**
Perichaetial lvs wider, 4–6 times as long as wide, acute, nerve ending in or below apex 2

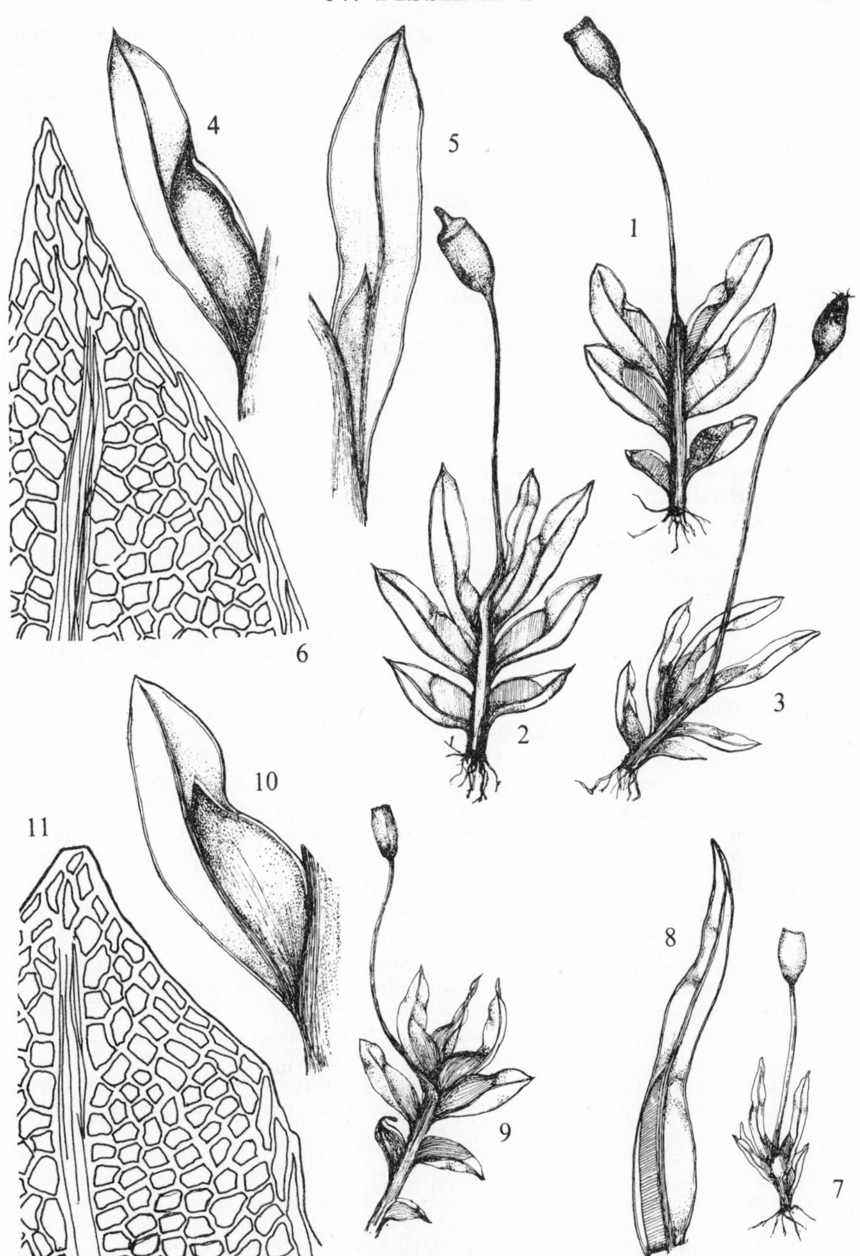

Fig. **83**. 1–6, *Fissidens viridulus* var. *viridulus*: plants from 1, limestone boulder in stream, 2, soil, 3, shaded calcareous rock; 4, stem leaf; 5, perichaetial leaf; 6, perichaetial leaf apex. 7–8, *F. viridulus* var. *tenuifolius*: 7, plant from shaded chalk; 8, perichaetial leaf. 9–11, *F. viridulus* var. *bambergeri*: 9, plant; 10, stem leaf; 11, stem leaf apex. Plants × 10, leaves × 40, cells × 450.

2 Plants usually autoecious or synoecious, border of perichaetial lvs extending to or nearly to apex, cells 8–14 μm wide var. **viridulus**
Plants synoecious, border of perichaetial lvs poorly developed and ending well below apex, cells 6–10 μm wide var. **bambergeri**

Var. **viridulus**

F. minutulus Sull., *F. pusillus* Wils., *F. viridulus* var. *lylei* Wils.

Dioecious or autoecious. Perichaetial lvs acute, 4–7 times as long as wide; border ending at or just below apex in perichaetial and upper stem lvs; nerve ending in or below apex; cells 8–14 μm wide. Perichaetial lvs not markedly differing in size. Antheridia in dwarf male shoot at base of female stem or on separate plant. $n=5$*, 10, 12. Dark green patches on soil in woods, on banks in shady places, and on shaded rocks especially by or in ponds and streams. Frequent or common in England, occasional in Wales and Ireland, occasional in S. Scotland, becoming rarer northwards. 98, H30, C. Europe, Siberia, China, Algeria, Madeira, N. America, Greenland, West Indies.

Var. **tenuifolius** (Boul.) A. J. E. Smith, J. Bryol., 1972

F. minutulus auct., *F. minutulus* Sull. var. *tenuifolius* (Boul.) Norkett

Autoecious. Fertile plants mostly 1.5–2.5(–3.5) mm with 4–8 or rarely more lvs. Lvs lanceolate to linear-lanceolate, acuminate, perichaetial lvs acuminate, narrower, longer, 7–9 times as long as wide with border reaching apex and often confluent with percurrent or excurrent nerve; cells 8–14 μm wide. $n=10$. Patches on shaded chalk or basic rocks or rarely on acidic sandstone. Frequent in S.E. England, rare elsewhere and extending north to Kintyre and E. Perth. 72, H7. Europe, Caucasus.

Var. **bambergeri** (Schimp. ex Milde) Waldh. in Weim, Fört. Skand. Växt. Moss., 1937

F. bambergeri Schimp. ex Milde

Synoecious. Lvs ± similar in shape to those of *F. viridulus* but 1 perichaetial lf often very small; border of stem lvs poorly developed and sometimes present only in sheathing laminae, border present in all laminae of perichaetial lvs but ending well below apex; nerve ending below apex; cells 6–10 μm wide. Patches on basic soil, sometimes mixed with var. *viridulus*, rare. Southern England, S. Wales, Yorks, Lancs, Stirling, S. Tipperary. 21, H1. Europe, Palestine, Egypt, Azores.

This species is very variable in size, the shape of the perichaetial lvs, the distribution of gametangia and habitat. The status of the various forms is unknown and requires experimental investigation. Small plants with long narrow perichaetial lvs tend to occur on dry rocks whilst with increasing dampness of the substrate the plants tend to be larger with less well differentiated perichaetial lvs.

Some forms of var. *viridulus* have an excurrent nerve and can only be separated from var. *tenuifolius* by the larger size and wider perichaetial lvs. Although var. *bambergeri* has been treated by some authorities as a species, the characters used to define it are sometimes found in various combinations in var. *viridulus*. Forms of var. *viridulus* may be encountered in which the border is poorly defined in the perichaetial lvs and poor or even absent in the stem lvs. Such plants may be distinguished from var. *bambergeri* by the autoecious or dioecious inflorescence, the larger cells and the perichaetial lvs of ± similar size. Occasional plants of var. *viridulus* have a synoecious inflorescence but are otherwise typical of the variety.

Saxicolous plants were considered to belong to a separate species, *F. minutulus* Sull., but are indistinguishable morphologically from terrestrial plants (see Smith, *Trans. Br. bryol. Soc.* **6**, 56–68, 1970). Large plants may be difficult to distinguish from small individuals of *F. crassipes* but that species has chlorophyllose cells outside the border of the sheathing laminae.

F. herzogii Ruthe ex Herzog has been recorded from 4 British localities: Land's End (Pierrot, *Rev. Bryol. Lichen.* **37**, 651–2, 1970); near Sunningwell, coll. Boswell, 1878;

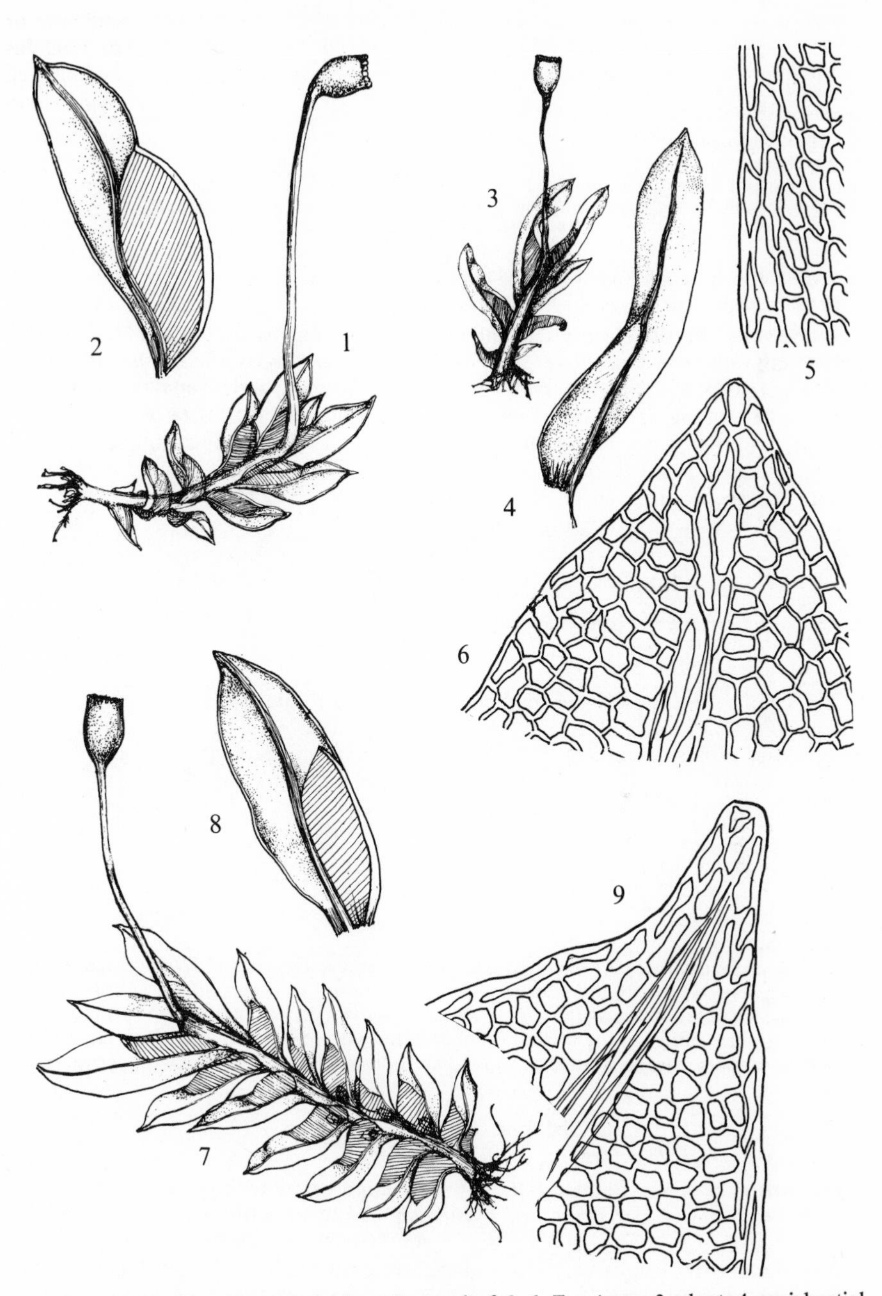

Fig. **84**. 1–2, *Fissidens incurvus*: 1, plant; 2, stem leaf. 3–6, *F. exiguus*: 3, plant; 4, perichaetial leaf; 5, margin of sheathing lamina; 6, leaf apex. 7–9, *F. bryoides*: 7, plant; 8, stem leaf; 9, stem leaf apex. Plants $\times 10$, leaves $\times 40$, cells $\times 450$.

Silverton, Devon, coll. Savery, 1908; between Malplash and Loscombe, Dorset, coll. Hill, 1966 (all Bruggeman-Nannenga, personal communication). This plant is said to differ from *F. viridulus* in the relatively wider and more shortly pointed lvs, areolation and the nerve reaching the apex (Potier de la Varde, *Ann. Crypt. Exot.* **4**, 161–5, 1931). I have seen the specimen from Dorset and slides of 2 of the other gatherings. All come within the range of the exceedingly variable *F. viridulus* and I can see no reason for treating the plants concerned as anything other than forms of *F. viridulus*. Some of the forms of *F. viridulus* are very distinctive and unless much material is examined might well be regarded as distinct species: the variation in *F. viridulus* is difficult to categorise and much is not amenable to orthodox taxonomic treatment.

3. F. incurvus Starke ex Röhl., Deutschl. Fl. Krypt., 1813
F. bryoides ssp. *incurvus* (Röhl.) Bertsch

Autoecious. Fertile shoots decumbent, 2–5(–10) mm with 5–12(–15) lvs. Lvs oblong-lanceolate to lanceolate, margin entire, all laminae bordered, border pale, usually confluent with percurrent nerve; cells 6–10 μm wide; perichaetial lvs narrower, more acute. Antheridia in dwarf shoot at base of female stem. Seta terminal, red; capsule cernuous or horizontal, often curved; spores 12–16 μm. Fr common, winter, spring. Patches or scattered plants on soil on arable and waste ground especially where calcareous, in open or shaded places, Frequent in the south-east, rare or occasional elsewhere, extending north to Stirling and Angus. 67, H7, C. Europe, Caucasus, Syria, N. Asia, N. America, Australia.

The cernuous or horizontal capsule is the only reliable character distinguishing *F. incurvus* from terrestrial forms of *F. viridulus*. *F. bryoides* has antheridia in bud-like axillary shoots and a ± erect capsule and is specifically distinct from *F. incurvus*.

4. F. bryoides Hedw., Sp. Musc., 1801

Autoecious. Fertile plants procumbent to ± erect, 3–20 mm with 6–20 lvs, sterile shoots sometimes longer, rhizoids few, brownish. Lvs oblong to lanceolate, border unistratose, usually confluent with nerve at apex; nerve percurrent or excurrent; cells 8–12(–14) μm wide; perichaetial lvs longer and more acute than stem lvs. Antheridia in dwarf, bud-like axillary shoots, rarely naked in lf axils. Seta terminal, red; capsule ± erect, ellipsoid; spores 10–14 μm. Fr common, winter, spring. $n=10$, 12. Dense, green patches on soil or more rarely on wood or stones in open places, arable fields, woods, on banks, etc.; common. 112, H32, C. Europe, Caucasus, Siberia, N. India, China, Canaries, Madeira, Cameroon, N. America, Mexico, New Zealand.

F. bryoides is usually easily distinguished from other small terrestrial *Fissidens* species by the bordered lvs with border confluent with the percurrent or excurrent nerve in the stem lvs and the small, bud-like, axillary male branches. Other species with such male shoots are *F. curnovii* which differs in the tomentum of deep-red rhizoids, *F. rivularis* which has a very thick border and occurs on wet rocks, and *F. monguillonii* which has very narrow perichaetial lvs.

5. F. curnovii Mitt., J. Linn. Soc. Bot., 1885
F. curnowii auct., *F. bryoides* var. *caespitans* Schimp.

Close to *F. bryoides*. Plants (4–)6–20 mm, matted below with tomentum of deep-red rhizoids. Nerve percurrent or excurrent, confluent with border at lf apex; Capsule inclined or occasionally erect; spores 16–20 μm. Deep green patches or scattered plants on soil or rocks in seepage areas, waterfalls and in rock crevices in and by streams. Western Britain from Cornwall to Shetland, occasional or rare, scattered localities in coastal counties of Ireland, rare. 27, H9, C. W. and C. Europe, Azores, N. Africa, Australia.

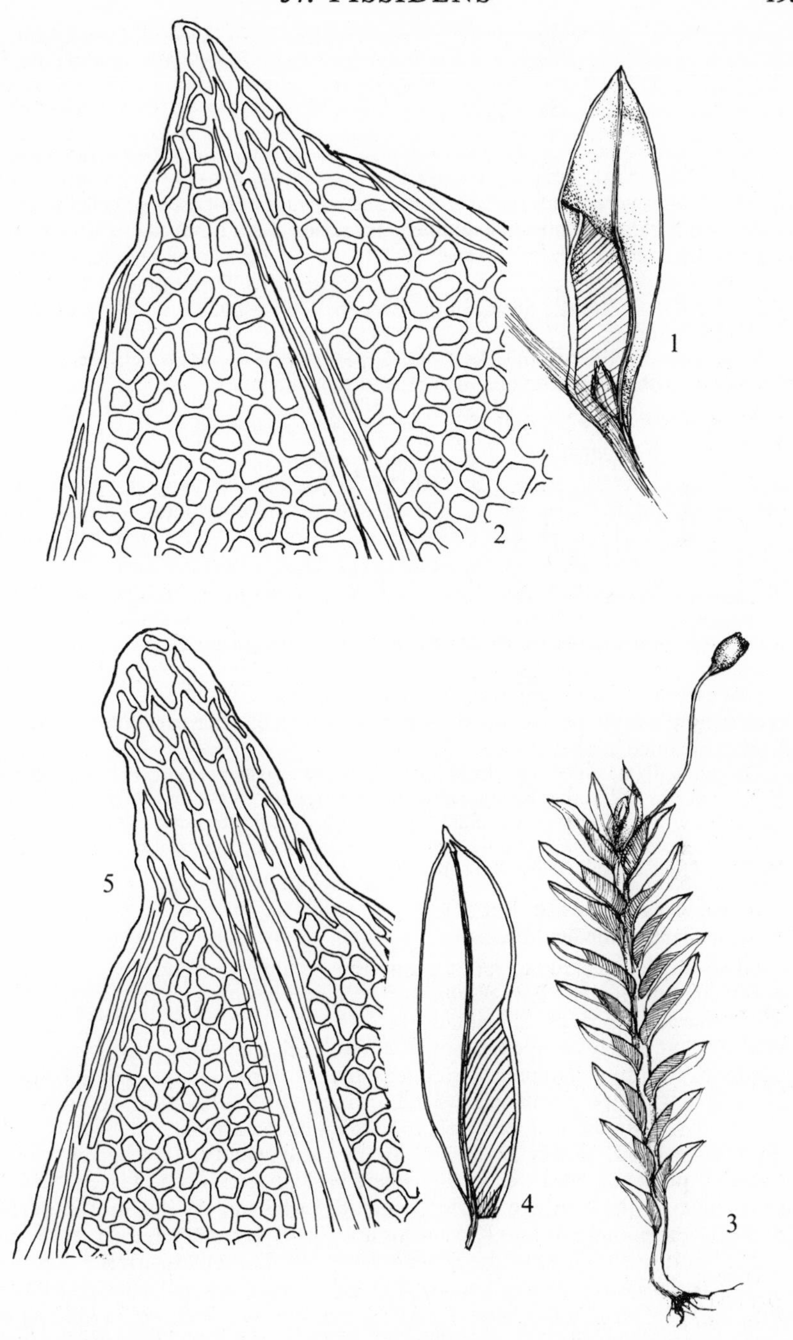

Fig. 85. 1–2, *Fissidens curnovii*: 1, leaf with axillary male branch; 2, leaf apex. 3–5, *F. rivularis*: 3, plant (× 5); 4, leaf; 5, leaf apex. Leaves × 20, cells × 450.

Likely to be confused with *F. rivularis*, *F. crassipes* and *F. rufulus*. It differs from all 3 in the deep red rhizoids, from *F. rivularis* in the thinner border and from the other 2 species in the border confluent with the nerve at the lf apex.

6. F. rivularis (Spruce) Br. Eur., 1851

Autoecious. Plants 7–20 mm, rhizoids brownish. Lvs numerous, lingulate to lingulate-lanceolate, acute to obtuse and mucronate, border very stout, yellowish, confluent with percurrent nerve at apex and forming stout mucro; cells (6–)8–10 μm wide. Antheridia in bud-like axillary branches. Seta terminal; capsule erect or slightly inclined, ellipsoid; spores 17–20 μm. Fr occasional, winter. Dark green patches or scattered plants on wet or submerged rocks. Occasional in S.W. England, rare elsewhere, Sussex, Monmouth, Brecon, Guernsey. 10, C. W. Europe, Italy, Tenerife, Madeira.

Readily distinguished from other *Fissidens* species by the very stout yellowish border which is confluent with the excurrent nerve, forming a stout mucro.

7. F. monguillonii Thér., Bull. Soc. Agri. Sc. Arts Sarte, 1899
F. rivularis var. *monguillonii* (Thér.) Podp.

Autoecious. Plants 4–15 mm with 15–20 lvs, rhizoids few, brownish. Stem lvs oblong to lingulate-lanceolate, obtuse and often mucronate, margin obscurely denticulate above, border well developed, 2–3-stratose, ± confluent with percurrent nerve; cells (8–)10–14 μm wide; perichaetial lvs very narrow, linear-lanceolate and markedly differing from preceding stem lvs. Antheridia in bud-like axillary shoots or terminal on short lateral branches. Setae terminal or on short lateral branches arising on ventral side of shoot; capsule ellipsoid, inclined; spores 14–18 μm. Fr occasional, winter. Dull green patches or scattered plants, usually silt-encrusted, on soil or exposed muddy tree roots in flood zone of rivers and on *Phragmites* stems in reed swamp, very rare. Devon, S. Somerset, Pembroke, Caerns, Fermanagh. 5, H1. France, N. and C. Africa.

The markedly narrower perichaetial lvs which are usually present will readily distinguish *F. monguillonii* from other large aquatic *Fissidens* species; if the perichaetial lvs are lacking the larger cells will separate the plant from *F. rivularis* and the border confluent with nerve at lf apex from *F. crassipes* and *F. rufulus*. For the occurrence of this plant in Britain see Norkett, *Trans. Br. bryol. Soc.* **2**, 11–14, 1952.

8. F. crassipes Wils. ex Br. Eur., 1849
F. mildeanus Schimp. ex Milde

Dioecious, autoecious or rarely synoecious. Plants ± decumbent, 7–20 mm with numerous lvs, rhizoids few, brownish. Lvs lanceolate to narrowly lanceolate, apex acute, margin obscurely denticulate above, border 1 or more stratose, colourless to reddish, ending below apex, a single row of chlorophyllose cells present outside border of sheathing laminae near base, border of dorsal lamina ending before base; nerve ending in apex; cells thin-walled to incrassate, (6–)10–14(–18) μm wide; perichaetial lvs ± similar in appearance to stem lvs. Seta terminal, red; capsule ellipsoid, erect or inclined; spores 18–28 μm. Fr frequent, winter. Dark green patches or scattered plants on rocks, especially where basic, in streams and rivers. Occasional to frequent in the southern half of England, Lancs and Yorks, rare in Wales and Scotland, extending north to Kincardine and Skye, rare in Ireland. 71, H16. Europe, Asia Minor, Kashmir, Madeira, Algeria, Morocco, Turkey, Australia.

In its typical form *F. crassipes* is quite distinct from *F. rufulus* in the narrower lf apex, cell size and the row of chlorophyllose cells outside border of sheathing lamina, but intermediate forms do occur, especially in N. Britain and Ireland, which cannot be named. *F. rufulus* might perhaps be better treated as a subspecies of *F. crassipes*.

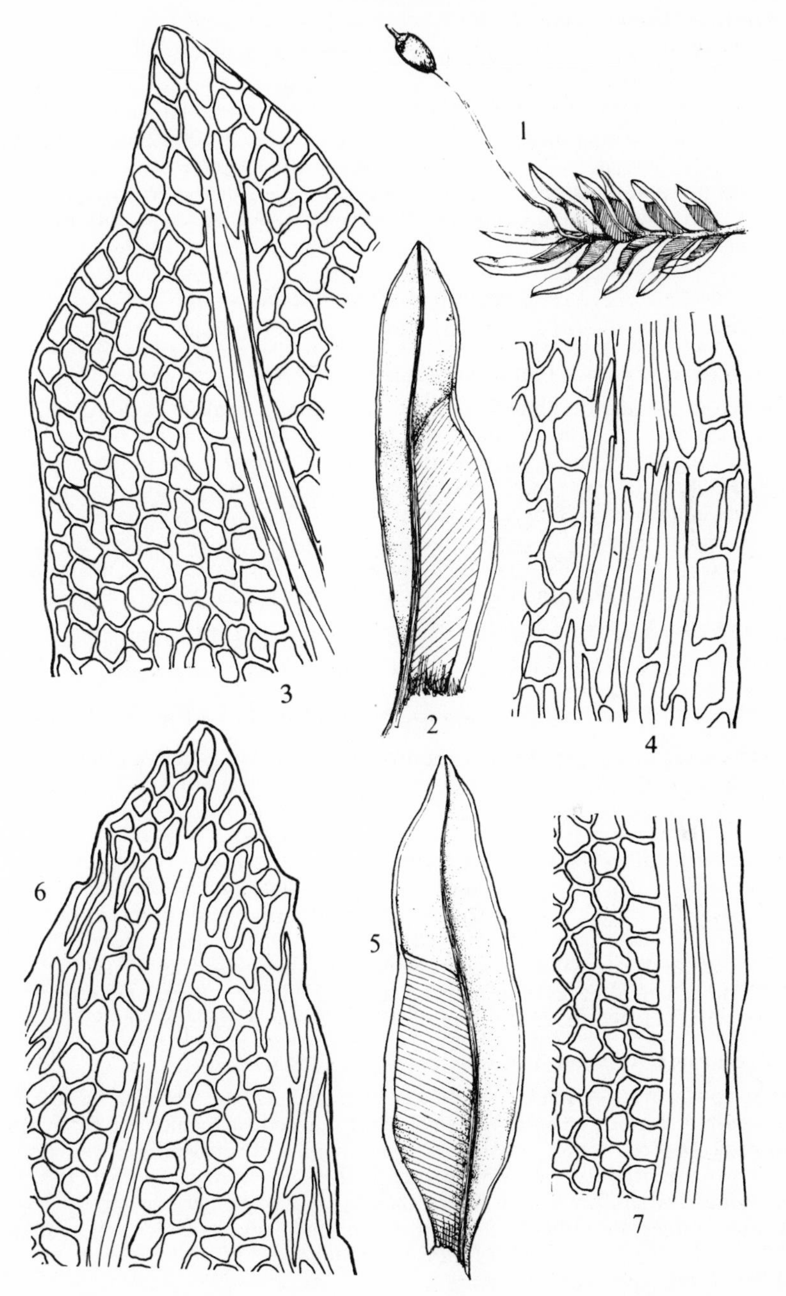

Fig. **86**. 1–4, *Fissidens crassipes*: 1, plant (×5); 2, leaf; 3, leaf apex; 4, margin of sheathing lamina near leaf base. 5–7, *F. rufulus*: 5, leaf; 6, leaf apex; 7, margin of sheathing lamina near leaf base. Leaves ×20, cells ×450.

9. F. rufulus Br. Eur., 1851

Close to *F. crassipes*. Dioecious. Plants 5–20 mm. Lvs oblong-lanceolate to lingulate-lanceolate, apex broad, ± obtuse; border strong, chlorophyllose cells outside border of sheathing lamina lacking, border of dorsal lamina extending to lf base; nerve and border orange to red, rarely paler; cells 8–10(–12) μm wide. Spores 18–22 μm. Fr occasional, winter. Basic rocks in streams and rivers. Occasional in S. Wales and the Pennines, very rare elsewhere in W. Britain, extending north to Kintyre and Angus, rare in Ireland. 18, H5. W. and C. Europe, Balearic Is., N. America.

Section II. *Pycnothallia*

A heterogeneous group of species with strongly papillose lf cells.

10. F. algarvicus Solms., Tent. Bryogeog. Algarv., 1868

Dioecious. Plants decumbent, 1.5–3.0 mm, fertile stems with 6–10 lvs, sterile stems with more. Lvs linear-lanceolate, gradually tapering to acuminate apex, all laminae with strong border, confluent with nerve at apex; cells in upper part of lf irregular, elongate-hexagonal, about twice as long as wide, 6–8 μm wide; cells in sheathing lamina larger, longer. Male plants bud-like, minute. Seta terminal; capsule erect, ovoid; spores *ca* 14 μm. Fr occasional, spring. Small patches or scattered stems on clayey soil on shaded or rocky banks, very rare and sporadic in appearance. Cornwall, Devon, Glos, Radnor, Pembroke, Caerns, S. Kerry. 7, H1. France, Spain, Portugal, Italy, Dalmatia, Sardinia, Madeira.

Readily recognised by the gradually tapering lvs and the elongated upper cells, but possibly overlooked because of its small size and frequent occurrence in small quantity.

Section III. *Aloma*

Lf margin crenulate, unbordered. Sporophyte terminal, peristome teeth spirally thickened and papillose. Minute plants on soil.

11. F. exilis Hedw., Sp. Musc., 1801

Dioecious or autoecious. Stems decumbent, 1.5–3.0 mm with 4–8 lvs. Lvs lingulate to narrowly lanceolate, apex acute, margin crenulate or crenulate-serrulate, unbordered; nerve ending in apex; cells 8–12(–14) μm wide. Male branches bud-like at base of female stems or as separate plants. Seta terminal; capsule erect, ellipsoid to sub-cylindrical; spores 10–12 μm. Fr common, autumn to spring. $n=12$. Patches on clayey soil on banks, edges of wood rides, arable fields, etc. Occasional to frequent in S. Britain, rare elsewhere. 75, H9, C. Europe, Kashmir, Algeria, N. and S. America.

12. F. celticus J. A. Paton, Trans. Br. bryol. Soc., 1965

Dioecious. Plants ± erect, occasionally branched, 2.5–4.5 mm with up to 28(–36) lvs, deep-red rhizoids at base. Lvs lingulate to lanceolate, acute, margin crenulate with projecting cell walls, unbordered; nerve with distinct bend halfway along its length, percurrent; cells ± hexagonal, marginal row 8–12 μm wide, others 12–20 μm; perichaetial lvs similar to but larger than stem lvs. Antheridia and sporophyte unknown. Patches or scattered plants on shaded soil banks in woods and by streams, occasional in England and Wales, rare in Scotland and Ireland but probably overlooked. 29, H4. Apparently endemic.

Easily recognised in the field by the ± erect stems with numerous lvs resembling miniature palm fronds. For the occurrence of this plant in Britain see Paton, *Trans. Br. bryol. Soc.* **4**, 780–4, 1965.

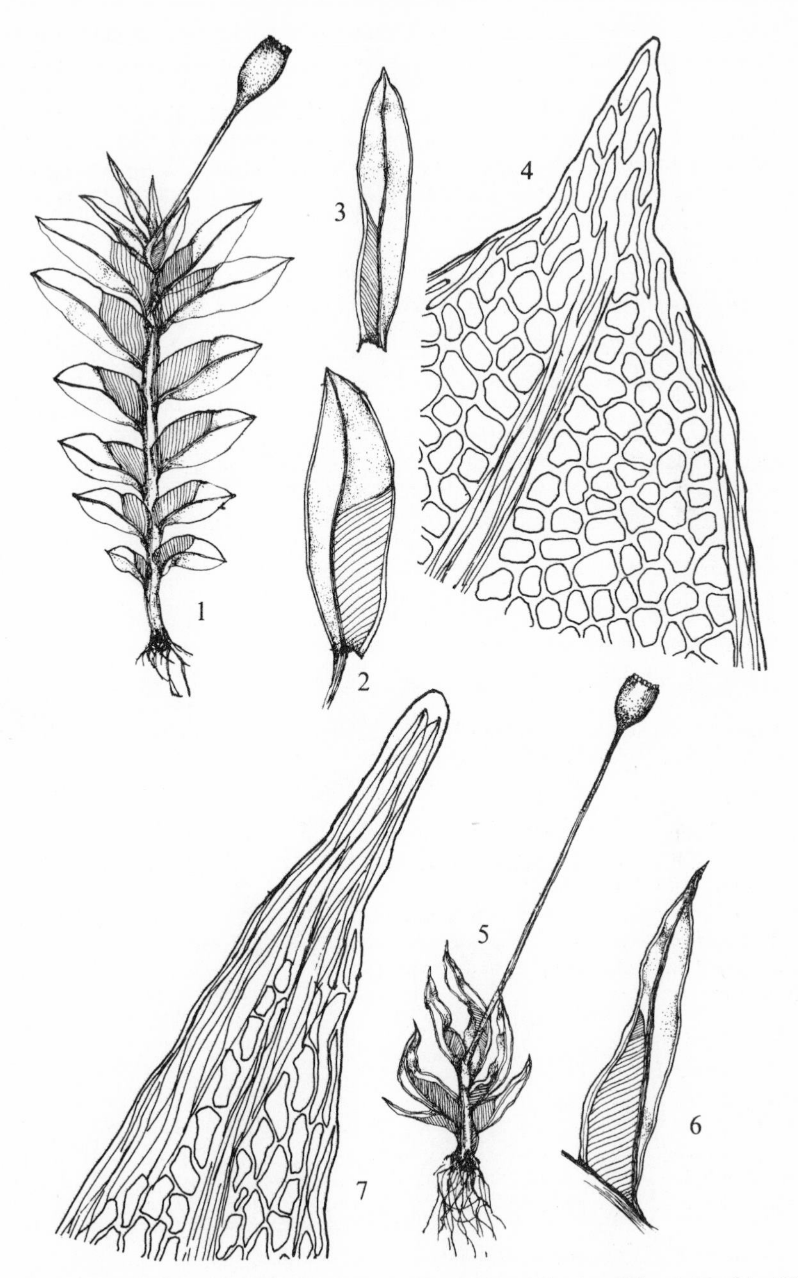

Fig. 87. 1–4, *Fissidens monguillonii*: 1, plant; 2, stem leaf (×20); 3, perichaetial leaf (×20); 4, stem leaf apex. 5–7, *F. algarvicus*: 5, plant; 6, leaf (×40); 7, leaf apex. Plants ×10, cells ×450.

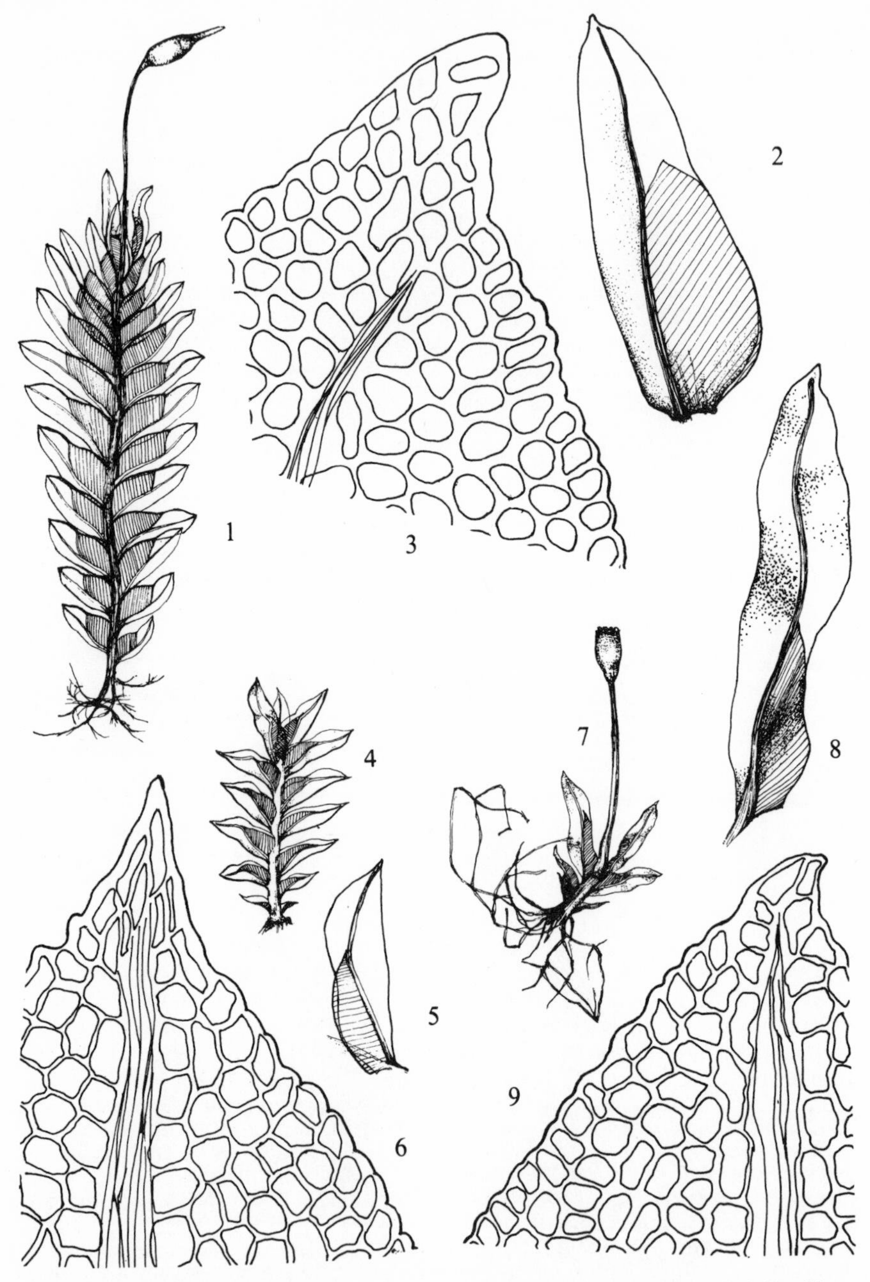

Fig. **88**. 1–3, *Fissidens osmundoides*: 1, plant (× 5); 2, leaf (× 30); 3, leaf apex. 4–6, *F. celticus*: 4, plant (× 10); 5, leaf (× 40); 6, leaf apex. 7–9, *F. exilis*: 7, plant (× 10); 8, leaf (× 40), 9, leaf apex. Cells × 450.

Section IV. *Serridium*

Lf margin crenulate to irregularly toothed, unbordered. Sporophyte terminal or lateral; peristome teeth papillose above. Usually large plants.

13. F. osmundoides Hedw., Sp. Musc., 1801

Dioecious. Plants erect, to 10 cm. Lvs numerous, oblong-lingulate to lingulate-lanceolate, acute to sub-obtuse and apiculate, margin regularly crenulate, unbordered; nerve ending below or in apex; cells incrassate, variable in size, 12–22 μm wide, marginal row smaller. Seta terminal, purple; capsule erect or inclined, ellipsoid; spores 18–24 μm. Fr occasional, winter. $n=11$, 16. Dense bright or dark green tufts in damp rock crevices and in flushes, mainly in montane areas. Frequent in western and northern Britain and in Ireland; Hants; absent from S.E. England and the Midlands. 77, H31. Europe, Faroes, Iceland, Siberia, C. Asia, Japan, Malaya, N. America, Greenland, southern S. America.

Distinguished from other large species with unbordered lvs by the purple seta, the relatively shorter and wider lvs and the usually wider cells. *F. taxifolius* has an excurrent or percurrent nerve and in the other species the lf margin toothed in varying degrees.

14. F. taxifolius Hedw., Sp. Musc., 1801

Autoecious. Plants procumbent to erect, branching from base, old stems rhizome-like, tomentose. Lvs oblong-lanceolate to linear-lanceolate, margin unbordered; nerve strong, percurrent to excurrent; cells 6–10 μm wide, marginal row smaller, quadrate, often pellucid. Seta from near base of main branches, red; capsule erect to horizontal, ellipsoid; spores 12–18 μm.

Ssp. **taxifolius**

Shoots mostly 1–2 cm long. Lvs lanceolate-lingulate, ± shortly tapering to acute to obtuse and apiculate apex, margin crenulate or serrulate; cells 6–10 μm. Dark brown to blackish rhizoidal gemmae, irregular in shape, 300–700 μm diameter, sometimes present. Capsule usually horizontal. Fr frequent, winter. $n=10$, 12*, 12+1*. Light green to reddish-brown patches, sometimes extensive, on soil in woods, fields, on banks, disturbed ground, etc., common except at high altitudes. 112, H40, C. Europe, Faroes, W. Asia, Siberia, Kashmir, Nepal, Japan, Tunisia, Macaronesia, N. and C. America, southern S. America.

Ssp. **pallidicaulis** (Mitt.) Mönk., Laubm., 1927
F. pallidicaulis Mitt.

Shoots 2–3 cm. Lvs mostly lanceolate to linear-lanceolate and tapering from end of sheathing laminae to acuminate apex, margin finely crenulate or finely serrulate; cells 7–9 μm wide. Capsule erect. Fr unknown in Britain. Small patches on damp soil on rocky stream banks, very rare. W. Inverness, Argyll, Kintyre, Mull, S. Kerry, S. Tipperary. 5, H2. Jura, Spain, Portugal, Dalmatia, Italy, Macaronesia.

For an account of ssp. *pallidicaulis* in Scotland and Ireland see Wallace, *J. Bryol.* **9**, 161–2, 1977. In its extreme forms, such as are found on Madeira, it is very distinctive with shoots to 6 cm long and the lvs up to 10 times as long as wide. The Scottish plants are less distinctive but still separable from ssp. *taxifolius* in the more longly tapering lf, the more finely crenulate or serrulate margin and the usually slightly smaller cells. The lvs are also more closely set on the stem. Potier de la Varde (*Rev. Bryol. Lichen.* **15**, 30–9, 1945) discusses the reasons for regarding the taxon as a subspecies of *F. taxifolius*. The narrower and more closely set lvs may lead to confusion with *F. polyphyllus* but the smaller cells and usually distinctly excurrent nerve will separate ssp. *pallidicaulis* from that species. Rhizoidal gemmae have only been found in sterile plants of ssp. *taxifolius* and the relationship between these and fertile plants requires investigation.

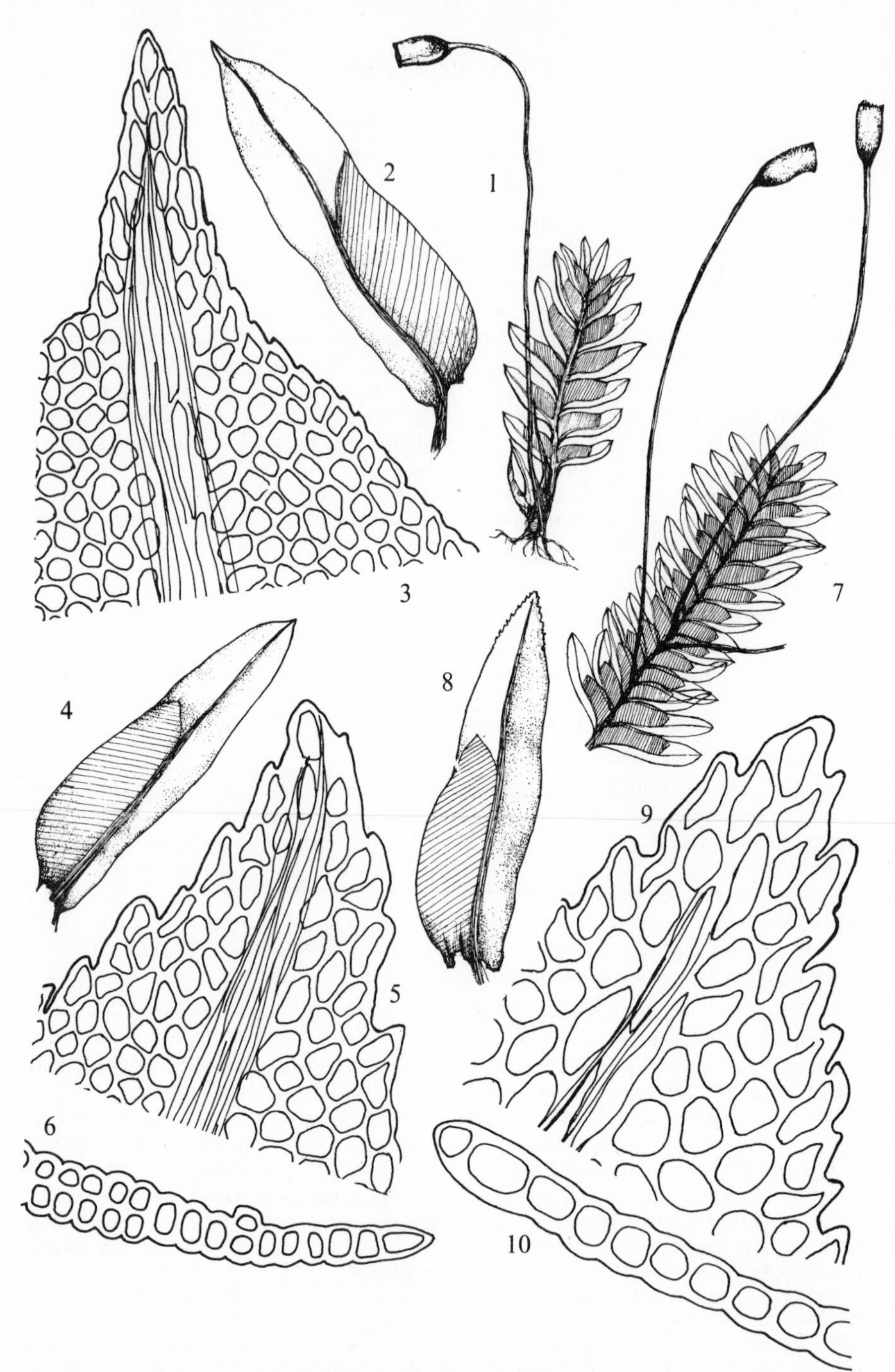

Fig. **89**. 1–3, *Fissidens taxifolius* ssp. *taxifolius*: 1, plant (×5), 2, leaf (×30); 3, leaf apex. 4–6, *F. cristatus*: 4, leaf (×15); 5, leaf apex; 6, section of apical lamina. 7–10, *F. adianthoides*: 7, plant (×4); 8, leaf (×15); 9, leaf apex; 10, section of apical lamina. Cells ×450.

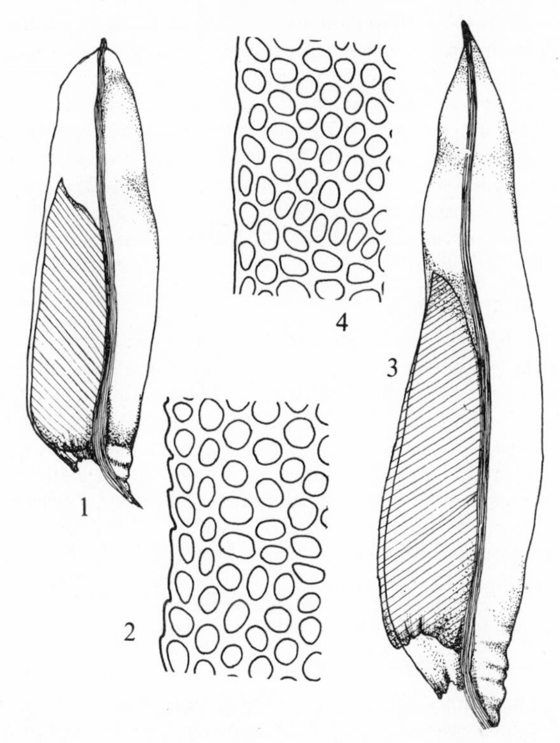

Fig. **90**. 1–2, *Fissidens taxifolius* ssp. *taxifolius*: 1, leaf; 2, cells from apical lamina. 3–4, *F. taxifolius* ssp. *pallidicaulis*: 3, leaf; 4, cells from apical lamina. Leaves ×25, cells ×415.

15. F. cristatus Wils. ex Mitt., J. Linn. Soc. Bot., Suppl., 1859
F. decipiens De Not.

Autoecious or dioecious. Plants ± erect, branched, older stems often with eroded lvs, tomentose below, branches to 2(–3) cm with numerous lvs. Lvs ovate-lanceolate to lanceolate, apex acute to obtuse and apiculate, margin crenulate-serrulate to irregularly toothed towards apex, unbordered; nerve ending below apex to excurrent; cells incrassate, opaque, bistratose in patches in upper part of lf, mostly 6–12 μm wide, 3–4 marginal rows more incrassate, pellucid, forming distinct pale band. Seta from middle or base of branch, pale red; capsule ± horizontal, ellipsoid; spores 10–16 μm. Fr occasional, winter, spring. $n=12$*, 13+2, 16. Dense tufts on shaded soil and rocks by streams, on cliffs and on soil in calcareous grassland, sand-dunes, occasionally on wood and in bogs. Common in wetter parts of Britain but usually limited to calcareous places elsewhere. 102, H33, C. Europe, Faroes, Caucasus, Manchuria, Japan, India, Java, Azores, N. America.

Close to *F. adianthoides* but usually distinguishable by the smaller lf cells and more conspicuous marginal band. Forms of *F. adianthoides* on dry ground may have smaller cells and may be difficult to distinguish but have the lf cells unistratose throughout. It seems possible that *F. adianthoides* is an autopolyploid derivative of *F. cristatus*.

16. F. adianthoides Hedw., Sp. Musc., 1801

Autoecious or dioecious. Plants decumbent to erect, branches to 5(–)7 cm with numerous lvs. Lvs broadly lingulate to lingulate-lanceolate, acute to obtuse and apiculate, margin crenulate below, sharply and irregularly toothed above, unbordered; nerve ending below or in apex; cells opaque, incrassate, unistratose throughout, 12–20 μm wide, 3–4 marginal rows more incrassate and pellucid, forming pale band. Seta or setae from middle of stem, reddish; capsule erect or inclined, straight or curved, ellipsoid; spores 18–24 μm. Fr frequent, autumn to spring. $n=24$*. Lax to dense, yellowish-green to dark green tufts up to 10(–25) cm high on wet rock ledges especially near streams and waterfalls, in seepage areas, fens and basic grassland, rarely on wood, common. 111, H40, C. Europe, Faroes, Iceland, Japan, Hong Kong, Algeria, Macaronesia, N. America, Tierra del Fuego, New Zealand.

17. F. serrulatus Brid., Sp. Musc., 1806

Dioecious. Plants procumbent to erect, simple or branched, to 7.5 cm tall. Lvs numerous, oblong-lanceolate to lingulate, apex acuminate to obtuse or obtuse and apiculate, margin crenulate below, irregularly toothed towards apex, unbordered; nerve ending in or below apex; cells incrassate, conically mamillose, partially bistratose, 10–16 μm wide, 3–4 marginal rows often more incrassate and pellucid, forming pale band. Seta terminal; capsule inclined, ovoid. Only male plants known in Britain and Ireland. Loose, green tufts or patches on soil and rocks by streams or rivers, very rare. W. Cornwall, S. Devon, Merioneth, W. Mayo. 3, H1. France, Spain, Portugal, Corsica, Macaronesia, Tunisia, Algeria.

Similar to large forms of *F. adianthoides* but with smaller, conically mamillose cells. *F. polyphyllus* has narrower lvs and lacks mamillose cells and pale marginal band. The mamillae are best seen in side view on a folded lf.

18. F. polyphyllus Wils. ex Br. Eur., 1851

Dioecious or autoecious. Plants procumbent, branches to 10 cm. Lvs numerous, narrowly lanceolate or narrowly lingulate, apex acuminate to obtuse or obtuse and apiculate, margin finely and obscurely denticulate towards apex, unbordered; nerve ending in apex; cells pellucid, smooth, 8–14 μm wide, single marginal row smaller but no pale marginal band present. Seta lateral, yellowish; capsule inclined, ovoid. Fr unknown in Britain or Ireland. Dark green patches on soil or rocks by streams and rivers and on flushed rocks, rare, sometimes locally abundant. Cornwall, S. Devon, Montgomery, Merioneth, Caerns, Argyll, Kintyre, Kerry, Cork, Waterford, Wicklow, Sligo. 8, H7. Norway, France, Spain, Portugal, Macaronesia.

38. Octodiceras Brid., Sp. Musc., 1806

Stem without central strand. Sheathing laminae $\frac{1}{4}$–$\frac{1}{3}$ total lf length. Capsule ± immersed, without stomata. About 15 species distributed through Europe, Asia and America.

1. O. fontanum (La Pyl.) Lindb., Bidr. Moss, Syn., 1863
O. julianum Brid., *Fissidens fontanus* (La Pyl.) Steud.

Autoecious. Plants to 5 cm. Lvs distant, spreading, linear to linear-lanceolate, gradually tapering from near base to blunt apex, margin entire, unbordered; nerve ending below apex; cells thin-walled to incrassate, *ca* 10 μm wide, larger towards margin and base. Seta lateral; capsule ellipsoid. Fr unknown in Britain. Floating or submerged patches on wood, stones or concrete in sluggishly flowing, often slightly polluted water, occasional but sometimes locally abundant in central and southern England. 16. N., W. and C. Europe, Madeira, Algeria, N. and C. America, Chile.

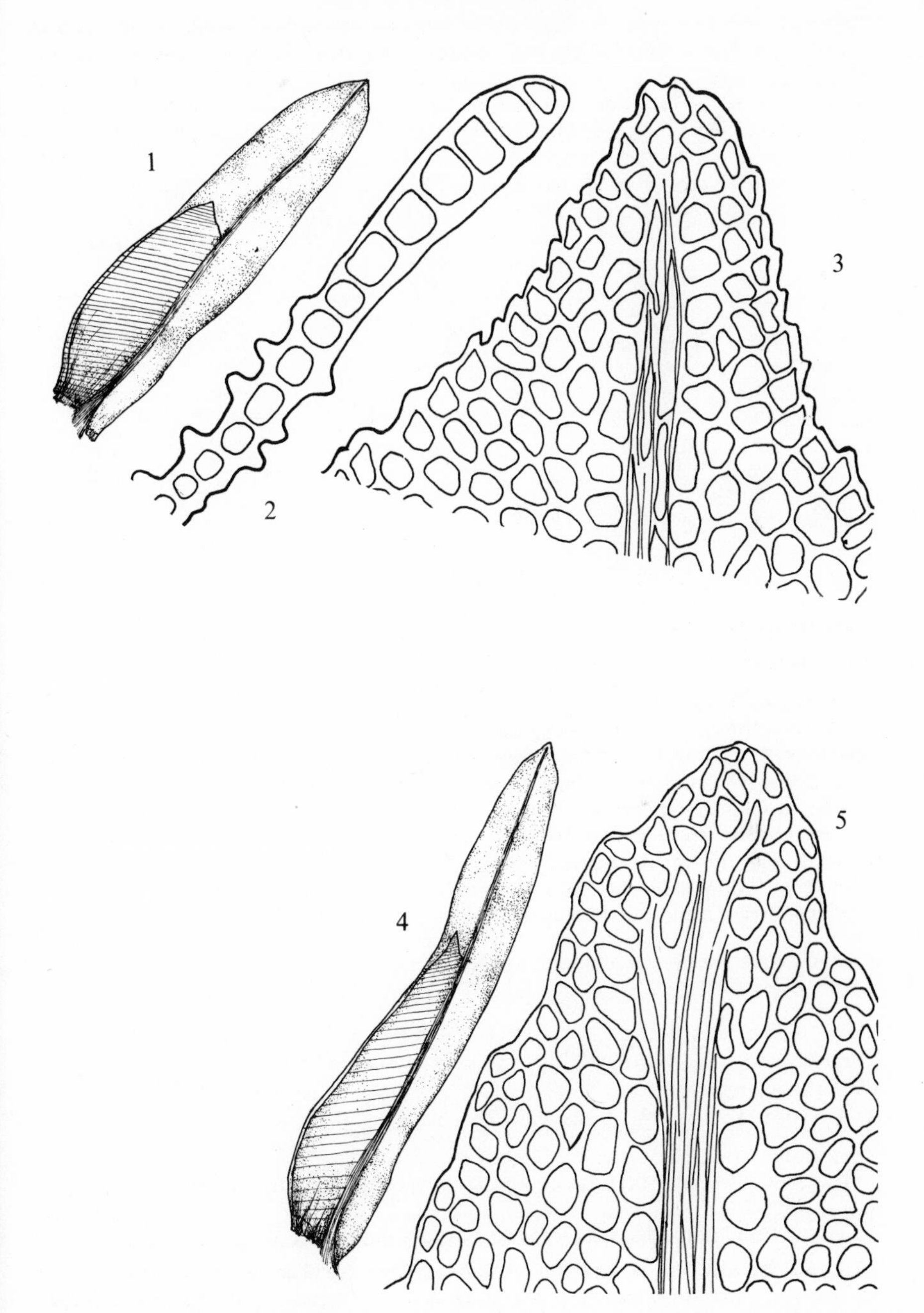

Fig. **91**. 1–3, *Fissidens serrulatus*: 1, leaf; 2, section of apical lamina; 3, leaf apex. 4–5, *F. polyphyllus*: 4, leaf; 5, leaf apex. Leaves ×15, cells ×450.

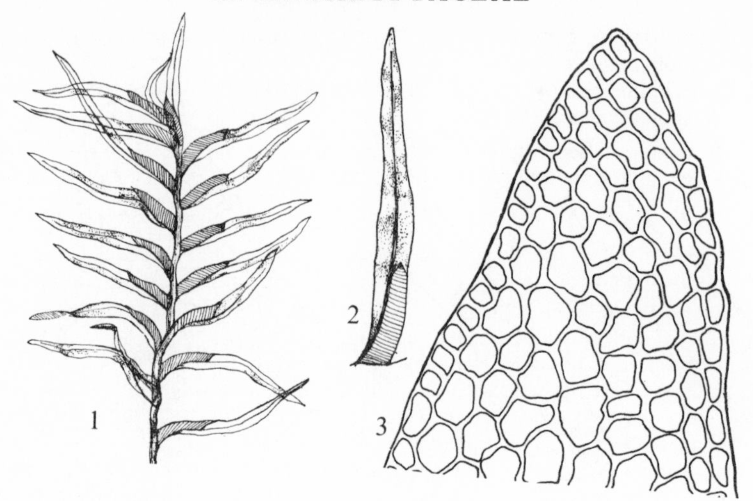

Fig. 92. *Octodiceras fontanum*: 1, plant (× 5); 2, leaf (× 10); 3, leaf apex (× 450).

9. ENCALYPTALES

With characters of the family.

12. ENCALYPTACEAE

Plants tufted. Lvs lingulate, spathulate or lanceolate, often broadly pointed, margin plane or reflexed, rarely recurved; nerve ending below apex to longly excurrent; basal cells narrowly rectangular, hyaline, smooth, marginal cells narrower, cells above hexagonal to quadrate, very papillose, obscure. Seta long, straight; capsule erect, cylindrical, smooth or striate; calyptra large, covering capsule, entire or fringed or erose at base; lid longly rostrate; peristome variable, well developed and double or single or rudimentary or absent. With only 1 genus.

On the basis of peristome structure this family is anomalous. The various species possess peristomes similar to or approaching those of the 3 major groupings of mosses based on peristome structure, Nematodonteae, Aplolepideae and Diplolepideae, recognised by older authorities (see Dixon, 1924).

39. Encalypta Hedw., Sp. Musc., 1801

With the characters of the family. About 35 species distributed through Europe, Asia, Africa, America, Australia, New Zealand, Oceania.

1 Lvs gradually narrowed from middle to apex, cells 8–10 μm wide in mid-lf, capsule smooth **1. E. alpina**
Lvs narrowed only near apex, cells 10–20 μm wide, capsule smooth or striate 2

2 Capsule spirally ribbed, lvs obtuse, often cucullate, nerve ending below apex **5. E. streptocarpa**
Capsule smooth or longitudinally furrowed, lvs acute to obtuse and apiculate, rarely obtuse, nerve usually ending in apex to longly excurrent 3

3 Calyptra fringed, capsule smooth, lf margin recurved at middle of lf, spores smooth or ridged **3. E. ciliata**
Calyptra entire or erose, capsule smooth or furrowed, margin plane at middle of lf, spores coarsely papillose 4

4 Capsule smooth, nerve smooth at back above **2. E. vulgaris**
Capsule longitudinally furrowed, nerve papillose at back above **4. E. rhaptocarpa**

1. E. alpina Sm., Eng. Bot., 1805
E. commutata Nees & Hornsch.

Autoecious. Plants to 6 cm. Lvs appressed, twisted, apices incurved when dry, reflexed, squarrose when moist, 2.0–3.5 mm long, gradually tapering to acute apex, very rarely ± obtuse and cucullate, margin plane or inflexed, ± denticulate; nerve stout, reddish, ending below apex to longly excurrent in hyaline arista, finely scabrous at back above; cells in mid-lf obscure, 8–10 μm wide. Seta 5–6 mm; capsule cylindrical, smooth when fresh, sulcate when dry and empty; calyptra irregularly torn at base, smooth towards apex; peristome lacking; spores variable in size, 28-40 μm. Fr common, autumn. $n=14$. Dull green tufts on dry, basic, montane rocks at altitudes of 350–1050 m, rare. Caerns, N.W. Yorks, Perth, Angus, Inverness, Argyll, W. Sutherland, Sligo. 10, H1. N. and montane parts of Europe, Caucasus, N. and C. Asia, Himalayas, China, Morocco, N. America.

Readily distinguished from other members of the genus by the usually gradually narrowed lvs and smaller cells.

2. E. vulgaris Hedw., Sp. Musc., 1801

Autoecious. Plants 0.5–2.0 cm. Lvs incurved, twisted when dry, erecto-patent to spreading when moist, mostly 2–4 mm long, lingulate or spathulate, acute to obtuse with or without apiculus, margin plane or inflexed; nerve ending below apex to excurrent, reddish below, smooth at back or with a few papillae at extreme apex; basal cells often with heavily thickened transverse walls, marginal cells narrow, sometimes extending up margin, cells above coarsely papillose, 14–20 μm wide in mid-lf. Seta reddish; capsule smooth at maturity, sulcate when dry and empty; calyptra entire or erose, scabrous towards apex; peristome teeth absent or rudimentary; spores coarsely papillose, irregular in size, 30–45 μm. Fr frequent, spring. $n=26$*, 39. Dull or yellowish-green patches or tufts on basic rocks, walls, cliffs and soil, frequent in basic areas, rare elsewhere. 87, H25. Europe, W. and C. Asia, Tibet, Kashmir, Himalayas, New Guinea, Algeria, Tunisia, Canaries, Madeira, N. America, Tasmania, New Zealand.

3. E. ciliata Hedw., Sp. Musc., 1801

Autoecious. Plants to 5 cm. Lvs incurved, ± crisped when dry, erecto-patent to spreading when moist, (2.5–)3.0–5.5 mm long, lingulate-spathulate to lingulate, acute to obtuse and apiculate, margin recurved at middle of lf; nerve percurrent to excurrent in a sometimes hyaline arista, reddish below, smooth at back above; cells 10–15 μm wide in mid-lf. Seta yellow to pale red; capsule smooth at maturity, contracted below mouth when dry and empty; calyptra fringed, smooth towards apex; peristome single, short; spores smooth or ridged, 30–38 μm. Fr frequent, late summer, autumn. $n=13$*. Bright green tufts or patches on damp, montane rocks and rock crevices particularly where basic, occasional in montane parts of Britain and Ireland. 39, H4. Faroes, N. and montane parts of Europe, Caucasus, N., C. and E. Asia, Kashmir, Himalayas, Algeria, Ethiopia, C. Africa, N. America, Australia, New Zealand, Hawaii.

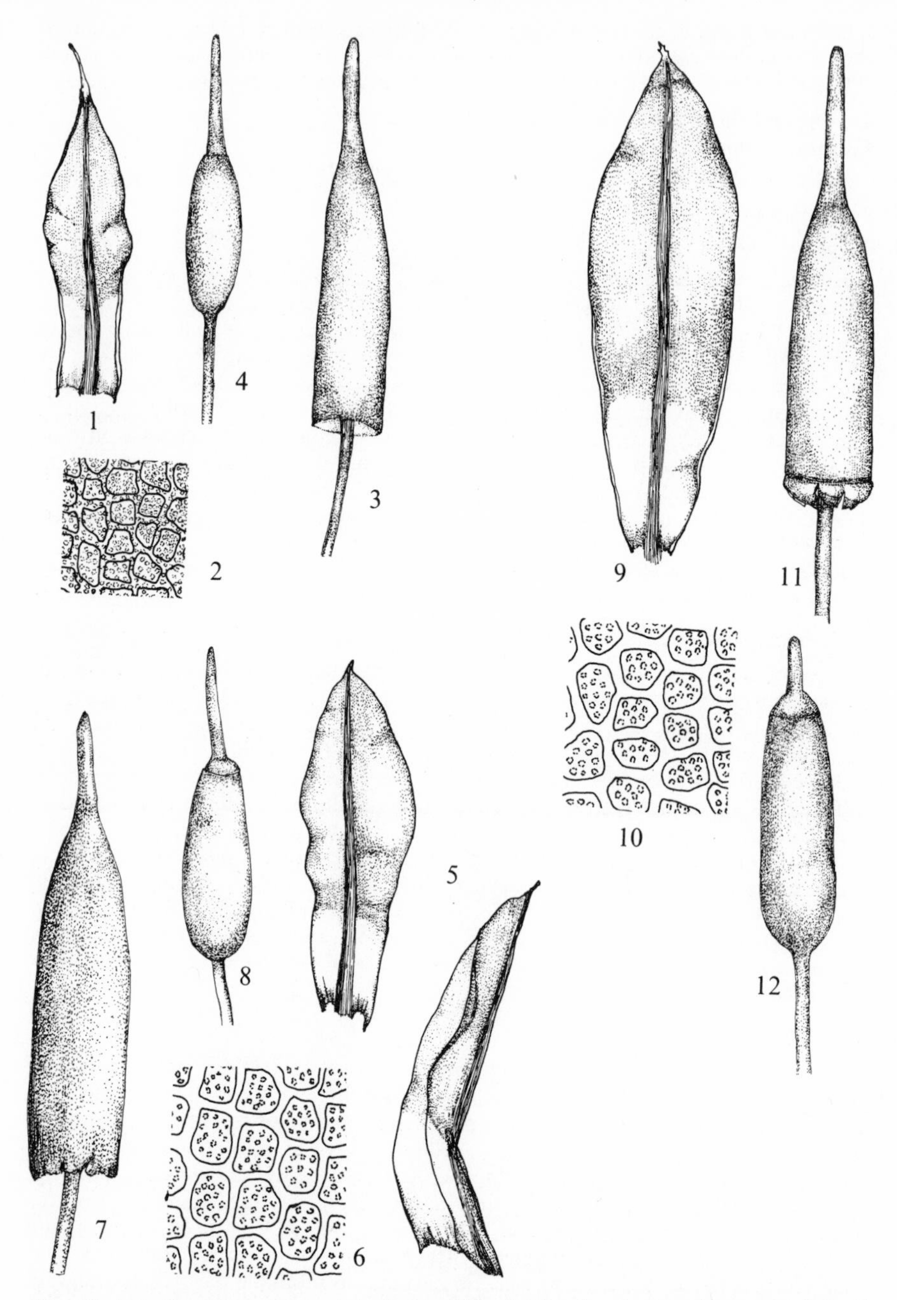

Fig. **93**. 1–4, *Encalypta alpina*: 1, leaf; 2, mid-leaf cells; 3, calyptra; 4, capsule. 5–8, *E. vulgaris*: 5, leaf; 6, mid-leaf cells; 7, calyptra; 8, capsule. 9–12. *E. ciliata*: 9, leaf; 10, mid-leaf cells; 11, calyptra; 12, capsule. Leaves ×15, cells ×415, capsules and calyptras ×10.

Easily recognised by the fringed calyptra. The fringe may be lost in old calyptras but the plant may be distinguished from *E. vulgaris* by the apex of the calyptra smooth, the smooth or ridged spores, the short peristome and the lf margin recurved at mid-lf.

4. E. rhaptocarpa Schwaegr., Suppl., 1811
E. rhabdocarpa auct.

Autoecious. Stems to 2(–5) cm. Lvs erect, incurved to crisped when dry, erecto-patent to spreading when moist, 2.0–3.5 mm long, lanceolate to lingulate-spathulate, acute to obtuse and apiculate, margin plane or recurved near base, inflexed above; nerve reddish, excurrent in acuminate, sometimes hyaline point, papillose at back above; basal marginal cells very narrow, extending up margin, cells above coarsely papillose, 10–16 μm wide in mid-lf. Seta red; capsule longitudinally striate at maturity, deeply furrowed when dry and empty; calyptra entire or erose, papillose towards apex; peristome single, short; spores coarsely papillose, variable in size, 34–50 μm. Fr frequent, summer, autumn. $n=26$. Dull green patches on basic rocks and in crevices in montane habitats. Caerns, Denbigh, Yorks and Westmorland northwards, rare to occasional, also on calcareous sand-dunes in E. and N. Scotland. 23, H4. Europe, Faroes, Caucasus, N., W. and C. Asia, Tibet, Himalayas, Kashmir, N. Africa, N. America, Greenland, Hawaii.

5. E. streptocarpa Hedw., Sp. Musc., 1801

Dioecious. Plants to 6 cm. Lvs incurved or crisped when dry, spreading when moist, lingulate or lingulate-spathulate, 3–7 mm long, apex ± obtuse, sometimes cucullate, margin plane, crenulate with papillae; nerve strong, reddish, ending in or below apex, scabrous at back above; basal cells rectangular, marginal narrower and not ascending up margin, cells above, quadrate, papillose, obscure, 10–14 μm wide in mid-lf. Capsule cylindrical, spirally ribbed; peristome double, outer teeth long, filiform; spores smooth, 10–14 μm wide; calyptra entire, scabrous above. Fr very rare, spring. Dull or yellowish-green tufts or patches on usually basic rocks and soil, mortar and old walls, often where shaded or damp, generally distributed, frequent or common. 104, H40. Europe, Caucasus, Siberia, Morocco, Canaries, N. America.

Although nearly always without capsules, readily distinguished from other *Encalypta* species by its larger size, the nerve ending below the apex and coarsely scabrous at the back. The other British species usually fruit freely and the capsules are either smooth or longitudinally furrowed.

10. POTTIALES

With the characters of the family.

13. POTTIACEAE

Usually small, acrocarpous or rarely cladocarpous plants. Lvs linear to spathulate or ovate, acuminate to rounded at apex; nerve usually strong; basal cells rectangular, ± hyaline, cells above hexagonal or quadrate, often papillose, sometimes obscure. Seta long or short, usually straight; capsule erect or inclined, cleistocarpous or stegocarpous, ovoid to cylindrical, straight or curved; peristome single, often spirally twisted and teeth divided into filiform segments or absent.

A diverse family the generic distinctions within which are sometimes somewhat arbitrary, as for example between *Pottia* and *Phascum*. The differences between the subfamilies, at least in the European genera, seem at least as good as those between the families of the Dicranales, and I consider that the taxa would probably be better treated as families.

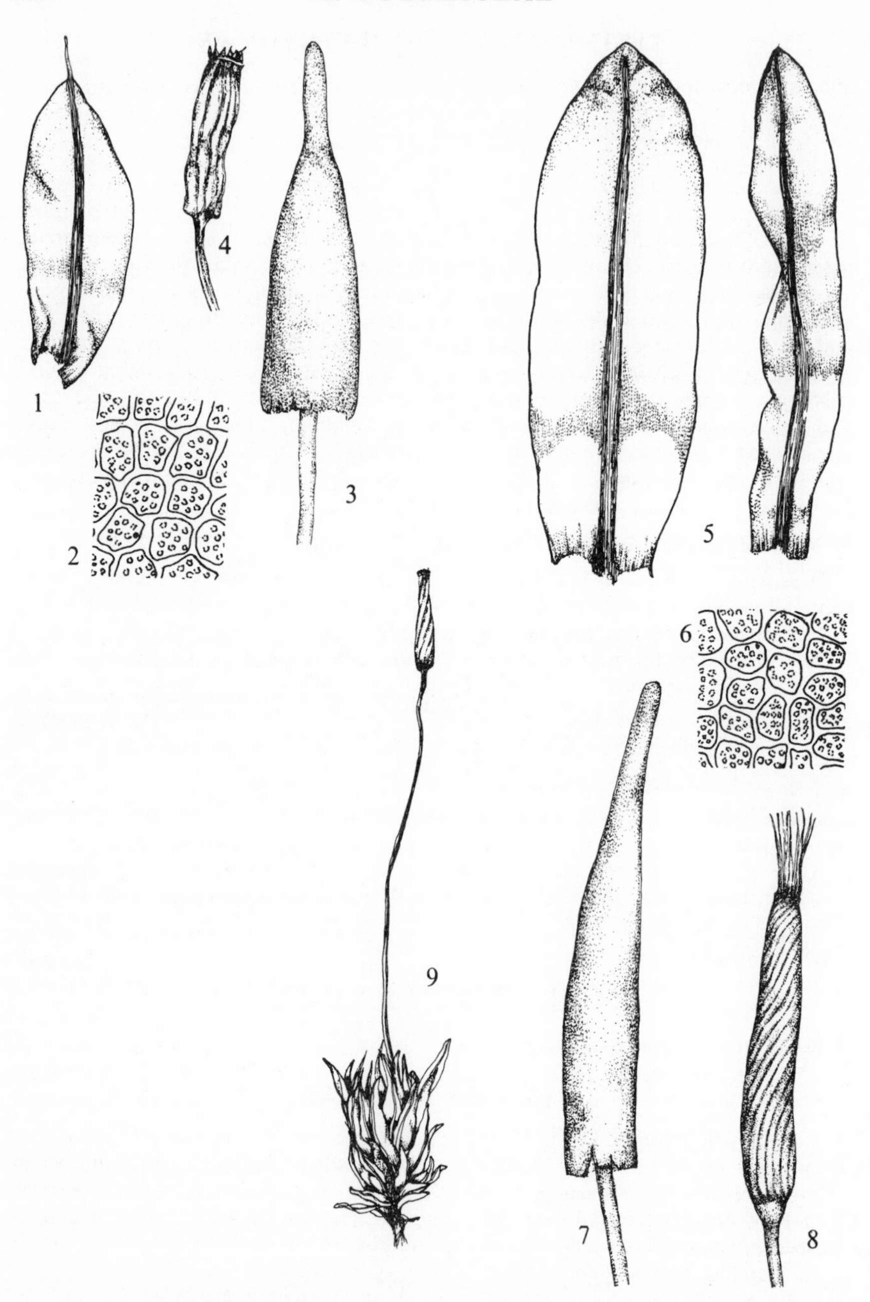

Fig. 94. 1–4, *Encalypta rhaptocarpa*: 1, leaf; 2, mid-leaf cells; 3, calyptra; 4, capsule. 5–9, *E. streptocarpa*: 5, leaves; 6, mid-leaf cells; 7, calyptra; 8, capsule; 9, plant (×2.5). Leaves ×15, cells ×415, calyptras and capsules ×10.

SUBFAMILY 1. POTTIOIDEAE

Plants acrocarpous. Lvs ovate or spathulate, rarely narrow; nerve often excurrent; basal cells rectangular, hyaline, cells above hexagonal, pellucid or obscure; peristome of 16 or 32 teeth sometimes divided into filiform segments, sometimes united into tube at base or absent.

40. TORTULA HEDW., SP. MUSC., 1801

Dioecious, autoecious, rarely synoecious or polyoecious. Plants forming greenish, sometimes hoary patches or cushions. Stems usually with central strand. Upper lvs often larger than lower, forming comal tuft, lingulate or spathulate, apex usually obtuse or rounded, margin plane or recurved, usually entire or papillose-crenulate; nerve usually excurrent, sometimes longly so, in a hyaline hair-point, in section with 2–4 large median guide cells, 1 to several rows of large ventral cells and 2 or more rows dorsal stereids; cells in lower part of lf rectangular, smooth, hyaline, above rounded-hexagonal or quadrate, papillose, chlorophyllose, marginal cells sometimes differentiated into yellowish border. Seta straight; capsule ± erect, ellipsoid or cylindrical, straight or slightly curved; lid longly rostrate; peristome teeth 32, spirally twisted, sometimes fused for up to $\frac{3}{4}$ their length into spirally twisted tube. A cosmopolitan but mainly temperate genus of about 280 species.

1 Nerve excurrent in hyaline or rarely reddish hair-point 2
Nerve ending below apex or excurrent in yellowish or greenish point 11

2 Lvs with incurved or involute margins, upper with 2–several-celled gemmae on upper surface of nerve **11. T. papillosa**
Lf margin plane or recurved, gemmae lacking or present at stem apex only 3

3 Hair-point denticulate to spinulose, plants (0.2–)1.0–10.0 cm 4
Hair-point ± smooth, plants 0.2–1.5(–3.0) cm 8

4 Hair-point reddish except sometimes at tip, margin recurved to $\frac{2}{3}$–$\frac{3}{4}$ way up lf **2. T. norvegica**
Hair-point hyaline except sometimes at base, recurvature of margin various 5

5 Lvs squarrose when moist, not or scarcely constricted at or below middle, margin recurved almost to apex **1. T. ruralis**
Lvs erecto-patent to spreading, constricted at or below middle, margin recurved to $\frac{1}{2}$–$\frac{2}{3}$ way up lf 6

6 Plants rarely more than 2 cm tall, cells in lower part of lf rectangular, 20–40 μm long, nerve in section with 1–2 rows of stereids **5. T. virescens**
Plants 1–4 cm tall, cells in lower part of lf narrowly rectangular, 50–80 μm long, nerve in section with several rows of stereids 7

7 Dioecious, lf cells 8–10(–12) μm wide at widest part of lf **3. T. intermedia**
Synoecious, cells (10–)12–20(–22) μm wide **4. T. princeps**

8 Margin plane, cells smooth, 16–24 μm wide at widest part of lf **14. T. cuneifolia**
Margin recurved, cells papillose, 10–16 μm wide 9

9 Lf margin revolute almost from base to apex, peristome teeth free almost to base **8. T. muralis**
Lf margin plane or recurved to $\frac{3}{4}$ way up lf, peristome teeth fused for $\frac{1}{2}$–$\frac{3}{4}$ their length 10

10 Plants 10–30 mm tall, lvs twisted when dry, constricted below middle **6. T. laevipila**
Plants 2–5 mm, lvs flexuose when dry, not or hardly constricted below middle **7. T. canescens**

11 Lvs with distinct border of elongated narrow cells, at least below 12
Lvs unbordered or if bordered then border cells of similar shape but of different appearance from other lamina cells 13

12 Lvs mostly 3.5–7.0 mm long, nerve ending in apex or excurrent in mucro, lower $\frac{3}{4}$ of peristome tubular **10. T. subulata**
Lvs less than 3 mm long, nerve excurrent in yellowish point to $\frac{1}{4}$ length of lamina, peristome teeth free almost to base **9. T. marginata**

13 Margin plane, cells smooth 14
Margin plane or recurved, cells papillose 15

14 Autoecious, nerve percurrent or excurrent in straight point, cell walls pale, rhizoidal gemmae absent **14. T. cuneifolia**
Dioecious, nerve excurrent in reflexed apiculus, cell walls brownish, rhizoidal gemmae present **17. T. rhizophylla**

15 Nerve excurrent 16
Nerve ending in or below apex 17

16 Margin revolute almost from base to apex, cells 8–16 μm wide at widest part of lf **8. T. muralis** var. **aestiva**
Margin plane or narrowly recurved near middle of lf, cells 16–20 μm wide **13. T. vahliana**

17 Lf apex rounded, emarginate or not, small ± spherical gemmae usually present on upper surface of lvs **12. T. latifolia**
Lf apex acute to obtuse, apiculate or not, gemmae not present on surface of lvs 18

18 Dioecious, lf margin recurved, rhizoidal gemmae abundant **16. T. amplexa**
Autoecious, lf margin plane, rhizoidal gemmae lacking **15. T. freibergii**

1. T. ruralis (Hedw.) Gaertn., Meyer & Scherb., Fl. Wetterau, 1802

Dioecious. Lvs appressed, ± twisted when dry, squarrose when moist, oblong-spathulate, not or hardly constricted at or below middle, apex rounded, sometimes emarginate, to acute, margin strongly recurved almost from base to apex, papillose-crenulate; nerve stout, brownish, excurrent in denticulate to spinulose hair-point to as long as lamina; cells in lower part of lf rectangular, hyaline, shorter and narrower towards margin, above smaller, hexagonal, strongly papillose, obscure, 12–16 μm wide at widest part of lf. Capsule cylindrical, slightly curved; lower $\frac{2}{3}$–$\frac{3}{4}$ of peristome tubular; spores 10–12 μm. Fr occasional, spring, early summer.

Ssp. **ruralis**

Plants 1–8 cm. Lf apex rounded, sometimes emarginate. Loose cushions, golden-green above, reddish below when moist, reddish-brown when dry on wall tops, stony ground, roofs, occasionally on sand-dunes, very rarely on trees. Frequent in calcareous districts, occasional elsewhere. 103, H33, C. Europe, Iceland, Asia, N. and southern Africa, N. America, Greenland, Patagonia, Australia, Oceania.

Ssp. **ruraliformis** (Besch.) Dix., Stud. Handb. Brit. Moss., 1896

T. ruraliformis (Besch.) Ingham, *T. ruralis* var. *arenicola* Braithw.

Plants mostly 1–4 cm but sometimes almost buried in sand. Lf apex tapering into hair-point. $n=13$*. Often extensive and sometimes locally dominant lax patches,

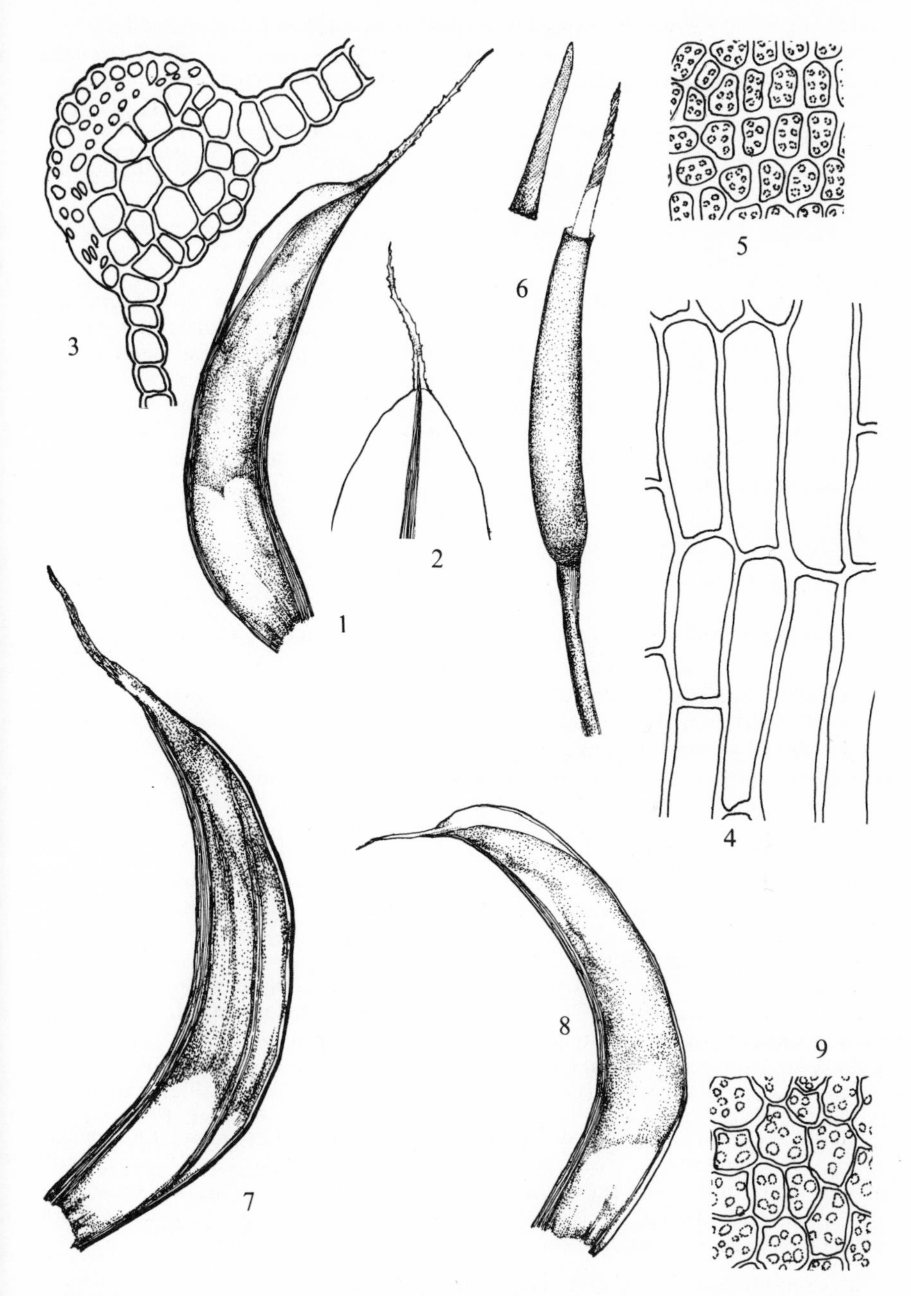

Fig. **95**. 1–6, *Tortula ruralis* ssp. *ruralis*: 1, leaf; 2, leaf apex (×100); 3, nerve section (×250); 4, basal cells; 5, cells from widest part of leaf; 6, capsule (×10). 7, *T. ruralis* ssp. *ruraliformis*: leaf. 8–9, *T. norvegica*: 8, leaf; 9, cells from widest part of leaf. Leave ×25, cells ×415.

yellowish-green to golden-brown when moist, brownish when dry on sand-dunes, particularly where calcareous and on sandy soil, very common on coastal sand-dunes, rare or occasional elsewhere. 76, H21, C. Europe, Faroes, W. Asia, Algeria, N. America.

Although usually distinct in ecology and lf morphology, intermediates between the two subspecies do sometimes occur. Ssp. *ruralis* is sometimes confused with *T. intermedia* but differs in the lvs squarrose when moist, the margin recurved to near the apex and the lvs not or scarcely constricted below the middle.

2. T. norvegica (Web. f.) Wahlenb. ex Lindb., Öfv. K. V. A. Förh., 1864
T. ruralis var. *alpina* Wahlenb.

Dioecious. Plants 1–10 cm. Lvs similar to *T. ruralis* but apex more tapering, margin recurved only to $\frac{2}{3}$–$\frac{3}{4}$ way up lf; hair-point denticulate, reddish throughout or hyaline only at apex; cells 12–16(–20) μm wide at widest part of lf. Capsule shortly cylindrical; spores 10–14 μm. Fr unknown in Britain. Lax cushions, yellowish-green above, reddish-brown below when moist, reddish-brown when dry, amongst boulders at high altitudes in the Scottish mountains, very rare. Mid Perth, S. Aberdeen, W. Inverness, 3. Montane parts of Europe, Corsica, Caucasus, N. Asia, Kashmir, Madeira, N. America, Greenland.

3. T. intermedia (Brid.) De Not., Syll., 1838
T. ruralis var. *crinita* De Not.

Dioecious. Plants 1–4 cm, Lvs appressed, incurved or twisted when dry, erecto-patent to spreading, rarely sub-squarrose when moist, broadly spathulate, constricted at or below the middle, apex rounded, sometimes emarginate, margin recurved to $\frac{1}{2}$–$\frac{2}{3}$ way up lf, papillose-crenulate; nerve strong, reddish, excurrent in denticulate, hyaline hair-point to $\frac{3}{4}$–1 length of lamina; cells in lower part of lf narrowly rectangular, hyaline, mostly 50–80 μm long, shorter and narrower towards margin, above hexagonal, papillose, obscure, 8–10(–12) μm wide at widest part of lf. Capsule shortly cylindrical, slightly curved or not, 2–3 mm long; lower $\frac{2}{3}$ of peristome tubular; spores 14–16 μm. Fr occasional, spring. $n=12$, 12+1*, 13*. Lax cushions, greenish when moist, greyish-brown, hoary when dry, on rocks, walls, roof-tops and occasionally on soil, frequent in but not restricted to calcareous habitats. Occasional to frequent throughout most of Britain and Ireland. 100, H30, C. Europe, W. Asia, Punjab, Azores, Madeira, southern Africa, N. America.

4. T. princeps De Not., Mem. R. Acc. Sc. Torino, 1838

Synoecious. Plants 1–4 cm. Lvs in interrupted tufts on stems, flexuose, twisted when dry, patent or recurved when moist, oblong-spathulate, constricted at or below middle, apex rounded, sometimes emarginate, margin recurved to $\frac{1}{2}$–$\frac{3}{4}$ way up lf; nerve stout, reddish-brown, excurrent in hyaline, denticulate hair-point; areolation as in *T. intermedia* but cells larger, (10–)12–20(–22) μm wide in widest part of lf. Capsule cylindrical, slightly curved, 3.5–4.5 mm long; spores 12–14 μm. Fr frequent, autumn. $n=12$, 24+1, 26, 36+2. Dense, spreading tufts or patches, green above, reddish-brown below, on basic rocks and cliffs in montane habitats, occasional. Shrops, N. Wales and Derby north to Sutherland, N. Ireland. 23, H3. Europe, Caucasus, Turkestan, Himalayas, Algeria, Morocco, Madeira, Canaries, N. and S. America, New Zealand, Hawaii.

5. T. virescens (De Not.) De Not., Musc. Ital., 1862
T. pulvinata (Jur.) Limpr.

Similar to *T. laevipila*. Dioecious. Plants 0.2–2.0(–3.0) cm. Lvs as in *T. laevipila* but

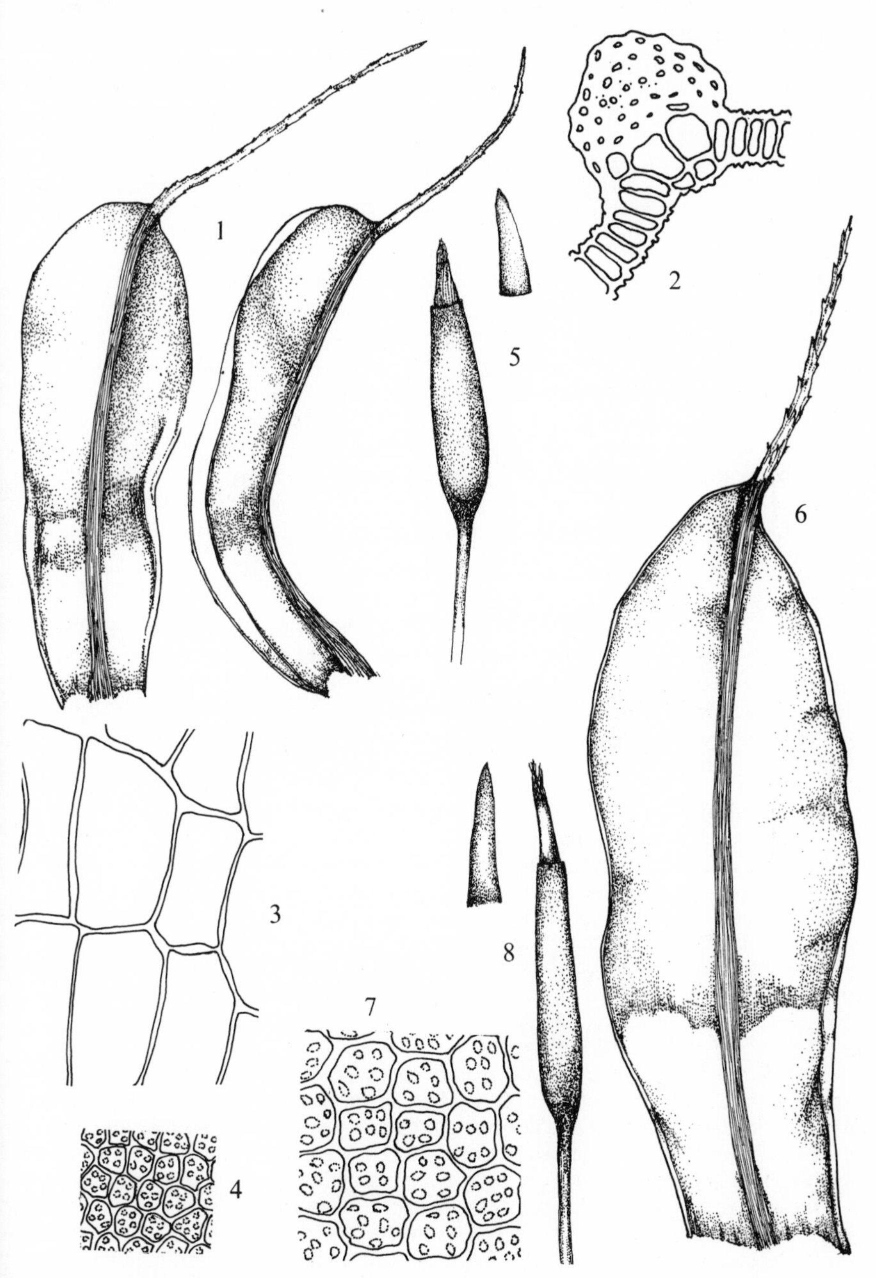

Fig. **96**. 1–5, *Tortula intermedia*: 1, leaves; 2, nerve section (×250); 3, basal cells; 4, cells from widest part of leaf; 5, capsule. 6–8, *T. princeps*: 6, leaf; 7, leaf cells; 8, capsule. Leaves ×25, cells ×415, capsules ×10.

margin crenulate with projecting cells; nerve in section with only 1–2 rows stereid cells, excurrent in hyaline, denticulate hair-point; cells in basal part of lf 20–40 μm long. Capsule similar to *T. laevipila*; spores 8–11 μm. Fr unknown in Britain. Dull green patches on bark where shade is not dense, rarely on rocks or soil, rare. Sussex, Cambridge and Hereford north to E. Ross. 12. Europe, N. Africa.

This plant is said to produce gemmae occasionally (Demaret & Castagne, 1964), but these have not been detected in British material. When this species occurs on rocks it may be confused with *T. intermedia* but differs in the smaller cells in the basal part of the lf. For the occurrence of this plant in Britain see Warburg & Crundwell, *Trans. Br. bryol. Soc.* **3**, 568–70, 1959.

6. T. laevipila (Brid.) Schwaegr., Suppl. 2, 1823

Plants mostly 1.0–1.5(–3.0) cm. Lvs twisted, incurved when dry, spreading to recurved when moist, broadly spathulate, constricted below middle, apex rounded, margin plane or recurved near middle of lf, papillose-crenulate; nerve stout, excurrent in hyaline, smooth to obscurely and sparsely denticulate hair-point, $\frac{1}{4}$–1 length of lamina, in section with several rows of stereid cells; cells in lower part of lf rectangular, narrower towards margin, above quadrate-hexagonal, papillose, 10–14 μm wide at widest part of lf, 0–4 marginal rows more translucent and incrassate, less papillose, forming yellowish marginal band. Capsule cylindrical, slightly curved; peristome tubular in lower half; spores 14–20 μm. Fr occasional to frequent, late summer, autumn.

Var. **laevipila**

Autoecious. Lvs with 0–4 rows of marginal cells forming yellowish marginal band. Plants without gemmae. Fr occasional to frequent. $n=12$, 15, 26*. Light or yellowish-green tufts or patches on bark of trees and shrubs where shade is not dense. Frequent throughout most of Britain and Ireland but rare in N. Scotland. 105, H35, C. Europe, W., C. and E. Asia, Algeria, Morocco, Canaries, Azores, N. America, southern S. America, Australia.

Var. **laevipiliformis** (De Not.) Limpr., Laubm., 1888

Dioecious. Lvs with *ca* 4 rows of cells forming yellowish marginal band. Gemmae in the form of minute lvs present at stem apices. Fr unknown. In similar habitats to the type but rarer, extending north to Stirling and Ayr. 31, H7. Europe.

The taxonomic status of var. *laevipiliformis* is not clear and there is some difference of opinion as to whether it is worth maintaining as a variety or whether it should be raised to specific status. *T. virescens* may be overlooked for *T. laevipila* but differs in the characters indicated under that species.

7. T. canescens Mont., Arch. Bot., 1833

Autoecious. Plants 2–5 mm. Lvs erect, flexuose, scarcely twisted when dry, erecto-patent to patent when moist, from broad base broadly ovate to lanceolate or spathulate, not or scarcely constricted at or below middle, apex ± obtuse, margin plane or recurved at middle of lf; nerve strong, excurrent in smooth, hyaline hair-point to $\frac{1}{2}$–$\frac{2}{3}$ length of lamina; cells in lower part of lf rectangular, above hexagonal, papillose, not incrassate, 12–16 μm wide at widest part of lf. Seta reddish; capsule cylindrical, peristome tubular for $\frac{1}{3}$–$\frac{1}{2}$ its length; spores 14–16 μm. Fr common, late winter, spring. $n=26$*. Dense patches or scattered stems on acidic soil in turf and on and amongst rocks, rare. Cornwall, Devon, Radnor, Pembroke, Montgomery, Merioneth, Kintyre, Jersey, Guernsey. 8, C. W., C. and S. Europe, Algeria, Canaries, Madeira, Caucasus.

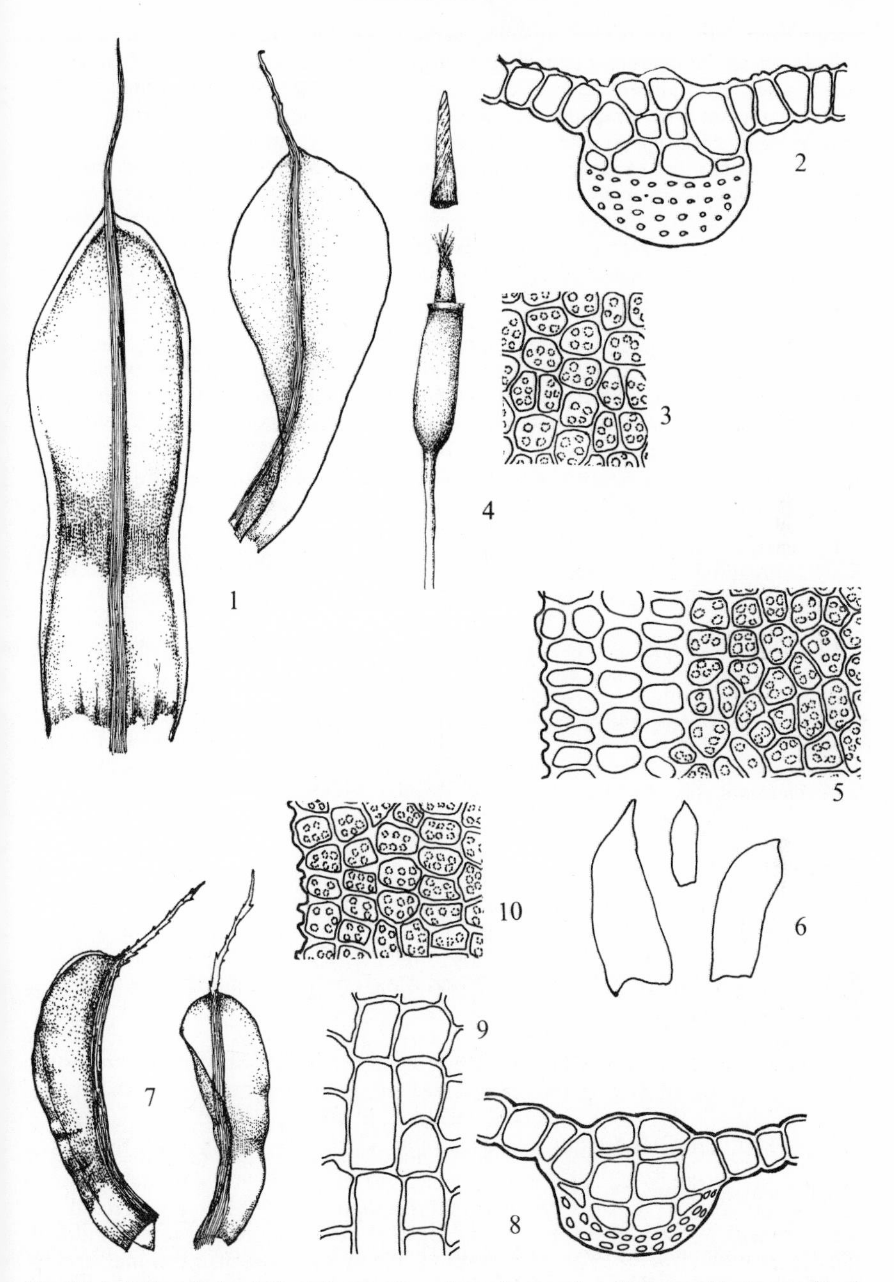

Fig. **97**. 1–4, *Tortula laevipila*: 1, leaves; 2, section of nerve; 3, cells from widest part of leaf; 4, capsule (×10). 5–6, *T. laevipila* var. *laevipiliformis*: 5, leaf cells; 6, gemmae (×62.5). 7–10, *T. virescens*: 7, leaves; 8, nerve section; 9, basal cells; 10, cells from widest part of leaf. Leaves ×25, sections ×250, cells ×415.

8. T. muralis Hedw., Sp. Musc., 1801

Autoecious. Plants rarely exceeding 1 cm. Lvs twisted or curved when dry, erecto-patent when moist, lingulate or lingulate-spathulate, apex obtuse or rounded, sometimes emarginate, occasionally sub-acute, margin strongly recurved almost from base to apex; nerve stout, often reddish, excurrent in ± smooth, hyaline hair-point to as long as lamina, rarely only shortly excurrent in yellowish point; cells in basal part of lf rectangular, smaller towards margin, above quadrate-hexagonal, incrassate, papillose, 8–16(–20) μm wide at widest part of lf, 1–2 marginal rows less papillose, more incrassate. Seta purple; capsule erect, cylindrical or narrowly ellipsoid, ± symmetrical; peristome teeth free almost to base; spores 7–12(–14) μm, Fr common, spring, summer.

Var. **muralis**

Nerve excurrent in ± smooth, hyaline hair-point to as long as lamina. $n=24$, 26*, 26+1*, 27*, *ca* 40, 48, 50*, 52*, 55*, 60, 66. Small tufts, patches or rounded hoary cushions, greyish when dry, green when moist, on mortar, concrete, basic rocks and walls, rarely on acidic rocks, very rarely on trees or hard-packed soil. Very common in man-made habitats, occasional elsewhere. 112, H40, C. Cosmopolitan.

Var. **aestiva** Hedw., Sp. Musc., 1801

Lvs narrowly lingulate-spathulate; nerve excurrent in short yellowish-green point. Short, spreading, bright green patches on shaded rocks and walls. Occasional in England from Devon and Kent north to Cheshire and Northumberland, very rare elsewhere, Denbigh, Wigtown, Berwick, Down. 36, H1. Europe, Asia, N. America.

Var. *aestiva* is distinctive in its appearance and is probably more than just a habitat form. *T. muralis* is variable in size but there is no constant correlation between chromosome number and morphology (see Newton, *Trans. Br. bryol. Soc.* **5**, 523–35, 1968). *T. canescens* has less strongly recurved lf margins, the peristome tubular for ⅓–½ its length and larger spores.

9. T. marginata (Br. Eur.) Spruce, Lond. J. Bot., 1845

Dioecious. Plants 2–3 mm. Lvs twisted when dry, erecto-patent when moist, narrowly lanceolate or lingulate to narrowly lingulate-spathulate, occasionally wider, spathulate, apex obtuse or acute, margin plane, bordered, sinuose with projecting cell walls; nerve yellowish, excurrent in yellowish point to ¼ length of lamina; cells in lower part of lf narrowly rectangular, above shortly rectangular to irregularly quadrate-hexagonal, papillose, not incrassate, 8–14 μm wide in widest part of lf, 2–4 marginal rows long, incrassate, forming distinct border extending almost from base to apex. Seta orange-red; capsule shortly cylindrical, ± symmetrical; peristome teeth free almost to insertion; spores *ca* 8 μm. Fr common, spring. Yellowish-green patches on damp, usually shaded, basic rocks and walls. Locally common in S. England, rare elsewhere, extending north to Cheshire and Durham, Argyll, Dumbarton, a few localities in Ireland. 43, H7, C. S. and W. Europe, Mediterranean Islands, W. Asia, India, Algeria, Morocco, Madeira, Azores.

10. T. subulata Hedw., Sp. Musc., 1801

Autoecious. Plants rarely more than 1 cm. Lvs incurved, twisted when dry, spreading when moist, very variable in shape, narrowly lanceolate or narrowly lingulate or spathulate or ± ovate, apex obtuse to acute or acuminate, margin recurved below, bordered at least below, entire or obscurely toothed; nerve stout, percurrent to excurrent in mucro; cells in lower part of lf rectangular, hyaline, at margin narrower with marginal band of narrow incrassate cells extending up margin

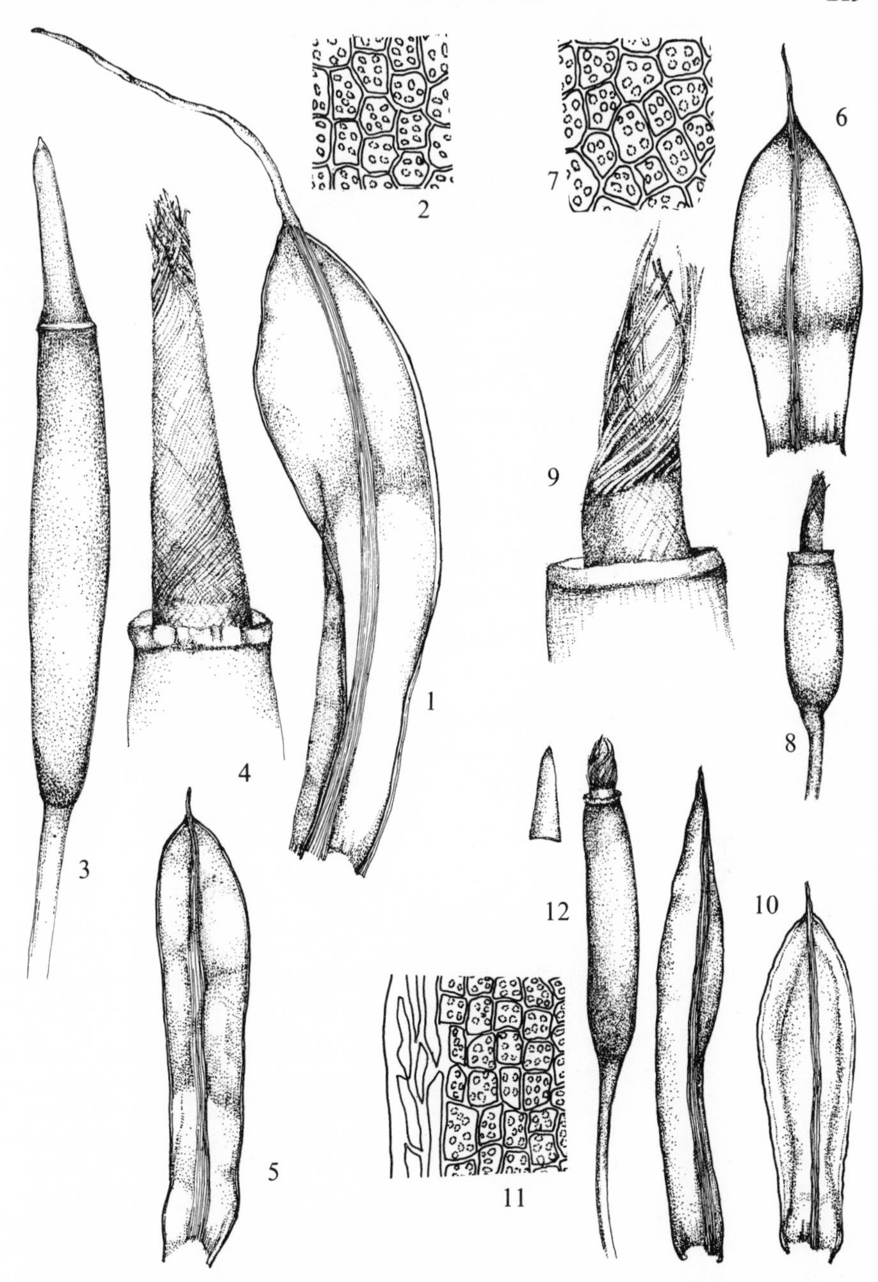

Fig. **98**. 1–4, *Tortula muralis* var. *muralis*: 1, leaf; 2, cells from widest part of leaf; 3, capsule; 4, peristome. 5, *T. muralis* var. *aestiva*: leaf. 6–9, *T. canescens*: 6, leaf; 7, cells from widest part of leaf; 8, capsule; 9, peristome. 10–12, *T. marginata*: 10, leaves; 11, leaf cells; 12, capsule. Leaves ×25, cells ×415, capsules ×15, peristomes ×62.5.

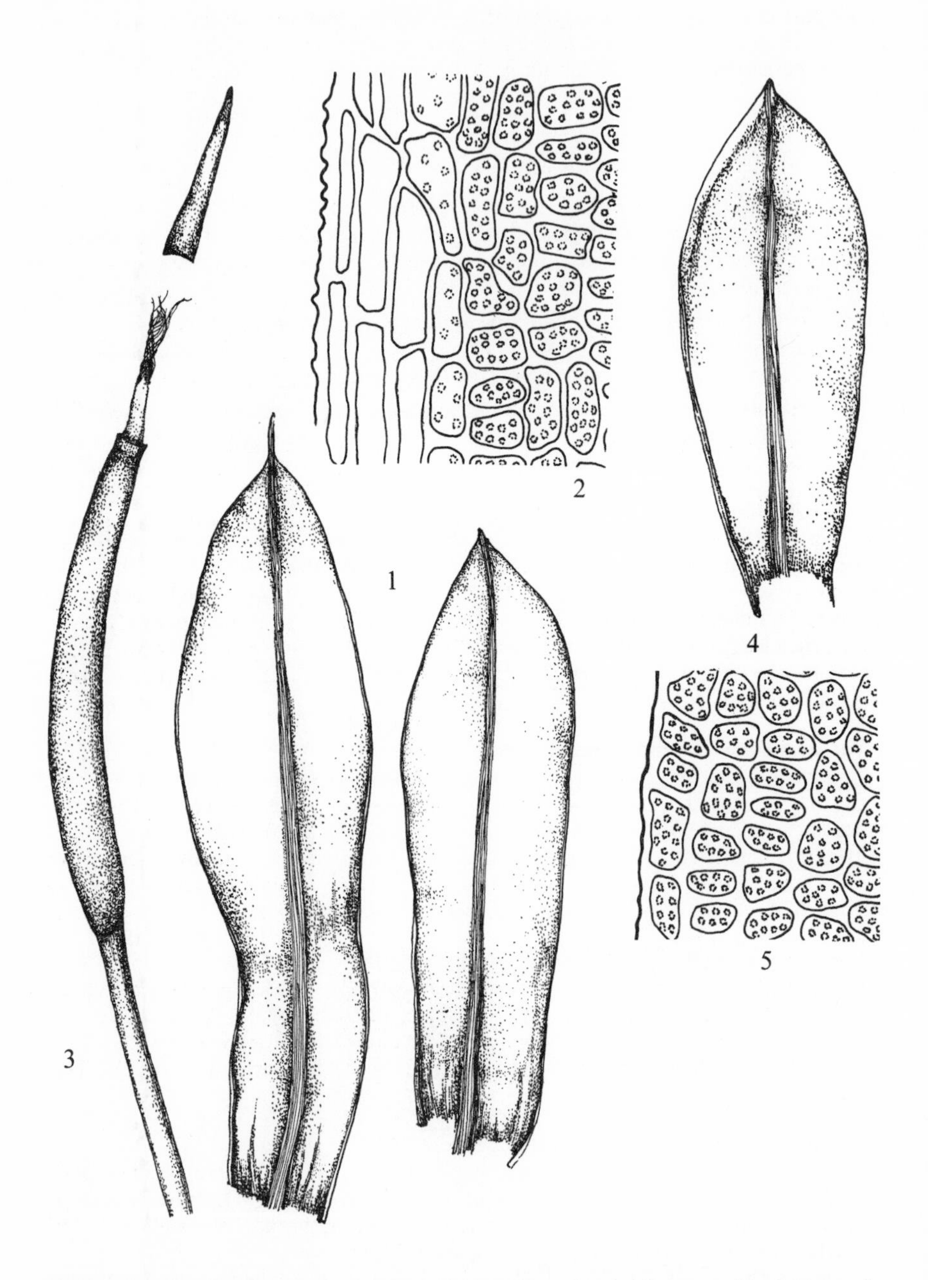

Fig. 99. 1–3, *Tortula subulata* var. *subulata*: 1, leaves; 2, cells at widest part of leaf; 3, capsule (×10). 4–5, *T. subulata* var. *subinermis*: 4, leaf; 5, cells from widest part of leaf. Leaves ×15, cells ×415.

to half way or more and sometimes forming conspicuous yellowish border, cells in upper part of lf irregularly hexagonal or ± quadrate, papillose, obscure, variable in size, 12–28 μm wide in widest part of lf. Seta reddish; capsule cylindrical, slightly curved; peristome tubular for about $\frac{3}{4}$ of its length; spores 10–28 μm. Fr common, spring.

Key to intraspecific taxa of *T. subulata*

1 Lf apex acute or acuminate, border poorly developed and barely reaching to mid-way, cells obscurely papillose, 16–28 μm wide at widest part of lf — var. **graeffii**
Lf apex rounded to acuminate, border variously developed, cells markedly papillose, (10–)12–20(–22) μm wide — 2
2 Lvs narrowly lingulate-lanceolate, acuminate, border strongly developed, extending almost to apex — var. **angustata**
Lvs wider, apex rounded to acuminate, border extending $\frac{1}{2}$–$\frac{3}{4}$ way up lf — 3
3 Lvs spathulate, apex rounded or obtuse and apiculate, border poorly developed, usually ending midway up lf — var. **subinermis**
Lvs usually lingulate, acute, border variable — var. **subulata**

Var. **subulata**
Lvs narrowly lingulate to spathulate or ovate-lanceolate, apex obtuse to acuminate, margin bordered to $\frac{1}{2}$–$\frac{3}{4}$ way up lf, entire or obscurely toothed; cells 12–20(–22) μm wide at widest part of lf. Spores 10–20 μm. $n = 13+1$*, 26*, $48+1$, 60. Patches or scattered plants amongst other mosses on light soil in open or shaded places, hedge-banks, rock ledges, rarely on wood, in acidic or basic habitats, generally distributed, occasional to frequent. 109, H28. Europe, Faroes, W. Asia, Kashmir, China, Algeria, Canaries, N. America.

Var. **angustata** (Schimp.) Limpr. in Rab., Krypt.-Flora, 1888
T. angustata Lindb.
Lvs narrowly lanceolate, tapering to acuminate apex, margin strongly bordered almost to apex, irregularly and obscurely toothed; cells strongly papillose, mostly 12–16 μm wide. Spores 10–12 μm. On light, usually acidic soil in shady places, rare. From Devon and Hants north to the Lothians, Orkney, Shetland, Down. 30, H1. Europe, S.W. Asia, Kashmir, China, Algeria, Canaries, N. America.

Var. **subinermis** (Brid.) Wils., Br. Brit., 1855
Lvs spathulate, apex obtuse or rounded, apiculate, border poorly developed, ceasing at about mid-lf; cells strongly papillose, 12–16 μm wide. Spores *ca* 16 μm. On soil, tree bases and exposed roots by streams and rivers, rare. Scattered localities from Dorset and Sussex north to Stirling, Fife and Mid-Perth, Leitrim. 24, H1. Spain, Algeria.

Var. **graeffii** Warnst., Krypt, Fl. Brandenb., 1904
Lvs acute to acuminate, border poorly developed, ending at about mid-lf; cells obscurely papillose, 16–28 μm wide. Spores 14–20 μm. On light calcareous soil in western and northern Britain, from Somerset, Monmouth and Derby north to W. Sutherland, Shetland, rare to occasional, Cavan. 25, H1.

A species exceedingly variable in lf characters and spore size as well as in chromosome number. The nature of this variation requires further investigation. Var. *graeffii* retains its characters in cultivation but the situation with the other varieties is unknown.

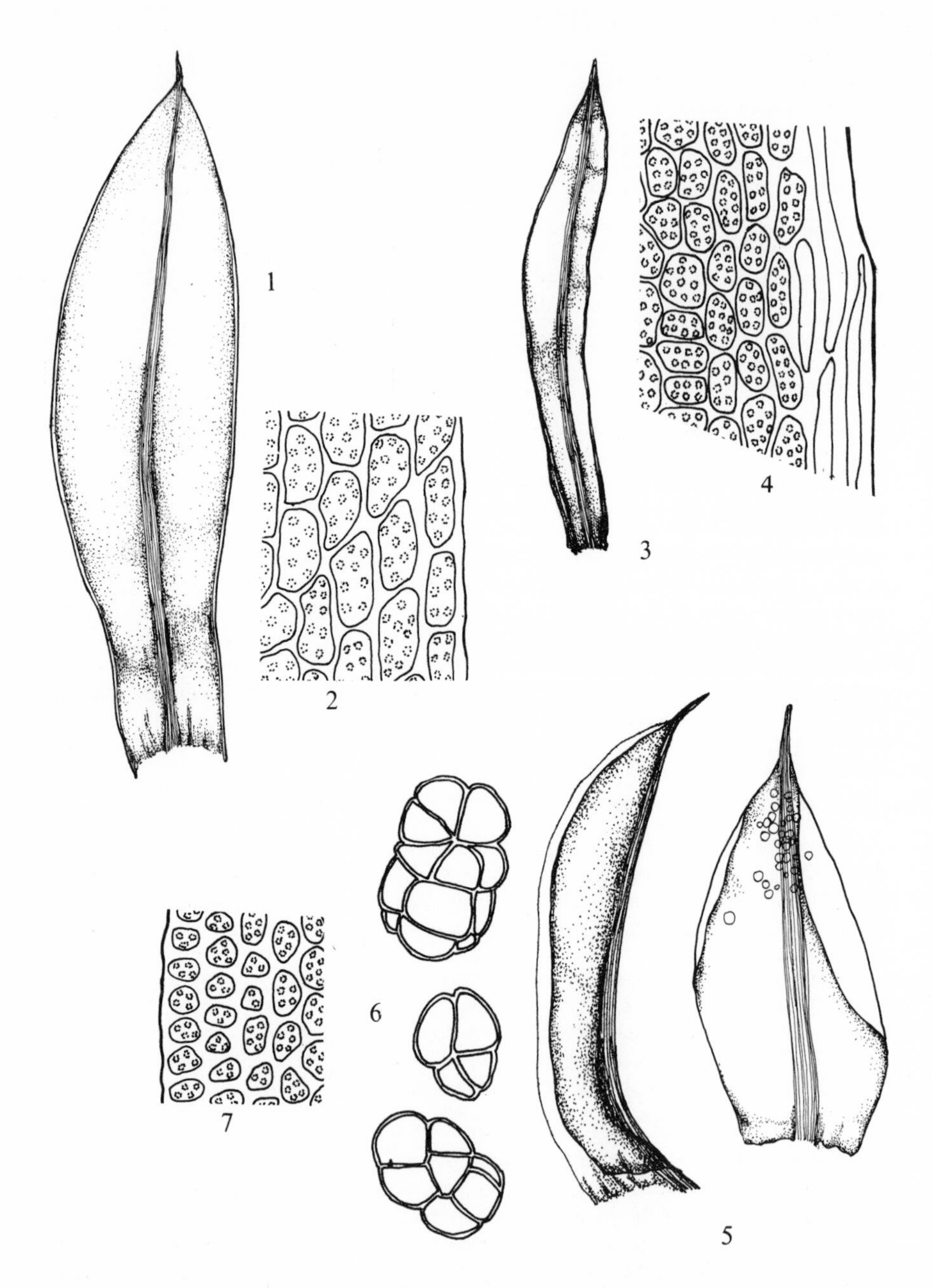

Fig. **100**. 1–2, *Tortula subulata* var. *graeffii*: 1, leaf (×15); 2, cells from widest part of leaf. 3–4, *T. subulata* var. *angustata*: 3, leaf (×15); 4, cells from widest part of leaf. 5–7, *T. papillosa*: 5, leaves (×25); 6, gemmae (×250); 7, cells from widest part of leaf. Cells ×415.

11. T. papillosa Wils. ex Spruce, Lond. J. Bot., 1845

Dioecious. Plants 2–10 mm. Lvs incurved when dry, spreading or recurved when moist, broadly obovate-spathulate, constricted below middle, apex rounded, margin inflexed above; nerve broad, strongly papillose at back above, excurrent in smooth, hyaline hair-point $\frac{1}{8}$–$\frac{1}{2}$ length of lamina; cells in lower part of lf rectangular, hyaline, cells above quadrate-hexagonal, papillose, 16–24 μm wide at widest part of lf. Irregularly spherical or ovoid, 2–several-celled gemmae occur on upper part of nerve in younger lvs. Lower third of peristome tubular. Fr known only from Australia, Tasmania and New Zealand. $n=12$, 24. Small, dark green patches or tufts on bark, particularly of mature trees, very rarely on rocks or walls, occasional and decreasing in areas subject to atmospheric pollution. 80, H11, C. Europe, southern Africa, N. America, Ecuador, Andes, Tierra del Fuego, Falkland Is., Australia, Tasmania, New Zealand.

12. T. latifolia Bruch ex Hartm., Skand. Fl., 1832
T. mutica Lindb.

Dioecious. Plants to 3 cm. Lvs ± appressed or incurved when dry, lower erecto-patent, upper crowded, soft, spreading when moist, broadly spathulate, apex rounded, sometimes emarginate, margin plane or narrowly recurved below, papillose-crenulate, often eroded; nerve strong, brownish, ending below apex to percurrent; cells in lower part of lf rectangular, smaller and shorter towards margin, above hexagonal, papillose, pellucid, 12–16 μm wide at widest part of lf. Small, ± spherical gemmae, 24–36 μm diameter, produced on upper surface of lvs. Capsule cylindrical, slightly curved; lower $\frac{1}{3}$ of peristome tubular; spores 10–12 μm. Fr very rare, spring. $n=12$. Dull or brownish-green, often silt-encrusted patches on tree bases, exposed roots and wood in flood zone of streams and rivers, very rarely on rocks or gravelly soil or away from water. Occasional to frequent in England and Wales, rare in Scotland, extending north to Perth and Banff, occasional in Ireland, 75, H11. N., W. and C. Europe, Caucasus, N. America.

13. T. vahliana (Schultz) Mont. in Gay, Hist. Fis. Polit. Chil. Bot., 1850

Autoecious. Plants to 5 mm. Lvs erect, twisted when dry, thin, soft, spreading when moist, narrowly lingulate to spathulate, apex rounded, margin plane or recurved at middle, papillose-crenulate; nerve excurrent in smooth, greenish point to *ca* 320 μm long; cells in lower part of lf rectangular, above hexagonal, thin-walled, papillose, 16–20 μm wide, 1–2 marginal rows ± quadrate, less papillose. Seta red; capsule narrowly cylindrical, slightly curved; peristome teeth free almost to base; spores 12–16 μm. Fr occasional, spring. Bright green patches or scattered stems on shaded, clayey soil, rare. Scattered localities in S. England extending north to Hereford and N. Lincs, Wicklow, Dublin, Leitrim, Antrim. 13, H4. S. and W. Europe, W. Asia, Algeria, Morocco, Canaries, southern S. America.

14. T. cuneifolia (With.) Turn., Musc. Hib., 1804

Autoecious. Plants to *ca* 5 mm. Lvs shrunken, appressed when dry, lower erecto-patent, upper thin, soft, crowded, patent to erecto-patent when moist, broadly spathulate, concave, abruptly narrowed to acute apex, margin plane, entire; nerve percurrent to excurrent in smooth, yellowish or sometimes hyaline point to $\frac{1}{3}$ length of lamina; cells in lower part of lf rectangular, above irregularly quadrate-hexagonal to shortly rhomboidal, thin-walled, smooth, 16–24 μm wide at widest part of lf. Seta red; capsule cylindrical to narrowly ellipsoid, straight; peristome teeth free nearly to base; spores 14–18 μm. Fr common, spring. $n=20$*. Green patches or scattered

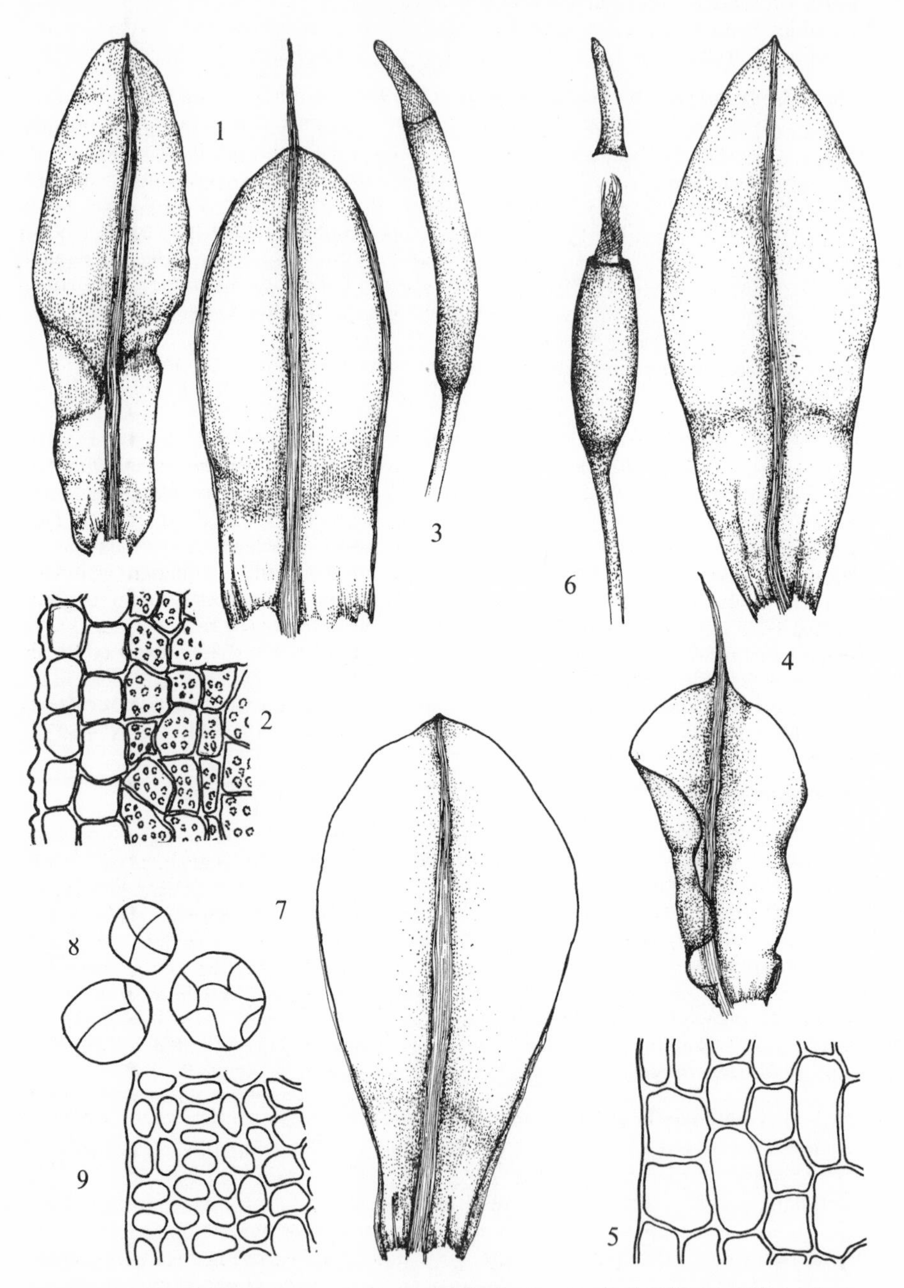

Fig. **101**. 1–3, *Tortula vahliana*: 1, leaves; 2, leaf cells; 3, capsule. 4–6, *T. cuneifolia*: 4, leaves; 5, leaf cells; 6, capsule. 7–9, *T. latifolia*: 7, leaf; 8, gemmae (×250); 9, leaf cells. Leaves ×25, cells ×415, capsules ×15.

stems on shaded rocks particularly near the coast. Rare to occasional in scattered localities from Cornwall east to Kent and north to Anglesey and Stafford, a few localities in Ireland. 18, H7, C. W. and S. Europe, Syria, Palestine, Algeria, Canaries.

15. T. freibergii Dix. & Loeske., Ann. Bryol., 1934

Autoecious. Plants to 2 mm. Lvs slightly twisted, incurved when dry, erecto-patent to spreading when moist, slightly concave, lingulate-lanceolate, apex acute to obtuse, margin plane, entire to crenulate; nerve ending below apex, rarely percurrent; cells in lower part of lf narrowly rectangular, above quadrate to rectangular, thin-walled, papillose but not obscure, 9–13 μm wide, smaller towards apex, cells near but not at margin often somewhat narrower and sometimes much elongated, incrassate. Seta pale yellow; capsule cylindrical, ± symmetrical; peristome teeth free almost to base; spores 8–11 μm. Fr common. Yellowish-green patches on vertical sandstone rocks and walls, very rare, Sussex. 1. Giglio Is., Italy.

For the occurrence of this plant in Britain see Crundwell & Nyholm, *J. Bryol.* **7**, 161–4, 1972.

16. T. amplexa (Lesq.) Steere in Grout, Moss Fl. N. Amer., 1939

Dioecious. Plants to 6 mm. Lvs ± twisted when dry, spreading when moist, slightly concave, lingulate-spathulate, apex obtuse, apiculate or not, margin recurved below, plane, entire above; nerve ending just below apex; basal cells rectangular, narrower towards margin, above irregularly quadrate, papillose, (13–)15–25(–28) μm wide, *ca* 4 marginal rows smaller, more incrassate, less papillose, forming yellowish border. Abundant rhizoidal gemmae always present, pale brown, of 1–3 thick-walled cells, 20–75 μm × 20–35 μm. Only female plants known in Britain. Dense green tufts, reddish-brown below, on disturbed clay, Leicester. 1. California, Washington, British Columbia.

Differs from *T. freibergii* in the less spathulate lvs with recurved margins, larger cells and rhizoidal gemmae. For the occurrence of this plant in Britain see Side & Whitehouse, *J. Bryol.* **8**, 15–18, 1974.

17. T. rhizophylla (Saki) Iwats. & Saito, Miscn. br. lichen, Nichinan, 1972
T. vectensis Warb. & Crundw.

Dioecious. Plants *ca* 3 mm. Lvs ± incurved when dry, sub-squarrose when moist, spathulate, ± rapidly narrowed into acute or acuminate apex, margin plane, sub-crenulate above; nerve excurrent in short reflexed apiculus; cells smooth, not incrassate, collenchymatous, shortly rectangular below, above irregularly hexagonal with brownish walls, very variable in size, 14–30 μm wide at widest part of lf, smaller towards apex, marginal cells often slightly transversely rectangular. Irregular, ± elongated brownish rhizoidal gemmae, 75–175 μm × 50–75 μm, abundant. Gametangia and capsules unknown in the wild. Green or brownish-green patches or scattered plants on basic soil in arable fields, I. of Wight. 1. Louisiana, Japan.

In cultivation plants from I. of Wight and Louisiana have proved to be female. Most likely to be mistaken for *Hyophila stanfordensis* or a *Pottia* species; differing from the former in the smooth cells and from the latter in the dioecious inflorescence and rhizoidal gemmae. For the occurrence of *T. rhizophylla* in Britain see Warburg & Crundwell, *Trans. Br. bryol. Soc.* **4**, 763–6, 1965.

41. ALOINA (C. MÜLL.) KINDB., BIH. K. SV. VET. AK. HANDL., 1882

Dioecious or polyoecious. Stems short. Lvs rigid, thick, usually incurved when dry, erect to spreading when moist, concave, frequently obtuse, margin inflexed, entire,

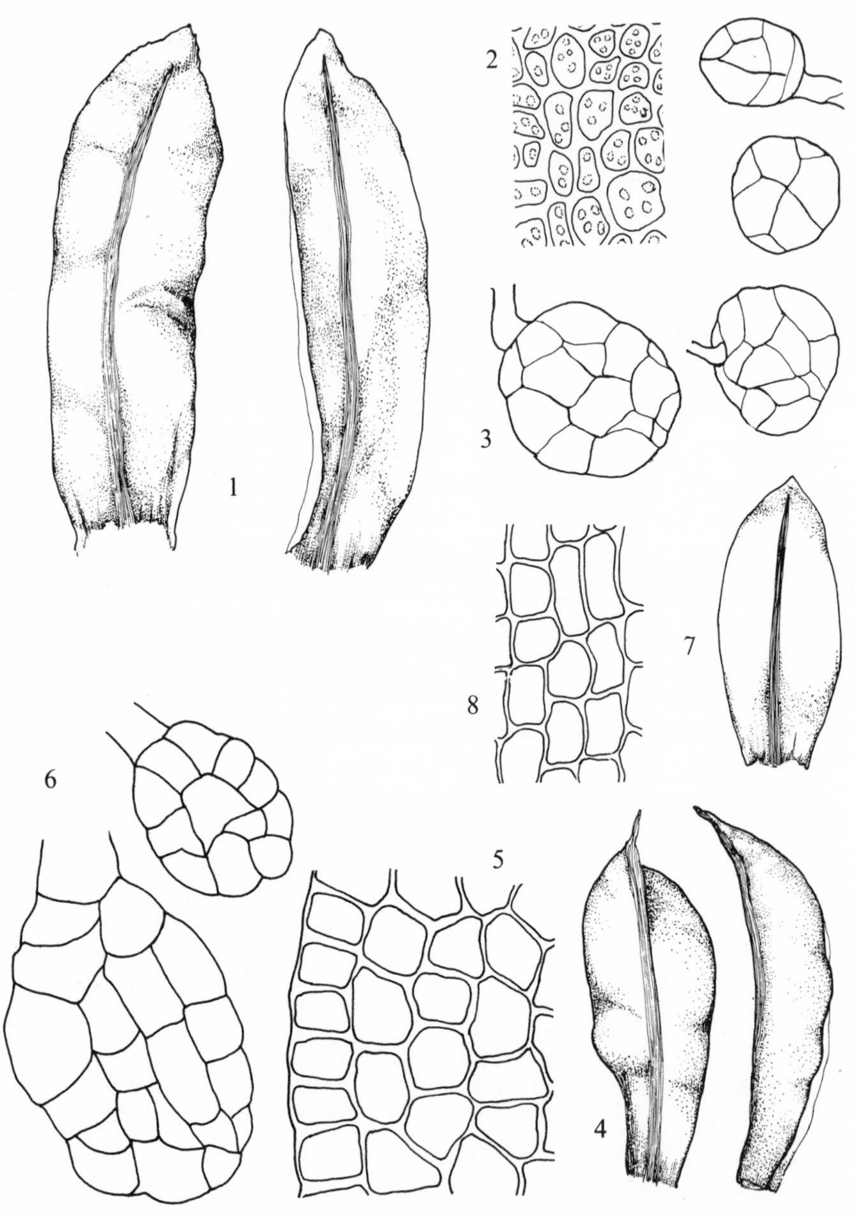

Fig. **102**. 1–3, *Tortula amplexa*: 1, leaves; 2, mid-leaf cells; 3, rhizoidal gemmae. 4–6, *T. rhizophylla*: 4, leaves; 5, marginal cells at margin of leaf; 6, rhizoidal gemmae. 7–8, *T. freibergii*: 7, leaf; 8, cells at widest part of leaf. Leaves ×25, cells ×415, gemmae ×250.

unbordered; nerve broad, usually ending in or below apex, with chlorophyllose, branched, granulose filaments on ventral surface of upper part; cells in lower part of lf rectangular, hyaline, above quadrate to transversely oblong or elliptical. Seta straight; capsule erect or inclined, ellipsoid to cylindrical, straight or curved; annulus persistent or not; peristome with short basal membrane and 32 spirally twisted teeth. A small ± world-wide genus of 9 species.

For a monograph of the genus see Delgadillo, *Bryologist*, **78**, 245–303, 1975.

1 Marginal cells near lf base elongate, hyaline, annulus separating 2
Marginal cells near lf base quadrate or rectangular, not hyaline, annulus persistent 3

2 Synoecious, lvs mostly 2.0–3.5 times as long as wide, spores 18–22 μm **1. A. brevirostris**
Dioecious, lvs mostly 4–6 times as long as wide, spores 14–16 μm **2. A. rigida**

3 Basal membrane of peristome not projecting above mouth of capsule, spores 18–25 μm **3. A. aloides** var. **aloides**
Basal membrane projecting above mouth of capsule, spores 12–16 μm **3. A. aloides** var. **ambigua**

1. A. brevirostris (Hook. & Grev.) Kindb., Bih. K. Sv. Vet. Ak. Handl., 1883
Tortula brevirostris Hook & Grev.

Synoecious and sometimes male only. Plants 1–2 mm. Lvs incurved when dry, erect to spreading when moist, very concave, lower lvs ± orbicular, upper broadly lingulate, 2.0–3.5 times as long as wide, lf base longer than lf blade, apex rounded or obtuse, cucullate; marginal cells near base elongate, hyaline. Capsule erect, narrowly ellipsoid, straight, 1.8–2.0 mm long; lid $\frac{1}{3}(-\frac{1}{2})$ length of capsule; annulus of large, separating, fugacious cells; spores finely papillose, 18–22 μm. Fr common, spring, summer. $n=28$*. Small, greenish or brownish patches on calcareous soil, very rare. Derby, Cheshire, Yorks, W. Lothian. 6 N., W. and C. Europe, Siberia, N. America, Greenland.

2. A. rigida (Hedw.) Limpr., Laubm., 1888
Tortula rigida (Hedw.) Schrad. ex Turn.

Dioecious. Plants *ca* 2 mm. Lvs incurved when dry, spreading when moist, mostly 4–6 times as long as wide, lingulate to lingulate-spathulate or lingulate-lanceolate, hyaline basal part shorter than rest of lf, apex rounded and cucullate or rarely obtuse and mucronate; nerve ending below apex or rarely excurrent in mucro. Capsule erect, narrowly ellipsoid, straight, *ca* 2–3 mm long; lid *ca* $\frac{1}{2}$ length of body of capsule; annulus of large, separating, fugacious cells; spores ± smooth. Fr common, spring. $n=24$. Small, dull green to brownish patches or scattered plants on calcareous soil, occasional in England and Wales, rare or very rare elsewhere and extending north to E. Perth and E. Ross. 54, H5. Europe south to northern Italy and the Pyrenees, Greece, Faroes, Turkestan, U.S.S.R., China, India, Madeira, N. America, Ecuador, Bolivia, Peru.

Delgadillo (*op. cit.*) reports var. *mucronulata* (Br. Eur.) Limpr. from Britain and Ireland. The variety differs from the type in the lingulate-lanceolate lvs with excurrent nerve. Even from the limited number of gatherings observed, however, there seem too many intermediates to make the variety worth maintaining. Demaret & Castagne (1964) comment similarly. *A. brevirostris* differs in the hyaline lf base as long as or longer than the rest of the lf, the shorter capsule lid, the synoecious inflorescence and larger spores 18 –22 μm.

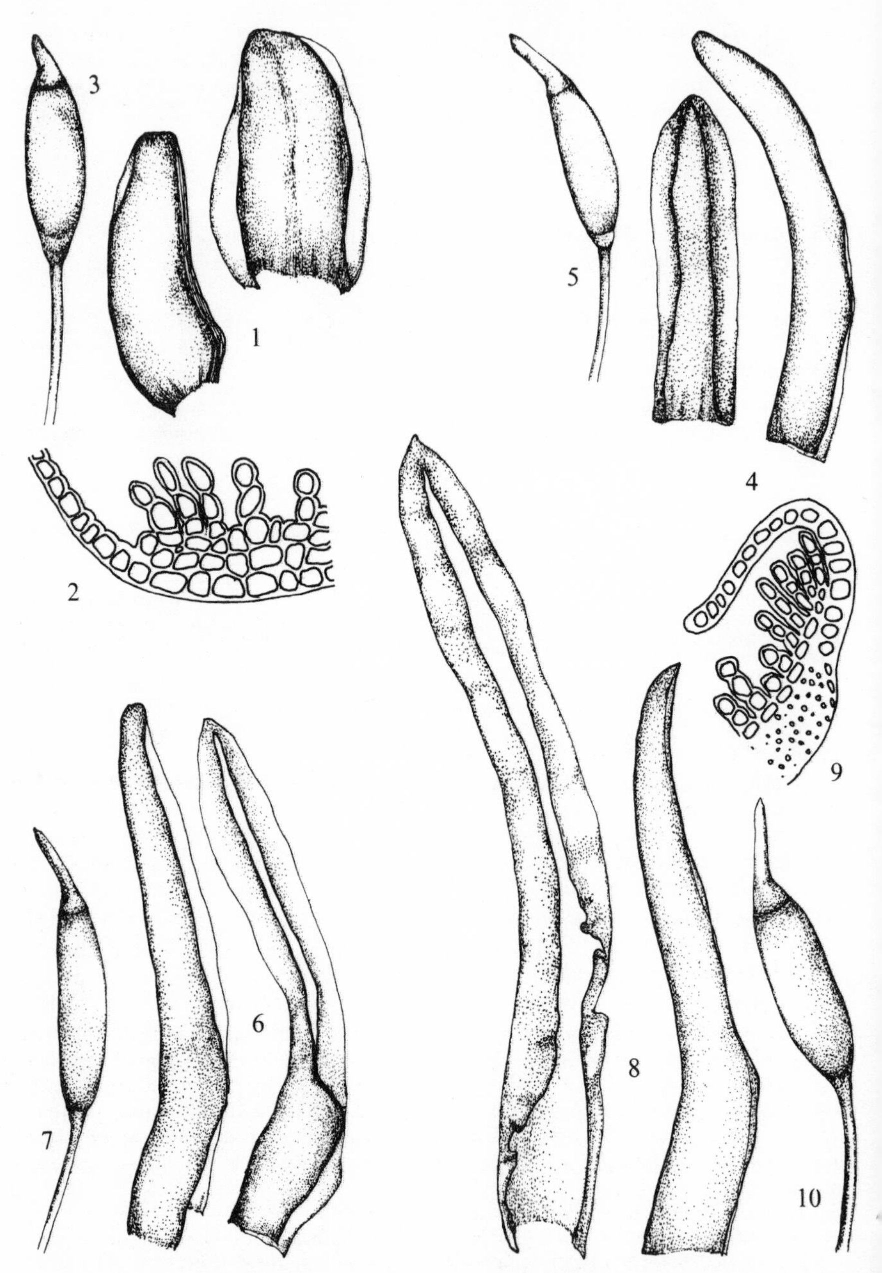

Fig. **103**. 1–3, *Aloina brevirostris*: 1, leaf; 2, leaf section; 3, capsule. 4–5, *A. rigida*: 4, leaves; 5, capsule. 6–7, *A. aloïdes* var. *ambigua*: 6, leaves; 7, capsule. 8–10, *A. aloides* var. *aloides*: 8, leaves; 9, leaf section; 10, capsule. Leaves ×25, sections ×165, capsules ×10.

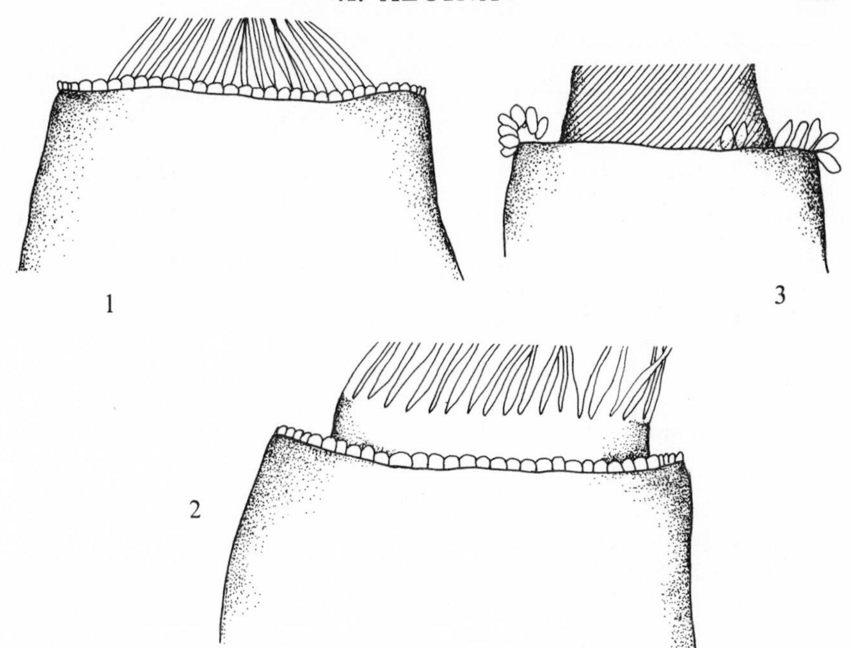

Fig. **104**. Mouth of capsules: 1, *Aloina aloides* var. *aloides*; 2, *A. aloides* var. *ambigua*; 3, *A. rigida*. All ×100.

3. A. aloides (Schultz) Kindb., Bih. K. Sv. Vet. Ak. Handl., 1883
Tortula aloides Schultz De Not.

Dioecious. Plants 2–5 mm. Lvs incurved when dry, erecto-patent to spreading when moist, apex cucullate or not, obtuse and apiculate to acute or acuminate; nerve ending in apex or shortly excurrent; marginal cells near lf base quadrate to rectangular, not hyaline. Capsule erect or inclined, narrowly ellipsoid to cylindrical, straight to strongly curved; annulus 1–2 rows persistent cells; lid obliquely rostrate. Fr common, autumn to spring. Dark green or reddish-brown patches or scattered plants on disturbed soil in open places, on banks, rock ledges and cliffs especially where calcareous and on mortar of old walls.

Var. **aloides**
Lvs erecto-patent when moist, apex cucullate or not, obtuse and apiculate to acute or acuminate, apiculus or acumen sometimes hyaline. Capsule erect or more usually inclined, narrowly ellipsoid to cylindrical, straight to strongly curved, 1.6–2.4(–2.8) mm long; basal membrane not projecting above mouth of capsule; spores ± smooth, 18–22(–25) μm. $n=26$*. Occasional to frequent in England and Ireland, rare in Scotland and extending north to Elgin and E. Ross. 84, H29, C. Europe, Lebanon, Siberia, Algeria, Morocco, Tunisia, Canaries.

Var. **ambigua** (Br. Eur.) Craig in Grout, Moss Fl. N. Amer., 1939
A. ambigua (Br. Eur.) Limpr., *Tortula ambigua* (Br. Eur.) Ångstr.
Lvs erecto-patent to spreading when moist, lanceolate-lingulate, apex cucullate, obtuse, sometimes with reflexed apiculus. Capsule erect or slightly inclined, cylindrical,

straight or slightly curved, (2.0–)2.2–3.2 mm long; basal membrane projecting above capsule mouth; spores ± smooth, (12–)14–16 μm. Fr common, winter, spring. $n=24$. Occasional in calcareous districts of England and Ireland, rare elsewhere, extending north to Elgin. 79, H21, C. Europe, Lebanon, Turkey, Iran, Siberia, Tunisia, Macaronesia, N. America, Mexico, Australia.

The 2 varieties are distinct in Britain but sometimes intergrade elsewhere; var. *aloides* has less spreading lvs frequently more acute than in var. *ambigua* and often markedly inclined capsules, but the only constantly reliable characters are basal membrane height and spore size. *A. aloides* var. *ambigua* is probably a rarer plant than the number of records suggests, as forms of var. *aloides* with erect capsules have been incorrectly named. *A. rigida* and *A. brevirostris* differ in the usually longer hyaline cells of the lf base and in the separating annulus.

42. Desmatodon Brid., Mant. Musc., 1819

Autoecious or paroecious. Lvs concave, ovate, obovate or spathulate, margin plane or recurved, entire or denticulate above, bordered or not; nerve sometimes enlarged in upper part but without appendages on ventral surface, usually excurrent in longer or shorter hair-point; cells quadrate, hexagonal or rhomboidal, smooth or papillose above. Seta long; capsule erect to pendulous, usually symmetrical, ovoid to cylindrical; lid rostrate; peristome mostly well developed, basal membrane short, teeth bi- or trifid into filiform, papillose, straight or spirally curved segments. About 40 mainly montane species distributed ± throughout the world.

1 Capsule cernuous or horizontal, lf margin bordered, denticulate **1. D. cernuus**
Capsule erect or sub-erect, lf margin unbordered, entire or papillose-crenulate 2

2 Lf margin entire, nerve thickened, very stout in upper half of lf **2. D. convolutus**
Lf margin papillose-crenulate, nerve not thickened above **3. D. leucostoma**

1. D. cernuus (Hüb.) Br. Eur., 1843
Tortula cernua (Hüb.) Lindb.

Plants 2–7 mm. Lvs erect, twisted when dry, erecto-patent or patent when moist, lanceolate, oblanceolate or spathulate, acuminate, margin bordered, recurved to middle or higher, denticulate above; nerve stout, excurrent; cells in lower part of lf rectangular, lax, hyaline, narrower at margin, above variable in shape and size, ± irregularly hexagonal, smooth or papillose, 10–20 μm wide at widest part of lf, 1–3 marginal rows longer and narrower, forming distinct border, bistratose below. Seta flexuose; capsule cernuous or horizontal, ovoid, curved; lid obliquely rostrate; peristome teeth ± straight; spores coarsely papillose, 36–40 μm. Fr common, autumn. $n=25, 26$*. Patches on calcareous soil, very rare. Notts, Cheshire, Yorks. 4. N. and C. Europe, Iceland, Siberia, C. Asia, Tibet, N. America, Greenland.

2. D. convolutus (Brid.) Grout, Moss. Fl. N. Amer., 1939
Tortula atrovirens (Sm.) Lindb.

Plants *ca* 2 mm. Lvs spirally twisted when dry, erecto-patent or patent when moist, concave, oblanceolate or obovate, broadly acute to obtuse and mucronate, margin usually strongly recurved, unbordered, entire; nerve stout, thickened in upper half of lf, excurrent in mucro; basal cells ± rectangular, hyaline, upper cells regularly quadrate-hexagonal, finely papillose, 10–12 μm wide in widest part of lf, not narrowed at margin. Capsule erect or slightly inclined, ovoid or ellipsoid, lid rostrate; peristome with short basal membrane, teeth straight; spores finely papillose, 24–28

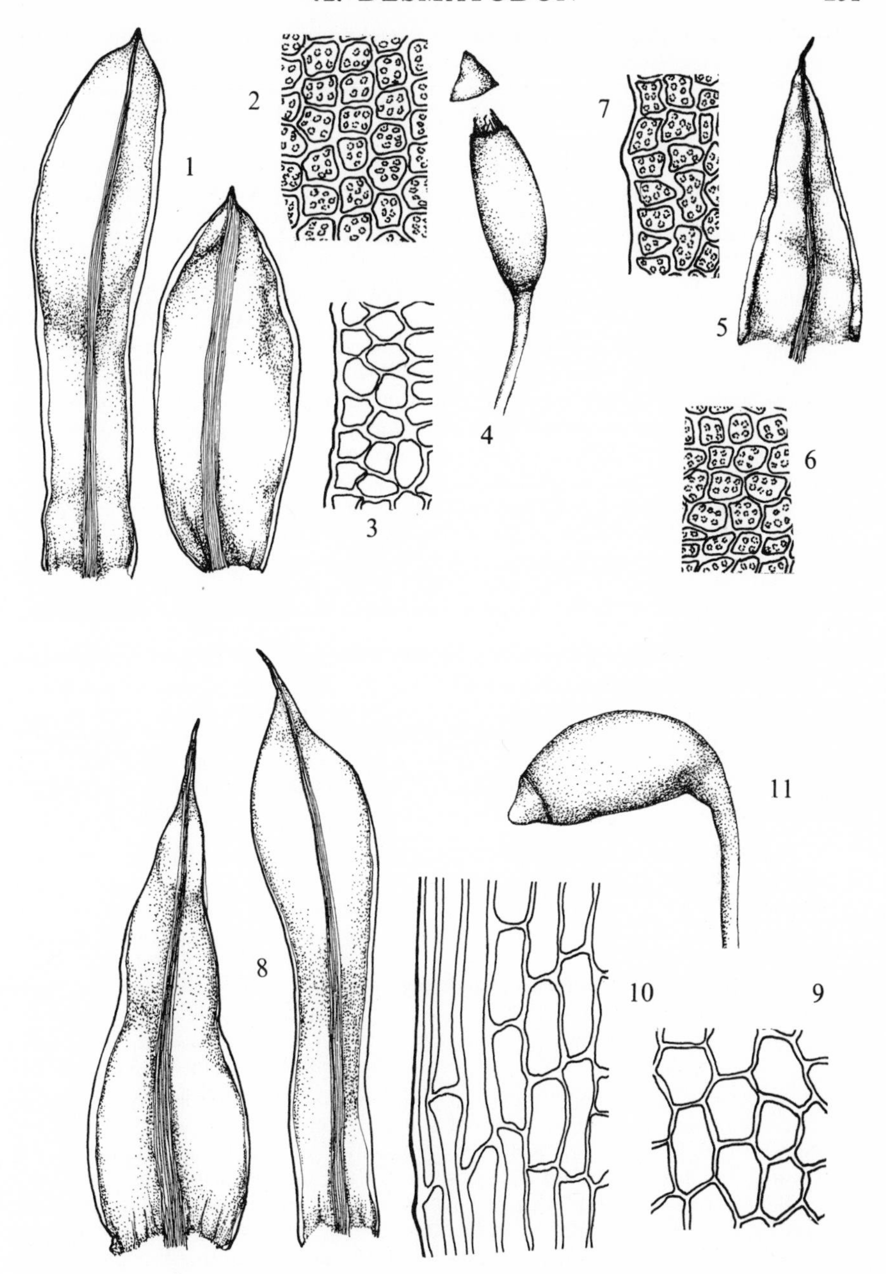

Fig. **105**. 1–4, *Desmatodon convolutus*: 1, leaves; 2, mid-leaf cells; 3, marginal cells; 4, capsule. 5–8, *D. leucostoma*: 5, leaf; 6, mid-leaf cells; 7, marginal cells. 8–11, *D. cernuus*: 8, leaves; 9, mid-leaf cells; 10, marginal cells; 11, capsule. Leaves ×25, cells ×415, capsules ×15.

μm. Fr common, winter, spring. $n=26$*. On soil on banks, rocks and cliffs, particularly near the sea. Occasional along the south coast, rare elsewhere, extending north to E. Ross and Dumbarton, rare in Ireland. 29, H10, C. W., C. and S. Europe, Caucasus, Iran, Turkestan, Kashmir, Africa, Azores, Tenerife, N. America, Tasmania, New Zealand, Oceania.

Likely to be mistaken for a species of *Pottia* or *Pterygoneurum* but differing from the former in the expanded nerve and the latter in the lack of lamellae on the upper surface of the nerve.

3. D. leucostoma (R. Br.) Berggr., Musci Spetzb., 1874
D. suberectus (Hook.) Limpr., *Tortula suberecta* Hook.

Plants 2–3 mm. Lvs erect when dry, erecto-patent when moist, broadly to narrowly triangular, to ovate-lanceolate or lanceolate, apex acute, margin revolute nearly to apex, unbordered, papillose-crenulate above; nerve stout, excurrent in short or long hair-point; basal cells rectangular, smooth, hyaline, becoming smaller, quadrate, papillose above, 8–12 μm wide in mid-lf, smaller and very obscure towards apex, cells towards margin less papillose, not obscure, forming pale band. Capsule erect or inclined, shortly cylindrical, straight or slightly curved; lid rostrate; peristome spirally twisted with tall basal membrane; spores coarsely papillose, 20–26 μm. Fr common, summer. Short-lived plants on montane calcareous rock ledges, very rare. E. Perth, S. Aberdeen. 2. Europe, N. and E. Asia, N. America, Greenland.

43. Pterygoneurum Jur., Laubmfl., 1882

Autoecious. Lvs concave, ovate or obovate, margin plane or narrowly recurved; nerve expanded above with 2–4 chlorophyllose lamellae on ventral surface, excurrent in frequently hyaline hair-point; cells rhomboidal, papillose above. Seta long or short, straight; capsule ± erect, symmetrical, ovoid to cylindrical; lid rostrate; peristome absent or imperfect and spirally curved. Eight species distributed through Europe, N. and W. Asia, N. and southern Africa, N. America, southern S. America and Australia.

Seta 2–3 mm, capsule ovoid or ovate-ellipsoid; cells of lid in straight rows, spores papillose, 20–36 μm **1. P. ovatum**
Seta 3–6 mm, capsule ellipsoid; cells of lid in spiral rows, spores smooth, 14–20 μm **2. P. lamellatum**

1. P. ovatum (Hedw.) Dix., Rev. Bryol. Lichen., 1934
Tortula pusilla Mitt.

Plants 1–2 mm. Lvs imbricate when dry, ± imbricate to erecto-patent when moist, very concave, ovate, obovate or obovate-spathulate, obtuse, margin entire; nerve stout, enlarged in upper part of lf with 2–4 chlorophyllose lamellae on upper surface, excurrent in short to long (to 1 mm), often hyaline, slightly denticulate hair-point; cells in lower part of lf rectangular, hyaline, above quadrate-hexagonal, incrassate, papillose, 10–18 μm wide in widest part of lf, smaller towards margin and apex. Seta 2–3 mm; capsule ovoid or ovate-ellipsoid, erect, straight, dark purplish-brown, sulcate when dry and empty; lid obliquely rostrate with cells in straight rows; peristome lacking; spores papillose, 20–36 μm. Fr common, autumn, winter. $n=26$. Small greenish or sometimes hoary patches on usually basic soil in turf, on rocks and on mud-capped walls, occasional, extending north to Inverness and E. Ross, Shetland, Mid Cork, Dublin. 64, H2. Europe, Caucasus, W. Asia, Algeria, Morocco, N. America, Madeira.

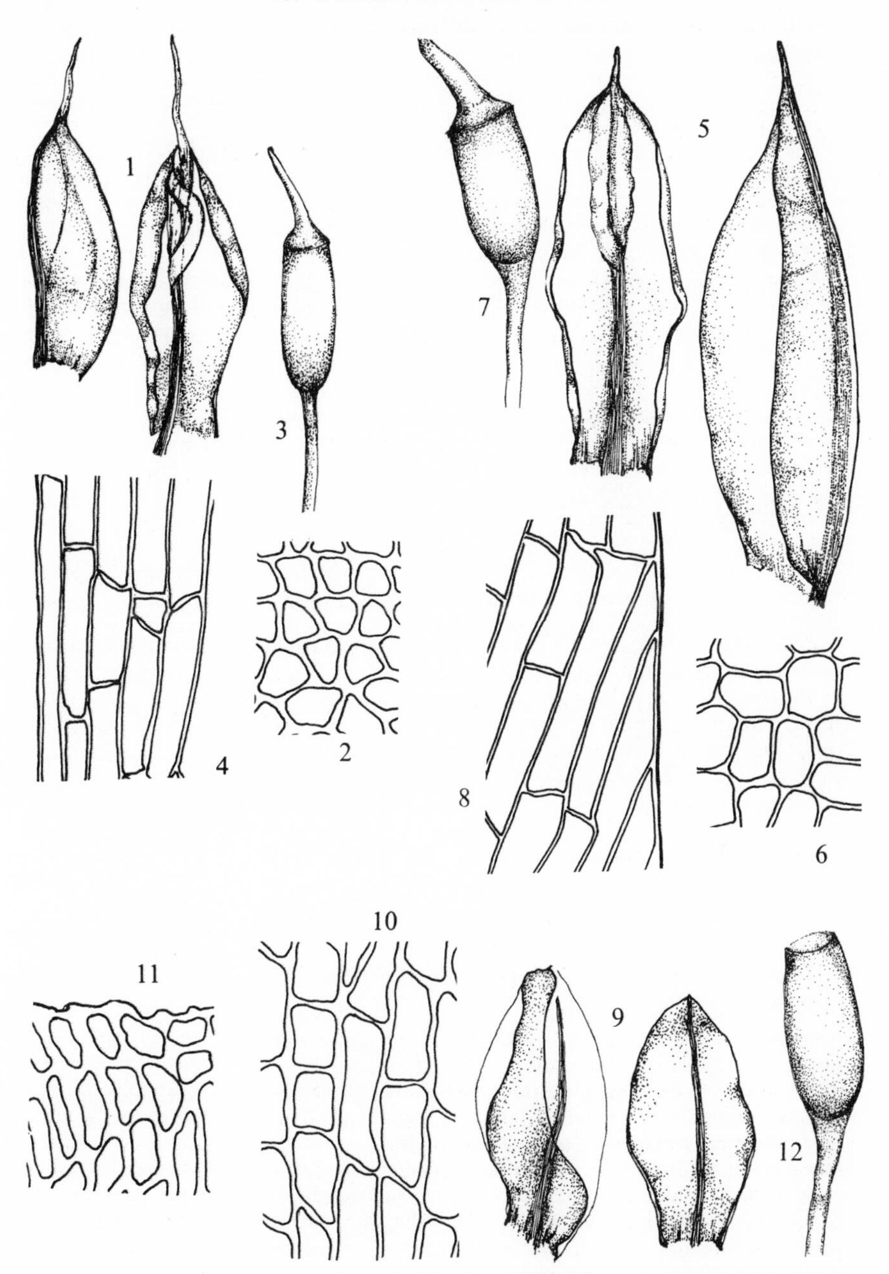

Fig. **106**. 1–4, *Pterygonerum ovatum*: 1, leaves; 2, mid-leaf cells; 3, capsule; 4, cells of lid. 5–8, *P. lamellatum*: 5, leaves; 6, mid-leaf cells; 7, capsule; 8, cells of lid. 9–12, *Stegonia latifolia*: 9, leaves; 10, mid-leaf cells; 11, cells from leaf apex; 12, old capsule. Leaves ×25, cells ×415, capsules ×15.

2. P. lamellatum (Lindb.) Jur., Laubmfl., 1882
Tortula lamellata Lindb.

Very close to *P. ovatum*. Hair-point of lf not exceeding 400 μm. Seta 3–6 mm; capsule ellipsoid, erect, straight or slightly curved, brown; lid obliquely rostrate with cells in spiral rows; peristome poorly developed and falling with lid; spores very finely papillose, 14–20 μm. Fr common, autumn, winter. On calcareous soil and mud-capped limestone walls, very rare and decreasing. Scattered localities from S. Devon and Sussex north to Derby and Yorks, Midlothian, Fife, Dublin, Down. 21, H2. N., W. and C. Europe, Italy, C. Asia, N. America.

Decreasing because of the destruction of the species' main habitats, mud-capped walls. *P. ovatum* differs in the sometimes longer hair-point, the larger spores and the cells of the lid in straight rows.

44. Stegonia Vent., Rev. Bryol., 1883

Autoecious. Lvs broad, very concave, imbricate giving plants bud-like appearance; nerve weak; cells smooth, weakly chlorophyllose. Seta long; capsule erect; peristome well developed. Two species.

1. S. latifolia (Schwaegr.) Vent. ex Broth., Laubm. Fennosk., 1923
Pottia latifolia (Schwaegr.) C. Müll.

Plants bud-like, *ca* 2 mm. Lvs imbricate when moist, hardly altered when dry, very concave, obovate to broadly obovate-spathulate, apex rounded or obtuse, margin plane, faintly denticulate above; nerve weak, ending below apex; cells in lower part of lf narrowly rectangular, hyaline, becoming smaller, narrowly rectangular to irregularly rhomboidal above, 8–16 μm wide in widest part of lf, smaller and more incrassate towards margin and apex. Capsule narrowly ellipsoid, erect or nearly so; peristomate; spores papillose, 40–50 μm. Fr common, summer autumn. $n=26$. Small, pale green patches or clusters on calcareous soil on rock ledges in the Scottish mountains, very rare. Perth, S. Aberdeen, Banff. 4. N., W. and C. Europe, Faroes, Caucasus, N. and C. Asia, China, N. America, Greenland.

The very concave lvs with nerve ending below apex will distinguish this plant from other British Pottiaceae. *Desmatodon leucostoma*, with which it cometimes occurs, has acuminate lvs with strongly recurved margins.

45. Pottia Fürnr., Flora, 1829

By D. F. Chamberlain

Plants small, in tufts or patches, stems usually unbranched, erect. Lvs narrowly lanceolate to elliptical, 2.0–4.5 times as long as wide; upper cells quadrate or hexagonal, usually lax, papillose or not; nerve percurrent or excurrent in long point. Seta always present; capsule cleistocarpous or stegocarpous; peristome absent, rudimentary or with 16 imperfect or bifid teeth; annulus present or absent; calyptra cucullate. A cosmopolitan genus with *ca* 35 species.

Pottia recta and *P. bryoides* form a link between *Phascum* and *Pottia* in having more or less cleistocarpous capsules. The separation of the 2 genera is somewhat arbitrary and it might be more satisfactory taxonomically if all the species were placed in only 1 genus.

1 Capsule cleistocarpous or with a persistent lid — 2
  Capsule with a deciduous lid — 3
2 Capsule ± spherical, seta 0.6–1.0 mm, lf cells 9.5–15.0 μm wide — **11. P. recta**
  Capsule ellipsoid, seta 2.0–5.5 mm, lf cells 16–22 μm wide — **10. P. bryoides**

3 Peristome with well developed teeth 4
Peristome absent or rudimentary, lacking well developed teeth 7

4 Lf cells incrassate, perichaetial lvs broad, concave, spores 15.5–19.0 μm **1. P. caespitosa**
Lf cells thin- or thick-walled but not incrassate, perichaetial lvs larger but same shape as lower lvs, spores 19–37 μm 5

5 Lid rostrate, annulus present, lf nerve excurrent 150–300 μm **6. P. lanceolata**
Lid conical to mamillate, annulus absent, lf nerve excurrent up to 150(–175) μm 6

6 Spores with a warty outline but not papillose **2. P. starkeana**
Spores with a regular outline, papillose **3. P. commutata**

7 Lvs sometimes toothed near apex, nerve usually percurrent, seta 5–9 mm, lid remaining attached to columella **9. P. heimii**
Lvs entire, nerve usually distinctly excurrent, seta 2–6 mm, lid deciduous 8

8 Lf margin plane, cells smooth, capsule turbinate, widest at mouth **8. P. truncata**
Lf margin recurved, cells usually papillose, capsule obloid to ellipsoid, widest below mouth or turbinate 9

9 Lid conical to mamillate, annulus absent **2. P. starkeana**
Lid rostrate, annulus present 10

10 Cells in upper part of lf 13–17 μm wide, strongly papillose, spores 19.0–26.5 μm **4. P. wilsonii**
Upper cells 17–24 μm wide, usually faintly papillose, rarely smooth, spores 25–34 μm 11

11 Lvs spathulate–oblong, apex rounded, nerve excurrent 100–1100 μm **5. P. crinita**
Lvs oblong–lanceolate, apex acute, nerve excurrent 100–300(–500) μm **7. P. intermedia**

1. P. caespitosa (Bruch ex Brid.) C. Müll., Syn. Musc., 1849
Trichostomum caespitosum (Bruch ex Brid.) Jur.

Autoecious. Lvs broadly lanceolate, 2–3 times as long as wide, margin plane, nerve excurrent 25–75 μm; upper cells irregularly rounded-quadrate, 7–11 μm wide, verrucose, incrassate. Perichaetial lvs distinctly broader and concave. Seta 1.7–3.0 mm; capsule obovoid, 0.9–1.0 mm × 0.55–0.70 mm, with 1 row thick-walled cells at the constricted mouth; lid rostrate; annulus present; peristome teeth 25–110 μm, imperfect, usually cleft; calyptra smooth; spores minutely papillose, 15.5–19.0 μm. Fr spring. Bare patches of soil on chalk, S. England, extending north to Hereford and Cambridge. 10. C. and S. Europe.

P. caespitosa is distinguished from the other British species of *Pottia* by the differentiated perichaetial lvs and the incrassate lf cells.

2. P. starkeana (Hedw.) C. Müll., Syn. Musc., 1849

Autoecious. Lvs ovate-lanceolate, 2.0–4.5 times as long as wide, margin recurved; nerve percurrent or excurrent to 280 μm; upper cells quadrate, 8.5–21.5 μm wide, papillose, thin- or thick-walled. Seta 1.0–4.3 mm; capsule ellipsoid to turbinate, 0.35–1.70 mm × 0.3–1.0 mm with 1–4 rows differentiated cells below mouth; lid conical to mamillate; annulus absent; peristome absent, rudimentary or with well-developed teeth up to 190 μm long; calyptra usually scabrous, occasionally smooth; spores with a regular to warty outline, with or without spiny or tuberculate projections, 19.0–41.5 μm. Fr common, winter, spring, rarely summer.

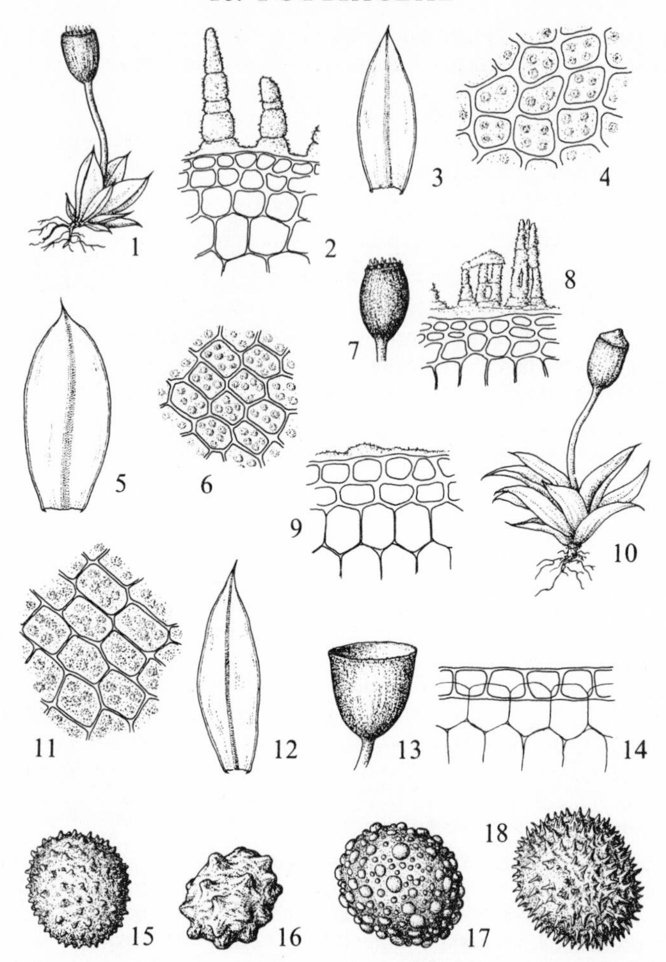

Fig. **107**. 1–4, 15, *Pottia commutata*: 1, plant; 2, capsule mouth; 3, upper leaf (×30); 4, cells from upper part of leaf; 15, spore. 5–8, 16, *P. starkeana* ssp. *starkeana*: 5, upper leaf (×30); 6, cells from upper part of leaf; 7, capsule; 8, capsule mouth; 16, spore. 9–10, 17, *P. starkeana* ssp. *conica*: 9, capsule mouth; 10, plant; 17, spore. 11–14, 18, *P. starkeana* ssp. *minutula*: 11, cells from upper part of leaf; 12, upper leaf (×15); 13, capsule; 14, capsule mouth; 18, spore. Plants ×15, leaf cells ×250, capsules ×15, spores ×600.

Key to intraspecific taxa of *P. starkeana*

1 Capsule turbinate, widest at mouth, usually with 1 row of unthickened cells below mouth, peristome absent — ssp. **minutula**

Capsule ellipsoid to obovoid with 1–4 rows thickened cells below constricted mouth, peristome absent, rudimentary or with well-developed teeth — 2

2 Spores (22.0–)27.5–34.0(–42.0) μm, outline regular with spiny, papillose or tuberculate projections, peristome absent or rudimentary and to 40 μm long — ssp. **conica**

Spores (19–)22–28(–32) µm, outline regular to warty, rarely ± smooth, projections absent, peristome absent, rudimentary or with teeth to 175 µm long
3 (ssp. **starkeana**)

3 Peristome teeth well developed, (60–)75–150(–190) µm long
ssp. **starkeana** var. **starkeana**

Peristome absent or rudimentary, without well developed teeth, basal membrane to 70(–155) µm high ssp. **starkeana** var. **brachyodus**

Ssp. **starkeana**

Lf cells (9–)10–12(–18) µm wide, strongly papillose; nerve percurrent or excurrent to 175 µm. Capsule ellipsoid, usually widest at about the middle with (1–)2–4 rows thickened cells below strongly constricted mouth; peristome absent to well developed; spores smooth or with a warty outline, projections lacking, 22–28(–32) µm. Disturbed soils on basic substrata, local. 43, H6, C. Europe, W. Asia, N. Africa, Canaries, Madeira, Australasia.

Var. **starkeana**

Peristome teeth well developed, (60–)75–150(–190) µm long.

Var. **brachyodus** C. Müll. ,Syn. Musc., 1849

Peristome absent or rudimentary, without well developed teeth, basal membrane to 70(–155) µm high.

Ssp. **conica** (Schleich. ex Schwaegr.) Chamberlain, Notes Roy. Bot. Gard. Edin., 1969

Lf cells (9.5–)13.0–16.0(–19.0)µm wide, moderately papillose; nerve percurrent or excurrent to 110(–280) µm. Capsule obloid, widest at about middle, with 1–3 rows thick-walled cells below mouth, mouth not or only slightly constricted; peristome absent or rudimentary, to 40 µm long; spores with smooth outline with spiny to ± irregularly papillose projections. Fields, disturbed soil, etc., on basic substrata. Fr spring, summer. $n=28$. Common in the south becoming very local in the north and absent from most of Scotland. 75, H10, C. Europe, W. Asia, W. and S. Africa, N. America, Australasia.

Ssp. **minutula** (Schleich. ex Schwaegr.) Chamberlain, Notes Roy. Bot. Gard. Edin., 1969

P. davalliana (Sm.) C. Jens., *P. minutula* (Schleich. ex Schwaegr.) Fürnr.

Lf cells (11.0–)13.5–19.0(–21.5) µm wide, moderately papillose; nerve excurrent (70–)100–200(–280) µm. Capsule turbinate, widest at mouth, with 1(–2) rows thin-walled differentiated cells below mouth; peristome absent; spores with smooth outline with spiny projections, (22–)31.5–39.5(–41.5) µm. Fr spring, summer. Basic soil, mostly on inland sites, local in England and Ireland, usually with ssp. *conica*. 30, H4. N. and C. Europe, Madeira.

P. starkeana ssp. *starkeana* and ssp. *minutula* are traditionally treated as separate species. Indeed, var. *starkeana* with its long peristome and warty spores is clearly distinct from ssp. *minutula* and ssp. *conica*. It is, however, often difficult to distinguish var. *brachyodus* from ssp. *conica* as both may have a rudimentary peristome, while the spores of the former may have a ± irregular outline and the spores of the latter very poorly developed projections. Ssp. *conica* is very variable and spans the gap between ssp. *starkeana* and ssp. *minutula*.

3. P. commutata Limpr., Laubm., 1888

P. davalliana ssp. *commutata* (Limpr.) Podp.

Autoecious. Lvs broadly ovate-lanceolate, 2–3 times as long as wide, margin recurved; nerve excurrent 10–100 µm; upper cells quadrate, (10–)13–17 µm wide, papillose. Seta 1.0–2.0(–2.5) mm; capsule broadly cylindrical, 0.35–1.00 mm × *ca*

0.50 mm, with 1–4 rows of thickened differentiated cells at slightly constricted mouth; lid conical to mamillate; peristome teeth (65–)90–100 μm, entire, blunt; calyptra papillose; spores papillose, (23.5–)28.0–32.0(–37.0) μm. Fr common, winter, spring. On bare soil, usually in turf near the sea, coasts of Cornwall, Devon, Dorset, I. of Wight, E. Sussex, very rare. 7. Norway (?), S.W. Europe, Mediterranean.

Closely resembling *P. starkeana* ssp. *conica* but much less variable and distinguished by the well developed peristome.

4. P. wilsonii (Hook.) Br. Eur., 1843
P. asperula Mitt.

Paroecious. Lvs obovate-lanceolate, 2.5–4.0 times as long as wide, apex acute to rounded, margin recurved; nerve excurrent 150–650 μm; upper cells quadrate, 15–17 μm wide, strongly papillose, walls moderately thick. Seta 2.0–3.5 mm; capsule obovoid, 0.8–1.7 mm × 0.7–1.0 mm, widest well below constricted mouth, with 3–5 rows differentiated cells below mouth; lid rostrate; annulus present; peristome absent or rudimentary and up to 75 μm long; calyptra smooth to scabrid; spores granulate-papillose, 19.0–26.5 μm. Fr common, spring. Disturbed ground, banks, etc. in coastal habitats, local and rare. Scottish records require confirmation. 21, H5, C. Europe, Mediterranean, Faroes.

At least some plants of the type specimen of *P. asperula* fall within the range of *P. wilsonii.*

5. P. crinita Wils. ex Br. Eur., 1843
P. wilsonii var. *crinita* (Br. Eur.) Warnst., *P. viridifolia* Mitt., *P. crinita* var. *viridifolia* (Mitt.) Kindb.

Autoecious. Lvs spathulate-oblong, apex rounded, (2.5–)3.0–4.0 times as long as wide, margin recurved; nerve excurrent 100–1100 μm; upper cells quadrate, 17–24 μm wide, papillose, thin-walled. Seta 2–4 mm; capsule obloid, 0.8–1.7 mm × 0.6–1.0 mm, widest at or slightly constricted towards mouth, with 2–4 rows thin-walled differentiated cells below mouth; lid rostrate; annulus present; peristome absent; calyptra usually ± roughened, occasionally scabrous; spores granulate-papillose, 25–35 μm. Fr common, spring. $n=48$*. Disturbed soil, rocks, etc., near the coast. Frequent in the south becoming very local in the north, occasional in Ireland. 38, H10, C. W. Europe, Morocco, Turkey.

Plants with shortly excurrent nerve have been described as var. *viridifolia* (Mitt.) Kindb., but as there is no other reliable difference nor a clear-cut discontinuity in the degree of excurrence of the nerve the variety is not recognised here.

6. P. lanceolata (Hedw.) C. Müll., Syn. Musc., 1849

Autoecious. Lvs ovate-lanceolate, 2.5–3.0 times as long as wide, margin recurved; nerve excurrent, 150–300 μm; upper cells quadrate, 13.0–17.0(–19.5) μm wide, papillose, thin-walled. Seta 2.8–5.5 mm; capsule ellipsoid, 1.00–1.75 mm × 0.6–0.9 mm with 2–4(–7) rows of thick-walled cells below constricted mouth; lid with long beak; annulus present; peristome teeth 230–450 μm long, imperfectly divided towards apex; calyptra smooth; spores densely and finely papillose, (19.0–)21.5–30.0 μm. Fr common, winter, early spring. Bare ground, wall tops, etc., on basic substrata, frequent in the south and becoming rarer in the north, rare in Ireland. 77, H5. Europe, N. Africa, Madeira, W. Asia, Japan, N. America.

7. P. intermedia (Turn.) Fürnr., Flora, 1829
P. littoralis Mitt., *P. intermedia* var. *littoralis* (Mitt.) Corb., *P. truncata* var. *major* (Web. & Mohr) Br. Eur.

Autoecious. Lvs oblong-lanceolate, 3–4 times as long as wide, margin plane or recurved; nerve excurrent, 100–300(–500) μm; upper cells quadrate, 17–22 μm,

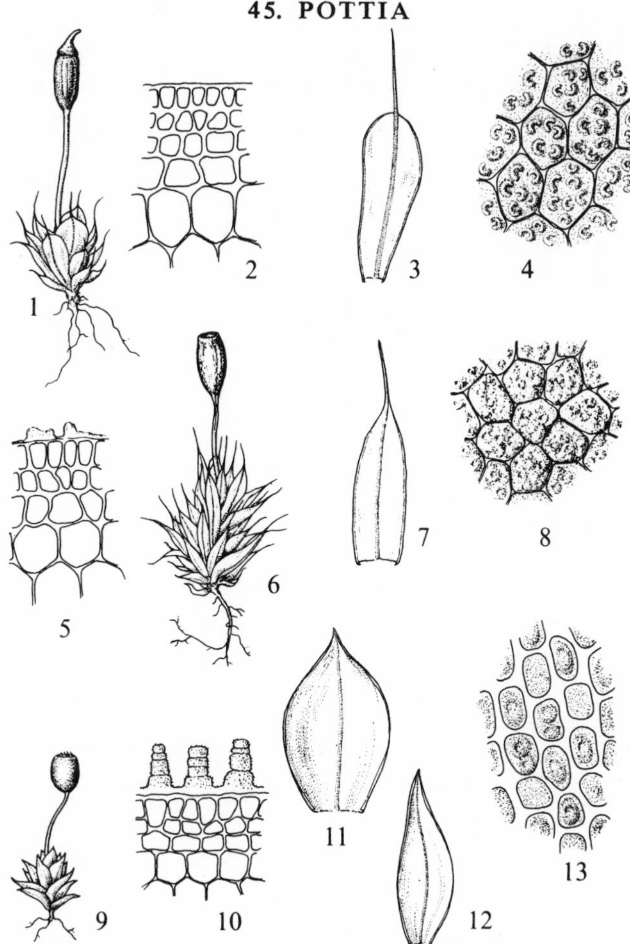

Fig. **108**. 1–4, *Pottia crinita*: 1, plant; 2, cells at mouth of capsule; 3, upper leaf (×15); 4, cells from upper part of leaf. 5–8, *P. wilsonii*: 5, cells at mouth of capsule; 6, plant; 7, upper leaf (×15); 8, cells from upper part of leaf. 9–13, *P. caespitosa*: 9, plant; 10, mouth of capsule; 11, perichaetial leaf (×30); 12, lower leaf (×30); 13, cells from upper part of leaf. Plants ×7.5, leaf cells ×325.

smooth or slightly papillose. Seta 4.0–6.2 mm; capsule obloid, 0.8–1.6 mm×0.6–1.0 mm with 3–5 rows thick-walled, differentiated cells below slightly constricted mouth; lid rostrate; annulus present; peristome absent or rudimentary, reduced to a basal membrane to 35 μm high; calyptra smooth or slightly scabrous; spores papillose, 27–34 μm. Fr common, mainly winter and spring. $n=52$. Disturbed ground, widespread but local. 83, H15, C. Europe, E. Asia, S. Ontario, Australia.

Intermediate between *P. truncata* and *P. lanceolata* and often found with one or both of these species and possibly a hybrid derivative of them. May be distinguished from *P. lanceolata* by the absent or rudimentary peristome and from *P. truncata* by the presence of an annulus and several rows of thick-walled differentiated cells below the capsule mouth. The plant referred to as var. *littoralis* by Dixon (1924), which in its extreme state is a well

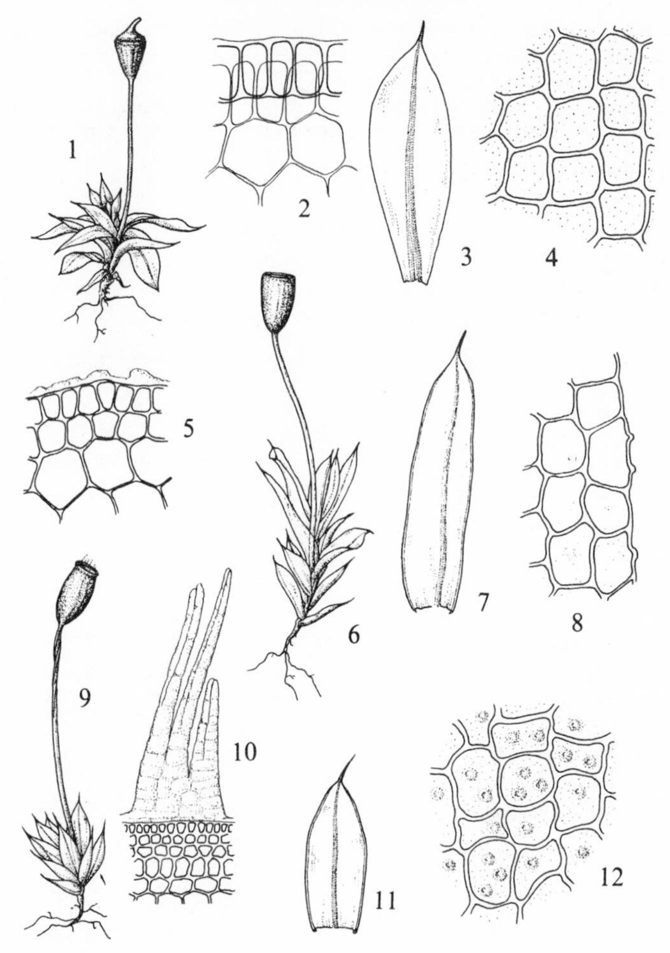

Fig. **109**. 1–4, *Pottia truncata*: 1, plant (×7.5); 2, cells at mouth of capsule; 3, leaf (×30); 4, cells from upper part of leaf. 5–8, *P. intermedia*: 5, cells at mouth of capsule; 6, plant (×7.5); 7, leaf (×15); 8, cells from upper part of leaf. 9–12, *P. lanceolata*: 9, plant (×7.5); 10, peristome teeth; 11, leaf (×15); 12, cells from upper part of leaf.

marked form with thick-walled, smooth lf cells and a relatively narrow capsule, may be of different origin from the typical form. On present evidence it is not sufficiently distinct from the type to be maintained as a variety.

8. P. truncata (Hedw.) Fürnr., Flora, 1829
P. truncatula (With.) Bus.

Autoecious. Lvs oblong-lanceolate, 3–4 times as long as wide, margin plane; nerve excurrent 50–300 μm; upper cells quadrate, 17–24 μm wide, smooth, thin-walled. Seta 2.0–3.5 mm; capsule turbinate, 0.9–1.2 mm×0.65–1.00 mm, widest at mouth, with 1–2 rows thin-walled differentiated cells below mouth; lid rostrate; annulus and peristome absent; calyptra smooth; spores papillose, 24–36 μm. Fr throughout year but more commonly in spring. $n=20^*$, 56^*. Fields, waste places, etc., on acid soil, abundant in the south, becoming less frequent in the north. 109, H34, C. Europe, Asia, N. Africa, N. and S. America.

Occasional gigas forms, about twice the size of typical plants with larger spores may be encountered either as solitary plants in a normal population or as a pure population.

9. P. heimii (Hedw.) Fürnr., Flora, 1829.
Desmatodon heimii (Hedw.) Mitt.

Autoecious, rarely synoecious. Lvs narrowly oblong-lanceolate, 3–5 times as long as wide, margin plane, toothed or entire towards apex; nerve percurrent to excurrent to 100 μm; upper cells quadrate, 13.0–21.5 μm wide, faintly papillose, thick-walled. Seta 5–9 mm; capsule obloid, 1.00–1.95 mm × 0.7–1.0 mm with 2–4 rows of thickened

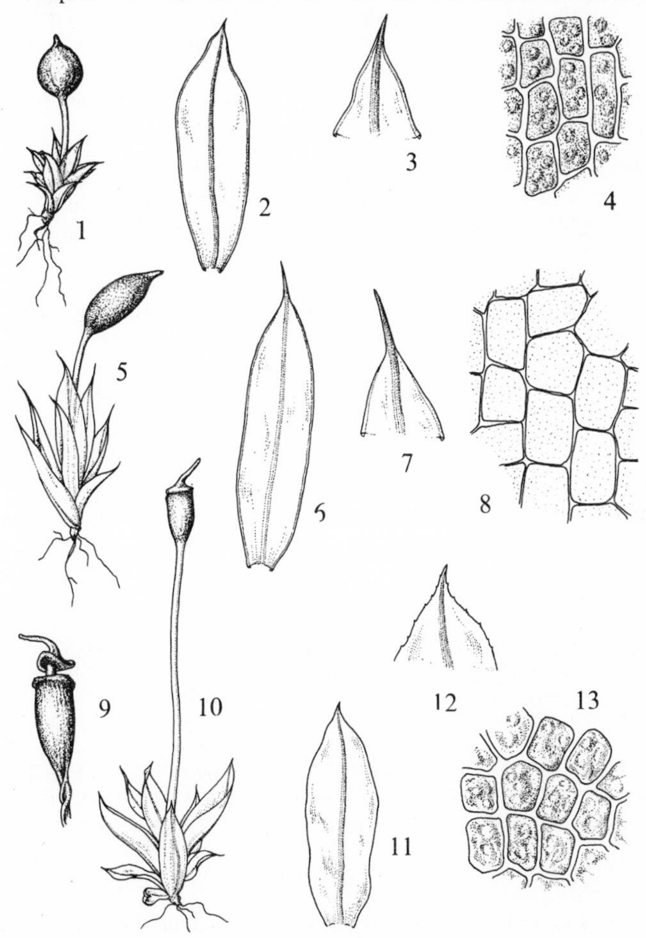

Fig. **110**. 1–4, *Pottia recta*: 1, plant (× 15); 2, upper leaf (× 30); 3, leaf apex; 4, cells from upper part of leaf. 5–8, *P. bryoides*: 5, plant (× 7.5); 6, upper leaf (× 15); 7, leaf apex; 8, cells from upper part of leaf. 9–13, *P. heimii*: 9, dry capsule (× 15); 10, plant (× 7.5); 11, upper leaf (× 15); 12, leaf apex; 13, cells from upper part of leaf.

cells below slightly constricted mouth; lid remaining attached to columella, persistent for some time after dehiscence, rostrate; annulus and peristome absent; calyptra smooth; spores minutely to coarsely papillose, 29–34 μm. Fr common, spring to early summer. $n=26$*. Earth-covered rocks, bare patches of soil, upper reaches of salt marshes, usually near the sea, frequent in coastal areas, very rare inland. 84, H19, C. Europe, Faroes, Asia, N. and S. America, Tasmania, New Zealand.

P. heimii may usually be recognised by the persistent columella and toothed lvs.

10. P. bryoides (Dicks.) Mitt., Ann. Mag. Nat. Hist., 1851

Autoecious. Lvs lanceolate, 2.5–4.0 times as long as wide, margin recurved; nerve excurrent 250–750 μm; upper cells quadrate, 16–22 μm wide, smooth or slightly papillose. Seta 2.0–5.5 mm; capsule ellipsoid, 1.0–1.8 mm×0.7–0.9 mm, cleistocarpous but with at least 1 row of differentiated cells at base of beak; peristome rudimentary; calyptra smooth; spores minutely papillose, 25.5–32.5 μm. Fr common, winter, spring. An ephemeral on exposed soil, by the sea or more rarely inland. Frequent in England, rare in Wales, very rare in Scotland and Ireland. 47, H2. Europe, W. Asia, western N. America.

11. P. recta (With.) Mitt., Ann. Mag. Nat. Hist., 1851

Paroecious. Lvs lanceolate, 2–3 times as long as wide, margin recurved; nerve excurrent 50–130 μm; upper cells quadrate, 9.5–12.5(–15.0) μm wide, strongly papillose. Seta 0.6–1.0 mm when mature; capsule ± spherical, 0.6–1.0 mm diameter, cleistocarpous or with lid semi-deciduous, with 1 row of thickened cells at base of short beak; peristome absent; calyptra rough; spores with long spines, 23–30 μm. $n=26$*. Fr common, winter, spring. An ephemeral on disturbed basic soil, frequent on the south coast of England, becoming rarer in the north and absent from most of Scotland, occasional in Ireland. 56, H9, C. C. and S. Europe, Caucasus, N. Africa.

46. Phascum Hedw., Sp. Musc., 1801

Paroecious, autoecious or synoecious. Small plants; stem without central strand. Lvs ovate to narrowly lanceolate, acute, margin plane or recurved, entire; nerve excurrent, in section with 2 large median cells, 2 large ventral cells and dorsal stereid band; basal cells rectangular, cells above quadrate, hexagonal or rhomboidal, papillose, sometimes strongly so. Seta short, straight or curved; capsule immersed or emergent, erect to pendulous, cleistocarpous, ovoid or sub-globose with blunt apiculus; columella present; calyptra cucullate or rarely mitriform. About 24 species.

The account of this genus is based upon unpublished notes provided by Mr Lewis Derrick, to whom I am much indebted.

1 Seta cygneous, 0.5–1.5 mm long, capsule laterally exserted **2. P. curvicolle**
Seta straight or curved, to 0.5 mm long, capsule immersed or slightly emergent 2
2 Plants greenish, (1–)2–9 mm tall, seta 0.1–0.5 mm long, spores 25–40(–44) μm **1. P. cuspidatum**
Plants reddish-brown, 0.6–1.2 mm, seta *ca* 0.07–0.10 mm, spores 15–25 μm **3. P. floerkeanum**

1. P. cuspidatum Hedw., Sp. Musc., 1801

Autoecious or paroecious. Plants tufted, (1–)2–9 mm. Lvs slightly twisted, appressed-flexuose when dry, lower patent, upper imbricate to convolute when moist, lower lvs ovate, acute, upper and perichaetial lvs larger, ovate to ovate-

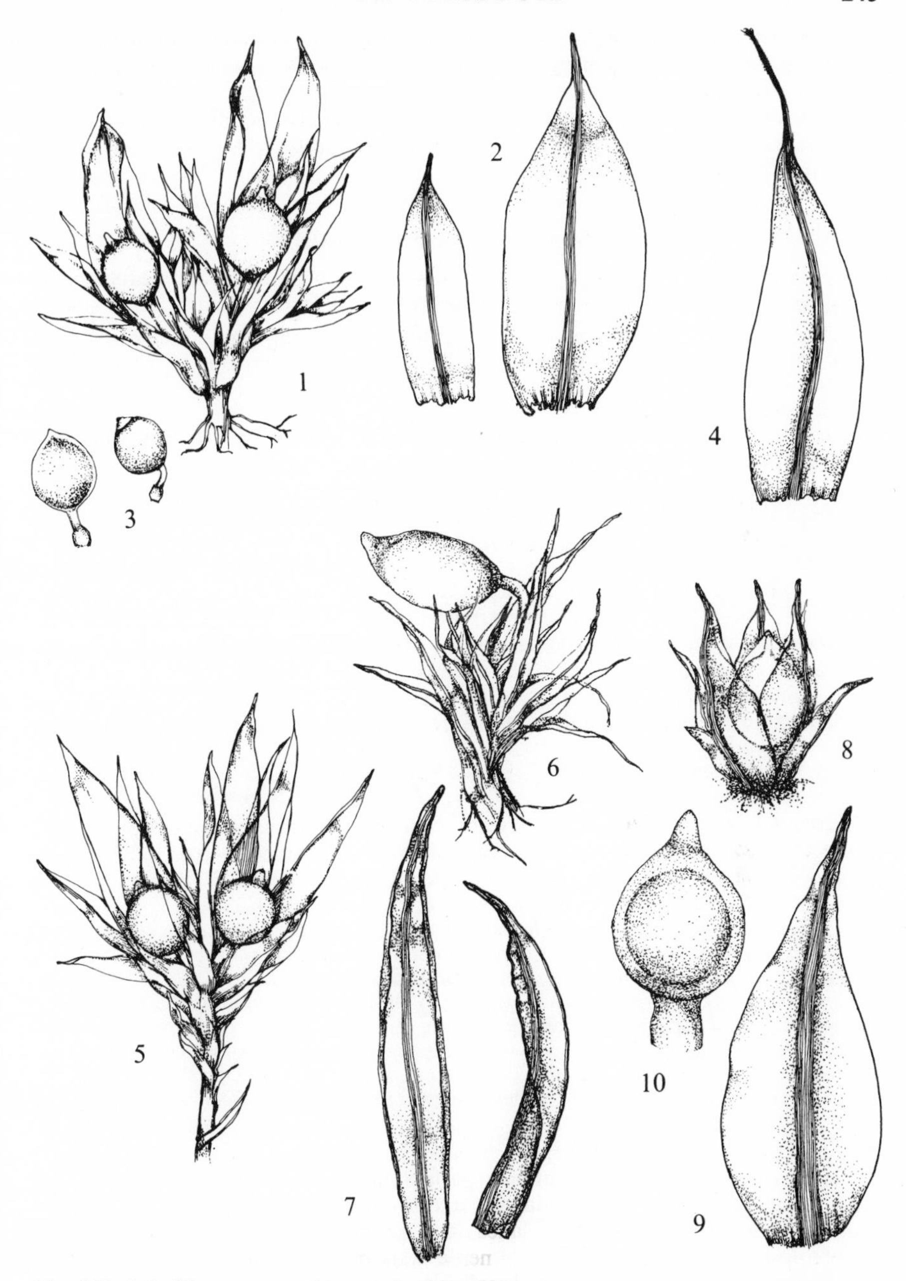

Fig. **111**. 1–3, *Phascum cuspidatum* var. *cuspidatum*: 1, plant (×10); 2, leaves (×20); 3, capsules (×10). 4, *P. cuspidatum* var. *piliferum*: leaf (×20). 5, *P. cuspidatum* var. *schreberanum*: plant (×10). 6–7, *P. curvicolle*: 6, plant (×25); 7, leaves (×40). 8–10, *P. floerkeanum*: 8, plant (×25); 9, leaf (×65); 10, capsule (×65).

lanceolate, acute, margin recurved, entire; nerve excurrent; cells very variable, lower lax, irregularly rectangular, upper rectangular to hexagonal, usually ± thick-walled, weakly to strongly papillose, (9–)14–24 μm wide. Seta short, 0.1–0.5 mm long, straight or curved; capsule cleistocarpous, immersed or slightly emergent, subglobose, bluntly apiculate; spores 25–40(–44) μm; calyptra conical, cucullate, rarely mitriform. Fr common, autumn to spring.

Key to the varieties of *P. cuspidatum*

1 Plants 4.5–9.0 mm tall, stems branched above and not at base — var. **schreberanum**
Plants rarely more than 4 mm, if branched then branched at base — 2
2 Nerve excurrent in cuspidate, yellowish point up to 0.6 mm long — var. **cuspidatum**
Nerve excurrent in slightly flexuose hyaline point 0.6–1.4 mm long — var. **piliferum**

Var. **cuspidatum**

Plants (1.0–)2.0–4.5 mm. Stems unbranched or forked at base. Lvs crowded; nerve excurrent in yellowish cuspidate point, (0.1–)0.2–0.6 mm long. $n=21$, 26*, 42, 52. Green tufts on bare soil in arable fields, gardens, roadsides, banks, quarries, etc., common in lowland habitats, occasional or rare elsewhere. 101, H29, C. Europe, Caucasus, Algeria, N. America, Ecuador (?).

Var. **piliferum** (Hedw.) Hook. & Tayl., Musc. Brit., 1818

As var. *cuspidatum* but nerve excurrent in slightly flexuose hyaline point, 0.6–1.4 mm long. Soil in turf on cliffs, banks, edges of paths, on stony ground, etc., particularly near the sea, rare. $n=27$. Scattered localities from Cornwall and Kent north to Angus and Kincardine, not recorded from Wales (except Glamorgan) or Ireland. 23, C. Europe (mainly in the south), Iran, Algeria, Morocco, N. America.

Var. **schreberanum** (Dicks.) Brid., Musc. Rec., 1806

P. cuspidatum var. *maximum* auct. non var. *maximum* Web. & Mohr

Plants 4.5–9.0 mm. Stems simple below, branched above. Lower lvs distant, upper crowded, erect. Arable fields, banks, waste places, etc., rare. Scattered localities from S. Devon, N. Somerset and Surrey north to Midlothian and Mid Perth. 13. Scandinavia, Germany, France, Italy, Switzerland.

A very variable species for which in the past some 3 subspecies and 22 varieties have been described. Much of the variation is, however, continuous and many of the taxa are based on arbitrary and unsatisfactory characters. In Britain var. *piliferum* and var. *schreberanum* appear sufficiently distinct to merit recognition, although experimental studies are required to establish their nature. The characters given in Dixon (1924) for the varieties of *P. cuspidatum* are unsatisfactory; the only reliable feature of var. *piliferum* is the longly excurrent nerve. Var. *curvisetum* is not worth maintaining as the main distinguishing feature, the curved seta, is not infrequently encountered in other forms of *P. cuspidatum.*

2. P. curvicolle Hedw., Sp. Musc., 1801

Paroecious or synoecious. Plants gregarious, not tufted, 1.5–3.0 mm tall. Lower lvs patent, upper and perichaetial lvs erecto-patent when moist, lower oblong-lanceolate, upper larger, narrowly lanceolate, acuminate, margin recurved, entire or sometimes papillose-crenulate above; nerve weak below, strong above, excurrent in cuspidate point 0.1–0.3 mm long; basal cells rectangular, cells in mid-lf quadrate-rectangular to quadrate-hexagonal, thick-walled, strongly papillose, obscure, 12–16 μm wide, towards apex irregularly quadrate-hexagonal or rounded hexagonal. Seta cygneous, 0.5–1.5 mm long; capsule laterally exserted, cleistocarpous, horizontal to pendulous, ovoid with stout blunt oblique apiculus; spores 24–28 μm; calyptra cucullate. Fr common, autumn to spring. Greenish to brownish, gregarious or solitary, winter ephemeral plants on bare soil in calcareous grassland, especially chalk, frequent in S. England, rare in N. England, very rare in Wales, Scotland and Ireland. 50, H2. C. and N. Europe, Spain, Greece.

3. P. floerkeanum Web. & Mohr, Bot. Taschenb., 1807

Paroecious. Plants gregarious, not tufted, 0.6–1.2 mm tall. Lower lvs spreading, upper and perichaetial lvs erecto-patent, concave, ovate, acuminate, margin plane or narrowly recurved, entire; nerve reddish-brown, weak below, strong above, excurrent in cuspidate point; basal cells lax, hexagonal-rectangular, cells above quadrate, hexagonal or rhomboidal, faintly papillose at back, 12–16 μm wide, towards apex smaller, ± hexagonal. Seta straight, very short, 70–100 μm; capsule immersed but not concealed by perichaetial lvs, cleistocarpous, sub-globose with blunt apiculus, symmetrical; spores 15–25 μm; calyptra sub-cucullate. Fr autumn to spring. Reddish-brown, gregarious or solitary, winter ephemeral plants on calcareous soil in grassland and arable fields, occasional in S. England, rare elsewhere in England and extending north to Cheshire, Yorks and Durham, not known in Scotland or Ireland. 29. Fennoscandia, W. and C. Europe, Algeria, S. Ontario.

47. Acaulon C. Müll., Bot. Zeit., 1847

Plants minute, bud-like, ephemeral. Lvs ± imbricate, convolute or not, concave, sometimes strongly keeled, ovate, margin plane or recurved, entire or toothed above; nerve thin, ending below apex to excurrent; basal cells rectangular, cells above various in shape, rectangular to rhomboidal or quadrate, smaller towards margin and apex. Seta very short, straight or arcuate; capsule immersed, globose, with or without small apiculus, cleistocarpous; calyptra small, conical. A small genus of 13 species occurring in Europe, W. Asia, N. Africa, America, Australasia, Hawaii.

1 Upper lvs convolute, plants rounded when viewed from above, seta straight 2
Upper lvs barely overlapping, plants usually ± triangular when viewed from above seta arcuate **3. A. triquetrum**

2 Lf margin toothed near apex, cells in mid-lf mostly 2–3 times as long as wide, spores 30–50 μm **1. A. muticum**
Margin entire, cells in mid-lf 1.0–1.5 times as long as wide, spores 20–26 μm **2. A. minus**

1. A. muticum (Brid.) C. Müll., Bot. Zeit., 1847

Autoecious or dioecious. Plants minute, to 2 mm. Lower lvs patent, upper erect, convolute, concealing capsule when viewed from above, concave, not keeled, ovate, obtuse with reflexed apiculus to acute, margin plane, entire below, toothed near apex; nerve ending below apex to excurrent in apiculus; basal cells irregularly rectangular, cells in mid-lf rhomboidal or narrowly hexagonal, mostly 2–3 times as long as wide, 14–30 μm wide, narrower towards margin, shorter near apex. Seta *ca* 120 μm, straight; capsule immersed, erect, globose, cleistocarpous with small blunt apiculus; spores ± ovoid, papillose, 30–50 μm. Fr common, autumn to spring. Pale green, scattered or gregarious ephemeral plants on soil in arable fields, turf, on paths and wood rides, occasional. 79, H7, C. Europe, Siberia, Algeria, Morocco, N. America.

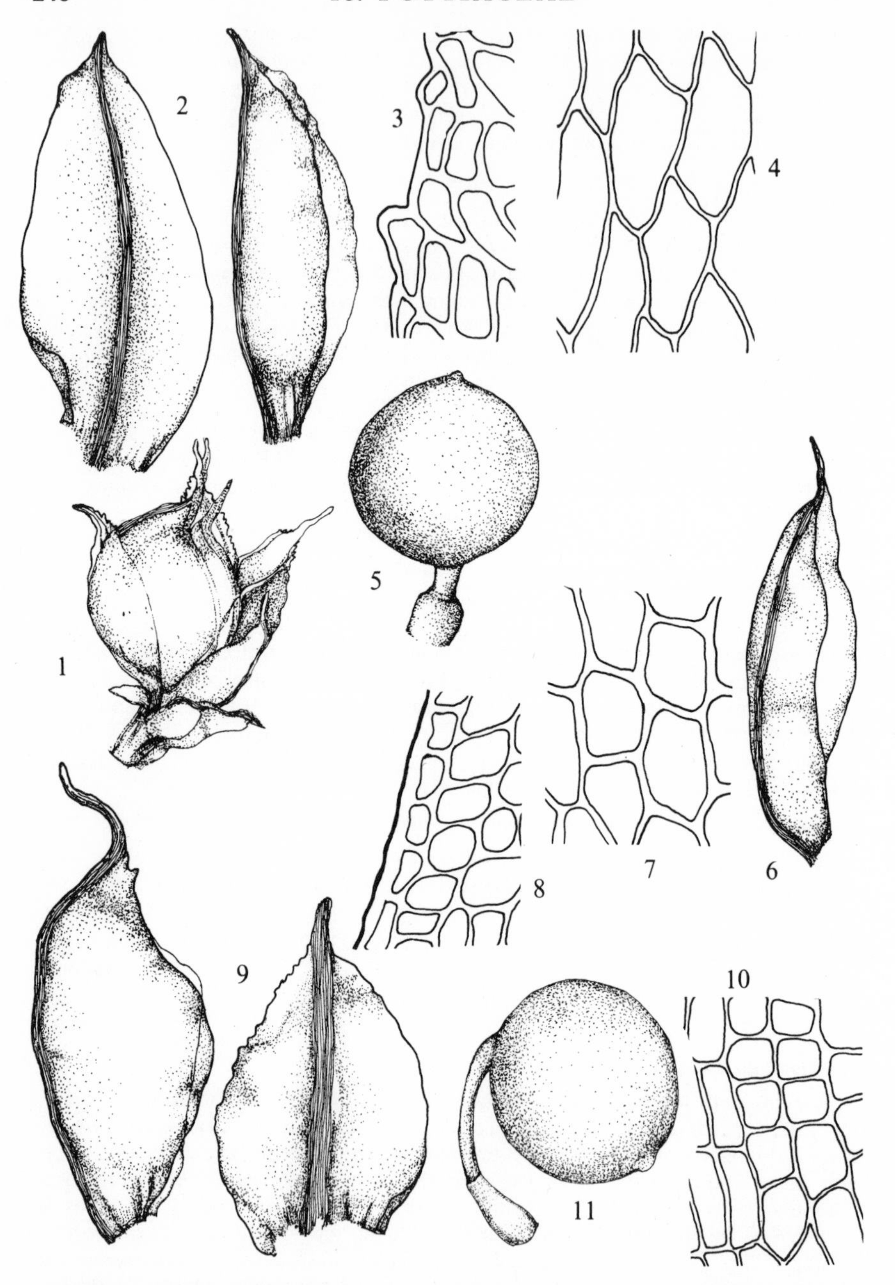

Fig. **112**. 1–5, *Acaulon muticum*: 1, plant (×20); 2, leaves (×40); 3, cells near leaf apex; 4, mid-leaf cells; 5, capsule. 6–8, *A. minus*: 6, leaf (×40); 7, cells near leaf apex; 8, mid-leaf cells. 9–11, *A. triquetrum*: 9, leaves (×65); 10, mid-leaf cells; 11, capsule. Capsules ×65, cells ×415.

2. A. minus (Hook. & Tayl.) Jaeg., Ber. S. Gall. Naturw. Ges., 1869
A. muticum var. *minus* (Hook. & Tayl.) Br. Eur.

Similar to *A. muticum* but lvs less convolute, hardly concealing capsule, lf margin entire; nerve strongly excurrent; cells in mid-lf 1.0–1.5 times as long as wide; spores ± spherical, finely papillose, 20–26 μm. Similar habitats to *A. muticum*, very rare. Cornwall, S. Devon, S. Somerset, W. Sussex, E. Norfolk, Anglesey, I. of Man, Guernsey. 8, C. Sweden, France, Algeria, Morocco.

3. A. triquetrum (Spruce) C. Müll., Bot. Zeit., 1847

Plants minute, to 1.5 mm. Upper lvs imbricate but scarcely overlapping, uppermost, usually 3, strongly keeled so that plant appears ± triangular when viewed from above, lvs concave, ovate, obtuse to acute, margin narrowly recurved above with a few sharp teeth near apex; nerve excurrent in reflexed apiculus; basal cells shortly rectangular, in mid-lf quadrate-rectangular to trapezoid, 12–20 μm wide, shorter towards apex. Seta arcuate, thin and fragile; capsule inclined, similar to that of *A. muticum*; spores finely papillose, *ca* 30 μm. Fr common, winter. Green to yellowish-brown, gregarious or solitary, ephemeral plants on soil on coastal cliffs, very rare. S. Devon, Dorset, I. of Wight, E. Sussex. 4. W. and C. Europe, W. Russia, Algeria, Morocco, N. America.

SUBFAMILY 2. TRICHOSTOMOIDEAE

Plants usually small, acrocarpous or rarely cladocarpous. Lvs lanceolate, acute to acuminate; basal cells quadrate to narrowly rectangular, hyaline or not, cells above quadrate, usually papillose, pellucid, or opaque. Capsule ovoid to cylindrical, cleistocarpous or stegocarpous; peristome of 16 teeth, cleft and spirally curved or not or lacking.

48. Hyophila Brid., Br. Univ., 1827

Dioecious. Lvs spathulate to longly lanceolate or linear, obtuse to acute, entire or toothed near apex; nerve ending below apex to excurrent; upper cells usually papillose. Capsule erect, ± cylindrical, sometimes ovoid; annulus fugacious; peristome lacking; calyptra cucullate. Rhizoidal or axillary gemmae sometimes present. A genus of *ca* 150 species occurring mainly in the warmer parts of the world.

1. H. stanfordensis (Steere) Smith & Whiteh., J. Bryol., 1974
Tortula stanfordensis Steere

Plants 2–7 mm. Lvs somewhat incurved when dry, spreading when moist, upper forming comal head, lingulate or spathulate or in stunted plants ovate, apex obtuse and apiculate, margin plane, bordered nearly to apex, irregularly toothed near apex; nerve strong, excurrent in small apiculus, in section with *ca* 4 rows of stereids; cells in lower part of lf rectangular, hyaline, above ± quadrate, very obscure, strongly papillose, 8–12 μm wide in widest part of lf, 2–4 rows marginal cells very narrow, incrassate, extending nearly to apex. Pale brown, irregularly ovoid rhizoidal gemmae, 30–60 μm × 50–100 μm, with protuberant cells always present. Seta erect, orange-red above, paler below, *ca* 2 mm long; capsule slightly inclined, narrowly ellipsoid, narrowing to mouth, *ca* 2 mm long; lid longly rostrate, falling with cucullate calyptra; peristome lacking; spores papillose, 20–22 μm. Fr very rare, summer. Yellowish-green or green patches or scattered plants on soil on footpaths, arable fields and river banks, rare. Cornwall, S.E. England, the Severn Basin, Mid W. Yorks. 9. Stanford, California.

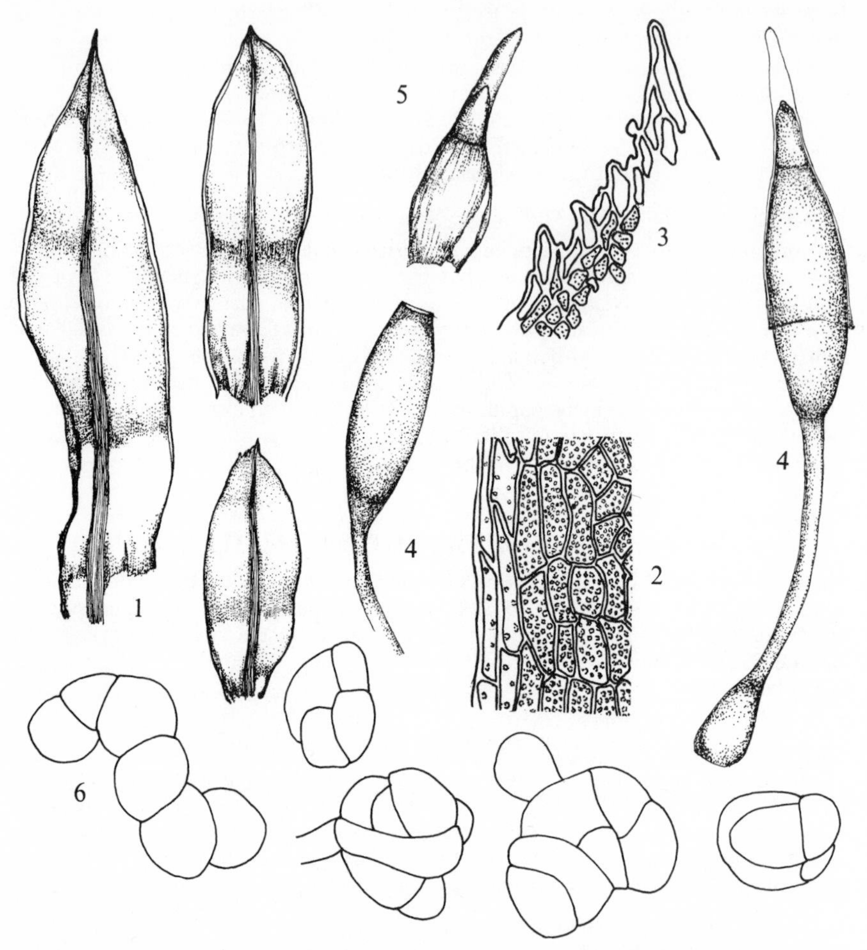

Fig. **113**. *Hyophila stanfordensis*: 1, leaves (×22.5); 2, marginal cells at widest part of leaf; 3, apical cells; 4, capsules (×15); 5, calyptra (×15); 6, gemmae (×250). Cells ×415.

This plant dies down in the summer and at the time when the fruits are ripe the gametophyte is much decayed. For the occurrence of this plant in Britain see Whitehouse, *Trans. Br. bryol. Soc.* **4**, 84–94, 1961.

49. Barbula Hedw., Sp. Musc., 1801

Usually dioecious. Plants forming tufts, cushions or short turfs, stems with central strand. Lvs mostly lanceolate, gradually tapering from below middle to acute apex, margin plane or recurved, entire or crenulate; nerve usually ending in apex or excurrent; basal cells rectangular, above quadrate or hexagonal, papillose or not, pellucid or obscure. Seta long; capsule erect, symmetrical, ellipsoid to cylindrical, straight or slightly curved; lid oblique, longly rostrate; peristome with very short basal membrane, long or short, teeth divided into 2–3 filiform, papillose, straight or twisted segments; calyptra cucullate. A cosmopolitan but mainly temperate genus of *ca* 500 species.

The genus *Barbula* as recognised here is a somewhat heterogenous group of plants and is divided by some authorities into 4 genera as follows: *Barbula* Hedw. (*B. unguiculata* to *B. maxima*, *B. vinealis* and *B. asperifolia*); *Didymodon* Hedw. (*B. spadicea*, *B. glauca*, *B. rigidula* to *B. tophacea*); *Bryoerythrophyllum* Chen (*B. recurvirostra* and *B. ferruginascens*); *Streblotrichum* P. Beauv. (*B. convoluta*). The separation of *Barbula* and *Didymodon* is based upon peristome characters and separates clearly closely related species. *Bryoerythrophyllum* and *Streblotrichum* are more distinct but if they are recognised then *B. unguiculata* should also be placed in an individual genus. It is more satisfactory and convenient to retain all the species in 1 genus.

1 Lvs oblong-lanceolate to narrowly lingulate, tapering in upper part only 2
Lvs broadly ovate to linear-lanceolate, gradually tapering from below middle 3

2 Seta yellowish, lf margin plane or narrowly recurved in lower part of lf, nerve ending below apex to shortly excurrent **1. B. convoluta**
Seta dark red or purplish, margin recurved except near apex, nerve excurrent **2. B. unguiculata**

3 Nerve excurrent 4
Nerve ending in or below apex 10

4 Lvs broadly ovate, nerve very stout, 70–100 μm wide near lf base, mid-lf cells 6–8 μm wide, axillary gemmae abundant **17. B. cordata**
Lvs linear-lanceolate to ovate, nerve narrower, gemmae lacking or if present then lf cells more than 8 μm wide 5

5 Margin revolute ± from base to apex, ± reaching nerve in upper part of lf 6
Margin plane in upper part of lf 7

6 Lvs twisted when dry, apex acute, cells 10–14 μm wide in mid-lf **3. B. hornschuchiana**
Lvs tightly incurved when dry, apex obtuse, cells 8–10 μm wide **4. B. revoluta**

7 Lvs tapering to thick obtuse point **14. B. rigidula**
Lvs tapering to acute or acuminate point 8

8 Upper lvs narrowly lanceolate to linear-lanceolate, not concave, cells in upper part of lf opaque **20. B. cylindrica**
Upper lvs ovate to lanceolate, concave, upper cells pellucid 9

9 Nerve shortly to longly excurrent, upper cells with ± rounded lumens, axillary gemmae sometimes present **5. B. acuta**
Lvs ending in long subulate point consisting mainly of longly excurrent nerve, upper cells with ± angular lumens, axillary gemmae lacking **6. B. icmadophila**

10 Cells over nerve (ventral cells) elongated at middle part of lf 11
Cells over nerve ± quadrate, hexagonal or shortly rectangular at middle of lf 15

11 Lvs patent when moist, upper part of lf ± lingulate with obtuse to acute apex **18. B. tophacea**
Lvs erecto-patent to strongly recurved when moist, upper part of lf gradually tapering to acute to obtuse apex 12

12 Lvs erecto-patent to recurved, cells usually slightly papillose 13
Lvs strongly recurved when moist, cells conspicuously papillose 14

13 Lvs usually ± recurved when moist, upper mostly 1.2–2.4 mm long, acuminate, peristome long (1.0–1.5 mm), spirally twisted **7. B. fallax**

Lvs patent or spreading when moist, upper mostly 1.8–4.0 mm long, apex acute to obtuse, peristome short (*ca* 0.5 mm), ± straight **10. B. spadicea**

14 Plants mostly to 2.5 cm high, lvs 0.8–1.8 mm, long, basal cells not incrassate **8. B. reflexa**
Plants 3–9 cm, lvs 2–4 mm, basal cells incrassate **9. B. maxima**

15 Plants reddish-brown ± throughout, plants to 6 cm 16
Plants green to olive-brown, darker or rusty-red below or not, robust or not 17

16 Lvs patent or recurved when moist, ovate, margin recurved ± from base to apex **12. B. asperifolia**
Lvs patent or spreading, margin recurved below, plane above **22. B. ferruginascens**

17 Plants rusty-red or reddish-brown below, margin often denticulate near apex, basal cells narrowly rectangular **21. B. recurvirostra**
Plants dull green. brownish or blackish below, margin entire near apex, basal cells quadrate to rectangular 18

18 Lf cells mamillosely protuberant, axillary gemmae abundant **13. B. mamillosa**
Lf cells not mamillosely protuberant, axillary gemmae present or not 19

19 Lf margin bistratose, axillary gemmae usually present, plants not glaucous 20
Lf margin unistratose, axillary gemmae lacking or if present then plants glaucous-green 21

20 Lvs lanceolate, cells 8–10 μm wide in mid-lf **14. B. rigidula**
Lvs ovate to ovate-lanceolate, mid-lf cells 6–8 μm wide **15. B. nicholsonii**

21 Upper lvs ovate-lanceolate to ovate, imbricate when dry **16. B. trifaria**
Upper lvs narrowly lanceolate to linear-lanceolate, flexuose to crisped when dry 22

22 Lf base ± erect with rectangular, ± hyaline cells, margin plane below, plane or narrowly recurved above, plants glaucous-green **11. B. glauca**
Lf base not erect, basal cells quadrate-rectangular, lf margin recurved ± from base to apex, plants not glaucous 23

23 Upper lvs flexuose or slightly twisted when dry, not flexuose when moist, mostly 1–3 mm long **19. B. vinealis**
Upper lvs curled when dry, flexuose when moist, mostly 2–5(–7) mm long **20. B. cylindrica**

1. B. convoluta Hedw., Sp. Musc., 1801
Streblotrichum convolutum (Hedw.) P. Beauv.

Dioecious. Plants 2–25 mm, stems tomentose below. Lvs strongly curved when dry, erecto-patent to spreading or recurved when moist, oblong-lanceolate or lingulate, obtuse to acute, margin plane or narrowly recurved below; crenulate-papillose; nerve strong, green to brownish, ending below apex to shortly excurrent in apiculus, ventral cells elongated; cells ± incrassate, basal irregularly rectangular, narrower towards margin, above ± quadrate or quadrate-hexagonal, strongly papillose, obscure, 8–10(–12) μm wide in mid-lf. Perichaetial lvs convolute, inner sheathing, nerveless. Brown, spherical or ovoid rhizoidal gemmae, 110–140 μm long, often present. Seta pale yellow; capsule erect, ellipsoid to sub-cylindrical, straight or slightly curved; lid longly rostrate; peristome teeth long, filiform, spirally twisted; spores 8–10 μm. Fr occasional, spring, summer.

Var. **convoluta**

Plants 2–5(–20) mm. Upper lvs 0.8–1.4 mm long, margin not undulate. $n=11$, 14*. Bright or yellowish-green tufts or patches, sometimes extensive, on soil on waste

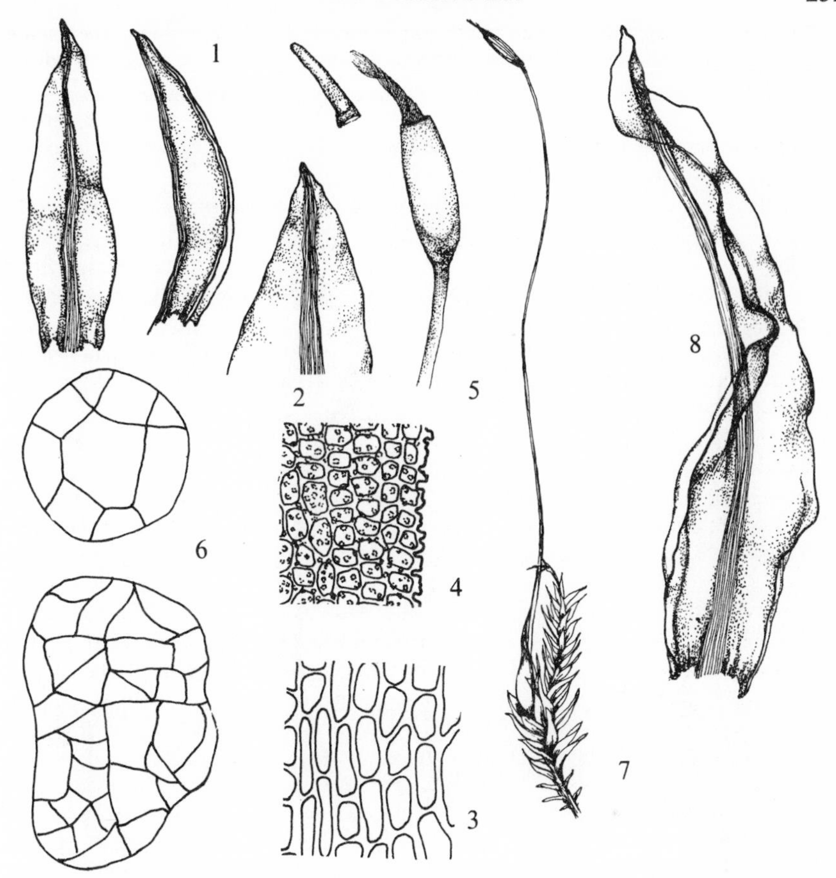

Fig. **114**. 1–7, *Barbula convoluta* var. *convoluta*: 1, leaves; 2, leaf apex (×100); 3, basal cells; 4, mid-leaf cells; 5, capsule (×15); 6, rhizoidal gemmae (×250); 7, plant (×5). 8, *B. convoluta* var. *commutata*: leaf. Leaves ×40, cells ×415.

ground, paths, in arable fields, crevices in footpaths, old buildings, etc., common. 111, H39, C. Europe, Caucasus, Siberia, China, Japan, Tunisia, Algeria, Azores, N. America, Mexico, Australia.

Var. **commutata** (Jur.) Husn., Fl. Mouss. N. Ouest, 1886
Var. *sardoa* Br. Eur.
Larger than the type. Plants 5–10(–25) mm. Upper lvs 1.2–1.4 mm long, lingulate, margin strongly undulate. Similar habitats to the type. Occasional in N. Britain, frequent elsewhere. 77, H31, C. S.W. Europe, Sardinia, Balearic Is., Cyprus, Morocco.

Cultivation experiments suggest that var. *commutata* is merely a more vigorous form resulting from higher nutrient status of the substrate.

2. B. unguiculata Hedw., Sp. Musc., 1801

Dioecious. Plants mostly 5–25 mm, stems orange-red, not tomentose. Lvs twisted

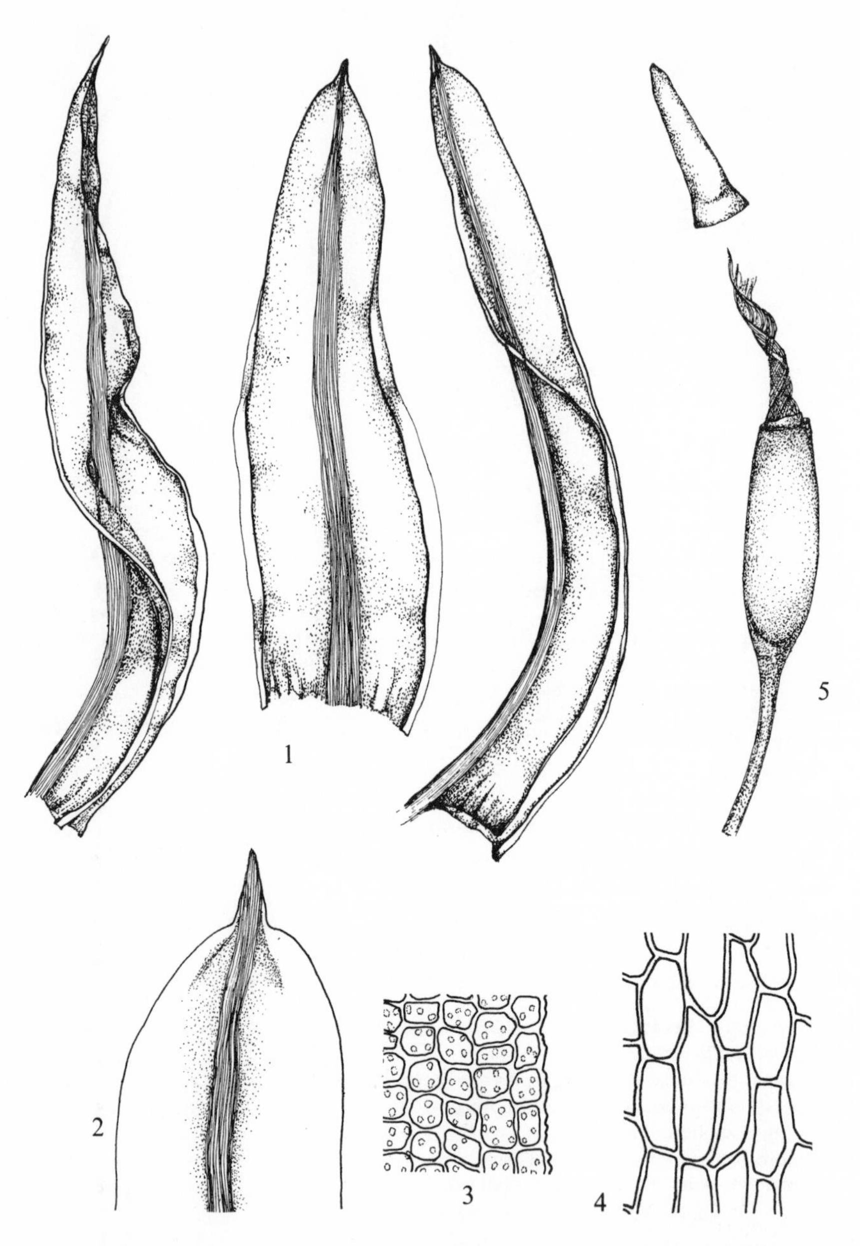

Fig. **115**. *Barbula unguiculata*: 1, leaves (× 40); 2, leaf apex (× 100); 3, mid-leaf cells; 4, basal cells; 5, capsule (× 15). Cells × 415.

or incurved when dry, erecto-patent to patent or recurved when moist, lingulate to narrowly lingulate or lingulate-lanceolate, apex obtuse or rounded, apiculate, occasionally acute or acuminate, margin recurved except towards apex, papillose-crenulate; nerve stout, green to brownish, excurrent in thick point, ventral cells elongate; basal cells rectangular or narrowly rectangular, shorter towards margin, above ± quadrate, incrassate, papillose, 10–14 μm wide in mid-lf. Perichaetial lvs larger with slightly sheathing base. Seta dark red to purplish; capsule erect, narrowly ellipsoid, straight; lid longly rostrate; peristome teeth filiform, spirally curved; spores 10–14 μm. Fr frequent, autumn to spring. $n=11^*$, $11+1$, 13^*, $13+1^*$, $13+2^*$, 14, $14+2$, 16^*, 24^*. Patches, bright green above, reddish-brown below, on soil on waste ground, in arable fields, on paths, walls, old buildings, etc., common. 112, H40, C. Europe, Faroes, Iceland, N., W. and C. Asia, China, Japan, Hong Kong, Algeria, N. America, Mexico, southern S. America, Australia.

A very variable species. A form with narrow lvs and more acute tapering apex, var. *cuspidata* (Schultz.) Brid., occurs in Britain but with the range of forms that exist it is impossible to delimit the variety satisfactorily and it is not recognised here. The species may be confused with *B. convoluta* which differs in the tomentose stems, pale yellow seta, the lf margin not or only narrowly recurved below and the nerve rarely excurrent.

3. B. hornschuchiana Schultz, Nov. Acta Acad. Leop., 1823

Dioecious. Plants mostly 3–15 mm. Lvs incurved, spirally twisted when dry, erecto-patent when moist and giving branches a stellate appearance when viewed from above, lanceolate, tapering from near base to apex, often recurved to nerve in upper part of lf; nerve stout, excurrent in acuminate point, ventral cells ± quadrate; basal cells rectangular, above quadrate-hexagonal, incrassate, papillose, pellucid, 10–14(–16) μm wide in mid-lf. Perichaetial lvs larger, longer, tapering to filiform apex. Seta orange-red; capsule narrowly ellipsoid; lid long, oblique; peristome teeth long, filiform, spirally twisted; spores 8–10 μm. Fr rare, winter, spring. Green patches, sometimes locally abundant on soil on waste ground, arable fields, banks, paths, walls, cliff ledges, etc., frequent or common. 94, H24, C. Europe, Syria, Iran, Azores, Morocco, Ethiopia, southern Africa.

Differs from *B. revoluta* in the lvs tapering to acuminate apex, less tightly incurved when dry. May be recognised in the field from the stellate appearance of the branches when viewed from above.

4. B. revoluta Brid. in Schrad., J. f. Bot., 1801

Dioecious. Plants 3–15 mm. Lvs tightly incurved when dry, erecto-patent when moist, narrowly lingulate to narrowly lanceolate, apex obtuse and apiculate, margin revolute almost from base to apex and recurved to nerve in upper part of lf; nerve very stout, usually excurrent, ventral cells ± quadrate; basal cells rectangular, incrassate, above quadrate, very incrassate, papillose, opaque, lumens rounded, 8–10 μm wide in mid-lf. Seta flexuose, orange-red; capsule narrowly ellipsoid; lid longly rostrate, oblique; peristome teeth filiform, spirally twisted, fragile; spores 10–14 μm. Fr occasional to rare, spring. Small, dense, dark green tufts or cushions on basic rocks, mortar of walls, old buildings etc., frequent. 109, H39, C. Europe, Iran, Madeira, Canaries, Tunisia, Morocco.

Readily recognised by the stout nerve, strongly revolute lf margin, the obtuse and apiculate apex and, when dry, the tightly incurved lvs.

5. B. acuta (Brid.) Brid., Mant. Musc., 1819
B. gracilis (Schleich.) Schwaegr.

Dioecious. Plants 3–20 mm. Lvs appressed, flexuose when dry, erecto-patent when

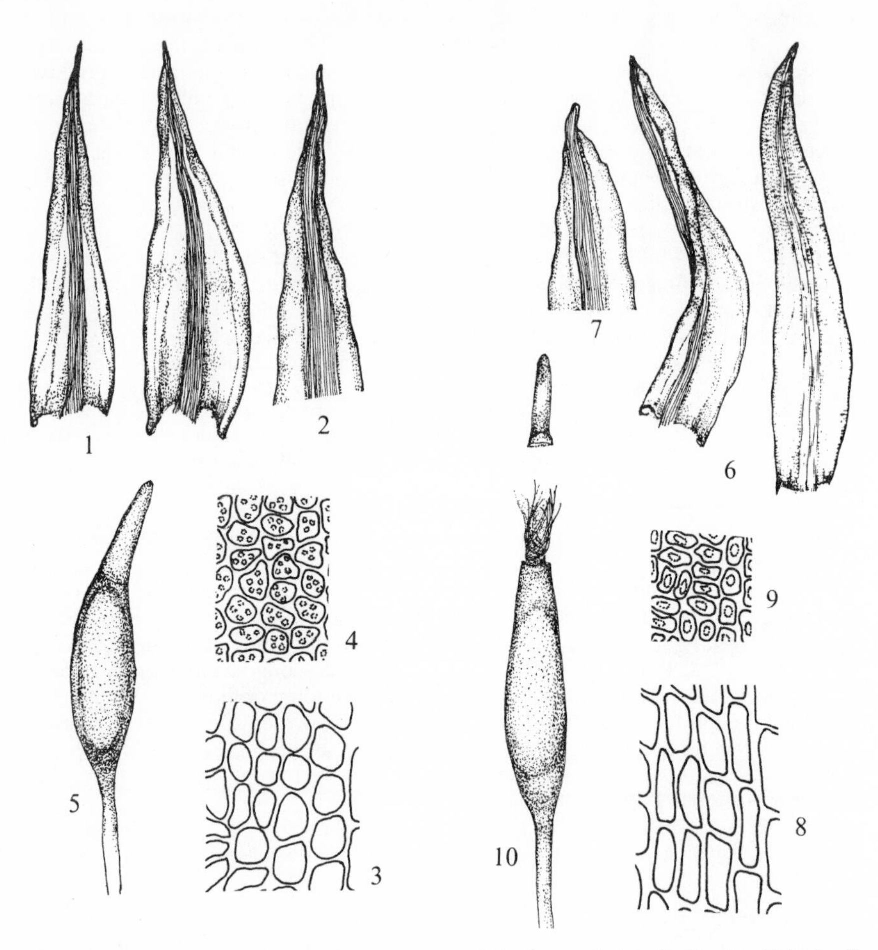

Fig. **116**. 1–5, *Barbula hornschuchiana*: 1, leaves; 2, leaf apex; 3, basal cells; 4, mid-leaf cells; 5, capsule. 6–10, *B. revoluta*: 6, leaves; 7, leaf apex; 8, basal cells; 9, mid-leaf cells; 10, capsule. Leaves ×40, apices ×100, cells ×415, capsules ×15.

moist, concave, ovate to lanceolate, gradually tapering to acute apex, margin narrowly recurved below, entire; nerve strong, occupying most of upper part of lf, excurrent in upper lvs, ventral cells hexagonal; basal cells rectangular, cells above hexagonal, incrassate with rounded lumens, smooth, pellucid, 8–10(–15) μm wide in mid-lf. Small spherical gemmae sometimes present in axils of upper lvs. Capsule ellipsoid or narrowly ellipsoid; lid obliquely rostrate; peristome teeth filiform, spirally twisted; spores 9–12 μm. Fr very rare, late winter, spring. Green to brownish patches or scattered plants on calcareous soil in turf and in sand-dunes and on montane rock ledges, occasional from Cornwall and Sussex north to Yorks and Westmorland, Angus, N. Aberdeen, rare in Ireland. 33, H9, C. Europe, Caucasus, Morocco, Madeira, Azores, N. America.

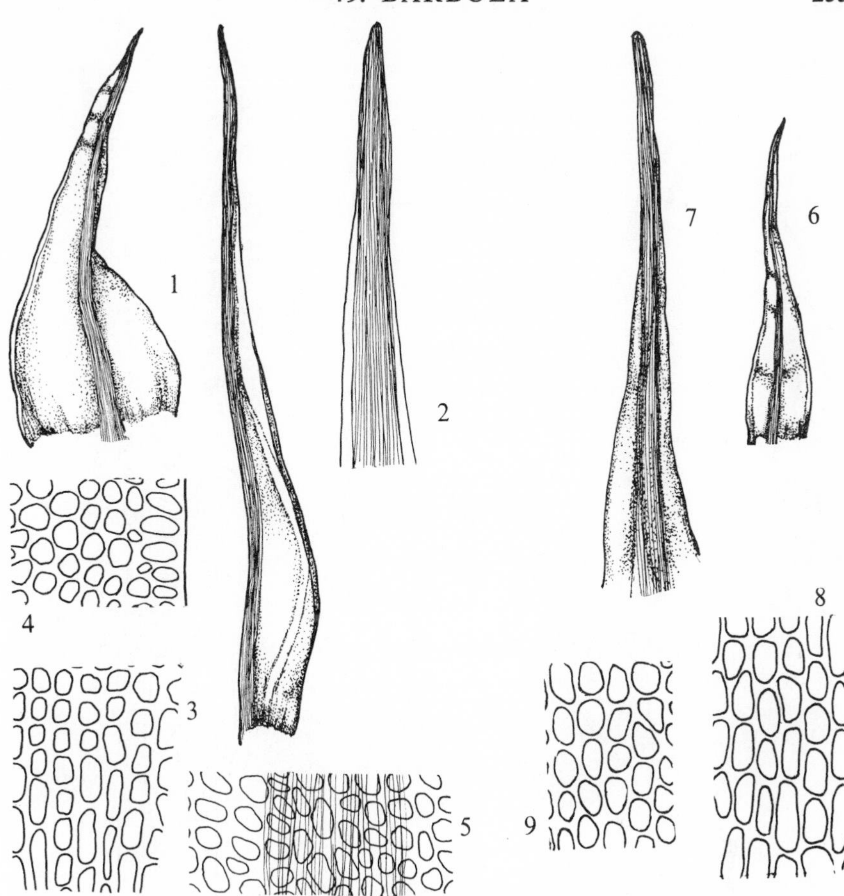

Fig. **117**. 1–5, *Barbula acuta*: 1, leaves; 2, leaf apex; 3, basal cells; 4, mid-leaf cells; 5, cells on ventral face of nerve. 6–9, *B. icmadophila*: 6, leaf; 7, leaf apex; 8, basal cells; 9, mid-leaf cells. Leaves ×40, apices ×100, cells ×415.

6. B. icmadophila Schimp. ex C. Müll., Syn., 1849
B. acuta ssp. *icmadophila* (C. Müll.) Amann

Dioecious. Plants to 2(–6) cm, shoots very slender. Lvs appressed, slightly flexuose when dry, patent when moist, concave, lanceolate, tapering to long, subulate apex composed mainly of nerve; margin narrowly recurved below, entire; nerve strong, excurrent, often longly so; basal cells shortly rectangular, above ± quadrate, slightly incrassate with angular lumens, *ca* 10 μm wide in mid-lf. Capsule narrowly ellipsoid. Fr unknown in Britain. Brownish-green tufts or patches on basic, montane rocks, very rare. Cumberland, Mid Perth, Banff, Skye. 4. Montane parts of Europe, Faroes, Iceland, Caucasus, Siberia, Kamchatka, Kashmir, N. America.

The tall, slender stems and longly excurrent nerve will distinguish this from any other British *Barbula* species except some forms of *B. acuta* with a longly excurrent nerve: these differ in the cells very incrassate with rounded lumens and frequent occurrence of axillary gemmae.

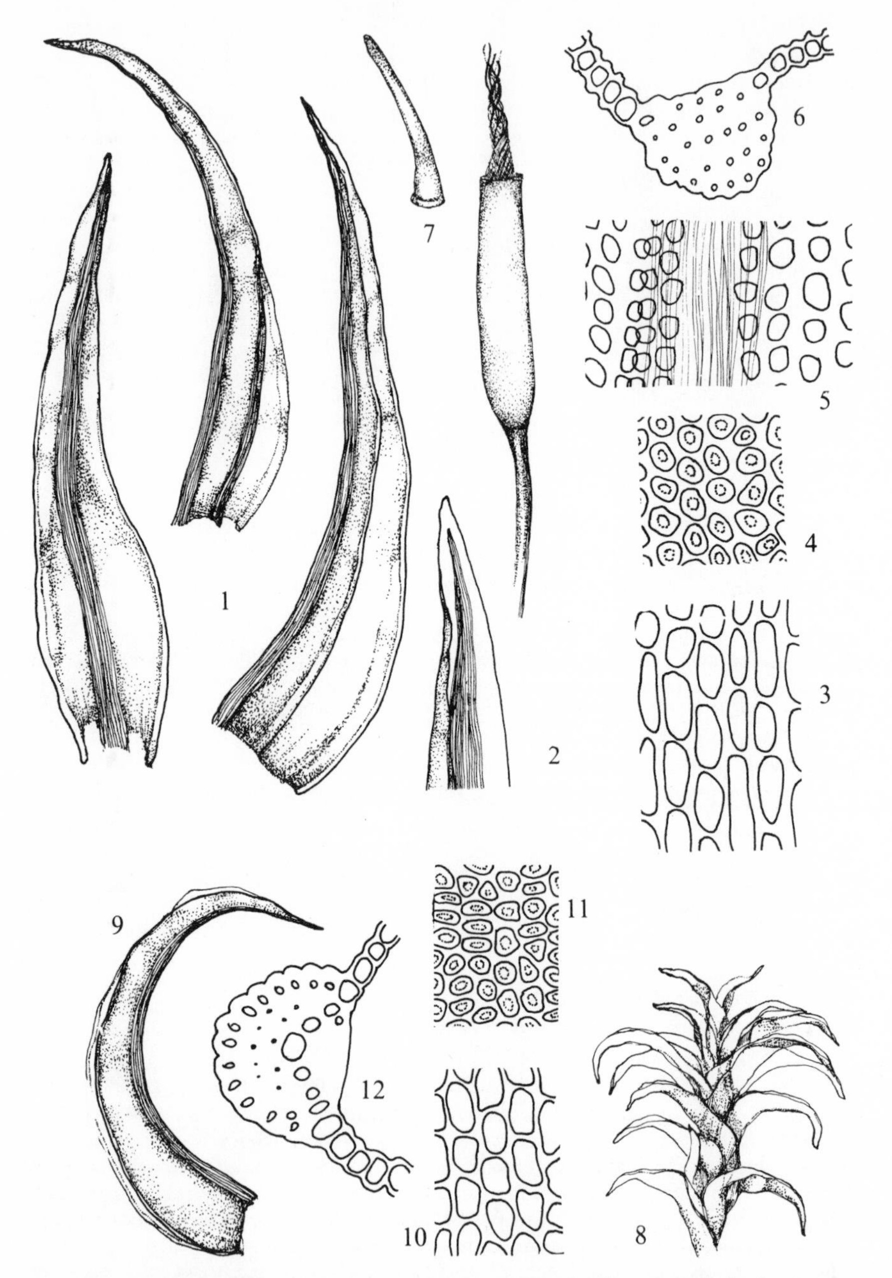

Fig. **118**. 1–7, *Barbula fallax*: 1, leaves; 2, leaf apex; 3, basal cells; 4, mid-leaf cells; 5, upper surface of nerve in middle of leaf; 6, section of nerve; 7, capsule. 8–12, *B. reflexa*: 8, shoot (×15); 9 leaf; 10, basal cells; 11, mid-leaf cells; 12, nerve section. Leaves ×40, apices ×100, cells ×415, sections ×250.

7. B. fallax Hedw., Sp. Musc., 1801

Dioecious. Plants to 15(–30) mm. Lvs ± distant, erect, flexuose to twisted when dry, erecto-patent to spreading or recurved when moist, 1.2–2.4 mm long, keeled, lanceolate to narrowly lanceolate from broad base, acuminate, margin recurved, at least below, entire; nerve strong, reddish-brown, ending below apex to percurrent, ventral cells elongated, in section cells ± uniform; basal cells rectangular, above irregularly hexagonal, incrassate with rounded lumens, papillose, pellucid, 8–12 μm wide in mid-lf. Seta flexuose, reddish-brown; capsule ellipsoid or narrowly ellipsoid, straight or slightly curved, lid oblique, longly rostrate; peristome teeth filiform, spirally twisted, 1.0–1.5 mm long; spores 12–16 μm. Fr occasional, winter, spring. $n=13$. Greenish-brown, often reddish-tinged, lax patches or scattered plants on soil in fields, waste ground, paths, wood rides, walls, old buildings, etc., common. 111, H40, C. Europe, Faroes, Asia, Madeira, N. Africa, N. America, Greenland.

The lvs are usually longer, narrower and with less conspicuous papillae than in *B. reflexa*, but forms with shorter lvs or more conspicuous papillae occur and may only be separated from *B. reflexa* by the less recurved lvs and the nerve uniform in section.

8. B. reflexa (Brid.) Brid., Mant. Musc., 1819
B. recurvifolia Schimp.

Dioecious. Plants to 2.5 cm. Lvs erect, flexuose when dry, recurved when moist, keeled above, lanceolate, tapering from broad base, acute, margin recurved below, entire; nerve ending below apex, papillose at back, ventral cells elongated, with 2 guide cells in section; basal cells shortly rectangular, above hexagonal, incrassate, lumens rounded, papillose with tall papillae, 6–10 μm wide in mid-lf. Fr similar to *B. fallax*, unknown in Britain. Green to reddish-brown, ± lax tufts or patches on calcareous rocks, soil and sand-dunes, generally distributed, occasional. 67, H22. Europe, Caucasus, Siberia, Himalayas, China, Tunisia, Algeria, N. America.

9. B. maxima Syed & Crundw., J. Bryol., 1973
B. reflexa var. *robusta* Braithw., *B. recurvifolia* var. *robusta* (Braithw.) Par.

Plants tall, 2–8 cm, stem with well developed central strand. Lvs 2–4 mm long, flexuose when dry, recurved when moist, narrowly lanceolate from broad basal part, gradually tapering to acute apex, margin recurved, entire; nerve ending in apex, ventral cells elongated; cells very incrassate, basal narrowly rectangular, cells above irregularly hexagonal with ± angular lumens, papillose, 10–12 μm wide in mid-lf. Fr unknown. Tall, yellowish-brown tufts on basic montane ledges, very rare. Sligo, Leitrim. H2. Endemic to Ireland.

Differs from *B. reflexa* in its larger size and relatively narrower and more longly tapering lvs and the angular cell lumens.

10. B. spadicea (Mitt.) Braithw., Brit. Moss Fl., 1887
Didymodon spadiceus (Mitt.) Limpr.

Dioecious. Plants mostly 10–30 mm. Lvs erect, flexuose to ± twisted when dry, patent to spreading when moist, (1.6–)1.8–4.0 mm long, from broad base tapering to acute to obtuse apex, margin recurved below, entire; nerve stout, ending below apex, in section with several guide cells, ventral cells elongated; extreme basal cells rectangular to narrowly rectangular, above irregularly hexagonal, incrassate, papillose, 8–10(–12) μm wide in mid-lf. Seta deep red below, paler above; capsule narrowly ellipsoid to cylindrical, straight or slightly curved; lid oblique, longly rostrate, peristome teeth short, *ca* 0.5 mm, filiform, ± straight; spores 10–16 μm. Fr occasional, autumn. Greenish-brown tufts on rocks and alluvial detritus by streams and rivers,

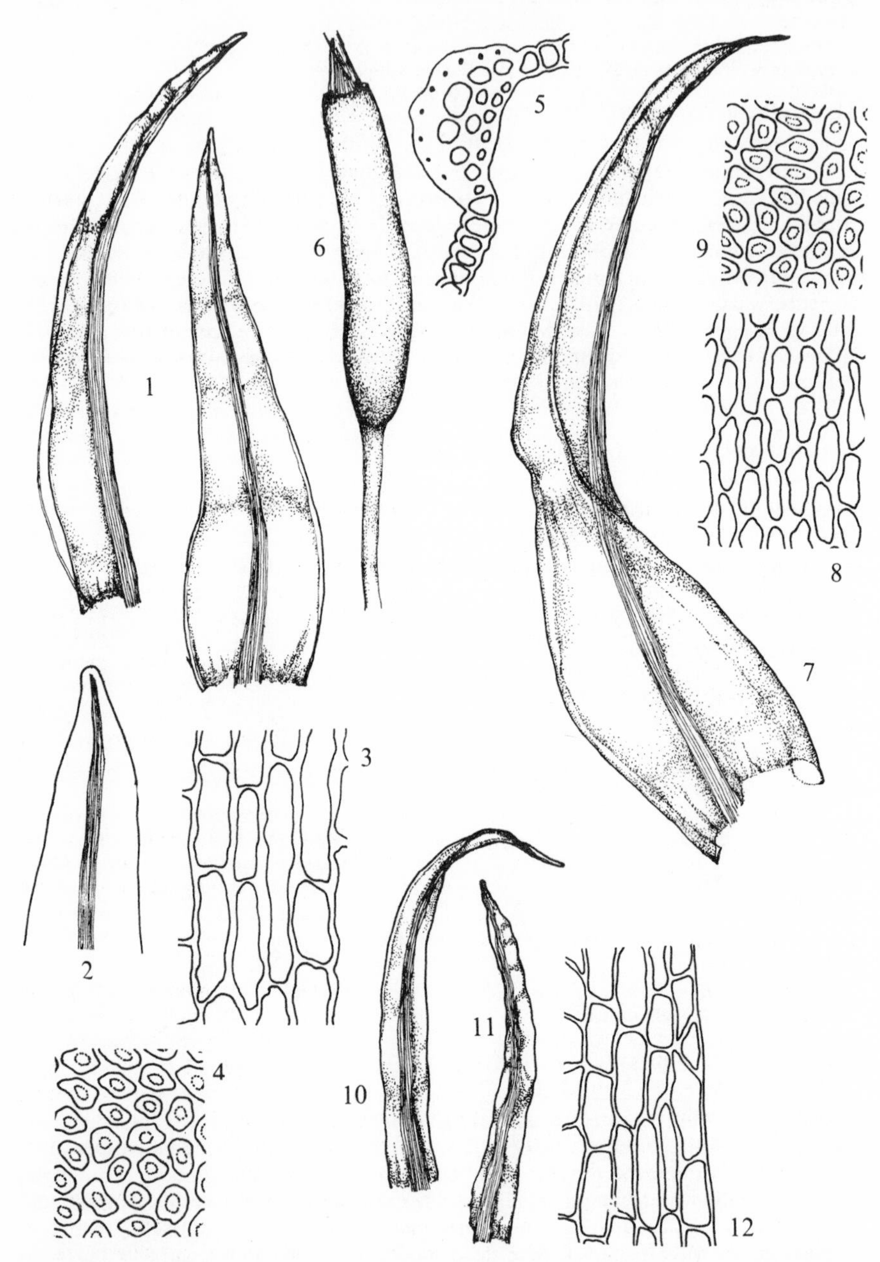

Fig. **119**. 1–6, *Barbula spadicea*: 1, leaves (×25); 2, leaf apex; 3, basal cells; 4, mid-leaf cells; 5, nerve section (×250); 6, capsule (×15). 7–9, *B. maxima*: 7, leaf (×25); 8, basal cells; 9, mid-leaf cells. 10–12, *B. glauca*: 10, leaf (×40); 11, leaf apex; 12, basal cells. Apices ×100, cells ×415.

especially where sheltered and humid. Occasional in W. and N. Britain, very rare elsewhere. 75, H24. Europe, Caucasus, N. America, Greenland.

Small forms may be difficult to distinguish from *B. fallax* but the lvs are less recurved, the nerve has several guide cells in section and the peristome is shorter.

11. B. glauca (Ryan) Möll., Bot. Not., 1907
Didymodon rigidulus var. *glaucus* (Ryan) Wijk & Marg.

Plants very small, 1–3 mm. Lvs incurved when dry, erecto-patent to recurved from ± erect base when moist, linear-lanceolate, tapering to long acute apex, margin plane or variously recurved, sometimes obscurely denticulate near base, slightly crenulate above; nerve ending below apex, ventral cells quadrate; basal cells rectangular, hyaline, above hexagonal, thin-walled, papillose, 8–10 μm wide in mid-lf. Occasional, irregular, several-celled gemmae, *ca* 80 μm long, sometimes present in axils of upper lvs. Fr unknown. Bright green patches in crevices of dry chalk or limestone, very rare. N. Wilts, W. Sussex, N.W. Yorks. 3. Norway, Czechoslovakia, Austria, Switzerland, Hungary, Germany, Sweden.

There has been confusion between *B. glauca* and *Trichostomopsis umbrosa*, but the latter differs in the expanded lf base with narrow marginal cells, the plane, strongly crenulate margin and the nerve ending a considerable distance below the apex. *T. umbrosa* occurs on damp shaded brickwork whilst *B. glauca* is found in crevices of dry chalk or limestone. The exact nature of *B. glauca* requires further study as the descriptions given by different authorities vary widely.

12. B. asperifolia Mitt., J. Linn. Soc. Bot., Suppl. I, 1859
B. rufa (Lor.) Jur.

Dioecious. Plants robust, to 6 cm, stems without or with thin central strand. Lvs erect, flexuose when dry, patent to recurved when moist, ovate, tapering to acute to obtuse apex, margin entire, recurved almost from base to apex; nerve stout, ending in apex, ventral cells quadrate; basal cells rectangular, above hexagonal, slightly papillose, 10–12 μm wide in mid-lf. Fr unknown. Dense or lax, reddish-brown tufts on basic, alpine rocks, reported from Ben Lawers. Europe, Iceland, C. Asia, Japan, C. Africa, arctic N. America.

There are two specimens of this plant collected by Stirton, apparently from Ben Lawers in 1867, but there is some doubt as to the authenticity of the locality; see Crundwell, *Trans. Br. bryol. Soc.* **3**, 174–9, 1957. It approaches *B. maxima* in stature but differs in the reddish-brown colour, the much wider lvs with blunter apex and the ventral cells of the nerve quadrate.

13. B. mamillosa Crundw., J. Bryol., 1976

Dioecious. Plants 4–8 mm, stems frequently branched. Lvs appressed, scarcely twisted and not at all crisped when dry, erect-spreading when moist, 0.5–1.2 mm long, ovate to lanceolate, channelled, tapering to acute, not or only slightly acuminate apex, margin narrowly recurved above; nerve ending shortly below apex or per-current, lamina cells continuous over ventral surface except in lower part of lf; cells in basal part of lf quadrate to shortly rectangular, cells above quadrate or wider than long, 8–10 μm wide, incrassate with round or oval lumens, mamillosely protuberant, obscurely papillose. Brown, spherical or ovoid gemmae, 30–45 μm diameter, abundant in axils of upper lvs. Only male plants known. Dense, compact, green and brown tufts on calcareous rock. W. Perth. 1. Endemic.

Known only from a single gathering. Closely related to *B. rigidula* but differing in the smaller size, the lvs straighter and more closely appressed when dry, relatively wider lvs and the margins less regularly recurved; the mamillose cells are also distinctive. For an account of the species see Crundwell, *J. Bryol.* **9**, 163–6, 1977.

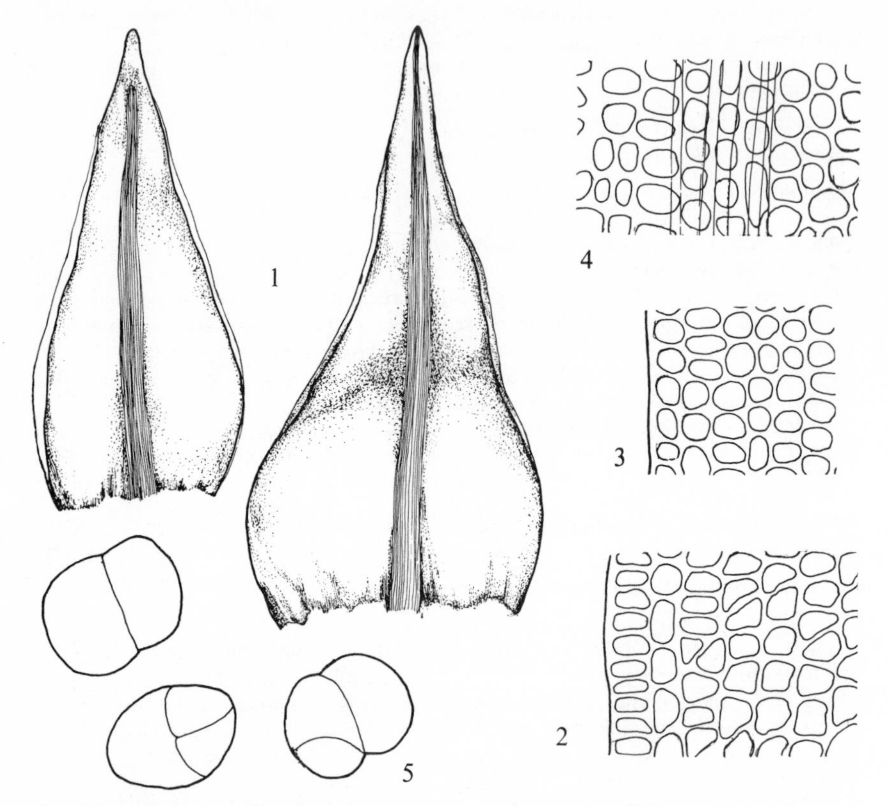

Fig. **120**. *Barbula mamillosa*: 1, leaves (×65); 2, basal cells; 3, mid-leaf cells; 4, cells over nerve at middle of leaf; 5, gemmae. Cells and gemmae ×415.

14. B. rigidula (Hedw.) Mitt., J. Bot., 1867
Didymodon rigidulus Hedw., *D. rigidulus* ssp. *andreaeoides* (Limpr.) Wijk & Marg., *Grimmia andreaeoides* Limpr.

Dioecious. Plants mostly 3–10 mm. Lvs appressed, flexuose, sometimes slightly twisted when dry, patent when moist, lanceolate or lingulate-lanceolate, gradually tapering to long, thick, ± subulate, obtuse apex, margin recurved or plane below, bistratose above; nerve ending well below apex to shortly excurrent, ventral cells ± hexagonal; cells in lower part of lf rectangular, above quadrate-hexagonal, incrassate, sometimes strongly so, slightly papillose, obscure, 8–10 μm wide in mid-lf, bistratose at margin. Globose several-celled gemmae, 25–80 μm diameter, frequently present in axils of upper lvs. Seta orange-red; capsule ellipsoid to cylindrical; lid oblique, longly rostrate; peristome teeth short, 320–480 μm long, filiform, ± straight; spores 10–12 μm. Fr occasional, autumn to spring. $n=12$*, 13*. Dark to yellowish-green tufts or cushions on walls, old buildings and rocks, usually where basic, frequent to common. 111, H38, C. Europe, Caucasus, C. Asia, Siberia, Kamchatka, Kashmir, China, Canaries, N. and C. America, Greenland, Falkland Is.

The plant originally described as *Grimmia andreaeoides* is a montane form of *B. rigidula*. Although apparently distinct in its lingulate-lanceolate lvs, short nerve and very incrassate

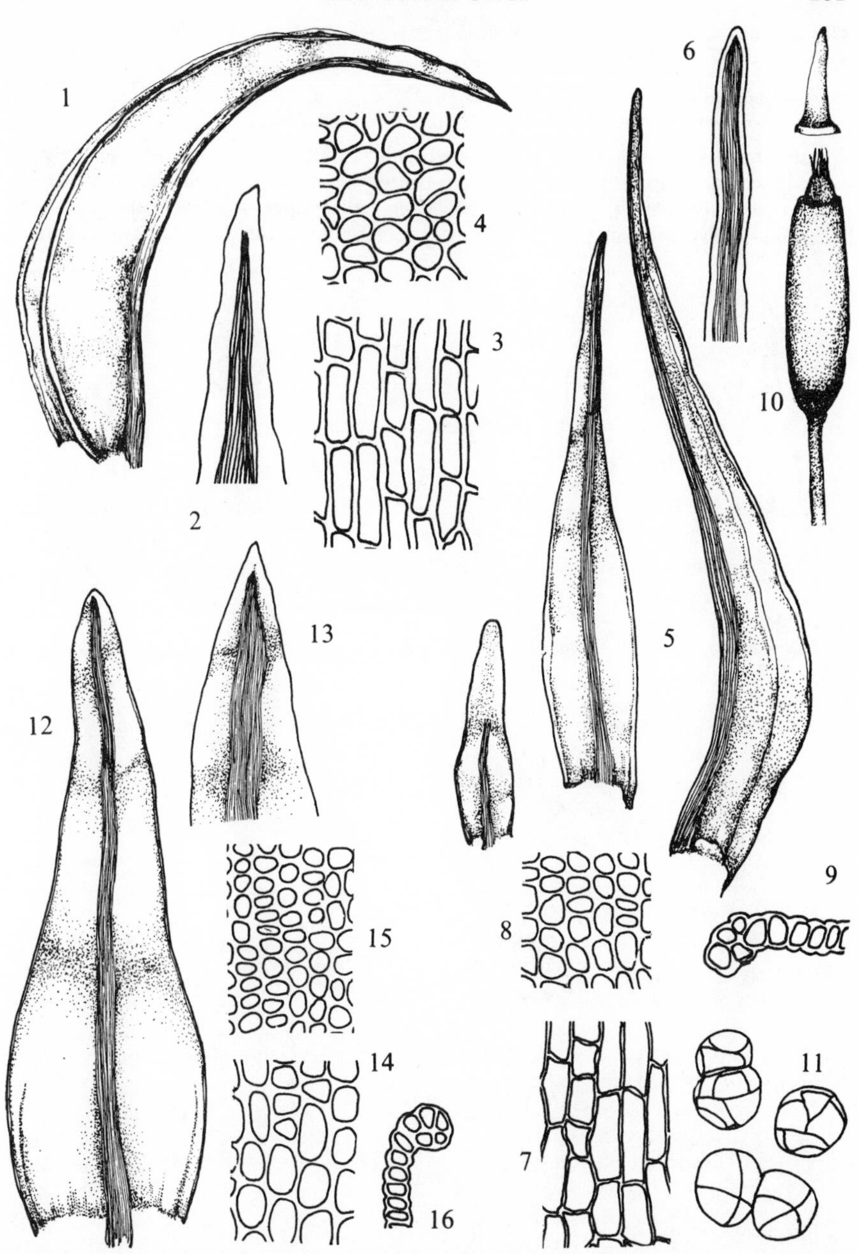

Fig. **121**. 1–4, *Barbula asperifolia*: 1, leaf; 2, leaf apex; 3, basal cells; 4, mid-leaf cells. 5–11, *B. rigidula*: 5, leaves, 6, leaf apex, 7, basal cells, 8, mid-leaf cells; 9, section of leaf margin (×250); 10, capsule; 11, axillary gemmae. 12–16, *B. nicholsonii*: 12, leaf; 13, leaf apex; 14, basal cells; 15, mid-leaf cells; 16, section of leaf margin (×250). Leaves ×40, cells × 415, apices × 100.

cells, intermediates linking it with typical *B. rigidula* occur (see Jones & Warburg, *Trans. Br. bryol. Soc.* **1**, 367–8, 1950).

15. B. nicholsonii Culm., Rev. Bryol., 1907
Didymodon trifarius var. *nicholsonii* (Culm.) Wijk & Marg.

Plants mostly 5–20 mm. Lvs flexuose when dry, patent, soft when moist, ovate to ovate-lanceolate, tapering to acute to sub-obtuse, sometimes slightly cucullate apex, margin recurved below, entire; nerve ending below apex, ventral cells quadrate-hexagonal; cells in lower part of lf shortly rectangular, above quadrate-hexagonal,

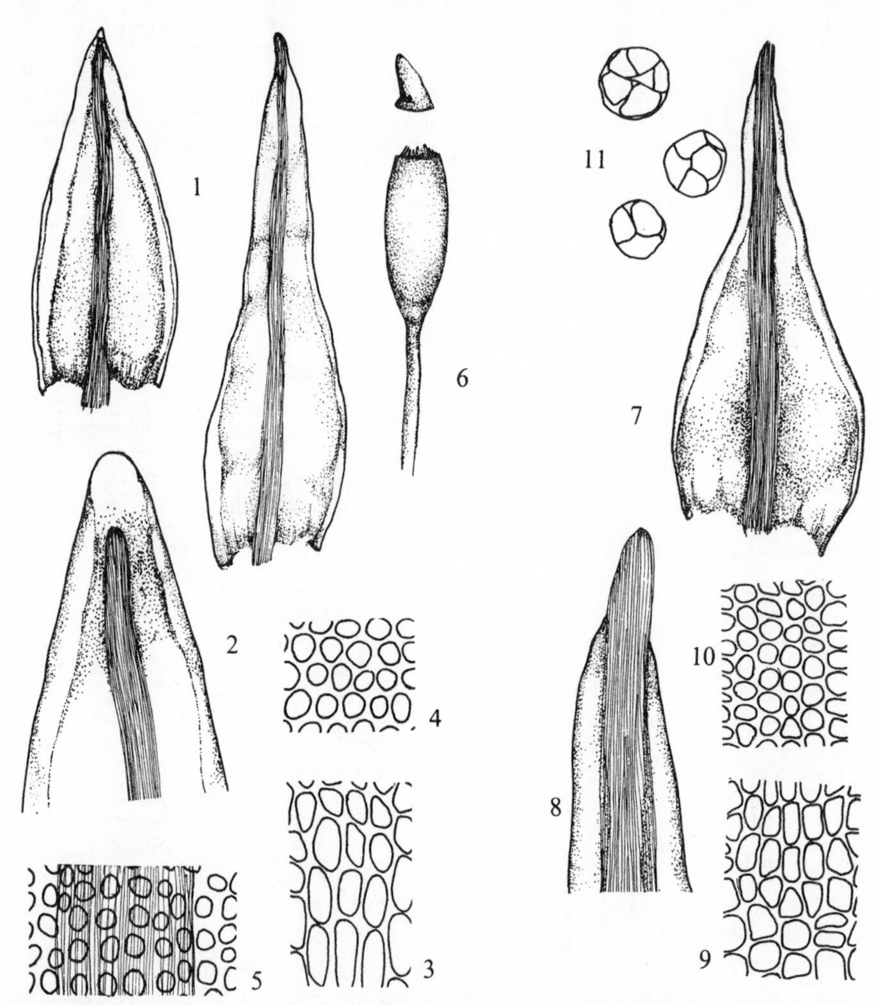

Fig. **122**. 1–6, *Barbula trifaria*: 1, leaves; 2, leaf apex; 3, basal cells; 4, mid-leaf cells; 5, ventral surface of nerve in middle of leaf; 6, capsule (×15). 7–11, *B. cordata*: 7, leaf; 8, leaf apex; 9, basal cells; 10, mid-leaf cells; 11, gemmae (×250). Leaves ×40, apices ×100, cells ×415.

slightly incrassate, papillose, bistratose at margin above, 6–8 μm wide in mid-lf. Globose, several-celled gemmae occasionally present in the axils of upper lvs. Fr unknown. Dull green tufts or patches, often covered with alluvial detritus on rocks and concrete below flood level of streams and rivers. Occasional in scattered localities from Cornwall and W. Sussex north to Berwick and Arran, N. Kerry, E. Cork, Wexford. 24, H3. Endemic.

Distinct from *B. rigidula* in the wider lvs and smaller cells. Van der Wijk & Margadant (*Taxon* 7, 289, 1958) reduced *B. nicholsonii* to a subspecies of *B. trifaria*. On morphological grounds this is unjustified, as the species, which is distinct, is closest to *B. rigidula*.

16. B. trifaria (Hedw.) Mitt., J. Linn. Soc. Bot., Suppl. I, 1859
B. lurida (Hornsch.) Lindb., *Didymodon trifarius* (Hedw.) Röhl.

Dioecious. Plants 5–30 mm. Lvs imbricate, not shrunken when dry, patent when moist, concave, broadly ovate to ovate-lanceolate, apex rounded to acute, base slightly decurrent, margin recurved below, entire; nerve stout, 40–60(–80) μm wide near base, ending in or below apex, ventral cells ± hexagonal; cells at extreme base shortly rectangular, elsewhere ± hexagonal, incrassate, smooth, pellucid, 6–10 μm wide in mid-lf. Seta purplish; capsule ellipsoid to sub-cylindrical; lid obliquely rostrate; peristome teeth short, *ca* 100 μm long; spores 12–16 μm. Fr occasional, winter to spring. $n=12$. Dark green or brownish tufts or patches on rocks, walls, old buildings, occasional on sand-dunes but not on soil. Frequent in S. Britain, extending north to Dumfries and the Lothians, Fife, Banff, occasional in Ireland. 75, H18, C. N., W. and C. Europe, Caucasus, Syria, Iran, Macaronesia, Algeria, Tunisia, N. America.

Recognised by the broad lvs, appressed but otherwise hardly altered when dry. Sometimes confused with *B. tophacea* but differs in the lvs imbricate when dry, lf shape, the ± hexagonal ventral cells of the nerve and the smooth cells, 6–10 μm wide. *B. cordata* has a broader nerve, the lvs slightly incurved and twisted when dry, a more strongly revolute margin and axillary gemmae.

17. B. cordata (Jur.) Braithw., Brit. Moss Fl., 1887
Didymodon cordatus Jur.

Dioecious. Plants *ca* 10 mm. Lvs slightly incurved and narrowed when dry, patent when moist, concave, broadly ovate, tapering to acuminate apex (in British plants), base slightly decurrent, margin entire, strongly recurved; nerve very stout, 70–100 μm wide near base, excurrent, basal cells rectangular, above quadrate-hexagonal, slightly incrassate, smooth, 6–8 μm wide in mid-lf. Numerous, spherical, several-celled gemmae, *ca* 30 μm diameter, in axils of upper lvs. Fr unknown, only female plants known in Britain. Dark green or brownish-green tufts or patches on exposed rocks, very rare. N. Devon. 1. C. and S. Europe, Caucasus.

18. B. tophacea (Brid.) Mitt., J. Linn. Soc. Bot., Suppl. I, 1859
Didymodon tophaceus (Brid.) Lisa

Dioecious. Plants 5–50 mm. Lvs erect, flexuose to slightly incurved, shrunken or not when dry, ± patent when moist, concave or not, variable in shape, basal part broadly ovate to narrowly lanceolate, above ± lingulate, apex rounded to acute, margin plane or recurved, entire; nerve ending below apex, ventral cells elongate; cells usually incrassate, basal rectangular, above irregular, hexagonal to rhomboidal in narrow-leaved forms, ± regularly quadrate in broad-leaved forms, slightly papillose, pellucid, 10–12 μm wide in mid-lf. Seta deep red; capsule ellipsoid; lid obliquely rostrate, long or short; peristome short, *ca* 250 μm long; spores 12–16 μm. Fr frequent, winter, spring. $n=12$*, 13. Olive green to brownish tufts or patches on

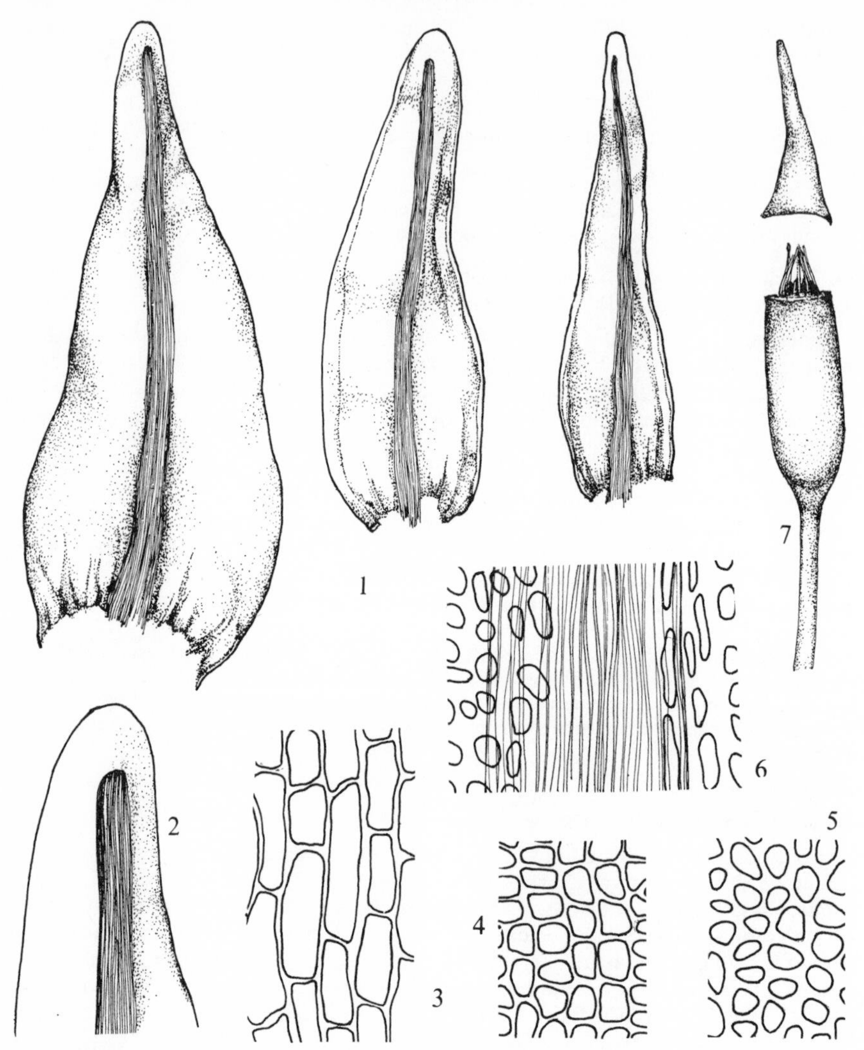

Fig. **123**. *Barbula tophacea*: 1, leaves (×40); 2, leaf apex (×100); 3, basal cells; 4, mid-leaf cells from broad-leaved form; 5, mid-leaf cells from narrow-leaved form; 6, ventral surface of nerve in middle of leaf; 7, capsule (×15). Cells ×415.

soil, walls, rocks, sand-dunes especially where damp or wet, often lime encrusted and forming tufa in wet calcareous habitats, frequent to common. 108, H34, C. Europe, Iceland, W. Asia, Tibet, Kashmir, China, Macaronesia, Tunisia, Morocco, N. America, Mexico, Bolivia.

19. B. vinealis Brid., Br. Univ., 1827

Dioecious. Plants mostly 5–20 mm. Lvs erect, flexuose or slightly twisted when dry, not flexuose, erecto-patent to patent when moist, upper 1–3 mm long, lanceolate

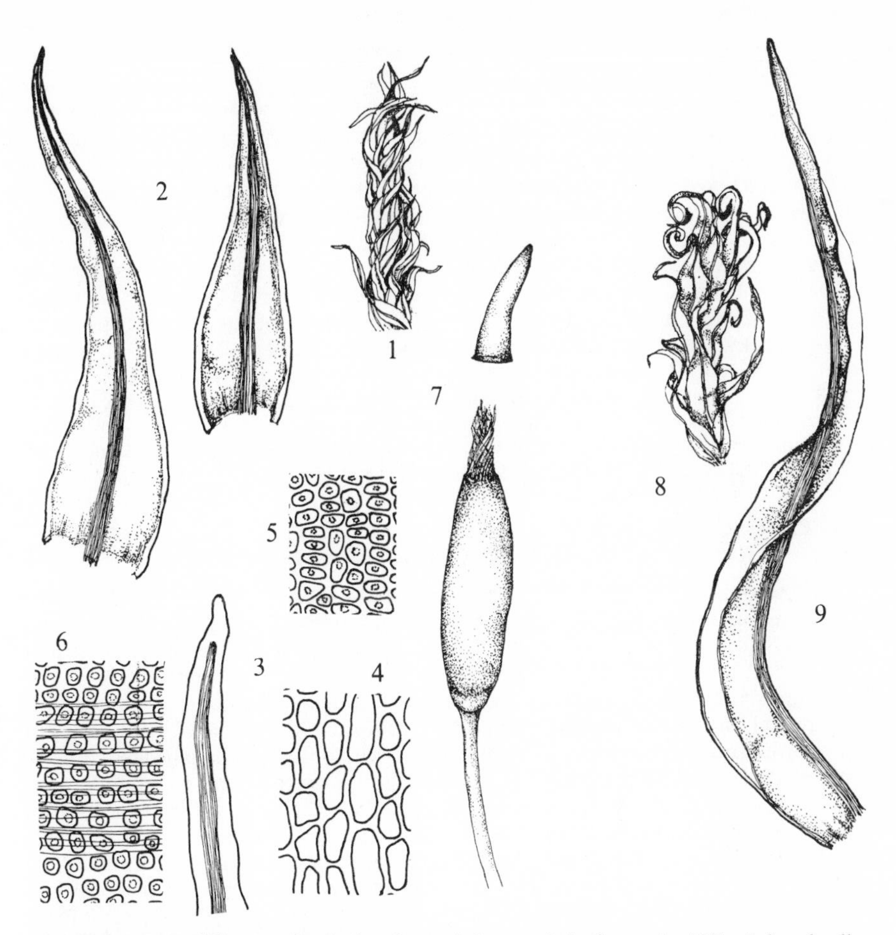

Fig. **124**. 1–7, *Barbula vinealis*: 1, dry shoot, 2, leaves; 3, leaf apex (×100); 4, basal cells; 5, mid-leaf cells; 6, ventral surface of nerve in middle of leaf; 7, capsule (×15). 8–9, *B. cylindrica*: 8, dry shoot; 9, upper leaf. Shoots ×10, leaves ×25, cells ×415.

to linear-lanceolate, tapering to acute apex, lower lvs shorter than upper, margin recurved from base almost to apex, entire; nerve ending in or below apex, ventral cells quadrate-hexagonal; basal cells quadrate to shortly hexagonal, above irregularly quadrate-hexagonal, papillose, opaque, 6–10 μm wide in mid-lf. Seta red below, orange above; capsule ellipsoid to narrowly ellipsoid; lid oblique, longly rostrate; peristome teeth filiform, spirally twisted, *ca* 500–600 μm long; spores *ca* 10 μm. Fr occasional, spring to autumn. $n=14$. Tall, dull green tufts on usually basic rocks, walls, old buildings and sand-dunes, never on soil or wood, frequent or common in the southern half of Britain, occasional or rare elsewhere. 77, H18, C. Europe, Faroes, Asia Minor, Nepal, China, Macaronesia, N. Africa, N. America, Mexico, Hawaii.

20. B. cylindrica (Tayl.) Schimp., Hedwigia, 1873
B. vinealis ssp. *cylindrica* (Tayl.) Podp., *B. vinealis* var. *cylindrica* (Tayl.) Boul., *B. vinealis* var. *flaccida* Br. Eur.

Dioecious. Plants mostly 5–50 mm. Upper and sometimes lower lvs crisped when dry, patent, flexuose and sometimes undulate when moist, upper 2–5(–7) mm long, linear-lanceolate, very longly tapering to acute apex, lower lvs shorter than upper, margin recurved from base almost to apex, entire; nerve ending below apex to excurrent, ventral cells quadrate-hexagonal; basal cells quadrate to shortly hexagonal, above irregularly quadrate-hexagonal, papillose, opaque, 6–10 μm wide in mid-lf. Seta red below, orange above; capsule cylindrical; lid oblique, longly rostrate; peristome teeth filiform, spirally twisted, *ca* 500–600 μm long; spores *ca* 10 μm. Fr rare, spring, summer. Green tufts or patches on damp walls, particularly the bases, rocks, tree boles and soil in shady places, common. 108, H39, C. Europe, Faroes, W. Asia, Tibet, China, Macaronesia, N. Africa, N. and C. America.

B. vinealis and *B. cylindrica* are usually quite distinct but occasional intermediates are encountered. As sporophytes are only rarely produced by *B. cylindrica* it is not certain how constant are the capsule differences between the 2 taxa. The general consensus of opinion in Britain is that they should be treated as separate species.

21. B. recurvirostra (Hedw.) Dix., Rev. Bryol. Lichen., 1933
B. rubella (Hüb.) Mitt., *Bryoerythrophyllum recurvirostre* (Hedw.) Chen.

Paroecious or synoecious. Plants to 5 cm. Lvs flexuose to curled when dry, patent to spreading when moist, lanceolate to narrowly lanceolate, tapering from expanded base to acute apex, margin strongly recurved, crenulate, often irregularly denticulate near apex; nerve strong, ending below apex to percurrent, ventral cells shortly rectangular; basal cells narrowly rectangular, hyaline, above quadrate to quadrate-hexagonal, incrassate, papillose, opaque, *ca* 10 μm wide in mid-lf. Seta deep red; capsule ellipsoid to cylindrical; lid oblique, conical, obtuse; peristome teeth filiform, straight, short, *ca* 250 μm long; spores 14–20 μm. Fr frequent to common, autumn to spring. $n=12+1$, 13*. Tufts or patches, green above, rusty-red or reddish-brown below, on basic rocks, walls, old buildings, soil and sand-dunes, common. 112, H39, C. Europe, Faroes, Iceland, Asia, N. Africa, N. America, Greenland, Tasmania.

A variable species usually readily recognised by the bright reddish-brown older parts, the lf margin denticulate near apex and the narrow, hyaline basal cells. *B. ferruginascens* is closely related but differs in the reddish-brown colour of the whole plant, the slender, tomentose stems and the rhizoidal gemmae.

22. B. ferruginascens Stirt., Ann. Scot. Nat. Hist., 1900
B. rubella var. *ruberrima* Ferg., *B. botelligera* Mönk., *Bryoerythrophyllum ferruginascens* (Stirt.) Giac.

Plants deep reddish-brown, to 5 cm, stems slender, tomentose. Lvs erect, slightly incurved when dry, patent to spreading from ± erect base when moist, lanceolate, tapering to ± acute apex, margin strongly recurved below, papillose-crenulate above; nerve stout, ending in apex; cells incrassate, basal narrowly rectangular, above irregularly quadrate, strongly papillose, obscure, 8–10(–12) μm wide in mid-lf. Elongate, reddish-brown rhizoidal gemmae, 200–400 μm×115–140 μm, often present. Fr unknown. Bright, reddish-brown tufts or patches on soil and rock ledges and crevices in montane habitats, frequent in base-rich areas, rare elsewhere. S. Somerset, Hereford, Staffs and Glamorgan northwards. 39, H12. N. and C. Europe, Iceland, Siberia, Greenland.

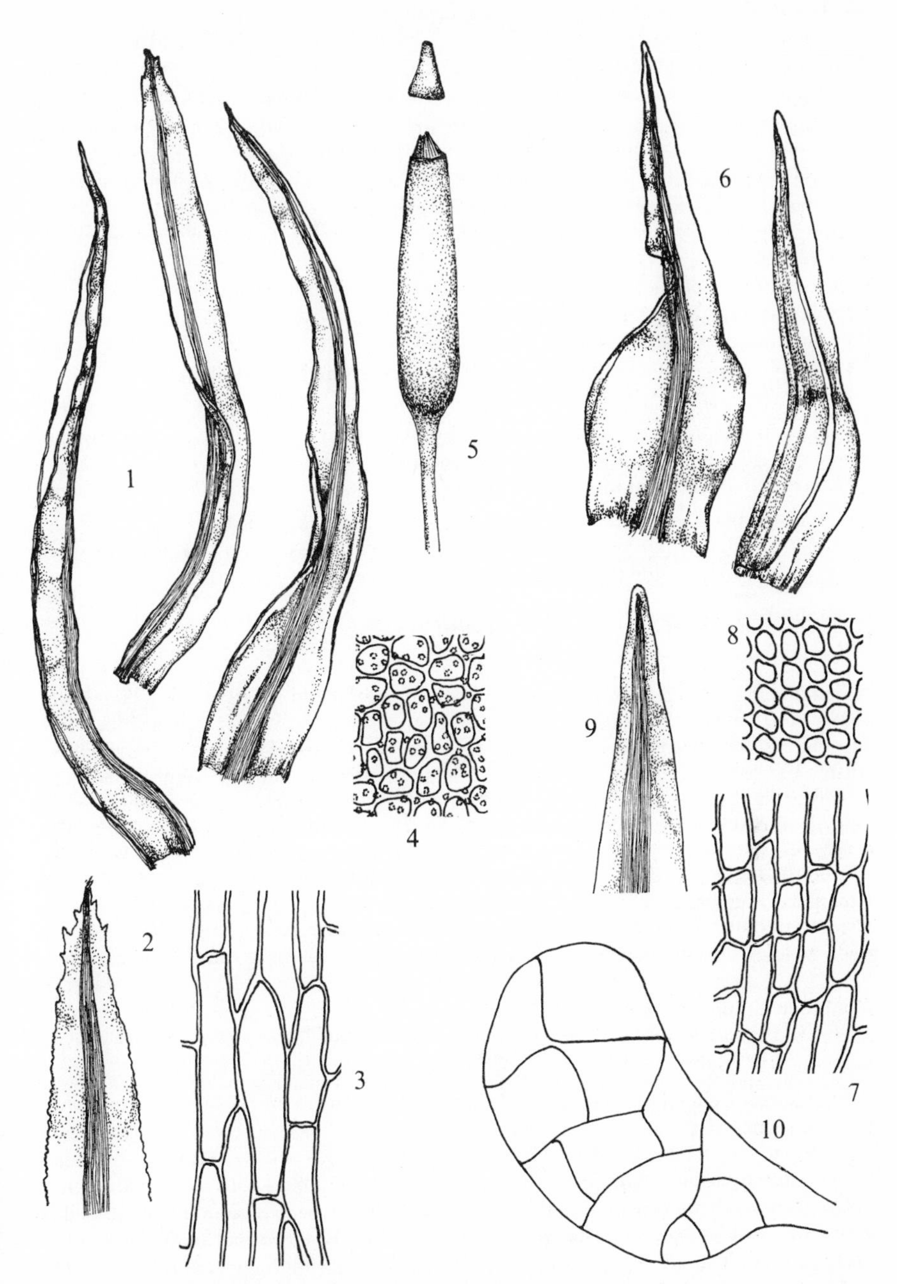

Fig. **125**. 1–5, *Barbula recurvirostra*: 1, leaves (×25); 2, leaf apex; 3, basal cells; 4, mid-leaf cells; 5, capsule (×15). 6–10, *B. ferruginascens*: 6, leaves (×40); 7, basal cells; 8, mid-leaf cells; 9, leaf apex; 10, rhizoidal gemma (×250). Cells ×415, apices ×100.

50. Gymnostomum Nees & Hornsch., Br. Germ., 1823

Dioecious. Plants forming lax or dense tufts. Lvs appressed, ± incurved when dry, erecto-patent to spreading when moist, linear-lanceolate to lanceolate, apex acute to obtuse, margin plane or recurved, entire or papillose-crenulate; nerve ending below apex; basal cells rectangular, smooth, above quadrate to rounded-hexagonal, papillose or smooth, obscure or pellucid. Seta long, straight; capsule ovoid or ellipsoid, symmetrical, smooth; lid longly rostrate, oblique; annulus of 1 row of small persistent cells; peristome absent; calyptra cucullate. A ± world-wide genus of some 55 species.

1 Lf margin recurved, at least below, cells in upper part of lf pellucid, spores papillose, 18–20 μm — 2
Lf margin plane, cells in upper part of lf obscure, spores smooth, 8–14 μm — 3

2 Lvs mostly 1.0–1.6 mm long, base not or only slightly expanded, not sheathing, basal cells not porose, nerve 30–40 μm wide near base — **3. G. recurvirostrum**
Lvs mostly 1.6–2.4 mm long, base expanded, ± sheathing, basal cells porose, nerve (35–)60–100 μm wide near base — **4. G. insigne**

3 Nerve 30–40(–60) μm wide near base, cells near apex 4–6 μm wide — **1. G. calcareum**
Nerve 60–110 μm wide near base, cells near apex (8–)10–14 μm wide — **2. G. aeruginosum**

1. G. calcareum Nees & Hornsch., Br. Germ., 1823
Weissia calcarea (Nees & Hornsch.), C. Müll.

Plants 1.5–5.0(–20.0) mm. Lvs flexuose or somewhat crisped but not neatly curled when dry, spreading to ± squarrose when moist, lower lvs lingulate, apex rounded or obtuse, upper larger, sometimes with expanded basal portion, apex obtuse to acute or rarely acuminate, margin plane, strongly papillose-crenulate almost to base; nerve 30–40(–80) μm wide, ending below apex, with dorsal stereids only in section; basal cells rectangular, to 36 μm long, cells above irregularly quadrate, papillose, opaque, towards apex 4–6(–8) μm wide. Perichaetial lvs wider obtuse to acute or rarely acuminate, margin plane, strongly papillose-crenulate with sheathing base. Seta *ca* 3 mm; capsule erect, ellipsoid; lid longly rostrate, oblique; peristome absent; spores smooth, 8–10 μm. Fr rare, summer. $n=13$. Bright green patches on damp, calcareous rocks and mortar on walls, rare to occasional. 40, H14. Europe, N., W. and C. Asia, China, Macaronesia, Tunisia, Algeria, Tristan da Cunha, N. America, Mexico, Chile, Australia, Tasmania, New Zealand.

Distinct from other species of the genus in its small size and bright green coloration. Small forms of *G. aeruginosum* may be distinguished by the lvs neatly curled when dry, the larger apical cells and the wider nerve with dorsal and ventral stereids in section. *Gyroweisia tenuis* commonly has fruit and the basal cells of the lf are longer.

2. G. aeruginosum Sm., Fl. Brit., 1804
Weissia rupestris (Schwaegr.) C. Müll.

Plants mostly 0.5–8.0 cm. Lvs twisted when dry, erecto-patent to patent when moist, variable in shape, ± linear to ligulate or linear-lanceolate, basal part sometimes expanded, apex sub-acute to acute, margin erect or incurved and papillose-crenulate above, plane, entire or obscurely toothed below; nerve stout, 60–100 μm wide near base, ending in or below apex, in section with dorsal and ventral stereids; cells incrassate, basal rectangular or narrowly rectangular, above irregularly

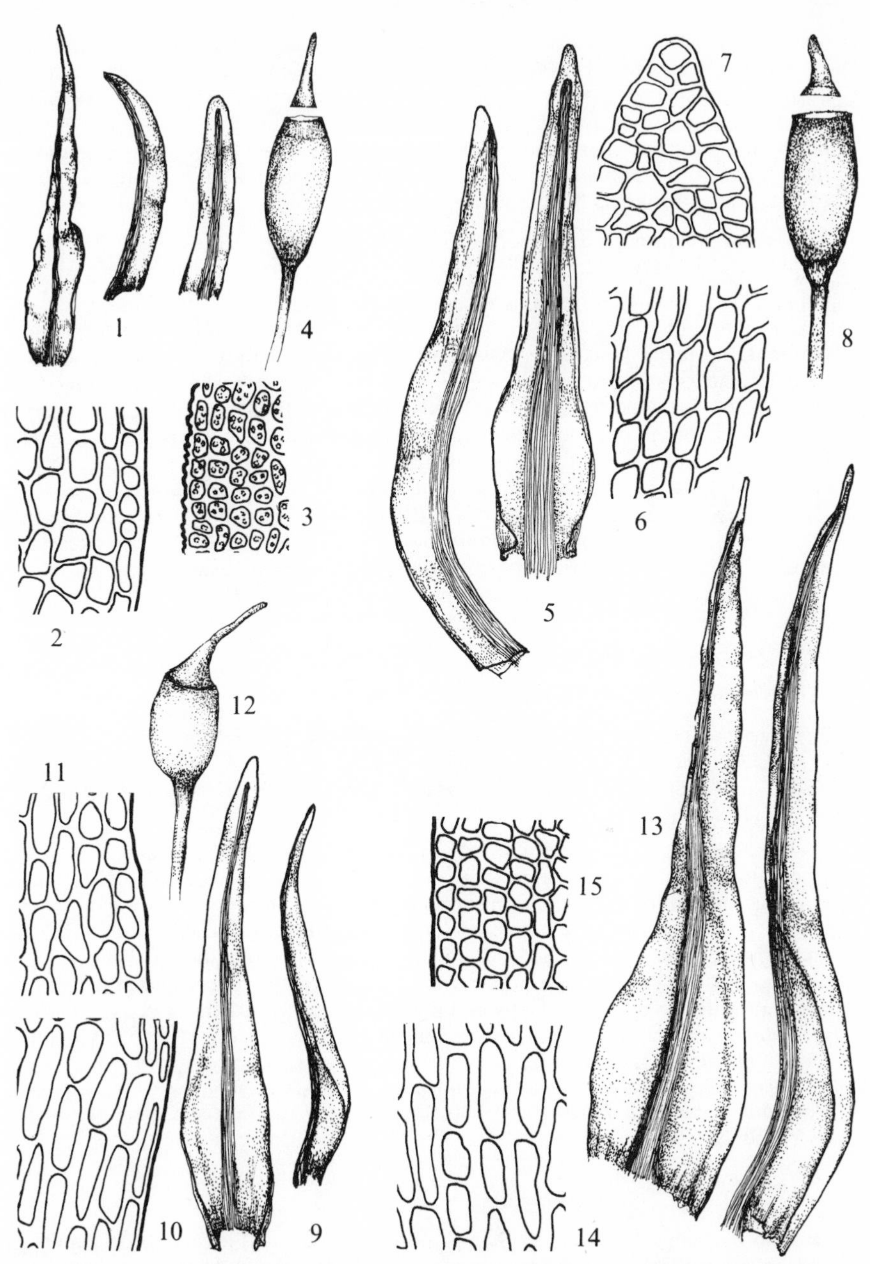

Fig. **126**. 1–4, *Gymnostomum calcareum*: 1, leaves; 2, basal cells; 3, cells near leaf apex; 4, capsule. 5–8, *G. aeruginosum*: 5, leaves; 6, basal cells; 7, cells near leaf apex; 8, capsule. 9–12, *G. recurvirostrum*: 9, leaves; 10, basal cells; 11, cells near leaf apex; 12, capsule. 13–15, *G. insigne*: 13, leaves; 14, basal cells; 15, cells near leaf apex. Leaves ×40, cells ×415, capsules ×20.

quadrate, papillose or strongly papillose, obscure, near apex (8–)10–14 μm wide. Perichaetial lvs larger with sheathing base. Seta 4–8 mm; capsule erect, ellipsoid; lid obliquely rostrate; peristome absent; spores smooth, 8–14 μm. Fr occasional, late summer, autumn. *n*=13*. Dull green tufts or cushions, reddish-brown below, on usually basic rocks and mortar of old buildings. Absent from S.E. England and the Midlands, occasional elsewhere. 71, H24. Europe, Faroes, Iceland, Caucasus, China, Japan, Canaries, N., C. and S. America, W. Indies.

3. G. recurvirostrum Hedw., Sp. Musc., 1801
Hymenostylium recurvirostrum (Hedw.) Dix., *Weissia curvirostris* (Ehrh.) C. Müll.

Plants (0.5–)1.0–10.0 cm. Lvs erect, appressed or flexuose, occasionally slightly twisted when dry, erecto-patent to patent when moist, mostly 1.0–1.6 mm long, lanceolate to linear-lanceolate, basal part not or only slightly expanded, apex acute to acuminate, margin recurved on one or both sides, papillose; nerve ending below apex, 30–60 μm wide towards base; cells incrassate, basal rectangular to narrowly rectangular, often with oblique transverse walls thinner than the non-porose longitudinal walls, cells above rectangular, smooth to strongly papillose, towards apex rectangular to ± quadrate, pellucid, 8–12 μm wide. Capsule erect, ovoid with long narrow lid attached to columella and sometimes persisting; columella lengthening at maturity, peristome lacking; spores papillose, 18–20 μm. Fr occasional, autumn. *n*=12+1, 13*. Green or dull green, dense tufts or cushions, often encrusted with calcareous matter below on usually basic rocks and walls, occasional in montane areas. Cornwall, Dorset, Wales, Derby and Yorks northwards. 56, H23. Europe, Asia, Algeria, N. America, Mexico, S. America, Greenland, W. Indies, Australasia.

Var. *cataractarum* Schimp. appears to be a dustbin taxon in which aberrant forms of *G. recurvirostrum* are placed, and cannot be maintained.

4. G. insigne (Dix.) A. J. E. Smith, J. Bryol., 1976
Weissia recurvirostris Hedw. var. *insignis* Dix., *G. recurvirostrum* var. *insigne* (Dix.) Rich. & Wallace, *Hymenostylium insigne* (Dix.) Podp., *Weissia curvirostris* var. *insigne* (Dix.) Amann.

Dioecious. Plants 4–12 cm. Lvs loosely incurved when dry, ± patent to spreading when moist, mostly 1.6–2.4 mm long, basal part erect, ± sheathing, limb gradually tapering to acuminate apex, margin recurved below, entire; nerve ending in or below apex, (35–)60–100 μm wide near base; cells incrassate, basal rectangular to narrowly rectangular, transverse walls ± oblique, thinner than longitudinal walls, longitudinal walls often porose, cells above irregularly rhomboidal to ± quadrate, smooth, 8–12 μm wide towards apex. Capsule ellipsoid. Fr very rare. Dark green to brownish-green tufts or patches on damp, calcareous, montane rocks, rare. Scottish Highlands, Sligo, Leitrim. 6, H2. Endemic.

Differs from *G. recurvirostrum* in the lvs loosely incurved when dry, more widely spreading when moist, the often porose basal cells, the expanded, ± sheathing basal part and the smooth cells.

51. GYROWEISIA SCHIMP., SYN., 2, 1876

Dioecious. Plants very small. Stems mostly simple, without central strand. Lvs appressed, incurved when dry, erecto-patent to recurved when moist, linear-lanceolate to lingulate, apex usually obtuse or rounded, margin plane, papillose-crenulate; nerve ending below apex; basal cells rectangular, cells above rounded-quadrate, incrassate, papillose or smooth. Seta straight; capsule erect, ovoid or ellipsoid; lid conical; peristome rudimentary or absent; annulus of 2–3 rows large cells; calyptra cucullate. Eleven species distributed through Europe, Asia, Africa, America, New Zealand, Oceania.

Peristome absent, upper and perichaetial lvs patent to reflexed **1. G. tenuis**
Rudimentary peristome present, upper and perichaetial lvs strongly reflexed to squarrose **2. G. reflexa**

1. G. tenuis (Hedw.) Schimp., Syn., 2, 1876
Weissia tenuis (Hedw.) C. Müll.

Plants 1.0–2.5 mm. Lvs erect to spreading, flexuose when dry, patent or spreading when moist, lower lingulate with rounded apex, upper larger, lingulate-lanceolate, apex rounded or obtuse, margin plane, papillose; nerve ending below apex, 20–30 μm wide towards base; basal cells narrowly rectangular, in lower lvs to 50 μm long, in upper to 90 μm long, cells above quadrate-rectangular, ± pellucid, towards apex irregularly quadrate, papillose, *ca* 8 μm wide. Perichaetial lvs patent to reflexed from sheathing base. Pale, ± ovoid or irregular rhizoidal gemmae, 50–70 μm long, sometimes present. Seta yellowish, 3–5 mm long; capsule narrowly ellipsoid to sub-cylindrical; lid rostrate, oblique; peristome lacking; spores 8–10 μm. Fr common, summer, autumn. Light green patches on damp, shaded, often vertical chalk, limestone, sandstone, old mortar and brickwork particularly by streams and pools. Occasional to frequent. 93, H17. N., W. and C. Europe, Syria, Iran, Madeira, Tunisia, N. America.

2. G. reflexa (Brid.) Schimp., Syn., 2, 1876
Weissia reflexa Brid.

Plants 1.5–2.0 mm. Lower lvs erecto-patent, upper and perichaetial lvs crowded, much larger, strongly recurved to squarrose when moist, ligulate, sometimes with expanded basal part, apex rounded to acuminate, margin plane, entire; nerve ending below apex; basal cells rectangular, 2–4 times as long as wide, above shortly rectangular, incrassate, papillose, towards apex ± quadrate, 6–8 μm wide. Perichaetial lvs with sheathing base, acuminate. Rhizoidal gemmae similar to those of *G. tenuis* often present. Seta flexuose, yellowish; capsule ovoid to ellipsoid, lid conical, oblique; 16 irregular, rudimentary peristome teeth present, soon falling; spores 10–14 μm. Fr winter. Sandstone quarry near Nuneaton, now extinct. 1. Mediterranean region, Madeira.

Some forms of *G. tenuis* approach this plant but the upper lvs are less crowded and not strongly recurved or squarrose. For the occurrence of this plant in Britain see *Rep. Brit. bryol. Soc.* **3**, (2), 118, 1933, under *Weissia tenuis*.

52. Anoectangium Schwaegr., Sp. Musc., Suppl. I, 1811

Dioecious, gametangia axillary. Plants tufted, stems tomentose. Lvs appressed when dry, erecto-patent when moist, margin papillose-crenulate; nerve ending in or below apex; basal cells rectangular, cells above rounded-quadrate, papillose, obscure. Seta long, lateral; capsule ovoid to ellipsoid, gymnostomous; calyptra cucullate. A world-wide genus of about 70 species.

Lvs acute with well defined, translucent apical cell **1. A. aestivum**
Lf apex rounded, apical cell not differing from other cells, obscure with papillae **2. A. warburgii**

1. A. aestivum (Hedw.) Mitt., J. Linn. Soc. Bot., 1869
A. compactum Schwaegr.

Plants 2–10 cm, stems slender, often with well defined annual growth zones,

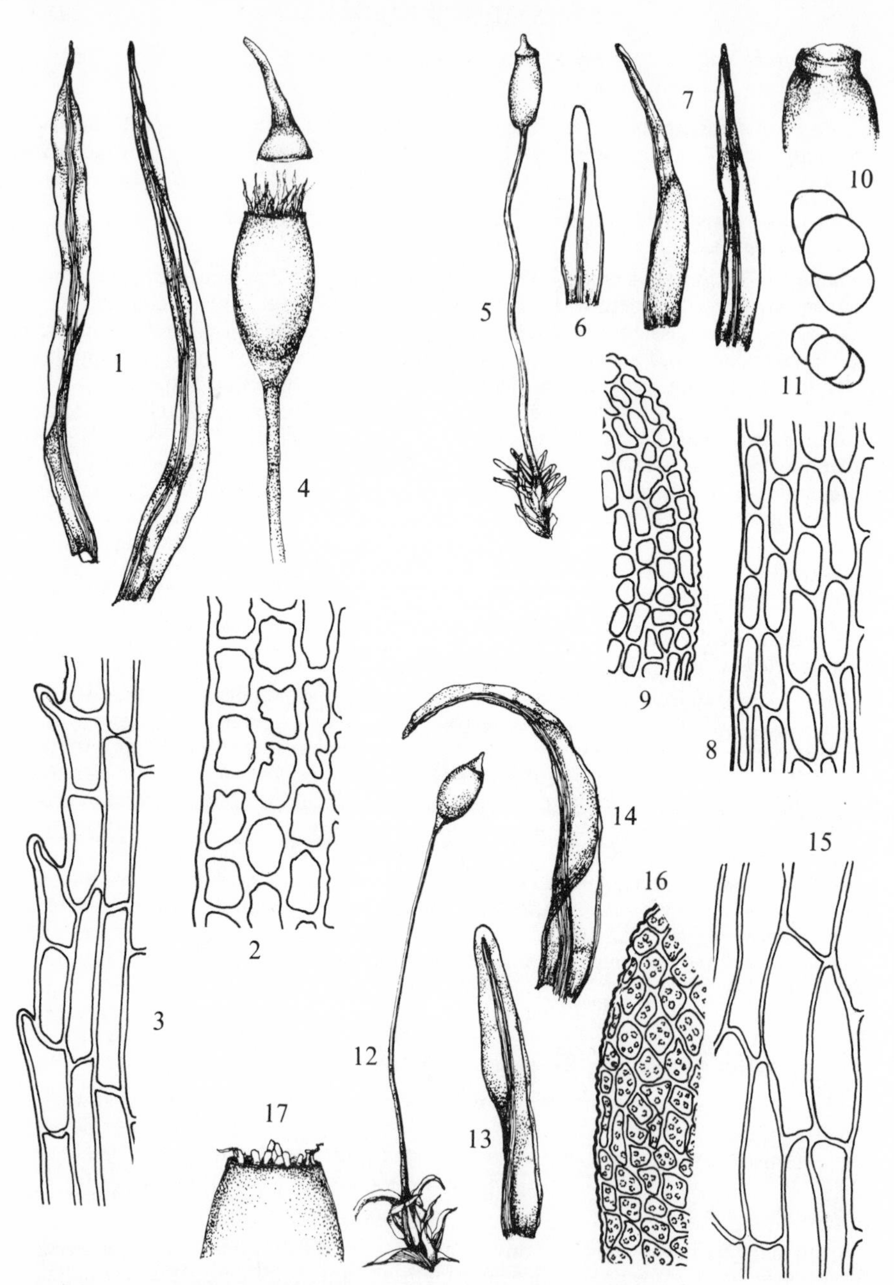

Fig. **127**. 1–4, *Eucladium verticillatum*: 1, leaves (×25); 2, mid-leaf cells; 3, margin near leaf base; 4, capsule (×10). 5–11, *Gyroweisia tenuis*: 5, plant; 6 and 7, lower and upper leaves (×40); 8, basal cells; 9, apical cells; 10, mouth of capsule (×40); 11, rhizoidal gemmae. 12–17, *G. reflexa*: 12, plant; 13 and 14, lower and upper leaves (×40); 15, basal cells; 16, apical cells; 17, mouth of capsule (×40); Plants ×10, cells ×415.

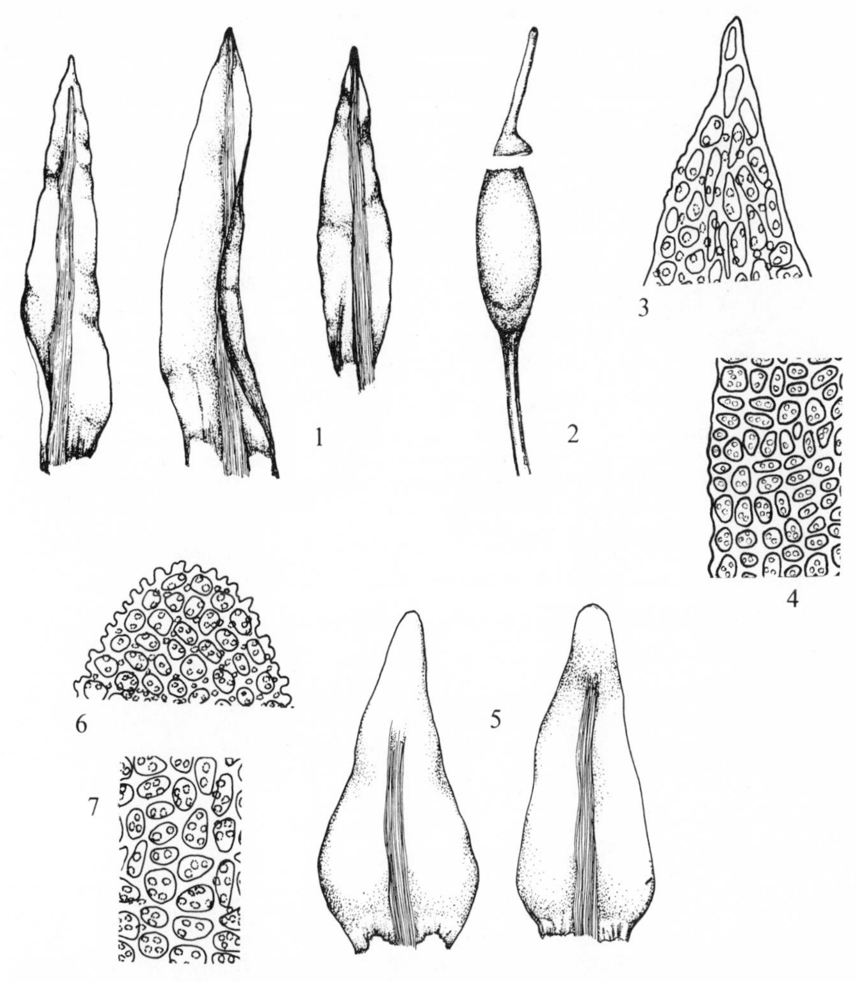

Fig. **128**. 1–4, *Anoectangium aestivum*: 1, leaves (×40); 2, capsule (×10); 3, leaf apex; 4, mid-leaf cells. 5–7, *A. warburgii*: 5, leaves (×65); 6, leaf apex; 7, mid-leaf cells. Cells ×415.

tomentose below with reddish-brown rhizoids. Lvs spirally curled when dry, patent when moist, keeled, ovate to lanceolate, acute, margin plane or narrowly recurved below, papillose-crenulate; nerve ending in apex to shortly excurrent, lamina cells not extending over nerve on upper surface; cells incrassate, basal rectangular, cells above rounded-hexagonal, obscure with papillae, apical cell clearly defined, translucent, cells 8–10 μm wide in mid-lf. Seta lateral, 7–10 mm long; capsule emerging 1–2 mm above lvs, ellipsoid, gymnostomous; lid with oblique, subulate beak; spores 12–16 μm. Fr rare, spring, summer. Dense, soft tufts or cushions, bright green above, light brown below, in rock crevices and on rock ledges, particularly where basic, frequent in montane areas. Staffs and Wales northwards. 51, H16. Montane parts of Europe, Iceland, Faroes, Caucasus, Himalayas, Japan, Philippines, Canaries, Madeira, Cameroon, N., C. and S. America, Greenland, West Indies, New Zealand.

2. A. warburgii Crundw. & Hill, J. Bryol., 1977

Plants slender, to 1 cm, stems often innovating from below, rhizoids deep reddish-brown. Lower lvs minute, upper larger, appressed when dry, erecto-patent to patent when moist, ovate to lanceolate, apex rounded, margin plane, papillose-crenulate; nerve broad, ending below apex, lamina cells extending over nerve in upper part of lf; extreme basal cells shortly rectangular, cells elsewhere ± quadrate, gradually becoming smaller towards apex, apical cells similar to other cells, obscure with papillae, cells in mid-lf *ca* 8 μm wide. Seta lateral; capsule gymnostomous. Lax, small to extensive, yellowish-green patches on moist, usually vertical, acidic to strongly basic rock in montane areas but descending to sea level in places, frequent in the central Scottish Highlands, rare elsewhere. Caerns, Perth north to Shetland. 16. Endemic.

Most likely to be confused with *Gymnostomum calcareum* which, however, occurs in drier and more exposed habitats, forms dense, bright green tufts and has less strongly papillose lvs with more extensive and markedly longer basal cells. *A. aestivum* is usually larger, but depauperate plants may be distinguished by the translucent well defined apical cell. For an account of this plant see Crundwell & Hill, *J. Bryol.* **9**, 435–40, 1977.

53. Eucladium Br. Eur., 1846

Dioecious. Plants forming dense tufts. Lvs appressed, incurved when dry, erecto-patent when moist, margin plane, toothed near base, papillose-crenulate above; nerve ending below apex to excurrent; basal cells rectangular, hyaline, cells above quadrate, papillose. Seta straight, long; capsule ovoid to cylindrical; lid rostrate, oblique; peristome present, well developed; calyptra cucullate. A small genus of 3 species occurring in Europe, Asia, N. and S. Africa, N. America and New Zealand.

1. E. verticillatum (Brid.) Br. Eur., 1846
Weissia verticillata Brid.

Plants mostly 0.5–3.0 cm. Lvs erect to erecto-patent, flexuose when dry, erecto-patent to spreading when moist, linear or linear-lanceolate from ± expanded basal part, acute, margin plane, toothed near base, finely papillose-crenulate above; nerve ending below apex to shortly excurrent; basal cells narrowly rectangular, hyaline, above rectangular to quadrate, incrassate, papillose, pellucid, *ca* 10 μm wide in mid-lf. Seta reddish; capsule erect, ovoid to ellipsoid; lid longly rostrate, oblique; peristome teeth 16, ± nodulose, irregularly cleft or perforated; spores 12–14 μm. Fr rare, spring. $n=13$*. Dense tufts, pale or glaucous green above, often whitish with calcareous matter below and sometimes forming tufa, on damp, shaded, basic rocks, damp walls and old buildings, occasional to frequent. 103, H29, C. Europe, Caucasus, Asia Minor, Kashmir, Macaronesia, Africa, N. and C. America.

54. Weissia Hedw., Sp. Musc., 1801

Usually autoecious. Small plants, frequently with much-branched stems. Lvs curled or crisped when dry, erecto-patent to spreading when moist, increasing in size up stem, perichaetial lvs similar to or larger than upper stem lvs, narrowly lanceolate to linear-lanceolate, acute or acuminate, margin plane or variously incurved or involute above, entire or papillose-crenulate, sometimes minutely denticulate near base of inner perichaetial lvs; nerve stout, excurrent, ventral cells quadrate or elongate; basal cells ± rectangular and hyaline, cells above rounded-quadrate, incrassate, very papillose, opaque or pellucid. Seta straight, long or short; capsule exserted or immersed, globose to cylindrical, stegocarpous or cleistocarpous; lid or beak usually longly rostrate, oblique; peristome of 16 imperfect teeth or absent; calyptra cucullate. A more or less world-wide genus of about 150 species.

Weissia as recognised here is divided by many authors into three genera, *Weissia* Hedw. (spp. 1–4), *Hymenostomum* R. Br. (spp. 5–8) and *Astomum* Hampe (spp. 9–13). The difference between *Weissia* and *Hymenostomum* is the presence or absence of peristome and hymenium, and in the presence of intermediates these differences can hardly be considered as worthy of generic status. *Astomum* differs in the immersed cleistocarpous capsule; *Hymenostomum rostellatum* is cleistocarpous, *Astomum levieri* is stegocarpous and *Weissia* (*Astomum*) *longifolia* var. *angustifolia* has a capsule which will dehisce under pressure. On this evidence *Astomum* cannot be considered a separate genus. Also the hybrids ***Weissia*** × ***Astomum*** and *Hymenostomum* × *Astomum* occur.

Species of *Weissia* and *Tortella* frequently occur mixed and the following hybrids have been recorded in Britain: ♀ *W. controversa* × ♂ *W. longifolia*, ♀ *W. longifolia* × ♂ *W. controversa*, ♀ *W. crispata* × ♂ *W. longifolia*, ♀ *W. longifolia* × ♂ *W. crispata*, ♀ *W. longifolia* × ♂ *W. microstoma*, ♀ *Tortella flavovirens* × ♂ *W. longifolia*.

1 Seta longer than capsule 2
  Seta not longer than capsule 10
2 Capsule with at least a rudimentary peristome, epiphragm lacking 3
  Capsule without peristome, epiphragm present, or capsule indehiscent 6
3 Margin of upper lvs plane or only narrowly incurved, spores 22–28 μm **4. W. rutilans**
  Margin of upper and perichaetial lvs involute, spores 16–20 μm 4
4 Cells on upper surface of nerve elongated in upper part of lf **3. W. perssonii**
  Cells on upper surface of nerve quadrate at least in patches in upper part of lf 5
5 Autoecious, widespread plant **1. W. controversa**
  Paroecious, alpine plant **2. W. wimmerana**
6 Seta 1.5–7.0 mm, capsule dehiscing in nature 7
  Seta 0.7–1.7 mm, capsule dehiscing or not 9
7 Nerve 60–120 μm wide near base, spores 14–20 μm **5. W. tortilis**
  Nerve to 55 μm wide near base, spores 20–34 μm 8
8 Lvs patent when moist, capsule wall tough, plants lacking distant-leaved innovations **6. W. microstoma**
  Lvs spreading to recurved, capsule wall fragile, readily rupturing, distant-leaved innovations arising from below perichaetium **7. W. squarrosa**
9 Perichaetial lvs similar to upper stem lvs, plants to 0.5 cm **8. W. rostellata**
  Perichaetial lvs distinct, larger than stem lvs, plants to 1.5 cm **9. W. mittenii**
10 Capsule dehiscent in nature, spores finely papillose **13. W. levieri**
  Capsule indehiscent in nature, spores finely or coarsely papillose 11
11 Fertile plants to 1.5 cm tall, spores finely papillose 12
  Fertile plants to 1.0 cm tall, spores coarsely papillose 13
12 Fertile stems hardly branched, perichaetial lvs distinct from stem lvs, seta 0.50–0.85 mm, rarely more than 1 capsule per stem **10. W. multicapsularis**
  Fertile stems usually with clustered short branches, perichaetial lvs similar to upper stem lvs, seta *ca* 0.4 mm, often 2 or more capsules per stem **11. W. sterilis**

13 Perichaetial lvs 3–6 mm long, margins plane or narrowly incurved, exothecial cells pale, capsule indehiscent **12. W. longifolia** var. **longifolia**
Perichaetia llvs 2.5–3.7 mm long, margins involute, exothecial cells yellowish, capsule dehiscing under pressure **12. W. longifolia** var. **angustifolia**

1. W. controversa Hedw., Sp. Musc., 1801
W. viridula Hedw.

Lvs strongly crisped when dry, patent to spreading when moist, from oblong-lanceolate basal part abruptly narrowed to linear-lanceolate upper part, apex acute,

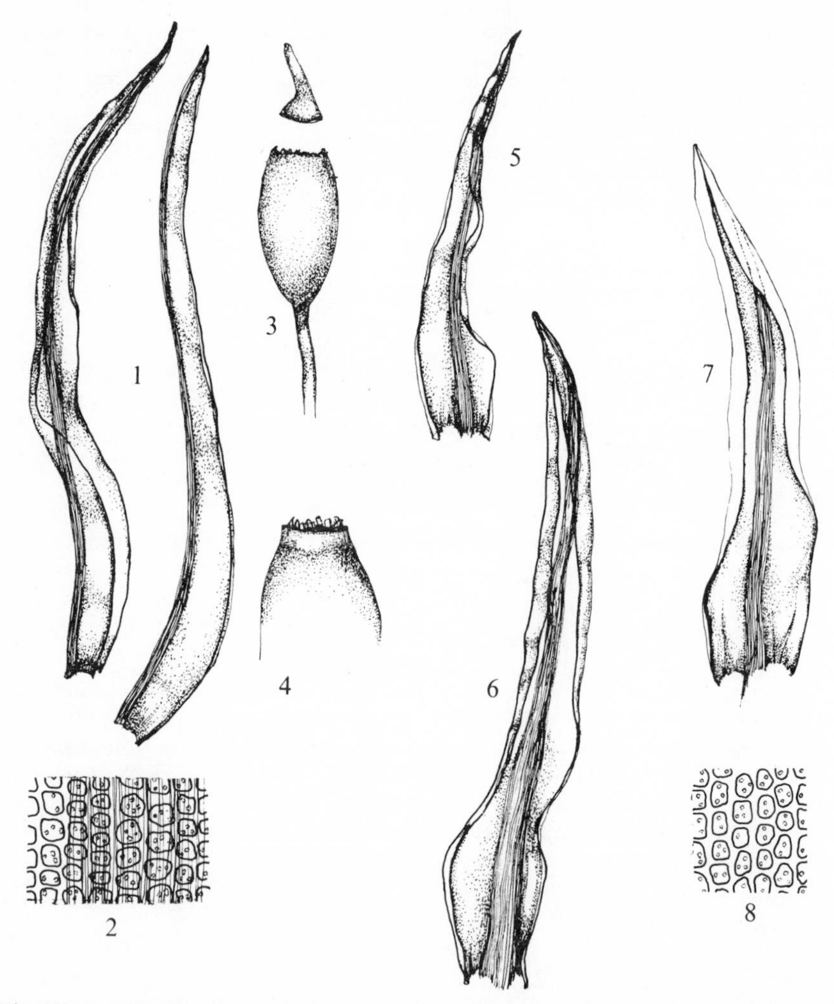

Fig. **129**. 1–4, *Weissia controversa* var. *controversa*: 1, leaves; 2, ventral side of nerve in mid-leaf; 3, capsule (×20); 4, capsule mouth (×40). 5, *W. controversa* var. *densifolia*: leaf. 6, *W. controversa* var. *crispata*: leaf. 7–8, *W. wimmerana*: 7, leaf; 8, mid-leaf cells. Leaves ×40, cells ×415

often ± mucronate, margin plane, entire below, involute above; nerve stout, excurrent, ventral cells in upper part of lf quadrate at least in patches; basal cells rectangular, hyaline, cells above quadrate-hexagonal, papillose, obscure. Seta yellowish; capsule exserted, erect or slightly inclined, ovoid to narrowly ellipsoid, symmetrical or slightly asymmetrical, narrowed at mouth, smooth or slightly furrowed when dry and empty; lid longly rostrate; peristome present, poorly developed, to 100 μm long or rudimentary; spores 16–20 μm. Fr common, winter, spring.

Key to intraspecific taxa of *W. controversa*

1 Nerve reddish, stout, (60–)70–100 μm wide near base — var. **crispata**
Nerve pale, 30–60(–65) μm wide near base — 2

2 Plants usually 3–10 mm tall, upper lvs larger, more crowded than lower, seta usually 3–6 mm long — var. **controversa**
Plants 15–40 mm, lvs ± uniform in size and distribution along stem, seta *ca* 1.5 mm long — var. **densifolia**

Var. **controversa**

Plants 3–10 mm. Lower lvs short, distant, upper longer, more crowded; nerve 30–60(–65) μm wide near lf base. Seta (1.5–)3.0–6.0 mm long. $n=8$, 11, 13*, 14. Dull or yellowish-green tufts or patches, sometimes extensive, on light soil on banks, walls, roadsides, cliffs, among rocks, on sand-dunes, etc., common. 112, H39, C. Europe, Madeira, Asia, N. and C. America, West Indies, Tasmania, New Zealand.

Var. **crispata** (Nees & Hornsch.) Nyholm, Moss Fl. Fenn., 1956

W. crispata (Nees & Hornsch.) C. Müll., *W. fallax* Sehlm.

Nerve reddish, stout, (60–)70–100 μm wide near lf base. Dull or yellowish-green patches on basic soil. Occasional in England, rare elsewhere. 45, H6. Europe, Caucasus, Asia Minor, Madeira, Canaries, Morocco, N. America.

Var. **densifolia** (Br. Eur.) Wils., Br. Brit., 1855

Plants 15–40 mm, stems frequently branched. Lvs of ± uniform size and distribution along stem. Seta *ca* 1.5 mm long. Very dense, dark green patches or tufts, sometimes extensive on soil, particularly in association with old heavy-metal mines, rare. W. Britain from Cornwall north to Cumberland, I. of Man, Berwick, W. Lothian, W. Perth, S. Kerry. 19, H1. Europe, N. America.

2. W. wimmerana (Sendtn.) Br. Eur., 1846

Very close to *W. controversa* but paroecious. Antheridia 1–3 together with paraphyses in the axils of lvs below the perichaetium. Peristome very short. Soil amongst rocks in alpine habitats, very rare. E. Inverness. 1. Pyrenees, Var, Jura, Alps, Scandinavia, Iceland, Caucasus, Kashmir.

For an account of this species in Britain see Warburg, *Trans. Br. bryol. Soc.* **3**, 171–3, 1957. The exothecial cell character given in that account has since been found to be unreliable. As the only difference between *W. controversa* and *W. wimmerana* is sexual, the latter would probably be better treated as a variety of the former.

3. W. perssonii Kindb., Bot. Not., 1896

W. controversa ssp. *perssonii* (Kindb.) Podp., *W. occidentalis* nom. nud.

Close to *W. controversa*. Nerve in lower part of lf (45–)50–70 μm wide, ventral cells elongated in upper part of lf; cells in mid-lf 9–10 μm wide. Capsule ovoid to ellipsoid, strongly contracted at mouth, smooth when dry and empty; peristome

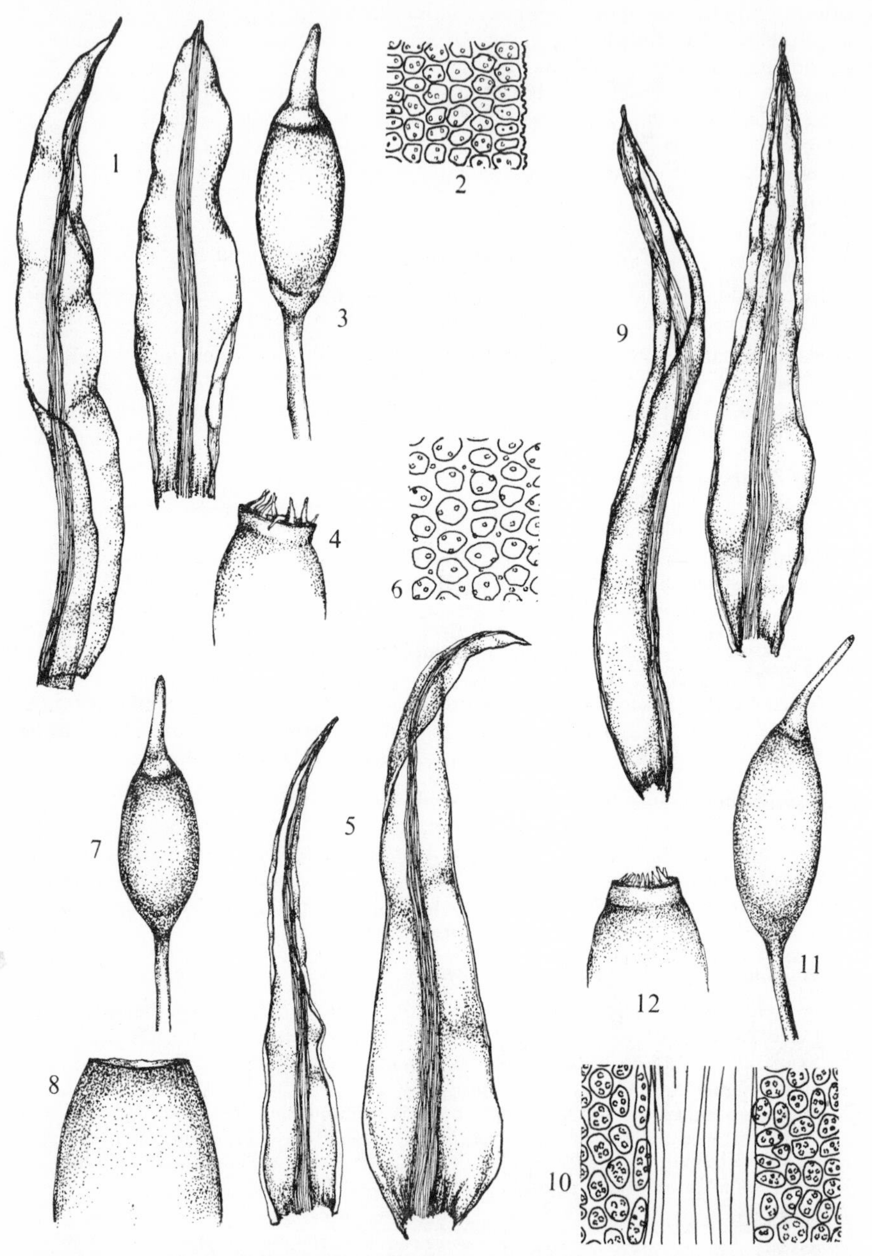

Fig. 130. 1–4, *Weissia rutilans*: 1, leaves (×40); 2, mid-leaf cells; 3, capsule; 4, capsule mouth. 5–8, *W. tortilis*: 5, leaves (×25); 6, mid-leaf cells; 7, capsule; 8, capsule mouth. 9–12, *W. perssonii*: 9, leaves (×40); 10, nerve and cells on ventral side in middle of leaf; 11, capsule; 12, mouth of capsule. Cells ×415, capsule figures ×20 and ×40.

teeth 20–60 μm long. Dull or yellowish-green patches on soil on banks, cliffs and amongst rocks, particularly near the sea, rare. Cornwall and Devon north to Skye and W. Ross, scattered localities round coast of Ireland, Jersey. 12, H6, C. Norway, Sweden.

Distinct from *W. controversa* in the wider nerve with elongated ventral cells in upper part of lf. Some gatherings of *W. crispata* are this plant. For information on *W. perssonii* see Crundwell, *Trans. Br. bryol. Soc.* **6**, 221–4, 1971.

4. W. rutilans (Hedw.) Lindb., Öfv. K. Vet. Ak. Förh., 1863
W. mucronata Br. Eur.

Plants 3–10 mm. Lvs crisped when dry, patent from ± erect base when moist, lower oblong-lanceolate, upper oblong-lanceolate to narrowly lanceolate, apex acute, margin plane or in uppermost lvs sometimes slightly incurved; nerve stout, excurrent in mucro, ventral cells ± quadrate in upper part of lf; mid-lf cells ± quadrate, papillose, obscure, 8–10 μm wide. Seta yellowish-green, 8–10 mm long; capsule ovate-ellipsoid to ellipsoid, straight or slightly curved, smooth or striate when dry and empty; peristome teeth rudimentary, fugacious, to 80 μm long; spores 22–28 μm. Fr common, autumn to spring. Dull or yellowish-green patches in similar habitats to *W. controversa*, occasional. 66, H4, C. Europe, Faroes, Morocco, Tunisia, Ethiopia.

5. W. tortilis (Schwaegr.) C. Müll., Syn., 1849
Hymenostomum tortile (Schwaegr.) Br. Eur.

Plants mostly 5–15 mm. Upper lvs crisped when dry, patent when moist, narrowly lanceolate, acute to obtuse and mucronate, margin plane below, involute above; nerve stout, 80–120 μm wide near base, ending in apex to excurrent in mucro, ventral cells ± quadrate in upper part of lf; cells in mid-lf obscure, 8–9 μm wide. Seta 3–4 mm long; capsule ellipsoid, symmetrical or slightly asymmetrical, mouth partially closed by fugacious membrane; peristome lacking; spores 14–20 μm. Fr frequent, spring. $n=13$. Loose, readily disintegrating patches, green above, brownish below, on calcareous soil, rocks and walls. Rare in S. England, very rare elsewhere, extending north to Flint, Westmorland, Berwick, Offaly. 16, H1. S. and W. Europe, Asia Minor, Caucasus, N. Africa, N. America, New Zealand.

A coarser plant than other *Weissia* species with the patches tending to become detached from the substrate. The very thick nerve and gymnostomous capsule are also characteristic.

6. W. microstoma (Hedw.) C. Müll., Syn., 1849
Hymenostomum microstomum (Hedw.) Br. Eur.

Plants 2–10 mm, stems erect. Lvs strongly crisped when dry, patent when moist, narrowly lanceolate, acute; nerve 35–50 μm wide near base, excurrent, ventral cells quadrate or elongated in upper part of lf; cells in mid-lf quadrate, papillose, obscure, 10–12 μm wide in mid-lf. Seta yellowish, 1.5–7.0 mm long; capsule ovate-ellipsoid to ellipsoid, symmetrical, brown, constricted at mouth, mouth covered by white membrane finally rupturing to release spores; exothecial cells not fragile; peristome absent. Fr common, winter, spring.

Var. **microstoma**
Lvs narrowly lanceolate, margin plane below, involute above. Capsule ovate-ellipsoid to ellipsoid; spores 20–28 μm. $n=13$*, 26. Dark green patches, sometimes extensive, on soil in turf, fields, banks, rock crevices, etc., especially where basic, frequent or common in Britain, occasional in Ireland. 90, H17, C. Europe, Caucasus, Algeria, Tunisia, N. America.

Var. **brachycarpa** (Nees & Hornsch.) C. Müll., Syn., 1849
Hymenostomum microstomum var. *brachycarpum* (Nees & Hornsch.) Hüb.
Lvs wider than in the type, margins plane. Capsule ovoid to ellipsoid; spores 24–34 μm. Damp arable fields, rare. Scattered localities north to Yorks, E. Lothian, Argyll. 23. Europe.

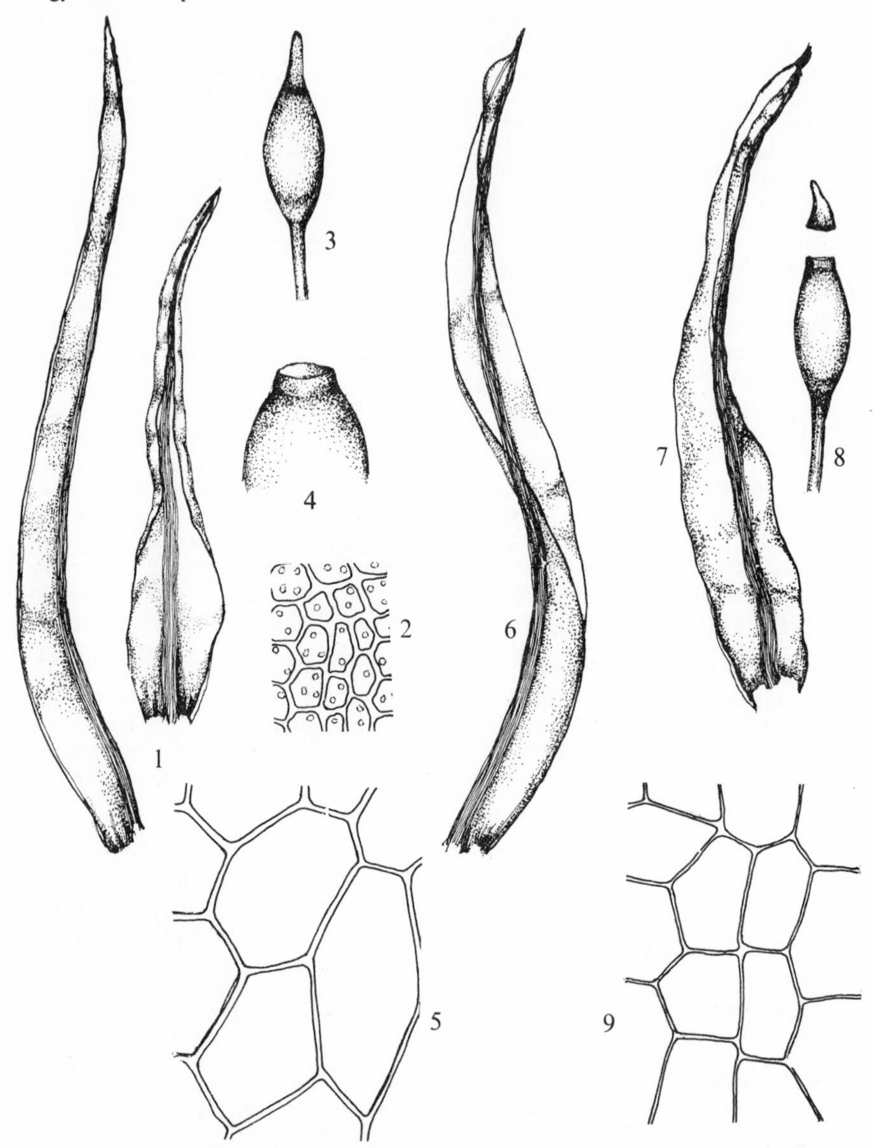

Fig. **131**. 1–5, *Weissia microstoma* var. *microstoma*: 1, leaves; 2, mid-leaf cells; 3, capsule; 4, capsule mouth (×40); 5, exothecial cells. 6, *W. microstoma* var. *brachycarpa*: leaf. 7–9, *W. squarrosa*: 7, leaf; 8, capsule; 9, exothecial cells. Leaves ×40, capsules ×20, cells ×415.

The name *W. microstoma* (Hedw.) C. Müll. is illegitimate but is used here because of uncertainty as to the correct name.

7. W. squarrosa (Nees & Hornsch.) C. Müll., Syn., 1849
Hymenostomum squarrosum Nees & Hornsch.

Close to *W. microstoma*. Stems becoming decumbent at maturity of capsules, producing distant-leaved innovations from below perichaetium. Lvs spreading to recurved when moist, margins of lower lvs plane, of upper lvs narrowly incurved above. Capsule dark greenish-brown at maturity, mouth covered by white membrane, exothecial cells thin, fragile, rupturing to release spores; spores 22–28 μm. Fr frequent, spring. Soil in damp fields, rare. Scattered localities north to Yorks, Westmorland, Fife. 24. Europe.

Distinguished from *W. microstoma* by the frequently decumbent stems with distant-leaved innovations and thin-walled capsules rupturing to release the spores; the membrane of the capsule does not appear to break down.

8. W. rostellata (Brid.) Lindb., Öfv. K. Vet. Ak. Förh., 1864
Hymenostomum rostellatum (Brid.) Schimp.

Plants to *ca* 5 mm. Lvs increasing in size up stem, perichaetial lvs similar to upper stem lvs, erecto-patent, 1.9–3.2 mm long, linear-lanceolate, acuminate, margin plane or occasionally narrowly incurved, entire or in innermost perichaetial lvs minutely toothed near base; nerve excurrent, *ca* 50 μm wide near base, ventral cells elongated; cells in mid-lf ± quadrate, pellucid, *ca* 10 μm wide. Seta 0.7–1.5 mm; capsule ovoid, indehiscent with oblique beak; spores (18–)22–27 μm. Fr frequent, autumn,

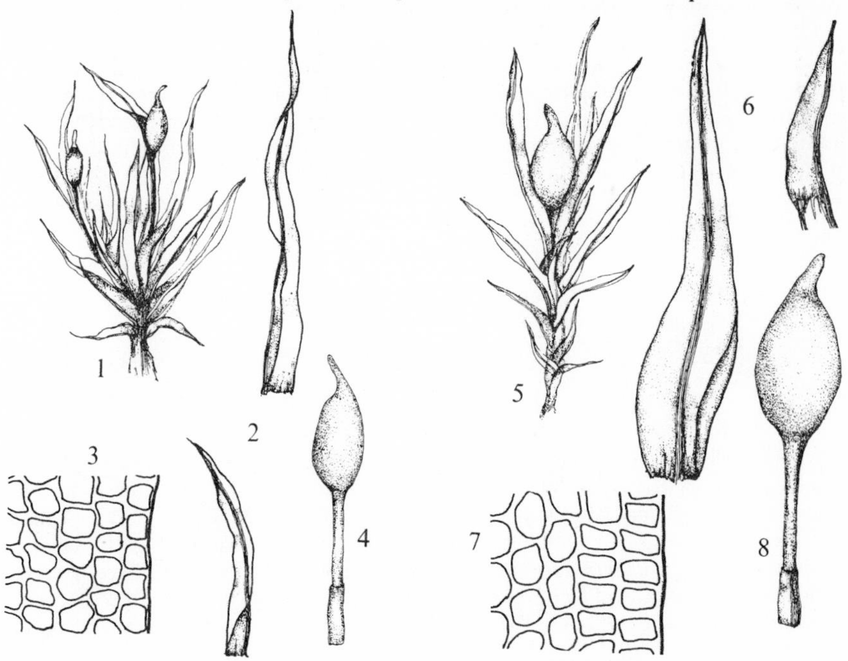

Fig. **132**. 1–4, *Weissia rostellata*: 1, plant; 2, leaves; 3, mid-leaf cells; 4, capsule. 5–8, *W. mittenii*: 5, plant; 6, leaves; 7, mid-leaf cells; 8, capsule. Plants ×10, leaves and capsules ×20, cells ×415.

winter. Loose patches on damp clay, especially by ponds and ditches, in arable fields, rare. Dorset and Kent north to Anglesey, Northumberland and Midlothian, Antrim. 18, H1. N., W. and C. Europe, Carpathians.

Likely to be confused with *W. squarrosa*, *W. mittenii* and forms of *W. microstoma* with short setae. Both *W. squarrosa* and *W. microstoma* have dehiscent capsules and *W. mittenii* has the perichaetial lvs markedly larger than most of the stem lvs. For a detailed account of *W. rostellata* and the following species of *Weissia* see Crundwell & Nyholm, *J. Bryol.*, 7, 7–19, 1972, upon which these descriptions are based.

9. W. mittenii (Br. Eur.) Mitt., Ann. Mag. Nat. Hist., 1851
Astomum mittenii Br. Eur.

Plants to 15 mm. Lower lvs spreading and ± uniform in size, only a few transitional to larger perichaetial lvs, perichaetial lvs 3–4 mm long, narrowly lanceolate, margin usually plane or occasionally very narrowly incurved, entire; nerve shortly to longly excurrent, ventral cells elongated; cells in mid-lf ± quadrate, pellucid, *ca* 10 μm wide. Seta 1.0–1.7 mm; capsule ± ovoid but often malformed, indehiscent; beak long or short, blunt or acute; spores variable in size (13–)18–28(–40) μm, often aborted. Fr spring. Clay banks, arable fields and wood rides, very rare. Sussex, Surrey, not seen since 1920. 3. Endemic.

The sporophyte resembles *W. rostellata* and the gametophyte *W. multicapsularis* except that the ventral cells of the nerve are elongated. The plant is possibly of hybrid origin although the parentage is obscure. It has not been found in mixed populations of the two species and hence its status is doubtful.

10. W. multicapsularis (Sm.) Mitt., Ann. Mag. Nat. Hist., 1851
Astomum multicapsularis (Sm.) Br. Eur.

Plants to 1.5 cm tall. Stem lvs spreading and ± uniform in size, only a few transitional to larger perichaetial lvs, perichaetial lvs 3.5–4.5 mm long, narrowly lanceolate, margin plane or occasionally very narrowly incurved, ± entire; nerve shortly excurrent, 60 μm wide near base, ventral cells quadrate; mid-lf cells irregularly quadrate, 9–10 μm wide. Seta 0.5–0.8 mm; capsule ovoid, indehiscent, 0.70–0.85 mm long; spores 16–21 μm, finely papillose. Fr occasional, spring. Loose patches on damp, clayey acidic soil in turf, fields and wood rides, rare. Cornwall, S. Devon, Sussex, W. Kent, Oxford, Cheshire. 9. Endemic.

11. W. sterilis Nicholson, J. Bot., Lond., 1903
Astomum crispum var. *sterilis* (Nicholson) Mönk.

Plants to 15 mm, stems usually branched, often with clustered branches and 2 or more capsules per perichaetium. Lvs gradually increasing in size up stem with perichaetial lvs similar to upper stem lvs, perichaetial lvs strongly crisped when dry, erecto-patent to spreading when moist, 3.0–3.7 mm long, narrowly lanceolate, margin erect or narrowly incurved, innermost perichaetial lvs with a few small teeth near base; nerve excurrent, 30–80 μm wide near base, ventral cells quadrate; cells in mid-lf *ca* 10 μm wide. Seta *ca* 0.4 mm; capsule ovoid or globose, *ca* 0.7 mm long, indehiscent; spores finely papillose, 18–20 μm. Fr common, winter. Small groups or loose patches on calcareous soil in south-facing chalk or limestone grassland slopes, rare. From S. Devon and Kent north to Worcs and Cambridge. 17. N.W. France.

12. W. longifolia Mitt., Ann. Mag. Nat. Hist., 1851

Plants to 10 mm. Lvs gradually increasing in size up stem to perichaetial lvs, strongly curled when dry, erecto-patent when moist, margin plane or incurved; nerve to 50(–80) μm wide, shortly to longly excurrent, ventral cells quadrate or not;

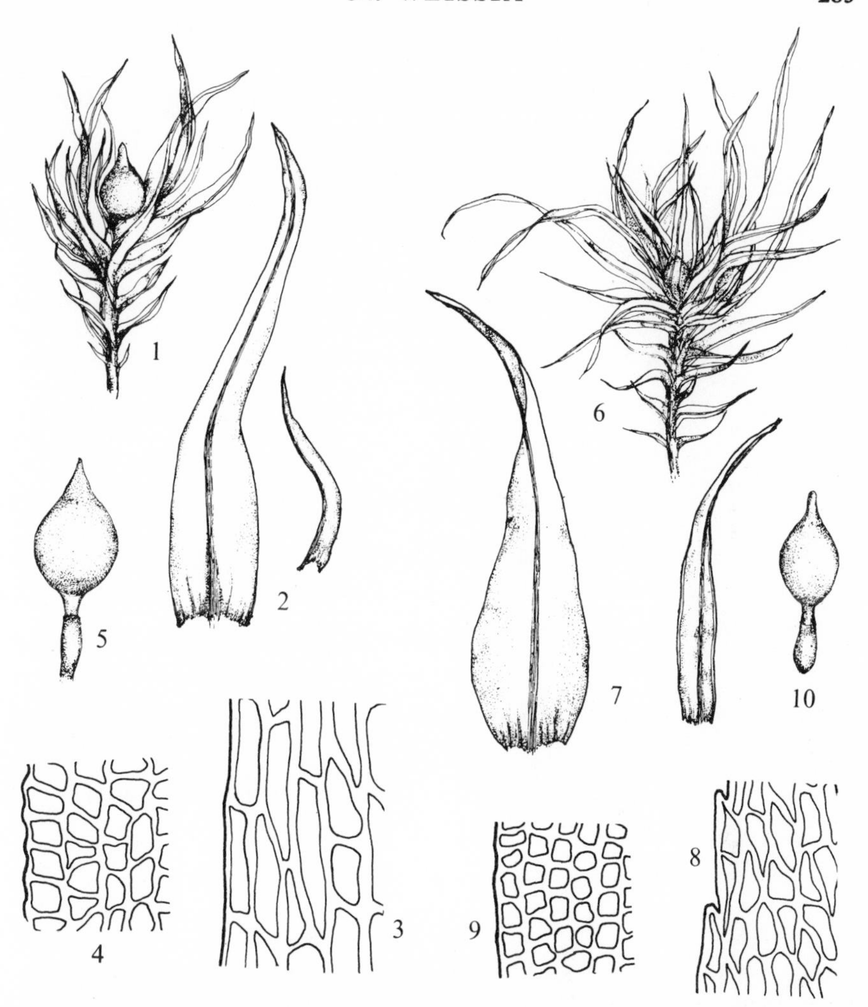

Fig. **133**. 1–5, *Weissia multicapsularis*: 1, plant; 2, leaves; 3, cells near perichaetial leaf base; 4, mid-leaf cells; 5, capsule. 6–10, *W. sterilis*: 6, plant; 7, leaves; 8, cells near perichaetial leaf base; 9, mid-leaf cells; 10, capsule. Plants × 10, leaves and capsules × 20, cells × 415.

basal cells thin-walled, hyaline, cells above ± quadrate, papillose, opaque, 9–10 μm wide. Seta 0.30–0.85 mm; capsule ovoid, 0.7–1.0 mm long, indehiscent in nature; beak variable; exothecial cells colourless to yellow; calyptra 0.5–1.2 mm long; spores coarsely papillose, 17–23 μm.

Var. **longifolia**

W. crispa (Hedw.) Mitt., *W. crispa* var. *aciculata* (Hedw.) Dix., *Astomum crispum* (Hedw.) Hampe, *Phascum crispum* Hedw.

Perichaetial lvs longer than stem lvs, conspicuous, 3–6 mm long, to 0.9 mm wide, from broad, ± colourless basal part linear-lanceolate, acuminate, margin plane or

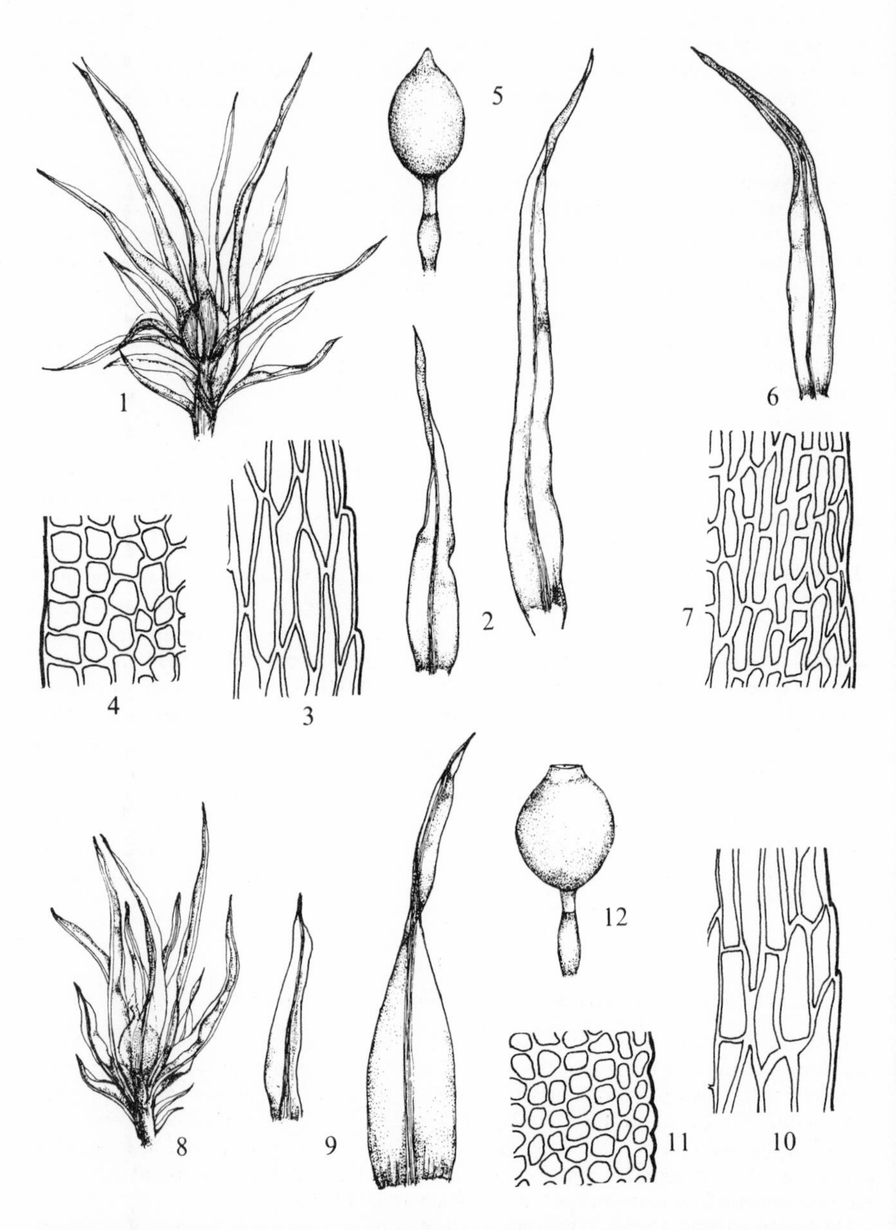

Fig. **134**. 1–5, *Weissia longifolia* var. *longifolia*: 1, plant; 2, leaves; 3, cells near perichaetial leaf base; 4. mid-leaf cells; 5, capsule. 6–7, *W. longifolia* var. *angustifolia*: 6, leaf; 7, cells near perichaetial leaf base. 8–12, *W. levieri*: 8, plant; 9, leaves; 10, cells near perichaetial leaf base; 11, mid-leaf cells; 12, capsule. Plants ×10, leaves and capsules ×20, cells ×415.

narrowly incurved, margin of inner perichaetial lvs often toothed near base; nerve excurrent to 130 μm, ventral cells quadrate or not. Seta 0.3–0.6 mm long; capsule ovoid, indehiscent, 0.7–1.0 mm long, beak short, 80–140(–210) μm long; exothecial cells pale; calyptra 0.5–0.7 mm. Fr common, winter, spring. Loose or compact patches on non-calcareous soil in arable fields, on banks and in limestone grassland, rare. From Cornwall and Kent north to Leics, Notts and Yorks. 23. Europe, S.W. Russia, N. Africa, Palestine.

Var. **angustifolia** (Baumg.) Crundw. & Nyh., J. Bryol., 1972
W. crispa auct angl. non *Phascum crispum* Hedw.

Lvs increasing in size up stem so that perichaetial lvs less conspicuous than in type, perichaetial lvs 2.5–3.7 mm long, to 0.8 mm wide, expanded basal part of lf yellowish, margin strongly incurved, usually entire in inner perichaetial lvs; nerve excurrent to 50–100 μm, ventral cells quadrate; basal cells incrassate, mostly yellowish. Seta 0.75–0.85 mm; beak of capsule 200–350 μm, capsule indehiscent in nature but lid coming off under pressure; exothecial cells deep yellow; calyptra 0.75–1.20 mm. Compact patches in chalk grassland and among limestone rocks. Common in southern Britain, extending north to Northumberland, a few localities in Ireland. 53, H5. Scattered localities in C. and S. Europe.

The type and var. *angustifolia* are quite distinct in Britain in the degree of incurving of lf margin, differences in form and structure of perichaetial lvs, seta length, capsule beak and calyptra and their ecology, but morphological intermediates occur elsewhere.

13. W. levieri (Limpr.) Kindb., Eur. N. Amer. Bryin., 1897
Astomum levieri Limpr.

Plants to 1 cm, hardly branched. Lvs gradually increasing in size up stem, perichaetial lvs not markedly distinct from upper stem lvs, strongly curled when dry, erect to erecto-patent when moist, 3.0–4.5 mm long, 0.6–0.8 mm wide, narrowly lanceolate, acuminate, margin plane or occasionally very slightly incurved, often slightly denticulate near base of innermost perichaetial lvs; nerve stout, 50–80 μm wide near base, prominent at back, excurrent to 130 μm, ventral cells quadrate; basal cells thin-walled, above quadrate, *ca* 9 μm wide in mid-lf. Seta 0.3–0.6 mm; capsule ovoid, 0.7–0.9 mm long, dehiscent, beak 150–200 μm long; spores finely papillose, 18–22(–24) μm. Fr common, late winter, early spring. Compact tufts on soil amongst limestone rocks, very rare. N. Somerset, Glamorgan. 2. France, Italy, Sicily, Yugoslavia, Ukraine, Crimea, Algeria.

55. Oxystegus (Lindb.) Hilp., Beih. Bot. Centralbl., 1933

Dioecious. Plants forming tufts or patches. Lvs variously incurved and crisped when dry, linear-lanceolate or ligulate, longly or shortly pointed, margin plane or narrowly recurved below, papillose-crenulate, sinuose, notched, often slightly toothed above; nerve ending in or below apex; basal cells rectangular, lax and hyaline or not, cells above hexagonal or rounded-hexagonal, incrassate or not, papillose. Capsule erect; peristome teeth irregularly perforated, cleft or divided, straight or spirally twisted. About 9 species found in Europe, Asia, Africa and America.

On morphological grounds *Oxystegus* produces a more satisfactory taxonomic situation than the more usually accepted procedure of placing *O. sinuosus* in *Barbula* or *Trichostomum*. *O. tenuirostris* and *O. hibernicus* show closer affinities with *O. sinuosus* than with species of *Trichostomum*.

1 Lvs without distinct hyaline base, cells in basal part of lf rectangular, not inflated **1. O. sinuosus**

Lvs with distinct hyaline base, cells in basal part of lf inflated 2

2 Basal part of lf hardly expanded, margin often sinuose or notched, mid-lf cells ± quadrate **2. O. tenuirostris**

Lf base expanded, margin not sinuose or notched, mid-lf cells often shortly rectangular **3. O. hibernicus**

1. O. sinuosus (Mitt.) Hilp., Beih, Bot. Centralbl., 1933
Barbula sinuosa (Mitt.) Grev., *Trichostomum sinuosum* (Mitt.) C. Müll.

Plants to 2 cm. Lvs curled when dry, patent, flexuose when moist, fragile, linear-lanceolate from slightly expanded basal part, tapering to narrowly lingulate to subulate apex, margin plane or narrowly recurved below, undulate, crenulate, sinuose and sometimes notched below or irregularly toothed towards apex; nerve ending below apex, ventral cells quadrate; basal cells rectangular, above ± quadrate-hexagonal, incrassate, papillose, 6–8 μm wide in mid-lf. Fr unknown, only female plants known in Britain. Green tufts or patches on damp, shaded, usually basic rocks, walls and old buildings. Frequent in southern Britain, rare to occasional elsewhere and extending north to Mull and Inverness. 69, H21, C. W. and C. Europe, Balearic Is., Caucasus, N. America.

A plant, resembling in certain respects *Oxystegus obtusifolius* Hilp., and likely to be mistaken for a small form of *O. sinuosus* has been found in Roscommon. A description is as follows:

Dioecious. Plants *ca* 6 mm. Lvs incurved and crisped when dry, patent to spreading when moist, lower small, upper larger, 1.0–1.6 mm long, ligulate, tapering to acute apex, margin plane, papillose-crenulate with occasional irregular projections but no distinct teeth; nerve ending below apex; basal cells rectangular or narrowly rectangular, cells above rounded-hexagonal, incrassate, unipapillose, 8–10 μm wide in mid-lf. Lax patches, green above, reddish-brown below, on shaded limestone (see Fig. **135**).

2. O. tenuirostris (Hook. & Tayl.) A. J. E. Smith, J. Bryol., 1977
O. cylindricus (Brid.) Hilp., *Trichostomum cylindricum* (Brid.) C. Müll., *T. tenuirostre* (Hook. & Tayl.) Lindb.

Plants to 7 cm. Lvs distant or crowded, often varying in size along stem, loosely incurved and curled when dry, erecto-patent to spreading, sometimes secund, often undulate when moist, 3–7 mm long, from ± erect but not expanded basal part ± linear-lanceolate, apex acute to sub-obtuse, margin plane, papillose-crenulate, often sinuose, notched or obscurely toothed above or with 2–3 teeth near apex; nerve ending below apex to slightly excurrent; cells in basal part inflated, thin-walled, hyaline, irregularly rectangular, cells above quadrate, papillose or not, 1–2 marginal rows sometimes more incrassate and distinct. Seta yellowish, flexuose; capsule cylindrical; peristome teeth short, straight, porose; spores 16–18 μm. Fr very rare, late spring, early summer. Lax tufts or patches on soil in rock crevices, on rocks and exposed tree roots by rivers, streams and waterfalls in shaded situations.

Var. **tenuirostris**

Lvs not crowded, spreading, sometimes secund when moist, narrowly linear-lanceolate; cells papillose, 8–10 μm wide in mid-lf. $n=13$. Lax tufts or patches, yellowish-green above, darker or blackish below. S.W. and W. England, E. Sussex, Wales, Shrops and Derby northwards, frequent or common. 59, H20. Europe, Faroes, Asia Minor, Siberia, Nepal, Sikkim, China, Korea, Sri Lanka, C. and southern Africa, Malagasay Republic, America, Greenland.

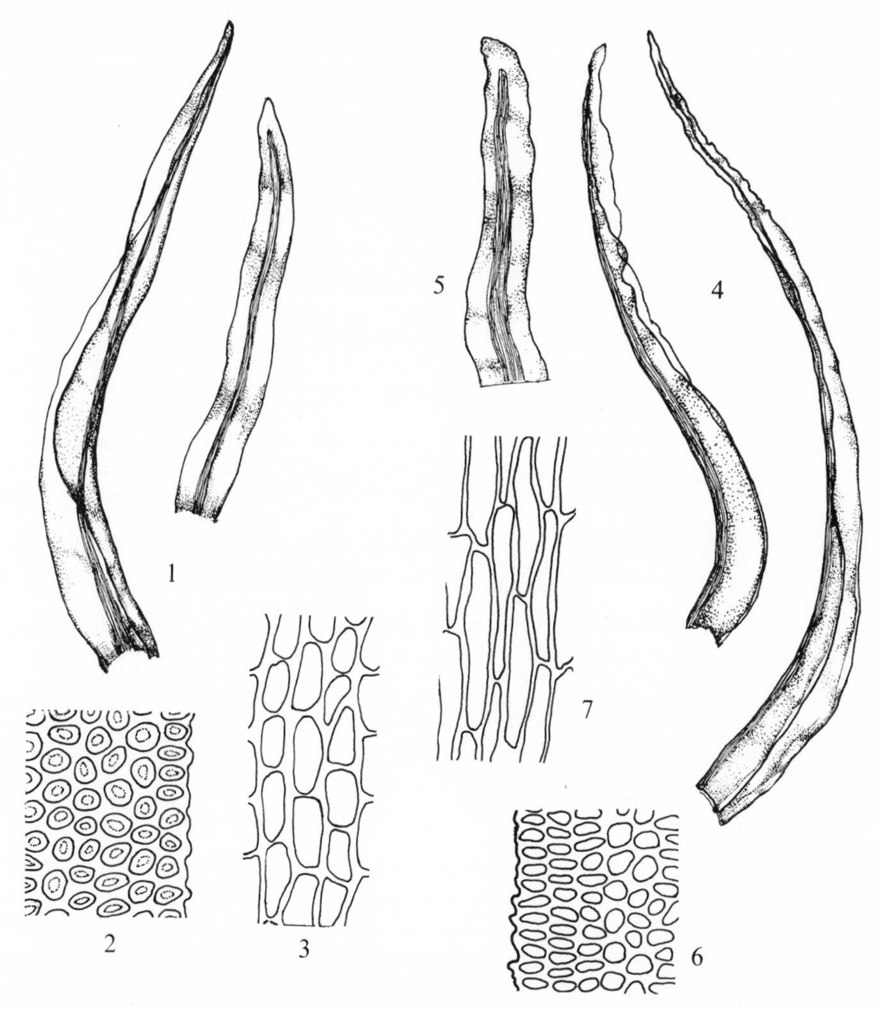

Fig. **135**. 1–3, plant from Roscommon: 1, leaves; 2, mid-leaf cells; 3, basal cells. 4–7, *Oxystegus sinuosus*: 4, leaves; 5, leaf apex (×100); 6, mid-leaf cells; 7, basal cells. Leaves ×25, cells ×415.

Var. **holtii** (Braithw.) A. J. E. Smith, J. Bryol., 1977

O. cylindricus var. *holtii* (Braithw.) Podp., *Trichostomum tenuirostre* var. *holtii* (Braithw.) Dix.

Lvs more crowded, ± erecto-patent, sometimes secund when moist, linear-lanceolate; cells less papillose to almost smooth, 7–8 μm wide in mid-lf. Plants greenish, on damp shaded rocks, rare. S. Somerset, N. Wales, S. Lancs and Durham north to Angus and Argyll, Inner Hebrides, Kerry. 16, H2. Norway.

The status of var. *holtii* is open to question as the characters attributed to it may occur independently in the type. *O. tenuirostris* has sometimes been confused with *O. hibernicus*,

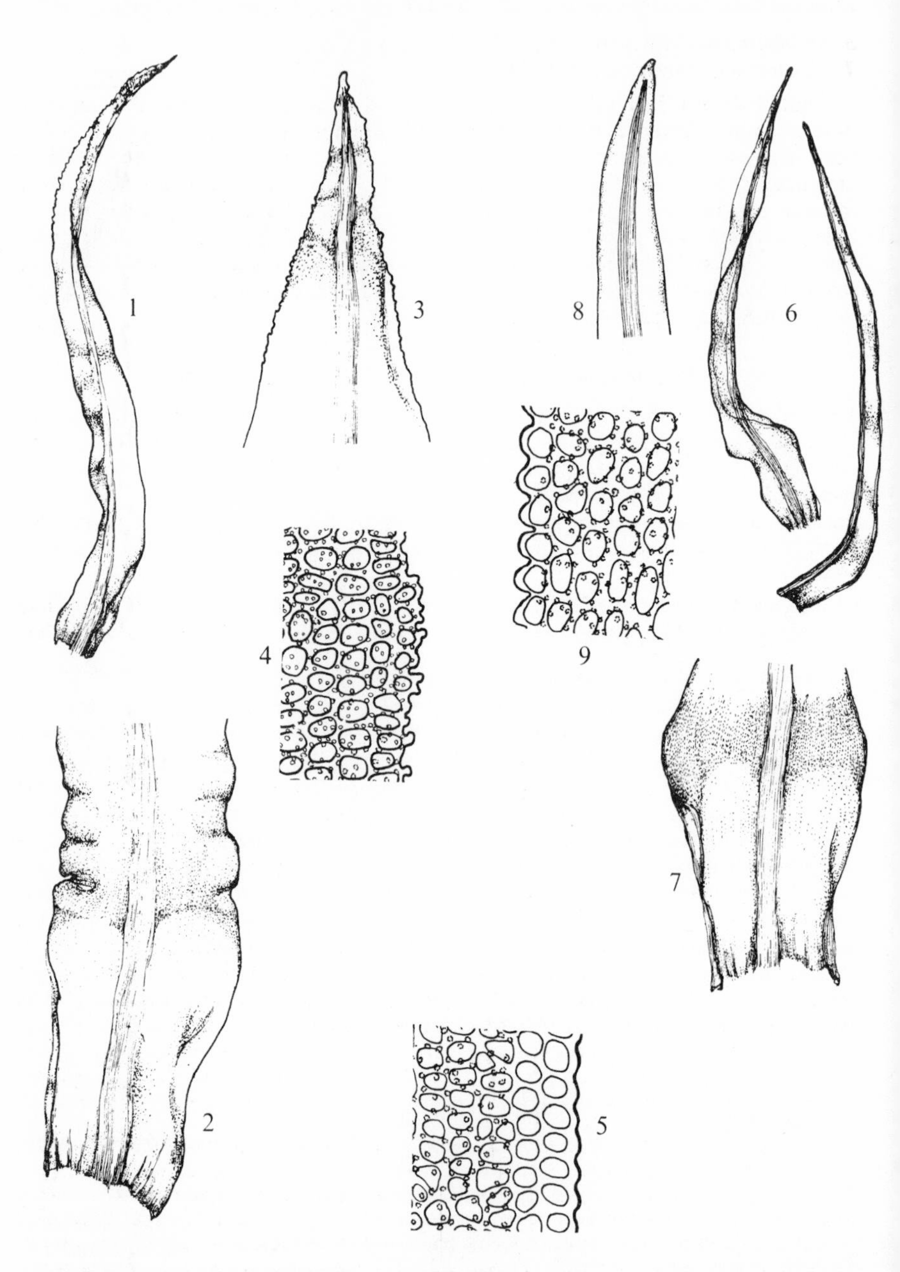

Fig. **136**. 1–4, *Oxystegus tenuirostris* var. *tenuirostris*: 1, leaf; 2, leaf base; 3, leaf apex; 4, mid-leaf cells. 5, *O. tenuirostris* var. *holti*: mid-leaf cells. 6–9, *O. hibernicus*: 6, leaves; 7, leaf base; 8, leaf apex; 9, mid-leaf cells. Leaves ×15, bases ×40, apices ×100, cells ×415.

but that species is distinct in the not or scarcely undulate lvs sharply reflexed from an expanded base, the margin not notched or toothed and the cells often shortly rectangular.

3. O. hibernicus (Mitt.) Hilp., Beih. Bot. Centralbl., 1933
Trichostomum hibernicum (Mitt.) Dix.

Plants 1–10 cm. Lvs not crowded, crisped and incurved when dry, from an erect base spreading, flexuose, usually not undulate when moist, from an expanded basal part, narrowly linear-lanceolate, tapering to acute apex, margin plane, finely crenulate, not toothed or notched; nerve ending in apex or slightly excurrent; cells in basal part inflated, hyaline, irregularly rectangular, above shortly rectangular to irregularly quadrate, papillose, 8–10 μm wide in mid-lf. Capsule cylindrical. Fr very rare. Lax, yellowish-green tufts on damp rocks and in rock crevices, rare. W. Scotland from Dumfries north to Ross, Arran and the Hebrides, Kerry, E. Cork, W. Galway, W. Mayo, W. Donegal. 11, H6. Endemic.

56. TRICHOSTOMUM HEDW., SP. MUSC., 1801

Usually dioecious. Stem with central strand. Lvs linear-lanceolate to lingulate or lanceolate, acute or obtuse, margin plane or incurved, papillose-crenulate; nerve ending below apex or excurrent, in section with 2 stereid bands; basal cells ± rectangular, hyaline or yellowish, cells above smaller, ± hexagonal, papillose, obscure. Seta long; capsule erect, ellipsoid or cylindrical; lid rostrate; peristome teeth divided, perforated or entire, straight; calyptra cucullate. A ± world-wide genus of 145 species.

Lf margin erect or incurved above, apex usually cucullate **1. T. crispulum**
Margin usually plane above, apex not cucullate **2. T. brachydontium**

1. T. crispulum Bruch, Flora, 1829

Plants 0.5–4.0(–8.0) cm. Lvs incurved, often crisped when dry, erecto-patent to patent, sometimes slightly incurved when moist, lingulate to lanceolate or narrowly linear-lanceolate, usually tapering to cucullate, apiculate apex, margin entire, erect or incurved above and forming with upturned nerve the cucullate apex, rarely plane above; nerve strong, ending below apex to excurrent; hyaline basal cells variable in extent, towards nerve incrassate, narrowly rectangular, towards margin thinner-walled, shorter and wider, cells above quadrate-hexagonal, papillose, obscure, 6–8(–10) μm wide in mid-lf. Seta reddish; capsule narrowly cylindrical; spores 16–18 μm. Fr rare or occasional, spring. Dense tufts or patches, yellowish-green, green or dark green, sometimes blackish below, in basic habitats on soil, in sand-dunes, on rocks, cliff ledges, mortar of old walls, etc. Generally distributed, frequent or common in basic areas, rare elsewhere. 97, H38, C. Europe, Caucasus, Asia Minor, Siberia, E. Asia, N. Africa, Madeira, N. America.

An exceedingly variable species usually easily recognised by the cucullate lf apex. Forms with a plane lf apex may be separated from *Oxystegus tenuirostris* by the lf margin not undulate; *T. brachydontium* has wider lvs and a more strongly excurrent nerve; species of *Tortella* differ in the hyaline basal cells ascending up the margin.

Four varieties of *T. crispulum* have been recognised in Britain; var. *brevifolium* (C. Müll.) Br. Eur.; var. *elatum* Schimp.; var. *nigroviride* (Braithw.) Dix.; var. *viridulum* (Bruch) Dix. These varieties are based on habit, colour, lf shape and size, form of apex and cell size. They are of doubtful status and are not recognised here as equally distinctive but undescribed plants occur and some of the varietal characters, especially those involving the lvs, may be found in otherwise typical plants. It seems likely that the species is subject to environmental modification and random genetic variations not amenable to orthodox taxonomic treatment. A somewhat parallel though less extreme situation occurs in *T. brachydontium*.

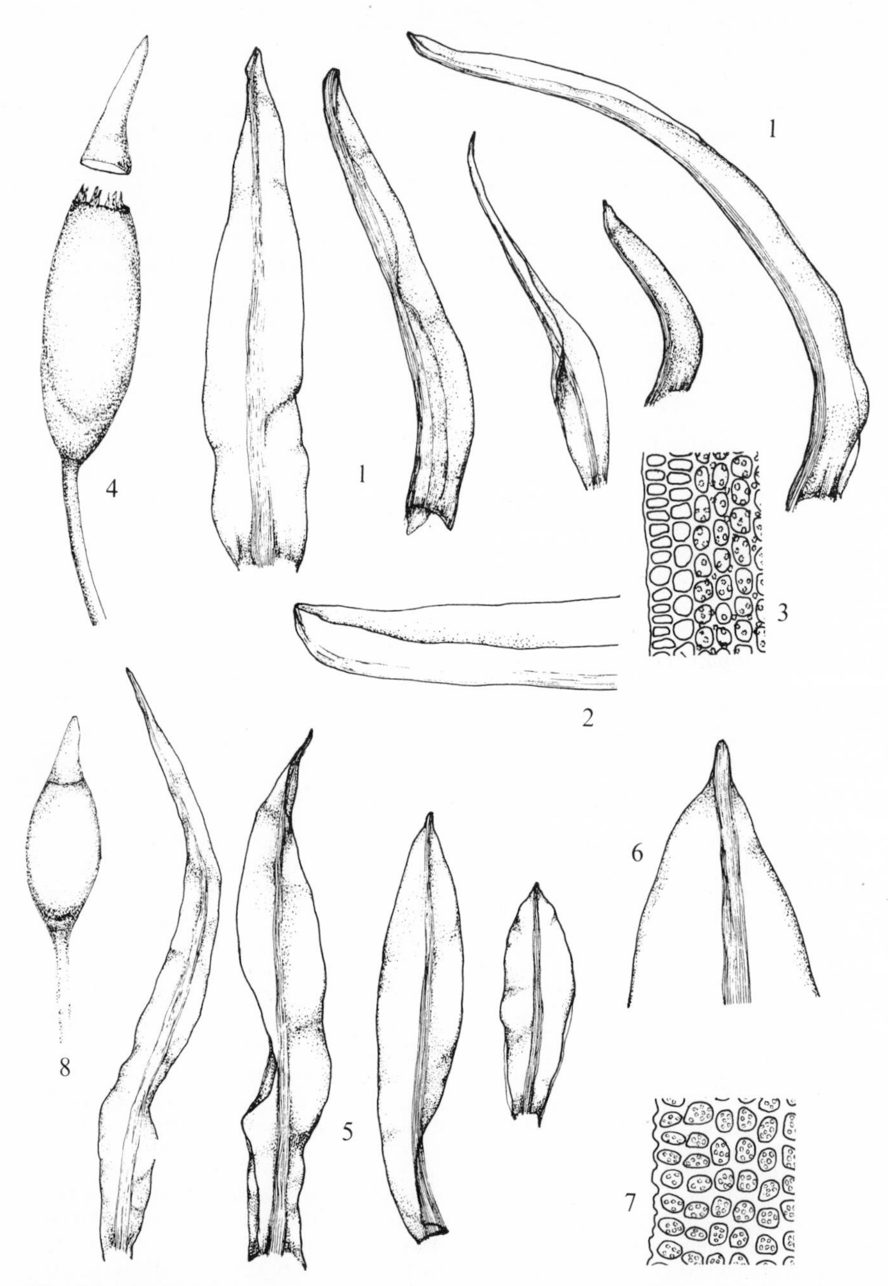

Fig. **137**. 1–4, *Trichostomum crispulum*: 1, leaves; 2, leaf apex; 3, mid-leaf cells; 4, capsule. 5–8, *T. brachydontium*: 5, leaves; 6, leaf apex; 7, mid-leaf cells; 8, capsule. Leaves ×25, apices ×100, cells ×415, capsules ×15.

2. T. brachydontium Bruch, Flora, 1829
T. mutabile Bruch

Plants (0.5–)1.0–4.0 cm. Lvs crisped and incurved when dry, erecto-patent to patent or spreading, sometimes recurved when moist, lingulate to narrowly lingulate-lanceolate or linear-lanceolate, apex acute to obtuse, rarely acuminate, not cucullate, margin plane or narrowly recurved, rarely narrowly incurved towards apex, papillose-crenulate, sometimes denticulate near base; nerve strong, usually excurrent in a stout mucro, rarely in a cuspidate point; basal cells irregularly rectangular or narrowly rectangular, hyaline or yellowish, above irregularly hexagonal, papillose, obscure, 6–8 μm wide in mid-lf. Seta yellow; capsule ellipsoid to narrowly ellipsoid; peristome teeth very short, fragile; spores 14–18 μm. Fr very rare. Yellowish-green to dark green patches or tufts, sometimes reddish below in usually but not necessarily basic habitats on soil in turf, on banks, cliffs, in rock and wall crevices, in shaded or exposed places, generally distributed, frequent. 87, H25, C. Europe, Caucasus, Syria, China, Japan, N. and S. Africa, Macaronesia, Juan Fernandez, New Zealand.

A variable species for which two varieties have been recorded from Britain, var. *littorale* (Mitt.) C. Jens. and var. *cophocarpum* (Schimp.) P. Cout., but both intergrade with the type to such an extent that they cannot be maintained.

57. Tortella Limpr., Laubm., 1888

Dioecious. Stems usually without central strand. Lvs crisped and curled when dry, erect to recurved when moist, usually fragile, lanceolate to linear-lanceolate, acute to obtuse and mucronate, margin plane or incurved above; nerve ending in apex or excurrent; all basal cells hyaline and hyaline cells ascending up margin, transition to chlorophyllose cells usually abrupt, upper cells hexagonal, opaque or pellucid. Perichaetial lvs ± similar to stem lvs. Capsule erect, ellipsoid or cylindrical, straight or slightly curved; peristome teeth filiform, usually spirally twisted; calyptra cucullate. A ± world-wide genus of about 60 mainly calcicole species.

1 Lvs ± straight or curved but not twisted or crisped when dry, cells in upper part of lf 6–8 μm wide, 2–3-stratose and forming with nerve long, trigonous, very fragile, setaceous point **1. T. fragilis**
Lvs usually strongly twisted, curled or incurved when dry, cells unistratose throughout, apex various but never setaceous 2

2 Lvs tightly incurved and imbricate with nerve conspicuous and glossy when dry, upper part of lf nearly always lost **8. T. nitida**
Lvs not tightly incurved nor nerve conspicuously glossy when dry, at least some lvs intact 3

3 Lamina cells continuous over upper (ventral) surface of nerve at least in middle part of lf 4
Lamina cells not continuous over upper surface of nerve 6

4 Lvs strongly curled and crisped when dry, flexuose and undulate when moist, longly tapering to acuminate apex **2. T. tortuosa**
Lvs lacking above combination of characters 5

5 Nerve excurrent, transition from basal to upper cells abrupt, plants on soil or sand **3. T. flavovirens**
Nerve ending in or below apex, transition from basal to upper cells not abrupt, plants saxicolous **7. T. inflexa**

6 Lvs gradually tapering to acuminate apex **6. T. densa**
Lvs shortly pointed, apex acute or sub-obtuse 7

7 Lvs patent or spreading when moist, upper cells strongly papillose, obscure, 8–10 μm wide **4. T. inclinata**
Lvs reflexed from ± erect base when moist, upper cells obscurely papillose, ± pellucid, 8–12 μm wide **5. T. limosella**

1. T. fragilis (Hook. f. & Wils.) Limpr., Laubm., 1888
Trichostomum fragile (Hook. f. & Wils.) C. Müll.

Plants 0.5–2.5 cm. Lvs rigid, lower ± straight, upper curved when dry, erecto-patent, not or scarcely flexuose or undulate when moist, narrowly linear-lanceolate, gradually tapering from hyaline basal portion to long subulate apex, usually broken except in young lvs, margin plane or inflexed above, papillose-crenulate; nerve extending to apex and forming with upper bistratose lamina cells a long fragile, setaceous apex, glossy at back when dry; basal cells linear, hyaline, extending up margin, transition to chlorophyllose cells abrupt, chlorophyllose cells small, ± hexagonal, papillose, obscure, 6–10 μm wide in upper part of lf, cells in upper part of lf bistratose and hardly distinguishable from nerve. Fr unknown in Britain. Yellowish-green to green patches on basic montane rocks and calcareous dune slacks, very rare. Fife, Mid Perth, E. Sutherland, Caithness, Outer Hebrides. 5. Europe, Asia, Morocco, N. America.

The fragile lf apices appear to provide a means of vegetative propagation. When dry the rigid, curved rather than strongly crisped lvs will distinguish this plant from *T. tortuosa*, some forms of which may have fragile lf apices. The setaceous lf apex if present is also distinctive.

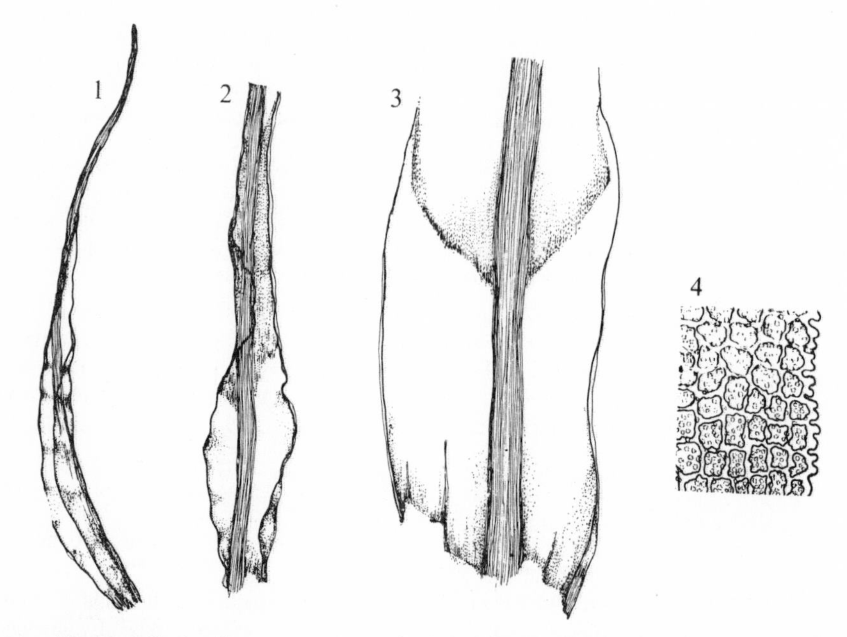

Fig. **138**. *Tortella fragilis*: 1, very young leaf; 2, older broken leaf; 3, leaf base (×40); 4, mid-leaf cells (×415). Leaves ×15.

2. T. tortuosa (Hedw.) Limpr., Laubm., 1888
Trichostomum tortuosum (Hedw.) Dix.

Plants 1–8 cm. Lvs strongly curled and contorted when dry, spreading, flexuose when moist, linear, gradually tapering from hyaline basal part to long acuminate apex, rarely narrowly lanceolate and more abruptly tapering, lamina sometimes torn or broken, margin plane, undulate, papillose-crenulate, denticulate or not near apex; nerve excurrent, lamina cells continuous over upper (ventral) surface in middle part of lf; basal cells narrowly rectangular or linear, hyaline, ascending up margin, transition to chlorophyllose cells ± abrupt, chlorophyllose cells quadrate-hexagonal, papillose, unistratose throughout, 8–10 μm wide in upper part of lf. Seta yellowish-red; capsule cylindrical, straight to arcuate; peristome teeth long, filiform, spirally twisted; spores 10–16 μm. Fr rare, summer. $n=13$. Yellowish to green cushions, tufts or patches on rocks, in rock crevices, soil in turf and in flushes where the substrate has at least some trace of base. Rare to occasional in lowland Britain, frequent to common in W. and N. Britain and Ireland in basic areas, very rare in acidic habitats. 102, H40, C. Europe, Faroes, Caucasus, Iran, Asia, Canaries, Madeira, Algeria, Morocco, N. America, Greenland.

The linear, longly tapering lvs, curled and contorted when dry, usually readily distinguish *T. tortuosa*. Short-leaved forms may be confused with *T. densa*, *T. inclinata* or *T. flavovirens*. The two latter have less contorted lvs when dry and in the former they are scarcely undulate when moist. In *T. densa* and *T. inclinata* the cells over the nerve are elongated. In the short-leaved forms of *T. tortuosa*, which occur particularly in Ireland and have also been mistaken for *T. fragilis*, the cells of the lamina are only patchily continuous over the nerve. Such plants may be worthy of varietal recognition and their nature requires investigation. *T. fragilis* differs in the lf apex.

3. T. flavovirens (Bruch) Broth., Nat. Pfl., 1902
Trichostomum flavovirens Bruch

Plants to 1.5 cm, stems often with central strand, not tomentose below. Lvs crisped when dry, lower small, erecto-patent, upper larger, spreading when moist, narrowly lanceolate to linear-lanceolate, shortly or abruptly tapering to acute to sub-obtuse, mucronate, frequently cucullate apex, margin plane below, usually erect or incurved above, papillose-crenulate; nerve excurrent, lamina cells continuous over upper (ventral) surface, at least in middle part of lf; basal cells linear to narrowly rectangular, hyaline, ascending up margin, transition to chlorophyllose cells abrupt, upper cells hexagonal, papillose. Seta purplish-red; capsule narrowly ellipsoid; peristome teeth filiform, straight, fugacious; spores 12–14 μm. Fr very rare.

Var. **flavovirens**
Cells in upper part of lf 8–10 μm wide. Yellowish-green to brownish-green tufts or patches in dune slacks and soil in maritime situations, frequent along the coast, absent inland. 55, H18, C. Atlantic coasts of Europe and N. America, Mediterranean coast, Macaronesia, China, Japan.

Var. **glareicola** (Christens.) Crundw. & Nyholm, Trans. Br. bryol. Soc., 1962
Cells in upper part of lf 10–14 μm wide. In similar habitats to the type, rare. 14, H3, C. Denmark, Sweden, Canada.

4. T. inclinata (Hedw. f.) Limpr., Laubm., 1888
Trichostomum inclinatum (Hedw. f.) Dix.

Plants to *ca* 1 cm, stems without central strand, tomentose below. Lvs curved inwards, not or scarcely crisped when dry, patent to spreading, slightly undulate

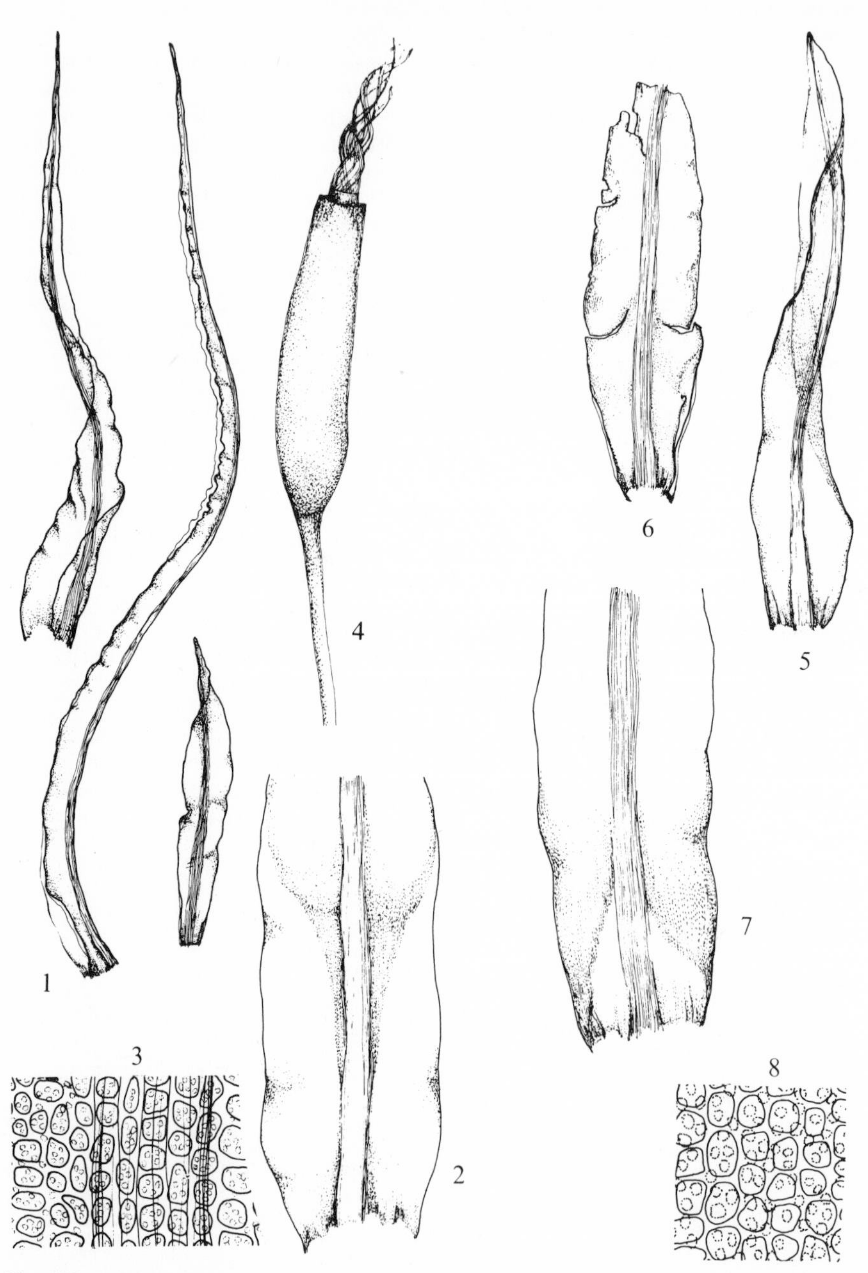

Fig. **139**. 1–4, *Tortella tortuosa*: 1, leaves (×15); 2, leaf base (×40); 3, cells over nerve in middle of leaf; 4, capsule (×15). 5–8, *T. nitida*: 5 and 6, young and old leaves (×25); 7, leaf base (×40); 8, mid-leaf cells. Cells ×415.

when moist, linear-lanceolate, shortly or abruptly tapering to slightly cucullate acute or sub-obtuse apex, margin plane, papillose-crenulate; nerve excurrent, cells on ventral surface elongated; basal cells linear to narrowly rectangular, hyaline, extending up margin, transition to rounded-hexagonal, papillose, obscure upper cells abrupt, cells 8–10 μm wide in upper part of lf. Small terminal, propaguliferous shoots sometimes present. Capsule ellipsoid, straight or slightly curved; peristome teeth filiform, spirally twisted, fugacious; spores 10–12 μm. Fr very rare, summer. Yellowish-green patches on shallow basic soil in turf and on sand-dunes, rare. Scattered localities from Surrey and W. Norfolk north to E. Sutherland, W. Mayo, Down. 19, H2. Europe, N. America.

Very similar in appearance to *T. flavovirens* but readily distinguished by the elongated ventral cells of the nerve.

5. T. limosella (Stirt.) Rich. & Wall., Trans. Br. bryol. Soc., 1950
T. flavovirens ssp. *limosella* (Stirt.) Podp., *Trichostomum limosellum* (Stirt.) Dix.

Plants 6–8 mm. Upper lvs longer than lower, loosely curled, hardly crisped when dry, reflexed from ± erect base when moist, lanceolate-lingulate, apex acute to sub-obtuse and apiculate, margin plane, crenulate-serrulate with scattered irregular poorly defined teeth; nerve excurrent, ventral cells elongated; basal cells hyaline, extending up margin, sharply differentiated from upper cells, upper cells thick-walled, quadrate-hexagonal, obscurely unipapillose, hardly obscure, 8–12 μm wide. Gametangia and sporophyte unknown. Dense patches on seashore west of Arisaig, 1906, not seen since. Endemic.

Dixon (1924) suggests that this plant is closely related to *T. flavovirens*, but the elongated ventral cells of the nerve indicate that its relationship is with *T. inclinata*. It differs from that species in the reflexed lvs and sometimes larger, unipapillose cells. The exact status of *T. limosella* is obscure but on present evidence there are no grounds for reducing it to an intraspecific taxon of any other *Tortella* species.

6. T. densa (Mol.) Crundw. & Nyholm, Trans. Br. bryol. Soc., 1962
T. tortuosa fo *curta* Albertson

Plants to 4 cm, stems not tomentose. Lower lvs straight or slightly curved, upper ± erect and curled when dry, erecto-patent, hardly flexuose when moist, narrowly lanceolate to linear-lanceolate, gradually tapering to acuminate apex, margin papillose-crenulate, hardly undulate; nerve ending in apex or shortly excurrent, cells on upper (ventral) surface elongated; lamina cells not extending over nerve, basal cells linear to narrowly rectangular, hyaline, ascending up margin, transition to ± hexagonal, papillose, obscure upper cells abrupt, cells in upper part of lf 7–10 μm wide. Fr unknown. Dull green patches or tufts on shallow soil on limestone rocks and in crevices, rare. E. Glos, Yorks, Durham, Westmorland, I. of Man, Argyll, W. Sutherland, Clare, Mayo, Sligo. 8, H4. N., W. and C. Europe.

7. T. inflexa (Bruch) Broth., Nat. Pfl., 1902

Plants 2–6 mm. Lvs strongly curled when dry, erect to spreading when moist, linear-lanceolate to linear, apex ± acute, often sub-cucullate, margin inflexed or incurved at least above, papillose-crenulate; nerve ending below apex, cells of lamina continuous over upper (ventral) surface of nerve; basal cells narrowly rectangular, hyaline, ascending up margin, transition to chlorophyllose cells gradual, upper cells quadrate-hexagonal, papillose, obscure, cells in upper part of lf 8–10 μm wide. Fr very rare. Small, bright green patches on chalk or oolitic limestone stones in open or shaded habitats in S. England from S. Wilts and E. Glos to E. Kent and Cambridge. 18. France, Portugal, Sardinia, Minorca, Malta, Crete, Israel, Algeria.

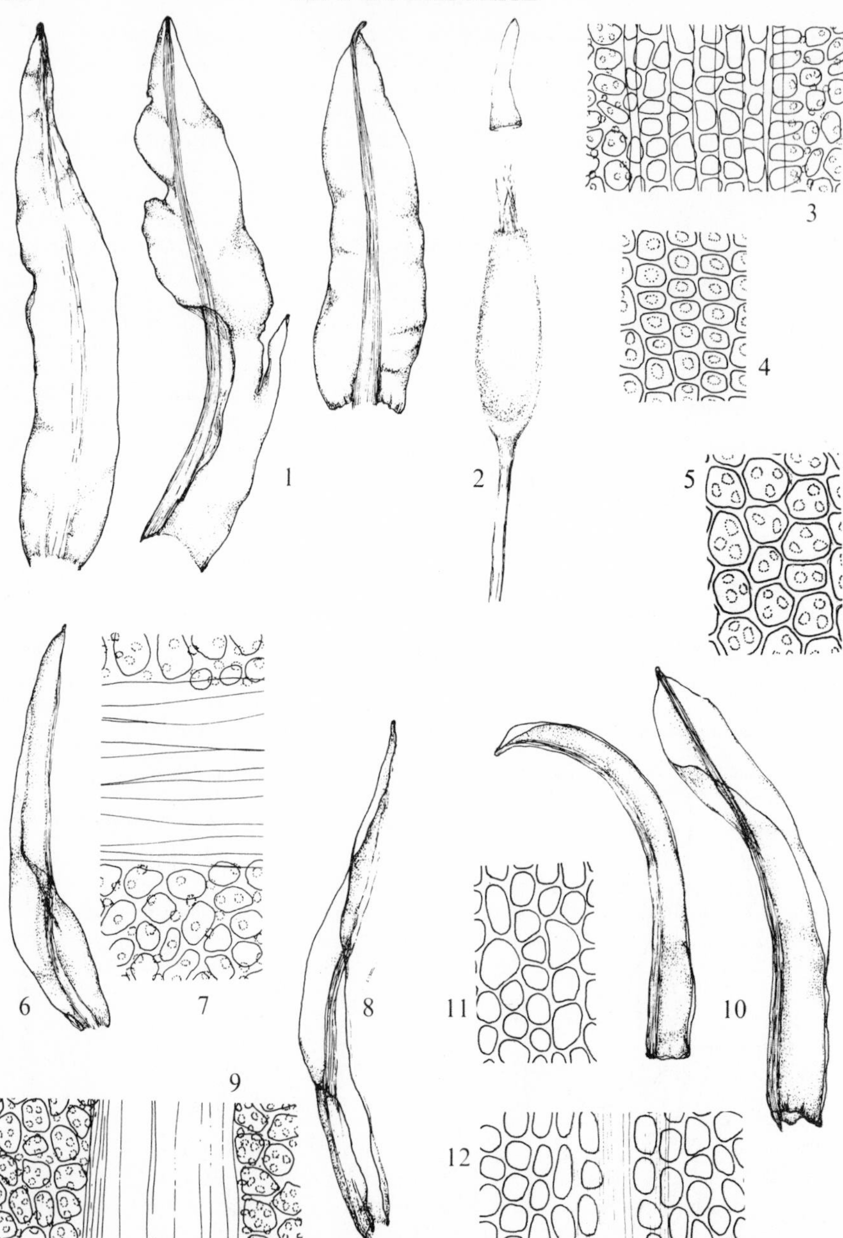

Fig. **140**. 1–4, *Tortella flavovirens* var. *flavovirens*: 1, leaves (×25); 2, capsule (×10); 3, ventral surface of nerve in middle of leaf; 4, mid-leaf cells. 5, *T. flavovirens* var. *glareicola*: mid-leaf cells. 6–7, *T. inclinata*: 6, leaf (×25); 7, ventral surface of nerve in middle of leaf. 8–9, *T. densa*: 8, leaf (×25); 9, ventral surface of nerve in middle of leaf. 10–12, *T. limosella*: 10, leaves (×40); 11, mid-leaf cells; 12, ventral surface of nerve in middle of leaf. Cells ×415.

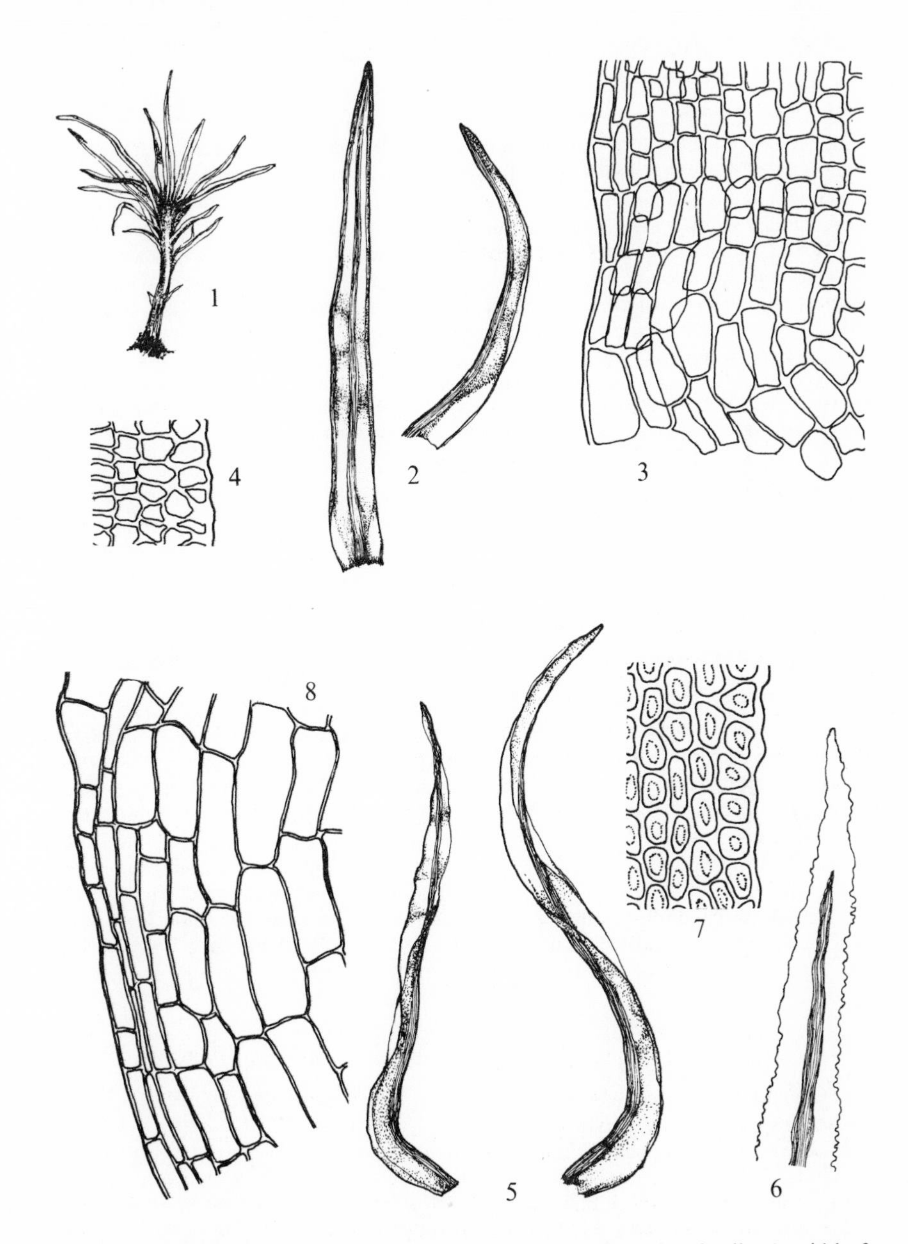

Fig. **141**. 1-4, *Tortella inflexa*: 1, plant (×10); 2, leaves (×40); 3, basal cells; 4, mid-leaf cells. 5–8, *Trichostomopsis umbrosa*: 5, leaves (×40); 6, leaf apex (×100); 7, mid-leaf cells; 8, basal cells. Cells ×415.

Possibly an introduction and apparently spreading in southern England. As fruiting material with immature capsules has only once been found here, it may spread by vegetative means, pairs of deciduous lvs having been seen to fall away from stem apices. For an account of this plant in Britain see Wallace, *J. Bryol.* 7, 153–6, 1972.

8. T. nitida (Lindb.) Broth., Nat. Pfl., 1902
Trichostomum nitidum (Lindb.) Schimp.

Plants 0.5–1.0 cm. Lvs tightly incurved when dry, erecto-patent to spreading, slightly undulate when moist, very fragile, upper part of lf usually missing, linear-lanceolate, acute to acuminate, margin plane, papillose-crenulate; nerve excurrent in short mucro, whitish and shining at back when dry; basal cells irregularly rectangular, ascending up margin, transition to chlorophyllose cells not very abrupt, chlorophyllose cells hexagonal, papillose, obscure, 6–10 μm wide in upper part of lf. Fr unknown in Britain. Light olive-green to yellowish-green tufts on basic rocks and walls in W. and S.W. Britain, W. Ireland and Jersey. 32, H12, C. Europe, S.W. Asia, N. Africa, N. America.

Recognisable by the lvs tightly incurved when dry, the neat, rounded tufts and the fragile lvs. Has been confused with *T. flavovirens* but differs in habit, habitat and in the transition from basal hyaline to chlorophyllose cells being not very abrupt.

58. Pleurochaete Lindb., Öfv. K. Vet. Ak. Förh., 1864

Dioecious, gametangia axillary. Lvs crisped when dry, squarrose from erect, sheathing base when moist, acute, margin toothed above; nerve percurrent or excurrent; basal cells rectangular, coloured except for marginal band, cells above hexagonal, papillose. Seta lateral, long; capsule erect, cylindrical, curved; peristome with basal membrane, 32 filiform, papillose, slightly twisted teeth; calyptra cucullate. A small genus of 5 species found in Europe, Asia, N. and C. Africa, America.

1. P. squarrosa (Brid.) Lindb., Öfv. K. Vet. Ak. Förh., 1864

Plants 1–7 cm. Upper lvs crisped when dry, squarrose from ± erect, sheathing base when moist, lanceolate to linear-lanceolate, acuminate, margin plane, denticulate below, crenulate and irregularly toothed above; nerve percurrent or very shortly excurrent, smooth at back; basal cells rectangular or narrowly rectangular, incrassate, coloured, 3–4 marginal rows thinner-walled, hyaline, ascending up margin and forming hyaline band in lower part of leaf, upper cells ± quadrate, papillose, obscure, 8–10 μm wide in mid-lf. Fr unknown in Britain. Lax yellowish-green tufts on dry, open, sandy or calcareous soil, often near the sea, occasional but sometimes locally abundant. Southern half of Britain, extending north to Yorks and Westmorland, Clare, Dublin, Wicklow. 30, H3, C. W., S. and C. Europe, Caucasus, Iran, Himalayas, China, Azores, Canaries, N. Africa, Kenya, Arizona.

59. Trichostomopsis Card., Rev. Bryol., 1909

Lvs incurved or crisped when dry, spreading when moist, lanceolate or narrowly lanceolate with usually expanded basal part, margin plane or narrowly recurved, bistratose in mid-part; nerve ending in or below apex; basal cells lax, hyaline, above hexagonal, often collenchymatous, papillose or not. Seta long; capsule ovoid to cylindrical; peristome teeth bifid almost to base, straight or slightly twisted. A small genus of 7 species occurring in W. Europe, S. Africa, America, Asia, Australasia.

1. T. umbrosa (C. Müll.) H. Robinson, Phytologia, 1970

Plants very small, 2–3 mm. Lvs spreading, flexuose or somewhat curled when dry,

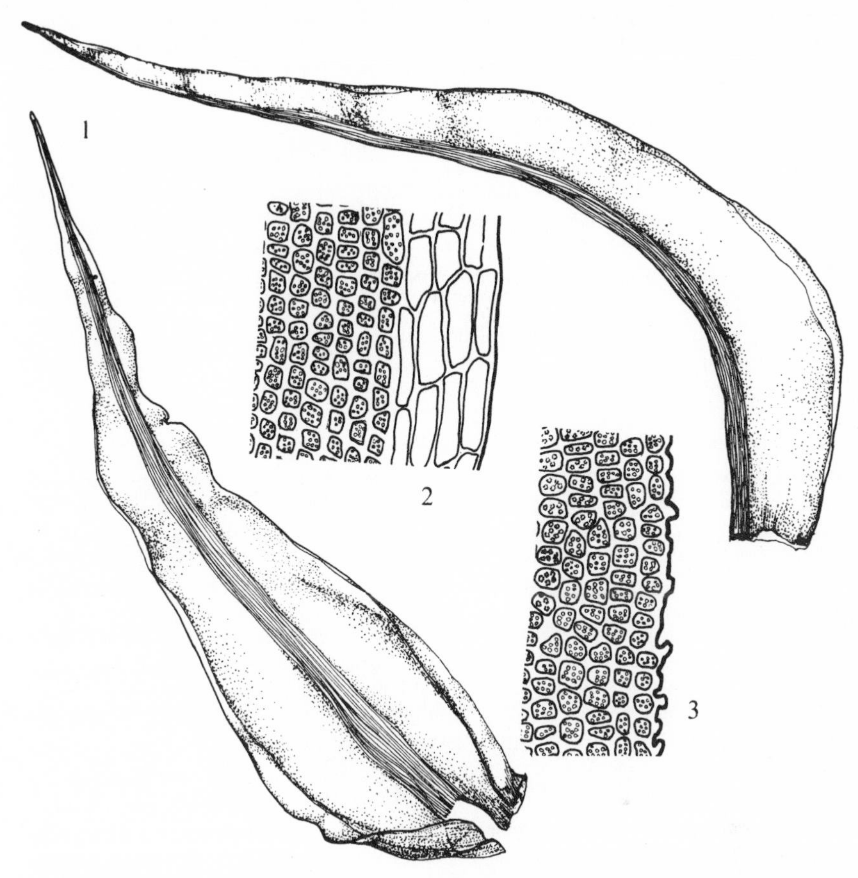

Fig. **142**. *Pleurochaete squarrosa*: 1, leaves (×25); 2, basal cells; 3, mid-leaf cells. Cells ×415.

spreading when moist, lower small, upper much longer, narrowly linear-lanceolate, tapering from expanded basal portion to acute to acuminate apex, margin plane, bistratose in middle part, crenulate; nerve thin, ending below apex; basal cells rectangular, hyaline, very narrow near margin, cells above hexagonal, thin-walled, *ca* 8 μm wide in mid-lf, marginal cells often wider than long. Fr unknown in Britain. Pale green patches, often encrusted with calcareous matter, on damp, shaded brick walls, very rare. S. Hants, E. Kent, Middlesex, Bucks, Cambridge. 5. Barcelona, California, Mexico, Uruguay, Argentina.

In England this plant seems restricted to damp walls such as support walls of bridges; in Barcelona it has been found on a number of walls of buildings; in America, where the plants are larger than in England (to 1 cm with lf cells 9–14 μm wide), it occurs on dry walls and soil.

60. LEPTODONTIUM (C. MÜLL.) HAMPE EX LINDB., ÖFV. K. VET. AK. FÖRH., 1864

Dioecious. Plants forming tufts or short turfs, stems without central strand. Lvs flexuose or crisped when dry, patent to squarrose when moist, oblong-lanceolate, obtuse to acute, margin plane or recurved below, coarsely toothed towards apex; nerve ending below apex to excurrent; basal cells rectangular, above hexagonal, papillose. Seta erect, flexuose; capsule ellipsoid to cylindrical; lid conical or rostrate; peristome without basal membrane, teeth 32, filiform, smooth or finely papillose; annulus present; calyptra cucullate. A ± world-wide genus of *ca* 120 species.

1 Lvs strongly curled when dry, ± undulate when moist, 1–2 marginal cells elongate, smooth, very incrassate **3. L. recurvifolium**
Lvs ± erect, flexuose when dry, not undulate when moist, marginal cells not elongate 2

2 Lf apex rounded to obtuse and apiculate, nerve ending below apex in upper lvs, not gemmiferous **1. L. flexifolium**
Lf apex acute, nerve excurrent with clusters of gemmae at apex **2. L. gemmascens**

1. L. flexifolium (With.) Hampe, Linnaea, 1847

Plants 2–15 mm. Lvs erect, flexuose when dry, soft, patent to spreading or recurved from erect base when moist, oblong-lanceolate to broadly lingulate or lingulate-spathulate, apex rounded to obtuse and apiculate, margin plane or recurved below, coarsely toothed above; nerve stout, ending below apex, smooth at back; basal cells shortly rectangular to rectangular, above hexagonal, thin-walled to strongly incrassate, smooth to papillose, 10–16(–20) μm wide in mid-lf, a single marginal row pellucid. Seta flexuose, yellowish; capsule narrowly ellipsoid to shortly cylindrical, straight or slightly curved; lid longly rostrate; peristome teeth erect, filiform; spores 12–14 μm. Fr occasional, spring, autumn. Dull green to yellowish or brownish-green patches on peaty soil, often on sites of fires, on heaths, moorland, wood rides, occasional. 86, H14. W. Europe, Italy, Iceland, Canaries, Cameroon, Fernando Po.

2. L. gemmascens (Mitt. ex Hunt). Braithw., Brit. Moss Fl., 1887

Plants 2–10 mm. Lvs erect, flexuose when dry, soft, spreading when moist, oblong-lanceolate to lanceolate, ± tapering to acute apex, margin plane or narrowly recurved below, coarsely toothed above; nerve stout, excurrent with clusters of gemmae on excurrent portion in upper lvs; basal cells shortly rectangular, upper hexagonal, thin-walled, 10–18 μm wide in mid-lf. Gemmae obovoid to shortly fusiform, 50–100 μm long. Fr unknown. Dull green patches on old thatch, rarely on trees, ephemeral, very rare and decreasing. A few localities from S. Devon, S. Somerset and Sussex north to E. Glos and Hereford. 10. Normandy, France, Marion Is., Indian Ocean.

3. L. recurvifolium (Tayl.) Lindb., Öfv. K. Vet. Ak. Förh., 1864
Bryoerythrophyllum recurvifolium (Tayl.) Zander

Plants 3–10 cm. Lvs strongly crisped when dry, patent to squarrose from erect base, ± undulate when moist, ovate to lanceolate, apex acute to obtuse and apiculate, margin plane, coarsely toothed above; nerve ending in apex to shortly excurrent; basal cells rectangular to narrowly rectangular, above quadrate-hexagonal, papillose, obscure, thin-walled to incrassate, 12–16 μm wide in mid-lf, 1–2 marginal rows elongated, more incrassate, smooth, forming narrow border. Fr unknown. Lax, green or yellowish-green tufts on base-rich cliffs, rock ledges, banks and rock crevices, often near waterfalls, rare. W. Britain from Merioneth and Caerns, Westmorland and Cumberland to W. Sutherland, W. Ireland. 15, H4. Spain.

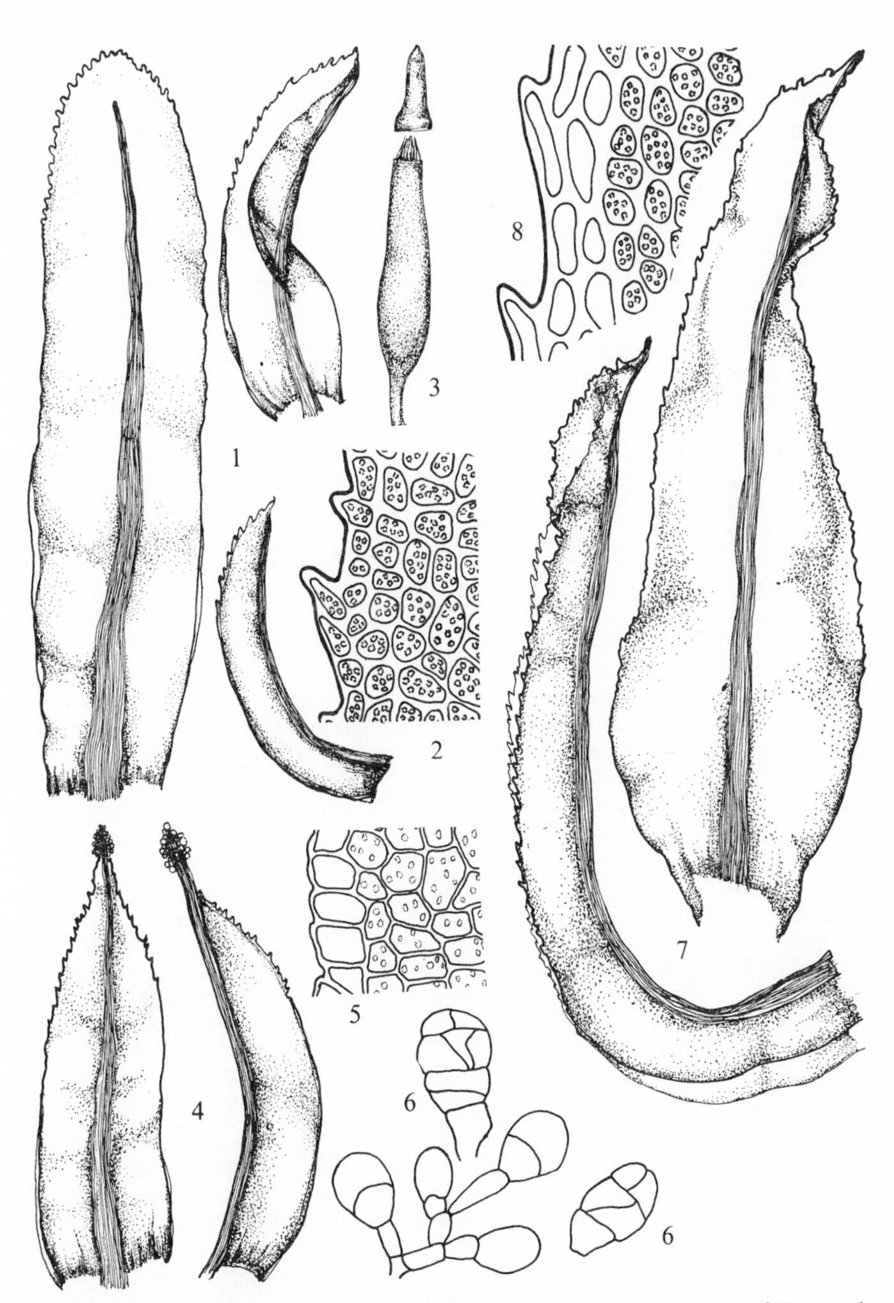

Fig. 143. 1–3, *Leptodontium flexifolium*: 1, leaves; 2, cells from upper part of leaf; 3, capsule (×15). 4–6, *L. gemmascens*: 4, leaves; 5, cells from upper part of leaf; 6, gemmae (×250). 7–8, *L. recurvifolium*: 7, leaves; 8, cells from upper part of leaf. Leaves ×40, cells ×415.

SUBFAMILY 3. CINCLIDOTOIDEAE

Plants robust, ± aquatic, dioecious, cladocarpous. Lvs lanceolate with several-stratose margin. Capsule immersed or exserted; peristome teeth filiform.

61. CINCLIDOTUS P. BEAUV., PRODR., 1805

European species dioecious. Robust, often aquatic plants, with lateral or terminal gametangia. Lvs erect to curled when dry, erecto-patent to spreading when moist, lanceolate to broadly lingulate, apex acute to rounded, margin plane or recurved, strongly thickened; nerve ending in apex; basal cells rectangular, above hexagonal, smooth or papillose. Seta short or long; capsule immersed or exserted, ellipsoid to cylindrical; peristome teeth 16, bifid above, ± united below, smooth or papillose, straight or twisted; calyptra cucullate. Ten species occurring in Europe, Asia, N. Africa, Macaronesia and N. America.

1 Capsule immersed, lvs narrowly lanceolate to oblong-lanceolate, margin plane, thickened, cells faintly papillose **1. C. fontinaloides**
Capsule exserted, lvs oblong-lanceolate or broadly lingulate, margin plane or recurved, thickened, cells faintly or strongly papillose 2

2 Lf cells faintly papillose, seta 3–6 mm long **2. C. riparius**
Lf cells strongly papillose, seta 8 mm or more **3. C. mucronatus**

1. C. fontinaloides (Hedw.) P. Beauv., Prodr., 1805

Plants floating or ± decumbent, 2–12 cm long, stems fastigiately branched, gametangia borne on short lateral branches. Lvs flexuose or contorted when dry, patent, often secund when moist, oblong-lanceolate to narrowly lanceolate, apex obtuse to acute, margin thickened, plane, entire or obscurely and irregularly toothed near apex, older lvs frequently eroded to nerve; nerve stout, ending in apex; cells at extreme base shortly rectangular, elsewhere regularly hexagonal, faintly papillose, 8–14 μm wide in mid-lf, margin of 3–6 layers of cells. Seta very short, *ca* 0.5 mm long; capsule immersed, ellipsoid; lid longly rostrate, oblique, red at maturity; peristome teeth divided into 2–3 long, filiform, papillose, purplish-red segments; spores coarsely papillose, 18–20 μm. Fr frequent, spring, summer. $n=13$*. Usually lax, olive green to blackish-green tufts on rocks and tree bases by waterfalls and in the flood zone of streams and rivers or tighter tufts on rocks in and by lakes and pools, usually in basic areas. Frequent and sometimes locally abundant in suitable habitats but rare in S.E. England and the Midlands. 100, H38. Europe, Asia Minor, Caucasus, Tibet, Madeira, Algeria, Morocco, Tunisia.

A very variable species, some forms of which resemble *C. riparius* (q.v.). In habit the plant may resemble *Schistidium alpicola* var. *rivularis*, but it has longer, narrower lvs with thickened margins, different areolation and capsule characters. *Orthotrichum rivulare* differs in the rounded lf apex, the recurved margin and the capsule with broad peristome teeth.

2. C. riparius (Web. & Mohr) Arnott, Mem. Soc. Linn. Paris, 1827
C. nigricans (Brid.) Loeske

Resembling compact forms of *C. fontinaloides*. Lvs oblong-lanceolate, obtuse and apiculate or rarely acute, margin thickened; cells smooth or faintly papillose, 10–14 μm wide in mid-lf. Seta 3–6 mm; capsule exserted, ellipsoid, straight or slightly curved; peristome teeth yellow, ± smooth; spores *ca* 20 μm. Fr occasional, spring,

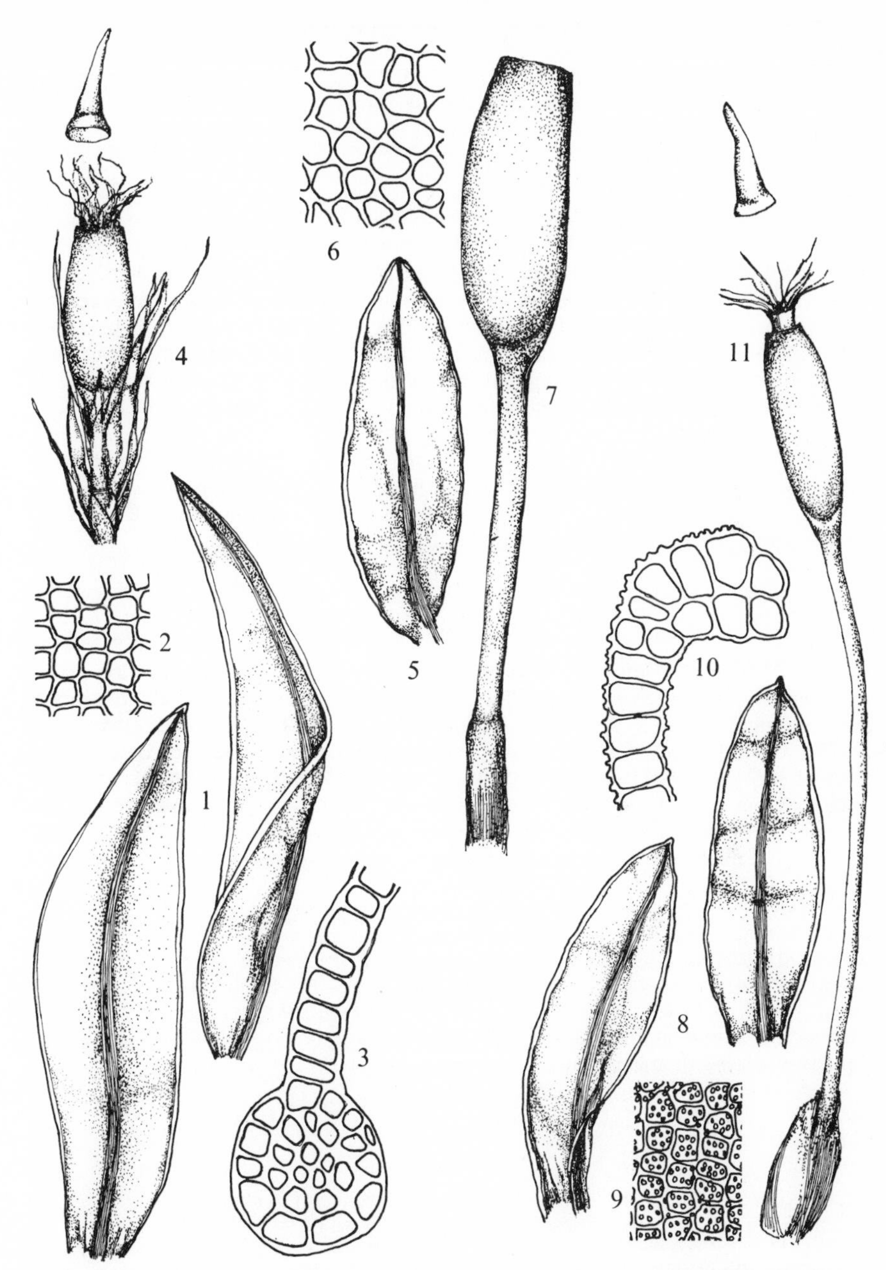

Fig. **144**. 1–4, *Cinclidotus fontinaloides*: 1, leaves; 2, mid-leaf cells; 3, section of leaf margin; 4, capsule. 5–7, *C. riparius*: 5, leaf; 6, mid-leaf cells; 7, capsule. 8–11, *C. mucronatus*: 8, leaves; 9, mid-leaf cells; 10, section of leaf margin; 11, capsule. Leaves ×15, cells ×415, capsules ×10.

summer. Dark green tufts on basic rocks in the flood zone of streams and rivers. Europe, Asia, N. America.

C. riparius has been recorded from Hereford, Worc, Shrops and Clare, but all the gatherings are sterile and *C. riparius* is indistinguishable from compact forms of *C. fontinaloides* when not in fruit. *C. mucronatus* differs in the papillose lf cells, the recurved margin and terminal seta. For a discussion of British material see Warburg, *Trans. Br. bryol. Soc.* **3**, 383–5, 1958.

3. C. mucronatus (Brid.) Mach., Cat. Descr. Briol. Portug., 1919
C. brebissonii (Brid.) Husn., *Dialytrichia mucronata* (Brid.) Broth.

Plants 0.5–3.0 cm, stems sparsely branched with terminal gametangia. Lvs flexuose or curled when dry, erecto-patent to spreading when moist, oblong-lanceolate or broadly lingulate, obtuse and mucronate, margin recurved, thickened, papillose-crenulate above; nerve stout, excurrent in mucro; basal cells narrowly rectangular, above irregularly quadrate-hexagonal, papillose, obscure, 8–12 μm wide in mid-lf, margin *ca* 3 cells thick above. Seta terminal, 8–10 mm; capsule erect, ellipsoid to cylindrical; lid longly rostrate, oblique; peristome teeth filiform, papillose, reddish; spores finely papillose, 14–16 μm. Fr occasional, spring. Dull or dark green tufts on rocks, walls and tree bases below and above flood zone of streams and rivers but also away from water. Occasional in S. Britain from Devon and Sussex north to Caerns and Durham, Roxburgh. 41. W., C. and S. Europe, Asia, Algeria.

11. GRIMMIALES

Acrocarpous or sometimes cladocarpous, mostly tufted or cushion-forming plants. Lvs ovate to linear, often with hyaline apices, upper cells small, usually quadrate or rounded, incrassate, ± sinuose, 1–4-stratose, papillose or not, lower cells longer, less incrassate, sometimes very sinuose or nodulose or thin-walled and hyaline. Seta long or short, straight or curved; capsule spherical to cylindrical, smooth or striate; peristome usually present, without basal membrane, single, similar to that of *Dicranum* but not vertically striate, entire or variously perforated or cleft; calyptra cucullate or mitriform. The majority of species is saxicolous.

14. GRIMMIACEAE

Lvs rarely crisped when dry, often with hyaline apices, margin usually entire; nerve ending below apex or excurrent, if apex hyaline then nerve ending below or excurrent into hyaline part; basal cells quadrate to rectangular or linear, smooth, sinuose or strongly nodulose-sinuose, towards margin shorter, sometimes pellucid or hyaline, upper cells 1–4-stratose. Capsule immersed or exserted.

62. Coscinodon Spreng., Einl. Krypt. Gew., 1804

Close to *Schistidium*. Calyptra campanulate, plicate, ± enveloping capsule; columella not attached to lid, persistent; peristome teeth cribrose. A small genus of 7 species occurring in the northern hemisphere, S. America and New Zealand.

1. C. cribrosus (Hedw.) Spruce, Ann. Mag. Nat. Hist., 1849

Dioecious. Plants to 1 cm. Lvs appressed when dry, erecto-patent when moist, ovate to lanceolate, apex broad, upper lvs longitudinally plicate near nerve above, margin plane, entire; hair-point of upper lvs $\frac{1}{4}$–1(–2) length of lamina, flattened at

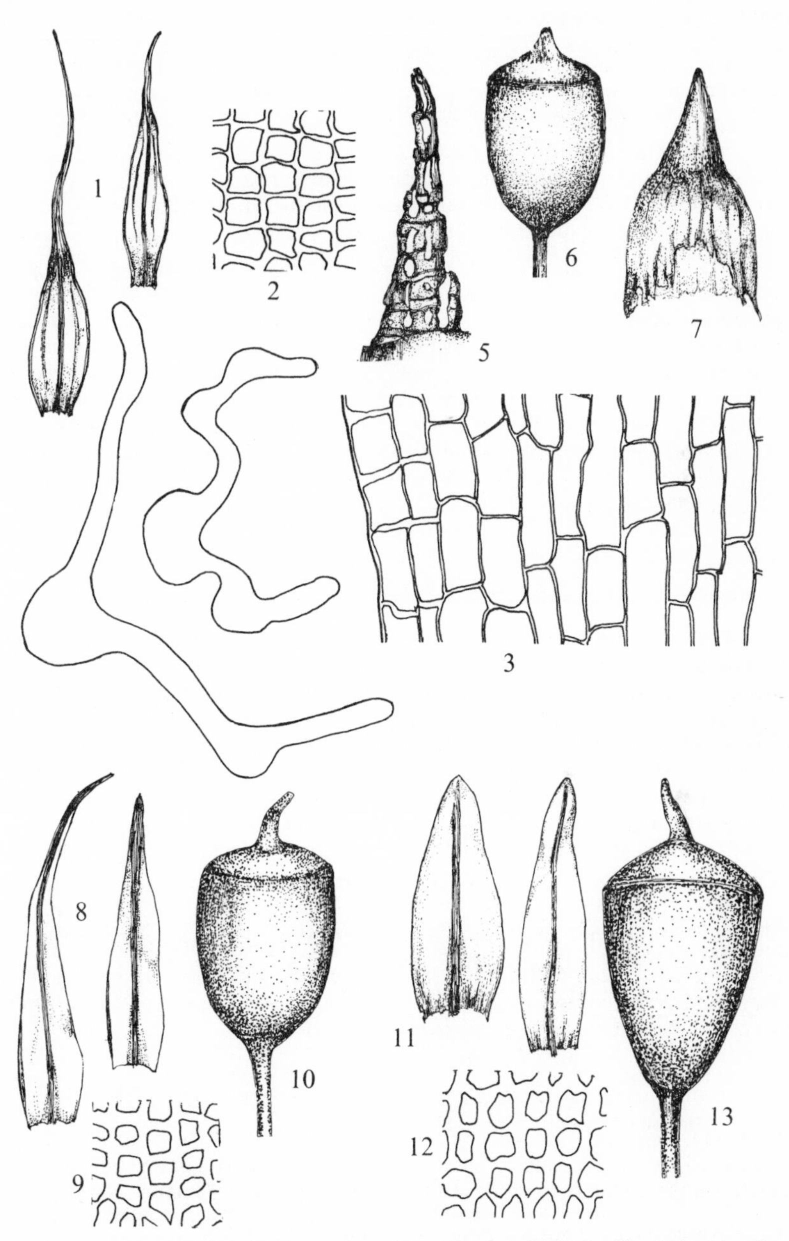

Fig. **145**. 1–7, *Coscinodon cribrosus*: 1, leaves (×15); 2, mid-leaf cells; 3, basal cells; 4, transverse sections of leaf (×250); 5, peristome tooth (×115); 6, capsules; 7, calyptra. 8–10, *Schistidium maritimum*: 8, leaves (×22.5); 9, mid-leaf cells; 10, capsule. 11–13, *S. agassizii*: 11, leaves (×22.5); 12, mid-leaf cells; 13, capsule. Cells ×450, capsules ×22.5.

base and sometimes decurrent down margin, terete above, smooth or occasionally slightly denticulate; basal cells rectangular, thin-walled, not sinuose, ± hyaline, shorter towards margin, cells above quadrate, not sinuose, often partially bistratose, 8–10(–12) μm wide in mid-lf, cells of plicae with thickened walls. Seta straight; capsule partially immersed, obloid; calyptra large, campanulate, plicate; peristome teeth strongly cribrose, fragile; spores 10–12(–14) μm. Fr rare, spring. Low, hoary, dense, often dust-filled tufts or patches in crevices and on cracks of acidic, especially slatey rocks, rare. E. Cornwall, N. Devon, Wales and a few localities from Yorks north to Banff. 13. Europe, Caucasus, Asia, Kashmir, Himalayas, N. America.

In fruit likely to be confused with *Schistidium* species and when sterile with *Grimmia* species; from the former it differs in capsule characters and from both in the lvs plicate above; the plane margin will also distinguish the plant from most forms of *Schistidium* species.

63. Schistidium Brid., Mant. Musc., 1819

Autoecious. Tufted or spreading plants. Lvs ± lanceolate, acuminate to obtuse, with or without hair-point; cells often bistratose above. Seta very short, straight; capsule ± immersed; columella attached to and falling with lid; peristome teeth entire, cleft or variously perforated; calyptra small, cucullate or mitriform. A cosmopolitan genus of about 20 species.

1 Spores 16–28 μm, lvs usually without hair-points — 2
  Spores 6–12(–14) μm, at least upper lvs usually with hair-points — 4
2 Spores 20–28 μm, nerve percurrent or excurrent, plants of acid rocks near the sea — **1. S. maritimum**
  Spores 16–20 μm, nerve ending in or below apex, plants of rocks in or by fresh water — 3
3 Lf margin recurved below, margin and upper part of lamina bistratose — **2. S. alpicola**
  Lf margin usually plane, lamina unistratose throughout — **3. S. agassizii**
4 Nerve smooth or with low opaque papillae at back towards apex — 5
  Nerve of younger lvs scabrous with colourless, conical papillae at back above — 7
5 Plants forming loose, procumbent to sub-erect, greenish to brownish patches or tufts, hair-points to 0.75–1.00 mm long usually present — **4. S. apocarpum**
  Plants forming dense, dark brown cushions or straggling blackish patches, hair-points very short (to 60 μm) or absent — 6
6 Plants forming dense brown cushions or tufts — **5. S. atrofuscum**
  Plants forming blackish ± flat, straggling patches — **8. S. trichodon**
7 Plants reddish or reddish-brown — **6. S. strictum**
  Plants blackish — **7. S. boreale**

1. S. maritimum (Turn.) Br. Eur., 1845
Grimmia maritima Turn.

Plants to *ca* 2 cm. Lvs ovate to lanceolate, tapering to blunt apex, margin recurved below on one or both sides; hair-point absent; nerve strong, sometimes thickened above, percurrent to excurrent in stout point; basal cells rectangular, pellucid, shorter towards margin, above quadrate, incrassate, opaque, slightly papillose, bistratose towards apex, 8–10 mm wide in mid-lf. Seta 1.0–1.4 mm; capsule ovate-obloid, turbinate when dry and empty; spores 20–26(–28) μm. Fr winter, common.

$n=13$*. Dark green to brownish tufts or cushions on acidic or very rarely basic maritime rocks. Frequent along rocky coasts of W. and N. Britain and Ireland, rarer in the east and absent from Dorset east and north to Yorks. 52, H21, C. Atlantic coasts of Europe and America.

Usually found only on coastal rocks beyond the spray zone and, except in the Scottish Islands or as a relic of silted estuaries, not found more than 400 m from the sea. Some gatherings from the Baltic have a hair-point, but such plants have not been found in Britain.

2. S. alpicola (Hedw.) Limpr., Laubm., 1889
Grimmia alpicola Sw. ex Hedw.

Plants to 10 cm. Lvs ovate-lanceolate to ovate, acute to obtuse, entire or bluntly toothed near apex, margin recurved on one or both sides below; hair-point lacking or rarely a few apical cells hyaline; nerve stout, ending in or below apex; basal cells rectangular, shorter towards margin, cells above quadrate, thin-walled to incrassate or collenchymatous, not or slightly sinuose, bistratose at margin and towards apex, 8–10 μm wide in mid-lf. Seta 0.4–0.5 mm; capsule sub-globose, hemispherical when empty; peristome teeth bright red; spores 16–20 μm. Fr common, autumn to spring.

Var. **alpicola**
Grimmia apocarpa var. *alpicola* (Hedw.) Röhl.

Plants to 3 cm. Lvs ovate, acute to sub-obtuse. Perichaetial lvs scarcely exceeding capsule. $n=13$*, 14. Small, dark tufts on usually basic rocks by water, rare. Hereford, Glos, Derby and Yorks northwards, Kildare. 20, H1. N., W. and C. Europe, Faroes, Siberia, Kashmir, Japan, C. Africa, N. America, Greenland.

Var. **rivulare** (Brid.) Limpr., 1889
Grimmia apocarpa var. *rivularis* Web. & Mohr, *G. alpicola* var. *rivularis* (Brid.) Wahlenb.

Plants much branched, to 10 cm, stems often denuded below. Lvs ovate-lanceolate, acute to obtuse; nerve percurrent. Perichaetial lvs overtopping capsule by *ca* $\frac{1}{3}$ their length. $n=13$*. Dark green, prostrate or floating tufts on rocks at and below flood level of streams and rivers. Frequent in N. and W. Britain, absent from S.E. England. 77, H17. N., W. and C. Europe, Iceland, Corsica, Siberia, Japan, N. America, Greenland, southern S. America.

S. agassizii differs from *S. alpicola* in the more obtuse lvs with usually plane margins and cells unistratose throughout lf and the longer seta. *S. apocarpum* occasionally occurs in similar habitats but differs in capsule shape, spore size and lvs usually with hair-points. For the differences between var. *rivulare* and *Cinclidotus fontinaloides* see under the latter.

3. S. agassizii Sull. & Lesq. in Sull., Musc. & Hep., 1856
Grimmia agassizii (Sull. & Lesq.) Jaeg.

Plants to *ca* 5 cm. Lvs lanceolate to narrowly lingulate, apex rounded, obtuse or sub-acute, margin entire or crenulate above, plane or rarely narrowly recurved below; hair-point lacking; nerve ending below apex; basal cells rectangular, shorter towards margin, above quadrate, unistratose throughout, 8–10 μm wide in mid-lf. Perichaetial lvs barely reaching mouth of capsule. Seta 1.0–1.2 mm; capsule shortly obloid, turbinate when empty; peristome teeth red; spores 16–20 μm. Fr common. Blackish tufts on basic rocks subject to flooding in fast-flowing streams and rivers, very rare. Durham, Mid Perth. 2. N., W. and C. Europe, Estonia, Iceland, N. America.

For the occurrence of this plant in Britain see Birks & Birks, *Trans. Br. bryol. Soc.* **5**, 215–17, 1967. Forms growing ± submerged in fast-flowing water tend to have narrower and sometimes more acute lvs than do plants less subject to submergence.

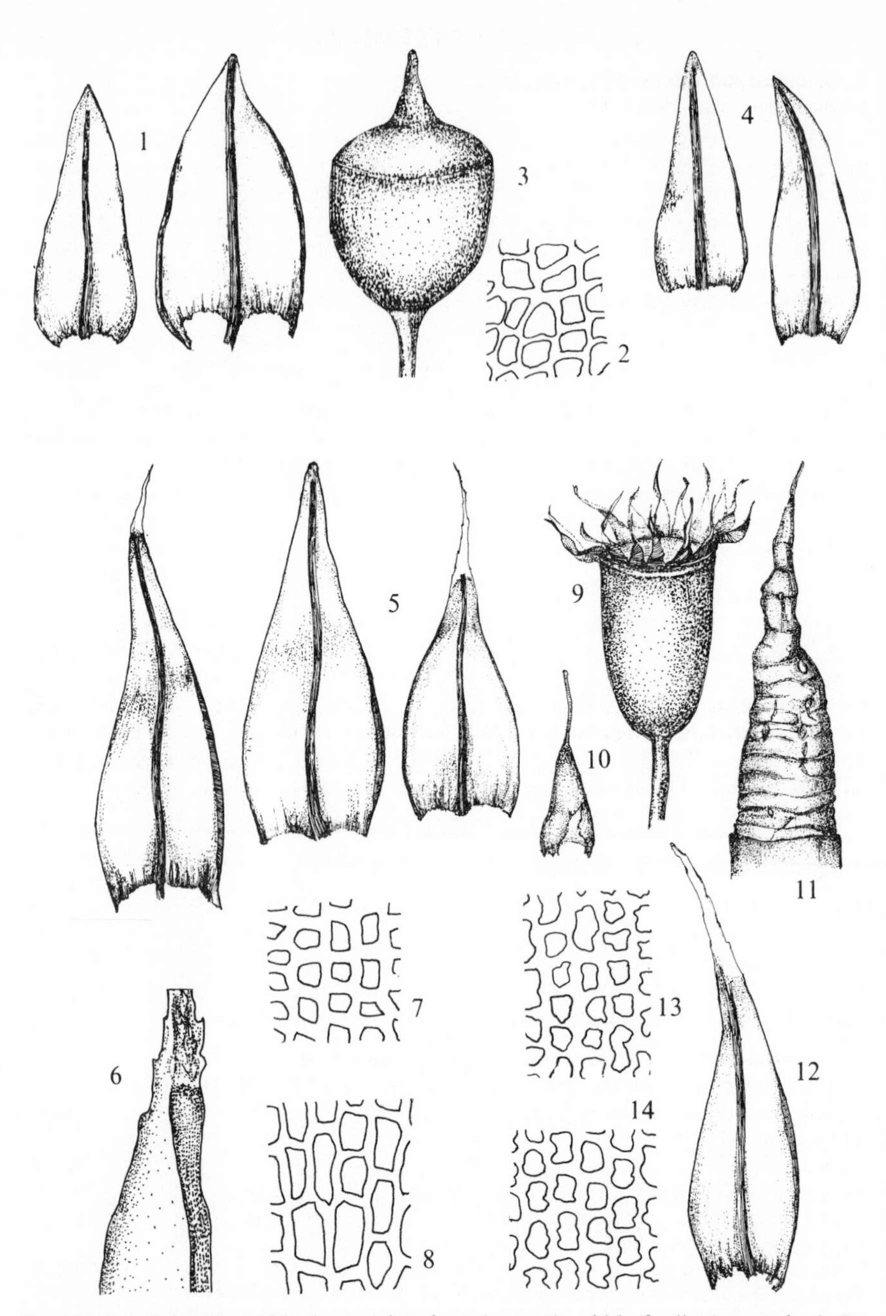

Fig. **146**. 1–3, *Schistidium alpicola* var. *alpicola*: 1, leaves; 2, mid-leaf cells; 3, capsule. 4, *S. alpicola* var. *rivularis*: leaves. 5–11, *S. apocarpum* var. *apocarpum*: 5, leaves; 6, leaf apex in side view (×115); 7, mid-leaf cells; 8, basal cells; 9, capsule. 10, calyptra; 11, peristome tooth (×115). 12–14, *S. apocarpum* var. *homodictyon*: 12, leaf; 13, mid-leaf cells. 14, basal cells. Leaves and capsules ×22.5, cells ×450.

4. S. apocarpum (Hedw.) Br. Eur., 1845
Grimmia apocarpa Hedw.

Stems ± procumbent to sub-erect. Lvs appressed when dry, erecto-patent to spreading or recurved when moist, ovate to lanceolate, tapering to acute apex, margin entire or occasionally bluntly toothed near apex, recurved on one or both sides below; hair-point usually present, smooth to denticulate or spinulose; nerve excurrent, smooth or with low, opaque papillae at back above; cells ± incrassate, basal quadrate to rectangular, quadrate towards margin, above quadrate, ± sinuose, smooth or with low papillae, 1–2-stratose at margin. Fr common, winter, spring.

Key to intraspecific taxa of *S. apocarpum*

1 Lf cells 6–8 μm wide in mid-lf, peristome teeth orange-red, strongly perforated var. **confertum**
Lf cells 8–10 μm wide, peristome teeth deep red, perforated or not 2
2 Hair-point ± smooth to denticulate, cells towards base of lf not or scarcely sinuose, peristome teeth papillose var. **apocarpum**
Hair-point spinulose, cells very sinuose throughout lf, peristome teeth smooth var. **homodictyon**

Var. **apocarpum**
Plants 0.5–6.0 cm. Hair-point usually present, 5–650(–1000) μm long, smooth to denticulate; basal cells rectangular, not or scarcely sinuose, mid-lf cells 8–10 μm wide. Seta *ca* 0.6 mm; capsule obloid; often widest at mouth when dry and empty; peristome teeth deep red, papillose, entire or perforated above; spores 6–14 μm. $n=13^*$, 14, 26. Green to brownish tufts or patches on wet or dry, usually basic rocks, walls, mortar and concrete, common in calcareous districts, rare elsewhere, rarely on trees by rivers. 112, H40, C. Cosmopolitan.

Var. **confertum** (Funck) Möll., Ark. f. Bot., 1934
S. confertum (Funck) Br. Eur., *Grimmia conferta* Funck
Plants to 1.5 cm. Lf cells 6–8 μm wide in mid-lf, otherwise areolation similar to that of type. Seta 0.3–0.5 mm; capsule ± globose to ovate-obloid; peristome teeth orange-red, strongly perforated; spores *ca* 10 μm. Fr common, winter. $n=13^*$, 26, 26+1. Small, tight tufts, usually on sub-alpine or alpine rocks, rare. S. Devon, Wales, Yorks and Westmorland northwards. 19, H6. Europe, Iceland, Caucasus, N. and C. Asia, Japan, Morocco, Ethiopia, N. America, Patagonia, Falkland Is.

Var. **homodictyon** (Dix.) Crundw. & Nyholm, Bot. Not., 1963
Grimmia homodictyon Dix., *G. apocarpa* var. *homodictyon* (Dix.) Crundw.
Hair-points in upper lvs to 1 mm long, spinulose; cells at extreme base near nerve rectangular, elsewhere shortly rectangular to quadrate, very sinuose, strongly incrassate. Perichaetial lvs often concealing capsule. Capsule hardly widened at mouth when dry and empty; peristome teeth smooth. Loose, readily disintegrating, hoary, brown tufts or patches on basic rocks in N. Britain, rare. Westmorland, Angus and Argyll northwards to Sutherland. 10. Sweden.

A variable species, some forms of which may be confused with the following 4 species of this genus. All differ from *S. apocarpum* in colour, *S. atrofuscum* in habit and sparse production of fruit, *S. trichodon* in its very prostrate habit and *S. strictum* and *S. boreale* in the papillose lvs. These taxa are probably best regarded as microspecies, and any plant resembling but not exactly agreeing with any one is best referred to *S. apocarpum*. *S. apocarpum* is usually readily recognised by the immersed capsules which are only very rarely completely absent. Sterile plants may be difficult to separate from *Grimmia* species; the combination of usually some hint of brown or reddish-brown coloration, the recurved lf margin, the relatively short, ± incrassate basal cells and pellucid upper cells will separate it from most species of that genus.

5. S. atrofuscum (Schimp.) Limpr., Laubm., 1889
Grimmia atrofusca Schimp., *G. apocarpa* var. *obscuriviridis* Crum

Stems to 3 cm, ± erect, forming dark brown to blackish tufts. Lvs lanceolate to broadly lanceolate, apex acute to obtuse, margin plane or recurved below; hair-point usually absent, occasionally present and up to 60 μm long; nerve percurrent, smooth at back; cells thin-walled to incrassate, sinuose or not above, 8–12 μm wide in mid-lf. Capsule shortly obloid; peristome teeth strongly papillose, *ca* 320 μm long; spores 10 μm. Fr occasional. Tight, dark brown to blackish, readily disintegrating tufts on basic montane rocks, very rare. Inverness, S. Aberdeen. 2. Norway, C. Europe, Spain, N. America.

For the occurrence of this plant in Britain see Warburg, *Trans. Br. bryol. Soc.* **3**, 172–3, 1957. It is possible that *S. atrofuscum* is only a variety of *S. apocarpum* as the 2 are said to intergrade in C. Europe.

6. S. strictum (Turn.) Loeske, Hedwigia, 1908
S. papillosum Culm., *Grimmia stricta* Turn., *G. apocarpa* var. *gracilis* (Schleich.) Mohr

Plants decumbent, reddish to reddish-brown, to 7 cm, occasionally more. Lvs spreading or sub-secund when moist, lanceolate, tapering to acute apex, margin toothed above; hair-point usually present, to 600 μm long, rough; nerve scabrous at back above with tall, conical papillae; cells incrassate, sinuose, papillose above, 8–10 μm wide. Capsule ovoid to obloid; peristome teeth 240–440 μm long; spores 12–14 μm. Fr frequent, winter. $n=13$*, 26. Reddish to reddish-brown patches on usually basic, sub-alpine or alpine rocks, occasional. S. Devon, Wales, Hereford, Worcs, Derby, Yorks northwards. 42, H15. Europe, Caucasus, Siberia, Himalayas, China, Japan, N. America.

7. S. boreale Poelt, Svensk. Bot. Tidskr., 1953
Grimmia borealis (Poelt) Crundw.

Plants decumbent, blackish, stems to 5 cm. Lf margin entire or with a few blunt teeth above; hair-point to 160 μm long; nerve in younger lvs scabrous at back above with tall, conical papillae; cells papillose above in younger lvs. Peristome teeth to *ca* 480 μm long, not filiform. Fr frequent. Blackish, straggling patches on basic, montane rocks, very rare. Angus. 1. Europe.

Similar to *S. trichodon* but differing in the presence of conical papillae on the back of the lf nerve and the shorter, wider peristome teeth. How reliable the lf papilla character is is open to question, as I have seen gatherings of *S. trichodon* from Yorks with the typical habit of that plant but with conical papillae on the youngest lvs. In *S. boreale* the papillae are usually on the older lvs. For the occurrence of this plant in Britain see Warburg, *Trans. Br. bryol. Soc.* **4**, 757–9, 1965.

8. S. trichodon (Brid.) Poelt, Svensk. Bot. Tidskr., 1953
Grimmia trichodon Brid.

Plants blackish, stems leafless below, to 8 cm. Lf margin entire, plane or recurved below; hair-point 0–60 μm long; nerve and cells smooth above or with low papillae. Peristome teeth (450–)500–650(–750) μm long, filiform, fragile; spores 8–10 μm. Fr common, winter. Blackish, straggling patches on basic, montane rocks, rare. Caerns, Yorks and Westmorland northwards. 14. Europe.

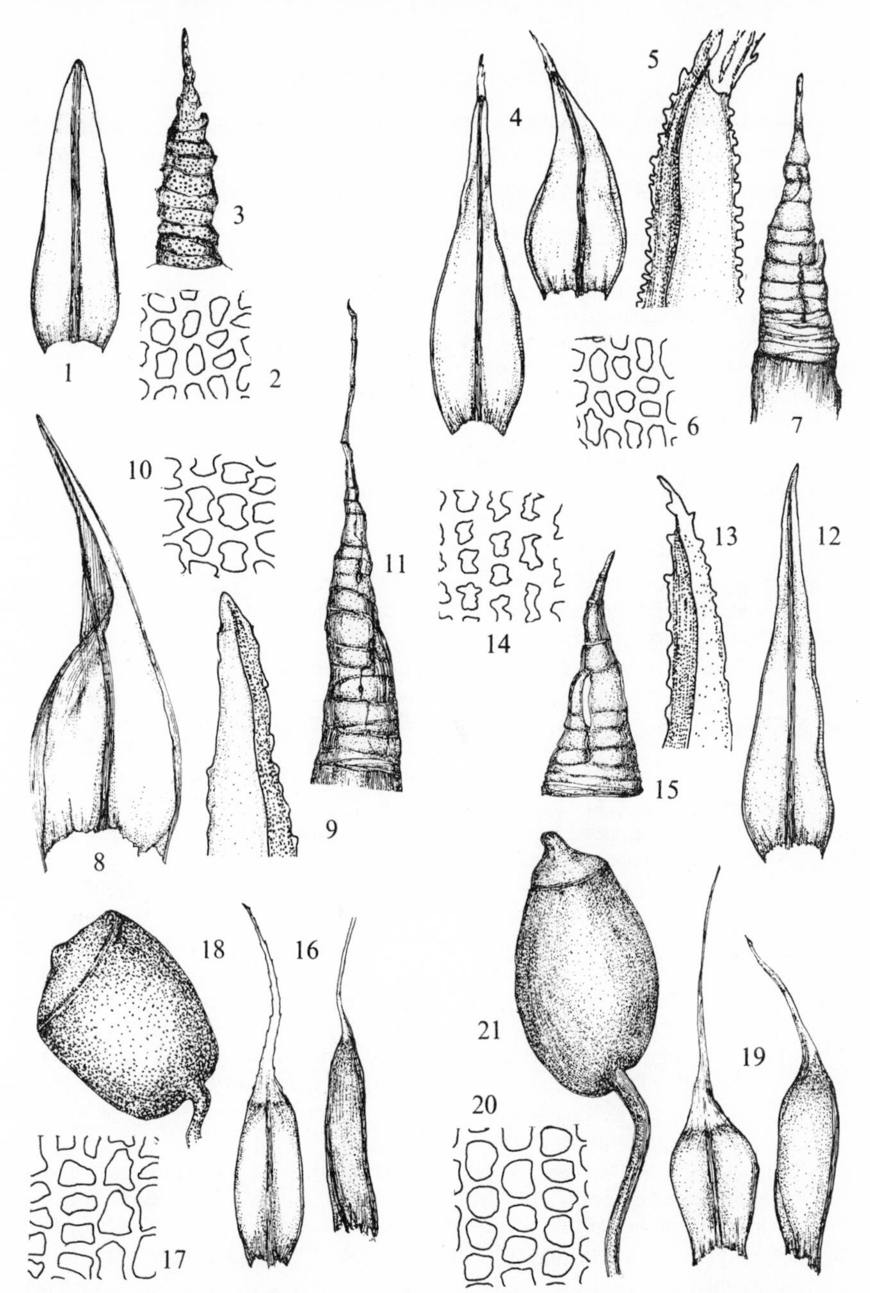

Fig. **147**. 1–3, *Schistidium atrofuscum*: 1, leaf; 2, mid-leaf cells; 3, peristome tooth. 4–7, *S. strictum*: 4, leaves; 5, leaf apex in side view; 6, mid-leaf cells; 7, peristome tooth. 8–11, *S. trichodon*: 8, leaf; 9, leaf apex in side view; 10, mid-leaf cells; 11, peristome tooth. 12–15, *S. boreale*: 12, leaf; 13, leaf apex in side view; 14, mid-leaf cells; 15, peristome tooth. 16–18, *Grimmia anodon*: 16, leaves; 17, mid-leaf cells; 18, capsule. 19–21, *G. crinita*: 19, leaves; 20, mid-leaf cells; 21, capsule. Leaves and capsules $\times 22.5$, apices and teeth $\times 115$, cells $\times 450$.

For the occurrence of this plant in Britain see Crundwell, *Trans. Br. bryol. Soc.* 3, 558–62, 1959.

64. Grimmia Hedw., Sp. Musc., 1801

Autoecious or dioecious, acrocarpous. Usually tufted or cushion-forming plants with short, dichotomously branched stems. Lvs lanceolate, acuminate, usually with hyaline hair-points; basal cells quadrate to linear, sinuose or not, above quadrate or rounded, 1–several-stratose, papillose or not. Seta straight or curved; capsule usually exserted; calyptra mitriform or cucullate, not papillose; columella persistent; peristome teeth 16, entire or variously perforated or divided to the base. A cosmopolitan genus of *ca* 170 almost exclusively saxicolous species.

1 Lvs without hair-points — 2
  At least upper lvs with hair-points — 3
2 Lf margin erect, apex cucullate — **10. G. unicolor**
  Lf margin narrowly recurved, apex flat — **11. G. atrata**
3 Lvs concave, obovate or oblong-lanceolate, margin plane, hair-point as long as lamina in upper lvs, seta 1 mm or less, curved, capsules ± immersed — 4
  Plants lacking above combination of characters — 5
4 Lf cells bistratose towards apex, capsule immersed, peristome lacking — **1. G. anodon**
  Lvs unistratose throughout, capsule partially immersed, peristome present — **2. G. crinita**
5 Lf margin usually plane or incurved, cells in upper part of lf opaque, bistratose, seta straight — 6
  Lf margin recurved, cells ± pellucid, unistratose except at margin and extreme apex, seta arcuate when moist — 12
6 Hair-point of upper lvs very short (to 150 μm long), lvs of ± uniform size throughout stem — **7. G. elongata**
  Hair-point usually $\frac{1}{4}$ or more length of lamina in upper lvs, lower lvs smaller than upper — 7
7 Lvs concave, triangular to lanceolate with broad base, apex obtuse, basal cells towards margin often wider than long — **3. G. laevigata**
  Lvs keeled or channelled, ovate-lanceolate to linear-lanceolate, apex ± acute, basal marginal cells longitudinally rectangular or quadrate — 8
8 Basal cells to 3 times as long as wide — 9
  Basal cells 4 or more times as long as wide — 10
9 Lvs linear-lanceolate to lanceolate, cells in mid-lf 8–10 μm wide, not or only slightly sinuose, spores *ca* 12 μm — **4. G. montana**
  Lvs ovate-lanceolate, cells 10–12 μm wide, spores 8–10 μm — **5. G. alpestris**
10 Cells 10–12 μm wide in mid-lf, walls of marginal cells near base ± uniformly thickened — **6. G. donniana**
  Cells 6–10 μm wide, transverse walls of marginal cells near base thicker than longitudinal walls — 11
11 Lf margin plane or incurved, cells 6–8 μm wide in mid-lf, 5–10 rows hyaline marginal cells forming conspicuous band at base of lf — **9. G. ovalis**
  Margin plane or slightly recurved, cells 8–10 μm wide, no conspicuous hyaline band of marginal cells at lf base — **8. G. affinis**

12 Plants dark green, lvs linear to linear-lanceolate, crisped when dry, hair-point very short, to 150 μm long **12. G. incurva**
Lvs wider, straight to twisted when dry or if spirally crisped then plants brownish, hair-point various 13

13 Upper lvs ovate-lanceolate or lanceolate, abruptly narrowed into hair-point 14
Upper lvs lanceolate or narrowly lanceolate, gradually tapering into hair-point 15

14 Basal cells 2–4 times as long as wide, capsule usually ellipsoid, lid rostrate **13. G. pulvinata**
Basal cells 4–8 times as long as wide, capsule ovoid, lid mamillate **14. G. orbicularis**

15 Lvs crisped or spirally curled when dry 16
Lvs straight or curved but not spirally so nor crisped when dry 17

16 Plants brownish, hair-points short, to 250 μm long, lvs crisped when dry **15. G. torquata**
Plants greyish, hair-point ½–1 length of lamina in upper lvs, lvs spirally curled when dry **16. G. funalis**

17 Hair-points short, to 320(–400) μm long in upper lvs 18
Hair-points to as long as lamina in upper lvs 19

18 Lvs often secund when moist, basal cells ± sinuose, clusters of gemmae often present at stem apex **18. G. hartmanii**
Lvs not secund, basal cells ± smooth, gemmae absent **19. G. retracta**

19 One margin of lf strongly recurved, other less so or plane, cells in upper part of lf opaque, bistratose **20. G. elatior**
Lf margins ± equally recurved, cells ± pellucid, unistratose except at margin 21

20 Hair-point smooth to denticulate, basal cells to 6(–8) times as long as wide, upper cells ± sinuose **17. G. trichophylla**
Hair-point spinosely denticulate, basal cells *ca* 10 times as long as wide, upper cells very sinuose **20. G. decipiens**

Section I. *Gasterogrimmia*

Seta short, arcuate; capsule sub-globose, gibbous, with or without peristome. Lvs unistratose or bistratose above.

1. G. anodon Br. Eur., 1845

Autoecious. Plants to 1.5 cm. Lvs appressed when dry, patent when moist, concave, oblong-lanceolate to oblong-obovate, obtuse, margin plane; hair-point to as long as lamina, smooth to denticulate, often decurrent down margin of lamina; basal cells rectangular, hyaline, above quadrate, incrassate, bistratose at margin and towards apex, 8–10 μm wide in mid-lf. Seta arcuate, very short, *ca* 0.4 mm; capsule immersed, inclined, ± globose, gibbous, wide-mouthed when empty; peristome lacking; spores 8–10 μm. Fr common, spring. $n=13$. Small, hoary cushions on dry, basic rocks, very rare, Westmorland, Midlothian, Argyll. 3. Europe, Caucasus, W. and C. Asia, Tibet, Algeria, Morocco, N. America, Greenland.

Distinct from *G. crinita* with stems not julaceous when moist, the smaller, partially bistratose lf cells, the shorter seta and lack of peristome. Not likely to be confused with any other species of the genus; *Schistidium* species differ in habit, lf shape, straight seta and symmetrical capsule.

2. G. crinita Brid., Sp. Musc., 1806

Autoecious. Plants to 1 cm, stems julaceous when moist. Lvs ovate-lanceolate to obovate, upper concave, abruptly narrowed at apex, margin plane; hair-point in upper lvs as long as or longer than lamina, flattened at base, terete above, smooth; cells at base narrowly rectangular, hyaline, shorter and wider towards margin, above shortly rectangular to quadrate, incrassate, sinuose, unistratose throughout, 10–12 μm wide in mid-lf. Seta curved, short, *ca* 1 mm; capsule ± immersed in hair-points of perichaetial lvs, ovoid, gibbous; peristome present; spores 10–12 μm. Fr common, spring. Low, grey patches on basic rocks and mortar. Recorded from near Hatton, Warwicks, where it was probably introduced, now extinct. C. and S. Europe, Caucasus, S.W. Asia, Tunisia, Algeria, Morocco.

Section II. *Guembelia*

Seta long, straight; capsule erect. Lvs with plane or incurved margins; cells above bistratose, opaque.

3. G. laevigata (Brid.) Brid., Br. Univ., 1826
G. campestris Burchell ex Hook., *G. leucophaea* Grev.

Dioecious. Plants 0.5–1.5 cm. Lvs appressed, straight when dry, erecto-patent when moist, concave, triangular to lanceolate with broad base; apex obtuse, margin plane or erect; hair-point in upper lvs as long as or longer than lamina, decurrent down margin of lamina, flattened, finely denticulate; nerve broad at base, narrower and obscure above, ending in apex; cells incrassate, not sinuose, bistratose except at base, basal cells rectangular, at margin and above quadrate or wider than long, 6–10 μm wide in mid-lf. Seta straight; capsule ellipsoid; spores 10–15 μm. Fr rare, spring. $n = 13$. Hoary, readily disintegrating tufts or patches on exposed, acidic rocks and roofing slates, rare. Scattered localities from Cornwall north to Perth and Argyll, Cork, Antrim. 25, H3, C. Europe, W. and C. Asia, Madeira, Canaries, Kilimanjaro, southern Africa, N. America, Australia, New Zealand, Hawaii.

4. G. montana Br. Eur., 1845

Dioecious. Plants to 1 cm, occasionally more. Lvs ± appressed, straight when dry, erecto-patent when moist, linear-lanceolate to lanceolate, keeled above, margin plane or incurved, hair-point $\frac{1}{2}$–$\frac{3}{4}$(–1) length of lamina in upper lvs, denticulate; nerve stout; cells at base incrassate 1.5–2.0(–3.0) times as long as wide, not sinuose, longer near nerve, shorter, more hyaline towards margin, transverse walls of marginal cells sometimes thickened, cells above irregularly quadrate, incrassate, slightly sinuose, bistratose, opaque, 8–10 μm wide in mid-lf. Seta straight; capsule ovoid; lid rostrate; spores *ca* 12 μm. Fr rare, spring. $n = 13$. Small, compact, sometimes hoary cushions on dry, acidic or basic rocks, rare. Scattered localities from S. Devon and Monmouth to Argyll and S. Aberdeen, Jersey. 14, C. Europe, Iceland, Morocco, N. America, Greenland.

The short basal cells will distinguish *G. montana* from *G. donniana*, *G. ovalis* and *G. affinis*.

5. G. alpestris (Web. & Mohr) Schleich. ex Hornsch. in Somm., Suppl. Fl. Lapp., 1826

Close to *G. montana*. Autoecious in Britain. Plants *ca* 1.5 cm. Lvs ovate-lanceolate, keeled above, margin incurved; hair-point $\frac{1}{2}$–$\frac{2}{3}$ length of lamina in upper lvs, denticulate; cells bistratose above, 10–12 μm wide in mid-leaf. Lid rostellate; spores 8–10 μm. Fr common, spring, summer. Montane rocks, very rare. Angus, Aberdeen. 3.

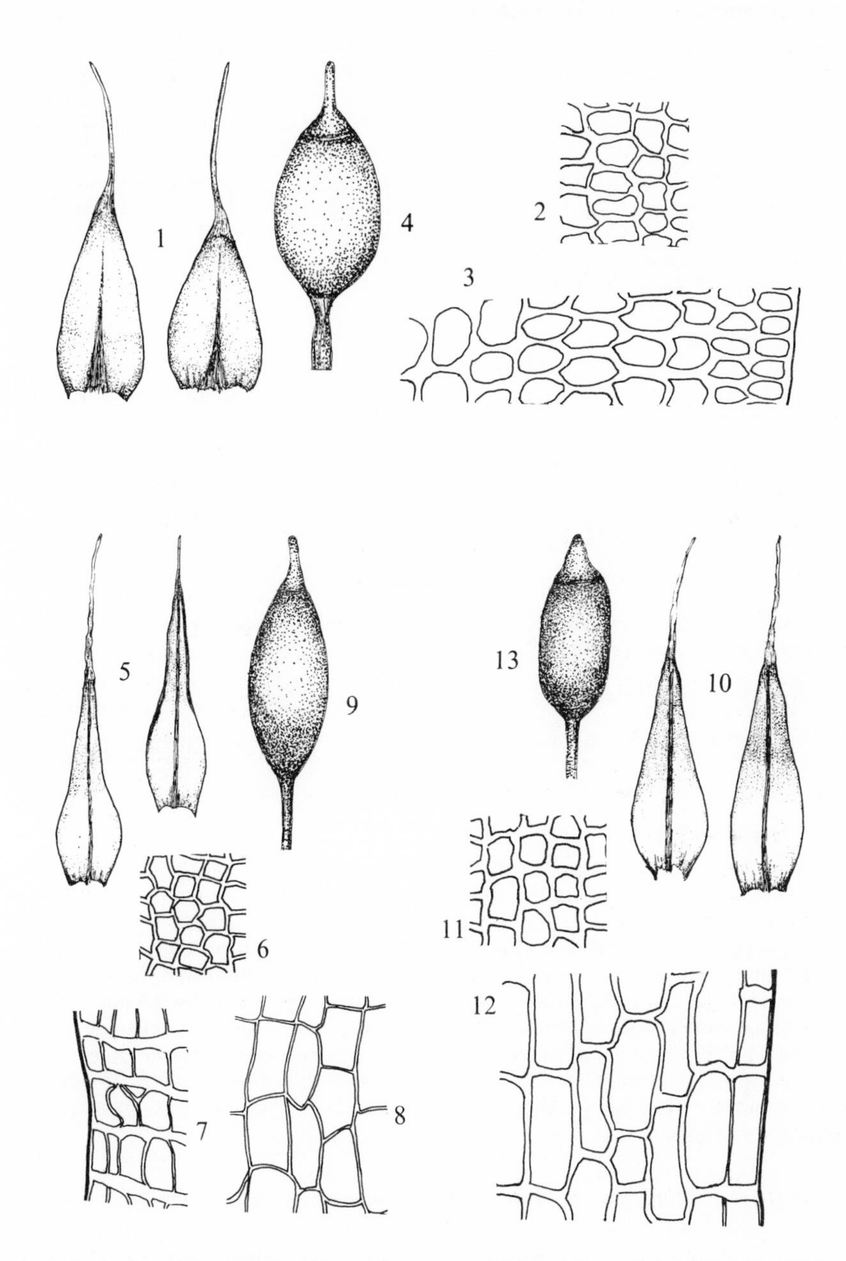

Fig. **148**. 1–4, *Grimmia laevigata*: 1, leaves; 2, mid-leaf cells; 3, basal cells; 4, capsule. 5–9, *G. montana*: 5, leaves; 6, mid-leaf cells; 7, marginal cells near leaf base; 8, basal cells midway between margin and nerve; 9, capsule. 10–13, *G. alpestris*: 10, leaves; 11, mid-leaf cells; 12, basal cells; 13, capsule. Leaves and capsules ×15, cells ×450.

N., W. and C. Europe, Iceland, Sardinia, Corsica, Caucasus, Altai, Kashmir, N. America.

6. G. donniana Sm., Eng. Bot., 1804

Autoecious. Plants to 1.5 cm. Lvs loosely appressed, straight when dry, erecto-patent when moist, from broad base lanceolate to narrowly lanceolate, margin plane or incurved; hair-point to as long as or longer than lamina in upper lvs, ± smooth to denticulate; nerve strong; basal cells rectangular, 3–6 times as long as wide, thin-walled, 3–4 rows marginal cells shorter with ± uniformly thickened walls, cells above quadrate, incrassate, ± sinuose, opaque, bistratose, 10–12 μm wide in mid-lf. Capsule ovoid, pale when empty; lid rostrate; spores 10–12 μm. Fr common, spring to autumn.

Var. **donniana**

Hair-points smooth to slightly denticulate, not homomallous. Seta straight or slightly curved. $n=13$*. Small, hoary, dark green tufts or scattered plants in crevices of hard, acidic rock, especially slate, in montane areas. Occasional in S.W. England, Scotland and Ireland, frequent in Mid and N. Wales and the Lake District. 65, H13. Europe, Caucasus, E. Asia, N. Africa, N. America, Greenland, southern S. America, Antarctica.

Var. **curvula** Spruce, Musc.Pyren., 1847

G. arenaria Hampe, *G. donniana* var. *arenaria* (Hampe) Loeske

Hair-points smooth to denticulate, often pointing in 1 direction giving tufts a brushed appearance. Seta arcuate; capsule scarcely emerging from hair-points of perichaetial lvs; peristome teeth often split into 2–3 coherent divisions. Fr common, spring. Hoary tufts in crevices of slatey rocks at low altitudes. Occasional in Merioneth very rare in Caerns and Denbigh. 3. Norway, Finland, C. Europe, N. Italy, Pyrenees.

The curved seta of the variety may lead to confusion with species of the section *Rhabdogrimmia*, but the incurved lf margin is distinctive. The usually numerous, pale capsules and hoary appearance of the tufts give *G. donniana* a distinctive appearance in the field.

7. G. elongata Kaulf. in Sturm., Deutsch. Fl., 1815

Dioecious. Stems to 2 cm. Lvs ± appressed and straight to slightly twisted when dry, erecto-patent when moist, of ± uniform length throughout stem, linear-lanceolate, keeled above, apex obtuse, margin plane or narrowly recurved; hair-point very short, to 150 μm long in upper lvs, smooth; nerve ending in or below apex; basal cells *ca* 3 times as long as wide, walls uniformly thickened, not sinuose except towards margin, cells above irregularly quadrate, incrassate, strongly sinuose, bistratose, opaque, 8 μm wide in mid-lf. Seta straight; capsule ovoid; spores 12–15 μm. Fr unknown in Britain. Brownish, erect or depressed tufts on damp or dry, acidic, montane rocks, very rare. Caerns, N. Northumberland, Cumberland, Westmorland, Angus, S. Aberdeen. 6. W. and C. Europe, Iceland, Caucasus, Sikkim, Siberia, Taiwan, Japan, N. America, Greenland, N. Africa.

8. G. affinis Hornsch. Flora 1819

G. ovalis auct. non *Dicranum ovale* Hedw., *G. ovata* Schwaegr.

Autoecious. Plants 1.0–1.5(–2.5) cm. Lvs ± appressed, straight when dry, erecto-patent when moist, lanceolate to narrowly lanceolate with broad base, keeled above, margin plane or slightly recurved; hair-point ¼–⅓(–1) length of lamina in upper lvs, smooth to slightly denticulate; nerve strong below, obscure above; basal cells elongate, 8 or more times as long as wide, longer near nerve, incrassate, ± sinuose,

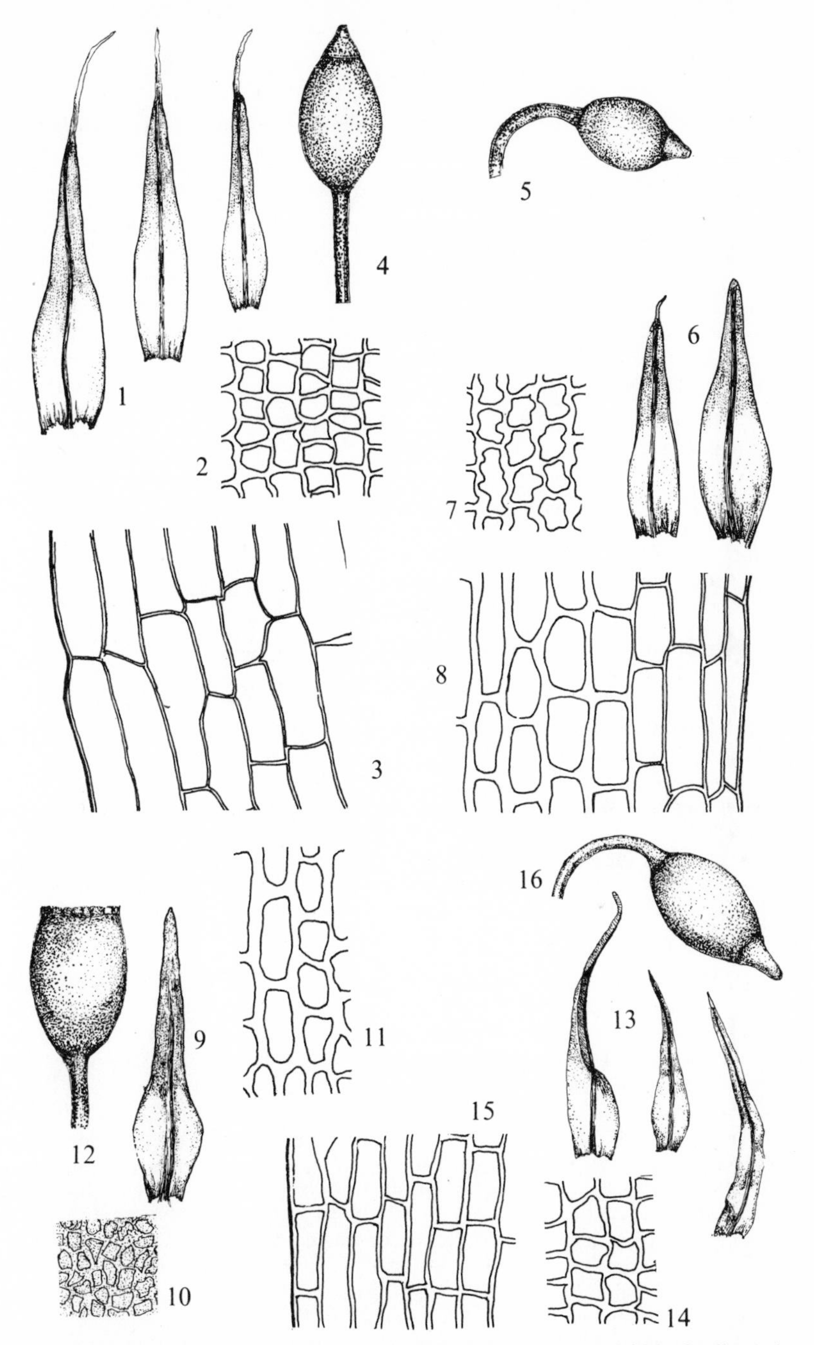

Fig. **149**. 1–4, *Grimmia donniana* var. *donniana*: 1, leaves (×15); 2, mid-leaf cells; 3, basal cells; 4, capsule. 5, *G. donniana* var. *curvula*: capsule. 6–8, *G. elongata*: 6, leaves (×15), 7, mid-leaf cells; 8, basal cells. 9–12, *G. unicolor*: 9, leaf (×25); 10, mid-leaf cells; 11, basal cells; 12, old capsule. 13–16, *G. incurva*: 13, leaves (×15); 14, mid-leaf cells; 15, basal cells; 16, capsule. Capsules ×15, cells ×450.

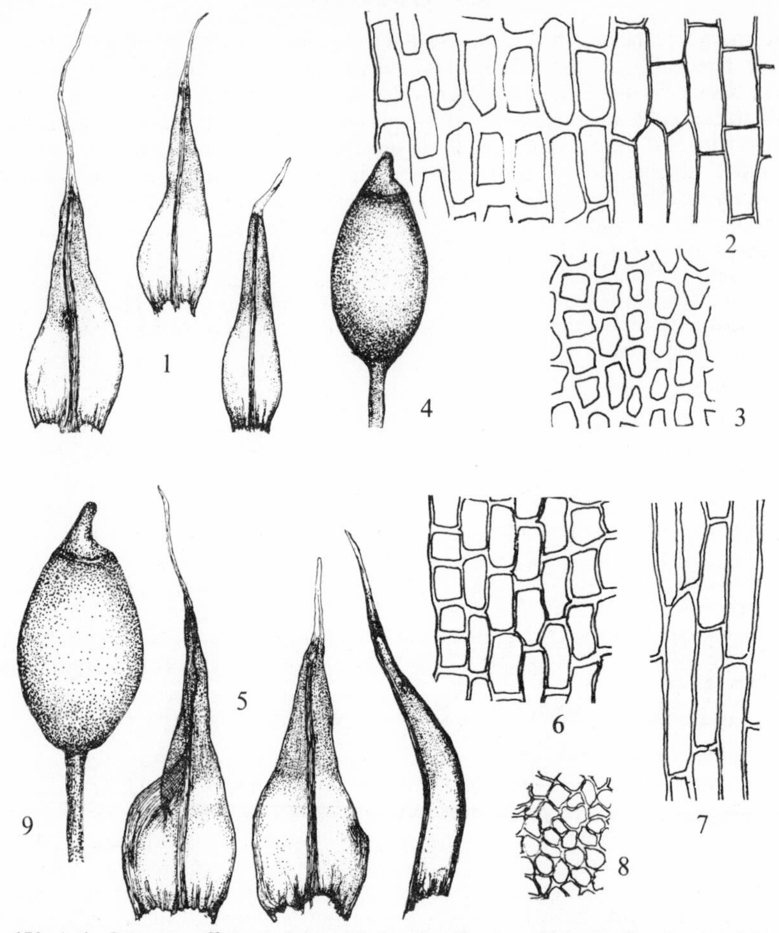

Fig. **150**. 1–4, *Grimmia affinis*: 1, leaves; 2, basal cells; 3, mid-leaf cells; 4, capsule. 5–9, *G. ovalis*: 5, leaves: 6, basal cells near margin; 7, basal cells between margin and nerve; 8, mid-leaf cells; 9, capsule. Leaves and capsules ×15, cells ×450.

marginal cells shorter, more hyaline with transverse walls thickened, cells above irregularly quadrate, more incrassate, sinuose, opaque, bistratose, 8–10 μm wide in mid-lf. Seta straight; capsule ovate-ellipsoid to ellipsoid with rostellate lid, rarely cylindrical with long, acute lid; spores 10–12 μm. Fr frequent, winter. $n=13$*. Small, compact cushions on dry, exposed, acidic or basic rocks, rare. Devon, Staffs, N. Wales, Durham and Westmorland northwards. 22, H5. Europe, Caucasus, Himalayas, Sri Lanka, N. and E. Asia, Borneo, N. America, Greenland, Andes, Morocco, C. Africa.

9. G. ovalis (Hedw.) Lindb., Rev. Fl. Dan., 1871
G. commutata Hüb., *G. ovata* Web. & Mohr

Dioecious. Plants (0.5–)1.0–4.0 cm. Lvs loosely appressed, straight when dry, erecto-patent when moist, concave, channelled above, lanceolate with very broad base, margin plane or incurved; hair-point to as long as lamina in upper lvs, flattened below, terete above, faintly denticulate; nerve wide below, obscure above, ending in apex; basal cells 4–8 times as long as wide, sinuose or not, 5–10 rows marginal cells with thickened transverse walls forming conspicuous pale marginal band below widest part of lf, upper cells irregularly quadrate, opaque, bistratose, 6–8 μm wide, in mid-lf, smaller above. Seta straight; capsule ovoid, narrowed at mouth; lid obliquely rostrate; spores 9–12 μm. Fr rare, spring. $n=13$. Hoary tufts on acidic or neutral rocks or roofing slates, rare. E. Sussex, Glos and Wales northwards. 19. Europe, Faroes, Asia, Africa, Madeira, Canaries, N. America.

Confused taxonomically and nomenclaturally with the previous species, from which it differs in its usually larger size, the very broad lf base with conspicuous marginal band of cells below and in the inflorescence. This plant is the *G. commutata* of Dixon (1924); *G. affinis* is his *G. ovata.*

10. G. unicolor Hook. in Grev., Scot. Crypt., 1825

Dioecious. Plants 2–5 cm, stems ± unbranched. Lvs loosely appressed when dry, patent when moist, lingulate with broad base, channelled above, margin erect, apex obtuse, cucullate; hair-point lacking; nerve strong below, obscure above, ending below apex; a few cells at base 2–5 times as long as wide, becoming quadrate-rectangular to quadrate towards margin, 0–3 marginal rows thin-walled, hyaline, cells above irregularly quadrate, very incrassate, sinuose, obscure, 2–4-stratose, *ca* 8 μm wide in mid-lf. Seta straight; capsule ovoid to ovate-ellipsoid; lid longly rostrate; spores 10–12 μm. Fr occasional, spring. Dark green tufts on acidic montane rocks, very rare. Angus. 1. Scandinavia, Alps, Spain, Corsica, Caucasus, N. America.

It is reported from C. Europe that if for some reason the substrate becomes dry *G. unicolor* develops hair-points.

11. G. atrata Mielichh. in Hoppe & Hornsch., Pl. Crypt. Sel., 1817–18

Dioecious. Plants 1–4(–7) cm, stems radiculose. Lvs somewhat incurved or ± twisted when dry, erect to erecto-patent when moist, narrowly lanceolate, apex obtuse, flat, margin narrowly recurved; hair-point lacking; nerve strong, prominent but not winged on dorsal side, ending in apex; basal cells quadrate-rectangular, not or only slightly sinuose, quadrate, smooth, hyaline towards margin, cells above quadrate-rectangular to quadrate, very sinuose, incrassate, 2–3-stratose at margin, 8–12 μm wide in mid-lf. Seta straight; capsule ovate-ellipsoid; lid rostellate; spores *ca* 14 μm. Fr occasional, autumn. $n=13$*. Greenish-black tufts on damp, acidic, often copper-bearing, montane rocks, rare. N. Wales, Lake District, Dumfries, Perth, Angus, S. Aberdeen, E. Inverness, Kintyre. 12. Europe, Japan.

Section III. *Rhabdogrimmia*

Seta long, arcuate when moist; capsule cernuous or pendulous. Lf margin recurved; cells usually unistratose above except at margin, pellucid.

12. G. incurva Schwaegr., Suppl. I, 1811

Dioecious. Plants 1.0–2.5 cm. Lvs crisped when dry, spreading when moist, linear from broad base or occasionally linear-lanceolate, apex acute, margin plane or slightly recurved; hair-point very short in upper lvs, to 120 μm long, lacking in lower lvs; nerve very strong, ending in or below apex; basal cells thin-walled, hyaline, not sinuose, *ca* 6 times as long as wide, ± uniform to margin, above quadrate, not sinuose, bistratose at margin and apex, 10–12 μm wide in mid-lf. Seta curved; capsule buried in perichaetial lvs, ovate-ellipsoid; spores *ca* 15 μm. Fr very rare, spring. Tufts or patches, greenish above, blackish below, on dry acidic montane rocks, rare.

Caerns, Shrops and Lake District north to Banff and Argyll. 12. N. and C. Europe, Belgium, Pyrenees, Iceland, Caucasus, N. Asia, N. America, Greenland.

When dry closely resembling *Dicranoweisia crispula* which, however, lacks hair-points, usually has abundant capsules on straight setae and has different areolation.

13. G. pulvinata (Hedw.) Sm., Eng. Bot., 1807

Autoecious. Plants to 3 cm. Lvs erect, appressed when dry, erecto-patent when moist, upper lanceolate to ovate-lanceolate, abruptly narrowed into hair-point, margin recurved, bistratose above; hair-point $\frac{1}{2}$–1(–2) length of lamina in upper lvs, rough; basal cells ± rectangular, sinuose, 2–4 times as long as wide, longer towards nerve but otherwise uniform throughout base, cells above quadrate to quadrate-rectangular, incrassate, sinuose, pellucid, 7–10 μm wide in mid-lf. Seta cygneous when moist, erect, flexuose when dry; capsule striate; calyptra mitriform; spores 8–12 μm. Fr common, spring.

Var. **pulvinata**

Capsule ovate-ellipsoid, lid rostellate or rostrate, acute. n=13*, 26*, 26+1*. Dense, rounded, hoary cushions on usually basic walls, rock, mortar, etc., very rarely on trees. Common throughout lowland Britain and Ireland. 112, H40, C. Europe, Faroes, Caucasus, W. and C. Asia, Macaronesia, Tunisia, Morocco, Algeria, Ethiopia, N. America, Greenland, Australia, New Zealand.

Var. **africana** (Hedw.) Hook. f., Fl. Nov. Zel., 1854

G. pulvinata var. *obtusa* (Brid.) Hüb.

Capsule ovoid; lid shortly conical. Similar habitats to the type, rare. Scattered localities from S. Somerset and W. Sussex north to Caerns and Derby. 9. Europe, southern Africa.

G. orbicularis differs in the laxer, darker tufts, shape of the basal cells and fruit. The capsule of var. *africana* is similar in shape to that of *G. orbicularis* but the basal cells of the lf are shorter and of more even size. *G. trichophylla* forms laxer, more untidy, readily disintegrating tufts, occurs on acidic substrata and rarely fruits. *G. decipiens* may resemble large plants of *G. pulvinata* but has long narrow basal cells and as with *G. trichophylla* the lamina gradually tapers into the hair-point.

14. G. orbicularis Bruch ex Wils., Eng. Bot. Suppl. 1844

Close to *G. pulvinata.* Plants to *ca* 4 cm. Lvs similar in shape to those of *G. pulvinata* but margin unistratose or only rarely bistratose near apex; basal cells narrowly rectangular, to 8 times as long as wide, slightly sinuose, shorter, more hyaline towards margin, mid-lf cells 10–12 μm wide. Capsule sub-spherical to ovoid, rarely ovate-ellipsoid; lid mamillate, obtuse; calyptra cucullate; spores 10–14 μm. Fr common, spring. n=26+1*. Dark, hoary cushions on basic rocks and walls, occasional but sometimes locally abundant in England and Wales, rare in Scotland and Ireland. 32, H4. Europe, Caucasus, W. Asia, Tunisia, Algeria, Morocco, N. America, Mexico, Patagonia, Tierra del Fuego.

15. G. torquata Hornsch. in Grev., Scot. Crypt., 1826

Dioecious. Plants 1–5 cm. Lvs spirally crisped when dry, erecto-patent when moist, narrowly oblong-lanceolate, acute, margin recurved on one or both sides; hair-point short, to 250 μm long in upper lvs, very rarely longer, smooth; nerve faint below, strong above, ending in or below apex; basal cells rectangular, *ca* 6 times as long as wide, very incrassate, sinuose, uniform except for 1–2 thin-walled, hyaline marginal cells, above quadrate-rectangular to quadrate, very incrassate, sinuose, pellucid, 8–10 μm wide in mid-lf, marginal cells in upper part of lf often shorter than other cells. Brown, multicellular gemmae, 40–120 μm diameter often present on back

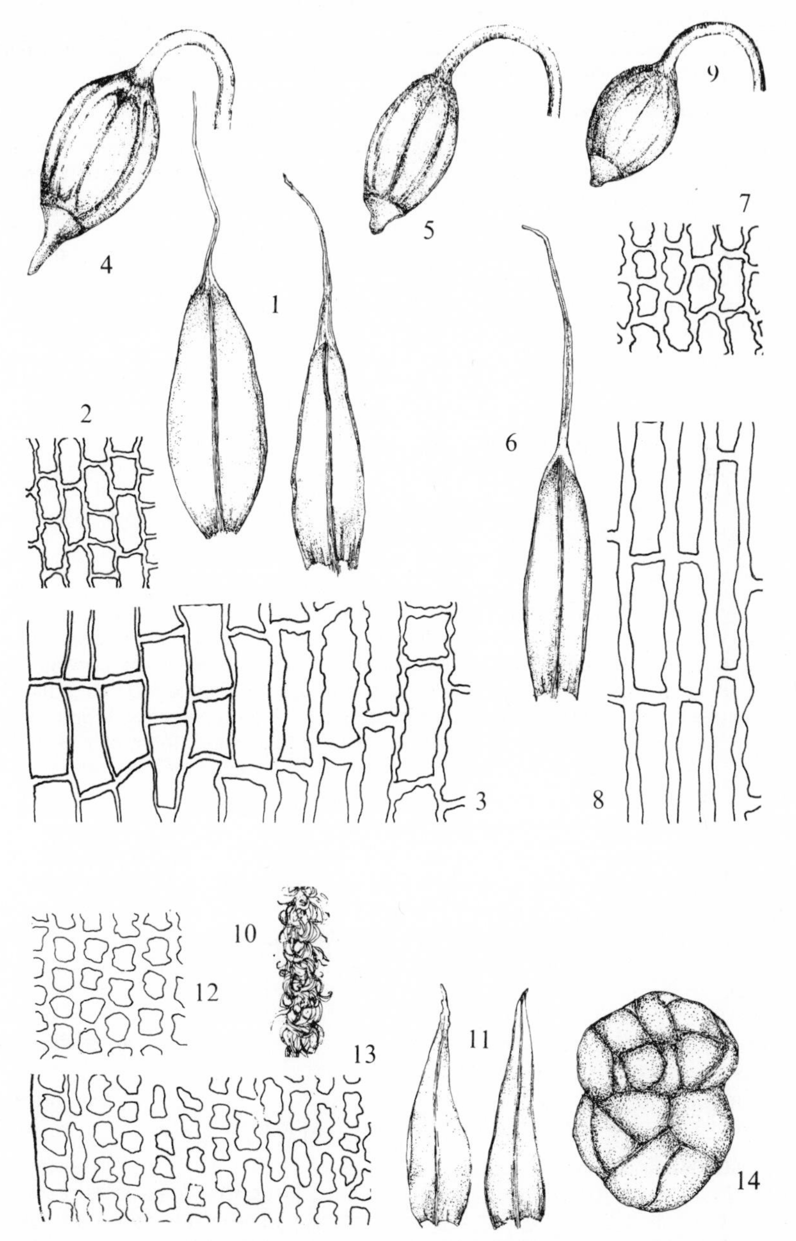

Fig. **151**. 1–4, *Grimmia pulvinata*: 1, leaves (×15), 2, mid-leaf cells; 3, basal cells; 4, capsule. 5, *G. pulvinata* var. *africana*: capsule. 6–9, *G. orbicularis*: 6, leaf (×15); 7, mid-leaf cells; 8, basal cells; 9, capsule. 10–14, *G. torquata*: 10, portion of dry shoot (×7.5); 11, leaves (×25); 12, mid-leaf cells; 13, basal cells; 14, gemma (×450). Capsules ×15, cells ×450.

of upper lvs. Seta arcuate; capsule cylindrical. Fr unknown in Britain. Brownish, often tumid tufts on shaded, usually basic montane rocks, occasional. Scattered localities in the mountains from Brecon northwards. 34, H6. N. and montane Europe, Iceland, Faroes, Madeira, N. America, Greenland, Hawaii.

16. G. funalis (Schwaegr.) Br. Eur., 1845

Dioecious. Plants (1–)2–4 cm, shoots string-like in appearance when dry. Lvs appressed, spirally curved when dry, erecto-patent when moist, lanceolate to linear-lanceolate, margin recurved; hair-point ½–1 times as long as lamina in upper lvs, slightly roughened; nerve faint below, stronger above, ending in apex; basal cells narrow, to 10 times as long as wide, sinuose, very incrassate, shorter by nerve at extreme base, towards margin quadrate-rectangular, 1–2 marginal rows elongated, hyaline, cells above quadrate-rectangular to quadrate, sinuose, pellucid, 10 μm wide in mid-lf. Seta arcuate; capsule obovoid, often ± concealed in hair-points of perichaetial lvs; spores 16–18 μm. Fr occasional, late summer. Dense, greyish, readily disintegrating tufts in crevices of montane, siliceous or basic rocks, occasional. 34, H11. Europe, Faroes, Iceland, Sardinia, Corsica, Altai, Japan, Canaries, Algeria, N. America, Greenland, Tasmania.

The neat, dark grey tufts with the lvs spirally curved when dry giving the shoots a string-like appearance make this an easily recognisable plant. *G. trichophylla* is a coarser plant of different aspect when dry. Stunted plants may be mistaken for a species of the section *Guembelia* but differ in the recurved lf margin and pellucid upper cells.

17. G. trichophylla Grev., Fl. Edin., 1804

Dioecious. Plants to 3.5 cm. Lvs loosely appressed, straight to slightly twisted when dry, erecto-patent to squarrose when moist, lanceolate to narrowly lanceolate, tapering to acute apex, narrowed or not at insertion, margin recurved on one or both sides below; hair-point to ¾ length of lamina in upper lvs, smooth to denticulate or spinulose, very rarely absent; basal cells rectangular, 2–6(–8) times as long as wide, smooth to sinuose, cells above ± quadrate, smooth to sinuose, incrassate, 8–10 μm wide in mid-lf. Irregular-shaped gemmae, up to 60 μm diameter, occasionally present on upper surface of lvs. Seta arcuate when moist, erect, flexuose when dry; capsule ovoid to obloid, striate; spores 10–14 μm.

Key to intraspecific taxa of *G. trichophylla*

1	Cells incrassate, sinuose throughout	var. **robusta**
	Basal cells less incrassate than other cells, smooth to slightly sinuose	2
2	Lvs squarrose when moist, cells not or only slightly sinuose	var. **subsquarrosa**
	Lvs not squarrose, upper cells sinuose or not	3
3	Basal cells to 6(–8) times as long as wide	var. **trichophylla**
	Basal cells to 3 times as long as wide	4
4	Hair-point ± smooth, short	var. **stirtonii**
	Hair-point denticulate or spinulose, to ½ length of lamina	var. **tenuis**

Var. **trichophylla**

Plants to 3 cm. Hair-point to ¾ length of lamina in upper lvs, smooth to slightly denticulate; basal cells rectangular, 2–6(–8) times as long as wide, smooth to slightly sinuose, ± incrassate, cells above sinuose, incrassate. Fr rare, spring. $n=13$, 26. Hoary, yellowish-green to blackish tufts on acidic rocks and walls, occasional. 81, H28, C. Europe, Asia Minor, Macaronesia, Algeria, Morocco, N. and C. America, southern S. America, Australia, Tasmania, New Zealand, Hawaii.

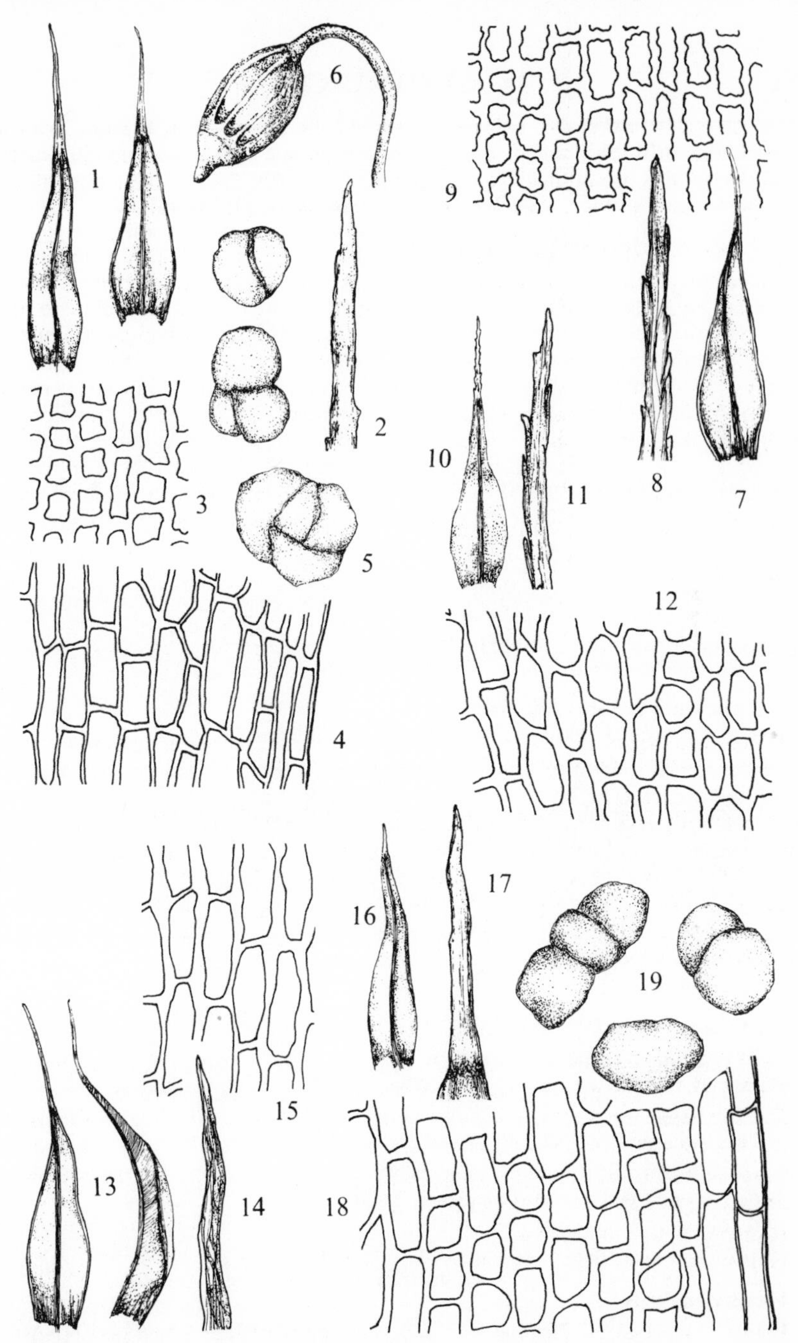

Fig. **152**. 1–6, *Grimmia trichophylla* var. *trichophylla*: 1, leaves; 2, hair-point; 3, mid-leaf cells; 4, basal cells; 5, gemmae; 6, capsule. 7–9, *G. trichophylla* var. *robusta*: 7, leaf; 8, hair-point; 9, basal cells. 10–12, *G. trichophylla* var. *tenuis*: 10, leaf; 11, hair-point; 12, basal cells. 13–15, *G. trichophylla* var. *subsquarrosa*: 13, leaves; 14, hair-point; 15, basal cells. 16–19, *G. trichophylla* var. *stirtonii*: 16, leaf; 17, hair-point; 18, basal cells; 19, gemmae. Leaves and capsules ×15, cells and gemmae ×450, hair-points ×115. All basal cells are from midway between margin and nerve.

Var. **subsquarrosa** (Wils.) A. J. E. Smith, J. Bryol., 1976
G. subsquarrosa Wils.
Plants forming lax tufts. Upper lvs squarrose when moist, often forming slight comal tuft, softer and wider than in type; hair-point to $\frac{1}{2}$–$\frac{2}{3}$ length of lamina in upper lvs, smooth or slightly denticulate; basal cells 2–3 times as long as wide, upper cells smooth or slightly sinuose. Fr very rare. Green to blackish-green tufts on acidic rocks, often near water, occasional. W. and N. Britain, Kerry, Limerick, Kilkenny. 26, H3, C. France.

Var. **stirtonii** (Schimp.) Möll., Ark. f. Bot., 1933
G. stirtonii Schimp.
Plants forming small tight tufts. Lvs spreading when moist, narrowly lanceolate, hair-point to $\frac{1}{4}$ length of lamina in upper lvs, smooth or slightly denticulate; basal cells smooth, 1–3 times as long as wide, upper cells smooth to sinuose. Fr unknown. Small dark grey tufts in crevices of dry, acidic rocks, occasional. 33, H4. Endemic.

Var. **tenuis** (Wahlenb.) Wijk & Marg., Taxon, 1959
G. muehlenbeckii Schimp.
Upper lvs with denticulate or spinulose hair-points; basal cells to 3 times as long as wide. Seta shorter, capsule smaller than in type. Fr unknown in Britain. $n = 14+2$, $28+4$. Dark grey tufts on acidic, montane rocks, very rare. Lanarks, S. Aberdeen, E. Inverness, Argyll. 4. Europe, Asia, N. America.

Var. **robusta** (Ferg.) A. J. E. Smith, J. Bryol., 1976
G. robusta Ferg., *G. decipiens* var. *robusta* (Ferg.) Braithw.
Plants 3.5 cm. Lvs erecto-patent when moist; hair-point to $\frac{1}{2}$(–$\frac{2}{3}$) length of lamina in upper lvs, smooth to slightly denticulate; basal cells 2–3 times as long as wide, sinuose except near nerve at extreme base, cells above incrassate, sinuose. Fr unknown. Hoary, dark tufts on acidic rocks, rare. W. Cornwall, S. Devon, N. Wales and N. Britain. 20, H8. Portugal.

An extremely variable species. The nature and status of the varieties is not clear; var. *stirtonii* and var. *tenuis* intergrade with one another and the type; var. *subsquarrosa* intergrades with the type. On the basis of inflorescence, areolation and the occurrence of intermediates, var. *robusta* clearly belongs here rather than under *G. decipiens* where it has been placed by earlier authors.

18. G. hartmanii Schimp., Syn., 1860

Dioecious. Plants (0.5–)1.0–4.0(–6.0) cm. Lvs slightly contorted and often with apices secund when dry, patent or recurved and frequently secund when moist, lanceolate, tapering to acute apex, one or both margins recurved below; hair-point short, 40–320(–500) μm long in upper lvs, very rarely longer; nerve strong, ending in apex; basal cells 3–6 times as long as wide, ± sinuose and incrassate, shorter towards margin, cells above quadrate, usually sinuose, incrassate, bistratose at margin above, 8–12 μm wide in mid-lf. Stem apices usually with clusters of brownish, mulberry-shaped gemmae, *ca* 250 μm diameter, borne on malformed lvs. Seta curved; capsule ovoid. Fr unknown in Britain. Yellowish-green patches on siliceous rocks, usually near water, rarely on trees, mostly at low altitudes. Occasional to frequent in W. and N. Britain. 47, H11. Europe, Corsica, Caucasus, N. Africa, N. America.

Usually distinguished by the presence of gemmae at the stem apices, but where these are lacking *G. hartmanii* may be separated from *G. retracta* by the secund lvs and longer, ± sinuose basal cells. Plants with longer hair-points to the lvs may be encountered and I have seen a specimen with the hair-point as long as the lamina. Such plants are exceedingly rare.

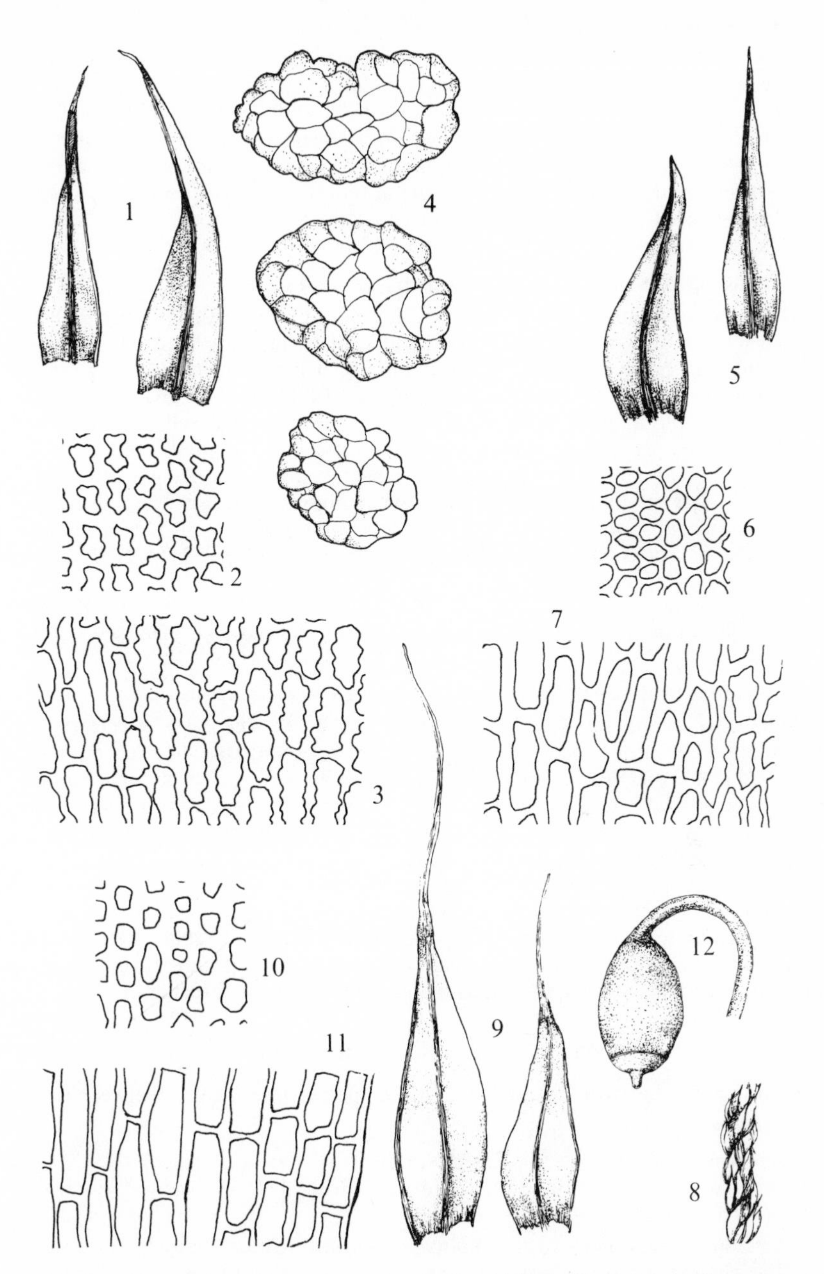

Fig. **153**. 1–4, *Grimmia hartmanii*: 1, leaves (× 15); 2, mid-leaf cells; 3, basal cells; 4, gemmae (× 180). 5–7, *G. retracta*: 5, leaves (× 15); 6, mid-leaf cells; 7, basal cells. 8–12, *G. funalis*: 8, portion of dry shoot (× 7.5); 9, leaves (× 25); 10, mid-leaf cells; 11, basal cells; 12, capsule (× 15). Cells × 450.

19. G. retracta Stirt., Scot. Nat., 1886

Plants 1.0–3.0(–3.5) cm. Lvs loosely appressed when dry, squarrose or recurved, not secund when moist, lanceolate or ovate-lanceolate, tapering to acute to obtuse apex, margin recurved below; hair-point short, 40–240(–500) μm in upper lvs; basal cells hyaline, 3–4 times as long as wide, not or scarcely sinuose, incrassate or not, shorter towards margin, cells above quadrate to wider than long, incrassate or not, bistratose at margin above, 8–10 μm wide in mid-lf. Gemmae lacking. Fr unknown. Greenish-black tufts on acidic rocks, often near water and sometimes embedded in alluvial sand, rare. Hereford, Wales and Yorks northwards. 19, H5. France(?).

20. G. decipiens (Schultz) Lindb. in Hartm., Scand. Fl., 1861

Autoecious. Plants 1.0–2.5(–4.0) cm. Lvs loosely appressed when dry, erecto-patent when moist, lanceolate to broadly lanceolate, tapering to acute apex, margin recurved; hair-point to as long as lamina in upper lvs, decurrent down margin at lf apex, strongly denticulate; nerve ending in apex; basal cells narrowly rectangular, to 10 times as long as wide, not or scarcely sinuose, marginal cells shorter, hyaline, with thickened transverse walls, forming conspicuous marginal band, cells above rectangular to quadrate, very sinuose, incrassate, bistratose at margin and near apex, 7–10 μm wide in mid-lf. Seta arcuate when moist, flexuose when dry; capsule ellipsoid, striate; spores 12–14 μm. Fr frequent. Hoary, readily disintegrating tufts on basic or acidic rocks. Occasional throughout Britain and Ireland. 44, H7, C. Europe, Algeria, Tenerife, N. America.

Resembling very coarse, large *G. pulvinata* but differing in lf shape and areolation. *G. trichophylla* has shorter basal cells, ± smooth or less denticulate hair-point and a dioecious inflorescence, and *G. trichophylla* var. *robusta* has short sinuose basal cells.

21. G. elatior Bruch ex Bals. & De Not., Mem. R. Acc. Sc. Torino, 1838

Dioecious. Plants to 7 cm. Lvs loosely appressed when dry, erecto-patent to spreading when moist, narrowly lanceolate from broad base, one margin strongly recurved, the other less so or plane; hair-point to ½ length of lamina, smooth to slightly denticulate; nerve ending in apex; basal cells rectangular to linear, sinuose, incrassate, towards margin rectangular to quadrate with thickened transverse walls, marginal row longer, cells above irregularly quadrate to rounded-hexagonal, very sinuose, incrassate, papillose, 2–3 stratose and opaque towards apex, 8–10 μm wide in mid-lf. Seta arcuate when moist, flexuose when dry; capsule ellipsoid; spores *ca* 15 μm. Fr unknown in Britain. *n* = 13. Dark tufts on montane rocks. Recorded from Angus and S. Aberdeen but not seen for over 100 years and probably extinct. Europe, Caucasus, Siberia, China, Manchuria, N. America, Greenland.

65. Dryptodon Brid., Br. Univ., 1826

A monotypic genus with the characters of the species, intermediate between *Grimmia* and *Racomitrium*.

1. D. patens (Hedw.) Brid., Bryol. Univ., 1826
Grimmia patens (Hedw.) Br. Eur., *G. curvata* (Brid.) J. L. de Sloover, *Racomitrium patens* (Hedw.) Hüb.

Dioecious. Plants (1–)2–10 cm, shoots procumbent to erect, branches often curved. Lvs loosely appressed when dry, patent, often sub-secund when moist, lanceolate, tapering to acute to sub-obtuse, entire or slightly toothed apex, margin recurved on one or both sides; hair-point absent; nerve stout, prominently 2-winged at back above; basal cells linear, incrassate, sinuose but not nodulose, at basal angles and

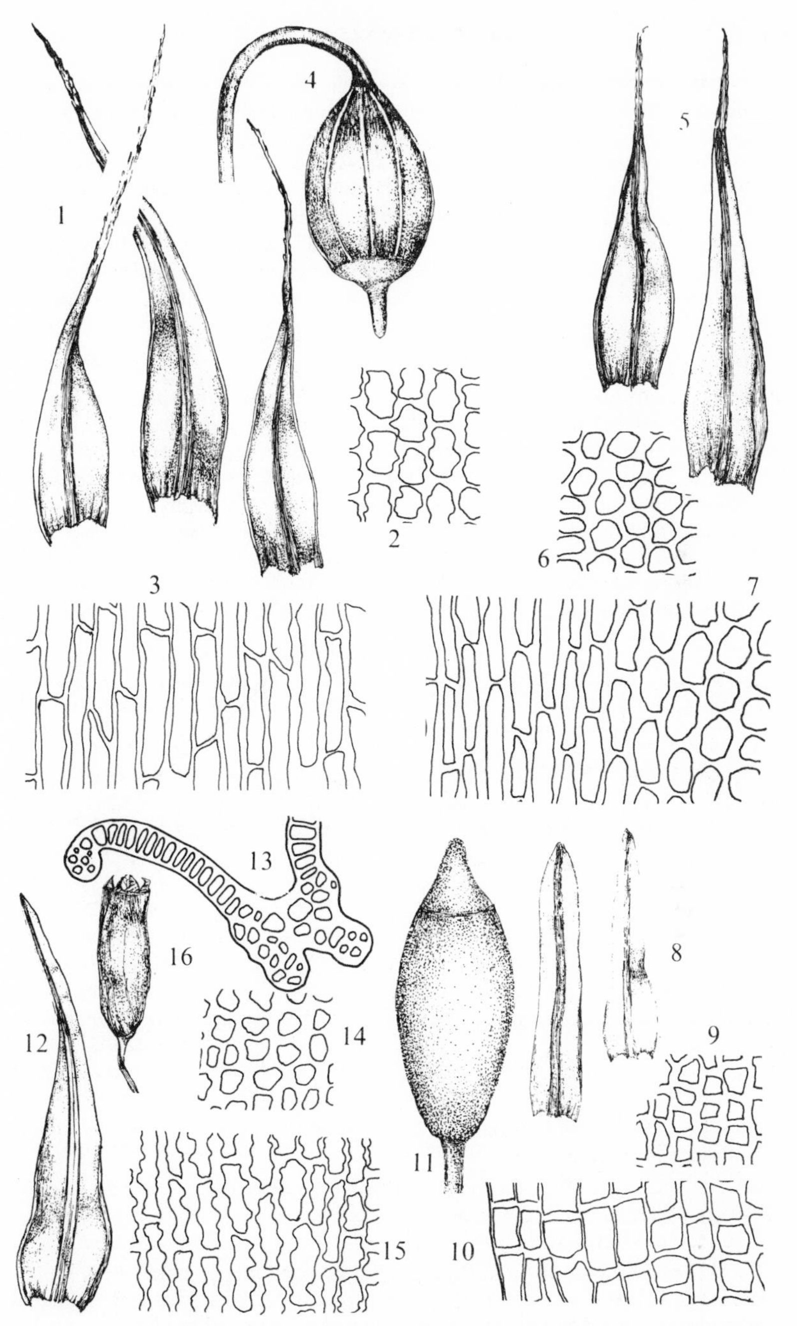

Fig. **154**. 1–4, *Grimmia decipiens*: 1, leaves; 2, mid-leaf cells; 3, basal cells; 4, capsule. 5–7, *G. elatior*: 5, leaves; 6, mid-leaf cells; 7, basal cells. 8–11, *G. atrata*: 8, leaves; 9, mid-leaf cells; 10, basal cells; 11, capsule. 12–16, *Dryptodon patens*: 12, leaf; 13, section of upper part of leaf ($\times 250$): 14, mid-leaf cells; 15, basal cells; 16, capsule. Leaves and capsules $\times 15$, cells $\times 450$.

margin rectangular, cells above irregularly quadrate or quadrate-rectangular, strongly incrassate and sinuose, bistratose at margin, in mid-lf 8–12 μm wide. Seta yellow, cygneous at maturity, flexuose when old; capsule ellipsoid, smooth, when old plicate; lid rostrate; peristome teeth divided to $\frac{1}{2}$ or $\frac{2}{3}$ into 2 subulate, partly adhering segments; calyptra mitriform; spores 12–16 μm. Fr rare, spring. n=22. Green or yellowish-green loose patches, blackish below, on moist rocks at medium to high altitudes, occasional to frequent in montane parts of Britain and Ireland. 40, H16. Montane and N. Europe, Sardinia, Iceland, Faroes, Altai, N. America, Greenland.

66. Racomitrium Brid., Mant. Musc., 1819
Rhacomitrium auct.

Plants cladocarpous; stems erect or elongated and prostrate, frequently with numerous short branches. Lvs lingulate to narrowly lanceolate, often ending in hyaline hair-point, margin recurved or not, usually entire; nerve ending in apex; basal cells linear, sinuose-nodulose, cells above quadrate to narrowly rectangular, usually strongly sinuose, unistratose or bistratose at margin and towards apex. Antheridia and archegonia borne laterally. Seta straight; capsule ovoid to shortly cylindrical, smooth; peristome teeth divided usually nearly to base into 2–3 filiform, papillose segments; calyptra mitriform, not plicate. A cosmopolitan genus of about 80 mainly saxicolous species.

1 Lf cells papillose, hair-point if present papillose and denticulate **9. R. canescens**
Lf cells smooth or only slightly papillose, hair-point when present smooth or if papillose then coarsely toothed 2
2 Lvs with hair-points 3
Lvs without hair-points 5
3 Hair-point strongly papillose, coarsely and irregularly toothed, seta papillose **8. R. lanuginosum**
Hair-point not papillose, denticulate, seta smooth 4
4 Cells towards lf apex rectangular, 3–4 times as long as wide except at margin, 1–2 marginal rows near base pellucid **7. R. microcarpon**
Cells towards lf apex 1.0–1.5(–2.0) times as long as wide, marginal cells at base not pellucid 9
5 Cells towards lf apex ± linear, stems with numerous short lateral branches **4. R. fasciculare**
Cells towards apex up to twice as long as wide, stems with or without short lateral branches 6
6 Lf apex rounded, often toothed **2. R. aciculare**
Apex acute to obtuse, entire 7
7 Cells in upper part of lf opaque, bistratose, capsule ovoid **1. R. ellipticum**
Cells in upper part of lf not opaque, unistratose or bistratose only at margin, capsule ovoid to shortly cylindrical 8
8 Lf apex obtuse, nerve stout, (70–)90–120 μm wide, cells ± papillose **3. R. aquaticum**
Apex acute or obtuse, nerve 20–80 μm wide, cells not papillose 9
9 Lvs (1.5–)2.2–3.6 mm long including hair-point, hair-point 0–45 % total lf length, seta 4–12 mm long, capsule narrowly ellipsoid to shortly cylindrical, 1.8–2.5 mm long (not including lid) **5. R. heterostichum**

Lvs 1.5–2.4(–2.6) mm long, hair-point 0–15 % total lf length, seta 2–4 mm long, capsule ovoid, 1.0–1.4 mm long (not including lid) **6. R. affine**

1. R. ellipticum (Turn.) Br. Eur., 1845

Plants tufted, 0.5–2.5 cm. Lvs ± straight, loosely appressed when dry, spreading when moist, lanceolate with broad base, apex obtuse, margin plane or recurved on one or both sides below; hair-point absent; nerve broad, ending below apex; cells sinuose, incrassate throughout, basal nodulose, ± linear, shorter towards margin, in mid-lf rectangular, at margin quadrate, bistratose, 8–10 μm wide, towards apex irregularly quadrate, bistratose, opaque. Seta straight or slightly curved; capsule ovoid, glossy, dark brown; spores 18–22 μm. Fr frequent, early summer. $n = 13$*. Dark green tufts on damp, siliceous, montane rocks, occasional in N. Wales, N. Britain and Ireland. 27, H14. Norway, Faroes.

2. R. aciculare (Hedw.) Brid., Mant. Musc., 1819

Plants tufted, to 6 cm. Lvs ± straight, appressed when dry, patent when moist, ovate to ovate-lanceolate, apex rounded, often toothed, margin recurved below; hair-point lacking; nerve ending below apex; cells sinuose, unistratose throughout, basal linear, nodulose, incrassate, at basal angles sometimes enlarged, in mid-lf rectangular to quadrate, 8–9 μm wide, towards apex irregularly quadrate. Capsule ellipsoid to shortly cylindrical, brown; lid longly rostrate; spores 16–20 μm. Fr common, spring, early summer. $n = 12$*, 13*. Green to greenish-black tufts on acidic rocks in or by water, common except in S.E. England and the Midlands. 80, H34. Europe, Iceland, Faroes, W. Asia, Madeira, southern Africa, N. America.

3. R. aquaticum (Schrad.) Brid., Mant. Musc., 1819
R. protensum (Braun) Hüb.

Plants 2–12 cm, branches mostly decumbent. Lvs straight, appressed when dry, patent or often secund when moist, lanceolate to narrowly lanceolate, tapering to obtuse apex, margin plane or recurved below, entire; hair-point absent; nerve ending below apex; cells sinuose, incrassate, unistratose throughout, basal nodulose, linear, shorter towards margin, in mid-lf rectangular, 8–10 μm wide, towards apex irregularly quadrate, slightly papillose. Capsule ellipsoid to shortly cylindrical; lid longly rostrate; spores 14–20 μm. Fr occasional, spring. $n = 12$, 13*. Yellowish-green to greenish-brown patches on moist, acidic, montane rocks, occasional in W. and N. Britain and Ireland. 61, H33. Europe, Caucasus, Japan, Azores, Madeira, N. America, Greenland, Tierra del Fuego, Kerguelen Is., New Zealand.

Small stunted forms may be difficult to separate from plants of *R. heterostichum* lacking hair-points; that plant is usually darker green with the narrower lvs bistratose at margin and not papillose (the papillae of *R. aquaticum* are most easily seen in transverse section of the lf). *Dryptodon patens* is superficially similar but has a winged nerve and hardly sinuose basal cells.

4. R. fasciculare (Hedw.) Brid., Mant. Musc., 1819

Plants to 10 cm, procumbent with numerous, fasciculate branches. Lvs loosely appressed to spreading or slightly crisped when dry, patent to spreading when moist, linear-lanceolate from broad base, apex acute to obtuse, margin recurved below, entire; hair-point absent; nerve ending below apex; cells sinuose, incrassate, unistratose throughout, basal linear, nodulose, shorter towards margin, in mid-lf narrowly rectangular, 8–12 μm wide, towards apex rectangular, not papillose. Capsule shortly cylindrical; lid longly rostrate; spores 13–15 μm. Fr frequent, late spring. $n = 12$*, 13*. Dense, golden-brown to yellowish-green or green patches on

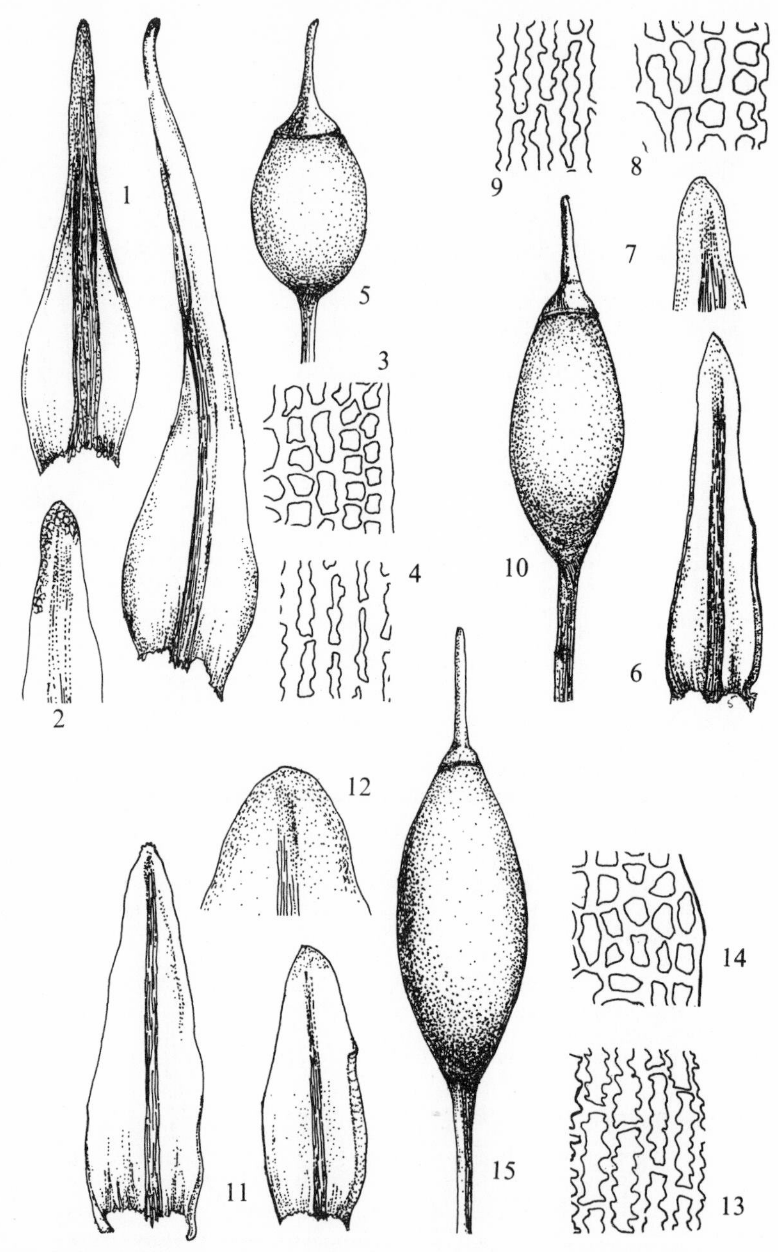

Fig. **155**. 1–5, *Racomitrium ellipticum*: 1, leaves (×30); 2, leaf apex; 3, cells from upper part of leaf; 4, basal cells; 5, capsule. 6–10, *R. aquaticum*: 6, leaf (×15); 7, leaf apex; 8, upper cells; 9, basal cells; 10, capsule. 11–15, *R. aciculare*: 11, leaves; 12, leaf apex; 13, basal cells; 14, upper cells; 15, capsule. Capsules ×15, apices ×115, cells ×450.

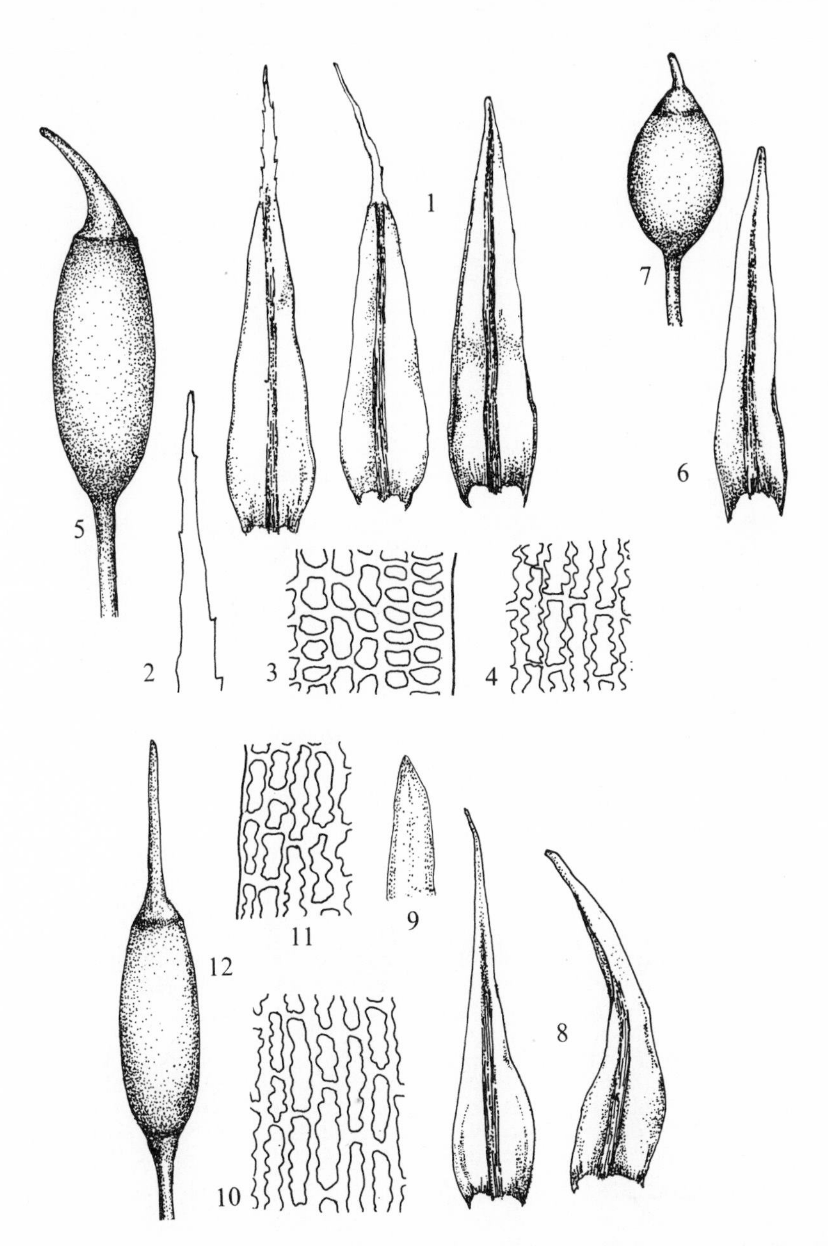

Fig. **156**. 1–5, *Racomitrium heterostichum*: 1, leaves; 2, apex of hair-point (×115); 3, cells from upper part of leaf; 4, basal cells; 5, capsule. 6–7, *R. affine*: 6, leaf; 7, capsule. 8–12, *R. fasciculare*: 8, leaves; 9, leaf apex (×115); 10, basal cells; 11, upper cells; 12, capsule. Leaves ×22.5, capsules ×15, cells ×450.

acidic rocks, frequent or common except in S.E. England and the Midlands. 90, H34. Greenland, southern S. America, New Zealand.

5. R. heterostichum (Hedw.) Brid., Mant. Musc., 1819

Plants with usually ± prostrate main stems to 5 cm long, branches usually ascending or erect and short. Lvs appressed to spreading or twisted when dry, spreading, sometimes secund when moist, upper lvs (1.5–)2.5–3.6 mm long (including hair-point when present), lanceolate to narrowly lanceolate from broad basal part, apex wide, obtuse when hair-point absent, margin recurved; hair-point (0–)15–45 % total lf length, flattened, denticulate, not papillose; basal cells narrowly rectangular to linear, sinuose-nodulose, incrassate, marginal cells shorter but not pellucid, cells in mid-lf, quadrate to rectangular, strongly sinuose, 6–10 μm wide, towards apex 1.0–1.5(–2.0) times as long as wide, not papillose, often bistratose at margin. Seta straight or curved, 4–8(–12) mm long; capsule narrowly ellipsoid to shortly cylindrical, 1.8–2.5 mm long (not including lid); lid longly rostrate; spores 10–18 μm. Fr occasional, spring. $n=13$*, 14. Yellowish-green to blackish, frequently hoary tufts or patches on acidic rocks, rarely on soil. Rare in S.E. England and the Midlands, common or very common elsewhere. 94, H36, C. Europe, Iceland, Faroes, Asia, Macaronesia, N. America, Greenland, southern Africa, Kerguelen Is., Tasmania, New Zealand.

R. heterostichum and *R. affine* have frequently been confused, as forms of the latter have been treated as varieties of the former. There is some intergradation in gametophyte characters between the 2 species though they are distinct in capsule characters. *R. heterostichum* is usually a coarser plant with larger lvs, but some small plants may be impossible to determine in the absence of capsules. *R. microcarpon* differs in the longer cells in the upper part of the lf and the marginal row of pellucid cells at the base.

6. R. affine (Web. & Mohr) Lindb., Acta Soc. Sc. Fenn., 1875
R. sudeticum (Funck) Br. Eur., *R. heterostichum* var. *gracilescens* Br. Eur., *R. heterostichum* var. *affine* (Web. & Mohr) Lesq., *R. heterostichum* var. *alopecurum* (Brid.) Hüb.

Plants with slender shoots up to 15 cm long, stems procumbent to erect, forked but usually without branches. Lvs appressed when dry, patent to spreading when moist, upper 1.5–2.2(–2.5) mm long (including hair-point when present), narrowly lanceolate, acute, margin recurved; hair-point 0–10(–15) % total lf length; aerolation as in *R. heterostichum*. Seta 2–4 mm; capsule ovoid, 1.0–1.2(–1.4) mm long (not including lid); lid rostrate. Fr occasional, spring. $n=13+1$*. Small, blackish tufts or cushions to dull green, extensive, flattish patches on acidic, usually montane rocks. Occasional to frequent in the mountains, very rare elsewhere. 67, H24, C. Europe, Iceland, Faroes, N., W. and C. Asia, Japan, N. America, Greenland.

A very variable species probably containing more than 1 taxon. The habit of the plants is exceedingly variable. Plants forming small cushions or extensive patches are readily distinguished from *R. heterostichum* in the absence of capsules. Plants of intermediate habit are more problematical but many may be recognised on the basis of the combination of slender stems, smaller lvs and shorter hair-points. Some of the plants referred to as *R. heterostichum* var. *alopecurum* belong here but others are slender green shade forms of *R. heterostichum*.

7. R. microcarpon (Hedw.) Brid., Mant. Musc., 1819
R. ramulosum Lindb.

Plants to *ca* 4 cm, ± decumbent, stems with numerous short branchlets. Lvs appressed when dry, patent to spreading when moist, lanceolate to narrowly lanceolate, margin recurved, entire; hair-point denticulate, not papillose, cells strongly sinuose, incrassate, unistratose throughout, basal linear, nodulose, 1–2 rows

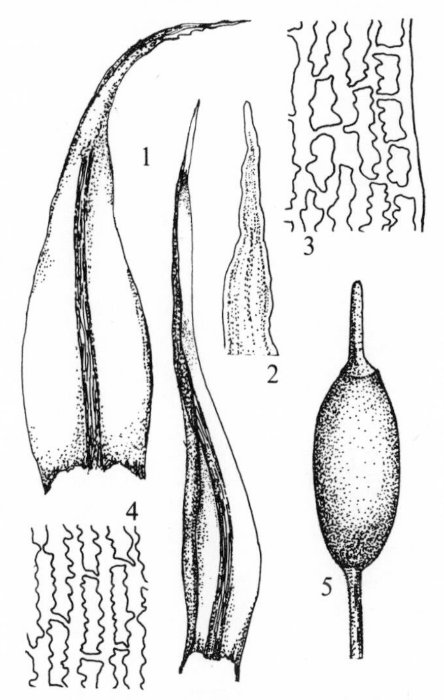

Fig. **157**. *Racomitrium microcarpon*: 1, leaves (×22.5); 2, apex of hair-point (×115); 3, cells from upper part of leaf; 4 basal cells; 5, capsule (×15). Cells ×450.

at margin ± quadrate, translucent, in mid-lf rectangular, 7–10 μm wide, towards apex rectangular, 3–4 times as long as wide, marginal quadrate. Capsule ellipsoid to shortly cylindrical; spores 12–16 μm. Fr frequent, spring. $n=14$. Greenish to yellowish patches on dry, acidic, montane rocks, rare. Scattered localities from Mid Perth and Inverness north to Caithness, S. Kerry. 8, H1. N., W. and C. Europe, N. America.

8. R. lanuginosum (Hedw.) Brid., Mant. Musc., 1819

Plants to 15 cm, occasionally more, stems with numerous branches. Lvs falcato-secund, particularly at stem tips, when dry appressed, when moist spreading, gradually narrowed from near base to long apex, margin recurved; hair-point to as long as lamina, coarsely toothed, strongly papillose, decurrent down margin of lamina, hyaline, in the dry state diaphanous, very rarely lacking; nerve strong, running into hair-point; cells strongly sinuose, incrassate, not papillose, unistratose throughout, basal sinuose-nodulose, linear, shorter towards margin, in mid-lf narrowly rectangular, *ca* 10 μm wide, towards apex rectangular. Seta coarsely papillose; capsule ovoid to ovate-ellipsoid; lid longly rostrate; spores 9–12 μm. Fr occasional, spring. $n=12$, 13*, 14. Dark green to yellowish-brown, hoary tufts or patches on exposed, acidic rocks and soil. Occasional in S.E. England and the Midlands, frequent to common elsewhere and sometimes abundant to dominant at high altitudes and forming *Racomitrium* heath. 110, H34, C. Cosmopolitan.

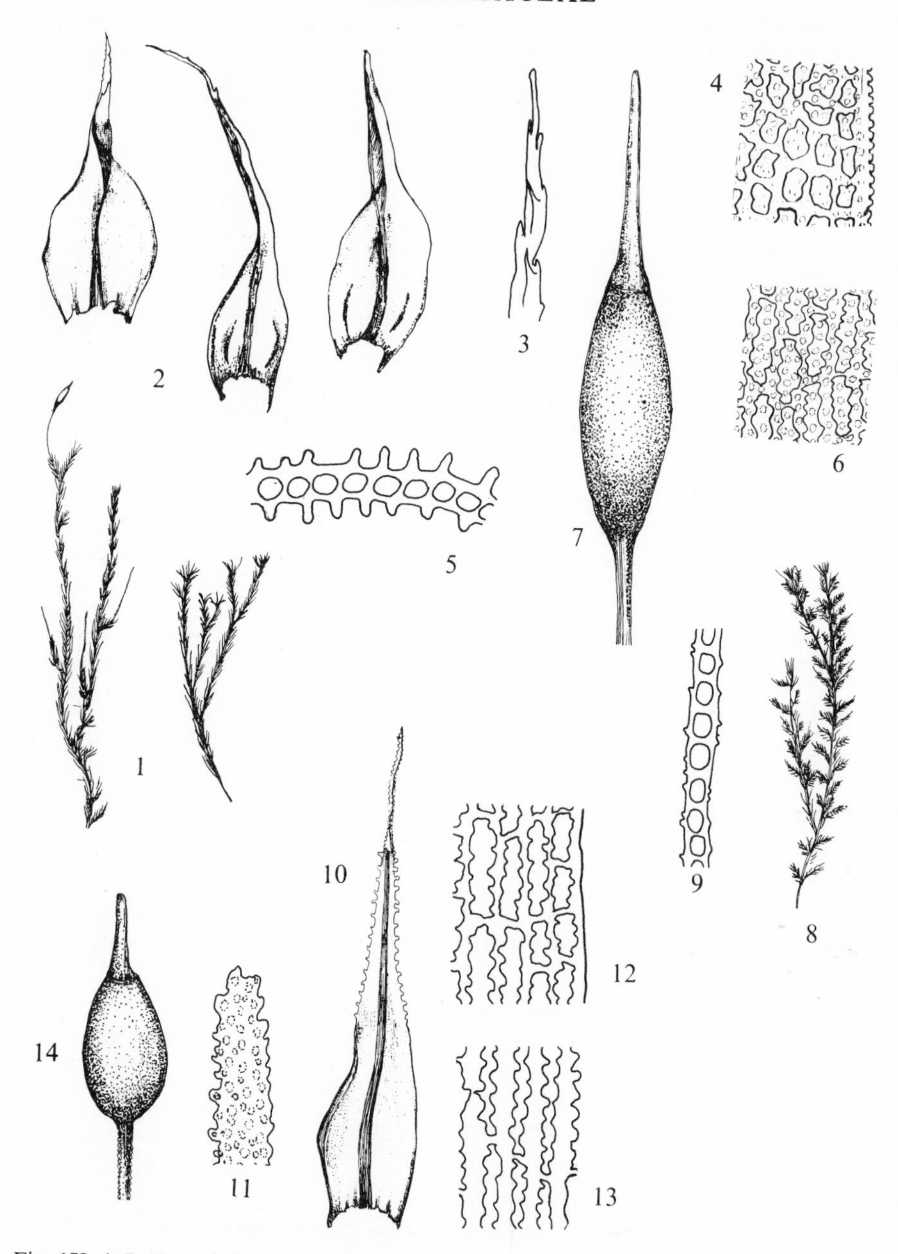

Fig. **158**. 1–7, *Racomitrium canescens* var. *canescens*: 1, habit sketch; 2, leaves; 3, apex of hair-point; 4 and 5, cells and transverse section of upper part of leaf; 6, basal cells; 7, capsule. 8–9, *R. canescens* var. *ericoides*: 8, habit sketch; 9, section of upper part of leaf. 10–14, *R. lanuginosum*: 10, leaf; 11, apex of hair-point; 12, upper cells; 13, basal cells; 14, capsule. Habits ×1, leaves ×15, apices ×115, cells ×415, capsules ×15.

Distinct in the falcate stem tips and the coarsely toothed, papillose hair-point. When growing vigorously it may form extensive, tumid patches. Occasional stunted forms may have lvs without hair-points but differ from *R. canescens* in the smooth cells and lack of small auricles, from *R. heterostichum* in the elongated upper cells and from that species and *R. microcarpon* in the falcate stem apices and generally larger size.

9. R. canescens (Hedw.) Brid., Mant. Musc., 1819

Shoots procumbent to erect, 2–6(–10) cm long, variously branched. Lvs loosely appressed, slightly curved when dry, patent to reflexed when moist, ovate to lanceolate, tapering to hair-point or to acute to obtuse apex, margin recurved, entire or with a few spinose teeth towards hair-point, papillose; hair-point 0–$\frac{1}{2}$ length of lf, spinosely dentate, papillose towards proximal end; nerve ending below apex; cells papillose throughout, unistratose, basal cells linear, sinuose-nodulose, cells at basal angles enlarged, rectangular, hyaline, not sinuose, $\pm$ forming small auricles, cells becoming shorter above, towards apex quadrate, obscure, *ca* 8 μm wide in mid-lf. Seta smooth; capsule ellipsoid; lid with very long beak, almost as long as capsule; spores 6–10 μm. Fr rare or occasional, winter.

Var. **canescens**

Branches few and of variable length in well grown forms, congested and to *ca* 5 mm long in small forms; male plants sometimes with numerous very short antheridial branches. Nerve extending $\frac{1}{2}$–$\frac{2}{3}$ way up lf; papillae, at least in upper part of lf and on back of nerve, 1–3 times as tall as wide, most conspicuous in congested forms. $n=12$. Green or dull green patches or tufts on silty, sandy or gravelly soil or thinly soil-covered rocks in open habitats on tracks, roadsides, by streams and rivers, in quarries, etc., rare in lowland Britain, very rare elsewhere. Distribution uncertain. Europe, Iceland, Faroes, Caucasus, Siberia, Sikkim, Nepal, Sri Lanka, China, Japan, Madeira, Azores, N. America.

Var. **ericoides** (Hedw.) Hampe, Flora, 1837

Stems $\pm$ pinnately branched, branches numerous, of $\pm$ uniform length, short, 2–4 mm long. Nerve extending $\frac{1}{2}$–$\frac{4}{5}$ way up lf; papillae $\frac{1}{2}$–1 times as tall as wide. $n=12$*. Green to yellowish-green tufts or patches in similar habitats to and sometimes with var. *canescens*, common. Europe, Caucasus, Siberia, Japan, N. America, Greenland.

An exceedingly variable plant some forms of which resemble other *Racomitrium* species, but easily distinguished by the papillose cells and small auricles. With the exception of *R. lanuginosum* it is the only British species which occurs on soil. Heinonen (*Ann. bot. Fenn.* **8**, 142–51, 1971) considers var. *ericoides* to be worthy of specific rank. Whilst there is a close correlation between habit and papilla height this is not absolute; also some plants are of indeterminate branching habit and I do not consider var. *ericoides* to be of specific rank. The 2 varieties can usually be distinguished on the basis of habit in the field; lf shape and length of hair-point quoted by some authorities as being of diagnostic value are of no significance. Male plants of var. *canescens* may have numerous antheridial branches which may cause confusion with var. *ericoides*, but these are shorter, barely exceeding the lvs in length. From the examination of herbarium specimens var. *ericoides* seems far commoner than var. *canescens*.

15. PTYCHOMITRIACEAE

Usually tuft- or cushion-forming mosses. Lvs usually crisped when dry, lanceolate to linear, acuminate, margin plane or recurved, entire to irregularly toothed; nerve ending below apex; basal cells rectangular to narrowly rectangular, cells above quadrate, rounded-quadrate or hexagonal. Seta long, straight or arcuate; capsule erect, smooth; annulus falling or not; peristome teeth single, entire, perforated or cleft, or in pairs, smooth or papillose; calyptra campanulate.

Most authorities place this family next to the Orthotrichaceae, but the meiotic chromosomes are totally different from those of that family and closely resemble those of the Grimmiaceae; also the peristome closely resembles that of the Grimmiaceae (S. R. Edwards, personal communication) and the leaf areolation differs from that of the Orthotrichaceae.

67. PTYCHOMITRIUM FÜRNR., FLORA, 1829

Autoecious. Lvs lanceolate, often crisped when dry, erecto-patent to spreading when moist, acuminate, recurved below, often bistratose above, upper cells quadrate or quadrate-rectangular, long and narrow at base. Perichaetial lvs similar to stem lvs. Seta straight; capsule ovoid to ellipsoid; calyptra campanulate, plicate; peristome teeth deeply bifid. A $\pm$ world-wide genus of *ca* 80 species.

1. P. polyphyllum (Sw.) Br. Eur., 1839

Plants to 4 cm. Lvs strongly crisped when dry, patent to spreading, flexuose when moist, narrowly lanceolate, longly acuminate; base plicate, margin recurved below, bistratose, coarsely toothed above; nerve stout, ending in or below apex; basal cells linear, longitudinal walls heavily thickened, above becoming quadrate-rectangular to quadrate, $\pm$ arranged in rows, uniformly thickened, smooth, 8–12 μm wide in mid-lf. Perichaetial lvs similar to stem lvs. Seta straight, 3–7 mm, sometimes 2 or more per perichaetium; capsule ellipsoid; peristome teeth deeply bifid; spores 10–14 μm. Fr common, late spring, early summer. $n=13$*. Dense, dull green, rounded cushions, blackish inside, on exposed, acidic rocks and wall-tops. Absent from S.E. England and the Midlands except W. Sussex, frequent to common elsewhere. 74, H37, C. N., W. and C. Europe, Macaronesia.

P. polyphyllum differs from *Glyphomitrium daviesii* in its larger size, capsule shape and erect peristome teeth. When sterile it may be distinguished by the plicate basal part of the leaf with cells with the longitudinal walls more heavily thickened than the transverse walls.

68. GLYPHOMITRIUM BRID., MUSC. REC., SUPPL., 1819

Close to *Ptychomitrium*. Perichaetial lvs with sheathing bases. Calyptra enclosing capsule; peristome teeth united in pairs, entire. A mainly northern hemisphere genus of 11 species.

1. G. daviesii (With.) Brid., Musc. Rec., Suppl., 1819

Autoecious. Plants to 1 cm. Lvs crisped when dry, spreading when moist, linear-lanceolate, acuminate, margin plane or slightly recurved below, entire; nerve ending in apex; basal cells rectangular, hyaline, above rounded-quadrate, in rows, bistratose at margin, 10–12 μm wide in mid-lf. Perichaetial lvs with sheathing bases. Seta straight, 2–3 mm; capsule ovoid; peristome teeth entire, united in pairs, reflexed when dry; spores 40–50 μm. Fr common, summer. Small, dense, dark green cushions or patches on acidic or basic rocks in coastal areas and basic rocks inland. Rare in S. Devon, Wales, Westmorland and Cumberland, frequent in W. Scotland and N.E. Ireland. 18, H10. Norway, Faroes, Madeira, Azores.

69. CAMPYLOSTELIUM BR. EUR., 1846

Plants minute, gregarious. Lvs crisped when dry, linear to linear-lanceolate; cells rounded-quadrate above, rectangular near base. Seta arcuate; calyptra mitriform; peristome teeth separate, cleft. Europe, Asia, Africa, America. Four species.

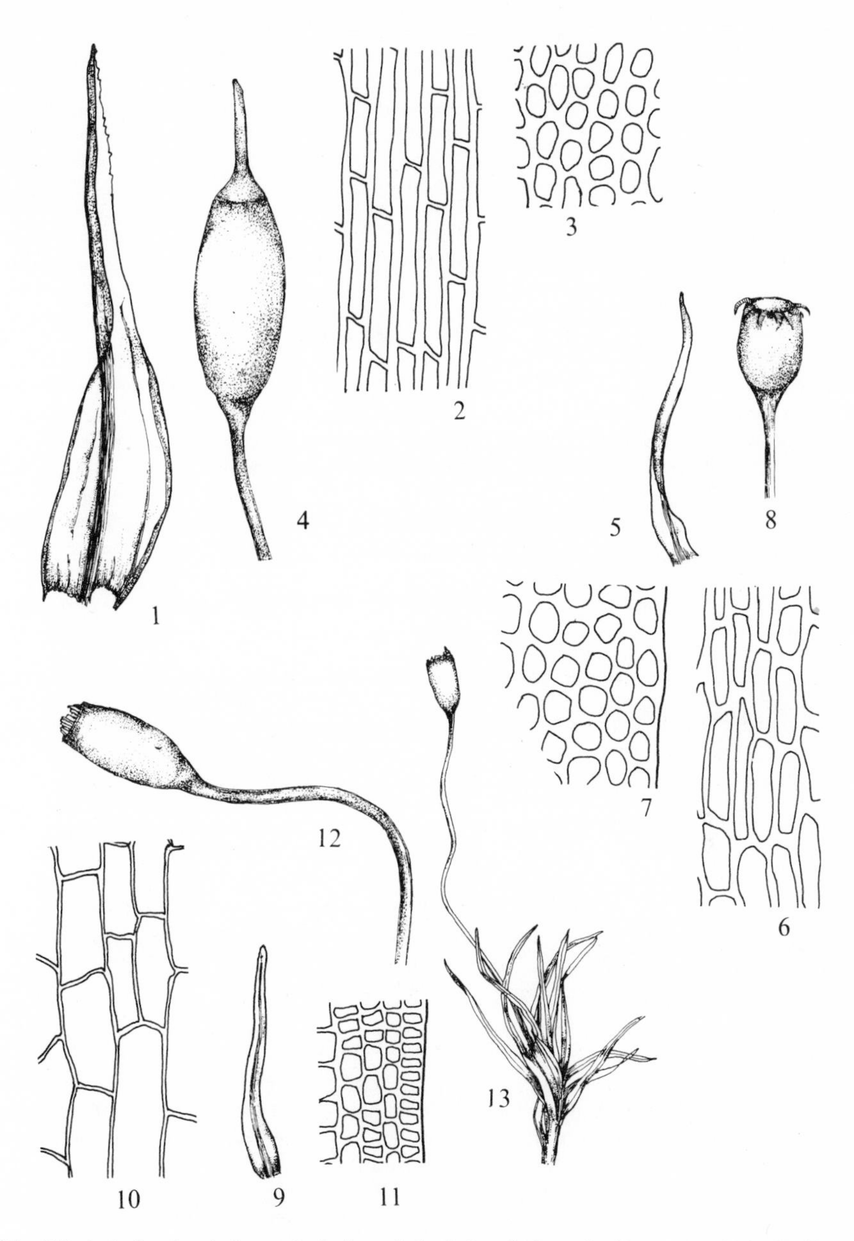

Fig. **159**. 1–4, *Ptychomitrium polyphyllum*: 1, leaf; 2, cells from leaf base; 3, mid-leaf cells; 4, capsule. 5–8, *Glyphomitrium daviesii*: 5, leaf; 6, cells from leaf base; 7, mid-leaf cells; 8, capsule. 9–13, *Campylostelium saxicola*: 9, leaf; 10, cells from leaf base; 11, mid-leaf cells; 12, capsule; 13, plant (× 10). Leaves × 30, cells × 415.

1. C. saxicola (Web. & Mohr) Br. Eur., 1846

Autoecious. Plants gregarious, 1–2 mm. Lvs flexuose-crisped when dry, erecto-patent when moist, lower linear-lanceolate, upper linear, apex acute to rounded, margin plane, entire; nerve ending in apex; basal cells rectangular, upper rounded-quadrate, 1–2-stratose, *ca* 8 μm wide in mid-lf. Seta cygneous when moist, flexuose when dry, 3–5 mm long; capsule narrowly ellipsoid, smooth. Fr common, winter. Patches on shaded sandstone or slatey rocks. Rare in S.E. England, rare or occasional in scattered localities in W. and N. Britain and Ireland. 27, H7. Europe, N. America.

Distinguished from similar species except *Seligeria recurvata* by the arcuate seta. *S. recurvata* and other *Seligeria* species differ in the elongated cells in the upper part of the lf. *Gyroweisia* species have papillose cells and rhizoidal gemmae.

12. FUNARIALES

Acrocarpous. Plants terrestrial. Lvs ovate-lanceolate; nerve thin; cells large, lax, rhomboid-hexagonal, thin-walled, smooth. Capsule stegocarpous or cleistocarpous, erect or inclined, symmetrical or asymmetrical, globose to ellipsoid or pyriform with distinct neck; peristome absent, single or double; calyptra large.

16. DISCELIACEAE

With the characteristics of the only species.

A monotypic family the systematic position of which is obscure. Cytologically it does not resemble either the Funariaceae or Catoscopiaceae with which it is variously classified.

70. Discelium Brid., Br. Univ., 1826

1. D. nudum (Dicks.) Brid., Br. Univ., 1826

Pseudo-dioecious, male and female plants produced from same persistent protonema. Plants minute, bud-like, gregarious. Lvs without chlorophyll, concave, ovate to lanceolate, acute, bluntly toothed above; nerve lacking or trace present in upper part of lf; cells lax, rhomboid-hexagonal, pellucid, narrower towards margin. Seta pale red, thin, to 2.5 cm; capsule ± horizontal, sub-globose; lid mamillate, obtuse; peristome single, teeth divided below, entire above; spores 20–24 μm. Fr common, autumn, winter. $n=13$*. Clayey banks of ditches and streams, dried-up beds of ponds and reservoirs. Frequent and sometimes locally abundant in the Pennines, very rare elsewhere, extending from E. Cornwall and Kent north to E. Ross, Cavan. 22, H1. Fennoscandia, E. Europe, Siberia, eastern N. America.

17. FUNARIACEAE

Plants mostly annual or biennial, short-stemmed. Lvs soft, concave, ovate-lanceolate, margin usually toothed above; nerve ending below apex to excurrent; cells large, lax, rhomboid-hexagonal, thin-walled, smooth. Seta long or short; capsule usually stegocarpous, erect or inclined, symmetrical or asymmetrical, usually pyriform with numerous stomata at base; peristome lacking, single or double; calyptra large, cucullate or mitriform, usually with long apex. Protonema not persisting.

71. Funaria Hedw., Sp. Musc., 1801

Autoecious (in Britain). Short-stemmed plants. Lvs often crowded at top of stem,

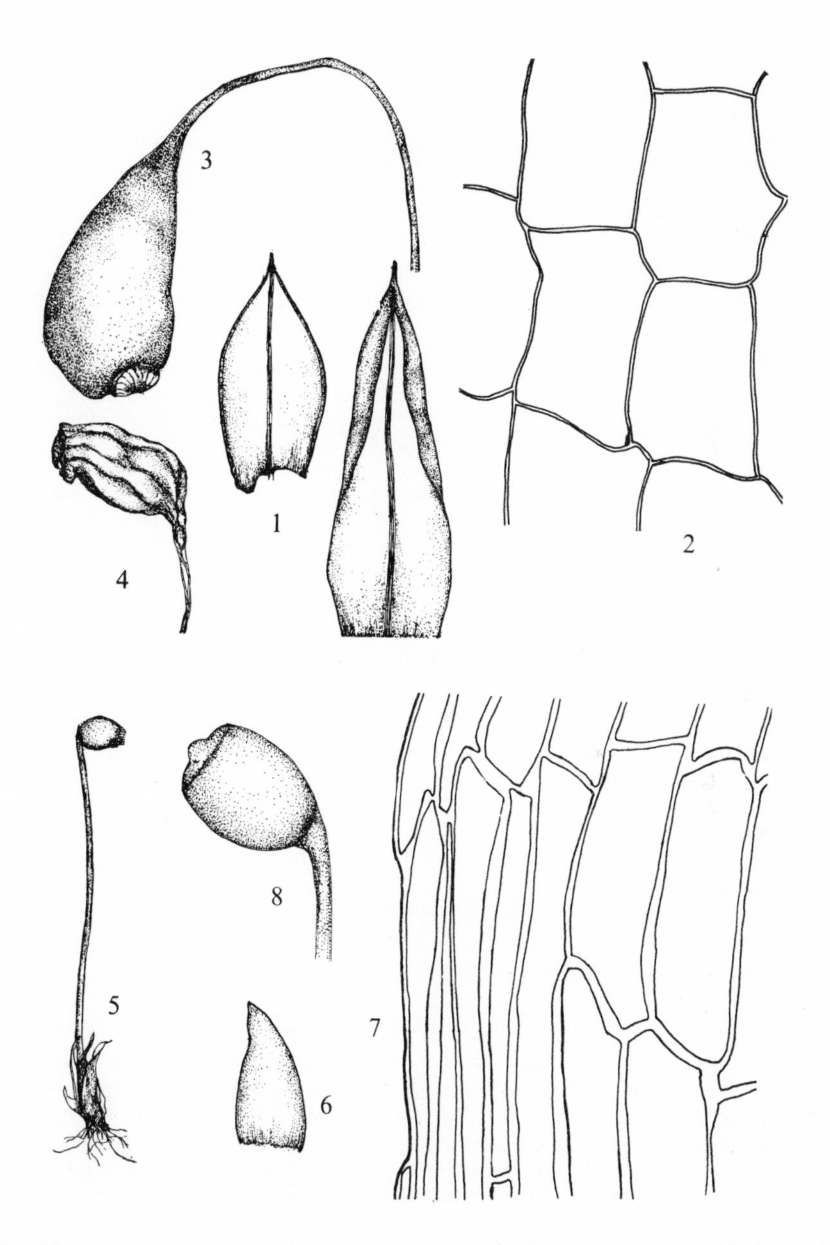

Fig. **160**. 1–4, *Funaria hygrometrica*: 1, leaves (×15); 2, cells from upper part of leaf; 3 and 4, mature and old capsules (×10). 5–8, *Discelium nudum*: 5, plant (×5); 6, leaf (×20); 7, leaf cells; 8, capsule (×15). Cells ×450.

lingulate to ovate, acute to acuminate; cells large, lax. Seta straight or arcuate; capsule pyriform, symmetrical or not; lid convex, rarely apiculate; peristome of 1 or 2 rows perfect teeth or rudimentary or absent; calyptra cucullate, oblique at maturity. A cosmopolitan genus of about 250 species.

Entosthodon Schwaegr. is not separated from *Funaria* here, as on morphological and cytological grounds the differences are not sufficient to merit generic distinction.

1 Capsule inclined, asymmetrical with oblique mouth 2
Capsule erect, ± symmetrical 4
2 Seta arcuate when young, flexuose with age, lvs shortly pointed **1. F. hygrometrica**
Seta straight when young, not flexuose when old, lvs with long apiculus 3
3 Lf margin toothed above, apical cell of lf to 450 μm long, spores coarsely papillose **2. F. muhlenbergii**
Lf margin ± entire, apical cell to 280 μm long, spores finely papillose **3. F. pulchella**
4 Lvs distinctly bordered with narrow cells, spores 30–38 μm **6. F. obtusa**
Lvs not obviously bordered, spores 24–30 μm 5
5 Capsule narrowly pyriform, neck $\frac{1}{2}$–$\frac{2}{3}$ length of theca, peristome present **4. F. attenuata**
Capsule obovoid or shortly pyriform, neck short, peristome rudimentary or absent **5. F. fascicularis**

1. F. hygrometrica Hedw., Sp. Musc., 1801

Plants to 3 cm, usually less, stems innovating from below. Lower lvs spreading to erecto-patent, upper imbricate, crowded, concave, ovate-lanceolate to lanceolate-spathulate, shortly pointed, entire or bluntly toothed; nerve ending below apex to percurrent; cells lax, thin-walled, basal rectangular, upper irregularly rectangular to quadrate-hexagonal, 35–50 μm wide in mid-lf, 1–2 rows marginal cells narrower. Seta arcuate when moist, flexuose, twisted when dry and old; capsule narrowly pyriform, gibbous, asymmetrical with oblique mouth, striate when fresh, sulcate when dry and empty; lid convex; peristome of 2 rows well developed teeth, outer row curved, joined at tips to central disc; spores finely papillose, 16–22 μm. Fr common, late winter to autumn. $n=14^*$, 21, 28*, 56. Yellowish-green or green patches or scattered stems on disturbed or cultivated ground, especially on the sites of fires, on old walls and buildings, etc., frequent and sometimes locally abundant. 112, H40, C. Cosmopolitan.

2. F. muhlenbergii Turn., Ann. Bot., 1805
F. calcarea Wahlenb., *F. mediterranea* Lindb.

Plants to 5 mm. Lower lvs spreading, upper ± erect, obovate to oblanceolate, abruptly narrowed to long fine apex, margin bluntly toothed above; nerve ending in or below apex; cells similar to those of *F. hygrometrica*, apical cell to 450 μm long. Seta straight at maturity, not flexuose when old, 7–11 mm long; capsule pyriform, gibbous, asymmetrical with oblique mouth, furrowed when dry, lid conical-mamillate, marginal cells of lid more strongly coloured than but not contrasting with other cells of lid; spores coarsely papillose, 18–28(–30) μm. Fr common, spring. Basic soil in turf and among rocks in open habitats, occasional in W. Britain and rare in the north and in Ireland. From S. Devon, and Dorset north to Arran and Shetland. 21, H2. Europe, Asia Minor, Caucasus, Syria, Morocco, Macaronesia, N. America.

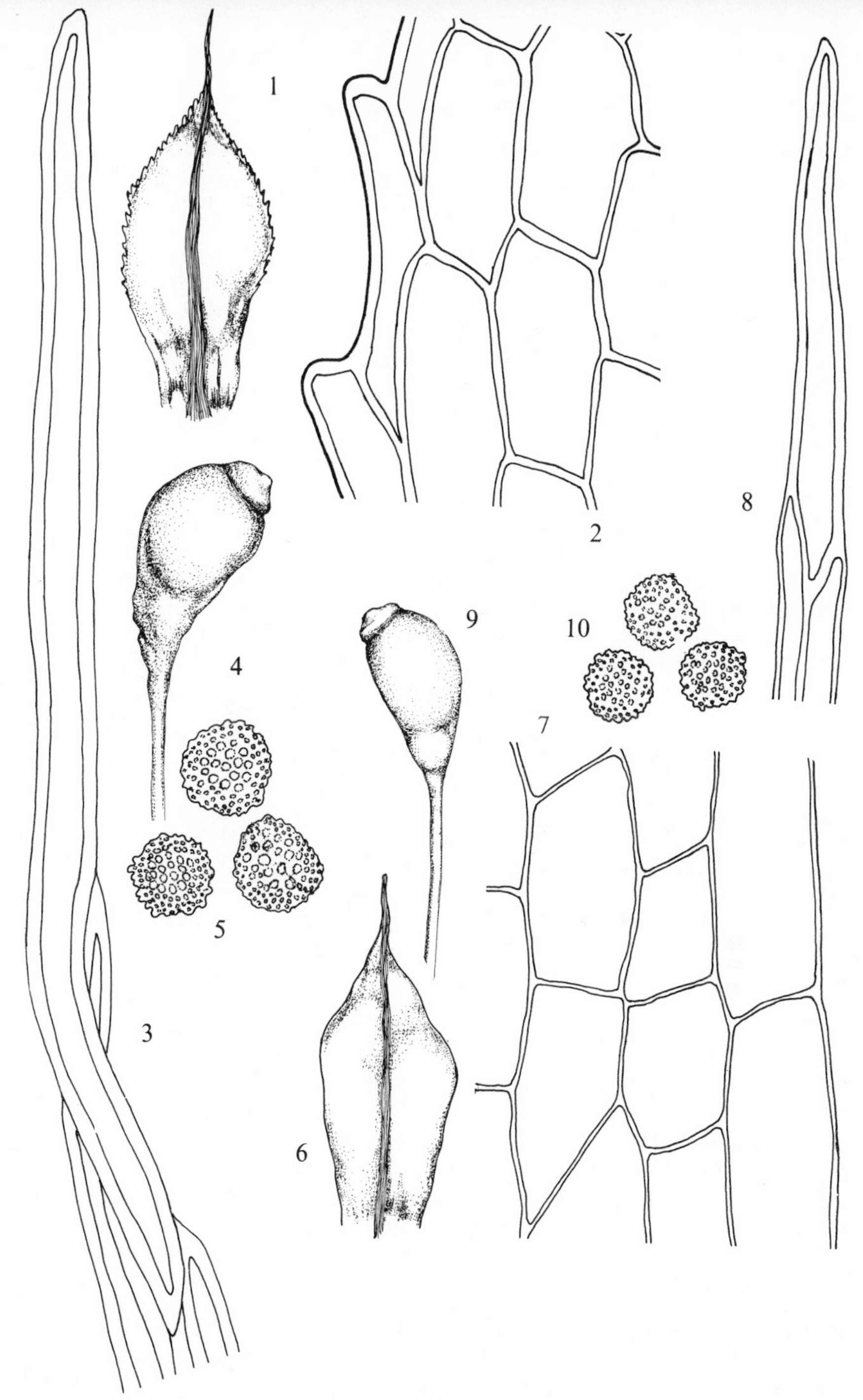

Fig. **161**. 1–5, *Funaria muhlenbergii*: 1, leaf; 2, cells at middle of leaf; 3, leaf apex,; 4 capsule; 5, spores. 6–10, *F. pulchella*: 6, leaf; 7, cells at middle of leaf; 8, leaf apex; 9, capsule; 10, spores. Leaves × 15, cells and spores × 415, capsules × 10.

3. F. pulchella Philib., Rev. Bryol., 1884

Close to *F. muhlenbergii*. Lvs ovate to obovate, ± gradually tapering to acuminate apex, margin entire or slightly sinuose, rarely obscurely toothed above; nerve ending well below apex, rarely percurrent; apical cell of lf to 280 μm long. Seta 5–8 mm long; capsule as in *F. muhlenbergii* but lid with margin of orange-red or red cells contrasting with other cells of lid; spores finely papillose, 20–28 μm. In similar habitats to *F. muhlenbergii* but more southerly, rare. E. Cornwall, S. Devon, N. Somerset, W. Sussex, W. Glos, Glamorgan, Pembroke, Anglesey, W. Lancs. 9. S., W. and C. Europe, Turkey, Pamirs, Egypt, Algeria, Morocco, Tenerife, Azores, Arizona.

Not distinguished previously from *F. muhlenbergii* in Britain but differing in the shorter lf apex, ± entire margin and finely papillose spores. For an account of this plant in Britain see Crundwell & Nyholm, *Lindbergia* **2**, 222–9, 1974.

4. F. attenuata (Dicks.) Lindb., Öfv. K. V. A. Förh., 1865
F. templetonii Sm., *Entosthodon attenuatus* (Dicks.) Bryhn

Plants to 5 mm with deep cerise rhizoids at base. Lvs erecto-patent to spreading, concave, lanceolate to ovate or obovate, acute to acuminate, bluntly toothed; nerve ending in or below apex; basal cells rectangular, cells above shorter, 24–40 μm wide in mid-lf, 1–2 marginal rows narrower but not forming distinct border. Seta straight, 5–15 mm; capsule pyriform, ± symmetrical, neck $\frac{1}{2}$–$\frac{2}{3}$ length of theca; peristome a single row of small fugacious teeth; spores slightly angular, smooth, 24–30 μm. Fr common, spring to autumn. $n=28$*. Damp soil by streams, on rock ledges and cliffs. Occasional in S.W. England and Wales, Cheshire and Yorks northwards, occasional. 54, H22, C. W., C. and S. Europe, Faroes, Corsica, Sardinia, Sicily, Syria, Egypt, Algeria, Morocco, Macaronesia, N. America.

5. F. fascicularis (Hedw.) Lindb., Öfv. K. V. A. Förh., 1865
Entosthodon fascicularis (Hedw.) C. Müll.

Plants to 1 cm. Lvs erecto-patent, lanceolate-spathulate to oblanceolate-spathulate, acuminate, toothed; nerve ending in or below apex; cells ± uniformly rectangular, 24–40 μm wide in mid-lf, 1–2 rows marginal cells narrower but not forming border. Seta straight; capsule pyriform, ± symmetrical; peristome rudimentary or absent; calyptra cucullate, oblique; lid convex, without an apiculus; spores coarsely papillose, 24–28 μm. Fr common, spring, summer. $n=26$*, 27. Patches or scattered plants on usually damp soil in fields, on waste ground, banks and paths. Rare in N. Scotland, occasional and generally distributed elsewhere. 78, H15, C. Europe, Caucasus, Algeria, Morocco.

Very likely to be confused with *Physcomitrium pyriforme*; for differences see under that species.

6. F. obtusa (Hedw.) Lindb., Nat. Sällsk. F. Fl. Fenn. Förh., 1870
F. ericetorum (Bals. & De Not.) Dix., *Entosthodon obtusus* (Hedw.) Lindb.

Plants to 5 mm with pale brown rhizoids at base of stem. Lvs lanceolate-spathulate to lanceolate, acuminate, ± entire to bluntly toothed; nerve ending below apex; basal cells rectangular, upper shortly rectangular to hexagonal, 2–3 marginal rows very narrow, incrassate, yellowish, forming distinct border. Seta straight, to 6 mm; capsule ovoid, symmetrical; peristome rudimentary or absent; spores papillose, 30–38 μm. Fr common, winter to spring. $n=26$*, 28*. Patches or scattered stems on damp, usually peaty soil on heaths, moorland, by streams, ditches, etc. Rare in S.E. England, occasional to frequent elsewhere. 83, H26, C. N., W. and C. Europe, Faroes, Iceland, Algeria, Tunisia.

The shorter capsule, larger spores, bordered lvs and brownish rhizoids distinguish this plant from *F. attenuata*.

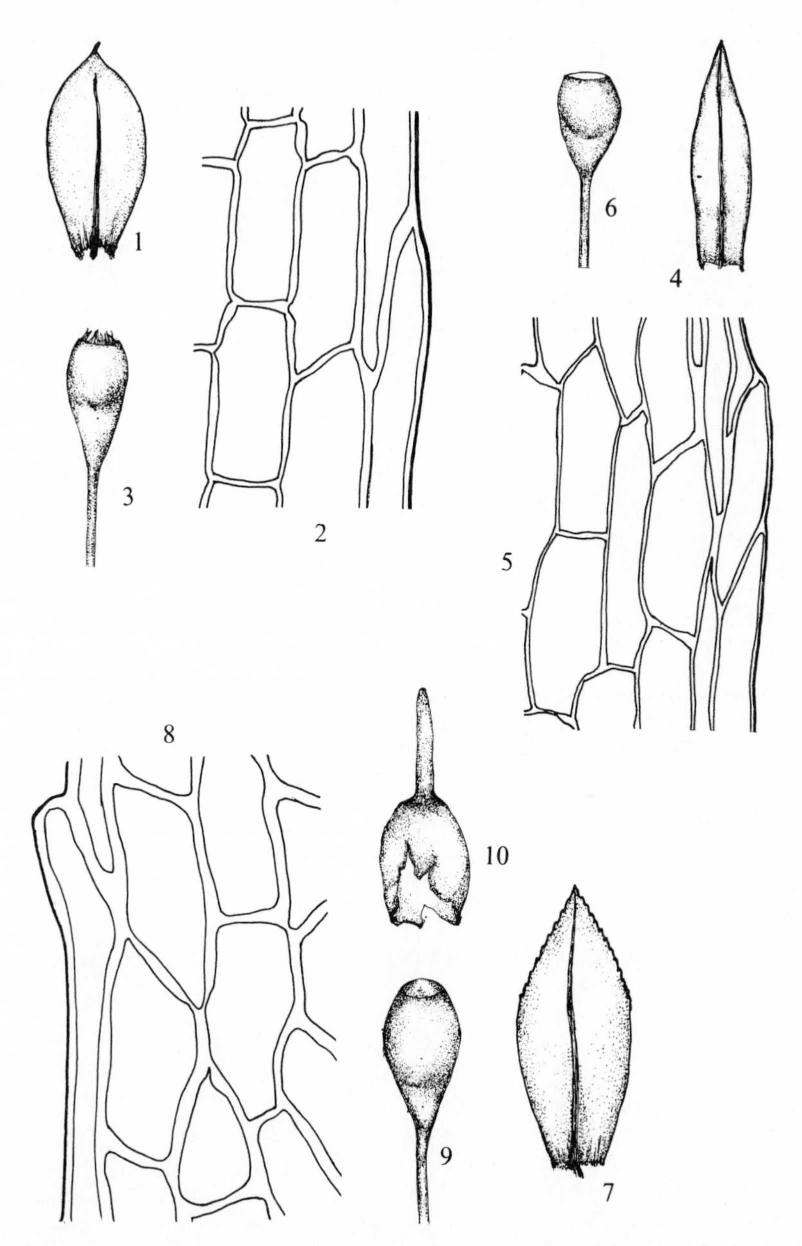

Fig. **162**. 1–3, *Funaria attenuata*: 1, leaf; 2, cells from upper part of leaf; 3, capsule. 4–6, *F. obtusa*: 4, leaf; 5, cells from upper part of leaf; 6, capsule. 7–10, *F. fascicularis*: 7, leaf; 8, cells from upper part of leaf; 9, capsule; 10, calyptra (×10). Leaves ×15, cells ×450, capsules ×10.

72. Physcomitrium (Brid.) Brid., Br. Univ., 1827

Autoecious. Similar to erect-fruited species of *Funaria*. Capsule erect, symmetrical, gymnostomous; lid apiculate to rostellate; calyptra symmetrical, mitriform. A cosmopolitan genus of about 100 species.

1 Capsule ± pyriform, about twice as long as wide, not hemispherical when empty, seta 5–15 mm long, lf margin toothed above **1. P. pyriforme**
Capsule spherical or turbinate, about as long as wide, ± hemispherical when empty, seta 1–5 mm long, lf margin entire or toothed 2
2 Lvs distinctly toothed above, seta 2–5 mm, spores 30–40 μm **2. P. eurystomum**
Lvs ± entire, seta 1–2 mm, spores 24–32 μm **3. P. sphaericum**

1. P. pyriforme (Hedw.) Brid., Br. Univ., 1827

Plants to 5 mm, rarely more. Lvs erecto-patent to spreading, concave, variable in shape, ovate to lancelolate, oblanceolate or spathulate, acute, margin toothed above; nerve ending in apex; basal cells narrowly rectangular to rectangular, cells above irregularly rectangular to hexagonal, 20–50 μm wide in mid-lf, marginal cells narrower. Seta 5–15 mm; capsule sub-globose to pyriform, narrowed at mouth, with short distinct neck; calyptra mitriform; lid variable, convex and apiculate to rostellate; spores papillose, 28–36 μm. Fr common, late winter to summer. n=9, 18, 26*, 27, 36, 52, 54, 72. Patches or scattered plants on damp soil in fields, waste places, by streams, ditches, pools, and on soil from cleared ditches, streams, etc., frequent except in the north of Scotland. 100, H23, C. Europe, Caucasus, Algeria, Morocco, Macaronesia, Australia.

The only reliable characters distinguishing *P. pyriforme* from *Funaria fascicularis* are the symmetrical mitriform calyptra and the apiculate to rostellate lid of the capsule. Also, in *P. pyriforme* the cells towards the middle of the lid are about the same size as those at the edge whilst in *F. fascicularis* the inner cells are about ½ the size of the outer.

2. P. eurystomum Sendtn., Denkschr., 1841

Plants to 3 mm. Lvs ovate to ovate-lanceolate, acuminate, bluntly toothed above; nerve ending in apex; areolation similar to that of *P. sphaericum* but cells slightly larger, 20–35 μm wide in mid-lf and 2–3 rows marginal cells narrower. Seta 2–5 mm; capsule turbinate with distinct neck; lid mamillate; spores 30–40 μm. Fr common, autumn, winter. On basic soil in dried-out ponds and reservoirs, very rare. Herts, W. Norfolk. 2. W. and C. Europe, Japan, China, Vietnam, Formosa, N. India, Sikkim.

For the occurrence of this plant in Britain see Ducker & Warburg, *Trans. Br. Bryol. Soc.* 4, 95–7, 1961.

3. P. sphaericum (Hedw.) Brid., Br. Univ., 1827

Plants to 4 mm, usually less. Lvs concave, ovate to ovate-lanceolate, acute to obtuse, entire or obscurely crenulate above; nerve ending below apex; lower cells rectangular, upper quadrate to hexagonal, 12–30 μm wide in mid-lf, marginal cells scarcely narrower. Seta 1–2 mm; capsule spherical or occasionally turbinate; lid mamillate; spores 24–32 μm. Fr common, autumn, winter. On mud of dried-out ponds and reservoirs, very rare and sporadic in appearance. E. Sussex, Surrey, Stafford, Derby, Cheshire, Yorks, Renfrew, Fife. 9. N., W. and C. Europe, Siberia, Amur, Japan, Korea, Formosa, China, C. India.

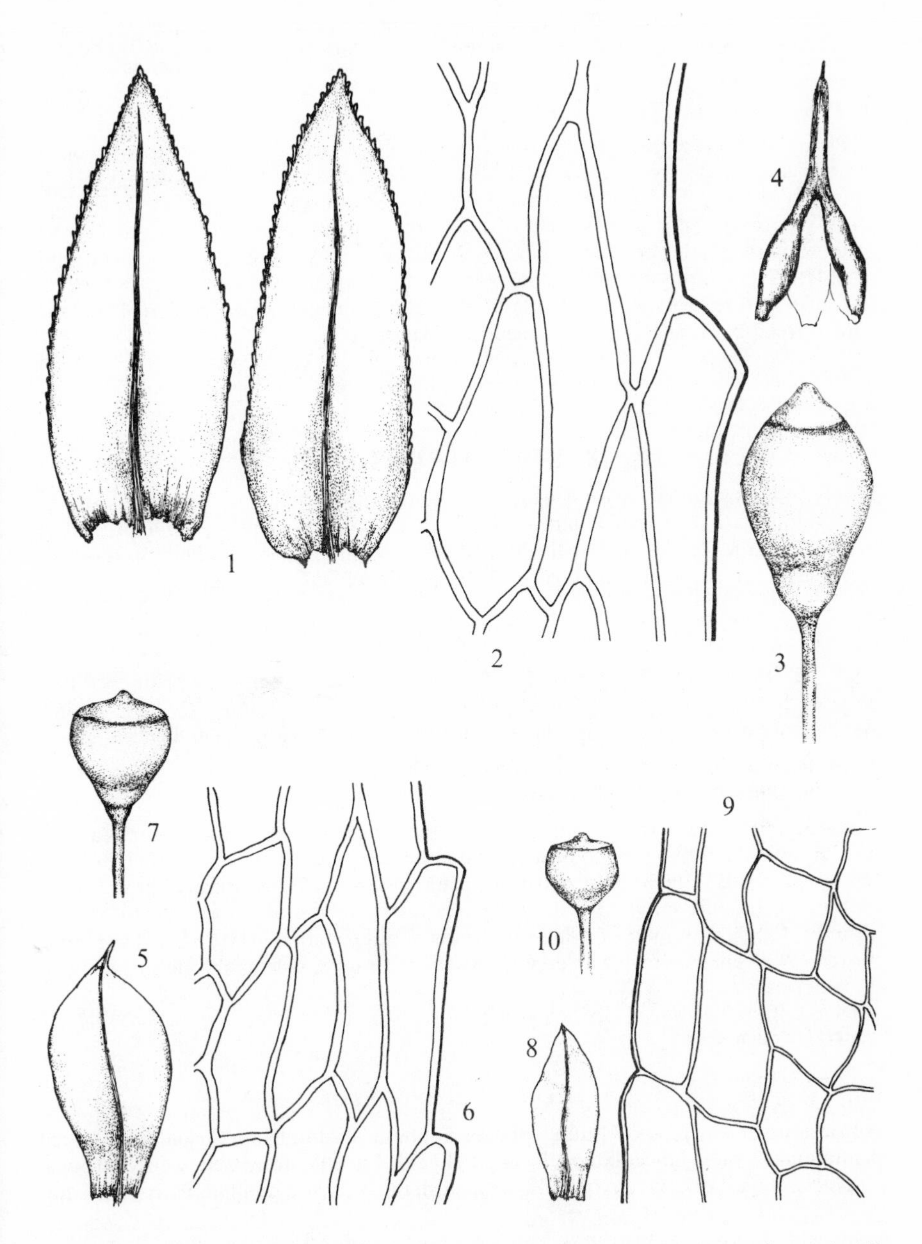

Fig. **163**. 1–4, *Physcomitrium pyriforme*: 1, leaves; 2, leaf cells; 3, capsule; 4, calyptra (× 10) 5–7, *P. eurystomum*: 5, leaf; 6, mid-leaf cells; 7, capsule. 8–10, *P. sphaericum*: 8, leaf; 9 mid-leaf cells; 10, capsule. Leaves × 15, cells × 450, capsules × 10.

73. Physcomitrella Br. Eur., 1849

Paroecious or synoecious. Plants minute, ephemeral, protonema not persistent. Seta very short; capsule cleistocarpous, globose, apiculate. Europe, Asia, N. America, Australia; 5 species.

1. P. patens (Hedw.) Br. Eur., 1849

Plants often gregarious, to 2.5 mm. Lvs erect or erecto-patent, ovate to lanceolate, shortly acuminate, bluntly toothed above; nerve ending in or below apex; cells narrowly rectangular below, shorter above, 15–20 μm wide in mid-lf, 1–2 marginal rows narrower. Seta very short, *ca* 100 μm; capsule immersed, cleistocarpous, globose with short blunt beak; spores 26–32 μm. Fr common, summer, autumn. $n = 14$*, 27. On damp soil or mud in fields, by streams, ditches, ponds, etc. Occasional in England, rare in Wales and Scotland, extending north to Selkirk and the Lothians, rare in Ireland. 61, H7. Europe, Yenesei, N. America.

Physcomitrella patens (♀) × *Physcomitrium pyriforme* (♂) has been recorded from Notts and *P. patens* (♀) × *Physcomitrium sphaericum* (♂) from Ches. (see Pettet, *Trans. Br. bryol. Soc.* **4**, 642–8, 1964).

18. EPHEMERACEAE

Plants minute, annual, arising from persistent protonema. Lvs forming a rosette, ovate-lanceolate to narrowly lanceolate, margin toothed or entire; nerve present or not; cells large, lax, ± narrowly hexagonal. Capsule immersed, cleistocarpous or stegocarpous and gymnostomous, ± globose, thin-walled; spores large.

74. Micromitrium Aust., Musci Appal., 1870

Close to *Ephemerum*. Capsule wall of 1 layer only, without stomata; ring of differentiated cells delimiting rudimentary lid present; calyptra minute. About 10 mainly American species with a few species also in Europe, Asia, Australia and Oceania.

1. M. tenerum (Bruch) Crosby, Bryologist, 1968
Nanomitrium tenerum (Bruch) Lindb.

Autoecious. Lower lvs spreading, upper erecto-patent, ovate-lanceolate to lanceolate, acuminate, sub-entire to obscurely denticulate; nerveless; cells lax, rhomboid-hexagonal, 20–25 μm wide in mid-lf. Capsule globose, cleistocarpous but dehiscing along ring of differentiated cells under pressure; spores papillose, 20–30 μm. Fr autumn. Patches on mud of dried-out ponds and reservoirs, very rare and sporadic. Sussex, W. Kent, Surrey, Anglesey. 5. Sweden, France, Germany, Silesia, E. Asia, N. America.

Differs from *Ephemerum* species in the capsule without stomata or apiculus and the scarcely denticulate lvs.

75. Ephemerum Hampe, Flora, 1837

Autoecious or dioecious. Minute ephemeral plants arising from a usually persistent protonema. Lvs ± lanceolate; cells lax, pellucid. Capsule immersed, cleistocarpous, ± globose, wall of 2 layers of cells, with stomata; calyptra campanulate. A cosmopolitan genus of about 27 species.

1 Lvs nerveless 2
Nerve present at least in upper part of lf 3

2 Lf margin entire or denticulate, cells in mid-lf 50–90 μm long, spores 40–50 μm **4. E. stellatum**
Lvs coarsely toothed, mid-lf cells 100–160 μm long, spores 40–70 μm **5. E. serratum**
3 Nerve very faint, in upper part of lf only **5. E. serratum**
Nerve present in both lower and upper parts of lf and strong in upper part 4
4 Upper lvs lanceolate, nerve ending in apex, mid-lf cells 50–80 μm long **3. E. cohaerens**
Upper lvs narrowly lanceolate to linear-lanceolate, nerve excurrent, mid-lf cells 10–50 μm long 5
5 Upper lvs recurved, capsule with oblique apiculus, stomata near base of capsule only **1. E. recurvifolium**
Lvs erecto-patent, capsule with straight apiculus, stomata scattered over whole of capsule **2. E. sessile**

1. E. recurvifolium (Dicks.) Boul., Musc. de l'Est, 1872
Ephemerella recurvifolia (Dicks.) Schimp.

Dioecious. Lvs erecto-patent, recurved above, narrowly lanceolate, acuminate, denticulate, sometimes with 1–2 coarse teeth near apex; nerve well defined, excurrent; cells in mid-lf 8–14(–20) μm × 10–40 μm. Capsule ± globose with oblique apiculus, stomata at base only; calyptra cucullate; spores ± smooth, 40–55 μm. Fr autumn, winter. Basic, clayey soil in fields, grassland, etc., occasional. Scattered localities from Devon and Somerset north to Cheshire and Durham. 28. Sweden, Finland, W. and C. Europe, Italy, Sardinia, N. Africa.

2. E. sessile (Br. Eur.) C. Müll., Syn., 1848
Ephemerella sessilis (Br. Eur.) Nyholm

Autoecious. Lvs erecto-patent, upper narrowly lanceolate, denticulate; nerve weak below, strong above, excurrent; cells in mid-lf 10–12 μm × 35–50 μm. Capsule ± globose with short, straight, apiculus, stomata scattered over whole surface; spores coarsely papillose, 60–80 μm long. Fr autumn, winter. Scattered plants or small patches on damp, compressed soil, rare. S. England, Cheshire, S. Northumberland, W. Galway, W. Donegal. 13, H2. Europe, Corsica, Morocco.

3. E. cohaerens (Hedw.) Hampe, Flora, 1837

Dioecious. Lvs patent or recurved, lanceolate, acuminate, toothed; in upper lvs nerve weak in lower part of lf, stronger above, ending in apex; cells in mid-lf 10–20 μm × 50–80 μm. Capsule ± globose, stomata scattered over whole surface, apiculus short, straight; spores coarsely papillose, 60–70 μm long. Fr autumn. $n = 27$. Scattered plants or small patches on fine, damp soil by ponds and lakes, very rare. W. Sussex, Herts, Leics, S.E. Galway. 3, H1. Germany, C. Europe, N. America.

4. E. stellatum Philib., Rev. Bryol., 1879

Autoecious or dioecious. Lower lvs spreading in stellate fashion, narrowly lanceolate, acute, upper ± straight, ligulate-lanceolate, acute to acuminate, margin entire to slightly denticulate; nerve absent; cells in mid-lf 10–16 μm × 50–90 μm, incrassate, basal larger and shorter, thin-walled. Capsule similar to but narrower than that of *E. serratum*; calyptra large, covering more than ½ the capsule; spores 40–50 μm. On compressed soil in open habitats, very rare. E. Sussex, W. Kent, S. Kerry. 2, H1. France, N. Africa.

Differs from *E. serratum* in the wider lf apex, smaller cells, larger calyptra and smaller

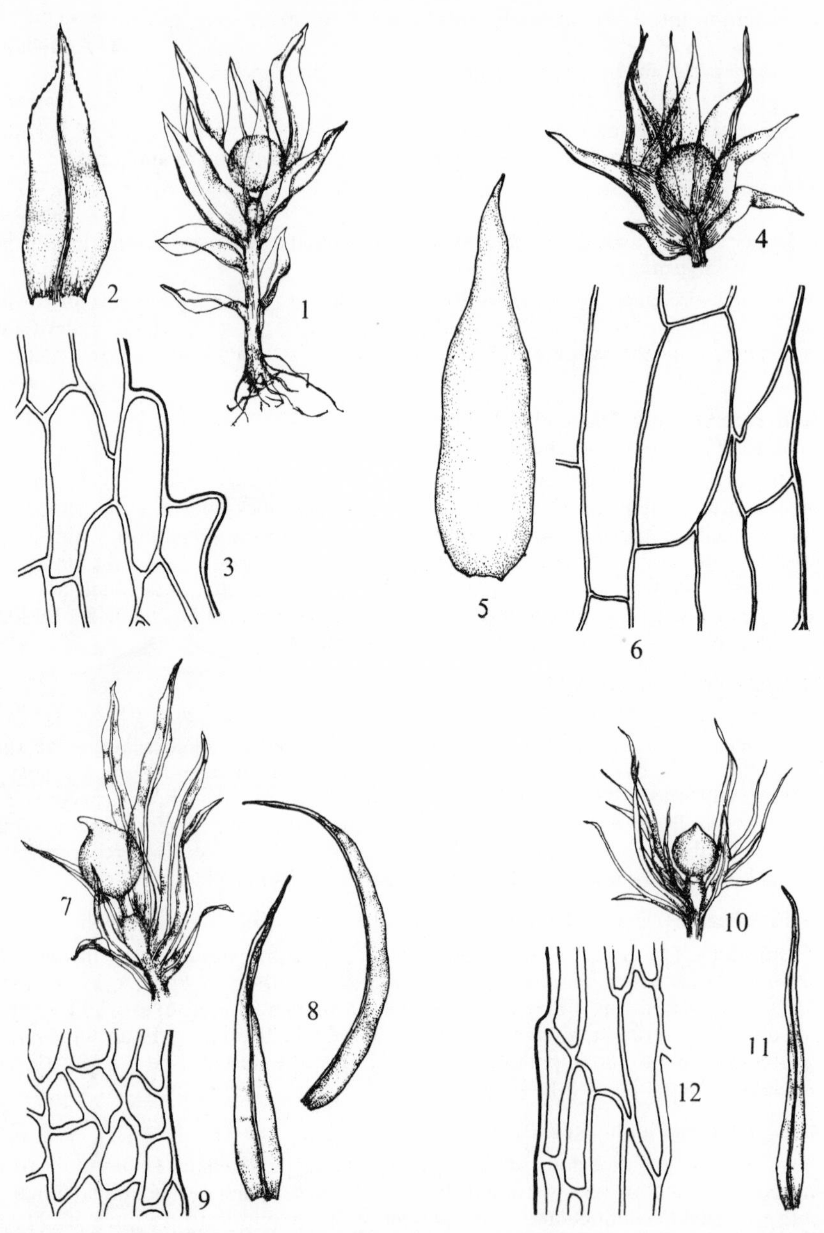

Fig. **164**. 1–3, *Physcomitrella patens*: 1, plant (×10); 2, leaf (×15); 3, cells from upper part of leaf. 4–6, *Micromitrium tenerum*: 4 plant (×30); 5, leaf (×60); 6, cells from upper part of leaf. 7–9, *Ephemerum recurvifolium*: 7, plant (×20); 8, leaves (×30); 9, cells from upper part of leaf. 10–12, *E. sessile*: 10, plant (×20); 11, leaf (×30); 12, cells from upper part of leaf. Cells ×415.

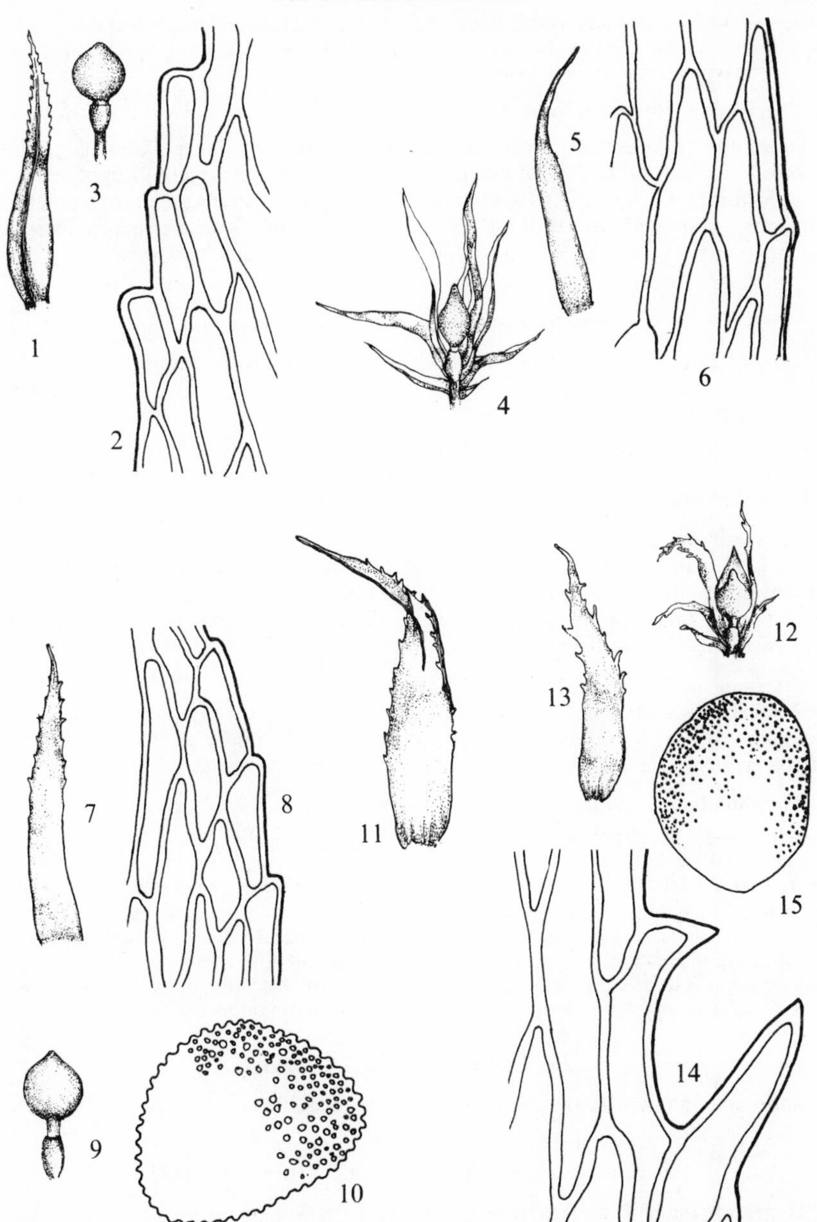

Fig. **165**. 1–3, *Ephemerum cohaerens*: 1, leaf; 2, cells from upper part of leaf; 3, capsule. 4–6, *E. stellatum*: 4, plant with young capsule; 5, leaf; 6, cells from near leaf apex. 7–10, *E. serratum* var. *serratum*: 7, leaf; 8, cells from upper part of leaf; 9, capsule; 10, spore. 11, *E. serratum* var. *praecox*: leaf. 12–15, *E. serratum* var. *minutissimum*: 12, plant; 13, leaf; 14, cells from upper part of leaf; 15, spore. Plants ×20, leaves and capsules ×30, cells and spores ×415.

spores. From *Micromitrium tenerum* it differs in the larger lf cells and spores and the apiculate capsule with stomata. Dixon (1924) says that there are 1 or more large oil bodies in the basal cells of the lvs but these are lost on drying.

5. E. serratum (Hedw.) Hampe, Flora, 1837

Dioecious. Lvs erecto-patent, lanceolate to narrowly lanceolate, apex often twisted, margin toothed; nerve absent or rarely faint nerve present in upper part of lf; cells in mid-lf 12–30 μm × 100–160 μm. Capsule with conical apiculus, stomata at base only; spores 40–70 μm long. Fr autumn to spring. Scattered plants or small patches on damp soil in fields, on wood rides, paths, etc. in lowland areas.

Key to intraspecific taxa of *E. serratum*

1 Lvs with faint nerve in upper part, cells in upper part 10–16 μm wide — var. **praecox**
Lvs completely nerveless, cells 16–24 μm wide above — 2
2 Lvs narrowly lanceolate, spores finely papillose and usually surrounded by hyaline membrane — var. **minutissimum**
Lvs lanceolate, spores coarsely papillose, membrane lacking — var. **serratum**

Var. **serratum**

Lvs lanceolate, teeth not recurved; nerve lacking. Spores coarsely papillose, without enveloping hyaline membrane, 40–70 μm long. Occasional in lowland areas and less common than var. *minutissimum*, generally distributed. 41, H8. N., W. and C. Europe, Morocco, southern Africa, N. America.

Var. **praecox** Walth. & Mol., Laubm. Oberfr., 1868
E. intermedium Mitt. & Braithw.
Similar to var. *serratum* but faint nerve present in upper part of lf and cells narrower. Very rare and not seen recently, W. Sussex. 1. Bavaria.

Var. **minutissimum** (Lindb.) Grout, Moss Fl. N. Amer., 1935
E. serratum var. *angustifolium* auct., *E. minutissimum* Lindb.
Lvs narrowly lanceolate, teeth recurved; nerve lacking. Spores finely papillose, often surrounded by hyaline membrane, 50–65 μm long. Fr autumn to spring. Occasional to frequent in lowland areas. 68, H14. N. and W. Europe, N. America.

Although var. *serratum* and var. *minutissimum* are usually distinct, occasional intermediates occur and Bryan & Anderson (*Bryologist* **60**, 67–102, 1957) are correct in considering the taxa to be varieties of one species. Var. *praecox* has been found in a few localities in W. Sussex, usually with the type, and its status is open to question.

19. OEDIPODIACEAE

A monotypic family with the characters of the species.

76. Oedipodium Schwaegr., Suppl. II, 1823

1. O. griffithianum (Dicks.) Schwaegr., Suppl. II, 1823

Autoecious or synoecious. Plants to 1 cm. Lvs soft, succulent, crowded, flaccid and shrivelled when dry, spreading to erecto-patent when moist, obovate-spathulate to ± orbicular with long narrow ciliate base, apex rounded, margin entire; nerve ending below apex; cells hexagonal, ± collenchymatous, 50–100 μm wide in mid-lf. Discoid or oval, stalked, multicellular gemmae, to 300 μm long, often present in lf

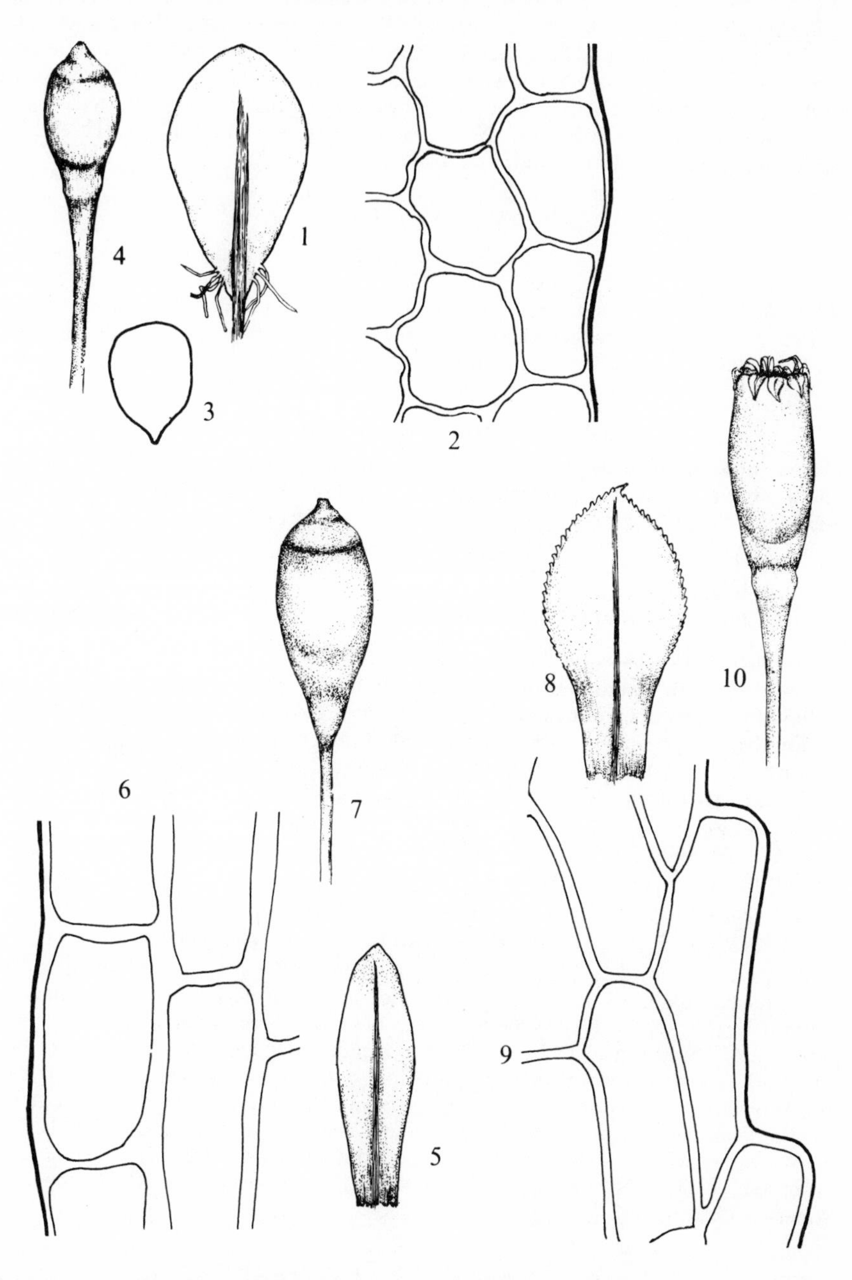

Fig. **166**. 1–4, *Oedipodium griffithianum*: 1, leaf; 2, mid-leaf cells; 3, axillary gemma (×115); 4, capsule. 5–7, *Tayloria lingulata*: 5, leaf; 6, mid-leaf cells; 7, capsule. 8–10, *T. longicollis*: 8, leaf; 9, mid-leaf cells; 10, capsule. Leaves and capsules ×10, cells ×415.

axils. Capsule erect, shortly ellipsoid, brown, with very long, yellowish apophysis with numerous narrow stomata, tapering into seta, apophysis and seta succulent; peristome absent; columella expanded above but not exserted after dehiscence; spores 24–30 μm. Fr occasional, summer. Green tufts or scattered plants on peaty soil in shaded rock crevices and boulder scree in montane areas, occasional. Wales, Yorks and Lake District north to W. Ross, S. Kerry, W. Donegal. 29, H2. Scandinavia, Japan, Alaska, Greenland, Falkland Is.

20. SPLACHNACEAE

Lvs soft, lanceolate to ± orbicular, margin entire or toothed; nerve thin; cells rectangular to hexagonal, lax, smooth. Seta long; capsule erect, theca of capsule ellipsoid to shortly cylindrical, apophysis similar to theca to greatly enlarged; peristome single; columella sometimes extending at maturity; spores often dispersed by flies. Plants frequently growing on decaying plant or animal matter.

77. Tayloria Hook., Jour. Sc. Arts., 1816

Autoecious or synoecious. Lvs soft, ovate-spathulate to lingulate, acute to obtuse, margin entire or toothed; nerve ending in or below apex; cells lax, rectangular to hexagonal. Seta thin, flexuose; capsule with pyriform apophysis narrower than theca; calyptra constricted at base. A ± cosmopolitan genus of about 65 species.

Lvs lingulate, obtuse, ± entire **1. T. lingulata**
Lvs broadly spathulate, acute, toothed **2. T. longicollis**

1. T. lingulata (Dicks.) Lindb., Musc. Scan., 1879

Plants to 5 cm. Lvs somewhat flexuose when dry, ± erect when moist, lingulate to ovate-lanceolate, apex obtuse to rounded, entire or obscurely toothed above; nerve ending below apex; cells ± rectangular, 30–50 μm wide in mid-lf. Seta bright red, slender, flexuose, to 3 cm; theca ovoid, wide-mouthed when empty, tapering into apophysis; peristome teeth erect when dry; spores 20–30 μm. Fr frequent, summer. Dense tufts, green above, blackish below, or scattered plants among other bryophytes in basic, montane flushes, very rare. Stirling, Perth, S. Aberdeen, Angus, Argyll, W. Sutherland. 8. Europe, N. Asia, N. America.

2. T. longicollis (Dicks.) Dix., Br. bryol. Soc. Rep., 1939
T. tenuis (With.) Schimp., *T. serrata* (Hedw.) Br. Eur. var. *tenuis* (With.) Br. Eur.

Plants to 2 cm. Lvs slightly shrunken when dry, erecto-patent when moist, obovate-spathulate to lanceolate-spathulate, acute or with acuminate apiculus, margin serrate; nerve ending below apex; basal cells rectangular, cells above ± hexagonal, 20–40 μm wide in mid-lf. Seta slender, flexuose, deep red, to 2.5 cm; theca of capsule ellipsoid, tapering into long narrow apophysis; peristome teeth reflexed when dry; spores *ca* 12 μm. Fr frequent, summer. On damp, decaying vegetable matter and soil in montane habitats. Very rare and recorded recently only from Caithness, but there are old records from Durham, Perth, S. Aberdeen, E. Inverness, Angus, N. Ebudes, W. Ross and Londonderry. 10, H1. Europe, N. America, Greenland.

78. Tetraplodon Br. Eur., 1844

Autoecious. Lvs obovate to lanceolate, acuminate, entire or toothed; cells lax, usually thin-walled. Capsule ± erect, apophysis of similar colour to and narrower to

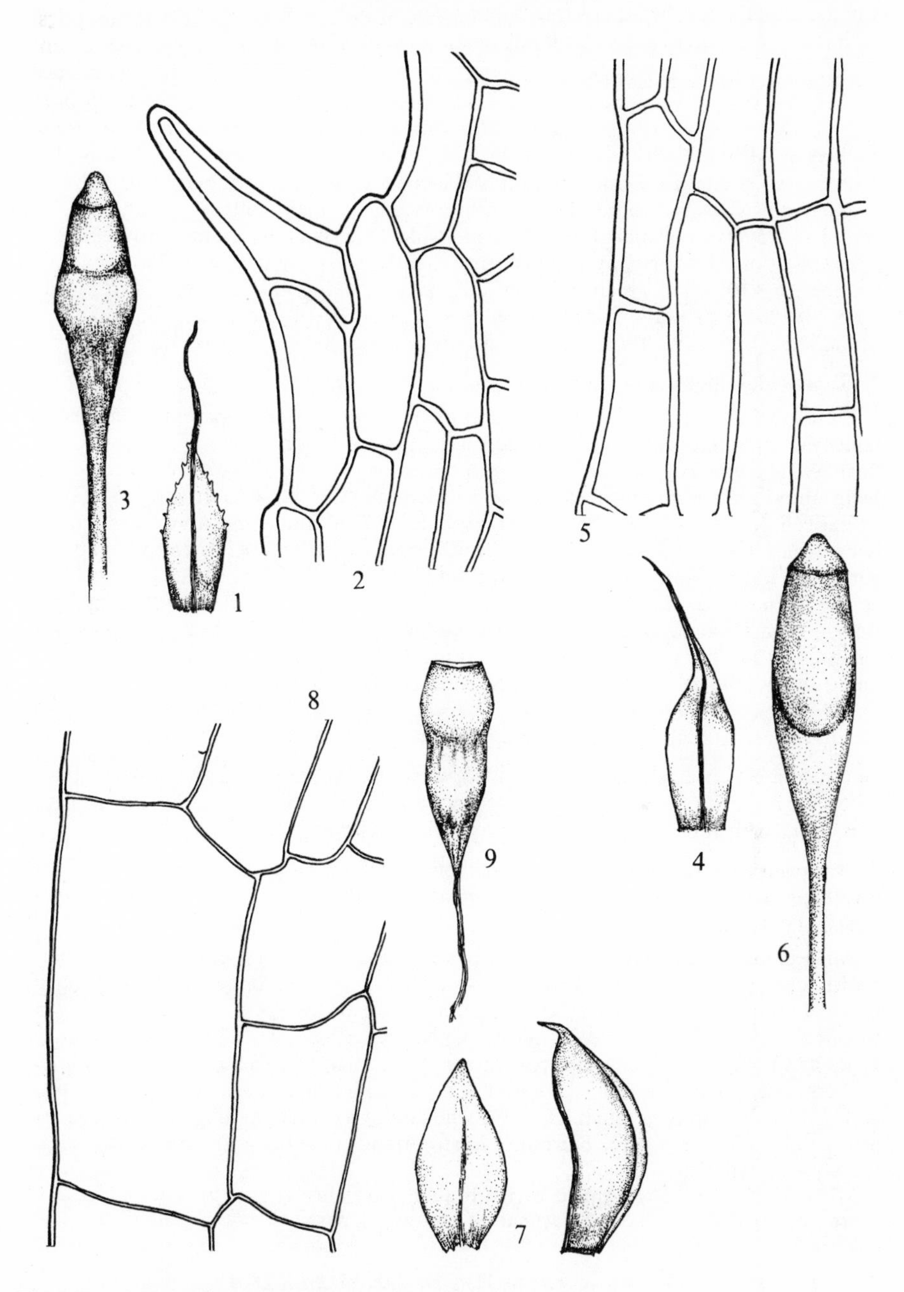

Fig. **167**. 1–3, *Tetraplodon angustatus*: 1, leaf; 2, mid-leaf cells; 3, capsule. 4–6, *T. mnioides*: 4, leaf; 5, mid-leaf cells; 6, capsule. 7–9, *Aplodon wormskjoldii*: 7, leaves; 8, mid-leaf cells; 9, capsule. Leaves and capsules ×10, cells ×450.

slightly wider than theca; columella not exserted after dehiscence; peristome teeth at first in fours then in pairs. A mainly northern hemisphere genus of 13 species.

Lvs ± entire, seta (7–)10–30 mm **2. T. mnioides**
Lvs toothed above, seta to 5 mm **1. T. angustatus**

1. T. angustatus (Hedw.) Br. Eur., 1844

Plants to 6 cm. Lvs lanceolate or oblanceolate, gradually tapering to subula, sparsely and sharply toothed above; nerve ending in apex; cells ± rectangular, in mid-lf 15–25 μm wide, apical cell 12–16 μm wide. Seta short, to 5 mm; capsule barely exserted above lvs, apophysis pyriform, slightly wider than theca. Fr common, summer. $n=10$. Light green tufts on dung and decaying animal remains in moist, open, montane habitats, rare. Caerns, S.E. Yorks, Perth and Angus north to W. Sutherland. 13. N., W. and C. Europe, Siberia, China, Japan, N. America.

2. T. mnioides (Hedw.) Br. Eur., 1844

Plants to 6 cm but usually less. Lvs shrunken when dry, erecto-patent when moist, concave, ovate-lanceolate to oblanceolate, gradually or abruptly narrowed to flexuose, sometimes long subula, margin entire or occasionally with a few obscure teeth above; nerve ending in apex; cells ± rectangular, 15–25 μm wide in mid-lf, apical cell 16–20 μm wide. Seta thick, succulent, 7–30 mm long; apophysis hardly wider than theca, the whole narrowly ellipsoid when moist, when dry capsule shrunken, apophysis pyriform, theca and apophysis of ± similar colour. Fr common, spring, summer. $n=8$, 11*, 19. Light green tufts on dung and decaying animal remains, occasional to frequent on open, boggy ground. Devon, Surrey, Wales, N. Lincs and Cheshire northwards, scattered localities in Ireland. 61, H18. Europe, N. and C. Asia, Sakhalin, China, Japan, N. America, Greenland, Australia.

79. Aplodon R. Br., Verm. Schr., 1825

Intermediate between *Tetraplodon* and *Splachnum*. Apophysis scarcely wider than theca at maturity; seta thin, hyaline; columella not exserted after dehiscence; peristome teeth at first in pairs then separate. One species.

1. A. wormskjoldii (Horn.) R. Br., Verm. Schr., 1825
Haplodon wormskjoldii auct., *Splachnum wormskjoldii* Horn., *Tetraplodon wormskjoldii* (Horn.) Lindb.

Autoecious. Plants slender, to 6 cm. Lvs shrunken when dry, erecto-patent when moist, concave, broadly ovate or obovate, lower obtuse, upper with frequently reflexed obtuse acumen, margin entire; nerve ending below apex; cells ± rectangular throughout, 20–40 μm wide in mid-lf, apical cell 30–50 μm × 30–80 μm, 1–2(–3) times as long as wide, marginal row larger than other cells. Seta flexuose, hyaline, to 2 cm; theca ovoid, apophysis obloid, scarcely wider than theca at maturity, theca turbinate and apophysis much narrower when old. Lax tufts on dung in boggy places at high altitudes, very rare. Durham, Westmorland, Cumberland, Mid Perth. 4. A mainly arctic species.

When fertile may be recognised by the hyaline seta; when sterile the much wider and shorter apical cell than other Splachnaceae and wider marginal row of cells of the lf is characteristic.

80. Splachnum Hedw., Sp. Musc., 1801

Gametophyte similar to that of *Tetraplodon* but lvs usually wider. Seta thin, flexuose, theca ovoid to shortly cylindrical, apophysis usually wider than theca, ovoid to

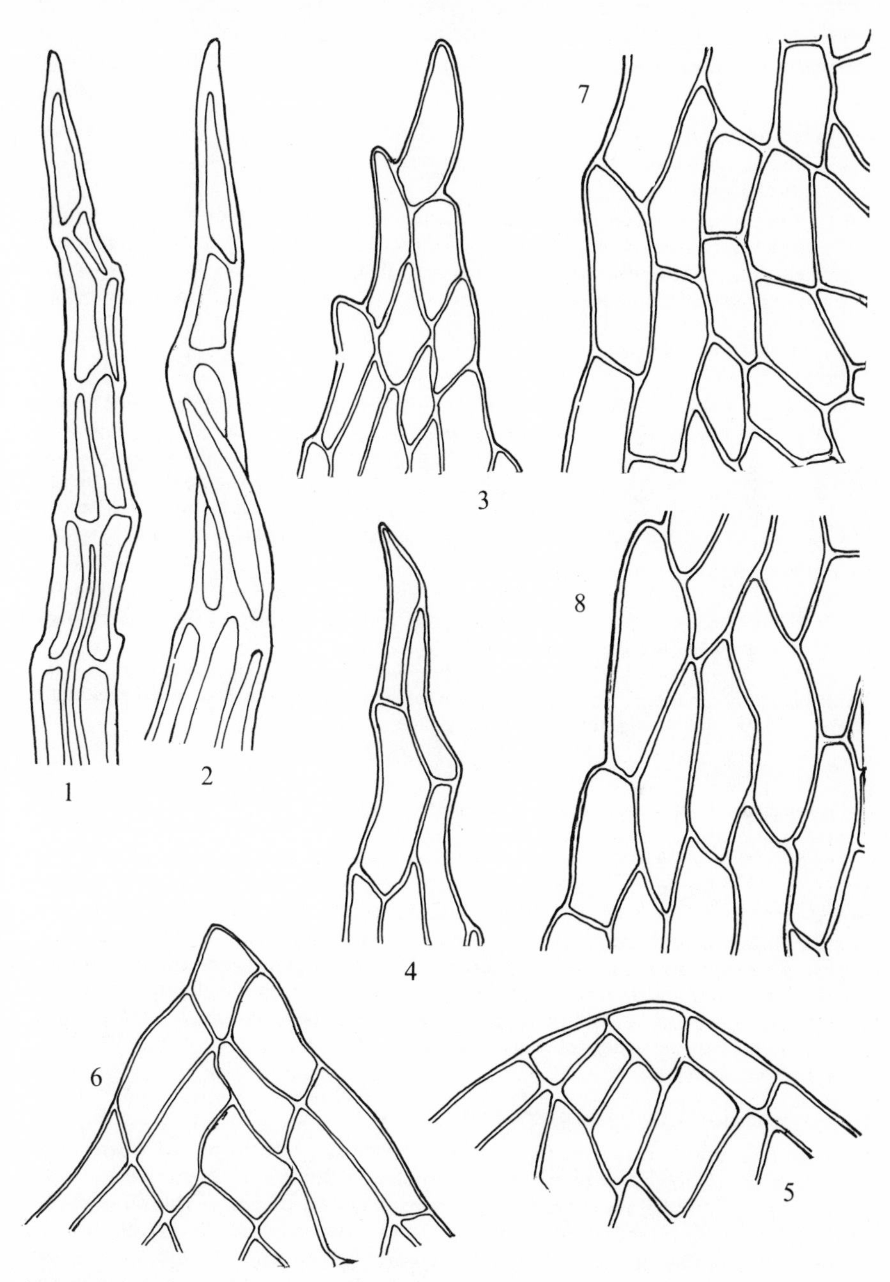

Fig. **168**. 1–6, leaf apices: 1, *Tetraplodon angustatus*; 2, *T. mnioides*; 3, *Splachnum sphaericum*; 4, *S. ampullaceum*; 5, *S. vasculosum*; 6, *Aplodon wormskjoldii*. 7–8, cells near leaf apex: 7, *Splachnum sphaericum*; 8, *S. ampullaceum*. All $\times 250$.

umbrella-shaped, of different colour and texture from theca; columella exserted and often inflated at apex after dehiscence; peristome teeth in pairs, reflexed when dry. A mainly northern hemisphere genus of 10 coprophilous species.

1 Lvs broadly ovate, lower obtuse, margin ± entire **3. S. vasculosum**
Lvs obovate or ovate to oblanceolate, lower acute to acuminate, margin entire or toothed 2
2 Lvs obscurely to strongly toothed, cells near lf apex 2–4 times as long as wide, apophysis much wider than theca of capsule **2. S. ampullaceum**
Lvs entire to obscurely toothed, cells near lf apex 1–2 times as long as wide, apophysis of similar diameter to theca **1. S. sphaericum**

1. S. sphaericum Hedw., Sp. Musc., 1801
S. ovatum Hedw.

Dioecious. Plants 1(–3) cm. Lvs shrunken when dry, erecto-patent when moist, obovate or ovate to lanceolate with narrow base, abruptly acuminate, entire or obscurely toothed, nerve ending in or below apex; cells in mid-lf rhomboid-hexagonal, 20–30 μm wide, cells near apex 1–2 times as long as wide, cells below apex 1–2 times as long as wide. Seta thin, flexuose, weak, reddish below, yellow above, to 4(–14) cm; theca of capsule obloid, apophysis deep purple, obloid, slightly wider to narrower than theca, rugose when dry; spores 8–12 μm. Fr common, spring to autumn. $n=9$. Light green tufts on dung on wet heaths, moorland and in bogs, occasional to frequent in western and northern Britain. Devon, Wales, Derby and Yorks northwards. 62, H20. Europe, N. Asia, N. America.

When sterile may be distinguished from *Tetraplodon* species by lf shape, from *Aplodon wormskjoldii* by the longer, narrower apical cell of the lf, and from *S. ampullaceum* by the shorter cells near lf apex and the ± entire margin.

2. S. ampullaceum Hedw., Sp. Musc., 1801

Autoecious. Plants to 3 cm. Lvs shrunken when dry, erecto-patent when moist, concave, obovate-lanceolate to oblanceolate, tapering to short to long acumen, base narrow, margin obscurely to spinosely toothed; nerve ending in or below apex; cells in mid-lf ± hexagonal, *ca* 25 μm wide, cells near apex 2–4 times as long as wide, apical cell 16–20 μm wide. Seta thin, flexuose, weak, deep red, to 6 cm; theca of capsule shortly cylindrical, apophysis pyriform, pale purple, much wider than theca, rugose when dry. Fr common, summer. $n = 10*$. Light green tufts on dung on wet heaths, moorland and in bogs. Rare in S.E. England and the Midlands, occasional to frequent elsewhere. 80, H27. Europe, Caucasus, Celebes, N. Asia, N. America.

When sterile may be distinguished from *Tetraplodon* species by the less attenuated lf apex with relatively shorter cells.

3. S. vasculosum Hedw., Sp. Musc., 1801

Dioecious. Plants to 7 cm. Lvs shrunken when dry, erecto-patent when moist, broadly ovate with narrow base, lower lvs obtuse, upper apiculate, margin ± entire; nerve ending in or below apex; basal cells rectangular, others ± rhomboidal, 2–3 marginal rows rectangular. Seta thin, to 2 cm; theca obloid, apophysis broadly globose, much wider then theca, rugose when dry. Fr frequent, summer. $n=9$. Loose patches on dung in flushes and wet places at high altitudes, rare but sometimes locally common. N. Northumberland and Westmorland north to E. Ross and E. Sutherland but mainly in the central Scottish Highlands. 14. A mainly circumboreal species.

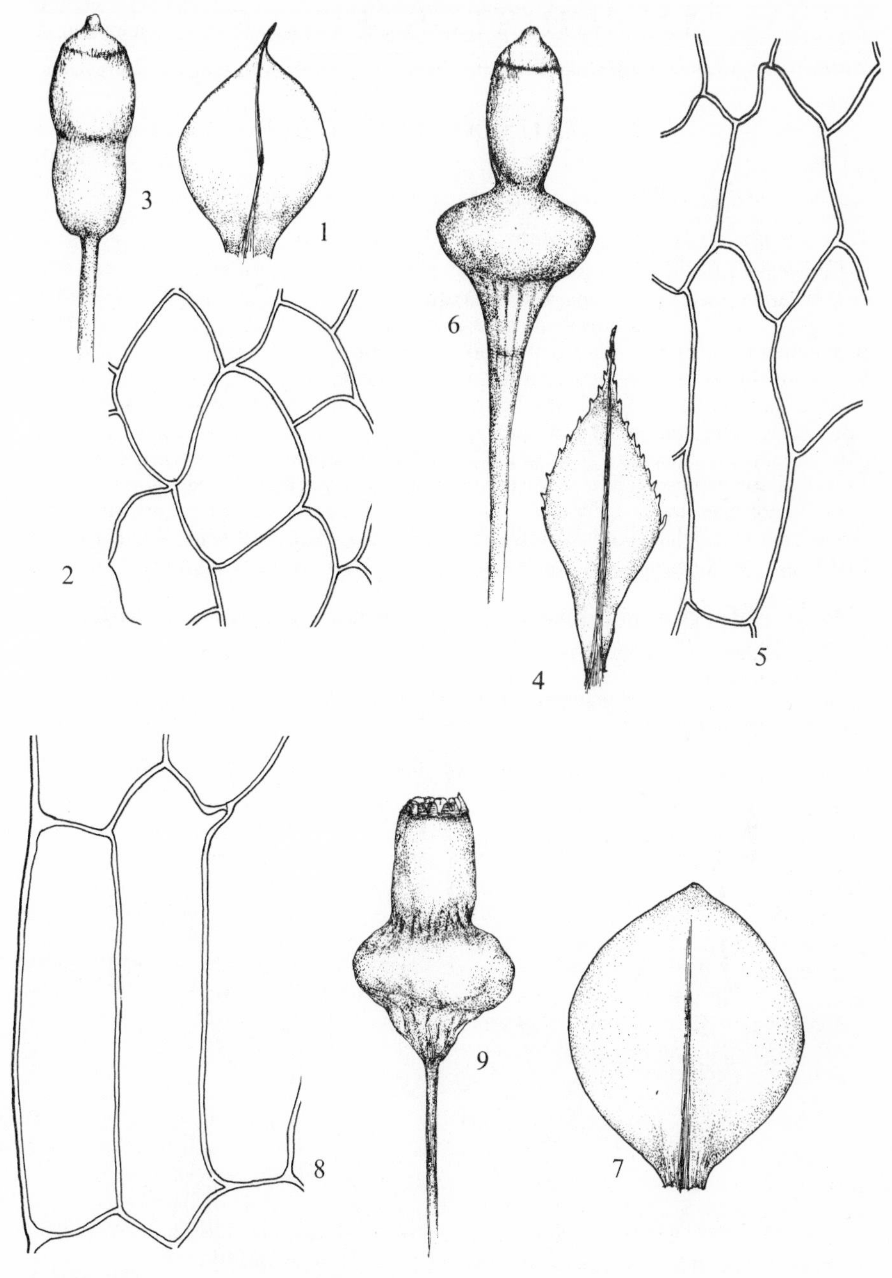

Fig. **169**. 1–3, *Splachnum sphaericum*: 1, leaf; 2, cells from near leaf apex; 3, capsule. 4–6, *S. ampullaceum*: 4, leaf; 5, cells from near leaf apex; 6, capsule. 7–9, *S. vasculosum*: 7, leaf; 8, mid-leaf cells; 9, capsule. Leaves × 10, cells × 415, capsules × 10.

13. SCHISTOSTEGALES

An order of uncertain affinities with the characters of the only species, *Schistostega pennata*.

21. SCHISTOSTEGACEAE

81. SCHISTOSTEGA MOHR, OBSER. BOT., 1803

1. S. pennata (Hedw.) Web. & Mohr, Ind. Musc. Pl. Crypt., 1803
S. osmundacea Mohr

Male and female plants separate but arising from same protonema. Plants variable in length, to *ca* 1.5 cm, stems flexuose, naked below, arising from persistent, thalloid protonema with light-refracting properties. Lvs on sterile stems in 2 ranks, ovate-lanceolate to lanceolate, acute to acuminate, entire, nerveless, confluent at base; fertile stems similar but with smaller lvs or only with rosette of lvs at apex; cells thin-walled, rhomboidal, 16–30 μm wide. Seta thin, to 4 mm; capsule ovoid, gymnostomous; lid convex; spores 8–12 μm. Fr occasional, spring, summer. $n = 11^*$, 14. Glaucous-green patches, reddish-brown below, on acidic, friable soil on shaded banks, entrances to caves, rabbit burrows, etc., particularly in sandstone areas. Occasional in England and Wales, very rare in Scotland, extending north to Mid Perth and W. Inverness, unknown in Ireland. 39, C. Europe, Amur, Japan, N. America.

Because of the light-refractive properties of the protonema this moss appears luminescent when growing in dark places.

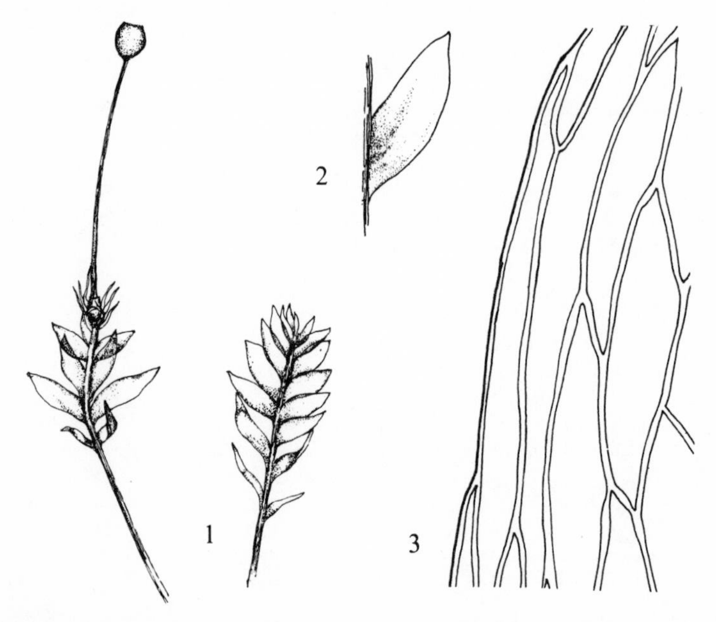

Fig. **170**. *Schistostega pennata*: 1, shoots (×10); 2, leaf (×25); 3, cells from upper part of **leaf** (×**450**).

14. BRYALES

Acrocarpous. Lvs bordered or not; cells isodiametric to linear, sometimes large, smooth or mamillose, rarely papillose. Seta long; capsule inclined to pendulous; peristome double, inner well developed and usually with cilia.

22. BRYACEAE

Usually erect tufted plants. Lvs ovate to lanceolate, often forming comal tuft, frequently bordered with narrow cartilaginous cells; cells smooth, ± rhomboidal, thin-walled. Seta long; capsule inclined to pendulous, ovoid to cylindrical; peristome usually double, often with well developed basal membrane, with or without cilia; calyptra cucullate.

SUBFAMILY 1. MIELICHHOFERIOIDEAE

Inflorescence apparently lateral; capsule usually erect; outer peristome sometimes lacking.

82. MIELICHHOFERIA HORNSCH., BR. GERM., 1831

Stems slender, densely tufted. Lvs ovate to lanceolate, margin entire or denticulate above; nerve ending in or below apex. Capsule erect or pendulous with distinct neck; outer peristome lacking. A ± cosmopolitan genus of *ca* 140 species which occur particularly on montane, copper-bearing rocks.

Lvs ovate to ovate-lanceolate, acute to obtuse, cells in mid-lf 10–20 μm wide **1. M. elongata**
Lvs lanceolate, acuminate, cells in mid-lf *ca* 10 μm wide **2. M. mielichhoferi**

1. M. elongata (Hoppe & Hornsch. ex Hook.) Hornsch., Br. Germ., 1831
Oreas mielichhoferi Brid. var. *elongata* (Hornsch.) Br. Eur.

Dioecious. Stems slender, to 1 cm. Lvs imbricate, ovate to ovate-lanceolate, acute to obtuse, margin plane, denticulate above; nerve weak or stout, ending well below apex, in section with 1 layer of large adaxial cells; cells thin-walled, 10–20 μm wide in mid-lf. Seta cygneous; capsule pyriform, horizontal or cernuous. Fr rare. Tumid, dull green patches, sometimes extensive, on permanently damp, shaded, highly acidic, heavy-metal-containing rocks, very rare. N.E. Yorks, S. Aberdeen. 2. N. and C. Europe, Pyrenees, Italy, E. Africa.

2. M. mielichhoferi (Hook.) Wijk & Marg., Taxon, 1961
M. nitida Nees & Hornsch.

Close to *M. elongata*. Shoots forming only small tufts. Lvs narrower, lanceolate, acuminate, margin sometimes recurved in mid-lf; nerve stout, extending almost to apex, in section with 2 rows large adaxial cells; cells slightly incrassate, to 10 μm wide in mid-lf. Seta flexuose; capsule ± erect. Fr unknown in Britain. Small, yellowish-green, glossy tufts in similar but less acid habitats than *M. elongata*, very rare. S. Aberdeen. 1. N. and C. Europe, Pyrenees, Spain, Caucasus, N. America.

The strength of the nerve in *M. elongata* is variable but it does not reach as high up the lf as in *M. mielichhoferi*; the shape of the basal cells is not a reliable character for distinguishing the two species. Although *M. elongata* and *M. mielichhoferi* appear reasonably distinct in Britain, some Continental gatherings are difficult to determine and *M. elongata* may possibly only be a variety of *M. mielichhoferi*. For the occurrence of this plant in Britain see Coker, *Trans. Br. bryol. Soc.* **5**, 448–51, 1968.

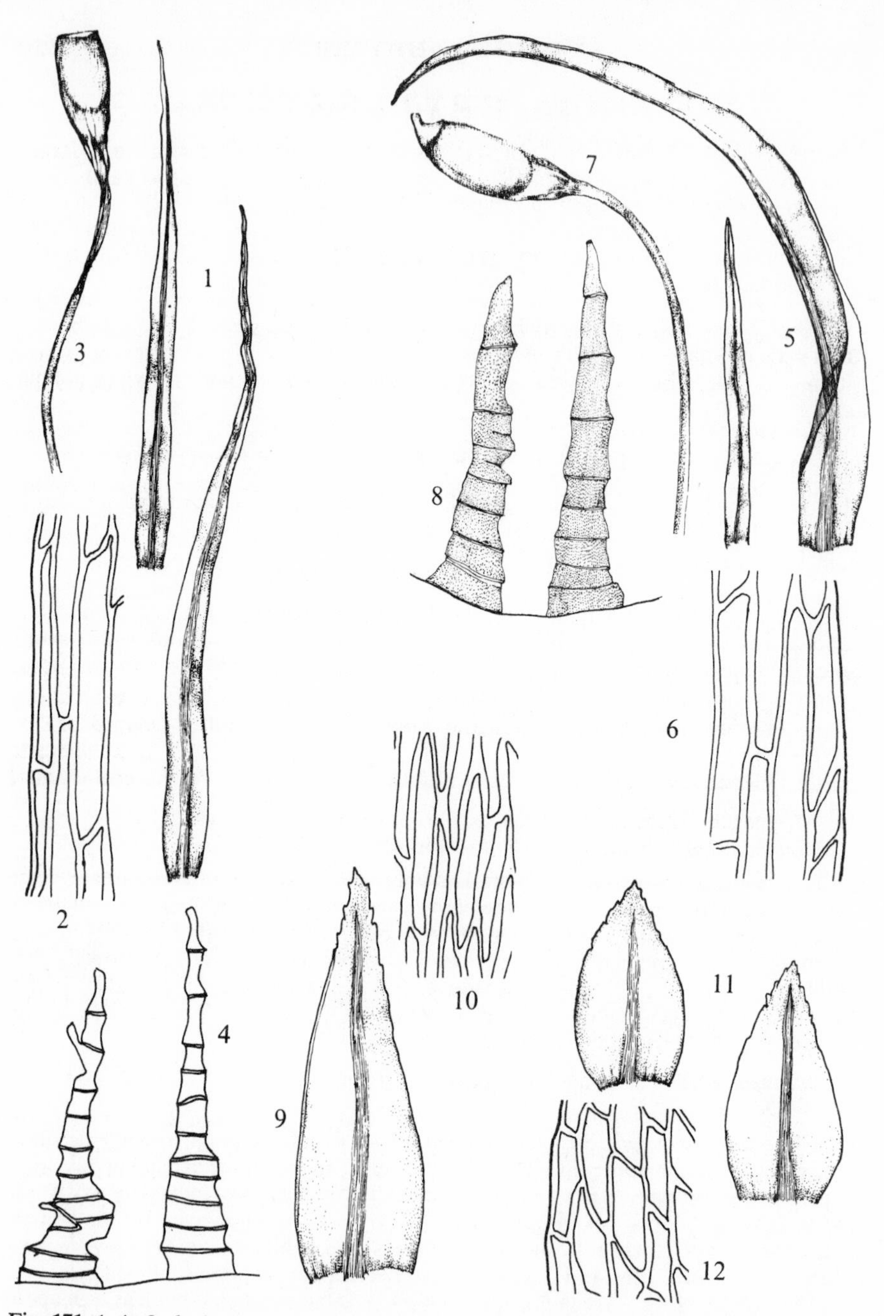

Fig. **171**. 1–4, *Orthodontium gracile*: 1, leaves (×25); 2, mid-leaf cells; 3, capsule; 4, outer peristome teeth. 5–8, *O. lineare*: 5, leaves (×25); 6, mid-leaf cells; 7, capsule; 8, outer peristome teeth. 9–10, *Mielichhoferia mielichhoferi*: 9, leaves (×65); 10, mid-leaf cells. 11–12, *M. elongata*: 11, leaves (×65); 12, mid-leaf cells. Cells ×415, capsules ×10, teeth ×250.

SUBFAMILY 2. **ORTHODONTOIDEAE**

Capsule ± erect; inner peristome teeth narrow, without basal membrane.

83. ORTHODONTIUM SCHWAEGR., SUPPL. II, 1827

Lvs linear-lanceolate, flexuose, unbordered; nerve narrow; cells ± linear except near base. Seta slender, flexuose; capsule erect or inclined, ovoid to cylindrical with tapering neck; peristome double, teeth narrow. A world-wide genus of 14 species.

Autoecious, synoecious or heteroecious, peristome teeth finely papillose **1. O. lineare**
Paroecious, peristome teeth smooth **2. O. gracile**

1. O. lineare Schwaegr., Suppl. 11., 1827
O. gracile var. *heterocarpum* W. Watson

Autoecious, synoecious or heteroecious. Plants to 1 cm, sometimes forming extensive patches. Lvs flexuose when dry, erect, falcate or recurved when moist, linear-lanceolate to linear, margin plane, obscurely denticulate above, unbordered; nerve ending in or below apex; basal cells rectangular, cells above ± linear-rhomboidal, 8–14 μm wide in mid-lf. Seta thin, flexuose; capsule erect or inclined, somewhat variable in shape, narrowly pyriform to clavate, sometimes gibbous, sulcate or not when dry; outer (except at base) and inner peristome teeth finely papillose, outer longer or shorter than inner; spores (10–)16–20 μm. Fr common, spring, summer. $n=20$*, 22*. Silky, dull green patches on acidic rocks, peat, rotting wood and tree boles in coniferous and deciduous woods and on shaded banks. Frequent to common in England and Wales except in the extreme west, and extending north to Perth and Angus, very rare in Ireland. 88, H4, C. W. Europe, Africa, southern S. America, Australia, New Zealand.

O. lineare more closely resembles a member of the Dicranaceae than of the Bryaceae in its general appearance but is readily distinguished by its capsule and leaf areolation. The species until its description from Laddy Rocks, Cheshire by Watson in 1922 was recorded only from the southern hemisphere. Since its original discovery in 1922 it has spread over most of England and Wales and is still extending its distribution in Scotland and also on the Continent. It is very probably an introduced species. *O. lineare* is very close to *O. gracile* and whilst the latter often has narrower lvs and fruit, the former is so variable that the only reliable diagnostic character is the inflorescence and papillosity of the peristome teeth. For a detailed account of European species of *Orthodontium* see Margadant & Meijer, *Trans. Br. bryol. Soc.* **1**, 266–74, 1950.

2. O. gracile Schwaegr. ex Br. Eur., 1844

Very similar to *O. lineare*. Paroecious. Lvs linear to linear-lanceolate. Capsule narrowly clavate; peristome teeth not papillose. Fr common. $n=12$. Dense, silky, bright green tufts on shaded acidic, often sandstone rock, rare and decreasing. S. Devon, E. Sussex, Stafford and Shrops north to Midlothian and Stirling. 16. Finistère (France), Madeira, California.

SUBFAMILY 3. **BRYOIDEAE**

Capsule usually inclined, cernuous or pendulous; inner peristome with distinct basal membrane.

84. LEPTOBRYUM (BR. EUR.) WILS., BR. BRIT., 1855

Synoecious or dioecious. Annual species with slender stems. Upper lvs setaceous,

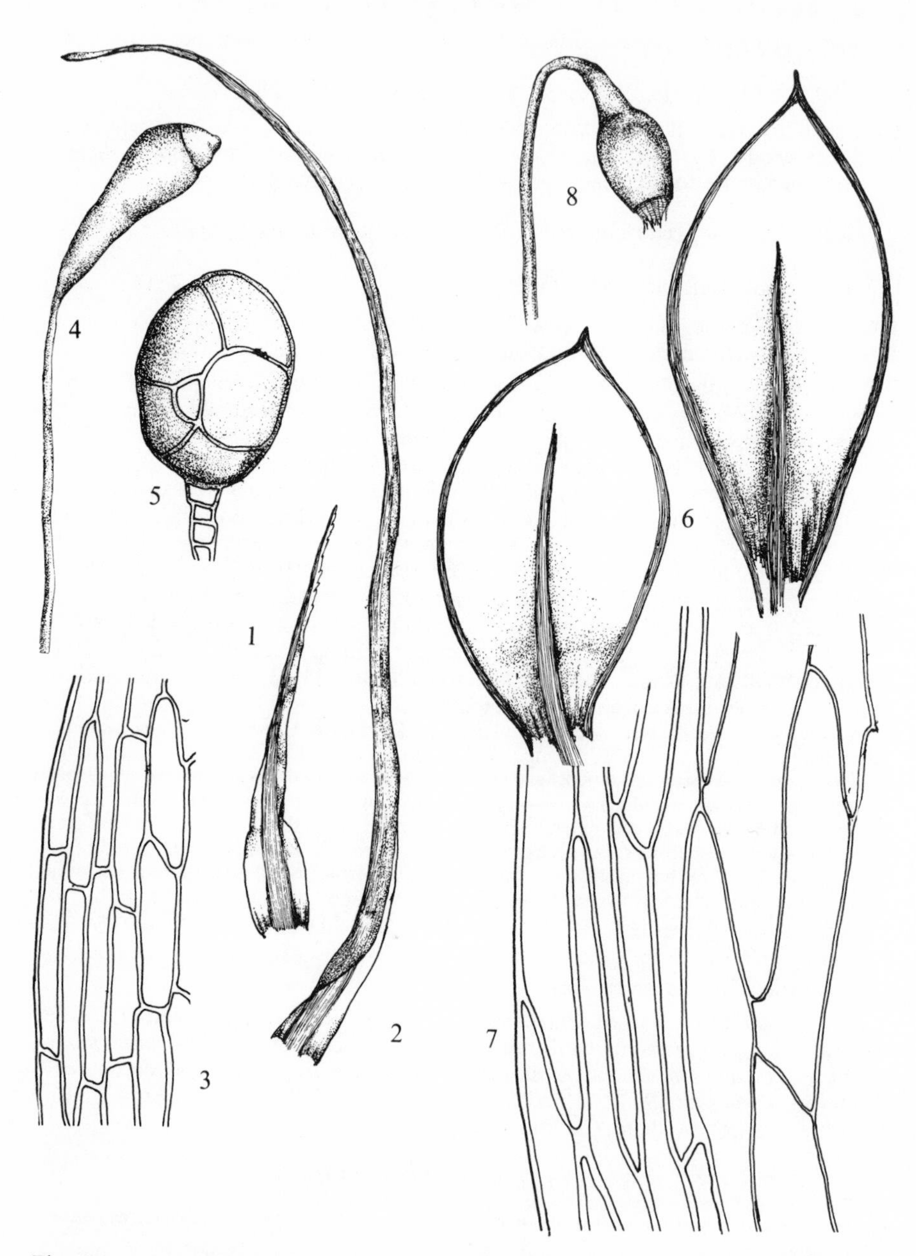

Fig. **172**. 1–5, *Leptobryum pyriforme*: 1, lower leaf; 2, upper leaf; 3, basal cells of leaf; 4, capsule; 5, rhizoidal gemma (×250). 6–8, *Epipterygium tozeri*: 6, leaves; 7, mid-leaf cells; 8, capsule. Leaves ×25, cells ×415, capsules ×10.

cells linear. Capsule pyriform, pendulous, glossy; peristome bryoid. Cosmopolitan genus of 5 species.

1. L. pyriforme (Hedw.) Wils., Br. Brit., 1855

Plants to 2 cm, stems slender, naked below. Lvs flexuose when dry, flexuose-spreading when moist, lower lanceolate, acuminate, upper and perichaetial forming comal tuft, ± abruptly narrowed from erect broad basal part to long setaceous subula composed mainly of nerve, margin unbordered, finely denticulate above; nerve broad below, occupying *ca* $\frac{1}{3}$ width of lf base, ending below apex in lower lvs, excurrent in upper; cells ± linear-rhomboidal throughout, basal larger than upper. Dark, brownish rhizoidal and occasionally axillary ellipsoid gemmae, 125–140 μm long frequently present. Seta thin, flexuose, variable in length, to 4 cm; capsule inclined or pendulous, pyriform; theca smooth, brown, glossy, neck sulcate; peristome teeth yellow; spores 14–16 μm. $n=20$*, 22, 24. Fr common, spring, summer. Lax, pale green tufts or patches on damp soil, arable fields, paths, roadsides, banks, peat, especially on sites of fires, throughout Britain and Ireland usually only occasional but common in greenhouses and flower pots. 92, H16, C. Cosmopolitan.

85. Pohlia Hedw., Sp. Musc., 1801

Dioecious, autoecious, paroecious, synoecious or polyoecious. Plants usually tufted. Lvs ovate to lanceolate, longer, narrower and more crowded towards stem apex, margin unbordered, usually denticulate above; nerve ending in or below apex; cells linear to narrowly hexagonal. Seta long; capsule erect to pendulous, ovoid to ellipsoid, neck short or long; annulus present or not; peristome double, inner with basal membrane, cilia present or sometimes rudimentary, appendiculate; calyptra small. A ± cosmopolitan genus of about 155 species.

1 Plants with axillary bulbils — 2
Plants without axillary bulbils — 11

2 Bulbils mostly 1–6 per axil on upper part of stem, ovoid or obconical, yellowish or reddish-brown — 3
Bulbils numerous, mostly 10–35 per axil on upper part of stem, ovoid, clavate or elongated, yellowish, green or orange, only reddish on older parts of stem — 7

3 Bulbils rarely more than 1 per axil, lf primordia not more than *ca* $\frac{1}{5}$ total length of bulbil or not restricted to apex of bulbil — 4
Bulbils usually 2–6 per axil, primordia mostly $\frac{1}{3}$–$\frac{1}{2}$ total length of bulbil and at apex of bulbil only — 5

4 Bulbils 350–500 μm long, yellowish, lf primordia restricted to apex of bulbil — **7. P. filum**
Bulbils 500–1000 μm long, reddish-brown, primordia extending to middle of bulbil or sometimes lower — **6. P. drummondii**

5 All except youngest bulbils reddish-brown — **8. P. rothii**
Bulbils yellowish-green — 6

6 Lf primordia of bulbils curved, forming dome-shaped cavity over apex of bulbil — **9. P. bulbifera**
Lf primordia ± erect, not incurved — **10. P. proligera**

7 Bulbils globose, ovoid or clavate, thread-like bulbils present or not — 8
Bulbils elongated, ± thread-like — 10

8 Apex of bulbils with small protuberances but without definite lf primordia, bulbils translucent **12. P. muyldermansii**
Apex of bulbils with definite lf primordia, bulbils opaque 9
9 Bulbils ovoid, lacking short, filamentous stalk, with 2–4 uni- or multicellular triangular to peg-like lf primordia **10. P. proligera**
Bulbils ± globose, usually with short filamentous stalk, with 2–3 unicellular lf primordia **11. P. camptotrachela**
10 Lf primordia to $\frac{1}{10}$ total length of bulbil **12. P. muyldermansii**
Lf primordia *ca* $\frac{1}{4}$ total length of bulbil **10. P. proligera**
11 Lf base decurrent **13. P. ludwigii**
Lf base not or hardly decurrent 12
12 Mid-lf cells 6–12 μm wide, plants without metallic sheen or rhizoidal gemmae, capsule long and narrow with neck $\frac{1}{3}$–1 length of theca 13
Mid-lf cells 10–30(–40) μm wide or if less than 10 μm then plants with metallic sheen or rhizoidal gemmae, capsule ovoid, ellipsoid or pyriform 14
13 Plants autoecious or paroecious, dull green, cells 8–12 μm wide in mid-lf **1. P. elongata**
Plants dioecious, glaucous-green, cells 6–8 μm wide **2. P. crudoides**
14 Lf cells lax, usually more than 15 μm wide, elongate-hexagonal or rhomboidal 15
Lf cells long and narrow, 6–14(–16) μm wide 16
15 Plants glaucous bluish-green, upper lvs ovate or ovate-lanceolate **17. P. wahlenbergii**
Plants pale green, not glaucous, upper lvs narrowly lanceolate **16. P. carnea**
16 Plants with metallic sheen, mid-lf cells 80–200 μm long, rhizoidal gemmae lacking **3. P. cruda**
Plants without metallic sheen, mid-lf cells to 100 μm long or rhizoidal gemmae present 17
17 Plants paroecious, 1–8 cm tall 18
Plants dioecious, 0.3–1.0 cm tall 19
18 Lower lvs plane, ovate-lanceolate, acute, base not decurrent **4. P. nutans**
Lower lvs ± concave, ± ovate, acute to obtuse, base slightly decurrent **5. P. obtusifolia**
19 Rhizoidal gemmae lacking, lf margin plane **17. P. wahlenbergii**
Rhizoidal gemmae present, lf margin usually recurved 20
20 Perichaetial lvs narrowly lanceolate, mid-lf cells 10–14 μm × *ca* 80 μm, rhizoidal gemmae pale brown, spherical to pyriform, not knobbly **15. P. lescuriana**
Perichaetial lvs linear-lanceolate, mid-lf cells 6–10 μm × 70–180 μm, rhizoidal gemmae yellowish, ellipsoid with knobbly outline **14. P. lutescens**

Section I. *Eupohlia*

Monoecious or dioecious. Comal lvs long and narrow; cells linear or linear-rhomboidal. Capsule usually horizontal to pendulous, cylindrical with long neck; exothecial cells rectangular; stomata superficial; annulus of large cells; basal membrane low, processes entire or narrowly perforated, cilia rudimentary or absent.

1. P. elongata Hedw., Spec. Musc., 1801
Webera elongata (Hedw.) Schwaegr.

Plants 0.5–5.0 cm. Lvs erect, straight to flexuose, somewhat shrunken when dry, erect or erecto-patent when moist, lower distant, lanceolate, acuminate, comal crowded, narrowly lanceolate to linear-lanceolate, acuminate, margin plane or recurved, denticulate above; nerve stout, ending in or below apex; cells except at extreme base linear, often ± vermicular, 8–12 μm×40–110 μm in mid-lf; capsule inclined to cernuous, narrowly pyriform to fusiform, sometimes curved; lid conical or mamillate, acute or obtuse; spores finely or coarsely papillose, 16–24 μm. Fr common, late summer, autumn.

Key to intraspecific taxa of *P. elongata*

1 Comal lvs 1.0–2.5 mm long, cells in mid-lf *ca* 40 μm long, capsule 1.5–3.5 mm long, neck less than ⅓ total length of capsule — ssp. **polymorpha**
Comal lvs 2–4 mm long, mid-lf cells 60–110 μm long, capsule 2.5–6.0 mm long, neck *ca* ½ total length of capsule — **2.** (ssp. **elongata**)
2 Paroecious — ssp. **elongata** var. **elongata**
Autoecious — ssp. **elongata** var. **acuminata**

Ssp. **elongata**
Autoecious or paroecious. Plants 0.5–5.0 cm. Comal lvs 2–4 mm long, mid-lf cells 60–110 μm long. Seta (1.0–)1.5–2.5 cm; capsule inclined to horizontal, very narrow, ± fusiform, 2.5–6.0 mm long, neck ± as long as theca; lid acute to obtuse; basal membrane *ca* ⅓ height of inner peristome; with or without cilia.

Var. **elongata**
Paroecious. $n=11$*. Tight, usually dull green tufts in rock crevices, ledges and on soil, rare to occasional in montane areas. N. Devon, S. Somerset, Wales, Shrops and Yorks northwards, scattered localities in Ireland. 43, H13. Europe, Faroes, N., W. and C. Asia, Sumatra, Macaronesia, Algeria, Kilimanjaro, Congo, Kerguelen Is., America, Greenland.

Var. **acuminata** (Hoppe & Hornsch.) Hüb., Musc. Germ., 1833
P. acuminata Hoppe & Hornsch., *Webera acuminata* (Hoppe & Hornsch.) Schimp.
Autoecious. Similar to var. *elongata*. Wales, Westmorland and Cumberland northwards, rare, very rare in Ireland, Kerry, S. Tipperary. 14, H3. N., W. and C. Europe, Caucasus, China, N. America.

Ssp. **polymorpha** (Hoppe & Hornsch.) Nyh., Ill. Moss Fl. Fenn. 3, 1958
P. minor Schwaegr., *P. polymorpha* Hoppe & Hornsch., *Webera polymorpha* (Hoppe & Hornsch.) Schimp.
Paroecious. Plants rarely more than 1 cm tall. Comal lvs usually 1.0–2.5 mm long; mid-lf cells *ca* 40 μm long. Seta rarely more than 1 cm; capsule cernuous, narrowly pyriform, 1.5–3.5 mm long, neck shorter than theca; lid conical or mamillate, obtuse; inner peristome with very short basal membrane, without cilia. In similar habitats to ssp. *elongata*. Caerns, Westmorland and Cumberland northwards, rare, very rare in Ireland, S. Kerry, S. Tipperary, W. Galway. 20, H3. N., W. and C. Europe, Faroes, Caucasus, Kashmir, Himalayas, China, N. America, Greenland.

A very variable species requiring experimental study. Although var. *acuminata* is regarded as a distinct species by many authorities, apart from the autoecious inflorescence the characters separating it from the type are inconstant. The plant is retained here pending further investigation; the largest plants of ssp. *elongata* are always paroecious and the smallest autoecious. Ssp. *polymorpha* is usually distinct in its smaller size and in its capsule shape but intermediate forms do occur very rarely. *P. cruda* differs in its metallic sheen and glaucous-green colour and the lf cells 80–120 μm long; *P. crudoides* has an erect or inclined ovoid capsule and is dioecious; *P. nutans* has wider lf cells.

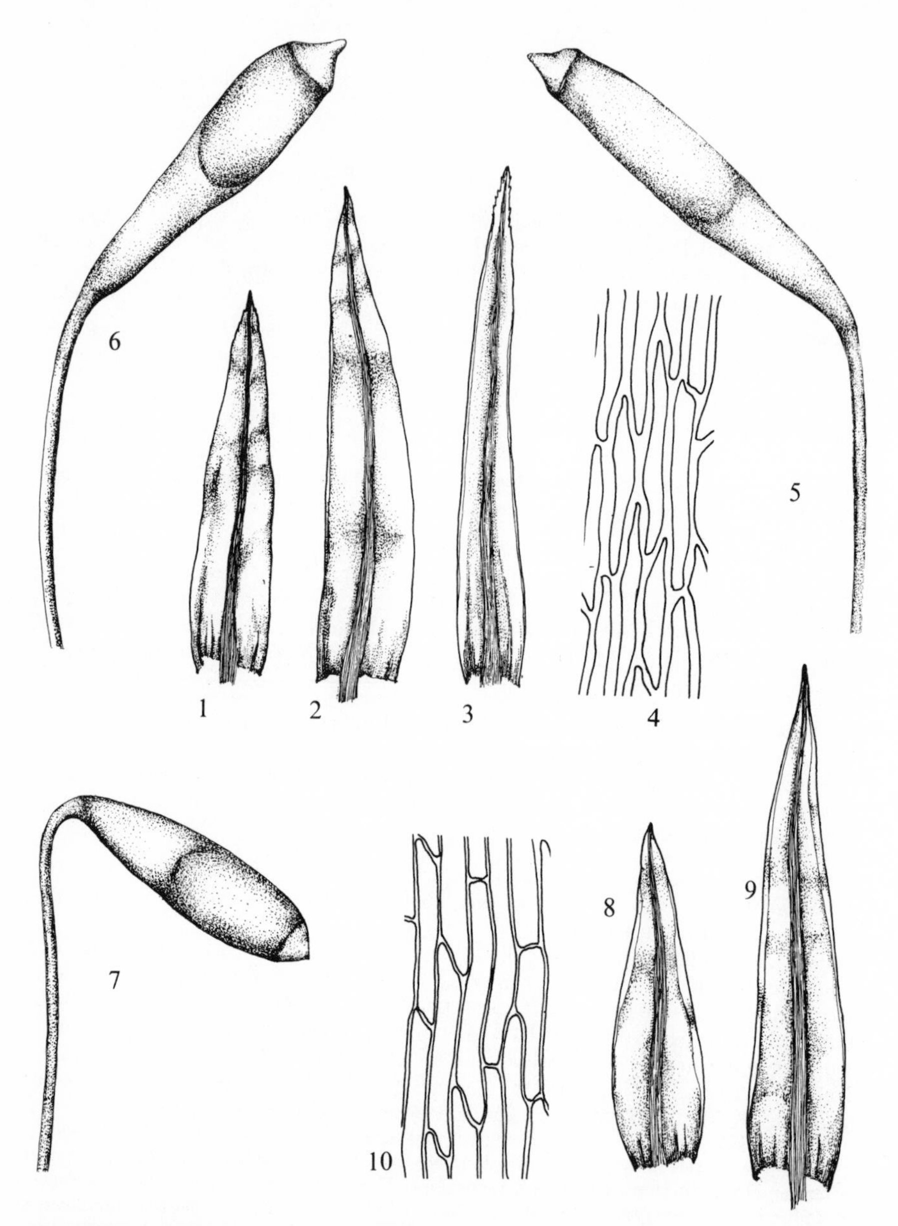

Fig. **173**. 1–5, *Pohlia elongata* var. *elongata*: 1 and 2, lower and upper leaves; 3, perichaetial leaf; 4, mid-leaf cells; 5, capsule. 6, *P. elongata* var. *acuminata*: capsule. 7, *P. elongata* ssp. *polymorpha*: capsule. 8–10, *P. crudoides*: 8 and 9, lower and perigonial leaves; 10, mid-leaf cells. Leaves ×25, cells ×415, capsules ×10.

2. P. crudoides (Sull. & Lesq.) Broth., Nat. Pfl., 1903

Dioecious. Plants *ca* 3.5 cm tall in Britain, stems red. Lvs erect, slightly shrunken when dry, erecto-patent when moist, lower lanceolate, upper narrowly lanceolate, acuminate, margin recurved, slightly denticulate near apex; nerve ending below apex in lower lvs, percurrent or excurrent in upper; cells linear, ± vermicular, 6–8 μm×40–100 μm in mid-lf. Capsule erect or inclined, ovate-ellipsoid. Only male plants known in Britain. Glaucous-green tufts, reddish below, in crevices on montane rocks, very rare. E. Inverness. 1. Arctic Europe and N. America, Greenland.

May be distinguished from *P. elongata*, *P. cruda*, *P. nutans* and *P. obtusifolia* by the very narrow cells and dioecious inflorescence. For the occurrence of this plant in Britain see Wallace, *J. Bryol.* 7, 157–9, 1972.

Section II. *Lamprophyllum*

Dioecious or paroecious. Comal lvs narrow; cells narrowly hexagonal to linear. Capsule usually horizontal to pendulous, narrowly ovoid or pyriform with short neck; exothecial cells long; stomata superficial; annulus of large cells; basal membrane of medium height, processes perforated, cilia usually perfect but not appendiculate.

3. P. cruda (Hedw.) Lindb., Musc. Scand., 1879
Webera cruda (Hedw.) Bruch

Dioecious, paroecious or synoecious. Plants to 4 cm, stems red. Lvs with metallic sheen, erect to erecto-patent when moist, hardly altered when dry, reddish at base, lower ovate-lanceolate, acute, margin entire, upper lanceolate and becoming narrower in coma, with acute apex, margin plane, denticulate; nerve reddish, ending below apex; extreme basal cells rectangular, elsewhere ± uniformly linear, 8–12 μm ×80–200 μm in mid-lf, slightly narrower at margin. Capsule inclined to subpendulous, ± ellipsoid, neck *ca* $\frac{1}{2}$ length of theca; lid mamillate; inner peristome ciliate; spores 18–24 μm. Fr occasional, summer. $n=10$, 10+4, 11*, 22*, 40. Tufts, glaucous-green with metallic sheen above, reddish-brown below, in crevices of acidic rocks, frequent in montane districts. Devon, Wales, Shrops and Derby northwards, scattered localities in Ireland. 64, H11. Europe, Faroes, Asia, Algeria, Azores, Kerguelen Is., America, Australasia, New Caledonia, Antarctica.

4. P. nutans (Hedw.) Lindb., Musc. Scand., 1879
Webera nutans Hedw.

Paroecious. Plants 1.0–7.5 cm, stems tomentose below. Lower lvs ± imbricate when dry, erect when moist, ovate-lanceolate, acute, margin plane or narrowly recurved, upper lvs shrunken, flexuose when dry, erecto-patent when moist, narrowly lanceolate, apex acuminate, margin narrowly recurved, entire to sharply denticulate; nerve reddish-brown, ending in or below apex; basal cells rectangular, incrassate, reddish, elsewhere narrowly to elongate-hexagonal, in lower lvs 8–14 μm ×25–80 μm, in upper lvs 10–16(–20) μm×60–100 μm in mid-lf. Seta flexuose, variable in length, 1–5 cm; capsule horizontal to pendulous, broadly pyriform to ovate-ellipsoid, neck conspicuous or not, less than $\frac{1}{2}$ length of theca; lid mamillate; spores 18–28 μm. Fr common, spring, summer. $n=11$, 21, 22*, 33*. Loose or dense tufts or patches, sometimes extensive, or scattered plants among other bryophytes, glossy, green or yellowish-green, brownish below, on peaty or gravelly soil, decaying wood, bark, rock crevices, on heaths, in bogs, woods, derelict industrial sites especially where heavy-metal-polluted, in shaded or open, wet or dry habitats, common

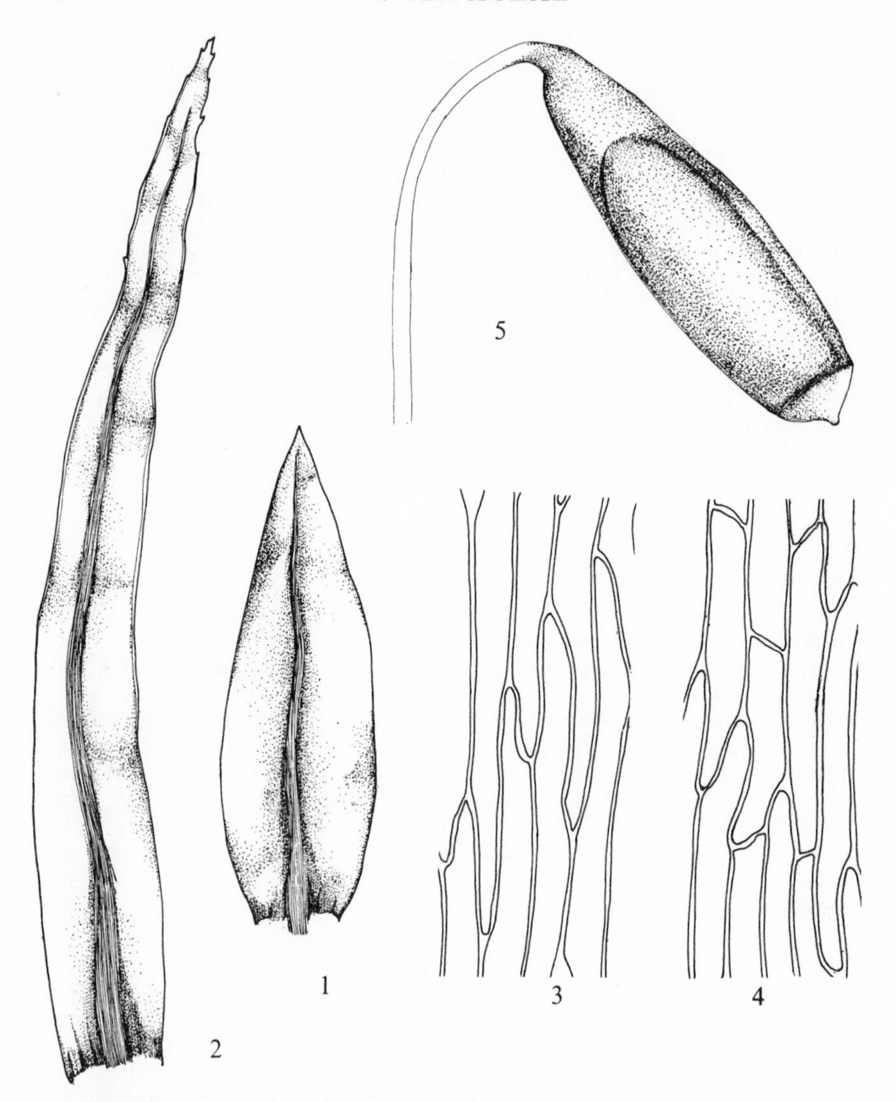

Fig. **174**. *Pohlia cruda*: 1, lower leaf; 2, perichaetial leaf; 3, leaf cells; 4, perichaetial leaf cells; 5, capsule (×10). Leaves ×25, cells ×415.

except in basic areas. 112, H38, C. Europe, Asia, southern Africa, Malagasy Republic, Kerguelen Is., N. America, Australia, New Zealand, Antarctica.

Very variable morphologically and cytologically. *P. cruda* differs in its glaucous-green colour, metallic sheen and narrower cells. A curious plant, reddish in colour and dioecious, has been found on several heaths in Berkshire. These plants in other respects come within the range of *Pohlia nutans* and it seems likely that they are habitat forms of that plant; the matter requires experimental investigation.

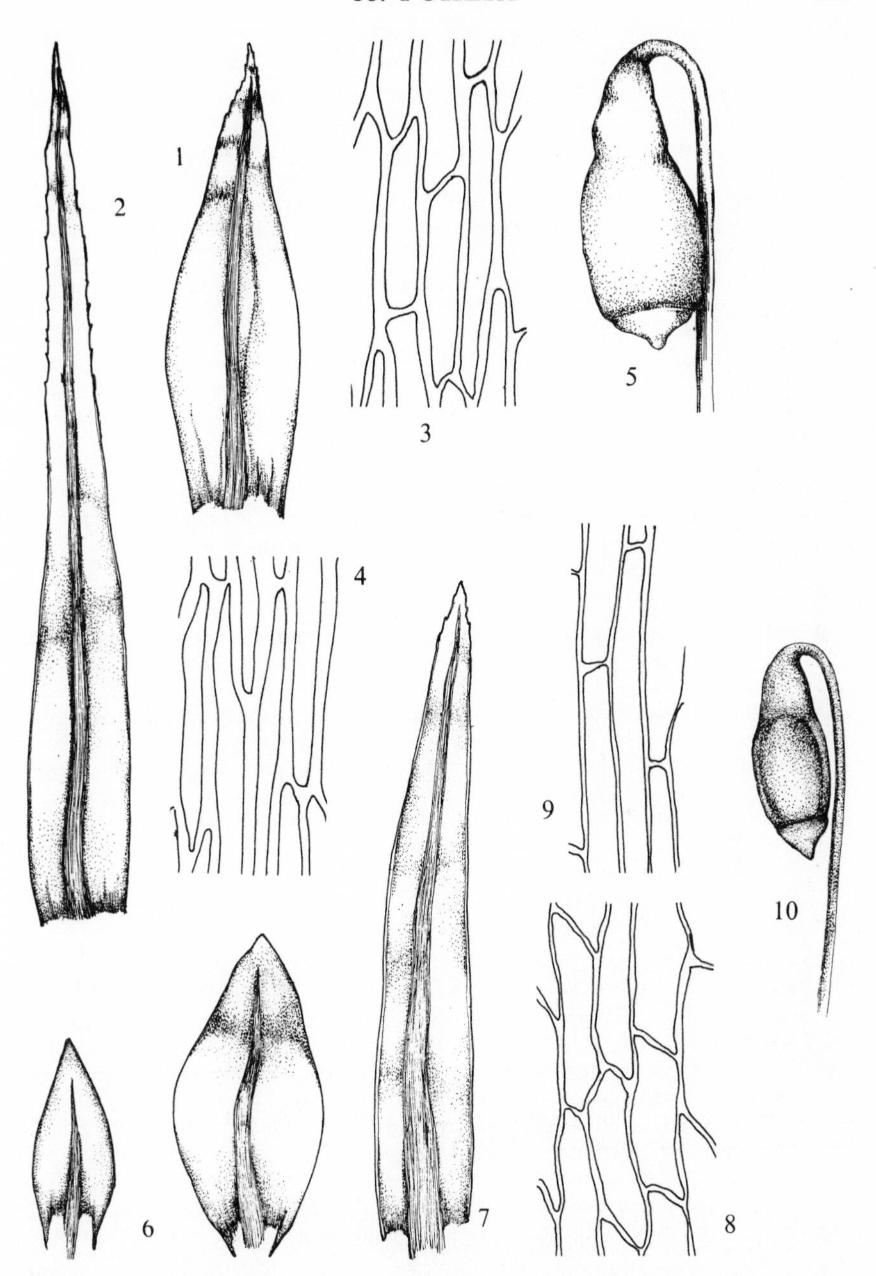

Fig. **175**. 1–5, *Pohlia nutans*: 1 and 2, stem and perichaetial leaves; 3 and 4, mid-leaf cells of stem and perichaetial leaves; 5, capsule. 6–10, *P. obtusifolia*: 6 and 7, stem and perichaetial leaves; 8 and 9, mid-leaf cells of stem and perichaetial leaves; 10, capsule. Leaves ×25, cells ×415, capsules ×10.

5. P. obtusifolia (Brid.) L. Koch, Leafl. West. Bot., 1950
P. cucullata (Schwaegr.) Bruch, *Webera cucullata* (Schwaegr.) Schimp.

Paroecious. Plants 1.0–2.5 cm, stems reddish. Lvs appressed when dry, erecto-patent when moist, ovate to ovate-lanceolate, base somewhat decurrent, apex acute to obtuse, comal lvs larger, lanceolate, acute, margin plane or narrowly recurved, entire or denticulate above; nerve stout, ending below apex in lower lvs, percurrent or excurrent in comal; cells in lower lvs elongate-hexagonal except at base, 12–24 μm ×40–80 μm, longer and narrower in comal lvs, 10–14 μm×70–130 μm in mid-lf. Capsule pendulous, pyriform, short-necked; lid mamillate or conical; spores 30–36 μm. Fr frequent, late summer. Pale green, glossy tufts in rock crevices at high altitudes, very rare. Mid Perth, E. Inverness, Arran, W. Ross. 4. Europe, N. and C. Asia, Japan, N. America, Greenland.

Similar in appearance to *P. nutans* but the lvs are relatively shorter and wider and less sharply pointed as well as being concave with a ± decurrent base; the spores are also larger. Also somewhat resembles *P. drummondii* but that species has axillary, reddish-brown bulbils and shorter lf cells.

Section III. *Pohliella*

Dioecious. Upper lvs not forming comal tuft and ± similar to lower lvs; cells narrowly hexagonal to linear-rhomboidal. Capsule horizontal to pendulous, ovoid or pyriform with short neck; exothecial cells short; stomata superficial; annulus of large cells; processes perforated, cilia present, not appendiculate.

6–12. P. annotina agg.

It has been shown that the following 7 species remain distinct in cultivation but the form of the axillary bulbils in any one species may vary depending upon the time of the year and the age of the plant. Whilst there are some differences in lf shape and cell size that are statistically significant, these are not of much practical use in the determination of species. Sporophytes are so rarely produced that they likewise are of little use. Bulbil descriptions are of bulbils from the upper $\frac{1}{3}$ of the stems; bulbils frequently become detached in dried specimens but may usually be found in the detritus at the bottom of the packet. The projections on the bulbils are, for convenience, referred to as lf primordia but they are not necessarily homologous with such structures. In all species the lf margin is plane, denticulate above and the nerve ends below the lf apex. For an account of *P. annotina* agg. see Lewis & Smith, *J. Bryol.* **9**, 539–56, 1977, and *J. Bryol* **10** (1), in press.

6. P. drummondii (C. Müll.) Andrews in Grout, Moss Fl. N. Amer., 1935
P. rothii auct. non *Webera rothii* Corr. ex Limpr., *Webera commutata* Schimp.

Plants 1–4(–10) cm. Lvs appressed when dry, erecto-patent when moist, lanceolate, apex acute; mid-lf cells 7–12 μm wide. Bulbils mostly 1 per axil but not regularly present in every axil, reddish-brown, ovoid, 500–1000 μm×150–400 μm, primordia (5–)6–8(–11), broad, green, arising at different levels on upper $\frac{1}{2}$ of body of bulbil and sometimes 1–2 small outgrowths from lower $\frac{1}{2}$; bulbils in tall montane forms may be longer and germinate *in situ*. Capsule cernuous to pendulous, narrowly pyriform; lid conical; spores 12–20 μm. Fr summer, occasional in Scotland, rare elsewhere. Lax or dense greenish tufts on wet sandy soil and gravel by streams, rivers, waterfalls and on muddy tracks, cuttings and in old gravel pits, generally distributed, occasional. 79, H15. Fennoscandia, probably N. and E. Asia and Patagonia.

Lowland forms of this plant have been named *P. rothii* but true *P. rothii* is rare in Britain. In *P. rothii* the lf primordia of the bulbils arise from the same level on the flattened apex and the bulbils are smaller and more numerous. In *P. filum*, which like *P. drummondii* usually has solitary bulbils, the bulbils are yellowish. Forms of *P. drummondii* occur in which bulbils are completely absent; the relationship between these and bulbiliferous forms requires study.

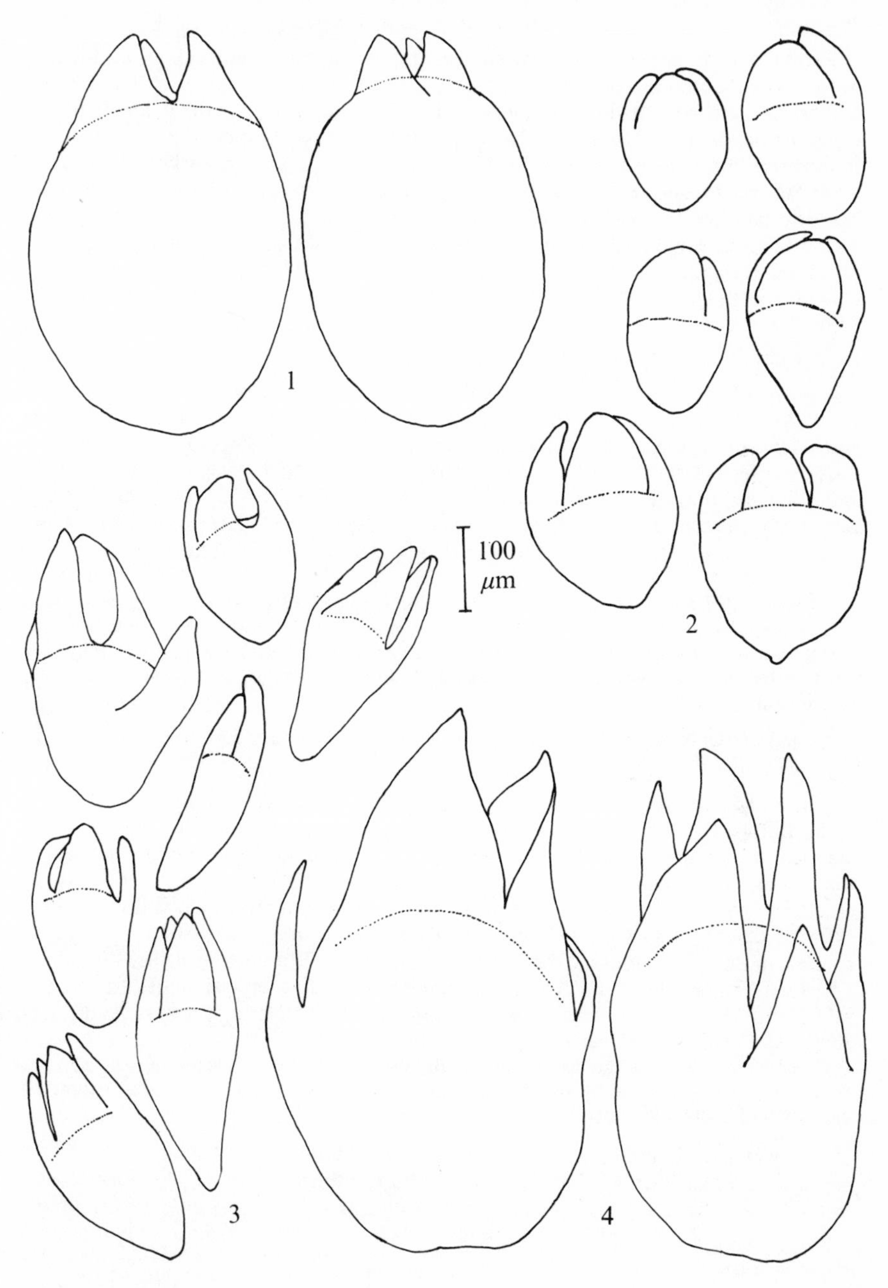

Fig. **176**. Axillary bulbils of *Pohlia* species, all × 100. 1, *P. filum*, 2, *P. bulbifera*; 3, *P. rothii*; 4, *P. drummondii*.

7. P. filum (Schimp.) Mårt., K. Svensk. Vet. Ak. Avh. Natursk., 1956
P. gracilis (Br. Eur.) Lindb., *P. schleicheri* Crum, *Webera gracilis* (Br. Eur.) De Not.

Plants to 4 cm. Lvs appressed when dry, erect when moist, lanceolate; mid-lf cells 8–14 μm wide. Bulbils solitary, very rarely 2 per axil, ovoid, yellowish, 350–500 μm ×200–350 μm, primordia 3–5, short and wide, triangular, crowded together and often overlapping, *ca* ⅕ total bulbil length. Capsule cernuous to pendulous, pyriform; lid convex to mamillate; spores 13–22 μm. Fr rare, late spring, early summer. Lax or dense, yellowish-green tufts, becoming reddish or brown with age, on wet sand and gravel on rocks and by streams and rivers in montane habitats, rare to occasional in the Scottish Highlands from Fife and Perth northwards. 14. C. and N. Europe, Pyrenees, Iceland, Siberia, N. America.

Differs from other related species in the erect lvs, bulbil morphology and the pyriform capsule.

8. P. rothii (Correns ex Limpr.) Broth., Nat. Pfl., 1903

Plants 1–3 cm. Lvs lanceolate; mid-lf cells 6–10 μm wide. Bulbils in upper part of shoot 3–6(–12) per axil, yellowish-green when young, soon becoming reddish-brown, obconical or ovoid with flattened apex, 200–300 μm×80–120 μm, primordia 3–5(–6), green, erect, broad, ± overlapping, arising at same level at apex, *ca* ⅓ total bulbil length, on older parts of stem bulbils fewer, larger, to 500 μm wide. Capsule horizontal to cernuous, oblong-ellipsoid; lid conical to rostellate; spores 14–24 μm. Fr unknown in Britain or Ireland. Lax, bright green tufts, sometimes extensive, on damp sandy soil, rare. Scattered localities from Cornwall to Dunbarton, S.W. Ireland, Waterford. 8, H3. Denmark, Norway, Finland, Germany.

This plant has only recently been discovered in Britain and Ireland. The name *P. rothii* has previously been erroneously applied to forms of *P. drummondii*. Young plants with yellowish-green bulbils may be confused with *P. proligera* but differ in the larger, broader, overlapping lf primordia of the bulbils.

9. P. bulbifera (Warnst.) Warnst., Krypt. Fl. Brandenb., 1904
Webera annotina var. *bulbifera* (Wanst.) Correns ex Dix.

Plants 0.5–2.0 cm. Lvs spreading when moist, lanceolate; mid-lf cells 7–11 μm wide. Bulbils (1–)3–5(–8) per axil, ovoid, yellowish-green, (150–)200–300 μm×120–200 μm, lf primordia usually 4, incurved, forming a hollow cover over the apex of the bulbil, *ca* ⅓ total bulbil length. Capsule cernuous to pendulous, obovoid; lid conical to rostellate; spores 14–24 μm. Fr rare, early summer. Lax, yellowish-green tufts on damp sandy soil and fine gravel by streams and lakes, often with related species, usually in montane habitats, E. Cornwall, Warwick, Carms, N. Wales, Cheshire, Westmorland and Northumberland north to Sutherland, scattered localities in Ireland. 32, H9. Fennoscandia, Belgium, Latvia, Iceland, Faroes, Azores, N. America, Greenland, N. Asia.

Readily recognised by the distinctive bulbils. When dried specimens are moistened an air bubble is usually trapped by the lf primordia over the apex of the bulbil, a feature not encountered in other species.

10. P. proligera (Kindb. ex Breidl.) Lindb. ex Arnell, Bot, Not., 1894
Webera annotina (Hedw.) Bruch, *W. proligera* Lindb ex Limpr., *P. annotina* (Hedw. emend Correns) Loeske, *P. annotina* var. *decipiens* Loeske, *P. annotina* var. *loeskei* Crum, Steere & Anderson, *P. camptotrachela* auct. non *Webera camptotrachela* Ren. & Card.

Plants 0.5–3.0 cm. Lvs erecto-patent to spreading when moist; cells (6–)8–10(–12) μm wide in mid-lf. Bulbils light green, very variable in shape and size depending on

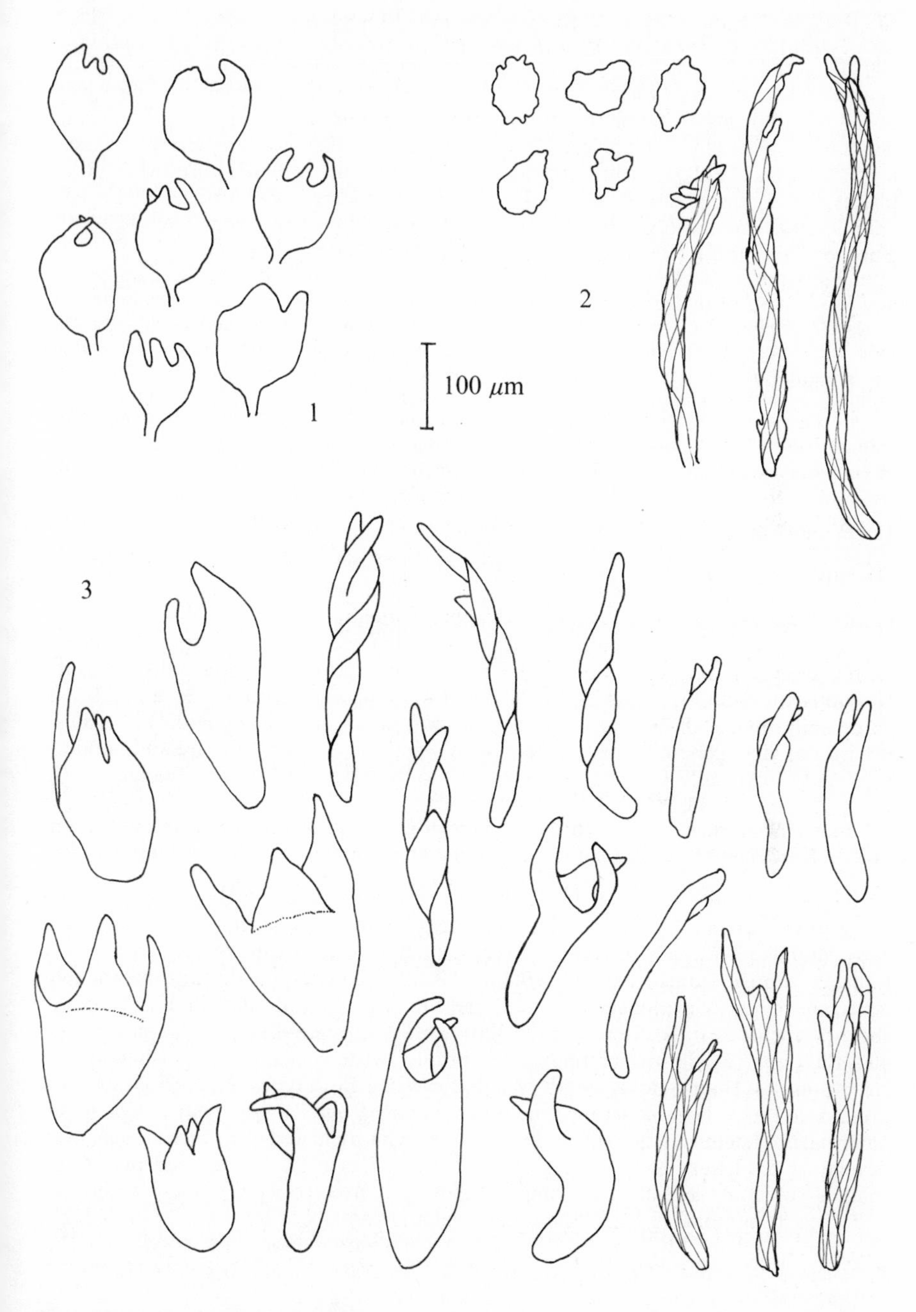

Fig. 177. Axillary bulbils of *Pohlia* species, all ×100. 1, *P. camptotrachela*; 2, *P. muyldermansii*; 3, *P. proligera*.

age of plant and position on stem; in upper part of stem 10–25(–30) per axil, below and in old plants (1–)3–10, ovoid to long and thread-like, size varying from (160–) 200–350 μm×(80–)100–150 μm in ovoid forms to (250–)300–450 μm×40–60(–80) μm in thread-like elongate forms, with 2–4(–5) uni- or multicellular, mostly narrowly triangular to prong-like lf primordia *ca* $\frac{1}{4}$ total length of bulbil. Capsule cernuous to pendulous, obovoid; lid conical; spores 14–22 μm. Fr very rare, early summer. $n=11$*. In lax, light green tufts to extensive patches, often mixed with other bryophytes, on disturbed soil in exposed situations, on damp rock surfaces and waste from copper mines, occasional to frequent, generally distributed. 102, H23, C. Europe, Asia Minor, Siberia, Azores, Madeira, N. America, Greenland.

Bulbil shape is exceedingly variable. They are most numerous and small in young plants and young parts of shoots; as bulbils are dislodged the remainder become larger, up to 4–5 times the original size; small narrow bulbils may be produced in the axils of new growth from old stems. Forms with only long narrow bulbils may be found on mud and wet soil in shaded situations in rock crevices, amongst boulders and in woods. Forms with few large bulbils may be mistaken for *P. drummondii*, *P. filum* or *P. rothii*. The first differs in the colour of the bulbils and the primordia arising at different levels on the bulbil, *P. filum* in the solitary bulbils, and *P. rothii* in bulbil colour and the larger, wider, overlapping lf primordia. *P. camptotrachela* and *P. muyldermansii* have smaller bulbils often with short filamentous stalks.

11. P. camptotrachela (Ren. & Card.) Broth., Nat. Pfl., 1903

Plants 0.5–2.0 cm. Lvs spreading when moist, lanceolate; mid-lf cells 7–14 μm wide. Bulbils 10–25(–30) per axil, yellowish or light green, rarely orange or in old plants deep red, ± globose, usually with short filamentous stalk, 80–140(–180) μm× 80–120 μm, with 2–4 unicellular inflexed apical projections, *ca* $\frac{1}{5}$ total length of bulbil, bulbils fewer, sometimes solitary and much larger in lower axils. Capsule cernuous to pendulous; spores 14–24 μm. Fr unknown in Britain. Small tufts on damp soil, often with *P. bulbifera* or *P. proligera* in damp fields, by paths and pools, occasional and generally distributed in W. and N. Britain and Ireland, under-recorded. Cornwall, Devon, Wales north to Shetland, Wexford, Wicklow, W. Donegal, Antrim. 38, H5. Sweden, N. America.

Distinct from other bulbilferous *Pohlia* species except *P. muyldermansii* in the smaller bulbils. *P. muyldermansii* has yellowish-green to orange bulbils without distinct lf primordia.

12. P. muyldermansii Wilcz. & Demar., Bull. Nat. Plant Belg., 1970

Plants 0.5–2.0 cm. Lvs erect when dry, spreading when moist; mid-lf cells 7–14 μm wide. Bulbils 15–25(–30) per axil, of 2 types: small ± clavate bulbils, yellowish-green to bright orange, translucent, 90–170 μm×50–80 μm, knobbly with protuberant cell walls, short stubby protuberances often present at apex but no definite lf primordia, short filamentous stalk often present; long, straight, thread-like bulbils, pale green, 400–1000 μm×25–50 μm, composed of spirally twisted cells, 2–4 finger-like apical structures, less than $\frac{1}{10}$ total length of bulbil present. Both types of bulbil sometimes present and mixed or one type or the other, but long bulbils very readily falling; on older parts of stem clavate bulbils may become cylindrical and 4–5 times larger, deep orange or reddish-brown. Fr very rare, autumn. Lax tufts or scattered plants on shallow soil, fragmenting rocks and gravel in sheltered, rocky places by waterfalls and streams in montane localities, occasional and probably very under-recorded. Fr very rare. Brecon, Merioneth, Caerns and Cumberland north to Argyll and W. Sutherland, S. Kerry, W. Cork, Waterford, S. Tipperary. 19, H4, Belgium, Holland.

Recognised by the dense clusters of small bulbils, yellowish-green to orange in colour. The thread-like bulbils may be mistaken for those of *P. proligera* but differ in the shorter apical outgrowths.

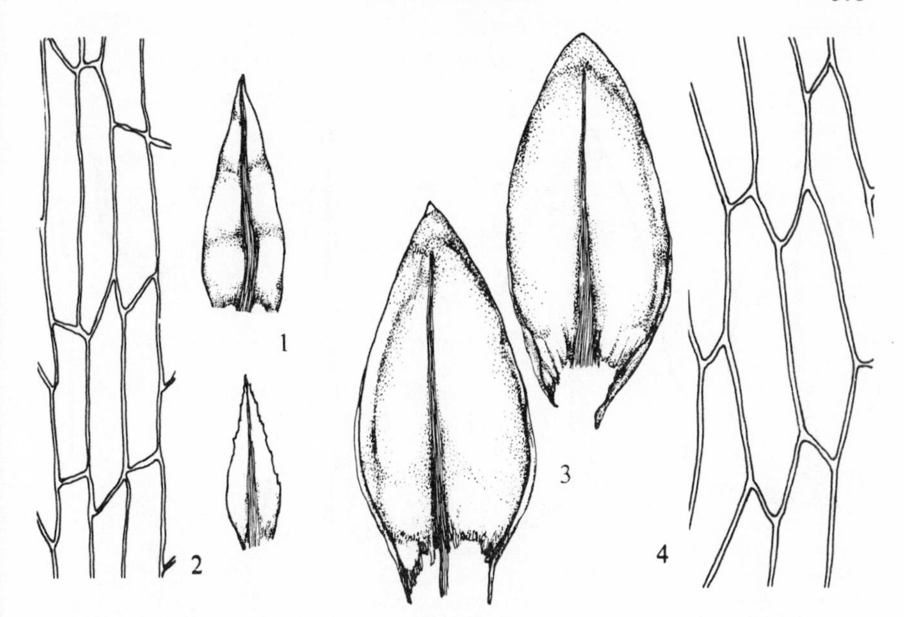

Fig. 178. 1–2, *Pohlia proligera*: 1, leaves; 2, mid-leaf cells. 3–4, *P. ludwigii*: 3, leaves; 4, mid-leaf cells. Leaves ×25, cells ×415.

13. P. ludwigii (Schwaegr.) Broth., Acta Soc. Sc. Fenn., 1892
Mniobryum ludwigii (Schwaegr.) Loeske, *Webera ludwigii* (Schwaegr.) Fürnr.

Dioecious. Plants (1–)2–10 cm. Lvs shrunken when dry, patent when moist, pink or red-tinged except uppermost, distant, concave, lower broadly ovate, obtuse to acute, upper ovate, acute, base longly decurrent, margin narrowly recurved, sinuose or faintly denticulate towards apex; nerve ending well below apex; cells lax, thin-walled, elongate-rhomboidal, 16–24 μm wide in mid-lf, narrower towards margin. Capsule ± pendulous, pyriform. Fr very rare, summer. Loose or dense tufts, often large, green above, reddish below, on damp montane soil and rock ledges, rare or occasional. N. Wales, N. England and Scottish Highlands. 21. N. and C. Europe, Faroes, Pyrenees, Caucasus, N. America, Greenland.

14. P. lutescens (Limpr.) Lindb. f., Acta Soc. Sc. Fenn., 1899
Mniobryum lutescens (Limpr.) Loeske

Dioecious. Plants 3–8 mm, stems brownish. Lvs flexuose, shrunken when dry, patent or spreading when moist, lanceolate to linear-lanceolate, acuminate, perichaetial lvs longer and narrower, margin plane or narrowly recurved below, denticulate, sometimes spinosely so, from about middle of lf; nerve ending below apex in lower lvs to longly excurrent in perichaetial lvs; cells narrowly rhomboidal to linear, in upper lvs 6–10 μm × 70–180 μm in mid-lf, wider in lower lvs. Rhizoidal gemmae, pale yellow, ± ellipsoid with markedly knobbly outline from projecting cell walls, 50–70 μm × 40–50 μm, always though sometimes sparsely present. Green, filamentous gemmae produced in lf axils during March. Male plants are rare in Britain. Small tufts or patches, glossy, yellowish-green, usually on clayey soil on banks, by paths, ditches streams, etc. Scattered localities from Cornwall north to C. Scotland and

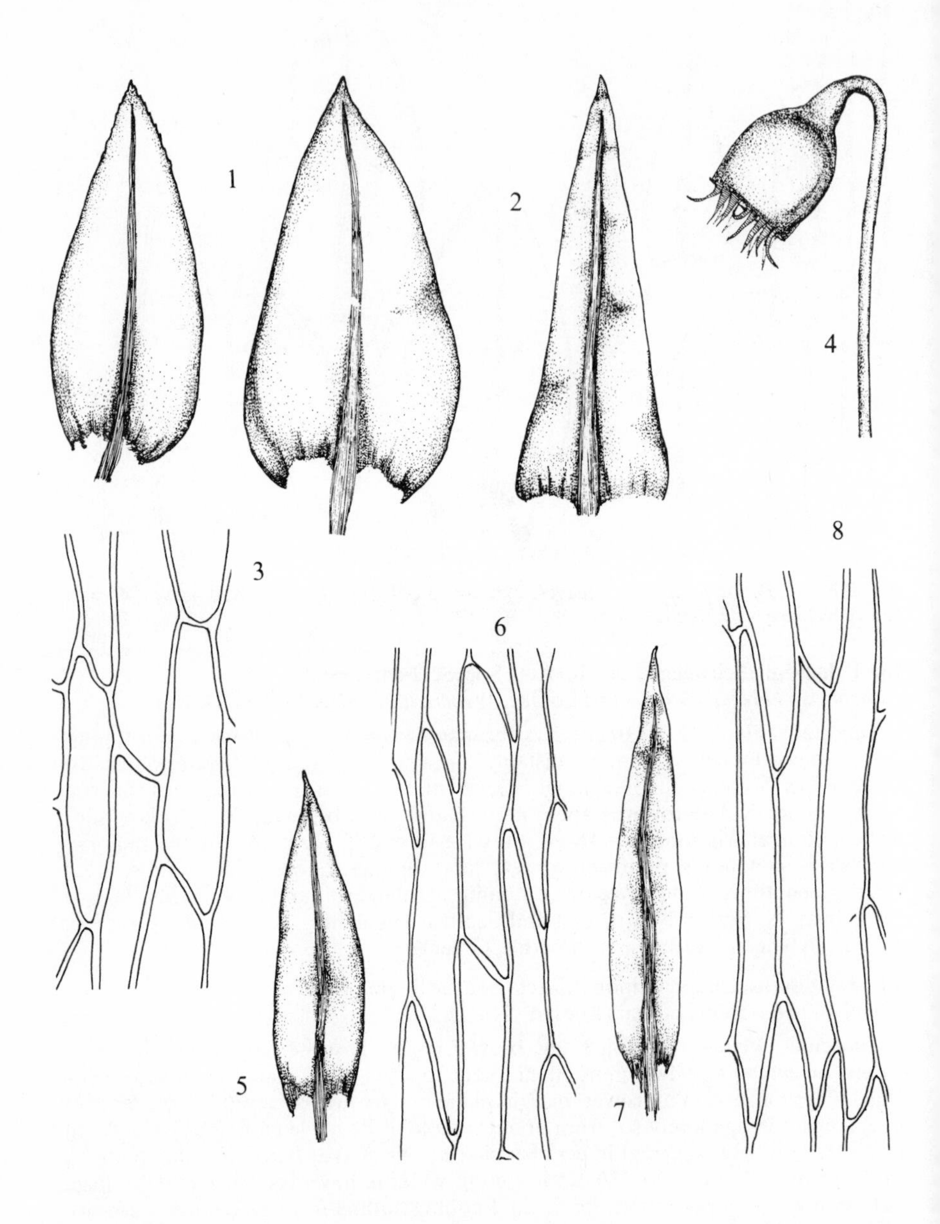

Fig. **179**. 1–4, *Pohlia wahlenbergii* var. *wahlenbergii*: 1, stem leaves; 2, perichaetial leaf; 3, mid-leaf cells; 4, capsule (×10). 5–8, *P. wahlenbergii* var. *calcarea*: 5 and 6, leaf and cells of broad-leaved form; 7 and 8, leaf and cells of narrow-leaved form. Leaves ×25, cells ×415.

Argyll, N. Kerry, occasional but previously overlooked as *P. carnea*. 33, H1, C. Europe, E. Asia.

Differs from *P. lescuriana* in the usually sharply denticulate upper lvs, narrower perichaetial lvs and narrower, longer cells. For an account of this plant in Britain see Watson, *Trans. Br. bryol. Soc.* **5**, 443–7, 1968.

15. P. lescuriana (Sull.) Andrews in Grout, Moss Fl. N. Amer., 1935
P. pulchella (Hedw.) Lindb., *Mniobryum pulchellum* (Hedw.) Schimp.

Dioecious. Plants *ca* 5 mm, stems brownish. Lvs flexuose when dry, spreading when moist, lower ovate-lanceolate, upper lanceolate, perichaetial lvs narrowly lanceolate, acuminate, margin plane or narrowly recurved below, obscurely toothed above; nerve ending below apex in vegetative lvs, percurrent to longly excurrent in perichaetial lvs; cells thin-walled, narrowly rhomboidal to linear-rhomboidal, in upper lvs 10–14 μm × *ca* 80 μm in mid-lf, wider in lower lvs. Rhizoidal gemmae, pale brown, spherical, ellipsoid or pyriform, 75–150 μm × 70–90 μm, always though sometimes sparsely present. Seta 1–2 cm; capsule pendulous, ovoid, neck short; exothecial cells ± isodiametric; lid conical, apiculate; inner peristome with nodulose cilia; spores 12–15 μm. Fr occasional. Dull or pale green tufts or patches, usually on clayey soil on banks, by streams, paths, ditches, etc., previously overlooked as *P. carnea*, occasional. Scattered localities from S. Devon and I. of Wight north to Sutherland and Caithness. 20, C. Europe, N. Asia, N. America.

For the occurrence of this plant in Britain see Warburg, *Trans. Br. bryol. Soc.* **4**, 760–2, 1965.

Section IV. *Mniobryum*

Dioecious. Capsule cernuous or pendulous, ovoid or pyriform, neck short; exothecial cells short; stomata immersed; basal membrane well developed, cilia simple or nodulose, processes perforated.

16. P. carnea (Schimp.) Lindb., Musc. Scand., 1879
P. delicatula (Hedw.) Grout, *Mniobryum delicatulum* (Hedw.) Dix., *Webera carnea* Schimp.

Dioecious. Plants 3–10 mm, stems red. Lvs flexuose when dry, patent when moist, frequently reddish at base, ovate to lanceolate, acute to acuminate, perichaetial lvs longer and narrower, margin plane, obscurely denticulate above; nerve ending below apex; cells thin-walled to slightly incrassate, ± rhomboidal, 16–30 μm × 120–200 μm in mid-lf, narrower at margin. Rhizoidal gemmae lacking. Seta short, *ca* 1 cm; capsule horizontal to pendulous, ovoid or pyriform, neck short; exothecial cells mostly about as long as wide; lid ± conical; inner peristome with appendiculate cilia; spores 14–22 μm. Fr occasional, winter, spring, Pale green to reddish, lax or dense tufts or patches on usually damp, clayey soil on banks, by paths, ditches, etc., frequent at low altitudes. 106, H30, C. Europe, N. and C. Asia, Japan, N. Africa, Azores, N. America.

P. lescuriana and *P. lutescens* differ in their narrower upper lvs and narrower cells and in the presence of rhizoidal gemmae. Lf shape and cell size are too variable in juvenile plants to allow identification unless gemmae are present. *P. wahlenbergii* differs in its wider lvs and glaucous-green colour.

17. P. wahlenbergii (Web. & Mohr) Andrews in Grout Moss. Fl. N. Amer., 1935
Mniobryum wahlenbergii (Web. & Mohr) Jenn., *P. albicans* (Wahlenb.) Lindb., *Webera albicans* (Wahlenb.) Schimp.

Dioecious. Plants 0.5–15.0 cm, stems bright red. Lvs shrunken when dry, erecto-patent to patent when moist, base slightly decurrent, narrowly lanceolate to ovate,

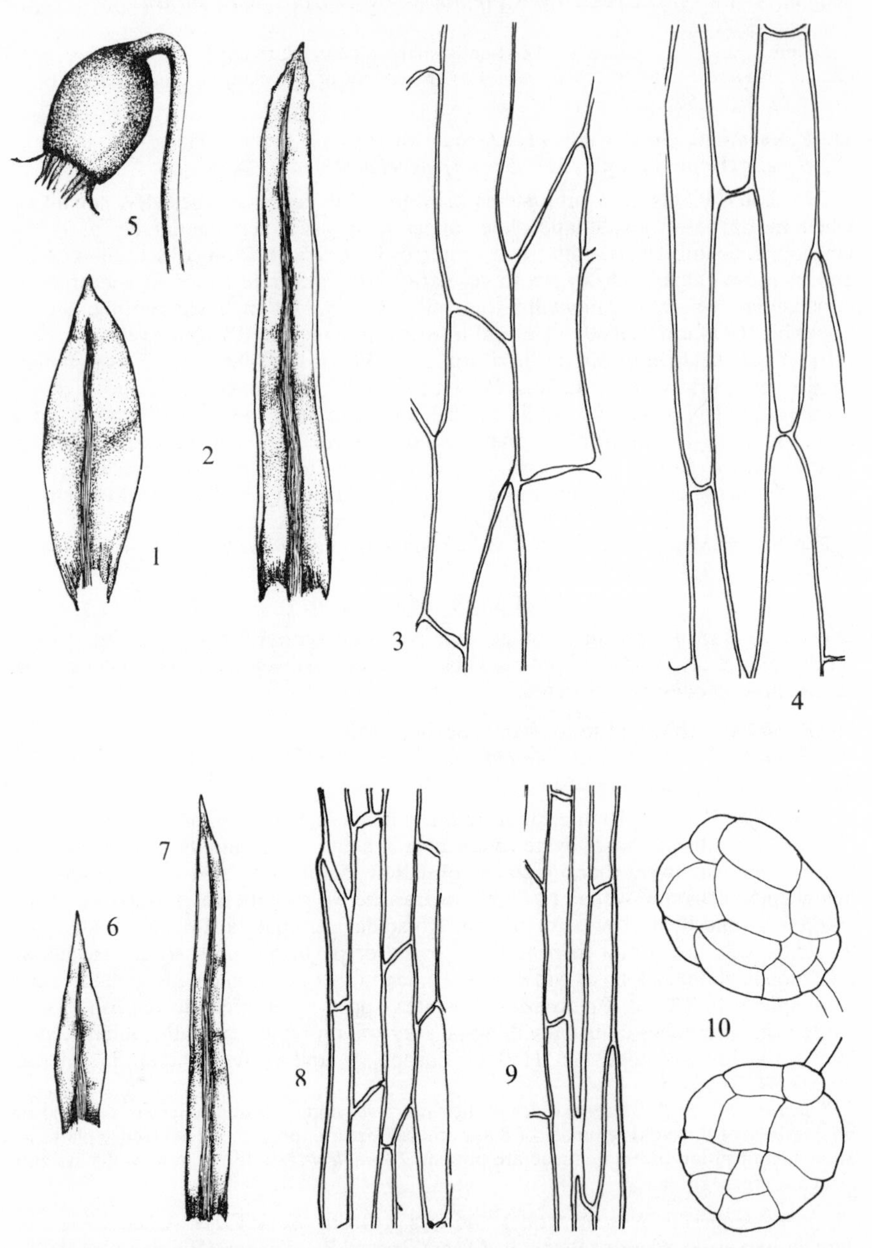

Fig. **180**. 1–5, *Pohlia carnea*: 1 and 2, stem and perichaetial leaves; 3 and 4, mid-leaf cells from stem and perichaetial leaves; 5, capsule (×10). 6–10, *P. lescuriana*: 6 and 7, lower and upper leaves; 8 and 9, mid-leaf cells from lower and upper cells; 10, rhizoidal gemmae (×250). Leaves ×25, cells ×415.

acute to acuminate, perichaetial lvs longer and gradually tapering from base to apex; margin plane, denticulate above; nerve reddish below, ending in or below apex or shortly excurrent in perichaetial lvs; cells elongate-hexagonal, narrower towards margin with 2–3 marginal rows very narrow. Seta pale red; capsule pendulous, pyriform, wide-mouthed when empty; cilia nodulose; spores 16–20 μm. Fr rare, summer.

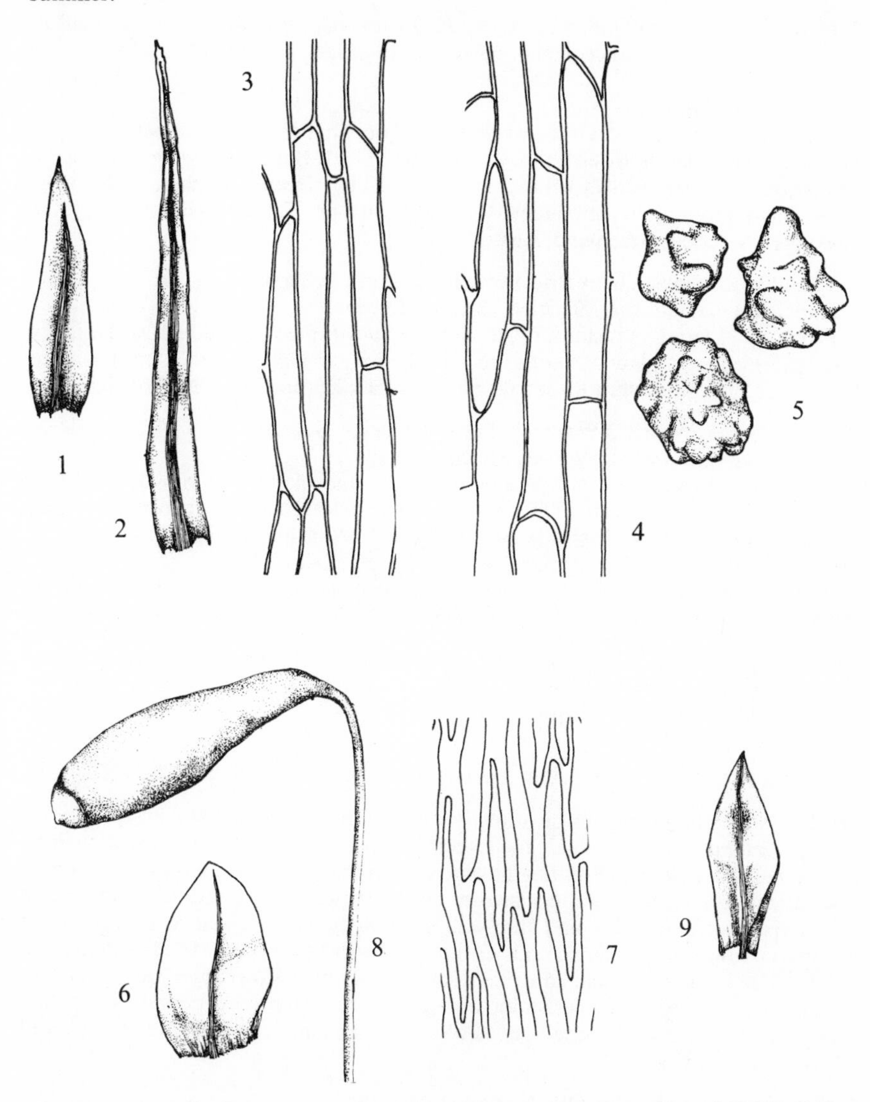

Fig. **181**. 1–5, *Pohlia lutescens*: 1 and 2, lower and upper leaves; 3 and 4, mid-leaf cells from lower and upper leaves; 5, rhizoidal gemmae (×250). 6–8, *Anomobryum filiforme* var. *filiforme*: 6, leaf; 7, mid-leaf cells; 8, capsule (×10). 9, *A. filiforme* var. *concinnatum*: leaf. Leaves ×25, cells ×415.

Key to intraspecific taxa of *P. wahlenbergii*

1 Plants pale, glaucous-green or whitish-green, mid-lf cells in upper lvs mostly 12–24 μm wide, lvs ovate or ovate-lanceolate var. **wahlenbergii**
Plants pale or dull green but not glaucous, lf cells mostly 20–30 μm wide or if less then lvs lanceolate or narrowly lanceolate 2

2 Plants to 15 cm, lvs ovate, cells mostly 20–30 μm wide in mid-lf var. **glacialis**
Plants to 1.5 cm, lvs lanceolate or narrowly lanceolate, cells mostly 10–20 μm var. **calcarea**

Var. **wahlenbergii**

Plants to 6 cm. Lvs ovate to ovate-lanceolate; mid-lf cells in upper lvs mostly 12–24 μm wide but in lower lvs to 40 μm wide. $n=11$. Soft, pale, glaucous-green or whitish-green tufts, reddish below, on damp soil by streams, ditches, paths, wood rides, waste ground, etc., frequent. 108, H39. Europe, Faroes, Asia, Algeria, Kerguelen Is., America, Greenland, Australasia.

Var. **glacialis** (Schleich. ex Brid.) Warburg, Trans. Br. bryol. Soc., 1962
P. albicans var. *glacialis* (Schleich. ex Brid.) Lindb.

Plants to 15 cm. Lvs distant, ovate; cells in mid-lf mostly 20–30 μm wide. Large, pale green but not glaucous tufts, reddish below, at high altitudes, rare. Caerns, Derby and Westmorland northwards, Sligo. 19, H1. Europe, Asia, N. America.

Var. **calcarea** (Warnst.) Warburg, Trans. Br. bryol. Soc., 1962
P. albicans var. *calcarea* (Warnst.) Rich. & Wall.

Plants to 1.5 cm. Lvs lanceolate or narrowly lanceolate; cells mostly 10–20 μm wide in mid-lf. Dull green tufts on chalky and calcareous soil in turf, very rare. Cornwall, Dorset, I. of Wight, E. Sussex, Surrey, W. Glos, Yorks, W. Galway, W. Donegal, Antrim. 10, H3. Germany, Hungary, Italy, Dalmatia.

Usually readily recognised by the glaucous, whitish-green soft tufts. The nature of the var. *calcarea* is obscure. Dixon & Nicholson (*J. Bot., Lond.* **63**, 126–7, 1925) describe 2 forms and this is borne out by observations on a very limited number of specimens, one form having lanceolate lvs with cells mostly 16–20 μm × 40–140 μm and the other with narrowly lanceolate lvs with cells mostly 14–16 μm × 100–200 μm. Further investigation is required. Var. *glacialis* is probably no more than a luxuriant habitat form.

86. Epipterygium Lindb., Öfv. K. V. A. Förh., 1863

Dioecious. Small, gregarious or tufted, terrestrial mosses. Lvs of sterile stems in 3–4 ranks, 2 lateral and 1–2 dorsal; the lateral lvs spreading, plane, broadly decurrent, from a narrow base ovate-oblong to obovate, shortly pointed, margin plane, entire or denticulate above; dorsal lvs smaller and narrower; nerve reddish, ending well below apex; cells very lax and thin-walled, elongate-rhomboidal to hexagonal, at margin narrower, prosenchymatous, forming distinct border; lvs of fertile shoots less differentiated. Capsule inclined or cernuous, ovoid with short thick neck; annulus broad, not persisting; inner peristome with tall basal membrane, processes perforated, cilia present, nodulose. A small genus of 18 species most of which occur in C. and S. America.

1. E. tozeri (Grev.) Lindb., Öfv. K. V. A. Förh., 1863
Webera tozeri (Grev.) Schimp.

Plants mostly *ca* 5 mm, stems reddish. Lvs distant, glossy, twisted when dry, spreading when moist, often of 2 distinct sizes, broadly ovate to ovate-lanceolate, obtuse and apiculate to acute, base very narrow, slightly decurrent, margin plane, entire or obscurely denticulate above, bordered; nerve reddish, ending well below apex; cells lax, narrowly hexagonal throughout except at margin, 20–40 μm wide in mid-lf, 3–6 marginal rows long, narrow, more incrassate, forming distinct, sometimes reddish border. Capsule horizontal to pendulous, pyriform or obovate, theca abruptly narrowed into neck, ± spherical when dry and empty; lid obtusely mamillate; inner peristome ciliate; spores 14–20 μm. Fr rare, spring. $n=11$. Small, pale green to reddish tufts or patches or as scattered plants on acidic, particularly clayey soil, on shaded banks in hedgerows, woods and by streams and rivers. Frequent in S.W. England, rare elsewhere and extending north to Shrops and Caerns, I. of Man, Argyll, Kintyre, east to Essex, rare in Ireland. 30, H6, C. France, Yugoslavia, Lebanon, Caucasus, Himalayas, Java, Morocco, Macaronesia, N. America.

Recognisable by the ± oval, bordered lvs with lax cells and nerve ending well below apex, these characters serving to distinguish the plant from species of *Pohlia* and *Bryum* with which it might be confused.

87. PLAGIOBRYUM LINDB., ÖFV. K. V. A. FÖRH., 1863

Dioecious. Plants tufted. Lvs imbricate or not, ovate to lanceolate, margin plane or recurved, unbordered; areolation lax, often thin-walled. Perichaetial and perigonial lvs longer, narrower, with narrower cells. Seta short, curved or cygneous; capsule gibbous, mouth oblique, neck long; peristome double, outer shorter than inner, cilia rudimentary. Seven species scattered through Europe, N. and E. Asia, Africa, N. America and New Zealand.

Tufts silvery or whitish above, reddish below, stem lvs imbricate, ovate, concave with nerve ending below apex **1. P. zieri**
Tufts reddish-brown, stem lvs not imbricate, lanceolate to ovate-lanceolate with excurrent nerve **2. P. demissum**

1. P. zieri (Hedw.) Lindb., Öfv. K. V. A. Förh., 1863

Plants to 2(–4) cm, shoots julaceous, stems brittle. Lvs imbricate, concave, hardly altered when dry, broadly ovate, acute to rounded or obtuse and apiculate, margin erect, unbordered, plane; nerve reddish, ending in or below apex; cells lax, thin-walled, elongate-hexagonal, 14–24 μm wide, 2–3 marginal rows narrower. Perichaetial and perigonial lvs longer, narrower. Seta curved, 6–10 mm long; capsule ± horizontal, ellipsoid, slightly asymmetrical with oblique mouth, neck 1–2 times length of theca; spores separating at maturity, 34–40 μm. Fr frequent, autumn. $n=10$*. Tufts silvery or whitish above, reddish below, on soil in crevices on basic rock, frequent in montane areas. Monmouth, Staffs, Wales and Derby northwards, occasional in Ireland. 54, H9. N., W. and C. Europe, Faroes, Iceland, Caucasus, N. and C. Asia, China, N. Africa, N. America, Greenland.

Readily recognised by the silvery or whitish shoots which are reddish below and, when fertile, by the asymmetrical, long-necked capsule. The reddish-tinged older parts distinguish the plant from *Anomobryum filiforme* and *Bryum argenteum*. *P. demissum* differs in the narrower lvs with excurrent nerve, the ± pendulous capsule, the spores adhering in tetrads and the reddish-brown coloration of the tufts.

2. P. demissum (Hook.) Lindb., Öfv. K. V. A. Förh., 1863

Shoots not julaceous, to 0.5 cm. Lvs ± erect, slightly concave, lanceolate to ovate-lanceolate, acuminate, margin plane or recurved; nerve reddish, excurrent in upper lvs; cells narrowly hexagonal to rectangular, thicker-walled than in *P. zieri*, 18–22 μm wide in mid-lf, 2–3 marginal rows narrower. Seta cygneous; capsule ± pendulous,

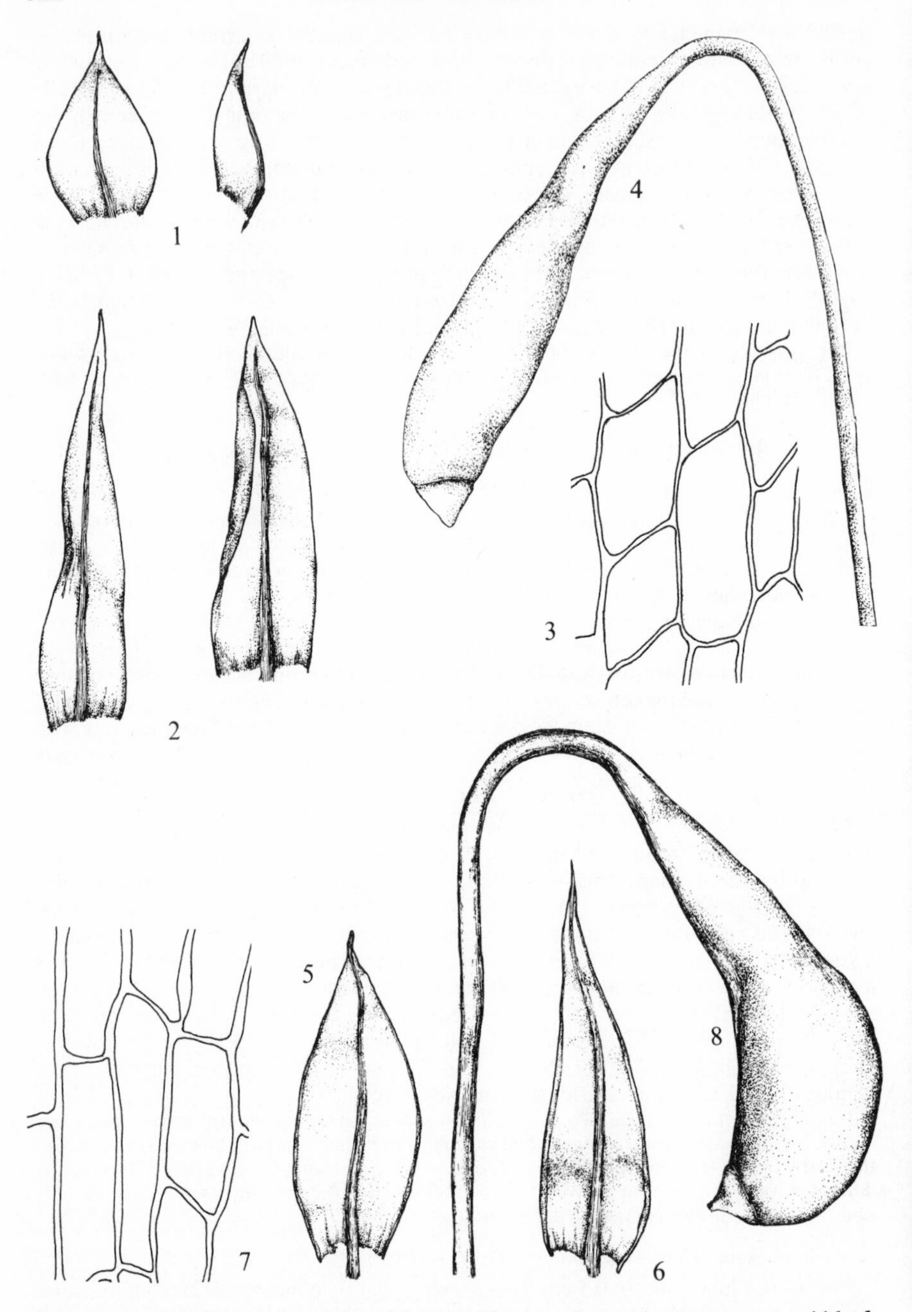

Fig. **182**. 1–4, *Plagiobryum zieri*: 1 and 2, leaves from sterile and fertile stems; 3, mid-leaf cells; 4, capsule. 5–8, *P. demissum*: 5 and 6, leaves from sterile and fertile stems; 7, mid-leaf cells; 8, capsule. Leaves ×25, cells ×415, capsules ×10.

very asymmetrical, mouth oblique, neck about same length as theca; spores adhering in tetrads, *ca* 40 μm diameter. Fr frequent, summer. Reddish-brown tufts on soil in calcareous rock crevices in the mountains, very rare. Mid Perth, Angus, Argyll. 3. N. and C. Europe, Iceland, Caucasus, C. Asia, China, N. America, Greenland.

88. Anomobryum Schimp., Syn., 1860

Shoots julaceous. Lvs concave, ± imbricate, lanceolate, obtuse to acute, margin plane, entire, unbordered; nerve ending below apex to excurrent; basal cells ± rectangular, upper narrowly hexagonal to vermicular. Seta long; capsule ± horizontal to pendulous, pyriform to narrowly pyriform with neck about ½ length of theca; cilia appendiculate. A ± world-wide genus of about 65 species.

1. A. filiforme (Dicks.) Solms-Laub. in Rabenh., Bryoth. Eur., 1873
A. julaceum (Sm.) Schimp., *Pohlia filiformis* (Dicks.) Andrews

Dioecious. Shoots slender, pale or yellowish-green above, pale brown below, to 5(–8) cm. Lvs imbricate, scarcely altered when dry, margin plane, entire or obscurely toothed above; nerve weak; basal cells rectangular, towards margin and above becoming narrower, linear-vermicular. Vegetative propagules in form of small, clavate axillary shoots sometimes present. Capsule inclined or pendulous, pyriform to narrowly pyriform; spore size varying from plant to plant, 10–18 μm.

Var. **filiforme**
Shoots julaceous. Lvs very concave, ovate, obtuse or obtuse and apiculate; nerve ending ½–¾ way up lf; cells in mid-lf 10–16 μm wide. Fr occasional, autumn. Dense, pale or yellowish-green tufts, pale brown below, usually on damp gravelly or stony soil and rocks near streams and waterfalls, frequent to common in montane areas. Cornwall, Monmouth, Wales and Derby northwards. 65, H22. Europe, Faroes, Iceland, Caucasus, China, Macaronesia, Cameroon, Ethiopia, Ruwenzori, Congo, America.

Var. **concinnatum** (Spruce) Loeske, Rev. Bryol. Lichen., 1933
A. concinnatum (Spruce) Lindb.
Shoots more slender than in the type, scarcely julaceous. Lvs slightly concave, ovate to lanceolate, acute; nerve extending to apex or nearly so; cells ± narrowly hexagonal. Cells 9–12 μm wide. Fr unknown. Similar habitats to the type, rare. Wales, Yorks and Westmorland northwards. 18, H7. N. and C. Europe, Pyrenees, Iceland, Caucasus, China, N. America, Greenland.

A. juliforme Solms-Laub., recorded from Cornwall, is said to differ from *A. filiforme* in its yellowish colour, more compact habit, lvs narrower with recurved apiculus and smaller capsule with orange peristome and smaller spores (*ca* 12 μm). In the absence of capsules the Cornish plant cannot be named, but it is very likely that *A. juliforme* is no more than a dry habitat form of *A. filiforme*, since the characters distinguishing it occur in various combinations in *A. filiforme*.

89. Bryum Hedw., Sp. Musc., 1801

Dioecious, synoecious or autoecious. Plants usually densely tufted, stems innovating from below the inflorescence. Lvs usually ovate or lanceolate, margin plane to revolute, entire to denticulate, rarely dentate above; nerve ending below apex to longly excurrent; cells ± rhomboid-hexagonal, rarely narrower, 1–several marginal rows often narrower forming obscure to well defined 1–3-stratose border. Seta long; capsule horizontal to pendulous, broadly pyriform to sub-cylindrical; lid conical to mamillate; calyptra small, cucullate, fugacious; peristome double, outer of 16 entire or occasionally perforated teeth with numerous thickened, transverse articulations, the inner of a thin basal membrane *ca* $\frac{1}{3}$–$\frac{1}{2}$ height of peristome bearing 16 perforated processes alternating with the outer teeth, 0–4 rudimentary, simple, nodulose or appendiculate cilia between processes. A very large cosmopolitan, taxonomically difficult genus of *ca* 1050 mainly terrestrial or saxicolous species. The arrangement of British species, with certain exceptions follows that of Nyholm (1954–69), which seems the most logical. In the classification given below the sections are somewhat arbitrary but group together mainly similar species, although intersectional hybridity may occur.

Notes on the identification of *Bryum* species

The key below is intended only as a guide to identification. Careful comparison of plants and descriptions should be made before naming. Some species of *Bryum* are very variable morphologically but the key caters only for 'typical' specimens: other species hybridise readily and it has to be accepted that not all gatherings can be named.

For determining the sex of plants, gametangia may be found within enlarged comal tufts of lvs; alternatively they may be sought at the base of setae of mature capsules, though in such situations old antheridia may readily be lost. The inner peristome is best observed by placing a ripe or dehisced capsule in a drop of water or dilute liquid detergent, cutting the top of the capsule longitudinally and then transversely immediately behind the mouth. This will provide 2 halves of peristome which may be flattened out under a cover-slip. Care should be taken not to confuse split processes (inner peristome teeth) with cilia, which are easily visible though fragile in species that possess them. Species with appendiculate or nodulose cilia have spores 8–20(–26) μm, those without or with rudimentary or simple cilia have spores (16–)20–45 μm.

1 Lf apices hyaline, shoots silvery-white when dry, julaceous with imbricate lvs **41. B. argenteum**
Lf apices not hyaline nor shoots silvery-white when dry, julaceous or not 2

2 Bulbils or filamentous or spherical gemmae present in lf axils at least of sterile shoots 3
Axillary bulbils or gemmae lacking, rhizoidal gemmae present or not 7

3 Plants with axillary bulbils 4
Plants with filamentous or spherical axillary gemmae 5

4 Lvs ovate to ovate-lanceolate, nerve to 80 μm wide near lf base (**B. bicolor** complex) 64
Lvs ovate-lanceolate to lanceolate, nerve 100 μm or more wide near lf base **32. B. gemmiparum**

5 Gemmae ± spherical, deep red (**B. erythrocarpum** complex) 56
Gemmae filamentous, greenish 6

6 Lf apex obtuse or rounded, lvs not reddish at base **13. B. cyclophyllum**
Lf apex acute, base reddish **23. B. flaccidum**

7 Lf cells very long and narrow, 8–12 μm × 60–80(–112) μm in mid-lf, lvs unbordered, plants usually purplish-red with metallic sheen **31. B. alpinum**
Plants lacking above combination of characters 8

8 Lf base longly decurrent, lvs pink-tinged, plants robust, to 12 cm tall **12. B. weigelii**
Lf base not decurrent, lvs pink-tinged or not, plants robust or not 9

9 Lf apex rounded or obtuse, apiculate or not 10
Lf apex acute or acuminate 15

10 Lvs with distinct border 11
Lvs unbordered or border ill-defined 12

11 Synoecious, lvs crowded, shoots bud-like, lf base not reddish **7. B. lawersianum**
Dioecious, lvs not crowded nor shoots bud-like, lf base reddish **29. B. neodamense**

12 Lvs not shrunken when dry, lf base reddish **33. B. muehlenbeckii**
Lvs shrunken when dry, base not reddish 13

13 Dioecious, lvs distant, obovate to orbicular **13. B. cyclophyllum**
Autoecious, lvs crowded, ovate 14

14 Capsule ± as long as wide, lf margin ± plane, very obscurely bordered, cells mostly 60–80 μm long in mid-lf **1. B. marratii**
Capsule longer than wide, margin of upper lvs recurved, ± bordered, cells mostly 40–60 μm long in mid-lf **4. B. calophyllum**

15 Plants without mature capsules 16
Plants with mature capsules 33

16 Lvs broadly ovate, very concave, margin plane, cells 20–40 μm wide, plants 5–10 cm tall **11. B. schleicheri** var. **latifolium**
Plants lacking above combination of characters 17

17 Basal cells of lvs of similar colour to cells above 18
Basal cells of lvs reddish to brownish, differing in colour from cells above, at least in older lvs 20

18 Lvs not pink-tinged, nerve longly excurrent, margin recurved, mid-lf cells 12–16 μm wide **30. B. caespiticium**
Plants lacking above combination of characters 19

19 Dioecious, plants pink to vinous-red, margin recurved, bistratose **9. B. pallens**
Plants lacking above combination of characters – it is not possible to proceed further without mature capsules

20 Lvs spirally twisted round stem when dry, widest above middle **20. B. capillare**
Lvs not spirally twisted round stem when dry, widest above middle or not 21

21 Lf margin plane, unbordered **36. B. dixonii**
Lf margin recurved at least below or bordered or both 22

22 Lf margin unbordered 23
Lvs with at least ill-defined border 25

23 Margin distinctly toothed above **25. B. canariense**
Margin entire or denticulate 24

24 Lf cells 20–24 μm wide, rhizoidal gemmae lacking **33. B. muehlenbeckii**
Lf cells 10–16 μm wide, rhizoidal gemmae present (**B. erythrocarpum** complex) 56

25 Lower lvs very concave, apex obtuse, sometimes cucullate **29. B. neodamense**
Lower lvs not as above 26

26 Lvs widest above middle 27
Lvs widest below middle 29

27 Shoots julaceous when moist, rhizoids coarsely papillose **21. B. elegans**
Shoots not julaceous, rhizoids smooth or finely papillose 28

28 Plants usually synoecious, lf cells not porose, rhizoidal gemmae present **24. B. torquescens**
Plants dioecious, lf cells porose, rhizoidal gemmae lacking **22. B. stirtonii**

29 Rhizoidal gemmae present 30
Rhizoidal gemmae lacking 31

30 Gemmae flattened, lobed **34. B. riparium**
Gemmae spherical or pyriform, not lobed but cells sometimes protuberant (**B. erythrocarpum** complex) 56

31 Upper lvs scarcely larger or more crowded than lower 32
Upper lvs larger and more crowded than lower forming comal tuft – it is not possible to proceed further without mature fruit

32 Plants 1–10 cm, lvs distinctly bordered, cells 14–30 μm wide in mid-lf **28. B. pseudotriquetrum**
Plants 0.5–1.5 cm, lvs obscurely bordered, cells 8–16 μm wide in mid-lf **35. B. mildeanum**

33 Outer peristome teeth with obvious vertical and oblique lines joining transverse articulations 34
Oblique and vertical lines joining transverse articulations absent or very obscure 35

34 Lf base reddish, spores 22–36 μm, mouth of capsule red **14. B. algovicum** var. **rutheanum**
Lf base not reddish, spores 28–45 μm, mouth of capsule yellow to orange **3. B. warneum**

35 Cilia of inner peristome simple, rudimentary or absent, spores (16–)20–42 μm 36
Cilia appendiculate or nodulose, spores 8–20(–24) μm 43

36 Cells at lf base not differing in colour from cells above 37
Basal cells reddish to brownish 41

37 Dioecious **9. B. pallens**
Autoecious or synoecious 38

38 Spores 30–42 μm 39
Spores 20–30 μm 40

39 Capsule pyriform, neck short, spores 36–42 μm, coastal plant **2. B. mamillatum**
Capsule narrowly pyriform, neck about ½ total length of capsule, spores 30–36 μm **5. B. purpurascens**

40 Plants deep red, border of lf partially bistratose, capsule symmetrical **6. B. arcticum**
Plants greenish, border 2 or more stratose, capsule mouth oblique **8. B. uliginosum**

41 Border weak, nerve ending below apex or percurrent **17. B. knowltonii**
Border well developed, nerve excurrent 42

42 Outer peristome teeth not perforated near base, exothecial cells near capsule mouth mostly 20–25 μm wide **16. B. inclinatum**
Outer peristome teeth frequently perforated near base, exothecial cells near mouth mostly 30–40 μm wide **15. B. salinum**

43 Lvs spirally twisted round stems when dry, widest above middle **20. B. capillare**
Lvs not spirally twisted round stem when dry, widest above middle or not 44

44 Basal cells of lvs of similar colour to cells above 45
Basal cells of lvs reddish or brownish at least in older lvs 47

45 Capsule turbinate when dry and empty, lf margin plane **10. B. turbinatum**
Capsule not turbinate when dry and empty, margin recurved 46

46 Lvs pinkish-tinged, border well developed, 2-stratose, cells 15–30 μm wide in mid-lf **9. B. pallens**
Lvs not pink-tinged, border poorly developed, unistratose, cells 12–16 μm wide in mid-lf **30. B. caespiticium**

47 Border very stout, 2–3-stratose, confluent with nerve at lf apex **19. B. donianum**
Border ± absent to well developed, unistratose, not confluent with nerve at lf apex 48

48 Lvs ± unbordered, distinctly toothed above **25. B. canariense**
Border poorly to well developed, margin entire or denticulate above 49

49 Dioecious 50
Autoecious or synoecious 52

50 Upper lvs larger and more crowded than lower, forming distinct comal tuft **30. B. caespiticium**
Upper lvs scarcely larger or more crowded than lower, not forming comal tuft 51

51 Plants robust, 1–15 cm tall, mid-lf cells 14–30 μm wide, rhizoidal gemmae lacking **28. B. pseudotriquetrum**
Plants to 1.5 cm tall, mid-lf cells 10–20 μm wide, rhizoidal gemmae present (**B. erythrocarpum** complex) 56

52 Lf border poorly developed, capsule mouth oblique, cilia nodulose, spores 18–24 μm **18. B. intermedium**
Border well developed, mouth oblique or not, cilia appendiculate, spores 12–20(–22) μm 53

53 Lvs mostly widest above middle, red rhizoidal gemmae present **24. B. torquescens**
Lvs mostly widest at or below middle, rhizoidal gemmae lacking 54

54 Upper lvs hardly forming comal tuft, nerve only slightly excurrent, processes with perforations about as wide as long **28. B. pseudotriquetrum** var. **bimum**
Upper lvs crowded into comal tuft, nerve markedly excurrent or if only shortly excurrent then process perforations 1½–2 times as long as wide 55

55 Spores 18–20(–22) μm, process perforations 1½–2 times as long as wide, plants sometimes autoecious **27. B. pallescens**
Spores 14–16 μm, process perforations about as long as wide, plants never autoecious **26. B. creberrimum**

Key to species of the *B. erythrocarpum* complex

56 Rhizoidal gemmae mostly less than 100 μm 57
Gemmae mostly more than 120 μm 59

57 Gemmae brown, pyriform, about twice as long as wide, 3–5 cells long, 2 cells wide **46. B. sauteri**
Gemmae reddish, ± spherical, at least 3 cells wide throughout 58

58 Rhizoids dull mauve to bright violet, cells of gemmae not protuberant **44. B. violaceum**
Rhizoids pale brown, gemmae with protuberant cells **45. B. klinggraeffii**

59 Rhizoids usually deep violet **43. B. ruderale**
Rhizoids paler, not violet 60

60 Lvs not or scarcely bordered, cells 10–16 μm wide 61
Lvs distinctly bordered, cells 14–20 μm wide 63

61 Nerve strong, longly excurrent, basal cells of lvs ± quadrate, rhizoids densely papillose, gemmae usually brownish, not contrasting with rhizoids, calcicole **42. B. radiculosum**
Nerve less strong, shortly excurrent, basal cells (except sometimes in lower lvs) rectangular, rhizoids not densely papillose, gemmae red or yellowish, contrasting strongly in colour with rhizoids 62

62 Gemmae yellowish, seldom more than 180 μm diameter **47. B. tenuisetum**
Gemmae red, frequently more than 200 μm diameter **48. B. microerythrocarpum**

63 Gemmae often more than 250 μm, usually on long rhizoids, never axillary, cells of gemmae protuberant or not, 40–70 μm diameter **49. B. bornholmense**
Gemmae rarely more than 250 μm diameter, clustered round base of stem, often axillary, cells of gemmae protuberant, 25–40(–50) μm diameter **50. B. rubens**

Key to bulbiliferous species of *B. bicolor* complex

64 Nerve in upper lvs longly excurrent, to $\frac{2}{3}$ length of lamina, bulbils solitary in lf axils, 480–640 μm long **40. B. dunense**
Nerve not or only shortly excurrent or if strongly excurrent then bulbils many per axil, bulbils 50–480 μm long 65

65 Bulbils (5–)20–30 per axil, 100–160(–200) μm long with distinct lf primordia, orange or reddish in colour **38. B. gemmiferum**
Bulbils not more than 5 per axil, 110–480 μm, with or without distinct lf primordia, green or yellowish in colour 66

66 Bulbils green, lf primordia about $\frac{1}{2}$ total length of bulbil **37. B. bicolor**
Bulbils yellowish, lf primordia rudimentary or indistinguishable **39. B. gemmilucens**

Section I

Lf of uniform colour throughout; border well defined or not, usually bistratose; cells usually thin-walled, 16–40 μm wide in mid-lf; cilia absent to long and appendiculate; spores 20–40 μm.

1. B. marratii Wils., Br. Brit., 1855

Autoecious. Plants 2–5 mm. Lower lvs distant, upper crowded, shrunken when dry, spreading when moist, not reddish at base, ovate, obtuse, margin plane or narrowly recurved below, entire or obscurely denticulate above, upper lvs obscurely bordered; nerve yellowish to brown, ending below apex; basal cells rectangular, above irregularly rectangular to rhomboid-hexagonal, 16–26 μm × 60–80(–90) μm in mid-lf, 2–3 marginal rows in upper lvs narrower forming very ill-defined border. Seta very slender; capsule pendulous, broadly pyriform, hardly longer than wide, neck short, abruptly narrowed into seta, mouth small, orange; lid mamillate with obtuse beak; inner and outer peristome partly fused, outer teeth reddish below, yellow above, processes coarsely papillose, narrowly perforated, cilia rudimentary; spores 26–32 μm. Fr occasional, summer. Patches in damp dune slacks, rare. Merioneth north to Ayr and Outer Hebrides, Lincs to Fife and Angus, W. Mayo, W. Donegal, Londonderry. 13, H3. Scandinavia, Latvia, Estonia, Faroes, Holland, France, Germany, N. America.

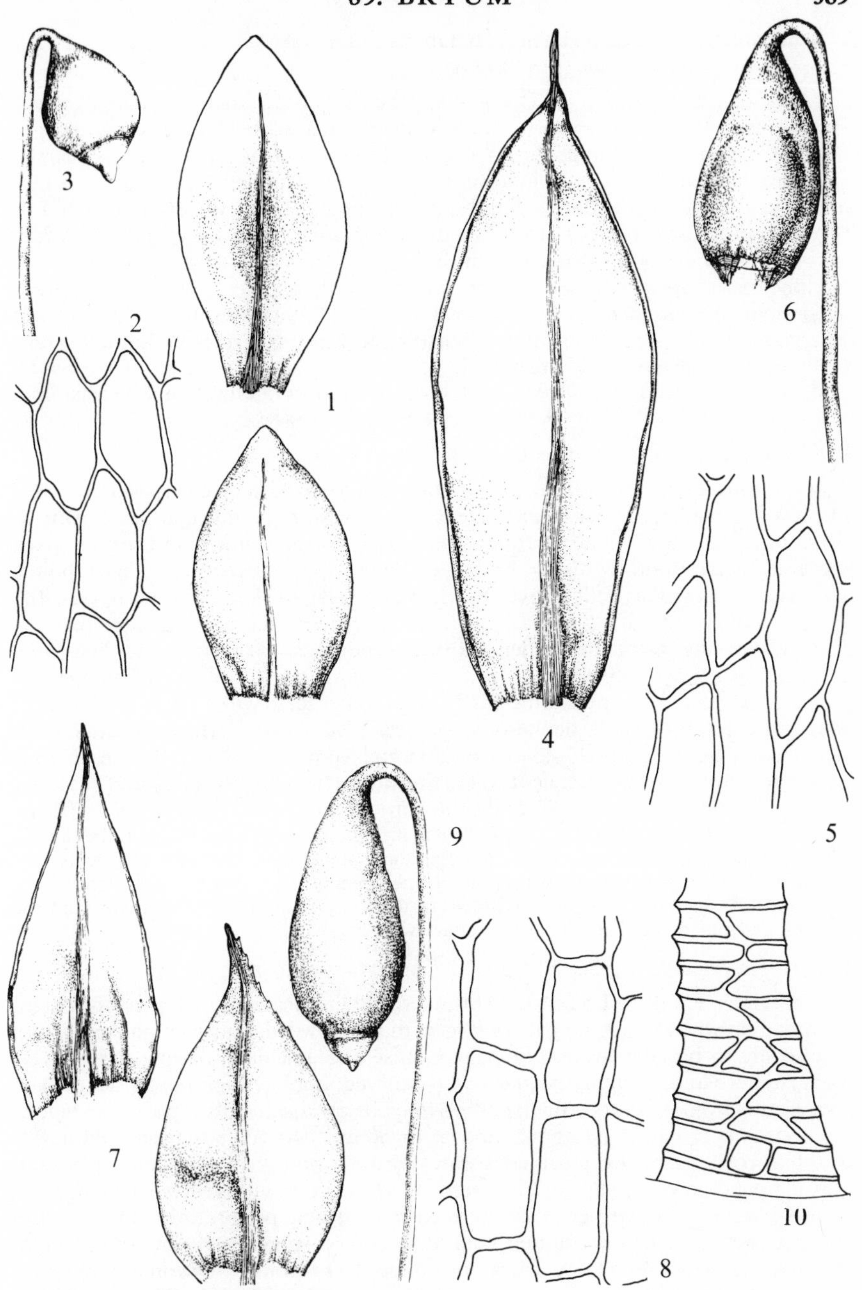

Fig. **183**. 1–3, *B. marratii*: 1, leaves; 2, mid-leaf cells; 3, capsule. 4–6, *B. mamillatum*: 4, leaf; 5, mid-leaf cells; 6, capsule. 7–10, *B. warneum*: 7, leaves; 8, mid-leaf cells; 9, capsule; 10, lower part of outer peristome tooth (× 250). Leaves × 25, cells × 415, capsules × 10.

2. B. mamillatum Lindb. in Hartm., Handb. Skand. Fl. 1864
B. warneum ssp. *mamillatum* (Lindb.) Podp.

Autoecious. Plants to *ca* 0.5 cm. Lvs appressed, twisted when dry, erecto-patent when moist, not reddish at base, ovate to lanceolate, acute to acuminate, margin recurved to about mid-lf in upper lvs, obscurely denticulate towards apex, bordered; nerve brownish, ending in apex in lower lvs, excurrent in upper; basal cells rectangular, above shortly rectangular to rhomboid-hexagonal, 12–20(–30) μm wide in mid-lf, 2–3 marginal rows narrow, forming distinct border. Capsule pyriform, wide-mouthed when dry and empty, mouth reddish-brown; lid convex, apiculate; outer and inner peristome partly fused, outer teeth yellowish, papillose, processes narrowly perforated, cilia rudimentary or absent; spores 36–42 μm. Fr spring. Sandy soil and dune slacks by the sea, very rare. W. Norfolk, N. Lincs, S. Lancs. 3. Baltic Islands, Svalbard Archipelago, Alps, Greenland.

Intermediate between *B. marratii* and *B. warneum*, distinct in capsule and lf shape from the former; for differences from *B. warneum* see under that species.

3. B. warneum (Röhl.) Bland. ex Brid., Br. Univ., 1826

Autoecious. Plants 0.5–1.0 cm. Lvs ± equally spaced on stem, erect, flexuose when dry, lower spreading, upper erecto-patent when moist, not reddish at base, ovate to ovate-lanceolate, acuminate, margin recurved, entire or denticulate towards apex, bordered; nerve stout, reddish to brown, ending in apex to excurrent in short point; basal cells rectangular, cells above shortly rectangular to rhomboid-hexagonal, 18–30 μm wide in mid-lf, 2–3 marginal rows narrower, more incrassate, forming distinct, partially bistratose border. Capsule pendulous, pyriform, *ca* twice as long as wide, neck gradually tapering into seta, mouth small, yellow to orange; lid conical; outer and inner peristome partly fused, outer teeth with vertical and oblique lines joining transverse articulations; processes narrowly perforated, cilia rudimentary or absent; spores 28–45 μm. Fr common, summer, autumn. Patches in dune slacks, sometimes locally frequent, rare. From N. Devon to W. Ross and N. Lincs to Nairn, Dublin. 21, H1. N., W. and C. Europe, Altai, Himalayas, southern Africa.

Differs from *B. marratii* and *B. calophyllum* in lf shape; from *B. mamillatum* in the capsule small-mouthed and peristome teeth with longitudinal and oblique lines joining articulations (the teeth of *B. mamillatum* have occasional lines joining the transverse articulations but these are not obvious). *B. algovicum* differs in the longer capsule, smaller spores and the more excurrent lf nerve. Cannot be determined without capsules.

4. B. calophyllum R. Br., Suppl. Append. Capt. Parry's Voyage, 1823

Autoecious. Plants to 1.5 cm. Lvs shrunken and imbricate when dry, erecto-patent to spreading when moist, comal lvs larger, more crowded, concave, not reddish at base, ovate to broadly ovate, obtuse or obtuse and apiculate, margin in lower lvs plane, unbordered, in upper lvs narrowly recurved, bordered, obscurely denticulate above; nerve yellowish to brown, ending in or below apex; cells rectangular below, rhomboid-hexagonal to hexagonal above, 16–30 μm × 40–60(–80) μm in mid-lf, 0–2 marginal rows narrower, more incrassate, forming poorly to well defined border. Capsule pendulous, ± ellipsoid to narrowly obovoid, mouth yellowish-orange; lid bluntly mamillate; outer peristome teeth with occasional perforations along median line; procseses narrowly perforated, cilia rudimentary; spores 26–40 μm. Fr common, autumn, winter. $n=40$. Damp, calcareous dune slacks from N. Devon and Durham north to Caithness, W. Mayo, rare. 16, H1. Europe, N. and E. Asia, Tibet, C. Africa, Arctic, N. America, Greenland.

Most likely to be confused with *B. marratii* or *B. cyclophyllum*. The latter differs in habitat and has broader, less concave lvs, the former differs in capsule shape. Sterile material

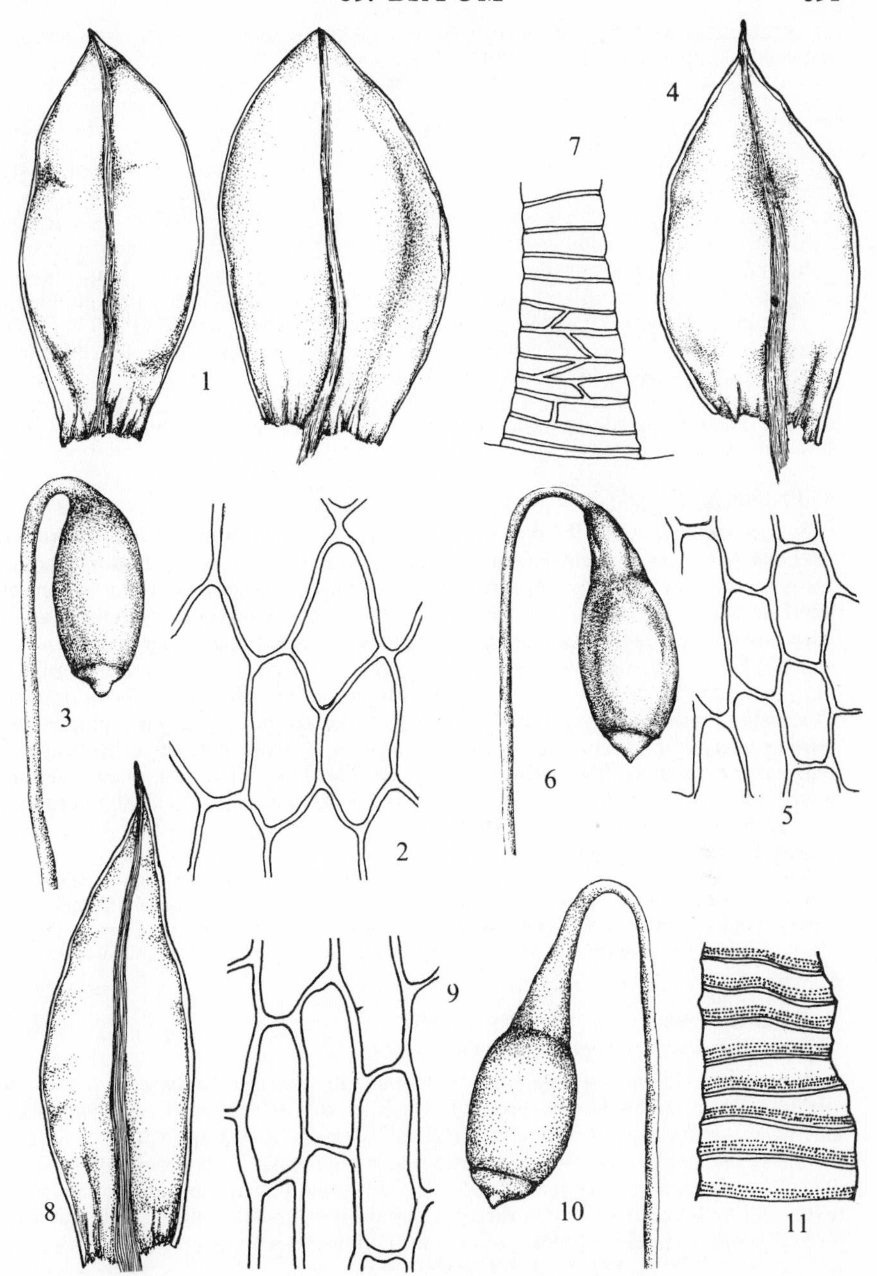

Fig. **184**. 1–3, *Bryum calophyllum*: 1, leaves; 2, mid-leaf cells; 3, capsule. 4–7, *B. arcticum*: 4, leaf; 5, mid-leaf cells; 6, capsule; 7, lower part of outer peristome tooth (×250). 8–11, *B. purpurascens*: 8, leaf; 9, mid-leaf cells; 10, capsule; 11, outer peristome tooth (×250). Leaves ×25, cells ×415, capsules ×10.

of *B. calophyllum* differs from that of *B. marratii* in the upper lvs wider with narrowly recurved margins with ± distinct border. Depauperate specimens may be difficult to separate but the lf cells of *B. calophyllum* are shorter in relation to their width than in *B. marratii*.

5. B. purpurascens (R. Br.) Br. Eur., 1846

Synoecious. Plants to 2 cm. Lvs erecto-patent when moist, ovate, acuminate, margin narrowly recurved, bordered, denticulate towards apex; nerve green to brownish, ending below apex to shortly excurrent; cells thin-walled, ± narrowly hexagonal, 16–30 μm wide in mid-lf, 2–3 marginal rows narrow, incrassate, forming yellowish border. Seta long, flexuose; capsule narrowly pyriform, neck long, about as long as theca; lid mamillate; outer peristome teeth opaque, with small papillae in horizontal rows between articulations, cilia poorly developed; spores 30–36 μm. Fr frequent, summer. Green or reddish-brown patches on damp sandy soil. Northern parts of Europe, Asia and N. America, Greenland.

Recorded by Whitehead from crevices of limestone rocks, Litton, Yorks, but in the absence of any specimens the record must be regarded as extremely doubtful in view of the low altitude and the habitat.

6. B. arcticum (R. Br.) Br. Eur., 1846

Synoecious. Plants *ca* 0.5 cm, red. Upper lvs crowded in comal tuft, appressed, flexuose when dry, erecto-patent or patent when moist, of ± uniform colour throughout, ovate-lanceolate, acuminate, margin recurved, entire or obscurely denticulate above, bordered; nerve reddish, excurrent; basal cells rectangular, above narrowly rectangular to hexagonal, 20–32 μm wide in mid-lf, 2–3 marginal rows very narrow, forming distinct border. Capsule cernuous or pendulous, pyriform, slightly asymmetrical, theca abruptly narrowed into neck, mouth small; exothecial cells thin-walled; lid shortly conical; outer peristome teeth with a few oblique lines joining transverse articulations on inside surface of lower part of teeth, processes narrowly perforated, cilia rudimentary; spores 20–28 μm. Fr common, late summer. $n=20$. Small, deep red patches or tufts on basic soil among rocks at high altitudes, very rare. Perth, Angus, S. Aberdeen. 4. N. and C. Europe, Siberia, Altai, Tibet, Korea, N. America, Greenland.

The lines joining the transverse articulations of the outer peristome teeth are faint and difficult to detect in British material. The plant is best recognised by its small stature and deep red coloration; it is only likely to be mistaken for alpine forms of *B. pallens*, especially its var. *fallax* which differs in the pink coloration, usually larger size and dioecious inflorescence.

7. B. lawersianum Philib., Rev. Bryol., 1899
B. pallens ssp. *lawersianum* (Philib.) Podp.

Synoecious. Plants bud-like. Lvs imbricate, not reddish at base, lower ovate, obtuse, comal much larger, broadly ovate to sub-orbicular, obtuse, innermost narrower, acute, margin recurved, bordered, towards apex plane with 1–2 obscure teeth; nerve orange-red, stout, shortly excurrent; basal cells rectangular, above irregularly narrowly hexagonal, 20–35 μm wide in mid-lf. Capsule narrowly pyriform with long neck about as long as theca; lid obtusely mamillate; inner peristome without cilia; spores finely papillose, 26–32 μm. Dull green patches on damp, micaceous soil at about 1070 m (3500 ft), Ben Lawers. 1. Endemic.

B. lawersianum is very distinctive in the bud-like plants with very broad lvs. It seems to be a good species and is probably related to *B. pallens* and *B. arcticum*. According to Dixon (1924) the lf base is reddish and the plant probably close to *B. inclinatum*, but both these points are incorrect. The original gathering was by Nicholson, Salmon & Dixon from Ben

Lawers on 27 July 1899 and there is one other by Stirling & Wheldon dated 9 September 1899, upon which the above description is based. The species has not been seen since.

8. B. uliginosum (Brid.) Br. Eur., 1839
B. cernuum (Hedw.) Lindb.

Autoecious. Plants to 3 cm. Lvs ± erect, slightly twisted when dry, erecto-patent when moist, not reddish at base, lower distant, ovate, acute, upper closer, lanceolate to narrowly lanceolate, acuminate, base slightly decurrent, margin recurved, ± entire, strongly bordered; nerve reddish-brown, percurrent or slightly excurrent; basal cells shortly rectangular, above rectangular to rhomboid-hexagonal, (10–)16–30(–40) μm wide in mid-lf, several marginal rows narrower, very incrassate, forming strong, 2–3-stratose border. Capsule cernuous or occasionally pendulous, narrowly ellipsoid, asymmetrical, mouth oblique, light brown, neck distinct, abruptly or gradually tapering into seta; lid mamillate; processes finely papillose with ± rectangular perforations, cilia short; spores 22–30 μm. Fr frequent, autumn. $n=10$. Greenish patches on damp soil by streams and in dune slacks, recorded from scattered localities from Berks, Worcs and N. Wales north to Caithness and in Ireland but now apparently very rare. 25, H9. N., W. and C. Europe, Faroes, Iceland, E. Siberia, Himalayas, N. America, Greenland.

This plant has decreased markedly in Britain over the past 100 years. It has been confused with *B. inclinatum*, *B. intermedium* and *B. pallens*, but differs from all these species in the autoecious inflorescence, from the first two in the lf base not reddish, and the last in the very stout border and short, simple cilia.

9. B. pallens Sw., Monthl. Rev. Lond., 1801

Dioecious. Plants usually pink to vinous-red, to 6 cm. Lvs more crowded above than below, successive comal tufts sometimes present, ± erect, rigid, flexuose or slightly curved when dry, erecto-patent to spreading when moist, base of similar colour to rest of lf, shortly decurrent, ovate to lanceolate, acute, margin recurved, entire or denticulate towards apex, bordered; nerve stout, deep pink to brownish, ending below apex to excurrent in short point; cells lax, basal rectangular, cells above variable, incrassate or not, rectangular to hexagonal, 15–30 μm wide in mid-lf, several marginal rows usually much narrower, very incrassate, bistratose, forming distinct border ending at or below apex. Capsule cernuous to pendulous, narrowly pyriform, ± straight or curved, slightly asymmetrical, processes with widely gaping perforations; spores 16–26 μm, often variable in size within a capsule. Fr rare, summer, autumn.

Var. **pallens**
Outer and inner peristomes not adherent, cilia usually well developed and appendiculate. $n=10$. Pink to vinous-red or greenish-red patches, tufts or scattered plants on soil, gravel, peat, etc., in damp places in turf, by roadsides, streams, in old quarries, flushes, on sand-dunes, etc., occasional in S.E. England, frequent or common elsewhere. 109, H34. Europe, W. and N. Asia, China, N. and C. Africa, N. America, Greenland, Ecuador, Andes.

Var. **fallax** Jur., Laubmoosfl. Oesterr. Ungarn, 1882
B. fallax Milde
Inner peristome adherent to outer below, cilia rudimentary or absent. Damp soil. Sussex, Caerns, S. Lancs, Mid W. Yorks, Mid Perth, S. Aberdeen, Wicklow. 6, H1. N., W. and C. Europe, C. Asia, Tunisia, Canada.

The only constant character for separating the type and var. *fallax* is the nature of the peristome and cilia. As some forms of the type also have poorly developed cilia the status

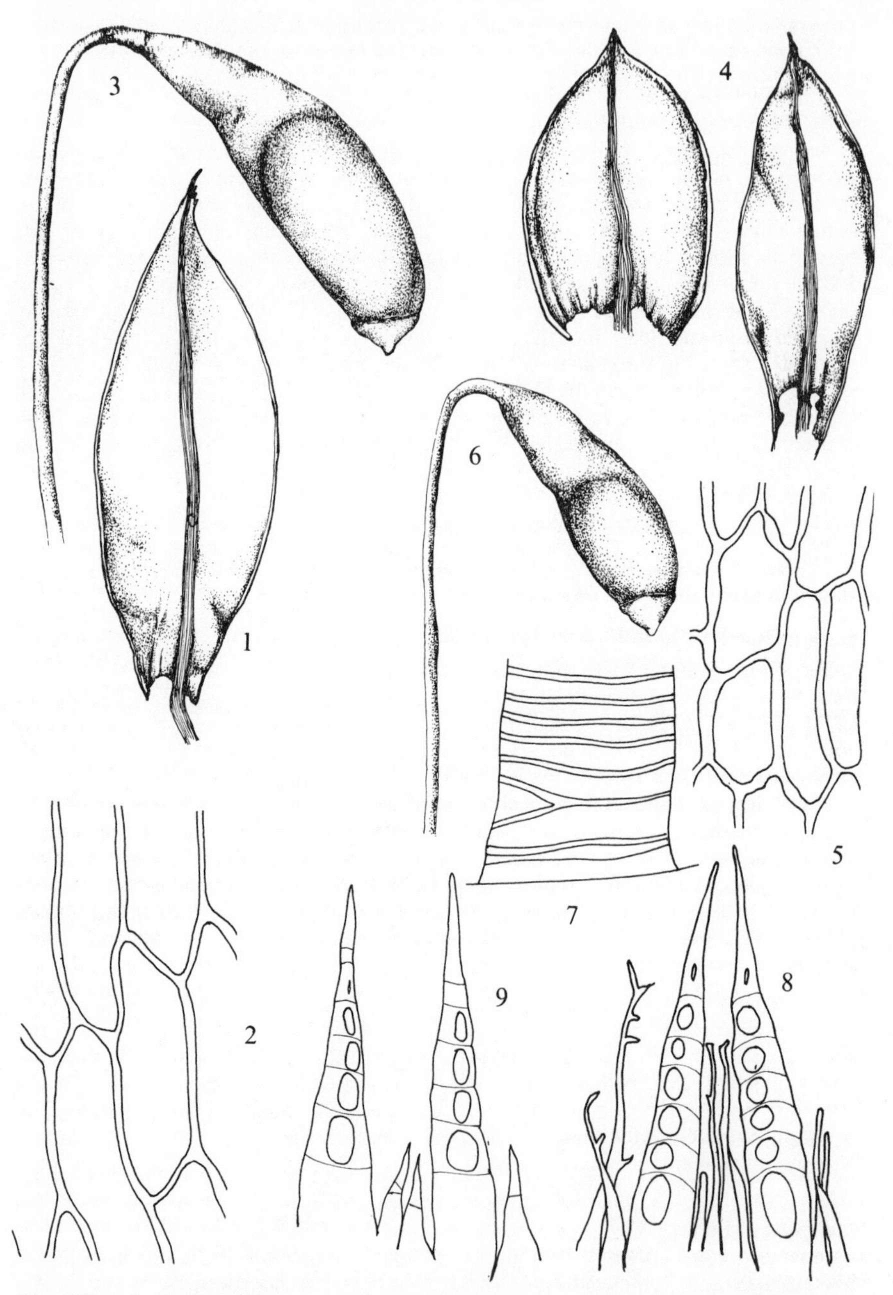

Fig. **185**. 1–3, *Bryum uliginosum*: 1, leaf; 2, mid-leaf cells; 3, capsule. 4–8, *B. pallens*: 4, leaves; 5, mid-leaf cells; 6, capsule; 7, basal part of outer peristome tooth (×250); 8, part of inner peristome. 9, *B. pallens* var. *fallax*: part of inner peristome. Leaves ×25, cells ×415, capsules ×10.

of the variety is very questionable. *B. pallens* is exceedingly variable but the pinkish or reddish, concolorous lvs will distinguish the species from most other British *Bryum* species. *B. alpinum* differs in lf shape, narrower cells and no border; *B. pseudotriquetrum* differs in the lvs with the base markedly reddish and unistratose border. *B. arcticum* has narrowly perforated processes, rudimentary cilia and synoecious inflorescence.

10. B. turbinatum (Hedw.) Turn., Musc. Hib., 1804

Dioecious. Plants 1–3 cm. Lvs erect, slightly twisted when dry, erecto-patent when moist, not reddish at base, ovate to ovate-lanceolate or narrowly triangular, acuminate, base slightly decurrent, margin plane, obscurely toothed above, bordered; nerve strong, reddish-brown, ending below apex to percurrent; basal cells shortly rectangular, cells above rectangular to rhomboid-hexagonal, 16–24(–30) μm wide in mid-lf, 2–3 marginal rows narrower, more incrassate, forming poorly defined border. Capsule pendulous, broadly pyriform, hardly longer than wide, small-mouthed, when dry markedly contracted below mouth both before and after dehiscence, when dry and empty turbinate; lid mamillate; processes widely perforated, cilia appendiculate; spores 16–20 μm. Fr spring, early summer, frequent. Green to pinkish patches on damp soil and dune slacks, formerly rare but sometimes locally abundant, but now very rare. W. Sussex, Oxford, Monmouth, Staffs, Merioneth, S. Lancs, S. Northumberland, E. Inverness, Skye. 9, H1. Europe, Iceland, Caucasus, C. and E. Asia, N. and C. Africa, N. America, Ecuador, Andes, Patagonia.

Readily recognised in fruit by the short capsule markedly contracted below the mouth when dry; the plane margin with poorly developed border is also a useful distinguishing character. Confused in the past with *B. pseudotriquetrum* which differs in capsule shape and recurved lf margin. Quite distinct from *B. schleicheri* var. *latifolium* in size, lf shape, areolation and habitat, but considered by Dixon (1924) to be linked by intermediate forms. Such forms, as for example that from Breconshire collected by Binstead & Dixon, on morphological grounds appear to be unrelated to either of the above plants but to be forms or hybrid derivates of *B. pseudotriquetrum*.

11. B. schleicheri DC. var. **latifolium** (Schwaegr.) Schimp., Syn. 2, 1876
B. turbinatum var. *latifolium* (Schleich.) Br. Eur.

Dioecious. Plants 5–10 cm. Lvs crowded, pinkish, appressed, straight when dry, erecto-patent when moist, very concave and frequently cucullate, reddish at base, broadly ovate, acute, sometimes apiculate, base not decurrent, margin plane or incurved below, denticulate above, bordered; nerve relatively thin, ending in apex or excurrent in short, reflexed apiculus; cells at extreme base lax, ± quadrate then rectangular or rhomboid, often reddish, cells above rhomboidal or rhomboid-hexagonal, gradually decreasing in size towards apex, 20–40(–60) μm wide in mid-lf, several marginal rows longer, narrower, forming distinct border. Capsule similar to that of *B. turbinatum*. Fr unknown in Britain. Tumid, yellowish-green tufts in montane flushes, very rare. Stirling, Mid Perth. 2. W. and C. Europe, Iran, Syria, C. and N. Asia, N. Africa.

12. B. weigelii Spreng., Mant. Prim. Fl. Halens., 1807

Dioecious. Plants to 12 cm. Lvs distant, ± uniformly spaced, shrunken when dry, patent when moist, pinkish throughout, ovate, obtuse and apiculate to acute or acuminate, base longly decurrent, margin plane or narrowly recurved below, entire or obscurely toothed above, bordered; nerve deep pink, ending below apex to shortly excurrent; basal cells rectangular, cells above irregularly rectangular to ± hexagonal, 12–24 μm wide in mid-lf, 2–3 marginal rows longer, narrower, more incrassate, forming distinct border. Fr unknown in Britain. Pinkish or greenish-pink patches in flushes and on sandy or gravelly soil by springs, usually at high altitudes,

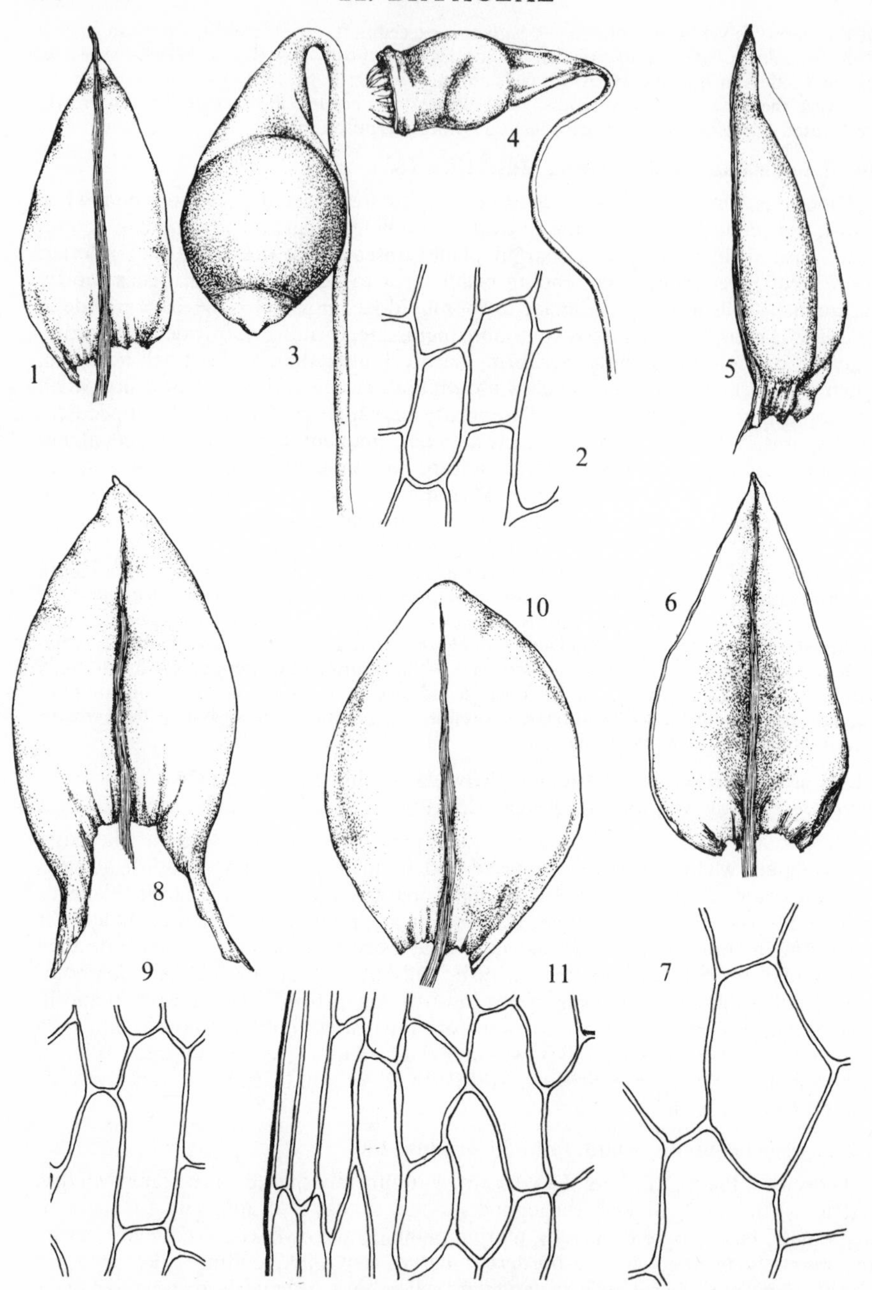

Fig. **186**. 1–4, *Bryum turbinatum*: 1, leaf; 2, mid-leaf cells; 3, mature capsule (×20); 4, dry empty capsule (×10). 5–7, *B. schleicheri* var. *latifolium*: 5 and 6, leaves; 7, mid-leaf cells; 8–9, *B. weigelii*: 8, leaf; 9, mid-leaf cells. 10–11, *B. cyclophyllum*: 10, leaf; 11, mid-leaf cells. Leaves ×25, cells ×415.

very rare in England, Wales and Ireland, occasional in Scotland. Shrops, Caerns, Yorks and Westmorland north to Orkney, Waterford. 29, H1. Europe, Caucasus, Siberia, E. Asia, Japan, N. America.

Only likely to be confused with *B. schleicheri* var. *latifolium* which has very concave, often cucullate lvs which have a usually reddish, non-decurrent base.

13. B. cyclophyllum (Schwaegr.) Br. Eur., 1846
B. tortifolium Brid.

Dioecious. Stems slender, often procumbent. Lvs distant, shrunken when dry, patent, soft when moist, not reddish at base, concave, obovate or ± orbicular, apex obtuse or rounded, base slightly decurrent, margin plane, entire, hardly bordered; nerve relatively thin, ending below apex; basal cells rectangular, cells above rectangular to rhomboid-hexagonal, 16–26 μm wide in mid-lf, 3–4 marginal rows somewhat narrower but not forming distinct border. Filamentous axillary gemmae often present. Fr unknown in Britain. $n=10$. Lax, straggling mats or patches on wet soil and marshy ground by streams and lakes very rare. Westmorland, Cumberland, Stirling, Argyll, Kintyre, W. Ross. 6. N.W. and C. Europe, Balkans, Siberia, Amur, Korea, Greenland.

Section II

Lf base red, differing in colour from rest of lf; border usually wide, unistratose, cells ± incrassate, 12-20(–24) μm wide; cilia mostly rudimentary; spores mostly 18–36 μm.

14. B. algovicum Sendtn. ex C. Müll. var. **rutheanum** (Warnst.) Crundw., Trans. Br. bryol. Soc., 1970
B. pendulum (Hornsch.) Schimp., *B. roellii* Phil., *B. angustirete* Kindb. ex Macoun

Synoecious. Plants to *ca* 1.5 cm. Lvs crowded above, sometimes in 2 or more successive comal tufts, appressed, flexuose when dry, lower erecto-patent, upper ± imbricate when moist, slightly concave, reddish at base, lanceolate to ovate, acuminate, margin recurved, entire or slightly denticulate above, bordered; nerve stout, reddish, excurrent in long cuspidate point; basal cells shortly rectangular, above ± rectangular to rhomboid-hexagonal or hexagonal, 16–28 μm wide in mid-lf, 3–4 marginal rows narrow, more incrassate, forming distinct border. Capsule pendulous, ± pyriform, symmetrical, mouth red; lid mamillate; outer teeth with oblique and vertical lines joining transverse articulations giving an irregular network appearance, basal membrane adherent to outer peristome, processes widely perforated, cilia absent or rudimentary; spores (18–)22–36 μm. Fr common, spring. $n=10$, 27*, 30*. Yellowish-green to dark green patches or tufts on sandy, especially basic soil in open places, sand-dunes, rock crevices and walls, frequent. 91, H30, C. Europe, Asia, southern Africa, N. America, New Zealand.

A somewhat variable species that can only be determined when capsules are present. The peristome teeth are of characteristic appearance and will readily separate *B. algovicum* var. *rutheanum* from all other British and Irish species except *B. warneum*; that species may be distinguished by the concolorous lvs, the small-mouthed capsule, the narrow perforations of the processes and the usually larger spores – the capsules are also later maturing. *B. algovicum* var. *algovicum*, which is dioecious, is not known in Britain or Ireland.

15. B. salinum Hagen ex Limpr., Laubm. Deutschl., 1892

Autoecious or synoecious. Close to *B. inclinatum*. Lf cells more pitted. Capsule of similar shape to that of *B. inclinatum* but neck tapering into seta, mouth red; 2–3

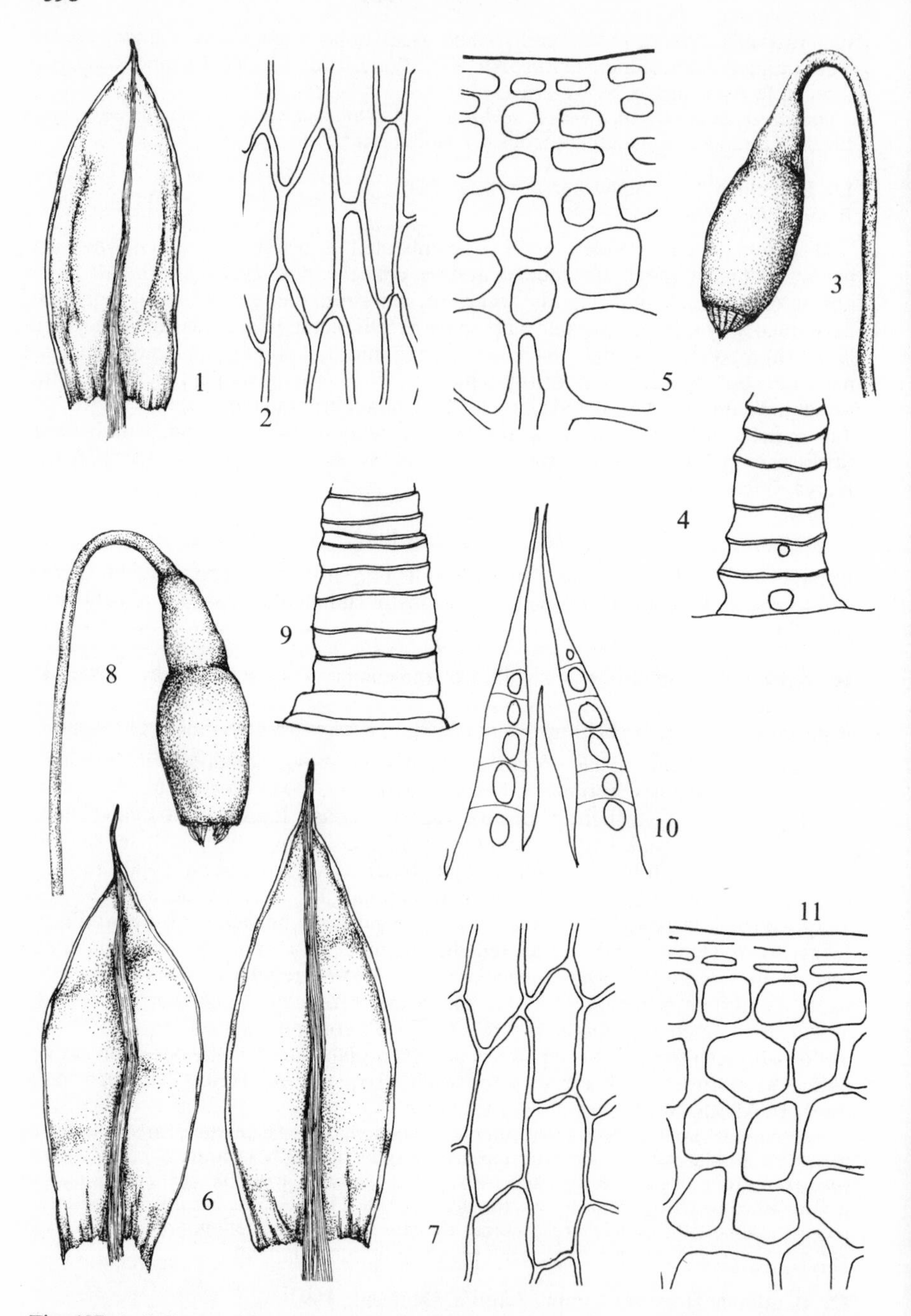

Fig. 187. 1–5, *Bryum salinum*: 1, leaf; 2, mid-leaf cells; 3, capsule; 4, lower part of outer peristome tooth; 5, exothecial cells from mouth of capsule. 6–11, *B. inclinatum*: 6, leaves; 7, mid-leaf cells; 8, capsule; 9, outer peristome tooth; 10, portion of inner peristome; 11, exothecial cells from mouth of capsule. Leaves ×25, cells ×415, capsules ×10, peristome ×250.

rows exothecial cells at mouth narrowly transversely rectangular, next cells ± quadrate-hexagonal, mostly 30–40 μm wide, becoming longitudinally rectangular below; outer peristome teeth straw-coloured with perforated longitudinal median groove on dorsal (outer) surface, processes uniformly finely papillose, cilia rudimentary; spores mostly 18–20 μm or 28–30 μm but in some capsules ranging in size from 18 to 36 μm. Fr frequent, spring. Patches on dune slacks and soil by the sea, very rare. S. Somerset, Ayr, Kincardine, S. Ebudes, E. Ross, W. Sutherland, N.E. Galway. 6, H1. W. and C. Europe, Spitzbergen, Alaska, Manitoba, Greenland.

16. B. inclinatum (Brid.) Bland., Übers. Mecklenb. Moos., 1809
B. stenotrichum C. Müll.

Synoecious. Plants to *ca* 1 cm. Lvs crowded above, sometimes in 2 or more successive comal tufts, appressed, straight or ± crisped when dry, erecto-patent when moist, reddish at base, ovate to lanceolate or oblong-lanceolate, acuminate to longly acuminate, margin recurved, entire or obscurely denticulate towards apex, bordered; nerve stout, yellowish to reddish-brown, excurrent, longly so in upper lvs; basal cells rectangular, in comal lvs angular cells slightly swollen, cells above ± rhomboid-hexagonal, 12–20(–28) μm wide in mid-lf, 3–4 marginal rows very narrow, more incrassate, forming distinct unistratose border. Capsule ± pendulous, narrowly ellipsoid or narrowly pyriform, symmetrical, neck abruptly narrowed into seta, mouth yellowish; 2–4 rows exothecial cells at mouth transversely rectangular, next cells ± quadrate-hexagonal, mostly 20–25 μm wide, then becoming longitudinally rectangular; lid ± mamillate; outer peristome teeth yellow above, reddish below, lacking perforations, processes with widely gaping perforations, more strongly papillose down middle than at edges, cilia rudimentary; spores 16–18 μm or 22–30(–34) μm. Fr common, spring, summer, autumn. $n=10^*$, 20*, 30. Patches on soil, rock crevices, walls, old buildings, sand-dunes, etc., particularly in basic habitats, frequent. 96, H26, C. Europe, Iceland, Asia, Morocco, Algeria, N. America, Greenland, Tierra del Fuego, Australia, Antarctica.

The relationship between spore size and chromosome number requires further investigation both in this plant and *B. salinum*. *B. salinum* differs in the perforated outer peristome teeth, the finely, uniformly papillose processes, colour of capsule mouth and peristome teeth and larger exothecial cells. *B. uliginosum* may be differentiated from both these species by the cernuous, asymmetrical capsule with oblique mouth and lvs with stronger, 2–3-stratose border and nerve at most only slightly excurrent. The name is illegitimate but is used here because of uncertainty as to the correct name.

17. B. knowltonii Barnes, Bot. Gaz., 1889
B. lacustre (Web. & Mohr) Bland.

Synoecious. Plants to 1 cm. Upper lvs more crowded than lower, sometimes in 2 or more successive comal tufts, erect, flexuose when dry, erecto-patent to spreading when moist, reddish at base, concave, lower ovate, obtuse, upper ovate-lanceolate acute, margin recurved, entire, ± bordered at least in upper lvs; nerve stout, reddish, ending below apex to percurrent, rarely excurrent; basal cells rectangular to narrowly rectangular, cells above narrowly rectangular to rhomboidal or hexagonal, 14–24 μm wide in mid-lf, 3–4 marginal rows narrow, forming ill-defined border. Capsule pendulous, theca ellipsoid, abruptly narrowed into neck, wide-mouthed when dry and empty; lid shortly conical; processes with oval perforations, cilia rudimentary; spores 20–26 μm. Fr frequent, late spring. Pale green to reddish patches on damp basic soil and dune slacks, rare. Scattered, mainly coastal localities from Devon and Essex north to Ross. 22. N., W. and C. Europe, Iceland, Faroes, Siberia, Himalayas, N. America, Greenland.

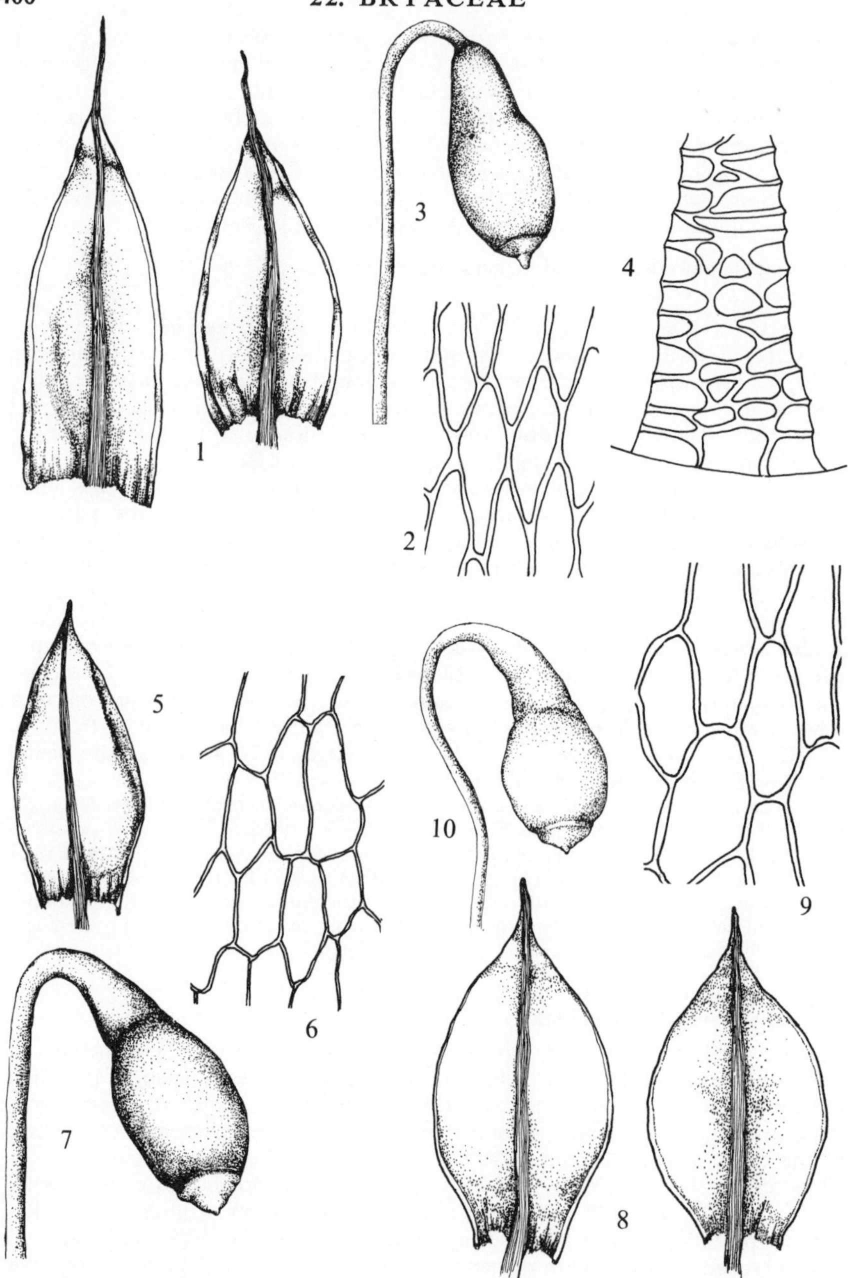

Fig. **188**. 1–4, *Bryum algovicum* var. *rutheanum*: 1, leaves; 2, mid-leaf cells; 3, capsule; 4, lower part of outer peristome tooth (×250). 5–7, *B. knowltonii*: 5, leaf; 6, mid-leaf cells; 7, capsule. 8–10, *B. lawersianum*: 8, leaves; 9, mid-leaf cells; 10, capsule. Leaves ×25, cells ×415, capsules ×10.

Most likely to be confused with *B. warneum*, from which it differs in the outer peristome teeth, spore size and the reddish lf base. It differs from allied species in the not or only slightly excurrent nerve.

18. B. intermedium (Brid.) Bland., Übers Mecklenb. Moos., 1809

Synoecious. Plants to 2.5 cm. Upper lvs in comal tuft, imbricate, straight when dry, erect when moist, reddish at base, lanceolate to ovate-lanceolate, acute to acuminate, margin revolute from base to apex, entire or nearly so, obscurely bordered; nerve stout, reddish, excurrent in short to long point; basal cells narrowly rectangular, angular cells in upper lvs swollen, cells above narrowly rectangular to rhomboid-hexagonal, 12–20 μm wide in mid-lf, 2–3 marginal rows narrower, forming ill-defined border. Capsule cernuous or pendulous, narrowly pyriform, slightly asymmetrical with oblique mouth, ± straight or incurved when dry; lid conical; peristome teeth brownish, processes narrowly perforated, cilia nodulose; spores 18–24 μm. Fr common, summer, autumn. n=20*, 24. Greenish tufts or patches on usually basic soil, on sand-dunes, hedgebanks, walls, old buildings, etc., occasional in England and Wales, rare in Scotland and Ireland. 76, H9. N., W. and C. Europe, Crete, Siberia, E. Asia, N. Africa, N. America, Greenland, Australia.

Recognised from other species of *Bryum* of similar lf shape, except *B. caespiticium*, by the lf margin with poorly developed border revolute from base to apex. *B. caespiticium* has smaller spores and a dioecious inflorescence. The slightly asymmetrical capsule will distinguish it from *B. inclinatum* and *B. algovicum*. *B. pallescens* and *B. creberrimum* differ in the appendiculate cilia as well as in lf character. A useful field character is that, unlike most other *Bryum* species, the capsules of any one plant ripen at different times and not all together.

Section III

Lf base red, border distinct, usually bistratose; cells ± incrassate, mostly 16–30 μm wide; cilia appendiculate; spores 10–20 μm.

19. B. donianum Grev., Trans. Linn. Soc., 1827

Dioecious. Plants to 1(–2) cm. Upper lvs forming tight comal tuft, shrunken and curved but not spirally twisted when dry, ± erecto-patent when moist, reddish at base, obovate, obovate-lanceolate or lanceolate, often widest above middle, acuminate, margin plane or recurved, obscurely toothed above, with strong yellowish border confluent with stout, shortly excurrent nerve at apex; basal cells rectangular, cells above rhomboid-hexagonal, incrassate, 14–22 μm wide in mid-lf, 4–5 marginal rows very long and narrow, 2–3-stratose, forming very distinct border. Capsule pendulous, narrowly pyriform, symmetrical or curved; lid mamillate; cilia appendiculate; spores 12–14 μm. Fr rare, spring, summer. Dark green tufts usually on light soil, in turf, edges of paths, soil crevices in walls, hedgebanks, etc., often near the coast, frequent or common in S.W. England, rare elsewhere. Channel Is., Cornwall and Devon east to Kent and north to Derby, S. Lancs, I. of Man, scattered localities in Ireland. 31, H10, C. S. Europe, E. Asia, Middle East, N. Africa, Macaronesia.

Readily recognisable by the lvs usually widest above the middle and with a very stout border confluent with the stout, shortly excurrent nerve. Plants of the next 5 species differ in the lvs spirally twisted when dry (except *B. torquescens*) and the border unistratose, as well as in the longly excurrent nerve which gives the dry tufts a characteristic hairy appearance.

20. B. capillare Hedw., Sp. Musc., 1801

Dioecious. Plants 1–5 cm. Lvs shrunken and twisted to strongly spirally twisted and with flexuose points when dry, when moist soft, ± patent and slightly curved in

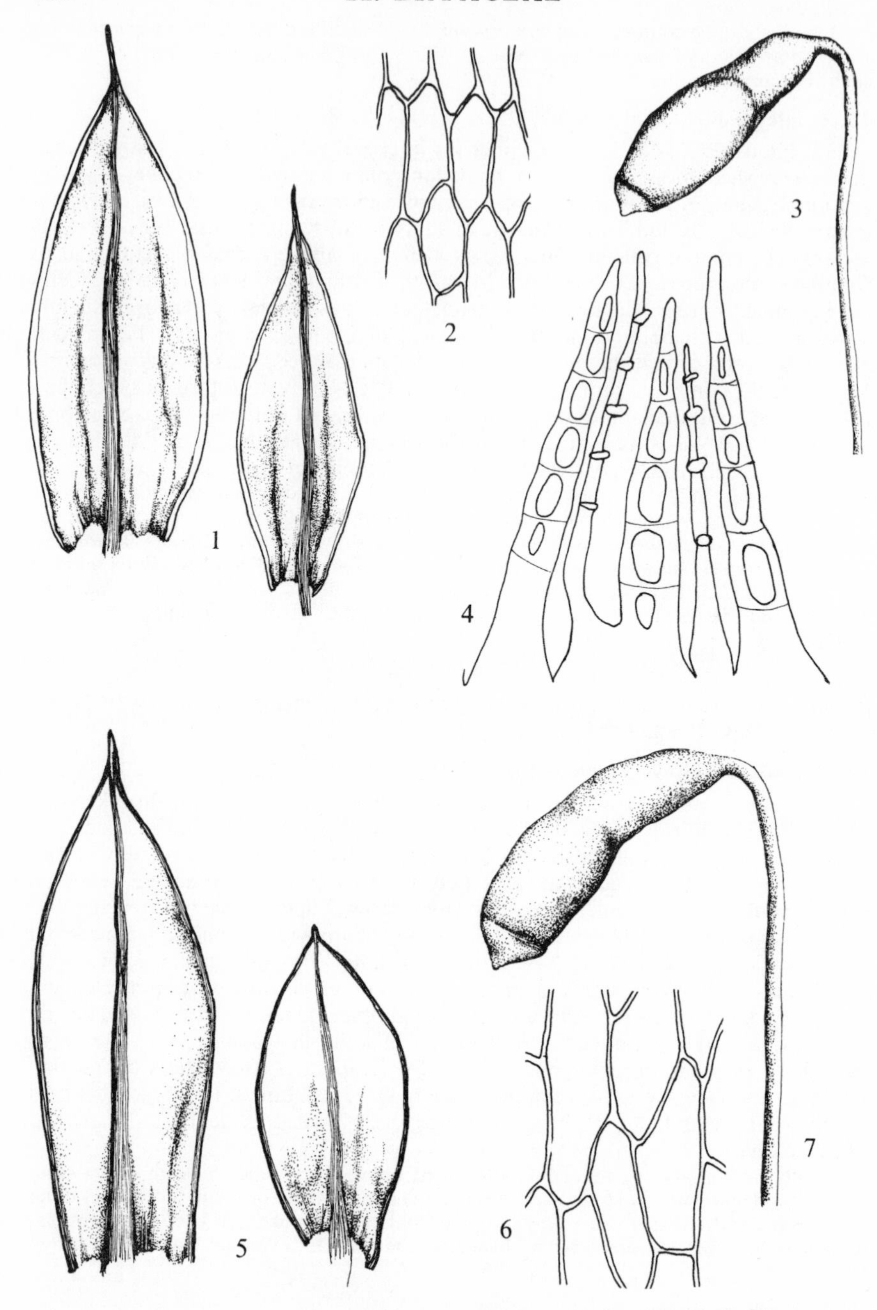

Fig. **189**. 1–4, *Bryum intermedium*: 1, leaves; 2, mid-leaf cells; 3, capsule; 4, portion of inner peristome. 5–7, *B. donianum*: 5, leaves; 6, mid-leaf cells; 7, capsule. Leaves ×25, cells ×415, capsules ×10.

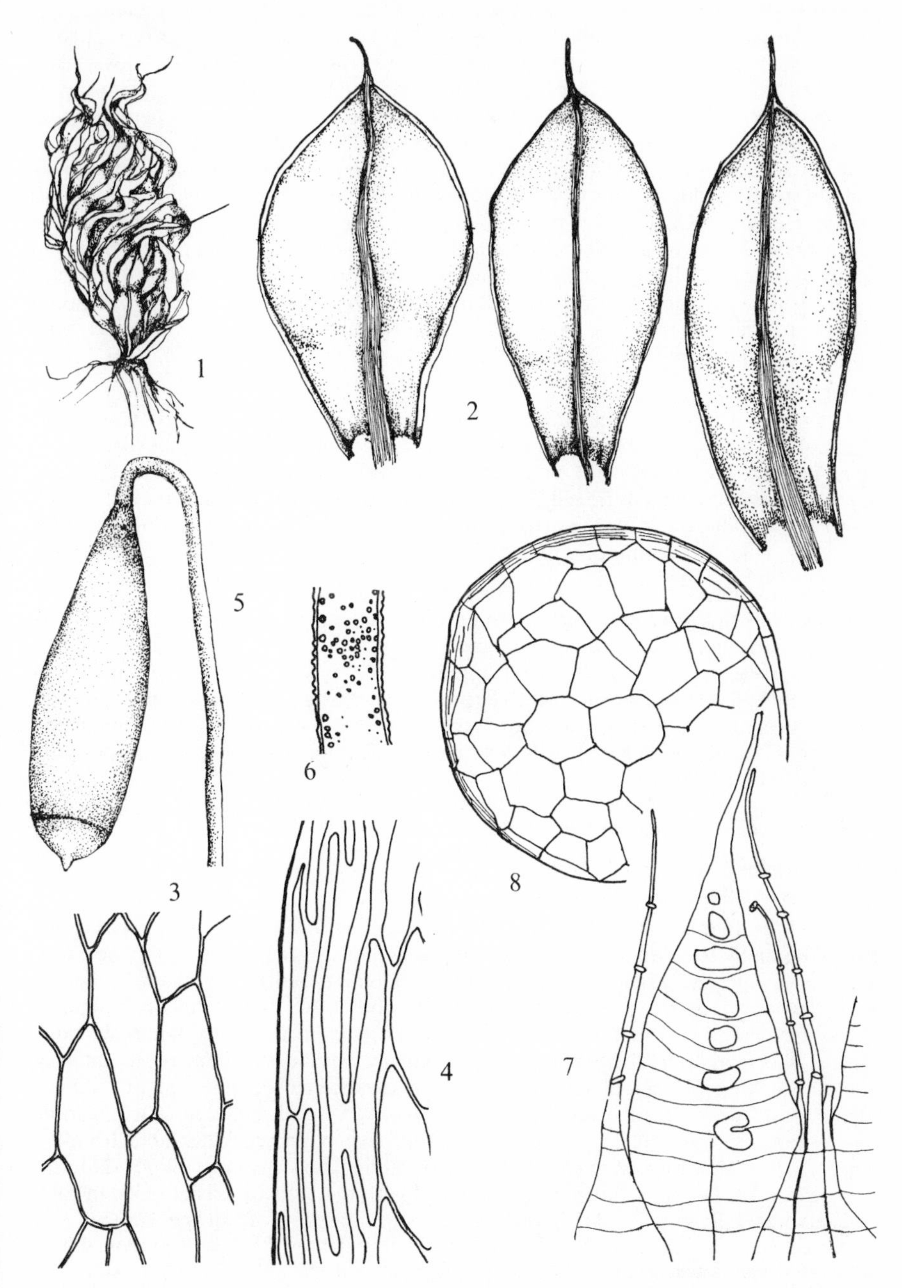

Fig. **190**. *Bryum capillare* var. *capillare*: 1, dry shoot (×10); 2, leaves (×25); 3 and 4, mid-leaf and marginal cells (×415); 5, capsule (×10); 6, portion of rhizoid (×415); 7, portion of inner peristome (×250); 8, rhizoidal gemma (×250).

direction of twisting, plane or concave, base reddish, narrowly lanceolate to oblong-spathulate, widest above middle, shortly or longly acuminate, not decurrent at base, margin narrowly recurved, bordered, denticulate or entire towards apex; nerve usually excurrent in piliferous or cuspidate point but sometimes ending below apex, usually colourless but sometimes red or brown; basal cells shortly rectangular, upper cells rhomboid-hexagonal, thin-walled, border of narrow, elongated, usually colourless but sometimes yellowish or reddish, unistratose cells. Perichaetial lvs narrowly lanceolate. Rhizoids brown to deep reddish-brown, papillose. Axillary filamentous gemmae absent. Rhizoidal gemmae usually on long rhizoids, similar in colour to rhizoids, scattered to abundant, spherical, 66–270(–440) μm in diameter, or ovoid or irregular, 65–250 μm × 105–350 μm, cells not protuberant. Capsule cernuous, cylindrical to pyriform, symmetrical, contracted below mouth when dry; exothecial cells at mouth not in rows; processes gradually narrowed to long fine apex, cilia appendiculate; spores (9–)12–15 μm.

Var. **capillare**

Lvs strongly spirally twisted when dry, obovate-spathulate, shortly acuminate; basal cells 16–31 μm × 32–88 μm, upper cells 16–25 μm × 38–60 μm, border of 3–5 rows narrow, elongated cells. Fr frequent, spring, summer. $n = 10$*. Dense or lax, green, sometimes reddish-tinged, rarely deep red or maroon tufts on rocks, in rock crevices, on walls, old buildings, tree trunks and branches, occasionally on soil, common. 112, H40, C. Cosmopolitan.

Var. **rufifolium** (Dix.) Podp., Acta Acad. Sci. nat. moravo-siles., 1950

Lvs shrunken or twisted, sometimes loosely or strongly spirally twisted when dry, narrowly lanceolate, longly acuminate but lower lvs sometimes like those of the type, border very wide; basal cells ± rectangular, 15–37 μm × 37–135 μm, upper cells 12–22 μm × 32–91 μm, border of 5–7 rows very narrow, incrassate cells. Fr rare. Dense, deep red, green or variegated tufts on dry, basic rocks, rare. Somerset, Radnor, Caerns, Denbigh, Argyll, Dunbarton, Rhum, W. Ross, W. Sutherland, Clare. 10, H1. Endemic.

The descriptions of *B. capillare* and the next 4 species are based on the account of *B. capillare* and related species by Syed, *J. Bryol.* 7, 265–326, 1973.

21. B. elegans Nees ex Brid., Bryol. Univ., 1826

B. capillare var. *elegans* (Brid.) Husn.

Dioecious. Plants 1–4 cm, with long slender julaceous branches. Lvs appressed, not shrunken or twisted, or sometimes loosely twisted at upper part of stem when dry, soft, concave, broadly ovate, not decurrent, margin plane, entire, narrowly bordered; nerve strongly excurrent in mucronate to piliferous point; basal cells rectangular, above hexagonal, distinctly porose, 13–18 μm wide, border of 2–3 narrow cells. Rhizoids brown to reddish-brown, very coarsely papillose. Filamentous axillary gemmae lacking. Rhizoidal gemmae on long rhizoids, brown, spherical, 90–200 μm diameter, cells not protuberant, only rarely present. Capsule sub-cylindrical, cernuous, symmetrical, hardly contracted below mouth when dry, neck distinct, shrunken when dry; exothecial cells near mouth in longitudinal rows; spores 12–15(–26) μm. Fr rare. $n = 10$. Dense, reddish or green tinged with red tufts on basic rocks, in rock crevices and walls, rare. 21, H1. Northern and montane parts of Europe, Iceland.

Distinctive in the julaceous shoots and very coarsely papillose rhizoids. *B. stirtonii* differs in the finely papillose rhizoids and shoots not julaceous.

22. B. stirtonii Schimp., Syn. 2, 1876

Dioecious. Plants 0.5–4.0 cm. Lvs hardly shrinking, incurved or closely appressed,

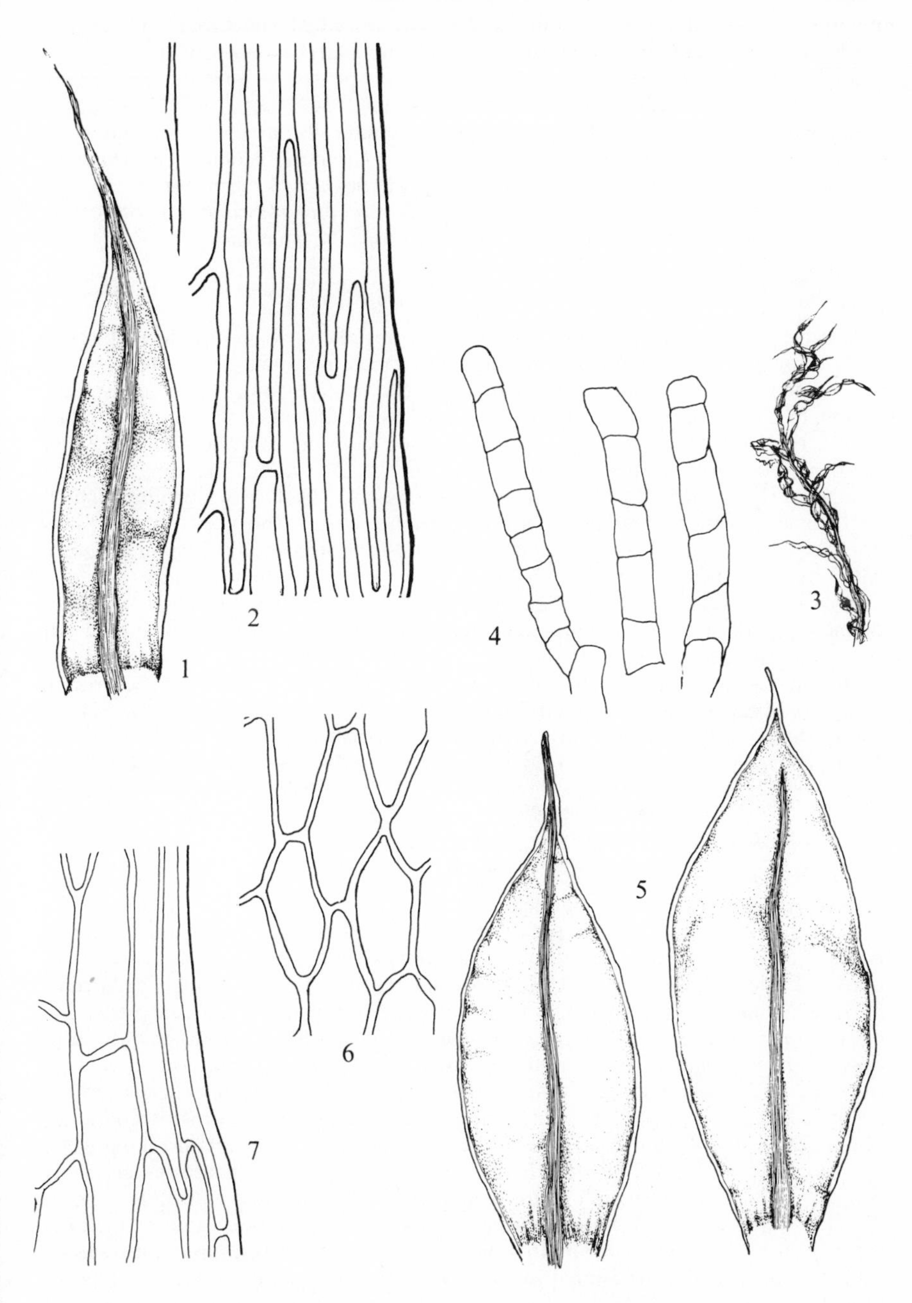

Fig. **191**. 1–2, *Bryum capillare* var. *rufifolium*: 1, leaf (×25); 2, marginal cells. 3–7, *B. flaccidum*: 3, dry shoot (×10); 4, axillary gemmae (×150); 5, leaf (×40); 6 and 7, mid-leaf and marginal cells. Cells ×415.

not spirally twisted when dry, soft, patent, concave when moist, ovate to broadly ovate, base decurrent, apex mucronate to cuspidate, margin slightly recurved, entire, indistinctly bordered; nerve ending below or in apex or excurrent in lvs round perichaetium; cell walls porose, basal rectangular, 15–25 μm × 56–67 μm, above hexagonal, 15–22 μm × 19–54 μm, 1–2 rows marginal cells narrow, forming obscure border. Rhizoids brown to reddish-brown, finely papillose. Axillary and rhizoidal gemmae absent. Capsule sub-cylindrical, symmetrical, neck shrunken when dry; spores 12–14(–19) μm. Fr rare. Dense, greenish tufts on soil, rarely on rocks, very rare. N. Northumberland, Perth. 4. N. and C. Europe, N. America.

23. B. flaccidum Brid., Br. Univ., 1826
B. capillare var. *flaccidum* (Brid.) Br. Eur., *B. laevifilum* Syed

Dioecious. Plants 0.5–4.0 cm. Lvs not much shrunken, twisted round themselves and not or only rarely round stem, sometimes spreading when dry, ovate or ovate-lanceolate to obovate, base shortly to longly decurrent, apex mucronate or cuspidate to longly cuspidate, margin plane or slightly recurved, entire or slightly toothed above, bordered; nerve ending below apex to excurrent, colourless or brown or reddish; cells rarely porose, basal rectangular, 12–25 μm × 25–113 μm, upper ± hexagonal, 12–19 μm × 22–47 μm, 1–3 rows marginal cells narrow, forming distinct border, yellowish in older lvs. Rhizoids light brown to brown, smooth or finely papillose. Filamentous, smooth or finely papillose, protonemal-like gemmae, green when young, brown at maturity, 15–35 μm wide, variable in length, branched or not, in lf axils. Brown, globose rhizoidal gemmae present on long rhizoids, 65–120 μm diameter, with non-protuberant cells. Capsule sub-cylindrical, symmetrical, or slightly asymmetrical, neck not shrunken when dry; spores 9–13 μm. Fr not known in Britain. Dense or loose, light green, soft tufts on tree trunks and branches, rotten logs, very rarely on soil, occasional. 34. Europe, N. America.

Close to *B. capillare* but distinctive in the filamentous axillary gemmae and decurrent lvs. *B. laevifilum* was distinguished on the basis of the smooth, not finely papillose, axillary gemmae and the non-decurrent lvs, but the two intergrade to such an extent that *B. laevifilum* cannot be maintained as a distinct taxon.

24. B. torquescens Bruch ex De Not., Syll., 1838
B. capillare var. *torquescens* (Bruch) Husn.

Usually synoecious, rarely autoecious or dioecious. Plants 1.0–2.5 cm. Lvs hardly shrinking, slightly twisted or spreading to closely appressed when dry, plane or concave, ovate, obovate or spathulate, mucronate or cuspidate, margin recurved, bordered, toothed, sometimes strongly so; nerve brown to red, strongly excurrent; basal cells narrowly rectangular, above narrowly hexagonal, 13–22 μm wide, 3–4 marginal rows elongate, incrassate, forming distinct border. Rhizoids bright red to brown, finely papillose. Globose, red gemmae usually abundant on long and short rhizoids, 75–255 μm diameter, walls usually non-protuberant. Filamentous axillary gemmae lacking. Capsule sub-cylindrical to cylindrical, cernuous or sub-pendulous, symmetrical, slightly constricted below mouth when dry, neck not shrinking when dry; exothecial cells below mouth in ± longitudinal rows; processes suddenly contracted at apex and ending in long projection, cilia appendiculate; spores (9–)11–15(–16) μm. Fr common, spring. $n=20$. Loose or dense, green, sometimes reddish-tinged tufts on basic soil in grassland, roadsides, banks, rarely on rocks or walls, never on trees. Occasional in England, very rare elsewhere, extending north to E. Ross, E. Donegal. 28, H1. Europe, W. Asia, Pakistan, Nepal, China, Azores, Madeira, N. and S. Africa, Kenya, N. America, Mexico, Chile, Australia, New Zealand.

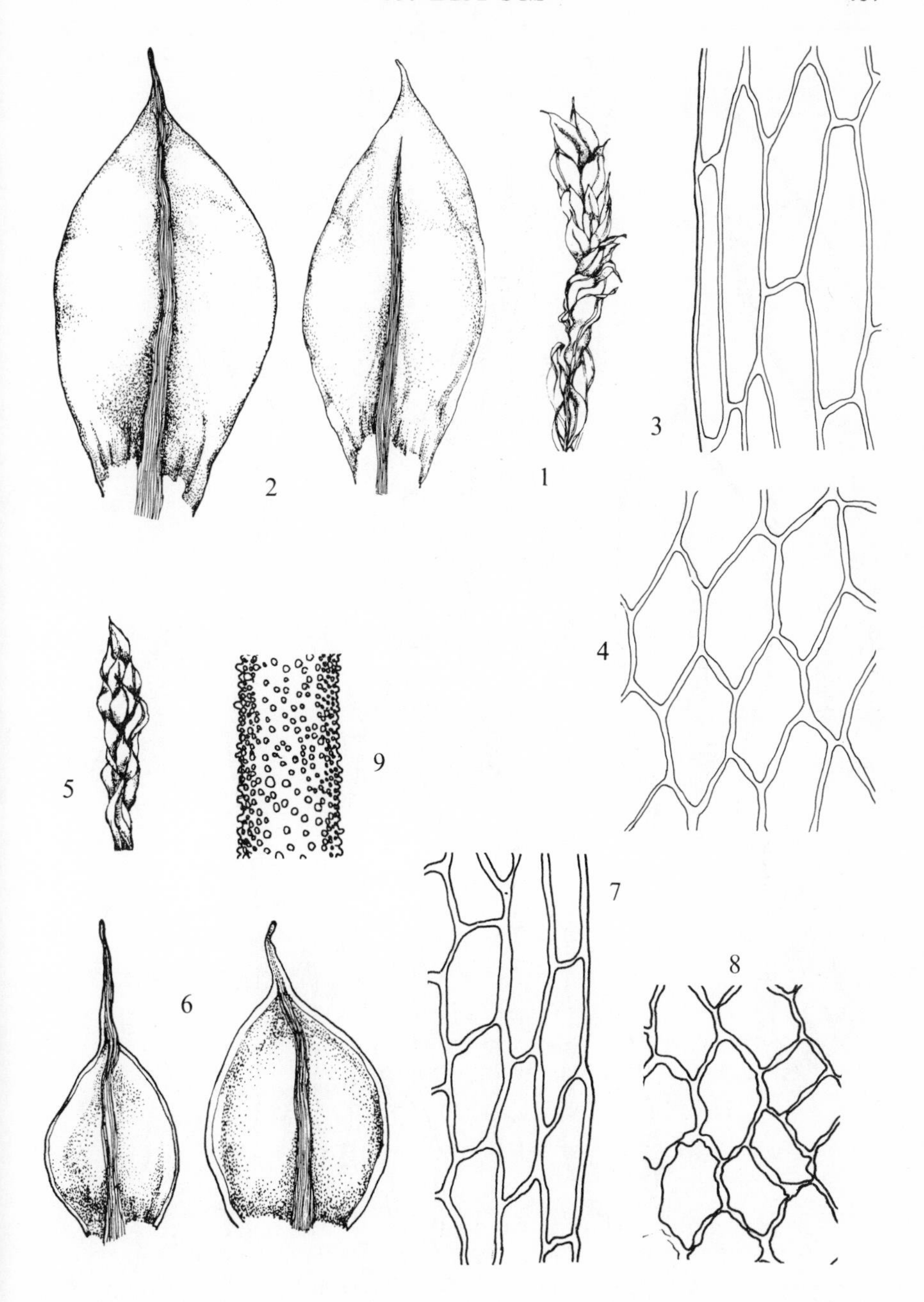

Fig. **192**. 1–4, *Bryum stirtonii*: 1, dry shoot; 2, leaves; 3 and 4, marginal and mid-leaf cells. 5–9, *B. elegans*: 5, dry shoot; 6, leaves; 7 and 8, marginal and mid-leaf cells; 9, portion of rhizoid (×415). Shoots ×10, leaves ×40, cells ×415.

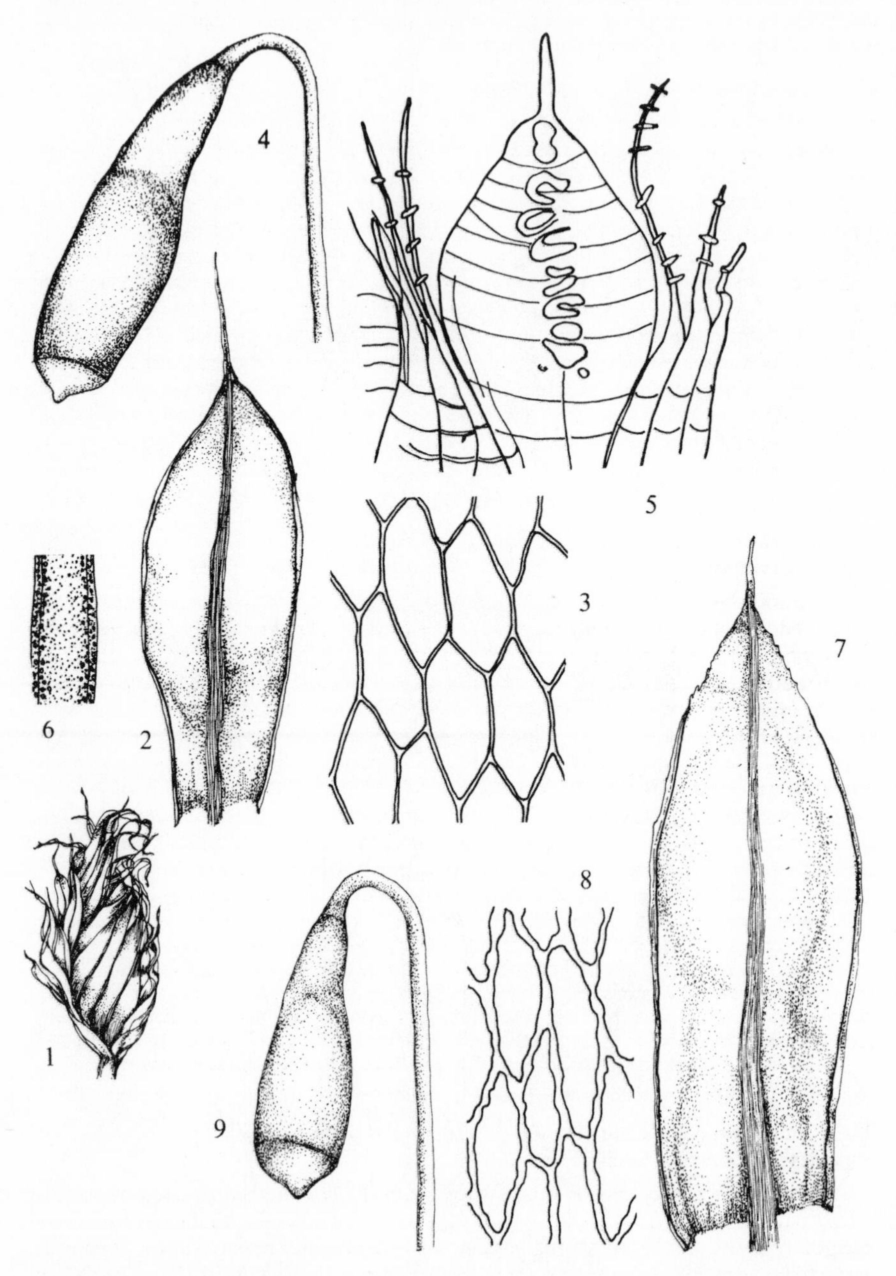

Fig. **193**. 1–6, *Bryum torquescens*: 1, dry shoot (×10); 2, leaf; 3, mid-leaf cells; 4, capsule; 5, portion of inner peristome (×250); 6, portion of rhizoid. 7–9, *B. canariense*: 7, leaf; 8, mid-leaf cells; 9, capsule. Leaves ×25, cells ×415, capsules ×10.

When synoecious the inflorescence will distinguish *B. torquescens* from other members of the group, but it is sometimes autoecious or dioecious. It may then be distinguished by the red rather than reddish-brown rhizoidal gemmae.

25. B. canariense Brid., Sp. Musc., 1817
B. canariense var. *provinciale* (Philib.) Husn., *B. provinciale* Philib.

Autoecious or occasionally synoecious. Plants 0.3–2.5 cm; stems matted with deep red rhizoids below. Lower lvs small, upper larger, crowded in comal tufts, 2 or more successive comal tufts sometimes present, appressed, ± straight when dry, erect or imbricate when moist, reddish at base, ovate to ovate-oblong or obovate-oblong, widest at or above middle, margin recurved below, distinctly toothed above, unbordered; nerve stout, reddish, excurrent in cuspidate point; basal cells rectangular, cells above narrowly hexagonal, 16–22 μm wide in mid-lf, a few marginal rows narrower but not forming border. Red, spherical, rhizoidal gemmae, 180–300 μm diameter, with non-protuberant cells usually present. Capsule pendulous, narrowly pyriform, ± symmetrical; lid mamillate; spores 13–16 μm. Fr occasional, spring, summer. Dull or reddish-green tufts or patches on soil in crevices of basic rocks, rarely on sand-dunes, rare. Scattered localities from S. Devon and E. Sussex north to N. Wales and W. Lancs. 17, C. Mediterranean coast, Atlantic coast of France, N. America.

26. B. creberrimum Tayl., Lond. J. Bot., 1845
B. affine Lindb. & Arnell

Synoecious. Plants 1–4 cm. Lvs slightly twisted when dry, erecto-patent when moist, reddish at base, ovate-lanceolate to lanceolate, often gradually tapering from widest point to acuminate apex, margin usually strongly recurved, entire or denticulate towards apex, bordered; nerve reddish, excurrent; basal cells rectangular, angular cells of comal lvs slightly turgid, above rhomboid-hexagonal, 14–24 μm wide in mid-lf, several marginal rows narrower, more incrassate, forming distinct border. Capsule inclined to pendulous, symmetrical, narrowly ellipsoid or narrowly pyriform; lid mamillate; process of inner peristome with perforations about as wide as long, cilia appendiculate; spores 14–16 μm. Fr common, summer. $n = 10$, 30. Tight tufts, green above, reddish-brown and matted with rhizoids below, on soil, in crevices in rocks and walls, occasional, extending north to E. Ross. 26. Europe, N. Asia, India, N. America, Australia, New Zealand.

B. creberrimum and *B. pallescens* are very close and have been confused in the past because of the statement in Dixon (1924) that *B. creberrimum* is synoecious and *B. pallescens* autoecious. The latter may also be synoecious. The smaller spores and widely perforated processes will usually separate *B. creberrimum* from *B. pallescens* but plants with spores of intermediate size occasionally occur. The gametophyte of *B. pallescens* is variable and some forms are indistinguishable from *B. creberrimum*, but plants with ± ovate, shortly pointed lvs are distinct. Robust plants may be mistaken for *B. pseudotriquetrum* var. *bimum* but differ in the more acuminate lvs with more longly excurrent nerve.

27. B. pallescens Schleich. ex Schwaegr., Sp. Musc. Suppl. I, 1816
B. obconicum Hornsch. ex Br. Eur.

Synoecious or autoecious. Plants 1.0–4.0(–7.5) cm. Lvs slightly twisted when dry, erecto-patent when moist, reddish at base, ovate to ovate-lanceolate, acuminate, margin recurved, sometimes strongly so, entire or denticulate towards apex, bordered; nerve red, excurrent; areolation as in *B. creberrimum*. Capsule as in *B. creberrimum*; processes of inner peristome with perforations, 1½–2 times as long as wide; cilia with poorly to well developed appendages; spores 18–20(–22) μm. Fr common, summer.

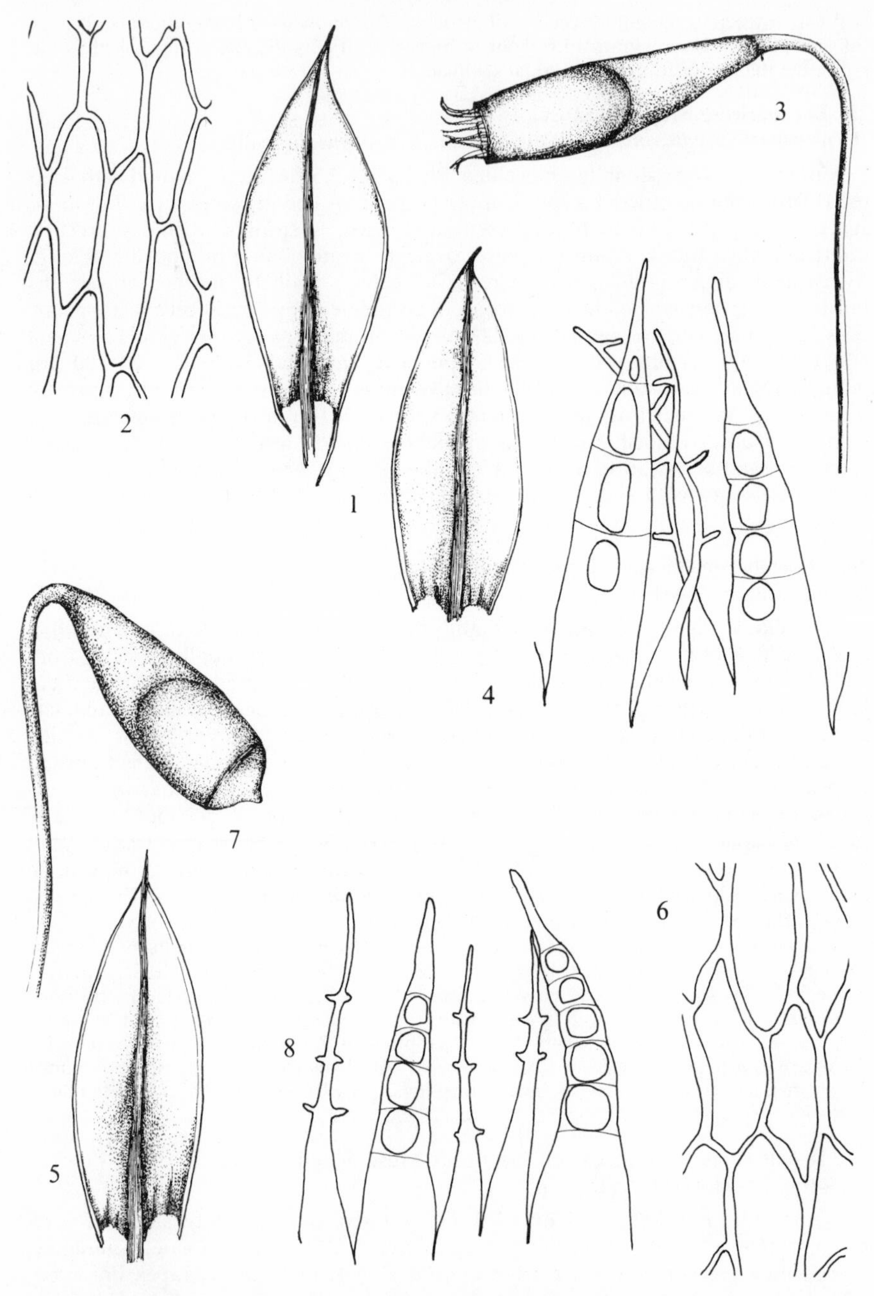

Fig. **194**. 1–4, *Bryum pallescens*: 1, leaves; 2, mid-leaf cells; 3, capsule; 4, portion of inner peristome. 5–8, *B. creberrimum*: 5, leaf; 6, mid-leaf cells; 7, capsule; 8, portion of inner peristome. Leaves ×15, cells ×415, capsules ×10, peristomes ×250.

n=10, 22. Tight tufts, green above, reddish-brown and matted with tomentum below, on soil, sand-dunes and in crevices in rocks and walls generally distributed, occasional. 42, H5, C. Europe, Asia, Africa, America.

The type specimen of *B. obconicum* is a specimen of *B. pallescens* and has usually been regarded as related to *B. capillare*. Many British gatherings named *B. obconicum* are *B. torquescens*.

28. B. pseudotriquetrum (Hedw.) Schwaegr., Spec. Musc. Suppl. 1, 1816

Plants glossy, green to reddish, 1–10(–15) cm, stems matted together below with brown tomentum. Lvs not much crowded towards stem apex, shrunken, flexuose when dry, erecto-patent when moist, basal cells reddish, more deeply coloured than

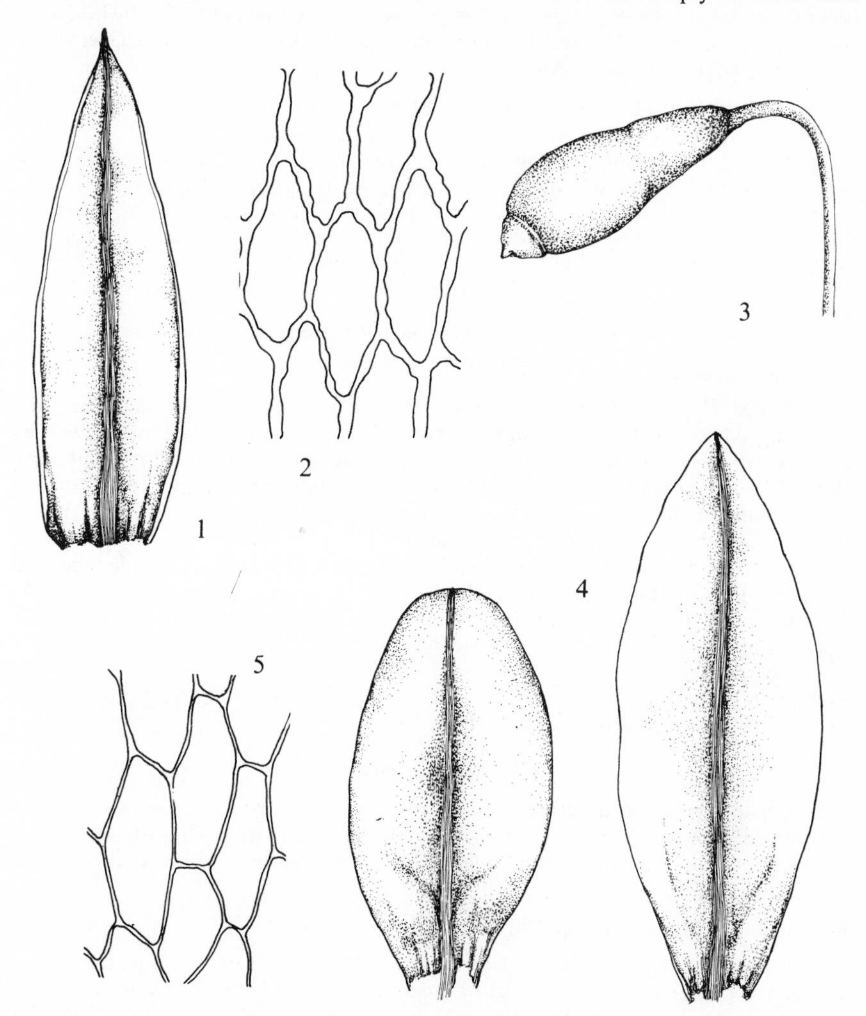

Fig. **195**. 1–3, *Bryum pseudotriquetrum*: 1, leaf; 2, mid-leaf cells; 3, capsule (×10). 4–5, *B. neodamense*: 4, leaves; 5, mid-leaf cells. Leaves ×25, cells ×415.

other cells, ovate to narrowly lanceolate, acuminate, margin slightly recurved, entire or denticulate towards apex, bordered; nerve stout, percurrent to shortly excurrent; basal cells shortly rectangular, cells above rhomboid-hexagonal, narrower towards margin, 16–30 μm wide in mid-lf, several marginal rows narrow, forming distinct unistratose border. Capsule large, inclined to ± pendulous, narrowly ellipsoid, symmetrical, wide-mouthed and narrowed below mouth when dry and empty; lid mamillate; cilia appendiculate; spores 12–18 μm, sometimes variable in size within a capsule.

Var. **pseudotriquetrum**

Dioecious. Fr occasional, summer, autumn. $n=10$*, 10+1, 20. Green to reddish, loose or compact tufts or patches on damp or wet soil, fens, wet heaths, flushes, dune slacks, etc., common. 106, H39, C. Europe, Asia, W. Africa, America, Greenland, Argentina, Tierra del Fuego, Australia.

Var. **bimum** (Brid.) Lilj., Utkast Svensk. Fl. ed. 3, 1816

B. bimum (Brid.) Turn., *B. pseudotriquetrum* ssp. *bimum* (Brid.) Hartm.

Synoecious. Fr common. $n=20$*, 22. Similar habitats to the type, occasional. 89, H25, C. Europe, E. Asia, Middle East, America, Australia, New Zealand.

The only constant difference between the type and var. *bimum* is the nature of the inflorescence, although there are possibly cytological differences which require further investigation. *B. pseudotriquetrum* is usually easily recognised by the robust, glossy, often reddish-tinged plants with rigid lvs. Stunted forms may be mistaken for *B. pallens* but the latter differs in the usually uniform pinkish tinge to the lvs which do not have reddish bases, the more shortly pointed lf apex, the margin strongly recurved and the border bistratose.

29. B. neodamense Itzigs. ex C. Müll., Syn., 1849

Dioecious. Plants 1–10 cm. Lvs very distant except at stem apex, curled when dry, spreading when moist, very concave, base slightly decurrent, reddish, lower lvs broadly ovate, apex rounded or obtuse, ± cucullate, nerve ending below or at apex, upper lvs ovate to lanceolate, obtuse to acute, nerve ending below apex to percurrent, margin plane or slightly recurved in upper lvs, ± entire, bordered; basal cells shortly rectangular, cells above shortly rectangular to rhomboid-hexagonal, 16–30 μm wide in mid-lf, 2–5 marginal rows narrow, forming distinct border. Capsule obovate-pyriform. Fr very rare, summer. Reddish-green tufts or patches in fens, on wet limestone and dune slacks, calcicole, very rare. Caerns, S. Lancs, N.W. Yorks, Angus, Caithness, W. Ireland. 5, H5. N., W. and C. Europe, Iceland, Siberia, Altai, N. America, Greenland.

Likely to be mistaken only for forms of *B. pseudotriquetrum* but recognised by the very concave, obtuse or rounded, ± cucullate lower lvs.

Section IV

Lf base usually red and differing in colour from rest of lf; border narrow or absent, unistratose; cells usually narrow, incrassate, 10–16(–24) μm wide; cilia appendiculate; spores 8–14 μm; vegetative propagules (axillary bulbils and/or rhizoidal gemmae) frequently produced.

30. B. caespiticium Hedw., Sp. Musc., 1801

Dioecious. Plants to 1(–2) cm. Upper lvs forming comal tuft, imbricate, slightly twisted when dry, imbricate to erect or erecto-patent when moist, reddish or not at base, ovate to ovate-oblong, acuminate, widest below middle, margin plane or strongly recurved, ± entire, obscurely bordered; nerve yellowish to reddish-brown, in upper lvs excurrent in long entire or slightly denticulate point; basal cells rectangular or shortly rectangular, angular cells swollen in comal lvs, above narrowly rhomboid-hexagonal or narrowly hexagonal, 12–16 μm wide in mid-lf, 2–3 marginal rows narrower, forming poorly defined border. Capsule ± pendulous, narrowly ellipsoid or narrowly pyriform, sometimes slightly asymmetrical, ± straight or up-curved, wide-mouthed and narrowed below mouth when dry and empty, mouth reddish-brown; lid mamillate; outer peristome teeth pale brown, cilia appendiculate; spores 10–14 μm. Fr common, summer.

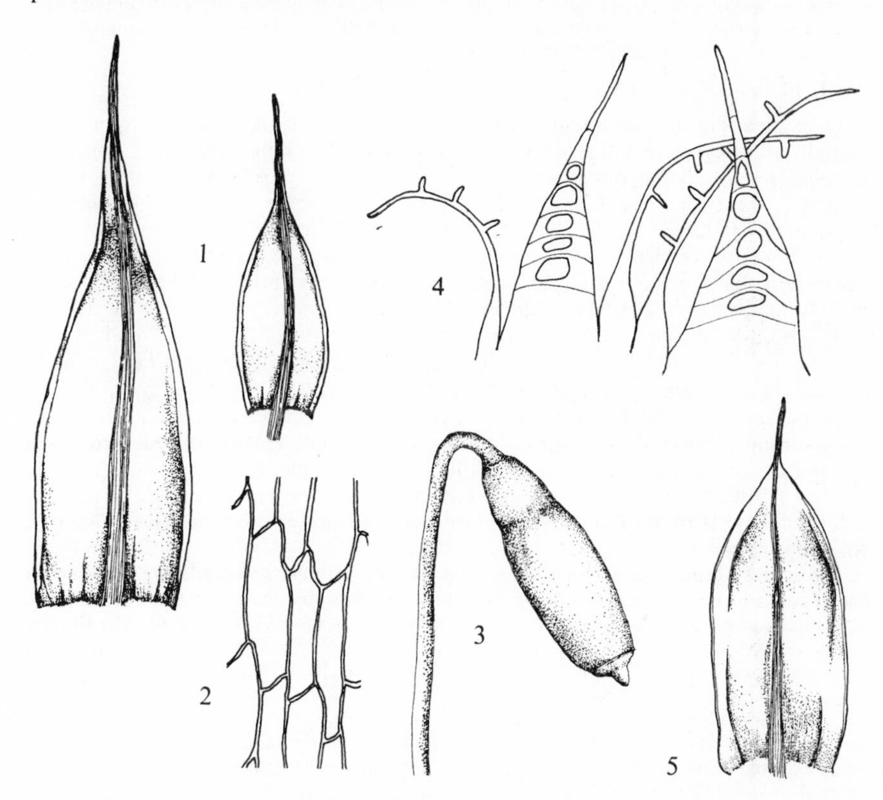

Fig. **196**. 1–4, *Bryum caespiticium* var. *caespiticium*: 1, leaves (× 25); 2, mid-leaf cells (× 415); 3, capsule (× 10); 4, portion of inner peristome (× 250). 5, *B. caespiticium* var. *imbricatum*: leaf (× 25).

Var. **caespiticium**

Lvs erect or erecto-patent when moist, ovate to ovate-oblong, tapering from below middle, margin strongly recurved. $n = 9+1$, 10*, $10+0$–1*, 20. Dull green tufts or patches on soil, dune slacks, waste ground, walls, rocks, etc., usually common. 108, H30, C. Europe, Asia, N. and C. Africa, America, Australia, New Zealand, Oceania.

Var. **imbricatum** Br. Eur., 1846

Var. *kunzei* (Hoppe & Hornsch.) Braithw.

Lvs imbricate, very concave, broadly ovate to ovate-oblong, ± abruptly narrowed to acuminate apex, margin plane. Similar habitats to the type, very rare. E. Sussex,

Hereford, Radnor, Derby, S.W. Yorks, Durham. 6. Europe, Middle East, Kashmir, N. Africa.

Distinct from the preceding species in the scarcely bordered lvs with narrower cells, although the plant cannot be reliably determined in the absence of fruit. The nature and status of var. *imbricatum* requires further investigation; Dixon (1924) regards the plant as a variety of *B. caespiticium*, Nyholm (1956–69) considers it may be a form of *B. capillare* and other authorities treat it as a species. The hybrid *B. caespiticium* × *B. algovicum* occurs occasionally and may be recognised by the *B. caespiticium* gametophyte bearing capsules resembling those of *B. algovicum* but differing in the smaller spores (*ca* 20 μm) and the lines on the outer peristome teeth less heavily thickened.

31. B. alpinum Huds. ex With., Syst. Arr. Brit. Pl., 1801

Dioecious. Plants with a metallic sheen, to 6 cm, stems rigid. Lvs ± equidistant, straight, imbricate when dry, erect to erecto-patent when moist, glossy, ± uniformly purplish-red, rarely paler or green, concave, lanceolate to narrowly lanceolate, widest at and ± tapering from below middle to obtuse to acute apex, margin plane or occasionally recurved or revolute below, entire or denticulate, rarely toothed above, unbordered; nerve stout or very stout, pale to purplish-red, ending below apex to percurrent; basal cells ± quadrate, cells above narrowly rhomboidal to vermicular, heavily incrassate in older lvs, thinner-walled and wider in younger, 8–12(–16) μm wide in mid-lf, marginal cells not or hardly narrower. Purplish-red to brownish-red, ± spherical, rhizoidal gemmae, 120–200 μm diameter, usually present in tomentum on older parts of stem. Capsule large, pendulous, narrowly ellipsoid to sub-cylindrical, symmetrical, neck short; lid mamillate; cilia appendiculate; spores 12–14 μm. Fr rare, summer. Purplish-red, red or, more rarely, bright green and red variegated, glossy tufts or patches on wet rocks, cliffs, in quarries, etc. Frequent or common in W. and N. Britain, particularly in montane areas, very rare in S.E. England and the Midlands. 81, H28, C. Europe, Asia, Morocco, C. and southern Africa, Canaries, Madeira, America.

The robust, usually purplish-red, glossy metallic-tinged plants with stiff lvs with narrow cells make this an easy plant to recognise. Somewhat variable in colour and degree of cell wall thickening and 2 varieties based on these characters have been recognised in Britain. Green plants with wider cells and denticulate lf apex have been called var. *viride* Husn., but the variety does not seem worth maintaining as the range in colour from green to purplish-red can occur in one tuft and red plants may sometimes have toothed lvs. Deeply coloured plants usually have lvs with very stout nerve and narrow, very incrassate cells, while paler or green plants have wider, thinner-walled cells. Var. *meridionale* Schimp. seems merely to be an extreme form with narrower cells and lvs, possibly a dry habitat variant.

32. B. gemmiparum De Not., Cronac. Briol. Ital., 1866
B. alpinum ssp. *gemmiparum* (De Not.) Kindb.

Dioecious. Plants to 3 cm. Lvs ± straight and imbricate when dry, erecto-patent when moist, not reddish at base, concave, ovate-lanceolate or lanceolate, acute, margin plane or slightly recurved, entire, ± unbordered; nerve strong, ending in or below apex; basal cells lax, ± quadrate, above rhomboidal to narrowly rhomboidal, 12–18(–28) μm wide in mid-lf, marginal cells narrower, occasionally forming very poorly defined border. Upper lf axils often with green to reddish, sessile bulbils, mostly 250–750 μm long, with rudimentary apical lvs. Orange or pink spherical rhizoidal gemmae, 100–160 μm diameter, sometimes present in tomentum on older part of stem. Capsule similar to *B. alpinum*, unknown in Britain. Dull, dense tufts, greenish above, reddish below, in crevices in rocks by streams or rivers, very rare. Devon, Monmouth, Brecon. 4. Europe, Caucasus, India, Yunnan, Tunisia, Algeria, Morocco, Macaronesia, America.

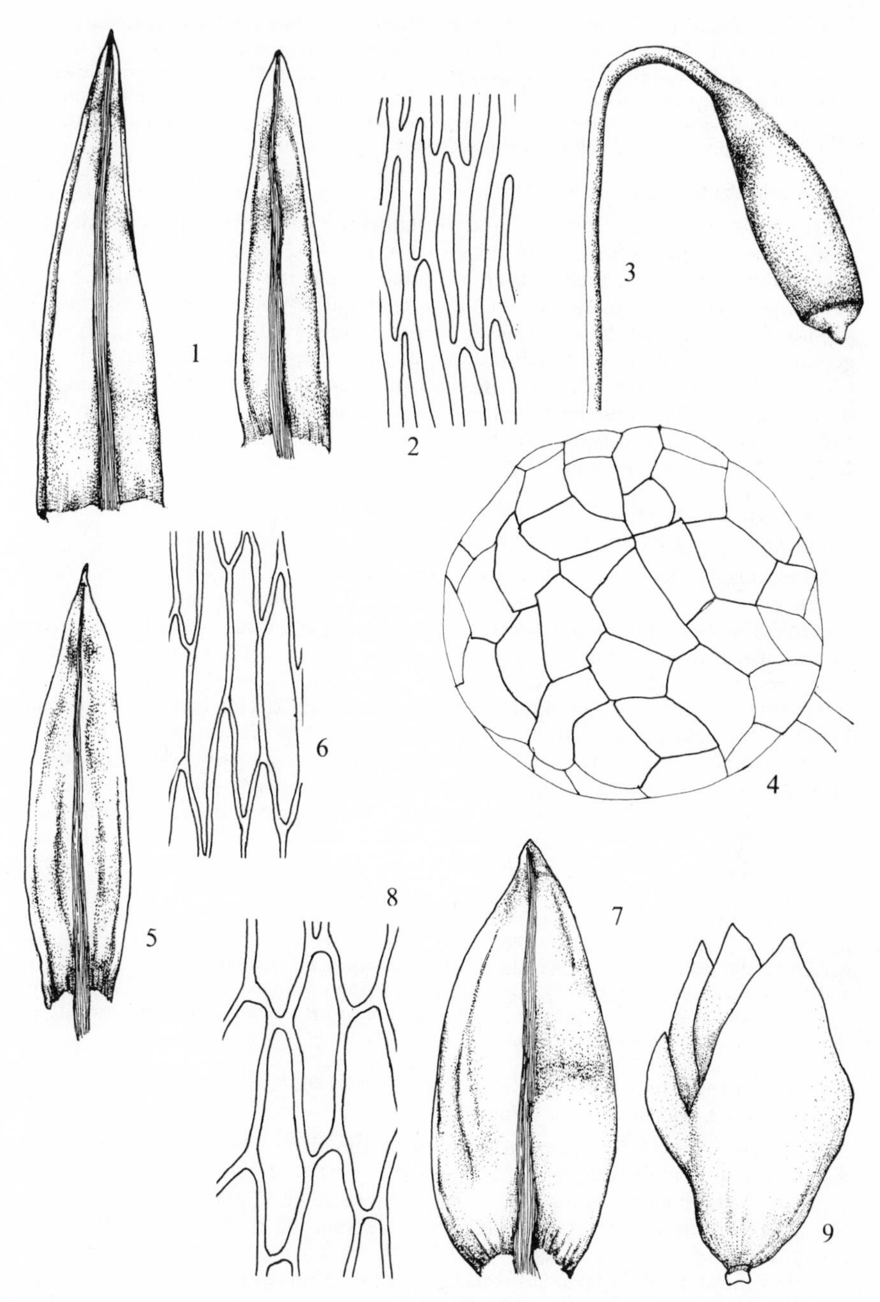

Fig. **197**. 1–6, *Bryum alpinum*: 1, leaves; 2, mid-leaf cells; 3, capsule (×10); 4, rhizoidal gemma (×250); 5 and 6, leaf and cells of green-coloured form. 7–9, *B. gemmiparum*: 7, leaf; 8, mid-leaf cells; 9, axillary bulbil (×100). Leaves ×25, cells ×415.

B. gemmiparum is intermediate in lf form and areolation between *B. alpinum* and *B. muehlenbeckii*. It is distinguishable from them by the axillary bulbils or, if these are lacking, by lf shape and areolation. *B. gemmiparum* has been confused in Britain with forms of *B. bicolor* agg., but these differ in their wider, more concave lvs which may be bordered, at least in older lvs, in the weaker nerve and ± hexagonal cells. The only definite material I have seen is from Wales; specimens from Somerset belong to *B. bicolor* agg.

33. B. muehlenbeckii Br. Eur., 1846

Dioecious. Plants to 7 cm. Lvs straight, imbricate when dry, erecto-patent when moist, glossy, sometimes pinkish-tinged, concave, base reddish, ovate, obtuse to acute, margin plane or recurved, entire, unbordered; nerve reddish, ending in or below apex; basal cells quadrate to quadrate-rectangular, upper rhomboid-hexagonal slightly incrassate, 20–24 μm wide in mid-lf. Orange red, ± spherical, rhizoidal gemmae, 160 μm diameter, sometimes present in tomentum on older parts of stem. Capsule similar to *B. alpinum*, unknown in Britain. Dark, greenish-red tufts or patches on acidic rocks by water in montane areas of northern Scotland, very rare. Mid Perth, S. Aberdeen, Inverness, Mid Ebudes, Ross, W. Sutherland. 8. Europe, Middle East, N. America, Greenland.

34. B. riparium Hagen, K. Norsk. Vid. Selsk. Skrift., 1908
B. alpinum ssp. *riparium* (De Not.) Kindb.

Dioecious. Plants to 3 cm, stem brownish, concealed by the closely placed lvs. Lvs imbricate, slightly twisted at apices when dry, erect to erecto-patent when moist, lanceolate to ovate-lanceolate, acuminate, abruptly narrowed into shortly excurrent nerve, base decurrent, margin plane or recurved, entire, bordered above at least in older lvs; nerve green or reddish-brown, distinctly excurrent; basal cells lax, rectangular or quadrate-rectangular, 16–40 μm wide at extreme base, reddish except in young lvs, cells above rhomboidal to narrowly rhomboidal, incrassate, (8–)12–16 (–20) μm wide in mid-lf, marginal cells narrow, forming ± distinct border. Reddish, flattened, lobed rhizoidal gemmae, 100–200 μm × 75–150 μm, present in tomentum on older part of stem. Fr unknown. Dense tufts, pale green above, blackish below, on soil, rocks and in rock crevices, especially by streams but also on arable soil. Occasional in W. and N. Britain from Mid and North Wales and the Lake District north to Sutherland and Outer Hebrides, very rare in Ireland. 26, H4. Norway.

Confused in the past with *B. mildeanum*, from which it differs in the more closely placed, narrower lvs concealing the brown, not reddish, stem, the distinct border, green or brownish nerve and the distinctive gemmae which also distinguish *B. riparium* from all other British and Irish species of the genus. For detailed comparison of the two species see Whitehouse, *Trans. Br. bryol. Soc.* **4**, 389–403, 1963. *B. mildeanum* and *B. riparium* are distinguished from the 3 preceding species by the excurrent nerve.

35. B. mildeanum Jur., Verh. Zool. Bot. Ges. Wien, 1862
B. alpinum var. *mildeanum* (Jur.) Podp.

Dioecious. Plants 0.5–2.5 cm, stems reddish, sometimes visible between somewhat distant lvs. Lvs imbricate, straight when dry, erect to erecto-patent when moist, concave, base reddish, ovate-lanceolate, acuminate, apex gradually tapering into shortly excurrent nerve; margin plane or recurved, entire or slightly denticulate above, obscurely bordered; nerve reddish at least at base, shortly excurrent; basal cells lax, quadrate-rectangular to rectangular, 14–25 μm wide at extreme base, cells above rhomboidal to narrowly rhomboidal, slightly incrassate, 8–16 μm wide in mid-lf, marginal cells narrower, forming obscure border. Gemmae lacking. Fr unknown in Britain. Tight, yellowish-green tufts on soil, rocks and rock crevices

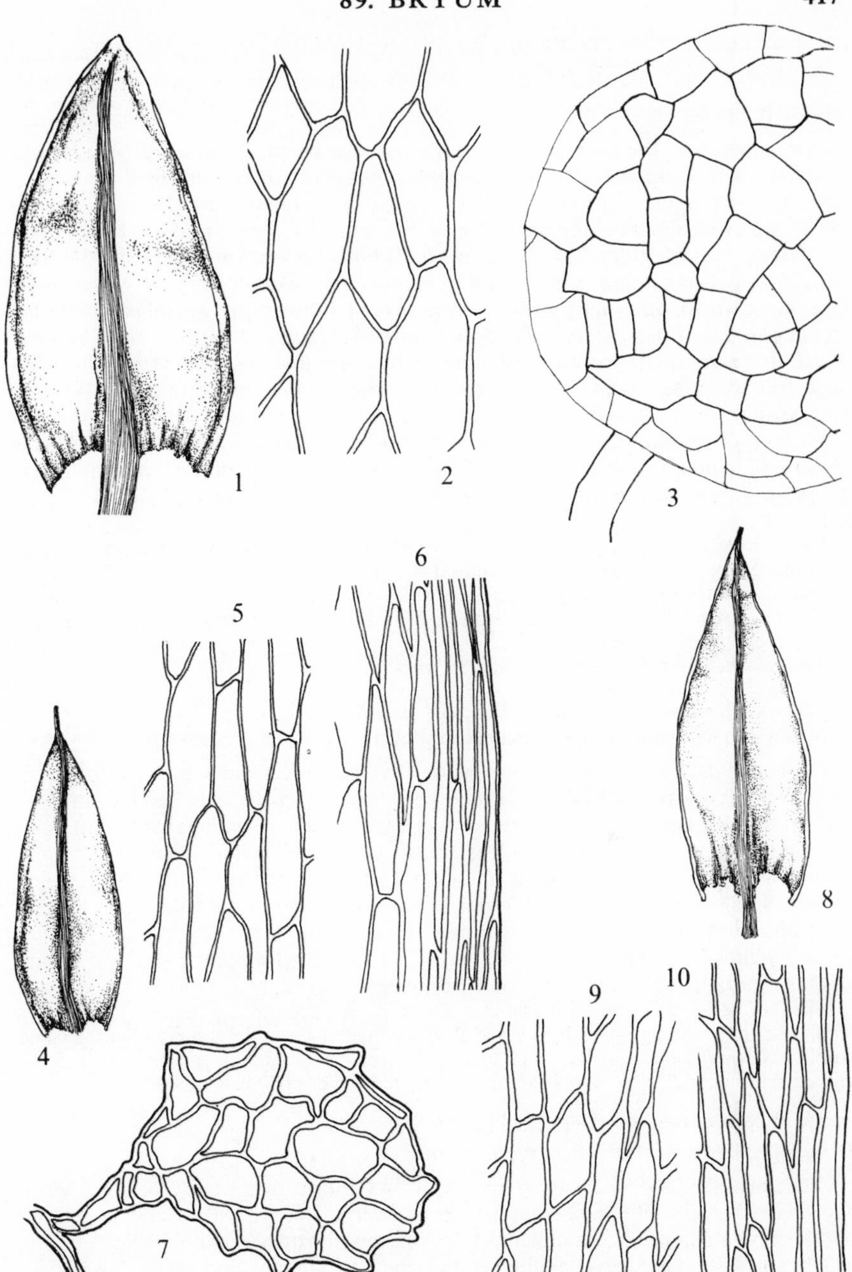

Fig. **198**. 1–3, *Bryum muehlenbeckii*: 1, leaf; 2, mid-leaf cells; 3, rhizoidal gemma. 4–7, *B. riparium*: 4, leaf; 5, cells at leaf base; 6, mid-leaf cells; 7, rhizoidal gemma. 8–10, *B. mildeanum*: 8, leaf; 9, cells from leaf base; 10, mid-leaf cells. Leaves ×25, cells ×415, gemmae ×250.

particularly by streams and on arable soil, rare. Hereford, Brecon and Carms north to E. Sutherland. 16. Europe, Asia, N. Africa.

36–40. B. bicolor agg.

The 5 species of the *B. bicolor* agg. have been shown by Dr H. L. K. Whitehouse to retain their distinctive characteristics when cultivated under uniform conditions Species 37–40 regularly produce axillary bulbils (and sometimes rhizoidal gemmae) except when fertile. The nomenclature of the bulbil-bearing species is based upon the descriptions of Wilczek & Demaret, *Bull. Jard. bot. nat. Belg.* **46**, 511–51, 1976 and Smith & Whitehouse, *J. Bryol.* **10**(1), in press. British material of *B. gemmilucens* does not correspond exactly with descriptions and illustrations of Belgian material. In the absence of more detailed and experimental studies of the *B. bicolor* agg., of which there are other variants not covered by the following descriptions, it seems best for the time being to assume that the British and Belgian taxa are conspecific.

Wilczek & Demaret report a fifth bulbil-producing species from Britain. This is *B. barnesii* Wood. I have not, however, been able to find any marked morphological discontinuity between *B. bicolor s.s.* and *B. barnesii* and have therefore not described the latter.

Plants of the *B. bicolor* agg. differ from bulbiliferous *Pohlia* species in lf shape, the narrowly recurved, entire margin and relatively shorter lf cells. When fruiting, plants of the complex are readily recognised by the ovoid capsule with neck abruptly narrowed into seta; when sterile the axillary bulbils provide adequate distinguishing characters from other *Bryum* species except *B. gemmiparum*. In *B. bicolor* agg. the nerve is rarely more than 60 μm wide at the base whereas in *B. gemmiparum* it is 80 μm or more.

36. B. dixonii Card. ex Nicholson, Rev. Bryol., 1901
B. bicolor ssp. *dixonii* (Card. ex Nicholson) Podp.

Plants to *ca* 1 cm, stems slender, brittle. Lvs imbricate when dry, erecto-patent when moist, concave, reddish at base, ovate, acute to broadly ovate and acuminate, margin plane, obscurely denticulate above, unbordered; nerve relatively thin, ending below apex to shortly excurrent; basal cells rectangular, above rhomboid-hexagonal, 10–16 μm wide in mid-lf, 2–3 marginal rows narrower but not forming border. Small, axillary, propaguliferous shoots sometimes present. Gametangia and fruit unknown. Dense patches or small tufts among other bryophytes, reddish below, yellowish-green above when moist, yellowish-white when dry, on damp, sloping montane rocks, very rare. W. Perth, Angus, Argyll, N. Ebudes. 4. Endemic.

Likely to be mistaken only for plants of the *B. bicolor* agg., *B. argenteum*, *Anomobryum filiforme* or *Plagiobryum* spp. Members of *B. bicolor* agg. differ in the recurved lf margin and bulbils, *B. argenteum* in colour and hyaline lf apex, *Anomobryum filiforme* in lf shape and areolation and *Plagiobryum* in aerolation and reddish or brownish tinge.

37. B. bicolor Dicks., Pl. Crypt. Brit., 1801
B. atropurpureum Br. Eur., *B. barnesii* Wood

Dioecious. Plants 0.5–1.5 cm; stems green. Lvs imbricate, scarcely shrunken when dry, erect to erecto-patent when moist, very concave, not reddish at base, ovate and obtuse to ovate-lanceolate and acuminate, margin narrowly recurved below, entire, usually unbordered; nerve ending below apex to shortly excurrent; basal cells quadrate-rectangular, cells above rhomboid-hexagonal, in mid-lf 10–18 μm wide, marginal cells narrower but only rarely forming poorly defined border. Bulbils 1–2 per axil, green, (125–) 200–480 μm long with lf primordia about $\frac{1}{2}$ total length of bulbil, primordia arising from upper $\frac{1}{2}$–$\frac{2}{3}$ of bulbil; rarely yellowish-brown

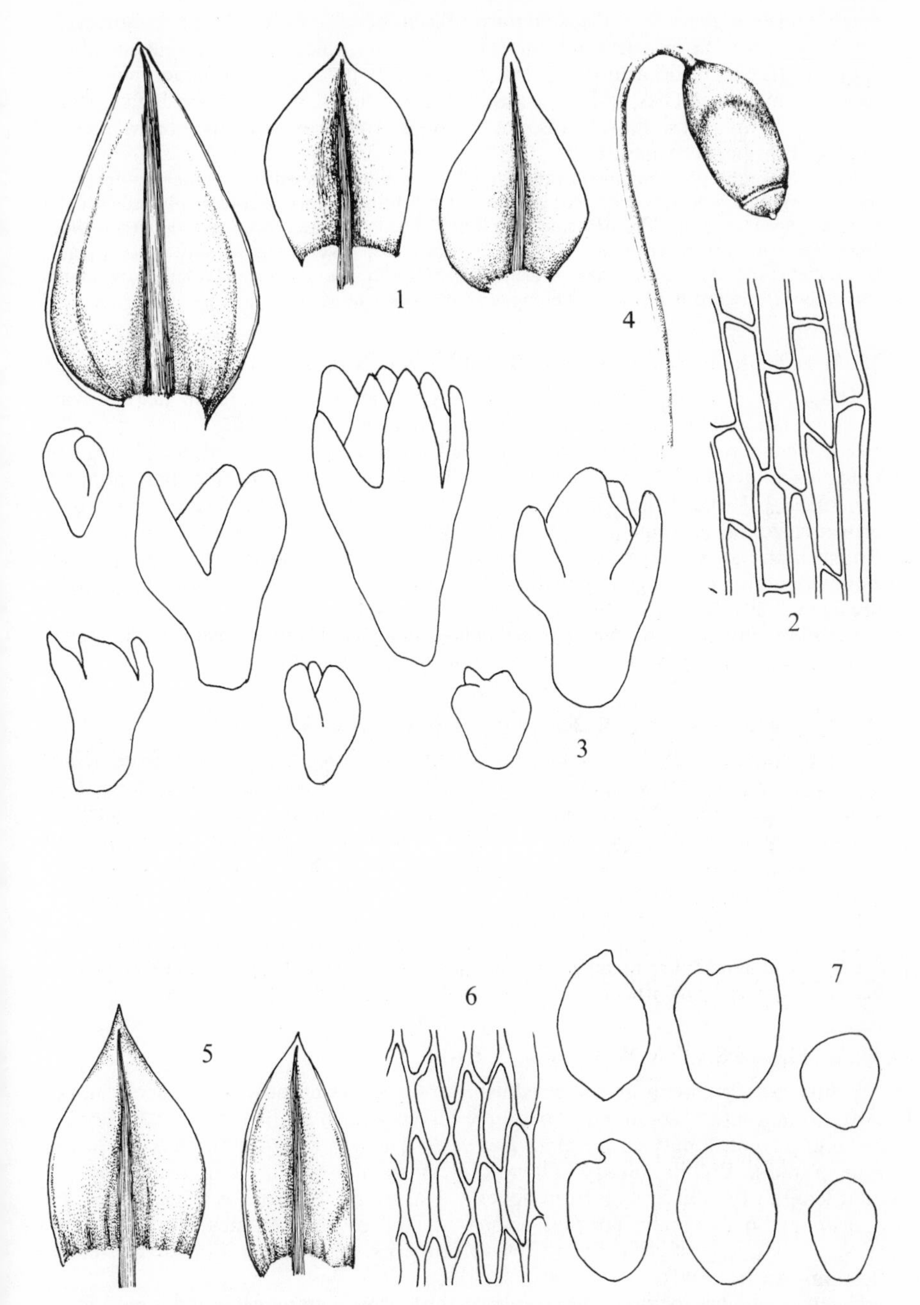

Fig. 199. 1–4, *Bryum bicolor*: 1, leaves; 2, mid-leaf cells; 3, axillary bulbils; 4, capsule (×10). 5–7, *B. gemmilucens*: 5, leaves; 6, mid-leaf cells; 7, axillary bulbils. Leaves ×50, cells ×415, bulbils ×100.

rhizoidal gemmae present. Capsule purplish-red, pendulous, ovoid, neck short, very abruptly narrowed into seta, lid conical and shortly apiculate; cilia appendiculate; spores 8–16 μm. Fr occasional, spring, summer. $n=10$*. Lax or compact, pale green to green tufts or patches, reddish below, on soil in fields, by roads, on banks, cliffs, on walls, in quarries, flower-pots and amongst rocks by streams, common. 111, H34, C. Europe, N. America.

By far the commonest member of the complex. *B. gemmiparum* differs in the broader nerve and the frequent presence of reddish rhizoidal gemmae. Very vigorous plants with the general appearance of *B. bicolor* and which have been mistaken for *B. gemmiparum* sometimes occur in rock crevices by rivers (e.g. by East Lyn River). The identity of such plants is not clear; the lower leaves have a distinct border and it is not certain whether the plants should be referred to *B. bicolor* or belong to some as yet undescribed taxon.

38. B. gemmiferum Wilcz. & Dem., Bull. Jard. Bot. Nat. Belg., 1976

Dioecious. Similar to *B. bicolor*, 0.5–1.5 cm. Stems often reddish. Leaves ovate to ovate-lanceolate; nerve distinctly excurrent in leaves of fertile stems and taller sterile stems; cells 16–22 μm wide in mid-lf. Sterile stems with usually orange or reddish axillary bulbils, (5–)20–30 per axil, 100–160(–200) μm long, base pointed, primordia $\frac{1}{4}$–$\frac{1}{3}$ total length of bulbil, restricted to apex of bulbil, brown or reddish-brown rhizoidal gemmae are reported in Belgian plants but have not been seen in British material. Capsule as in *B. bicolor*. Fr occasional, spring, summer. $n=10$*. Tufts or patches on damp soil by ditches, pools, gravel pits, etc., and in greenhouses, occasional. 29. Holland, Belgium, Lanzarote (Canary Is.).

Usually distinguished by the very small bulbils, though old bulbils remaining on the stem may approach those of *B. bicolor* in size but differ in the shorter primordia.

39. B. gemmilucens Wilcz. & Dem., Bull. Jard. Bot. Nat. Belg., 1976

Plants similar to *B. bicolor*, 0.5–1.0 cm; stems orange-red. Leaves ovate, acute; nerve ending in or below apex; cells in mid-lf 10–16 μm wide. Bulbils *ca* 5 per axil, yellow, glossy, 120–200 μm long, base rounded, primordia very rudimentary or indistinct with bulbils appearing ± similar at both ends, primordia where present restricted to bulbil apex. Gametangia not known in Britain; capsules unknown. Small patches on non-calcareous soils in arable fields and roadsides in S. Britain, rare. Belgium, France, Hungary, California.

Recognised by the lf primordia of the bulbils so poorly developed that the bulbils often appear ovoid and similar in appearance at either end. Wilczek & Demaret (1976) describe the bulbils as being pyriform or obovoid and in this respect British material differs.

40. B. dunense Smith & Whitehouse, J. Bryol., 1978

Plants *ca* 1 cm; stems green to reddish. Lower leaves broadly ovate, acute, upper ovate to lanceolate, acuminate, margin recurved below, slightly denticulate above; nerve excurrent, longly so in upper leaves (to $\frac{2}{3}$ length of lamina); cells 12–20 μm wide in mid-lf. Bulbils solitary in lf axils, greenish, 400–600 μm long, primordia $\frac{1}{2}$–$\frac{3}{4}$ total length of bulbil arising from upper $\frac{1}{2}$–$\frac{2}{3}$ of bulbil. Capsule as in *B. bicolor*. Fr occasional. Small, greenish or pink-tinged tufts on soil and sand near the coast, rare. W. Norfolk and west coast from Cornwall to Ross. 12. Sweden, France, Germany, Belgium, Austria, Italy.

Differs from other species of the complex in the longly excurrent nerve and is more likely to be mistaken for a species such as *B. caespiticium* but may be distinguished by the axillary bulbils.

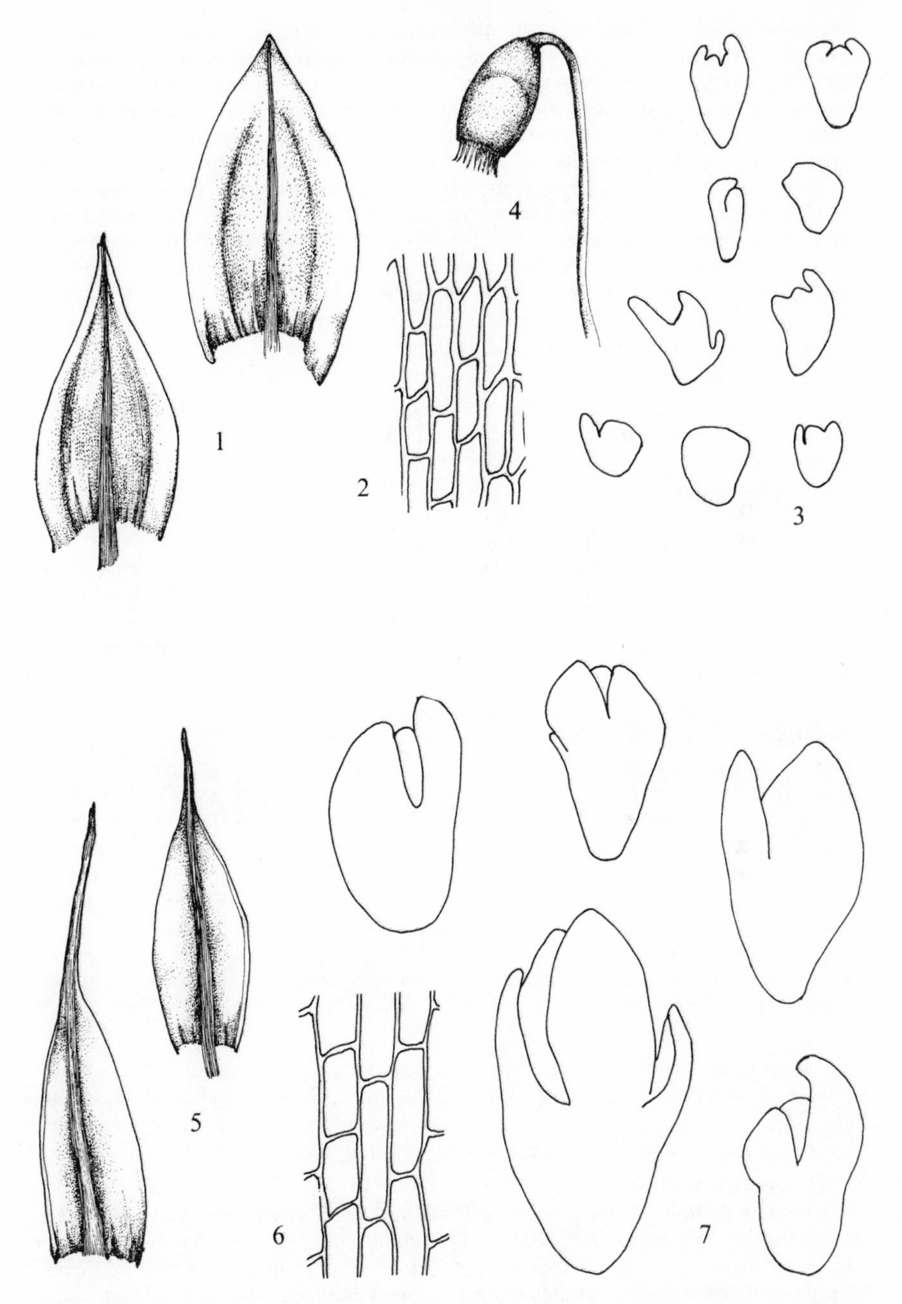

Fig. **200**. 1–4, *Bryum gemmiferum*: 1, leaves; 2, mid-leaf cells; 3, axillary bulbils; 4, capsule (×10). 5–7, *B. dunense*: 5, leaves; 6, mid-leaf cells; 7, axillary bulbils. Leaves ×50, cells ×415, bulbils ×100.

41. B. argenteum Hedw., Sp. Musc., 1801

Dioecious. Plants to 1.5 cm, shoots slender, fragile, julaceous. Lvs imbricate, scarcely altered when dry, reddish at base, ovate to broadly ovate, variously narrowed to short to long acuminate apex, ± hyaline in upper part, base decurrent, margin plane, entire, unbordered; nerve relatively thin; basal cells reddish, quadrate, quadrate or rectangular cells extending to *ca* ⅓ way up margin, cells elsewhere rhomboid-hexagonal to narrowly rhomboid-hexagonal, sometimes very incrassate, 10–16 μm wide in mid-lf, cells in upper part of lf pellucid with colourless walls, causing the whitish or silvery appearance of the shoots, 1–2 marginal rows in upper part of lf narrow but not forming border. Seta short, capsule small, cernuous or pendulous, ellipsoid to pyriform, wide-mouthed, contracted below mouth when dry and empty, neck short; lid mamillate; processes narrowly perforated; cilia appendiculate; spores 8–14 μm. Fr occasional, autumn to spring.

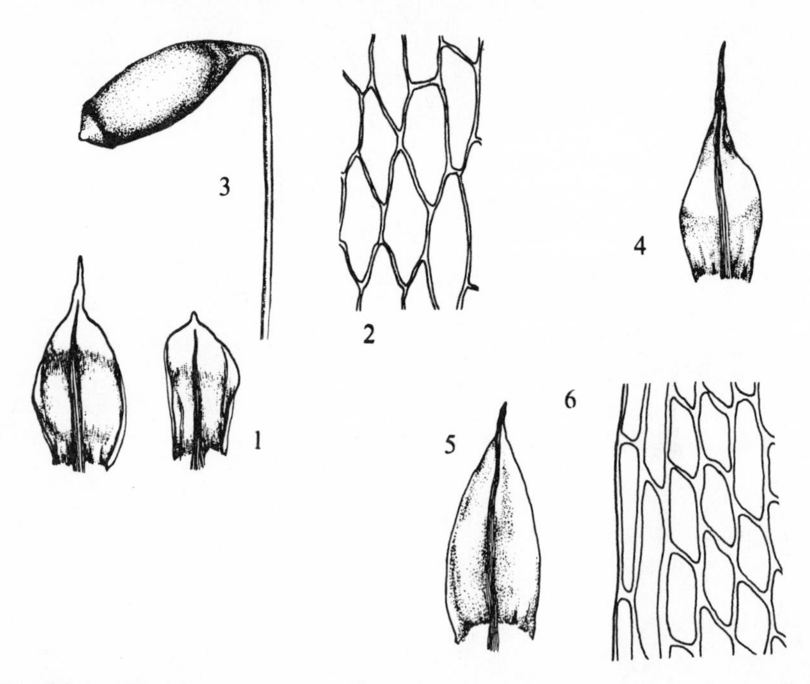

Fig. **201**. 1–3, *Bryum argenteum* var. *argenteum*: 1, leaves (×25); 2, mid-leaf cells; 3, capsule (×10). 4, *B. argenteum* ver. *lanatum*: leaf (×25). 5–6, *B. dixonii*: 5, leaf (×40); 6, mid-leaf cells. Cells ×415.

Var. **argenteum**

Plants hardly hoary. Upper part of lf hyaline, ± shortly tapering to apex; nerve not reaching lf apex. $n=10$*, 10+1, 11. Small, tight tufts or patches to dense, spreading mats or scattered shoots amongst other bryophytes, whitish or silvery, glossy above, brownish-green below, in pavement crevices, on damp concrete, edges of roads, wall crevices, hard-packed soil, etc., common in man-made habitats, rare elsewhere. 112, H40, C. Cosmopolitan.

Var. **lanatum** (P. Beauv.) Hampe, Linnaea, 1839

Plants ± hoary. Lvs with long hyaline point, whitish when dry, more gradually tapering to apex than in type; nerve excurrent. $n=12$. Hoary, whitish tufts in similar but usually drier habitats than the type and often ascending to higher altitudes. Occasional to frequent. 78, H15, C. Cosmopolitan.

Var. *lanatum* may only be a dry ground variant of the type, but as it sometimes grows mixed with the type its status requires further investigation. The capsule is similar in appearance to that of *B. bicolor* agg. but the slender, whitish or silvery stems will readily distinguish the plant from other species of *Bryum*. *Plagiobryum* species differ in their reddish or brownish-red tint and *Anomobryum filiforme* in its lf shape and areolation.

42–50 B. erythrocarpum complex

Usually dioecious in Britain. Plants 0.3–1.5 cm. Lvs ovate-lanceolate or lanceolate to triangular, erect or slightly spreading, slightly shrunken when dry, erect to erect-spreading when moist, base reddish, margin recurved in upper lvs, denticulate towards apex, bordered or not; nerve strong, ± excurrent; cells in mid-lf elongate-hexagonal, shorter and wider towards base. Rhizoidal gemmae constantly present. Capsule cylindrical to ovoid, symmetrical or slightly curved; cilia appendiculate. Plants loosely or densely tufted or as scattered groups of stems among other mosses, ± tinged with red.

The following nine species are characterised by the possession of rhizoidal and sometimes axillary gemmae. For a detailed account of the species see Crundwell & Nyholm, *Trans. Br. bryol. Soc.* **4**, 597–637, 1964. For information on the occurrence of rhizoidal gemmae in these and other moss species see Whitehouse, *Trans. Br. bryol. Soc.* **5**, 103–16, 1966.

For naming to be meaningful, identification of the plants of the *B. erythrocarpum* complex should be taken to the species level.

42. B. radiculosum Brid., Sp. Musc., 1817

B. murale Wils. ex Hunt, *B. murorum* (Schimp.) Berk., *B. atrovirens* var. *radiculosum* (Brid.) Wijk & Marg.

Plants densely tufted, 3–10 mm. Lvs not or scarcely bordered; nerve very strong, longly excurrent, yellow, sometimes reddish when old; mid-lf cells 10–12 μm × 40–60 μm, slightly longer and narrower at margin. Rhizoids yellowish to brown, coarsely papillose. Gemmae constantly present but sometimes few, never axillary, brown to bright red, spherical, 120–180 μm diameter, cells not protuberant. Capsule variable in shape and size, ± narrowly ellipsoid; spores 10–14 μm. Fr frequent, early summer. Dense tufts on old mortar, limestone rocks and hard calcareous soil, calcicole. Frequent in lowland areas of England and Wales, very rare in Scotland, occasional in Ireland. 72, H19, C. S., C. and W. Europe, Macaronesia, Bermuda.

Most likely to be confused with *B. microerythrocarpum* but is a calcicole species with a stronger, yellowish nerve and narrower cells. Sometimes mistaken for *B. bicolor* agg. but differs in the narrower lf cells and rhizoidal gemmae.

43. B. ruderale Crundw. & Nyh., Bot. Not., 1963

Dioecious. Plants 5–8 mm. Nerve shortly excurrent; mid-lf cells 10–14 μm × 30–60 μm, usually narrower at the margin but hardly forming a border. Rhizoids distinctly papillose, deep bright violet, sometimes purple, rarely reddish and paler. Gemmae on long rhizoids, never axillary, bright or pale purplish-red, occasionally orange, spherical, 125–180(–200) μm diameter, cells scarcely protuberant. Capsule pyriform; spores 9–11 μm. Fr unknown in Britain. On strongly basic to slightly acid soil in arable fields, on earthy banks, roadsides, etc., sand-dunes. Frequent to common especially in basic lowland areas. 70, H17, C. Europe, Azerbaydzhan, Egypt, N. America.

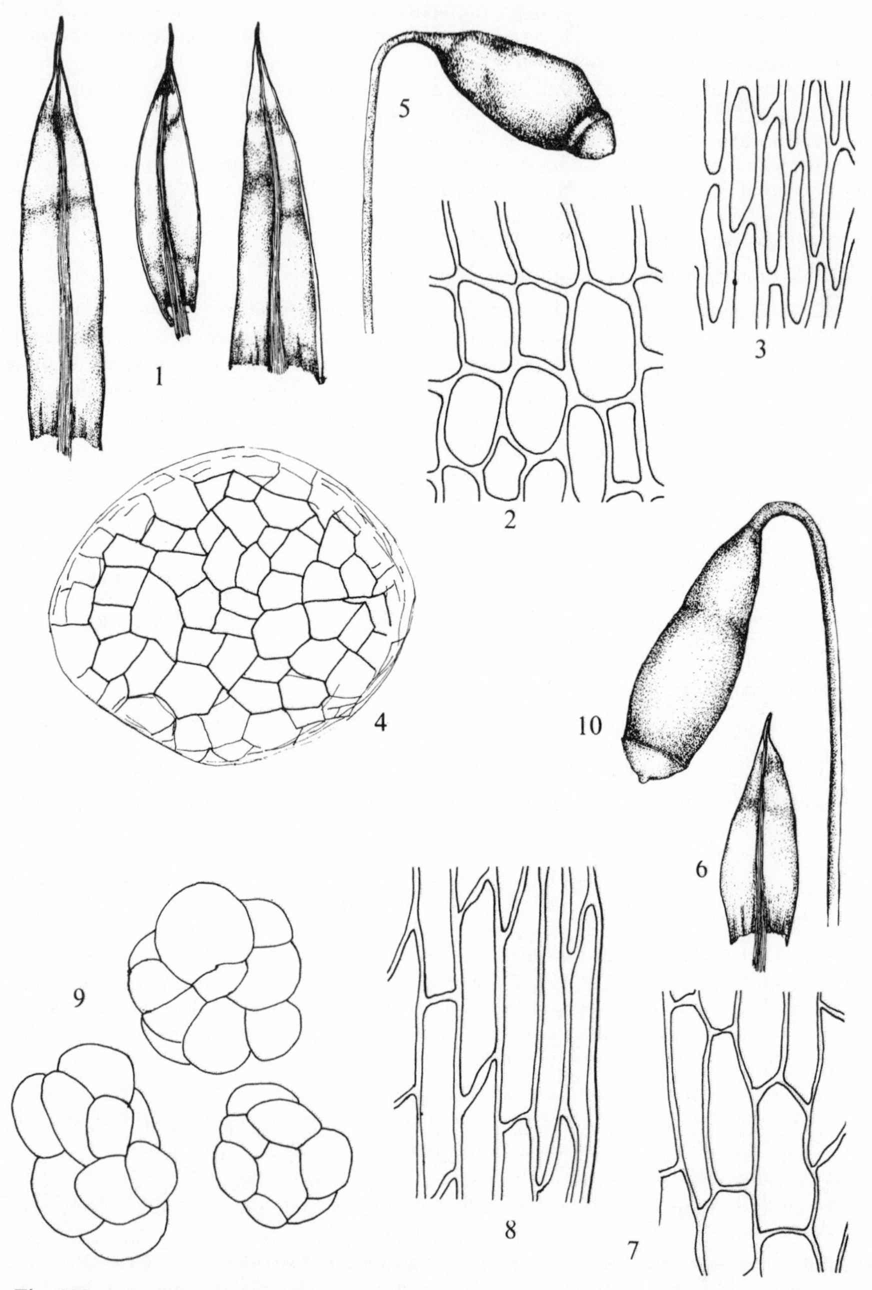

Fig. **202**. 1–5, *Bryum radiculosum*: 1, leaves; 2, basal cells; 3, mid-leaf cells; 4, gemma; 5, capsule. 6–10, *B. klinggraeffii*: 6, leaf; 7, basal cells; 8, mid-leaf cells; 9, gemma; 10, capsule. Leaves ×25, cells ×415, gemmae ×250, capsules ×10.

Very close to *B. violaceum* but differing in the larger, less regularly spherical gemmae and the distinctly papillose rhizoids. The usually deep violet, markedly papillose rhizoids render the plant easily recognisable.

44. B. violaceum Crundw. & Nyh., Bot. Not., 1963

Very close to *B. ruderale*. Rhizoids smooth, rarely finely papillose, dull mauve to violet. Gemmae numerous on long rhizoids, never axillary, bright or pale purplish-red, occasionally orange, regularly spherical, 60–90(–110) μm diameter, cells not protuberant. Fr unknown in Britain. On calcareous to slightly acid soil in arable fields, earthy banks etc. Generally distributed in Britain, rare in Scotland and Ireland. 50, H4. C. and N. Europe, Kashmir, N. America, Patagonia.

45. B. klinggraeffii Schimp. in Klinggr., Höh. Crypt. Pruess., 1858

Plants 2–5 mm. Nerve shortly excurrent; mid-lf cells 10–14 μm × 45–60 μm, usually longer and narrower at margin but hardly forming a border. Rhizoids finely papillose to almost smooth, pale yellowish-brown. Gemmae usually abundant, never axillary, bright crimson, irregularly spherical, (50–)60–100(–115) μm diameter, cells protuberant. Capsule broadly pyriform, strongly contracted below mouth when dry and empty; spores 8–12 μm. Fr unknown in Britain. On highly calcareous to slightly acid soils in arable fields, etc., also at margins of ponds and reservoirs. Frequent in England and Wales, occasional to rare in Scotland and Ireland. 66, H14, C. W. and N. Europe, China, Japan, N. America, Patagonia.

Likely to be confused only with *B. sauteri* or *B. violaceum*; the former has smaller, pyriform gemmae, browner and the same colour as the rhizoids, the latter has gemmae of a similar size but of purplish-red colour and without protuberant cells, and the rhizoids are nearly always violet.

46. B. sauteri Br. Eur., 1846

Dioecious in Britain. Plants *ca* 4 mm. Nerve very stout, excurrent in short point; mid-lf cells *ca* 14 μm × 70 μm, incrassate, slightly narrower towards margin. Rhizoids red-brown, finely papillose. Gemmae usually abundant, never axillary, brown to red-brown, pyriform, 40–60 μm × 60–100 μm, cells not or slightly protuberant. Capsule narrowly pyriform, distinctly contracted just below mouth; spores 16–20 μm. Fr unknown in Britain. On basic or acidic soils (but not extremes of either) in arable fields, on earthy banks, molehills, etc., in lowland habitats, generally distributed, occasional. 36 H9, C. Europe, Caucasus, New Zealand.

47. B. tenuisetum Limpr., Jahresber. schles. Ges. vaterl. Kult., 1897

Dioecious with occasional synoecious inflorescences. Plants 2–10 mm. Nerve strong, excurrent, becoming dark purple with age; mid-lf cells 12–14 μm × 50–60(–100) μm, incrassate, usually somewhat longer, narrower and more incrassate at margin but hardly constituting a border. Rhizoids pale, usually yellow, papillose. Gemmae usually abundant, never axillary or clustered at base of stem but mostly on long rhizoids, spherical to slightly angled, yellowish, 120–180(–220) μm diameter, cells distinctly protuberant. Capsule narrowly ellipsoid; spores 12–16 μm. Fr frequent. In open places on damp sandy or peaty soils, very rarely in arable fields, calcifuge, generally distributed, rare to occasional. 21, C. C. and N. Europe, N. America.

48. B. microerythrocarpum C. Müll. & Kindb. in Macoun, Cat. Can. Pl., 1892

Plants 4–10 mm. Nerve strong, excurrent; mid-lf cells 10–16 μm × 45–70 μm, usually somewhat narrower, longer and more incrassate at the margin but hardly forming a border. Rhizoids brownish, papillose. Gemmae mostly on long rhizoids

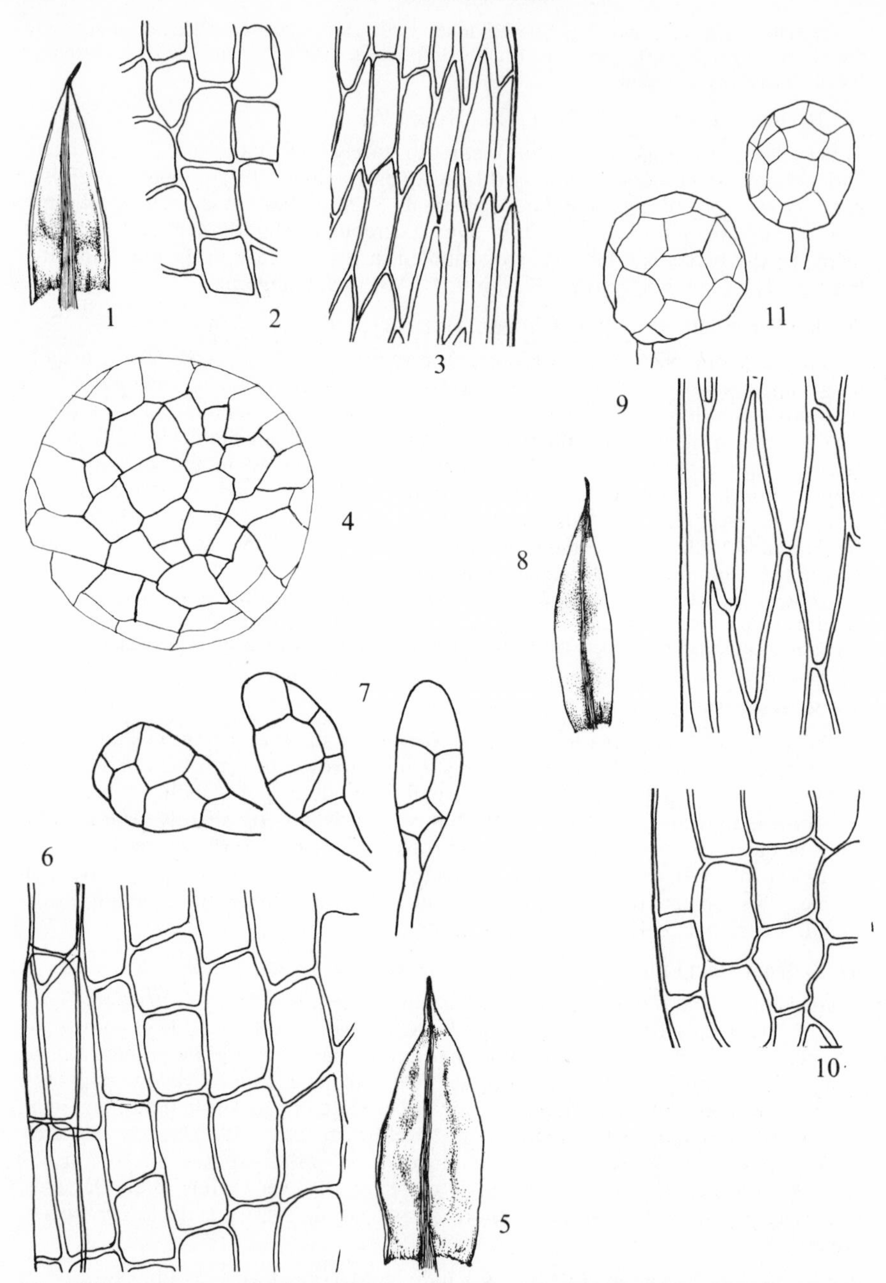

Fig. **203**. 1–4, *Bryum ruderale:* 1, leaf; 2, basal cells; 3, mid-leaf cells; 4, rhizoidal gemma. 5–7, *B. sauteri*: 5, leaf; 6, mid-leaf cells; 7, rhizoidal gemmae. 8–11, *B. violaceum*: 8, leaf; 9, basal cells; 10, mid-leaf cells; 11, rhizoidal gemmae. Leaves ×25, cells ×415, gemmae ×250.

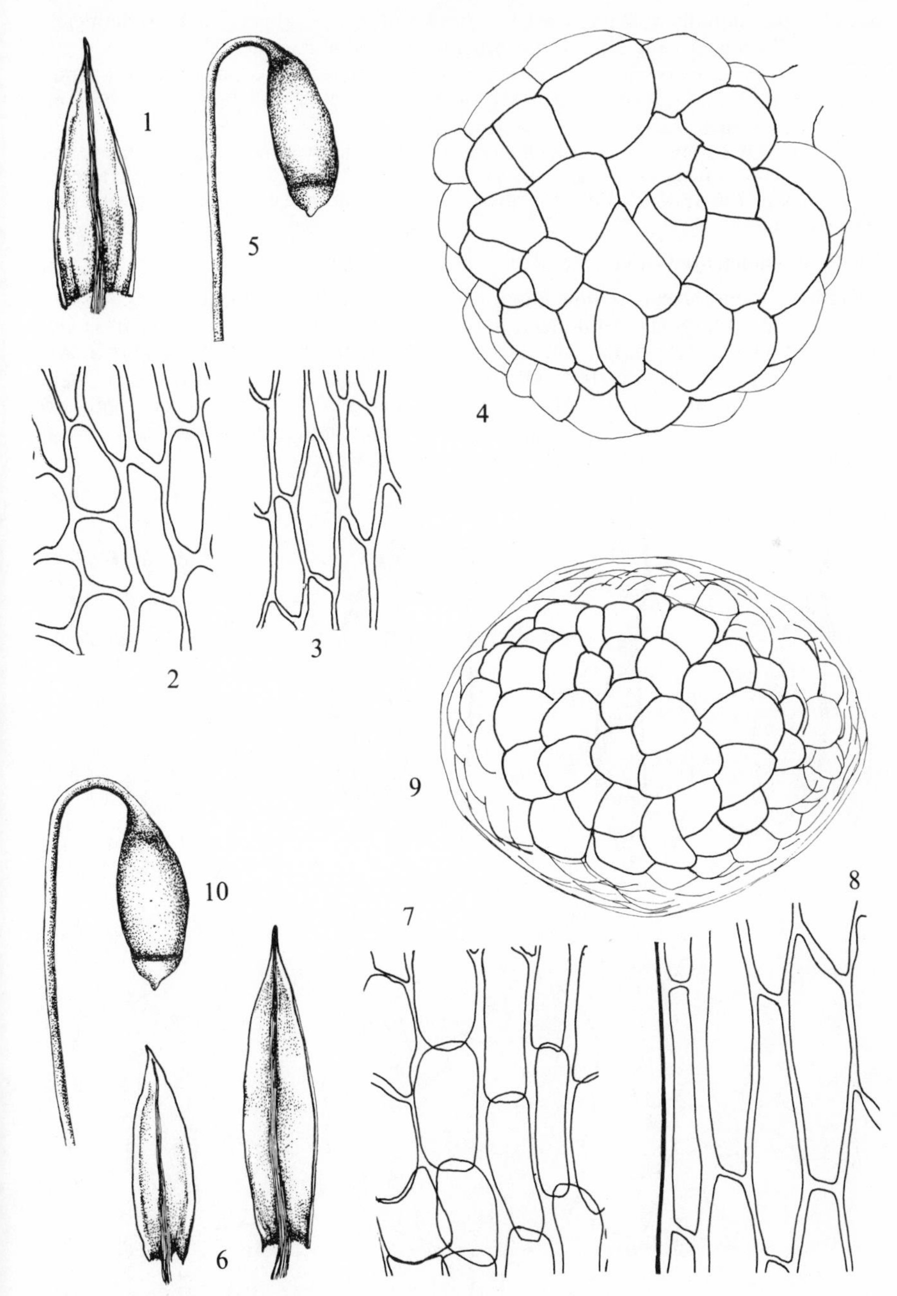

Fig. **204**. 1–5, *Bryum tenuisetum*: 1, leaf; 2, basal cells; 3, mid-leaf cells; 4, gemma; 5, capsule. 6–10, *B. microerythrocarpum*: 6, leaves, 7, basal cells; 8, mid-leaf cells; 9, gemma; 10, capsule. Leaves $\times 25$, cells $\times 415$, capsules $\times 10$, gemmae $\times 250$.

but also occasionally axillary, scarlet or brick-red, spherical, 190–260 μm diameter, cells usually not or only slightly protuberant. Capsule narrowly ovate-cylindrical; spores 10–12 μm. Fr occasional, late spring. On mildly acidic, usually peaty or sandy soils in disturbed places in arable fields, by paths, on cliffs, molehills, etc., common. 75, H6, C. Europe, N. America, New Zealand.

An extremely variable species which might possibly consist of more than one taxon and may contain derivatives of hybrids with *B. tenuisetum* and *B. rubens*. Most likely to be confused with *B. rubens* which differs in the lvs distinctly bordered and gemmae with very protuberant cells.

49. B. bornholmense Winkelm. & Ruthe, Hedwigia, 1899

Plants 3–15 mm. Nerve strong, longly excurrent; mid-lf cells 14–20 μm × 60–80 μm, incrassate, 2–3 marginal rows longer, narrower, with more incrassate and darker cell walls forming distinct border. Rhizoids brown, papillose. Gemmae on long rhizoids, never axillary, ± spherical, 160–300 μm diameter, cells not protuberant to strongly so. Capsule pyriform; spores 12–14 μm. Fr occasional, spring. On very acidic to highly calcareous soils in ± permanent habitats and sandy soil by the sea, never in arable fields, generally distributed, frequent. 49, H8, C. Europe, N. America.

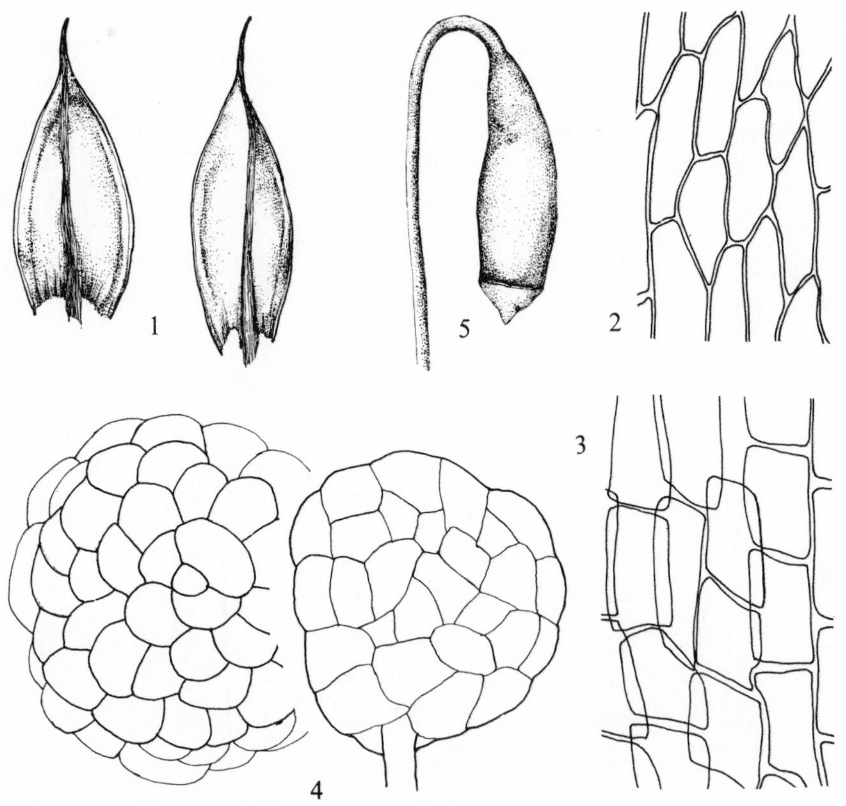

Fig. **205**. *Bryum bornholmense*: 1, leaves (×25); 2, mid-leaf cells; 3, basal cells; 4, rhizoidal gemmae (×250); 5, capsule (×10); Cells ×415.

Morphologically and ecologically very variable and possibly consisting of more than one taxon. The bordered lvs will distinguish *B. bornholmense* from all other related species except *B. rubens*, from which it may be distinguished by the larger size of all its parts and in the gemmae never being axillary. Gemmiferous forms of *B. capillare* have the nerve more longly excurrent, the lvs of a different shape and spirally twisted when dry.

50. B. rubens Mitt., Kew J. Bot., 1856

Dioecious. Plants 2–15 mm. Nerve shortly excurrent; mid-lf cells 16–20 μm×40–60 μm, ± thin-walled, 2–3 marginal rows longer, narrower with more incrassate and deeply coloured walls forming a distinct border. Rhizoids brown, papillose. Gemmae usually abundant, often on short rhizoids and in clusters at base of stem or solitary in the lf axils, bright crimson, sometimes darker in basic soils, ± spherical, 180–260 μm diameter, walls strongly protuberant. Capsule cylindrical; spores 8–10 μm. Fr frequent, spring. On slightly acid to highly calcareous disturbed soils in temporary habitats, common. 101, H26, C. Europe, Azerbaydzhan, India, Japan, New Zealand, possibly N. America.

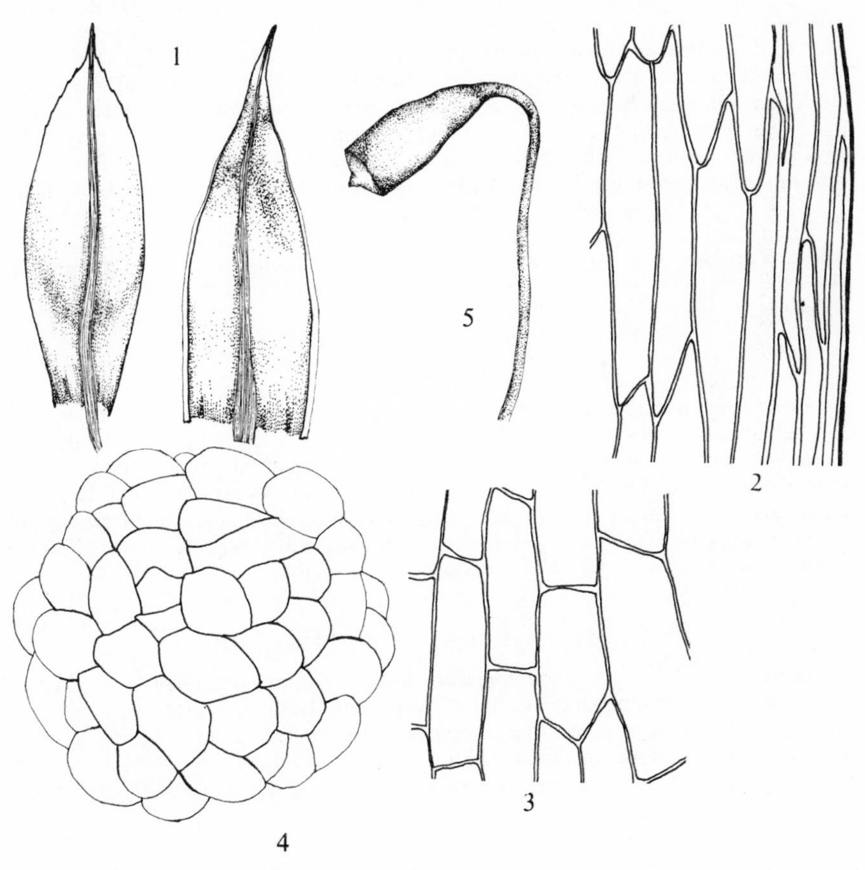

Fig. **206**. *Bryum rubens*: 1, leaves (×25); 2, mid-leaf cells; 3, basal cells; 4, rhizoidal gemma (×250); 5, capsule (×10). Cells ×415.

Likely to be confused only with *B. bornholmense* or *B. microerythrocarpum*, for the distinctions from which see under these species.

90. Rhodobryum (Schimp.) Limpr., Laubm., 1895

Plants with underground rhizomatous stems, branches erect. Lvs in terminal rosette, obovate or spathulate, margin unistratose, usually recurved below, dentate above; nerve in section with few or no stereids. Capsule similar to that of *Bryum*. A small north temperate genus of 4 species, but it is possible that additional species occur in subtropical or tropical regions.

1. R. roseum (Hedw.) Limpr., Laubm., 1895
Bryum roseum Hedw.

Dioecious. Shoots 5 cm or more, simple or with a single branch from below perichaetium, arising from underground rhizomatous stem with scale lvs. Shoots with 16–21 lvs in comal tuft, lower lvs scale-like, comal lvs spreading when moist, somewhat shrunken but not twisted when dry, outer lvs obovate, acute, middle spathulate, acute, innermost spathulate, acuminate, margin recurved to about widest part of lf, dentate above, unbordered; nerve ceasing below apex; cells in lower part of lf rectangular, at widest part and above rhomboidal, 24–36 μm wide at widest part of lf. Seta red, long; capsule pendulous, ellipsoid with short neck; lid conical; spores 16–24 μm. Fr very rare, autumn. $n=11$. Green patches on soil of usually reasonable nutrient status, in turf, on banks, on soil-capped walls and rock ledges, dune slacks, generally distributed, occasional. 98, H22. A mainly north temperate species distributed through Europe, Asia and N. America.

A second species, *R. ontariense* (Kindb.) Kindb., has been recorded from Sweden (Iwatsuki & Koponen, *Acta bot. fenn.* **96**, 1–22, 1972). It is distinguished from *R. roseum* by the more numerous comal lvs (18–52) which when dry are strongly twisted and turned upwards.

23. MNIACEAE

Dioecious or synoecious, inflorescence bud-like or discoid. Stem with central strand. Lvs pellucid, usually small below, larger above, not crowded except at apex of fertile stems, ovate to narrowly lanceolate, apex acute to rounded, margin usually bordered, with single or double teeth; nerve with stereid band in section; cells large, ± hexagonal to quadrate, smooth. Seta long, 1 to several per perichaetium; capsule cernuous to pendulous, ovoid to cylindrical; peristome double; calyptra cucullate.

The treatment of this family follows Koponen, *Ann. bot. fenn.* **5**, 117–51, 1968.

91. Mnium Hedw., Spec. Musc., 1801

Sub-apical branching rare, stem epidermis unistratose, cell walls heavily thickened; sterile shoots erect; reddish coloration often present. Lvs mostly ovate or lanceolate, margin with 2 to several stratose border and double teeth; nerve ending in apex, toothed at back above, in section with 2 stereid bands; cells ± isodiametric, not much smaller near margin than near nerve, rarely in rows. A northern hemisphere genus of *ca* 12 species.

1 Lvs unbordered **6. M. stellare**
Lvs bordered 2

2 Lf base hardly decurrent, nerve ending below apex, lid mamillate **1. M. hornum**
Lf base distinctly decurrent, nerve ending in apex or excurrent at least in upper lvs of fertile stems, lid rostrate 3

3 Nerve lacking teeth at back above, plants usually synoecious **4. M. marginatum**
At least some lvs with nerve toothed at back above, plants dioecious 4

4 Cells in ± regular rows radiating from nerve **5. M. spinosum**
Cells not in regular radiating rows 5

5 Cells mostly 18–28 μm wide in mid-lf, collenchymatous **3. M. ambiguum**
Cells mostly 14–18 μm wide in mid-lf, not collenchymatous **2. M. thomsonii**

1. M. hornum Hedw., Sp. Musc., 1801

Dioecious. Plants 2–10 cm, stems red to brown, with brown tomentum below. Lvs erect, twisted when dry, patent when moist, lower small, distant, upper larger, crowded, lanceolate to narrowly lanceolate, base slightly decurrent, apex acute to acuminate, margin plane, bordered, with spinose double teeth from below middle; nerve ending below apex with a few single teeth at back above; basal cells rectangular, above irregularly hexagonal with uniformly thickened walls, variable in size, (14–)18–24(–30) μm wide in mid-lf. Seta red; capsule horizontal to pendulous, ovoid to ovate-ellipsoid; lid mamillate; spores 26–35 μm. Fr frequent, spring. $n=6$*, 7, 12. Dark green tufts or patches, paler in younger parts, on acidic soil in woods, on banks, sides of streams, on logs, decaying wood, rocks and in rock crevices, common and sometimes abundant except in basic habitats and mainly at low altitudes. 112, H40, C. Europe, Faroes, Japan, Algeria, N. America.

2. M. thomsonii Schimp., Syn., 1876
M. orthorhynchum auct. non Brid., *M. lycopodioides* ssp. *orthorrhynchum* (Hartm.) Wijk & Marg.

Dioecious. Plants 1–6 cm, stems red. Lvs twisted and incurved when dry, patent when moist, distant below, more crowded, larger above, ovate to ovate-lanceolate on sterile stems, ovate-lanceolate to lanceolate on fertile stems, acute to acuminate, base decurrent, margin with strong reddish border and double spinose teeth to below halfway; nerve reddish-brown, excurrent in upper lvs, ending below apex in lower, toothed at back above; basal cells rectangular, above quadrate-hexagonal, not or scarcely collenchymatous, variable in size, (12–)14–18(–20) μm wide in mid-lf. Seta red; capsule cernuous to horizontal, ellipsoid; lid rostrate; spores 25–36 μm. Fr very rare, summer. $n=6$. Dull green or pale green, dense tufts or patches, reddish below, in crevices and on ledges of basic rocks at 700–1190 m, occasional. Caerns, Pennines, Lake District, Dumfries, Scottish Highlands, very rare in Ireland, Clare, W. Galway, Sligo, Leitrim. 18, H4. Europe, Faroes, Caucasus, C. Asia, Siberia, China, N. America, Greenland.

3. M. ambiguum H. Müll., Verh. Bot. Ver. Brandenburg, 1866
M. lycopodioides auct. non Schwaegr.

Dioecious. Plants 3–7 cm, stems red or brownish. Lvs twisted when dry, patent when moist, crowded, ovate-oblong, acute or apiculate, lower shorter and wider, base decurrent, margin with strong reddish-brown border with double, spinose teeth from below midway; nerve percurrent to excurrent at least in upper lvs, usually toothed at back above; basal cells rectangular, above ± hexagonal, collenchymatous, variable in size, (15–)18–28(–32) μm wide in mid-lf. Seta red; capsule cernuous to horizontal, ellipsoid to narrowly ellipsoid; lid rostrate; spores 22–34 μm. Fr very rare, summer.

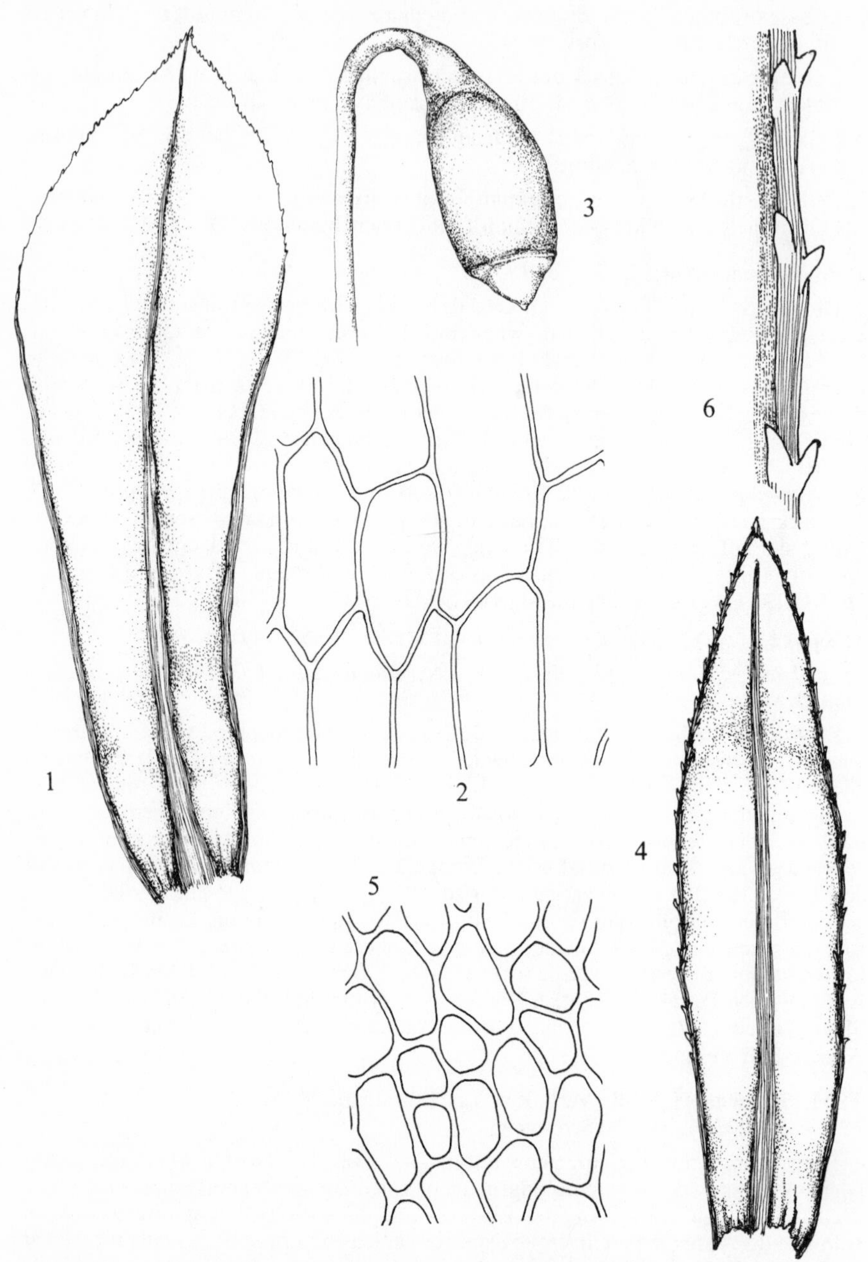

Fig. **207**. 1–3, *Rhodobryum roseum*: 1, leaf (×10); 2, cells at widest part of leaf; 3, capsule (×10). 4–6, *Mnium ambiguum*: 4, leaf (×15); 5, mid-leaf cells; 6, leaf margin (×100). Cells ×415.

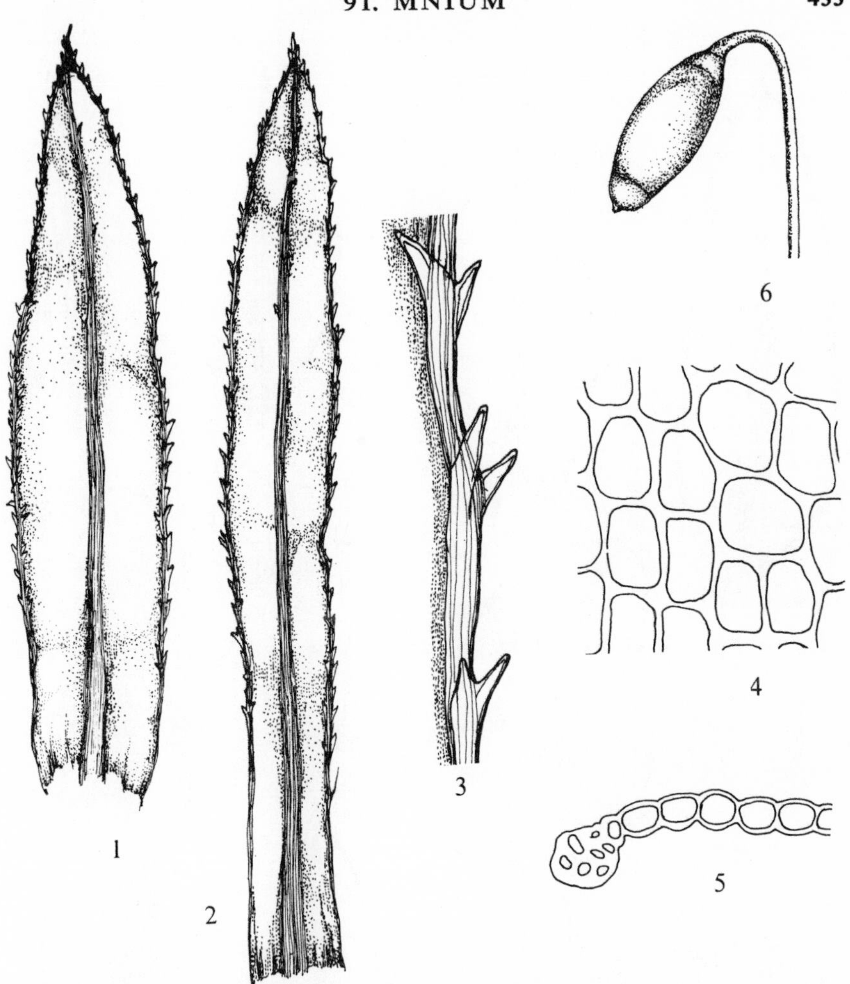

Fig. **208**. *Mnium hornum*: 1, and 2 lower and upper leaves (× 15); 3, leaf margin (× 100); 4, mid-leaf cells (× 415); 5, section of leaf margin (× 415); 6, capsule (× 5).

n = 6. Dark green tufts, reddish below, in shaded crevices and on ledges of damp basic montane rocks, very rare. Mid Perth, Angus. 2. Scandinavia, mountains of C. Europe, Faroes, Pyrenees, Himalayas, China, N. America.

Closely allied to *M. marginatum* but differing in the spinulose nerve and dioecious inflorescence. The lvs are usually more crowded, more sharply toothed and less conspicuously reddish-tinted than *M. marginatum*, but these characters are not always constant.

4. M. marginatum (With.) P. Beauv., Prodr., 1805
M. serratum Schrad. ex Brid.

Plants 2–4(–6) cm, reddish-brown in older parts. Lvs twisted when dry, patent when moist, ovate to broadly ovate, acute to obtuse and apiculate on sterile stems,

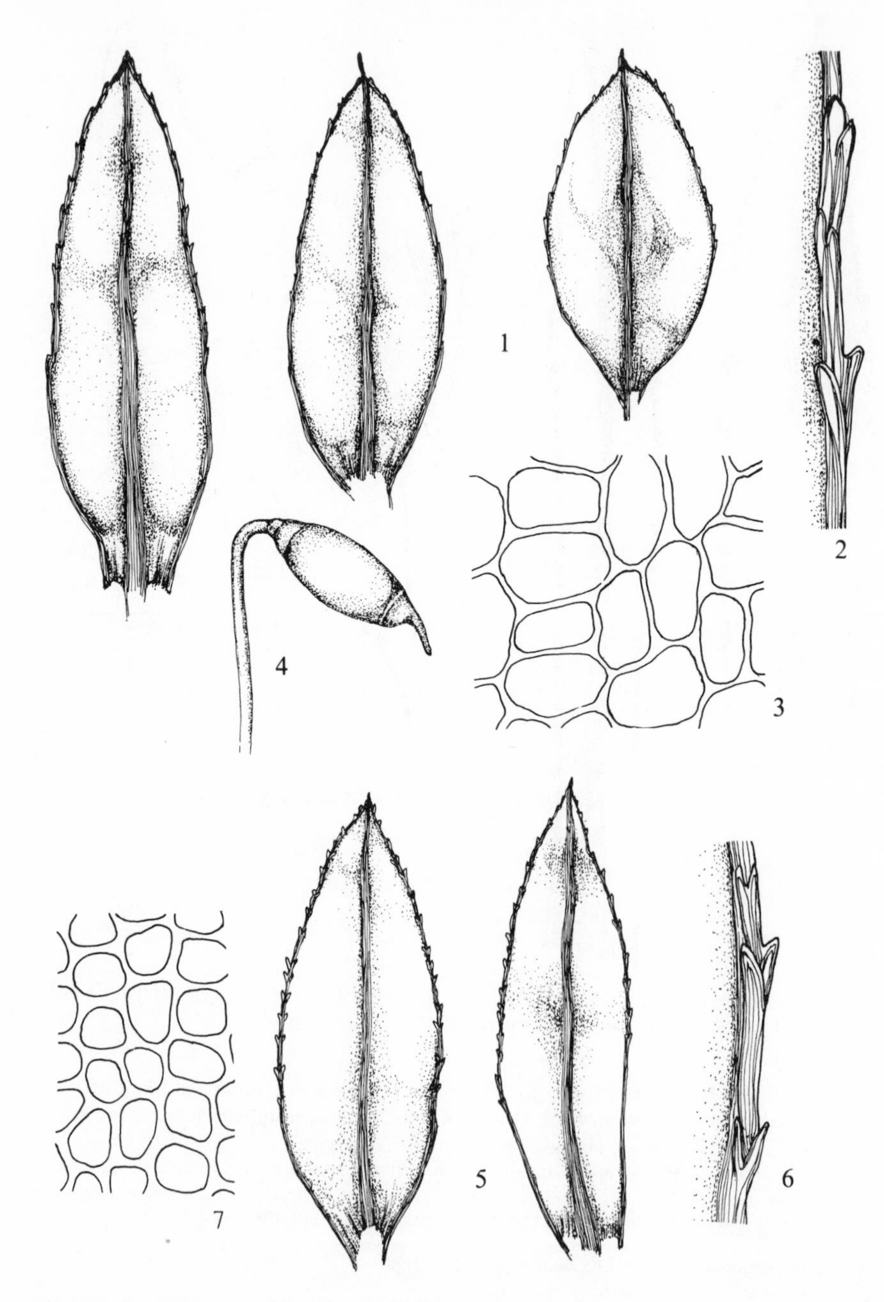

Fig. **209**. 1–4, *Mnium marginatum*: 1, leaves; 2, leaf margin; 3, mid-leaf cells; 4, capsule (×5). 5–7, *M. thomsonii*: 5, leaves; 6, leaf margin; 7, mid-leaf cells. Leaves ×10, margins ×100, cells ×415.

lanceolate, acute on fertile stems, base decurrent, margin with strong border, reddish-brown in older lvs, toothed from halfway or below with small, double, blunt or sharp teeth or sometimes ± entire in lower lvs; nerve ending in apex of upper lvs of fertile stems, below apex in other lvs, rarely toothed at back above, reddish-brown in older lvs; cells ± hexagonal, collenchymatous, not in rows, (14–)18–32(–36) μm wide in mid-lf. Seta pale red; capsule sub-pendulous, ellipsoid; lid rostrate; spores 24–34 μm.

Var. **marginatum**

Synoecious. Fr frequent, spring, early summer. $n=12$. Dark green or light green tufts or patches, deep red below, on moist soil and rocks and in rock crevices in woods and shaded, usually basic habitats. Rare in southern England and Ireland, occasional elsewhere. 69, H10. Europe, Turkey, Caucasus, Siberia, N. America, Greenland.

Var. **dioicum** (H. Müll.) Crundw., Trans. Br. bryol. Soc., 1968

M. dioicum H. Müll., *M. riparium* Mitt., *M. marginatum* var. *riparium* (Mitt.) Husn.

Dioecious. Fr unknown in Britain. Scattered stems and small patches on damp basic soil and rocks in woods and by streams, rare. W. Sussex, Hereford, Denbigh, Yorks, Cumberland, Mid Perth, E. Mayo, Armagh, Antrim. 8, H3. Belgium, C. Europe, Caucasus, Himalayas, Siberia, China, N. America (?).

The only difference between the 2 varieties is the inflorescence and it is uncertain whether var. *dioicum* should be maintained as a variety or merely be regarded as a dioecious form of a normally synoecious species.

5. M. spinosum (Voit.) Schwaegr., Suppl., 1816

Dioecious. Plants 1.5–8.0 cm, reddish or reddish-brown in older parts. Lower lvs distant, scale-like, appressed, upper large, crowded, twisted, undulate when dry, spreading when moist, similar on fertile and sterile stems, ovate-lanceolate to ovate or obovate, acute or sub-acute and apiculate, base narrow, longly decurrent, margin with strong border, spinosely double-toothed from or below halfway; nerve excurrent in upper lvs, toothed at back above, reddish; basal cells rectangular, cells above in radiating rows, irregularly hexagonal, sometimes somewhat elongated, especially near nerve, not collenchymatous, 16–40 μm wide in mid-lf. Seta red or orange, 1 to several per perichaetium; capsule horizontal or pendulous, narrowly ellipsoid; lid with curved beak; spores 20–28 μm. Fr very rare, late summer. $n=6$. Lax, dull or dark green tufts or patches in turf, rock crevices and on rock ledges in shaded, highly basic habitats between 600 and 1070 m, rare. Perth, Angus, W. Inverness. 4. Pyrenees, C. and N. Europe, Iceland, Urals, Caucasus, Siberia, China, N. America.

6. M. stellare Hedw., Spec. Musc., 1801

Dioecious. Plants 1–10 cm with brownish tomentum below. Lvs small, distant below, larger, more crowded above, slightly curled and undulate when dry, patent when moist, similar in sterile and fertile stems, broadly ovate to ovate-lanceolate, acute to obtuse and apiculate, base decurrent, margin unbordered, irregularly dentate, at least above, with single or double blunt teeth; nerve ending well below apex, not toothed at back above; cells hexagonal, not in rows, ± collenchymatous, 20–34 μm wide in mid-lf, 1–2 marginal rows often narrower, more incrassate but not forming border. Seta reddish; capsule cernuous to horizontal, ellipsoid; lid convex. Fr very rare, spring. $n=7$. Dense tufts or patches, pale or bright green above, dark green to brownish below, sometimes bluish-tinged, on damp shaded rocks and soil in woods, rock crevices and ledges in basic habitats. Rare in S. England, N. Scotland

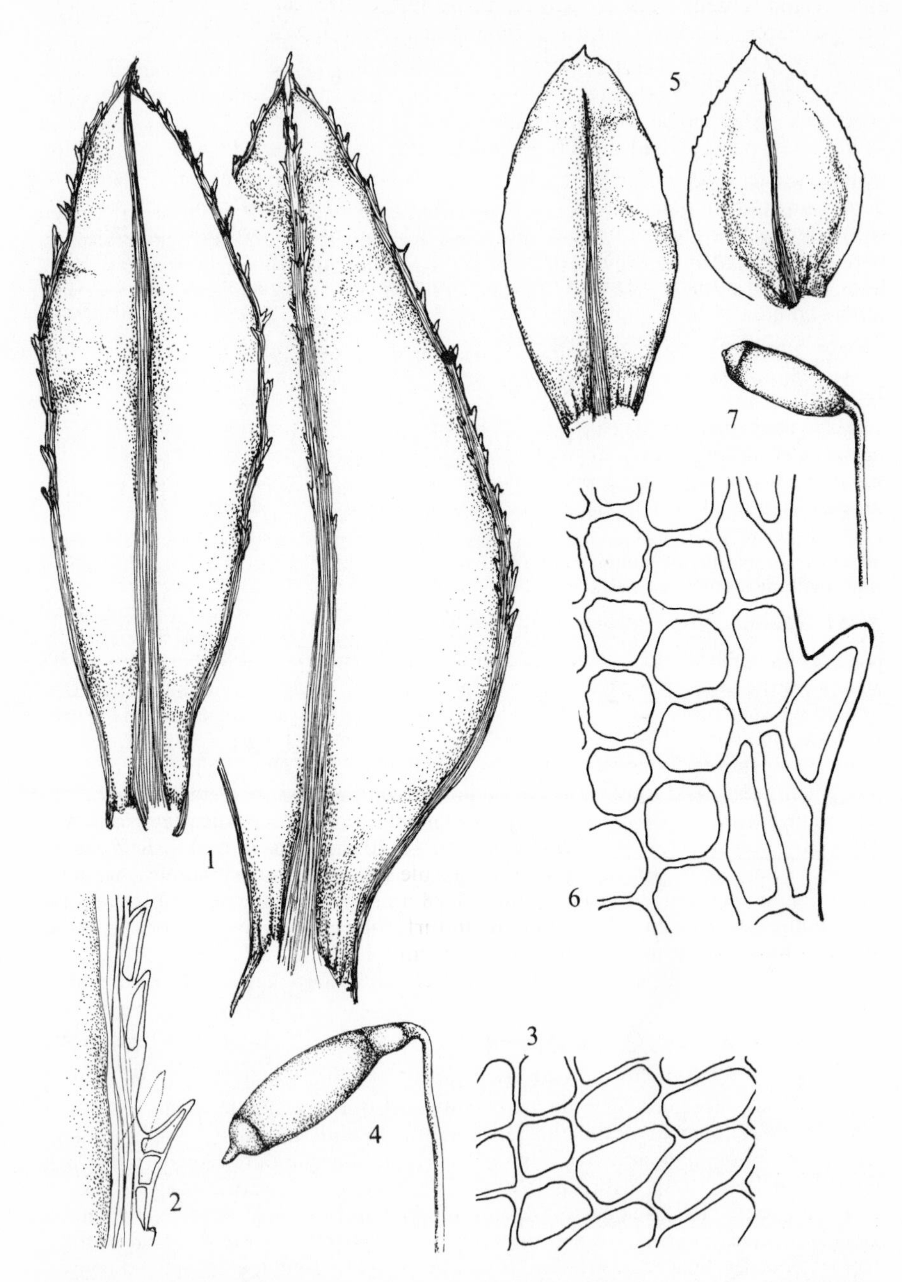

Fig. **210**. 1–4, *Mnium spinosum*: 1, leaves; 2, leaf margin (×100); 3, mid-leaf cells; 4, capsule. 5–7, *M. stellare*: 5, leaves; 6, mid-leaf cells; 7, capsule. Leaves ×15, cells ×415, capsules ×10.

and Ireland, absent from E. Anglia, occasional or frequent elsewhere. 93, H13. Europe, Faroes, Turkey, Caucasus, Himalayas, Siberia, Japan, N. America.

Likely only to be passed over for *M. hornum* but differing in its lighter, paler green coloration and unbordered lvs.

92. Cinclidium Sw., Schrad. J. f. Bot., 1803

Synoecious or dioecious. Sterile stems erect, sub-apical branching rare; stem epidermis unistratose, cells heavily thickened, micronemata and macronemata present; strong reddish coloration present. Lvs elliptical to orbicular or obovate, attenuate but not decurrent at base, margin entire, border 1–2-stratose; cells elongate-hexagonal, in radiating rows. Capsule pendulous with distinct neck; lid conical; teeth and cilia of inner peristome fused into conical structure with irregular openings. A mainly boreal genus of 4 species.

1. C. stygium Sw., Schrad. J. f. Bot., 1803

Synoecious. Plants 3–8 cm, stems reddish to reddish-black and covered with dense brown tomentum except in youngest parts. Lvs shrunken when dry, patent to spreading when moist, small below, larger, more crowded above, concave, elliptical to orbicular or obovate, apiculate, base slightly decurrent, margin with strong, reddish, unistratose border, entire; nerve percurrent, red to blackish; cells in radiating rows, rectangular-hexagonal, 15–35 μm wide in mid-lf. Seta long, red; capsule pendulous, ovoid; lid conical; spores 25–55 μm. Fr rare, summer. $n = 14$. Lax, densely tomentose, reddish-brown tufts or patches, in fens, basic springs and flushes. E. Anglia, Merioneth, Lancs and Yorks northwards, rare, very rare in Ireland. 23, H3. N. and C. Europe, Mongolia(?), Kamchatka(?), northern N. America, Greenland.

93. Rhizomnium (Broth.) Kop., Ann. Bot. Fenn., 1968

Plants with reddish coloration, stem epidermis unistratose, cells thick-walled, micronemata present or absent. Lvs ellipsoid, orbicular or obovate, base narrowly decurrent, apex rounded, sometimes emarginate and with an apiculus; margin entire, border 1 to several stratose; nerve ending well below apex to percurrent, without stereids in section; cells in rows radiating from nerve, elongate-hexagonal, decreasing in size from nerve to margin. Ten species distributed through Europe, N. Asia, N. America, Greenland.

1 Stems without micronemata (see fig. 333), border cells 3–4-stratose, 12–20 μm wide at middle of lf, prosenchymatous at apex **1. R. punctatum**
 Stems with micronemata (see fig. 333), border cells usually unistratose, 15–40 μm wide at middle of leaf, with flat end walls near apex 2
2 Dioecious, lvs often with small apiculus, capsule ovoid or ovate-ellipsoid **2. R. magnifolium** (p. 687)
 Synoecious, lvs never apiculate, capsule sub-globose **3. R. pseudopunctatum**

1. R. punctatum (Hedw.) Kop., Ann. Bot. Fenn., 1968
Mnium punctatum Hedw.

Dioecious. Plants 1–10 cm, stems red or brown, without micronemata, covered with reddish-brown or brown tomentum below. Lvs shrunken when dry, spreading when moist, distant below, larger, more crowded above, concave, elliptical to

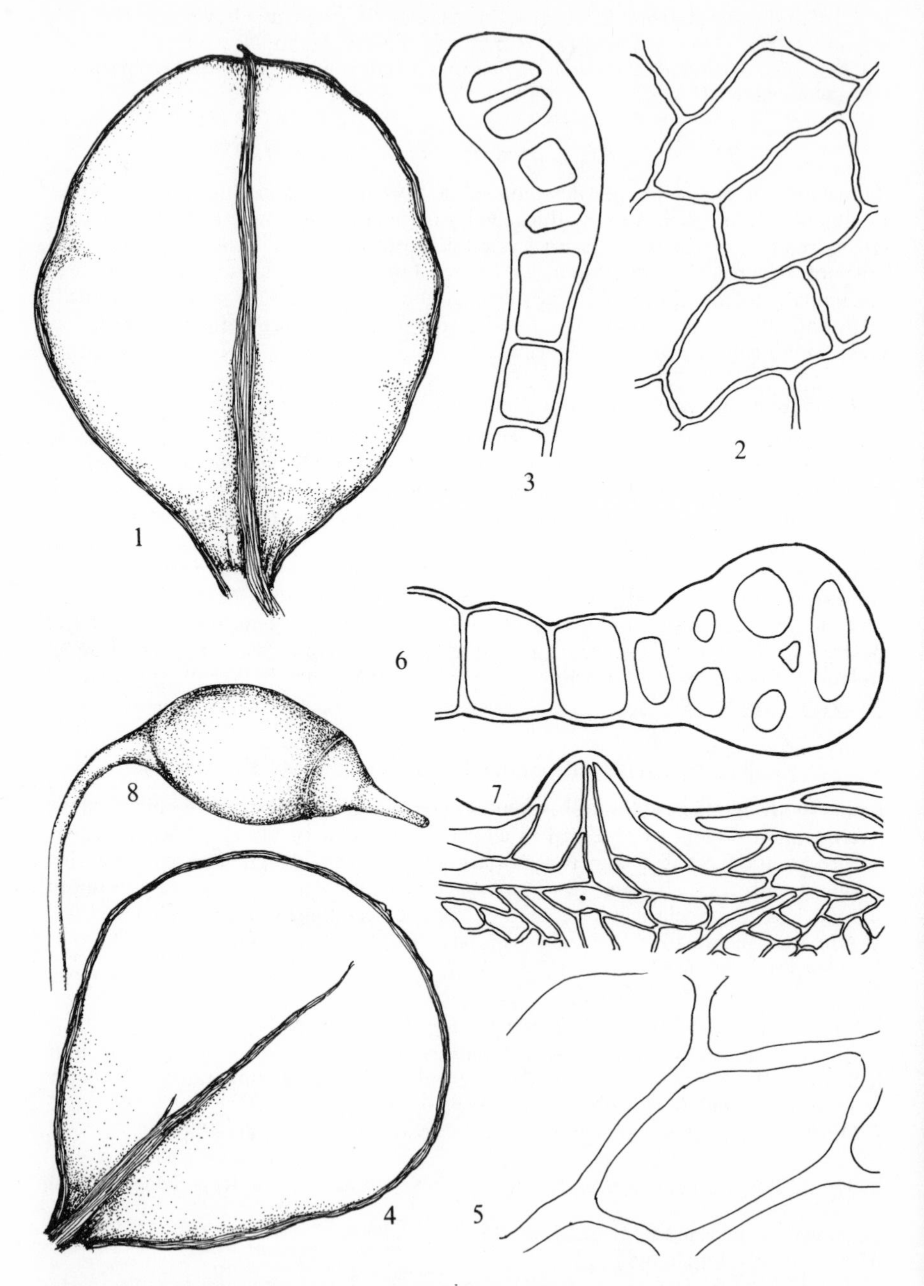

Fig. **211**. 1–3, *Cinclidium stygium*: 1, leaf (× 15); 2, mid-leaf cells; 3, section of leaf margin. 4–7, *Rhizomnium punctatum*: 4, leaf (× 10); 5 mid-leaf cells; 6, section of leaf margin; 7, leaf apex (× 165); 8, capsule (× 10). Cells and sections × 415.

orbicular or broadly ovate, base very narrowly decurrent, apex rounded or emarginate, usually apiculate, margin entire with strong reddish to brownish, 3–4-stratose border; nerve ending below apex to percurrent; cells in radiating rows, elongate-hexagonal, 35–50 μm × 70–120 μm in mid-lf, cells of border 3–4-stratose, 12–20 μm wide at middle of lf, at apex long and narrow. Seta orange-red; capsule cernuous, ovoid or ovate-ellipsoid; lid rostrate; peristome teeth pale yellow; spores 25–50 μm. Fr frequent, autumn, winter. $n=7$*. Scattered plants, tufts or patches, dark green above, blackish below, in fens and marshes, on soil and rocks by streams in woods and on mountains or as persistent protonema with depauperate shoots on damp rocks and wood in woods and by streams, common. 112, H39, C. Europe, Faroes, Asia, Madeira, N. America, Greenland.

2. R. magnifolium (Horik.) Kop.
See p. 687.

3. R. pseudopunctatum (Br. Eur.) Kop., Ann. Bot. Fenn., 1968
Mnium pseudopunctatum Br. Eur., *M. subglobosum* Br. Eur.

Synoecious. Plants 4–12 cm, stems black or brown, densely tomentose below, micronemata present. Lvs shrunken when dry, patent to spreading when moist, ± orbicular to broadly obovate, base very narrowly decurrent, apex rounded, sometimes emarginate, rarely with small blunt apiculus; margin entire, border pale or yellowish-brown, unistratose except at base; nerve ending well below apex; cells in indistinct rows, elongate-hexagonal, 35–50 μm × 60–140 μm in mid-lf, border cells unistratose at middle of lf 20–30 μm wide, at apex marginal cells shortly rectangular or trapezoid. Seta orange-red; capsule sub-pendulous to pendulous, sub-globose; lid with short thick beak; peristome brownish; spores 30–50 μm. Fr occasional, winter, spring. $n=13$, 14. Lax tufts or patches, dark green above, blackish below, often tinged with red, in fens and marshes, basic flushes and springs, generally distributed, occasional or rare. 71, H15. Europe, Faroes, Siberia, Korea, Formosa, northern N. America, Greenland.

94. Plagiomnium Kop., Ann. Bot. Fenn., 1968

Sterile shoots arcuate, fertile shoots erect; stem epidermis 2 to several stratose, cells not much thickened; micronemeta present, sometimes to stem apex; red coloration lacking. Lf margin with unistratose border, toothed; nerve not toothed at back above, with 1 stereid band in section. Lid usually conical. A cosmopolitan genus of 20 or so species.

1 Upper lvs narrowly lingulate, strongly undulate, lower lvs shorter and wider, cells 12–25 μm wide in mid-lf **6. P. undulatum**
Lvs broad, ± ovate, not or scarcely undulate, cells mostly 20–50 μm wide in mid-lf 2

2 Margin sharply toothed from about middle of lf, cells 20–24 μm wide in mid-lf, not arranged in divergent rows **1. P. cuspidatum**
Lvs bluntly or sharply toothed ± from near base, cells in ± divergent rows or if not then 25–50 μm wide in mid-lf 3

3 Lid rostrate, lf cells hardly porose, hardly enlarged towards nerve **7. P. rostratum**
Lid conical, lf cells distinctly porose, becoming larger towards nerve 4

4 Cells in mid-lf elongate-hexagonal, 1.5–3.0 times as long as wide, lf base decurrent* 5
Cells in mid-lf mostly ± isodiametric, lf base decurrent or not 6
5 Lf base narrowly decurrent, cells 1.5–2.0 times as long as wide in mid-lf, sterile shoots arcuate **2. P. affine**
Lf base broadly decurrent, cells 2–3 times as long as wide in mid-lf, sterile shoots erect **3. P. elatum**
6 Lf base not decurrent, cells in divergent rows, 22–43 μm wide in mid-lf **5. P. ellipticum**
Lf base decurrent, cells not in rows, 30–50 μm wide in mid-lf **4. P. medium**

1. P. cuspidatum (Hedw.) Kop., Ann. Bot. Fenn., 1968
Mnium cuspidatum Hedw.

Synoecious. Plants 1.5–4.0 cm, fertile stems erect, sterile stems arcuate or erect. Lower lvs distant, upper larger, more crowded, crisped when dry, patent to spreading when moist, obovate to broadly ovate, base decurrent, apex sharply pointed, margin with unistratose yellowish border, 2–4 cells wide, sharply toothed from about halfway; nerve ending below apex to percurrent; cells ± regularly hexagonal, collenchymatous, not in rows, 20–24(–28) μm wide in mid-lf. Seta reddish; capsule horizontal to sub-pendulous, ovoid; lid shortly conical; spores 18–36 μm. Fr frequent, spring, early summer. $n=6$, 12. Dark green tufts or patches with paler young growths, on damp soil, walls, rocks and tree boles in shaded habitats, generally distributed, frequent. 97, H23. Europe, Faroes, Caucasus, Iran, Himalayas, N., E. and C. Asia, N. America.

2. P. affine (Funck) Kop., Ann. Bot. Fenn., 1968
Mnium affine Funck

Dioecious. Fertile branches erect, 0.5–6.0 cm, angled in section, sterile stems arcuate, rooting at tips, often complanate, to 10 cm long. Lvs crisped when dry, spreading when moist, broadly ovate or obovate, base narrowly and longly decurrent, apex obtuse or sub-obtuse, apiculate, margin with unistratose border, 2–4 cells wide, toothed nearly to base, teeth usually sharp, 1–3(–4) cells long, at least some teeth ± perpendicular to margin at middle part of lf; nerve ending in or below apex; cells elongate-hexagonal, in rows, slightly collenchymatous, porose, mostly 1.5–2.0 times as long as wide and 30–45 μm wide in mid-lf (22–57 μm × 45–100 μm). Seta pale red, mostly 1–2(–3) per perichaetium; capsule sub-pendulous, ellipsoid, neck not distinct; lid shortly conical; spores 18–24 μm. Fr very rare, spring. $n=6$. Loose, spreading, green patches or scattered shoots on damp soil in woods and in turf in basic habitats. Generally distributed and probably frequent in England, rare elsewhere. 49, H6. Europe, Faroes, western Asiatic Russia, Turkey, Iran, Madeira.

This and the next 5 species are likely to be confused. *P. medium* is distinct in the lf cells not arranged in rows and the usually present synoecious inflorescence. In *P. ellipticum* the sterile stems are usually erect and the lf bases not or scarcely decurrent, the teeth blunt, small or even lacking and the cells ± isodiametric. *P. elatum* usually has erect sterile shoots, the lvs with widely decurrent bases and the cells smaller and relatively longer than in typical *P. affine*. Poorly developed forms of the latter may, however, have smaller cells and it is uncertain if such plants are forms of *P. elatum* growing in unfavourable habitats. *P. rostratum* differs from *P. affine* and related species in the smaller, hardly porose lf cells, the rostrate beak of the capsule and the stomata scattered over the capsule instead of concentrated near the base. For a monograph of these species see Koponen, *Ann. Bot. Fenn.* **5**, 213–24, 1968.

* Decurrent lf bases are usually lost from detached lvs and are best viewed *in situ* on the stem.

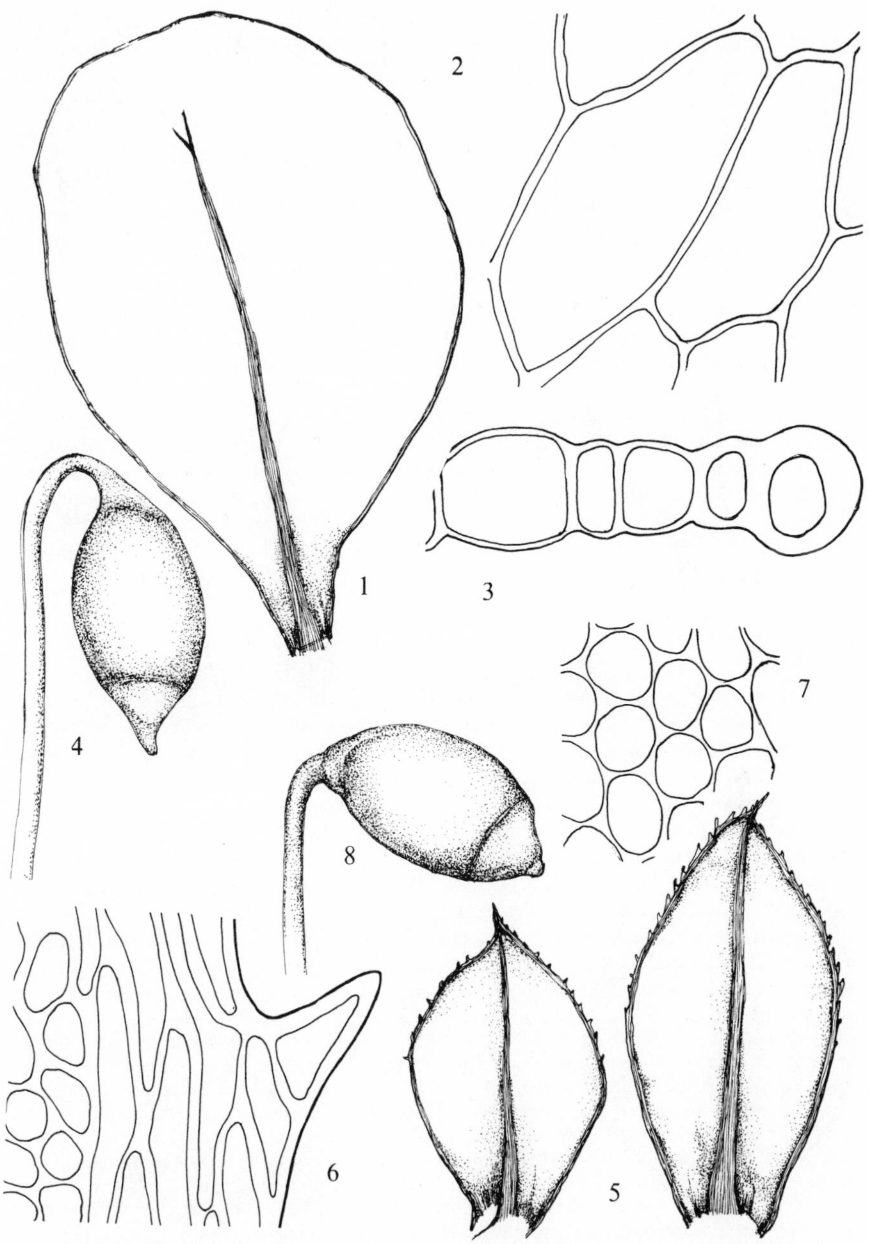

Fig. **212**. 1–4, *Rhizomnium pseudopunctatum*: 1, leaf (×10); 2, mid-leaf cells; 3, section of leaf margin; 4, capsule. 5–8, *Plagiomnium cuspidatum*: 5, leaves (×15); 6, marginal cells $\frac{3}{4}$ way up leaf; 7, mid-leaf cells; 8, capsule. Capsules ×10, cells ×415.

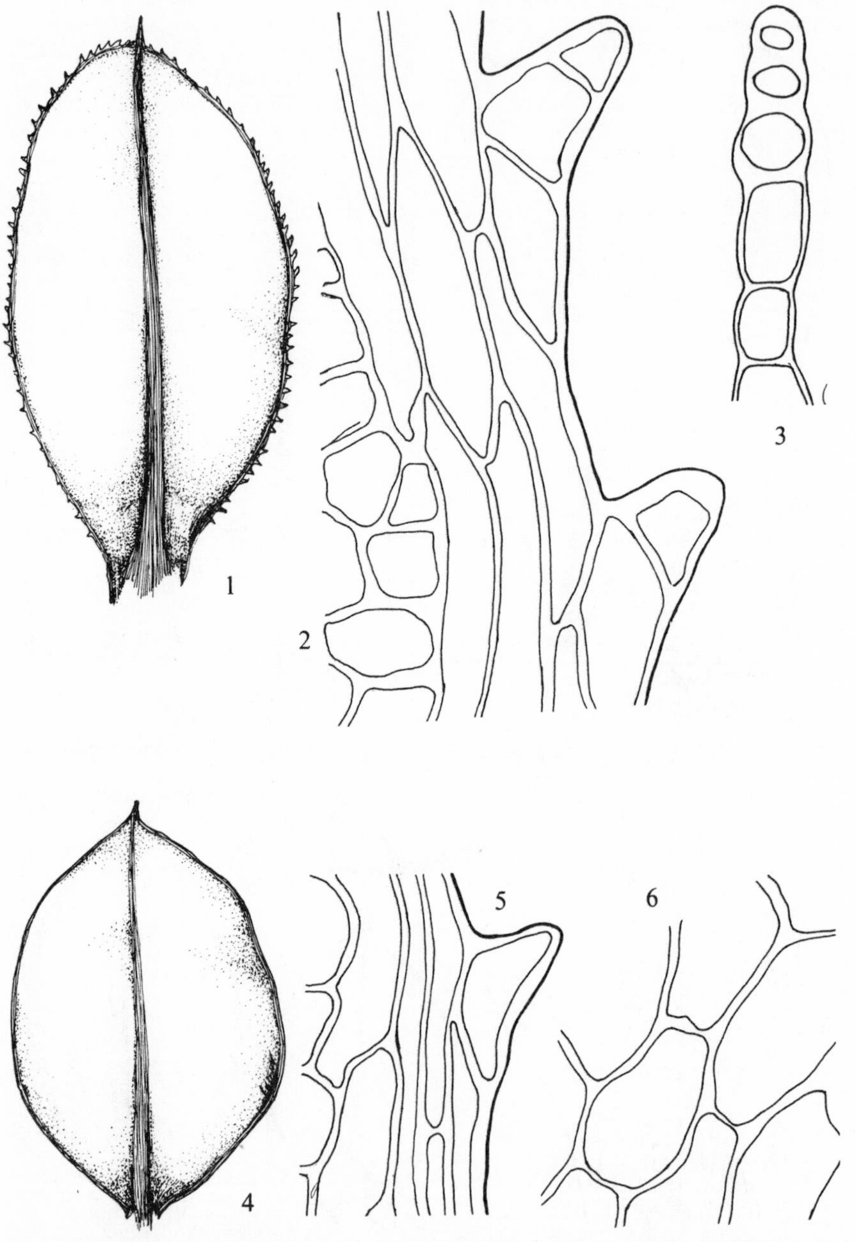

Fig. 213. *Plagiomnium affine*: 1 and 2, leaf and marginal cells at middle of leaf from erect stem; 3, section of leaf margin; 4 and 5, leaf and marginal cells at middle of leaf from prostrate stem; 6, mid-leaf cells. Leaves × 10, cells × 415.

3. P. elatum (Br. Eur.) Kop., Ann. Bot. Fenn., 1968
Mnium seligeri Jur. ex Lindb., *M. affine* var. *elatum* Br. Eur.

Dioecious. Fertile branches erect, 4–10 cm, sterile stems erect or sometimes arcuate and rooting at tips, to 10 cm, angular in section, old stems matted with rhizoids. Lvs crisped when dry, spreading when moist, broadly ovate to obovate or elliptical, apex obtuse to sub-obtuse, apiculate, base broadly decurrent in erect stems, margin with unistratose border, 2–4 cells wide, toothed ± from base, teeth of 1–3 cells, at middle of lf at an angle to margin; nerve percurrent or excurrent; cells in

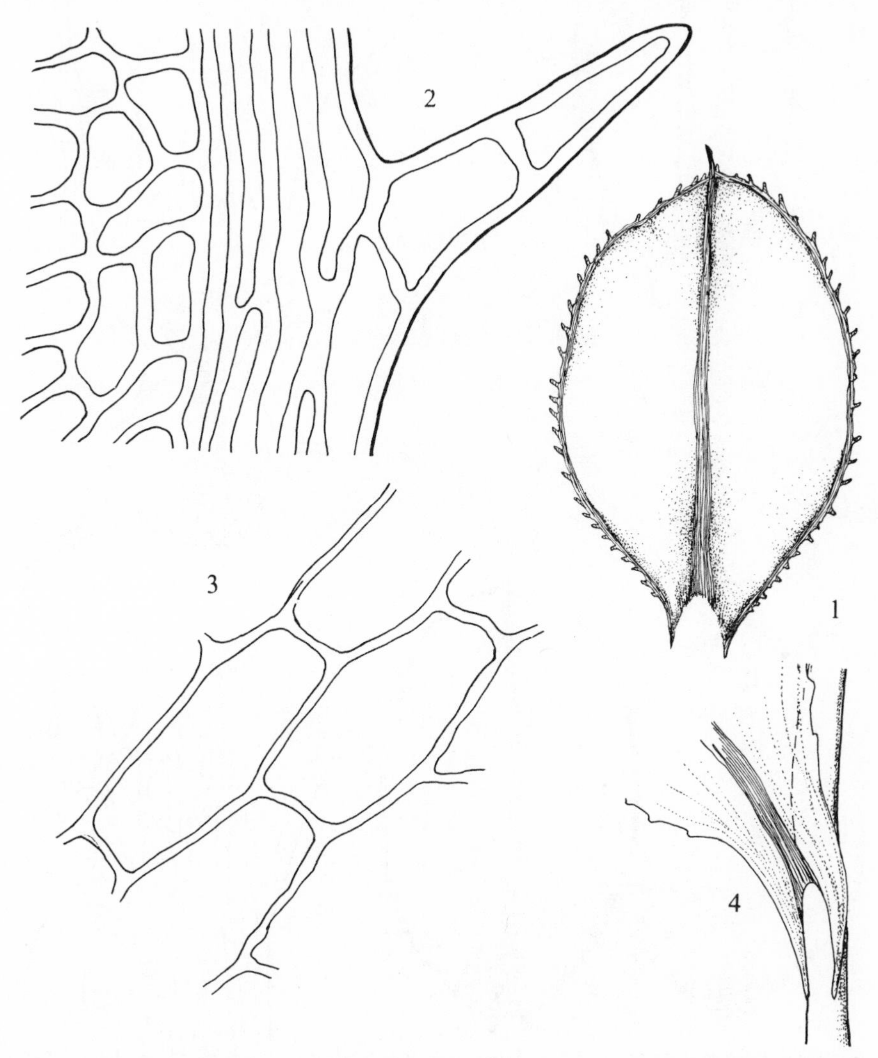

Fig. **214**. *Plagiomnium elatum*: 1, leaf (×10); 2, marginal cells at middle of leaf; 3, mid-leaf cells (×415); 4, leaf base (×15).

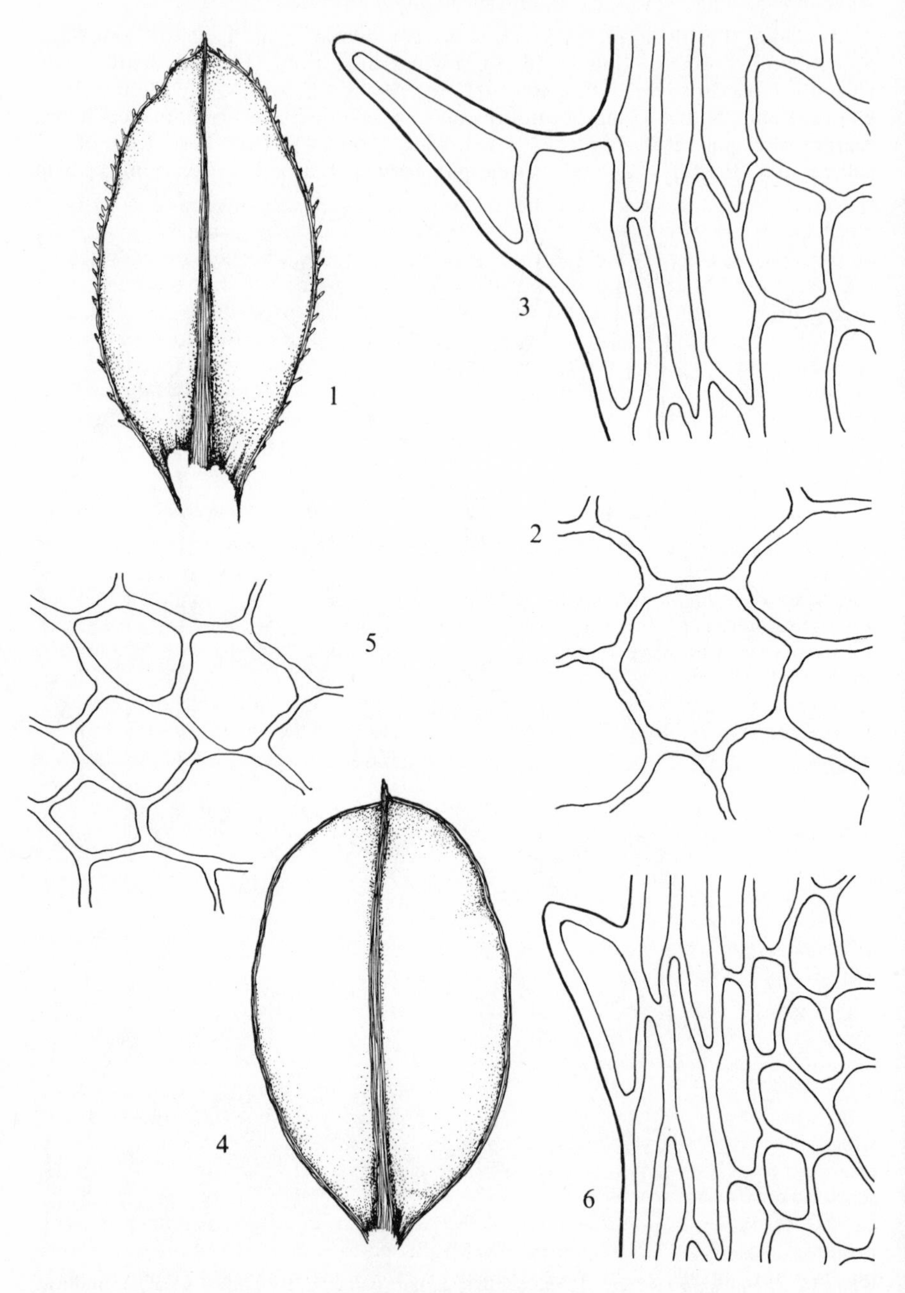

Fig. **215**. 1–3, *Plagiomnium medium*: 1, leaf; 2, mid-leaf cells; 3, margin at middle of leaf. 4–6, *P. ellipticum*: 4, leaf; 5, mid-leaf cells; 6, margin at middle of leaf. Leaves ×10, cells ×415.

rows, elongate-hexagonal or ± rectangular, porose, not collenchymatous, mostly 2–3 times as long as wide, 15–35 μm×40–70 μm in mid-lf. Sporophyte as in *P. affine*. Fr spring but probably unknown in Britain. Yellowish-green tufts or patches in fens, fen carr, by streams, springs and in seepage areas in basic habitats. Generally distributed and probably frequent. 91, H22. Europe, Iceland.

4. **P. medium** (Br. Eur.) Kop. Ann. Bot. Fenn., 1968
Mnium medium Br. Eur.

Synoecious. Fertile branches erect, 3–9 cm, sterile stems few, arcuate, not rooting at apex complanate, stems angled in section. Lvs crisped when dry, spreading when moist, ovate to ovate-lanceolate, base broadly decurrent, apex acute, sub-obtuse or obtuse apiculate, margin with unistratose border, 3–5 cells wide, dentate, teeth sharp, 1–2 cells long, set at an angle to margin at middle part of lf; nerve excurrent or ending below apex; cells hexagonal, not in rows, collenchymatous, porose, 1.0–1.5(–2.0) times as long as wide, mostly 30–50 μm wide in mid-lf (25–70 μm×45–102 μm). Setae 1.5(–8) per perichaetium; capsule sub-pendulous, ellipsoid; spores 24–34 μm. Fr unknown in Britain. Loose, green patches or tufts in basic, montane flushes, very rare. Mid Perth, Argyll. 2. Europe, Faroes, Iceland, Caucasus, Siberia, Chitral, Punjab, Kashmir, Japan, Korea, Morocco, N. America, Greenland.

5. **P. ellipticum** (Brid.) Kop., Ann. Bot. Fenn., 1968
P. rugicum (Laur.) Kop., *Mnium rugicum* Laur., *M. affine* var. *rugicum* (Laur.) Br. Eur.

Dioecious. Fertile stems erect, 1–10 cm, sterile stems usually erect, sometimes arcuate and complanate, to 9 cm long, stems rounded or obscurely angled in section. Lvs crisped when dry, spreading when moist, orbicular to ovate, base not or scarcely decurrent, apex acute, sub-obtuse or obtuse, apiculate, margin with unistratose border 3–4 cells wide, toothed ± from base, teeth blunt, of 1(–2) cells, sometimes lacking; nerve percurrent to excurrent; cells ± hexagonal, in rows, not collenchymatous, porose, ± 1.0–1.5 times as long as wide and 20–35 μm wide in mid-lf. Setae 1–3 per perichaetium; capsule pendulous, ellipsoid; spores 24–40 μm. Fr very rare, summer. $n=6$. Light green tufts or patches in fens, marshes, flushes, wet places in fields, dune slacks, by streams, rivers and lakes. Generally distributed and probably frequent. 66, H19. N., W. and C. Europe, Faroes, Iceland, N. Asia, Caucasus, China, Japan, N. America, Greenland, Chile, Argentina, Australia.

Differs from the preceding 3 species in the not or scarcely decurrent lf base and the small, blunt teeth.

6. **P. undulatum** (Hedw.) Kop., Ann. Bot. Fenn., 1968
Mnium undulatum Hedw.

Dioecious. Fertile stems erect, 2–11 cm, sterile stems erect or arcuate and frequently complanate, to 15 cm, branching frequently sub-apical. Lvs crisped when dry, spreading when moist, strongly transversely undulate, lower distant, ovate to ovate-oblong, upper more crowded, much larger, narrowly lingulate, base decurrent, apex acute to obtuse or rounded and apiculate, margin with unistratose border 3–5 cells wide, toothed ± from base, teeth spinose, 1(–2) cells long; nerve percurrent or excurrent; cells hexagonal to elongate-hexagonal, not arranged in rows, ± incrassate, scarcely porose, mostly 10–16 μm wide in mid-lf. Setae 1–10 per perichaetium; capsule cernuous, ellipsoid; spores 24–32 μm. Fr rare but often abundant when produced, spring. $n=6$*. Pale green, loose tufts, patches or scattered plants in damp places in turf, on banks, in woods, boggy ground, dune slacks, etc., common. 112, H40, C. Europe, Faroes, Asia, N.W. Africa, Macaronesia.

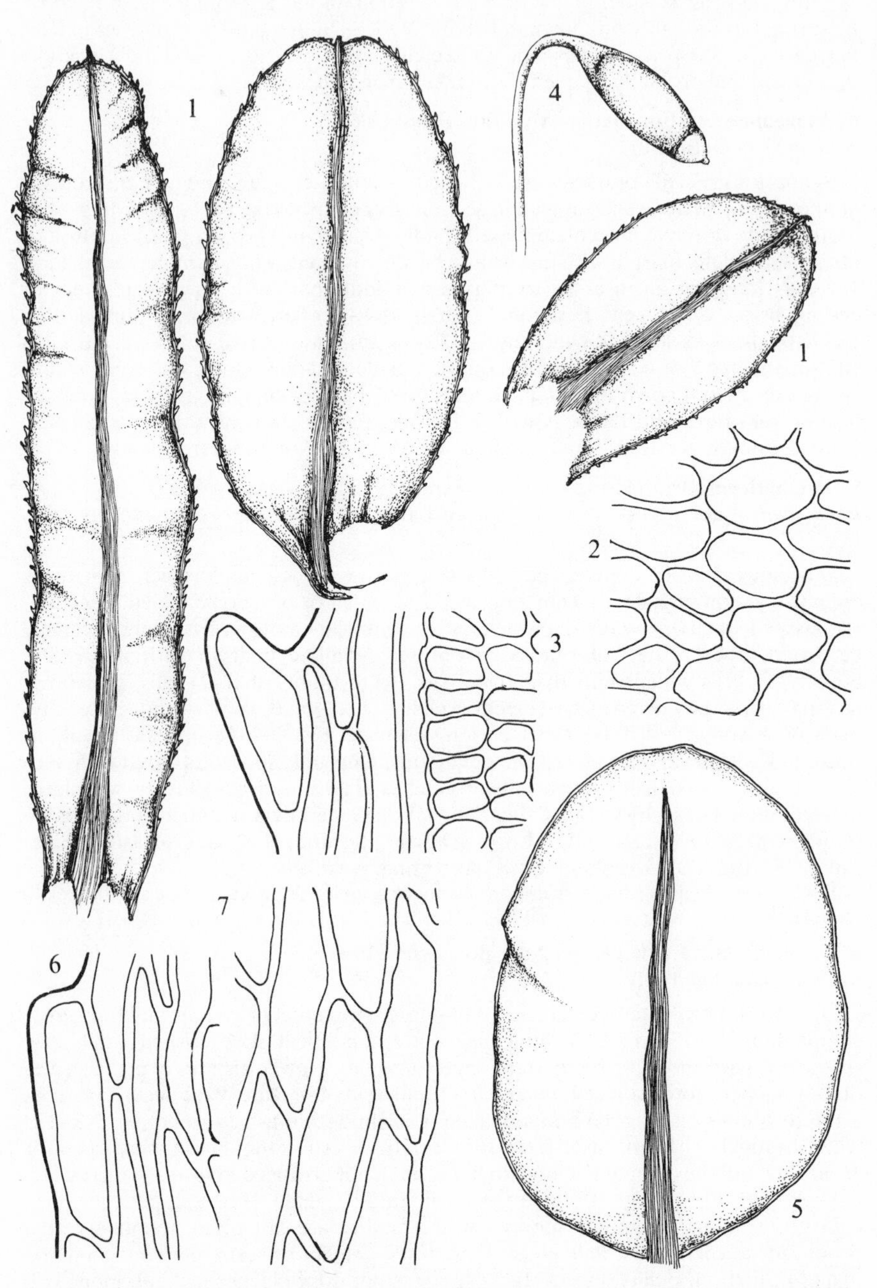

Fig. **216**. 1–4, *Plagiomnium undulatum*: 1, leaves (× 20); 2, mid-leaf cells; 3, margin at middle of leaf; 4, capsule (× 5). 5–7, *Pseudobryum cinclidioides*: 5, leaf (× 15); 6, marginal cells; 7, mid-leaf cells. Cells × 415.

Stunted plants sometimes found in turf with shorter, hardly undulate lvs may be mistaken for other species of *Plagiomnium*. *P. affine* and its allies differ in their larger lf cells. *P. rostratum* has larger, more collenchymatous cells and smaller, blunter teeth. *Atrichum undulatum* has acute lvs with double teeth and lamellae on the nerve.

7. P. rostratum (Schrad.) Kop., Ann. Bot. Fenn., 1968
Mnium rostratum Schrad., *M. longirostrum* Brid.

Synoecious. Fertile stems ± erect, 2–5 cm, sterile stems arcuate, often complanate. Lvs crisped when dry, spreading when moist, broadly ovate to ovate or ovate-oblong, not or hardly decurrent at base, apex obtuse or sub-obtuse, apiculate, margin with yellowish, unistratose border 2–5 cells wide, toothed from below middle, teeth small, blunt, usually unicellular, sometimes obsolete; nerve ending in or below apex; cells irregularly hexagonal, not markedly larger near nerve, collenchymatous, sometimes strongly so, not or hardly porose, 20–35(–40) μm wide in mid-lf. Setae 1–5 per perichaetium; capsule cernuous, ellipsoid, stomata scattered over surface of capsule; lid rostrate; spores 18–30 μm. Fr common, spring. $n=7, 12$. Lax, green or dark green patches on soil, rocks and wood in woods, on banks, by streams and in marshes and fens. Generally distributed, occasional to frequent. 110, H25, C. Cosmopolitan.

95. Pseudobryum (Kindb.) Kop., Ann. Bot. Fenn., 1968

Dioecious. All stems erect. Lvs orbicular to ovate-oblong, unbordered; nerve without stereids; cells elongate-hexagonal or rhomboidal. Lid conical. Chromosomes longer and thinner than in preceding genera of Mniaceae. Two species only, the second, *P. speciosum* (Mitt.) Kop., occurring in C. Asia.

1. P. cinclidioides (Hüb.) Kop., Ann. Bot. Fenn., 1968
Mnium cinclidioides Hüb.

Stems erect, 3–10(–15) cm, with brownish tomentum below. Lvs shrunken when dry, spreading, often slightly undulate when moist, ovate, obovate-oblong or orbicular, apex rounded, emarginate or apiculate, base not decurrent, margin unbordered, entire or obscurely denticulate above; nerve stout, green to brown, ending well below apex; cells in rows, elongate-hexagonal, not collenchymatous, porose, 3–4 times as long as wide, 20–30 μm × 70–125 μm in mid-lf, 2–3 marginal rows longer, narrower but not forming border. Seta orange-red; capsule pendulous, ovoid; lid conical; spores 24–36 μm. Fr very rare, early summer. $n=6$. Light, bright green patches or deep tufts, brownish or blackish below, on lake-side waterlogged ground, peaty marshes and carr, rare. N. Wales, N. England, scattered localities in Scotland. 27. Europe, Siberia, Mongolia, Kamchatka, northern N. America, Greenland.

24. AULACOMNIACEAE

Lvs ovate to lanceolate, margin unbordered, usually denticulate; nerve ending below apex; extreme basal cells enlarged or elongated, elsewhere uniformly rounded-hexagonal, incrassate, papillose. Seta long; capsule erect to horizontal, ± ellipsoid, striate, sulcate when old; peristome double, inner ciliate; calyptra cucullate.

96. Aulacomnium Schwaegr., Suppl., III., 1827

Dioecious (in British species) or autoecious. Tufted plants. Lvs ovate to lanceolate, cells usually with single central conical papilla on both surfaces. Stems sometimes terminating in gemmae-bearing pseudopodia. Capsule straight or curved, striate. Europe, Asia, Africa, America, Australasia. Eight species.

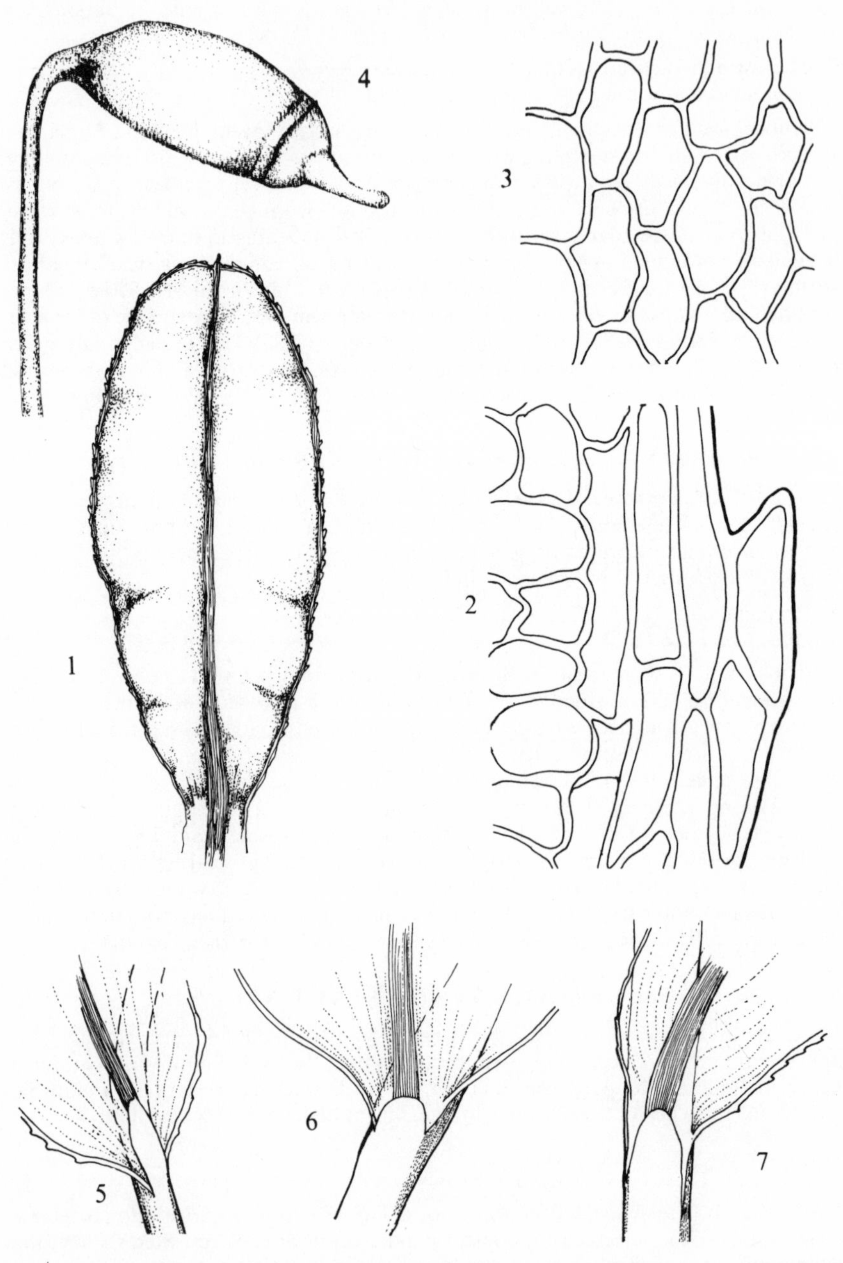

Fig. **217**. 1–4, *Plagiomnium rostratum*: 1, leaf (×10); 2 and 3, marginal and mid-leaf cells (×415); 4, capsule (×10). 5–7, leaf bases: 5, *P. affine*; 6, *P. ellipticum*; 7, *P. medium* (all ×**15**).

1 Lvs imbricate when dry, entire, cells with star-shaped lumens, not or scarcely papillose **2. A. turgidum**
Lvs crisped or twisted when dry, entire or toothed, cells with rounded lumens, strongly papillose 2
2 Plants to 2.5 cm, stems not densely tomentose, basal cells narrow, rectangular **3. A. androgynum**
Plants to 10 cm or more, stems matted with tomentum, basal cells inflated **1. A. palustre**

1. A. palustre (Hedw.) Schwaegr., Suppl., III., 1827

Plants to 10 cm or more, stems matted with dense brown tomentum. Lvs erecto-patent when moist, narrowly lanceolate to lingulate, apex acuminate to rounded, margin recurved on one or both sides below, entire to denticulate above; nerve ending below apex, shining white at back when dry; extreme basal cells enlarged, 2–3-stratose, cells elsewhere unistratose, irregularly rounded to quadrate, incrassate, with rounded lumens and conical papillae, 8–14 μm wide in mid-lf. Stems occasionally producing terminal pseudopodia with apical, ensiform, flattened gemmae to 400 μm long. Fr occasional, summer.

Var. **palustre**

Lvs variously crisped or twisted when dry, narrowly lanceolate, tapering to acuminate to obtuse apex, margin entire to dentate above. $n=10$*, 12*. Yellowish-green or green, matted tufts or patches in bogs, acidic flushes, wet places on heaths, moors, etc. Common and sometimes locally abundant in suitable habitats. 111, H40, C. Europe, Faroes, Asia, N. and C. Africa, N. America, Patagonia, Australia, Tasmania, New Zealand.

Var. **imbricatum** Br. Eur., 1841

Lvs ± straight and imbricate when dry, often lingulate with obtuse or rounded apices, entire. Dense tufts on wet ground, very rare. W. Suffolk, I. of Wight, Hereford, Caerns, N. Lincs, Mid W. Yorks, Durham, Ayr. 7. Europe, Siberia, N. America.

A very variable species some small forms of which approach *A. androgynum* in size. The var. *imbricatum* is probably no more than a drier ground form of the type. The densely tomentose stems and cells with conical papillae and rounded lumens will distinguish imbricate-lvd forms from *A. turgidum*. *A. palustre* occasionally produces pseudopodia but may be distinguished from *A. androgynum* by the basal cells enlarged and 2–3-stratose.

2. A. turgidum (Wahlenb.) Schwaegr., Suppl., III, 1827

Stems turgid when moist, radiculose, to 10 cm. Lvs imbricate, concave, ovate or obovate to lanceolate, apex obtuse to rounded, flat to cucullate, margin recurved to above middle, entire; nerve ending below apex; basal cells swollen, 2–3-stratose, elsewhere unistratose, rounded-hexagonal with stellate lumens, not or scarcely papillose, 10–16 μm wide in mid-lf. Capsule ± curved. Fr unknown in Britain. $n=12$. Yellowish to green tufts on montane heaths and grassland at altitudes of 750–1050 m. Occasional on the higher Scottish mountains from Perth north to Sutherland with an old record from Yorks. 9. Northern and montane parts of Europe, Faroes, Asia and N. America, C. Africa.

3. A. androgynum (Hedw.) Schwaegr., Suppl., III, 1827

Stems tomentose below, slender, to 2.5(–3.5) cm. Lvs appressed, crisped when dry, patent when moist, ovate-lanceolate to narrowly lanceolate, acuminate, margin recurved or not below, irregularly denticulate above; nerve ending below apex; cells

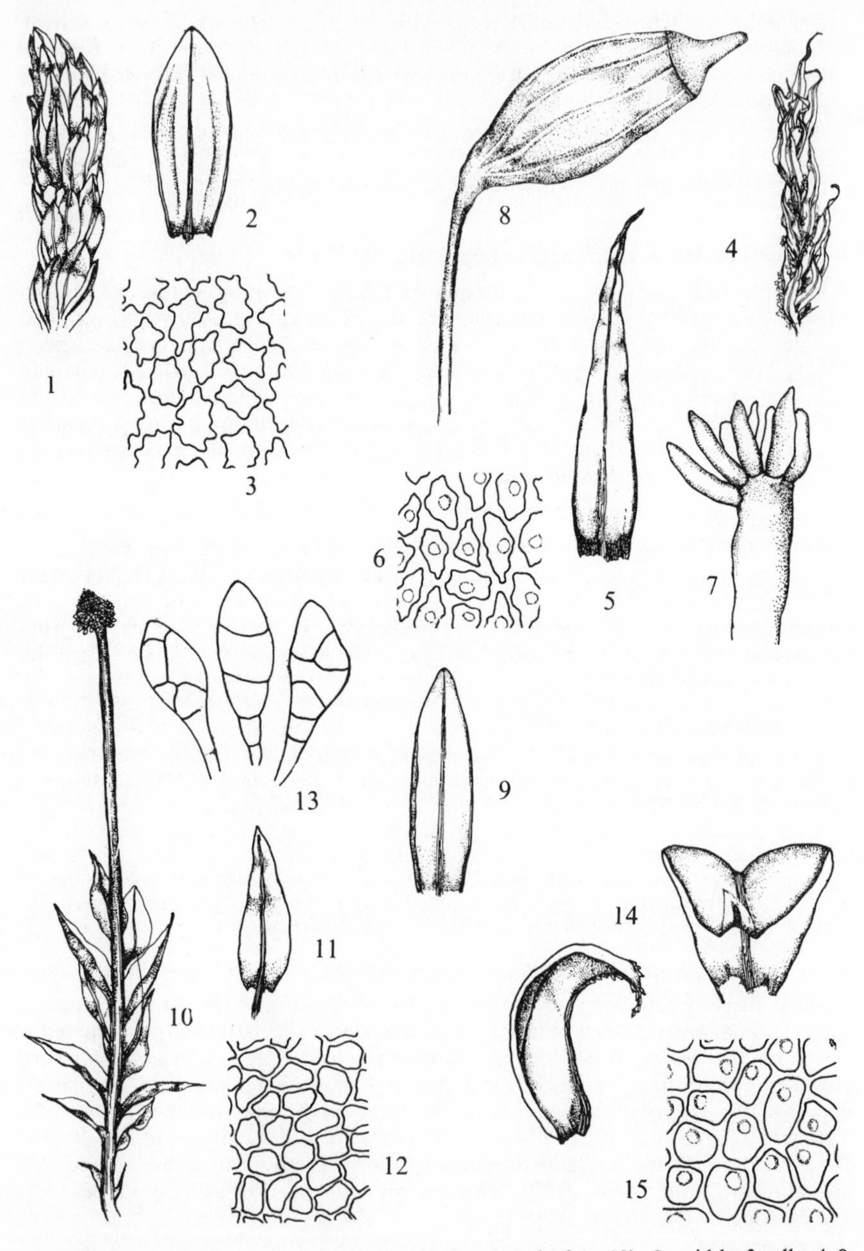

Fig. **218**. 1–3, *Aulacomnium turgidum*: 1, dry shoot; 2, leaf (×10); 3, mid-leaf cells. 4–8, *A. palustre* var. *palustre*: 4, dry shoot; 5, leaf (×10); 6, mid-leaf cells; 7, gemmae-bearing pseudopodium (×40); 8, capsule (×10). 9, *A. palustre* var. *imbricatum*: leaf (×10). 10–13, *A. androgynum*: 10, shoot with pseudopodium (×15); 11, leaf (×20); 12, mid-leaf cells; 13, gemmae (×250). 14–15, *Paludella squarrosa*: 14, leaves (×15); 15, mid-leaf cells. Cells ×415.

incrassate, unistratose throughout, rectangular at base, ± uniformly rounded-hexagonal with conical papillae elsewhere, 10–12 μm wide in mid-lf. Stems frequently terminating in a pseudopodium bearing club-shaped head of stalked spherical to ovoid gemmae to 55 μm long. Capsule ± erect, becoming inclined to horizontal with age, narrowly ellipsoid. Fr very rare, spring, summer. $n=11$–12, 12. Pale to dark green tufts or patches on damp rocks, peat, decaying wood, humus, in knot holes, crevices in bark, etc. Frequent to common in drier parts of Britain, rare in the west and north. 90, H8. Europe, Japan, Canaries, N. America, Patagonia.

25. MEESIACEAE

Large tufted plants. Lvs erecto-patent or squarrose; nerve stout or thin, ending in or below apex; cells rectangular or hexagonal, smooth or mamillose. Seta long, thin; capsule asymmetrical, ± pyriform or clavate, curved, smooth, with distinct neck; lid conical; peristome double, outer of 16 short (long in *Paludella*), obtuse, articulate teeth, inner of 16 longer processes, cilia short to absent.

The genera show wide divergence in gametophyte characters but close similarity in the sporophyte.

97. Paludella Brid., Sp. Musc., 1817

A monotypic genus with the characters of the species.

1. P. squarrosa (Hedw.) Brid., Sp. Musc., 1817

Dioecious. Plants to 15 cm, stems tomentose below. Lvs in 5 ranks, strongly squarrose-recurved, base decurrent, ovate, acute, margin narrowly recurved below, serrate above; nerve ending below apex; basal cells narrowly rectangular, becoming ± rounded-hexagonal above, coarsely mamillose, *ca* 10 μm wide in mid-lf. Seta to 10 cm; capsule ellipsoid, curved; inner and outer peristome teeth about same length. Fr unknown in Britain. Dense, matted, light green tufts in fens, recorded from Cheshire, S.E. and N.E. Yorks, but now extinct in Britain. N. and C. Europe, N. Asia, N. America, Greenland.

98. Meesia Hedw., Sp. Musc., 1801

Lvs with thick nerve; cells small, narrowly rectangular to shortly rectangular, smooth or slightly mamillose. Seta long, capsule asymmetrical, narrowly pyriform, curved; outer peristome teeth shorter than inner, cilia rudimentary. Europe, N. and S. Asia, C. Africa, America, Australia, New Zealand. Twelve species.

Lvs erecto-patent, obtuse, nerve ½ or more width of lf at base	**1. M. uliginosa**
Lvs squarrose, acute, nerve to ¼ width of lf at base	**2. M. triquetra**

1. M. uliginosa Hedw., Sp. Musc., 1801 CHECK
M. trichoides Spruce

Autoecious or synoecious. Stems radiculose below, to 5 cm. Lvs appressed, straight when dry, erecto-patent when most, ligulate to linear-lanceolate, apex obtuse or rounded, margin recurved, entire; nerve very stout, occupying ½–¾ lf base, ending below apex; cells narrowly rectangular below, rectangular above. Seta thin, variable in length, 1–4 cm; capsule asymmetrical, narrowly pyriform, curved; spores 40–50 μm. Fr frequent, late spring, summer. $n=13$, 14, 14+1*. Dark green tufts in flushes and rock crevices in montane habitats or as scattered plants or patches in dune slacks, calcicole, rare or occasional. Anglesey, Lancs and Yorks north to Caithness.

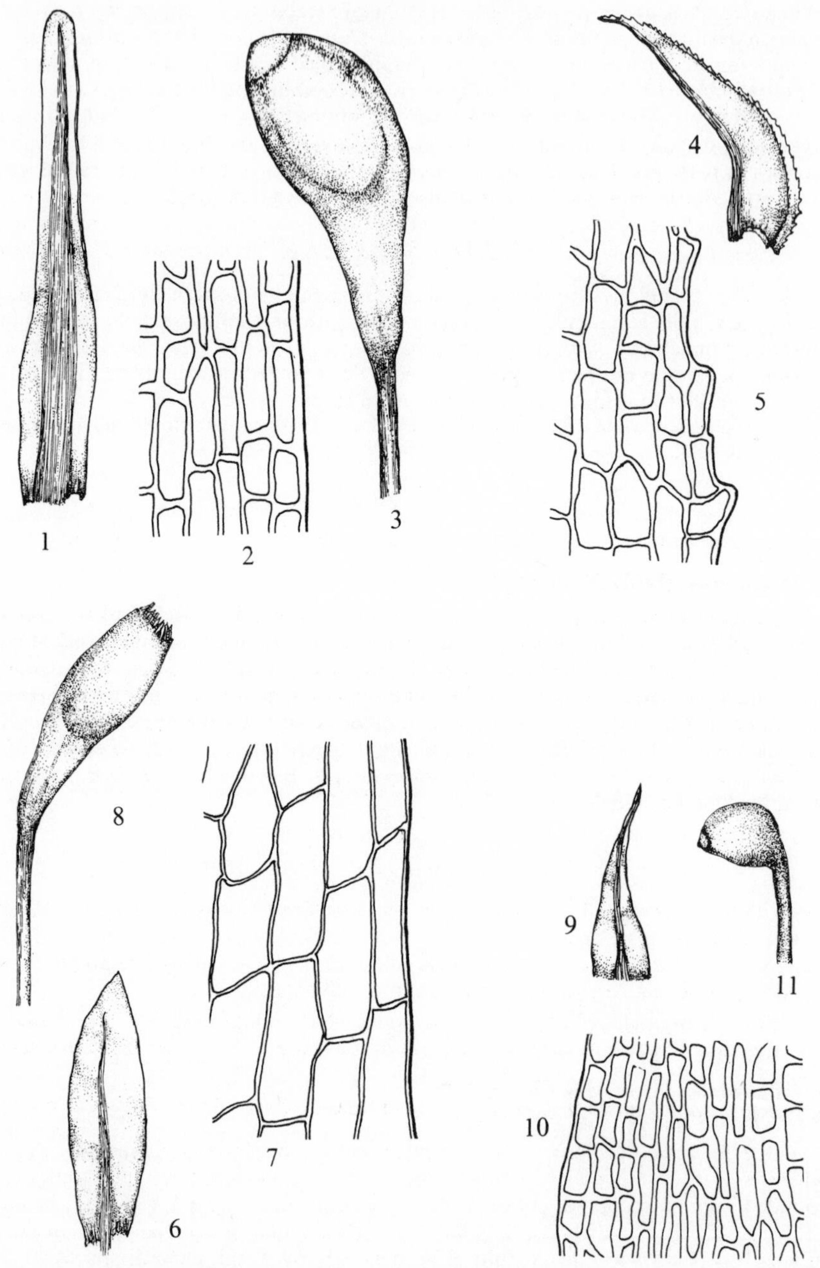

Fig. **219.** 1–3, *Meesia uliginosa*: 1, leaf; 2, mid-leaf cells; 3, capsule (×15). 4–5, *M. triquetra*: 4, leaf; 5, mid-leaf cells. 6–8, *Amblyodon dealbatus*: 6, leaf; 7, mid-leaf cells; 8, capsule (×7.5). 9–11, *Catoscopium nigritum*: 9, leaf; 10, mid-leaf cells; 11, capsule (×15). Leaves ×15, cells ×415.

29. Europe, Iceland, Caucasus, N. and C. Asia, Himalayas, China, N. America, Greenland.

2. M. triquetra (Hook. & Tayl.) Ångstr., Nov. Act. Soc. Ups., 1844
M. trifaria Crum, Steere & Anderson, *M. tristicha* Br. Eur.

Dioecious. Plants 3–5 cm. Lvs trifarious, squarrose, from erect sheathing basal part narrowly triangular, acute, margin plane, serrate almost from base; nerve strong, to $\frac{1}{4}$ width of lf at base, ending below apex; cells pellucid, in basal part rectangular, above smaller, ± shortly rectangular, smooth. Capsule similar to that of *M. uliginosa* but larger. Fr unknown in Ireland. $n=10$. Large, lax, bright green tufts in basic flushes, very rare. W. Mayo. H1. Europe, N. Asia, N. America, Greenland.

Most likely to be confused with *Dicranella palustris* or *Dichodontium pellucidum*. It differs from the former in the serrate lf margin, from the latter in the smooth cells, and from both in the trifarious lvs. For an account of the occurrence of the species in Ireland see Warburg, *Trans. Br. bryol. Soc.* **3**, 378–81, 1958.

99. AMBLYODON BR. EUR., 1841

A monotypic genus with the gametophytic characters of the Funariaceae and the sporophytic characters of *Meesia*. On cytological grounds this genus clearly belongs near *Meesia*.

1. A. dealbatus (Hedw.) Br. Eur., 1841

Autoecious or polyoecious. Plants to 2 cm. Lvs appressed, shrunken when dry, erecto-patent when moist, oblong-lanceolate, acute, margin plane, entire or denticulate above; nerve stout, ending below apex; cells thin-walled, lower rectangular, above narrowly hexagonal, 12–30 μm wide in mid-lf. Seta thin, 1.0–4.5 cm long; capsule narrowly pyriform, curved, asymmetrical, smooth; spores *ca* 40 μm. Fr frequent, summer. $n=14$*, 18, 20*. Tufts in montane flushes and rock crevices or in dune slacks, usually in basic habitats. Occasional from Derby and Yorks northwards, very rare elsewhere, N. Devon, Lancs, Wales. 44, H16. Europe, Caucasus, Iran, C. Asia, N. America, Greenland.

26. CATOSCOPIACEAE

A monotypic family of uncertain affinity.

100. CATOSCOPIUM BRID., BR. UNIV., 1826

1. C. nigritum (Hedw.) Brid., Br. Univ., 1826

Dioecious. Plants to 4(–7) cm, slender. Lvs ± erect when dry, erecto-patent, lanceolate, gradually tapering to acuminate apex, margin recurved below, entire; nerve stout, ending in apex; cells incrassate, smooth, quadrate-rectangular, narrower near nerve and at base, 8–10 μm wide in mid-lf. Seta very slender, rigid, to 1 cm; capsule minute, hardly 1 mm long, globose, cernuous, smooth, glossy, black; lid conical; outer peristome teeth poorly developed, inner rudimentary or lacking; spores 40–45 μm. Fr frequent, autumn. $n=13$*, 14. Dense, bright to dark green tufts, blackish below, in montane flushes or in dune slacks, calcicole. Rare in sand-dunes from Anglesey northwards and northern coast of Ireland, rare inland from Yorks and Durham northwards. 19, H4. Europe, Caucasus, N. and C. Asia, N. America, Greenland.

27. BARTRAMIACEAE

Plants usually tufted, stems tomentose. Lvs ovate-lanceolate to linear-lanceolate, toothed, nerve ending below apex to excurrent; cells ± hexagonal, mamillose. Seta long, usually straight; capsule erect or inclined, globose, symmetrical or not with mouth often oblique, usually longitudinally furrowed when dry; peristome usually double, teeth 16, processes divided, cilia rudimentary or absent; calyptra minute, cucullate.

101. Plagiopus Brid., Br. Univ., 1826

Gametophyte similar to that of *Bartramia* but lf cells very finely papillose-striate and not mamillose. Seta long, flexuose; capsule ± globose, striate when moist, furrowed when dry; peristome double. Only 2 species, the second of which occurs in New Zealand.

1. P. oederi (Brid.) Limpr., Laubm., 1893
Bartramia oederi Brid.

Synoecious. Plants to 6 cm, stems tomentose below. Lvs flexuose to crisped when dry, erecto-patent when moist, narrowly lanceolate, acuminate, base not sheathing; margin recurved below, serrate above; nerve ending in apex or excurrent; basal cells linear, at basal angles and short distance up margin quadrate-rectangular, above ± rectangular, not mamillose but faintly papillose, 8–10 μm wide in mid-lf. Seta 5–15 mm, straight, capsule ± globose, asymmetrical with oblique mouth, smooth, sulcate and curved when dry; spores 18–25 μm. Fr frequent, spring, summer. $n = 7$*. Dense, green tufts in crevices and on ledges of shaded, basic, usually montane rocks, occasional. W. Cornwall, Glos, Glamorgan and Derby north to Inverness and Skye, S. Kerry, Leitrim, Antrim. 41, H3. Europe, Caucasus, N. and C. Asia, N. India, Nepal, China, Japan, Java, N. America, Greenland, Hawaii.

102. Bartramia Hedw., Sp. Musc., 1801

Autoecious or synoecious. Plants often robust, stems tomentose below. Lvs erect to spreading, long and narrow, sometimes with sheathing base, margin usually toothed, teeth sometimes double; cells quadrate to narrowly rectangular, mamillose. Capsule ± globose, striate; lid short; peristome usually double. A world-wide genus of *ca* 175 species.

1 Lvs with white sheathing base, cells narrowly rectangular in upper part of lf, spores 34–40 μm **4. B. ithyphylla**
Lvs without whitish sheathing base or if base sheathing then upper cells quadrate-rectangular, spores 20–32 μm 2

2 Lvs ± erect when moist, cells rectangular to narrowly rectangular, ± similar throughout, capsule symmetrical **1. B. stricta**
Lvs erecto-patent to spreading when moist, cells quadrate to rectangular above, narrowly rectangular towards base, mouth of capsule oblique 3

3 Upper lvs flexuose to ± crisped when dry, plants usually glaucous green, seta straight, 5–25 mm long, capsule exserted **2. B. pomiformis**
Lvs ± crisped when dry, plants green or brownish-green, seta curved, 2–3 mm long, capsule usually concealed by lvs **3. B. hallerana**

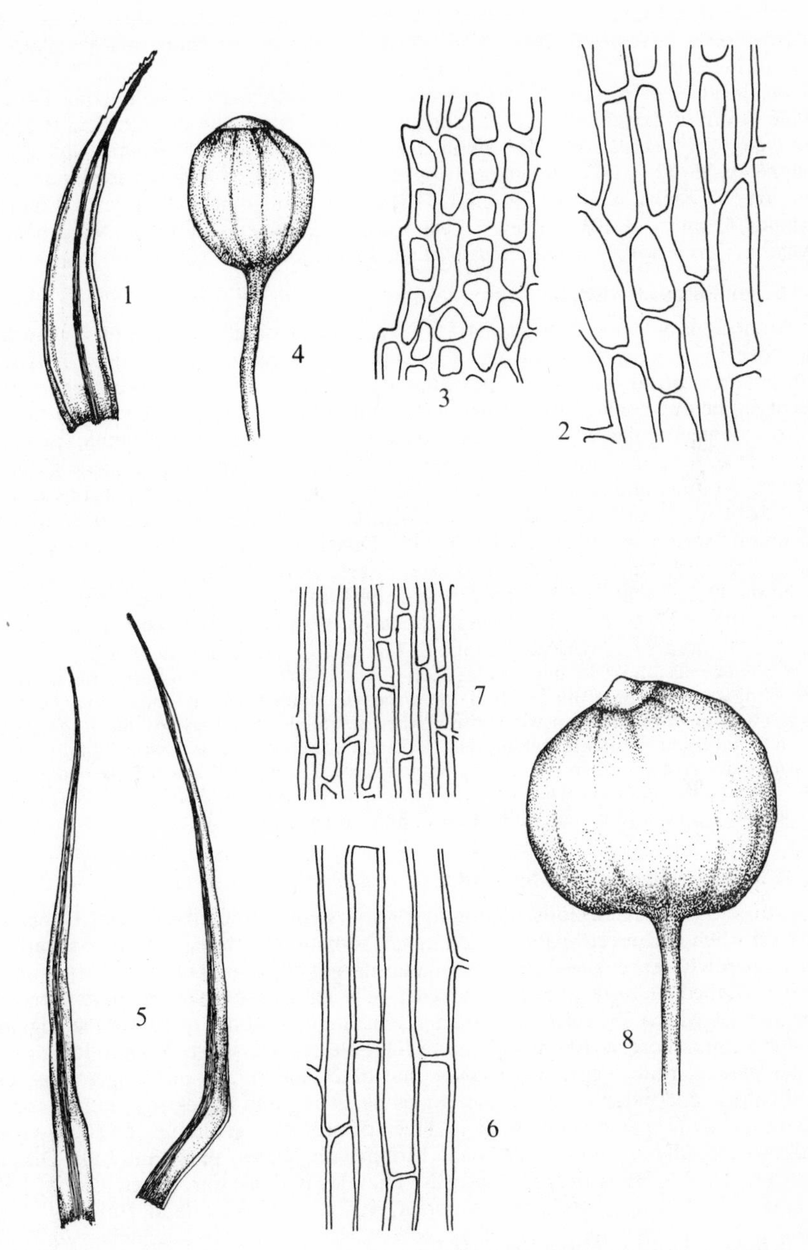

Fig. **220**. 1–4, *Plagiopus oederi*: 1, leaf, 2, basal cells; 3, mid-leaf cells; 4, capsule. 5–8, *Bartramia ithyphylla*: 5, leaves; 6, basal cells; 7, mid-leaf cells; 8, capsule. Leaves ×15, capsules ×10, cells ×415.

1. B. stricta Hedw., Sp. Musc., 1801

Synoecious. Plants to 3 cm, tomentose below. Lvs erect, rigid when dry, ± erect when moist, narrowly lanceolate, acuminate, base not sheathing, margin plane or narrowly recurved below, toothed above; nerve excurrent; cells ± similar throughout lf except at base, rectangular to narrowly rectangular, mamillose, opaque 4–8 μm wide in mid-lf, basal longer, ± hyaline, cells at basal angles shorter. Seta straight, *ca* 1 cm long; capsule globose, symmetrical, otherwise as in *B. pomiformis*; spores papillose, 26–32 μm. Dense, glaucous-green tufts on soil in rock crevices and ledges and on soil banks, very rare. Radnor, Pembroke, Montgomery, Mid Perth, Channel Islands, formerly also W. Sussex, 5, C. S. Europe, Turkey, Syria, Libya, Macaronesia, Algeria, Cameroon, N. America, Australia, Tasmania.

2. B. pomiformis Hedw., Sp. Musc., 1801

Autoecious or synoecious. Plants 0.5–8.0 cm, stems tomentose below. Lvs crowded, flexuose to ± crisped when dry, erecto-patent to patent when moist, narrowly lanceolate to linear-lanceolate, apex subulate, base not or scarcely sheathing, margin recurved below, dentate above; nerve toothed at back above, excurrent; basal cells narrowly rectangular, above quadrate-rectangular, quadrate or hexagonal, pellucid, mamillose, 4–8 μm wide in mid-lf. Seta straight, 5–25 mm long; capsule globose, striate, mouth oblique, ripe capsules globose, shiny when dry, immature capsules ovate-obloid, sulcate when dry, spores with large rounded papillae, 20–26 μm. Fr common, spring, summer. $n=8$*, $8+1$*. Dense, glaucous-green or rarely green tufts, brownish below. on soil in rock crevices, stone walls, banks, etc., in acidic habitats. Occasional in S.E. England and the Midlands, frequent or common elsewhere. 105, H27, C. Europe, Faroes, Caucasus, N. and C. Asia, Himalayas, China, Japan, Madeira, N. America, Greenland, Tierra del Fuego, New Zealand.

B. pomiformis is a somewhat variable species, particularly in relation to size and degree of crowding and crisping of the lvs. In damp or heavily shaded habitats the lvs may be laxer, longer and narrower and somewhat crisped when dry. Such forms may be difficult to separate from *B. hallerana* if fruit are lacking. This form has been regarded as a variety (var. *elongata* Turn.) but would appear to be merely a habitat form. The lvs of *B. stricta* have opaque cells and are straight and rigid when dry; the capsule is symmetrical. *B. ithyphylla* may be distinguished in the field by the white lf bases and microscopically by the very long, narrow cells.

3. B. hallerana Hedw., Sp. Musc., 1801

Autoecious or synoecious. Plants to 15 cm, stems tomentose below. Upper lvs crisped when dry, erecto-patent when moist, sometimes sub-secund, from sheathing base narrowly lanceolate, longly subulate, margin recurved below, dentate above; nerve toothed at back above, excurrent; cells incrassate, in basal part narrowly rectangular with a few cells in basal angles rectangular, above quadrate-rectangular, pellucid, mamillose, 6–10 μm wide in mid-lf. Seta curved short, 2–3 mm long, often 2 per perichaetium; capsule similar to that of *B. pomiformis* but longer when dry and empty, concealed by lvs, old capsules persisting and appearing lateral; spores 20–24 μm. Fr frequent, summer. Silky green or brownish-green tufts on shaded rock ledges, especially in woods, occasional. Monmouth, Wales, Yorks and Lake District northwards. 37, H9. Europe, Caucasus, Himalayas, Kashmir, Tibet, China, Kilimanjaro, N. America, southern S. America, Australia, New Zealand, Hawaii.

4. B. ithyphylla Brid., Musc. Rec., 1803

Synoecious. Stems tomentose below, 0.5–4.0 cm. Lvs erect, rigid to flexuose or crisped when dry, spreading when moist, blade narrowly lanceolate, subulate from

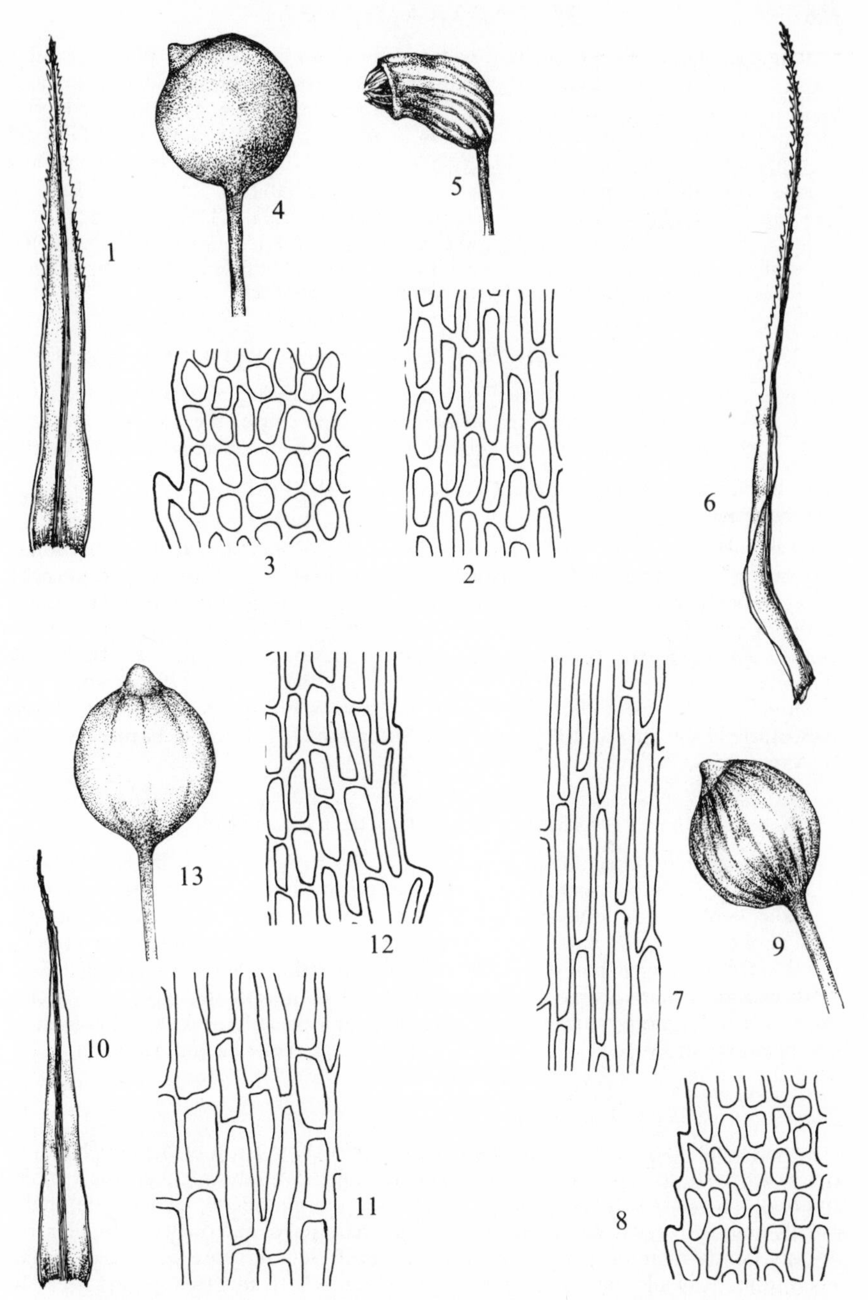

Fig. **221**. 1–5, *Bartramia pomiformis*: 1, leaf; 2, basal cells; 3, mid-leaf cells; 4 and 5, mature and dry capsules. 6–9, *B. hallerana*: 6, leaf; 7, basal cells; 8, mid-leaf cells; 9, mature capsule. 10–13, *B. stricta*: 10, leaf; 11, basal cells; 12, mid-leaf cells; 13, mature capsule. Leaves $\times 15$, capsules $\times 10$, cells $\times 415$.

rectangular, white, sheathing base widened at shoulders, margin plane, serrulate above; nerve excurrent; cells in sheathing base narrowly rectangular, hyaline, in blade much smaller, narrowly rectangular, opaque, mamillose, 4–6 μm wide in mid-lf. Seta straight, 1–3 cm; capsule as in *B. pomiformis*; spores papillose, 34–40 μm. Fr frequent, summer. *n*=12*. Silky, glaucous-green tufts on soil in rock crevices, ledges, banks and turf, particularly in acidic, montane habitats. Rare in S.E. England and the Midlands, occasional to frequent elsewhere. 71, H15. Europe, Faroes, Iceland, N. and C. Asia, China, N. and E. Africa, N. America, Greenland, Patagonia.

Readily distinguished from other members of the genus in the field by the white, sheathing leaf bases and under the microscope by the narrow opaque leaf cells.

103. Conostomum Sw. in Web. & Mohr, Nat. Reise Schwed., 1804

Stems fastigiately branched, densely tufted. Lvs imbricate in 5 rows; cells ± rectangular. Capsule ± obloid, cernuous, striate; peristome single, the 16 teeth united at apex. With the exception of *C. tetragonum*, a southern hemisphere genus of 15 species.

1. C. tetragonum (Hedw.) Lindb., Öfv. K. V. A. Förh., 1863
C. boreale Sw.

Dioecious. Plants to 2(–12) cm. Lvs imbricate, 5-ranked, narrowly lanceolate, acuminate, margin plane, denticulate above; nerve excurrent; cells variable in shape, ± rectangular to narrowly rectangular, towards apex ± quadrate to elliptical, narrower at margin, 8–13 μm wide in mid-lf. Seta to 12 mm; capsule cernuous, ± obloid, asymmetrical; spores *ca* 40 μm. Fr occasional, summer. *n*=16. Dense, glaucous to dark green tufts on peaty soil at altitudes of 600 to 1200 m, sometimes locally abundant, particularly in areas of late snow-lie. Caerns (not seen recently). Westmorland, Stirling and Perth north to Sutherland. 21. Europe, Faroes, N. Asia, N. America, Greenland.

104. Bartramidula Br. Eur., 1846

Synoecious, rarely dioecious or autoecious. Slender plants, stems branching below inflorescence. Lvs erecto-patent or secund, not plicate, lanceolate, acuminate, margin ± plane, denticulate; nerve ending below apex to excurrent; basal cells rectangular, cells above shortly to narrowly rectangular, smooth or with mamillae at ends. Perichaetial lvs longer than stem lvs, base partly sheathing. Setae thin, often 2–3 per perichaetium, usually cygneous; capsule inclined or horizontal, globose-pyriform, short-necked, smooth, sulcate when dry; peristome usually absent; lid convex. A small genus with 25 species occurring in Europe, Asia, America and Australia.

1. B. wilsonii Br. Eur., 1846
Philonotis wilsonii (Br. Eur.) Mitt.

Synoecious. Plants 2–5 mm, shoots slender, branching from below perichaetium. Lvs erect-flexuose when dry, erecto-patent, sometimes sub-secund when moist, lanceolate to narrowly lanceolate, acuminate, margin plane, denticulate; nerve usually ending in apex; basal cells rectangular, cells above narrowly rectangular, not or only faintly mamillose, *ca* 8 μm wide in mid-lf. Seta flexuose when dry, arcuate when moist; capsule inclined to pendulous, globose with shortly tapering neck; lid convex; peristome absent; spores 28–40 μm. Fr common, summer, autumn. Pale green patches or scattered plants on peaty soil in open habitats in montane areas, very rare. Merioneth, Angus, W. Inverness, Argyll, W. Ross, Outer Hebrides, S. Kerry, W. Cork. 6, H2. Spain, Yunnan, Fernando Po.

105. Philonotis Brid., Br. Univ., 1827

British species dioecious except for *P. rigida*. Plants very variable phenotypically, often tall, stems innovating from below perichaetium or perigonium. Lvs ovate to lanceolate, acute to longly acuminate, margin plane or recurved, toothed, teeth single or double; at least upper cells usually mamillose. Fertile male stems in some species (species 4–8) with smaller, more appressed and shortly pointed lvs than sterile or female stems. Capsule globose, inclined, striate, when dry sulcate, inner peristome with well developed cilia. A world-wide genus of about 240 species of mainly damp habitats.

It is important that lvs produced late in the current season or from the upper part of the previous years growth are examined, as those produced early in the season may be abnormal and lead to misidentification.

1 Autoecious, androecia bud-like, lvs rigid, mamillae apparently on end-walls of cells **1. P. rigida**
Dioecious, androecia disciform, lvs soft, mamillae near end-walls of cells or absent 2
2 At least upper cells of lvs with only distal mamillae or if cells smooth then plants very slender with lanceolate, ± tapering lvs 3
Mamillae proximal and sometimes also distal 4
3 Most cells with mamillae near upper ends (i.e. distal), plants 1–5 cm tall **2. P. marchica**
Cells in lower ½ of lf smooth, upper cells smooth or with distal mamillae, plants slender, 0.5–1.0 cm tall **3. P. arnellii**
4 Lvs of mature growth with plane margin and single teeth, not plicate, cells ± rectangular throughout **4. P. caespitosa**
Lvs of mature growth with margin recurved and teeth double, plicate at least near base, upper cells much narrower than lower 5
5 Lvs spirally imbricate, nerve coarsely papillose at back ± from base **7. P. seriata**
Lvs not spirally imbricate, nerve smooth or only slightly papillose at back 6
6 Lvs 1.8–3.0 mm long, nerve (200–)320–600 μm wide near base, cells near nerve at widest part of lf mostly 48–88 μm long, perigonial lvs acute, plants ± robust **8. P. calcarea**
Lvs 0.5–1.5(–2.0) mm long, nerve 160–320 μm wide, cells 24–40 μm long, perigonial lvs obtuse or if acute then plants slender 7
7 Lvs ± ovate, acute to acuminate, nerve not or scarcely excurrent, inner perigonial lvs obtuse **5. P. fontana**
Lvs narrowly lanceolate to ovate-lanceolate, acuminate to cuspidate, nerve excurrent to longly excurrent, inner perigonial lvs acute **6. P. tomentella**

1. P. rigida Brid., Br. Univ., 1827

Autoecious. Plants to 10(–12) mm, stems radiculose. Lvs crowded, appressed, straight when dry, erect to erecto-patent, rigid when moist, lanceolate, acuminate, margin plane or narrowly recurved, toothed, teeth single; nerve strong, excurrent; basal cells rectangular, shorter towards margin, above narrowly rectangular, mamillose, mamillae distal but apparently on end-walls of cells, cells 5–8 μm wide in mid-lf. Androecia bud-like below perichaetia. Seta red, to 2 cm; capsule globose, asymmetrical, striate; spores ovoid, 28–32 μm long. Fr frequent, spring. $n=6^*$, 12. Green to brownish-green tufts or patches in crevices on cliffs, on damp soil by streams, especially near the coast, rare. Cornwall, N. Devon, I. of Wight, Radnor, N. Wales,

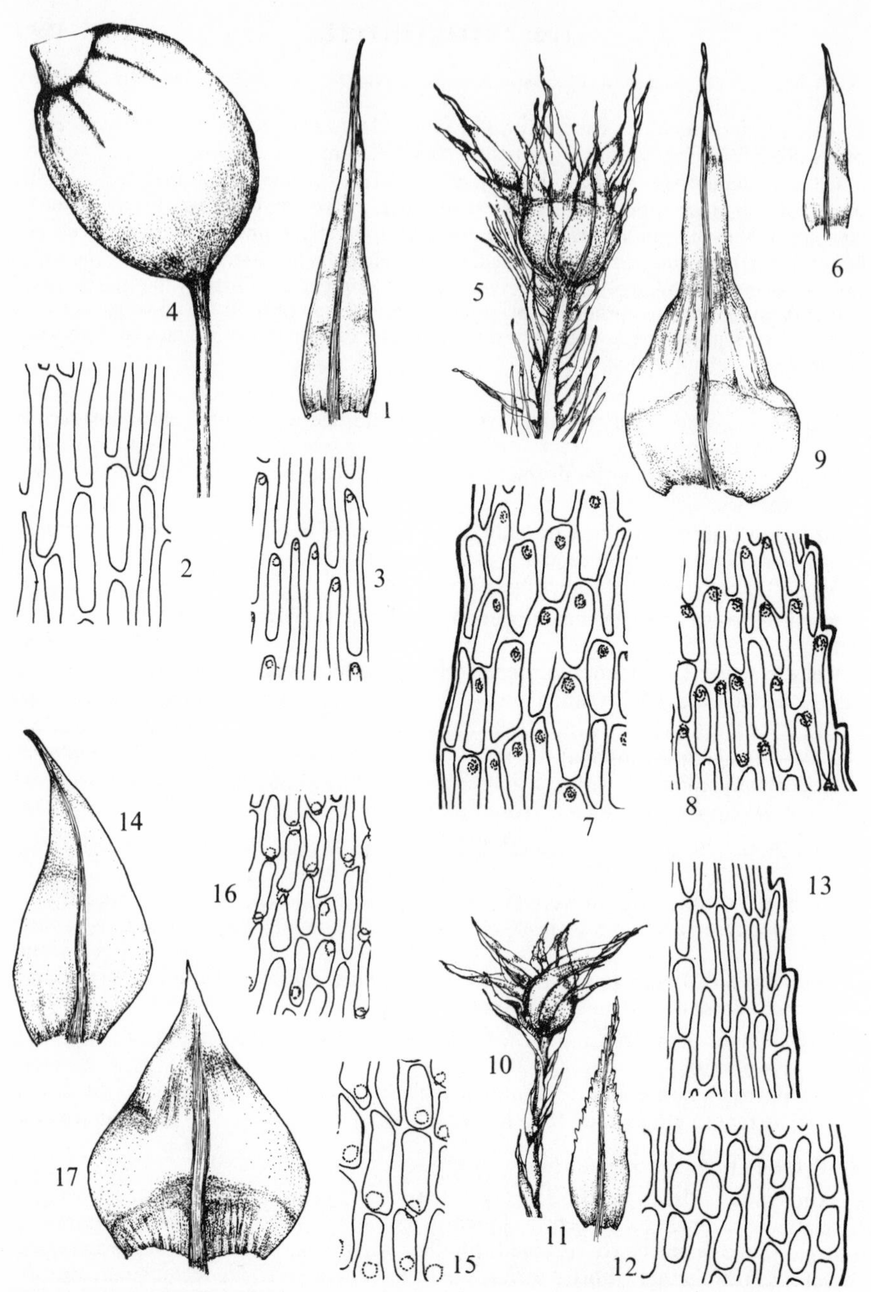

Fig. **222**. 1–4, *Philonotis rigida*: 1, leaf (×25); 2, basal cells; 3, upper cells; 4, capsule (×10). 5–9, *P. marchica*: 5, male stem; 6, leaf (×25); 7, basal cells; 8, upper cells; 9, perigonial leaf (×25). 10–13, *P. arnellii*: 10, male stem; 11, leaf (×40); 12, basal cells; 13, upper cells. 14–17, *P. caespitosa*: 14, leaf (×25); 15, basal cells; 16, upper cells; 17, perigonial leaf (×25). Cells ×415.

I. of Man, Ayr, W. Perth, coastal counties of Ireland. 11, H7, C. Belgium, Switzerland, Portugal, Italy, Sicily, Caucasus, Algeria, Macaronesia.

2. P. marchica (Hedw.) Brid., Br. Univ., 1827

Plants slender or robust, 1–5(–9) cm. Lvs erecto-patent in slender plants to falcato-secund in robust plants, narrowly lanceolate to ovate-lanceolate, acute to acuminate, not plicate near base, margin plane, toothed, teeth single, blunt or sharp; nerve ending below apex to shortly excurrent; basal cells rectangular, 8–16 μm wide, above narrowly rectangular, 6–10 μm wide in mid-lf, most cells except at extreme base with distal mamillae. Perigonial lvs ± erect, apex long, acuminate. Capsule as in *P. fontana* but smaller. Fr very rare, unknown in Britain. Pale green tufts on damp soil and in rock crevices, very rare. I. of Wight, N.W. Yorks. 2. N., W. and C. Europe, W. Asia, Japan, Korea, N. America, Algeria, Madeira.

Whilst well grown specimens are distinct from *P. arnellii* in their larger size and secund lvs, the only reliable character for identifying small gatherings which are sterile is the presence of mamillae on the cells in the lower part of the lf. For the occurrence of this plant in Britain see Smith, *J. Bryol.* **8**, 5–8, 1974.

3. P. arnellii Husn., Musc. Gall., 1890
P. capillaris Lindb., *P. fontana* var. *capillaris* (Lindb.) Braithw.

Plants slender, mostly 0.5–1.0 cm. Lvs erect or slightly secund when dry, patent, sometimes slightly secund when moist, 0.6–0.9 mm long, mostly 3–5 times as long as wide, lanceolate, apex acuminate to longly acuminate, not plicate at base, margin plane or narrowly recurved, toothed, teeth single; nerve excurrent; lower cells rectangular, smooth, 8–10 μm wide, cells above narrower, smooth or with distal mamillae. Perigonial lvs squarrose, acute. Capsule as in *P. fontana* but smaller. Fr very rare, summer. Small tufts or patches, often in small quantity in rock crevices, on ditch and stream banks and damp soil in shaded places. Occasional throughout Britain, rare in Ireland. 51, H5. N., W. and C. Europe, Faroes, Iceland, Asia Minor, Greenland.

In both *P. arnellii* and *P. marchica* depauperate specimens may be difficult to determine, but careful study of well developed lvs of the latter will usually reveal distal mamillae on the lower cells. All the succeeding species except *P. caespitosa* have double teeth on the margins of mature lvs but immature or poorly developed lvs may only have single teeth. Petit (*Bull. Jard. bot. nat. belg.* **46**, 221–6, 1976) points out a character which might be useful in the separation of *P. arnellii* and *P. marchica*. *P. marchica* produces ellipsoid axillary gemmae, 0.5–1.0 mm long, whilst *P. arnellii* has axillary propagules, *ca* 1 mm long, resembling flagelliform shoots.

4. P. caespitosa Wils. ex Milde, Br. Siles., 1869

Plants slender, to 5 cm. Lvs ± imbricate, straight or sub-secund when dry, erecto-patent to falcato-secund when moist, 0.6–1.7 mm long, 1.5–3.0(–4.0) times as long as wide, ovate and acute to ovate-lanceolate and shortly acuminate, not plicate near base, margin plane or rarely slightly recurved, toothed, teeth single; nerve ending below apex to shortly excurrent; cells thin-walled to incrassate, ± rectangular throughout, 10–20 μm wide in lower part of lf, mamillae proximal. Perigonial lvs acute. Capsule as in *P. fontana*. Fr not known in Britain. Tufts or patches on stream banks, damp tracks, marshes and flushed rocks. Occasional in England and Wales, very rare in Scotland and Ireland. 42, H4, C. N., W. and C. Europe, N., W. and C. Asia, N. America, Greenland.

Differs from the following species in the lvs not plicate below, the cells of ± similar shape throughout, the margin plane or only narrowly recurved and with single teeth. Small forms of *P. fontana* may be mistaken for this plant but have smaller cells in the upper part of the lf.

5. P. fontana (Hedw.) Brid., Br. Univ., 1827

Plants slender to robust, (0.5–)1.0–10.0(–15.0) cm, older stems tomentose. Lvs imbricate when dry, erecto-patent or, especially in female plants, falcate and sub-secund or secund, 0.8–1.5(–2.0) mm long, 2–3(–4) times as long as wide, straight to falcate, ovate, tapering to acute to acuminate apex, plicate below, margin recurved with double teeth at least in mature lvs; nerve stout, 160–320(–400) μm wide near base, ending below apex to shortly excurrent; cells mamillose, mamillae proximal and sometimes also distal, cells in lower part of lf ± rectangular, at widest part of lf towards nerve 8–16 μm×(16–)24–40 μm, cells becoming narrower above, towards apex narrowly rectangular. Inner perigonial lvs reflexed from sheathing base, obtuse, toothed. Seta to 5 cm long; capsule ± globose, asymmetrical, inclined, striate when fresh, when dry ovoid, furrowed and horizontal; spores ovoid, 26–30 μm long. Fr occasional, summer. $n=6$*. Yellowish-green, bright green or brownish-green tufts or patches in springs, flushes and wet places, or small tufts or patches on damp woodland rides, wet gravelly soil and tracks. Frequent to common except in S.E. England and the Midlands. 112, H38, C. Europe, Asia, Algeria, E. Africa, N. America, Mexico, Greenland.

A very variable species some forms of which may be confused with the previous species and others with *P. calcarea* or *P. tomentella*. Young lvs may have plane margins with simple teeth and abnormally large cells. For this reason it is essential in determining *Philonotis* species that lvs at the end of the previous year's growth, taken from amongst the tomentum, are examined. The first 3 species differ in the position of the cell mamillae and narrower more tapering lvs. *P. caespitosa* has wider, ± uniformly shaped cells. *P. tomentella* is close to *P. fontana* but has more slender stems with narrower lvs with long, sometimes subulate points and acute perigonial lvs. These differences appear constant and the plant is worthy of specific status. *P. seriata* differs in the spirally arranged lvs and the nerve scabrous with papillae at back; the basal cells in that species are also more heavily and coarsely mamillose. *P. calcarea* may sometimes be difficult to distinguish and occasional *Philonotis* plants are encountered that cannot be named in the absence of perigonial lvs. *P. calcarea* usually has longer lvs (1.8–3.0 mm, whereas in *P. fontana* they are 0.8–2.0 mm), the cells are frequently but not always wider (10–24 μm, in *P. fontana* 8–16 μm) and the longest cells in *P. fontana* (40 μm) do not reach the length of the longer cells in *P. calcarea* (48–88 μm), though it must be noted that in *P. calcarea* there are also scattered shorter cells. Cell length seems to be the most satisfactory character and I have seen only 2 gatherings in which, in the last resort, this has not served to name the plants.

6. P. tomentella Mol. in Lor., Moost., 1864
P. fontana var. *tomentella* (Mol.) Dix., *P. fontana* var. *pumila* (Turn.) Brid.

Plants slender, to 5(–9) cm, stems densely tomentose except at apices. Lvs imbricate when dry, erecto-patent to sub-secund when moist, 0.5–1.5 mm long, 3–4 times as long as wide, narrowly lanceolate to ovate-lanceolate, tapering to acuminate to subulate apex, margin plane or recurved, toothed, teeth single or double; nerve excurrent to longly excurrent; areolation as in *P. fontana*. Perigonial lvs acute. Seta 1.5–2.0 cm; capsule as in *P. fontana* but smaller; spores 24–26 μm long. Fr very rare, autumn. Densely tomentose tufts or patches or straggling shoots in damp rock crevices, on damp shaded soil by streams, usually in montane regions, rare. N. Wales, Yorks and Westmorland northwards, Louth, W. Donegal. 19, H2. Europe, N., W. and C. Asia, N. America, Greenland.

7. P. seriata Mitt., J. Linn. Soc. Bot., Suppl., 1869

Plants 3–12 cm, older stems tomentose. Lvs spirally imbricate, mostly 1–2 mm long, 1.5–3.0(–3.5) times as long as wide, straight to falcate, plicate, broadly ovate and acute to ovate and shortly acuminate, margin narrowly recurved, teeth single or

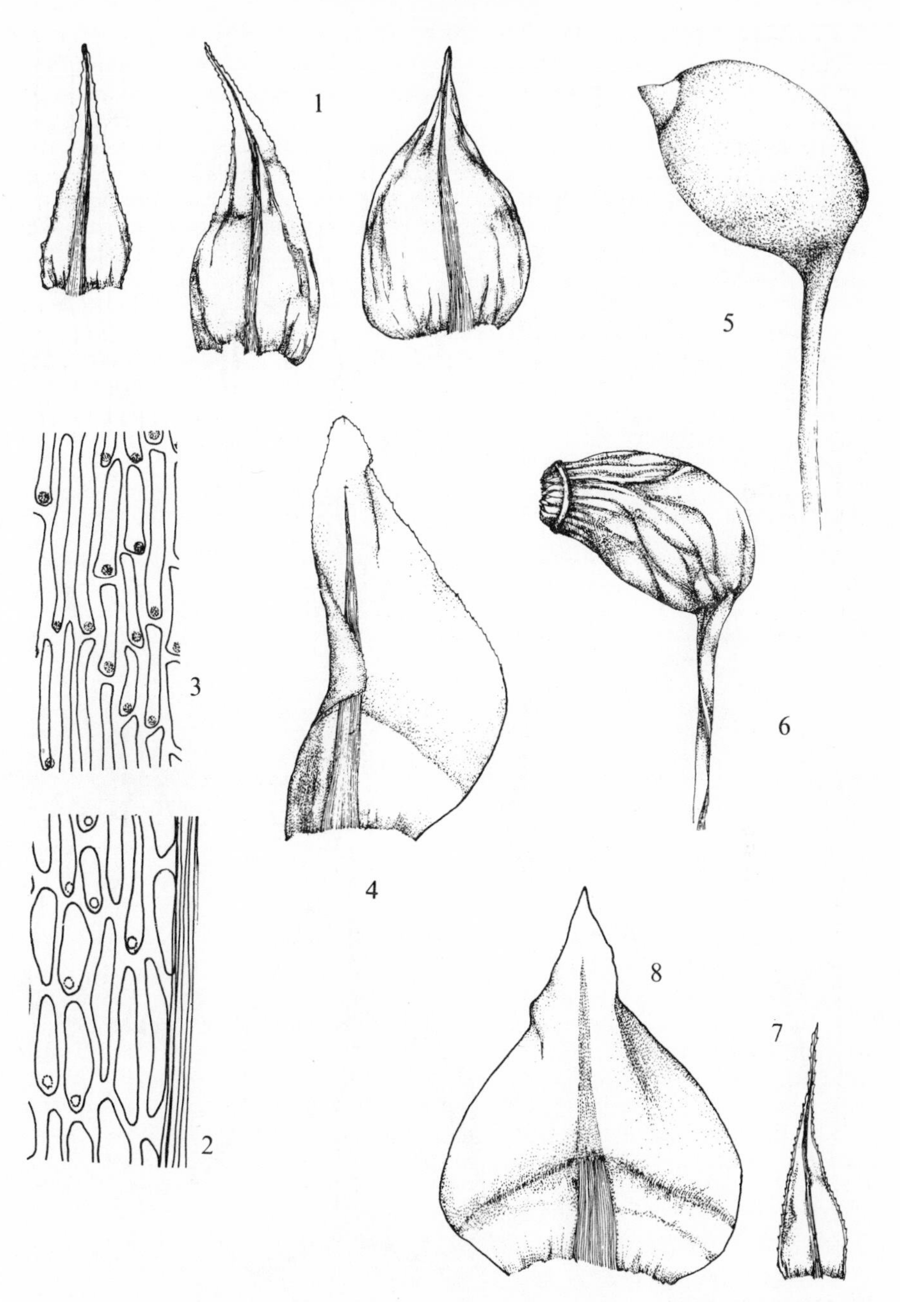

Fig. **223**. 1–6, *Philonotis fontana*: 1, leaves; 2, cells near nerve at widest part of leaf; 3, cells in upper part of leaf; 4, perigonial leaf; 5 and 6, fresh and dry capsules (×10). 7–8, *P. tomentella*: 7, leaf; 8, perigonial leaf. Leaves ×25, cells ×415.

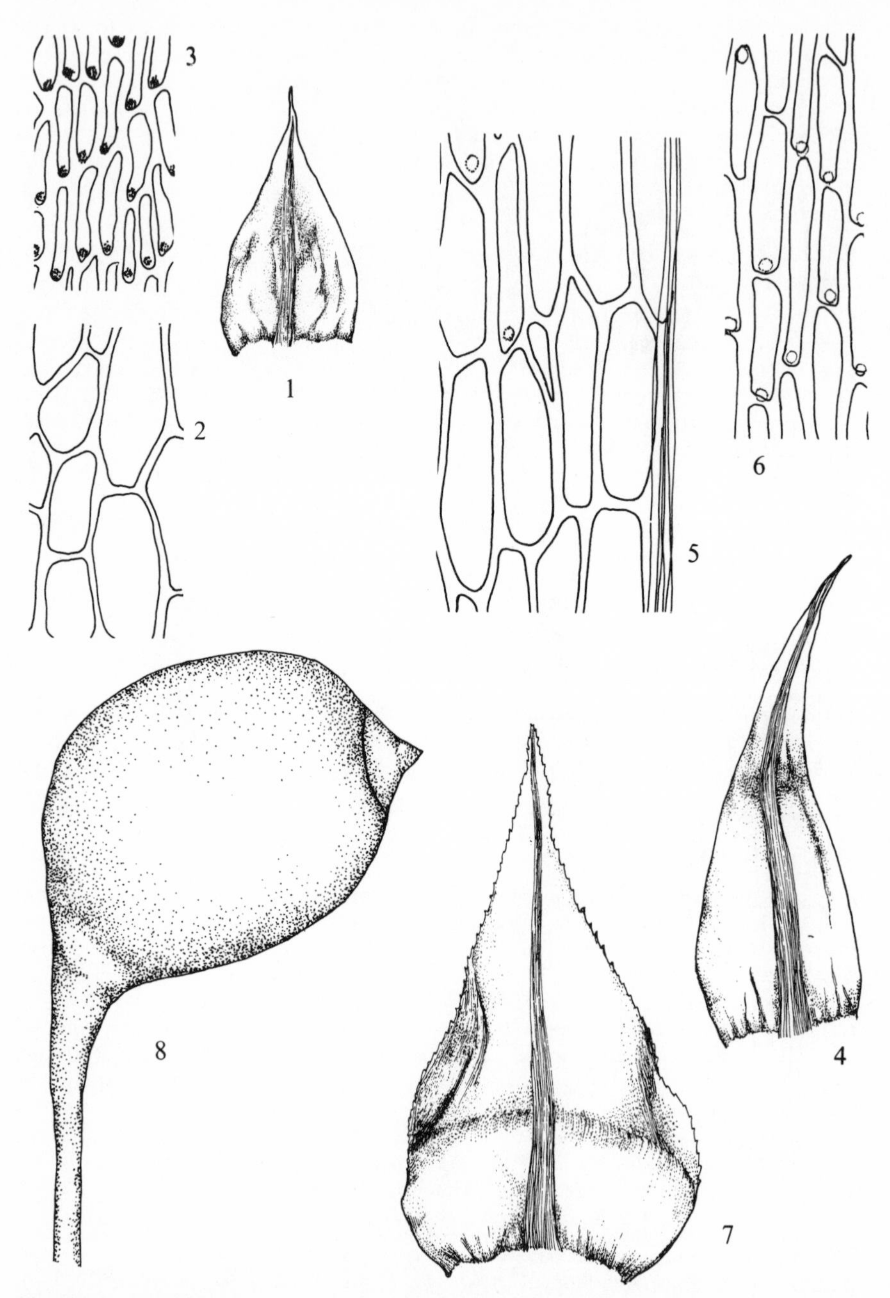

Fig. **224**. 1–3, *Philonotis seriata*: 1, leaf; 2, cells near nerve at widest part of leaf; 3, cells from upper part of leaf. 4–8, *P. calcarea*: 4, leaf; 5, cells near nerve at widest part of leaf; 6, upper cells; 7, perigonial leaf; 8, capsule ($\times 10$). Leaves $\times 25$, cells $\times 415$.

double; nerve stout, 240–480 μm wide near base, ending below apex to shortly excurrent coarsely papillose at back; cells with proximal mamillae, in lower part of lf rectangular to elongate-hexagonal, cells near nerve at widest part of lf 8–18 μm × 16–56 μm, towards apex narrowly rectangular. Capsule as in *P. fontana*. Fr very rare, summer. $n=6$. Green to yellowish-green tufts or patches in acidic springs and flushes and on wet ground at high altitudes, rare but sometimes locally frequent in the Scottish Highlands, Caerns. 14. Europe, Faroes, Caucasus, Iran, Siberia, Himalayas, N. Africa, Greenland.

8. P. calcarea (Br. Eur.) Schimp., Coroll., 1856

Plants ± robust, older stems tomentose. Lvs appressed, usually secund when dry, erecto-patent to falcato-secund when moist, 1.8–3.0 mm long, 2.5–4.0 times as long as wide, usually falcate, plicate, lanceolate to ovate-lanceolate, rarely ovate, tapering to acuminate apex, margin recurved with double teeth in mature lvs; nerve stout, (200–)320–600 μm wide near base; cells with proximal mamillae, areolation variable, in lvs produced early in season lower cells large, lax, ovate-rectangular, thin-walled, in lvs produced later in season narrower, ± narrowly rectangular, thick-walled, cells towards nerve at widest part of lf 10–24(–32) μm × (32–)48–88(–108) μm, upper cells narrowly rectangular to ± linear. Perigonial lvs narrowly triangular from ± erect base, acute. Seta to 4 cm; capsule as in *P. fontana*; spores *ca* 25 μm. Fr rare, summer. $n=6$*. Pale to bright or yellowish-green tufts or patches in fens, basic springs and flushes and dune slacks, calcicole, generally distributed, frequent. 83, H29. Europe, Faroes, N. and W. Asia, Himalayas, Tibet, Algeria, N. America.

106. Breutelia (Br. Eur.) Schimp., Coroll., 1856

Dioecious. Plants robust, stems tomentose. Lvs spreading, plicate; cells narrow, mamillose above. Seta usually flexuose-arcuate; capsule pendulous, ovoid, sulcate when dry; peristome double, rudimentary or absent. A world-wide genus of about 140 species.

1. B. chrysocoma (Hedw.) Lindb., Öfv. K. V. A. Förh., 1863
B. arcuata Schimp.

Shoots decumbent to ascending, densely tomentose with brownish rhizoids except at apex, to 10(–15) cm long. Lvs spreading when moist, somewhat scarious but otherwise unaltered when dry, longitudinally plicate, reflexed from sheathing base, lanceolate, acuminate, margin plane or slightly recurved above, slightly denticulate above; nerve ending below apex; basal cells linear to narrowly rectangular, at angles wider and shorter, cells above rectangular to narrowly rectangular, slightly sinuose, mamillose. Seta cygneous; capsule usually hidden in lvs, ovoid to spherical, striate. Fr rare. Shoots yellowish-green above, brownish below, in lax tufts, patches or scattered shoots amongst rocks, on rock ledges particularly near streams and waterfalls, on damp soil on heaths, banks, etc. Frequent to common in montane areas, occasional elsewhere. S.W. England, Wales, Hereford and Staffs northwards. 73, H31. W. Europe, Faroes, Switzerland, Corsica, N. America.

28. TIMMIACEAE

Autoecious or dioecious. Robust plants resembling *Polytrichum* in habit. Lvs with ± erect sheathing base and lanceolate to narrowly lanceolate, spreading limb, margin toothed above; nerve ceasing in or below apex, papillose or not at back, sometimes toothed at back above; cells in basal part linear, narrower at margin, smooth or papillose, cells in limb quadrate or hexagonal, mamillose at back, papillose or not. Seta long; capsule inclined to pendulous, ovate-ellipsoid, furrowed when dry; outer peristome teeth articulated, papillose, inner with tall basal membrane and appendiculate, free or united cilia; calyptra large, cucullate.

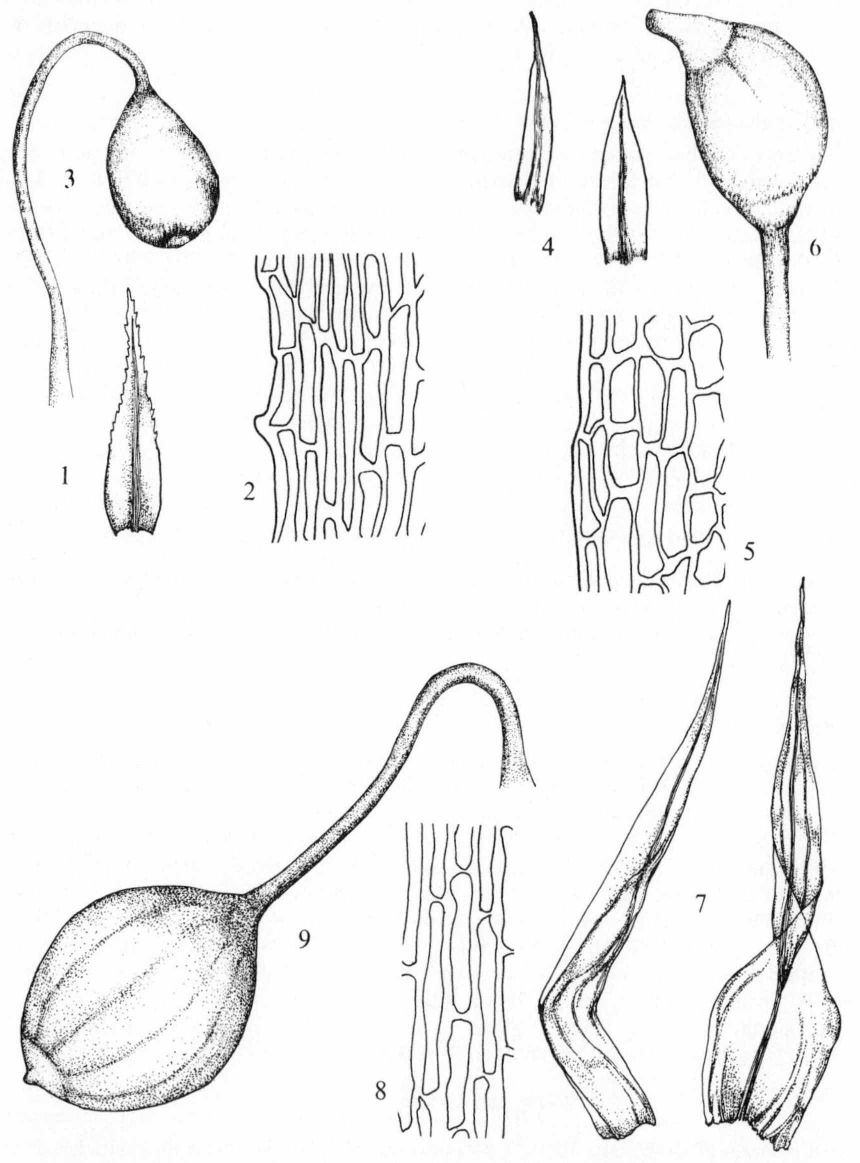

Fig. **225**. 1–3, *Bartramidula wilsonii*: 1, leaf (×40); 2, mid-leaf cells; 3, capsule (×15). 4–6, *Conostomum tetragonum*: 4, leaves (×15); 5, mid-leaf cells; 6, capsule (×15). 7–9, *Breutelia chrysocoma*: 7, leaves (×25); 8, mid-leaf cells; 9, capsule (×10). Cells ×415.

107. Timmia Hedw., Sp. Musc., 1801

The only genus, with the characters of the family. Nine northern hemisphere species.

Nerve not papillose at back, with a few teeth at back above, cells smooth **1. T. austriaca**

Nerve papillose at back, not toothed, cells papillose **2. T. norvegica**

1. T. austriaca Hedw., Sp. Musc., 1801

Dioecious. Plants 4–9 cm, stems tomentose below. Lvs of ± similar size along stem, lower appressed, upper curled when dry, patent when moist, differentiated into sheathing base and narrowly lanceolate limb, tapering to blunt to acute apex, margin plane or inflexed, obscurely to coarsely toothed above; nerve strong, ending in apex, not papillose at back but with a few teeth at back near apex; cells not papillose, in basal part ± linear, narrower towards margin, in limb irregularly quadrate, 8–15 μm wide in mid-lf. Fr unknown in Britain. Dark green tufts or patches on soil at high altitudes, very rare. Perth, Angus, Sligo. 3, H1. Northern and montane Europe, Siberia, C. Asia, Himalayas, Yunnan, Szechwan, Japan, N. America, Greenland.

The 2 British species of *Timmia* may be distinguished in the field by their general appearance. *T. austriaca* superficially resembles a species of *Polytrichum*, is dark green in colour and has ± uniformly sized lvs. *T. norvegica* is lighter in colour, more closely resembles a *Dicranum* species in general aspect and has the lvs variable in length.

2. T. norvegica Zett., Öfv. K. V. A. Förh., 1862

Dioecious. Plants 2–8 cm, stems tomentose below. Upper lvs larger than lower, lower curved, upper strongly curled when dry, patent when moist, differentiated or not into sheathing base and limb, narrowly lanceolate, acute, margin plane, coarsely toothed above; nerve stout, ending in apex, papillose on both surfaces but not toothed at back near apex; basal cells narrowly rectangular to linear, narrower towards margin, sometimes papillose, cells above hexagonal or quadrate-hexagonal, mamillose with projecting cell walls and papillose, 9–18 μm wide in mid-lf. Fr unknown in Britain. Green tufts or patches on soil at high altitudes, very rare. Perth, Angus, W. Inverness, Argyll, Sligo. 5, H1. Northern and montane Europe, Iceland, Caucasus, Siberia, Altai, N. America, Greenland.

15. ORTHOTRICHALES

Plants acrocarpous but sometimes apparently pleurocarpous because of development of branches from beneath perichaetium. Lvs ovate to lanceolate; nerve with or without large, median guide cells or absent; basal cells elongate, upper cells rounded-hexagonal, often papillose, unistratose. Seta straight, very short to long; capsule erect, immersed or exserted; peristome double, outer of 16 usually papillose or striate teeth, inner thin or absent or capsule gymnostomous. Basic chromosome numbers of the order almost entirely $x=6$ or $10+1$ or 11.

29. ORTHOTRICHACEAE

Plants usually tufted. Lvs lanceolate or narrowly lanceolate, acute, margin usually recurved; nerve present, ending below apex to excurrent; basal cells elongate, upper cells rounded-hexagonal, often thick-walled, papillose. Seta long or short; capsule immersed to exserted, ± ellipsoid, often with long neck, striate at maturity, sulcate when dry and empty; peristome double, single or, rarely, absent; calyptra mitriform or cucullate, often with coarse, erect hairs. Mainly saxicolous or corticolous plants.

108. Amphidium Schimp., Coroll., 1856

Gametophyte similar to *Zygodon* but stem irregular in section and papillae of upper cells rounded and of lower cells oval to linear; nerve with median guide cells. Sporophyte similar to *Zygodon* but seta short and capsule more longly tapering into seta, striate, sulcate when dry; peristome lacking; calyptra cucullate. Twelve species with a ± world-wide distribution.

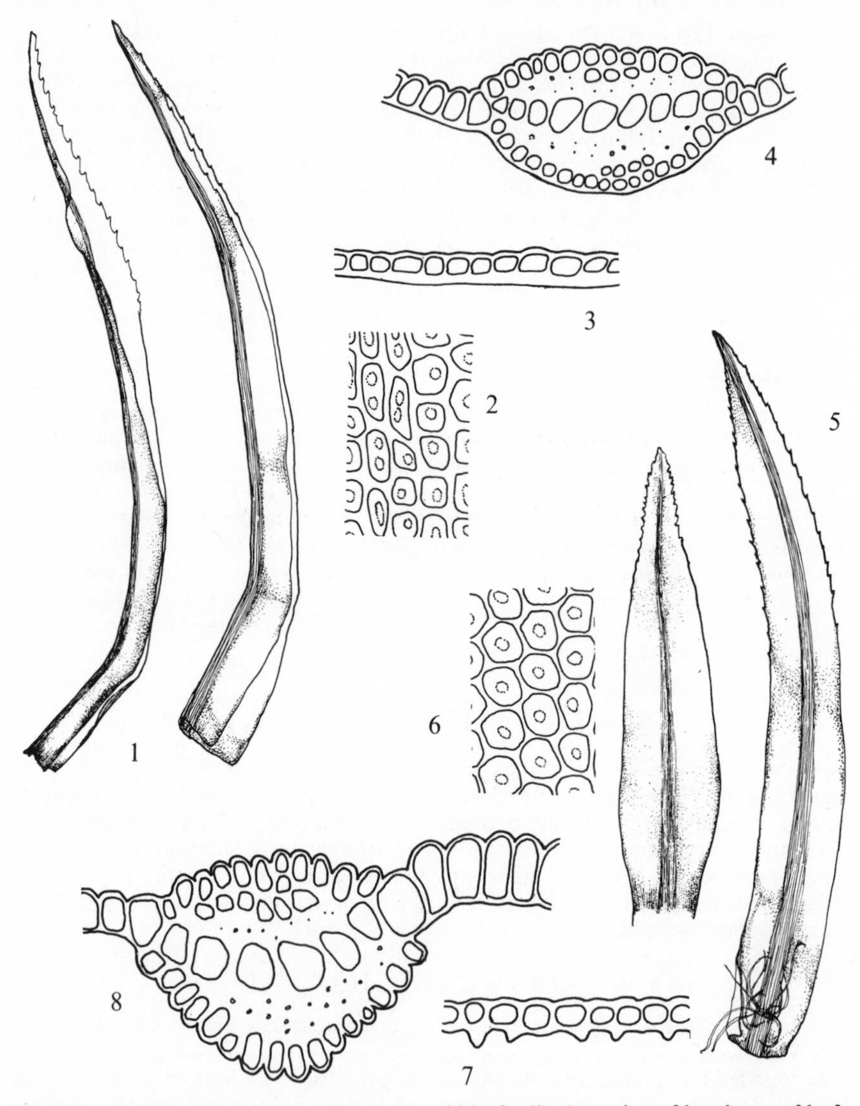

Fig. **226**. 1–4, *Timmia austriaca*: 1, leaves; 2, mid-leaf cells; 3, section of basal part of leaf; 4, nerve section. 5–8, *T. norvegica*: 5, leaves; 6, mid-leaf cells; 7, section of basal part of leaf; 8, nerve section. Leaves ×10, cells ×415.

The taxonomic position of *Amphidium* is obscure, being morphologically similar to *Zygodon* but cytologically showing affinities with *Grimmia* and *Ptychomitrium* and clearly not belonging to the Orthotrichaceae. As the position of the genus is obscure it is placed here as this seems the most convenient pigeon-hole.

Upper lf cells strongly papillose, basal cells thin-walled, hyaline, capsules common, just emergent above lvs **1. A. lapponicum**
Upper lf cells faintly papillose, basal cells incrassate, only hyaline in old lvs, capsules very rare, exserted **2. A. mougeotii**

1. A. lapponicum (Hedw.) Schimp., Coroll., 1856
Zygodon lapponicus Br. Eur.

Autoecious. Plants to 5 cm. Lvs crisped when dry, patent, flexuose when moist, ligulate to linear, acute to acuminate, margin plane or slightly recurved below, papillose-crenulate above; nerve ending in or below apex; basal cells thin-walled, hyaline, cells above rounded-hexagonal, strongly papillose, 10–14 μm wide in mid-lf. Seta short, 1.5–2.5 mm; capsule just emergent above lvs, narrowly pyriform, striate, when dry and empty plicate; spores smooth, 8–12 μm. Fr common, summer. $n=13$*, 16. Dark green tufts, reddish-brown below, on damp, shaded, basic rocks, occasional in montane areas. Mid and N. Wales, Yorks and Westmorland northwards, absent from Ireland. 19. Europe, Faroes, Asia, N. Africa, N. America.

2. A. mougeotii (Br. Eur.) Schimp., Coroll., 1856
Zygodon mougeotii Br. Eur.

Dioecious. Plants to 8 cm. Lvs flexuose-curled when dry, erecto-patent, flexuose when moist, narrowly linear-lanceolate, acuminate, margin recurved, entire or faintly

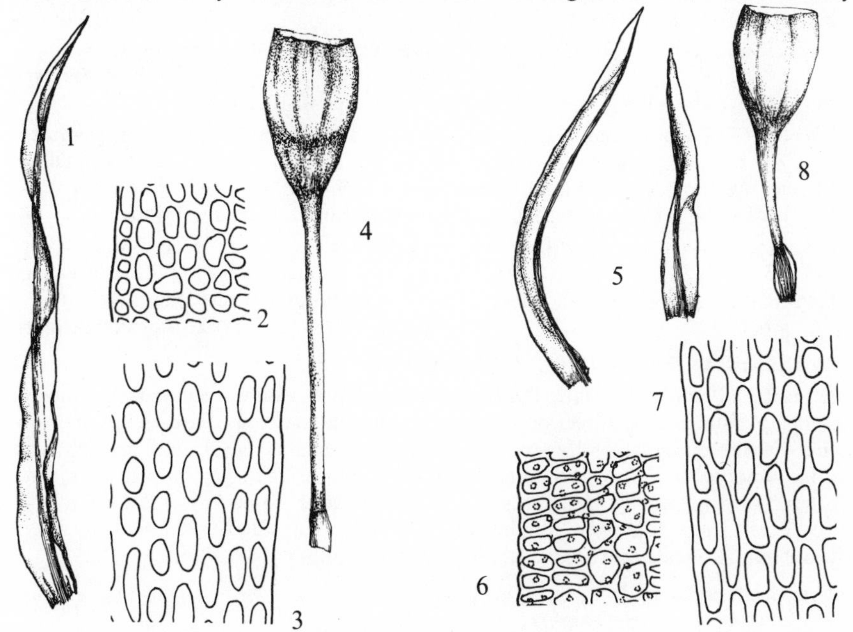

Fig. **227**. 1–4, *Amphidium mougeotii*: 1, leaf; 2, mid-leaf cells; 3, basal cells; 4, capsule. 5–8, *A. lapponicum*: 5, leaves; 6, mid-leaf cells; 7, basal cells; 8, capsule. Leaves ×25, cells ×415, capsules ×10.

papillose-crenulate to denticulate towards apex; nerve ending in or below apex; basal cells rectangular, incrassate, cells above incrassate, rounded-hexagonal, faintly papillose, 8–10 μm wide in mid-lf. Seta *ca* 3 mm; capsule shortly exserted, narrowly pyriform, striate, when dry and empty plicate; spores 10–12 μm. Fr very rare, summer. Yellowish-green tufts, brownish below, on damp rocks and in rock crevices in wooded valleys, quarries, on cliffs and in the mountains. W. England and Wales, Notts, Derby and Yorks northwards, frequent. 75, H31. Europe, Faroes, Caucasus, Madeira, Azores, N. America, Greenland.

A. lapponicum, which is frequently fertile, may be distinguished from *A. mougeotii* by its shorter, broader, more papillose lvs with thin-walled hyaline basal cells. *Anoectangium aestivum*, which may be mistaken for *A. mougeotii*, differs in its vivid green colour, the slender stems and the shorter, wider lvs with more strongly papillose cells.

109. ZYGODON HOOK. & TAYL., MUSC. BRIT., 1818

Autoecious or dioecious. Lvs flexuose or twisted when dry, linear-lanceolate to ovate, margin plane or recurved below, entire or toothed towards apex; nerve ending below apex to excurrent; basal cells rectangular, hyaline, smooth, upper cells rounded to hexagonal, often obscure, papillose. Stem gemmae frequently present. Seta long; capsule ovoid to ellipsoid or pyriform, with 8 longitudinal striae; calyptra smooth, cucullate; peristome double, single or absent, teeth when present united in pairs. A ± world-wide genus of about 90 species the majority of which occur in Africa and America.

1 Plants 2–6 cm, at least upper lvs toothed towards apex **4. Z. gracilis**
Plants to 1(–2) cm, lvs entire 2

2 Capsule narrowly pyriform, lf cells 14–24 μm wide in mid-lf, not papillose **5. Z. forsteri**
Capsule ovoid to ellipsoid, lf cells 6–12 μm wide in mid-lf, papillose 3

3 Lf cells 10–12 μm wide in mid-lf, lvs patent when moist, spores 18–20 μm, gemmae 7–8 cells long without longitudinal walls **3. Z. conoideus**
Lf cells mostly 7–9 μm wide in mid-lf, lvs spreading-recurved when moist, spores 14–16 μm, gemmae 4–6 cells long with or without longitudinal walls 4

4 Gemmae 30–40 μm wide with some longitudinal walls, plants green to dark green **1. Z. viridissimus**
Gemmae mostly 20–30 μm wide, without longitudinal walls, plants yellowish-green **2. Z. baumgartneri**

1. Z. viridissimus (Dicks.) R. Br., Trans. Linn. Soc., 1812

Dioecious. Plants to 1(–2) cm. Lvs twisted when dry, spreading-recurved when moist, tending to point in one direction, narrowly lanceolate to lanceolate, acute, margin plane, papillose; nerve ending below apex to excurrent; basal cells rectangular, hyaline, smooth, cells above rounded, papillose, obscure, 7–9(–12) μm wide in mid-lf. Scattered stem gemmae usually present, clavate to ovoid, 30–40 μm wide at maturity, (4–)5–6 cells long with some longitudinal cell walls. Capsule ovoid, smooth, when dry and empty plicate; peristome lacking; spores papillose, 14–16 μm.

Var. **viridissimus**

Z. viridissimus var. *occidentalis* (Correns) Malta

Nerve ending below apex, not widened above. Fr rare, spring. $n=12$*. Green or dark green tufts or patches on trees, rocks and walls, frequent. 109, H40, C. Europe, Faroes, Caucasus, Algeria, Madeira, N. and C. America.

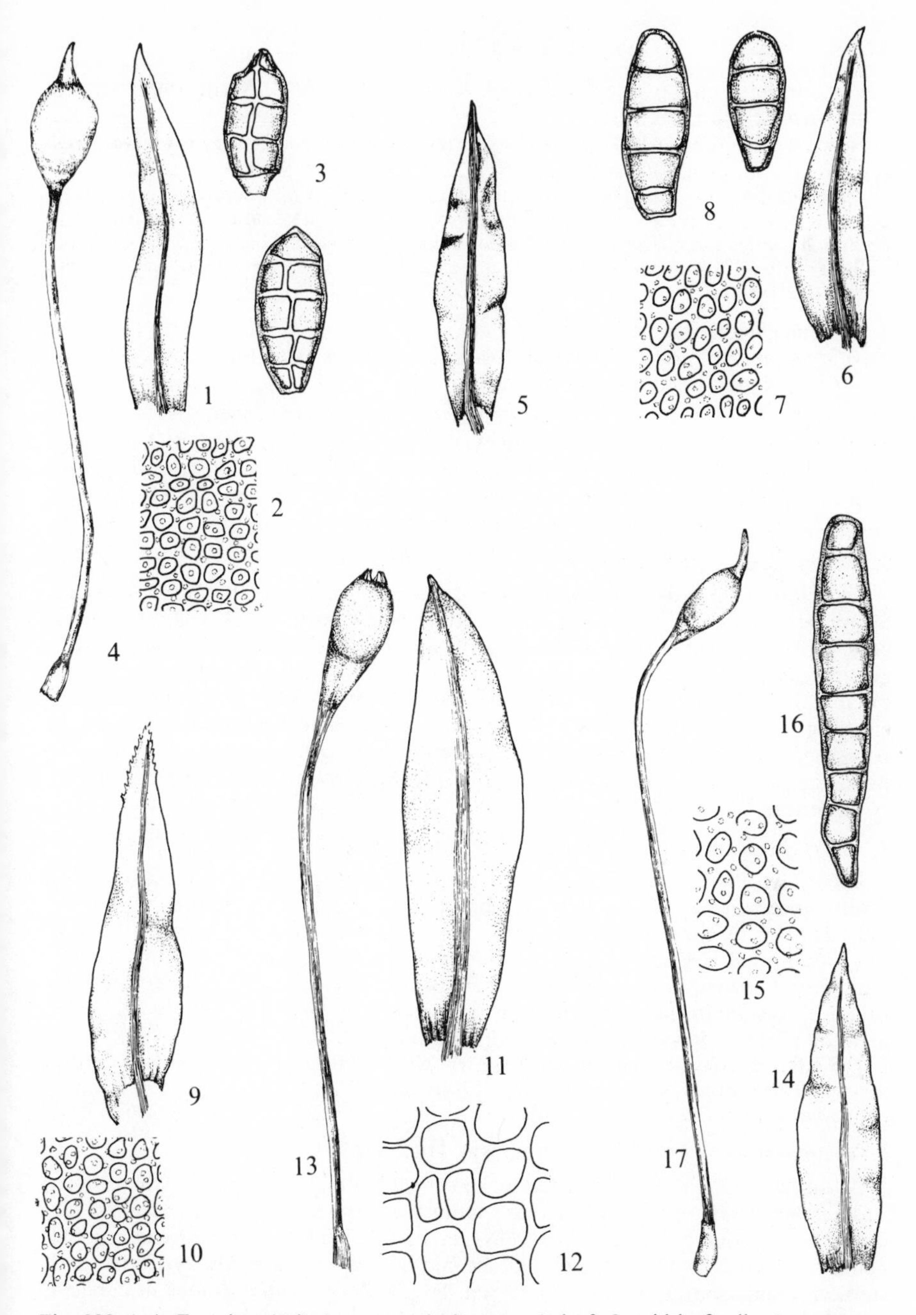

Fig. **228**. 1-4, *Zygodon viridissimus* var. *viridissimus*: 1, leaf; 2, mid-leaf cells; 3, gemmae; 4, capsule. 5, *Z. viridissimus* var. *stirtonii*: leaf. 6–8, *Z. baumgartneri*: 6, leaf; 7, mid-leaf cells; 8, gemmae. 9–10, *Z. gracilis*: 9, leaf; 10; mid-leaf cells. 11–13, *Z. forsteri*: 11, leaf; 12, mid-leaf cells; 13, capsule. 14–17, *Z. conoideus*: 14, leaf; 15, mid-leaf cells; 16, gemmae; 17, capsule. Leaves $\times 25$, cells $\times 415$, gemmae $\times 250$, capsules $\times 10$.

Var. **stirtonii** (Schimp. ex Stirt.) Hagen, K. Norsk. Vid. Selsk. Skrift, 1908
Z. stirtonii Schimp. ex Stirt.

Nerve widened towards lf apex and excurrent. Tufts on rocks, very rarely on trees, occasional and generally distributed. 73, H22, C. N. and W. Europe.

Although the only distinguishing feature of var. *stirtonii* is the excurrent nerve, this is usually a distinct character and the plant seems worthy of varietal rank. The lvs tending to point in one direction, the green or dark green colour and the gemmae with longitudinal cell walls separate *Z. viridissimus* from *Z. baumgartneri*, and the spreading-recurved lvs with smaller cells from *Z. conoideus.*

2. Z. baumgartneri Malta, Acta Univ. Latv., 1924
Z. viridissimus var. *vulgaris* Malta, *Z. viridissimus* var. *rupestris* Lindb. ex C. Hartm., *Z. vulgaris* Nyholm., *Z. rupestris* (C. Hartm.) Milde.

Close to *Z. viridissimus.* Lvs not homomallous, linear-lanceolate to lanceolate. Gemmae 20–30(–40) μm wide at maturity, 4–5 cells long without longitudinal walls, walls brownish. Peristome absent or rudimentary. Yellowish-green spreading tufts or patches on bark, occasional in W. and N. Britain, rare elsewhere. 28, H4. Europe, Macaronesia, Tunisia, Middle East, Caucasus, N. America.

3. Z. conoideus (Dicks.) Hook, & Tayl., Musc. Brit., 1818

Dioecious. Plants to 5 mm. Lvs appressed when dry, patent when moist, lanceolate, acuminate, margin entire, plane; nerve ending below apex; basal cells rectangular, cells above rounded, papillose, 10–12 μm wide in mid-lf. Scattered, fusiform stem gemmae usually present, *ca* 30 μm wide at maturity, 7–8 cells long, without longitudinal walls, walls ± colourless. Capsule ovoid to ellipsoid, striate, when dry and empty plicate; peristome double, fugacious; spores papillose, (16–)18–20 μm. Fr occasional, spring, summer. $n = 10 + 1^*$. Green or yellowish-green tufts or patches on bark, occasional and generally distributed. 61, H24, C. N. and W. Europe, Canaries, N. America.

4. Z. gracilis Wils. ex Berk., Handb. Brit. Mosses, 1863

Dioecious. Plants 2–6 cm. Lvs incurved when dry, spreading to recurved when moist, lanceolate, tapering to acute to obtuse apex, margin plane, entire in lower lvs, toothed towards apex in upper; nerve ending in or below apex; cells papillose, 8–10 μm wide in mid-lf. Capsule shortly cylindrical, slightly inclined; peristome double. Only found once in fruit in Britain in 1866. Brownish-green tufts on limestone walls or rarely on limestone rocks. Known only from the Malham and Ingleborough areas of Yorks. 1. Bavaria, Tirol, Switzerland, Italy.

The large size and lvs incurved when dry will differentiate *Z. gracilis* from other British species of the genus. The lvs in stunted specimens and the lower lvs in larger plants may be entire. For comments on *Z. gracilis* in Britain see Shaw, *Trans. Br. bryol. Soc.* **4**, 206–8, 1962.

5. Z. forsteri (With.) Mitt., Ann. Mag. Nat. Hist., 1851

Autoecious. Plants to 5 mm. Lvs flexuose when dry, erecto-patent when moist, ovate to lanceolate or slightly obovate-spathulate, acute to acuminate, margin plane, entire; nerve stout, excurrent; basal cells rectangular, hyaline, cells above hexagonal, pellucid, smooth, 14–20(–24) μm wide in mid-lf. Capsule narrowly pyriform; peristome present; spores smooth, *ca* 10 μm. Small, dark green cushions in crevices, knot holes and seepage areas of trunks and exposed roots of *Fagus sylvatica*, very rarely on other tree species. Very rare and seen recently only from the New Forest Forest and Burnham Beeches, but there are old confirmed records from Epping Forest and Worcs. 4. France, Germany, Switzerland, Italy, Dalmatia.

For an account of this plant in Britain see Proctor, *Trans. Br. bryol. Soc.* **4**, 107–10, 1961.

110. ORTHOTRICHUM HEDW., SP. MUSC., 1801

Autoecious or dioecious. Lvs usually straight, imbricate when dry, erecto-patent to spreading when moist, lanceolate, apex acute to obtuse or rounded, margin usually strongly recurved; nerve ending below apex; basal cells elongate, shorter towards margin, cells above rounded or hexagonal, papillose, papillae simple or bi- or trifid. Seta short, vaginula glabrous or hairy, ochrea present; capsule immersed to shortly exserted with 8 or 16 ribs or striae; stomata superficial or immersed; peristome of 16 outer teeth which may be fused in pairs, 0, 8 or 16 inner processes, capsule rarely gymnostomous. A large, world-wide genus of about 250 saxicolous or corticolous species.

Some species show considerable morphological variation. Whether this is environmental genotypic or due to hybridity is not known. Many species are highly susceptible to atmospheric pollution and have decreased markedly this century. All the species with superficial stomata examined cytologically have $x = 6$ and those with immersed stomata have $x = 10 + 1$ or 11.

Classification of the species follows Vitt, *Nova Hedwigia* **21**, 683–711, 1971.

1 Lvs with toothed, hyaline apices **17. O. diaphanum**
  Lvs without hyaline apices 2
2 Lf margin plane or strongly incurved, apex obtuse to rounded 3
  Lf margin recurved at least below, apex various 4
3 Lf margin strongly incurved, cells with 2–3 papillae on each face **1. O. gymnostomum**
  Lf margin plane or slightly incurved, cells with 1 papilla on each face **2. O. obtusifolium**
4 Lvs with numerous, 2- to several-celled gemmae, plants dioecious, capsules very rare **5. O. lyellii**
  Gemmae absent or few, plants autoecious, capsules common 5
5 Capsule smooth or with faint striae just below mouth at maturity, smooth when dry and empty **3. O. striatum**
  Capsules striate, at least some sulcate when dry and empty 6
6 Stomata superficial 7
  Stomata immersed 9
7 Calyptra hairy, peristome teeth erect or spreading when dry, spores 14–20 μm, plants usually saxicolous **7. O. rupestre**
  Calyptra glabrous or hairy, at least some capsules with reflexed peristome teeth, spores 18–26 μm, plants usually corticolous 8
8 Lower part of capsule smooth, striate below mouth only, calyptra hairy, seta *ca* 2 mm, spores 24–26 μm **4. O. speciosum**
  Capsule striate from mouth to base, calyptra glabrous or sparsely hairy, seta 0.4–1.2 mm, spores 18–24 μm **6. O. affine**
9 Lf apex broad, usually rounded 10
  Lf tapering to acute to obtuse apex 11
10 Plants 1–3 cm, lf cells 10–14 μm wide in mid-lf **8. O. rivulare**
  Plants 0.5–1.0 cm, cells (14–)16–24 μm **9. O. sprucei**
11 Plants 0.5–5.0 cm, usually saxicolous, peristome teeth erect to spreading when dry, not in pairs, stems with or without reddish-brown tomentum 12
  Plants mostly 0.5–1.0 cm, usually corticolous, at least some capsules with peristome

teeth reflexed when dry, in 8 pairs, stems without conspicuous reddish-brown tomentum 14

12 Capsule exserted, sporophyte reddish **10. O. anomalum**
Capsule emergent, pale 13

13 Stems without reddish-brown tomentum, capsules erect, vaginula naked or with a few sparse hairs* **11. O. cupulatum**
Stems with reddish-brown tomentum, capsules erect or laterally emergent, vaginula with long hairs **12. O. urnigerum**

14 Capsule exserted, peristome teeth orange-red, lvs twisted when dry **18. O. pulchellum**
Capsule immersed to emergent, peristome teeth rarely orange, lvs ± straight when dry 15

15 Calyptra conspicuously dark-tipped, vaginula with numerous long hairs* **14. O. stramineum**
Calyptra not dark-tipped, vaginula glabrous or with a few short hairs 16

16 Capsule orange-brown, narrowly ellipsoid to sub-cylindrical, stomata restricted to neck of capsule **16. O. tenellum**
Capsule pale, ellipsoid, stomata scattered at least in lower ½ of capsule 17

17 Calyptra smooth, capsule abruptly narrowed into seta, stomata ± obscured by overlying exothecial cells **13. O. schimperi**
Calyptra plicate, capsule gradually tapering into seta, stomata not obscured **15. O. pallens**

Subgenus I: ORTHOPHYLLUM

Lf margin plane, incurved or inrolled; stomata superficial; peristome present or absent. $x = 6$.

1. O. gymnostomum Bruch ex Brid., Br. Univ., 1826
Nyholmiella gymnostoma (Bruch ex Brid.) Holmen & Warncke, *Stroemia gymnostoma* (Bruch ex Brid.) Hagen

Dioecious. Plants to 0.3 cm (in Britain). Lvs imbricate when dry, spreading when moist, lanceolate to ovate, apex obtuse to rounded, sometimes slightly cucullate, margin strongly incurved; cells 12–14 μm wide in mid-lf, with 2–3 papillae on each face. Two- to several-celled simple or branched gemmae sometimes on both surfaces of lvs. Seta *ca* 0.3 mm; capsule immersed or slightly emergent, ellipsoid, ± abruptly narrowed into seta; stomata superficial; peristome absent; spores 18–21 μm. Fr unknown in Britain. Small, yellowish-green tufts on bark, very rare. E. Inverness. 1. Europe, Ukraine, Japan, Newfoundland.

For the occurrence of this plant in Britain see Perry & Dransfield, *Trans. Br. bryol. Soc.* **5**, 218–21, 1967. Distinguished from *O. obtusifolium* by the strongly incurved lf margin and cells with 2–3 papillae on each face.

2. O. obtusifolium Brid., Musc. Rec., 1801
Stroemia obtusifolia (Brid.) Hagen

Dioecious. Plants *ca* 0.5 cm (in Britain). Lvs patent when moist, ovate, apex rounded, margin plane to slightly incurved; cells 10–14 μm wide in mid-lf, unipapillose. Two- to several-celled gemmae scattered over surface of lvs. Seta *ca* 0.4 mm; capsule slightly emergent, ellipsoid, gradually tapering into seta; peristome present; stomata superficial; spores 16–20 μm. Fr unknown in Britain. $n = 6$.

* Not to be confused with paraphyses.

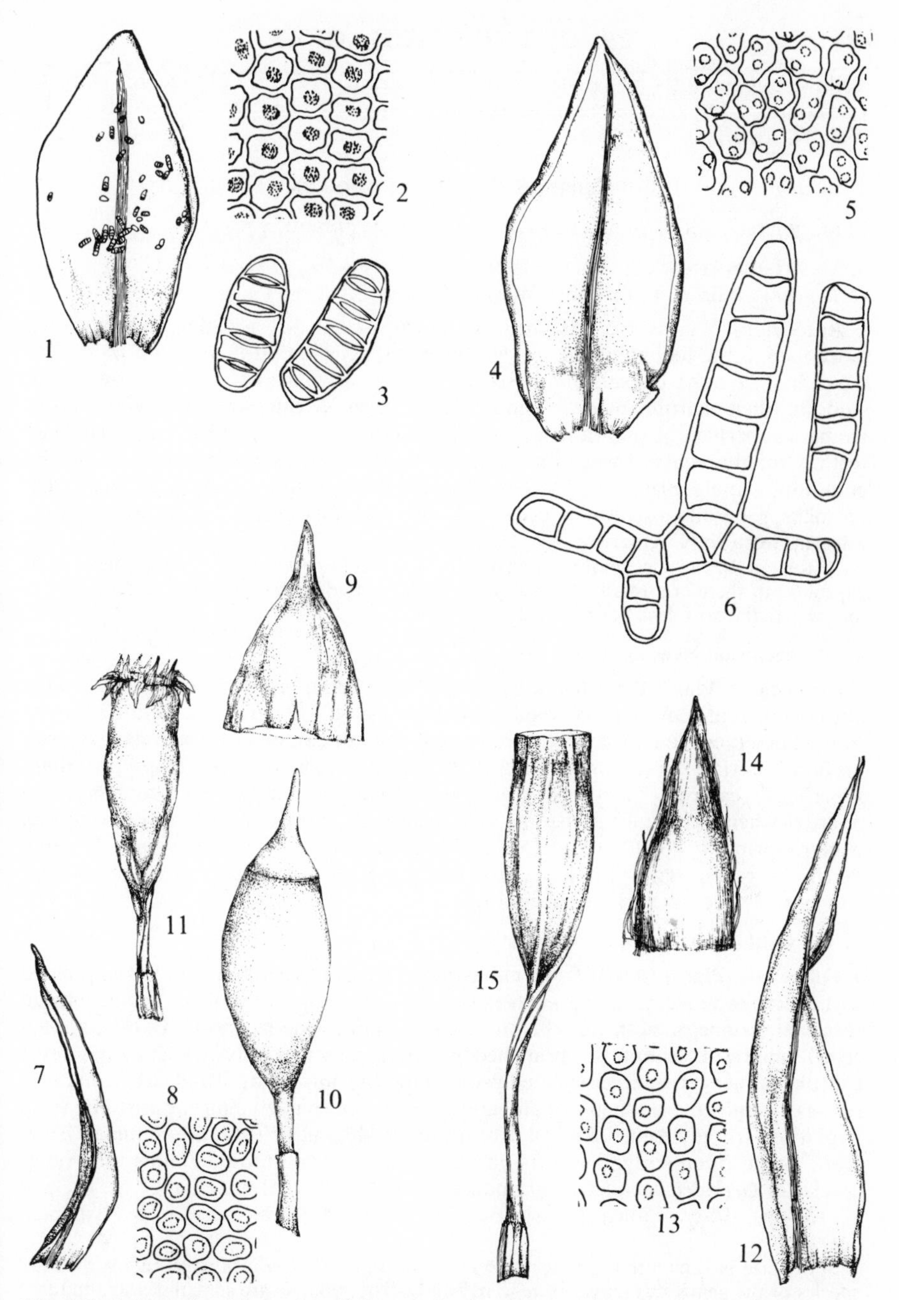

Fig. **229**. 1–3, *Orthotrichum obtusifolium*: 1, leaf (×25); 2, mid-leaf cells; 3, gemmae. 4–6, *O. gymnostomum*: 4, leaf (×25); 5, mid-leaf cells; 6, gemmae. 7–11, *O. striatum*: 7, leaf (×15); 8, mid-leaf cells; 9, calyptra; 10 and 11, mature and dry empty capsules. 12–15, *O. speciosum*: 12 ,leaf (×15); 13, mid-leaf cells; 14, calyptra; 15, dry empty capsule. Calyptras and capsules ×15, cells × 415, gemmae × 250.

Yellowish-green tufts or patches on bark, very rare. There are old records from 13 vice-counties, but seen in recent years only in Angus. C. and N. Europe, Siberia, C. Asia, N. America.

Subgenus II: PHANEROPORUM

Lf margin recurved; stomata superficial; peristome present. $x=6$.

3. O. striatum Hedw., Sp. Musc., 1801
O. leiocarpum Br. Eur., *O. shawii* Wils. ex Schimp.

Autoecious. Plants 0.5–3.0 cm. Lvs erecto-patent to spreading when moist, lanceolate, acute to acuminate, margin recurved; cells in mid-lf *ca* 10 μm wide. Seta 0.5–1.5(–2.2) mm; capsule immersed to emergent, ellipsoid, tapering into seta, smooth when mature, pale, ± smooth and never sulcate when dry and empty; stomata superficial; calyptra hairy; peristome double, outer of 16 strongly papillose teeth, strongly recurved when dry, inner of 16 erect processes; spores *ca* 30 μm. Fr common, winter, spring. $n=6$*. Usually small, dark green tufts on bark, very rarely on rocks, generally distributed, occasional. 80, H23, C. Europe, Caucasus, Kashmir, Algeria, western N. America.

Distinct in the capsules smooth when dry and empty. *O. speciosum* may have some smooth capsules but there are usually also sulcate capsules present in the same tuft and the seta is longer with the capsules often exserted.

4. O. speciosum Nees in Sturm, Deutschl. Fl., 1819

Autoecious. Plants 0.5–4.0 cm. Lvs spreading when moist, lanceolate to narrowly lanceolate, acute, margin recurved; cells in mid-lf 10–12 μm wide. Seta *ca* 2 mm; capsule emergent to exserted, ellipsoid to sub-cylindrical, tapering into seta, smooth or faintly striate at maturity, smooth to sulcate when dry and empty; stomata superficial; calyptra hairy; peristome teeth papillose, united in pairs, erect to reflexed when dry; inner peristome of 8 processes; spores 34–36 μm. Fr common, winter, spring. $n=6$, 12, 12+1. Yellowish-green tufts on bark, very rare. Stirling, Perth, Aberdeen, Banff, Elgin, E. Inverness and formerly from Sussex, N.E. Yorks, Angus and Kincardine. 12. Circumboreal.

5. O. lyellii Hook. & Tayl., Musc. Brit., 1818

Dioecious. Plants (0.5–)1.0–4.0 cm, shoots often curved when dry. Lvs appressed to flexuose when dry, spreading when moist, narrowly lanceolate, acute, margin recurved on one or both sides below; cells 8–10 μm wide in mid-lf. Two- to many-celled uniseriate, simple or branched gemmae present on lvs. Seta *ca* 1.4 mm; capsule immersed to emergent, ellipsoid, tapering into seta, striate when mature, narrower and sulcate when dry and empty; calyptra sparsely hairy; peristome teeth papillose, strongly recurved when dry; spores 26–40 μm. Fr very rare, summer. $n=6$. Dark green, often straggling tufts on bark, often in less humid habitats than other species of *Orthotrichum*, frequent and generally distributed. 98, H26, C. N., W. and C. Europe, Italy, Caucasus, Algeria, Morocco, Madeira, western N. America, Mexico.

Spore size is sometimes very variable and as fruiting *O. lyellii* often occurs with other species of the genus this may be due to hybridity. The gemmae are sometimes so abundant as to give the lvs a dusty, brownish appearance.

6. O. affine Brid., Musc. Rec., 1801

Autoecious. Plants usually 0.5–2.5 cm. Lvs erecto-patent to spreading when moist, broadly lanceolate to lanceolate, acute to obtuse and apiculate, margin recurved; cells

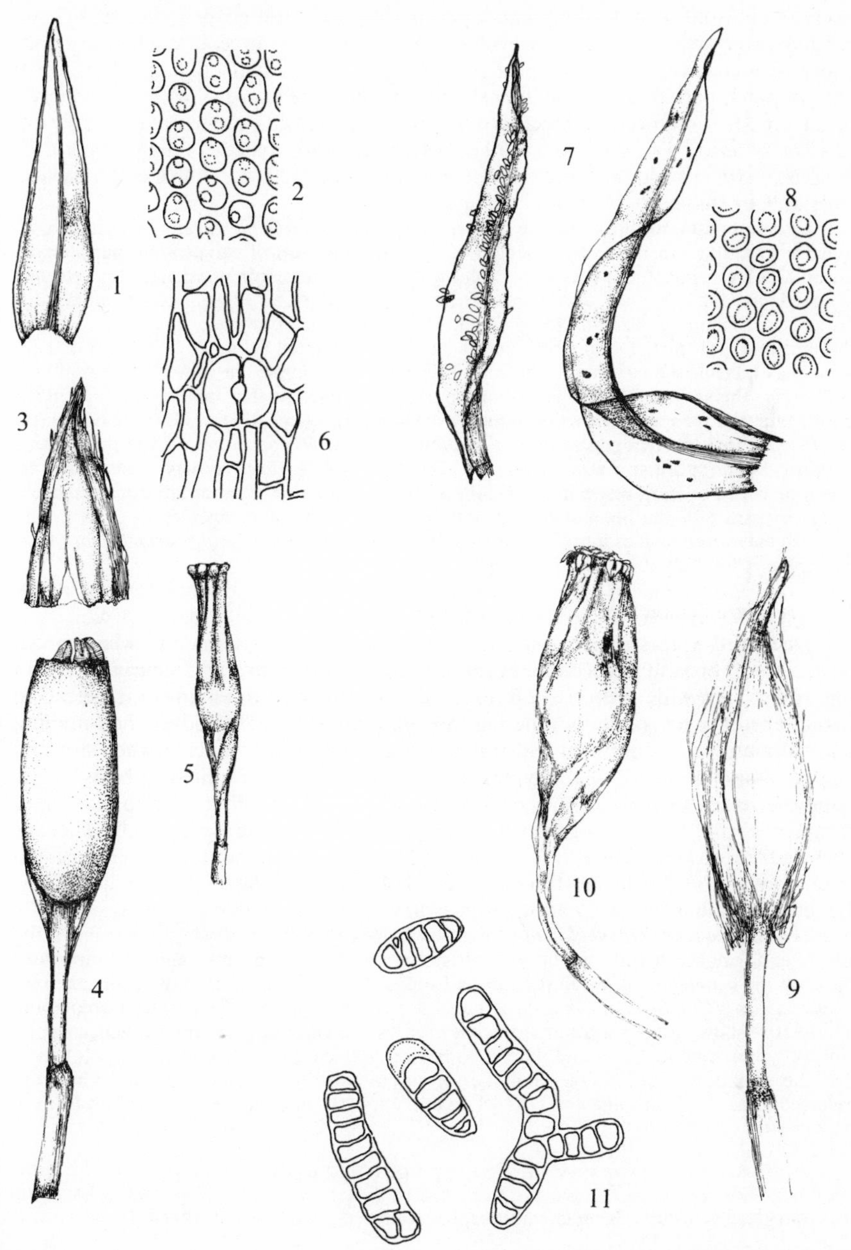

Fig. **230**. 1–6, *Orthotrichum affine*: 1, leaf; 2, mid-leaf cells; 3, calyptra; 4 and 5, mature and dry empty capsules; 6, stoma. 7–11, *O. lyellii*: 7, leaves; 8, mid-leaf cells; 9, calyptra; 10, dry empty capsule; 11, gemmae (×165). Leaves, calyptras and capsules ×15, cells ×415.

in mid-lf 6–10 μm wide. Seta 0.4–1.2 mm; capsule greenish, emergent to slightly exserted, ellipsoid to sub-cylindrical, tapering into seta, striate when mature, sulcate and narrower and of ± uniform width throughout or ± urceolate when dry and empty; stomata superficial; calyptra greenish, glabrous to sparsely hairy; peristome teeth in pairs, papillose, sharply reflexed at least in some capsules when dry; spores 18–24 μm. Fr common, summer. $n=6$*. Dark green tufts on bark, especially of *Sambucus*, *Salix*, *Fraxinus* and *Betula* in humid, shady places and by pools and streams, very occasionally on rock, common. 110, H39, C. Europe, Caucasus, Siberia, Kamchatka, N. Africa, N. America.

The commonest and most variable of corticolous species of *Orthotrichum* in Britain and Ireland. The shape of the capsule, length of seta and hairiness of calyptra are particularly variable. A form, shorter and more compact with shorter, more immersed capsules, referred to in Dixon (1924) as var. *fastigiatum* (Brid.) Hüben., does not appear to be sufficiently distinct in Britain to warrant recognition. In Scandinavia, however, this form is given specific status (*O. fastigiatum* Brid.) and it has been suggested (Nyholm, 1954–69) that *O. affine* may have arisen as a hybrid between this and *O. speciosum*. In view of the variability of *O. affine* this is a possibility. The situation clearly requires further investigation. *O. affine* is distinguished from most other corticolous species of the genus by the superficial stomata. *O. striatum* may be recognised by the smooth capsules, *O. speciosum* by the more hairy calyptra, larger spore size and longer seta. *O. affine* and *O. rupestre* may occasionally be found on rock and bark respectively. *O. affine* may be recognised by the glabrous or sparsely hairy calyptra and the peristome teeth reflexed at least in some capsules when dry. The greenish calyptra provides a useful field character for separation from *O. stramineum* which has a pale yellow calyptra with a dark apex.

7. O. rupestre Schleich. ex Schwaegr., Suppl. I., 1811

Autoecious. Plants mostly 1–4 cm. Lvs erecto-patent to spreading when moist, lanceolate to broadly lanceolate, acute to obtuse, margin strongly recurved, cells in mid-lf 8–10 μm wide. Seta 0.8–2.0 mm; capsule immersed to emergent, ellipsoid to ovoid, abruptly to gradually tapering into seta, faintly striate above when mature, narrower and of ± uniform width throughout and usually sulcate when dry and empty; stomata superficial; calyptra hairy; peristome teeth united in pairs, finely papillose, erect to spreading when dry; spores 14–20 μm. Fr common, winter to summer. $n=6$*, 12. Dark or brownish-green tufts on dry, exposed, usually siliceous rocks, rarely on trees. Rare in W. England, N. Wales and Ireland, rare to occasional from Lancs and Yorks northwards. 63, H14. Europe, Faroes, Iceland, Turkey, Caucasus, India, Algeria, Canaries, Madeira, western N. America.

Distinguished from *O. cupulatum* by the very hairy calyptra, the dry empty capsule being of ± uniform width throughout, the peristome teeth fused in pairs and the superficial stomata. *O. anomalum* differs in the exserted capsules and longer seta. May be mistaken for muticous forms of *Schistidium* species but differs in capsule and lf shape and areolation. *O. rupestre* shows some variation in thickness of lvs and capsule shape (plants with capsules abruptly narrowed into seta and thicker lvs having been called var. *sturmii* (Hornsch.) Jur.), but the variation is continuous and the forms are probably habitat variants. The plant referred to as var. *franzonianum* (De Not.) Vent. appears merely to be a corticolous form of *O. rupestre*.

Subgenus III: Rivularium

Lf margin recurved; lf apex obtuse to rounded; stomata immersed; peristome present. $x=10+1, 11$.

8. O. rivulare Turn., Musc. Hib., 1804

Autoecious. Plants 1–3 cm, shoots sometimes denuded below. Lvs erecto-patent

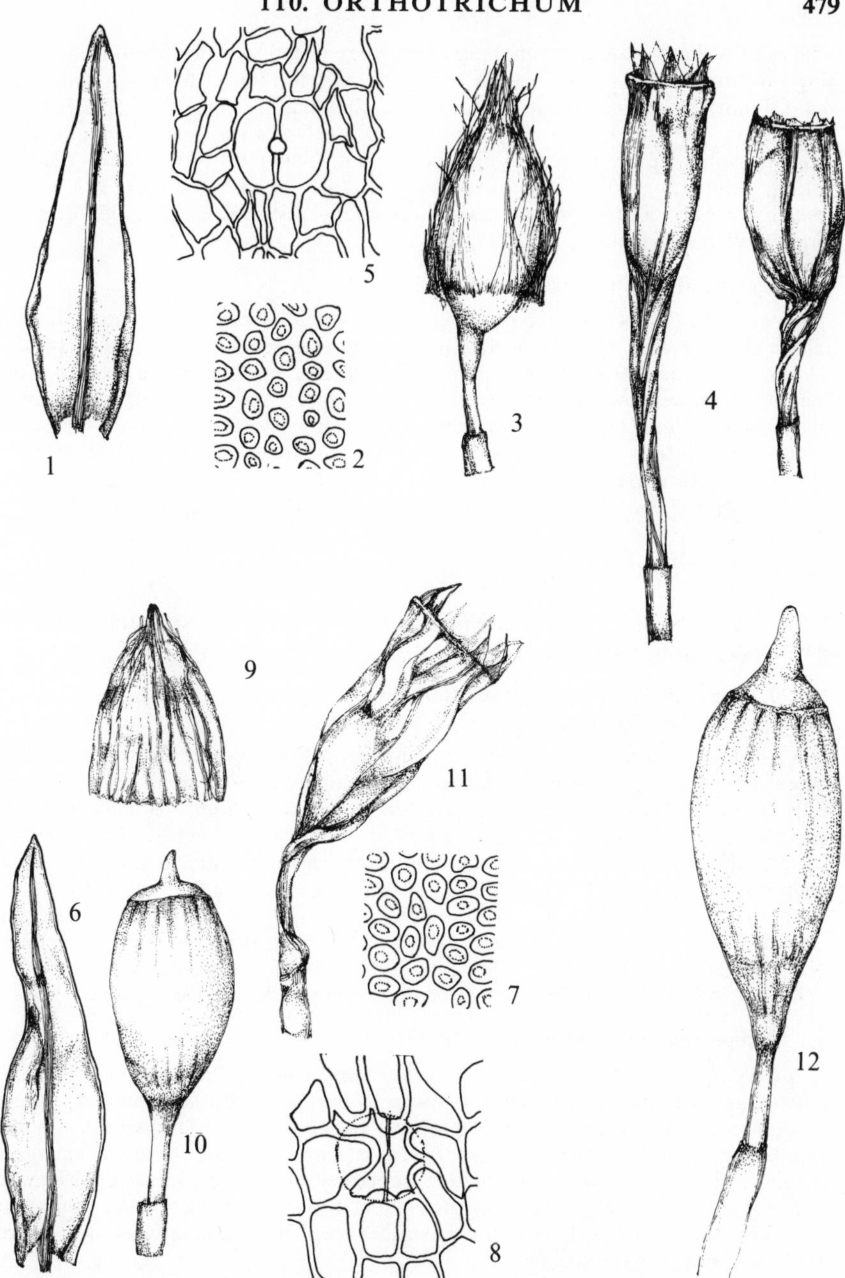

Fig. **231**. 1–5, *Orthotrichum rupestre*: 1, leaf; 2, mid-leaf cells; 3, calyptra; 4, dry empty capsules; 5, stoma. 6–11, *O. cupulatum* var. *cupulatum*: 6, leaf; 7, mid-leaf cells; 8, stoma; 9, calyptra; 10 and 11, mature and dry empty capsules. 12, *O. cupulatum* var. *riparium*: mature capsule. Leaves, calyptras and capsules ×15, cells ×415.

when moist, lanceolate to ovate-lanceolate, apex broad, obtuse to rounded, sometimes toothed or cucullate, margin recurved; cells in mid-lf 10–14 μm wide. Seta 0.4–1.5 mm; capsule emergent, ovoid, striate when mature, narrower, urceolate and sulcate when dry and empty; stomata immersed; calyptra glabrous; peristome teeth fused in pairs, papillose, erect to reflexed when dry; spores 14–20 μm. Fr common, spring, summer. $n=10+1$*, 11. Lax, dark green tufts on bark and rocks below flood level by streams and rivers, occasional, generally distributed. 67, H13. Belgium, France, western Germany, N. America.

9. O. sprucei Mont. in Spruce, Lond. J. Bot., 1845

Autoecious. Plants 0.5–1.0 cm. Lvs erecto-patent when moist, lingulate to lanceolate or oblanceolate, apex rounded, entire, margin recurved; cells in mid-lf (14–)16–24 μm wide. Seta 0.6–0.8 mm; capsule emergent, ovate-ellipsoid, striate when mature, narrower and sulcate when dry and empty; stomata immersed; calyptra glabrous; peristome teeth papillose, reflexed when dry; spores 14–16 μm. Fr frequent, summer. Small tufts, often encrusted with alluvial matter on bark by streams, rivers and ditches. Rare to occasional in scattered localities from E. Cornwall and W. Kent north to Stirling and Mid Perth, very rare in Ireland. 41, H4. Belgium, France.

Subgenus IV: Orthotrichum

Lvs ± straight when dry, margins recurved, apex acute; stomata immersed; peristome present. $x=10+1$, 11.

10. O. anomalum Hedw., Sp. Musc., 1801
O. anomalum var. *saxatile* auct.

Autoecious. Plants 0.5–5.0 cm, stems not conspicuously tomentose. Lvs erecto-patent to spreading when moist, lanceolate, acute, margin recurved at least below; cells in mid-lf 8–10 μm wide. Seta reddish, (1.5–)2.0–4.0 mm, vaginula glabrous; capsule reddish, exserted, ovoid to ellipsoid, striate when mature, narrower and sulcate when dry and empty; stomata immersed; calyptra sparsely hairy; peristome teeth not united in pairs, longitudinally striate, erect to spreading when dry; spores 10–14 μm. Fr common, spring, early summer. $n=10+1$*, $20+2$. Dark green tufts on usually basic rocks and walls, very rarely on trees, frequent. 110, H40, C. Europe, N., W. and C. Asia, Morocco, Algeria, N. and C. America, Greenland.

Readily recognised in the field by the reddish sporophyte and exserted capsules.

11. O. cupulatum Brid., Musc. Rec., 1801

Autoecious. Plants 0.5–2.0 cm, stems not conspicuously tomentose. Lvs erecto-patent to spreading when moist, lanceolate to broadly lanceolate, acute to obtuse, margin recurved, cells in mid-lf 8–10 μm wide. Seta pale green to brown, 1.0–1.6 mm, vaginula glabrous or with a few sparse hairs; capsule straw-coloured, emergent, ovoid to ellipsoid, smooth when mature, narrowed, urceolate and sulcate when dry and empty; stomata immersed; calyptra glabrous or very sparsely hairy; peristome teeth not fused in pairs, longitudinally striate, erect or spreading when dry; spores 14–16 μm. Fr common, spring.

Var. **cupulatum**

Capsule ovoid to ellipsoid, ± abruptly narrowed into seta; calyptra sparsely hairy; pre-peristome poorly developed. $n=10+1$*, 11. Dark green to blackish tufts on usually basic rocks and walls, very rarely on trees. Occasional to frequent in basic

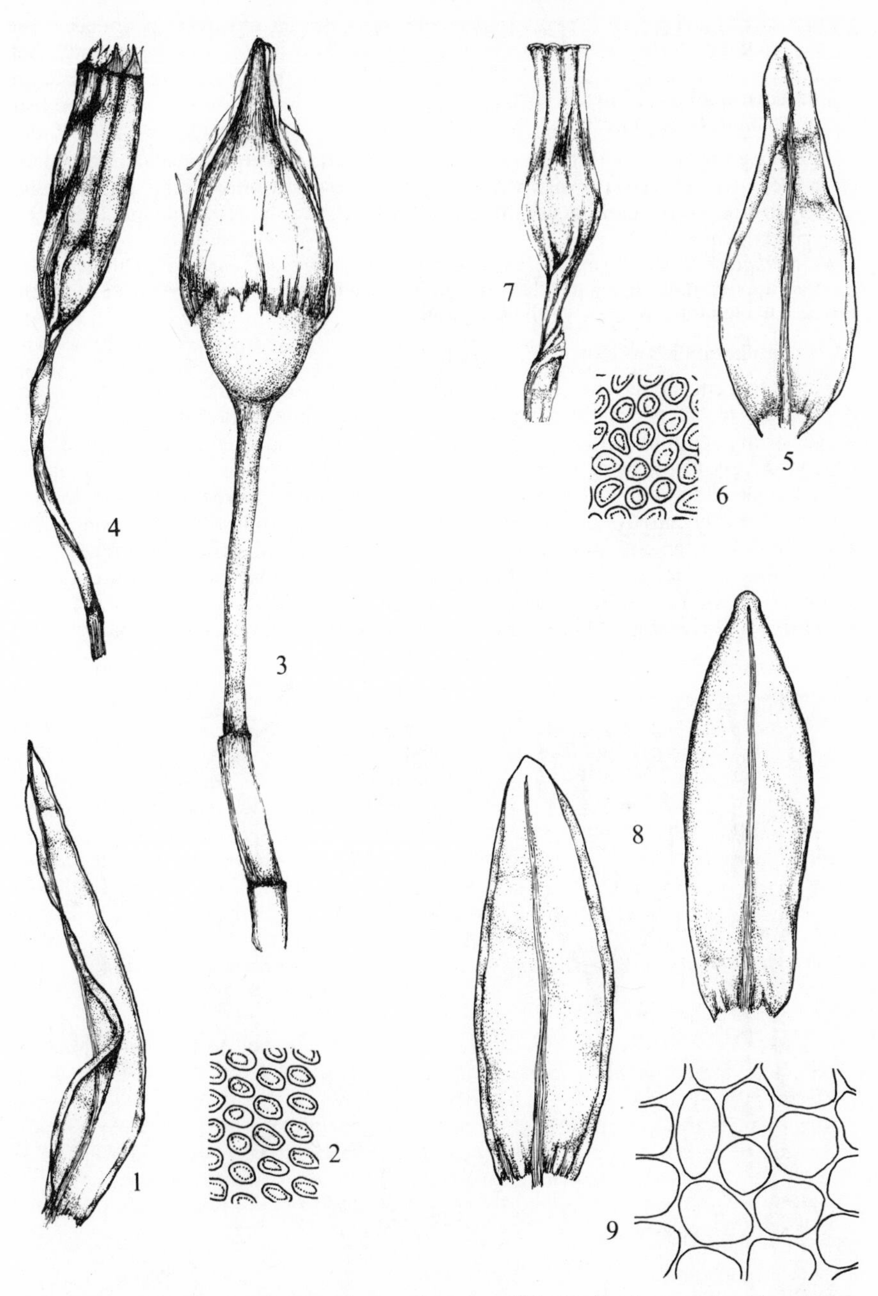

Fig. **232**. 1–4, *Orthotrichum anomalum*: 1, leaf; 2, mid-leaf cells; 3, capsule with calyptra; 4, dry empty capsule. 5–7, *O. rivulare*: 5, leaf; 6, mid-leaf cells; 7, dry empty capsule. 8–9, *O. sprucei*: 8, leaves; 9, mid-leaf cells. Leaves and capsules × 15, cells × 415.

areas, rare elsewhere. 88, H24. Europe, Turkey, N., W. and C. Asia, Algeria, Morocco, N. America.

Var. **riparium** Hüb., Musc. Germ., 1833
Var. *nudum* (Dicks.) Braithw.

Capsule ellipsoid, tapering into seta, less immersed; calyptra glabrous; peristome teeth more strongly striate with well developed pre-peristome. $n=11$. Basic rocks, usually in streams or pools, generally distributed but rare. 44, H11. Europe, U.S.S.R., Algeria, N. America.

O. cupulatum shows a similar range in capsule shape to *O. rupestre*. Although var. *riparium* appears more distinctive than the parallel form in *O. rupestre*, it is likewise probably merely a habitat form and not worth distinguishing.

12. O. urnigerum Myr., Coroll. Fl. Upsal., 1833

Plants to 2 cm, stems with tomentum of reddish-brown rhizoids. Lvs ± erect when dry, patent to erect-spreading when moist, ± lanceolate, acute, margin recurved; nerve strong, ending below apex; basal cells rectangular, cells above quadrate-hexagonal, 10–12 μm wide in mid-lf. Seta *ca* 2 mm long, vaginula with long hairs; capsule pale, erect and emergent or inclined and laterally emergent, ovate-ellipsoid, striate, abruptly narrowed into seta, narrower, sulcate and somewhat contracted above middle when dry and empty; stomata immersed; calyptra with a few coarse hairs; peristome teeth ± erect when dry, strongly papillose below, striate above; spores *ca* 14 μm. Fr common, summer. Lax tufts, green above, reddish-brown below, very rare. W. Inverness. 1. N., W. and C. Europe, Caucasus, C. Asia, Sinkiang.

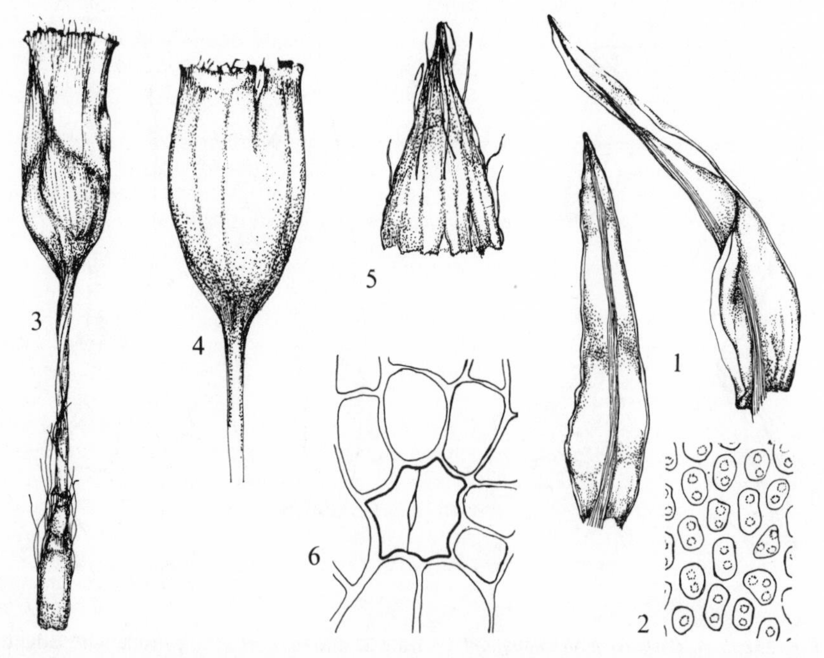

Fig. **233**. *Orthotrichum urnigerum*: 1, leaves; 2, mid-leaf cells; 3 and 4, dry and moist capsules; 5, calyptra; 6, stoma. Leaves and capsules ×15, cells ×415.

This species usually occurs on shaded basic or acidic rocks but in the only known British locality it is unusual in growing on bark. The plant is distinctive in the reddish-brown tomentum, the longly hairy vaginula and the erect or spreading peristome teeth (normally corticolous species have reflexed teeth and normally saxicolous species a ± glabrous vaginula). For an account of this plant in Britain see Appleyard, *J. Bryol.* **9**, 159–60, 1976.

13. O. schimperi Hammar, Mon. Orthotr. Sp. Suec., 1852

Autoecious. Plants *ca* 0.5 cm. Lvs erecto-patent when moist, lanceolate, acute to obtuse, margin recurved; cells in mid-lf 10–14 μm wide. Seta 0.5 mm, vaginula glabrous; capsule emergent, ellipsoid, tapering into seta, striate when mature, narrower and sulcate when dry and empty; stomata immersed, obscured by overlying exothecial cells; calyptra smooth, glabrous; peristome teeth fused in pairs, papillose, reflexed when dry; spores 12–16 μm. Fr common, spring. $n=11$. Small, tight, pale green tufts on bark, very rare. S. Devon, W. Suffolk, E. Norfolk, Northants, Angus, Kincardine. 6. Europe, Caucasus, N. Asia, Japan, Algeria, Morocco.

Very close to *O. pumilum* Sw. and distinguished on characters considered by Dixon (1924) as unimportant. Nyholm (1954–69) suggests it is probably a drier habitat form of *O. pumilum*. Distinguished by its small size, compact habit and smooth calyptra.

14. O. stramineum Hornsch. ex Brid., Br. Univ., 1827

Autoecious. Plants 0.5–1.0 cm. Lvs erecto-patent when moist, lanceolate, acute, margin recurved, cells in mid-lf 8–12 μm wide. Seta 0.7–1.0(–1.5) mm, vaginula with numerous long hairs; capsule emergent, ellipsoid, gradually tapering into seta, striate when mature, narrower, often markedly constricted below mouth and sulcate when dry and empty, ribs of 4–5 rows thick-walled cells; stomata restricted to lower ½ of capsule, immersed, partially obscured by overlying exothecial cells; calyptra yellowish with dark apex, plicate, glabrous or sparsely hairy; peristome teeth ± erect to reflexed when dry; spores 12–14 μm. Fr common, summer. Light or yellowish-green tufts on bark, rare in southern England, occasional to frequent elsewhere. 80, H8. Europe, Asia Minor, N. Africa.

The capsule markedly constricted below mouth when dry and empty and the dark-tipped calyptra provide useful field characters for identifying this species, although very old capsules are of uniform width throughout. The hairy vaginula, partially obscured guard cells and the shape of the empty capsule will separate *O. stramineum* from *O. pallens*.

O. patens Bruch ex Brid. (as *O. stramineum* var. *patens* (Brid.) Vent.) has been reported from 4 localities in Britain, but all the specimens I have seen are *O. affine*.

15. O. pallens Bruch ex Brid., Br. Univ., 1827

Autoecious. Plants 0.5–1.0 cm. Lvs erecto-patent when moist, lanceolate, acute to obtuse, margin recurved; cells in mid-lf 12–18 μm wide. Seta 0.5–0.8 mm, vaginula without hairs; capsule emergent, ellipsoid, gradually tapering into seta, striate when mature, narrower and of ± uniform width throughout when dry and empty; stomata immersed, hardly obscured by surrounding exothecial cells; calyptra plicate, glabrous; peristome teeth yellow or orange, papillose, erect to reflexed when dry; spores 10–16 μm. Fr common, summer. Small tufts on bark, rare, in a few localities from Essex and Derby north to E. Ross, rare in Ireland. 11, H6. Europe, Caucasus, N. America.

16. O. tenellum Bruch ex Brid., Br. Univ., 1827

Autoecious. Plants to 0.5 cm. Lvs appressed, straight when dry, erecto-patent to spreading when moist, lanceolate to oblong-lanceolate, acute to obtuse or obtuse and apiculate, margin recurved, cells in mid-lf 10–12(–14) μm wide. Seta 0.5–1.0(–1.6) mm,

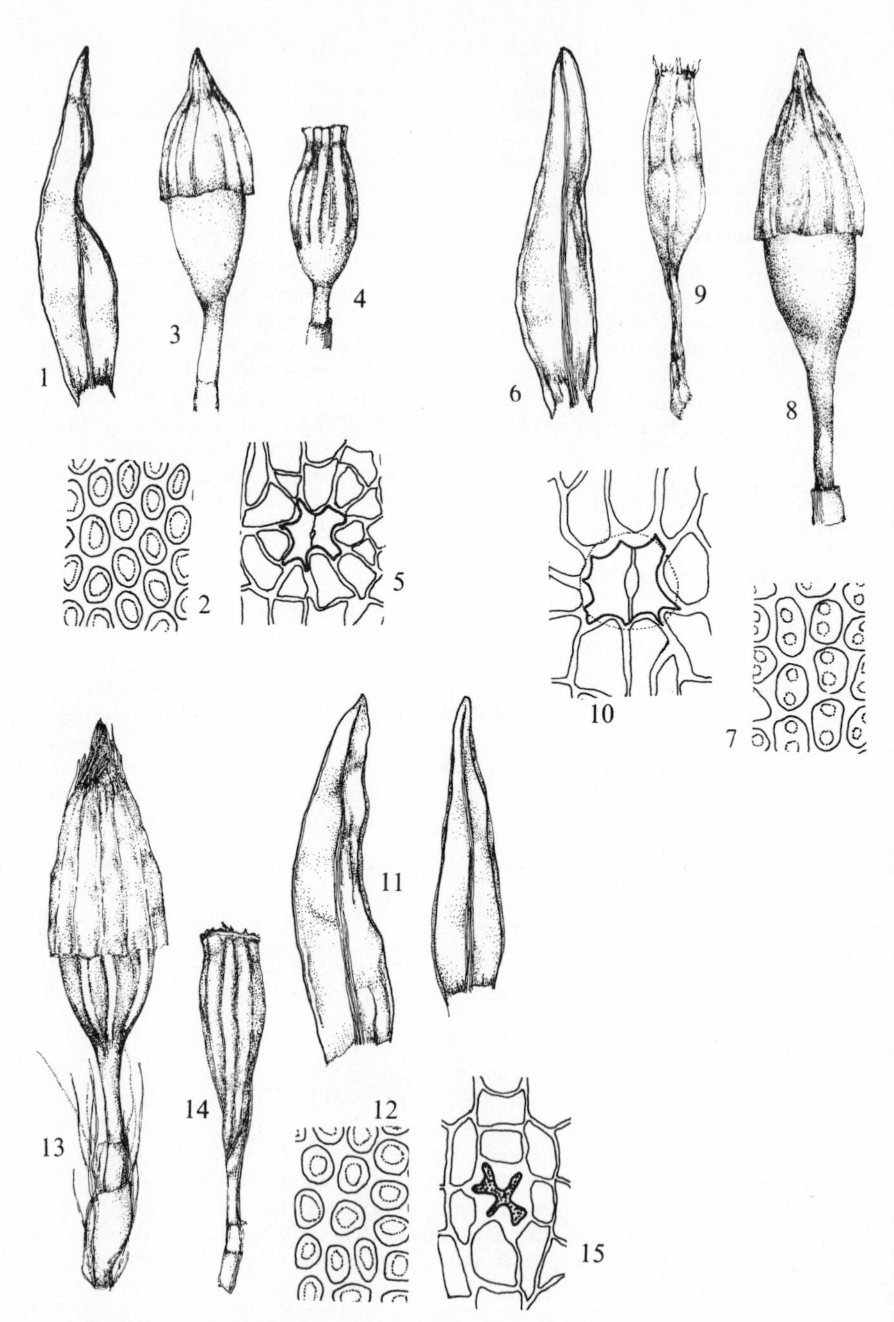

Fig. **234**. 1–5, *Orthotrichum schimperi*: 1, leaf; 2, mid-leaf cells; 3, capsule with calyptra; 4, dry empty capsule; 5, stoma. 6–10, *O. pallens*: 6, leaf; 7, mid-leaf cells; 8, capsule with calyptra; 9, dry empty capsule; 10, stoma. 11–15, *O. stramineum*: 11, leaves; 12, mid-leaf cells; 13, capsule with calyptra; 14, dry empty capsule; 15, stoma. Leaves and capsules ×**15**, cells ×**415**.

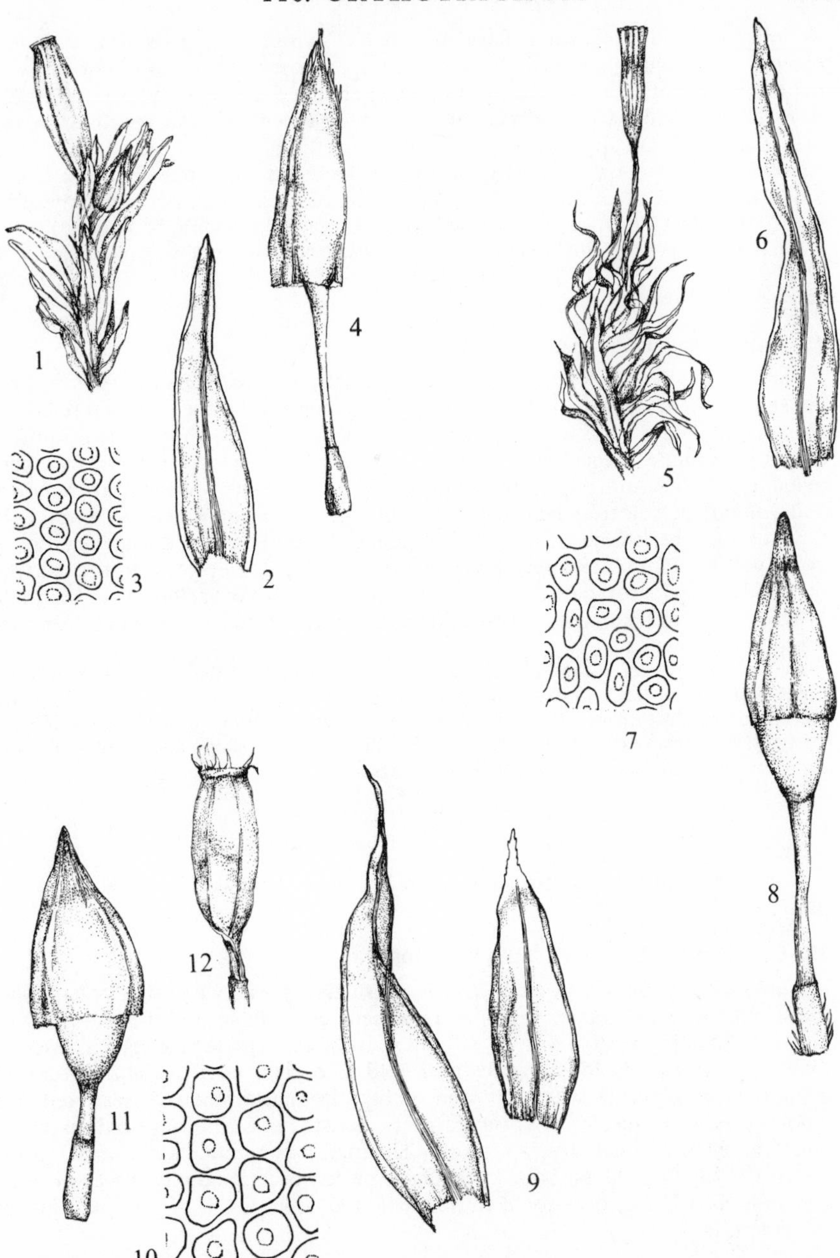

Fig. **235**. 1–4, *Orthotrichum tenellum*: 1, dry shoot (×10); 2, leaf; 3, mid-leaf cells; 4, capsule with calyptra. 5–8, *O. pulchellum*: 5, dry shoot (×10); 6, leaf; 7, mid-leaf cells; 8, capsule with calyptra. 9–12, *O. diaphanum*: 9, leaves; 10, mid-leaf cells; 11, capsule with calyptra; 12, dry empty capsule. Leaves and capsules ×15, cells ×415.

vaginula glabrous or with a few short hairs; capsule orange-brown, emergent, narrowly ellipsoid to sub-cylindrical, ± abruptly narrowed into seta, striate when mature, narrower, of ± uniform width throughout and sulcate when dry and empty; stomata immersed, partially obscured by overlying exothecial cells, restricted to neck of capsule; calyptra sparsely hairy; peristome teeth light brown, papillose, erect to reflexed when dry; spores 10–12, 14–18 μm. Fr common, summer. $n=10+1$, 11. Small tufts, usually on *Sambucus*, *Salix*, *Fraxinus*, occasional and generally distributed. 74, H22, C. Europe, N. Asia, Morocco, Canaries, Madeira, N. America.

Similar in appearance to small forms of *O. affine* but recognisable in the field by the orange-brown capsules. *O. pulchellum* differs in the lvs flexuose when dry, the exserted capsule and the orange peristome teeth.

17. O. diaphanum Brid., Musc. Rec., 1801

Autoecious. Plants to *ca* 1 cm. Lvs soft, slightly twisted when dry, patent when moist, ovate-lanceolate with acuminate, hyaline, denticulate apex, margin recurved; cells in mid-lf 10–16 μm wide, rounded-hexagonal. Two- to several-celled simple or branched gemmae sometimes present on lvs. Seta 0.4–0.6 mm; capsule emergent, ovoid to ellipsoid, abruptly narrowed into seta, faintly striate when mature, narrower and smooth or sulcate when dry and empty; stomata immersed; calyptra hairy towards apex; peristome teeth coarsely papillose, whitish, recurved when dry; spores 14–18 μm. Fr common, winter, spring. $n=10+1$*. Small, grey-green tufts or patches on wood, damp fences, bark, rocks and stones and detritus of various types, common. 112, H38, C. Europe, Faroes, Siberia, Algeria, Madeira, Canaries, N. and C. America, southern S. America.

Occurs on a greater variety of substrata than other species of the genus and may sometimes be found on such unlikely materials as old linoleum, corrugated iron and old tar-macadam. Distinct from other British species of *Orthotrichum* in the hyaline lf apex. Distinguished from species of *Schistidium* by the soft lvs, lf shape and areolation, hairy calyptra, striate capsule and the pale, double peristome.

Subgenus V: Pulchellum

Lvs crisped when dry, margin recurved; apex acute; stomata immersed; peristome present. $x=10+1$, 11.

18. O. pulchellum Brunton in Sm., Eng. Bot., 1807

Autoecious. Plants to *ca* 1 cm. Lvs flexuose and curved when dry, patent when moist, narrowly lanceolate, acute, margin recurved; cells in mid-lf 8–12 μm wide. Seta 1.2–2.0 mm, vaginula with or without short hairs; capsule pale brown, exserted, ovoid to ovate-ellipsoid, abruptly narrowed into seta, striate when mature, ± cylindrical and sulcate when dry and empty; stomata immersed, scattered over whole surface of capsule; calyptra glabrous; peristome teeth orange-red, papillose, erect to reflexed when dry; spores 16–22 μm. Fr common, late winter, spring. $n=10+1$*. Small tufts on bark in sheltered places, occasionally on rocks or walls, generally distributed, occasional to frequent. 100, H30, C. N., W. and C. Europe, N. America.

111. Ulota Mohr in Brid., Musc. Rec., Suppl., 1819

Autoecious, rarely dioecious. Marginal branches of tufts often decumbent and sometimes creeping. Lvs linear to linear-lanceolate, usually curved or crisped when dry, erecto-patent to spreading when moist, margin plane or recurved, entire; nerve ending in or below apex; cells incrassate, basal often narrow or vermicular except towards margin, cells above ± rounded, papillose. Vaginula glabrous or hairy, ochrea small or absent; capsule exserted, with long neck, 8-striate, usually narrowed and sulcate when dry and empty; calyptra hairy, peristome double or inner lacking. About 65 species mainly in temperate parts of the globe.

Species frequently grow mixed together, this often causing difficulty with identification. It is possible that hybridisation occurs between *U. drummondii* and *U. crispa*, as occasional gametophytes occur with sporophytes of an intermediate nature. As with *Orthotrichum* the species of *Ulota* are susceptible to the effects of atmospheric pollution.

1 Nerve excurrent, upper lvs with fusiform gemmae at apices, dioecious, fr very rare **6. U. phyllantha**
Nerve ending in or below apex, gemmae absent, autoecious, fr frequent 2
2 Dry empty capsules inflated, whitish, smooth except immediately below mouth **1. U. coarctata**
Dry empty capsules fawn to brownish, narrowed, strongly furrowed 3
3 Lvs stiff, imbricate and ± straight when dry, cells very heavily incrassate with small lumens, saxicolous plants **5. U. hutchinsiae**
Lvs variously curved or crisped when dry, cells incrassate but lumens not very small, plants usually but not exclusively corticolous 4
4 Calyptra with a few sparse hairs, lvs strongly crisped when dry, basal part of lvs plicate, band of rectangular basal cells extending up margin **4. U. calvescens**
Calyptra very hairy, lvs curled to strongly crisped when dry, not plicate in basal part, band of rectangular cells not ascending up margin 5
5 Lvs curled when dry, capsule with numerous stomata in lower part (10–15 usually visible in flattened capsule), peristome teeth whitish, erect to spreading when dry **2. U. drummondii**
Lvs moderately to strongly crisped when dry, capsule with few stomata (usually 1–6 visible in flattened capsule), peristome teeth pale brown, reflexed when dry **3. U. crispa**

1. U. coarctata (P. Beauv.) Hammar, Mon. Orth. Sp. Suec., 1852
U. ludwigii Brid.

Plants *ca* 1 cm. Lvs slightly twisted when dry, patent when moist, lanceolate, basal part ± inflated, concave, apex acute to obtuse, margin plane or recurved; nerve ending below apex; basal cells narrowly rectangular to linear, towards margin quadrate-rectangular, cells above ovate or rounded, 10–12 μm wide in mid-lf. Seta 4.0–4.5 mm; capsule 2.0–2.5 mm long, narrowly pyriform, very pale brown, smooth except immediately below small mouth, when dry and empty whitish, inflated, smooth except immediately below very small mouth; calyptra hairy; peristome teeth whitish, erect when dry; spores finely papillose, 18–24 μm. Fr common, autumn. $n=9+1$. Small, dull green tufts on small trees and bushes by streams and in wet places, very rare except in parts of western Scotland. S. Somerset, S. Hants, W. Sussex, E. Norfolk, Merioneth and Selkirk, Angus and Argyll north to Ross, Kerry. 16, H2. Europe, N. America.

2. U. drummondii (Hook. & Grev.) Brid., Br. Univ., 1827

Plants 0.5–1.0 cm high, marginal branches creeping. Lvs curved when dry, erecto-patent when moist, narrowly lanceolate to lanceolate, basal part sometimes expanded but not plicate, apex acute to acuminate, margin plane or recurved below; nerve

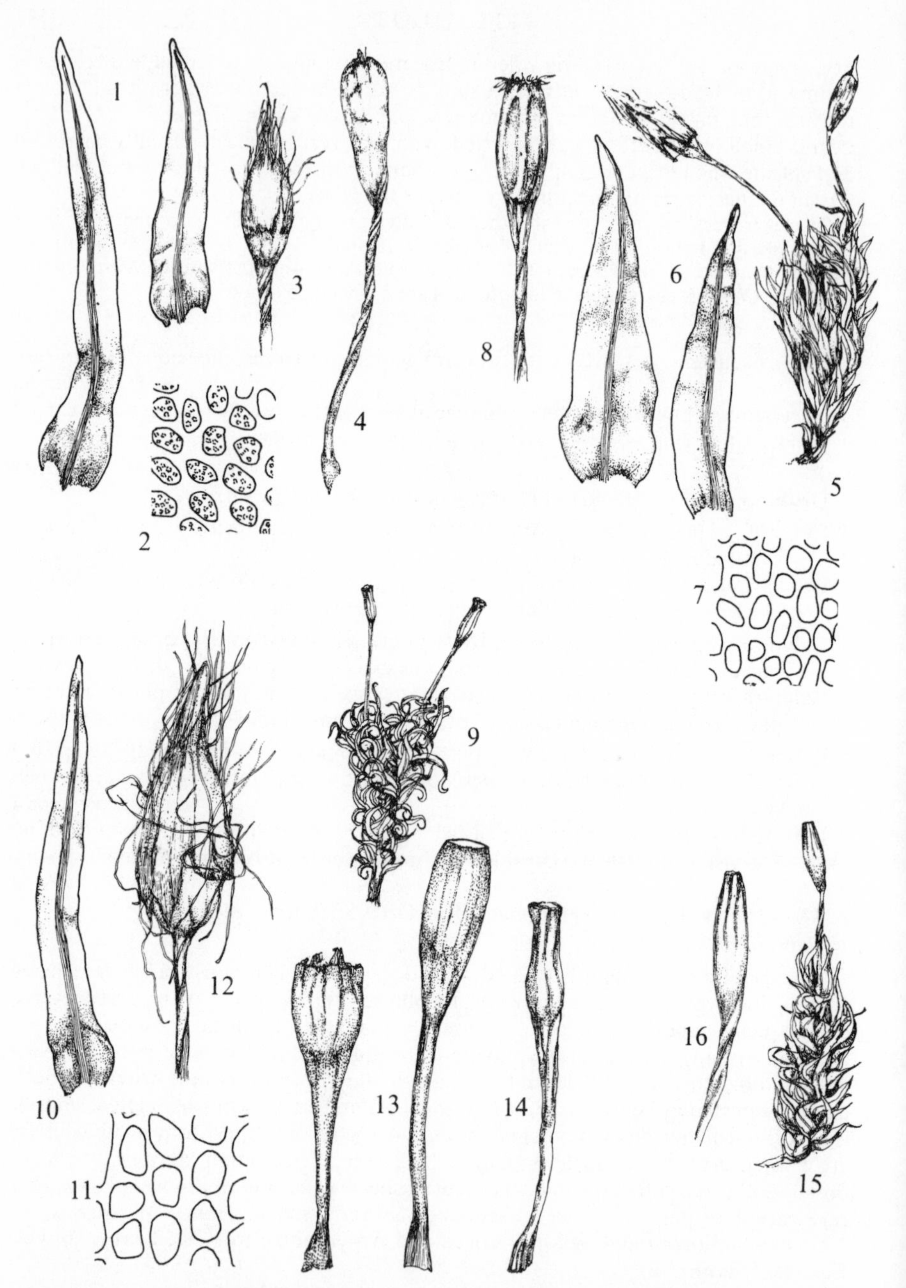

Fig. **236**. 1–4, *Ulota coarctata*: 1, leaves; 2, mid-leaf cells; 3, young capsule; 4, dry empty capsule. 5–8, *U. drummondii*: 5, dry shoot; 6, leaves; 7, mid-leaf cells; 8, dry empty capsule. 9–14, *U. crispa* var. *crispa*: 9, dry shoot; 10, leaf; 11, mid-leaf cells; 12, young capsule; 13 and 14, moist and dry empty capsules. 15–16, *U. crispa* var. *norvegica*: 15, dry shoot; 16, dry empty capsule. Shoots ×5, leaves ×15, cells ×415, capsules ×10.

ending below apex; basal cells linear, vermicular, 4–10 marginal rows shorter, ± rectangular, cells above rhomboidal to rounded, papillose, 8–10(–14) μm wide in mid-lf. Seta 3–5 mm; capsule 2–3 mm, pale brown, ovoid to ellipsoid with long neck, striate at maturity, when dry and empty narrowly ellipsoid to cylindrical, narrowed at mouth and sulcate; stomata *ca* 40, 28–36 μm diameter; peristome teeth white, erect to spreading when dry; spores 18–24 μm. Fr common, autumn. Tufts or spreading patches, often tinged with red, on branches of trees and shrubs in humid situations. Rare in N. Wales and N. England, occasional to frequent in Scotland and Ireland. 43, H10. N., W. and C. Europe, Japan, N. America.

May be confused with forms of *U. crispa* but differs in the lvs which are often wider and curved, not crisped when dry, and in the white peristome teeth and numerous stomata. The reddish tinge and creeping branches are also often characteristic.

3. U. crispa (Hedw.) Brid., Musc. Rec., Suppl., 1819
U. nicholsonii Culmann, *U. crispula* Brid.

Plants mostly 0.5–2.0 cm. Lvs moderately to strongly crisped when dry, patent to spreading when moist, lanceolate to linear-lanceolate, tapering to acute to acuminate apex, basal part concave, not plicate, margin plane or recurved below; nerve ending in or below apex; basal cells narrowly rectangular to vermicular, 4–10 marginal rows rectangular, cells above ovate or rounded, 8–14 μm wide in mid-lf. Capsule ellipsoid, abruptly narrowed to longly tapering into seta, striate at maturity, sulcate when dry and empty; stomata 4–15 in number, 24–28 μm diameter; peristome teeth brownish, reflexed when dry; spores 20–26 μm.

Var. **crispa**

Lvs strongly crisped when dry, upper lvs 2.0–3.5 mm long, cells in mid-lf 8–11 μm wide. Seta 2–4(–5) mm; capsule 1–3 mm, total length of sporophyte 3.0–6.0(–7.5) mm, capsule abruptly narrowed to longly tapering into seta, dry empty capsule wide-mouthed, narrowly urceolate to shortly cylindrical; stomata 24–42 μm wide. $n=10+1$*, $10+2$*, $19+2$, $20+2$*. Fr common, throughout the year but most frequently late summer and autumn. Small yellowish-green tufts or patches on branches of trees and shrubs, especially *Corylus*, *Fraxinus*, *Salix* and *Sambucus* in damp or humid open woodland and by water, rarely on rocks. Occasional in S.E. England and the Midlands, frequent to common elsewhere. 109, H40, C. Europe, Caucasus, Asia Minor, Amur, Sachalin, Canaries, N. America, Tasmania.

Var. **norvegica** (Grönvall) Smith & Hill, J. Bryol, 1975
U. bruchii Hornsch. ex Brid.

Lvs moderately crisped when dry, upper lvs 2.6–4.0 mm long; cells in mid-lf (9–)10–14 μm wide. Seta (3.0–)3.5–5.5 mm; capsule 2.0–3.2 mm, total length of sporophyte 5–8 mm, capsule tapering into seta, when dry and empty narrowed at the mouth, fusiform; stomata 32–48 μm wide. Fr common, autumn, winter. $n=9+1$, $19+1$*, $20+2$*. Loose, yellowish-green tufts or patches, sometimes tinged with red, in similar habitats to var. *crispa* but less frequent especially in drier areas. 92, H31, C. Europe.

For the reasons for reducing *U. crispula* to synonomy with *U. crispa* var. *crispa* and *U. bruchii* to a variety of *U. crispa*, see Smith & Hill, *J. Bryol.* **8**, 423–33, 1975. All N. American specimens I have seen named *U. bruchii* have proved to be *U. curvifolia* (Wg.) Brid.

4. U. calvescens Wils. in Rabenh., Bryoth. Eur., 1862
U. vittata Mitt.

Plants 0.5–1.5 cm. Lvs strongly crisped when dry, erecto-patent when moist,

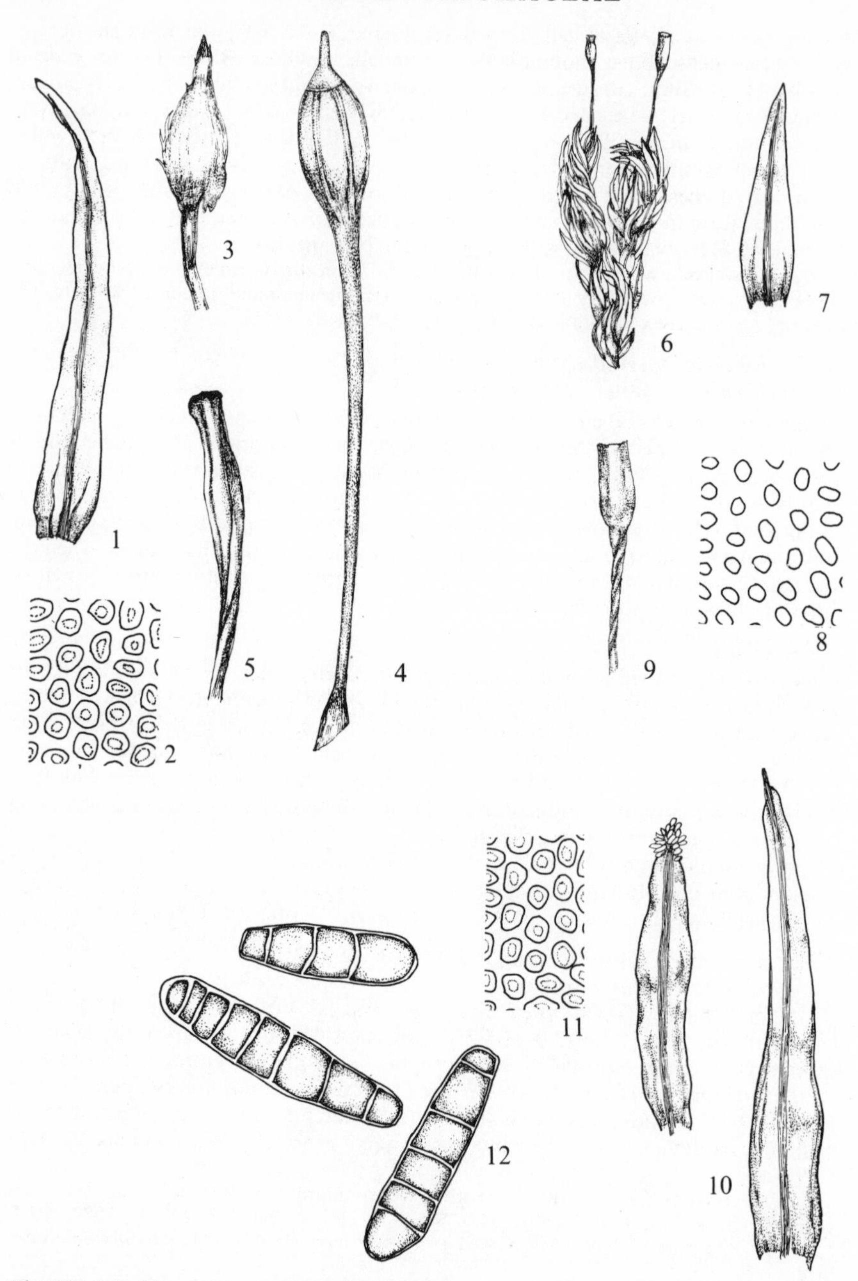

Fig. **237**. 1–5, *Ulota calvescens*: 1, leaf; 2, mid-leaf cells; 3 and 4, young and mature capsules; 5, dry empty capsule. 6–9, *Ulota hutchinsiae*: 6, plant ($\times 5$); 7, leaf; 8, mid-leaf cells; 9, dry empty capsule. 10–12, *Ulota phyllantha*: 10, leaves; 11, mid-leaf cells; 12, gemmae ($\times 250$). Leaves $\times 15$, cells $\times 415$, capsules $\times 10$.

narrowly lanceolate to linear-lanceolate, basal part with 2 plicae, not inflated, apex acute to acuminate, margin plane or recurved; nerve ending below apex; basal cells narrowly rectangular to linear, towards margin shorter and wider, quadrate-rectangular, a band of rectangular cells extending some way up margin, cells above rounded-quadrate, papillose, 8–10 μm wide in mid-lf. Seta 4–6 mm; capsule 1.5–2.5 mm, narrowly pyriform, striate at maturity, cylindrical and sulcate when dry and empty; calyptra glossy with a few sparse hairs; spores 22–26 μm. Fr common, autumn. Small tufts on branches of trees and shrubs, especially *Betula*, *Corylus* and *Sorbus*, often in calcareous habitats. Occasional and sometimes locally common in W. Scotland and W. Ireland, Merioneth. 13, H14. Portugal, Spain, Madeira, Canaries.

Recognisable in the field by the sparsely hairy calyptra and the capsule short in relation to the length of the seta when compared with other species of *Ulota*. Sterile plants may be recognised from the basal areolation and plicae of the lvs.

5. U. **hutchinsiae** (Sm.) Hammar, Mon. Orthotr. Suec., 1852
U. americana (P. Beauv.) Schimp.

Plants 1–2(–3) cm. Lvs rigid, imbricate, straight or slightly curved when dry, erecto-patent when moist, lanceolate, basal part not concave, apex acute to obtuse, margin recurved; nerve very strong, reddish or brownish; basal cells narrowly rectangular to linear, upper cells rounded, sometimes very heavily incrassate, papillose, 8–10 μm wide in mid-lf. Seta 3.5–4.0 mm; capsule *ca* 2 mm long, clavate, striate at maturity, when dry and empty ellipsoid to sub-cylindrical, hardly contracted at mouth, sulcate; calyptra hairy; peristome teeth light brown, spreading when dry; spores coarsely papillose, 16–18 μm. Fr common, summer. Dark green to brownish tufts or patches on rocks in and by streams, very rarely on trees. Very rare in S.W. England, rare in N. Wales, occasional from Westmorland and Cumberland northwards and in Ireland. 36, H20. Europe, Faroes, Japan, N. America.

6. U. **phyllantha** Brid., Musc. Rec., Suppl., 1819

Dioecious. Plants 0.5–2.0 cm. Lvs strongly incurved and crisped when dry, erecto-patent when moist, variable in shape, narrowly lanceolate to linear-lanceolate or ligulate, basal part sometimes slightly enlarged, not plicate, apex variable, often obtuse and apiculate or mis-shapen or malformed, margin plane or recurved below; nerve excurrent; basal cells narrowly rectangular, 4–10 marginal rows shorter, rectangular, cells above elliptical to rounded, 6–10 μm wide in mid-lf. Clusters of fusiform, brownish gemmae borne on the tips of the uppermost lvs. Capsule ellipsoid; calyptra sparsely hairy. Fr very rare, autumn. Yellowish-green to brownish-green tufts, reddish to brown below, on rocks, walls, trees and wood, particularly near the coast. Rare in S.E. England and the Midlands, occasional to frequent elsewhere and often common along the W. coast of Britain. 86, H40, C. N. and W. Europe, Faroes, Iceland, southern Africa, N. America.

30. HEDWIGIACEAE

Dioecious or autoecious, rarely synoecious. Stems without central strand, branching irregular to pinnate. Lvs ± spreading, when dry imbricate, ovate or ovate-lanceolate, concave, sometimes longitudinally plicate; nerveless; cells papillose, angular enlarged or not. Archegonia terminal on main or lateral branches. Perichaetial lvs long, erect. Seta very short, vaginula hairy, ochrea lacking; capsule immersed to exserted, erect, asymmetrical; stomata superficial; peristome present or lacking; calyptra very small or large, cucullate or mitriform, never plicate. Six genera.

112. Hedwigia P. Beauv., Mag. Enc., 1804

Autoecious. Main stems without stolons, branching irregular. Lvs concave, not plicate, ovate-lanceolate, acuminate; cells papillose, linear, porose in centre of lf, ± quadrate-rounded towards apex and margin. Seta short; capsule immersed to emergent; peristome lacking; calyptra small, cucullate, fugacious. Cosmopolitan; 10 species.

Lvs with spinosely toothed hyaline apices, cells with branched papillae, perichaetial lvs with long, hyaline cilia **1. H. ciliata**
Stem and perichaetial lvs entire, without hyaline apices, cells with simple papillae **2. H. integrifolia**

1. H. ciliata (Hedw.) P. Beauv., Prodr., 1805

Shoots ± decumbent, irregularly branched. Lvs imbricate or sometimes falcato-secund, with spreading apices when dry, giving shoots a bristly appearance, patent,

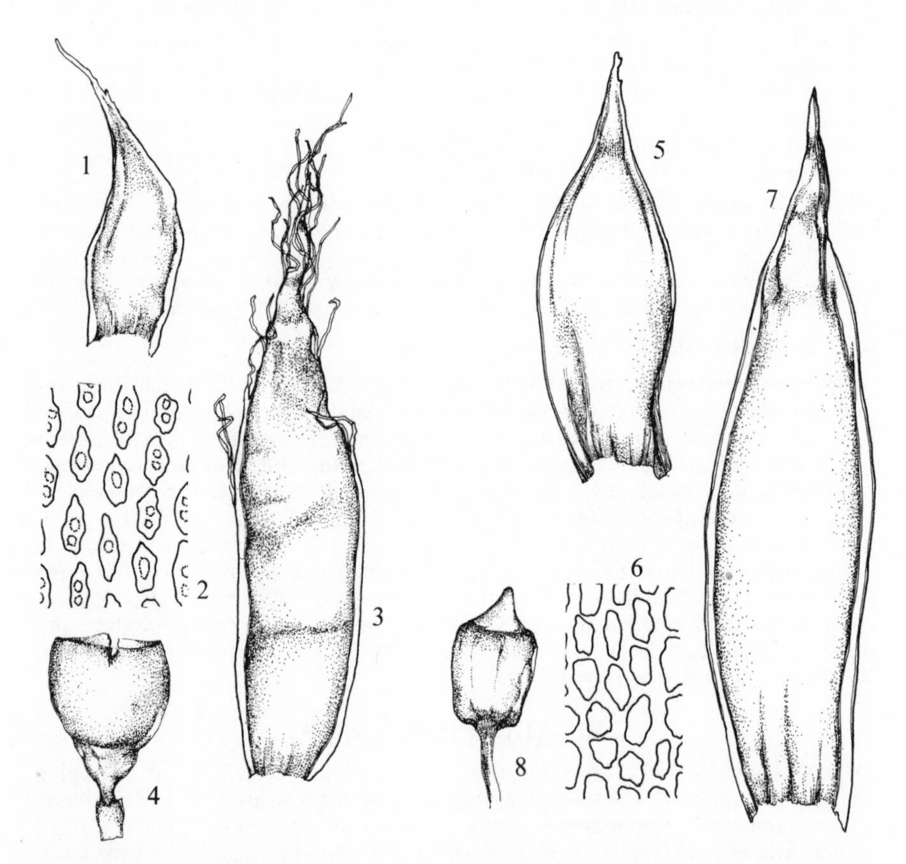

Fig. **238**. 1–4, *Hedwigia ciliata*: 1, leaf; 2, mid-leaf cells; 3, perichaetial leaf; 4, capsule. 5–8, *H. integrifolia*: 5, leaf; 6, mid-leaf cells; 7, perichaetial leaf; 8, capsule. Leaves ×25, cells ×415, capsules ×10.

sometimes sub-secund when moist, ovate, tapering to acuminate, spinosely toothed, hyaline apex, concave, margin recurved below; nerveless; cells incrassate throughout, coarsely papillose with branched papillae above, basal narrowly rectangular, extending up middle of lf, above and towards margin ovate or rounded-quadrate, 10–12 μm wide in mid-lf. Perichaetial lvs narrower, margins with long, hyaline cilia above. Capsule sub-sessile, obloid; spores 26–28 μm. Fr occasional to frequent, spring. $n=10$*, 10+1, 11, 21, 22. Hoary, yellowish-green to glaucous-green tufts or patches on dry, exposed, acidic rocks, walls and roofing slates. Rare in S.E. England and the Midlands, frequent to common elsewhere. 74, H34, C. Cosmopolitan.

2. H. integrifolia P. Beauv., Prodr., 1805
H. imberbis (Sm.) Spruce, *Hedwigidium integrifolium* (P. Beauv.) Dix.

Shoots yellowish-green above, brownish below, irregularly branched, sometimes with flagelliform branches with minute lvs. Lvs ± imbricate when dry, erecto-patent when moist, ovate, acute or acuminate, concave, margin strongly recurved from base to apex, entire; nerveless; cells incrassate, with simple papillae, basal cells narrowly rectangular and extending up lf, cells above and towards margin shorter, quadrate-rectangular or rounded-quadrate, sinuose, 6–9 μm wide in mid-lf. Perichaetial lvs longer and narrower, entire. Seta *ca* 1 mm; capsule scarcely emergent, obloid, faintly striate when dry and empty; spores *ca* 24 μm. Fr rare, early summer. Brownish or greenish patches on dry, exposed, acidic to slightly basic rocks at low to moderate altitudes. Frequent in N. Wales and the Lake District, occasional in W. Scotland, rare elsewhere, extending north to Skye and W. Ross. 19, H5. S. and W. Europe, Sri Lanka, Cameroon, E. Africa, Australia, Tasmania, New Zealand.

16. ISOBRYALES

Pleurocarpous. Primary stems usually creeping, irregularly branched, secondary stems procumbent to erect, branching irregular to pinnate or dendroid; paraphyllia present or not. Lvs in 3 or more ranks or complanate, branch lvs similar to or smaller than stem lvs, usually ovate to lanceolate, acute; nerve single, double or lacking; cells ± quadrate to linear, smooth or papillose. Seta long or short, straight; capsule immersed or exserted, erect or inclined, ovoid to cylindrical; peristome double, teeth of outer usually papillose or horizontally striate, inner usually well developed; calyptra cucullate or mitriform. Chromosome number usually $n=10+1$ or 11.

A heterogeneous order composed mainly of families of doubtful affinities. Those such as the Cryphaeaceae, Leucodontaceae and Thamniaceae may be related to members of the Orthotrichales, but the relationship of others such as the Fontinalaceae and Climaciaceae are obscure.

31. FONTINALACEAE

Dioecious or synoecious. Slender to robust, aquatic or amphibious plants; stems irregularly or pinnately branched. Lvs ± 3-ranked, ovate to lanceolate, tapering to acute apex, plane, concave or keeled, margin plane; nerve single or absent; cells narrow, marginal sometimes forming ill-defined border, cells at basal angles ± inflated, sometimes forming auricles. Perichaetial lvs convolute, forming tubular perichaetium. Capsule immersed or exserted, ovoid to cylindrical; inner peristome teeth and cilia united by transverse strands into conical lattice; calyptra cucullate or mitriform.

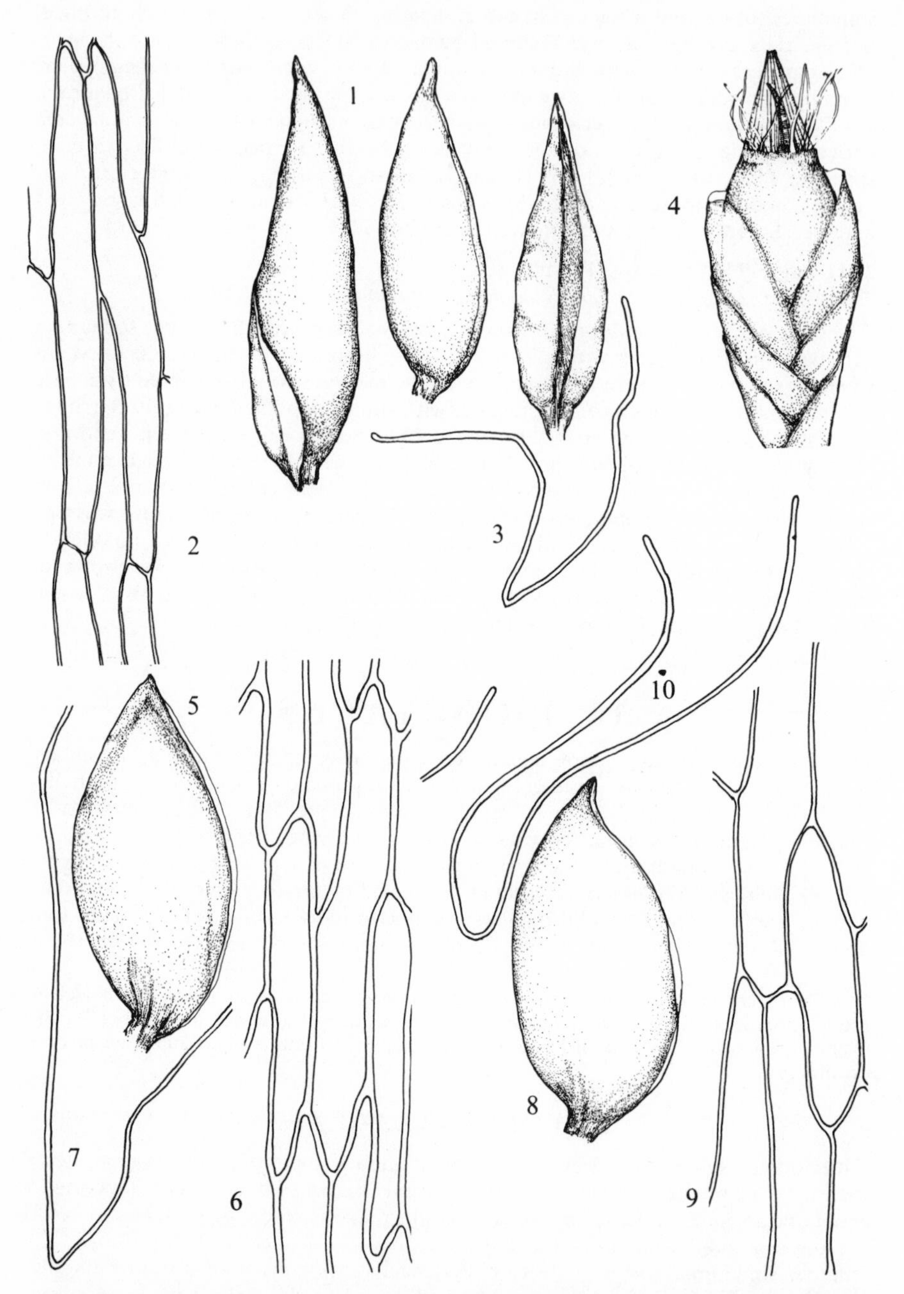

Fig. **239**. 1–4, *Fontinalis antipyretica* var. *antipyretica*: 1, leaves from main stem; 2, mid-leaf cells; 3, transverse section of leaf; 4, capsule (×10). 5–7, *F. antipyretica* var. *gigantea*: 5, leaf from main stem; 6, mid-leaf cells; 7, section of leaf. 8–10, *F. antipyretica* var. *cymbifolia*: 8, leaf from main stem; 9, mid-leaf cells; 10, section of leaf. Leaves ×10, cells ×415, sections ×65.

113. FONTINALIS HEDW., SP. MUSC., 1801

Plants usually aquatic, robust; stems long and irregularly branched. Stem and branch lvs of similar shape, keeled, concave or plane, ovate or lanceolate, acute, margin plane, entire; nerveless; angular cells usually enlarged and forming ± distinct auricles, cells above linear. Seta very short, straight; capsule immersed or emergent; peristome double, outer teeth entire or perforated, teeth of inner filiform, united into latticed cone by transverse bars; calyptra minute, smooth. About 20 mainly north temperate species.

At least lvs of main stems keeled or if not keeled then deeply channelled and widely spreading **1. F. antipyretica**
Lvs concave, not keeled or deeply channelled **2. F. squamosa**

1. F. antipyretica Hedw., Sp. Musc., 1801
F. dolosa Cardot

Plants to 80 cm, stems flexuose, irregularly branched. Lvs 3-ranked, imbricate to spreading and often giving shoots a triquetrous appearance, soft or rigid, keeled or deeply channelled, ovate to ± orbicular when flattened, scimitar to half-moon shaped in side view, apex acute to ± rounded, margin plane or reflexed near base, entire or obscurely denticulate towards apex; nerveless; cells thin-walled, enlarged, quadrate-hexagonal at basal angles and sometimes forming distinct auricles, elsewhere linear-rhomboid to linear, 10–19 μm wide in mid-lf. Capsule ± immersed; lid with stout conical beak; outer peristome teeth bright red; spores 10–12 μm. Fr rare, spring, summer. Green or dark green to golden-brown glossy or dull tufts on wet or submerged rocks and tree bases and roots.

Key to intraspecific taxa of *F. antipyretica*

1 Lvs distant, widely spreading, deeply channelled but not keeled var. **cymbifolia**
Lvs distant or not, imbricate to spreading, keeled at least on main stems 2
2 Lvs from mid part of main stem 2–3 times as long as wide when folded in half var. **gigantea**
Lvs from mid part of main stem 3–7 times as long as wide when folded in half 3
3 Lvs on main stem mostly 3.5–5.5 mm long, 2.0–2.6 mm wide (when flattened), mid-lf cells 12–19 μm wide var. **antipyretica**
Lvs mostly 2.4–3.6 mm long, 0.9–1.4 mm wide (when flattened), mid-lf cells 10–12 μm wide var. **gracilis**

Var. **antipyretica**
Stem and branch lvs keeled or apparently conduplicate when folded in half, lvs from mid part of main stems (3.0–)3.5–5.5(–8.0) mm long, 2.0–2.6(–4.0) mm wide (½ this width if folded), broadly ovate (lanceolate when folded), apex acute; cells 12–19 μm wide in mid-lf. Tufts on rocks, tree boles and exposed roots in and by rivers, streams and ponds, submerged or subject to submergence, and in seepage areas on cliffs and in quarries, common. 112, H40, C. Europe, Faroes, Iceland, N. and W. Asia, Japan, N. Africa, La Palma, Azores, Madeira, N. America, Greenland.

Var. **gracilis** (Hedw.) Schimp., Syn., 2, 1876
Plants to 40 cm. Lvs from mid-part of main stem 2.4–3.6(–4.0) mm × 0.9–1.4(–2.2) mm (½ that width when folded, 2–3 times as long as wide (twice these dimensions if folded in half), keeled, lanceolate, acute, narrowly lanceolate when folded; cells linear, mostly 10–12 μm wide in mid-lf. Usually on rocks or stones in fast-flowing

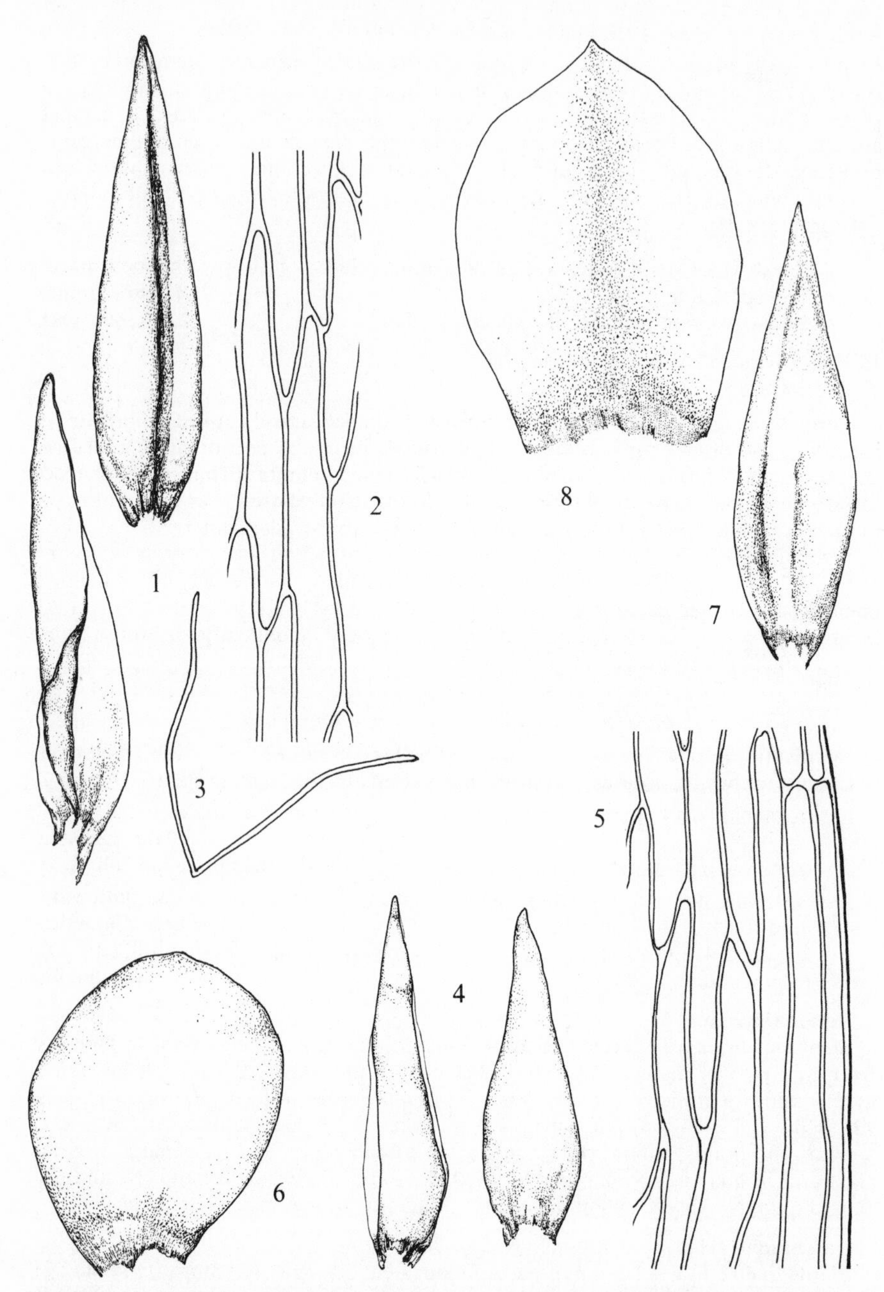

Fig. **240**. 1–3, *Fontinalis antipyretica* var. *gracilis*: 1, leaves; 2, mid-leaf cells; 3, leaf section (×65). 4–6, *F. squamosa* var. *squamosa*: 4, leaves; 5, cells from widest part of leaf; 6, perichaetial leaf. 7–8, *F. squamosa* var. *curnowii*: 7, leaf; 8, perichaetial leaf. Leaves ×10, cells ×415.

montane streams, rarely in ponds and lowland habitats. Generally distributed but rare. 45, H7. Europe, Siberia, Japan, Korea.

Var. **gigantea** (Sull.) Sull., Icon. Musc., 1864
Plants to 150 cm, shoots usually conspicuously triquetrous. Lvs from mid-part of main stems 3.6–7.0(–8.0) mm × 3.2–6.0 mm (½ that width when folded), mostly 1.0–1.5 times as long as wide (twice these dimensions if folded), keeled, ovate-lanceolate to broadly ovate when folded in half, obtuse; cells in mid-lf 15–18 μm wide. Rocks in streams, rivers and waterfalls, generally distributed but rare. 22, H11. W. and C. Europe, Algeria, N. America.

Var. **cymbifolia** Nicholson, J. Bot., Lond., 1901
Lvs distant, widely spreading, deeply channelled but not keeled, lvs from middle part of main stem 3.0–5.5 mm × 1.5–4.5 mm, 1.0–2.2 times as long as wide, broadly ovate in side view. Dark green tufts in sluggishly flowing water. A few localities in England, Clare. 9, H1. Endemic.

A species subject to considerable morphological variation. Vars. *cymbifolia*, *gigantea* and *gracilis* are of uncertain nature but may be more than mere habitat variants as they are not necessarily confined to one type of habitat. Thus var. *gracilis* may be found in fast-flowing montane streams and also montane pools. Welch (1960) considers vars. *gigantea* and *gracilis* worthy of recognition but reduces var. *cymbifolia* to synonomy with *F. antipyretica* – but it is a much more distinctive looking plant than the other varieties. *F. dolosa* Cardot reported from Lembury, Beds, appears merely to be a form of var. *antipyretica* and is not recognised here.

2. F. squamosa Hedw., Sp. Musc., 1801

Plants to 40 cm, stems flexuose or rigid, often fasciculately branched, frequently denuded below. Stem lvs similar to or larger than branch lvs, ± imbricate or sometimes patent, concave, ovate-lanceolate, tapering to acute to obtuse apex, base decurrent, margin plane or inflexed, entire or slightly denticulate towards apex; nerveless; cells ± incrassate, 1–2(–3) rows marginal cells narrower, more incrassate, yellowish, sometimes forming slight border, cells at basal angles inflated, forming ± distinct decurrent auricles. Capsule immersed. Glossy tufts in fast-flowing streams and rivers.

Key to intraspecific taxa of *F. squamosa*

1 Plants brownish, stem and branch lvs of similar size, lf base and auricles yellowish to brown — 2
Young parts of plants reddish-brown, stem lvs larger and wider than branch lvs, auricles orange-brown — var. **dixonii**
2 Perichaetial lvs with obtuse or truncate apices — var. **squamosa**
Perichaetial lvs with apiculate apices — var. **curnowii**

Var. **squamosa**
Stem and branch lvs not differing markedly in size, 2.0–4.2 mm long, base of similar colour to rest of lf or brownish; auricles yellowish to brownish. Perichaetial lvs with rounded or truncate apices. Spores 20–28 μm. Fr rare, spring. $n=11$*. Glossy, brownish tufts on rocks and stones in streams and rivers. S.W. England, Wales, Shrops and Derby northwards, frequent. 62, H21.

Var. **curnowii** Cardot, Mon. Font., 1892
Similar to var. *squamosa* but perichaetial lvs with apiculate apex, spores 18–22 μm. $n=10$*. Stones in streams, very rare. Cornwall, S. Devon, Cheshire. 4. France, Norway.

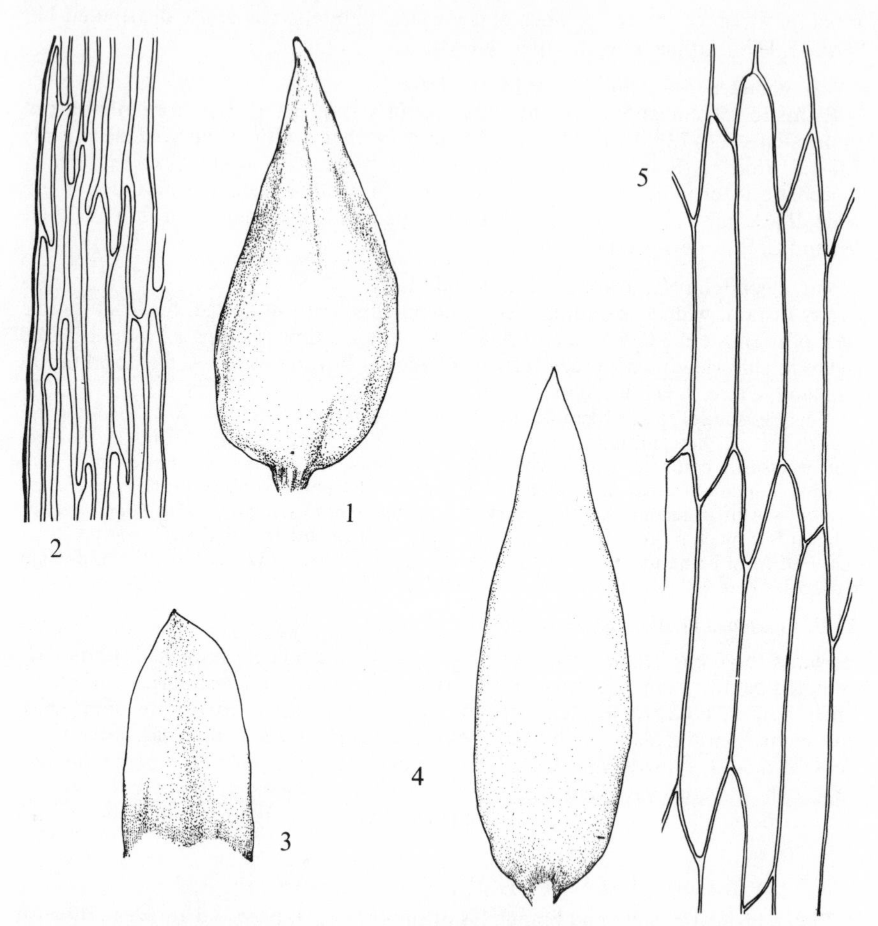

Fig. **241**. 1–3, *Fontinalis dalecarlica*: 1, leaf; 2, marginal cells from widest part of leaf; 3, perichaetial leaf. 4–5, *F. hypnoides*: 4, leaf; 5, mid-leaf cells. Leaves × 15, cells × 415.

Var. **dixonii** (Cardot) A. J. E. Smith, J. Bryol., 1976

F. dixonii Cardot, *F. squamosa* ssp. *dixonii* (Cardot) Dix.

Younger branches often with distinct reddish tint, older parts reddish-brown or golden-brown. Stem lvs 3.5–4.5 mm long, larger than branch lvs, branch lvs 2.5–3.5 mm long; lf base and auricles orange-brown. Fr unknown. Reddish-brown tufts in fast-flowing streams and rivers, rare. Merioneth, Caerns, Derby, Cumberland, W. Inverness, Hebrides, Fermanagh. 7, H1. Not known outside Britain and Ireland.

Welch (1960) reduces *F. dixonii* to synonomy with *F. squamosa* but, whilst in Wales occasional intermediates are found, the plant is distinctive in appearance and is at least as good as the varieties of *F. antipyretica*. Slender or depauperate forms of *F. squamosa* may be mistaken for *F. dalecarlica* or *F. hypnoides*.

F. dalecarlica Br. Eur. has been confirmed from several localities in Britain and Ireland by Welch (1960) but careful examination of the specimens concerned revealed that at least a few lvs had the border of *F. squamosa* and reports of this species from Britain and Ireland cannot be accepted. Typical plants of *F. dalecarlica* are up to 90 cm long with slender, flexuose stems; the lvs are similar in shape to those of *F. squamosa* but with 3–4 rows of marginal cells narrower and of similar colour to the other cells; the perichaetial lvs taper to an apiculate apex. Small forms of the 2 species may be green and where the lvs lack the border of *F. squamosa* the specimens cannot be named in the absence of fruit. However small greenish plants with lvs mostly of an indeterminate nature but perichaetial lvs of *F. squamosa* occur in Britain. The incorrectly named specimens of *F. squamosa* are of this type. *F. hypnoides* Hartm. has been reported from the River Wye but the specimens are *F. squamosa* (see Smith, *J. Bryol.* **9**, 279–80, 1977).

32. CLIMACIACEAE

Dioecious. Primary stems rhizomatous with scale lvs, secondary stems robust, erect, with scale lvs and central strand, usually branched in a dendroid fashion. Branches and upper parts of secondary stems with normal lvs. Lvs plicate; nerve single, ending below apex; cells smooth, rhomboidal above, rhomboidal to linear, porose towards base, alar cells hyaline. Perichaetial lvs elongated, nerveless, appressed. Seta long, twisted, vaginula cylindrical, usually glabrous; capsule erect or arcuate; stomata superficial; annulus not differentiated; peristome double, teeth papillose or not; calyptra cucullate. Two genera.

114. Climacium Web. & Mohr, Nat. Reise Schwed., 1804

Capsule erect, symmetrical; peristome teeth papillose; basal membrane very short; processes long, widely perforated; calyptra large. Four species. Europe, N.E. and S.E. Asia, N. America, New Zealand.

1. C. dendroides (Hedw.) Web. & Mohr, Nat. Reise Schwed., 1804

Primary stems rhizomatous, secondary stem erect to 2–10 cm tall, branching at top in dendroid fashion with crowded branches. Lvs of secondary stems appressed, scale-like, branch lvs imbricate, scarcely altered when dry, plicate, lanceolate to lingulate, obtuse to acute, margin plane or narrowly recurved below, coarsely toothed above; nerve strong below, faint above, extending nearly to apex; cells incrassate, narrowly rhomboidal, 5–6 μm wide in mid-lf, angular cells irregularly quadrate. Seta thin, flexuose, to 4 cm; capsule erect, cylindrical, slightly curved; columella extending beyond mouth of capsule after dehiscence; spores very variable in size with many abortive, 16–24 μm. Fr very rare but often abundant when produced, winter. $n=11$. Yellowish-green or green, usually numerous, dendroid plants in damp, marshy places, wet grassland and dune slacks, generally distributed, frequent. 109, H40, C. Europe, Faroes, N., C. and E. Asia, N. America, Greenland, New Zealand.

C. dendroides shows considerable phenotypic variation, the drier the habitat, the shorter the erect secondary stem and the more tightly compact the branches.

33. CRYPHAEACEAE

Autoecious. Primary stems creeping, stoloniform; secondary stems without or with rudimentary central strand, ± elongated, erect or procumbent, occasionally pendulous, irregularly, pinnately or fastigiately branched, leafy, paraphyllia lacking, pseudoparaphyllia present. Lvs imbricate when dry, spreading when moist, ovate or ovate-lanceolate, apex acute to longly acuminate, longitudinally plicate or not; nerve single; cells mostly smooth, with oval or elliptical lumens, towards margin oblique, elongated and porose near nerve. Perichaetial lvs erect. Seta usually very short; capsule immersed, symmetrical with a few superficial stomata; annulus differentiated; peristome double, single or absent, basal membrane when present very short; calyptra small, mitriform, ± papillose. Eleven genera.

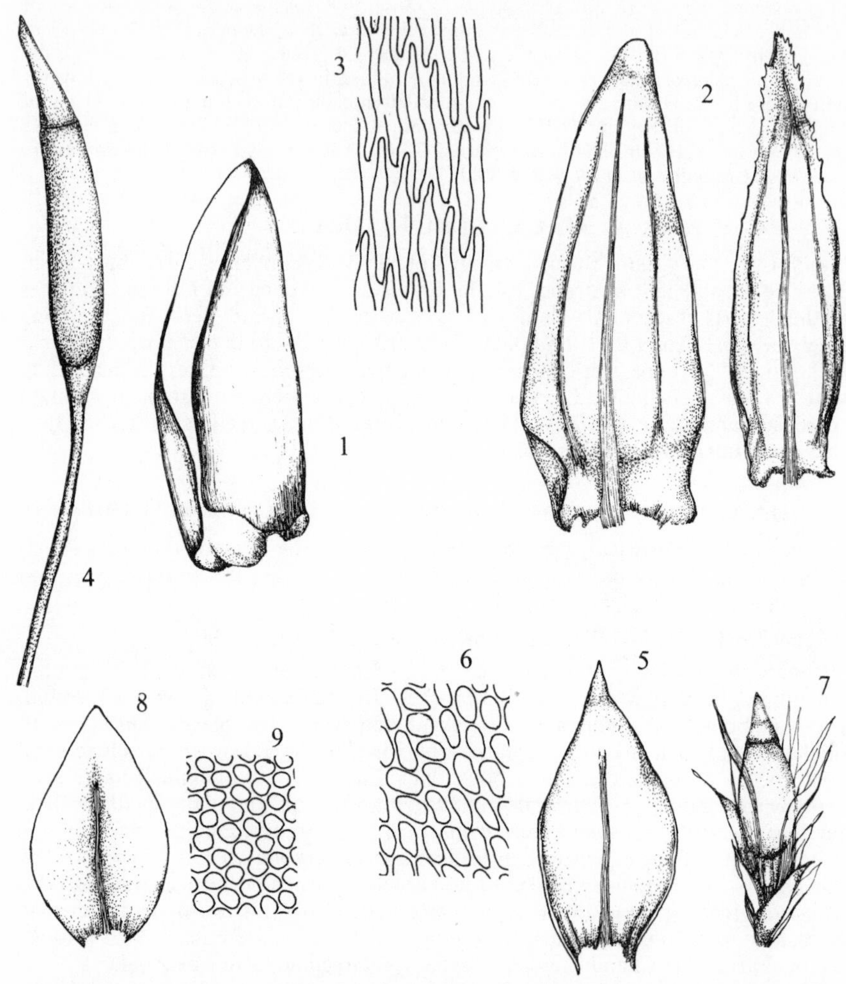

Fig. **242**. 1–4, *Climacium dendroides*: 1, stem leaf; 2, branch leaves; 3, mid-leaf cells of branch leaf; 4, capsule. 5–7, *Cryphaea heteromalla*: 5, leaf; 6, mid-leaf cells; 7, capsule (×10). 8–9, *C. lamyana*: 8, leaf; 9, mid-leaf cells. Leaves ×25, cells ×415.

115. Cryphaea Mohr in Web., Tab. Syn. Musc., 1814

Secondary stems erect or ascending, paraphyllia lacking. Upper lf cells with rounded or oval lumens. Perichaetial lvs long. Capsules borne unilaterally, ellipsoid to cylindrical; peristome double, outer teeth finely papillose; calyptra mitriform. About 70 spp. Europe, C. and E. Asia, Africa, Malagasy Republic, America, Australia, New Zealand, Oceania.

Primary stems with numerous branches forming tight, procumbent patches, lvs gradually tapering to acute to acuminate apex, margin narrowly recurved **1. C. heteromalla**

Primary stems sparsely branched, forming lax patches, lvs shortly pointed, obtuse, margin plane **2. C. lamyana**

1. C. heteromalla (Hedw.) Mohr in Web., Tab. Syn. Musc., 1814

Primary stems procumbent with numerous, ± erect branches with dwarf fertile shoots usually borne unilaterally. Branch lvs imbricate when dry, erecto-patent when moist, ovate, tapering to acute or acuminate apex, concave below, not plicate, margin narrowly recurved below, entire; nerve stout, extending about $\frac{3}{4}$ way up lf; cells smooth, incrassate, basal elliptical, above ± oval, 1–1½ times as long as wide, 8–12 μm wide in mid-lf. Perichaetial lvs larger, tapering gradually to abruptly into longly excurrent nerve; cells longer and narrower. Capsule immersed, ellipsoid; spores *ca* 18 μm. Fr common, late summer, autumn. $n = 10 + 1^*$, $10 + 2$. Dull or dark green patches on bark especially near water, rarely on rocks. Frequent in England, Wales and Ireland, occasional in Scotland. 72, H38, C. W., C. and S. Europe, Canaries, Azores, N. America.

2. C. lamyana (Mont.) C. Müll., Linnaea, 1845
C. heteromalla var. *aquatilis* (C. Müll.) Wils.

Close to *C. heteromalla* but stems longer, stouter and less frequently branched. Lvs ovate, broadly pointed, obtuse, margin plane. $n = 10 + 1^*$. Lax patches on rocks and exposed tree roots below flood level by streams and rivers, very rare. E. Cornwall, Devon, 3. S.W. Europe, N. Africa.

Dixon (1924) describes this plant as having denticulate perichaetial lvs and floating in water. Specimens seen recently had perichaetial lvs similar to *C. heteromalla* and would be floating only at times of high flood.

34. LEUCODONTACEAE

Dioecious. Stem nearly circular in section without or with rudimentary central strand. Primary stem stoloniform, rarely rhizomatous; secondary stems numerous, robust, without rhizoids, erect or ascending, rarely pendulous, branched or not, paraphyllia absent or present. Lvs often longitudinally plicate, ovate or lanceolate with long or short acumen, unbordered; nerve single, double or absent; cells incrassate, usually smooth, rhomboidal above, below and towards middle of lf longer, narrower. Perichaetial lvs longer. Seta very short to long; capsule usually erect, symmetrical, ovoid to cylindrical, usually without stomata or when present superficial; annulus usually undifferentiated; peristome double, usually papillose, basal membrane very short; calyptra cucullate, glabrous or hairy. Seven genera.

116. Leucodon Schwaegr., Suppl. I, 1816

Lvs imbricate when dry, longitudinally plicate or not; nerve lacking; cells smooth, upper with narrow lumens. Inner peristome very short. Thirty-five species. Eurasia, Africa, America, Oceania.

1. L. sciuroides (Hedw.) Schwaegr., Suppl. I, 1816

Primary stem stoloniform, branches decumbent to erect, straight or curved. Lvs appressed when dry, patent or secund when moist, longitudinally plicate, lanceolate to broadly ovate, acute to acuminate, margin plane, entire; nerveless; cells smooth, incrassate, narrow below, shorter and wider above, towards margin rounded. Deciduous, axillary shootlets sometimes produced in abundance at ends of branches. Seta long, reddish-brown; capsule ovoid and curved to ellipsoid or cylindrical and straight or curved; lid rostrate; spores 20–26 μm.

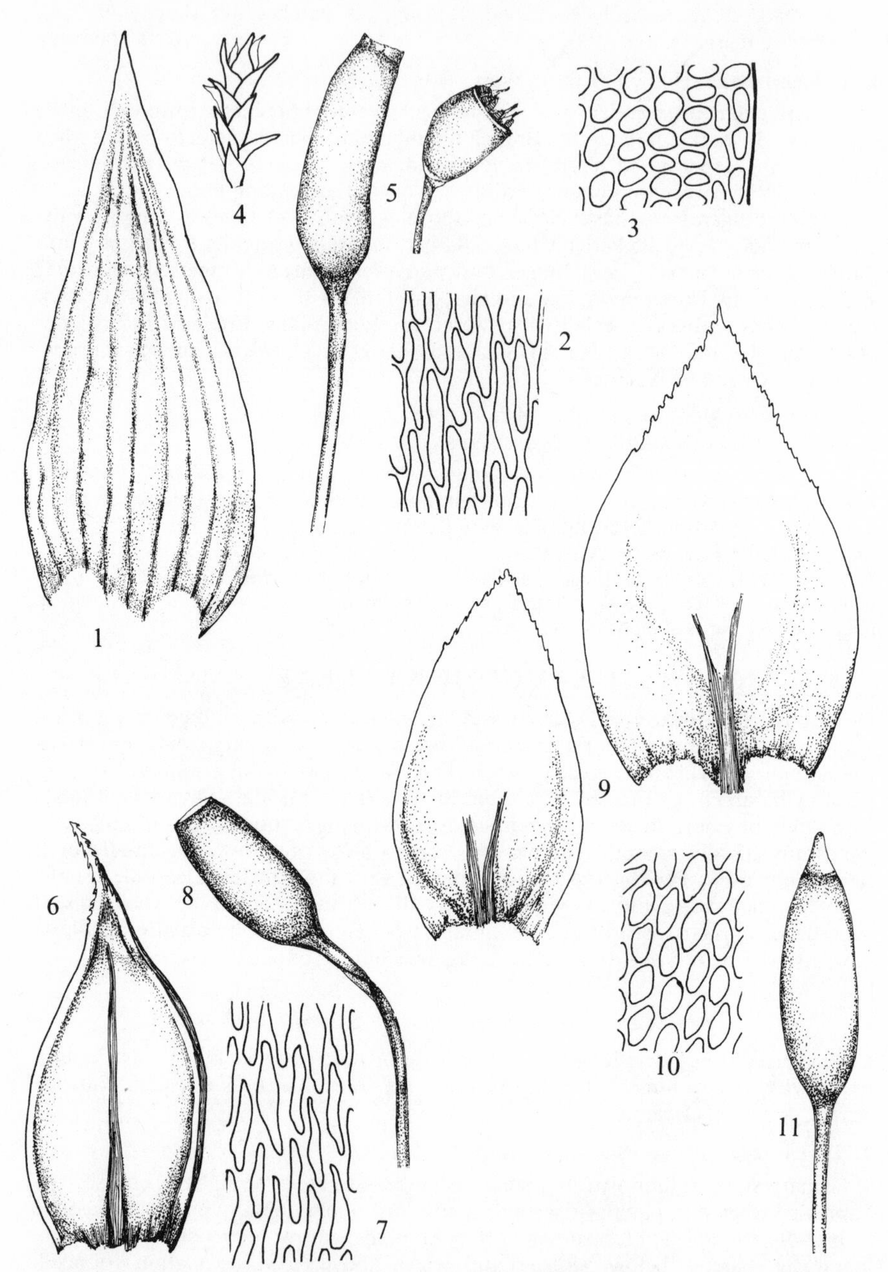

Fig. **243**. 1–5, *Leucodon sciuroides*: 1, leaf (×25); 2, mid-leaf cells; 3, marginal cells; 4, propaguliferous shoot (×40); 5, capsules. 6–8, *Antitrichia curtipendula*: 6, leaf (×25); 7, mid-leaf cells; 8, capsule. 9–11, *Pterogonium gracile*: 9, leaves (×40); 10, mid-leaf cells; 11, capsule. Cells ×415, capsules ×10.

Var. **sciuroides**

Branches 0.5–2.0(–2.5) cm long, 0.4–1.0 mm diameter when dry. Capsule ovoid to cylindrical, straight or curved. Fr very rare. *n*=11. Yellowish-green to green patches on walls, rocks, tree trunks in open habitats, generally distributed, occasional to frequent in S. Britain, rare to occasional elsewhere. 95, H14, C. Eurasia, Algeria, Madeira.

Var. **morensis** (Limpr.) De Not., Syll., 1838

Larger, branches (1.5–)2.0–5.0(–7.0) cm long, 1.2–1.8 mm diameter when dry. Capsule ellipsoid to cylindrical, straight or slightly curved. Fr occasional. *n*=10–11. Patches on basic rocks and walls, very rare. Dorset, Surrey, Kirkcudbright, Perth, Angus, E. Inverness. 6. S. and W. Europe, Turkey, Ukraine.

117. Antitrichia Brid., Mant. Musc., 1819

Secondary stems elongated, irregularly and sparsely branched. Lvs plicate; nerve single; cells smooth, lumens narrow above. Perichaetial lvs nerveless. Inner peristome well developed but lacking basal membrane. Four species. Europe, parts of Asia, Africa, America.

1. A. curtipendula (Hedw.) Brid., Mant. Musc., 1819

Primary stems short, stoloniform, secondary stems long, to 20 cm, irregularly pinnately branched, branches erect to spreading. Lvs sub-secund when dry, patent when moist, concave, plicate, ovate with long acuminate apex with spinose, often recurved teeth, margin narrowly recurved; nerve broad, branching below, thin above and extending to ¾ way up lf; cells incrassate, ellipsoid, 2–3 times as long as wide, 6–10 μm wide in mid-lf, ± rounded at base, basal angles and lower part of margin. Perichaetial lvs longer, sheathing, nerveless. Seta flexuose or arcuate; capsule inclined, ellipsoid; spores 34–36 μm. Fr rare, spring. Dull or yellowish-green, coarse wefts on rocks, walls, bark at various altitudes. Very rare in S.E. and C. England, occasional elsewhere. 79, H12. Europe, Faroes, Caucasus, Madeira, Canaries, Ethiopia, Morocco, C. and southern Africa, N. America, Greenland, Patagonia.

Possibly to be confused with *Rhytidiadelphus loreus* which differs in the more distant and longly pointed, less imbricate lvs.

118. Pterogonium Sm., Monthl. Rev., 1801

Lvs not plicate; nerve double or forked; cells papillose, lumens elliptical to ovate. Peristome double. Five species, Europe, Middle East, Africa, N. America.

1. P. gracile (Hedw.) Sm., Eng. Bot., 1802

Primary stems stoloniform, secondary stems erect, to 4 cm long, with numerous unidirectionally curved branches above, flagelliform shoots often present. Branch lvs imbricate when dry, spreading when moist, concave, not plicate, broadly ovate, acute to obtuse, scabrous at back above with papillae, margin plane, dentate above; nerve very short, single or double, extending about ¼ way up lf; cells beyond nerve elliptical, becoming oval above and towards margin, at extreme base transversely rectangular, at basal angles and extending up margin rounded, cells in mid-lf towards margin *ca* 1½ times as long as wide, 10 μm wide. Perichaetial lvs longer, sheathing, with single nerve. Capsule slightly inclined, narrowly ellipsoid; spores *ca* 30 μm. Fr autumn, very rare. Dense, dark or brownish-green patches on shaded rocks or tree bases, frequent in basic areas, rare elsewhere. 70, H18, C. S., W. and C. Europe, Madeira, Canaries, Africa, Malagasay Republic, Réunion, N. America.

Readily recognised by the curved branches of the secondary stems pointing ± in the same direction with the lvs imbricate when dry.

35. MYURIACEAE

Dioecious. Stems circular in transverse section, without central strand. Primary stem stoloniform with scale lvs; secondary stems stout, usually erect or ascending, simple or branched, leafy, paraphyllia lacking. Lvs symmetrical, ovate to lanceolate, ± longly acuminate; nerve lacking; cells smooth, angular cells differentiated. Perichaetial lvs small. Seta long, capsule symmetrical, ellipsoid, smooth, stomata superficial; annulus not differentiated; peristome double; calyptra cucullate. Two genera.

119. MYURIUM SCHIMP., SYN., 1860

Lvs very concave, not plicate. Seta long, smooth; capsule ellipsoid; peristome teeth not papillose, basal membrane short, processes lacking. Fourteen species, Europe, Asia, Macaronesia, Australia, Oceania.

1. M. hochstetteri (Schimp.) Kindb., Ottawa Nat., 1900
M. hebridarum Schimp.

Secondary stems crowded, erect, julaceous, to 6 cm long. Lvs ± imbricate, glossy, hardly altered when dry, not plicate, ovate-oblong, very concave, abruptly narrowed into long apiculus, margin plane below, inflexed above, denticulate; nerveless or nearly so; cells incrassate, porose, narrow, sinuose, *ca* 6 times as long as wide, 6 μm wide in mid-lf, cells at base and basal angles shorter. Gametangia unknown. Glossy, golden-yellow, dense tufts on soil. Frequent in Outer Hebrides, occasional in Inner Hebrides, and on coast of W. Inverness and W. Sutherland. 5. Macaronesia.

36. NECKERACEAE

Dioecious, autoecious or synoecious. Primary stem stoloniform or rhizome-like, secondary stems decumbent or erect, pinnately or bipinnately branched, complanate, without central strand, often with paraphyllia. Lvs complanate, usually asymmetrical, ovate-oblong to lingulate; nerve absent or short and faint, single or double; cells linear-vermicular to rhomboidal, never papillose. Capsule immersed, occasionally emergent or exserted, erect, symmetrical, with or without stomata; calyptra cucullate, with or without hairs; annulus usually lacking; peristome double or (in *Leptodon*) inner rudimentary.

120. LEPTODON MOHR, OBSERV. BOT., 1803

Secondary stems arising from rhizome-like primary stem, pinnately or bipinnately branched, stems and branches strongly inrolled when dry; small-leaved innovations often present. Lvs oval or shortly lingulate, obtuse; nerve slender, single; cells oval. Seta short; capsule exserted; calyptra hairy; inner peristome rudimentary. A small genus of 5 species distributed through Europe, W. Asia, Africa, N. America, Chile, Australasia.

1. L. smithii (Hedw.) Web. & Mohr, Ind. Musc. Pl. Crypt., 1803

Dioecious. Secondary stems decumbent to ascending, to *ca* 2.5 cm long, pinnately or bipinnately branched, branches curved when moist, inrolled when dry. Lvs

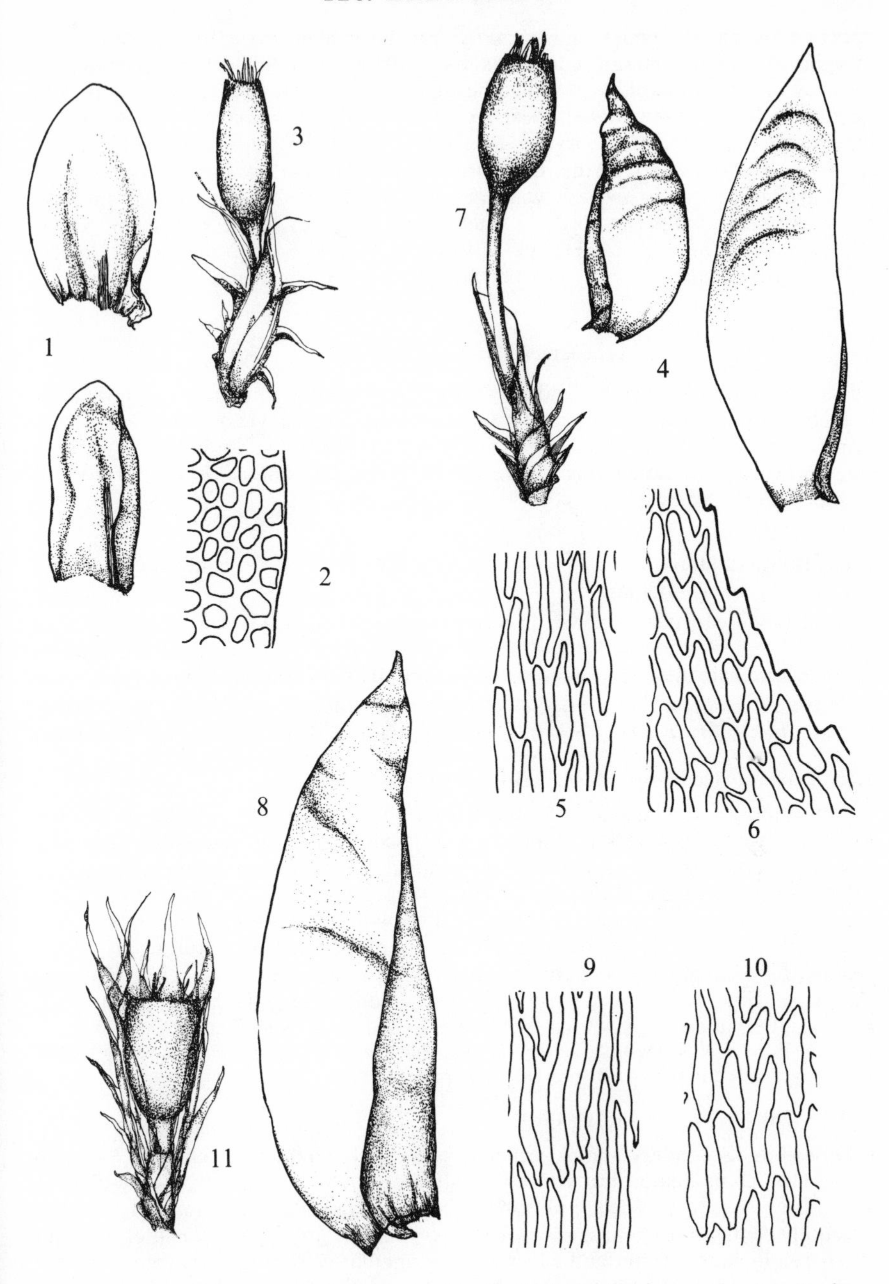

Fig. **244**. 1–3, *Leptodon smithii*: 1, leaves (×40); 2, mid-leaf cells; 3, capsule; 4–7, *Neckera pumila*: 4, leaves (×25); 5, mid-leaf cells; 6, cells from leaf apex; 7, capsule. 8–11, *N. pennata*: 8, leaf (×25); 9, mid-leaf cells; 10, cells from leaf apex; 11, capsule. Cells ×415, capsules ×10.

appressed when dry, spreading when moist, often asymmetrical, ovate, apex rounded, margin recurved on one side, entire or sub-crenulate; nerve single, faint, extending to $\frac{1}{4}$–$\frac{1}{2}$ way up lf; cells ovoid to rounded-hexagonal, 10–14 μm wide, basal cells longer, angular cells not differentiated. Perichaetial lvs longer, acuminate, outer squarrose, inner erect. Seta straight or curved, very short, 1.5–2.0 mm; capsule ellipsoid; spores *ca* 16 μm. Fr occasional, spring. Dark green patches on bark of trees, particularly in basic areas, very rarely on basic walls or rocks. Occasional in S. and S.W. England from Cornwall and Kent north to Caerns and Anglesey, Cumberland, W. Cork, Waterford, Channel Is. 25, H2, C. S. Europe, Middle East, Madeira, N. Africa, Kilimanjaro, southern Africa, S. America, Australia, New Zealand.

121. Neckera Hedw., Sp. Musc., 1801

Dioecious or autoecious. Secondary stems complanately, pinnately or bipinnately branched. Lvs complanate, often sub-falcate; nerve faint, single or double or absent. Capsule immersed or exserted, erect, symmetrical; peristome double, inner with 16 short processes and no cilia. A ± world-wide genus with about 120 mainly tropical or subtropical species.

1	Lvs not undulate	**4. N. complanata**
	Lvs transversely undulate	2
2	Autoecious, capsules common, immersed, lvs ± gradually tapering to apex	**1. N. pennata**
	Dioecious, capsules rare, exserted, lvs ± abruptly narrowed to apex	3
3	Secondary stems 4–20 cm long, lf cells 8–10 μm wide	**2. N. crispa**
	Secondary stems 1–4 cm long, lf cells 5–6(–8) μm wide	**3. N. pumila**

1. N. pennata Hedw., Sp. Musc., 1801

Autoecious. Secondary stems 5–10 cm, decumbent, irregularly pinnately branched. Lvs complanate, transversely undulate, sub-falcate, ovate to lanceolate, gradually tapering to acute apex, margin incurved, faintly denticulate towards apex; nerve very short, faint; cells very incrassate, narrow, vermicular, *ca* 8 μm wide, shorter and wider at extreme base and basal angles, at apex 4–8 times as long as wide. Seta very short, *ca* 1 mm; capsule immersed; spores *ca* 22 μm. Fr common, autumn and winter. Yellowish-green patches on tree trunks, Angus, not seen since 1823. Europe, Caucasus, Siberia, C. Asia, China, Japan, Hong Kong, Himalayas, Canaries, Madeira, Fernando Po, S. Africa, N. America, Tasmania, New Zealand.

Likely to be confused with *N. crispa* but distinguished by the ± gradually tapering lf apex with longer cells and the usually present, immersed capsules.

2. N. crispa Hedw., Sp. Musc., 1801

Dioecious. Secondary stems decumbent to erect, pinnately branched, 4–20 cm long. Lvs complanate, transversely undulate, often sub-falcate, ovate-oblong, abruptly tapered to rounded or obtuse or obtuse and apiculate apex, margin incurved, obscurely denticulate towards apex; nerve very short, faint; cells incrassate, vermicular, 8–10 μm wide, shorter and wider at base, angular cells irregularly rounded, very incrassate, towards apex 2–4 times as long as wide. Seta slightly flexuose, 8–12 mm; capsule exserted, ellipsoid; spores 24–30 μm. Fr rare, spring. $n=10+1$. Glossy, yellowish-green to golden, lax patches, sometimes very extensive, on basic rocks, especially where damp, stony calcareous soil, occasionally on trees. Occasional to

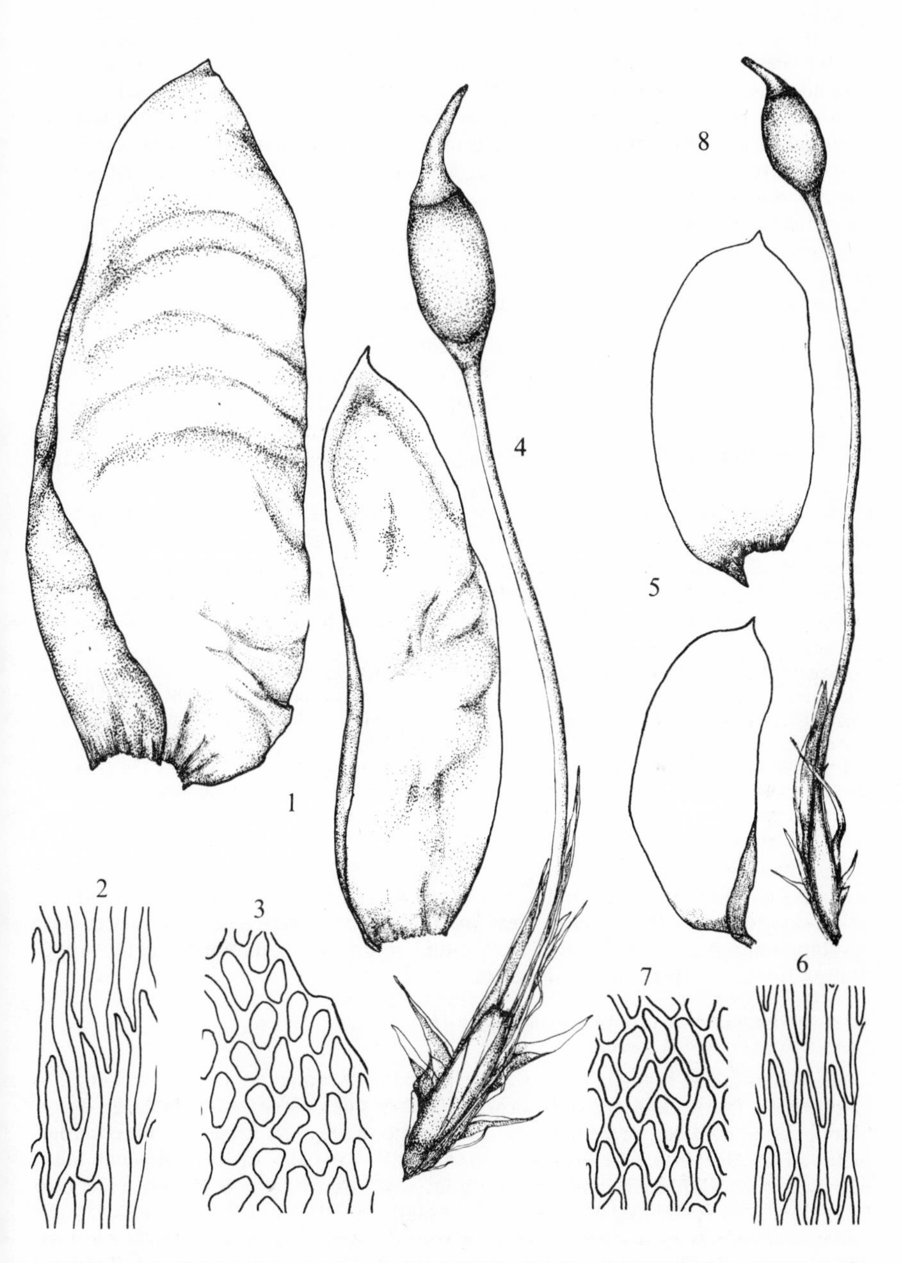

Fig. **245**. 1–4, *Neckera crispa*: 1, leaves; 2, mid-leaf cells; 3, cells from leaf apex; 4, capsule. 5–8, *Neckera complanata*: 5, leaves; 6, mid-leaf cells; 7, cells from leaf apex; 8, capsule. Leaves ×25, cells ×415, capsules ×10.

common in basic areas, rare elsewhere. 91, H30, C. Europe, Caucasus, Madeira, Canaries.

3. N. pumila Hedw., Sp. Musc., 1801

Dioecious. Secondary stems decumbent, 1–4 cm long. Lvs complanate, transversely undulate, ovate, acute to acuminate or obtuse with long apiculus, one margin incurved below, the other narrowly reflexed, finely denticulate towards apex; nerve short, faint; cells incrassate, vermicular, 5–6(–8) μm wide, basal shorter and wider, angular not distinct, near to apex 2–6 times as long as wide. Short flagelliform branches with minute lvs often present. Seta 2–4 mm; capsule exserted, ellipsoid; spores 15–20 μm. Fr occasional, winter. Glossy, pale green patches on trees, very rarely on rocks or walls. Occasional to frequent in England, Wales and Ireland, rare in Scotland. 62, H34, C. Europe, Madeira, Canaries.

4. N. complanata (Hedw.) Hüb., Musc. Germ., 1833

Dioecious. Secondary stems decumbent, complanate, to 5 cm long. Lvs complanate, not undulate, sub-falcate, lanceolate-oblong to ovate-oblong, acute to obtuse and apiculate, one margin slightly incurved below, margin finely denticulate above; nerve short, faint, double; cells incrassate, narrow, vermicular, 6–8 μm wide, shorter at base, angular cells ± distinct, cells near apex 1–3 times as long as wide. Flagelliform shoots sometimes present and abundant. Seta flexuose, 8–10 mm; capsule exserted, ovate-ellipsoid; spores very variable in size, 14–26 μm. Fr occasional, spring. $n=10$, $10+1$*, 12*. Glossy, pale or yellowish-green patches on usually shaded rocks, walls, soil banks, tree trunks, especially in basic areas, common. 112, H40, C. Europe, Caucasus, Iran, Kashmir, Algeria, Madeira, Canaries, C. Africa, N. America.

Flagelliform shoots may sometimes be so abundant as to give the plants a wispy appearance and occasionally constitute the whole plant (as in the var. *tenella* Schimp. of Dixon, 1924). Differs from *N. pumila* in the non-undulate lvs with ± plane margin and is most likely to be confused with *Homalia trichomanoides* which has a stronger nerve extending beyond the middle of the lf and shorter cells and may be recognised in the field by the lvs more strongly curved down on either side of the stem which is less regularly branched.

122. Homalia (Brid.) Br. Eur., 1850

Autoecious. Gametophyte similar to that of *Neckera* but stems never with paraphyllia or flagelliform branches. Seta long; peristome well developed, inner with basal membrane about ⅓ length of teeth, rudimentary cilia sometimes present. About 25 mainly tropical or subtropical species.

1. H. trichomanoides (Hedw.) Br. Eur., 1850
Omalia trichomanoides auct.

Secondary stems sparsely and irregularly pinnately branched, to *ca* 6 cm long. Lvs complanate, not undulate, curved down on either side of stem, asymmetrical, ovate-oblong, apex rounded or rounded and apiculate, margin denticulate towards apex, incurved on one side; nerve single, extending to ½–¾ way up lf; cells near nerve narrowly rhomboidal, becoming rhomboid towards margin, 6–10 μm wide near margin, cells in upper part of lf narrowly hexagonal. Seta 1.0–1.5 cm; capsule shortly cylindrical; spores 14–16 μm. Fr occasional to frequent, winter. $n=11$*, $11+1$. Lax, depressed patches on damp, shaded rocks, walls, soil banks and tree trunks. Frequent or common except in N.W. Scotland. 105, H35, C. Europe, Caucasus, Siberia, China, Japan, Madeira, N. America.

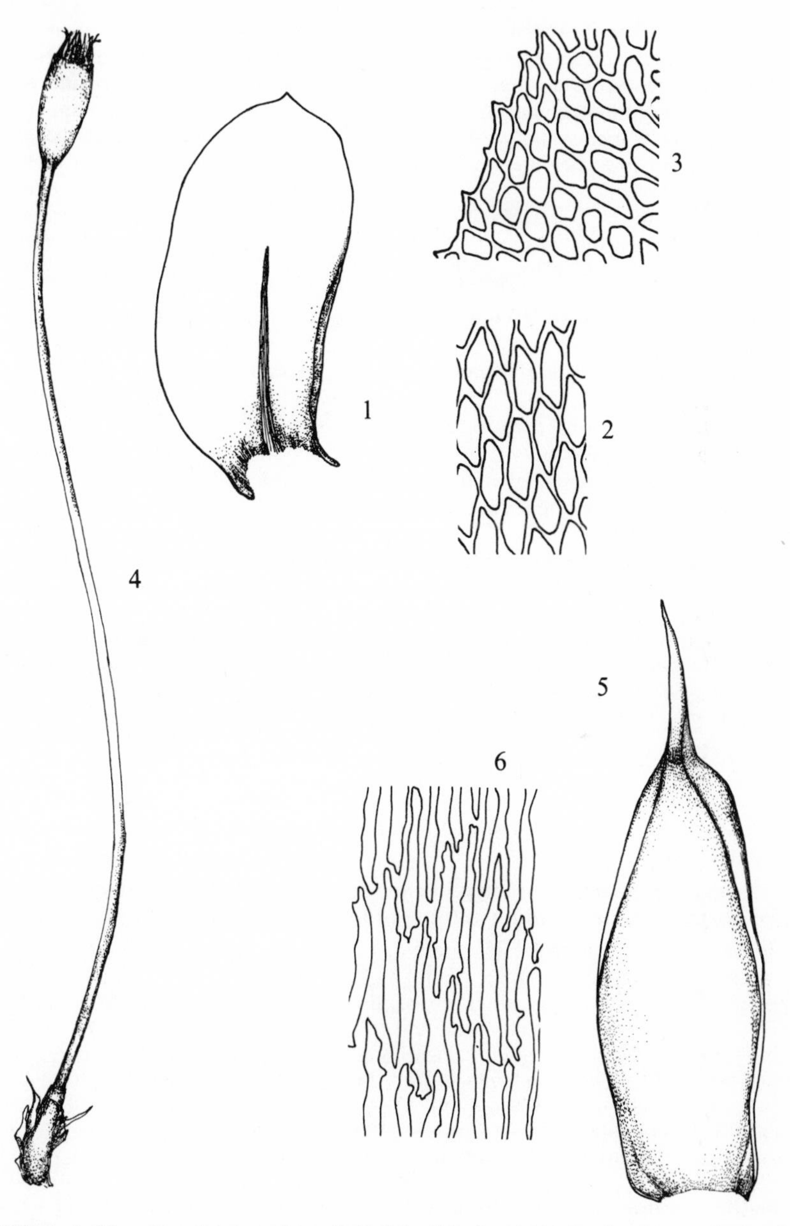

Fig. **246**. 1–4, *Homalia trichomanoides*: 1, leaf (×40); 2, mid-leaf cells; 3, marginal cells towards apex; 4, capsule (×10). 5–6, *Myurium hochstetteri*: 5, leaf (×25); mid-leaf cells. Cells ×415.

37. THAMNIACEAE

Primary stems stoloniform, secondary erect, often dendroid, sometimes stoloniform, central strand present. Lvs acute to obtuse, bordered with elongated cells or not, entire to strongly toothed; nerve single, ending below apex; cells mostly short, smooth or papillose. Seta long or short; capsule symmetrical or not; peristome well developed; calyptra cucullate, smooth or hairy.

123. THAMNOBRYUM NIEUWL., AM. MIDL. NAT., 1917

Dioecious. Usually robust, dendroid plants, secondary stems with numerous, often curved, homomallous branches above. Stem lvs broadly triangular, branch lvs ovate, toothed; cells short, papillose. Capsule inclined, arcuate; peristome teeth papillose throughout or horizontally striate below. A ± world-wide genus of about 40 species.

Lvs of branches ovate to lanceolate, nerve stout, occupying less than ¼ of lf base
1. T. alopecurum

Branch lvs ligulate to lanceolate, nerve thin, wide, occupying ⅓–½ lf base
2. T. angustifolium

1. T. alopecurum (Hedw.) Nieuwl., Am. Midl. Nat., 1917
Porotrichum alopecurum (Hedw.) Dix., *Thamnium alopecurum* (Hedw.) Br. Eur.

Dioecious. Secondary stems stout, erect, dendroid, unbranched below, densely branched near apex, branches irregularly pinnate or bipinnate, spreading, secund or arcuate, short to flagelliform. Stem lvs distant, appressed, broadly triangular, lower scarious, upper chlorophyllose; branch lvs imbricate when dry, patent when moist, concave or not, ovate, tapering from below middle to acute apex, margin plane, sub-entire to coarsely toothed, lvs on ultimate branches narrower, lanceolate; nerve single, stout, occupying less than ¼ lf base, ending shortly below apex; cells papillose, thin-walled to incrassate, in stem lvs elongated except near basal angles, in branch lvs elliptical or rounded-quadrate, 8–12 μm wide in mid-lf. Seta short, 1.0–1.5 cm; capsule inclined, asymmetrical, curved, ovoid to narrowly ellipsoid; spores 10–12 μm. Fr occasional, winter, spring. $n=11$*. Lax, green to dark green or brownish-green tufts or patches, sometimes extensive, on usually moist, shaded rocks, tree roots and boles in and by streams and damp places, rarely on dry rocks, common in most parts of Britain and Ireland. 110, H40, C. Europe, Faroes, Asia, N. Africa, Macaronesia, N. America.

2. T. angustifolium (Holt) Crundw., Trans. Br. bryol. Soc., 1971
Thamnium angustifolium Holt, *Porotrichum angustifolium* (Holt) Dix.

Secondary stems more slender, more distantly branched than in *T. alopecurum*, to *ca* 4 cm long, branches slender. Stem lvs narrowly triangular, branch lvs loosely imbricate when dry, patent when moist, ligulate to lanceolate, acute, margin very coarsely toothed above; nerve wide and thin, occupying ⅓–½ lf base. Fr unknown. Greenish patches, encrusted with calcareous matter below, on damp vertical shaded limestone, very rare. Apart from a doubtful record from Antrim the only known British site for this plant is in Derby. 1. Madeira.

Distinctive in appearance and with no intermediates this plant is certainly worthy of specific rank distinct from *T. alopecurum*.

17. HOOKERIALES

Pleurocarpous. Lvs frequently complanate and often asymmetrical, sometimes with

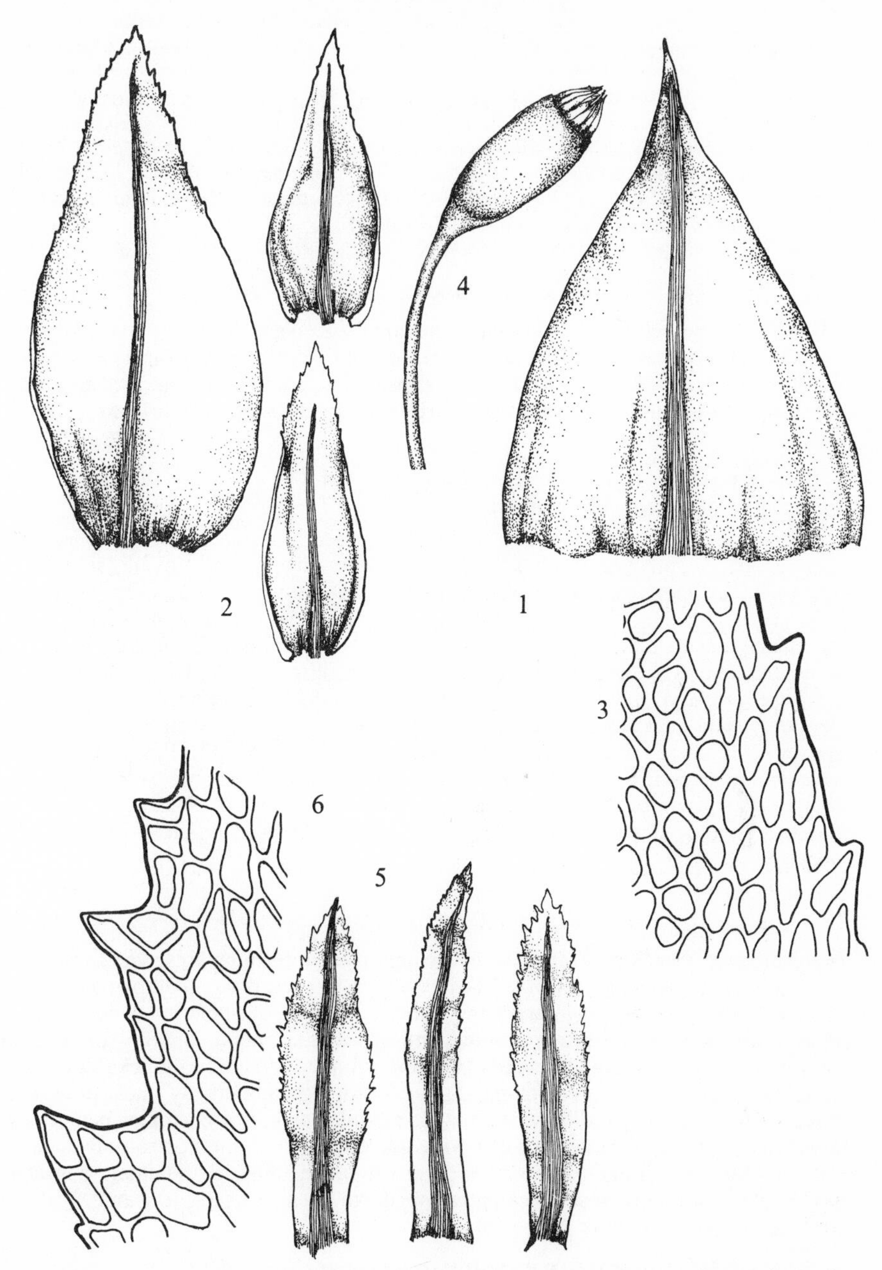

Fig. **247**. 1–4, *Thamnobryum alopecurum*: 1, stem leaf; 2, branch leaves; 3, branch leaf cells; 4, capsule. 5–6, *T. angustifolium*: 5, branch leaves; 6, branch leaf cells. Leaves ×25, cells ×415.

border of elongated cartilaginous cells; nerve single or double, long or short or absent. Peristome double, teeth usually with lamellae, cilia usually absent; calyptra conical, often fringed.

38. HOOKERIACEAE

Sporophyte relatively robust; seta usually smooth; exothecial cells rectangular below, quadrate-hexagonal and collenchymatous above; outer peristome teeth horizontally striate, inner peristome with high basal membrane, processes keeled, finely papillose, cilia sometimes present; calyptra naked, lobed at base.

124. HOOKERIA SM., TRANS. LINN. SOC. LOND., 1808

Autoecious or dioecious. Lvs complanate, large, succulent, ovate, ± asymmetrical, ± obtuse, margin plane, entire, not bordered; nerveless; cells very large, hexagonal, marginal scarcely differing. Seta stout; capsule ellipsoid; peristome perfect, inner with tall basal membrane; calyptra mitriform. Distribution ± world-wide, 20–25 species.

1. H. lucens (Hedw.) Sm., Trans. Linn. Soc. Lond., 1808
Pterygophyllum lucens (Hedw.) Brid.

Autoecious. Plants to 6 cm long, shoots decumbent, stems sparsely branched. Lvs complanate, slightly shrunken and undulate when dry, broadly ovate, sometimes asymmetrical, apex rounded or obtuse, margin plane, entire, unbordered; nerveless; cells very large, thin-walled, translucent, interspersed with occasional smaller cells, irregularly hexagonal to elongate-hexagonal, 60–100 μm wide in mid-lf, longer in mid-base, marginal cells undifferentiated or slightly narrower. Uniseriate, chlorophyllose, propaguliferous filaments sometimes produced from small cells in upper part of lf. Seta very stout, *ca* 2 cm long; capsule blackish, ± horizontal, ellipsoid; spores 12–16 μm. Fr occasional, late autumn to spring. $n = 6^*, 9^*$. Glossy, pale green, succulent patches on damp shaded soil and rocks in woods, on banks, sides of ditches, rare in drier parts of Britain, frequent elsewhere. 93, H36, C. W. and C. Europe, Faroes, Poland, Croatia, Caucasus, Asia Minor, Tunisia, Madeira, western N. America.

125. ERIOPUS C. MÜLL., BOT. ZEIT., 1847

Autoecious or dioecious. Plants slender to very robust, forming loose and sometimes extensive turfs; stems ascending. Lvs in 6 ranks but usually ± complanate, lower distant, upper crowded and larger, asymmetrical, often variable in shape on same stem, ovate or obovate, shortly pointed, entire, bordered; nerve forked with unequal branches, short or rarely long; cells very lax, rounded to rhomboid-hexagonal, 2–5 marginal rows very narrow, forming distinct border. Seta relatively thick, prickly or rarely only with low papillae, not twisted; capsule small, horizontal or pendulous, ovoid; annulus of large cells falling with lid; outer peristome horizontally striate, with median line, inner peristome with tall basal membrane and finely papillose keeled processes, cilia absent; calyptra mitriform. A mainly tropical and southern hemisphere genus of about 35 species.

1. E. apiculatus (Hook. f. & Wils.) Mitt., J. Linn. Soc. Bot., 1869

Dioecious. Plants to 1.5–3.0 cm, stems sub-erect, hardly branched, Lower lvs small, upper larger, twisted and slightly shrunken when dry, complanate when moist,

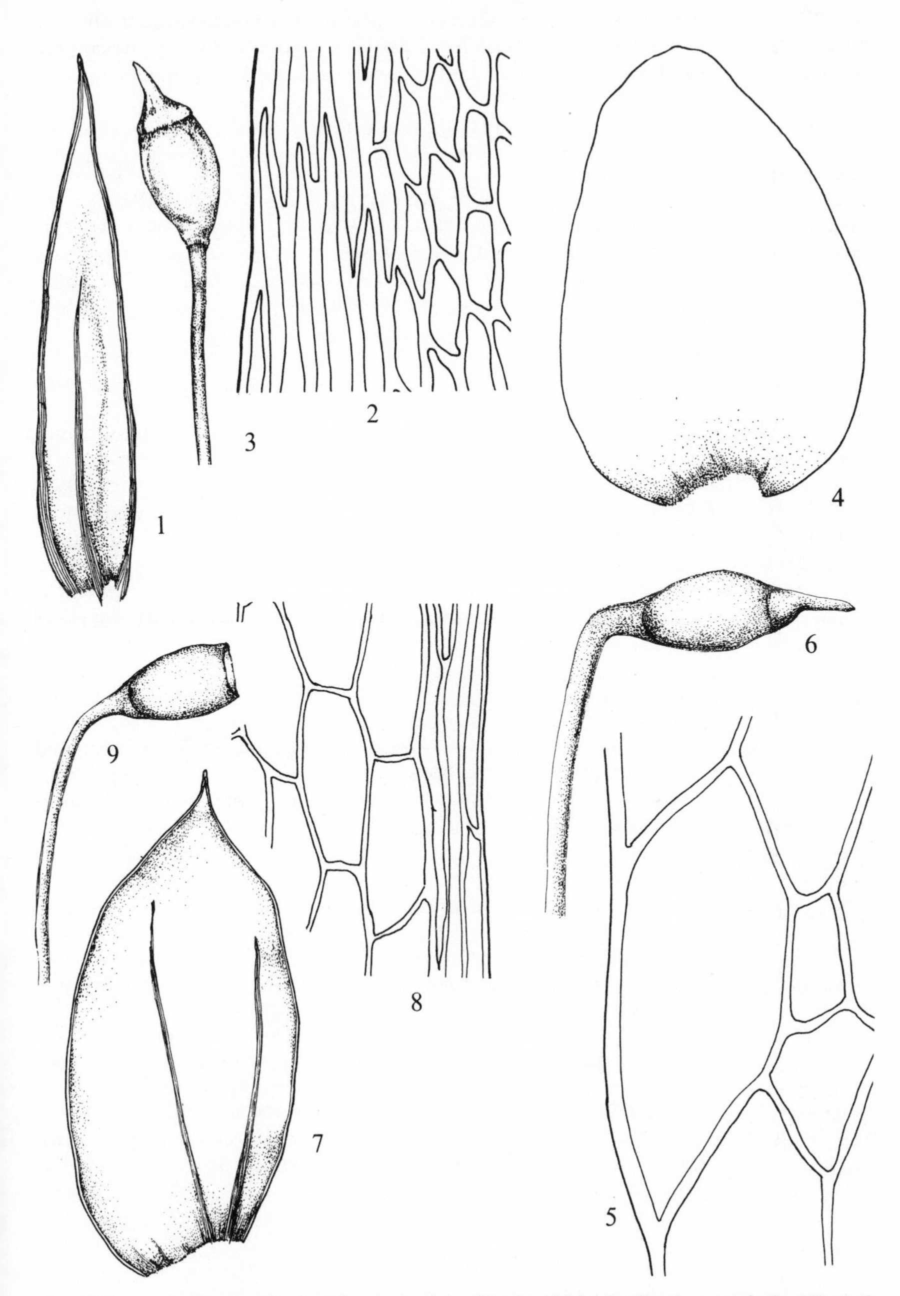

Fig. **248**. 1–3, *Daltonia splachnoides*: 1, leaf (×40); 2, mid-leaf cells; 3, capsule (×15). 4–6, *Hookeria lucens*: 4, leaf (×15); 5, mid-leaf cells; 6, capsule (×10). 7–9, *Cyclodictyon laetevirens*: 7, leaf (×25); 8, mid-leaf cells; 9, capsule (×10). Cells ×415.

broadly ovate or obovate, symmetrical or not, apex apiculate, margin plane, widely bordered, ± toothed above; nerve very short, single or forked; cells large, hexagonal, incrassate, smooth, 25–40 μm wide in mid-lf, larger towards base, smaller towards apex, 5–10 marginal rows narrow and elongated, forming strongly marked border. Only female plants known in Britain. Glossy, green or yellowish-green patches on shaded soil, very rare. Scilly Is. 1. Chile, Tierra del Fuego, Australia, Tasmania, New Zealand.

Almost certainly introduced from the southern hemisphere with horticultural plants. For an account of this plant in Britain see Paton, *Trans. Br. bryol. Soc.* **5**, 460–2, 1968. There is a second British record, from Sussex (see Wallace, *Trans. Br. bryol. Soc.* **6**, 327–8, 1971). Whether this is a young form of *E. apiculatus* or of another species is uncertain. The Sussex plant, which occurs on shaded sandstone, is smaller, the lf cells thin-walled and the border only about 3 cells wide (see Fig. **249**).

126. Cyclodictyon Mitt., J. Linn. Soc. Bot., 1864

Synoecious, autoecious or dioecious. Slender to moderately robust, turf-forming, soft, glossy plants; stems sparsely to densely tomentose. Lvs not crowded, in 5–8 ranks, ± complanate, variable in size, asymmetrical, concave, usually ovate-oblong or oblong, abruptly tapering to short to subulate point; nerve double, extending *ca* halfway up lf; cells very lax, rounded-hexagonal, marginal linear, forming 1–2-stratose border. Seta smooth; capsule inclined to horizontal, ovoid to obloid; annulus of single row of fugacious cells; lid longly rostrate; calyptra mitriform. A largely tropical or subtropical genus of about 110 species occurring particularly in C. and S. America and to a lesser extent in Africa.

1. C. laetevirens (Hook. & Tayl.) Mitt., J. Linn. Soc. Bot., 1864
Hookeria laetevirens Hook. & Tayl.

Stems irregularly pinnately branched, branches decumbent, to 8 cm long. Lvs complanate, glossy, shrunken when dry, broadly ovate, acute to obtuse with large apiculus, margin recurved near base, finely denticulate near apex, bordered; nerve double, thin, extending about $\frac{3}{4}$ way up lf; cells large, thin-walled, hexagonal, 15–30 μm wide in mid-lf, larger towards base, 3–4 marginal rows very narrow, forming distinct border. Seta stout; capsule horizontal, ovoid; spores 12–16 μm. Fr rare, autumn. Soft, glossy, dark green patches on wet shaded rocks, in ravines, caves and crevices near waterfalls, very rare. Cornwall (extinct), Jura, Kerry, Cork, Waterford, W. Mayo, Leitrim. 2, H7. Madeira, Azores, Africa.

Only likely to be confused with *Eriopus apiculatus* which differs in its very short nerve, wider border and its habitat.

39. DALTONIACEAE

Sporophyte not robust; seta usually papillose above; exothecial cells quadrate to rectangular with uniformly thickened walls (except *Daltonia*); outer peristome teeth papillose, inner peristome with very low or no basal membrane, processes papillose, cilia absent; calyptra often hairy, fringed or ciliate at base.

127. Daltonia Hook. & Tayl., Musc. Brit., 1818

Autoecious and dioecious. Small plants with decumbent or ascending stems with short ascending branches. Lvs erecto-patent, keeled, base decurrent, from ovate basal part lingulate-lanceolate to lanceolate or linear-lanceolate; nerve stout, ending

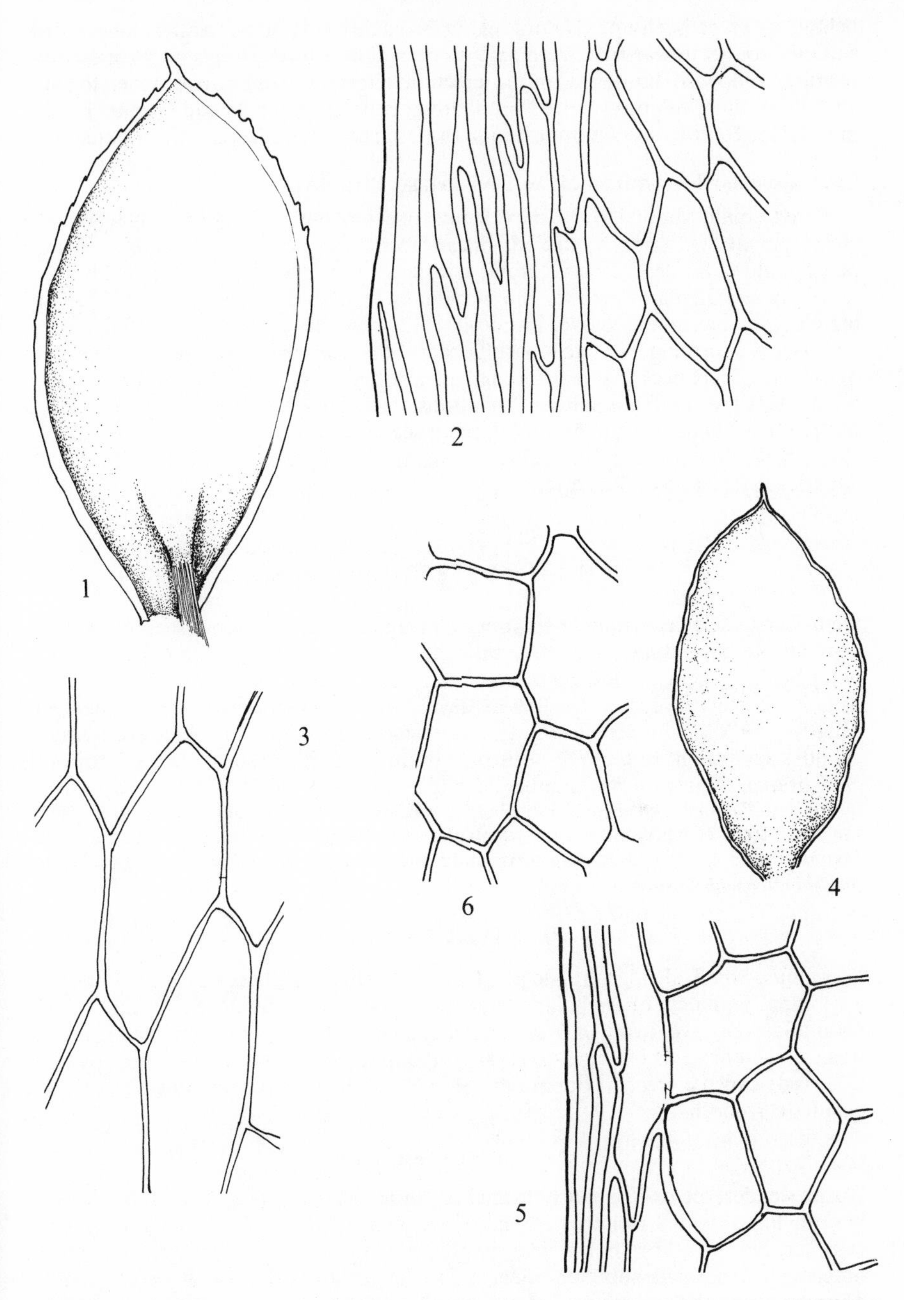

Fig. **249**. 1–3, *Eriopus apiculatus*: 1, leaf; 2, marginal cells; 3, cells from middle of leaf. 4–6, *Eriopus* species from Sussex: 4, leaf; 5, marginal cells, 6, cells from centre of leaf. Leaves ×25, cells ×415.

below apex; cells rhomboidal or elongate-hexagonal, rarely rounded-hexagonal, smooth, longer towards base, marginal cells unistratose, elongated, thick-walled, forming yellowish border. Capsule erect or slightly inclined, ovoid or ellipsoid; annulus absent; calyptra fringed or ciliate at base. A ± world-wide genus but with most of the 60 or so species occurring in the tropics or subtropics.

1. **D. splachnoides** (Sm.) Hook. & Tayl., Musc. Brit., 1818

Plants small, shoots to 10 mm, stems procumbent. Lvs appressed when dry, erecto-patent when moist, keeled, linear-lanceolate, acute to acuminate, margin plane, entire, bordered; nerve stout, ending below apex; cells variable in shape, narrowly rectangular, narrowly rhomboidal or trapezoid in lower part of lf, becoming elongate-hexagonal above, in mid-lf 6–12 μm wide, several marginal rows long and narrow, forming stout yellowish border. Seta deep red, papillose; capsule erect, ovoid with short neck; lid rostrate; calyptra fringed; spores *ca* 16 μm. Fr common. Small, dark green tufts or patches, on humus, decaying wood or rocks by streams and on tree boles in areas of high rainfall from sea level to 500 m, very rare. Argyll, W. Ross, S.W. Ireland, Dublin, Leitrim, Cavan. 2, H6. Madeira, Fernando Po, N. America, Mexico, New Zealand.

18. THUIDIALES

Pleurocarpous. Plants minute to robust, primary stems creeping, secondary stems prostrate to ascending, irregularly branched to regularly 1–3 pinnately branched; paraphyllia present or not. Stem and branch lvs similar or stem lvs larger, ovate, usually acute; nerve usually ± well developed or not, double or wanting; cells usually papillose, rounded to linear. Seta long; capsule erect or inclined, ovoid to cylindrical, straight or curved; peristome double, usually perfect. Plants form wefts or tight mats. Chromosome number usually $n = 10$, $10+1$ or 11.

The families included in this order bear a greater similarity to one another than to the families of the Hypnobryales with which they are usually included. The Thuidiales and Hypnobryales as recognised here form more natural units than when all the families are included in a single order.

40. THELIACEAE

Dioecious. Small, slender, fragile plants forming thick patches; stems decumbent or ascending, pinnately or irregularly branched, branches ± terete. Stems and branch lvs similar, concave, ovate or broadly ovate, acute to obtuse; nerve single or double, weak or absent; cells elliptical to rhomboidal, papillose at back. Seta short or long, thin, red; capsule erect to horizontal, ± symmetrical; peristome double; calyptra cucullate, smooth.

128. Myurella Br. Eur., 1851

Plants slender, stems irregularly branched, branches julaceous, ascending or erect; paraphyllia absent. Lvs imbricate, concave, obtuse; nerve faint, double or absent; cells lax, hexagonal or rhomboidal, papillose. Capsule ± erect; peristome perfect, outer teeth lanceolate-subulate, inner with cilia; calyptra minute. A small genus of 4 species occurring in arctic and alpine regions of the northern hemisphere.

Plants bluish-green, lvs obtuse or with thin, straight apiculus **1. M. julacea**
Plants pale or yellowish-green, lf apex with reflexed apiculus **2. M. tenerrima**

1. M. julacea (Schwaegr.) Br. Eur., 1851

Shoots to 10(–15) mm long, very slender, fragile, older parts with sparse, long, thin rhizoids. Lvs closely imbricate, hardly altered when dry, very concave, broadly ovate, apex rounded or obtuse, sometimes with long, straight apiculus, margin ± entire to finely denticulate above; nerve double or single and forked, very indistinct; cells ± elliptical in mid-lf, 1.5–3.0 times as long as wide, 4–6 μm wide in mid-lf. Seta curved; capsule obovoid; lid conical, obtuse; spores *ca* 10 μm. Fr very rare. summer. Scattered shoots among other bryophytes, or small dense patches with younger parts bluish-green and older pale reddish-brown, on damp or dry, basic soil among rocks or in rock crevices at high altitudes.

Var. **julacea**
Lf margin denticulate below, apex not apiculate; cells with low, obscure, rounded papillae. Rare, Caerns, Yorks, Scottish Highlands, Waterford, W. Galway, Londonderry. 16, H3. Europe, N. and C. Asia, Kashmir, China, N. America, Greenland.

Var. **scabrifolia** Lindb. ex Limpr., Laubm., 1895
Lf margin spinosely denticulate below, apex with long fine apiculus; cells, especially in young lvs, with conical papillae at back. Very rare, Mid Perth, Leitrim. 1, H1. Sweden, Finland, N. America.

Whether var. *scabrifolia* is worth maintaining is open to question. Nyholm (1954–69) suggests it is not distinct. *M. tenerrima* may be distinguished by the differences listed under that species.

2. M. tenerrima (Brid.) Lindb., Musc. Scand., 1879
M. apiculata (Hüb.) Br. Eur.

Differing from *M. julacea* in the lvs less closely imbricate when dry and somewhat spreading when moist, lf margin crenulate-denticulate above, apex with recurved apiculus. Pale or yellowish-green plants among other bryophytes and as small patches on basic soil among rocks at high altitudes, very rare. Mid Perth, Angus. 2. W. and C. Europe, Iceland, Caucasus, N. and C. Asia, N. America, Greenland.

41. FABRONIACEAE

Dioecious or autoecious. Mostly slender plants, primary stems creeping, regularly branched or divided, stems usually without central strand or paraphyllia. Stem and branch lvs similar in shape, ovate to lanceolate, apex acute to acuminate, margin entire or denticulate; nerve usually present, single, short; cells rhomboidal in mid-lf, shorter towards margin and basal angles, smooth. Seta elongate; capsule ± erect and symmetrical; annulus usually of small, persistent cells; peristome single or double; lid conical; calyptra cucullate.

129. Myrinia Schimp., Syn., 1860

Slender, patch-forming plants; branching irregular. Lvs ovate or ovate-lanceolate, obtuse to shortly acute; nerve single, extending about ½ way up lf; cells small, rhomboidal, marginal and angular cells quadrate. Perichaetial lvs sheathing. Annulus absent; peristome double. Two species occurring in Europe, N. Asia and N. America, and a third species from Brazil.

1. M. pulvinata (Wahlenb.) Schimp., Syn., 1860
Helicodontium pulvinatum (Wahlenb.) Lindb.

Autoecious. Stems procumbent with ± erect branches. Lvs soft, appressed when

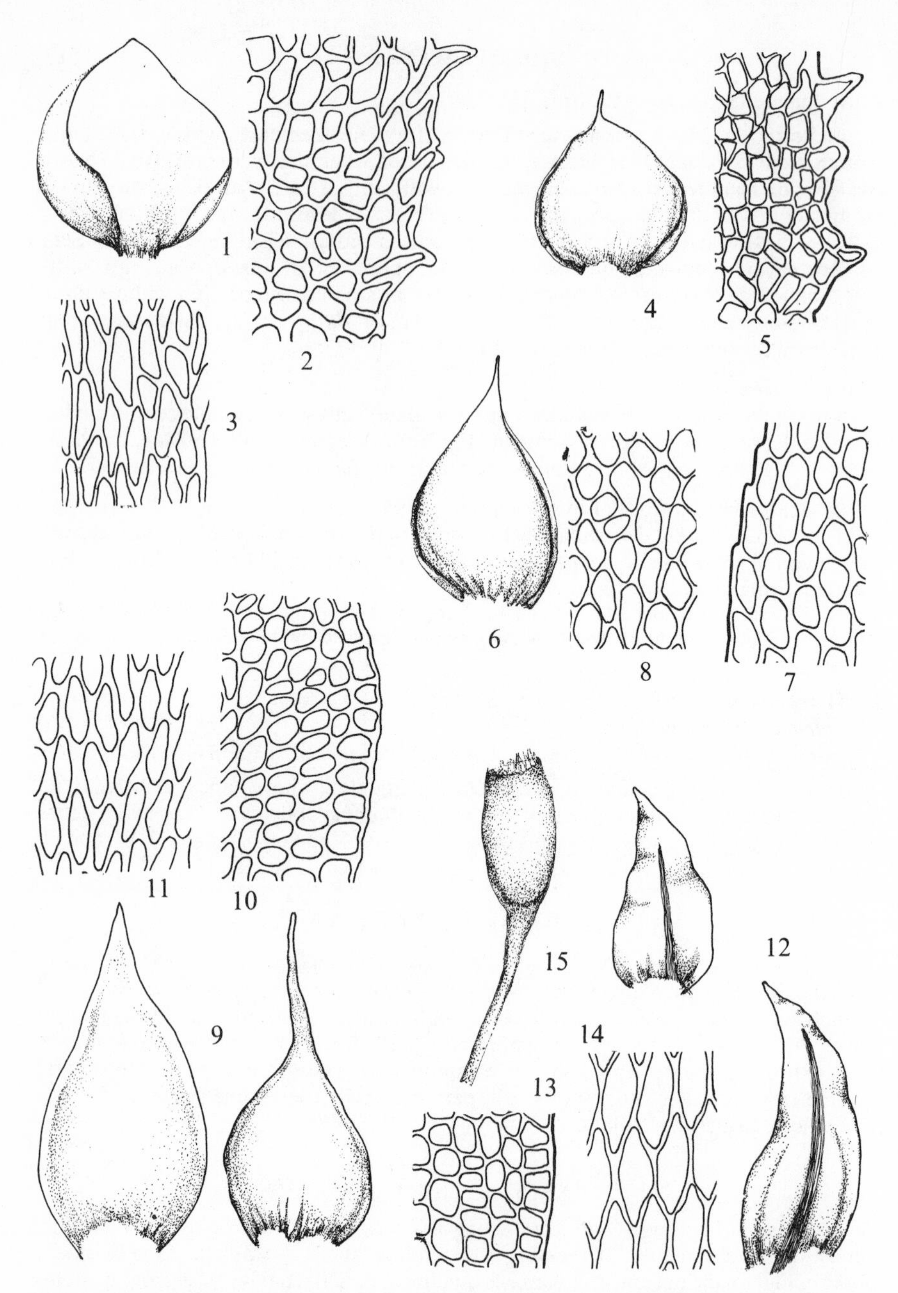

Fig. **250**. 1–3, *Myurella julacea* var. *julacea*: 1, leaf (×65); 2, marginal cells near leaf base; 3, cells from centre of leaf. 4–5, *M. julacea* var. *scabrifolia*: 4, leaf (×65); 5, marginal cells near leaf base. 6–8, *M. tenerrima*: 6, leaf (×65); 7, marginal cells near leaf base; 8, cells from centre of leaf. 9–11, *Habrodon perpusillus*: 9, leaves (×65); 10, marginal cells at widest part of leaf; 11, cells at centre of leaf. 12–15, *Myrinia pulvinata*: 12, leaves (×40); 13, marginal cells at widest part of leaf; 14, mid-leaf cells; 15, capsule (×15). Cells ×415.

dry, patent, sometimes secund when moist, ovate to ovate-lanceolate, apex acute to obtuse, margin plane, entire; nerve faint, single, extending to $\frac{1}{3}(-\frac{2}{3})$ way up lf, sometimes forked above; cells smooth, in mid-lf rhomboidal, 2–3 times as long as wide, 12–16 μm wide, shorter towards margin, ± rectangular or quadrate at basal angles. Capsule erect, slightly asymmetrical, ellipsoid; lid conical; spores 12–20 μm. Fr common, summer. Soft, dull green patches, often silt-encrusted on wood and bark in flood zone of streams, rivers and ditches in widely scattered localities from Dorset and E. Sussex north to Perth, rare. 21. France, Italy, N. Europe, Urals, Siberia, C. Asia, Canada.

Often accompanied by *Leskea polycarpa* but may be distinguished in the field by its smaller size, darker colour, lvs appressed when dry, faint nerve and shorter capsule.

130. HABRODON SCHIMP., SYN., 1860

Dioecious. Plants very slender, primary stems creeping, branches ± erect. Lvs ovate, tapering to long or short acuminate apex, margin entire or crenulate; nerve short and single or absent; cells smooth, narrowly elliptical in centre of lf, ± quadrate at margin. Annulus large, separating; inner peristome absent. Three mainly northern hemisphere species.

1. H. perpusillus (De Not.) Lindb., Öfv. K. V. A. Förh., 1863
H. notarisii Schimp.

Branches ± procumbent and homomallous. Lvs appressed with flexuose apices when dry, patent to spreading when moist, ± concave, ovate, tapering to short or long acuminate apex, margin plane, entire or finely crenulate; nerve extending up to $\frac{3}{4}$ way up lf or absent; cells smooth, incrassate, in mid-lf rhomboidal, 1–2 times as long as wide, *ca* 10 μm wide, shorter towards margin and at basal angles, longer towards middle of lf. Ovoid to shortly fusiform, (1–)2–5-celled brownish gemmae, 20–50 μm long, sometimes produced on younger parts of branch stems. Seta short, 3–4 mm; capsule narrowly ellipsoid. Fr very rare, spring. Small, pale green patches or growing through other bryophytes, on tree trunks, rare. Scattered localities from Devon, Somerset and Glamorgan north to Bute and Argyll. 15. Europe, Algeria.

42. LESKEACEAE

Primary stems usually stoloniform, irregularly or pinnately branched, branches erect or procumbent, paraphyllia often present. Lvs erect to spreading when moist, appressed when dry, concave, sometimes plicate, ovate to lanceolate, obtuse to acuminate, margin plane or recurved, entire or denticulate; nerve usually present, single or double, usually ending well below apex; cells smooth or papillose, usually 1–2 times as long as wide, quadrate or wider than long at basal angles. Seta long; capsule erect or inclined, straight or curved; peristome usually double; calyptra cucullate.

131. PSEUDOLESKEELLA KINDB., EUR. N. AMER. BRYIN., 1896

Dioecious. Lvs ± patent, when moist, ovate or lanceolate, obtuse to longly acuminate; nerve single, forked or double; mid-lf cells rounded or elongate, smooth or with papillae on distal walls. Seta elongate, capsule ovoid to cylindrical, straight or slightly curved; annulus usually separating; lid conical or rostrate. Eight or 9 mainly arctic or alpine northern hemisphere species.

Lf apex acute or if acuminate then mid-lf cells mostly 3–4 times as long as wide, nerve ceasing about $\frac{2}{3}$ way up lf **1. P. catenulata**
Lf apex acuminate to ± filiform, mid-lf cells 1–2 times as long as wide, nerve reaching acumen **2. P. nervosa**

1. P. catenulata (Brid.) Kindb., Eur. N. Amer Bryin., 1896
Pseudoleskea catenulata (Brid.) Br. Eur.

Stems irregularly pinnately branched, paraphyllia sparse. Lvs appressed when dry, patent when moist, slightly concave, from broad base ovate, tapering to acuminate to sub-obtuse apex, margin plane or narrowly recurved below; nerve extending to $\frac{2}{3}$ way up lf; cells incrassate, not or hardly papillose, cells towards basal angles transversely rectangular. Capsule erect, narrowly ellipsoid to sub-cylindrical; lid rostrate. Fr not known in Britain.

Var. **catenulata**
Stems and branches ± julaceous when dry. Lvs ovate, acute; cells in mid-lf rounded or oval, 1–2 times as long as wide, 12–14 μm long, 8–12 μm wide, cells near apex rounded. Dark or olive-green to brownish, dense patches on basic montane rocks in scattered localities from Caerns and Yorks north to W. Sutherland, rare. 17. N., W. and C. Europe, N. and C. Asia, Japan, Caucasus, China.

Var. **acuminata** (Culmann) Amann, Fl. Mousse Suisse, 1919
Pseudoleskea catenulata var. *acuminata* (Culmann) Dix.
Stems and branches slightly wispy with flexuose lf apices when dry. Lvs often sub-secund, ovate with ± long, acuminate apex, margin ± denticulate towards apex; cells in mid-lf elliptical to narrowly elliptical, (2–)3–4 times as long as wide, 20–30 μm long, 5–8 μm wide, cells near apex elliptical. Green or yellowish-green patches in similar habitats to the type, very rare. Westmorland, Perth, Angus, W. Inverness, W. Ross, W. Sutherland. 7. Europe, N. America.
Although completely distinct in Britain, var. *catenulata* and var. *acuminata* intergrade in continental Europe. See Crundwell, *Trans. Br. bryol. Soc.* **2**, 278–82, 1953, for an account of var. *acuminata*.

2. P. nervosa (Brid.) Nyholm, Moss Fl. Fenn., 1960
Leskea nervosa (Brid.) Myr., *Leskeella nervosa* (Brid.) Loeske

Stems irregularly branched. Lvs appressed when dry, patent when moist, concave, from a broad base ovate, tapering to long acuminate to filiform apex, acumen often turned to one side, margin plane, entire; nerve reaching acumen, stout; cells incrassate, slightly papillose, in mid-lf rounded to oval, 1–2 times as long as wide, 8–10 μm wide, towards basal angles wider than long. Dwarf, axillary, propaguliferous branches often present in upper parts of branches. Capsule erect, obovoid, slightly curved. Fr unknown in Britain. Patches on rocks or scattered shoots amongst other bryophytes on rocks or soil in basic, montane habitats. Recorded from Mid Perth, Aberdeen. 2. Europe, Iceland, Caucasus, N. and C. Asia, N. America.

132. Leskea Hedw., Sp. Musc., 1801

Autoecious. Primary stem creeping, branches erect or decumbent. Lvs ± ovate, obtuse to acute, margin entire or denticulate, recurved below; nerve usually ceasing below apex; cells ± quadrate or hexagonal, with central papillae. Seta long, arising from primary stem; capsule cylindrical, straight or curved; peristome double, without cilia. About 75 species occurring mainly in Europe, Asia and N. America.

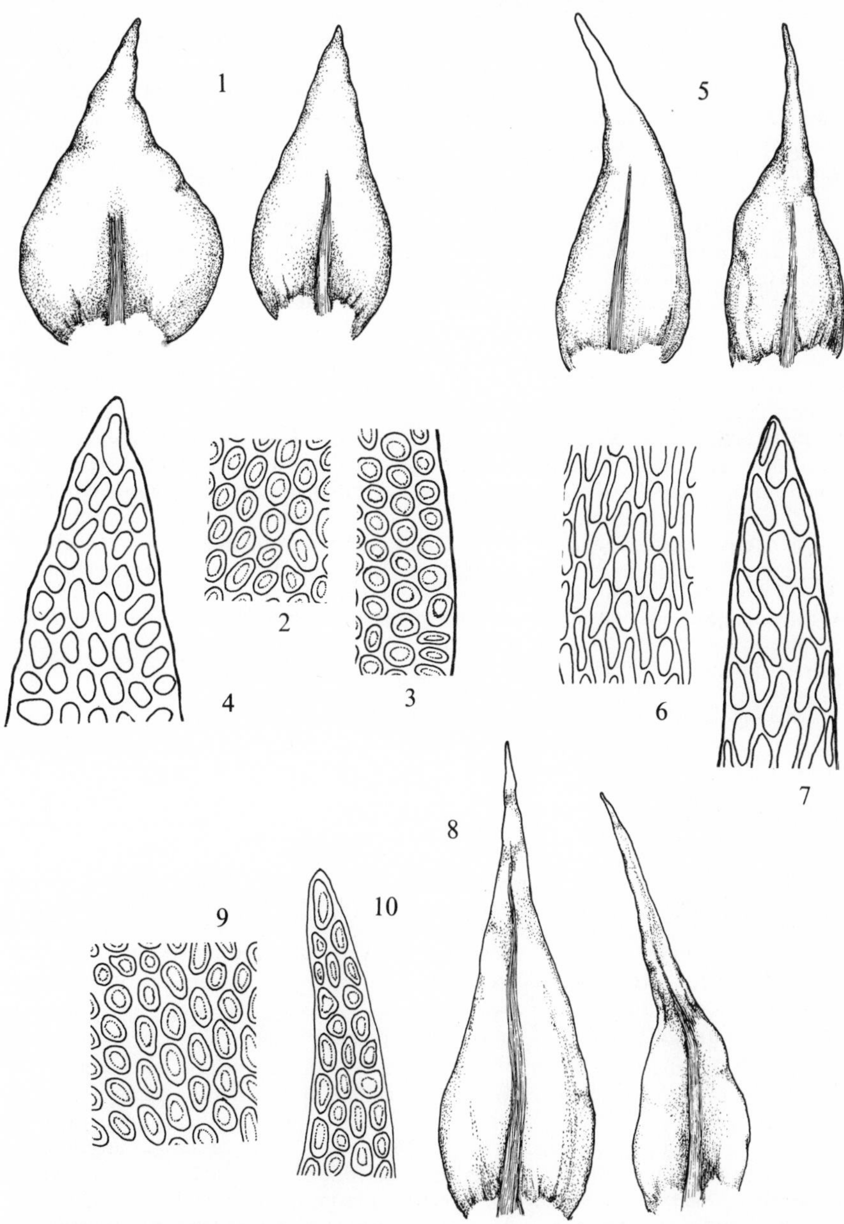

Fig. **251**. 1–4, *Pseudoleskeella catenulata* var. *catenulata*: 1, leaves; 2, mid-leaf cells; 3, marginal cells at widest part of leaf; 4, leaf apex. 5–7, *P. catenulata* var. *acuminata*: 5, leaves; 6, mid-leaf cells; 7, leaf apex. 8–10, *P. nervosa*: 8, leaves; 9, mid-leaf cells; 10, leaf apex. Leaves ×65, cells ×415.

1. L. polycarpa Hedw., Sp. Musc., 1801

Branches decumbent to erect. Lvs patent, spreading or sub-secund when moist, not much altered when dry, ovate or broadly ovate, obtuse to acute, margin recurved below, entire; nerve strong below, ceasing well below apex; cells ± uniform throughout lf, irregularly quadrate or quadrate-hexagonal, with central papilla, in mid-lf about as long as wide, 8–11 μm wide. Seta reddish-brown; capsule erect or slightly inclined, straight or curved, cylindrical; lid conical; spores 12–16 μm. Fr common, summer. $n=10^*$, $10+1$, 11, $11+2^*$. Dull, dark or brownish-green patches, often silt-encrusted, on tree roots, trunks and branches, occasionally on rocks, rarely on soil by streams and rivers, often in the flood zone, and by pools, common in lowland areas and extending north to Argyll and Angus, W. Sutherland. 94, H27, C. Europe, Caucasus, C. Asia, N. America.

For the differences from *Myrinia pulvinata* see under that species.

133. Lescuraea Br. Eur., 1851

Dioecious. Stems creeping, branched, without or with poorly developed central strand, usually with rhizoids and simple paraphyllia. Lvs lanceolate to ovate, acute to acuminate, plicate, margin plane or recurved, entire or denticulate above; nerve single, usually ceasing below apex; cells ± isodiametric to elongate-rhomboidal, papillose or not, at basal angles quadrate or transversely rectangular. Seta long; capsule erect and straight or curved and inclined; annulus poorly developed; peristome double, outer fused below into yellow or brown band; calyptra cucullate. About 10 arctic or alpine northern hemisphere species.

I have followed Lawton, *Bull. Torrey bot. Cl.* **84**, 281–307, 337–55, 1957, in the treatment of this genus.

1 Cells 1(–2) times as long as wide in mid-lf, with centrally placed papillae **1. L. patens**
  Lf cells 5–8 times as long as wide, or if shorter then lvs mostly falcato-secund, with papillae on end-walls 2
2 Cells in mid-lf short, less than 20 μm long **2. L. incurvata**
  Cells in mid-lf elongate, 30–55 μm long 3
3 Lvs strongly plicate, basal cells with pitted walls **3. L. plicata**
  Lvs slightly plicate, walls of basal cells not pitted **4. L. saxicola**

1. L. patens (Lindb.) Arn. & C. Jens., Naturw. Unt. Sarekgeb., 1910
Pseudoleskea patens (Lindb.) Limpr.

Stems irregularly branched, branches decumbent, straight. Lvs appressed when dry, uniformly spreading, rarely slightly secund when moist, plicate, ovate to broadly ovate, tapering to acuminate to filiform apex, margin recurved on one or both sides below and sometimes above, crenulate-denticulate above; nerve ending well below apex; cells with central conical papilla, irregularly quadrate-hexagonal, in mid-lf 1.0(–2.0) times as long as wide, 8–12 μm wide, 10–14(–16) μm long, marginal cells near base wider than long, basal cells about as wide as long. Seta brownish; capsule erect or inclined, ellipsoid, slightly curved; lid conical; spores 12–14 μm. Fr rare, spring, summer. Dark green to brownish patches on basic rocks, or wefts in rock crevices, usually at high altitudes, very rarely on trees, very rare. Mid Perth, Angus, S. Aberdeen, Inverness, Argyll, E. Ross. 7. Finland, Norway, Sweden, C. Europe, Corsica, Faroes, Iceland, N. America.

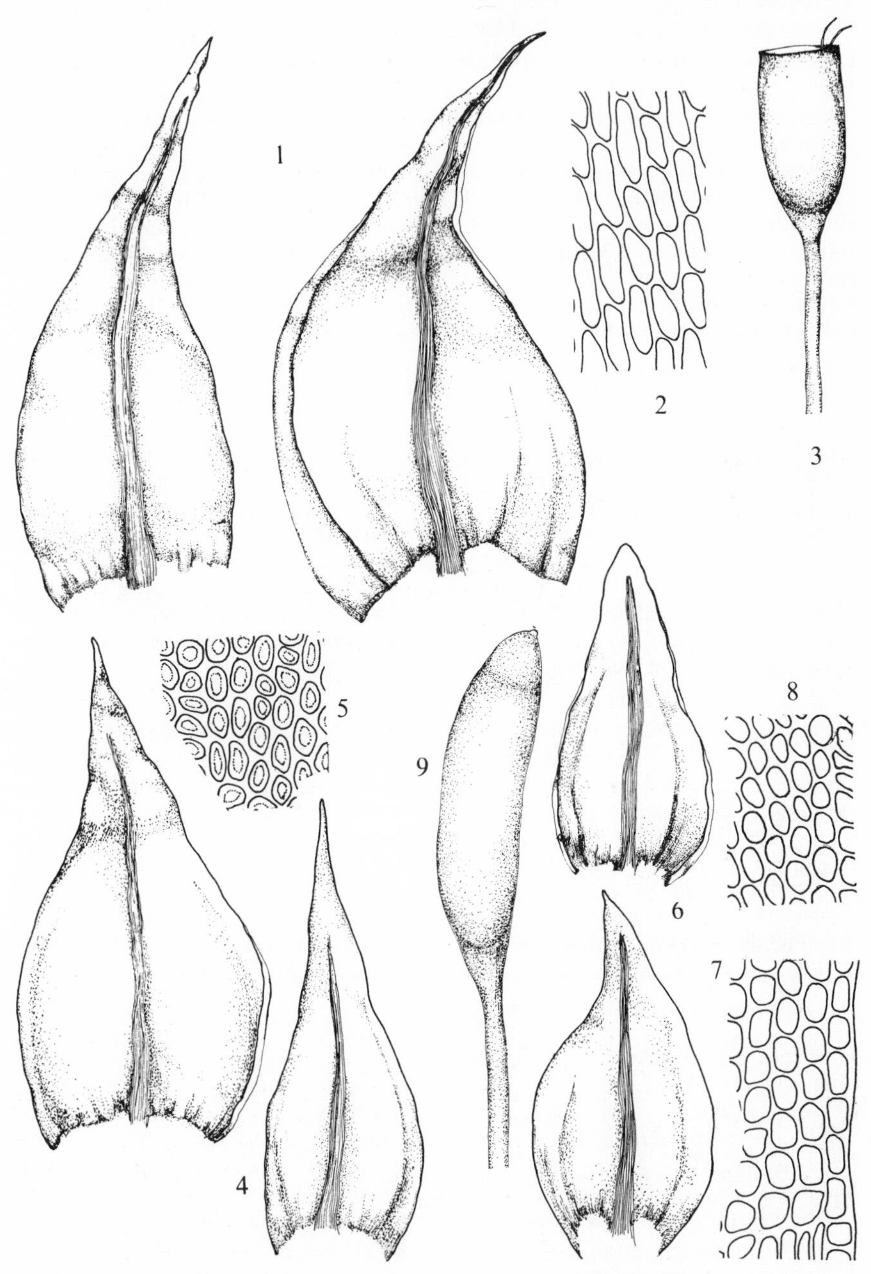

Fig. **252**. 1–3, *Lescuraea incurvata*: 1, leaves (×65); 2, mid-leaf cells; 3, capsule. 4–5, *L. patens*: 4, leaves; 5, mid-leaf cells. 6–9, *Leskea polycarpa*: 6, leaves (×40); 7, marginal cells at widest part of leaf; 8, mid-leaf cells; 9, capsule. Cells ×415, capsules ×15.

Confused in the past with *L. incurvata*, but distinguished by the ± uniformly isodiametric cells with thinner walls and centrally placed papillae.

2. L. incurvata (Hedw.) Lawton, Bull. Torrey Bot. Cl., 1957
Pseudoleskea atrovirens Br. Eur., *P. incurvata* (Hedw.) Loeske

Stems irregularly branched, stem and branch apices curved or hooked. Lvs appressed when dry, falcato-secund when moist, plicate, ovate, tapering to acuminate apex, margin recurved, denticulate towards apex; nerve ending in apex or percurrent, papillose at back above; cells incrassate, papillae near end-walls, in mid-lf variable in length, 1–3 times as long as wide, 6–8 μm wide, 10–16(–20) μm long, basal cells about as wide as long. Sporophyte similar to that of *L. patens*. Fr very rare, spring, Dark green patches, often blackish in older parts, on dry basic montane rocks, rare. Roxburgh, Fife, Perth, Angus, S. Aberdeen, Banff, Inverness, Argyll. 10. Europe, Iceland, N. and C. Asia, Sakhalin, Japan, N. America, Greenland.

3. L. plicata (Schleich. ex Web. & Mohr) Lawton, Bull. Torrey Bot. Cl., 1957
Brachythecium plicatum (Web. & Mohr) Br. Eur., *Ptychodium plicatum* (Web. & Mohr) Schimp.

Stems creeping, ± pinnately and much branched, branches prostrate to sub-erect. Lvs appressed when dry, patent, sometimes slightly secund when moist, strongly plicate, ovate-lanceolate to broadly ovate, ± abruptly narrowed to long or short acuminate apex, margin recurved, entire or obscurely denticulate towards apex; nerve reaching acumen; cells smooth, incrassate, basal cells shortly rectangular, above elongate, 5–8 times as long as wide, 5–8 μm wide, 35–55 μm long in mid-lf. Capsule ± horizontal, curved; spores 15–24 μm. Lax, yellowish-green to brownish patches on basic soil and rocks at high altitudes, very rare. Perth, Angus, W. Inverness. 4. Montane and arctic regions of Europe, Caucasus.

4. L. saxicola (Br. Eur.) Milde, Br. Siles., 1869
L. mutabilis var. *saxicola* (Br. Eur.) Hagen, *Pseudoleskea striata* var. *saxicola* Br. Eur.

Stem ± pinnately branched, branches decumbent, often curved or with curved tips. Lvs appressed when dry, sub-secund when moist, plicate, lanceolate to ovate, tapering to acuminate apex, margin plane or recurved, denticulate towards apex; nerve stout, ending in or below apex; cells incrassate with papillae at ends, in mid-lf elongate, 6–8 times as long as wide, 30–40 μm long, 6–8 μm wide, basal cells mostly ± 3 times as long as wide, at basal angles quadrate. Capsule erect, symmetrical, oblong-ellipsoid; lid conical; spores *ca* 16 μm. Fr unknown in Britain. Green or brownish-green, glossy patches on basic montane rocks. Recorded from Mid Perth in 1880, not seen since. Europe, Caucasus, N. and C. Asia, N. America.

134. Pterigynandrum Hedw., Sp. Musc., 1801

Dioecious. Usually slender plants, stems creeping or ascending, branches crowded, often secund or flagelliform. Lvs ± concave, imbricate to erecto-patent, ovate, acute, obtuse or obtuse and apiculate, margin plane or recurved below, entire or denticulate above; nerve slender, single or double, extending about ½ way up lf; cells about 2–6 times as long as wide with large, solitary papillae on dorsal side, angular cells ± quadrate. Capsule erect, cylindrical, straight, annulus falling with lid; lid rostrate; inner peristome without basal membrane or cilia; calyptra cucullate. Only 4 species, distributed through Europe, Asia, N. Africa, Macaronesia, N. and C. America.

1. P. filiforme Hedw., Sp. Musc., 1801

Plants slender or very slender, stems prostrate, branches numerous, procumbent to ascending, often short or flagelliform, often curved and pointing in one direction.

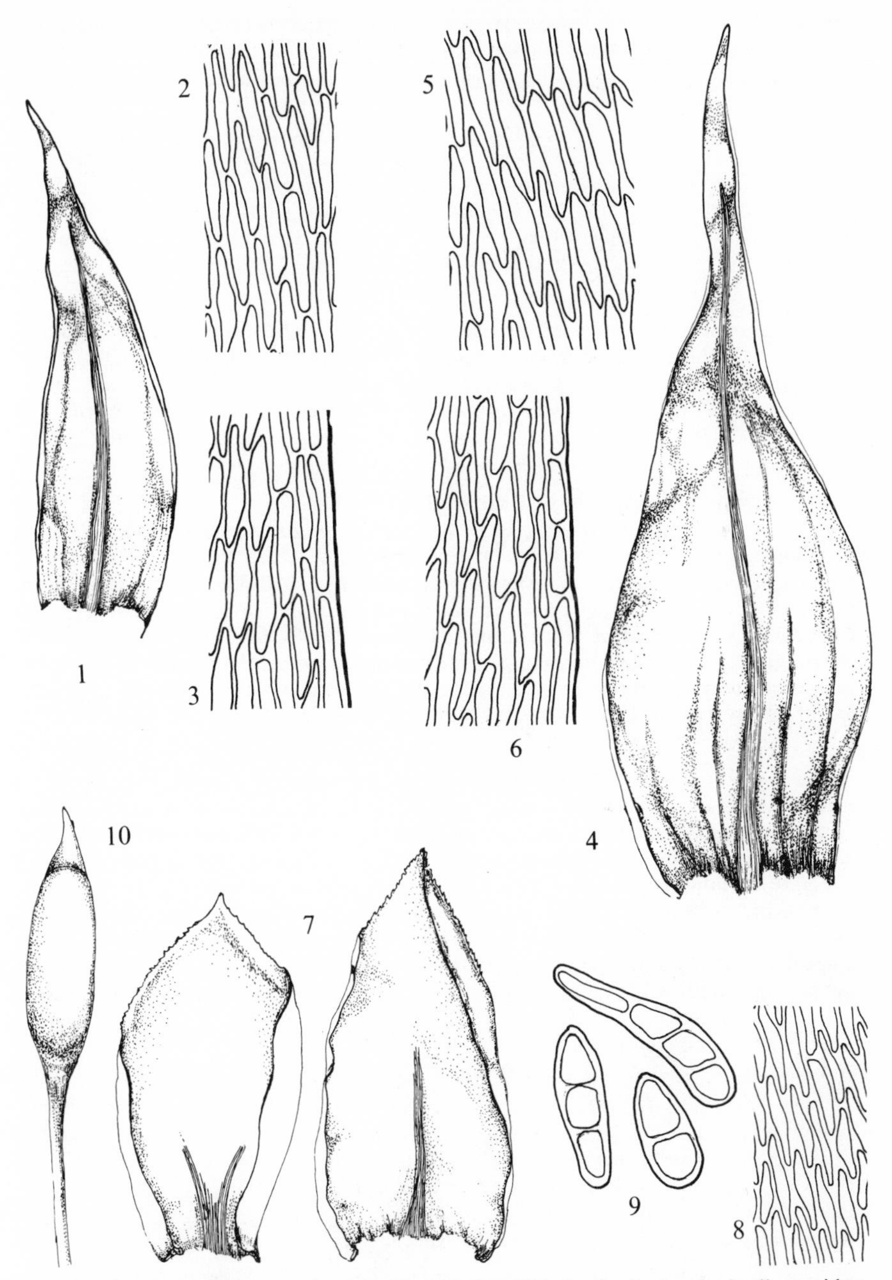

Fig. 253. 1–3, *Lescuraea saxicola*: 1, leaf (× 63); 2, mid-leaf cells; 3, marginal cells at widest part of leaf. 4–6, *L. plicata*: 4, leaf (× 40); 5, mid-leaf cells; 6, marginal cells at widest part of leaf. 7–10, *Pterigynandrum filiforme*: 7, leaves (× 63); 8, mid-leaf cells; 9, gemmae; 10, capsule (× 10). Cells × 415.

Lvs appressed when dry, imbricate to erecto-patent, sometimes sub-secund when moist, stem and branch lvs ± similar, concave at base, ovate to broadly ovate, acute to obtuse or obtuse and apiculate, margin plane or inflexed above, denticulate; nerve single, weak, extending about ½ way up lf or double and shorter; cells incrassate, basal narrowly rhomboidal, becoming rectangular to quadrate at basal angles, cells above linear-rhomboidal, radiating from nerve, towards margin and apex shorter, cells unipapillose with large distal, apically directed papilla on dorsal side, cells in mid-lf 5–8 μm×16–28 μm, 2.5–6.0 times as long as wide. Uniseriate, 2–5-celled gemmae, 50–80 μm long, often present on stems. Seta reddish; capsule erect, cylindrical; lid obliquely rostrate, acuminate; spores *ca* 12 μm. Fr very rare, summer.

Var. **filiforme**

Branches very slender, acute at tips, straight or slightly curved, (120–)160–240 (–320) μm diameter when dry. Lvs straight or sub-secund. Yellowish-green to green or brown mats on basic rocks and tree branches and roots in the mountains, occasional in the Scottish Highlands, very rare in N. Wales, N. England, Dumfries and Ireland. 24, H4. Europe, Iceland, Faroes, Caucasus, Siberia, Kashmir, Japan, Korea, Algeria, Madeira, Canaries, N. America, Greenland.

Var. **majus** (De Not.) De Not., Cronac. Briol. Ital., 1867

P. filiforme var. *heteropterum* (Brid.) Br. Eur., *P. filiforme* var. *decipiens* (Web. & Mohr) Limpr.

Branches stouter, obtuse, curved, especially at tips, 320–480 μm diameter when dry. Lvs sub-secund. Rocks and boulders by montane lakes, rare. Caerns, Mid Perth, Angus, S. Aberdeen, E. Inverness, E. Ross, Down. 6, H1. N. and C. Europe, Pyrenees, Caucasus, Kashmir, Punjab, Indo-China, Japan, N. America.

The main differences between the 2 varieties are indicated above and I have not been able to find any correlated lf characters. The varieties are poorly defined with numerous intermediates, but in the absence of more detailed study I have retained them.

43. THUIDIACEAE

Dioecious or autoecious. Primary stems creeping, secondary stems procumbent to erect, distantly and irregularly branched to 1- to 3-pinnate, sometimes producing stoloniform innovations, central strand absent or rudimentary; paraphyllia often present and numerous, various in form. Stem and branch lvs usually differing, stem lvs often concave, plicate or not, with broad, often decurrent base, ovate, ovate-lanceolate or lingulate, acute or occasionally obtuse, margin usually denticulate; nerve well developed, single, rarely short and double; cells short at least towards margin, sometimes elongated or linear towards nerve, papillose on one or both sides, rarely smooth. Seta elongated, reddish; capsule erect and symmetrical to horizontal and curved; annulus differentiated; lid conical or convex and apiculate; peristome double, inner with well developed basal membrane, processes and cilia; calyptra cucullate.

135. Heterocladium Br. Eur., 1852

Dioecious. Primary stems stoloniform, secondary stems irregularly branched, paraphyllia few. Stem lvs ovate, acuminate, base cordate, excavate, decurrent, margin entire or denticulate; nerve short and double or longer and single; cells unipapillose, median elongate or not, towards margin and apex quadrate-hexagonal to rounded-quadrate. Branch lvs smaller, more bluntly pointed. Perichaetial lvs with sheathing base and long, narrow, spreading to reflexed apex. Seta long; capsule inclined or horizontal, ellipsoid, straight or curved; annulus of large cells, separating; peristome perfect, inner with cilia. A small genus of 8 species occurring in Europe, N. and E. Asia, Macaronesia, N. America, Brazil.

Stem lvs patent or spreading, tapering to acute to acuminate apex **1. H. heteropterum**

Stem lvs squarrose or squarrose-recurved, abruptly narrowed to filiform to acuminate apex **2. H. dimorphum**

1. H. heteropterum (Bruch ex Schwaegr.) Br. Eur., 1852

Plants slender or very slender, primary stems stoloniform, secondary stems procumbent to ascending, irregularly branched, branches often homomallous. Stem lvs patent or spreading, often sub-secund, excavate at base, ovate, acute to acuminate, margin plane, denticulate; nerve short and double or longer and single; cells papillose, median, ± elliptical or quadrate-hexagonal, shorter towards margin and apex. Branch lvs smaller, relatively wider, often secund. Perichaetial lvs from broad expanded basal part ± linear, entire. Seta purplish, erect or decumbent; capsule inclined to horizontal, narrowly ellipsoid; lid with stout, blunt, straight or curved beak; spores 14–16 μm.

Var. **heteropterum**

Plants slender. Stem lvs mostly 0.8–1.2 mm long, ovate, acute to acuminate; nerve short and double or longer and single; median cells elliptical, 5–7 μm × 12–24 μm, 2–4 times as long as wide. Branch lvs 0.3–1.0 mm long, ovate, acute. Fr very rare, spring. Dull or bright green, rarely yellowish-green, dense patches on damp shaded rocks, sometimes locally abundant, or wefts among other bryophytes on soil and rocks, in ravines and by streams and rivers. Rare in lowland Britain, frequent or common in suitable habitats elsewhere. 72, H32. N., W. and C. Europe, Faroes, Macaronesia.

Var. **flaccidum** Br. Eur., 1852

H. heteropterum var. *fallax* Milde

Plants very slender. Stem lvs scarcely larger than branch lvs, 0.24–0.40 mm ovate-lanceolate to lanceolate, acute to acuminate; nerve very short, double; cells ± quadrate-hexagonal throughout lf, median cells 5–8 μm × 8–16 μm, 1–2 times as long as wide. Branch lvs 0.15–0.32 mm long, ovate-lanceolate to lanceolate, acute. Fr unknown. Thin patches in similar, though usually more heavily shaded habitats to var. *heteropterum*, very rare in lowland Britain, occasional elsewhere. 63, H13, C. Norway, Sweden, Germany, France, Silesia.

A very variable species, the most distinctive form of which that occurs in Britain being var. *flaccidum*. A N. American species, *H. macounii* Best, has been reputed to occur in Britain and other parts of Europe; it differs from *H. heteropterum* in its larger size, yellowish-green colour and the terminal cell of ultimate branch lvs having 2 papillae (only 1 papilla in *H. heteropterum*). Warburg (*Trans. Br. bryol. Soc.* **3**, 126–7, 1956) has shown that this plant does not occur in Britain and indeed suggests that it is not distinct from the N. American species *H. heteropteroides* Best. *H. dimorphum* differs in the squarrose stem lvs with a wider band of short cells on either side of the median band of elongated cells and the more bluntly pointed branch lvs.

2. H. dimorphum Br. Eur., 1852

H. squarrosulum (Voit.) Lindb.

Similar to *H. heteropterum* in habit. Stem lvs squarrose or squarrose-reflexed, broadly ovate, abruptly narrowed to usually long acuminate to filiform apex, base excavate, decurrent, margin sharply denticulate; nerve short, double; cells papillose, median narrowly rectangular, 5–8 μm × 20–32 μm, 3–5 times as long as wide, towards margin ± abruptly rectangular, trapezoid or quadrate-hexagonal. Branch lvs concave, ovate, apex rounded to acute. Fr unknown in Britain. Dull green patches on shaded montane rocks, very rare. Stirling, Perth, Angus, E. Inverness, Dunbarton. 6. Montane and northern Europe, Iceland, Caucasus, N. America, Greenland, Brazilian Andes.

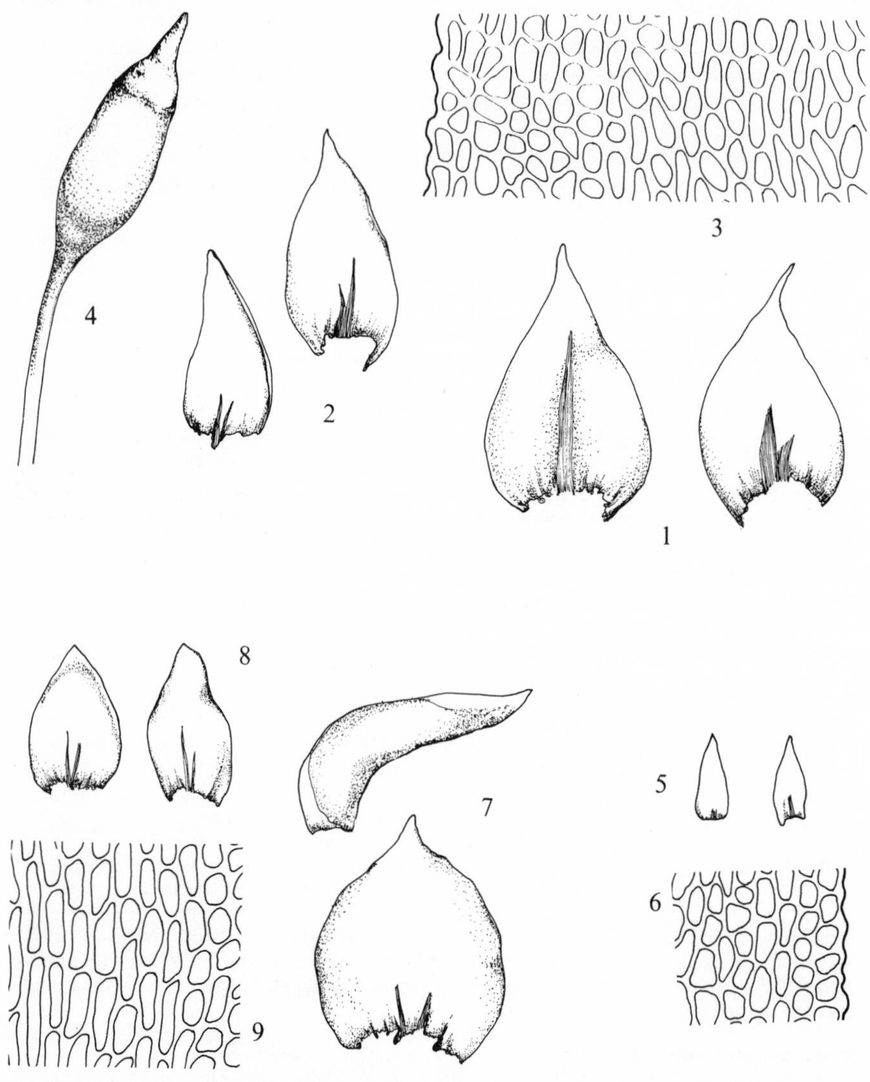

Fig. **254**. 1–4, *Heterocladium heteropterum* var. *heteropterum*: 1, stem leaves; 2, branch leaves; 3, cells at middle of stem leaf; 4, capsule (× 15). 5–6, *H. heteropterum* var. *flaccidum*: 5, stem leaves; 6, cells at middle of stem leaf. 7–9, *H. dimorphum*: 7, stem leaves; 8, branch leaves; 9, mid-leaf cells. Leaves × 40, cells × 415.

136. Anomodon Hook. & Tayl., Musc. Brit., 1818

Dioecious. Primary stems stoloniform with small lvs differing in shape from those of secondary stems, secondary stems ± erect, irregularly branched, branches sometimes flagelliform, paraphyllia lacking. Lvs erecto-patent to spreading, sometimes secund, ovate to lanceolate or lingulate, apex rounded to acute, margin papillose-crenulate, sometimes toothed above; nerve single, ending below apex; cells incrassate, rounded-hexagonal except at base, usually multipapillose. Perichaetial lvs with sheathing base and reflexed apex. Seta long; capsule erect, cylindrical; cilia of inner peristome rudimentary or absent; calyptra cucullate. About 30 species distributed through Europe, Asia, N. Africa, America, Australasia.

1 Lvs acuminate, cells unipapillose **1. A. longifolius**
Lf apex usually rounded or obtuse, apiculate or not, cells with 2–3 papillae 2
2 Plants slender, lvs 1.0–2.2 mm long, margin with a few teeth near apex **2. A. attenuatus**
Plants robust, lvs 2.0–3.5 mm long, margin without teeth near apex **3. A. viticulosus**

1. A. longifolius (Brid.) Hartm., Handb., Skand. Fl., 1838

Plants slender, primary stems stoloniform, secondary stems irregularly branched, sometimes flagelliform. Lvs loosely appressed when dry, patent when moist, 0.9–2.4 mm long, ovate to lanceolate, tapering to acuminate apex, base decurrent, sub-sheathing, margin plane, crenulate below, slightly denticulate towards apex; nerve ending below apex; cells incrassate, unipapillose, at basal angles slightly enlarged, marginal cells for some distance up lamina transversely rectangular, cells elsewhere rounded-hexagonal, conically papillose, 8–10 μm wide in mid-lf. Capsule ellipsoid. Fr unknown in Britain. $n=11$. Yellowish-green patches on sheltered basic rocks, very rare and decreasing. N. Somerset, W. Glos, Monmouth, Hereford, N.W. Yorks, Durham, Mid Perth, Angus. 8. N., W. and C. Europe, Caucasus, N. and C. Asia.

A. attenuatus and *A. viticulosus* differ in lf shape, obtuse or rounded lf apices and cells with 2–3 papillae. *A. viticulosus* is usually a much larger plant with secondary stems up to 12 cm long.

2. A. attenuatus (Hedw.) Hüb., Musc. Germ., 1833

Plants slender, primary stems stoloniform, secondary stems irregularly branched, some branches flagelliform. Lvs loosely appressed when dry, patent to spreading, sometimes secund when moist, 1.0–2.2 mm long, basal part obtuse, lingulate or broadly lingulate above, apex obtuse, apiculate or not, base decurrent, sub-sheathing, margin plane, papillose, with a few teeth near apex; nerve ending below apex; cells incrassate with 2–3 papillae, basal elongated, at basal angles not enlarged, marginal cells near base slightly wider than long, elsewhere rounded-hexagonal, 8–12 μm wide in mid-lf. Fr unknown in Britain. Yellowish-green patches on damp, shaded logs and rocks. Recorded from Mid Perth and Angus but not seen for more than 50 years. N., W. and C. Europe, Caucasus, Iran, Kashmir, Japan, N. America, Mexico, Guatemala, Cuba, Jamaica.

3. A. viticulosus (Hedw.) Hook. & Tayl., Musc. Brit., 1818

Plants robust, primary stems stoloniform, secondary stems sparsely branched, 2–12 cm long. Lvs loosely appressed or incurved and curled when dry, spreading, sometimes secund, somewhat undulate when moist, 2.0–3.5 mm long, from ovate base lingulate-lanceolate, apex obtuse or rounded, base decurrent, sub-sheathing,

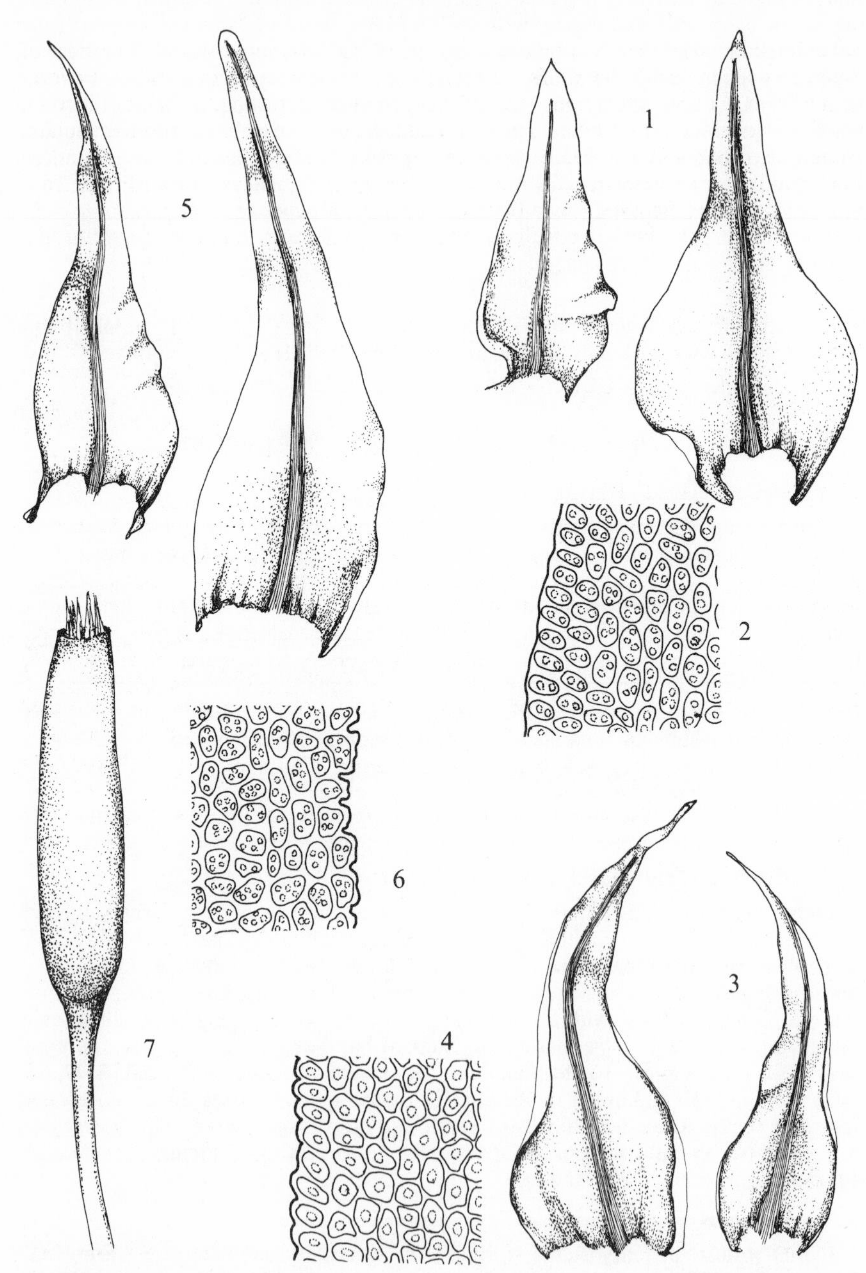

Fig. **255**. 1–2, *Anomodon attenuatus*: 1, leaves; 2, mid-leaf cells. 3–4, *A. longifolius*: 3, leaves; 4, mid-leaf cells. 5–7, *A. viticulosus*: 5, leaves; 6, mid-leaf cells; 7, capsule (×15). Leaves ×25, cells ×415.

margin plane or narrowly recurved, papillose-crenulate but not toothed; nerve ending below apex; cells incrassate, with 2–3 papillae, basal rectangular, angular cells not enlarged, marginal cells near base wider than long, cells above rounded-hexagonal 7–10 μm wide in mid-lf. Seta long; capsule ± cylindrical; spores *ca* 16 μm. Fr rare, winter. $n=11$. Large, lax, bright green or yellowish-green tufts or patches, dull green when dry, on usually ± vertical rocks, rock faces, walls, tree bases and tree trunks. Generally distributed, common and sometimes locally abundant in basic habitats at low altitudes, rare elsewhere. 102, H38, C. Europe, Caucasus, Asia Minor, Iran, Siberia, Kashmir, Nepal, China, Canaries, Algeria, N. America.

137. Thuidium Br. Eur., 1852

Dioecious. Primary stems stoloniform, secondary stems usually arcuate, irregularly 1–3 pinnately branched, paraphyllia present. Stem lvs plicate, concave, broadly ovate or ovate-triangular with long or short acumen, margin crenulate or denticulate; nerve ending in or below apex; cells unipapillose, short, to 3 times as long as wide. Perichaetial lvs narrowly lanceolate, with longly tapering acuminate apex, margin entire or ciliate. Seta long; capsule inclined or horizontal, curved; annulus present, peristome perfect, inner with cilia. A world-wide genus of about 240 species.

I have not recognised *Abietinella* C. Müll. as a separate genus as the only distinguishing character is unipinnate as opposed to 2- to 3-pinnate branching.

1 Stems simply pinnate — **1. T. abietinum**
  Stems 2- or 3-pinnate — 2
2 Apex of apical cell of ultimate branch lvs acute, not papillose — **2. T. tamariscinum**
  Apex of apical cell of ultimate branch lvs obtuse, crowned with 2–3 papillae — 3
3 Stem lf apex not reflexed when dry, apical cell of acumen elongated or not, other cells of acumen similar to lower cells, not uniseriate, perichaetial lvs ciliate — **3. T. delicatulum**
  Stem lf apex reflexed when dry, apex filiform, composed of uniseriate row of cells or cells of acumen longer than lower cells, perichaetial lvs not ciliate — 4
4 Cells beyond end of nerve of stem lvs similar to other cells of lf, paraphyllia with median or sub-apical papillae — **4. T. philibertii**
  Cells beyond end of nerve elongated and differing from other lamina cells, paraphyllia with terminal papillae — **5. T. recognitum**

1. T. abietinum (Hedw.) Br. Eur., 1852
Abietinella abietina (Hedw.) Fleisch.

Stems ± arcuate, simply pinnate, branches in 4 rows, not complanate. Stem lvs patent, often curved, plicate, broadly ovate with broad base, tapering to long acuminate apex. Branch lvs concave, broadly ovate to lanceolate, obtuse to acuminate; nerve extending about $\frac{3}{4}$ way up lf; cells rounded to elliptical, incrassate, papillose, shorter towards margin, 8–10 μm wide. Perichaetial lvs not ciliate. Paraphyllia irregularly branched, papillose. Capsule sub-erect, cylindrical, curved; lid conical, acute.

Ssp. abietinum

Branches ± terete with appressed lvs when dry. Stem lvs mostly 1.0–1.4 mm long, tapering to acuminate apex. Branch lvs mostly 0.6–1.0 mm long, ovate, acute to obtuse, margin entire; cells in mid-lf 8–16 μm long, 1.0–1.5(–2.0) times as long as wide. Fr very rare. $n=11$. Lax, yellowish-green to brownish wefts in turf, especially

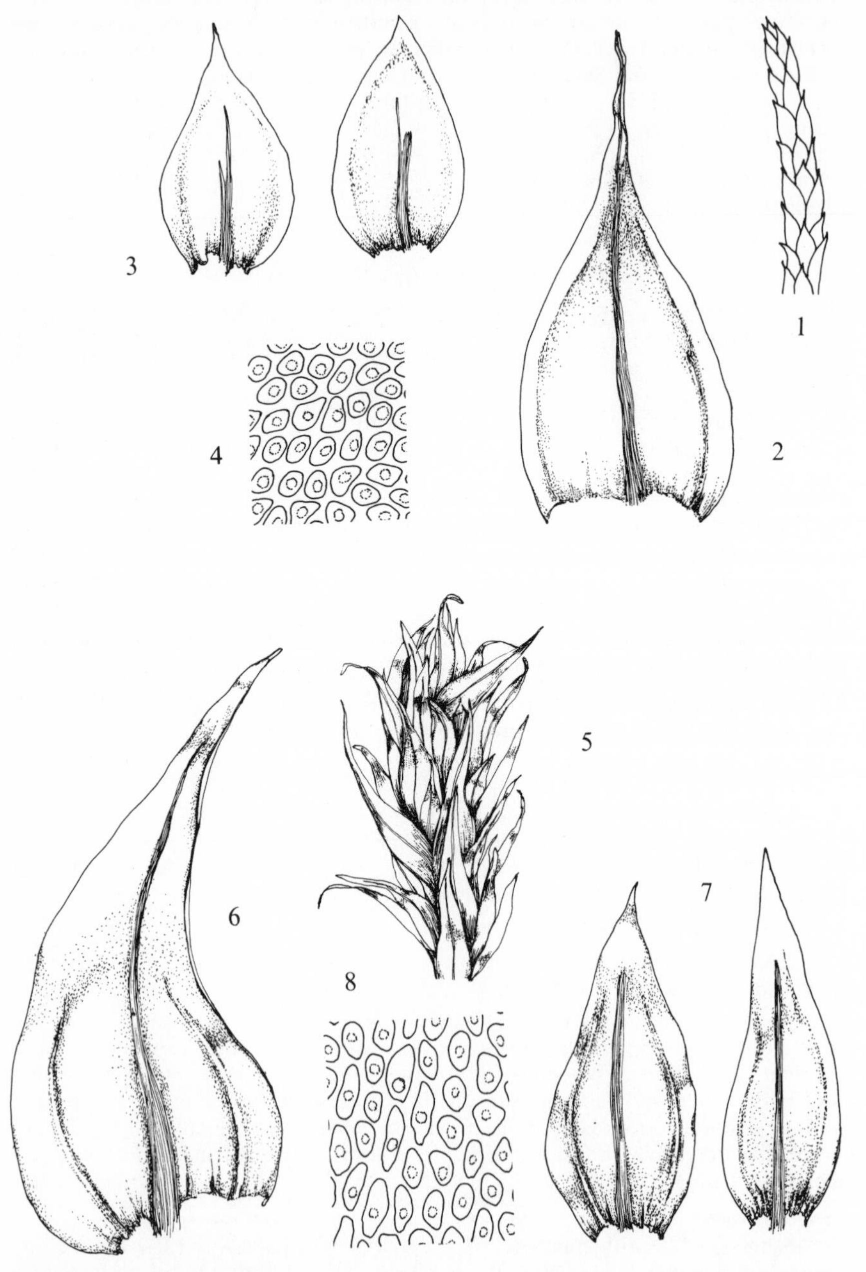

Fig. **256**. 1–4, *Thuidium abietinum* ssp. *abietinum*: 1, branch; 2, stem leaf; 3, branch leaves; 4, mid-leaf cells of stem leaf. 5–8, *T. abietinum* ssp. *hystricosum*: 5, branch; 6, stem leaf; 7, branch leaves; 8, mid-leaf cells of stem leaf. Branches ×15, leaves ×40, cells ×415.

in chalk grassland, sand-dunes and scree, usually where basic. Occasional in England and Wales, rare in Scotland and Ireland, extending from W. Cornwall and E. Kent north to Perth, Angus, Banff and E. Ross. 35, H7, C. Europe, Iceland, Caucasus, N. and C. Asia, Kashmir, China, Japan, N. America, Greenland.

Ssp. **hystricosum** (Mitt.) Kindb. Eur. N. Am. Bryin., 1897
T. hystricosum Mitt., *A. histricosa* (Mitt.) Broth.
Branch lvs ± patent when dry, branches not terete. Stem lvs mostly 1.5–2.0 mm long, tapering to longly acuminate apex. Branch lvs sometimes secund, mostly 0.9–1.3 mm long, ovate to lanceolate, shortly or longly tapering to acute to acuminate apex, margin entire to denticulate; cells in mid-lf 12–20 μm long, 1.5–3.0 times as long as wide. Fr unknown in Britain or Ireland. Yellowish-green to brownish wefts, occasional in chalk grassland, on chalk banks and chalk quarries, very rarely in other types of basic habitat. Dorset and Kent north to Bucks and Beds, Yorks, Mid Perth, Sligo, Donegal, Derry. 17, H4. France, Spain, Czechoslovakia, Switzerland, Italy, China.

Although ssp. *hystricosum* is usually distinct in its appearance, with the lvs patent when dry and the narrower, more longly tapering branch lvs with longer cells, occasional forms intermediate between it and ssp. *abietinum* do occur.

2. T. tamariscinum (Hedw.) Br. Eur., 1852

Stems ± arcuate, (2–)3-pinnate, often elegantly so, branches complanate, branches sometimes interrupted, stems brownish with paraphyllia. Lvs loosely appressed when dry, erecto-patent to patent when moist, stem lvs from broad base widely ovate, rapidly narrowed to lingulate to acuminate apex, margin recurved below, crenulate to denticulate above, nerve strong, ending below apex; cells thin-walled to incrassate, unipapillose, 8–10 μm wide in mid-lf. Lvs of ultimate branches concave, ovate, acute. Inner perichaetial lvs with ciliate margin. Capsule inclined, cylindrical, curved; lid

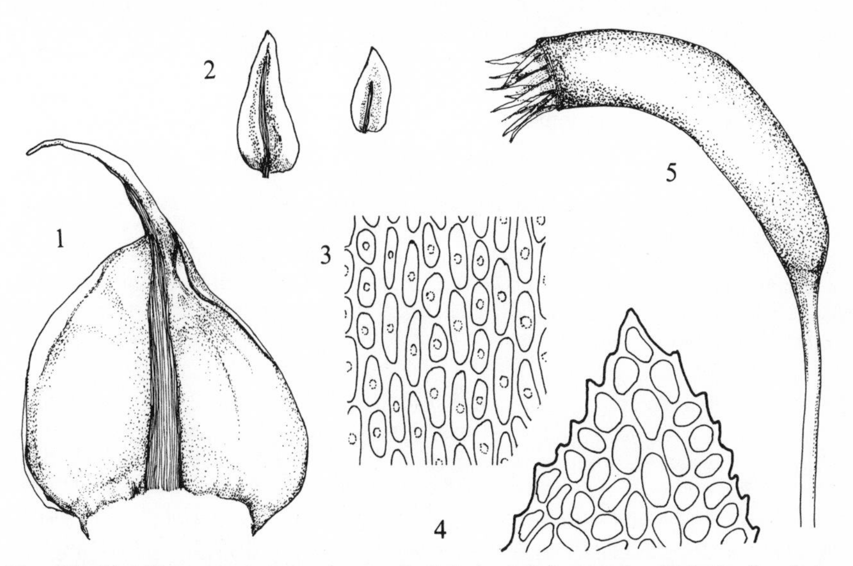

Fig. 257. *Thuidium tamariscinum*: 1, stem leaf; 2, branch leaves; 3, mid-leaf cells of stem leaf; 4, apex of ultimate branch leaf; 5, capsule (×15) Leaves ×40, cells ×415.

with oblique beak; spores 12–20 μm. Fr occasional in W. and N. Britain, very rare elsewhere, winter, spring. *n*=10+1, 11*. Bright green to rich golden-brown wefts, sometimes extensive, on damp ground, tree boles, rotting logs and rocks in woods, in rock crevices and in damp turf, frequent to common. 112, H40, C. Europe, Faroes, Japan, Azores, Madeira, N. America, Jamaica.

Usually easily recognised by the handsome 3-pinnate, arcuate shoots. Stunted forms may be confused with *T. delicatulum* but in that species the branches are more congested and hardly complanate.

3. T. delicatulum (Hedw.) Mitt., J. Linn. Soc. Bot., 1869
T. recognitum var. *delicatulum* (Hedw.) Warnst.

Stems ± arcuate, (2–)3-pinnate, branches sometimes congested and hardly complanate; stems brownish with paraphyllia. Stem lvs appressed to spreading, acumen rarely reflexed when dry, ± patent when moist, plicate, from broad base ovate-triangular, gradually or abruptly tapering to long or short acumen, margin recurved, crenulate-denticulate; nerve ending below acumen; cells of acumen ± similar to cells of upper part of lamina, not incrassate, mostly 1–2 times as long as wide, 30–50 μm long except apical cell which may be much elongated. Lvs of ultimate branches concave, oval, acute; cells unipapillose, apical cell obtuse with 2–3 terminal papillae. Perichaetial lvs ciliate. Paraphyllia with median or sub-apical papillae, papillae never at distal walls. Capsule inclined, cylindrical, curved; spores 16–20 μm. Fr occasional, winter. *n*=11*. Green wefts on shaded, siliceous rocks in woods or shaded montane habitats, in bogs and on heaths, occasionally on sand-dunes. Occasional in W. and N. Britain and Ireland. 45, H26. Europe, Iceland, Faroes, Caucasus, N. and C. Asia, China, Japan, N. America, Mexico, Jamaica, Haiti, St Vincent, Andes.

Although *T. delicatulum* and *T. philibertii* are reduced to intraspecific taxa of *T. recognitum* by some authorities, the differences between them are sufficient for the maintenance of specific status. *T. philibertii* is usually distinct from *T. delicatulum* in the gametophytic characters listed under the former and in the non-ciliate perichaetial lvs. *T. recognitum* differs from both in the elongated cells of the lf apex continuing beyond the end of the nerve and the position of the papillae of the paraphyllia. For a discussion of these 3 species see Tallis, *Trans. Br. bryol. Soc.* **4**, 102–6, 1961.

4. T. philibertii Limpr., Laubm., 1895
T. recognitum ssp. *philibertii* (Limpr.) Dix.

Close to *T. delicatulum* but differing in the following characters. Stem lvs appressed to patent with apex reflexed or recurved when dry, apex filiform; nerve ending in acumen; ultimate cells of acumen incrassate, cells beyond end of nerve similar to upper cells of lamina, 1–2 times as long as wide. Paraphyllia as in *T. delicatulum*. Perichaetial lvs not ciliate. Fr not known in Britain or Ireland. Yellowish-green wefts in chalk and limestone grassland, on basic sand-dunes and on damp basic rocks in woods and montane habitats, generally distributed, occasional. 87, H16. Europe, Iceland, Siberia, Altai, Kashmir, China, N. America.

5. T. recognitum (Hedw.) Lindb., Not. Sällsk. F. Fl. Fenn. Förh., 1874

Close to *T. delicatulum* and *T. philibertii* but differing from one or other in the following characters. Stem lvs with reflexed or recurved apices when dry, apex filiform; nerve extending almost to apex of acumen, cells of acumen incrassate, cells beyond end of nerve elongated. Paraphyllia with papillae against distal walls of cells. Perichaetial lvs not ciliate. Fr not known in Britain or Ireland. Yellowish-green wefts on limestone or basic rocks in woods or in shaded habitats, especially in Derby, Mid W. Yorks, Lake District and central Scottish Highlands, very rarely in

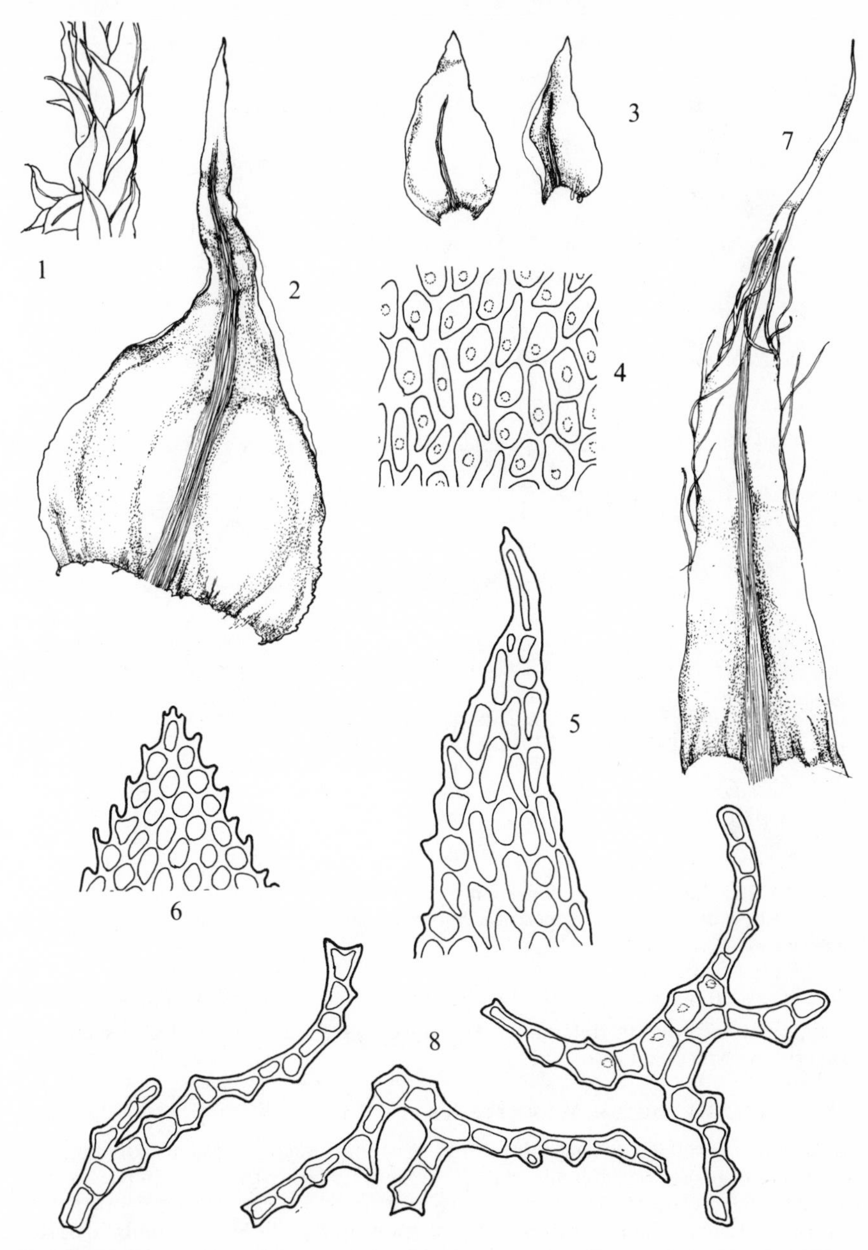

Fig. **258**. *Thuidium delicatulum*: 1, portion of stem (×15); 2, stem leaf; 3, branch leaves; 4, mid-leaf cells of stem leaf; 5, stem leaf apex; 6, apex of ultimate branch leaf; 7, perichaetial leaf; 8, paraphyllia. Leaves ×40, cells ×415.

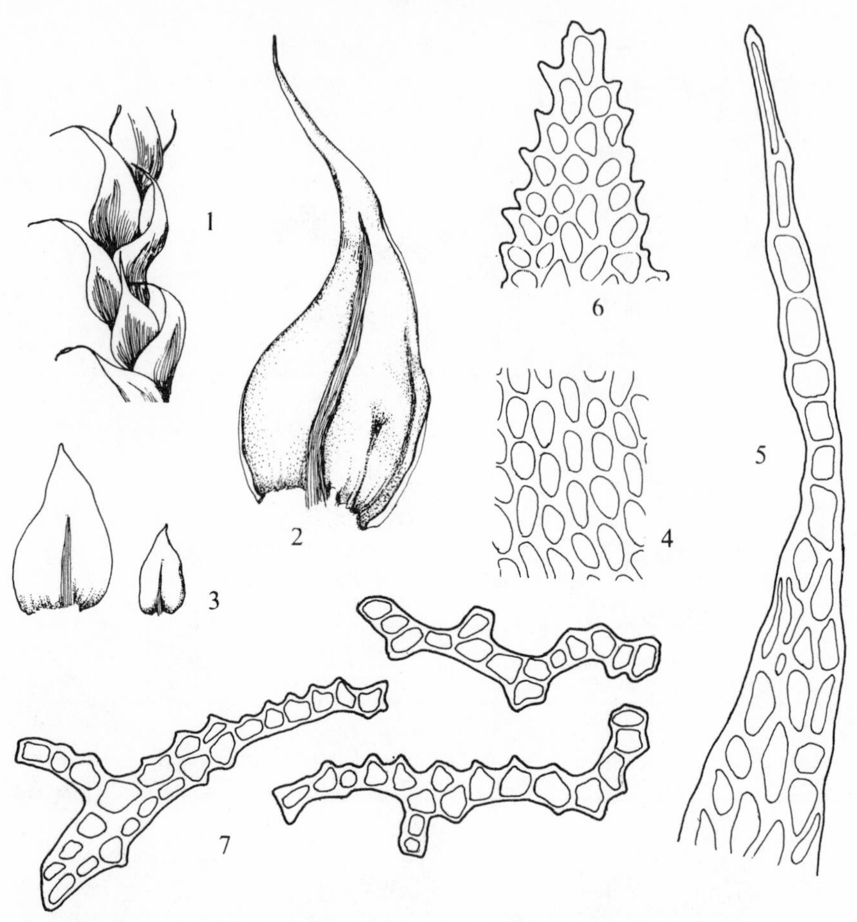

Fig. **259**. *Thuidium philibertii*: 1, portion of stem ($\times 15$); 2, stem leaf; 3, branch leaves; 4, mid-leaf cells of stem leaf; 5, apex of stem leaf; 6, apex of ultimate branch leaf; 7, paraphyllia. Leaves $\times 40$, cells $\times 415$.

sand-dunes. Extending from Somerset, Glos, N. Wales and Yorks north to Sutherland, a few localities in Ireland. 20, H5. Europe, Caucasus, N. and C. Asia, Japan, Algeria, N. America, Greenland.

138. HELODIUM WARNST., KRYPT. FL. BRANDENBURG, 1905

Autoecious. Primary stems stoloniform, secondary stems erect, usually regularly pinnately branched, paraphyllia abundant. Stem lvs larger than branch lvs, concave, plicate, broadly ovate, acuminate, margin recurved at least below, denticulate or entire; nerve extending $\frac{1}{2}$ way up lf to just below apex; cells linear-rhomboidal, thin-walled or incrassate, unipapillose on dorsal surface. Perichaetial lvs plicate, lanceolate, acuminate, not ciliate. Seta elongate; capsule inclined to horizontal, cylindrical, curved; annulus differentiated; lid conical, acute or convex and apiculate; peristome

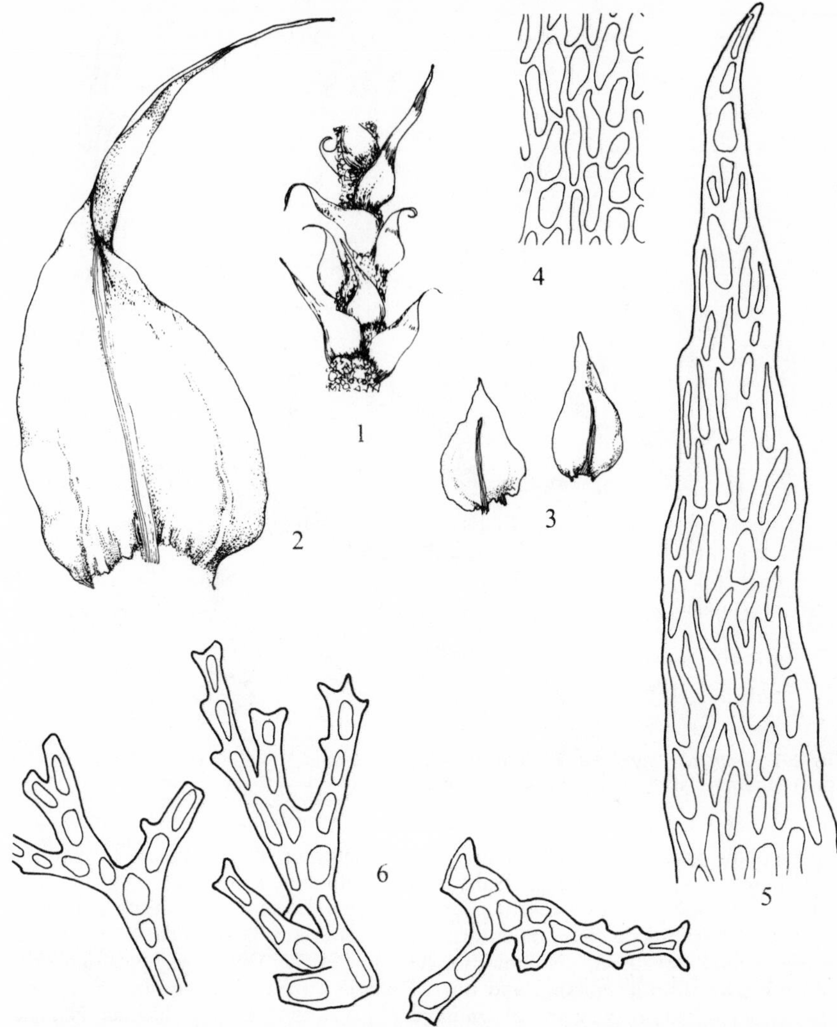

Fig. **260**. *Thuidium recognitum*: 1, portion of stem (×15); 2, stem leaf; 3, branch leaves; 4, mid-leaf cells of stem leaf; 5, apex of stem leaf; 6, paraphyllia. Leaves ×40, cells ×415.

perfect, inner with well developed basal membrane and cilia; calyptra cucullate. A small genus of 4 species of usually marshy habitats; circumboreal.

1. H. blandowii (Web. & Mohr) Warnst., Krypt. Fl. Brandenburg, 1905
Thuidium blandovii (Web. & Mohr) Br. Eur.

Primary stems stoloniform, secondary stems erect, regularly pinnately branched, stems and branches densely clothed with brownish paraphyllia. Lvs loosely appressed when dry, patent when moist, stem lvs crowded, with paraphyllia at basal angles and back of nerve, concave, plicate, broadly ovate, shortly tapering to acuminate apex,

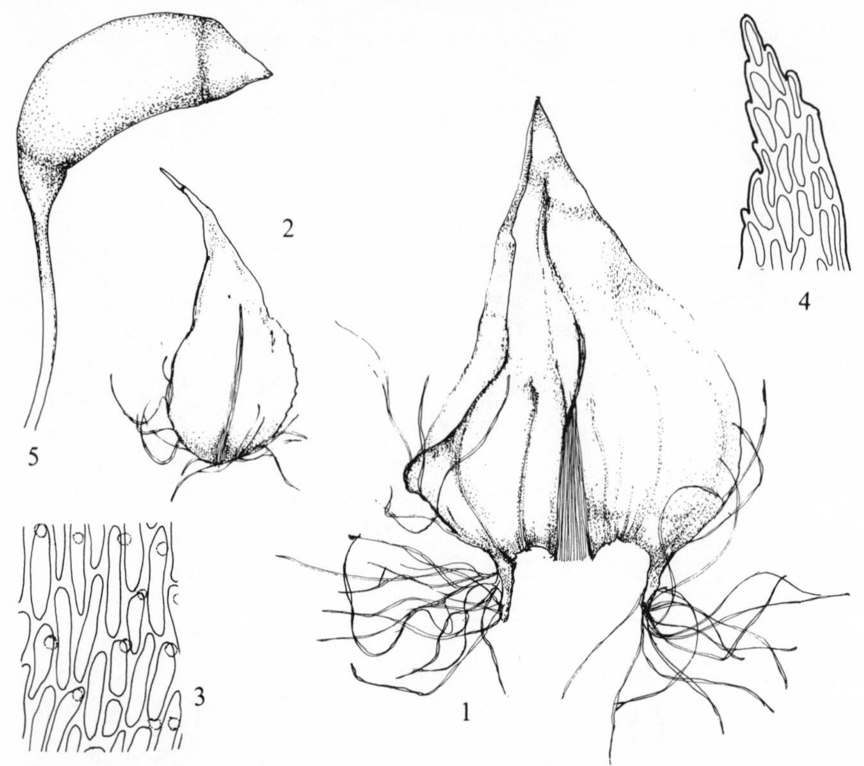

Fig. **261**. *Helodium blandowii*: 1, stem leaf; 2, branch leaf; 3, mid-leaf cells of stem leaf; 4, apex of branch leaf; 5, capsule (×15). Leaves ×40, cells ×415.

margin plane or recurved below, sharply denticulate; nerve extending $\frac{1}{2}$–$\frac{3}{4}$ way up lf; cells elongate, unipapillose on dorsal surface, papillae on distal walls, cells in mid-lf 6–9μm × 28–44 μm, 5–7 times as long as wide. Branch lvs smaller, ovate or lanceolate, acuminate. Paraphyllia long, much branched, uniseriate. Seta inclined, cylindrical, curved; spores 16–20 μm. Fr common, summer. $n=12$. Dense, yellowish-green to golden-brown tufts in marshes and bogs. Recorded from Cheshire and Yorks but now extinct in Britain as a result of drainage. N., W. and C. Europe, Iceland, N. Asia, Japan, N. America, Greenland.

19. HYPNOBRYALES

Pleurocarpous, slender to robust plants. Primary stems variously branched; paraphyllia and pseudoparaphyllia present or not. Lvs linear-lanceolate to orbicular; nerve single, double or absent; cells narrowly rhomboidal to linear-vermicular, usually smooth, angular cells differentiated or not. Seta long, smooth or papillose; capsule inclined to horizontal, rarely erect, ovoid to cylindrical; lid concave to longly rostrate or subulate; peristome double, outer papillose, inner with or without tall basal membrane, cilia present or not.

Some of the families in this order such as the Plagiotheciaceae, Hypnaceae and Hylocomiaceae are very poorly defined and, whilst maintaining with misgivings the Plagiotheciaceae, I have treated the Hylocomiaceae as a sub-family of the Hypnaceae. The disposition of some of the genera is also controversial, again because the characters used are vague and unsatisfactory.

44. AMBLYSTEGIACEAE

Autoecious to dioecious, antheridia and archegonia borne in dwarf bud-like branches on main stems. Slender or robust, lax or densely matted plants, stems irregularly to pinnately branched; stolons lacking; paraphyllia usually absent. Lvs not usually complanate, symmetrical, of one kind only although branch lvs usually smaller with weaker nerve, imbricate to spreading, sometimes sub-secund to falcato-secund; nerve well developed, single, rarely double or faint or absent; cells prosenchymatous and hexagonal to long and narrow, smooth, angular cells frequently differentiated and distinct. Perichaetial lvs differentiated. Seta reddish, smooth; capsule inclined or horizontal, ellipsoid to cylindrical, curved, frequently contracted below mouth when dry and empty; peristome double, outer teeth often distinctly articulated, inner with well developed basal membrane, processes broad, cilia entire; lid shortly conical with sharp or mamillose apex; calyptra cucullate, smooth.

A natural family the species of which usually occur in moist, wet or aquatic habitats. Although the chromosome numbers both within and between species may be very variable, the meiotic complements of those species studied are of characteristic appearance, differing from those of other members of Hypnobryales.

139. CRATONEURON (SULL.) SPRUCE, CAT. MUSCI AMAZ. AND., 1867

Dioecious. Usually robust plants forming lax tufts, often glossy when dry; stems procumbent to erect, often matted with rhizoids, paraphyllia mostly numerous, branching ± regularly pinnate. Stem lvs larger and wider than branch lvs, secund or falcato-secund, from wide decurrent base cordate-triangular, gradually or abruptly narrowed to channelled acumen, margin plane, entire or denticulate below; nerve strong or very strong, reaching at least to base of acumen; cells narrow, prosenchymatous, smooth, at basal angles enlarged, thin- or thick-walled, hyaline or coloured, forming distinct auricles. Inner perichaetial lvs erect, pale, lanceolate, acutely pointed. Seta reddish; capsule inclined or horizontal, curved, symmetrical or gibbous, often contracted below mouth when dry and empty; annulus differentiated; peristome perfect; lid conical. A ± world-wide genus of about 19 species.

1 Stem lvs not plicate, mid-lf cells 2–4(–6) times as long as wide, cells smooth **1. C. filicinum**
 Stem lvs plicate, mid-lf cells (6–)9–15 times as long as wide or if shorter then cells papillose 2
2 Stem and branch lf cells (6–)9–15 times as long as wide in mid-lf, smooth **2. C. commutatum**
 Stem and branch lf cells (2.5–)3.0–6.0 times as long as wide in mid-lf, papillose at back at least in young lvs **3. C. decipiens**

1. C. filicinum (Hedw.) Spruce, Cat. Musci Amaz. And., 1867
Amblystegium filicinum (Hedw.) De Not.

Shoots procumbent, ascending or erect, stems irregularly to pinnately branched, rarely unbranched, usually with dense tomentum of rhizoids, paraphyllia few to

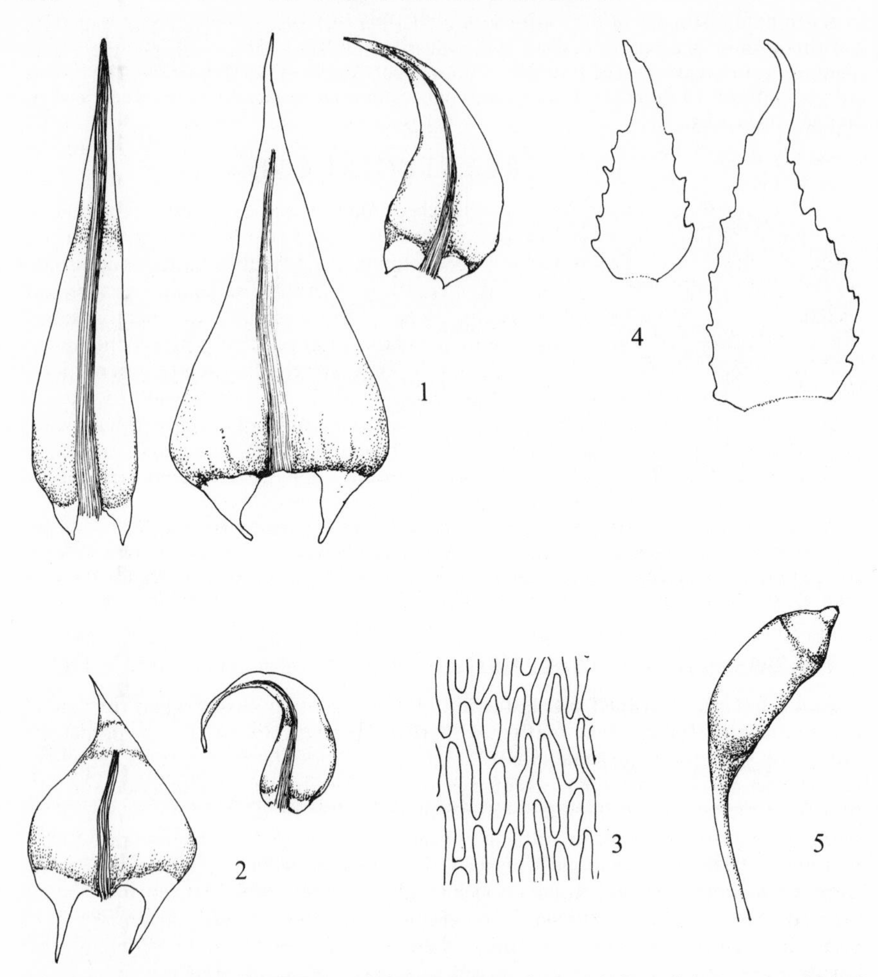

Fig. **262**. *Cratoneuron filicinum* var. *filicinum*: 1, stem leaves; 2, branch leaves; 3, mid-leaf cells (×415); 4, paraphyllia (×165); 5, capsule (×10). Leaves ×40.

many. Stem lvs erecto-patent to spreading, not plicate, scarcely altered when dry, rarely falcato-secund, cordate-triangular, acuminate, margin denticulate throughout with teeth most prominent at widest part of lf; nerve stout, ending below acumen to excurrent; angular cells enlarged, forming decurrent auricles extending almost to nerve, basal cells irregularly rectangular, cells above shorter, ± elliptical, smooth, 2–6 times as long as wide in mid-lf. Branch lvs similar but smaller and narrower. Capsule inclined, ellipsoid, curved; lid conical, obtuse; spores 12–14 μm.

Var. **filicinum**

Stem and branch apices with hooked appearance from strongly falcate lvs; paraphyllia usually present. Stem lvs concave, tapering from widest part to acuminate apex; branch lvs ± straight to strongly curved, gradually tapering from widest part to acuminate apex; nerve very stout; angular cells incrassate, cells in mid-lf mostly 2–4 times as long as wide, 6–8 μm×16–24 μm. Fr rare, spring. Tight patches with regularly pinnately branched stems to lax, erect tufts with sparsely branched stems, golden-yellow to bright green, on damp soil on banks, rocks, cliffs, in dune slacks, marshy places, flushes, by streams, on waste ground, etc., frequent or common, especially in basic habitats. 112, H40, C. Europe, Iceland, Faroes, N., W., C. and E. Asia, Madeira, Algeria, N. America, Mexico, Ecuador, New Zealand.

Var. **curvicaule** (Jur.) Mönk., Hedwigia, 1911
Amblystegium curvicaule (Jur.) Lindb.
Stem and branch apices not or scarcely hooked; paraphyllia absent. Stem and branch lvs concave, straight or only slightly curved, ± abruptly narrowed from broad base to fine acumen; nerve less stout than in var. *filicinum*, ending below acumen; angular cells ± thin-walled, cells in mid-lf 3–6 times as long as wide. Fr unknown. On basic rocks at high altitudes, very rare. Mid Perth, Skye. 2. N. and C. Europe, Spain, Iceland, Siberia, Yunnan.

An extremely variable species with a range in form from small, slender plants to lax, coarse tufts 5–7 cm high. A number of varieties has been described but intergrade to such an extent that with the exception of var. *curvicaule* they are not recognised here. Coarse plants with erect stems differ markedly in appearance from the more usual procumbent plants with pinnately branched stems but the stem and branch apices generally have the characteristic hooked appearance of *C. filicinum*. The plant referred to as var. *fallax* (Brid.) Roth (*Amblystegium filicinum* var. *vallisclausiae* auct.) is merely a phenotypic variant, as I have seen plants with the characteristics of that variety organically attached to typical *C. filicinum*.

2. C. commutatum (Hedw.) Roth, Hedwigia, 1899

Plants robust, rarely slender, stems procumbent, ascending or erect, regularly or irregularly and frequently plumosely branched; stems tomentose with rhizoids and paraphyllia or not. Stem and branch lvs falcate or falcato-secund, sometimes strongly so, rarely ± straight. Stem lvs plicate, mostly gradually tapering from near base to acumen, not abruptly narrowed at insertion, margin plane or recurved, denticulate below and not or obscurely so above; nerve stout; cells at basal angles enlarged, yellowish to brownish, forming decurrent auricles extending ± to nerve, other cells linear, smooth or with solitary distal papillae in lower part of lf, cells in mid-lf (6–)9–15 times as long as wide. Capsule ± horizontal, narrowly ellipsoid or sub-cylindrical, curved; lid with straight, stout, acute point; spores 16–20 μm. Fr rare, spring.

Key to intraspecific taxa of *C. commutatum*

1 Stem lvs ± straight or rarely slightly curved, nerve percurrent to excurrent — var. **virescens**
  Stem lvs falcate to falcato-secund, nerve ending below apex — 2
2 Plants slender or robust, nerve extending ¾ way or more up lf, mid-lf cells mostly 9–15 times as long as wide — 3
  Plants slender, nerve reaching *ca* ½ way up lf, mid-lf cells mostly 6–8 times as long as wide — var. **sulcatum**
3 Stems regularly pinnately branched, stem lvs cordate-triangular, stems tomentose with brown rhizoids — var. **commutatum**
  Stems irregularly or sub-pinnately branched, stem lvs ovate to lanceolate, stems with few or no rhizoids — var. **falcatum**

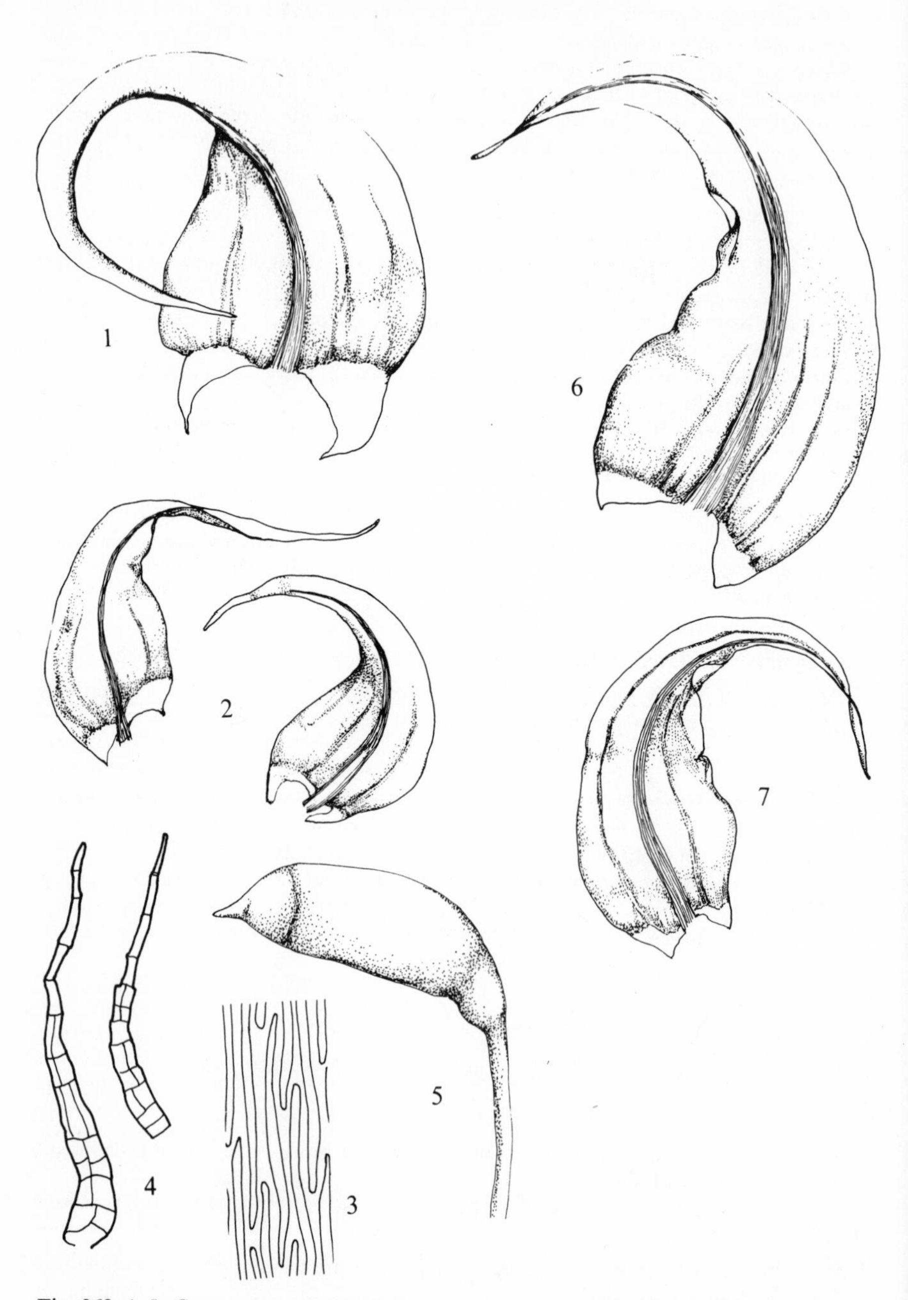

Fig. **263**. 1–5, *Cratoneuron commutatum* var. *commutatum*: 1, stem leaf; 2, branch leaves; 3, mid-leaf cells (×415); 4, paraphyllia (×165); 5, capsule (×10). 6–7, *C. commutatum* var. *falcatum*: 6, stem leaf; 7, branch leaf. Leaves ×40.

Var. **commutatum**
Hypnum commutatum Hedw.
Stems regularly complanately pinnately branched, with paraphyllia and tomentum of brownish rhizoids, branches without tomentum. Stem lvs cordate-triangular, sometimes widely so, gradually tapering from near base to acumen, branch lvs ovate-lanceolate; nerve extending *ca* ¾ way up lf; cells in mid-lf 5–7 μm × 55–100 μm, 9–15 times as long as wide. $n=10$*. Dense or lax, dull or bright green tufts or patches, often encrusted with calcareous matter below, in streams, flushes, on wet basic rocks, in dune slacks, frequent or common in suitable habitats in basic areas, rare elsewhere. 105, H34, C. Europe, Iceland, Faroes, W. Asia, Kashmir, Tibet, Japan, Korea, Kamchatka, Madeira, N. Africa, N. America, Greenland.

Var. **falcatum** (Brid.) Mönk., Hedwigia, 1911
Hypnum falcatum Brid., *C. falcatum* (Brid.) Roth
Plants more robust than in var. *commutatum*, stems irregularly to sub-pinnately branched, with few or no rhizoids. Lvs narrower than in var. *commutatum*, tapering from ovate or lanceolate basal part but otherwise similar. $n=7$, 10. Dense or lax, greenish or more usually yellowish-green to orange-brown tufts or patches, often encrusted with calcareous matter below, on wet rocks by streams, waterfalls, on cliffs, in flushes, fens and dune slacks, frequent or common in suitable habitats in basic areas, rare elsewhere. 100, H31. Montane parts of Europe, Iceland, Faroes, W. and C. Asia, Punjab, Kashmir, Szechwan, N. Africa, N. America.

Var. **virescens** (Schimp.) Rich. & Wall., Trans. Br. bryol. Soc., 1950
Hypnum falcatum var. *virescens* Schimp.
Stems sparsely branched, without rhizoids. Lvs ± straight or only curved at stem tips; nerve very stout, percurrent to excurrent. Dull green tufts on submerged basic rocks in streams, rare. In a few scattered localities from Northants, Worcs and Lincs north to W. Sutherland. 16. Europe, Iceland, Caucasus, Turkestan.

Var. **sulcatum** (Lindb.) Mönk., Hedwigia, 1911
Hypnum sulcatum Lindb.
Plants slender, stems regularly pinnately branched, stems with sparse paraphyllia and rhizoids. Stem lvs abruptly or gradually tapering to long fine acumen; nerve less stout than in var. *commutatum* and only extending about ½ way up lf; cells in mid-lf 5–7 μm × 30–40 μm, 6–8 times as long as wide. Fr unknown. Golden-brown patches on basic rocks at high altitudes, very rare. Mid Perth, W. Inverness, Argyll, Mid Ebudes. 4. Montane and northern Europe, Greenland, Alaska.

A very variable species the status of the varieties of which is open to question. The relationship between var. *commutatum* and var. *falcatum* is controversial. Both Dixon (1924) and Nyholm (1954–69) say that plants of var. *commutatum* occur with peripheral branches with the characters of var. *falcatum*. Nyholm also reports of individual shoots, var. *commutatum* below, var. *falcatum* above. On the other hand Bell & Lodge (*J. Ecol.* **51**, 113–22, 1963) say that the 2 varieties remain distinct in culture and that var. *falcatum* will tolerate water much lower in bases than will var. *commutatum*. I have retained var. *falcatum* pending further investigation. Var. *virescens* is probably merely a form of the type occurring on submerged rocks. Var. *sulcatum* is a more distinctive plant in the shorter nerve and cells and seems to possess reasonably distinctive features even for a species as polymorphic as *C. commutatum*.

3. C. decipiens (De Not.) Loeske, Moosfl. Harz, 1903
Hypnum decipiens De Not.

Plants slender, procumbent, stems pinnately branched and sometimes forked, with tomentum of rhizoids and paraphyllia; stem and branch tips sometimes hooked. Lvs falcato-secund, plicate, concave, stem lvs cordate-triangular, sometimes broadly

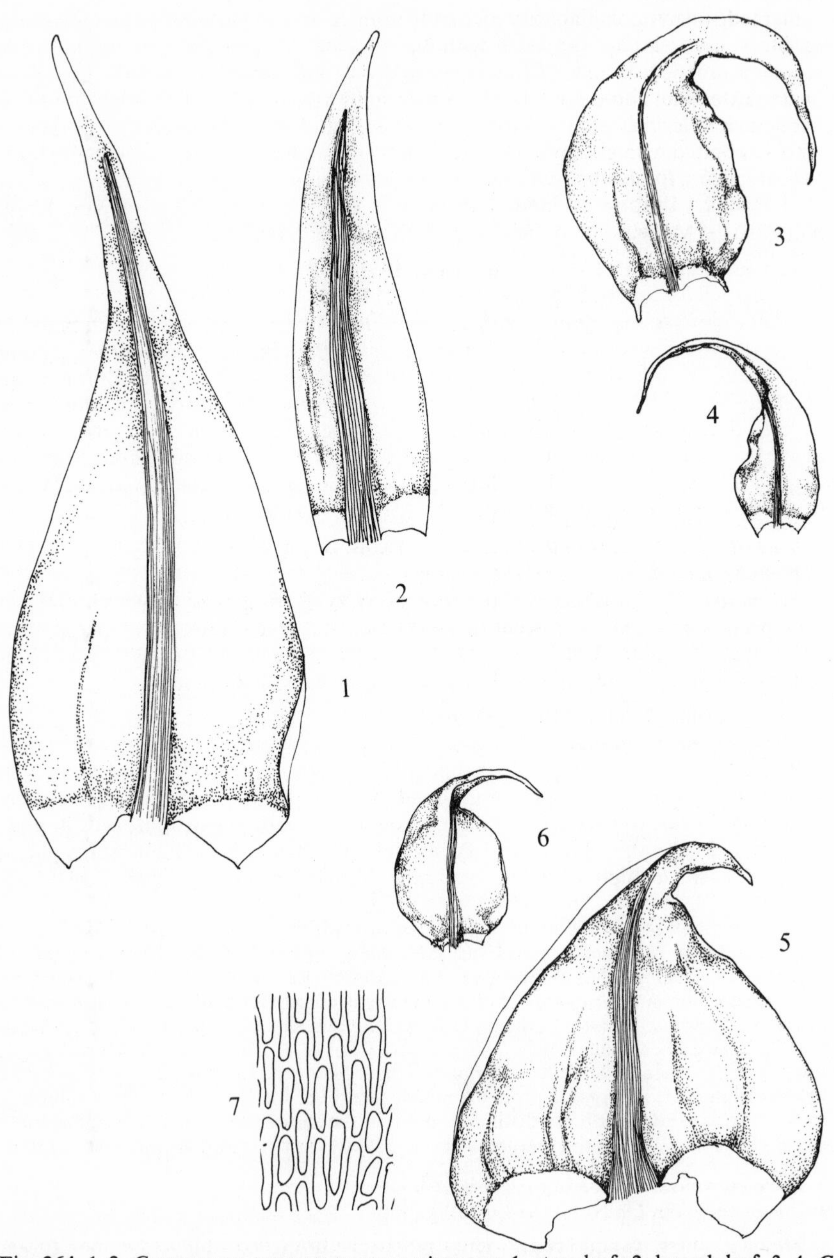

Fig. **264**. 1–2, *Cratoneuron commutatum* var. *virescens*: 1, stem leaf; 2, branch leaf. 3–4, *C. commutatum* var. *sulcatum*: 3, stem leaf; 4, branch leaf. 5–7, *C. decipiens*: 5, stem leaf; 6, branch leaf; 7, mid-leaf cells. Leaves ×40, cells ×415.

so, much narrowed at insertion, rapidly tapering to long acumen, margin plane or recurved, denticulate below; nerve extending $\frac{2}{3}$–$\frac{3}{4}$ way up lf; cells at basal angles enlarged, forming distinct auricles extending ± to nerve, other cells narrow, with solitary conical papillae at distal ends at least in younger lvs, cells in mid-lf (5–)6–9 μm×20–35 μm, (2.5–)3.0–6.0 times as long as wide. Branch lvs concave; nerve extending about $\frac{1}{2}$ way up lf; areolation as in stem lvs. Fr unknown in Britain. Bright or yellowish-green patches in basic springs and flushes at high altitudes, rare. Yorks, Scottish Highlands. 11. N., W. and C. Europe, Iceland, W. Asia, Yenisei, Japan, Canada, Alaska.

Likely to be passed over as a small form of *C. commutatum* but differing in the more abruptly tapered stem lvs, markedly narrowed at insertion, and in the presence of conical papillae best seen on the back of folded lvs.

140. Campylium (Sull.) Mitt., J. Linn. Soc. Bot., 1869

Autoecious or dioecious. Slender to robust, often yellowish-green or golden-yellow plants, stems irregularly or pinnately branched, without paraphyllia, with or without tomentum of brown rhizoids. Lvs patent to spreading or squarrose, from broad basal part gradually or abruptly tapering to channelled acumen, base somewhat decurrent, margin plane, entire or denticulate; nerve absent or short and double or longer and single; cells linear-vermicular with pointed ends, angular cells quadrate and incrassate or inflated and hyaline and forming auricles. Perichaetial lvs with sheathing bases, inner abruptly narrowed to squarrose acumen. Seta reddish; capsule inclined, ellipsoid to cylindrical, curved; annulus present, lid conical; peristome perfect, inner with processes and nodulose cilia; calyptra cucullate. A ± world-wide genus of about 30 species found mainly in moist or wet habitats.

1 Nerve single, extending to $\frac{1}{2}$ way up lf or beyond in all or most of lvs — 2
  Nerve absent or short, single or double usually ceasing well below middle of lf — 4
2 Nerve usually extending into acumen, margin obscurely denticulate — **4. C. elodes**
  Nerve reaching $\frac{1}{2}$–$\frac{3}{4}$ way up lf, margin entire or obscurely denticulate near base only — 3
3 Lvs 1–2 mm long, angular cells hardly forming auricles, cells in mid-lf 5–8 times as long as wide, dioecious — **2. C. chrysophyllum**
  Lvs 1.6–3.0 mm long, angular cells forming distinct auricles, cells in mid-lf 12–18 times as long as wide, autoecious or synoecious — **3. C. polygamum**
4 Usually robust plants, lvs 1.2–3.6 mm long, margin entire or obscurely denticulate at widest part of lf only — **1. C. stellatum**
  Slender plants, lvs mostly 0.4–1.0 mm long, margin denticulate at least below — 5
5 Lvs spreading to sub-squarrose, angular cells forming small auricles, cells in mid-lf mostly 4–7 times as long as wide — **5. C. calcareum**
  Lvs squarrose from ± erect base, angular cells not forming auricles, cells in mid-lf 8–12 times as long as wide — **6. C. halleri**

1. C. stellatum (Hedw.) J. Lange & C. Jens., Medd. Groenland, 1887
Hypnum stellatum Hedw.

Dioecious. Plants procumbent, ascending or erect, stems pinnately to sparsely and irregularly branched, older parts with tomentum of brown rhizoids. Lvs patent to spreading or squarrose, occasionally sub-falcate, concave, ± plicate, base somewhat decurrent, from broad or very broad cordate basal part gradually or abruptly

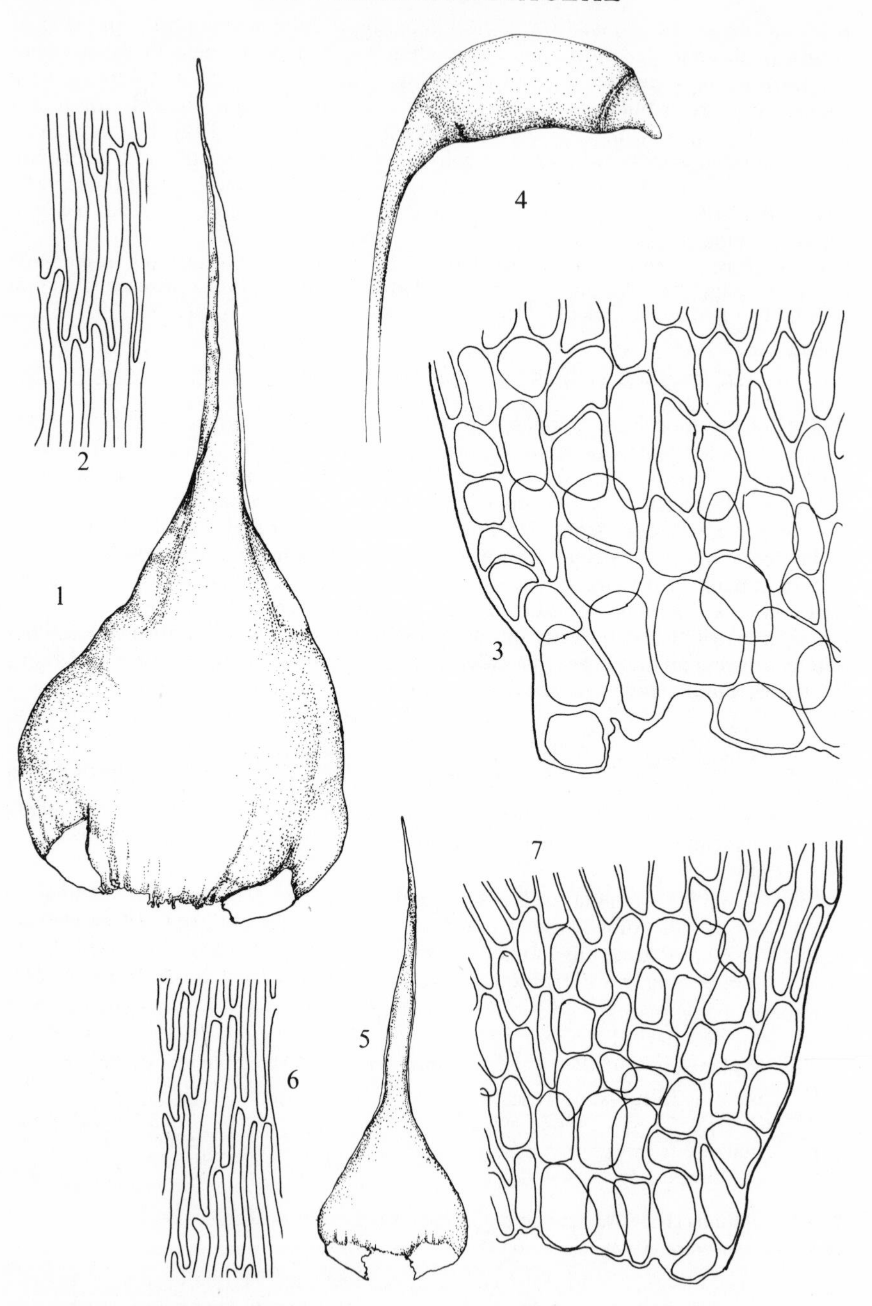

Fig. **265**. 1–4, *Campylium stellatum* var. *stellatum*: 1, stem leaves; 2, mid-leaf cells; 3, angular cells; 4, capsule (×10). 5–7, *C. stellatum* var. *protensum*: 5, stem leaf; 6, mid-leaf cells; 7, angular cells. Leaves ×40, cells ×415.

tapering to fine, channelled acumen, margin plane, entire or obscurely denticulate at widest part of lf; nerve absent or short and single or double; angular cells enlarged, quadrate to rectangular, thin-walled in young lvs, incrassate in old, other cells linear-vermicular, incrassate, porose in lower part of lf, in mid-lf 5–8 μm × 36–80 μm, (6–)8–15 times as long as wide. Capsule inclined, sub-cylindrical, curved; lid conical with small, acute apiculus; spores smooth, 12–18 μm.

Var. **stellatum**

Plants decumbent to erect, usually robust, stems irregularly and sparsely branched. Lvs usually spreading, rarely falcate, 2.0–3.6 mm long, from broad basal part gradually tapering to acumen or if abruptly narrowed then basal part constituting ⅓ or more of total lf length; angular cells forming distinct auricles. Fr rare, late spring, early summer. *n* = 10, 18 + 2, 22. Yellowish-green to greenish-brown patches, tufts or turfs to 10 cm tall on wet soil in grass-sward, on tracks, by streams, in flushes, fens, marshes, dune slacks, on cliffs, etc., especially where basic. Common in suitable habitats. 109, H40, C. Europe, Iceland, Faroes, Caucasus. N., C. and E. Asia, N. America, Greenland, New Zealand.

Var. **protensum** (Brid.) Bryhn, K. Norsk Vid. Selsk. Skrift, 1893
C. protensum (Brid.) Kindb., *Hypnum stellatum* var. *protensum* (Brid.) Röhl.

Plants less robust, usually decumbent, more closely and often pinnately branched. Lvs spreading or frequently squarrose, 1.2–2.0 mm long, gradually or abruptly narrowed above broad basal part, where abruptly narrowed basal part constituting less than ⅓ total lf length, acumen relatively longer and narrower than in var. *stellatum*; angular cells enlarged but hardly forming distinct auricles. Fr unknown in Britain. Green or dull green patches on soil in turf, on banks and on basic rocks in usually drier habitats than var. *stellatum*. Frequent in suitable habitats. 95, H24. Europe, Iceland, Faroes, Caucasus, Siberia, Altai, Kashmir, Punjab, N. America.

Although var. *protensum* tends to grow in drier habitats and is a more marked calciole than var. *stellatum*, the two are not always morphologically distinguishable and var. *protensum* cannot be treated as a distinct species.

2. C. chrysophyllum (Brid.) J. Lange, Nomencl. Fl. Dan., 1887
Hypnum chrysophyllum Brid.

Dioecious. Plants slender, stems procumbent, rarely ascending or erect, irregularly to sub-pinnately branched, branches usually ascending to erect. Lvs patent, often secund, rarely straight or falcato-secund, 1–2 mm long, from broad ovate-cordate basal part ± rapidly narrowed to long, channelled acumen, margin plane, entire or obscurely denticulate at widest part of lf; nerve usually single, extending ½–¾ way up lf; angular cells irregularly quadrate, incrassate, opaque, hardly forming auricles, cells elsewhere in basal part narrowly rhomboidal, 5–7 μm × 24–44 μm, (4–)5–8 times as long as wide. Capsule sub-erect to inclined, cylindrical, curved; lid conical; spores smooth, *ca* 14 μm. Fr rare, spring. *n* = 9, 20. Greenish to golden-yellow patches or more rarely tufts on soil in grassland, on sand-dunes, rocks, cliffs and old buildings, frequent or common in moist or dry basic habitats, very rare elsewhere. 95, H33, C. Europe, Iceland, Caucasus, C. Asia, Himalayas, China, Korea, Japan, N. Africa, N. America, Mexico, Guatemala, Greater Antilles, Patagonia.

Similar to small forms of *C. stellatum* especially those in which lvs have a single nerve reaching nearly ½ way up lf. *C. chrysophyllum* differs, however, in the somewhat shorter cells. Occasional lvs of *C. chrysophyllum* may have the nerve lacking or very short but there are always some lvs with well developed nerve. Dixon (1924) suggests that *C. chrysophyllum* should be treated as a subspecies of *C. stellatum*, but the shorter cells and the less strongly

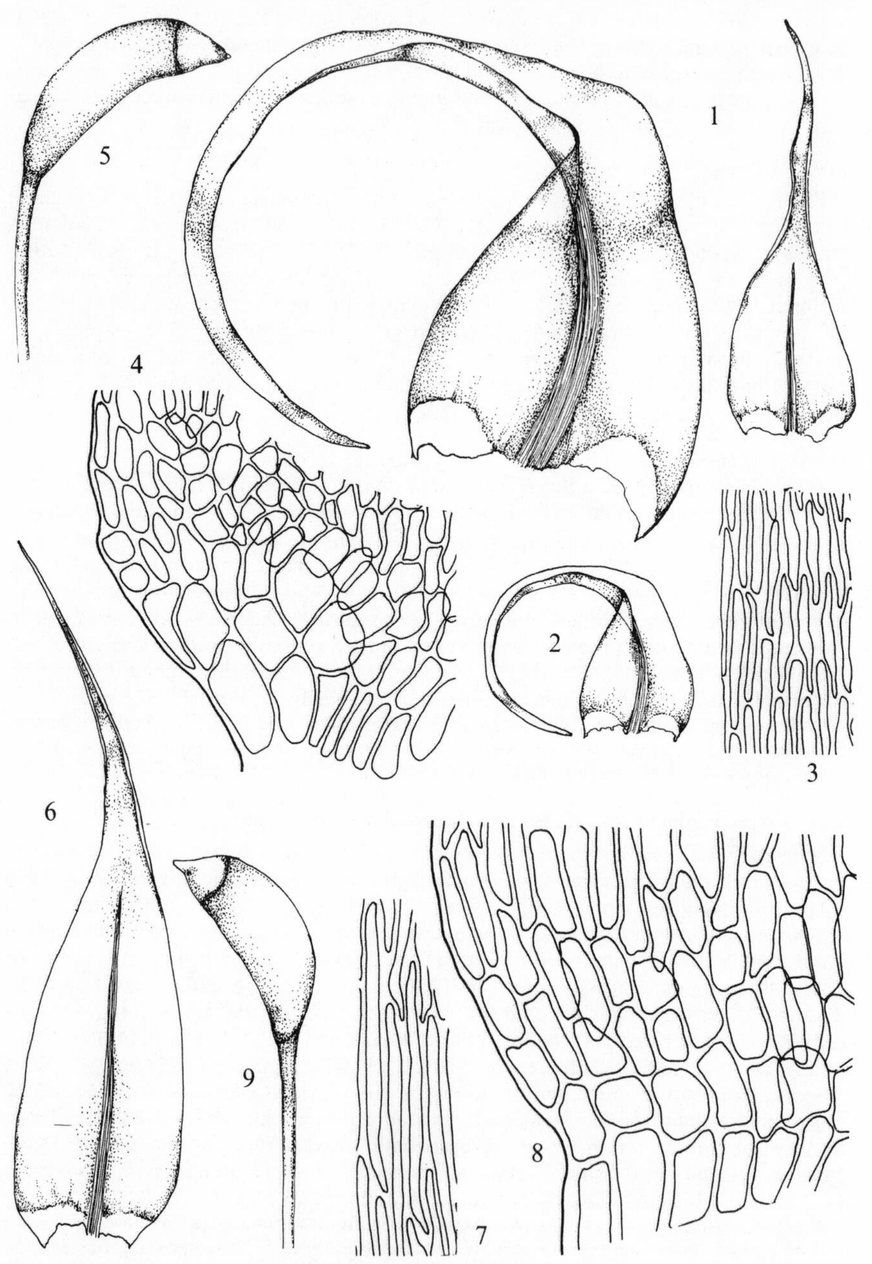

Fig. 266. 1–5, *Campylium chrysophyllum*: 1, stem leaves; 2, branch leaf; 3, mid-leaf cells; 4, angular cells; 5, capsule. 6–9, *C. polygamum*: 6, stem leaf; 7, mid-leaf cells; 8, angular cells; 9, capsule. Leaves ×40, cells ×415, capsules ×10.

inclined capsule with lid without an acute apiculus seem sufficiently definitive characters for maintaining specific rank.

3. C. polygamum (Br. Eur.) J. Lange & C. Jens., Medd. Groenland, 1887
Hypnum polygamum Br. Eur.

Autoecious or synoecious. Plants slender to moderately robust, stems procumbent, ascending or scrambling, occasionally erect, irregularly branched. Lvs patent to spreading, sometimes sub-secund but not squarrose, 1.6–3.0 mm long, narrowly to broadly lanceolate, basal part neither expanded nor ovate-cordate, tapering to long, fine, channelled acumen; nerve thin, extending $\frac{1}{2}$–$\frac{3}{4}$ way up lf; cells in basal angles enlarged, quadrate to rectangular, forming decurrent auricles extending almost to nerve, other cells linear, incrassate, porose in lower part of lf, in mid-lf 5–8(–12) μm × 100–130 μm, 12–18 times as long as wide. Capsule inclined, ellipsoid to sub-cylindrical; lid with short acute apiculus; spores papillose, 20–24 μm. Fr occasional to frequent, summer. $n = 11^*$, 20^*. Yellowish-green to golden-yellow patches or coarse wefts in wet turf, fens, dune slacks, occasional in basic habitats, very rare elsewhere, generally distributed. 66, H31. N. and C. Europe, Portugal, Iceland, Faroes, C. Asia, Manchuria, Japan, N. Africa, N. and S. America, Greenland, Australia, New Zealand.

Similar in appearance to some forms of *C. stellatum* but differing in the narrower lf base and nerve extending $\frac{1}{2}$–$\frac{3}{4}$ way up lf; the cells are also longer.

4. C. elodes (Lindb.) Kindb., Canad. Rec. Sc., 1894
Hypnum elodes Spruce

Dioecious. Plants slender, stems procumbent to ascending, irregularly pinnately branched, branches ascending to erect. Lvs patent to spreading, sometimes falcate or falcato-secund at stem and branch apices, branch lvs smaller and narrower than stem lvs, stem lvs 1–2 mm long, concave, broadly lanceolate to lanceolate, gradually to abruptly tapering to long, fine, channelled acumen, margin plane, obscurely denticulate; nerve usually reaching to acumen; angular cells enlarged, ± rectangular, forming distinct, decurrent auricles, other cells narrow, elongated, in mid-lf 5–8 μm × 48–100 μm, (7–)8–18 times as long as wide. Capsule inclined, ellipsoid, curved, lid with acute apiculus; spores 12–17 μm. Fr rare. $n = 11^*$. Lax or dense, green to yellowish patches or straggling shoots in fens, marshes, wet dune slacks and moist turf, occasional in basic habitats, very rare elsewhere, generally distributed. 54, H22. Europe, Himalayas.

The relatively long nerve and obscurely denticulate lf margin will distinguish this plant from related species.

5. C. calcareum Crundw. & Nyh., Trans. Br. bryol. Soc., 1962
C. sommerfeltii auct. eur., *Hypnum hispidulum* Brid. var. *sommerfeltii* auct. eur.

Dioecious. Plants slender, procumbent, irregularly branched, branches usually long. Lvs spreading or sub-squarrose, basal part concave, broadly ovate-cordate or cordate-triangular, abruptly tapering to long, fine, channelled acumen, margin plane, denticulate below, sinuose to sharply and sparsely denticulate above, nerve absent or short and double; angular cells slightly enlarged, quadrate to rectangular, ± forming auricles, other cells narrowly elliptical, variable in size, in mid-lf 5–8 μm × 24–56 μm, 4–7(–9) times as long as wide. Capsule inclined, ellipsoid to sub-cylindrical, curved; lid conical; spores 10–16 μm. Fr rare, summer. Dense, dull or yellowish-green patches on basic soil, rocks, tree boles and logs, occasional in basic habitats, generally distributed. 58, H6. W. and C. Europe.

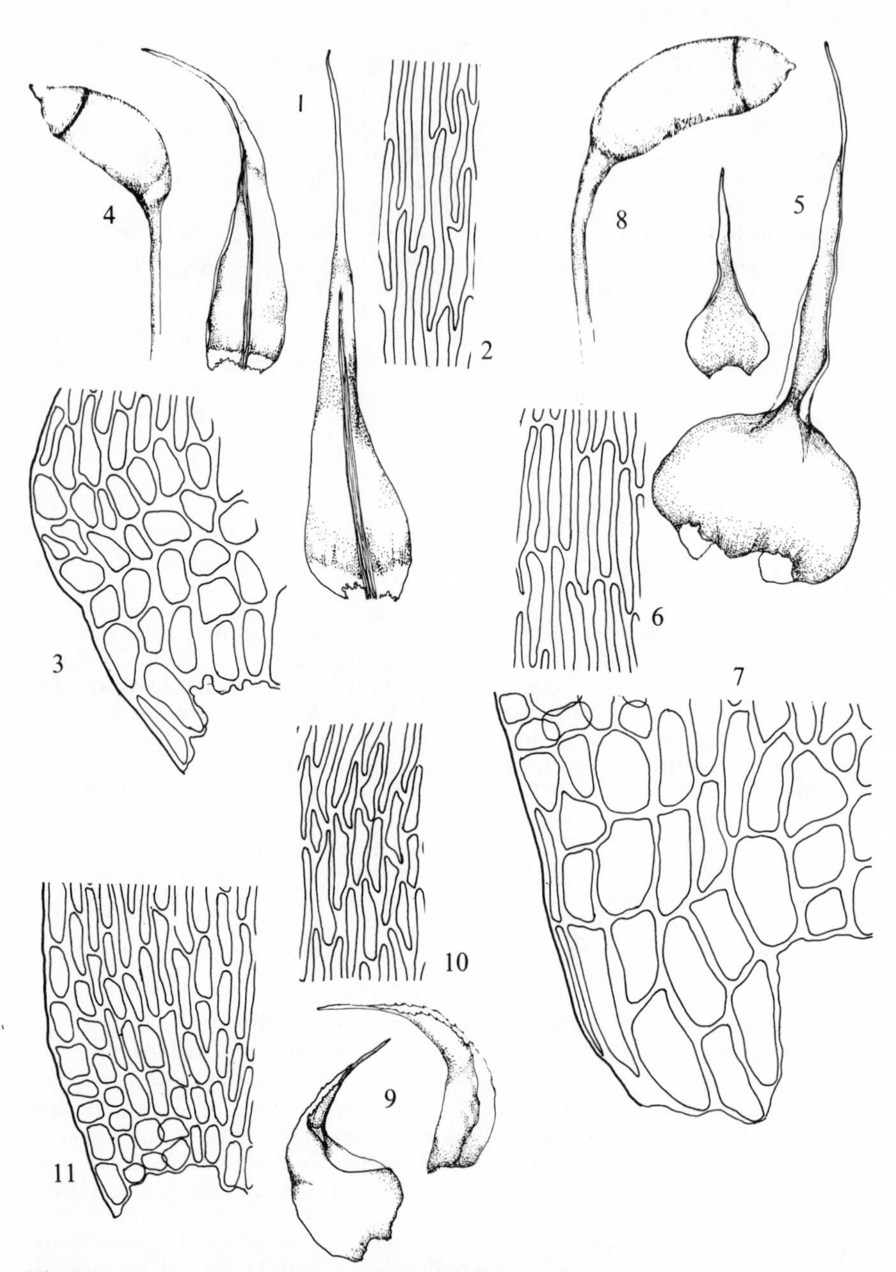

Fig. **267**. 1–4, *Campylium elodes*: 1, stem leaves; 2, mid-leaf cells; 3, angular cells; 4, capsule. 5–8, *C. calcareum*: 5, stem leaves; 6, mid-leaf cells; 7, angular cells; 8, capsule. 9–11, *C. halleri*: 9, leaves; 10, mid-leaf cells; 11, angular cells. Leaves ×40, cells ×415, capsules ×10.

6. C. halleri (Hedw.) Lindb., Musci Scand., 1879
Campylohypnum halleri (Hedw.) Fleisch., *Hypnum halleri* Hedw.

Autoecious. Plants slender, stems procumbent, ± pinnately branched, branches erect or ascending. Lvs squarrose, 0.4–1.0 mm long, basal part oblong to broadly ovate, ± abruptly narrowed to fine channelled acumen, margin plane, denticulate; a few angular cells, ± quadrate, other cells linear, in mid-lf 4–5 μm × 32–52 μm, 8–12 times as long as wide. Capsule inclined, sub-cylindrical, slightly curved; spores 10–13 μm. Fr unknown in Britain. Dense, golden patches on damp basic rocks at high altitudes, very rare. Mid Perth, Angus, W. Inverness. 3. Montane parts of Europe, Caucasus, Himalayas, Yunnan, N. America.

Very distinct in the dense patches with strongly squarrose lvs, the plants resembling miniature patches of *Rhytidiadelphus squarrosus*.

141. Amblystegium Br. Eur., 1853

Autoecious, rarely dioecious. Very slender or slender, rarely robust plants; stems procumbent, irregularly branched, branches few to numerous, decumbent to erect. Lvs erecto-patent to spreading, sometimes sub-secund, rarely sub-complanate, broadly ovate to lanceolate, gradually to abruptly tapering to acute to longly acuminate, rarely obtuse apex, margin plane, entire or denticulate; nerve usually thin, extending to $\frac{1}{2}$–$\frac{3}{4}$ way up lf, rarely absent or strong and ending in apex or excurrent; angular cells scarcely differentiated and not forming auricles, other cells rhomboid-hexagonal to narrowly rhomboid, rarely linear, smooth. Capsule inclined, ellipsoid to cylindrical, curved, when dry often horizontal and contracted below mouth; annulus present; lid conical, sometimes apiculate; peristome perfect, inner with processes and nodulose or appendiculate cilia. A world-wide genus of about 95 species occurring mainly in moist or wet habitats.

I have followed Nyholm (1954–69) in including *Hygroamblystegium* Loeske and *Leptodictyum* (Schimp.) Warnst. in *Amblystegium*. The only difference between *Hygroamblystegium* and *Amblystegium* is the somewhat larger size and stouter nerve; *Leptodictyum* also differs in larger size and, in some species, narrower cells. These characters alone are not sufficient to justify generic distinction.

1 Cells in mid-lf* linear-rhomboidal, mostly 7–14 times as long as wide, margin entire — **7. A. riparium**
Cells in mid-lf rhomboidal to narrowly rhomboidal, mostly 2–7 times as long as wide, margin entire or denticulate — 2

2 Nerve very stout, 40 μm or more wide near lf base, ± reaching apex — 3
Nerve thinner, not reaching 40 μm wide near lf base, reaching apex or not — 4

3 Stem ± pinnately branched, plants forming intricate patches, lf apex ± acute — **3. A. tenax**
Stems irregularly branched, stems and branches forming ± elongated tufts, lf apex obtuse — **2. A. fluviatile**

4 Nerve ± reaching apex — 5
Nerve extending to $\frac{3}{4}$ way up lf — 6

5 Lvs ± entire, cells 2–4 times as long as wide, gemmae lacking — **4. A. varium**
Lvs strongly denticulate, especially at widest part, cells 5–6 times as long as wide, uniseriate gemmae often present at lf apices — **8. A. compactum**

* N.B. Lvs referred to in *Amblystegium* key are stem lvs, not branch lvs.

6 Lf base longly decurrent **6. A. saxatile**
Lf base not or scarcely decurrent 7
7 Lvs distant, markedly narrowed at insertion, 1.2–2.2 mm long **5. A. humile**
Lvs crowded, not markedly narrowed at insertion, mostly 0.3–1.0 mm long **1. A. serpens**

1. A. serpens (Hedw.) Br. Eur., 1853
A. juratzkanum Schimp.

Autoecious. Plants slender, stems procumbent, irregularly branched, branches decumbent to erect. Lvs appressed to patent or spreading when dry, erecto-patent to spreading when moist, stem lvs ovate to lanceolate, tapering to ± long acuminate apex, margin plane, entire to denticulate; nerve almost absent to extending ¾ way up lf; basal cells quadrate or rectangular, cells above ± rhomboidal, prosenchymatous, marginal cells from widest part of lf to base quadrate to rectangular. Branch lvs smaller, narrower than stem lvs but otherwise similar. Capsule sub-erect to inclined, narrowly ellipsoid to cylindrical, curved; lid conical with blunt apiculus; spores 8–15 μm, calyptra whitish.

Var. **serpens**
Plants slender to moderately slender. Lvs appressed to spreading when dry, mostly 0.5–1.0 mm long; nerve extending ½–¾ way up lf; cells in mid-lf (6–)7–12 μm × 28–40(–48) μm, 3–5(–6) times as long as wide. Capsule inclined. Fr common, spring to autumn. $n=10+1$, 11, $11+1$, 12, $12+1$, 13, 19*, $19+1$*, $19+2$, 20*, $20+1$*, 21, 24, 28. Greenish, dense, intricate patches on soil, rocks, tree boles, logs, in turf, hedgebanks, woods, bases of walls, etc., in usually damp, shaded situations, common. 111, H40, C. Europe, Faroes, Caucasus, Siberia, Tibet, Chitral, Kashmir, Japan, Azores, Algeria, N. America, Greenland, Peru, New Zealand.

Var. **salinum** Carr., Trans. Bot. Soc. Edin., 1863
Plants very slender. Lvs appressed when dry, (0.24–)0.34–0.56 mm long; nerve ± absent to extending to ½ way up lf; cells in mid-lf 8–10 μm × 16–24 μm, 2–3 times as long as wide. Capsule sub-erect. Fr occasional. Dense yellowish-green patches on sand or soil, occasional or frequent in dune slacks and on soil in maritime habitats, generally distributed round the coast. 34, H15. Apparently endemic.

I have treated *A. juratzkanum* as a synonym of *A. serpens* as it is quite impossible to separate the 2 taxa. *A. juratzkanum* is described as having spreading lvs, a longer nerve than in *A. serpens* and rectangular basal marginal cells. Having examined a considerable number of gatherings of both I have found there is no correlation between spreading lvs and rectangular marginal cells, and nerve length is variable. Several of the gatherings could well be small forms of *A. varium* but cannot be named with certainty. *A. varium* usually differs in the larger, wider stem lvs with nerve extending to the acumen. The plants are also somewhat softer in texture and the lf cells are relatively shorter than in larger forms of *A. serpens*. Not all sand-dune plants of *A. serpens* belong to var. *salinum*, which is distinctive in its shorter lf cells and sub-erect capsules. Sand-dune plants without capsules can only be determined microscopically.

2. A. fluviatile (Hedw.) Br. Eur., 1853
Hygroamblystegium fluviatile (Hedw.) Loeske

Autoecious. Plants slender to moderately slender, stems sparsely and irregularly branched, branches long. Lvs appressed when dry, erecto-patent to patent when moist, concave, ovate to lanceolate, sometimes slightly curved, apex obtuse, margin plane or sinuose with projecting cell ends above; nerve stout, brownish, 40–70 μm wide near lf base, ending in apex to shortly excurrent; basal cells quadrate-rectangular, cells above rhomboidal to narrowly rhomboidal, in mid-lf 8–10 μm × (16–)

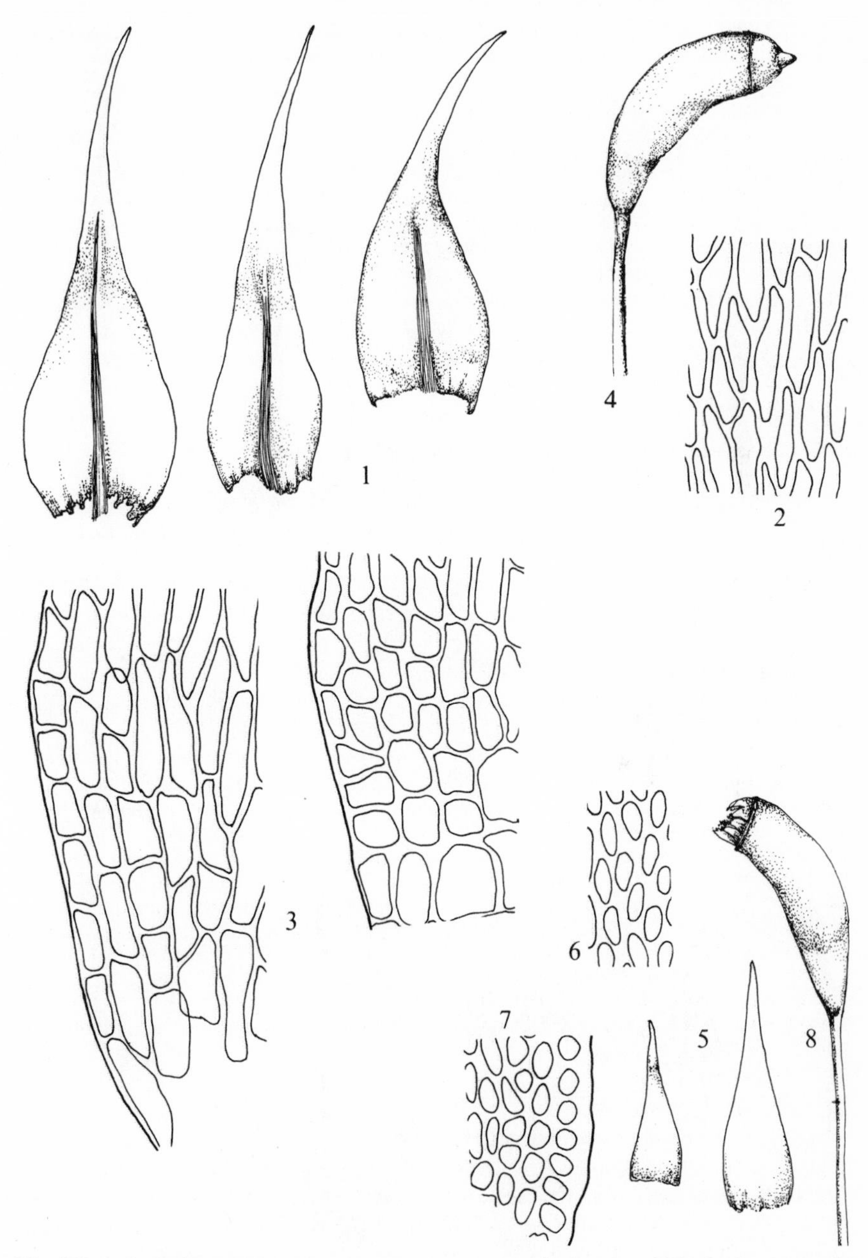

Fig. **268**. 1–4, *Amblystegium serpens* var. *serpens*: 1, stem leaves; 2, mid-leaf cells; 3, angular cells from different plants; 4, capsule. 5–8, *A. serpens* var. *salinum*: 5, leaves; 6, mid-leaf cells; 7, angular cells; 8, capsule. Leaves ×63, cells ×415, capsules ×10.

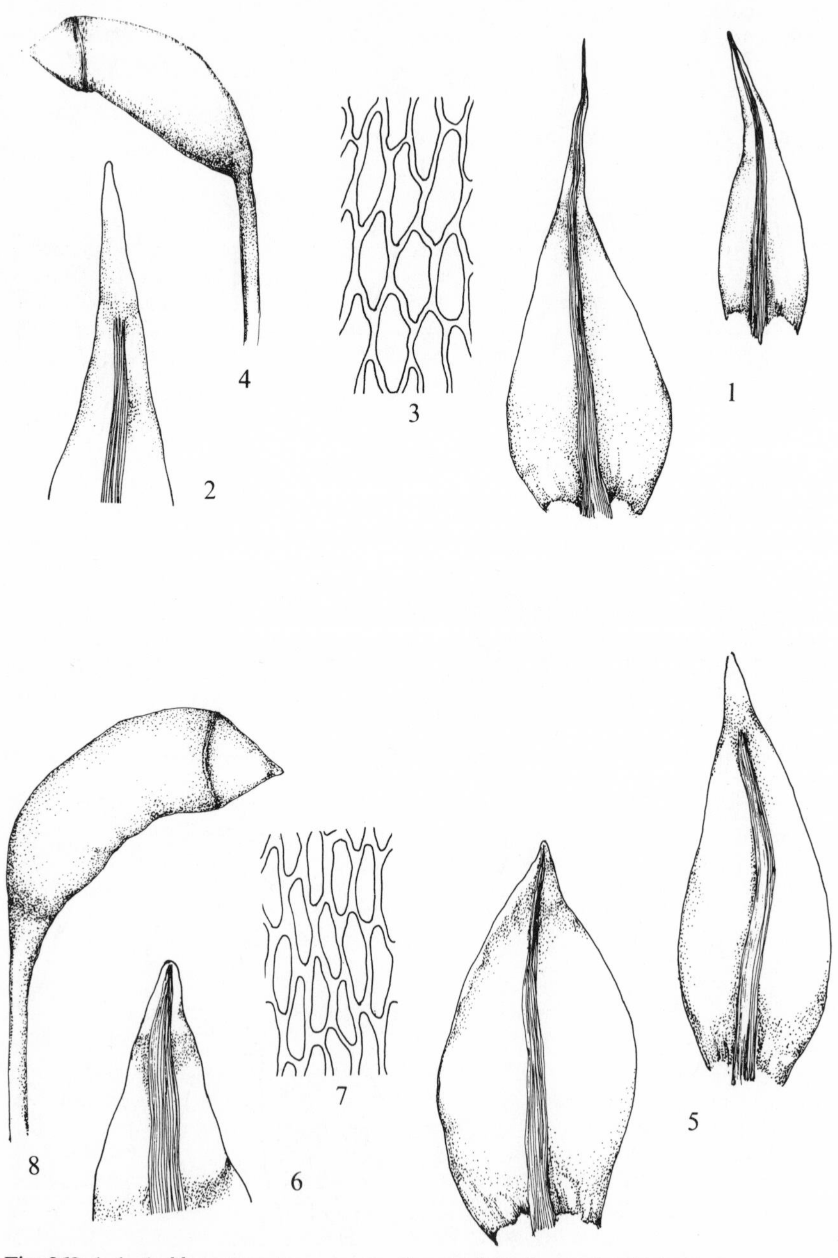

Fig. **269**. 1–4, *Amblystegium tenax*: 1, stem leaves; 2, leaf apex; 3, mid-leaf cells; 4, capsule. 5–8, *A. fluviatile*: 5, stem leaves; 6, leaf apex; 7, mid-leaf cells; 8, capsule. Leaves ×40, apices ×100, cells ×415, capsules ×10.

20–44 μm, 2–4(–6) times as long as wide. Capsule inclined, cylindrical, curved; lid conical, acute; spores 12–18 μm. Fr rare, summer. $n=20$*. Dark green tufts or patches on rocks and tree bases subject to submergence in streams and rivers. Absent from S.E. England, occasional elsewhere. 63, H11, C. Europe, Iceland, Madeira, N. and C. America, Andes.

3. A. tenax (Hedw.) C. Jens., Skand. Bladmoosfl., 1939
A. irriguum (Wils.) Br. Eur., *Hygroamblystegium tenax* (Hedw.) Jenn.

Autoecious. Plants slender, stems wiry in texture, denuded below, irregularly pinnately branched, branches rather short. Lvs rigid, loosely appressed when dry, erecto-patent, sometimes sub-secund when moist, ± concave, ovate to lanceolate, sometimes curved, tapering to ± acute apex, margin plane, entire or denticulate above; nerve stout, brownish, ending just below apex to shortly excurrent, usually 50–60 μm wide near lf base; basal cells quadrate-rectangular, cells above rhomboid-hexagonal, in mid-lf 8–11 μm × 16–28(–40) μm, 2–4 times as long as wide. Capsule inclined, ellipsoid to cylindrical, curved; lid conical, acute; spores 16–20 μm. Fr occasional, spring, summer. $n=12$, 20*, 30*. Dark green, rough, intricate patches on rocks and exposed tree bases subject to submergence in and by streams and rivers, occasional, generally distributed. 80, H20, C. Europe, Caucasus, Siberia, Altai, Madeira, Algeria, N. America.

A. fluviatile is a softer plant, the patches less intricate, the stems and branches forming elongated tufts; the lvs are also relatively shorter with obtuse or rounded apices. *A. varium* has a thinner nerve, the lvs are ± spreading when dry and the whole plant is softer in texture.

4. A. varium (Hedw.) Lindb., Musci Scand., 1870

Autoecious. Plants slender, stems irregularly branched, stems and branches ± procumbent. Lvs patent to spreading when moist, hardly altered when dry, stem lvs mostly 1.0–1.4 mm long, ovate, tapering to long acumen, margin plane, ± entire, nerve reaching ± to apex; basal cells shortly rectangular, cells above rhomboidal, in mid-lf 8–12 μm × 16–32(–44) μm, (1.3–)2.0–4.0(–4.5) times as long as wide, marginal cells from widest part of lf to base quadrate to rectangular. Branch lvs smaller, narrower; nerve extending to $\frac{3}{4}$ way or more up lf. Capsule erect or inclined, cylindrical, curved; lid with obtuse apiculus; spores 10–16 μm. Fr occasional, summer. $n=10$, 10+1, 12+1, 19+1, 20, 40. Lax patches on wet rocks, wood and soil by streams and pools and in fens and marshes, occasional in England, Wales and Ireland, rare in Scotland. 70, H17, C. N., W. and C. Europe, Caucasus, Iran, Siberia, China, Madeira, Azores, Morocco, N. America, Mexico, Haiti.

5. A. humile (P. Beauv.) Crundw., MS.
A. kochii Br. Eur., *Leptodictyum kochii* (Br. Eur.) Warnst.

Autoecious. Plants slender to moderate in size; stems irregularly branched, stems and branches procumbent or branches ascending. Lvs distant, patent to spreading both when moist and dry, stem lvs 1.2–2.2 mm long, ovate, narrowed at insertion and not decurrent, tapering to acuminate apex, margin entire; nerve extending $\frac{1}{2}$–$\frac{3}{4}$ way up lf; basal cells rectangular, cells above ± rhomboidal, in mid-lf 8–12 (–18) μm × 24–48(–72) μm, 2.5–5.0 times as long as wide. Branch lvs smaller, ovate-lanceolate but otherwise similar. Capsule inclined, narrowly ellipsoid, curved. Fr rare, summer. Dull green patches or straggling shoots on soil, rocks and tree bases by streams and in moist turf, rare. Scattered localities from Devon and Kent north to Yorks, Fife, Angus, Caithness, Antrim, Londonderry. 32, H2. N.W. and C. Europe, Caucasus, Iran, Siberia, C. Asia, N. America, Mexico.

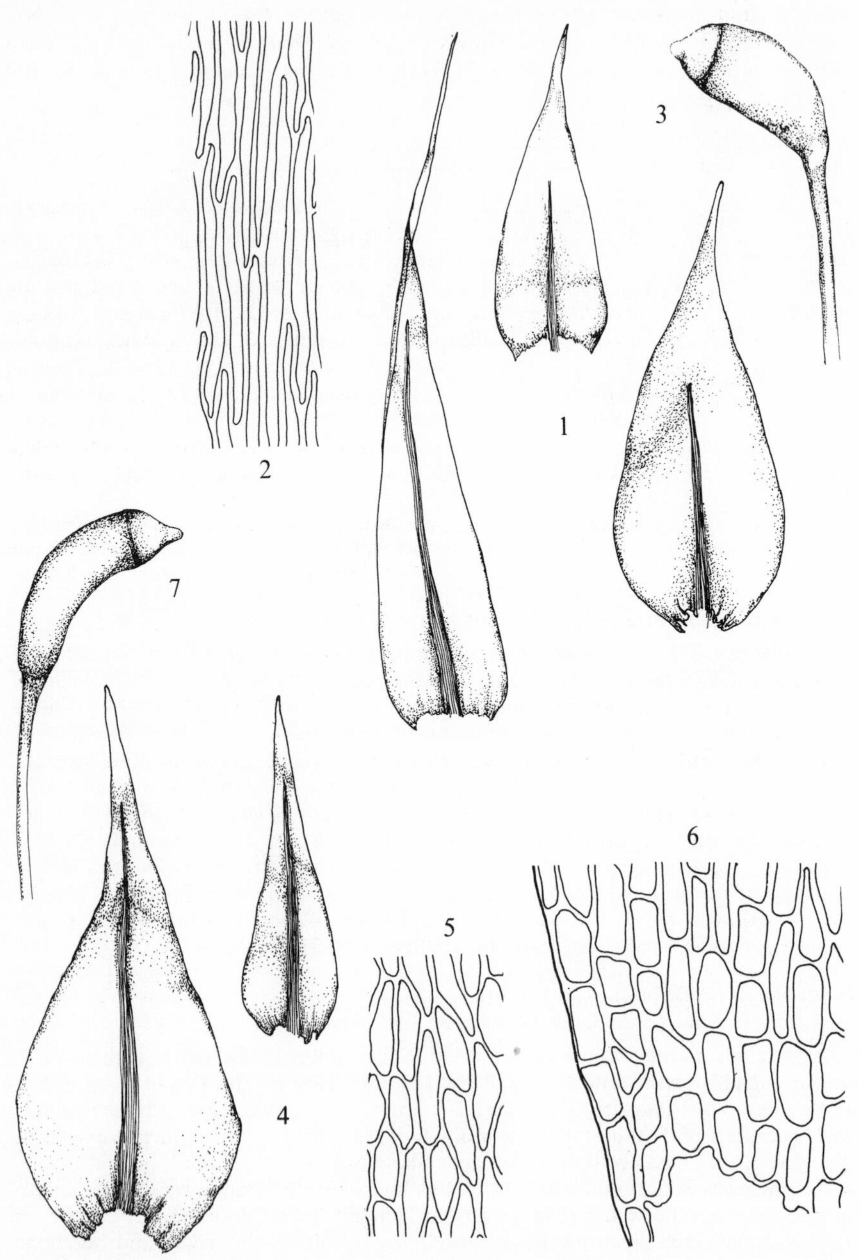

Fig. **270**. 1–3, *Amblystegium riparium*: 1, stem leaves (×25); 2, mid-leaf cells; 3, capsule. 4–7, *A. varium*: 4, leaves (×63); 5, mid-leaf cells; 6, angular cells; 7, capsule. Cells ×415, capsules ×10.

Differs from *A. serpens* and *A. varium* in the larger stem lvs markedly narrowed at insertion, from the former also in the less crowded somewhat spreading lvs and from the latter in the shorter nerve. *A. riparium* often has the lvs sub-complanate and the lf cells are longer.

6. A. saxatile Schimp., Syn., 1860
Campylium radicale (P. Beauv.) Grout

Autoecious. Plants of moderate size, stems irregularly pinnately branched. Lvs patent to spreading both when dry and moist, stem lvs 0.8–1.4 mm long, rapidly tapering to long channelled acumen from ovate-cordate basal part, base longly decurrent, margin sinuose with projecting cell ends; nerve extending from ½ way up lf to base of acumen; angular cells enlarged, rectangular but not forming auricles, cells above narrowly ellipsoid, in mid-lf 6–9 μm × 32–64 μm, 4–7(–10) times as long as wide. Branch lvs smaller and narrower. Capsule ± horizontal, ellipsoid, curved; lid conical, sub-acute; spores 10–18 μm. Fr common, June. Lax, straggling, yellowish-green patches on marshy ground, very rare. E. Cornwall. 1. Norway, Sweden, France, Germany, Poland, Czechoslovakia, Switzerland, Jugoslavia, Crimea, N. America, Ecuador.

This plant tends to occur in somewhat unpleasant habitats on swampy ground amongst decaying vegetation. It may be confused with *A. humile* but capsules are frequently produced and the seta is conspicuously long and flexuose; the decurrent lf base is distinctive. For an account of this plant in Britain see Crundwell & Nyholm, *Trans. Br. bryol. Soc.* **4**, 638–41, 1964.

7. A. riparium (Hedw.) Br. Eur., 1853
Hypnum riparium Hedw., *Leptodictyum riparium* (Hedw.) Warnst.

Autoecious. Plants slender to moderately robust, stems sparsely and irregularly to sub-pinnately branched, procumbent or rarely ascending. Lvs close or distant, divergent, sometimes falcato-secund towards stem and branch apices, usually sub-complanate, hardly altered when dry, typically lanceolate or ovate-lanceolate, gradually tapering to long acuminate apex but varying from narrowly lanceolate with long filiform acumen to ovate-triangular and acuminate, margin entire or sometimes sinuose with projecting cell ends, occasionally faintly denticulate near apex; nerve weak or strong, extending from ⅔ way up lf to almost to apex; basal cells enlarged, quadrate-hexagonal, pellucid but not forming auricles, cells above linear-rhomboidal, in mid-lf 5–12 μm × 40–120 μm, (5–)7–15 times as long as wide. Capsule inclined, narrowly ellipsoid to shortly cylindrical, curved; lid conical, obtuse; spores 12–16 μm. Fr frequent, spring, summer. $n = 10$, 12, *ca* 19*, 20*, 24, 36, 38 + 2*, 40*, 48, 96. Bright or yellowish-green patches on soil, stones, fallen logs, tree boles and exposed roots and amongst vegetable detritus in fens, marshes and carr, by streams, pools and in damp habitats, frequent and generally distributed. 101, H32, C. Europe, Siberia, Tibet, Tonkin, Japan, Macaronesia, Kerguelen Is., Algeria, southern Africa, N. America, Mexico, Guatemala, Cuba, Haiti, Australia.

8. A. compactum (C. Müll.) Aust., Musci Appal., 1870
Rhynchostegiella compacta (C. Müll.) Loeske, *Conardia compacta* (C. Müll.) Robins.

Dioecious in Europe. Plants very slender, procumbent, stems irregularly branched. Lvs appressed when dry, patent to spreading, often slightly secund when moist, stem lvs 0.3–0.7 mm long, lanceolate, ± acute, margin plane, strongly denticulate below, less so above, teeth at widest part of lf often recurved, nerve poorly defined, reaching acumen to percurrent; basal cells rectangular, cells above narrowly rhomboidal, in mid-lf 5–10 μm × 28–48 μm, mostly 5–6 times as long as wide. Branch lvs smaller but

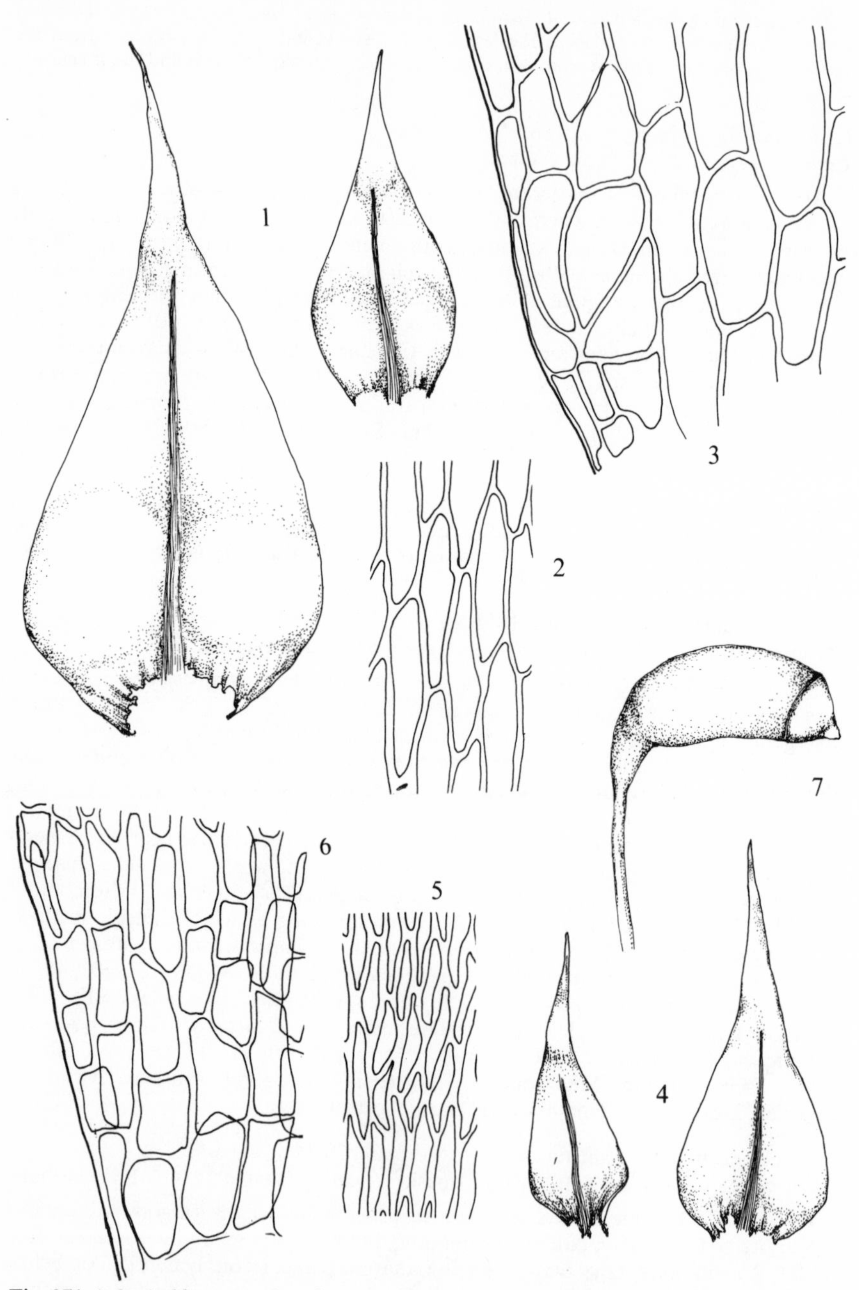

Fig. **271**. 1–3, *Amblystegium humile*. 1, stem leaves; 2, mid-leaf cells; 3, angular cells. 4–7, *A. saxatile*: 4, stem leaves; 5, mid-leaf cells; 6, angular cells; 7, capsule (×10). Leaves ×40, cells ×415.

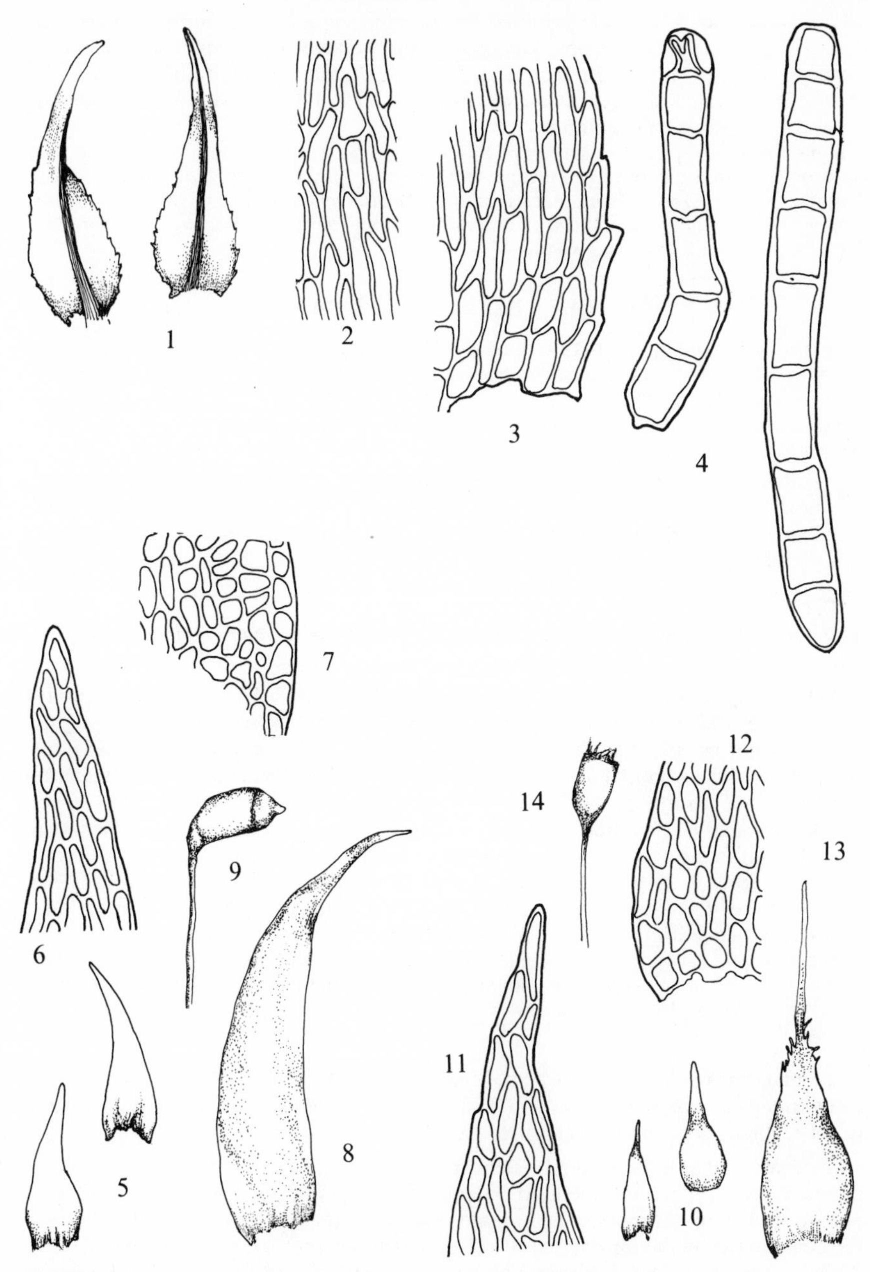

Fig. **272**. 1–4, *Amblystegium compactum*: 1, leaves; 2, mid-leaf cells; 3, angular cells; 4, gemmae. 5–9, *Platydictya confervoides*: 5, leaves; 6, leaf apex; 7, angular cells; 8, perichaetial leaf; 9, capsule. 10–14, *P. jungermannioides*: 10, leaves; 11, leaf apex; 12, angular cells; 13, perichaetial leaf; 14, capsule. Leaves $\times 63$, cells $\times 415$, capsules $\times 10$.

otherwise similar. Uniseriate gemmae, to 100 μm long, often on dorsal surface near lf apex. Fr unknown in Europe. Dense, green or yellowish-green patches on shaded basic rocks, in W. and N. Britain, rare. From S. Devon, Hereford and Derby north to W. Sutherland, Sligo, not known in Wales. 19, H1. N., W. and C. Europe, Iceland, N. America, Greenland.

Some authorities place this species in *Rhynchostegiella*, but the grounds for so doing are no stronger than those for retaining it in *Amblystegium*. The lvs, sharply toothed below, will distinguish *A. compactum* from other species of *Amblystegium*. In Europe *A. compactum* is rare, is not known to fruit and occurs on basic rocks and soil; in N. America it is common, fruits frequently and occurs on wood.

142. Platydictya Berk., Handb. Brit. Moss., 1863

Like *Amblystegium* but plants exceedingly slender, stems without a central strand and lvs with nerve very short and double or absent. A genus of 10 species found in Europe, Asia and N. America.

Very close to *Amblystegium* and I see no sound reason for separating the 2 genera as some slender forms of *A. serpens* approach very closely *Platydictya* species.

Autoecious, lvs entire, perichaetial lvs entire or obscurely denticulate, capsule ± horizontal **1. P. confervoides**

Dioecious, at least well developed lvs denticulate, perichaetial lvs spinosely denticulate above, capsule inclined **2. P. jungermannioides**

1. P. confervoides (Brid.) Crum, Mich. Bot., 1964
Amblystegiella confervoides (Brid.) Loeske, *Amblystegium confervoides* (Brid.) Br. Eur.

Autoecious. Plants exceedingly slender, stems irregularly branched. Stem lvs minute, 0.10–0.25(–0.42) mm long, appressed when dry, erecto-patent, slightly curved or not when moist, lanceolate, acute, margin plane, entire or sinuose; nerve absent; basal cells shortly rectangular, at basal angles ± quadrate and extending short distance up margin, cells above rhomboidal, in mid-lf 5–10 μm×12–24 μm, 1.5–4.0 times as long as wide, cells near lf apex 2–3 times as long as wide. Perichaetial lvs larger with long acumen, entire or finely denticulate. Capsule ± horizontal, ovoid, gibbous; spores 8–12 μm. Dull green confervoid patches on shaded calcareous rocks, rare. Scattered localities from Cornwall and Kent north to Yorks and Cumberland, Angus, a few localities in Ireland. 22, H5. Europe, Caucasus, N. America.

Somewhat similar to *P. jungermannioides* but differing in the characters given in the key. The angular cells in *P. confervoides* are more distinctive and the cells near the apex shorter. Also, in *P. jungermannioides* the lvs are not appressed when dry and have longer, finer apices.

2. P. jungermannioides (Brid.) Crum, Mich. Bot., 1964
Amblystegiella jungermannioides (Brid.) Giac., *Amblystegiella sprucei* (Bruch) Loeske, *Amblystegium sprucei* (Bruch) Br. Eur.

Dioecious. Plants exceedingly slender, stems irregularly branched. Lvs erecto-patent when dry, erecto-patent to patent, often sub-falcate when moist, 0.12–0.32 (–0.36) mm long, lanceolate, acuminate, margin plane, entire or in well developed lvs denticulate; nerve absent; basal cells shortly rectangular, angular cells scarcely differentiated, cells above rhomboidal, in mid-lf 6–9 μm×16–32 μm, 2.5–4.0 times as long as wide, cells near apex 3–4 times as long as wide. Perichaetial lvs larger, margin with spinose denticulations above. Axillary gemmae, 100–320 μm long, sometimes produced. Capsule ± inclined, ellipsoid; spores *ca* 14 μm. Fr very rare, summer. Fine, yellowish-green patches on shaded, basic soil on rocks and in rock crevices. Rare in S.W. England, Wales, N.W. Scotland and Ireland, occasional in Derby and the Pennines. 33, H6. Europe, Iceland, Caucasus, N. and C. Asia, Yunnan, N. America, Greenland.

143. Drepanocladus (C. Müll.) Roth, Hedwigia, 1899

Dioecious or autoecious. Plants slender to robust, procumbent to ascending, usually forming lax patches; stems irregularly to pinnately branched, rarely simple. Lvs usually falcato-secund, sometimes circinate, rarely straight, concave, plicate or not, ovate to narrowly lanceolate, shortly to longly tapering to acuminate apex, margin entire or denticulate; nerve single, ending at middle of lf to shortly excurrent; basal cells incrassate, often porose, at basal angles shorter or sometimes enlarged or inflated and hyaline and forming distinct auricles, cells above linear, ± vermicular, incrassate or not. Seta long, flexuose; capsule sub-erect to horizontal, sub-cylindrical, curved; lid conical, apiculate; peristome perfect, inner with nodulose cilia. A world-wide but mainly temperate genus of some 35 species occurring in mainly wet habitats.

1 Lvs plicate when moist 2
  Lvs not plicate when moist 3
2 Lvs tapering from about ¾ way up lf, margin entire, auricles absent **7. D. vernicosus**
  Lvs tapering almost from base, margin finely denticulate, auricles present **8. D. uncinatus**
3 Lf margin denticulate 4
  Lf margin entire or at most slightly sinuose 5
4 Nerve extending to ¾ way up lf, plano-convex in section, largest angular cells not more than 70 μm long and hardly forming distinct auricles **4. D. fluitans**
  Nerve reaching acumen, biconvex in section, largest angular cells 70–150 μm long, forming distinct auricles **5. D. exannulatus**
5 Lvs with angular cells differentiated into auricles 6
  Angular cells not differentiated into auricles or 2–3 cells inflated, hyaline, very fragile and fugacious 7
6 Nerve extending to ¾ way up lf, 40–60 μm wide near base, angular cells enlarged, hyaline **1. D. aduncus**
  Nerve extending into acumen, 60–120 μm wide near base, angular cells not much enlarged, yellowish to orange-brown **2. D. sendtneri**
7 Lvs falcate, outer layer of cells of stem cortex not enlarged **3. D. lycopodioides**
  Lvs strongly falcate to circinate, outer layer of cells of stem cortex enlarged **6. D. revolvens**

1. D. aduncus (Hedw.) Warnst., Beih. Bot. Centralbl., 1903
Hypnum aduncum Hedw.

Dioecious. Plants of medium size, soft in texture, shoots to 30(–45) cm long, stems irregularly to pinnately branched. Lvs patent, straight to falcate, rarely falcato-secund, not plicate, stem lvs ovate to ovate-lanceolate, rarely lanceolate, rapidly to longly tapering to long acuminate apex, base cordate, slightly decurrent, margin entire or slightly sinuose; nerve extending ½–¾ way up lf, weak or strong, 40–60 μm wide near base; basal cells irregularly rectangular, incrassate, not porose, angular cells inflated, hyaline to yellowish, forming distinct auricles not or hardly reaching

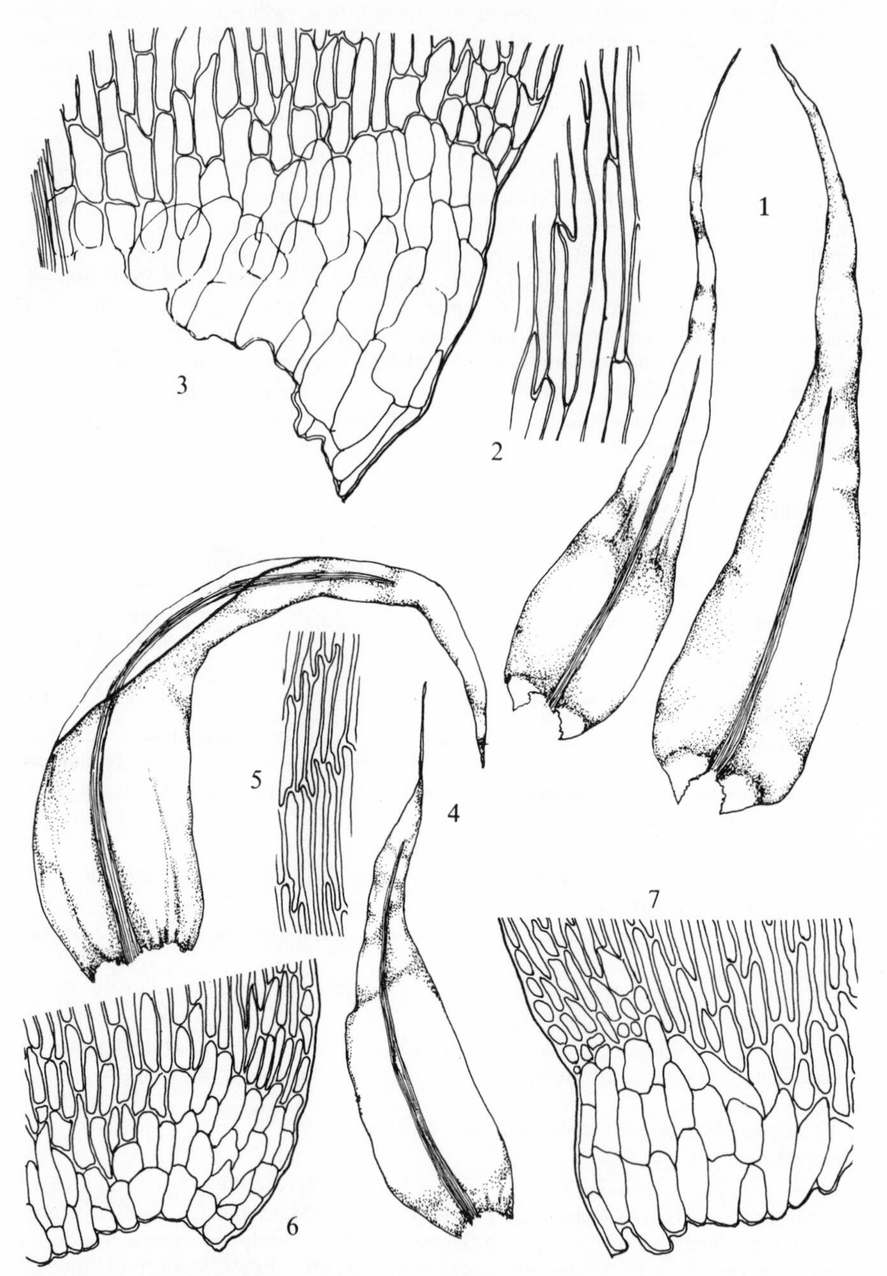

Fig. 273. 1–3, *Drepanocladus aduncus*: 1, leaves; 2, mid-leaf cells (×415); 3, angular cells (×250). 4–7, *D. sendtneri*: 4, leaves; 5, mid-leaf cells (×415); 6 and 7, angular cells from slender and robust plants (×250). Leaves ×25.

nerve, cells above linear-rhomboidal to linear, in mid-lf 5–10 μm×40–90(–120) μm, (7–)9–12(–14) times as long as wide. Capsule inclined, cylindrical, curved; annulus separating; spores *ca* 16 μm. Fr very rare, summer. $n=12$. Dull green to yellowish-green or brownish patches, sometimes extensive on wet ground, often submerged in and by pools, marshes, upper reaches of salt marshes, dune slacks, ditches, damp tracks, sometimes in calcareous habitats, occasional to frequent, usually at low altitudes, generally distributed. 100, H27, C. Europe, Faroes, Caucasus, Szechwan, Algeria, N. and C. Asia, N. America, Greenland, Mexico, Peru, Australia, New Zealand, Kerguelen Is.

Differs from *D. fluitans* and *D. exannulatus* in the entire lvs and the auricles scarcely reaching the nerve. *D. sendtneri* has smaller, less distinct, more opaque auricles, a stronger nerve extending further up the lf, and the basal cells are porose.

2. D. sendtneri (Schimp. ex H. Müll.) Warnst., Beih. Bot. Centralbl., 1903
D. sendtneri var. *wilsonii* (Schimp. ex Lorb.) Warnst., *Hypnum sendtneri* Schimp. ex H. Müll., *H. wilsonii* Schimp. ex Lorb.

Dioecious. Plants of moderate size to robust, shoots to 20 cm long, stems ± simple to pinnately branched. Lvs patent, secund to strongly falcato-secund, stem lvs 2.4–4.4 mm long, ovate-lanceolate to lanceolate, gradually tapering to long acuminate apex, base not decurrent, margin entire; nerve strong, extending into acumen, 60–100(–120) μm wide near base; basal cells rhomboidal, incrassate, porose, angular cells shorter, wider, incrassate, orange to yellowish-brown, differentiated into small auricles, not reaching nerve, distinct in small forms, poorly defined in robust forms, cells above linear, 5–7 μm×(44–)56–80(–104) μm, 9–16 times as long as wide. Capsule sub-erect, cylindrical, curved. Fr very rare, summer. Green to greenish-brown, reddish or golden-brown patches in and at edges of pools on sand-dunes, fens and marshes, generally distributed, rare to occasional. 46, H14. Europe, Iceland, Siberia, Szechwan, N. America, Greenland, Kerguelen Is.

The plant referred to as var. *wilsonii* is a large form with larger lvs and smaller auricles, but the range of intermediates between this taxon and more typical plants is such as to make discrimination impossible. Large forms may be difficult to separate from *D. lycopodioides*. That species has relatively wider lvs, a thinner nerve and hardly differentiated angular cells.

3. D. lycopodioides (Brid.) Warnst., Beih. Bot. Centralbl., 1903
Hypnum lycopodioides Brid., *Scorpidium lycopodioides* (Brid.) Paul

Dioecious. Plants robust, shoots to 25 cm, stems sparsely branched. Lvs crowded, falcate, concave, stem lvs 2.6–5.0 mm long, ovate, tapering to acuminate apex, margin entire or obscurely denticulate below; nerve thin, 40–60 μm wide near base, ending well below apex; basal cells rhomboidal, incrassate, porose, angular cells ± quadrate, incrassate, not forming auricles, cells above linear, in mid-lf 5–7 μm× 56–104 μm, 9–15(–20) times as long as wide. Capsule sub-erect, cylindrical, curved. Fr very rare. Greenish-brown to yellowish-brown patches, usually submerged, in pools on heaths and sand-dunes, generally distributed, rare. 33, H10. Scandinavia, C. Europe, Siberia, Miquelon, Greenland.

4. D. fluitans (Hedw.) Warnst., Beih. Bot. Centralbl., 1903
Hypnum fluitans Hedw.

Autoecious. Plants of moderate size, to 15 cm long, usually soft in texture, stems irregularly branched, rarely pinnate. Lvs straight to sub-falcate, rarely falcate, stem lvs narrowly lanceolate to ovate-lanceolate, gradually tapering to acuminate to longly acuminate apex, margin denticulate; nerve extending $\frac{1}{2}$–$\frac{3}{4}$ way up lf, 40–70 μm wide near base; angular cells enlarged and differentiated from basal cells but not or

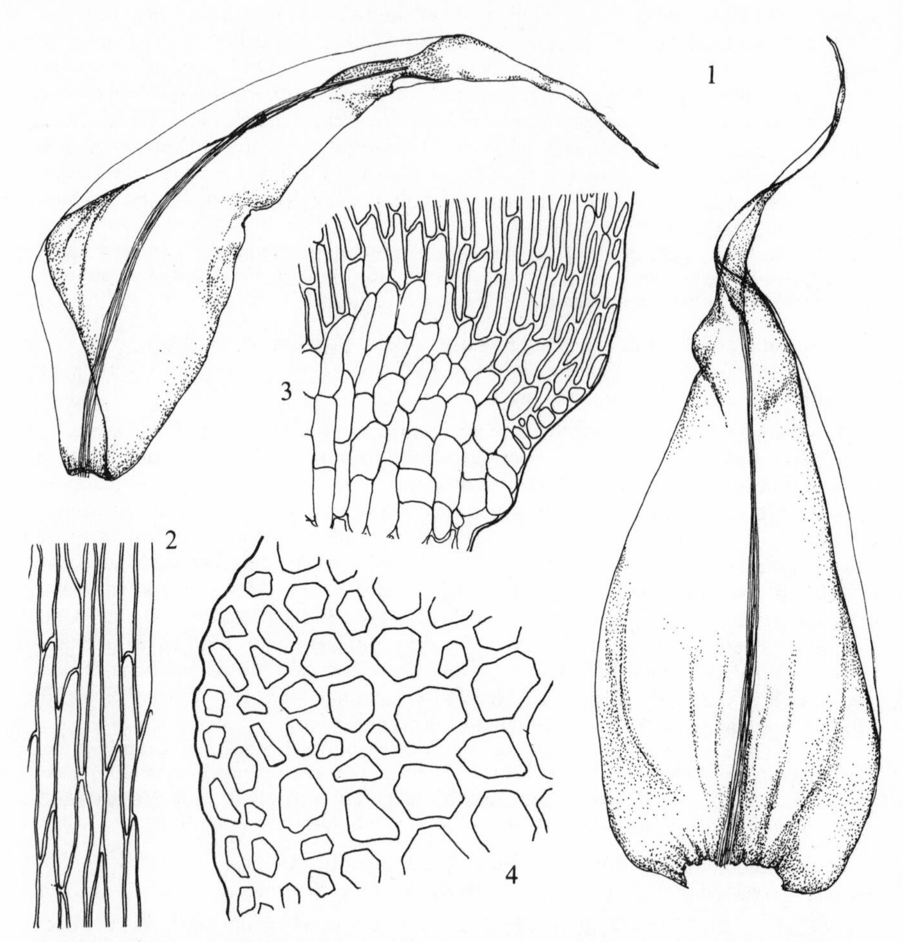

Fig. **274**. *Drepanocladus lycopodioides*: 1, leaves (×25); 2, mid-leaf cells (×415); 3, angular cells (×250); 4, stem section (×415).

scarcely forming auricles, mostly to 50(–60) μm long, extending to nerve, cells above linear, in mid-lf 5–10 μm × 80–160 μm, 10–20 times as long as wide. Capsule inclined or horizontal, shortly cylindrical, curved; annulus lacking; spores 16–24 μm. Fr rare, summer, autumn. $n = 22$, $22 + 2$, 24. Dull green to greenish-brown or reddish-brown tufts or patches on acidic marshy, boggy and peaty ground, in shallow pools, waterlogged turf, often submerged, usually at low altitudes.

Var. **fluitans**

Nerve plano-convex in section (plane on ventral side, convex on dorsal side), angular cells in 3–4 tiers extending from margin to nerve, thin or thick-walled, colourless. Seta 1–7 cm. $n = 20$*. Occasional to frequent in suitable habitats. 78, H12. Europe, Iceland, Faroes, N. Asia, Korea, Japan, Azores, Canaries, southern Africa, Kerguelen Is., N. America, Ecuador, Tasmania, New Zealand.

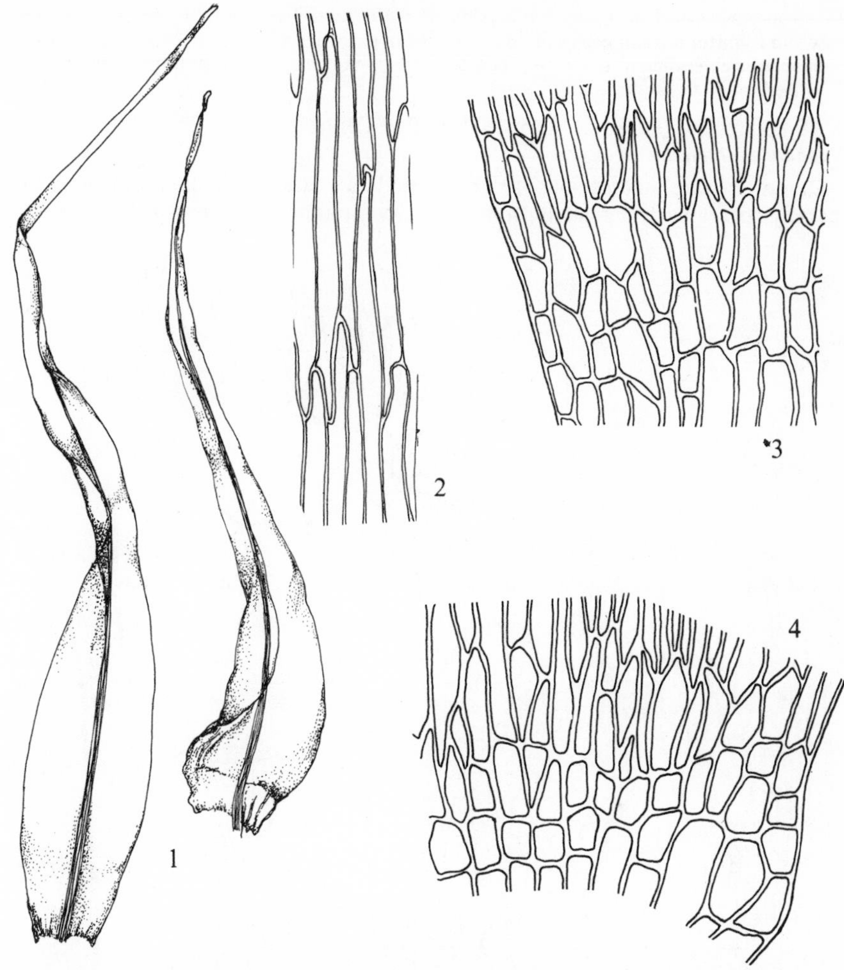

Fig. **275**. 1–3, *Drepanocladus fluitans* var. *fluitans*: 1, leaves (×25); 2, mid-leaf cells (×415); 3, angular cells (×250). 4, *D. fluitans* var. *falcatus*: angular cells (×250).

Var. **falcatus** (Sanio ex C. Jens.) Roth, Eur. Laubm., 1904

D. schulzei Roth, *D. fluitans* var. *uncatus* Crum, Steere & Anderson

Nerve stouter, often biconvex in section, angular cells in 3–4 tiers at margin, 1 tier at nerve, incrassate, yellowish to orange-brown, often pitted. Seta 5–12 cm. In similar habitats to but less frequent than var. *fluitans*. 65, H6. Europe, N. America.

A variable species, the only distinctive form of which that has been shown to have a genetic basis being var. *falcatus* (for an account of the variation in *D. fluitans* and *D. exannulatus* see Lodge, *Svensk bot. Tidskr.* **54**, 368–86, 1960). The angular cells of *D. fluitans* extend to the nerve and may be mistaken for basal cells but are considerably shorter and wider than the basal cells. Forms of *D. fluitans*, especially var. *falcatus* with incrassate angular cells, may be confused with forms of *D. exannulatus* with smaller than usual auricles. The angular cells of *D. fluitans* are never as long as those of *D. exannulatus* which usually has pronounced auricles which are lacking in *D. fluitans*. The latter species has a weaker nerve and shorter angular cells and tends to grow in more acidic habitats. The angular cells of both species extending to the nerve and the denticulate lf margin will distinguish them from *D. aduncus* and *D. sendtneri*.

5. D. exannulatus (Br. Eur.) Warnst., Beih. Bot. Centralbl., 1903
Hypnum exannulatus Br. Eur.

Dioecious, rarely autoecious. Plants of moderate size to 20(–30) cm long, stems sparsely to regularly pinnately branched. Lvs straight or more usually falcate to

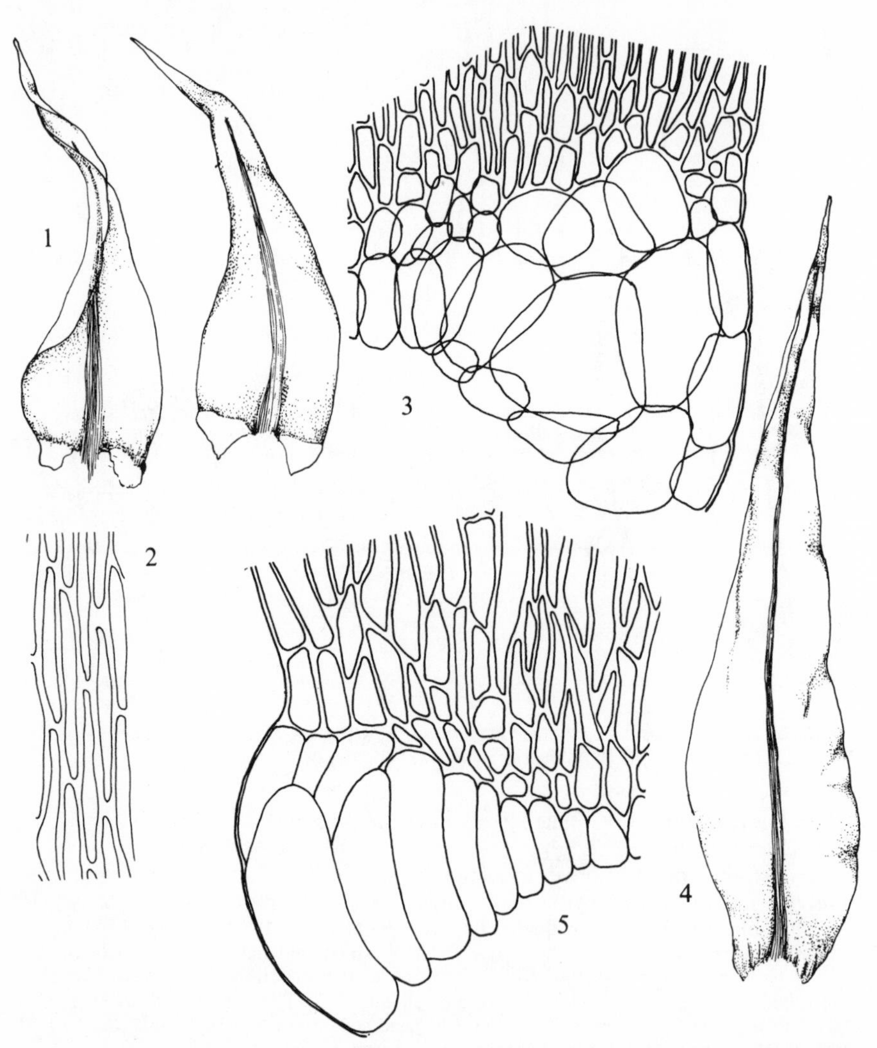

Fig. 276. 1–3, *Drepanocladus exannulatus* var. *exannulatus*: 1, leaves; 2, mid-leaf cells (×415); 3, angular cells (×250). 4–5, *D. exannulatus* var. *rotae*: 4, leaf; 5, angular cells (×250). Leaves ×25.

falcato-secund, stem lvs ovate to narrowly lanceolate, tapering to long acumen, margin denticulate; nerve strong, biconvex, usually reddish or brownish, extending into acumen, 50–100 μm wide near base; basal cells rectangular, angular cells inflated, largest 70–150 μm long, extending to nerve, cells above linear-rhomboidal, in mid-lf 4–8 μm × 40–80 μm, (6–)8–12(–17) times as long as wide. Capsule inclined, shortly cylindrical, curved; spores 16–20 μm. Fr rare, summer. Dull green to brownish or deep red tufts on wet heath, edges of pools, in springs, marshes and bogs in both lowland and montane habitats.

Var. **exannulatus**

Nerve 50–80 μm wide near base; angular cells strongly inflated, thin-walled, ± colourless, in 2–3 tiers extending from margin to nerve, forming distinct auricles. $n=11$*. Plants green to brownish, rarely reddish. Frequent or common in suitable habitats, generally distributed. 70, H25. Europe, Iceland, Caucasus, Altai, Kashmir, Yunnan, Japan, N. America, Greenland, Falkland Is., New Zealand.

Var. **rotae** (De Not.) Loeske, Mitteil. Bayer. Bot. Ges., 1910

Nerve 50–100 μm wide; angular cells inflated, elongated, walls thickened, reddish to brownish, in 2–3 tiers at margin, 1 tier at nerve. Plants usually brownish to deep red, only occasionally greenish. In similar habitats to but less common than var. *exannulatus*. 51, H3. Europe, N. Asia, N. America.

Var. *rotae* of *D. exannulatus* parallels but is less distinctive than var. *falcatus* of *D. fluitans*. Although shown to be genetically distinct, the difficulty experienced in naming a considerable proportion of gatherings raises doubts about the propriety of recognising var. *rotae* as a distinct taxon.

6. D. revolvens (Sw.) Warnst., Beih. Bot. Centralbl., 1903

D. revolvens var. *intermedius* (Lindb.) Grout, *D. intermedius* (Lindb.) Warnst., *Hypnum revolvens* Sw., *H. intermedium* Lindb.

Dioecious or autoecious. Plants slender to robust, procumbent to ascending, to 10 cm long; stems distantly pinnately branched, in section with single outer layer of large cells. Lvs strongly falcato-secund to circinate, unaltered or somewhat crisped at tips when dry, stem lvs ovate, tapering to long channelled acumen, base not decurrent, margin ± entire; nerve extending *ca* ¾ way up lf; basal cells very incrassate, porose, at basal angles shorter, slightly wider and less thick-walled, with 2–3 inflated, hyaline, very fragile cells, usually lost except in young lvs, cells above linear-vermicular, incrassate, in mid-lf 4–8 μm × 48–100(–120) μm, (7–)9–20 times as long as wide. Capsule inclined, shortly cylindrical, curved; annulus present; spores 12–16 μm. Fr rare, summer. $n=10+1$*, 20*, $22+1$. Light green to pinkish to deep purplish, sometimes variegated, glossy patches, sometimes extensive, in marshes, fens, springs, flushes, on wet shaded rocks and cliffs, usually where at least some base present, occasional to frequent, generally distributed. 97, H36. Europe, Faroes, Iceland, N. and C. Asia, Japan, New Guinea, Greenland, N. America, Colombia, Antarctica, New Zealand.

I have not recognised var. *intermedius* (Lindb.) Grout as a distinct taxon as I am unable to find any discontinuity in variation within the species. Var. *intermedius* is described as being small, green and dioecious and plants so named sometimes resemble superficially *Hypnum cupressiforme*; var. *revolvens* is a larger plant with reddish lvs, robust forms approaching *Scorpidium scorpioides* in appearance. Examination of a large number of gatherings has shown that there is no morphological discontinuity between the two extremes. Small green plants are dioecious, larger red-tinged ones autoecious, and the former tend to occur in more basic habitats than the latter. There may be some correlation between

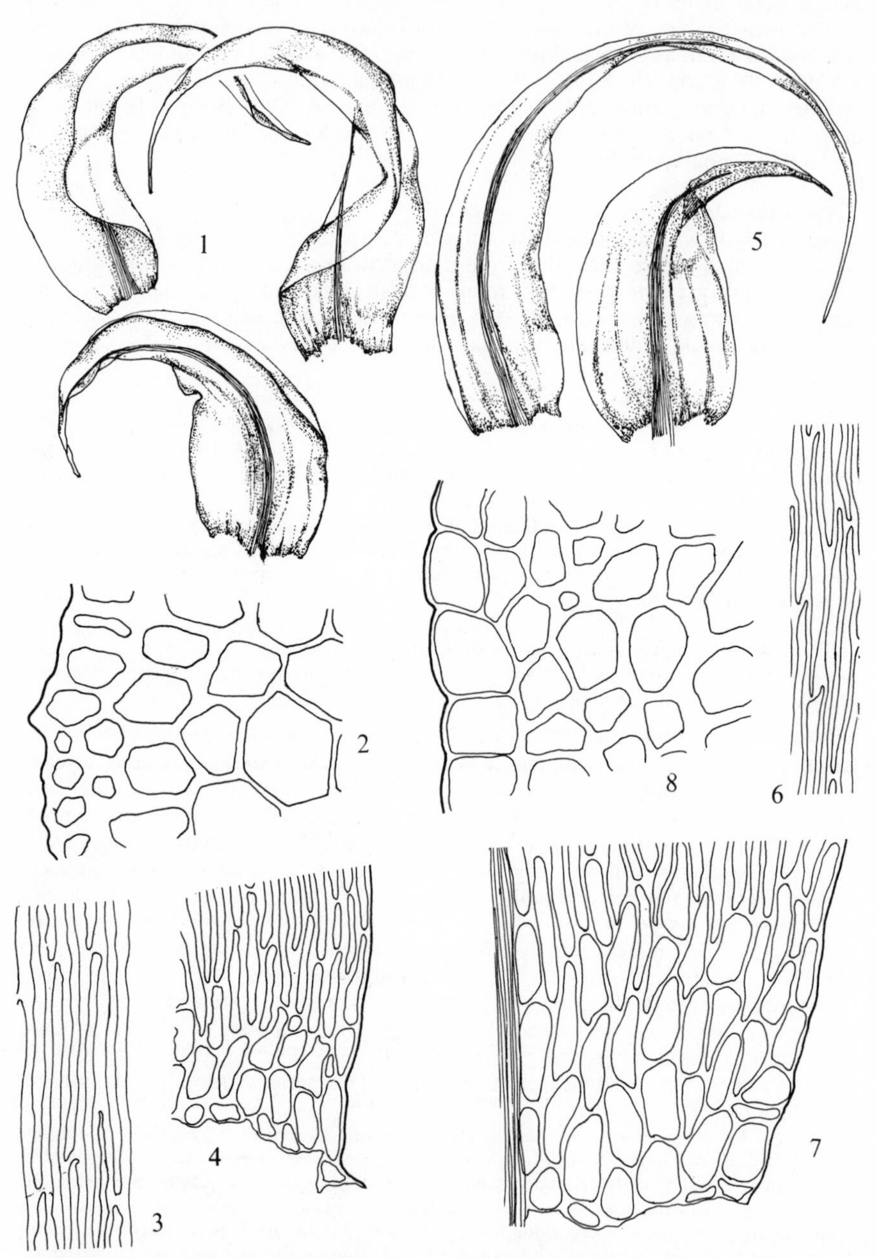

Fig. **277**. 1–4, *Drepanocladus vernicosus*: 1, leaves; 2, section of stem; 3, mid-leaf cells; 4, angular cells (×250). 5–8, *D. revolvens*: 5, leaves; 6, mid-leaf cells; 7, angular cells; 8, section of stem. Leaves ×25, cells ×415.

chromosome number and sex. The nature of variation in *D. revolvens* requires experimental investigation.

D. revolvens differs from *D. vernicosus* in the relatively longer, narrower, non-plicate lvs, the basal cells not brownish and the large epidermal cells of the stem. A few of these epidermal cells may tear off with the leaf base and form a useful distinguishing character from *D. vernicosus*, some forms of which may lack plicate lvs or brownish basal cells.

7. D. vernicosus (Mitt.) Warnst., Beih. Bot. Centralbl., 1903
Hypnum vernicosum Mitt., *Scorpidium vernicosum* (Mitt.) Tuom.

Dioecious. Plants of moderate size, to 10 cm long, procumbent to ascending; stems pinnately branched, outermost layer of cells of cortex small, incrassate. Lvs strongly falcato-secund, usually plicate both when wet and dry, stem lvs ovate, tapering from $\frac{1}{4}$–$\frac{1}{3}$ way up lf to acuminate apex, base not decurrent, margin entire; nerve extending $\frac{1}{2}$–$\frac{3}{4}$ way up lf; basal cells incrassate, porose, 1–2 basal rows brown, sometimes conspicuously so; angular cells not differentiated, cells above linear, incrassate or not, in mid-lf (3–)4–6 μm $\times$ 32–72 μm, (8–)10–20 times as long as wide. Capsule inclined, shortly cylindrical, curved; annulus present. Fr very rare, summer. Green to yellowish-green or brownish patches in bogs, fens and flushes, rarely by streams, generally distributed, occasional. 46, H20. N., W. and C. Europe, Faroes, Siberia, China, Japan, N. America, Greenland.

8. D. uncinatus (Hedw.) Warnst., Beih, Bot. Centralbl., 1903
Hypnum uncinatum Hedw.

Autoecious. Plants of medium size, to 10 cm long, stems pinnately branched. Lvs falcato-secund to circinate, plicate, from broad base gradually and longly tapering to $\pm$ filiform acumen, margin finely denticulate; nerve extending far into acumen; basal cells incrassate, not porose, a few angular cells enlarged, hyaline, forming auricles, cells above linear, incrassate, in mid-lf 5–7 μm $\times$ 48–72 μm, 10–14 times as long as wide. Capsule inclined, shortly cylindrical, curved; annulus present; spores 12–16 μm. Fr frequent to common, spring, summer. $n=10$*, 11, 12, 20. Pale to yellowish-green patches on tree boles, logs, soil and rocks by streams and in fens and carr and on flushed rocks on cliffs and in ravines, in basic or acidic habitats, generally distributed, rare in lowland areas, occasional to frequent in montane regions. 101, H28. Montane and northern Europe, Faroes, Iceland, N. and C. Asia, Kashmir, China, N. America, Greenland, Mexico, Brazil, Patagonia, Tierra del Fuego, Australia, New Zealand.

144. Hygrohypnum Lindb., Acta Soc. Sc. Fenn., 1872

Autoecious or dioecious. Slender to robust, decumbent to erect plants, stem sometimes denuded and with or without sparse rhizoids below. Lvs usually crowded, uniformly imbricate to spreading, sometimes secund or falcato-secund, $\pm$ concave, plicate, ovate-lanceolate and acuminate to $\pm$ orbicular, margin plane, entire or denticulate; nerve short or long, single, forked above or double, sometimes absent; cells at basal angles shorter and wider than elsewhere, sometimes enlarged or inflated, hyaline or coloured, thin-walled or incrassate, other cells narrow, blunt-ended, smooth, shorter near apex. Seta red; capsule inclined to horizontal, ovoid to cylindrical, curved, often contracted below mouth when dry and empty; annulus differentiated; peristome perfect; lid conical. A small genus of about 38 species distributed through Europe, Asia, C. Africa and America.

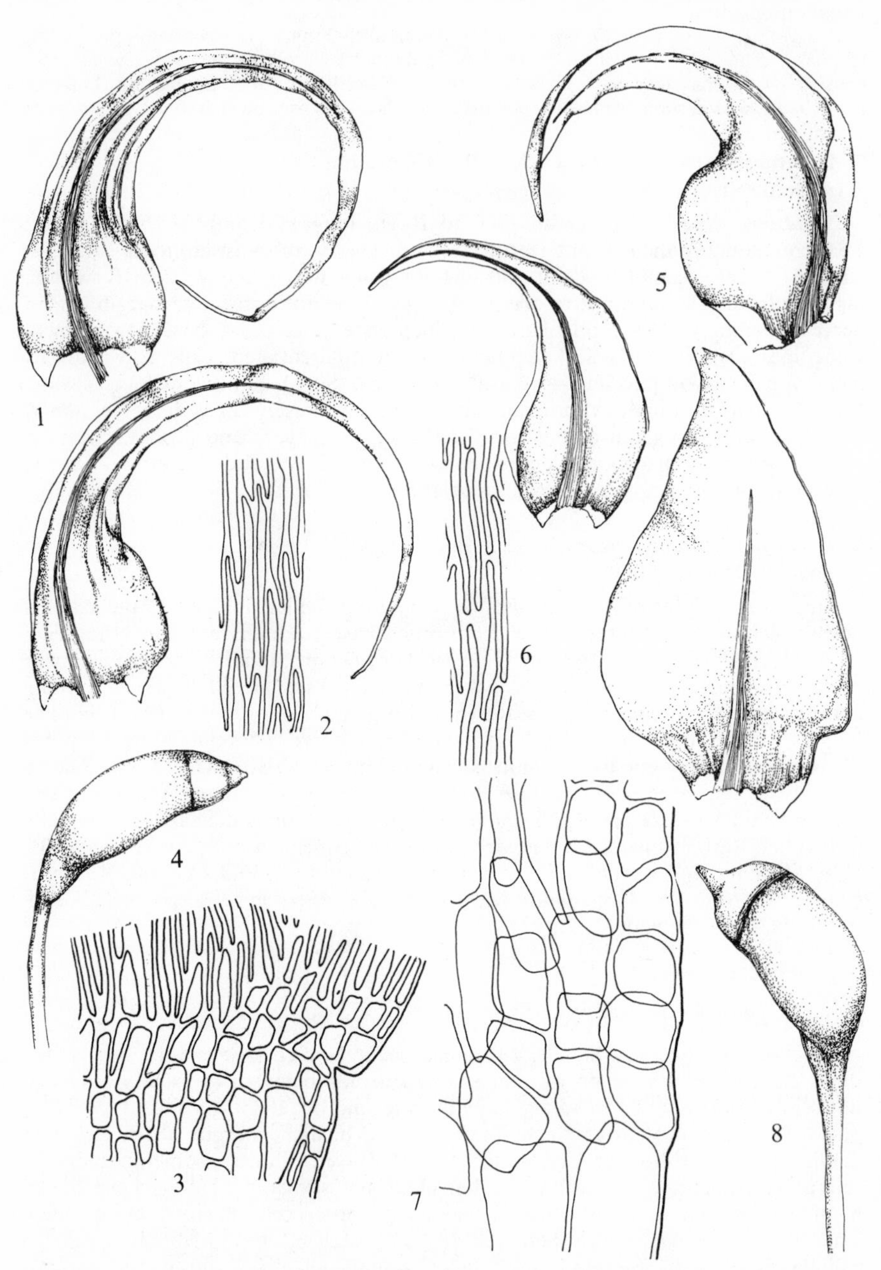

Fig. **278**. 1–4, *Drepanocladus uncinatus*: 1, leaves (×25); 2, mid-leaf cells (×415); 3, angular cells (×250); 4, capsule. 5–8, *Hygrohypnum ochraceum*: 5, leaves (×40); 6, mid-leaf cells (×415); 7, angular cells (×415); 8, capsule. Capsules ×10.

1 Lvs ovate-oblong or ovate to ovate-lanceolate, gradually or abruptly tapering to apex 2
Lvs broadly ovate to orbicular, apex wide 5
2 Angular cells thin-walled, inflated, hyaline and forming distinct auricles*, nerve single or double, extending ½–¾ way up lf **1. H. ochraceum**
Angular cells inflated or not, usually coloured at least in older lvs and not forming auricles, nerve various 3
3 Nerve single, ending in apex or percurrent **2. H. polare**
Nerve single or double, not extending more than ¾ way up lf or sometimes absent 4
4 Nerve extending to (½–)¾ way up lf, angular cells with granulose contents **3. H. luridum**
Nerve rarely extending more than ¾ way up lf, angular cells without granulose contents **4. H. eugyrium**
5 Nerve single or sometimes forked above, only occasional lvs with double nerve, cells in mid-lf 4–7 times as long as wide, lvs not sub-secund, entire except at apex **5. H. smithii**
Nerve double, cells in mid-lf 7–14 times as long as wide, lvs sub-secund or if not then margin denticulate from middle of lf or below 6
6 Lvs not sub-secund, denticulate from middle or below **6. H. molle**
Lvs sub-secund, margin entire except near apex **7. H. dilatatum**

1. H. ochraceum (Turn. ex Wils.) Loeske, Moosfl. Harz, 1903
Hypnum ochraceum Turn. ex Wils.

Dioecious. Plants medium-sized to robust, shoots 1–8 cm long, soft, usually procumbent; stems irregularly branched, branches usually parallel to stems. Lvs shrunken, glossy when dry, straight or more usually falcato-secund, sometimes slightly undulate when moist, concave, ovate-oblong to ovate-lanceolate, apex rounded to obtuse or sub-acute, margin erect or inflexed above, entire or minutely denticulate towards apex; nerve single or double, extending ½(–¾) way up lf; basal cells narrowly rhomboidal, angular cells inflated, thin-walled, hyaline, forming distinct decurrent auricles, cells above ± linear, in mid-lf 5–7 μm × 40–64 μm, 9–13 times as long as wide. Capsule inclined, ellipsoid, curved; lid conical, acute; spores 16–23 μm. Fr very rare. $n=10$, 11. Green to yellowish-green or brownish-green patches on rocks in fast-flowing streams and rivers and by waterfalls, frequent or common in montane areas, very rare elsewhere and absent from S.E. England and E. Anglia. 70, H21. Europe, Iceland, Faroes, Caucasus, C. Asia, Korea, Japan, N. America, Greenland.

Very variable and often resembling other *Hygrohypnum* species but distinct in the large hyaline angular cells. *H. ochraceum* is softer in texture than the next 3 species.

2. H. polare (Lindb.) Loeske, Verh. Bot. Ver. Brandenburg, 1905

Dioecious. Medium-sized or robust plants, to 3.5 cm long; stems irregularly branched, branches ascending to erect. Lower lvs imbricate, towards stem and branch tips falcato-secund, ovate, apex obtuse to rounded, margin inflexed above, entire; nerve stout, ending near apex or percurrent; basal cells narrowly rectangular, angular cells quadrate to rectangular, hyaline or, in older lvs, brownish, forming small decurrent auricles, cells above linear, in mid-lf 5–7 μm × 35–56 μm, 5–10 times as long as wide. Fr unknown. Green patches or tufts, brown below, on rocks by lakes, very rare, W. Ross. 1. Iceland, arctic Europe, Asia and N. America, Greenland.

* Often remain on stem when lvs are detached.

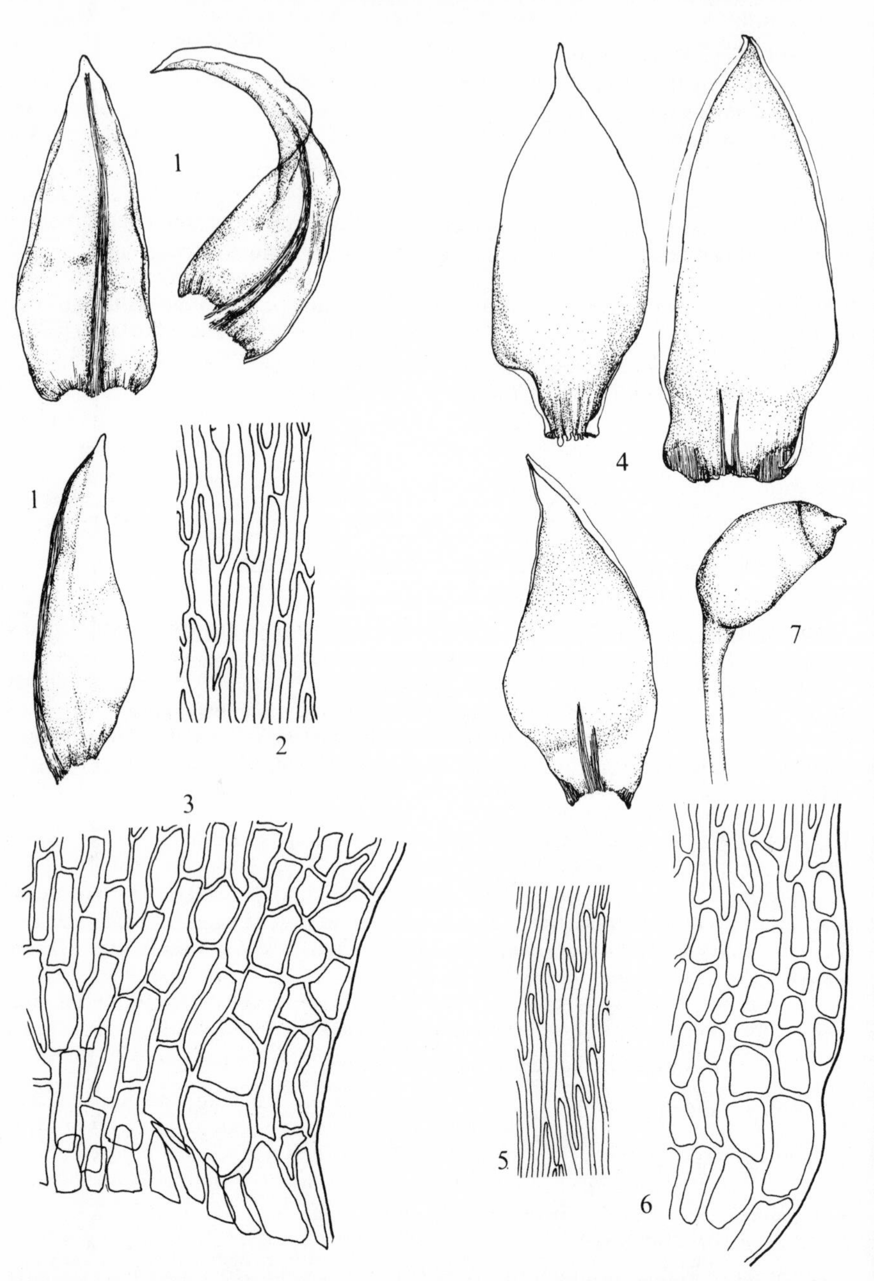

Fig. **279**. 1–3, *Hygrohypnum polare*: 1, leaves; 2, mid-leaf cells; 3, angular cells. 4–7, *H. eugyrium*: 4, leaves; 5, mid-leaf cells; 6, angular cells; 7, capsule (×10). Leaves ×40, cells ×415.

Distinct from other British species of *Hygrohypnum* in the strong nerve extending to lf apex. For the occurrence of this plant in Britain see Wallace, *J. Bryol.* 7, 157–9, 1972. The Scottish plants appear to have more strongly falcato-secund lvs than material from northern Europe.

3. H. luridum (Hedw.) Jenn., Man. Moss. West Pennsylv., 1913
Hypnum palustre Huds.

Autoecious. Medium-sized plants, stems procumbent, irregularly branched, branches ascending, ± parallel to stems. Lvs imbricate to strongly falcato-secund, concave, ovate-lanceolate to ovate-oblong, gradually or abruptly tapering to ± acute apex, margin erect or inflexed, minutely denticulate towards apex; nerve single, usually forked, extending ($\frac{1}{2}$–)$\frac{3}{4}$ way up lf and sometimes almost to apex; cells at base narrowly rhomboidal, angular cells ellipsoid to quadrate, not inflated, incrassate, opaque with granular contents, brownish in older lvs, cells elsewhere linear, in mid-lf 4–7 μm × 40–56 μm, 7–14 times as long as wide. Capsule inclined, ovoid to ellipsoid, curved; lid bluntly mamillate; spores 16–21 μm.

Var. **luridum**
Lvs straight to sub-falcate; nerve single, forked or double, extending ($\frac{1}{2}$–)$\frac{3}{4}$ way up lf. Capsule ellipsoid, curved. Fr common, summer. $n=10$, $10+1$*, 11*. Yellowish-green to green or brownish-green, sometimes red-tinged patches on rocks in fast-flowing streams and rivers, rare in lowland areas, frequent or common elsewhere. 98, H35. Europe, Iceland, Faroes, Caucasus, Siberia, Kashmir, Tibet, Yunnan, Japan, N. America.

Var. **subsphaericarpon** (Schleich. ex Brid.) C. Jens in Podp., Consp., 1956
H. palustre var. *subsphaericarpon* (Schleich. ex Brid.) Br. Eur.
Lvs falcato-secund; nerve single, extending almost to apex. Capsule ovoid or shortly ellipsoid, gibbous. In similar habitats to the type, rare in W. and N. England, very rare in Scotland, N. Kerry. 20, H1. Baltic region, Spain, Caucasus, Tibet, Japan.
Differs from *H. eugyrium* in the stronger nerve, granulose angular cells and slightly wider upper cells. *H. polare* has a longer nerve and non-granulose angular cells.

4. H. eugyrium (Br. Eur.) Broth., Nat. Pfl., 1908
Hypnum eugyrium (Br. Eur.) Schimp.

Autoecious. Medium-sized plants, shoots 0.5–2.0 cm long, stems procumbent, irregularly branched, branches procumbent to erect. Lvs falcate to falcato-secund, concave, ovate to ovate-lanceolate, tapering to short, obtuse to acuminate apex, margin erect or inflexed above, entire or denticulate towards apex; nerve thin, double, hardly extending to $\frac{1}{4}$ way up lf or absent; angular cells enlarged, incrassate, hyaline in young lvs, orange-brown and opaque in older lvs, other cells linear, 3–5 μm × 35–52 μm, 8–16 times as long as wide. Capsule inclined, shortly ellipsoid, curved; lid conical, acute; spores *ca* 20 μm. Fr common, spring. $n=8$*, 11. Yellowish-green to brownish-green patches on rocks in streams and rivers, occasional in montane areas. Devon, Wales, Yorks and Westmorland, northwards. 46, H15. N., W. and C. Europe, Faroes, E. Asia, Japan, N. America.

5. H. smithii (Sw.) Broth., Nat. Pfl., 1908
Hypnum arcticum Sommerf.

Autoecious. Plants forming tufts, 3–12 cm long. Lvs loosely imbricate when dry, spreading when moist, not secund, 0.7–1.2 mm long, concave, broadly ovate to orbicular with obtuse apex, margin plane, entire; nerve single, sometimes forked

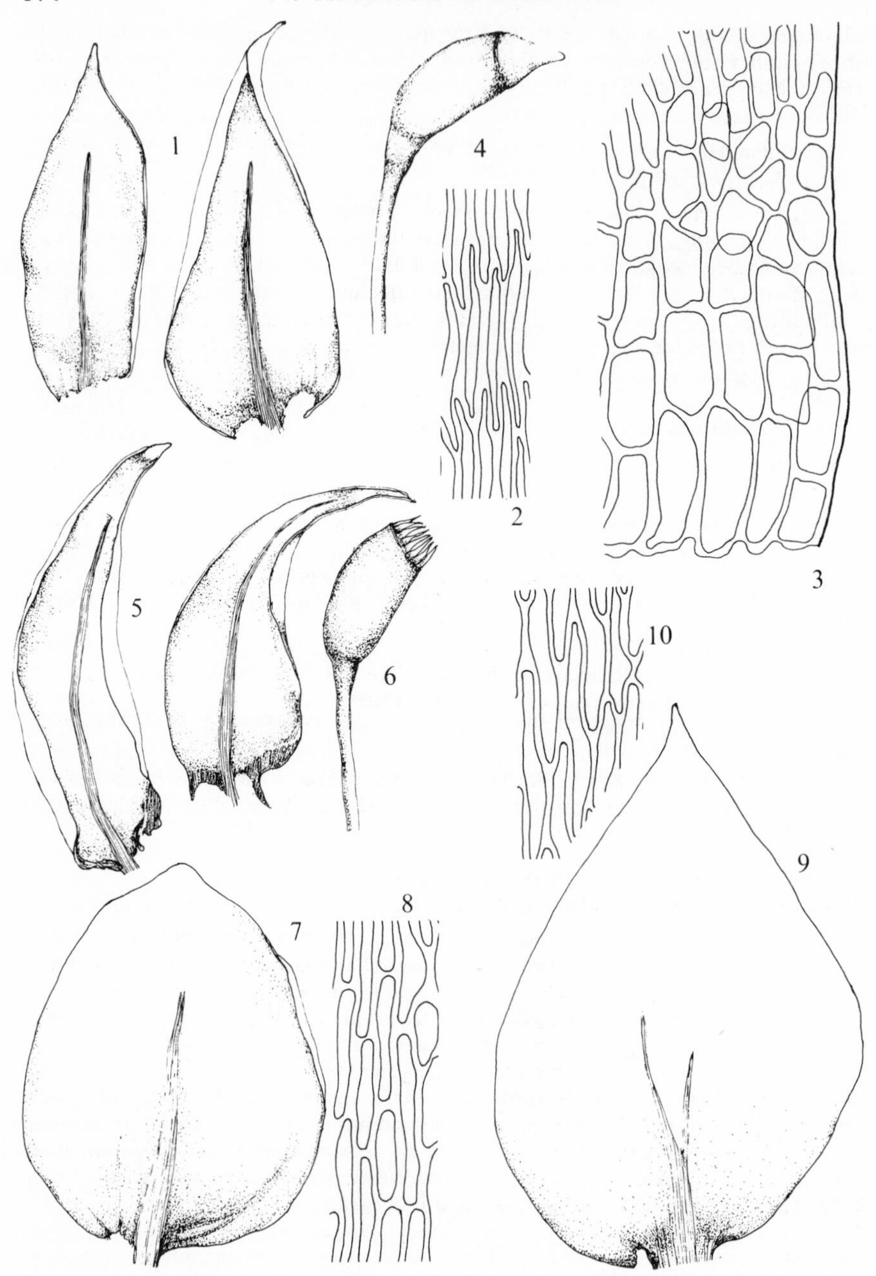

Fig. **280**. 1–4, *Hygrohypnum luridum* var. *luridum*: 1, leaves; 2, mid-leaf cells; 3, angular cells; 4, capsule. 5–6, *H. luridum* var. *subsphaericarpon*: 5, leaves; 6, capsule. 7–8, *H. smithii*: 7, leaf; 8, mid-leaf cells. 9–10, *H. molle*: 9, leaf; 10, mid-leaf cells. Leaves ×40, cells ×415, capsules ×10.

above, extending $\frac{1}{2}$–$\frac{3}{4}$ way up lf; basal cells narrowly rectangular, at basal angles shorter and wider, yellowish, not forming auricles, cells above ± ellipsoid or narrowly ellipsoid, in mid-lf 6–10 μm×28–56 μm, 4–7 times as long as wide. Perichaetial lvs narrowly lanceolate; nerve extending to apex. Capsule inclined, ellipsoid; spores 14–20 μm. Fr very rare. Brownish tufts on rocks in fast-flowing streams and by waterfalls at high altitudes, very rare. Perth, Angus, S. Aberdeen, W. Inverness. 5. N. and C. Europe, Faroes, Pyrenees, Silesia, Bulgaria, Iceland, Faroes, N. America, Greenland.

6. H. molle (Hedw.) Loeske, Moosfl. Harz., 1903
Hypnum molle Hedw.

Autoecious. Plants forming tufts, 2–8 cm long, not denuded below. Lvs imbricate, soft, 1–2 mm long, concave, broadly ovate, apex rounded or obtuse or occasionally ± acute, margin plane, denticulate; nerve double, extending $\frac{1}{3}$–$\frac{1}{2}$ way up lf; angular cells only slightly enlarged, hyaline but not forming distinct auricles, cells above linear, in mid-lf 4–7 μm×40–52 μm, (6–)7–12 times as long as wide. Capsule inclined, ovoid. Fr unknown in Britain. Soft, greenish tufts in fast-flowing streams at high altitudes, very rare. S. Aberdeen, Inverness. 3. N., W. and C. Europe, Faroes, Siberia, Montana, Greenland.

This species has been confused in the past with *H. dilatatum* and many older gatherings so named belong to the latter. *H. dilatatum* differs in the wider, sub-secund lvs with entire margins and more distinctive angular cells. *H. smithii* has smaller, wider lvs with a usually single nerve and wider cells.

7. H. dilatatum (Wils. ex Schimp.) Loeske, Moosfl. Harz, 1903
Hypnum dilatatum Wils. ex Schimp.

Dioecious. Plants tufted, shoots 2–6 cm long; stems irregularly branched, stems and branches sometimes denuded below. Lvs sub-secund, 1.0–1.8 mm long, concave, ovate-orbicular, apex obtuse to rounded, margin plane, entire or faintly denticulate near apex; nerve double, extending $\frac{1}{3}$–$\frac{1}{2}$ way up lf; angular cells somewhat enlarged, incrassate, orange or brownish at least in older lvs, cells above linear, in mid-lf 4–7 μm×44–72 μm, (7–)9–14 times as long as wide. Capsule inclined, ellipsoid, curved; lid conical, obtuse; spores *ca* 20 μm. Fr very rare, summer. $n=11$. Green or yellowish-green tufts in fast-flowing montane streams and rivers, rarely at low altitudes, rare but sometimes locally abundant. N. Wales, N. England, Scottish Highlands, Kerry. 15, H2. Montane and northern Europe, Faroes, Caucasus, N. and C. Asia, Kashmir, China, Japan, N. America, Greenland.

145. SCORPIDIUM (SCHIMP.) LIMPR., LAUBM., 1899

Dioecious. Robust plants, stems procumbent to ascending, julaceous or tumid with imbricate lvs. Lvs large, very concave, broadly ovate or ovate-oblong, apex acute to obtuse, apiculate or not, margin entire, erect or inflexed above; nerve absent or short and double, rarely longer and single; basal cells incrassate, porose, angular cells differentiated, forming distinct auricles or not, cells above linear or linear-rhomboidal. Seta long; capsule inclined, cylindrical, curved; annulus present, separating; inner peristome with cilia. Three species. Europe, N. and E. Asia, N. and S. America.

Lvs falcato-secund, angular cells inflated, hyaline, forming small auricles, cells in mid-lf 10–20 times as long as wide **1. S. scorpioides**
Lvs ± straight, angular cells neither inflated nor forming auricles, cells in mid-lf 5–8 times as long as wide **2. S. turgescens**

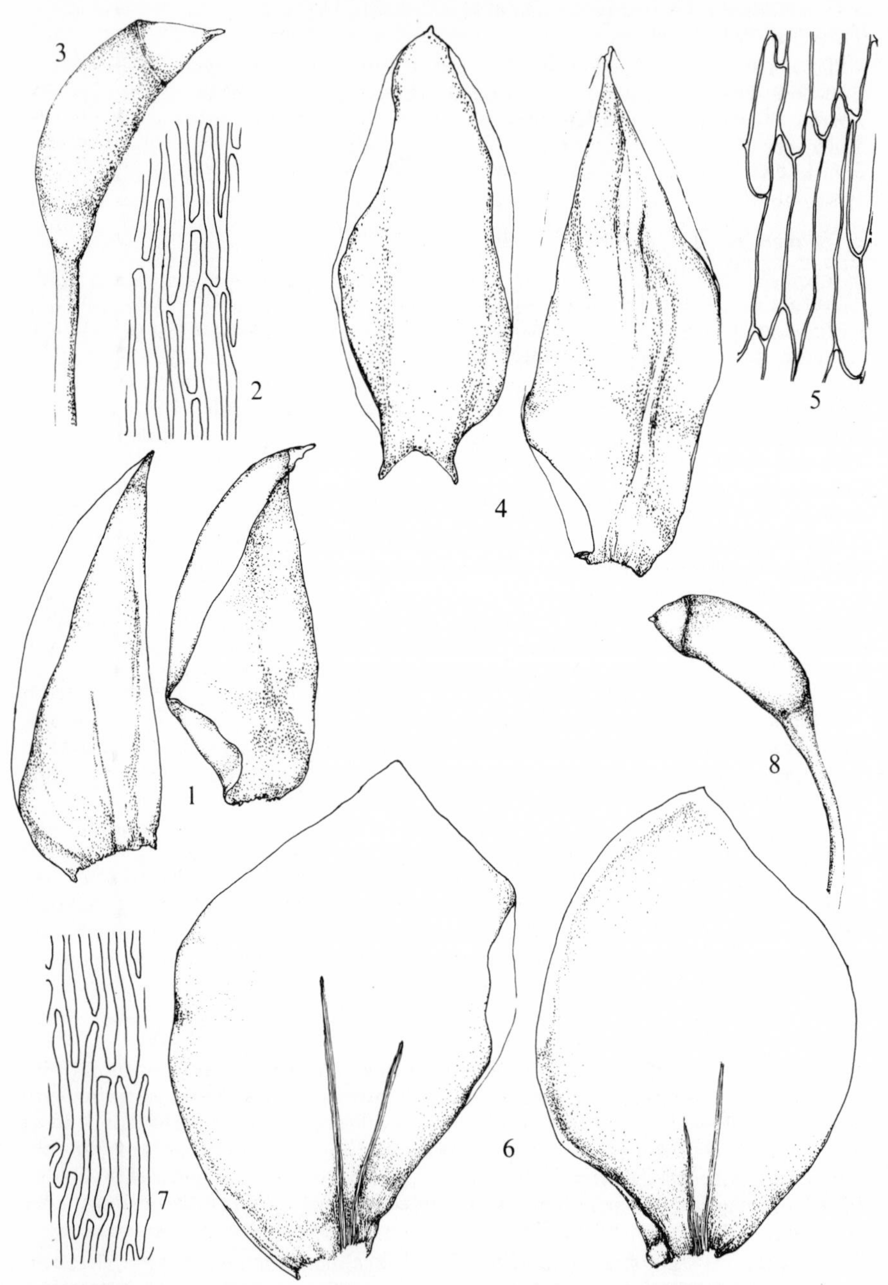

Fig. **281**. 1–3, *Scorpidium scorpioides*: 1, leaves (×25); 2, mid-leaf cells; 3, capsule. 4–5, *S. turgescens*: 4, leaves (×25); 5, mid-leaf cells. 6–8, *Hygrohypnum dilatatum*: 6, leaves (×40); 7, mid-leaf cells; 8, capsule (×10). Cells ×415.

1. S. scorpioides (Hedw.) Limpr., Laubm. Deutschl., 1899
Hypnum scorpioides Hedw.

Plants robust, to 15(–25) cm long, shoots tumid with imbricate lvs, sparsely branched. Lvs imbricate, falcato-secund, concave, often rugose, ovate, shortly tapered to obtuse or rounded apex, to more longly tapering acute apex, margin entire; nerve absent or very short and double, very rarely long and single; a few angular cells inflated, thin-walled to incrassate, forming small decurrent auricles, cells above linear-vermicular, incrassate, in mid-lf 4–7 μm×48–120 μm, 10–20 times as long as wide. Capsule sub-erect to inclined, cylindrical, curved; spores 18–22 μm. Fr rare, spring, summer. $n=8$, 11*. Dark green, to yellowish-brown, reddish-brown or purplish patches, often mud-encrusted and worm-like in appearance, on mud, sometimes submerged, by pools, in fens, marshes, bogs and flushes in acid or basic habitats. Rare in lowland areas, occasional to frequent elsewhere. 87, H34. Europe, Faroes, Iceland, N. Asia, N. America, Greenland, Andes.

2. S. turgescens (T. Jens.) Loeske, Verh. Bot. Ver. Prov. Brandenb., 1905
Hypnum turgescens T. Jens.

Plants robust, to 25 cm long, stems sparsely branched, tumid with imbricate lvs. Lvs straight, not secund or only very slightly so at stem and branch tips, very concave, broadly ovate-lanceolate, apex obtuse, with or without recurved apiculus, sometimes cucullate, margin entire, inflexed above; nerve short, single or double; basal cells incrassate, porose, angular cells quadrate to rectangular, not forming auricles, cells above linear-rhomboidal, in mid-lf 7–10 μm×48–64 μm, 5–8 times as long as wide. Capsule inclined, cylindrical, curved. Fr unknown in Britain. Green to greenish-brown patches, sometimes locally abundant in mires at high altitudes, very rare. Mid Perth. 1. Northern and montane Europe, Siberia, Yunnan, N. America, Greenland, Ecuador, Bolivia.

Although very rare in Britain the plant may be locally abundant where it does occur (see Birks & Dransfield, *J. Bryol.* **6**, 129–32, 1970). A few stems of *S. turgescens* were said to have been found amongst other bryophytes collected on Cader Idris, Merioneth, in 1922 but the plant has not been seen there since and the source of the stems is questionable.

146. Calliergon (Sull.) Kindb., Canad. Rec. Sc., 1894

Autoecious or dioecious. Mostly ± robust plants, procumbent to ascending or erect, stems ± simple to regularly pinnately divided, stem and branch tips sometimes cuspidate, paraphyllia few. Lvs imbricate to spreading, concave, stem lvs sub-orbicular to lanceolate, apex rounded to obtuse, apiculate or not, often sub-cucullate, margin entire, plane below, often erect above and slightly inflexed at apex; nerve single usually extending at least ½ way up lf, rarely absent or short and double; basal cells narrowly rectangular, angular cells thin-walled to incrassate, hyaline or not, decurrent down stem, sometimes inflated and forming distinct auricles, cells above ± linear-vermicular, shorter near apex. Branch lvs usually smaller and narrower. Seta long, flexuose, reddish; capsule inclined or horizontal, gibbous or curved; annulus absent except in *C. trifarium* and *C. cuspidatum*; lid conical, obtuse or acute, sometimes apiculate; outer peristome teeth coarsely papillose above, inner with appendiculate cilia. A ± world-wide genus of 15 species occurring mainly in moist or wet habitats.

1 Lvs nerveless or with very short, faint double nerve **6. C. cuspidatum**
Lvs with single nerve extending ½ way or more up lf 2

2 Lvs imbricate both when moist and dry, stems not or sparsely branched 3
Lvs erecto-patent to spreading except sometimes at stem and branch tips, stems usually branched 4
3 Lvs ovate to ovate-lanceolate, angular cells forming distinct decurrent auricles **1. C. stramineum**
Lvs sub-orbicular to ovate-orbicular, angular cells not forming distinct auricles **2. C. trifarium**
4 Lvs oblong-lanceolate, plants usually deep purplish-red **5. C. sarmentosum**
Lvs ovate-triangular or broadly ovate to ovate-lanceolate, plants yellowish-green to brownish 5
5 Lvs without distinct auricles, branches few **3. C. cordifolium**
Lvs with distinct decurrent auricles, branches numerous, crowded **4. C. giganteum**

1. C. stramineum (Brid.) Kindb., Canad. Rec. Sc., 1894
Acrocladium stramineum (Brid.) Rich. & Wall., *Hypnum stramineum* Brid.

Plants moderately slender, shoots procumbent or ascending, to 15 cm long, stems simple or sparsely branched. Lvs loosely imbricate when dry and moist, concave, ovate to ovate-lanceolate, widest near base, apex obtuse or rounded, often sub-cucullate, margin entire, plane below, erect or sometimes slightly inflexed above; nerve extending $\frac{3}{4}$ way or more up lf; basal cells narrowly rectangular, angular cells inflated, hyaline, forming distinct decurrent auricles, cells above linear, incrassate, in mid-lf 5–8 μm $\times$ 52–80 μm, (6–)9–14 times as long as wide. Capsule inclined, cylindrical, curved; lid with blunt apiculus; spores 12–16 μm. Fr rare. $n=11$*. Pale, yellowish-green tufts, patches or scattered shoots among other bryophytes on wet ground by streams and pools, in flushes, marshes and bogs, often submerged, occasional to frequent in suitable habitats, generally distributed. 103, H25. Europe, Iceland, Faroes, N. and C. Asia, Japan, N. America, Greenland, Australia.

2. C. trifarium (Web. & Mohr) Kindb., Canad. Rec. Sc., 1894
Acrocladium trifarium (Web. & Mohr) Rich. & Wall., *Hypnum trifarium* Web. & Mohr

Dioecious. Plants of moderate size, procumbent, shoots julaceous, 15 cm long, stems sparsely branched. Lvs imbricate when moist and dry, concave, sub-orbicular to ovate-orbicular, apex rounded, sometimes sub-cucullate, margin entire; nerve extending $\frac{1}{2}$ way or more up lf; angular cells enlarged, thin-walled to incrassate, not forming auricles, cells above linear-vermicular, $\pm$ incrassate, in mid-lf 6–9 μm $\times$ 36–80 μm, 6–9 times as long as wide. Capsule inclined, ellipsoid, curved. Fr unknown in Britain. Yellowish-brown to brownish patches in basic montane flushes and fens, rare. Scottish Highlands, Clare. 18, H1. N. and C. Europe, Iceland, Faroes, C. Asia, N. America, Greenland.

3. C. cordifolium (Hedw.) Kindb., Canad. Rec. Sc., 1894
Acrocladium cordifolium (Hedw.) Rich. & Wall., *Hypnum cordifolium* Hedw.

Autoecious. Plants of moderate size, shoots to 15 cm long, procumbent or ascending, stems irregularly to sparsely pinnately branched, branches short, branch and stem tips cuspidate. Lvs $\pm$ spreading when dry, patent to spreading when moist, concave, broadly ovate to ovate-lanceolate, apex rounded, margin entire; nerve extending almost to apex; basal cells enlarged, angular cells enlarged, decurrent down stem but not forming distinct auricles, cells above linear, incrassate, in mid-lf 8–12(–16) μm $\times$ (64–)80–160 μm, (5–)7–20 times as long as wide. Capsule horizontal,

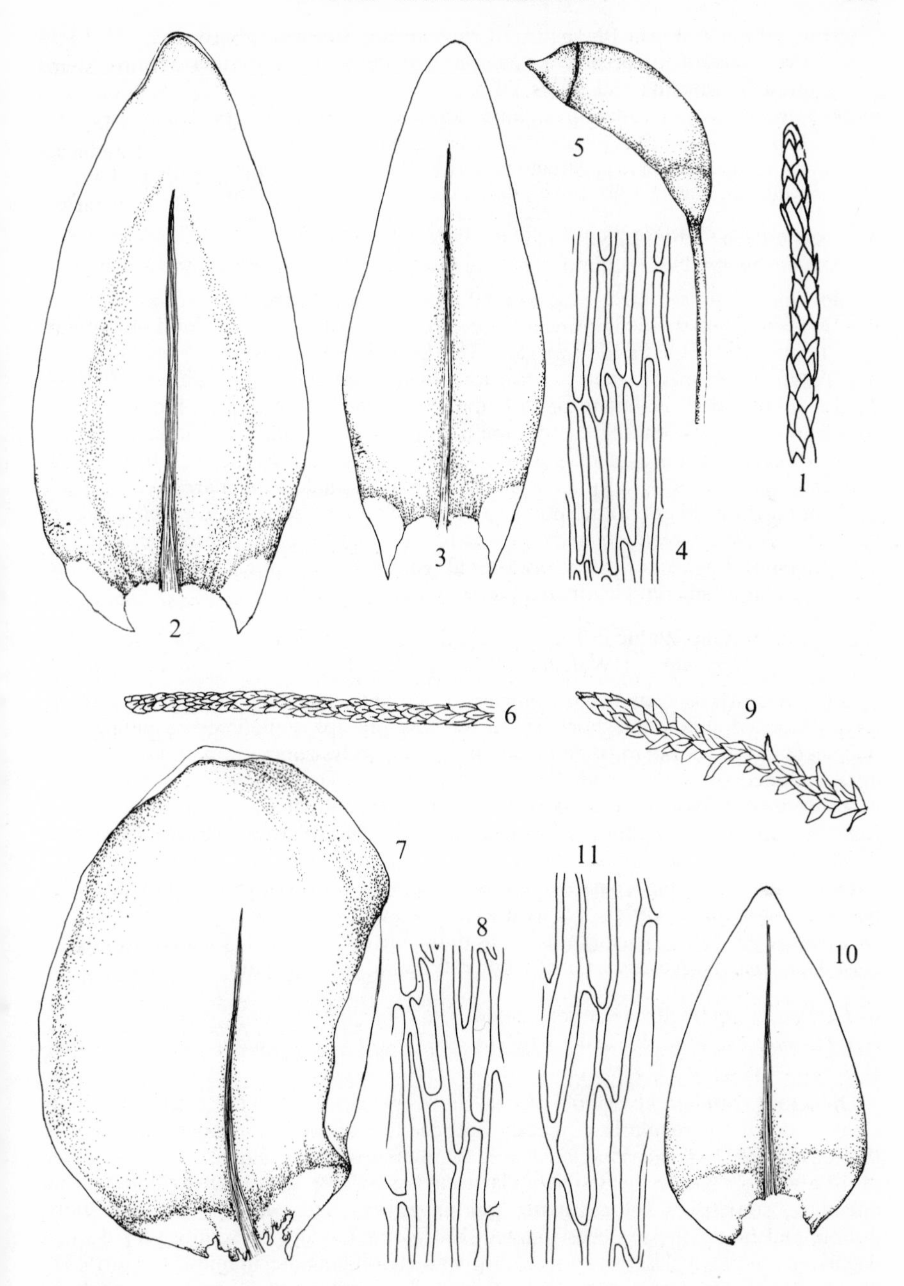

Fig. **282**. 1–5, *Calliergon stramineum*: 1, shoot; 2 and 3, stem and branch leaves (×25); 4, mid-leaf cells; 5, capsule (×10). 6–8, *C. trifarium*: 6, shoot; 7, stem leaf (×40); 8, mid-leaf cells. 9–11, *C. cordifolium*: 9, shoot; 10, stem leaf (×25); 11, mid-leaf cells. Shoots ×2, cells ×415.

ellipsoid, gibbous; spores 10–16 μm. Fr rare, spring, summer. *n*=10, 21*, 22*. Light or yellowish-green tufts or patches or scattered shoots among other bryophytes on wet ground by streams and pools, in fens, marshes, bogs and flushes, often submerged, frequent or common in suitable habitats. 110, H29. Europe, Iceland, Faroes, N. and C. Asia, Japan, N. America, Greenland, New Zealand.

C. stramineum differs in the julaceous stems and branches and the well defined auricles. In *C. cordifolium* the angular cells scarcely differ from the surrounding inflated basal cells.

4. C. giganteum (Schimp.) Kindb., Canad. Res. Sc., 1894
Acrocladium giganteum (Schimp.) Rich. & Wall., *Hypnum giganteum* Schimp.

Dioecious. Plants robust, shoots to 20 cm long, procumbent to ascending, stems closely pinnately branched, branches short, of varying lengths, stem and branch tips usually cuspidate. Lvs, except at stem and branch tips, patent to spreading when dry, patent when moist, stem lvs concave, ovate-triangular to broadly ovate, apex obtuse or rounded, apiculate or not, margin entire; nerve strong, reddish-brown below in older lvs, ending below apex; angular cells inflated, hyaline, forming distinct decurrent auricles, cells above linear, incrassate, in mid-lf 5–10 μm×52–100 μm, 7–16 times as long as wide. Branch lvs smaller and narrower. Capsule horizontal, ellipsoid, gibbous; spores *ca* 16 μm. Fr very rare, summer. Dense tufts or patches, yellowish-green, green or brownish, on wet ground by streams and pools, in fens, marshes, bogs and flushes, occasional, generally distributed. 91, H29. Europe, Faroes, Iceland, Siberia, N. America, Greenland.

5. C. sarmentosum (Wahlenb.) Kindb., Canad. Res. Sc., 1894
Acrocladium sarmentosum (Wahlenb.) Rich. & Wall., *Hypnum sarmentosum* Wahlenb.

Dioecious. Moderately robust, purplish-red plants, shoots to 12 cm, stems irregularly branched, branches short, stem and branch tips sometimes cuspidate. Lvs imbricate to erecto-patent when moist and dry, stem lvs concave, oblong-lanceolate, apex rounded to obtuse, sometimes with reflexed apiculus; nerve extending $\frac{3}{4}$ or more way up lf; basal cells rectangular, angular cells inflated, hyaline in young lvs, reddish in older lvs, forming distinct decurrent auricles, cells above linear, incrassate, in mid-lf 5–7 μm×44–80 μm, 8–14 times as long as wide. Branch lvs small and narrow. Capsule inclined, narrowly ellipsoid, curved; spores *ca* 16 μm. Deep purplish-red tufts, sometimes variegated with green or yellow, in montane flushes, rare to occasional. 53, H10. Northern and montane Europe, Iceland, Faroes, N. and C. Asia, C. Africa, N. America, Greenland, Patagonia, New Zealand, Antarctica.

6. C. cuspidatum (Hedw.) Kindb., Canad. Res. Sc., 1894
Acrocladium cuspidatum (Hedw.) Lindb., *Calliergonella cuspidata* (Hedw.) Loeske, *Hypnum cuspidatum* Hedw.

Dioecious. Moderately robust plants, shoots procumbent to erect, to 12 cm long, stems reddish, ± regularly pinnately branched, branches short, stem and branch tips cuspidate. Lvs except at stem and branch tips erecto-patent to patent when moist and dry, stem lvs ovate-triangular to ovate-oblong, apex obtuse or obtuse and apiculate, sometimes sub-cucullate, margin entire; nerve lacking or very short, double and faint; angular cells inflated, hyaline or coloured, forming very distinct decurrent auricles, cells above linear, thin-walled or incrassate, in mid-lf 4–7 μm×52–88 μm, 10–18 times as long as wide. Branch lvs smaller and narrower. Capsule ± horizontal, cylindrical, curved; spores 16–24 μm. Fr occasional, spring. *n*=11*, 12*. Tufts, patches or sometimes scattered shoots mixed with other bryophytes, yellowish-green, green or greenish-brown, in fens, marshes, bogs, flushes, moist turf, by pools

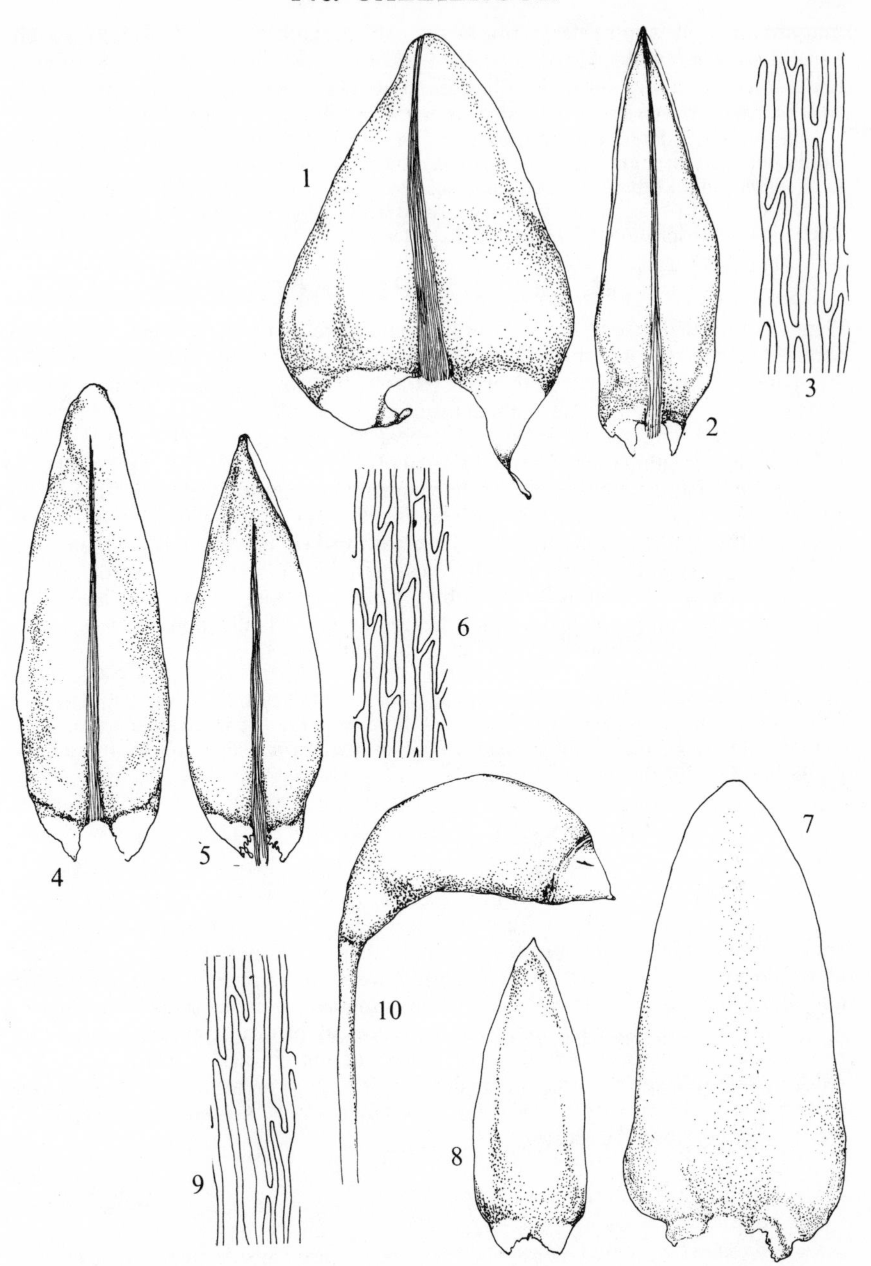

Fig. **283**. 1–3, *Calliergon giganteum*: 1 and 2, stem and branch leaves; 3, mid-leaf cells. 4–6, *C. sarmentosum*: 4 and 5, stem and branch leaves; 6, mid-leaf cells. 7–10, *C. cuspidatum*: 7 and 8, stem and branch leaves; 9, mid-leaf cells; 10, capsule (×10). Leaves ×25, cells ×415.

and streams, on wood rides, damp walls and in chalk grassland, usually but not necessarily where the substrate is moist and basic, sometimes submerged, common and sometimes abundant or locally dominant. 112, H40, C. Europe, Faroes, Asia, Macaronesia, N. Africa, N. America, Jamaica, Argentina, Australia, New Zealand.

I have followed Karczmarz (*Monogr. Bot.* **34**, 1972) in placing this species in *Calliergon* rather than in the monotypic genus *Calliergonella* Loeske. The differences between *C. cuspidatum* and the other species of *Calliergon* are not sufficient to merit generic delimitation. Robust forms of *C. cuspidatum* may be confused with *C. giganteum* but the nerve of the lf in the latter is a conspicuous distinguishing character.

45. BRACHYTHECIACEAE

Slender to robust plants, usually forming mats, patches or wefts; stems creeping to ascending, sparsely and irregularly to closely and pinnately branched, branches straight or curved, complanate to erect; paraphyllia usually absent. Lvs imbricate to spreading, rarely complanate, stem and branch lvs similar or not, plane or concave, plicate or not, broadly ovate to narrowly lanceolate, apex obtuse to longly acuminate; nerve single, sometimes terminating in small projection at back of lf, rarely forked or double; cells elongate-hexagonal to linear-vermicular, usually smooth, basal cells shorter, often porose, angular cells differentiated or not, forming auricles or not. Seta long, usually reddish, smooth or papillose with projecting cell ends; capsule erect, inclined or horizontal, ovoid to cylindrical, straight, gibbous or curved; annulus differentiated or not; peristome double, outer teeth horizontally striate below, papillose above, inner with tall basal membrane and usually well developed cilia; lid conical, with or without beak; calyptra cucullate, naked.

The limits between some of the genera (e.g. *Cirriphyllum*, *Eurhynchium*, *Rhynchostegium* and *Brachythecium*) are obscure and this makes for considerable difficulties for the beginner, especially in the absence of sporophytes. I have adopted a conservative treatment of *Eurhynchium* (see under that genus) because of the various unsatisfactory taxonomic arrangements of the species.

147. Isothecium Brid., Br. Univ., 1827

Dioecious. Primary stems creeping, secondary procumbent to erect, sparsely to closely branched, branches sometimes secund. Lvs imbricate to erecto-patent, concave, stem lvs usually differing from branch lvs, broadly ovate to cordate-triangular, apex obtuse or obtuse and apiculate to longly acuminate, margin denticulate at least above; branch lvs ovate to lanceolate, obtuse to acuminate; nerve single or occasionally forked above or double; basal cells rectangular to rhomboidal, angular cells incrassate, opaque, forming indistinct yellowish to brownish auricles, other cells rhomboidal to narrowly rhomboidal, shorter at margin and towards apex. Seta reddish, smooth; capsule erect or inclined, ellipsoid, straight or curved; peristome perfect, inner with or without cilia; lid obliquely rostrate. A mainly northern hemisphere genus of about 20 species.

1 Lvs longitudinally plicate **4. I. striatulum**
Lvs not plicate 2

2 Stem lvs shortly pointed, denticulate only near apex, capsule erect **1. I. myurum**
Stem lvs with long or short acumen, denticulate almost from base, at least obscurely so, capsule inclined 3

3 Lvs tapering to short or long fine acumen **2. I. myosuroides**
Lvs gradually or abruptly tapering to acute apex **3. I. holtii**

1. I. myurum Brid., Br. Univ., 1827
Eurhynchium myurum (Brid.) Dix.

Medium-sized to robust plants; secondary stems procumbent, irregularly branched, branches straight or curved, often pointing in same direction. Lvs imbricate, concave, stem lvs ovate to broadly ovate, shortly acute to obtuse or obtuse and apiculate, margin inflexed at least above middle of lf, entire below, denticulate towards apex; nerve extending $\frac{1}{2}$–$\frac{2}{3}$ way up lf, single, occasionally forked above; basal cells rhomboidal, at basal angles incrassate, opaque, forming small auricles, cells above narrowly rhomboidal, shorter at margin and towards apex, in mid-lf 5–8 μm×20–40 μm, 4–6 times as long as wide. Capsule ± erect and symmetrical, ellipsoid to shortly cylindrical; lid rostrate; spores 12–16 μm. Fr occasional to frequent, autumn, winter. n=10, 11, 11+1*, 12. Lax, pale to yellowish or greyish-green tufts or patches on logs, tree roots and boles, stumps and rocks in woods, by streams and rivers, on cliffs and shaded rocks and scree, generally distributed. 112, H40, C. Europe, Faroes, Caucasus, Canaries, Azores.

2. I. myosuroides Brid., Br. Univ., 1827
Eurhynchium myosuroides (Brid.) Schimp.

Plants slender to medium-sized, primary stems creeping, secondary erect and sub-dendroid to procumbent and irregularly branched. Lvs of secondary stems concave, ovate to cordate-triangular, tapering to long or short fine acumen, margin plane or inflexed above, obscurely to strongly denticulate; nerve single, sometimes forked or double, rarely absent; basal cells rectangular, angular cells shortly rectangular, opaque, forming small auricles, cells above linear-rhomboidal, in mid-lf 5–9 μm×28–48 μm, 4–8 times as long as wide. Branch lvs differing from stem lvs or not. Seta curved; capsule inclined, ellipsoid, not curved; lid obliquely rostrate; spores 16–24 μm.

Var. **myosuroides**
Secondary stems ascending to erect, sub-dendroid with branches crowded near ends, branches short or long, often curved and pointing in same direction. Lvs imbricate when dry, erecto-patent when moist, stem lvs ovate to cordate-triangular, tapering to long, sometimes fine apex, margin denticulate to dentate from near base. Branch lvs smaller, lanceolate to ovate, tapering to long, sometimes twisted acumen. Fr frequent, winter. n=11*, 11+1*. Pale or yellowish-green patches, sometimes extensive or locally dominant, on rocks, walls, tree boles and branches and logs in shaded habitats, especially woods, frequent to common in suitable habitats. 110, H40, C. Europe, Faroes, Algeria, Canaries, Madeira.

Var. **brachythecioides** (Dix.) Braithw., Brit. Moss Fl., 1905
Plants to 10 cm long, secondary stems long, procumbent, irregularly and distantly branched, branches procumbent, straight. Lvs imbricate, stem lvs ovate to broadly ovate, tapering to long or short acumen, margin obscurely denticulate; branch lvs similar but with more pronounced denticulations. Yellowish-green to green patches on cliff ledges and in turf on slopes in montane habitats, rare or occasional. Caerns, Dumfries, Scottish Highlands. 24, H12. Norway.

Var. *brachythecioides* in its typical form is a very different looking plant from var. *myosuroides*, resembling a *Brachythecium* species but differing in areolation. It is, however, linked to the type by a whole series of intermediates as, for example, on the Shetlands.

3. I. holtii Kindb., Rev. Bryol., 1895
I. myosuroides var. *rivulare*, Holt, ex Limpr., *Eurhynchium myosuroides* var. *rivulare* (Limpr.) Paris

Plants medium-sized, secondary stems ascending, long, irregularly to sub-pinnately

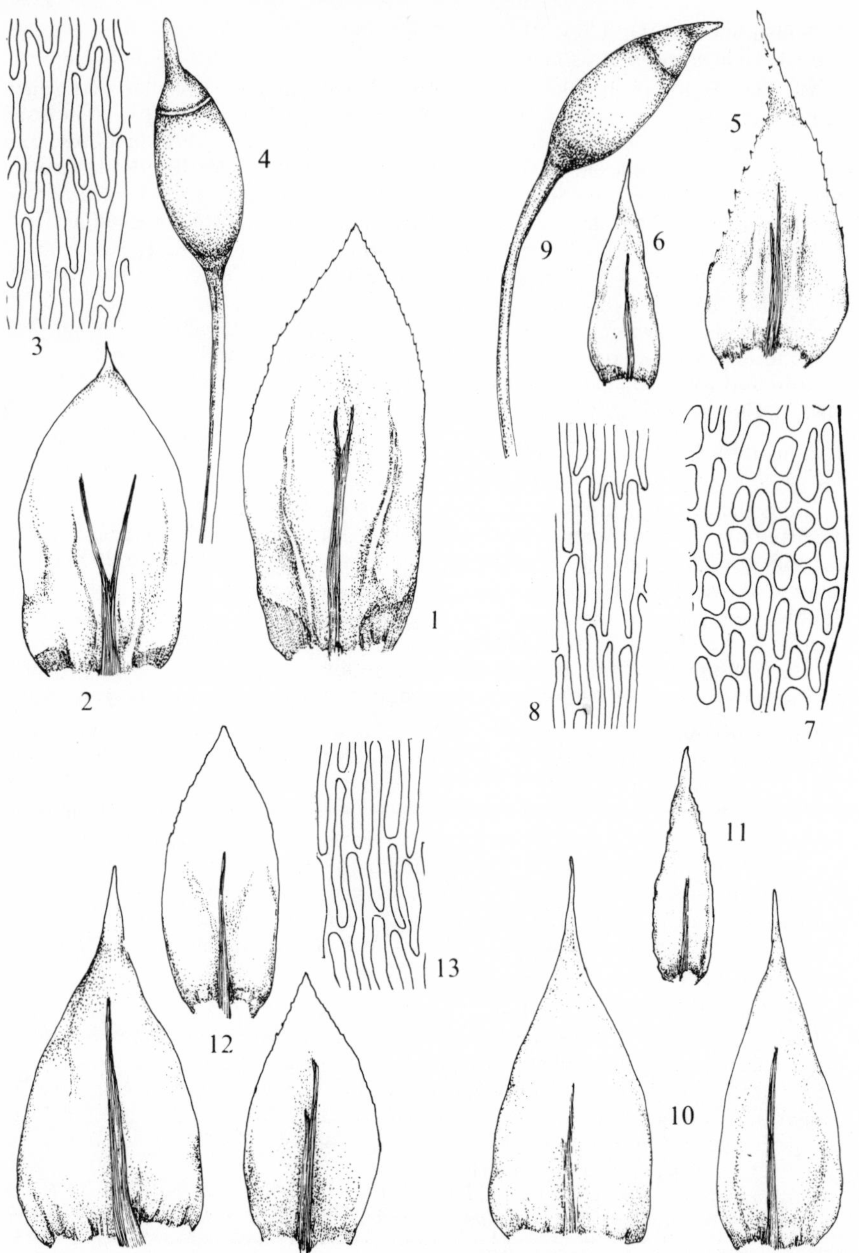

Fig. **284**. 1–4, *Isothecium myurum*: 1 and 2, stem and branch leaves; 3, mid-leaf cells; 4, capsule. 5–9, *I. myosuroides* var. *myosuroides*: 5 and 6, stem and branch leaves; 7, angular cells; 8, mid-leaf cells; 9, capsule. 10–11, *I. myosuroides* var. *brachythecioides*: 10, stem and branch leaves. 12–13, *I. holtii*: 12, leaves; 13, mid-leaf cells. Leaves ×25, cells ×415, capsules ×10.

branched, branches long, sub-complanate. Lvs imbricate, stem lvs ovate to cordate-triangular, tapering gradually or abruptly to acute apex, margin inflexed above, obscurely denticulate; nerve strong, extending almost to apex; basal cells rhomboidal, angular cells opaque, forming small auricles, cells above narrowly rhomboidal, in mid-lf 5–8 μm×28–48 μm, 4–8 times as long as wide. Capsule inclined, ellipsoid; spores 16–21 μm. Fr rare, winter. Yellowish-green to orange-brown patches on shaded rocks, boulders and tree roots by fast-flowing streams and rivers in W. Britain from Cornwall, Devon and Wales northwards, occasional but sometimes locally frequent. 28, H7. Norway, Harz.

The specific status of this plant is doubtful, as intermediates between it and *I. myosuroides* are said to occur and the situation requires further study. Superficially resembling small forms of *Thamnobryum alopecurum* which however, has a stronger nerve and short papillose cells; it differs from *I. myosuroides* in the longer branches, the more shortly pointed stem lvs and the frequent orange tint.

4. I. striatulum (Spruce) Kindb., Canad. Rec. Sc., 1894
Eurhynchium striatulum (Spruce) Br. Eur., *Plasteurhynchium striatulum* (Spruce) Fleisch.

Plants slender to medium-sized, primary stems creeping, secondary ascending to erect, branching sub-dendroid to irregularly pinnate, branches often curved and pointing in one direction. Lvs imbricate to erecto-patent, plicate, ± concave, stem lvs cordate-triangular, rapidly tapering to very long channelled acumen in large forms, to broadly ovate, acuminate in small forms, margin denticulate; nerve extending *ca* ¾ way up lf; basal cells shortly rectangular, angular cells smaller, opaque, forming auricles extending to nerve, cells above rhomboidal to narrowly rhomboidal, in mid-lf 5–8 μm×20–48 μm, 4–8 times as long as wide. Branch lvs narrowly lanceolate to ovate-lanceolate, acute, denticulate to dentate. Capsule inclined, ellipsoid, gibbous to slightly curved; lid with long thin oblique beak; spores 14–20 μm. Fr rare, winter. Yellowish-green patches on shaded basic rocks, stones and walls, rare. From Devon and Somerset east to Surrey and Sussex and north to N. Wales, Derby, Mid W. Yorks and Westmorland, Argyll, scattered localities in Ireland. 22, H10. Europe, Caucasus, Canaries, Algeria.

The taxonomic position is debatable. It has the habit, appearance and lf areolation of *Isothecium* but the nerve ending in the spine-like projection characteristic of species of *Eurhynchium*. There is a greater number of characters attributable to *Isothecium* than to *Eurhynchium* so the species is retained in the former genus.

Large forms have been confused with *Eurhynchium meridionale* but differ in the basal areolation and shorter cells of the lvs which are not widely spreading as in that species.

148. Scorpiurium Schimp., Syn. ed. 2, 1876

Dioecious. Relatively slender plants, stems creeping with tufts of rhizoids, secondary stems ascending to erect with numerous short, curved branches. Lvs of secondary stems distant, weakly plicate, ovate-triangular, acuminate, base decurrent; nerve extending to apex; cells rhomboidal to narrowly rhomboidal, at base and basal angles quadrate-hexagonal. Branch lvs ovate or ovate-lanceolate; nerve extending to apex. Seta purple, smooth; capsule inclined, ellipsoid, curved; lid with long oblique beak; peristome perfect, inner with tall basal membrane and nodulose cilia. Three species: Europe, Middle East, E. Asia, N. Africa, Macaronesia.

1. S. circinatum (Brid.) Fleisch. & Loeske, Allg. Bot. Zeitschr., 1907
Eurhynchium circinatum (Brid.) Br. Eur.

Plants slender, primary stems creeping, secondary stems erect, curved with

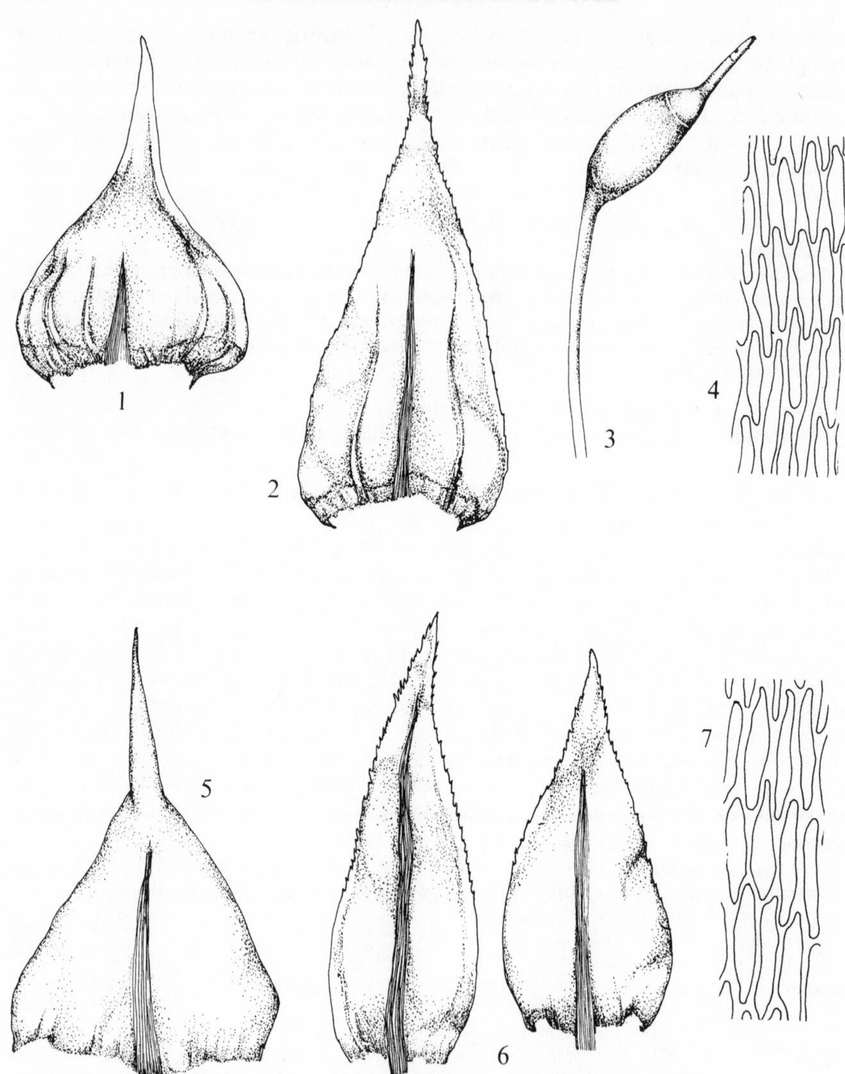

Fig. **285**. 1–4, *Isothecium striatulum*: 1 and 2, stem and branch leaves (×40); 3, capsule (×10); 4, mid-leaf cells. 5–7, *Scorpiurium circinatum*: 5 and 6, stem and branch leaves (×65); 7, mid-leaf cells. Cells ×415.

branches crowded at ends, branches short, curved. Lvs of secondary stems erect when dry, erecto-patent to spreading when moist, triangular to ovate-triangular, rapidly narrowed to long acuminate apex, margin plane, entire or denticulate; nerve reaching acumen; basal and angular cells ± hexagonal, opaque, cells above rhomboidal to narrowly rhomboidal, in mid-lf 5–8 μm×20–36 μm, 4–6 times as long as wide. Branch lvs larger, narrower, imbricate when dry, erecto-patent when moist, curved, ovate to lanceolate, acuminate, margin denticulate or dentate; nerve reaching nearly to apex; areolation as in stem lvs. Seta smooth; capsule inclined, ellipsoid, curved; lid with long, acute, oblique beak. Fr unknown in Britain or Ireland. Tight, green to yellowish-green patches on shaded or exposed calcareous rocks, walls and soil, occasional in southern England and S. Wales, rare in Anglesey and S. and W. Ireland. 21, H8, C. Belgium, Mediterranean region, Crimea, Asia Minor, Iran, Azores, Madeira.

149. HOMALOTHECIUM BR. EUR., 1851

Dioecious. Medium-sized to robust plants, stems procumbent to erect, radiculose to tomentose or not, irregularly to sub-pinnately branched. Lvs appressed or erect when dry, erect or erecto-patent when moist, deeply plicate, stem lvs similar to or larger than branch lvs, lanceolate-triangular, tapering from base to long acumen, margin entire or denticulate, plane or narrowly recurved; nerve single, extending about $\frac{3}{4}$ way up lf; cells $\pm$ uniformly linear-vermicular except at basal angles, porose or not. Outer perichaetial lvs recurved, inner erect. Seta smooth or papillose; capsule erect to horizontal, ovoid to cylindrical, straight or curved; lid conical, obtuse or acute; inner peristome with or without cilia. A small natural genus of about 11 species distributed through Europe, Asia, Africa and America.

1 Stems with tomentum of long brown rhizoids, seta smooth **3. H. nitens**
Stems without tomentum, with or without short rhizoids, seta papillose 2
2 Stems procumbent, attached to substrate by rhizoids for most of their length **1. H. sericeum**
Stems erect or ascending, not attached to substrate except at base **2. H. lutescens**

1. H. sericeum (Hedw.) Br. Eur., 1851
Camptothecium sericeum (Hedw.) Kindb.

Plants moderately robust, stems creeping, attached to substrate for most of their length by rhizoids, closely branched, branches erect, often curved, especially when dry. Lvs appressed when dry, erecto-patent when moist, strongly plicate, stem lvs lanceolate-triangular, tapering from base to long filiform acumen, margin plane or narrowly recurved, denticulate at base, entire or faintly denticulate above; nerve extending *ca* $\frac{3}{4}$ way up lf; angular cells larger, incrassate, forming small decurrent auricles, other cells $\pm$ uniformly linear-vermicular, not porose, cells in mid-lf 4–6 μm × 56–90 μm, 10–16 times as long as wide. Branch lvs smaller, narrower. Seta reddish, papillose; capsule erect, cylindrical, tapering from base to mouth, straight or slightly curved; lid rostrate, obtuse; spores 11–22 μm. Fr occasional, winter. $n=8$*, 9*, 10*, 10+1*, 11*, 11+2*, 12. Dense, yellowish-green to golden-brown mats or patches, sometimes extensive on dry, exposed, often vertical basic rocks, walls, on hard-packed soil, tree trunks and sand-dunes, common especially in basic habitats and sometimes abundant. 112, H40, C. Europe, Iceland, Faroes, Caucasus, Middle East, C. Asia, Kashmir, N. Africa, Macaronesia, Newfoundland.

2. H. lutescens (Hedw.) Robins., Bryologist, 1962
Camptothecium lutescens (Hedw.) Br. Eur.

Plants moderately robust, stems ascending, with few rhizoids, not attached to substrate except at base, irregularly to sub-pinnately branched, branches erect or $\pm$ directed forwards, not curved. Lvs erect when dry, erecto-patent when moist, strongly plicate, stem lvs lanceolate-triangular, tapering from base to long filiform acumen, margin plane or narrowly recurved, sinuose to denticulate; nerve extending *ca* $\frac{3}{4}$ way

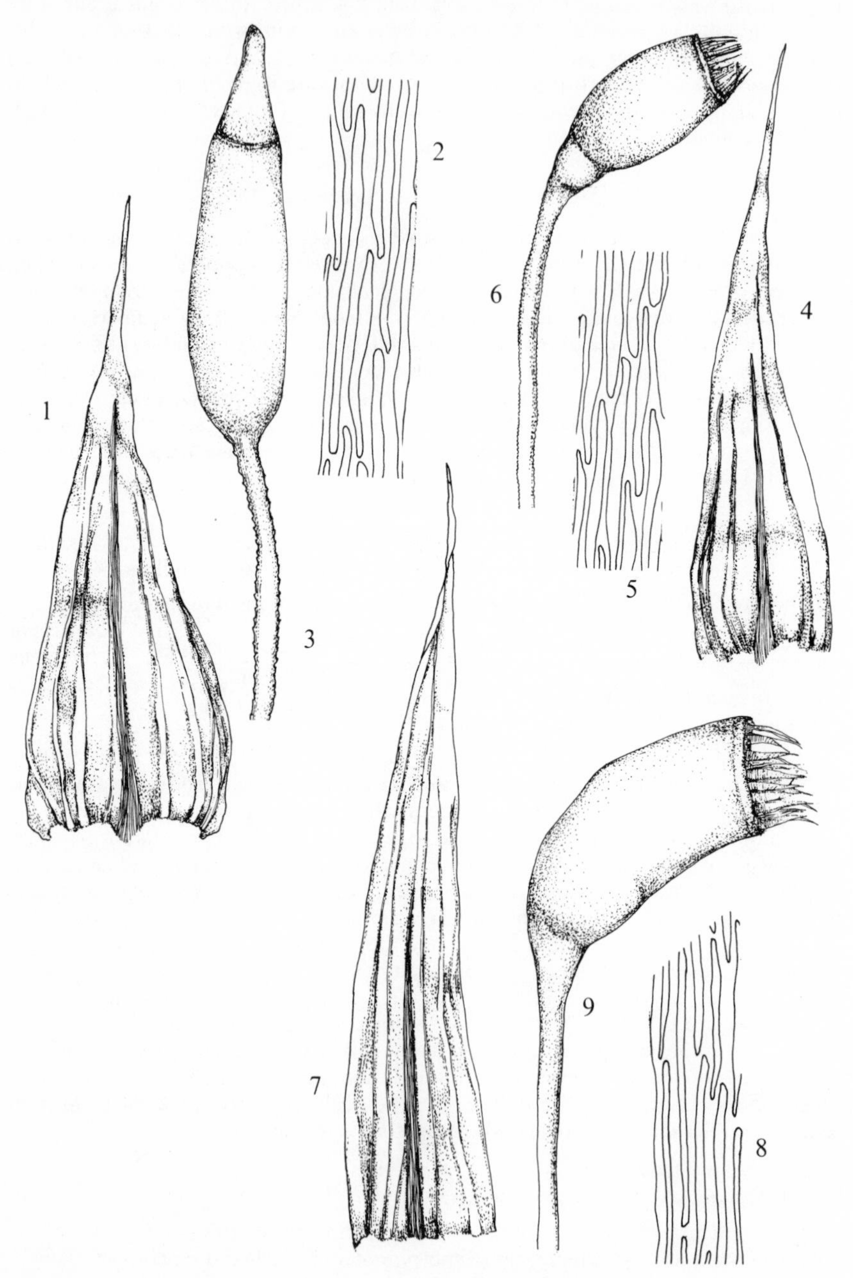

Fig. **286**. 1–3, *Homalothecium sericeum*: 1, leaf; 2, mid-leaf cells; 3, capsule. 4–6, *H. lutescens*: 4, leaf; 5, mid-leaf cells; 6, capsule. 7–9, *H. nitens*: 7, leaf; 8, mid-leaf cells; 9, capsule. Leaves ×25, capsules ×15, cells ×415.

up lf; basal cells incrassate, porose, angular cells larger, incrassate, forming small decurrent auricles, other cells linear-vermicular, in mid-lf 5–7 μm × 48–80 μm, *ca* 10 times as long as wide. Branch lvs narrower but otherwise similar. Seta purple, papillose; capsule erect or inclined, sub-cylindrical, straight or slightly curved; lid conical, acute; spores 16–20 μm. Fr rare, winter. $n=8$, 10, 11*, 12, 14. Yellowish-green to golden-brown irregular patches or spreading tufts, sometimes extensive, in turf, on soil, rocks, cliffs and sand-dunes in exposed, well-drained, often basic habitats, occasional to common. 105, H38, C. Europe, Iceland, Caucasus, Iran, Canaries.

3. H. nitens (Hedw.) Robins., Bryologist, 1962
Camptothecium nitens (Hedw.) Schimp., *Tomenthypnum nitens* (Hedw.) Loeske

Plants moderately robust, stems ascending or erect, densely tomentose with brown rhizoids, irregularly branched, branches erect or ascending, straight. Lvs erect, strongly plicate, stem and branch lvs similar, lanceolate-triangular, tapering ± from base to long acuminate apex, margin plane or narrowly recurved, entire or sinuose; nerve extending about $\frac{3}{4}$ way up lf; basal cells incrassate, porose, a few angular cells larger but not forming decurrent auricles, cells elsewhere linear-vermicular, *ca* 6 μm × 48–90 μm, 9–18 times as long as wide. Seta purple, smooth; capsule horizontal, cylindrical, curved; lid conical; spores 16–20 μm. Fr very rare, summer. $n=12$. Yellowish-green to golden-brown patches or scattered stems in basic flushes and at margins of fens, rare, decreasing in England. E. Anglia, N. Wales, Shrops and scattered localities north to Banff, Elgin and Inverness, Kildare, Westmeath, W. Mayo, Fermanagh. 34, H4. Europe, Iceland, Caucasus, N. and C. Asia, N. America, Greenland.

150. BRACHYTHECIUM BR. EUR., 1853

Dioecious, autoecious, rarely synoecious. Slender to robust plants, stems creeping, ascending or erect, irregularly to pinnately branched, branches short or long, straight or curved. Lvs imbricate to spreading, stem lvs similar to or differing from branch lvs, concave or plane, smooth or plicate, cordate-triangular to ovate-lanceolate, acute to longly acuminate, base decurrent or not, margin plane or recurved, entire or denticulate; nerve single, extending $\frac{1}{2}$–$\frac{3}{4}$ way up lf or to apex, basal cells rhomboidal or rectangular, porose or not, angular cells usually shorter, sometimes enlarged and forming auricles, cells above narrowly or linear-rhomboidal. Seta deep red, smooth or papillose; capsule inclined to horizontal, ovoid to cylindrical, symmetrical, gibbous or curved; annulus separating; lid conical; peristome perfect, inner with tall basal membrane and nodulose or appendiculate cilia. A large cosmopolitan genus of about 300 species.

Although a straightforward genus in Britain and Ireland, in continental Europe the distinctions between some of the species are very blurred. The genus has been divided into sections by several authors but there is little agreement between them and as the limits of the sections are so unsatisfactory they are not recognised here.

1 Lvs longitudinally plicate, margin entire or denticulate, seta smooth — 2
  Lvs not or only slightly plicate, margin denticulate, seta papillose at least above — 6
2 Stems with tufts of long reddish rhizoids, stem lvs curved — **3. B. erythrorrhizon**
  Stems without tufts of reddish rhizoids, stem lvs ± straight — 3
3 Lvs imbricate when dry giving shoots a string-like appearance — **1. B. albicans**
  Lvs erect or appressed-flexuose when dry, shoots not string-like — 4

4 Dioecious, stems creeping, stem lvs with long, twisted, filiform acumen **2. B. glareosum**
Autoecious, stems ascending or creeping, stem lvs with shorter, flat acumen 5

5 Lvs strongly plicate, mid-lf cells 6–9 μm wide **4. B. salebrosum**
Lvs moderately plicate, mid-lf cells 7–12 μm wide **5. B. mildeanum**

6 Nerve reaching acumen and sometimes almost to lf apex 7
Nerve rarely extending more than $\frac{3}{4}$ way up lf 8

7 Stem lvs distant, flexuose-spreading when dry, base longly decurrent **10. B. reflexum**
Stem lvs crowded, appressed when dry, base scarcely decurrent **12. B. populeum**

8 Lvs secund when moist, seta papillose above, smooth below **13. B. plumosum**
Lvs not secund, seta papillose throughout 9

9 Angular cells forming distinct auricles 10
Angular cells not forming distinct auricles 12

10 Dioecious, plants robust, stem lvs broadly ovate to ovate-oblong, acute to obtuse and apiculate **7. B. rivulare**
Autoecious, plants slender, stem lvs cordate-triangular to broadly ovate with long acumen 11

11 Margin plane or slightly recurved at base, angular cells rectangular **8. B. starkei**
Margin strongly recurved near base, angular cells quadrate **9. B. glaciale**

12 Plants slender, stem lvs lanceolate-triangular, tapering from near base **11. B. velutinum**
Plants medium-sized to robust, stem lvs ovate or ovate-cordate or ovate-lanceolate 13

13 Stem lvs denticulate, seta coarsely papillose **6. B. rutabulum**
Stem lvs entire, seta smooth **5. B. mildeanum**

1. B. albicans (Hedw.) Br. Eur., 1853

Dioecious. Slender to medium-sized plants to 8 cm long, shoots often string-like in appearance, stems ascending, irregularly branched with few rhizoids. Lvs imbricate, plicate both when wet and dry, concave, stem lvs ovate-lanceolate to ovate or ovate-triangular, tapering rapidly to long filiform acumen, acumen not twisted, margin recurved below, entire or obscurely denticulate above; nerve extending $\frac{1}{2}$–$\frac{2}{3}$ way up lf; basal cells shortly rectangular, angular cells quadrate, less translucent, decurrent, cells above linear-rhomboidal, in mid-lf 6–9 μm×48–80 μm, 7–11 times as long as wide. Branch lvs narrower, ovate-lanceolate to lanceolate, cells longer and narrower. Seta smooth; capsule inclined, ovoid, gibbous; lid conical; spores 12–14 μm. Fr rare, autumn, winter. n=7, 9. Pale, yellowish-green or whitish-green patches in turf, on soil and gravel, in old quarries and sand-dunes in exposed neutral or acidic habitats, occasional to frequent. 112, H22, C. Europe, Iceland, Faroes, Caucasus, Azores, Madeira, N. America, Greenland, Australia, New Zealand.

Easily recognised in the field by the pale, silky, string-like branches with imbricate, plicate lvs, and most likely to be mistaken for *Camptothecium lutescens* which differs in its larger size and the lvs tapering from the base; *C. lutescens* is a calcicole whereas *B. albicans* is usually a calcifuge.

2. B. glareosum (Spruce) Br. Eur., 1853

Dioecious. Plants medium-sized, shoots to 8(–20) cm, stems creeping, irregularly pinnately branched, branches procumbent to ascending. Lvs appressed-flexuose when dry, erecto-patent when moist, plicate, concave, stem lvs ovate to ovate-triangular,

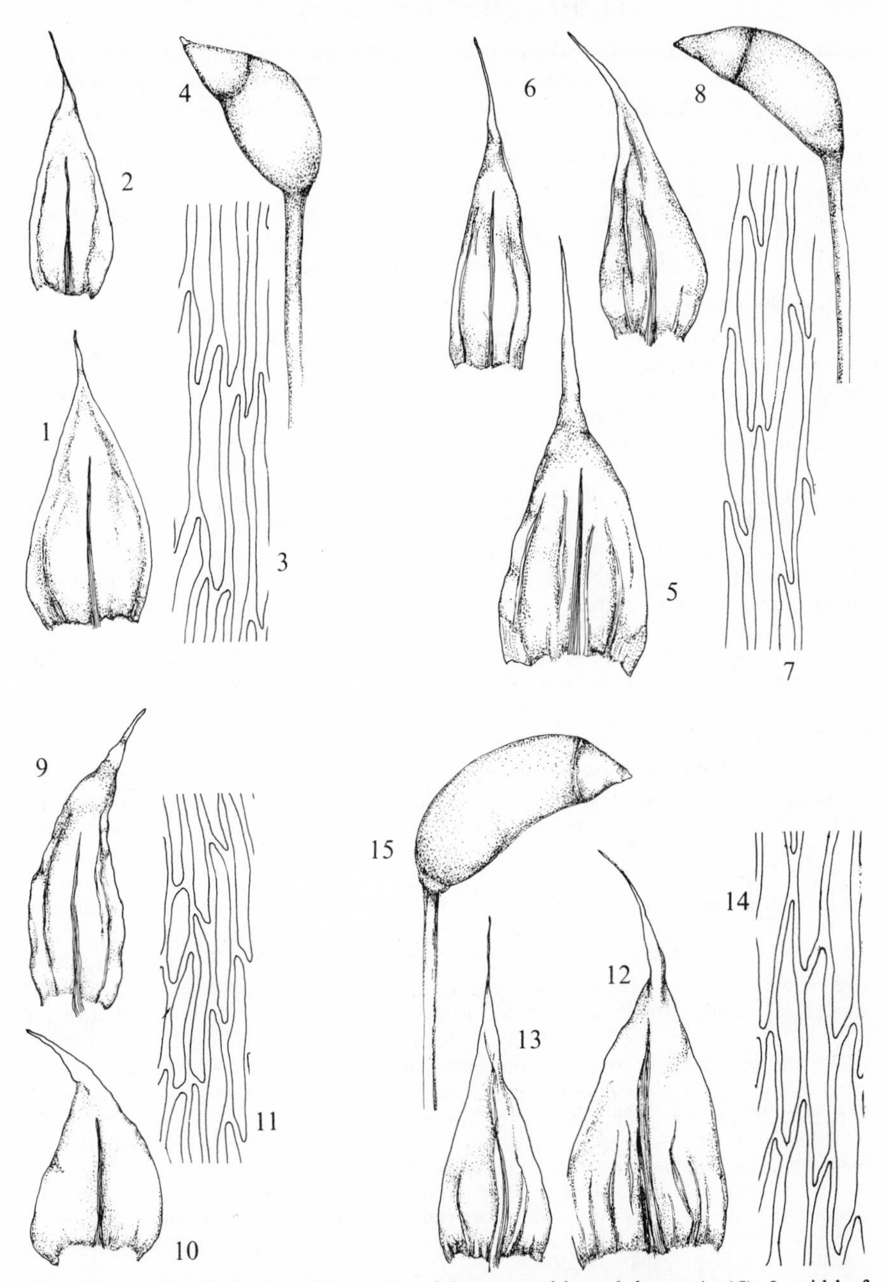

Fig. **287**. 1–4, *Brachythecium albicans*: 1 and 2, stem and branch leaves (×15); 3, mid-leaf cells; 4, capsule. 5–8, *B. glareosum*: 5 and 6, stem and branch leaves (×15); 7, mid-leaf cells; 8, capsule. 9–11, *B. erythrorrhizon*: 9 and 10, stem and branch leaves; 11, mid-leaf cells. 12–15, *B. salebrosum*: 12 and 13, stem and branch leaves; 14, mid-leaf cells; 15, capsule. Cells ×415, capsules ×10.

tapering to long, twisted, filiform acumen, margin plane or recurved, entire or finely denticulate above; nerve extending $\frac{1}{2}$–$\frac{2}{3}$ way up lf; basal cells rhomboidal, angular cells shortly rectangular to oval or rounded-quadrate, cells above narrowly rhomboidal, in mid-lf 6–10 μm×64–90 μm, 8–11(–13) times as long as wide. Seta smooth; capsule inclined, ovate-ellipsoid, gibbous or slightly curved; lid conical; spores 14–20 μm. Fr rare, winter. n=6, 9, 9+2, 10, 14. Silky, yellowish-green to whitish-green, straggling patches on rocks and soil in turf, on banks and in open woodland in basic habitats, occasional to frequent in England, rare to occasional elsewhere. 86, H21, C. Europe, Iceland, Caucasus, Siberia, Yunnan, Korea, Japan, N. America.

3. B. erythrorrhizon Br. Eur., 1853

Dioecious. Plants slender in Britain, stems procumbent, with tufts of reddish-brown rhizoids on older parts, pinnately branched, branches procumbent to ascending. Lvs erect when dry, stem lvs patent, sub-secund, plicate, branch lvs erecto-patent to patent, straight to sub-secund when moist, stem lvs curved, ovate, gradually or abruptly tapering to long acumen, base decurrent, margin plane or recurved below, finely denticulate above; nerve extending *ca* $\frac{3}{4}$ way up lf; basal cells narrowly linear-rhomboidal, angular cells irregularly hexagonal, decurrent cells rectangular, cells above linear, in mid-lf 6–10 μm×48–72 μm, 6–10 times as long as wide. Branch lvs smaller, ovate-lanceolate, more shortly pointed but otherwise similar to stem lvs. Seta ± smooth; capsule inclined, ellipsoid; spores *ca* 16 μm. Fr unknown in Britain. Small, glossy, yellowish-green patches on sand-dunes amongst *Dryas*, very rare, W. Sutherland. 1. Scandinavia, eastern Alps, Siberia, N. America.

This plant which was found by J. J. Barkman (see *Trans. Br. bryol. Soc.* **2**, 568–70, 1955) in 1948 has only been seen once since. Scottish material is very depauperate but differs from very slender forms of *B. albicans* in the frequently curved stem lvs and tufts of reddish-brown rhizoids.

4. B. salebrosum (Web. & Mohr) Br. Eur., 1853

Autoecious, rarely synoecious. Plants medium-sized to robust, stems procumbent to ascending, irregularly branched, branches ascending to erect. Lvs erect-flexuose, sometimes sub-secund when dry, erecto-patent when moist, strongly plicate, ± concave, stem lvs ovate-lanceolate, tapering to long flexuose acumen, margin recurved near base, entire or denticulate; nerve extending $\frac{1}{2}$–$\frac{2}{3}$ way up lf; basal cells rhomboidal, incrassate, porose, angular cells shortly rectangular, a few enlarged, ± hyaline, forming small decurrent auricles, cells above narrowly to linear-rhomboidal, 6–9 μm×56–100 μm, 9–13 times as long as wide. Branch lvs narrower. Seta smooth; capsule inclined, shortly cylindrical, curved; lid conical; spores 12–18 μm. Fr occasional, autumn. n=13. Yellowish-green or green patches on logs, branches, rocks and soil, usually in woods or shaded habitats, generally distributed, rare to occasional. 58, H19. Europe, Iceland, Caucasus, N. and C. Asia, Japan, Morocco, Macaronesia, N. America, Greenland, Tasmania, Kerguelen Is.

Although *B. salebrosum* and *B. mildeanum* are ecologically distinct, morphological intermediates occur and the status of the 2 taxa is uncertain. *B. salebrosum* is intermediate between *B. glareosum* and *B. rutabulum*; it differs from *B. glareosum* in habit, the more shortly pointed stem lvs and the autoecious inflorescence. It differs from *B. rutabulum* in its silky appearance and the more longly pointed, plicate lvs and smooth seta; it may, however, be readily passed over for that species.

5. B. mildeanum (Schimp.) Milde, Bot. Zeit., 1862
B. salebrosum var. *palustre* Schimp.

Autoecious. Plants medium-sized to robust, stems procumbent, irregularly branched, branches spreading to ascending. Lvs erect-flexuose when dry, erecto-

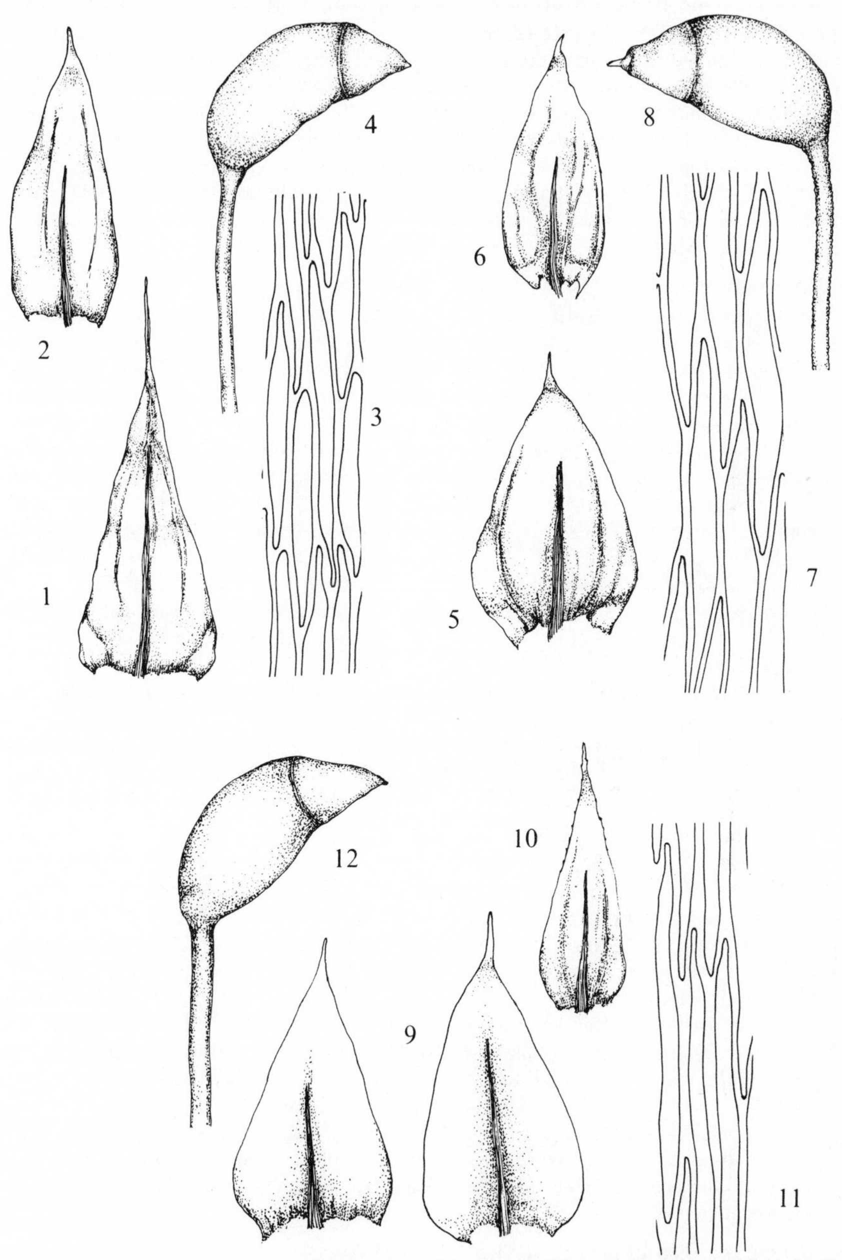

Fig. **288**. 1–4, *Brachythecium mildeanum*: 1 and 2, stem and branch leaves; 3, mid-leaf cells; 4, capsule. 5–8, *B. rivulare*: 5 and 6, stem and branch leaves; 7, mid-leaf cells; 8, capsule. 9–12, *B. rutabulum*: 9 and 10, stem and branch leaves; 11, mid-leaf cells; 12, capsule. Leaves ×15, cells ×415, capsules ×10.

patent when moist, plicate but sometimes only slightly so, plane or concave, stem lvs narrowly triangular to ovate-lanceolate, tapering to fine acuminate apex, margin entire or sinuose; nerve extending $\frac{1}{2}$–$\frac{2}{3}$ way up lf; basal cells rhomboidal, incrassate, porose, angular cells shortly rectangular, a few enlarged, ± hyaline, cells above narrowly to linear-rhomboidal, 7–12 μm × 80–120 μm, 9–18 times as long wide. Seta smooth; capsule inclined, shortly cylindrical, curved; lid conical; spores 12–18 μm. Fr occasional, autumn. $n=14$. Yellowish-green patches on damp soil and tracks, on marshy ground in open habitats and sometimes locally abundant in dune slacks, generally distributed, rare to occasional. 71, H14, C. Europe, Iceland, Faroes, N. Asia, Azores, Madeira, N. America.

6. B. rutabulum (Hedw.) Br. Eur., 1853

Autoecious. Plants medium-sized to robust, shoots to 12 cm, stems creeping, irregularly branched, branches erect, ascending or arcuate, often curved. Lvs ± erecto-patent, smooth, ± plicate or rugose when dry, erecto-patent, sometimes slightly plicate when moist, concave, stem lvs ovate or ovate-cordate, abruptly or gradually tapering to acute to acuminate apex, base decurrent, margin plane or recurved, denticulate; nerve extending $\frac{1}{2}$–$\frac{2}{3}$ way up lf; basal cells rhomboidal, porose, angular cells rectangular, decurrent but not forming distinct auricles, cells above linear-rhomboidal, in mid-lf 6–10 μm × 56–100 μm, 8–14 times as long as wide. Branch lvs ovate to ovate-lanceolate, acute to acuminate. Seta coarsely papillose; capsule inclined, ellipsoid, gibbous or curved; lid conical; spores 16–24 μm. Fr common, late autumn to spring. $n=5$, 6, 10, 10+1, 11*, 12*, 13, 20*, 22. Glossy, green or yellowish-green, lax tufts or patches or scattered shoots on soil, tree boles, logs and rocks in woods, in turf, on roadsides, by streams and rivers, on walls and cliffs, particularly in damp shaded habitats, common or very common. 112, H40, C. Europe, Iceland, Faroes, Asia, Algeria, Macaronesia, America, Tasmania, New Zealand, Hawaii.

Very variable in habit and size but usually recognised by the lvs erecto-patent both when wet and dry and the coarsely papillose seta. It cannot always be distinguished from *B. rivulare* in the field but that species differs in the distinct auricles and dioecious inflorescence.

7. B. rivulare Br. Eur., 1853

Dioecious. Plants robust, shoots to 12 cm long, stems creeping, irregularly and distantly to closely branched, branches long, erect or ascending. Lvs erect or imbricate when dry, imbricate to patent when moist, concave, stem lvs broadly ovate to ovate-oblong, acute to obtuse and apiculate or shortly acuminate, base decurrent, margin sinuose to denticulate; nerve extending $\frac{3}{4}$ way or more up lf; basal cells rhomboidal, angular cells enlarged, lax, hexagonal, hyaline or brownish, forming distinct decurrent auricles, cells above linear-rhomboidal, in mid-lf 6–10 μm × 64–128 μm, 9–20 times as long as wide. Branch lvs ovate to ovate-lanceolate. Seta coarsely papillose; capsule inclined, ovoid, gibbous; lid conical, obtuse and apiculate; spores 16–20 μm. Fr occasional, autumn to spring. Pale green to green or yellowish-green patches on rocks in fast-flowing streams and rivers, on soil, rocks, logs and tree boles by streams, in deep woodland, in flushes and wet turf, frequent or common. 111, H35, C. Europe, Iceland, Faroes, Asia, Madeira, Azores, N. America, Greenland, Kerguelen Is.

8. B. starkei (Brid.) Br. Eur., 1853

Autoecious. Plants slender, stems irregularly branched. Lvs patent, ± plicate below both when moist and dry, stem lvs plane, cordate-triangular, narrowed to long or short acumen, longly decurrent at base, margin plane or slightly recurved at base,

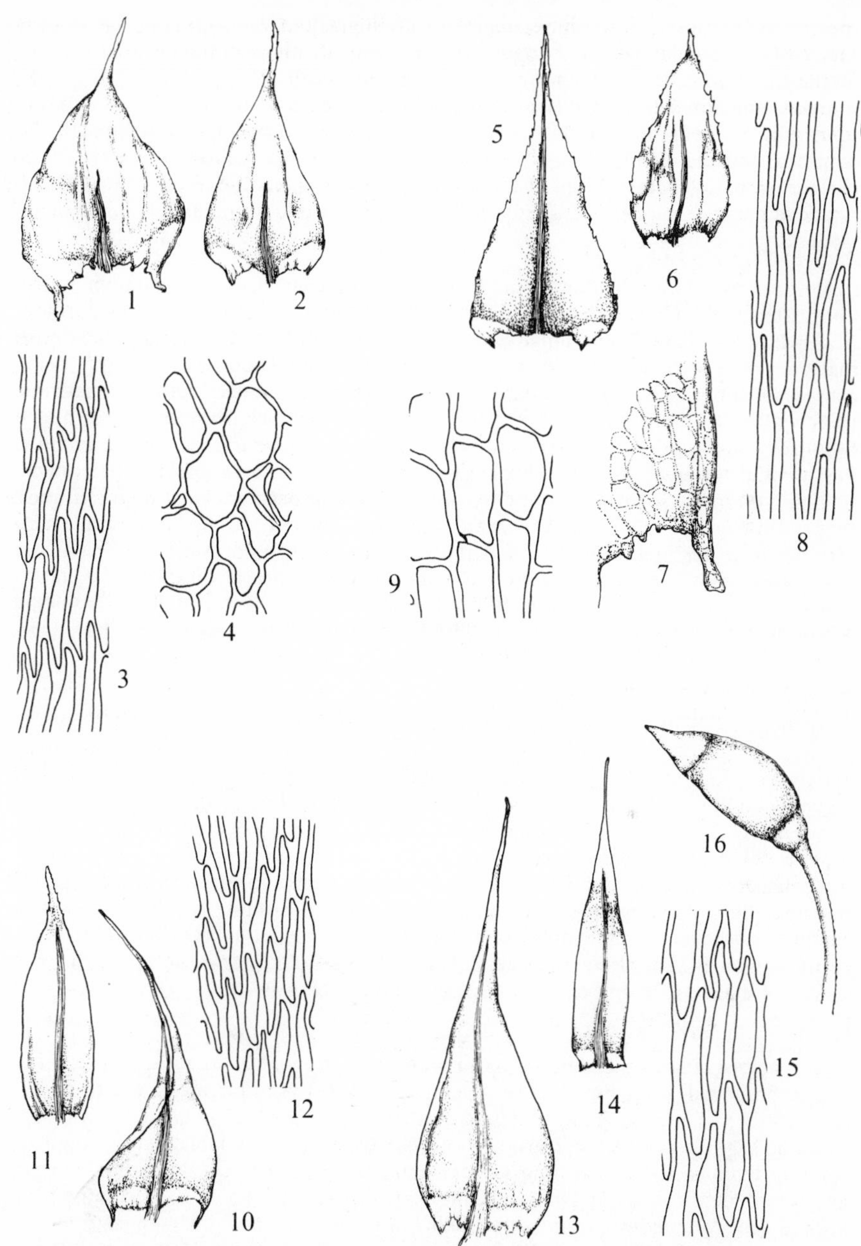

Fig. **289**. 1–4, *Brachythecium glaciale*: 1 and 2, stem and branch leaves; 3 and 4, mid-leaf and angular cells. 5–9, *B. starkei*: 5 and 6, stem and branch leaves; 7, basal angle of stem leaf (×165); 8 and 9, mid-leaf and angular cells. 10–12, *B. reflexum*: 10 and 11, stem and branch leaves; 12, mid-leaf cells. 13–16, *B. populeum*: 13 and 14, stem and branch leaves; 15, mid-leaf cells; 16, capsule (×10). Leaves ×25, cells ×415.

spinosely denticulate from base; nerve extending about ½ way up lf; basal cells rhomboidal, angular cells rectangular, thin-walled, hyaline, forming narrow, very fragile auricles, cells above linear-rhomboidal, in mid-lf 5–7 μm×44–80 μm, 7–14 times as long as wide. Branch lvs ± concave, ovate-lanceolate, acute. Perichaetial lvs squarrose. Seta slightly papillose; capsule inclined to horizontal, ovate-ellipsoid; lid conical; spores 12–15 μm. Fr unknown in Britain. $n=10, 20$. Yellowish-green tufts or patches in basic rock crevices at high altitudes, very rare. Mid Perth, W. Ross. 2. Arctic and alpine Europe, Caucasus, N. and C. Asia, Yunnan, Japan, N. America.

9. B. glaciale Br. Eur., 1853

Autoecious. Plants slender, stems irregularly branched. Lvs imbricate when dry, imbricate to patent when moist, concave, plicate or not below, stem lvs cordate-triangular to broadly ovate, narrowed to ± filiform acumen, base longly decurrent, margin strongly recurved near base, denticulate from about widest part; nerve extending about ½ way up lf; basal cells rhomboidal, angular cells ± quadrate, thick-walled, forming decurrent auricles, cells above narrowly rhomboidal, in mid-lf 6–10 μm×44–64 μm, 6–8 times as long as wide. Branch lvs ovate, apex acuminate. Perichaetial lvs erect. Seta papillose; capsule inclined, ovate-ellipsoid; lid conical. Fr rare, summer. Green patches or wefts in crevices of basic rocks at high altitudes, rare. Mid Perth, Angus, S. Aberdeen, Inverness, Argyll, Ross. 8. Montane and arctic Europe, Iceland, N. and C. Asia, N. America, Greenland.

B. glaciale and *B. starkei* have been confused in Britain. *B. glaciale* differs from *B. starkei* in the lf margin strongly recurved near the base and the ± quadrate angular cells, the margin less strongly denticulate, the shorter cells and straight perichaetial lvs. *B. reflexum* has a much longer stem lf apex and the nerve reaching the acumen.

10. B. reflexum (Starke) Br. Eur., 1853

Autoecious. Plants slender, stems irregularly branched. Stem lvs distant, flexuose-spreading when dry, patent to reflexed when moist, decurrent, cordate-triangular, narrowed to long, channelled, sometimes filiform acumen, margin plane, denticulate; nerve extending into acumen; basal cells rhomboidal, angular cells quadrate, cells above narrowly rhomboidal, in mid-lf 5–9 μm×32–60 μm, 5–9 times as long as wide. Branch lvs more crowded, patent when dry, patent to spreading when moist, lanceolate, acuminate, denticulate to dentate. Seta papillose; capsule horizontal, ellipsoid; lid conical; spores *ca* 16 μm. Fr unknown in Britain. $n=11, 20$. Greenish patches on rocks and tree boles or wefts in rock crevices at high altitudes, rare. Perth, Angus, S. Aberdeen, Inverness, Argyll, Ross. 9. Montane and arctic Europe, Iceland, Caucasus, N. and C. Asia, Kashmir, Japan, N. America.

11. B. velutinum (Hedw.) Br. Eur., 1853

Autoecious. Plants slender, stems procumbent, irregularly to sub-pinnately and closely branched, branches erect or ascending, short. Lvs imbricate to erecto-patent to spreading when dry, erecto-patent to patent, sometimes sub-secund when moist, plane or slightly concave, plicate or not near base, stem lvs lanceolate-triangular, tapering from near base to long slender acumen, base scarcely decurrent, margin sinuose to denticulate; nerve extending ½–⅔ way up lf; basal cells rhomboidal, angular cells rounded-quadrate, other cells linear, in mid-lf 5–7 μm×48–80 μm, 9–18 times as long as wide. Branch lvs lanceolate to narrowly lanceolate, apex filiform. Seta papillose; capsule inclined, ellipsoid, symmetrical or gibbous; lid conical with stout obtuse beak; spores 13–16 μm. Fr frequent, winter, spring. $n=10^*, 10+2, 11, 12^*$. Green or yellowish-green patches on shaded tree boles, stumps, rocks, rock crevices, walls and hedgebanks. Frequent or common in England and Wales,

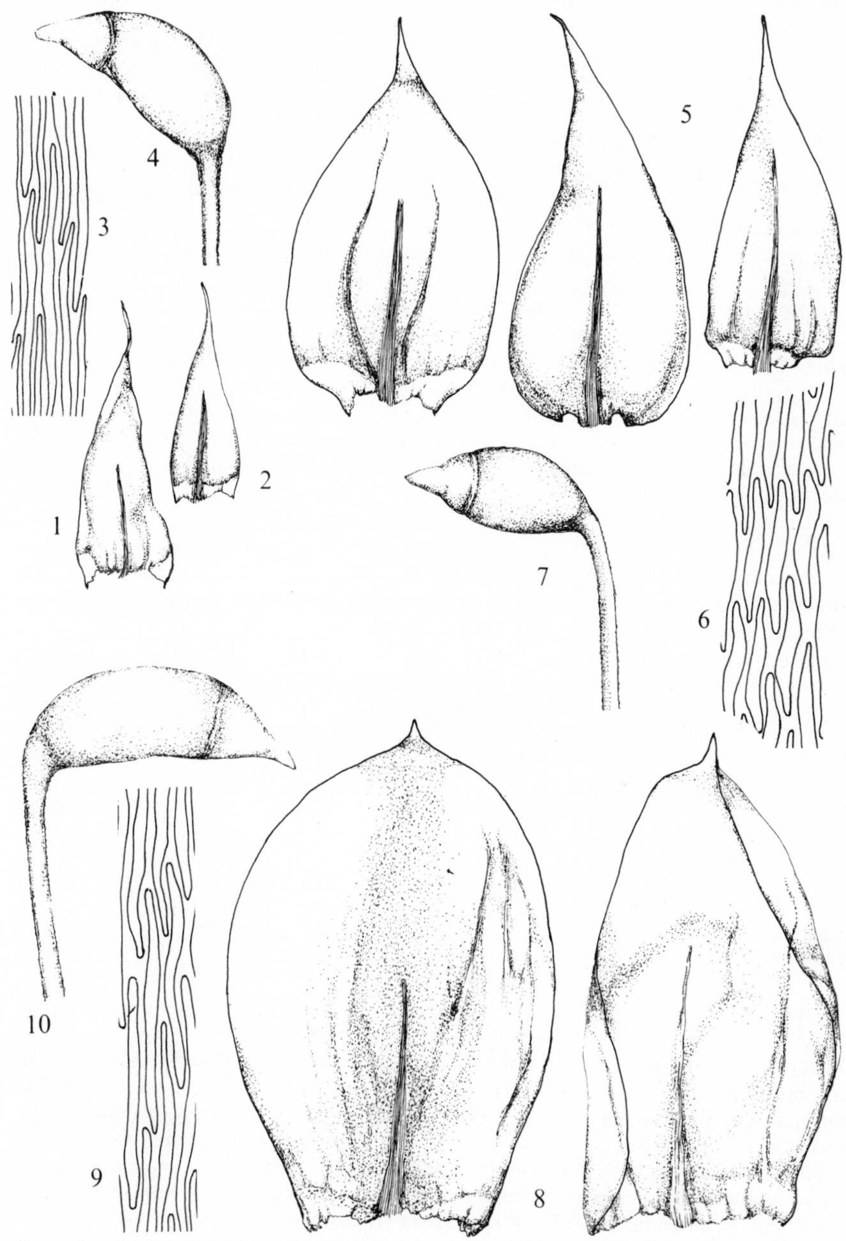

Fig. **290**. 1–4, *Brachythecium velutinum*: 1 and 2, stem and branch leaves; 3, mid-leaf cells; 4, capsule. 5–7, *B. plumosum*: 5, leaves; 6, mid-leaf cells; 7, capsule. 8–10, *Pseudoscleropodium purum*: 8, leaves; 9, mid-leaf cells; 10, capsule. Leaves $\times 25$, cells $\times 415$, capsules $\times 10$.

occasional in Scotland and Ireland. 106, H28, C. Europe, Iceland, Caucasus, Iran, N. Asia, Japan, Macaronesia, Algeria, Morocco, N. America.

A variable species, forms of which may be confused with *B. populeum*. That species has the lvs appressed when dry whereas *B. velutinum* often has them patent or spreading, it has a more sharply pointed lid to the capsule, a longer nerve and shorter cells.

12. B. populeum (Hedw.) Br. Eur., 1853

Autoecious. Plants slender, stems procumbent, pinnately branched, branches ascending to erect. Lvs appressed when dry, erecto-patent when moist, stem lvs ovate-triangular, tapering to long filiform or subulate apex, base scarcely decurrent, margin plane or recurved, sinuose to denticulate; nerve extending to acumen and sometimes almost to apex; basal cells rhomboidal, angular cells rectangular to rounded-quadrate, other cells linear-rhomboidal, in mid-lf 5–8 $\mu m \times 32$–56 μm, 5–9 times as long as wide. Branch lvs narrowly lanceolate, longly acuminate. Seta papillose above; capsule inclined, ellipsoid, symmetrical or gibbous; lid conical, acute; spores 12–20 μm. Fr frequent, autumn to spring. $n=9$, 10*, 10+1, 11. Tight, yellowish-green to green sometimes brownish-tinged patches in shaded habitats on rocks, walls, tree boles and stumps and soil banks, frequent to common. 93, H33, C. Europe, Iceland, Faroes, Caucasus, Iran, N. Asia, Kashmir, Japan, Algeria, Morocco, Azores, N. America.

13. B. plumosum (Hedw.) Br. Eur., 1853

Autoecious. Plants medium-sized to robust, stems procumbent, closely branched, branches erect or ascending. Lvs ± imbricate to erecto-patent when dry, erecto-patent to patent, usually secund when moist, very concave, stem lvs curved or straight, broadly-ovate to ovate-lanceolate, acuminate, base decurrent, margins recurved near base, one margin sometimes inflexed above, denticulate; nerve extending about $\frac{3}{4}$ way up lf; basal cells rectangular, incrassate, sometimes opaque or brownish, angular cells quadrate-rectangular, ± opaque, cells above linear-rhomboidal, in mid-lf 6–8 $\mu m \times 48$–72 μm, 7–11 times as long as wide. Branch lvs straight or curved, ovate-lanceolate to oblong-lanceolate, acuminate. Seta papillose above, smooth below; capsule inclined, ovoid to ellipsoid, gibbous; lid conical; spores 16–24 μm. Fr frequent, winter. $n=10$*, 10+1, 11. Yellowish-green to brownish green or golden-brown patches on rocks and tree roots subject to submergence in and by fast-flowing streams and rivers, rocks and pools and on flushed rocks and rock ledges, common in montane areas, occasional elsewhere. 92, H34, C. Almost cosmopolitan.

151. Pseudoscleropodium (Limpr.) Fleisch. in Broth., Nat. Pfl., 1925

Dioecious. Robust plants, stems procumbent or ascending, without rhizoids, stems and branches tumid, julaceous with ± well defined annual growths, branches complanate, obtusely pointed. Lvs very concave, decurrent at base, broadly ovate, apex obtuse or rounded with reflexed apiculus; nerve single, extending to mid-lf or double and shorter; cells linear-vermicular, at base rhomboidal, porose, at basal angles quadrate to rectangular. Seta red, smooth; capsule horizontal, ellipsoid to cylindrical, gibbous or curved; lid rostrate; peristome perfect, inner with tall basal membrane and appendiculate cilia. A genus of 3 species, the most widespread of which is *P. purum*.

Placed by some authorities in the Entodontaceae but on cytological grounds obviously very closely related to *Brachythecium*.

1. P. purum (Hedw.) Fleisch. in Broth., Nat., Pfl., 1925
Brachythecium purum (Hedw.) Dix., *Scleropodium purum* (Hedw.) Limpr.

Robust plants, shoots to 15 cm, stems and branches tumid, julaceous, stems procumbent to ascending, complanately pinnately branched. Lvs imbricate, very concave, plicate both when dry and moist, stem lvs rounded-ovate to broadly ovate, apex rounded or obtuse with acuminate apiculus, margin plane or recurved near base, finely denticulate above; nerve single, extending about $\frac{1}{3}$ way up lf; basal cells rhomboidal, incrassate, porose, angular cells larger, quadrate to rectangular, more pellucid, cells above linear-vermicular, in mid-lf 4–6 μm × 56–100 μm, 14–25 times as long as wide. Branch lvs ± similar. Seta smooth; capsule horizontal, cylindrical, gibbous or curved; lid rostrate; spores 12–16 μm. Fr rare, autumn, winter. $n = 7, 11$*. Pale green to brownish-green or off-white, lax patches or coarse wefts on soil and in turf in grassland, on heath, banks, roadsides, cliffs, in marshes, quarries, woods, etc., mainly at low altitudes, very common and sometimes abundant. 112, H40, C. Europe, Iceland, Faroes, Caucasus, Iran, N. Asia, Japan, Algeria, Macaronesia, St Helena, Tristan da Cunha, N. America, Jamaica, New Zealand.

152. SCLEROPODIUM BR. EUR., 1853

Dioecious. Slender to medium-sized plants, secondary stems ascending, closely branched, branches spreading or ascending, long or short. Lvs very concave, stem lvs broadly ovate, gradually to abruptly narrowed to acute to obtuse apex, apiculate or not, margin plane, entire or denticulate above; nerve extending $\frac{1}{2}$–$\frac{3}{4}$ way up lf, single or forked; cells narrowly rhomboidal to linear-vermicular, basal cells shorter, angular cells quadrate to rectangular. Seta deep red, papillose; capsule erect to horizontal, cylindrical to ovate-ellipsoid; annulus separating; lid conical or rostrate; peristome perfect, inner with tall membrane and nodulose or appendiculate cilia. A small genus of 9 species found in Europe, W. and E. Asia, N. Africa, Macaronesia, N. and C. America.

Stem lvs broadly ovate, cells 4–12 times as long as wide, capsule erect or slightly inclined **1. S. cespitans**
Stem lvs ovate-oblong to cordate-triangular, cells 10–20 times as long as wide, capsule much inclined to horizontal **2. S. tourettii**

1. S. cespitans (C. Müll.) L. Koch, Leafl. West Bot., 1950
S. caespitosum (Wils.) Br. Eur., *Brachythecium caespitosum* (Wils.) Dix.

Medium-sized plants, primary stems creeping, secondary stems ascending with numerous erect, frequently curved branches. Lvs imbricate when dry, erecto-patent when moist, very concave, stem lvs broadly ovate, acute or obtuse or abruptly narrowed to apex of varying length, margin plane, entire or denticulate above; nerve single, sometimes forked above, broad below, extending $\frac{1}{2}$–$\frac{2}{3}$ way up lf; basal and angular cells quadrate or shortly rectangular, cells above narrowly to linear-rhomboidal, in mid-lf 5–7 μm × 28–68 μm, (4–)6–12 times as long as wide. Branch lvs ovate, obtuse to acute, margin denticulate; cells longer. Seta papillose; capsule erect or slightly inclined, narrowly ellipsoid to cylindrical, straight or curved; lid rostrate; spores 12–24 μm. Fr rare, winter. Light or whitish-green patches on tree trunks and roots, rocks and soil, usually near water but also in woods. Occasional in England and Wales, very rare in Scotland, rare in Ireland. 53, H9, C. Holland, Belgium, France, Spain, Portugal, Madeira, Corsica, N. America.

Differs from *S. tourettii* in its larger size, laxer habit, longer, curved branches, shorter nerve and shorter cells, as well as in fruiting characters. *S. tourettii* also has more longly tapering lvs.

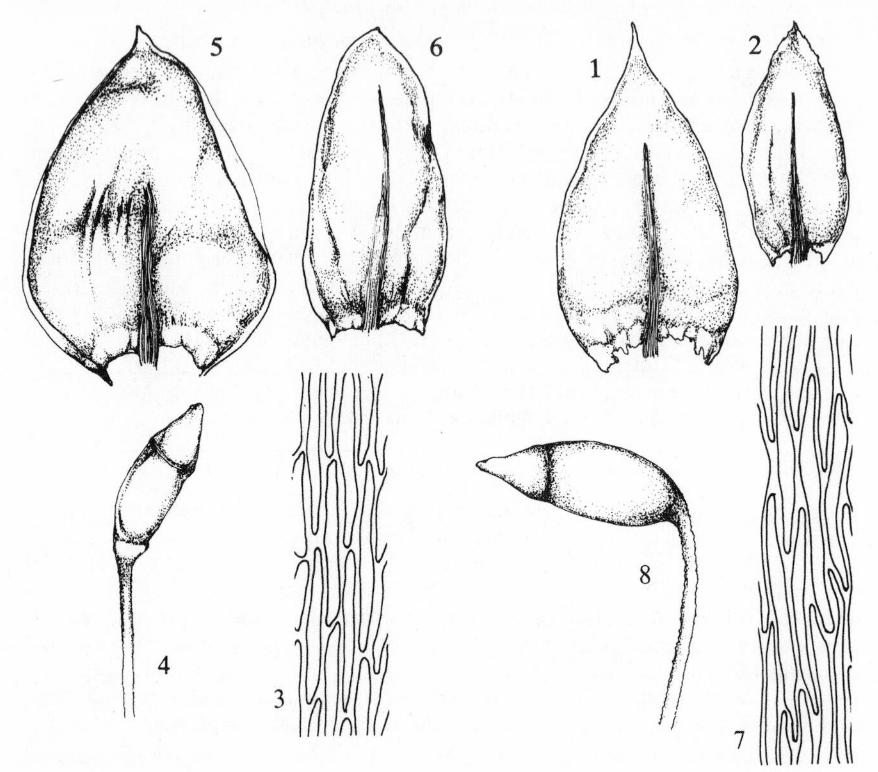

Fig. **291**. 1–4, *Scleropodium cespitans*: 1, stem leaf; 2, branch leaf; 3, mid-leaf cells; 4, capsule. 5–8, *S. tourettii*: 5, stem leaf; 6, branch leaf; 7, mid-leaf cells; 8, capsule. Leaves ×40, cells ×415, capsules ×10.

The 2 species of *Scleropodium* may be confused with *Rhynchostegium murale* or *Cirriphyllum crassinervium*. Both these species are autoecious with capsule with a longly rostrate beak; in *R. murale* the seta is smooth. In *C. crassinervium* the leaf tapers more gradually at the apex and the cells are shorter.

2. S. tourettii (Brid.) L. Koch, Rev. Bryol. Lichen., 1949
S. tourretii auct., *S. illecebrum* auct., *Brachythecium illecebrum* auct.

Slender or medium-sized plants, primary stems rhizomatous, secondary stems ascending, pinnately branched, branches short, spreading. Lvs imbricate, very concave, stem lvs ovate-oblong to cordate-triangular, acute to obtuse with or without blunt apiculus, margin plane, sinuose to finely denticulate above; nerve single, extending about $\frac{3}{4}$ way up lf or forked and shorter; basal cells rhomboidal, angular cells rectangular, pellucid, cells above linear-vermicular, in mid-lf 4–7 μm × 52–80 μm, 10–15(–20) times as long as wide. Branch lvs ovate, acute to obtuse, apiculate or not. Seta papillose; capsule strongly inclined to horizontal, ovate-ellipsoid; lid rostrate, obtuse; spores 12–15 μm. Fr rare. Dull green patches on soil in turf and rock crevices and on sunny banks, especially on coastal cliffs but also inland, locally common along the south coast, extending north to Anglesey and Derby, I. of Man, Mid

Perth, rare in Ireland. 52, H11, C. Denmark, Belgium, Luxemburg, France, Germany, Mediterranean region, Macaronesia, N. America.

153. CIRRIPHYLLUM GROUT, BULL. TORR. BOT. CL., 1898

Dioecious. Medium-sized to robust plants, stems creeping or ascending, irregularly or pinnately branched, branches erect or spreading. Lvs concave, sometimes strongly concave, stem lvs ovate to ovate-oblong, gradually or abruptly narrowed to acute to acuminate apex or apex obtuse to rounded with piliferous acumen, base decurrent, margin plane or recurved near base, entire to denticulate; nerve single or forked above, extending $\frac{1}{3}$–$\frac{3}{4}$ way up lf; cells rhomboidal to linear-rhomboidal, angular cells quadrate to rectangular, forming decurrent auricles. Branch lvs similar or narrower. Seta deep red or brownish red, strongly papillose; capsule inclined, ovoid to ellipsoid, gibbous or curved; annulus separating; lid conical with long subulate beak; peristome perfect, inner with tall basal membrane and nodulose or appendiculate cilia. About 20 species occurring mainly in temperate regions.

1	Lvs not very concave, apex acuminate	**3. C. crassinervium**
	Lvs very concave, apex obtuse or rounded with long piliferous acumen	2
2	Stems pinnately branched, lowland plants	**1. C. piliferum**
	Stem irregularly branched, high alpine plants	**2. C. cirrosum**

1. C. piliferum (Hedw.) Grout, Bull. Torr. Bot. Cl., 1898
Brachythecium piliferum (Hedw.) Kindb., *Eurhynchium piliferum* (Hedw.) Br. Eur.

Plants robust, to 15 cm long, stems prostrate, complanately pinnately branched. Lvs shrunken or not, erect with flexuose apices when dry, loosely imbricate when moist, very concave, plicate or not, stem lvs ovate-oblong, apex rounded or obtuse with long piliferous acumen, base decurrent, margin recurved near base, sinuose or finely denticulate; nerve extending about $\frac{3}{4}$ way up lf; basal cells elongate-rhomboidal, angular cells enlarged, rectangular, forming decurrent auricles, cells above linear-rhomboidal, in mid-lf 6–10 μm × 60–100 μm, 8–12 times as long as wide. Branch lvs similar or narrower. Seta strongly papillose; capsule inclined, ovate-ellipsoid to ellipsoid, gibbous or curved; lid conical with long subulate beak; peristome teeth yellowish; spores 12–16 μm. Fr rare, autumn, winter. $n = 12$–14. Light green, lax patches, coarse wefts or scattered shoots in moist turf and on soil in woods and on shaded banks and roadsides, especially where basic, frequent or common at low altitudes. 109, H35, C. Europe, Iceland, Faroes, Caucasus, N. and C. Asia, Japan, Morocco, N. America, Greenland.

2. C. cirrosum (Schwaegr.) Grout, Bull. Torr. Bot. Cl., 1898
Brachythecium cirrosum (Schwaegr.) Schimp., *Eurhynchium cirrosum* (Schwaegr.) Husn.

Plants robust, shoots to 10 cm long, stems prostrate, irregularly branched. Lvs erect with flexuose apices when dry, loosely imbricate when moist, very concave, plicate or not, stem lvs ovate-oblong, apex obtuse or rounded with long piliferous acumen, base decurrent, margin recurved near base, sinuose or finely denticulate; nerve extending $\frac{1}{3}$–$\frac{1}{2}$ way up lf; basal cells narrowly rhomboidal, angular cells quadrate to rectangular, forming decurrent auricles, cells above linear-rhomboidal, incrassate, in mid-lf 5–8 μm × 56–84 μm, 8–13 times as long as wide. Branch lvs ± similar. Capsule similar to that of *C. piliferum* but peristome teeth brown. Fr unknown in Britain. Lax or tight, pale or whitish-green patches on basic rock ledges

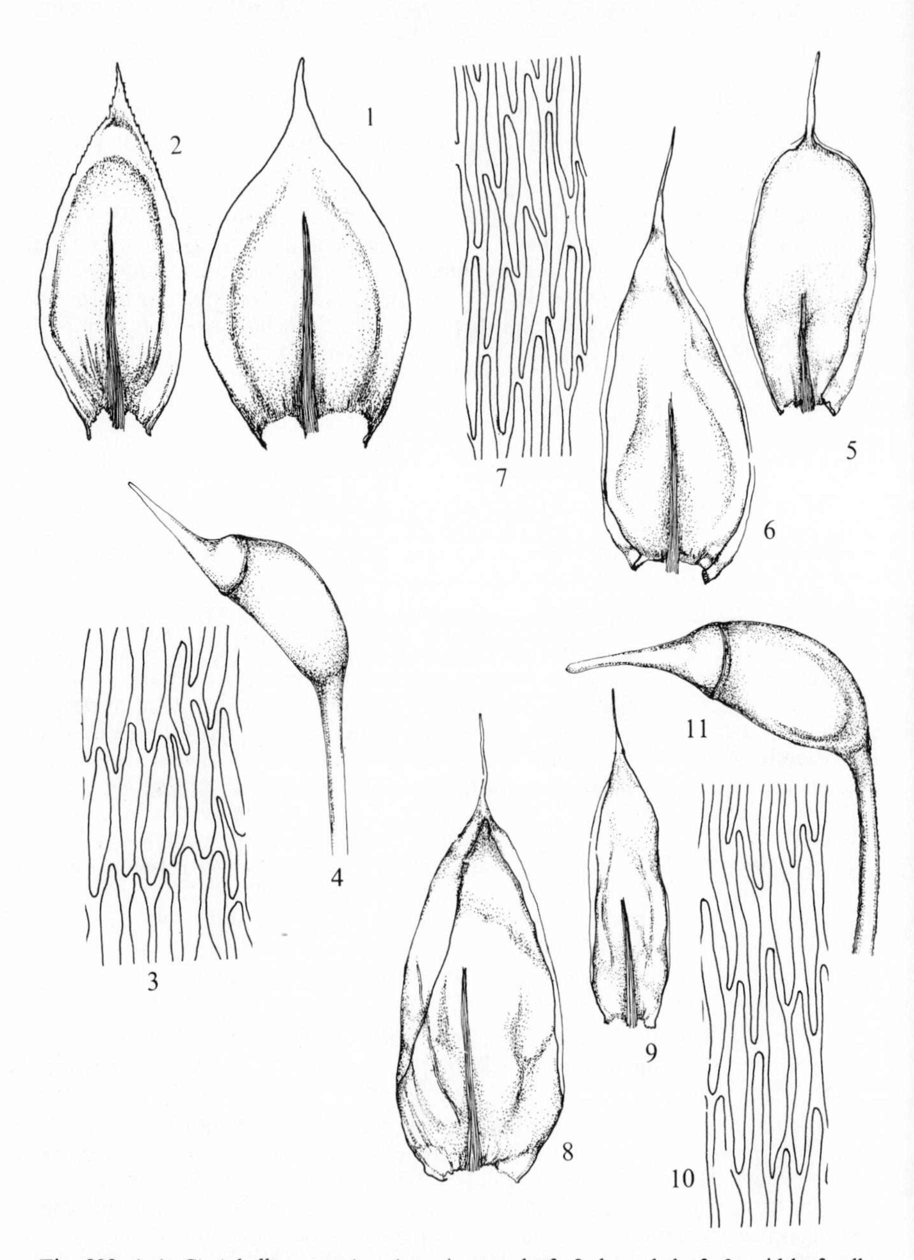

Fig. **292**. 1–4, *Cirriphyllum crassinervium*: 1, stem leaf; 2, branch leaf; 3, mid-leaf cells; 4, capsule. 5–7, *C. cirrosum*: 5, stem leaf; 6, branch leaf; 7, mid-leaf cells. 8–11, *C. piliferum*: 8, stem leaf; 9, branch leaf; 10, mid-leaf cells; 11, capsule. Leaves $\times 25$, cells $\times 415$, capsules $\times 10$.

at *ca* 1200 m, very rare. Mid Perth, Argyll, W. Ross. 3. Iceland, montane and arctic Europe and Asia, Caucasus, Atlas Mountains, N. America, Greenland.

3. **C. crassinervium** (Tayl.) Loeske & Fleisch., Allg. Bot. Zeitschr., 1907
Eurhynchium crassinervium (Tayl.) Br. Eur.

Medium-sized plants, stems creeping or ascending, closely branched, branches short, straight or curved, often pointing in one direction. Lvs erect-flexuose when dry, patent when moist, concave, stem lvs ovate, rapidly narrowed to acuminate apex, base decurrent, margin plane, denticulate; nerve extending $\frac{1}{2}$–$\frac{2}{3}$ way up lf, basal cells narrowly hexagonal, angular cells shortly rectangular, forming decurrent auricles, cells above rhomboidal to narrowly rhomboidal, in mid-lf 6–8 μm × 36–56 μm, 5–8 times as long as wide. Branch lvs similar or narrower. Seta strongly papillose; capsule inclined, ovate-ellipsoid to ellipsoid, gibbous or curved; lid conical with long subulate beak; spores 16–22 μm. Fr occasional, winter. Green or bright green dense tufts or patches on shaded rocks, soil and tree boles, usually in basic habitats, frequent except in N. Scotland. 90, H33, C. Europe, Caucasus, Japan, Algeria, Madeira, Canaries.

154. Rhynchostegium Br. Eur., 1852

Autoecious. Plants slender to robust; stems procumbent, ± irregularly branched; paraphyllia absent. Stem and branch lvs ± similar, usually concave, ovate-lanceolate to broadly ovate, acute to obtuse, margin entire or denticulate; nerve extending $\frac{1}{2}$ way or more up lf, not projecting at back of branch lvs; cells narrowly rhomboidal to linear. Seta reddish, smooth; capsule as in *Eurhynchium*. A mainly northern hemisphere genus of 195 species.

Intermediate between *Eurhynchium* and *Brachythecium*, differing from the former in the nerve not ending in a spine-like projection and the autoecious inflorescence, and from the latter in the rostrate lid of the capsule.

1 Lvs soft, broadly ovate, mid-lf cells 10–16 μm wide, 3–5 times as long as wide **6. R. rotundifolium**
  Lvs various, mid-lf cells 4–10 μm wide, 6–17 times as long as wide 2
2 Lvs shortly pointed, apex obtuse to acute 3
  Lvs tapering to acute to acuminate apex 5
3 Lf margin ± entire or only denticulate above, angular cells forming poorly defined auricles **3. R. murale**
  Lf margin denticulate ± from base to apex, angular cells not forming auricles 4
4 Lvs imbricate to spreading, mid-lf cells 6–11 μm wide **1. R. riparioides**
  Stems julaceous with closely imbricate lvs, cells 4–8 μm wide **2. R. lusitanicum**
5 Plants slender, stems attached to substrate for most of their length by rhizoids, lf apex plane **4. R. confertum**
  Plants medium-sized, stems attached to substrate at base only, lf apex twisted **5. R. megapolitanum**

1. **R. riparioides** (Hedw.) C. Jens., Skand. Bladmossfl., 1939
R. rusciforme Br. Eur., *Eurhynchium riparioides* (Hedw.) Rich., *E. rusciforme* (Br. Eur.) Milde, *Platyhypnidium riparioides* (Hedw.) Dix., *P. rusciforme* (Br. Eur.) Fleisch.

Autoecious. Plants robust, rigid or soft, to 15 cm long, stems prostrate or pendulous, often denuded below, irregularly branched, branches short or long, sometimes

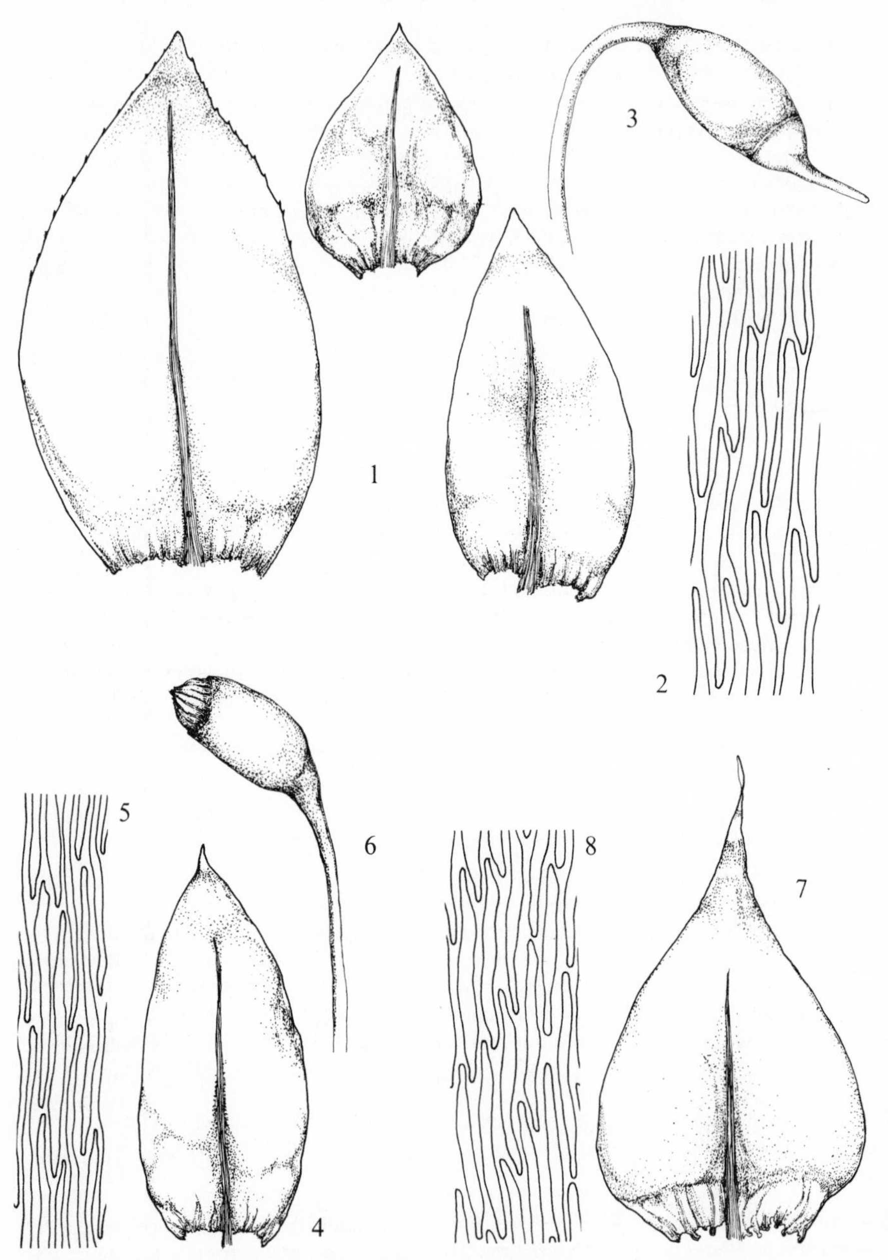

Fig. **293**. 1–3, *Rhynchostegium riparioides*: 1, leaves (×15); 2, mid-leaf cells; 3, capsule. 4–6, *R. lusitanicum*: 4, leaf (×25); 5, mid-leaf cells; 6, capsule. 7–8, *R. megapolitanum*: 7, leaf (×20); 8, mid-leaf cells. Cells ×415, capsules ×10.

attenuated. Lvs slightly shrunken or not when dry, imbricate to spreading, occasionally sub-secund, concave, ovate to broadly ovate, acute to obtuse, margin plane, denticulate all round; nerve strong, extending *ca* $\frac{3}{4}$ way up lf; basal cells narrowly rhomboidal, angular cells rectangular, cells above linear, in mid-lf 6–11 μm×48–100(–136) μm, (6–)8–13 times as long as wide. Seta smooth; capsule inclined to cernuous, ovate-ellipsoid to ellipsoid; lid with subulate beak, spores 16–22 μm. Fr frequent, autumn, winter. $n=8^*$, 10, 11*, 20. Bright or dark green to brownish-green tufts or patches, blackish below, glossy with metallic sheen when dry, on rocks and tree roots submerged or subject to submergence in and by streams and rivers and in wet places on cliffs and in quarries, very common in montane regions, occasional elsewhere. 111, H40, C. Europe, Iceland, Lebanon, Caucasus, Tibet, Kashmir, Nepal, China, Manchuria, Japan, Algeria, Morocco, Macaronesia, N. America, Mexico, Guatemala, northern S. America.

A very variable species, small plants tending to grow in drier habitats, large plants submerged or floating. May be confused with *Hygrohypnum* and aquatic *Brachythecium* species. Differs in lf shape and sharply denticulate lf margin from the former. *Brachythecium rivulare* has distinct, decurrent auricles and *B. plumosum* differs in lf shape. *R. lusitanicum* is usually brownish in colour, the stems julaceous with closely imbricate lvs with narrower lf cells.

2. R. lusitanicum (Schimp.) A. J. E. Smith, J. Bryol., 1977
Eurhynchium alopecuroides (Brid.) Rich. & Wall., *E. alopecurum* err. orthogr. pro *E. alopecuroides*, *E. rusciforme* var. *alopecuroides* (Brid.) Dix., *Hygrophynum lusitanicum* (Schimp.) Corb.

Plants medium-sized, prostrate or floating, stems sparsely branched, julaceous with closely imbricate lvs. Lvs closely imbricate when dry, erect when moist, very concave, ovate, acute to obtuse, margin denticulate all round; nerve reaching $\frac{3}{4}$ way or more up lf; basal cells narrowly rhomboidal, angular cells quadrate-rectangular or rectangular, cells above linear-vermicular, in mid-lf 4–8 μm×56–112 μm, 12–17 times as long as wide. Capsule ellipsoid; spores 16–20 μm. Fr very rare, winter. Dense brownish-green to dark brown patches on submerged rocks in fast-flowing streams and rivers, rare in montane areas. 24, H7. France, Spain, Portugal.

3. R. murale (Hedw.) Br. Eur., 1852
Eurhynchium murale (Hedw.) Milde

Autoecious. Plants medium-sized, stems creeping, irregularly pinnately branched, branches short, obtuse, erect or ascending. Lvs imbricate to erecto-patent, very concave, ovate, obtuse or obtuse and apiculate, margin plane or inflexed above, entire below, obscurely denticulate above; nerve extending $\frac{1}{2}$–$\frac{3}{4}$ way up lf; basal cells rhomboidal, angular cells enlarged, shortly rectangular, forming poorly defined auricles, cells above narrowly to linear-rhomboidal, in mid-lf 5–8 μm×48–72 μm, 7–12 times as long as wide. Perichaetial lvs entire, nerveless. Seta short, smooth; capsule inclined or horizontal, shortly cylindrical, curved; lid with long subulate beak; spores 12–16 μm. Fr common, winter, spring. $n=10$, 10+1, 11*. Light green patches on shaded rocks, walls, stony ground and tree boles, frequent in basic areas, rare or occasional elsewhere. 100, H27, C. Europe, Iceland, Caucasus, Syria, Japan, Algeria, Madeira, Canaries.

4. R. confertum (Dicks.) Br. Eur., 1852
Eurhynchium confertum (Dicks.) Milde

Autoecious. Plants slender, stems attached to substrate by rhizoids, irregularly branched, branches short, spreading to erect. Stem and branch lvs ± similar, not

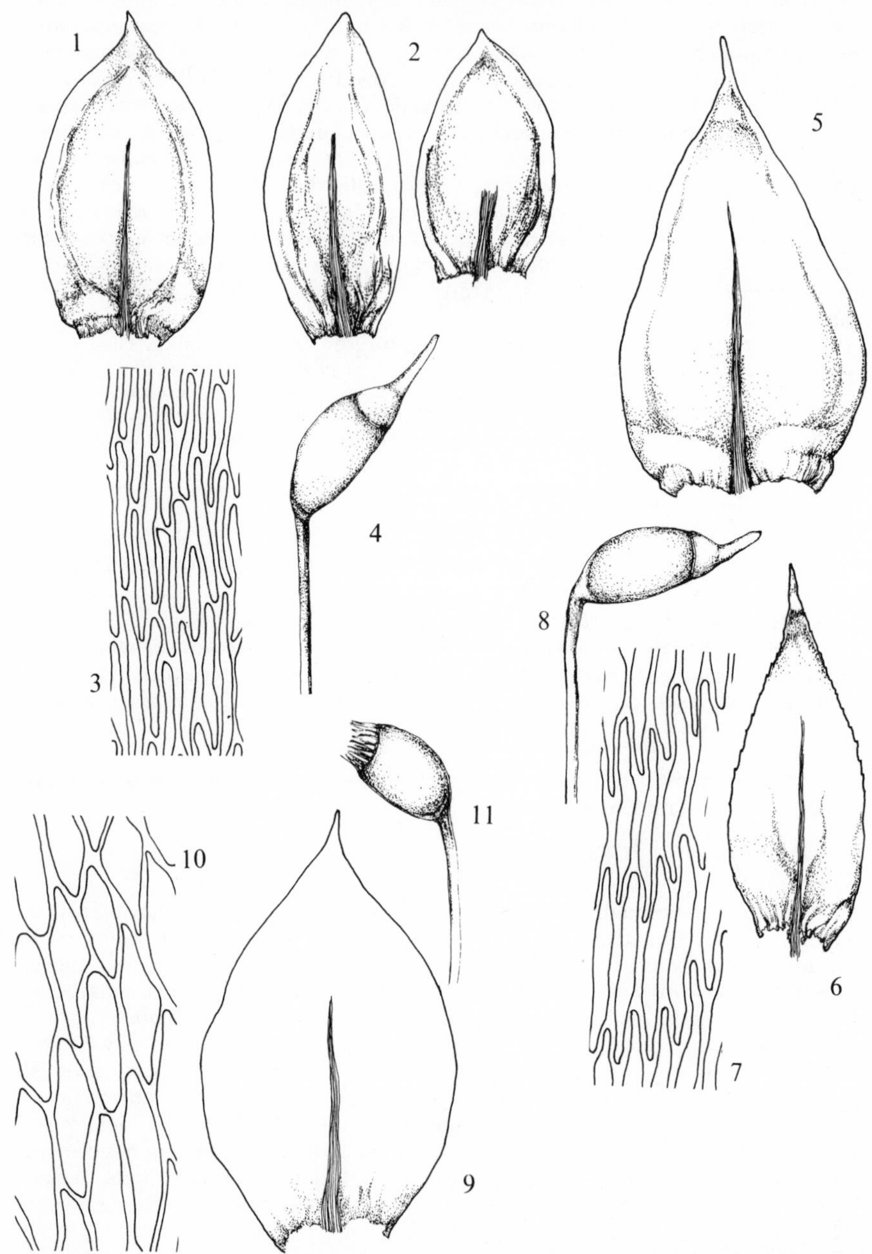

Fig. **294**. 1–4, *Rhynchostegium murale*: 1, stem leaf; 2, branch leaves; 3, mid-leaf cells; 4, capsule. 5–8, *R. confertum*: 5, stem leaf; 6, branch leaf; 7, mid-leaf cells; 8, capsule. 9–11, *R. rotundifolium*: 9, leaf; 10, mid-leaf cells; 11, capsule. Leaves ×40, cells ×415, capsules ×10.

crowded, spreading, sometimes sub-secund or complanate, ovate or ovate-lanceolate, acute, rarely acuminate, margin plane, denticulate ± all round; nerve extending $\frac{2}{3}$–$\frac{3}{4}$ way up lf; basal cells rhomboidal, angular cells shortly rectangular, not forming auricles, cells above linear, in mid-lf 5–7 μm × 48–60 μm, 9–11(–13) times as long as wide. Perichaetial lvs denticulate, nerved. Seta short, smooth; capsule inclined, ellipsoid; lid with long subulate beak; spores *ca* 15 μm. Fr common, winter. $n = 10$*, 10 + 1, 11*, 12*. Tight, dull green patches on rocks and stones, damp walls and bark in shaded habitats, common at low altitudes. 107, H39, C. N., W. and C. Europe, Caucasus, China, Algeria, Macaronesia.

A somewhat nondescript moss distinguished from small *Brachythecium* species by the wider lvs, from *R. murale* by the spreading, hardly concave, acute lvs and from *Eurhynchium pumilum*, which occurs on soil, in the longer, narrower lf cells and the nerve not ending in a small dorsal projection in the branch lvs.

5. R. megapolitanum (Web. & Mohr) Br. Eur., 1852
Eurhynchium megapolitanum (Web. & Mohr) Milde

Autoecious. Plants medium-sized, stems procumbent to ascending, irregularly branched. Stem and branch lvs ± similar, somewhat shrunken when dry, erect to patent when moist, ovate to broadly ovate, gradually to abruptly tapering to acuminate apex, apex twisted, margin plane, denticulate ± all round; nerve extending *ca* $\frac{3}{4}$ way up lf; basal and angular cells shortly rectangular or basal cells rhomboidal, cells above linear, 6–10 μm × 56–128 μm, 8–13 times as long as wide. Seta long, flexuose, smooth; capsule inclined, sub-cylindrical; lid with subulate beak; spores 12–16 μm. Fr frequent, autumn, winter. Pale or yellowish-green patches or wefts on sandy soil in turf, on banks and in sand-dunes. A mainly southern species frequent in coastal areas sand occasional elsewhere in England and Wales, very rare in Scotland and Ireland. 51, H5, C. W. and C. Europe, Caucasus, Iran, Syria, Lebanon, Algeria, Morocco, Macaronesia.

6. R. rotundifolium (Brid.) Br. Eur., 1852
Eurhynchium rotundifolium (Brid.) Milde

Autoecious. Plants medium-sized, stems creeping, irregularly branched, branches erect or ascending, short. Lvs soft, not crowded, shrunken and flexuose when dry, patent to spreading when moist, slightly concave, broadly ovate, acute, margin plane, bluntly denticulate above; nerve extending *ca* halfway up lf; cells narrowly hexagonal to rhomboidal, shorter at base, in mid-lf 10–16 μm × 48–60 μm, 3–5 times as long as wide. Seta short, smooth; capsule inclined, ovoid; spores 14–16 μm. Fr common, winter. $n = 10 + 1$. Dull green patches on tree and shrub boles in hedgerows, very rare. Recorded from N. Somerset, E. Glos and E. Sussex but seen recently only in Sussex. 3. France, Germany, Austria, Czechoslovakia, Switzerland, Italy, Caucasus, Japan.

155. Eurhynchium Br. Eur., 1854

Usually dioecious. Plants slender to robust, forming mats, patches or wefts, often glossy when dry; stems creeping, procumbent, arcuate or ascending, irregularly to pinnately or rarely bipinnately branched, branches spreading to erect, paraphyllia seldom present. Stem and branch lvs ± similar or not, imbricate to spreading, sometimes plicate, stem lvs broadly cordate-triangular to lanceolate, apex obtuse to acute or longly acuminate, base decurrent or not, margin denticulate; nerve extending $\frac{1}{2}$ way or more up lf; basal cells narrowly rhomboidal, angular cells usually shorter and wider, sometimes enlarged and forming auricles, cells above narrowly rhomboidal to linear. Branch lvs ovate to lanceolate; nerve ending in small projection from back of lf. Seta reddish, smooth or papillose; capsule inclined to horizontal, ovoid to shortly cylindrical, curved or gibbous or not; annulus present; lid with long, subulate beak; peristome perfect, inner with tall basal membrane and nodulose or appendiculate cilia; calyptra cucullate. A genus of about 85 species.

The genus *Eurhynchium* as recognised here is split into 2 or more genera by some authors. These are *Eurhynchium*, *Oxyrrhynchium* Warnst., *Plasteurynchium* Fleisch., and *Stokesiella* (Kindb.) Robins. Some of these genera are based on trivial characters and the limits of others are controversial, and for these reasons are not recognised here.

1 Lvs strongly plicate, seta smooth — 2
Lvs not or only slightly plicate, seta papillose (except in *E. pulchellum*) — 3

2 Plants forming coarse wefts, branches not crowded, angular cells rectangular — **1. E. striatum**
Plants forming dense patches, branches crowded, angular cells quadrate, very incrassate — **2. E. meridionale**

3 Stem lvs cordate-triangular, longly decurrent, branch lvs very different in shape, ovate to lanceolate — **5. E. praelongum**
Stem and branch lvs ± similar in shape, differing in size or not — 4

4 Plants very small, lvs not more than 0.5 mm long, cells in mid-lf 2–5 times as long as wide — **4. E. pumilum**
Plants and lvs larger, cells in mid-lf longer, 6–16 times as long as wide — 5

5 Primary stems subterranean, rhizomatous, mid-lf cells 3–5 μm wide, lf apex often twisted through 180°, branches short, crowded — **7. E. schleicheri**
Primary stems not subterranean, cells 6–9 μm wide, lf apex plane, branches of various lengths, not crowded — 6

6 Seta smooth, at least some branch lvs concave, — **3. E. pulchellum**
Seta papillose, branch lvs acute, not concave — 7

7 Dioecious, plants usually yellowish or yellowish-green, mid-lf cells 48–60 μm long — **6. E. swartzii**
Autoecious or synoecious, plants usually dark green, cells 40–90 μm long — **8. E. speciosum**

1. E. striatum (Hedw.) Schimp., Coroll., 1856

Dioecious. Plants medium-sized to robust, rigid, stems procumbent, ascending or arcuate, irregularly to sub-pinnately branched, branches not crowded, spreading to erect, obtuse, stout or somewhat slender. Lvs erecto-patent to spreading, strongly plicate both when moist and dry, stem lvs cordate-triangular to lanceolate-triangular, tapering from near base to acuminate apex, base shortly decurrent, margin narrowly recurved below, denticulate to dentate ± throughout; nerve extending to ¾ way or more up lf; basal cells narrowly rhomboidal, angular cells larger, rectangular, forming poorly defined auricles, cells above linear, in mid-lf 4–6 μm×36–72 μm, 10–14 times as long as wide. Branch lvs smaller, ovate to lanceolate, acute. Seta smooth; capsule inclined, sub-cylindrical, slightly curved; lid with long subulate beak; spores *ca* 14 μm. Fr occasional, autumn, winter. $n=6$, 11*, 12. Glossy, green, light green or yellowish-green, coarse wefts on soil, rocks and logs in woods, on banks, by streams, in damp turf and amongst rocks and boulders in limestone grassland, common and sometimes abundant. 112, H40, C. Europe, Caucasus, Siberia, Altai, Japan, Algeria, Azores.

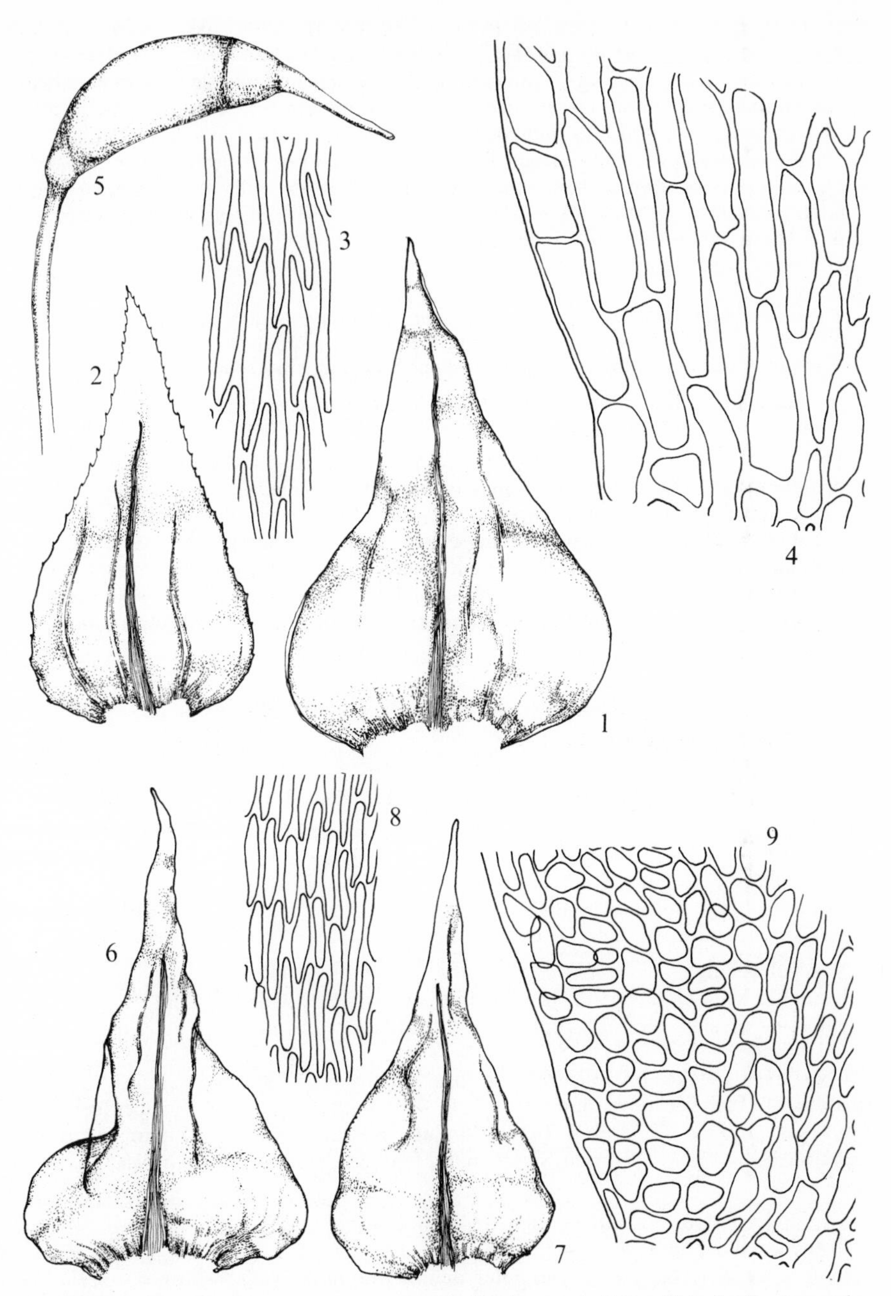

Fig. **295**. 1–5, *Eurhynchium striatum*: 1, stem leaf; 2, branch leaf; 3, mid-leaf cells; 4, angular cells; 5, capsule (×10). 6–9, *E. meridionale*: 6, stem leaf; 7, branch leaf; 8, mid-leaf cells; 9, angular cells. Leaves ×25, cells ×415.

2. E. meridionale (Br. Eur.) De Not. in Picc., Conn. Critt. Ital., 1863
Plasteurhynchium meridionale (Br. Eur.) Fleisch.

Dioecious. Plants medium-sized, stems creeping, closely branched, branches erect or ascending. Lvs spreading, flexuose when dry, patent when moist, plicate and undulate, stem lvs cordate-triangular, apex acuminate, margin plane, denticulate ± throughout; nerve reaching about $\frac{3}{4}$ way up lf; basal cells elongate-rhomboidal, incrassate, porose, angular cells quadrate, very incrassate, cells above linear-rhomboidal, incrassate, in mid-lf 5–6 μm × 48–56 μm, *ca* 10 times as long as wide. Fr unknown in Britain. Bright yellowish-green, glossy, dense patches on limestone rocks, very rare. Dorset. 1. Mediterranean region of Europe, Algeria, Madeira, Azores.

The lvs are similar in shape to those of *E. striatum* but differ in the ± quadrate, incrassate angular cells. The habit is also different and the plant is more likely to be confused with *Isothecium striatulum*, for the differences from which see under that species.

3. E. pulchellum (Hedw.) Jenn., Man. Moss. West Pennsylv., 1913
E. strigosum (Web. & Mohr) Br. Eur.

Dioecious. Plants small, male plants minute, bud-like on female plants, stems procumbent, closely, often sub-pinnately branched, branches erect, straight. Lvs imbricate to erecto-patent both when wet and dry, concave, ± plicate or not, stem lvs cordate-triangular to narrowly triangular, apex acute to acuminate, margin plane, denticulate; nerve slender, extending *ca* $\frac{3}{4}$ way up lf; basal cells rhomboidal, incrassate, angular cells oval to rounded-quadrate, incrassate, cells above linear-rhomboidal to linear, incrassate. Branch lvs ovate, obtuse to acute. Seta smooth; capsule ovoid to ellipsoid, curved. Fr unknown in Britain.

Key to intraspecific taxa of *E. pulchellum*

1 Stem and branch lvs lanceolate — Plant from Somerset
Stem and branch lvs wider — 2
2 Lvs not crowded, stem lvs not complanate, branches 3–10 mm long, branch lvs patent — var. **pulchellum**
Lvs crowded, stem lvs slightly complanate, branches 1.5–5.0 mm, branch lvs imbricate — 3
3 Montane plant, mid-lf cells 36–56 μm long — var. **diversifolium**
Lowland plant, mid-lf cells 56–80 μm long — var. **praecox**

Var. **pulchellum**
Branches slender, acute, 3–10 mm long. Lvs not crowded, patent, sometimes slightly plicate, stem lvs not complanate, ± triangular to ovate-triangular, apex acute to acuminate; mid-lf cells 6–8 μm × 48–90 μm, 6–11(–16) times as long as wide. $n = 10$. Green patches on soil, rocks or logs, usually in shaded basic situations, very rare and not seen for more than 40 years. E. Cornwall, S. Devon. 2. Europe, Iceland, Caucasus, Siberia, C. and E. Asia, Punjab, Kashmir, Chitral, Japan, Algeria, Canaries, N. America, Mexico, Guatemala, Ecuador, northern S. America, Greenland.

Var. **diversifolium** (Br. Eur.) C. Jens., Skand. Bladmfl., 1939
Branches short, obtuse, 1.5–5.0 mm long. Lvs crowded, stem lvs slightly complanate, cordate-triangular to triangular, acuminate; mid-lf cells 5–7 μm × 36–56 μm, 8–12 times as long as wide. Branch lvs imbricate, acute to obtuse. Basic soil or rocks in montane habitats, very rare and not seen for nearly 90 years. Angus (1889). 1. Montane and arctic Europe, Iceland, Caucasus, arctic Asia, N. America, Greenland.

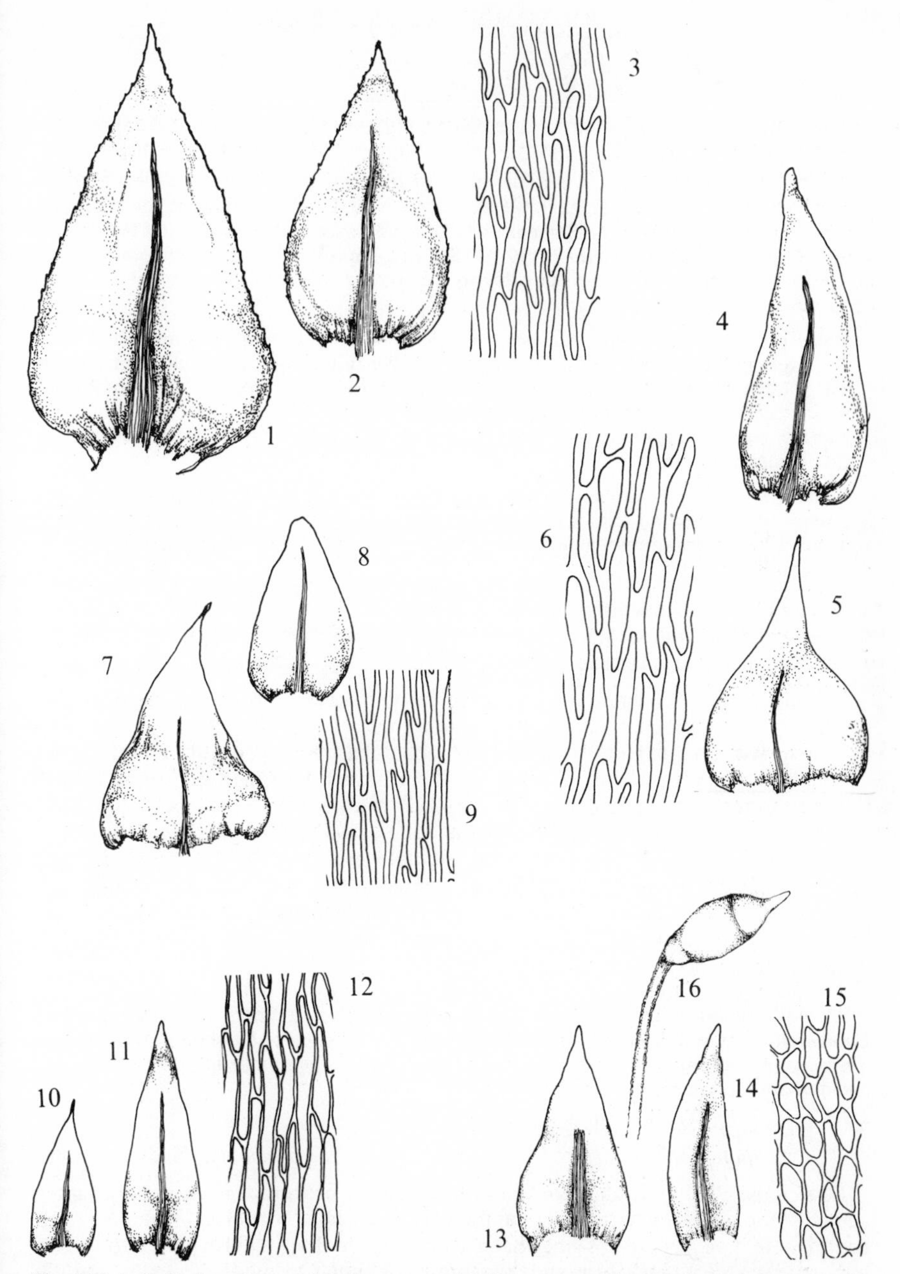

Fig. **296**. 1–3, *Eurhynchium pulchellum* var. *pulchellum*: 1, stem leaf; 2, branch leaf; 3, mid-leaf cells. 4–6, *E. pulchellum* var. *praecox*: 4, branch leaf; 5, stem leaf; 6, mid-leaf cells. 7–9, *E. pulchellum* var. *diversifolium*: 7, stem leaf; 8, branch leaf; 9, mid-leaf cells. 10–12, *E. pulchellum* from Somerset: 10, branch leaf; 11, stem leaf; 12, mid-leaf cells. Leaves ×40, cells ×415. 13–16, *E. pumilum*: 13 and 14, stem and branch leaves (×100); 15, mid-leaf cells (×415); 16, capsule (×10).

Var. **praecox** (Hedw.) Dix., Rev. Bryol. Lichen., 1934

Branches short, acute to obtuse, 2–5 mm long. Lvs crowded, stem lvs patent, slightly complanate, triangular to narrowly triangular with long acuminate apex, cells in mid-lf 5–7 μm × 56–80 μm, 10–13 times as long as wide. Branch lvs imbricate. $n=12$. Green or yellowish-green patches on soil in heathland, very rare. W. Suffolk. 1. Fennoscandia, C. Europe, Italy, Caucasus, C. Asia, Kashmir, Algeria, N. America.

A curious plant occurs in N. Somerset on sheltered limestone rocks and soil (see Appleyard, *Trans. Br. bryol. Soc.* **6**, 36–7, 1970). This has been referred to *E. pulchellum* but I do not think it belongs here or even possibly to the genus *Eurhynchium*. A description of the Somerset plant is as follows:

Stems procumbent, closely branched, branches to *ca* 7 mm long. Lvs appressed when dry, erecto-patent when moist, concave, stem lvs not complanate, lanceolate, acute, margin denticulate; nerve strong, extending *ca* ¾ way up lf; basal cells rhomboidal, angular cells shorter but not much differentiated, cells above narrowly rhomboidal, in mid-lf 5–6 μm × 24–40 μm, 4–8 times as long as wide. Branch lvs lanceolate, acuminate. Gametangia unknown.

Apart from the above Somerset plant, only var. *praecox* is still known to be extant. The nature of the 3 named varieties is not clear but Nyholm (1954–69) suggests that var. *praecox* and var. *diversifolium* are not genetically distinct and might merely be dry habitat forms of the type variety.

4. E. pumilum (Wils.) Schimp., Coroll., 1856

Oxyrrhynchium pumilum (Wils.) Loeske, *Rhynchostegiella pumila* (Wils.) E. F. Warb., *R. pallidirostra* (Brid.) Loeske

Dioecious. Plants slender, stems irregularly branched, branches filiform. Lvs spreading, often sub-complanate, stem and branch lvs of ± similar shape, 2–3 times as long as wide, ovate to ovate-lanceolate, acute, base not decurrent, margin plane, denticulate; nerve extending ½–⅘ way up lf, in branch lvs projecting from back of lf in small tooth; cells rhomboidal or narrowly rhomboidal, ± similar throughout, in mid-lf 6–8 μm × 16–36 μm, 2–5 times as long as wide. Seta papillose; capsule ± horizontal, ovoid to ellipsoid; lid with subulate beak; spores 12–15 μm. Fr rare, autumn, winter. $n=10+1$, 11*. Small, dense, dull green mats on shaded soil, particularly where clayey, in woods and on hedgebanks, usually where basic, frequent or common in England, occasional elsewhere. 101, H28, C. Europe, Algeria, Macaronesia.

This plant is sometimes placed in the genus *Rhynchostegiella*, but on morphological and cytological grounds is probably better placed in *Eurhynchium*. It differs from other species of *Eurhynchium* in the shorter cells, from *Rhynchostegiella curviseta* in the shorter, wider lvs and dioecious inflorescence, and from *Amblystegium serpens* in the shorter lvs, the papillose seta and dioecious inflorescence.

5. E. praelongum (Hedw.) Br. Eur., 1854

Oxyrrhynchium praelongum (Hedw.) Warnst., *Stokesiella praelonga* (Hedw.) Robins.

Dioecious. Plants slender, to 12 cm long, stems procumbent or arcuate, interruptedly pinnately or sometimes bipinnately branched, branches ± curved, sub-complanate. Lvs ± patent, somewhat shrunken when dry, stem lvs patent to spreading or squarrose when moist, widely cordate-triangular to ovate-triangular, rapidly or abruptly narrowed to usually long acumen, base longly decurrent, margin plane, denticulate; nerve extending ½–¾ way up lf; basal cells narrowly rhomboidal, at basal angles rectangular, forming poorly defined auricles, cells above becoming longer and narrower, towards acumen linear. Branch lvs patent to spreading when moist, much narrower than stem lvs, ovate to lanceolate, acute to acuminate. **Seta**

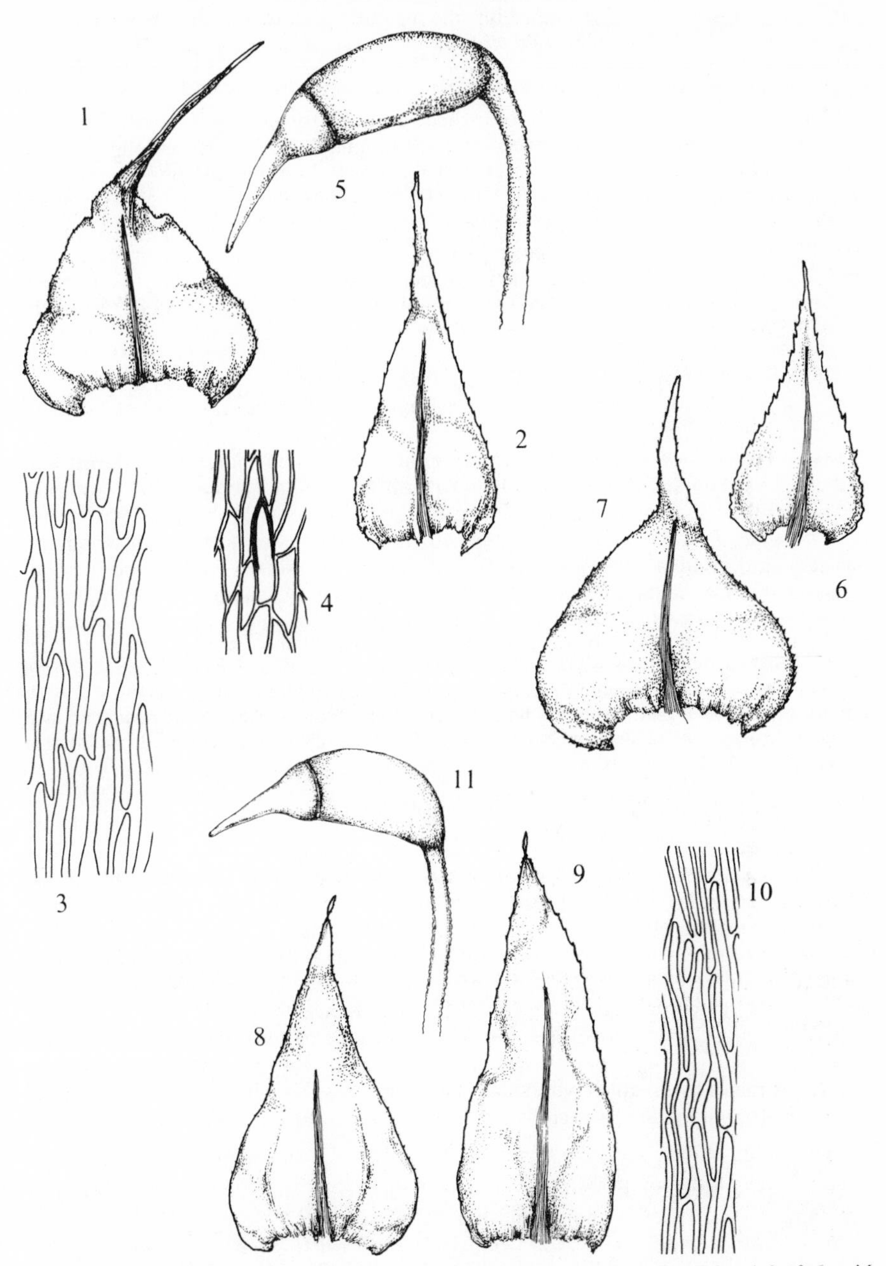

Fig. **297**. 1–5, *Eurhynchium praelongum* var. *praelongum*: 1, stem leaf; 2, branch leaf; 3, mid-leaf cells; 4, end of nerve at back of branch leaf; 5, capsule. 6–7, *E. praelongum* var. *stokesii*: 6, branch leaf; 7, stem leaf. 8–11, *E. schleicheri*: 8, branch leaf; 9, stem leaf; 10, mid-leaf cells; 11, capsule. Leaves ×40, cells ×415, capsules ×10.

papillose; capsule ± horizontal, ellipsoid to sub-cylindrical, curved, at maturity green with small black blotches; lid with long subulate beak; spores 11–33 μm.

Var. **praelongum**

Plants slender, stems pinnately branched, paraphyllia lacking. Lvs not crowded, branch lvs lanceolate to ovate-lanceolate, acuminate, 2.7–5.0 times as long as wide; cells in mid-lf of branch lvs 5–7 μm×40–60 μm, 8–12 times as long as wide. Fr frequent, winter. *n*=6, 7, 8*, 10, 10+1*, 11*. Light or dull green to yellowish-green or brownish-green wefts or lax patches on soil, rocks, litter, stumps and logs in woods, copses, shaded hedgebanks, stream and ditch sides, marshy ground and meadows, etc., common and sometimes locally abundant. 112, H40, C. Europe, Iceland, Faroes, Caucasus, N. Asia, Yunnan, Japan, Tunisia, Morocco, Macaronesia, N. America.

Var. **stokesii** (Turn.) Dix., Stud. Handb. Brit. Moss., 1896

E. stokesii (Turn.) Br. Eur., *Oxyrrhynchium praelongum* var. *stokesii* (Turn.) Podp., *Stokesiella praelonga* var. *stokesii* (Turn.) Crum.

Plants larger, stems pinnately and bipinnately branched, paraphyllia sometimes present. Lvs crowded, branch lvs ovate to ovate-triangular, acute, 1.6–2.6 times as long as wide; cells in mid-lf of branch lvs 5–7 μm×24–48(–52) μm, 5–8(–9) times as long as wide. Fr occasional, winter. *n*=10, 12. Lax patches or wefts in woodland habitats and shaded places by streams. Rare or occasional in lowland Britain, occasional to frequent elsewhere. 82, H30, C. Europe, N. America.

Although var. *stokesii* is treated as a separate species by some authors, intermediates between it and the type variety are frequent. It seems to be little more than just a large form of var. *praelongum* but does have wider branch lvs with shorter cells and paraphyllia often present. *E. swartzii* differs from *E. praelongum* in its usually more strongly marked yellowish colour and the stem and branch lvs hardly differing in shape. Some authors regard the two as synonymous but I have seen no intermediates. Dixon (1924) discusses this matter at length.

6. E. swartzii (Turn.) Curn. in Rabenh., Bryoth. Eur., 1862

Oxyrrhynchium swartzii (Turn.) Warnst.

Dioecious. Plants slender to medium-sized, stems prostrate, distantly to closely branched, branches straight, spreading or erect. Lvs patent, ± shrunken when dry, patent to spreading when moist, complanate or not, stem lvs somewhat wider and larger than branch lvs but otherwise not differing, stem lvs ovate to broadly ovate, acute to acuminate, base shortly decurrent, margin plane, denticulate to dentate; nerve strong, extending *ca* ¾ way up lf; basal cells narrowly rhomboidal, angular cells rectangular, cells above linear. Branch lvs ovate to ovate-lanceolate, acute to acuminate. Seta strongly papillose; capsule inclined, ellipsoid, gibbous or slightly curved, at maturity greenish with small black blotches; lid with long subulate beak; spores 12–16 μm. Fr rare, winter.

Var. **swartzii**

Stems distantly and irregularly branched, branches spreading. Lvs ± distant, complanate or sub-complanate; mid-lf cells of stem lvs 6–8 μm×48–80 μm, 6–11 times as long as wide. *n*=7*, 10. Yellowish or yellowish-green, occasionally green, lax patches or straggling shoots on damp soil in fields, by streams, shaded roadsides, banks and in woodland, occasionally on rocks or walls, frequent or common in lowland habitats. 112, H40, C. Europe, Iceland, Caucasus, Lebanon, C. Asia, Japan, Algeria, Madeira, Azores, St Helena, N. America.

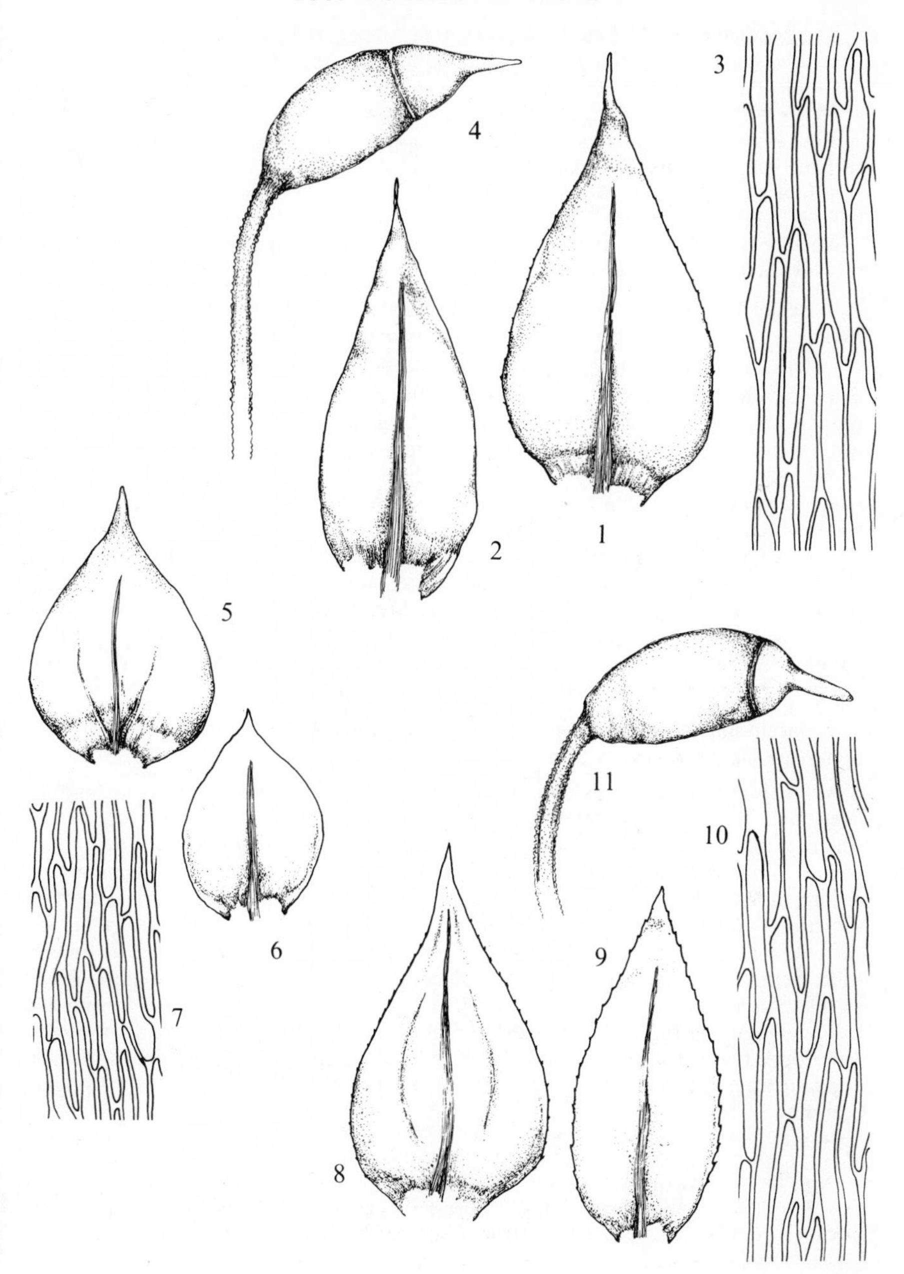

Fig. **298**. 1–4, *Eurhynchium swartzii* var. *swartzii*: 1, stem leaf; 2, branch leaf; 3, mid-leaf cells; 4, capsule. 5–7, *E. swartzii* var. *rigidum*: 5, stem leaf; 6, branch leaf; 7, mid-leaf cells. 8–11, *E. speciosum*: 8, stem leaf; 9, branch leaf; 10, mid-leaf cells; 11, capsule. Leaves ×25, cells ×415, capsules ×10.

Var. **rigidum** (Boul.) Thér., Bull. Acad. Inst. Geogr. Bot., 1901
Eurhynchium praelongum var. *rigidum* Boul., *Oxyrrhynchium swartzii* var. *rigidum* (Boul.) Barkm.

Stems with crowded, erect branches. Lvs crowded, not complanate; mid-lf cells of stem lvs 6–9 μm × 28–56 μm, 4–8(–10) times as long as wide. Yellowish patches in turf in open habitats, especially where basic. Rare in England and Scotland, very rare in Ireland, not known from Wales. 34, H4. W., C. and S. Europe, Tunisia, Algeria, Madeira, Azores.

Whether var. *rigidum* is worth treating as a distinct taxon is doubtful as numerous intermediates occur between it and the type.

7. E. schleicheri (Hedw. f.) Milde, Br. Siles., 1869
E. abbreviatum (Turn.) Brockm., *Oxyrrhynchium schleicheri* (Hedw. f.) Röll

Dioecious. Plants medium-sized with ± subterranean stolons, secondary stems creeping, closely branched, branches erect or ascending, short, obtuse. Stem and branch lvs ± similar, erect, somewhat shrunken when dry, erecto-patent to patent when moist, not complanate, ± concave, irregularly plicate, ovate-oblong to ovate, apex acute, usually twisted through 180°, margin plane or recurved below, denticulate; nerve extending *ca* $\frac{2}{3}$ way up lf; basal cells narrowly rhomboidal, angular cells shortly rectangular, cells above linear, ± vermicular or not, in mid-lf 3–5 μm × 32–80 μm, 10–16 times as long as wide. Seta papillose; capsule inclined to horizontal, ellipsoid to shortly cylindrical, gibbous or curved; lid with long subulate beak; spores 10–14 μm. Fr rare, winter. $n=7$. Dense yellowish-green tufts or patches on soil on banks in woodland and by roadsides, often but not necessarily in calcareous habitats, rare but sometimes locally frequent in England, extending north to Roxburgh and Berwick, W. Sutherland, Glamorgan. 30. Europe.

8. E. speciosum (Brid.) Jur., Verh. Zool. Bot. Ges. Wien, 1863
Oxyrrhynchium speciosum (Brid.) Warnst.

Autoecious or synoecious. Plants medium-sized, stems procumbent or ascending, irregularly branched, branches spreading or ascending. Lvs ± distant, spreading, when dry shrunken, stem lvs not complanate, ± similar to or wider than complanate or sub-complanate branch lvs, ovate to broadly ovate, apex acute to acuminate, base decurrent, margin sharply denticulate; nerve strong, extending to $\frac{4}{5}$ way up lf or sometimes nearly to apex; basal cells rhomboidal, angular cells rectangular, decurrent, cells above linear, in mid-lf 6–9 μm × 40–90 μm, (6–)9–16 times as long as wide. Seta papillose; capsule inclined or horizontal, ellipsoid, slightly gibbous; lid with subulate beak; spores *ca* 16 μm. Fr frequent, winter. Dull or dirty green, lax, straggling patches on muddy ground and tree bases by streams, rivers and pools, on waterlogged soil and wet, muddy rock ledges, occasional in suitable habitats in England and Wales, rare elsewhere. 61, H17, C. W. and C. Europe, Italy, Iran, Canaries.

Readily identified when well grown by the dull or dirty green colour, complanate and distant branch lvs and the coarsely toothed lf margin; in *Brachythecium* species with which it might be confused the lvs are not complanate, are more finely toothed and the nerve of branch lvs does not terminate in a small projection from the back of the lf. Depauperate specimens may be difficult to separate from *E. swartzii* but that plant is dioecious and usually has shorter cells.

156. Rhynchostegiella (Br. Eur.) Limpr., Laubm., 1896

Autoecious, rarely dioecious. Slender plants, stems creeping, with tufts of rhizoids, irregularly branched. Lvs small, linear-lanceolate to oblong-lanceolate or lanceolate, apex acute to acuminate, base not decurrent, stem and branch lvs ± similar or branch lvs smaller; nerve single; angular cells little differentiated, cells above narrowly rhomboidal to linear. Seta deep red, usually papillose, somewhat curved; capsule inclined or horizontal, ovoid or ellipsoid; annulus differentiated; lid rostrate; peristome perfect, inner with tall basal membrane and nodulose or shortly appendiculate cilia. Distribution world-wide. Fifty species.

1 Lvs linear-lanceolate, tapering to long fine acumen, 6–10 times as long as wide, seta usually smooth **1. R. tenella**
Lvs ovate to lanceolate, acute to sub-obtuse, 3–5 times as long as wide, seta papillose 2
2 Nerve extending $\frac{1}{2}$–$\frac{2}{3}$ way up lf, mid-lf cells 40–72 μm long, 8–10 times as long as wide **2. R. curviseta**
Nerve extending $\frac{3}{4}$ or more way up lf, mid-lf cells 32–48 μm long, 5–7 times as long as wide **3. R. teesdalei**

1. R. tenella (Dicks.) Limpr., Laubm., 1896
Eurhynchium tenellum (Dicks.) Milde

Autoecious. Slender plants, stems creeping with numerous erect branches. Lvs patent or erecto-patent, sometimes sub-secund, linear-lanceolate, tapering from near base to long acuminate apex, 5–10 times as long as wide, margin sinuose; nerve extending to acumen; basal cells narrowly rhomboidal, a few angular cells rectangular, cells above ± linear, in mid-lf 5–8 μm × 56–100 μm, 10–20 times as long as wide. Branch lvs with narrower bases. Capsule ± horizontal, ovoid; spores 10–15 μm. Fr common, autumn to spring. $n = 11$*, 22*. Yellowish-green to olive-green silky mats or tufts.

Var. **tenella**
Plants very silky. Lvs with long acumen; nerve extending into acumen. Seta smooth. On usually shaded, basic rocks and walls, rarely on tree stumps, common in basic habitats, rare or occasional elsewhere. 94, H38, C. N., W. and C. Europe, Lebanon, Sinai, Caucasus, N. Africa, Macaronesia.

Var. **litorea** (De Not.) Rich. & Wall., Trans. Br. bryol. Soc., 1950
Eurhynchium tenellum var. *scabrellum* Dix.
Less silky, branches more spreading. Lvs more shortly pointed; nerve varying in extent from $\frac{1}{2}$ way up lf to reaching acumen. Seta sparsely to moderately papillose. On tree stumps or trunks, occasionally on stones, rare. Devon and Kent to Oxford and Cambridge. 12. France, Elba, Giglio Is., Genoa, Sardinia, Sicily.

2. R. curviseta (Brid.) Limpr., Laubm. Deutschl., 1896
Eurhynchium curvisetum (Brid.) Husn.

Autoecious. Plants slender; stems creeping, irregularly branched, branches ± procumbent. Lvs patent to spreading, 3–5 times as long as wide, lanceolate to oblong-lanceolate, acute to sub-obtuse, margin denticulate above; nerve extending $\frac{1}{2}$–$\frac{2}{3}$ way up lf; cells narrowly to linear-rhomboidal, in mid lf (4–)5–7(–8) μm × 40–72 μm, (6–)8–10(–13) times as long as wide. Perichaetial lvs sub-erect. Seta strongly papillose; capsule ± horizontal, ovoid; lid with long subulate beak; spores 13–16 μm. Fr frequent, winter. Small, dense, green mats on soft rocks and stones, especially near water, very rarely on trees, rare or occasional. Cornwall and Kent north to Northants and Warwick, N. E. Yorks, Westmorland, S. Kerry, Leitrim, Fermanagh, Antrim.

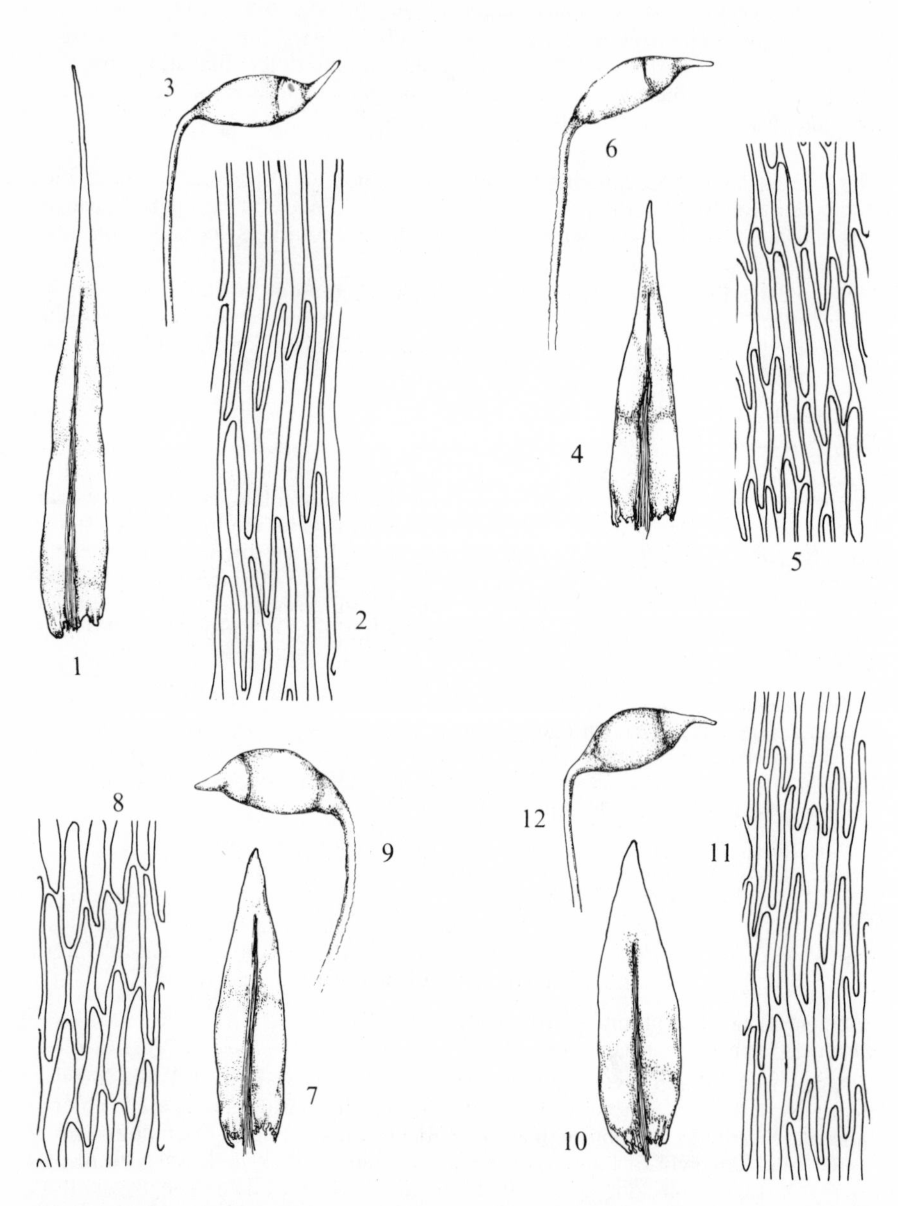

Fig. **299**. 1–3, *Rhynchostegiella tenella* var. *tenella*: 1, leaf; 2, mid-leaf cells; 3, capsule. 4–6, *R. tenella* var. *litorea*: 4, leaf; 5, mid-leaf cells; 6, capsule. 7–9, *R. teesdalei*: 7, leaf; 8, mid-leaf cells; 9, capsule. 10–12, *R. curviseta*: 10, leaf; 11, mid-leaf cells; 12, capsule. Leaves ×40, cells ×415, capsules ×10.

29, H4, C. Norway, France, Bulgaria, Germany, Bohemia, Switzerland, Italy, Madeira, Canaries, Morocco, N. America.

Although usually distinct there occur in S.W. England plants intermediate between *R. curviseta* and *R. teesdalei*. It is not clear whether such plants are intermediates or individuals of hybrid origin, but it is possible that the 2 species might be better treated as 2 subspecies of a single species.

3. R. teesdalei (Br. Eur.) Limpr., Laubm., 1896
Eurhynchium teesdalei (Br. Eur.) Milde

Autoecious. Plants slender; stems creeping, irregularly branched, branches ± procumbent. Lvs patent to spreading, 3–5 times as long as wide, lanceolate to oblong-lanceolate, apex sub-obtuse or occasionally acute; nerve extending $\frac{3}{4}$ way up lf or beyond, margin denticulate above; mid-lf cells 5–8 μm × 32–48 μm, (4–)5–7(–8) times as long as wide. Perichaetial lvs spreading. Seta strongly papillose; capsule ± horizontal, ovoid; lid with long or short subulate beak; spores 13–16 μm. Dense, dull or dark green mats on soft rocks or stones, usually near water, rare in Scotland, occasional elsewhere. 66, H15, C. Sweden, Belgium, France, Bohemia, Giglio Is., Iberian peninsula, Caucasus, Algeria, Madeira, Canaries.

46. ENTODONTACEAE

Slender or robust plants, stems creeping to ascending, irregularly to pinnately branched, branches long or short. Stem and branch lvs ± similar, usually appressed or imbricate, at least when dry, ovate-oblong to lanceolate, obtuse to longly acuminate, margin plane or recurved below, entire or denticulate above; nerve thin, single, double or absent; cells narrowly ellipsoid to linear-vermicular, differentiated at basal angles or not. Seta long, smooth; capsule erect, ellipsoid to cylindrical, straight or curved; annulus present or absent; lid conical, usually rostrate; outer peristome teeth papillose, striate or smooth, inner peristome absent or present with short or no basal membrane, cilia rudimentary or absent; calyptra cucullate, naked.

A heterogeneous family of very uncertain limits. Brotherus (1924–5) includes *Pseudoscleropodium* and *Pleurozium* here but on morphological and cytological grounds these are more satisfactorily placed in the Brachytheciaceae and Hypnaceae respectively.

157. Orthothecium Br. Eur., 1851

Dioecious. Plants slender to robust, primary stems creeping, secondary stems and branches prostrate to ascending. Lvs erect to erecto-patent or secund, lanceolate or lanceolate-triangular with long acuminate apex, margin plane or recurved, ± entire; nerve lacking or very poorly developed; cells ± uniformly linear-vermicular, shorter at extreme base, angular cells not differentiated. Seta reddish, smooth; capsule erect or slightly inclined, ellipsoid to cylindrical, straight or slightly curved; annulus separating; basal membrane of inner peristome long or short, cilia present or absent; calyptra cucullate. A small genus of 11 species occurring in Europe, Asia, N. Africa and N. America.

Plants robust, lvs 2–4 mm, plicate **1. O. rufescens**
Plants slender, lvs 0.5–2.0 mm, not plicate **2. O. intricatum**

1. O. rufescens Br. Eur., 1851

Plants robust, stems and branches procumbent to ascending. Lvs ± erect when dry, erecto-patent, sometimes slightly secund, when moist, strongly plicate, 2–4 mm long, narrowly lanceolate-triangular, tapering ± from base to long acuminate apex,

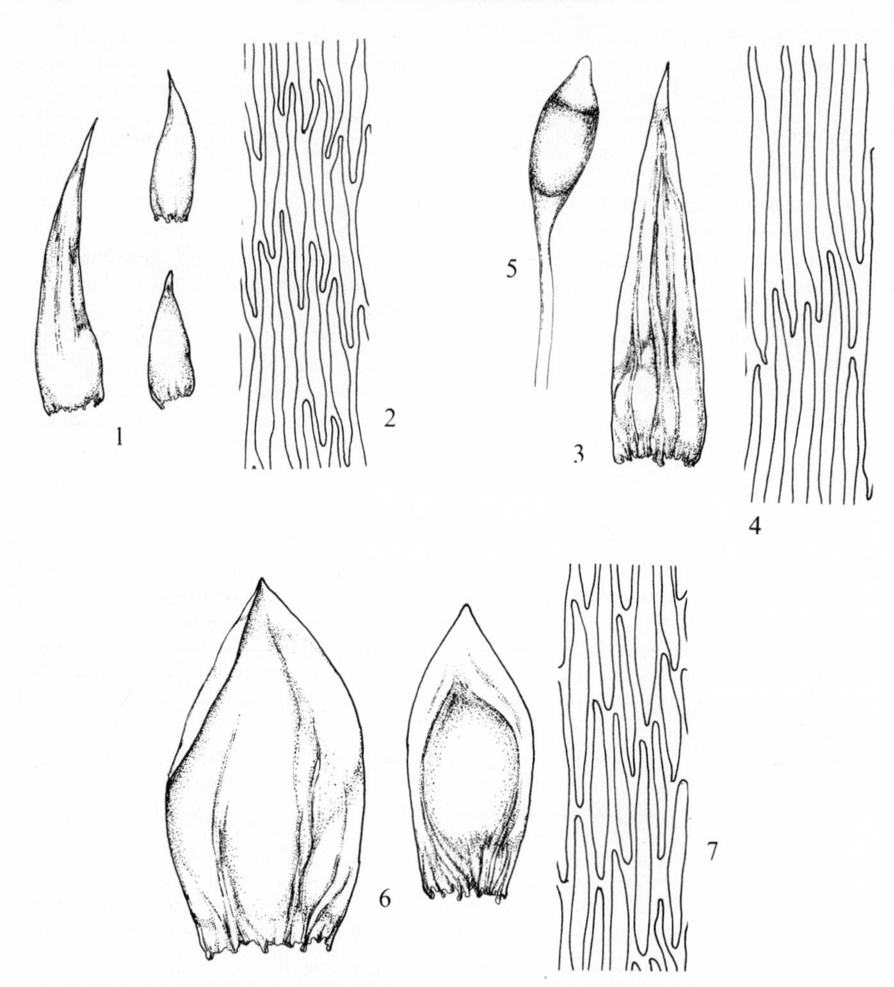

Fig. **300**. 1–2, *Orthothecium intricatum*: 1, leaves; 2, mid-leaf cells. 3–5, *O. rufescens*: 3, leaf; 4, mid-leaf cells; 5, capsule. 6–7, *Entodon concinnus*: 6, leaves; 7, mid-leaf cells. Leaves × 25, cells × 415.

margin recurved, entire; ± nerveless; cells at extreme base rhomboidal, strongly incrassate, porose, cells above linear, ± vermicular, 5–8 μm × 80–128 μm, (10–)12–20 times as long as wide. Seta long, pale red; capsule erect or very slightly inclined, ellipsoid to sub-cylindrical; lid with short, obtuse beak; spores *ca* 12 μm. Fr rare, summer. Dense, glossy, greenish-red to vinous-red patches, sometimes extensive, on damp or flushed, usually ± vertical, sheltered basic rock faces, rare or occasional in montane areas. Caerns, Yorks, Westmorland, Scottish highlands, Clare, Leitrim, Sligo, W. Donegal. 20, H4. N., W. and C. Europe, Iceland, Siberia, Japan, Canada, Greenland.

2. O. intricatum (Hartm.) Br. Eur., 1851

Plants slender, stems and branches procumbent to ascending. Lvs appressed when dry, secund when moist, 0.5–2.0 mm long, concave, not plicate, lanceolate or ovate-lanceolate with longly acuminate apex or, in stunted forms ovate, acute, margin plane, entire; ± nerveless; cells narrowly rhomboidal to linear-vermicular, long in well grown plants, short in stunted forms, in mid-lf 4–8 μm × 32–88 μm, (4–)6–18 times as long as wide. Capsule ovate-ellipsoid. Fr very rare, summer. Yellowish-green to reddish, glossy, small or extensive patches on dry or damp, usually vertical sheltered rock faces and in crevices, occasional in W. and N. Britain and Ireland. 61, H21. N., W. and C. Europe, Faroes, Caucasus, N. Asia, Punjab, N. Africa, N. America, Greenland.

Juvenile lvs are shorter, wider, less sharply pointed than mature lvs and have markedly shorter cells. Poorly developed plants may have lvs intermediate in form and the plant referred to as var. *abbreviatum* Dix. appears to be a very stunted state with lvs of juvenile type.

158. Entodon C. Müll., Linnaea, 1845

Autoecious, rarely dioecious. Plants slender to robust, forming mats, patches or tufts, stems creeping to ascending, irregularly to pinnately branched, branches short. Lvs imbricate, sometimes ± complanate, concave, ovate-oblong, margin plane or recurved below, entire or denticulate above; nerve very short and double or absent; cells linear-vermicular, porose, shorter at base, at basal angles ± quadrate, hyaline or not, forming distinct group. Seta yellow or red; capsule erect, cylindrical, straight or curved; annulus present or absent; lid with oblique beak; outer peristome teeth papillose, striate or smooth, inner peristome without basal membrane, cilia absent or very rudimentary. A world-wide genus of some 190 species, concentrated particularly in E. Asia.

1. E. concinnus (De Not.) Paris, Ind. Bryol., 1904
E. orthocarpus (Brid.) Lindb., *Cylindrothecium concinnum* (De Not.) Schimp.

Dioecious. Plants medium-sized or large, shoots to *ca* 10 cm long, procumbent to ascending, stems yellowish to pale brown, irregularly divided, pinnately branched, branches crowded, ± complanate, short, cuspidate. Lvs loosely imbricate, stem lvs broadly ovate or ovate-oblong, apex obtuse or obtuse and bluntly apiculate, often ± cucullate, margin erect or inflexed above, entire; nerve very short and double or absent; cells linear-vermicular, incrassate, shorter and wider at base, at basal angles quadrate, extending short way up margin and forming poorly defined auricles, cells at apex shorter and wider in mid-lf 5–7 μm × 48–82 μm, 7–12 times as long as wide. Branch lvs narrower, ovate-oblong. Capsule erect, cylindrical. Fr unknown in Britain. Yellowish-green or brownish-green patches or scattered shoots on well drained basic soil in turf, amongst rocks and in scree, occasional in chalk and limestone grassland, very rare elsewhere, generally distributed. 66, H19. Europe, Iceland, Faroes, Caucasus, Iran, N. and C. Asia, N. America.

Similar in appearance to *Pleurozium schreberi* which is, however, a calcifuge, has deep red stems and larger, hyaline angular cells. Some forms of *Pseudoscleropodium purum* approach *E. concinnus* but differ in the nerved lvs.

47. PLAGIOTHECIACEAE

Autoecious or dioecious. Plants slender to robust, primary stems usually procumbent, branches procumbent to erect; pseudoparaphyllia present or not. Lvs strongly complanate to imbricate, when dry slightly to very shrunken, symmetrical or slightly to strongly asymmetrical, ovate to lanceolate, shortly to longly tapering to acute to acuminate apex, margin entire or denticulate; nerve double, rarely absent; angular cells often enlarged, sometimes decurrent and forming distinct auricles or not, cells elsewhere linear-vermicular, rarely shorter and wider. Vegetative propagation by axillary or lf-tip gemmae of various types frequent. Seta long, straight, usually reddish; capsule erect or inclined, ellipsoid to cylindrical, straight or curved; annulus deciduous or persistent; lid conical to rostrate; peristome double, teeth of outer yellow, striate below, papillose above, inner with or without basal membrane and cilia; calyptra cucullate.

I have followed other European authors in treating this as a separate family but it is poorly defined and Ireland (1969) treats it, with good reasons, as a subfamily of the Hypnaceae. Recent monographs of the Plagiotheciaceae in Japan and Denmark have been published by Iwatsuki (*J. Hattori bot. Lab.* **33**, 331–80, 1970) and Lewinsky (*Lindbergia* **2**, 185–217, 1974) respectively. Studies on the family in Belgium have been carried out by Lefebvre, whose most recent paper is in *Bull. Soc. roy. bot. Belg.* **103**, 63–70, 1970. Ireland (*Nat. Mus. Canad. Nat. Sci. Publs. Bot.* **1**, 1–118, 1969) has monographed North American representatives of the genus *Plagiothecium* as recognised by Grout; this includes species of *Plagiothecium*, *Herzogiella*, *Isopterygium* and *Taxiphyllum*.

159. PLAGIOTHECIUM BR. EUR., 1851

Plants slender to robust, stems simple or branched, branches procumbent to erect; epidermal cells large, thin-walled; pseudoparaphyllia lacking. Stem and branch lvs similar, imbricate to strongly complanate, apices often directed downwards, concave or not, symmetrical or asymmetrical, ovate to lanceolate, apex obtuse to acuminate, margin plane or recurved, entire or denticulate near apex; nerve double or sometimes ± lacking; basal cells rhomboidal, angular cells decurrent, cells elsewhere linear-rhomboidal to linear-vermicular, rarely wider. Fusiform axillary gemmae often present. Capsule erect or inclined, ellipsoid to cylindrical, straight or curved; peristome perfect. Plants often forming glossy patches. A world-wide genus of about 110 species.

All British species come within the subgenus *Plagiothecium* except *P. latebricola* which is placed in the subgenus *Plagiotheciella* Broth. For an account of the species of the *P. denticulatum–nemorale* (*P. denticulatum–sylvaticum*) group see Greene, *Trans. Br. bryol. Soc.* **3**, 181–90, 1957.

1 Lvs whitish-green, strongly undulate, 2–5 mm long **11. P. undulatum**
 Lvs yellowish-green to dark green but not whitish, not or only slightly undulate, rarely more than 3.2 mm long 2
2 Lf apex abruptly narrowed to long filiform acumen **2. P. piliferum**
 Lvs gradually tapering to obtuse to acuminate apex 3
3 Most mid-lf cells less than 10 μm wide 4
 Most mid-lf cells more than 10 μm wide 6
4 Lvs symmetrical **1. P. latebricola**
 Most lvs strongly asymmetrical with one side ± flat 5
5 Lf apices curved downwards, capsule inclined to horizontal, curved **5. P. curvifolium**
 Lf apices not curved downwards, capsule ± erect and straight **6. P. laetum**
6 Lvs mostly asymmetrical or symmetrical and obtuse, decurrent angular cells* rounded to rounded rectangular, forming distinct auricles 7

* The decurrent angular cells usually remain on the stem when the lvs are removed and are best viewed *in situ* on a portion of stem from which the lvs have been stripped

Lvs mostly symmetrical, acute to acuminate, decurrent angular cells rectangular, not rounded, not or hardly forming distinct auricles 8

7 Most lvs with both sides curved, apex often denticulate **3. P. denticulatum**
Most lvs with one side ± flat, apex usually entire **4. P. ruthei**

8 Branches julaceous with imbricate concave lvs **8. P. cavifolium**
Branches not julaceous, lvs complanate, not or only slightly concave 9

9 Lvs sharply denticulate near apex, with group of thin-walled cells often eroded away near apex **7. P. platyphyllum**
Lvs entire or denticulate near apex, apical cells not differentiated 10

10 Lvs tapering to narrow, acuminate, entire apex, cells 10–22 μm wide, 100–200 μm long, not in transverse rows **9. P. succulentum**
Lvs shortly tapered to entire or denticulate, acute apex, cells 16–22 μm × 80–120 μm in ± transverse rows **10. P. nemorale**

1. P. latebricola Br. Eur., 1851
Plagiotheciella latebricola (Br. Eur.) Fleisch. in Broth.

Dioecious. Plants slender. Lvs complanate, when dry somewhat shrunken, symmetrical, ovate-lanceolate, tapering to long acuminate apex, margin plane, entire or faintly denticulate near apex; nerve extending up to $\frac{1}{3}$ way up lf; angular cells enlarged, decurrent in single row of elongated cells, cells elsewhere linear, in mid-lf 6–8 μm × 80–150 μm, 11–17 times as long as wide. Fusiform gemmae, 30–100 μm long, often present in lf axils and on lf tips. Capsule erect, narrowly ellipsoid, smooth when dry; lid conical-rostellate; spores *ca* 12 μm. Fr rare, spring. Glossy, pale or yellowish-green, flattish patches on decaying logs, fern bases or rarely on soil in damp shaded habitats. Occasional in S. England, rare elsewhere, extending from Cornwall and Kent north to Stirling and W. Perth, W. Ross, Leitrim. 52, H1. Europe, eastern N. America.

2. P. piliferum (Sw. ex Hartm.) Br. Eur., 1851
Isopterygium piliferum (Sw. ex Hartm.) Loeske, *Plagiotheciella pilifera* (Sw. ex Hartm.) Fleisch. in Broth.

Autoecious. Plants small. Lvs complanate, when dry somewhat shrunken, very concave, ovate, abruptly narrowed to long filiform acumen, margin entire; nerve extending up to $\frac{3}{4}$ way up lf, angular cells enlarged, hyaline, decurrent, cells elsewhere linear, in mid-lf 4–8 μm × 48–100 μm, 10–15 times as long as wide. Gemmae lacking. Capsule erect, cylindrical, straight; lid conical; spores 13–15 μm. Fr common, summer. $n=11$. Glossy, light green patches on shaded, acidic rock ledges at high altitudes, very rare. Mid Perth, Angus. 2. N. and C. Europe, Pyrenees, Corsica, Sardinia, N. Asia, western N. America.

3. P. denticulatum (Hedw.) Br. Eur., 1851

Autoecious. Plants medium-sized, branches prostrate. Lvs strongly complanate or rarely julaceous, often curving downwards when moist, when dry slightly shrunken, 1.4–2.4 mm long, slightly to strongly concave, mostly asymmetrical with both sides rounded, ovate, shortly tapering to acute to obtuse apex, margin entire or slightly denticulate near apex; nerve extending up to $\frac{1}{3}$–$\frac{1}{2}$ way up lf, rarely wanting; angular cells inflated, decurrent, forming distinct auricles with cells rounded to rounded-rectangular, rarely tapering with rectangular cells, cells elsewhere linear-vermicular, in mid-lf 10–16 μm × 96–140 μm, 8–12 times as long as wide. Axillary fusiform gemmae, 70–180 μm long, sometimes present. Capsule inclined, shortly cylindrical,

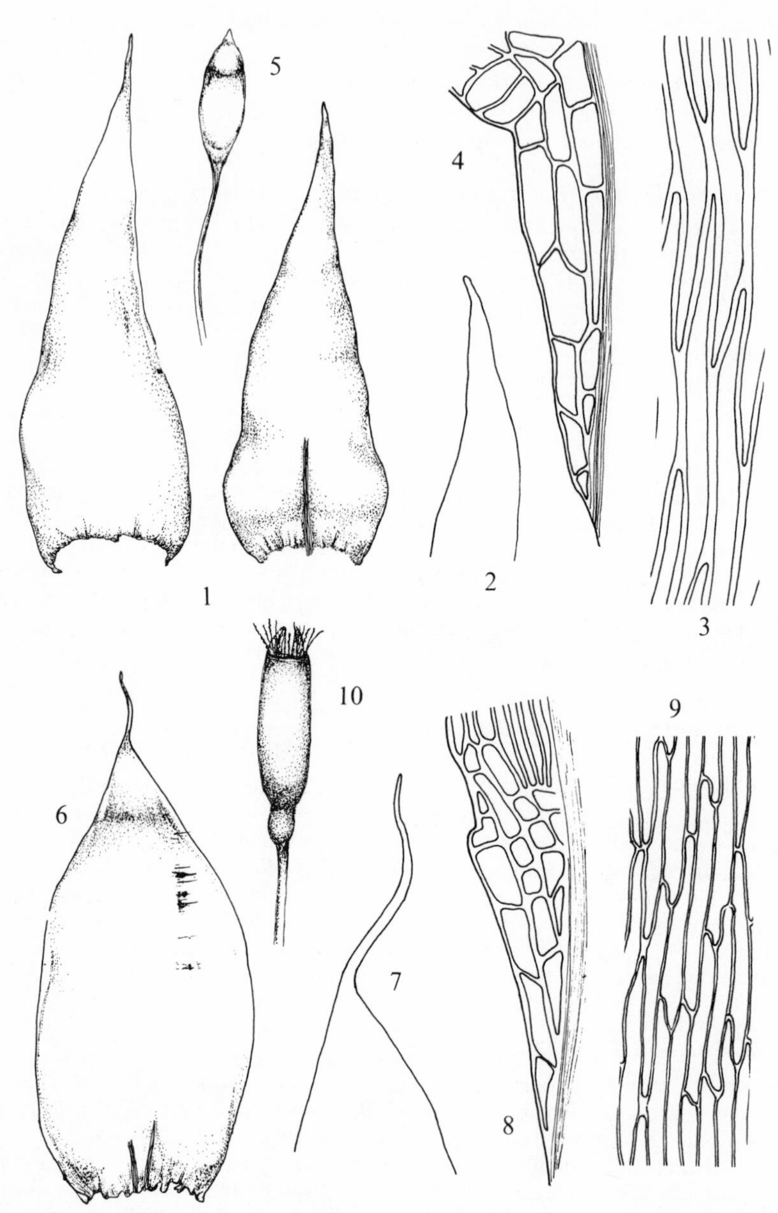

Fig. **301**. 1–5, *Plagiothecium latebricola*: 1, leaves; 2, leaf apex; 3, mid-leaf cells (×415); 4, decurrent angular cells (×250); 5, capsule. 6–10, *P. piliferum*: 6, leaf; 7, leaf apex; 8, decurrent angular cells (×250); 9, mid-leaf cells (×415); 10, capsule. Leaves ×40, apices ×100, capsules ×10.

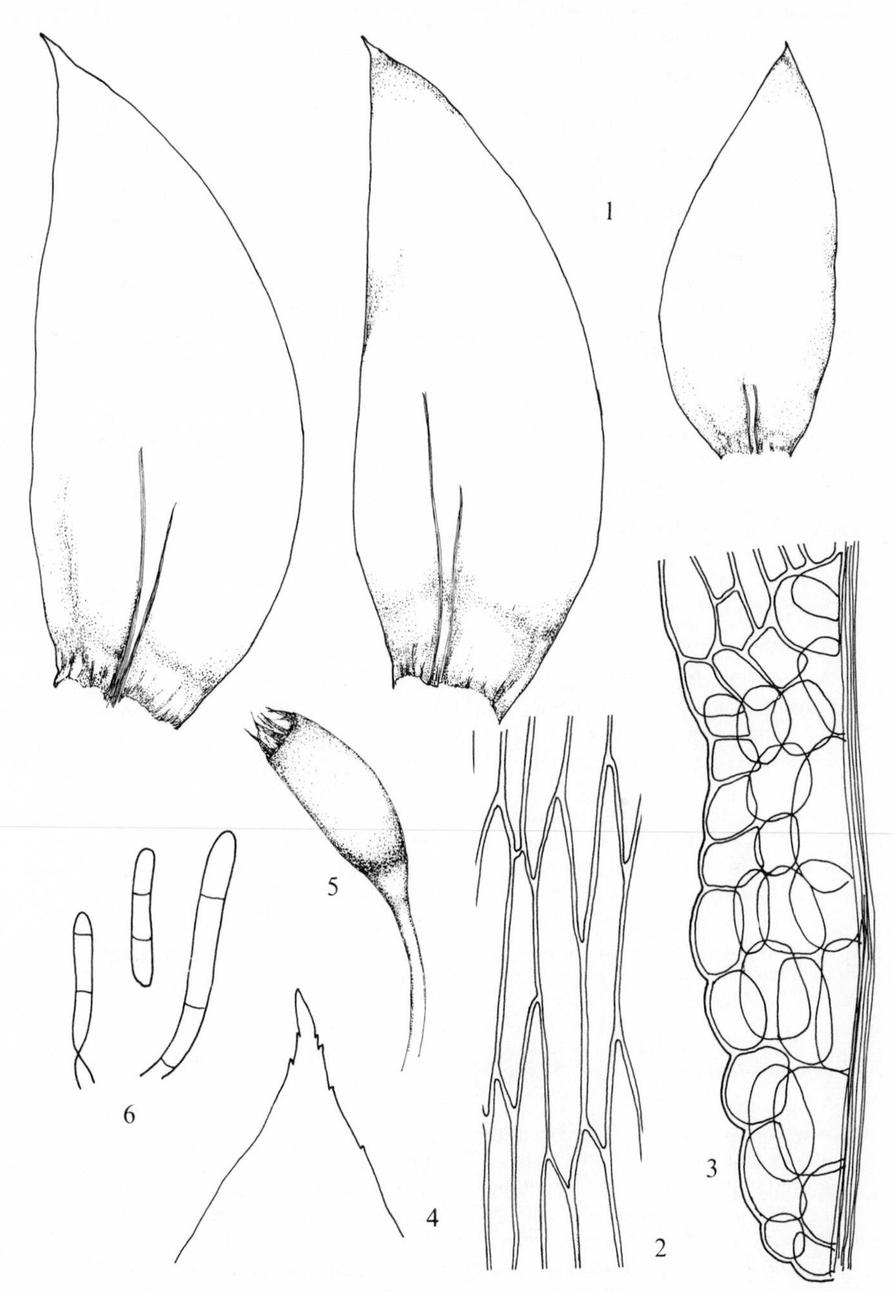

Fig. 302. *Plagiothecium denticulatum* var. *denticulatum*: 1, leaves (×40); 2, mid-leaf cells (×415); 3, decurrent angular cells (×250); 4, leaf apex (×100); 5, capsule (×10); 6, gemmae (×250).

curved, longitudinally furrowed when dry; lid conical-rostellate; spores 8–12 μm. Fr common, late spring, summer.

Var. **denticulatum**

Lvs strongly complanate, lf apex acute. $n=10$, 11*, 20, 25. Glossy, pale to dark green patches on soil, on banks, decaying logs and rocks in woods, on marshy ground at low altitudes and among boulders and on cliff ledges in montane habitats, common. 102, H13, C. Europe, Faroes, Caucasus, N., C. and E. Asia, Macaronesia, N. America, Greenland, Peru, Australia.

Var. **obtusifolium** (Turn.) Moore, Proc. Roy. Irish Acad. Sci., 1873

Lvs strongly complanate and asymmetrical or imbricate, symmetrical and strongly concave, lf apex obtuse. Glossy, pale to dark green patches among boulders or in rock crevices in montane habitats, usually above 600 m, occasional. E. Cornwall, Wales, Yorks and Westmorland northwards. 25, H5. N. and C. Europe, Sikkim, N. America, southern S. America, Tasmania, New Zealand.

Distinct from other *Plagiothecium* species except *P. ruthei* in the decurrent, round to rounded-rectangular angular cells forming distinct auricles. Usually differs from *P. ruthei* in that both sides of the lf are rounded whereas in *P. ruthei* in many lvs one side is ± flat. It is, however, uncertain whether *P. ruthei* is a distinct species or merely a habit form of *P. denticulatum*. Ireland (1969) regards them as synonymous whereas Lewinsky (op. cit.), says there are no intermediates. Var. *obtusifolium* requires further study as there are 2 forms, one with complanate asymmetrical lvs, the second with imbricate, symmetrical, concave lvs. This latter may be mistaken for *P. cavifolium* but has obtuse lvs and the angular cells characteristic of *P. denticulatum*.

4. P. ruthei Limpr., Laubm. Deutschl., 1897

Autoecious. Plants medium-sized, branches prostrate. Lvs strongly complanate, spreading, sometimes transversely undulate when moist, slightly shrunken when dry, 2.0–2.6 mm long, mostly strongly asymmetrical with one side ± straight, ovate to ovate-lanceolate, tapering to acute to acuminate apex, margin usually entire; nerve extending up to $\frac{1}{3}$ way up lf; angular cells inflated, decurrent, forming distinct auricles with rounded to rounded-rectangular cells, cells elsewhere linear-rhomboidal, in mid-lf 10–17 μm × 80–176 μm, 7–11(–14) times as long as wide. Fusiform axillary gemmae sometimes present. Capsule inclined, narrowly ellipsoid, curved, longitudinally furrowed when dry; lid rostellate; spores *ca* 10 μm. Fr occasional, spring. $n=11$. Glossy, pale green patches on damp soil, decaying vegetation, rocks and plant bases in reed swamps, marshes, carr and by streams. Occasional in England and Wales, rare in Scotland, not recorded from Ireland. 40. Fennoscandia, C. Europe, Japan, N. America.

5. P. curvifolium Schlieph. in Limpr., Laubm. Deutsch., 1897
P. denticulatum var. *aptychus* (Spruce) Lees

Autoecious. Plants small, branches prostrate. Lvs strongly complanate, curving downwards when moist, scarcely altered when dry, 1.0–2.6 mm long, mostly strongly asymmetrical with one side ± straight, lanceolate, shortly tapering to acute to acuminate apex, margin entire or slightly denticulate near apex; nerve extending up to $\frac{1}{4}$–$\frac{1}{2}$ way up lf; angular cells enlarged, decurrent in 1–4 rows of narrowly rectangular cells, not forming auricles, cells elsewhere linear-vermicular, in mid-lf 6–8(–10) μm × 80–140 μm, 12–19 times as long as wide. Fusiform axillary gemmae sometimes present. Capsule inclined, narrowly ellipsoid to cylindrical, curved, longitudinally furrowed when dry; lid conical; spores *ca* 12 μm. Fr occasional, late spring, early summer. $n=7$, 11*. Glossy, pale green patches on tree stumps, logs and litter in

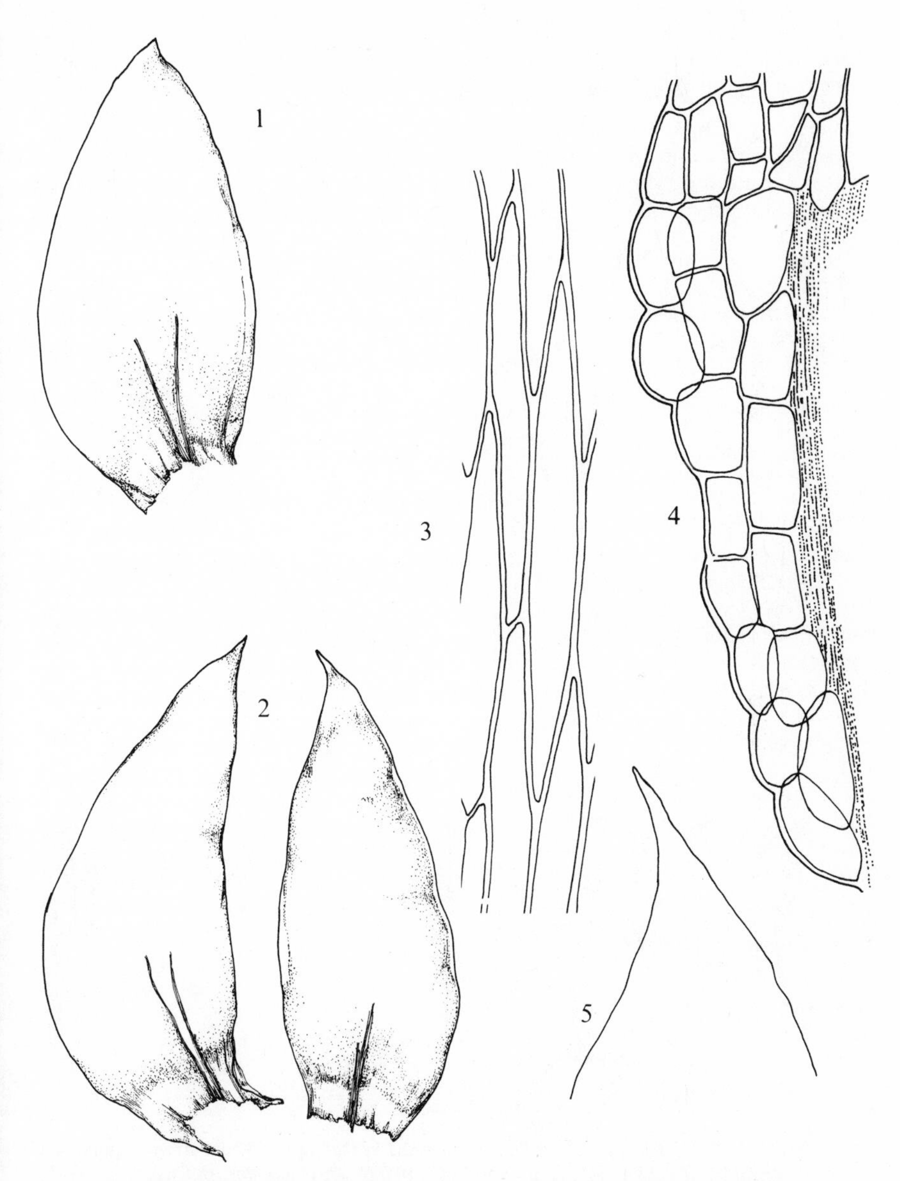

Fig. **303**, 1, *Plagiothecium denticulatum* var. *obtusifolium*: leaf. 2–5, *P. ruthei*: 2, leaves; 3, mid-leaf cells (×415); 4, decurrent angular cells (×250); 5, leaf apex (×100). Leaves ×25.

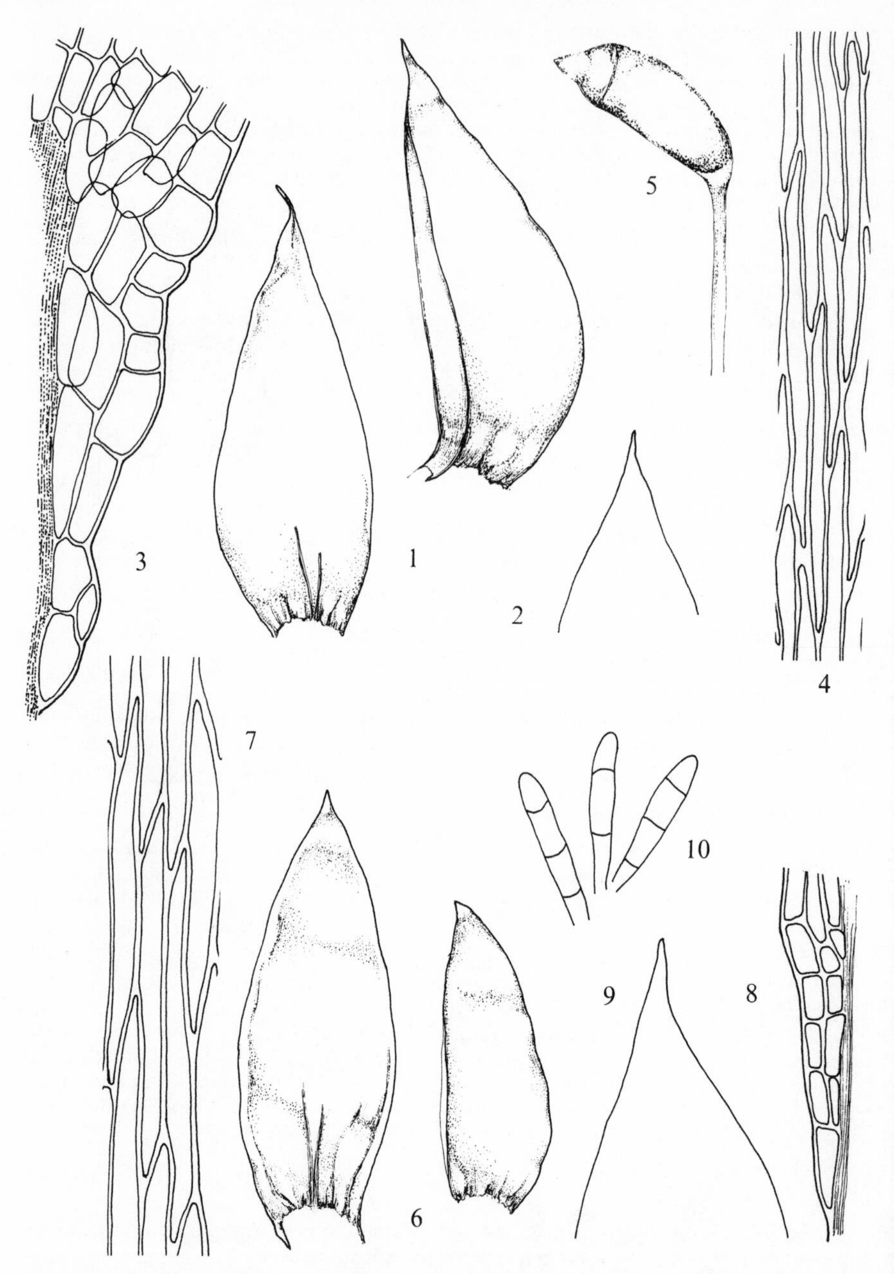

Fig. **304**. 1–5, *Plagiothecium curvifolium*: 1, leaves; 2, leaf apex; 3, decurrent angular cells (×250); 4, mid-leaf cells (×415); 5, capsule (×10). 6–10, *P. laetum*; 6, leaves; 7, mid-leaf cells (×415); 8, decurrent angular cells (×250); 9, leaf apex; 10, gemmae (×250). Leaves ×25, apices ×100.

woodland, generally distributed in Britain, occasional except in the north where it is rare, very rare in Ireland, Leitrim, Fermanagh. 75, H2. Europe, N. America, Japan.

P. curvifolium and *P. laetum* are distinct from all other *Plagiothecium* species in the narrow lf cells and from all except *P. ruthei* in the presence of numerous flat-sided asymmetrical lvs. Ireland (1969) regards *P. curvifolium* and *P. laetum* as synonymous and the 2 taxa are certainly not easy to distinguish. In *P. curvifolium* the lvs are usually larger and the tips curved down when moist. The form of the capsule, inclined and curved in *P. curvifolium* and ± erect and straight in *P. laetum*, is often quoted as a good distinguishing character, but even from the limited number of specimens I have examined this does not seem very reliable.

6. P. laetum Br. Eur., 1851

Autoecious. Plants small, branches prostrate. Lvs complanate, spreading, apices usually not curved downwards when moist, when dry sometimes undulate, 1.0–2.0 mm long, asymmetrical, many with one side ± flat, ovate-lanceolate, acute, margin entire or obscurely denticulate near apex; nerve extending up to $\frac{1}{3}$ way up lf; angular cells enlarged, decurrent in 1–3 rows of rectangular cells, not forming auricles, cells elsewhere linear-vermicular, in mid-lf 6–8(–10) μm×76–144 μm, 11–17 times as long as wide. Fusiform gemmae sometimes present in lf axils and on lf tips. Capsule ± erect, narrowly ellipsoid, ± straight; lid conical-rostellate; spores *ca* 10 μm. Fr frequent. $n=8, 10, 11$. Glossy, pale green patches on tree stumps, logs and soil in woodland and, in montane habitats, on soil among boulders, rare to occasional. Merioneth, W. England, Scotland, Dublin. 22, H1. N. and C. Europe, Spain, Caucasus, Siberia, N. America. Greenland.

7. P. platyphyllum Mönk., Laubm. Eur., 1927
P. denticulatum var. *majus* of Dix.

Autoecious. Plants large, branches prostrate to erect. Lvs complanate to sub-secund, spreading, when dry shrunken, 2.0–2.5 mm long, ovate to ovate-lanceolate, tapering from about halfway to acute apex, margin sharply denticulate near apex; nerve extending up to $\frac{1}{2}$ way up lf; angular cells enlarged, decurrent in 2–4 rows rectangular cells, ± forming auricles, cells elsewhere linear-rhomboidal, patch near apex very thin-walled and often eroded away, in mid-lf 10–15 μm×80–160 μm, 7–14 times as long as wide. Gemmae lacking. Capsule inclined, narrowly ellipsoid, curved, longitudinally furrowed when dry; lid conical-rostellate; spores *ca* 12 μm. Fr frequent, summer. Glossy or dull, pale to dark green patches in flushes, rock crevices by waterfalls and by streams in montane habitats, rare. Caerns and scattered localities from Westmorland and Cumberland north to Sutherland, Dublin. 15, H1. Europe, Turkey.

Differs from *P. nemorale* and *P. succulentum* in the narrower lf cells and lf shape. Sometimes resembles *P. denticulatum* in the angular cells forming auricles but the cells of these are not rounded. *P. platyphyllum* is, however, most distinct in the patch of cells near the lf apex with very thin walls. These cells are often eroded away so that there is a hole at the apex. The margin is also sharply denticulate above.

8. P. cavifolium (Brid.) Iwats., J. Hattori Bot. Lab., 1970
P. roeseanum Br. Eur.

Dioecious. Plants medium-sized, branches ascending to erect. Lvs imbricate to sub-complanate, when dry scarcely shrunken, not undulate, 1.2–2.2 mm long, very concave, ± symmetrical, ovate to broadly ovate, shortly tapering to acute, often reflexed apex, margin usually entire; nerve extending up to $\frac{1}{2}$–$\frac{2}{3}$ way up lf; angular cells enlarged, narrowly decurrent in 1–3 rows rectangular cells but not forming auricles, cells elsewhere linear-rhomboidal, in mid-lf 10–16 μm×76–144 μm, 6–13

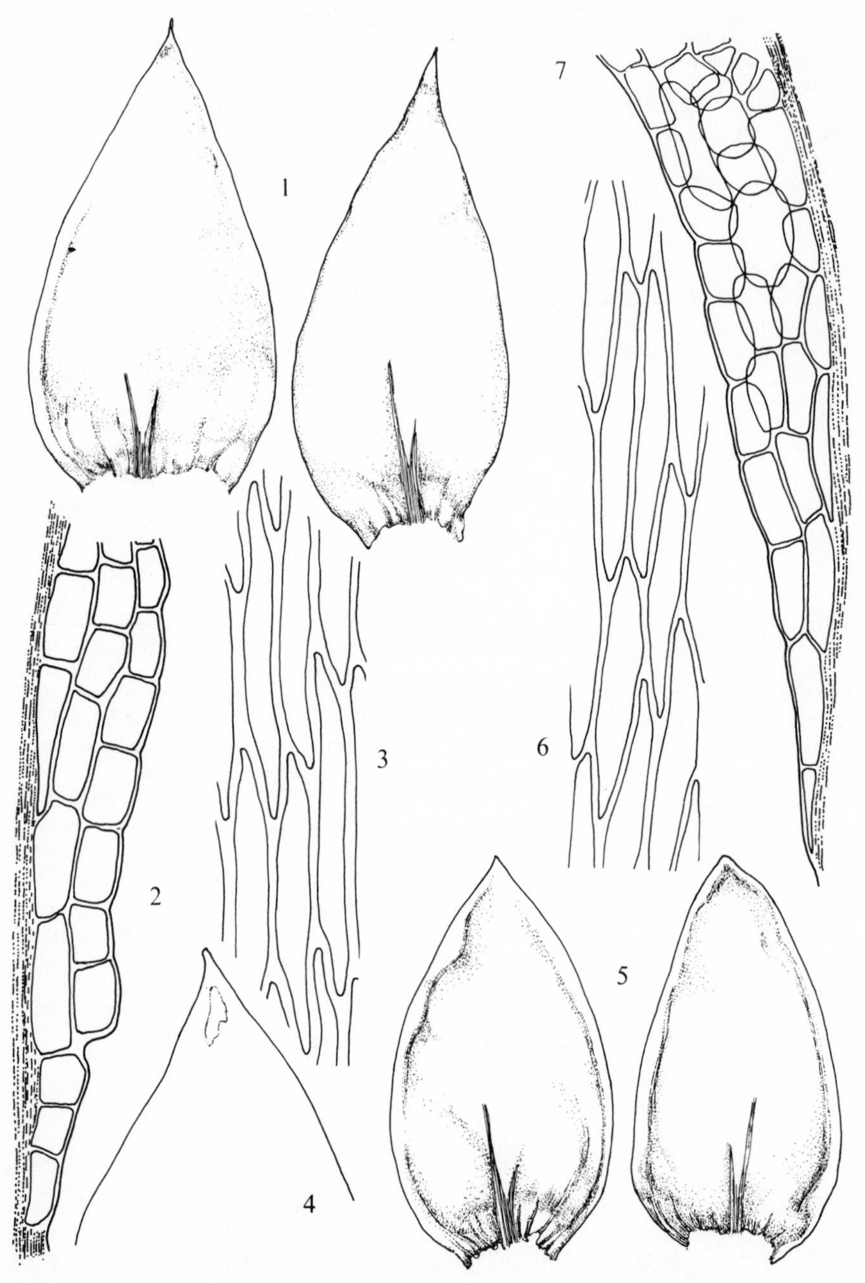

Fig. **305**. 1–4, *Plagiothecium platyphyllum*: 1, leaves; 2, decurrent angular cells (×415); 3, mid-leaf cells (×415); 4, leaf apex (×100). 5–7, *P. cavifolium*: 5, leaves; 6, mid-leaf cells (×415); 7, decurrent angular cells (×250). Leaves ×25.

times as long as wide. Fusiform axillary gemmae occasionally present. Capsule erect or inclined, cylindrical, straight or curved; lid rostrate; spores *ca* 12 μm. Fr rare. Glossy, yellowish-green patches on moist, basic rock ledges above 600 m in montane habitats, rare. Pembroke, Cards, Caerns, W. Lancs and N.E. Yorks northwards, S. Kerry, E. Donegal 24, H2. N. and C. Europe, Pyrenees, Faroes, Iceland, N. Asia, Japan, N. America, Greenland, Falkland Is.

9. P. succulentum (Wils.) Lindb., Bot. Not., 1865

Dioecious or autoecious. Plants medium-sized, branches prostrate to ascending. Lvs ± complanate, spreading, when dry moderately to strongly shrunken, 2.0–3.2 mm long, ± symmetrical, ovate to ovate-lanceolate, tapering to narrow acuminate apex, margin usually entire; nerve extending up to $\frac{1}{2}$ way up lf; angular cells enlarged, decurrent in band of 1–3 rows rectangular cells but not forming auricles, cells elsewhere linear-rhomboidal, overlapping, not in transverse rows, in mid-lf

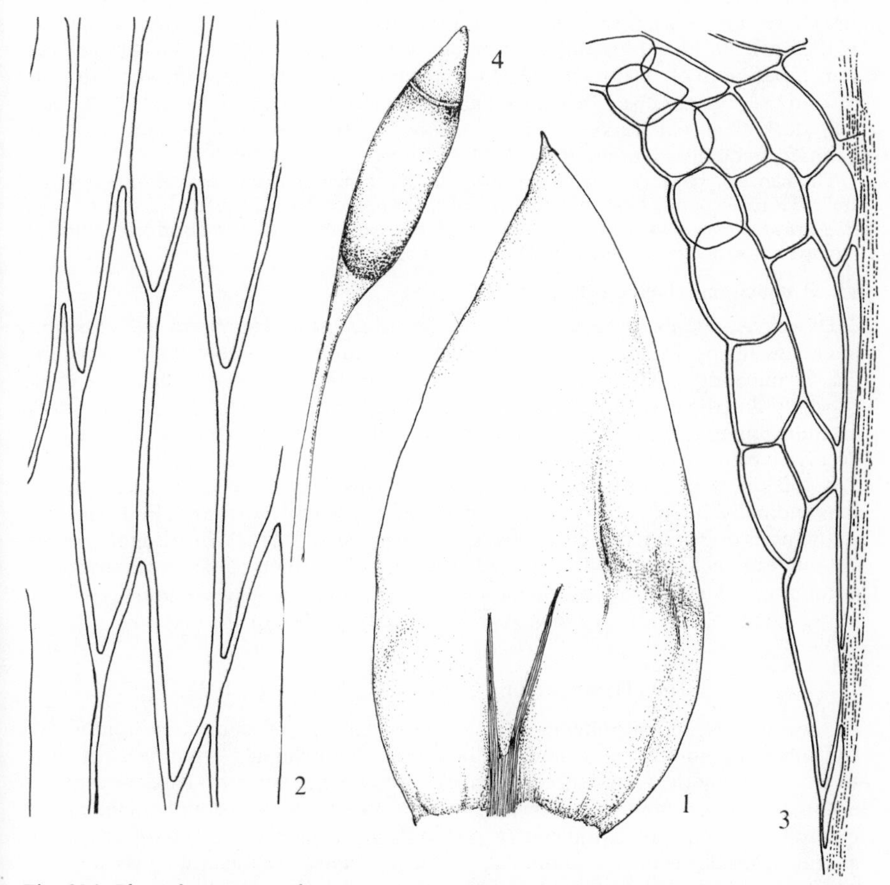

Fig. **306**. *Plagiothecium succulentum*: 1, leaf (×25); 2, mid-leaf cells (×415); 3, decurrent angular cells (×250); 4, capsule (×10).

10–22 μm×(80–)100–208 μm, 6–10 times as long as wide. Fusiform axillary gemmae sometimes present. Capsule inclined, shortly cylindrical, curved, smooth when dry; lid conical; spores 12–14 μm. Fr occasional, summer. Very glossy, usually golden-green patches on soil banks in woods, hedgerows, quarries, by streams, etc., in lowland habitats and amongst boulders in the mountains, frequent or common. 103, H27, C. Europe, Faroes, Madeira.

P. succulentum and *P. nemorale* cannot be distinguished in the field as there are forms of each species that resemble those of the other. In *P. succulentum* the lf apex is, however, longer and under the microscope the longer, overlapping cells are distinctive. In both species the lvs shrink more strongly on drying than is usual in other *Plagiothecium* species.

10. P. nemorale (Mitt.) Jaeg., Ber. S. Gall. Naturw. Ges., 1878
P. neglectum Mönk., *P. sylvaticum* auct. non *Hypnum sylvaticum* Brid.

Dioecious. Plants medium-sized, branches prostrate to ascending. Lvs complanate, spreading, when dry shrunken, (1.4–)1.8–3.0 mm long, ± symmetrical, ovate, acute, margin entire or denticulate near apex; nerve extending up to ½ way up lf; angular cells enlarged, narrowly to broadly decurrent in 1–3 rows rectangular cells, cells elsewhere narrowly hexagonal, in ± transverse rows, scarcely overlapping, in mid-lf 16–22 μm×80–120(–140) μm, 4–6 times as long as wide. Fusiform axillary gemmae sometimes present. Capsule inclined, cylindrical, curved, smooth when dry; lid rostrate; spores *ca* 12 μm. Fr occasional, summer, autumn. $n=8$, 10, 11*, 12. Usually dull, dark green patches on soil in woods, hedgebanks, streamsides, etc. in lowland habitats, frequent or common. 96, H22, C. Europe, E. Asia, Madeira.

The name *P. sylvaticum* by which this plant has previously been known in Britain and Ireland is based upon *Hypnum sylvaticum* Brid., the nomenclatural type of which is, however, *P. denticulatum* (see Iwatsuki, *op. cit.*). I have not cited *P. sylvaticum* (Brid.) Br. Eur. as a synonym of *P. denticulatum* as this will cause needless confusion.

11. P. undulatum (Hedw.) Br. Eur., 1851

Dioecious. Plants robust, stems and branches prostrate, branches sometimes attenuate at tips. Lvs complanate, transversely undulate, scarcely altered when dry, 2.4–4.0 mm long, ± symmetrical to slightly asymmetrical, narrowly decurrent, ovate, shortly pointed, acute to obtuse, margin plane, denticulate near apex; nerve double, extending up to ⅓ way up lf; basal and angular cells enlarged, ± rhomboid-hexagonal, decurrent cells narrowly rectangular, cells elsewhere linear, in mid-lf 7–10 μm×(80–)120–160 μm, (8–)13–18 times as long as wide. Capsule inclined, cylindrical, curved, longitudinally furrowed when dry; lid rostrate; spores 10–15 μm. Fr frequent in wetter parts of Britain, rare elsewhere, spring, summer. $n=11$*. Whitish-green patches on soil and rocks in woods, on shaded hedgebanks, ditch-banks, streamsides, on moorland under heather, etc., on montane slopes, etc., calcifuge, common. 110, H35, C. N., W. and C. Europe, Faroes, Asia Minor, Siberia, N. America.

160. HERZOGIELLA BROTH., NAT. PFL., 1925

Autoecious. Pseudoparaphyllia absent; epidermal cells of stem large, thin-walled. Lvs spreading, sometimes secund, not or scarcely complanate, ± symmetrical, ovate to ovate-oblong or ovate-lanceolate, apex acuminate or filiform, base decurrent or not, margin denticulate in upper half; nerve usually short and double; angular cells quadrate to rectangular, inflated or not, cells above linear. Vegetative propagules absent. Capsule inclined, cylindrical or ovoid, curved; annulus not persistent; lid conical; peristome perfect, inner ciliate; calyptra cucullate. A small genus of 5 or 6 species occurring in Europe, Asia, N. Africa, America.

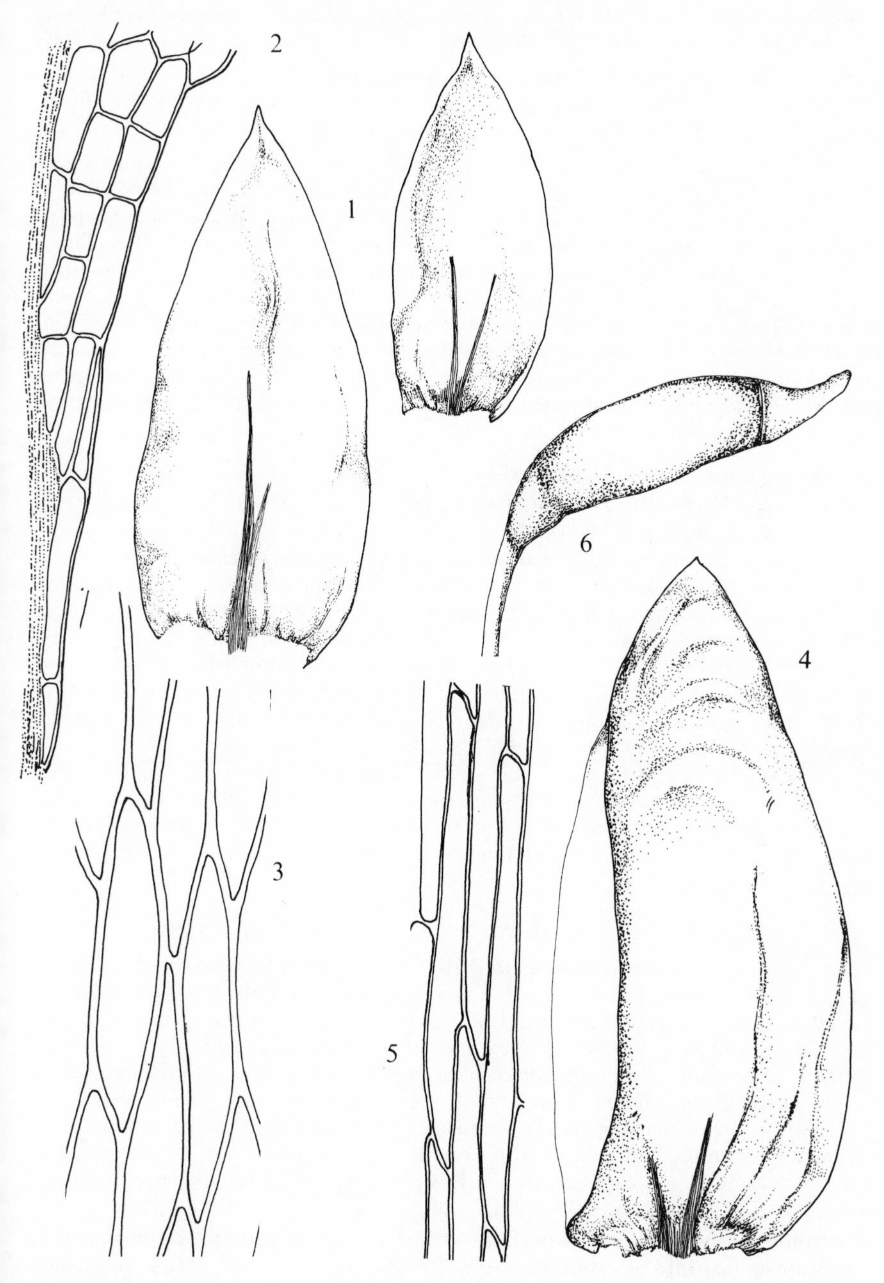

Fig. **307**. 1–3, *Plagiothecium nemorale*: 1, leaves; 2, decurrent angular cells (×25); 3, mid-leaf cells (×415). 4–6, *P. undulatum*: 4, leaf; 5, mid-leaf cells (×415); 6, capsule (×10). Leaves ×25.

Angular cells inflated, decurrent **1. H. striatella**
Angular cells not inflated or decurrent **2. H. seligeri**

1. H. striatella (Brid.) Iwats., J. Hattori Bot. Lab., 1970
Dolichotheca striatella (Brid.) Loeske, *Plagiothecium muehlenbeckii* Br. Eur., *P. striatellum* (Brid.) Lindb., *Sharpiella striatella* (Brid.) Iwats.

Autoecious. Plants slender, branches ascending. Lvs spreading, ± complanate, often secund with apices upturned, when dry somewhat shrunken and flexuose, concave, ovate-lanceolate to ovate-triangular, tapering to long fine acumen, margin plane, denticulate ± throughout; nerve short, faint, double; angular cells inflated, hyaline, decurrent, forming distinct auricles, cells elsewhere linear-rhomboidal, in mid-lf 5–7(–8) μm×40–64(–72) μm, 6–11 times as long as wide. Capsule inclined, sub-cylindrical, curved, irregularly furrowed when dry; lid conical; spores 10–12 μm. Fr common, summer. Glossy, green to yellowish-green patches on soil in crevices amongst boulders, about tree bases in montane areas, usually at high altitudes, rare. I. of Man, Stirling and Perth northwards. 17. Northern and montane Europe, N. Asia, N. America, Greenland.

2. H. seligeri (Brid.) Iwats., J. Hattori Bot. Lab., 1970
Dolichotheca seligeri (Brid.) Loeske, *Isopterygium seligeri* (Brid.) Dix., *Plagiothecium silesiacum* auct., *Sharpiella seligeri* (Brid.) Iwats.

Autoecious. Plants small, procumbent, stems irregularly branched; epidermal cells of branches thick-walled, 8–12 μm wide. Lvs spreading, sub-secund, often upturned at branch tips, hardly complanate when moist, less spreading, flexuose when dry, ± symmetrical, lanceolate to ovate-lanceolate or lanceolate-triangular, tapering to long fine apex, margin plane, denticulate from near base to apex; nerve short, double; angular cells rectangular, not inflated or decurrent, green, cells elsewhere linear, in mid-lf 5–9 μm×56–100(–112) μm, (8–)10–16 times as long as wide. Capsule inclined, cylindrical, curved; lid shortly conical; spores 9–12 μm. Fr common, summer. $n=10+1$, 11. Pale or yellowish-green, glossy patches on rotten logs and tree stumps in shaded habitats, especially *Castanea sativa* coppice. Occasional and apparently increasing in parts of southern England, extending west to N. Somerset and north to Norfolk, Yorks. 18. Europe, Caucasus, Kashmir, Hunan, Japan, N. America.

161. Isopterygium Mitt., J. Linn. Soc. Bot., 1869

Autoecious or dioecious. Pseudoparaphyllia lacking in British species, epidermal cells of stem small and thick-walled. Lvs spreading to complanate, tips often up-curved or down-curved, ± symmetrical, not plicate, ovate to lanceolate, gradually tapering to acuminate to filiform apex, base not decurrent, margin plane, entire or toothed; nerve short and double or absent; angular cells quadrate to rectangular or hardly distinct, not inflated or decurrent, green, cells elsewhere linear. Vegetative propagation by axillary gemmae frequent. Seta straight; capsule sub-erect to inclined, ovoid to cylindrical; lid with or without beak; annulus of large separating cells; peristome ± perfect. A large world-wide genus of about 270 species mostly occurring in the tropics and subtropics.

Autoecious; lvs not complanate, tapering ± from base to apex, margin entire, gemmae fusiform or cylindrical, to 60 μm long **1. I. pulchellum**
Dioecious; lvs complanate, widest at about $\frac{1}{4}$–$\frac{1}{3}$ way up lf, margin denticulate above, gemmae much larger, to 1.5 mm, flagelliform with minute lvs **2. I. elegans**

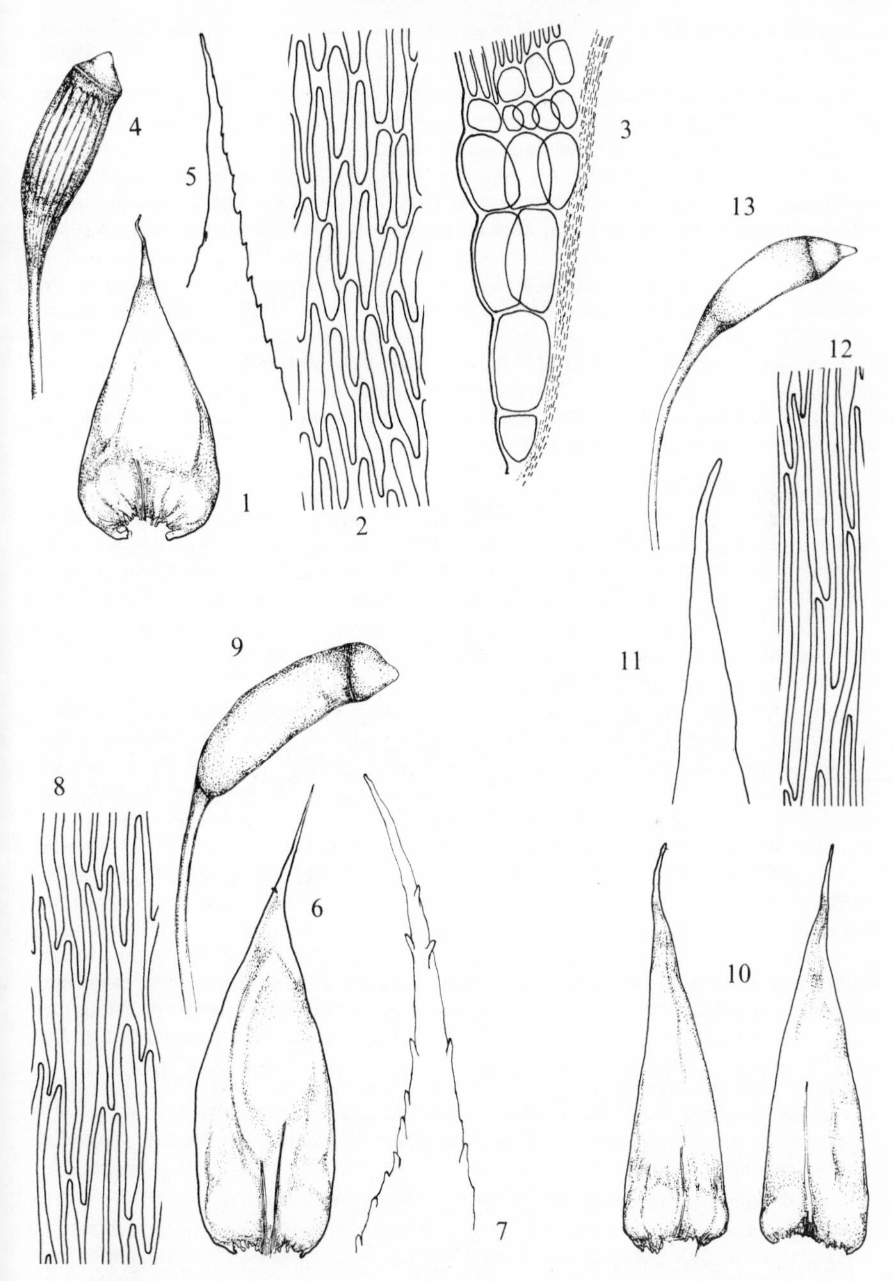

Fig. **308**. 1–5, *Herzogiella striatella*: 1, leaf (×25); 2, mid-leaf cells; 3, decurrent angular cells (×250); 4, capsule; 5, leaf apex. 6–9, *H. seligeri*: 6, leaf (×25); 7, leaf apex; 8, mid-leaf cells; 9, capsule. 10–13, *Isopterygium pulchellum*: 10, leaves (×40); 11, leaf apex; 12, mid-leaf cells; 13, capsule. Cells ×415, apices ×100, capsules ×10.

1. I. pulchellum (Hedw.) Jaeg., Ber. S. Gall. Naturw. Ges., 1876–77
Plagiothecium pulchellum (Hedw.) Br. Eur.

Autoecious. Plants slender, stems creeping with numerous branches, branches erect or occasionally spreading; epidermal cells of branches thick-walled, 7–12 μm wide. Lvs spreading, usually secund, rarely complanate when moist, flexuose but otherwise little altered when dry, narrowly triangular or narrowly lanceolate-triangular, tapering ± from base to filiform apex, margin plane, entire throughout or with 1–2 small blunt teeth at basal angles; nerve short and double or absent; angular cells rectangular, not inflated or decurrent, green, cells elsewhere linear, in mid-lf 5–8 μm × 56–120 μm, 10–20 times as long as wide. Fusiform or cylindrical axillary gemmae, 2–5 cells long, occasionally present. Capsule slightly inclined, ovate-ellipsoid to ellipsoid, straight or slightly curved; lid conical, with or without short beak; spores 8–10 μm. Fr common, late spring, early summer. $n = 10+2$, 11, 22*. Pale or yellowish-green patches or scattered shoots amongst other bryophytes on damp shaded ledges and in rock crevices in montane habitats and ravines, frequent but usually in small quantity. S. Devon, Monmouth and Derby northwards. 61, H17. Europe, Faroes, Iceland, Caucasus, N. Asia, Kashmir, N. America, Greenland, New Zealand.

A somewhat variable species, a variety of which, var. *nitidulum* (Wahlenb.) Broth. has been recorded in Britain. But I have followed Ireland (1969) in not recognising what appears to be a minor variant, the characters delimiting which do not necessarily correlate. *Isopterygiopsis muelleranum* differs in the long stems and branches with larger, thin-walled epidermal cells and the entire lvs with narrower cells.

2. I. elegans (Brid.) Lindb., Not. Sallsk. F. Fl. Fenn. Förh., 1874
Plagiothecium elegans (Brid.) Schimp.

Dioecious. Plants slender to medium-sized, stems procumbent, branches numerous, usually procumbent and all pointing in one direction; epidermal cells of stem thick-walled, 10–12 μm wide. Lvs complanate, tips often pointing downwards when moist, hardly altered when dry, ovate to ovate-oblong, abruptly to gradually tapering to filiform apex, margin plane, denticulate near apex or rarely entire; nerve short and double; angular cells hardly differentiated, not inflated or decurrent, green, cells elsewhere linear, in mid-lf 5–7 μm × 76–112 μm, 13–23 times as long as wide. Axillary, filiform propagules with minute lvs often present and sometimes so abundant as to render plant fluffy in appearance. Capsule inclined, ovoid; lid conical. Fr very rare, spring. $n = 11$. Glossy, pale green patches, sometimes extensive on shaded soil in woods, on banks, tree boles, rocks and in rock crevices, calcifuge, common. 110, H38, C. N., W. and C. Europe, Faroes, Iceland, N. America, Madeira.

162. ISOPTERYGIOPSIS IWATS., J. HATTORI BOT. LAB., 1970

Dioecious. Pseudoparaphyllia lacking; epidermal cells of stem large, thin-walled. Lf base not decurrent. Axillary gemmae, 60–100 μm, sometimes present. Annulus present. One species only.

1. I. muelleranum (Schimp.) Iwats., J. Hattori Bot. Lab., 1970
Isopterygium muelleranum (Schimp.) Jaeg., *Plagiothecium muelleranum* Schimp.

Dioecious. Plants slender or small, procumbent, stems and branches long, to 3 cm, sometimes ± stoloniform; epidermal cells of stem and branches thin-walled, 16–30 μm wide. Lvs distant or not, ± complanate, directed forwards and sometimes upwards, hardly altered when dry, ovate to ovate-oblong, rapidly contracted to narrow or filiform apex, margin plane, entire; nerve short and double or absent;

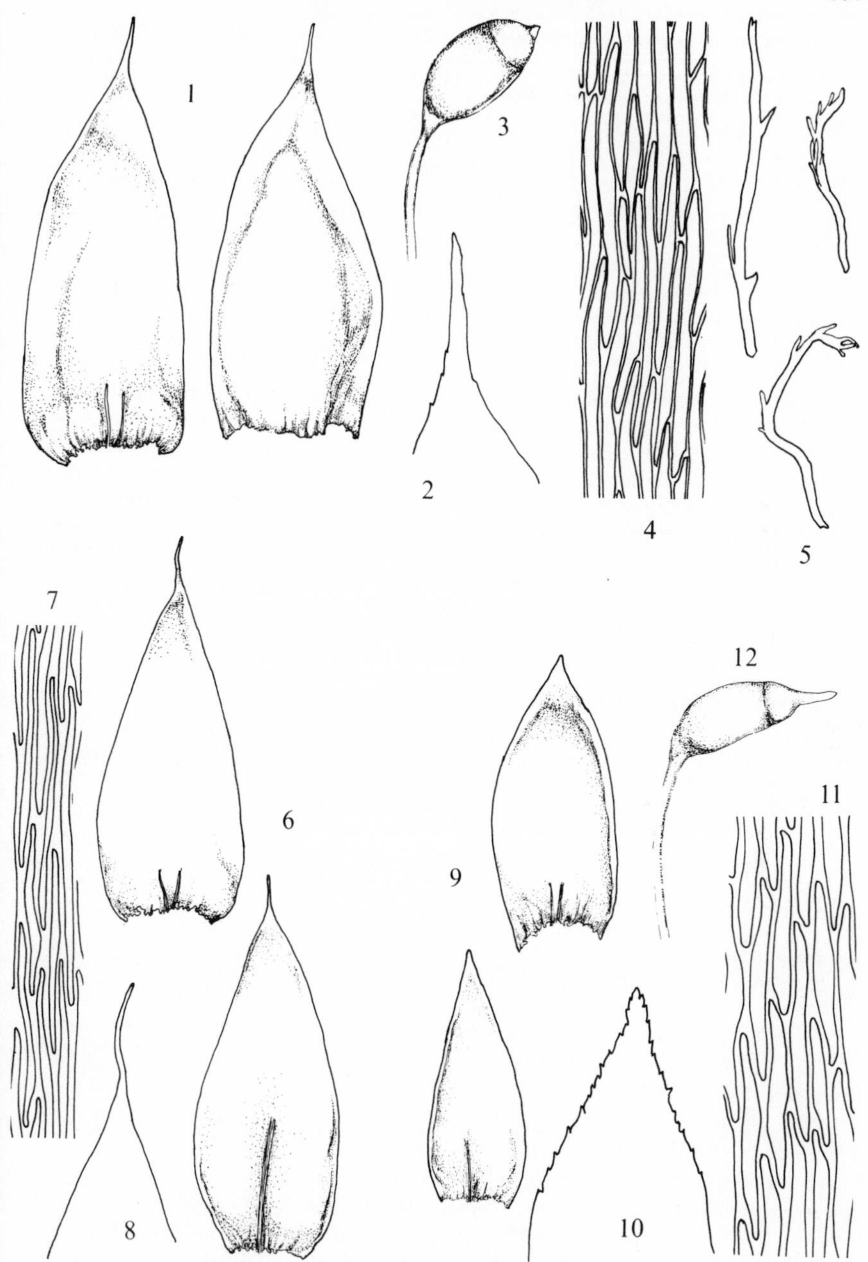

Fig. **309**. 1–5, *Isopterygium elegans*: 1, leaves; 2, leaf apex; 3, capsule; 4, mid-leaf cells; 5, propagules. 6–8, *Isopterygiopsis muelleranum*: 6, leaves; 7, mid-leaf cells; 8, leaf apex. 9–12, *Taxiphyllum wissgrillii*: 9, leaves; 10, leaf apex; 11, mid-leaf cells; 12, capsule. Leaves $\times 40$, cells $\times 415$, apices $\times 100$, capsules $\times 10$.

1–2 angular cells quadrate, not enlarged or decurrent, green, cells elsewhere linear, in mid-lf 4–6 μm×(56–)72–100 μm, 14–20 times as long as wide. Clusters of axillary gemmae, 60–100 μm long, occasionally present. Fr unknown in Britain or Ireland. Glossy, pale or yellowish-green patches or scattered shoots amongst boulders and in rock crevices in basic habitats at high altitudes, rare. Caerns, Scottish Highlands, S. Kerry. 11, H1. N. and C. Europe, Pyrenees, Italy, Caucasus, China, Japan, N. America, New Zealand.

May be mistaken for a slender form of *I. elegans*, but differing in the larger cells of the stem epidermis and the entire lvs with very narrow cells. Dixon (1924) reports the presence of flagelliform branchlets in *I. muelleranum* but I have seen these only in misnamed forms of *I. elegans* from montane habitats.

163. Taxiphyllum Fleisch., Musc. Fl. Buiten., 1923

Dioecious. Leafy pseudoparaphyllia present round branch primordia and branch bases; epidermal cells of branches small, thick-walled. Lvs usually complanate, not secund, ± symmetrical, ovate to oblong-lanceolate, apex acute to acuminate, base not decurrent, margin denticulate from middle or below; nerve short and double or absent; angular cells quadrate to rectangular, not inflated, green, cells elsewhere ± linear. Vegetative propagules unknown. Capsule erect to horizontal, ovoid, straight or curved; annulus persistent; lid with long beak; peristome perfect, inner ciliate; calyptra cucullate. A genus of about 36 species mostly occurring in eastern Asia but also found in N. and C. Africa, N. and S. America, Australia and Hawaii.

1. T. wissgrillii (Garov.) Wijk & Marg., Taxon, 1960
T. depressum (Brid.) Reim., *Isopterygium depressum* (Brid.) Mitt., *I. wissgrillii* (Garov.) Gillet-Lefebvre, *Plagiothecium depressum* (Brid.) Spruce

Dioecious. Plants small to medium-sized, procumbent, stems pinnately branched; epidermal cells of stem thick-walled, 10–15 μm wide. Lvs spreading, complanate, slightly secund, hardly altered when dry, ovate, narrowed to short acute apex, margin plane, denticulate above and sometimes ± to base; nerve short, double or absent; angular cells quadrate to shortly rectangular, not inflated or decurrent, green, cells elsewhere linear, in mid-lf 6–10 μm×(52–)64–80 μm, 8–10 times as long as wide. Capsule inclined, ovate-ellipsoid; lid with long beak; spores 10–12 μm. Fr very rare. $n=11$. Pale green, glossy, flat patches on soil, tree roots and bases, stones and porous rocks in shaded habitats, calcicole. Occasional to frequent in England and Wales, rare or occasional in Scotland and Ireland. 71, H12. Europe, Iceland, Caucasus.

Likely to be confused with *Isopterygium elegans* which is, however, calcifuge, has more longly tapering lvs, longer, narrower cells and often produces filiform propaguliferous shoots not found in *T. wissgrillii*.

48. SEMATOPHYLLACEAE

Plants slender to robust, stems creeping, procumbent or ascending, irregularly to pinnately branched; paraphyllia usually lacking. Stem and branch lvs usually similar, sometimes secund or complanate, usually symmetrical, ovate to lanceolate, acute to acuminate; nerve short and double or absent; angular cells enlarged or inflated, cells above linear or linear-rhomboidal. Seta long, smooth or papillose; capsule erect or more usually inclined to pendulous, ovoid to cylindrical, straight or curved; lid rostrate; peristome perfect or not; calyptra cucullate or rarely mitriform. Plants usually glossy, often yellowish-tinged, forming loose or tight mats.

164. SEMATOPHYLLUM MITT., J. LINN. SOC. BOT., 1864

Dioecious or autoecious. Stems creeping or procumbent, paraphyllia lacking. Lvs erecto-patent or patent, sometimes sub-secund or sub-complanate, ovate to lanceolate, acute to acuminate; a few angular cells enlarged or inflated forming small auricles, cells above linear to linear-rhomboidal. Seta smooth or papillose; capsule relatively small, erect to horizontal; lid with a long beak. A world-wide genus with about 160 species.

Lvs lanceolate, ± entire	1. **S. demissum**
Lvs ovate, denticulate above	2. **S. micans**

1. **S. demissum** (Wils.) Mitt., J. Linn. Soc. Bot., 1864

Autoecious. Plants slender, stems creeping, ± pinnately branched, branches short, procumbent, often curved. Lvs erecto-patent to patent, sometimes sub-secund, when dry slightly shrunken, 0.9–1.2 mm long, concave, sometimes slightly asymmetrical, lanceolate, tapering to acuminate apex, base narrow, not decurrent, margin plane or recurved below, ± entire; nerve very short, single or double or wanting; basal cells often orange, a few hyaline angular cells inflated, forming small auricles, a few marginal cells above rectangular, cells elsewhere linear to linear-rhomboidal, in mid-lf (5–)6–8 μm×44–94 μm, 7–16 times as long as wide. Seta reddish, flexuose; capsule inclined to horizontal, narrowly ellipsoid, slightly curved; lid longly rostrate; spores 12–18 μm. Fr common, autumn. Tight, glossy, yellowish-green to golden patches on lightly shaded rocks in humid habitats at low altitudes, very rare. Merioneth, Caerns, Cumberland, Kerry, W. Cork. 3, H3. Norway, W. and C. Europe, Asia Minor, Japan, N. America.

2. **S. micans** (Mitt.) Braithw., Brit. Moss. Fl., 1902
S. novae-caesareae (Aust.) Britt.

Dioecious. Plants very slender, stems procumbent, pinnately branched, branches long, sometimes filiform, procumbent. Lvs patent, sometimes sub-secund, when dry slightly shrunken, 0.4–0.8(–1.0) mm long, concave, sometimes slightly asymmetrical, ovate, acute to acuminate, base not decurrent, margin recurved below, denticulate above; nerve double, extending to $\frac{1}{3}$–$\frac{1}{2}$ way up lf; a few angular cells enlarged, forming small auricles, a few marginal cells above rectangular, cells elsewhere linear-rhomboidal, in mid-lf 5–7 μm×44–72 μm, 7–12 times as long as wide. Seta flexuose; capsule ellipsoid; lid rostrate. Fr unknown in Britain or Ireland. Tight yellowish-green patches or mixed with other bryophytes on lightly to moderately shaded rocks, up to *ca* 500 m altitude, rare. Merioneth, Caerns, Cumberland, western Scotland from Stirling and Dunbarton north to Skye and W. Ross, S.W. Ireland, W. Galway. 10, H4. Germany, N. America.

49. HYPNACEAE

Autoecious or dioecious. Slender to robust pleurocarpous mosses forming mats, tufts or lax, coarse wefts; stems procumbent to erect, stems usually forked, frequently pinnately branched with complanate branches; paraphyllia and pseudoparaphyllia present or not. Lvs frequently falcate and secund, smooth or plicate, stem lvs larger than branch lvs, similar in shape or not, lanceolate to cordate-triangular, abruptly to gradually tapering to usually acuminate apex, margin mostly denticulate at least towards apex; nerve double or absent, very rarely single; basal cells often porose, angular cells mostly differentiated, sometimes enlarged or inflated, forming auricles, cells above narrowly ellipsoid to linear. Seta long, smooth or rarely papillose;

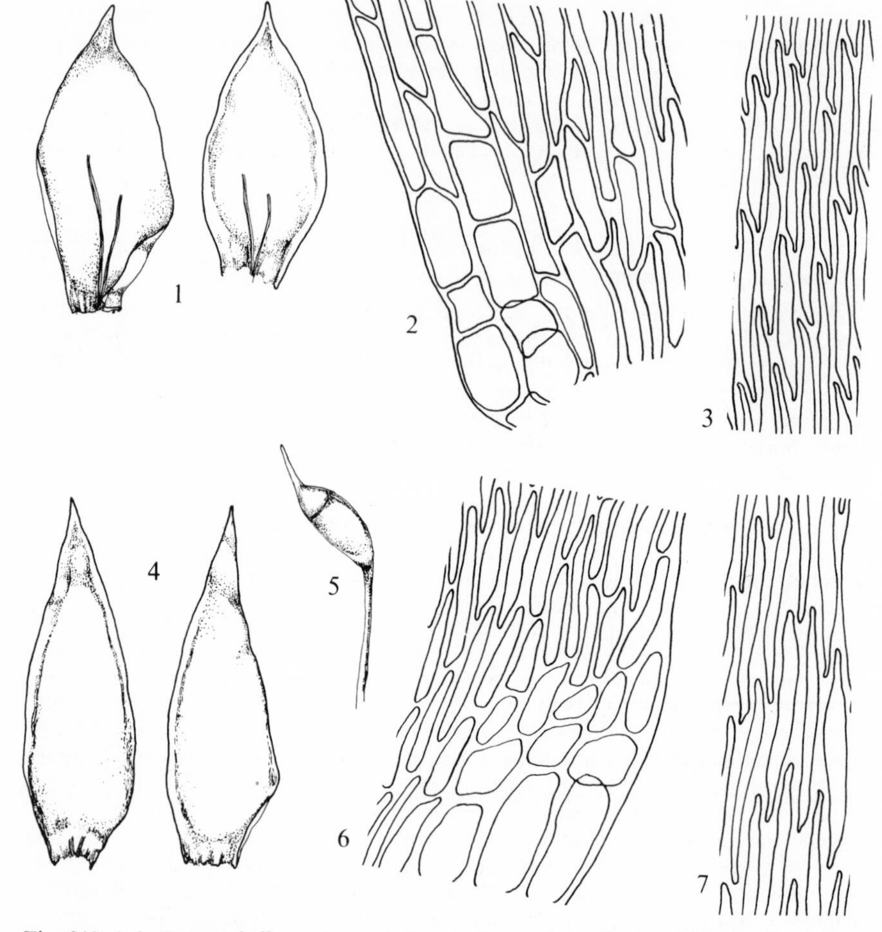

Fig. **310**. 1–3, *Sematophyllum micans*: 1, leaves; 2, angular cells; 3, mid-leaf cells. 4–7, *S. demissum*: 4, leaves; 5, capsule (×10); 6, angular cells; 7, mid-leaf cells. Leaves ×40, cells ×415.

capsule erect to horizontal, ovoid to cylindrical, symmetrical to curved; annulus usually separating; lid conical to mamillate or with rostellate to subulate beak; peristome double, outer teeth 16, vertically point-striate below, transversely point-striate or papillose above, sometimes with pale border, inner peristome usually with tall basal membrane, keeled processes and nodulose cilia; calyptra cucullate.

A large family, the status, limits and contents of the subdivisions of which are controversial. Many authorities divide the European members of the Hypnaceae into 6 subfamilies or families: Pylaisioideae (*Platygyrium*, *Pylaisia*, *Homomallium*), Hypnoideae (*Hypnum*), Ctenidioideae (*Ptilium*, *Ctenidium*, *Hyocomium*), Rhytidioideae (*Rhytidium*), Hylocomioideae (*Hylocomium*, *Pleurozium*, *Rhytidiadelphus*) and Climacioideae (*Climacium*). I consider on morphological and cytological grounds that *Climacium* should be placed in the Isobryales. I can find no satisfactory combinations of characters for discriminating the other subfamilies but there are good grounds for placing *Hylocomium* in a subfamily of its own, with the remaining genera being placed in a single subfamily, the Hypnoideae (see Noguchi, *J. Hattori Bot. Lab.* **35**, 156–68, 1972) and I have followed this course.

SUBFAMILY 1. HYPNOIDEAE

Stems with no paraphyllia, pseudoparaphyllia present or not.

165. PYLAISIA BR. EUR., 1851

Autoecious. Plants slender, stems creeping, irregularly to pinnately branched, branches erect or ascending. Stem and branch lvs similar, ovate, tapering to long acuminate apex, margin plane, ± entire; nerve short and double or absent; angular cells quadrate, cells above linear. Seta smooth, reddish; capsule erect, shortly cylindrical; exothecial cell walls uniformly thickened; annulus present or absent; outer peristome teeth transversely striate below, papillose above, bordered, inner peristome with well developed basal membrane, cilia rudimentary; calyptra naked. A mainly northern hemisphere genus of about 40 species.

1. P. polyantha (Hedw.) Br. Eur., 1851
Pylaisiella polyantha (Hedw.) Grout

Plants slender, stems creeping, pinnately branched, branches erect or ascending, dwarf fertile branches usually abundant. Lvs appressed, flexuose when dry, erect or sub-secund and upturned when moist, lanceolate to ovate-lanceolate, narrowed to long fine acumen, margin plane, entire; nerve very short and double or absent; basal cells thin-walled, not porose, narrowly rectangular, frequently with flat ends, intergrading with angular cells, angular cells ± quadrate and intergrading with cells above, forming ill-defined group of cells about $1\frac{1}{2}$ times as long as wide, cells above linear-rhomboidal, in mid-lf 5–8 μm × 56–92 μm, 8–12(–14) times as long as wide. Capsule erect, narrowly ellipsoid, straight; exothecial cells with uniformly thickened walls; lid conical, acute, 1.0–1.2 mm long; spores 13–16 μm. Fr common, 2 generations often present, autumn, winter. $n=11$. Glossy, yellowish-green or green patches on bark in hedgerows and open woodland, rare in lowland Britain, very rare elsewhere and extending north to Perth and Angus, Kerry, E. Cork. 42, H3, C. Europe, Caucasus, Siberia, Kashmir, Japan, N. America, Greenland.

Superficially similar to *Hypnum cupressiforme* var. *resupinatum* but differing in the shape of the lid, the frequent occurrence of 2 generations of capsules, the numerous fertile branches and the rectangular basal cells often with flat ends. The exothecial cell walls of the *Hypnum* have heavily thickened longitudinal walls. *Homomallium incurvatum* occurs on basic rocks and has an inclined, curved capsule; *Platygyrium repens* is dioecious, with sparse fertile branches, has recurved lf margins, the angular cells extend further up the margin and has axillary gemmiferous branchlets.

166. PLATYGYRIUM BR. EUR., 1851

Dioecious. Stems creeping, closely pinnately branched, branches erect or ascending, pseudoparaphyllia lacking. Stem and branch lvs similar, straight or slightly curved, ovate or ovate-lanceolate, tapering to acuminate apex, margin recurved, ± entire; nerve short and double or absent; angular cells quadrate or rectangular, cells above linear-rhomboidal. Vegetative reproduction by axillary branchlets frequent. Seta smooth, reddish; capsule erect, cylindrical, straight; exothecial cells thin-walled, annulus persisting; outer peristome teeth transversely striate below, papillose above,

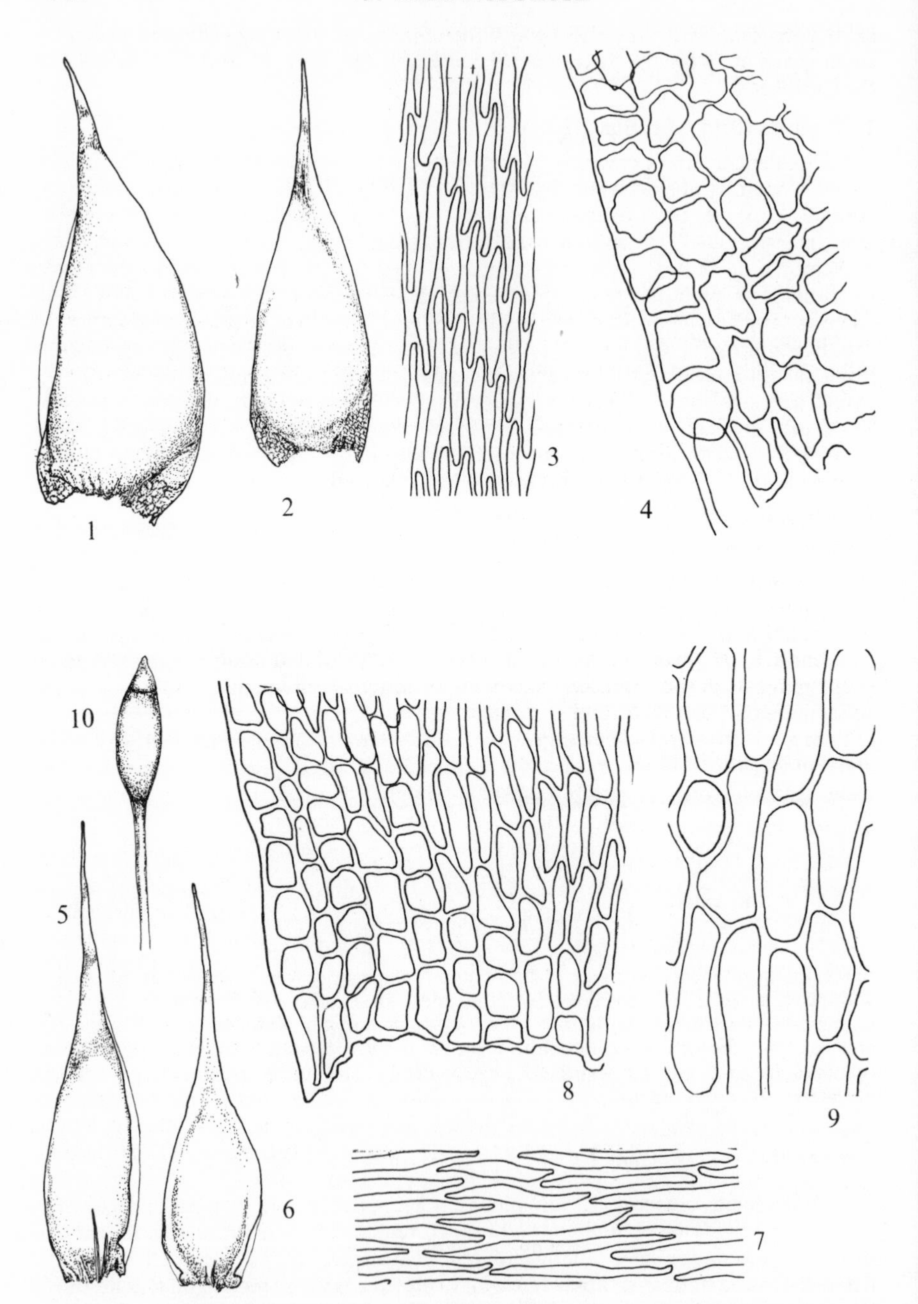

Fig. **311**. 1–4, *Platygyrium repens*: 1, stem leaf; 2, branch leaf; 3, mid-leaf cells; 4, angular cells. 5–10, *Pylaisia polyantha*: 5, stem leaf; 6, branch leaf; 7, mid-leaf cells; 8, angular cells; 9, exothecial cells; 10, capsule (×10). Leaves ×40, cells ×415.

inner peristome with very short basal membrane, cilia lacking; calyptra naked. A small genus of 10 species distributed through Europe, Asia, N. and C. America and the Caribbean.

1. P. repens (Brid.), Br. Eur., 1851

Plants slender, stems creeping, closely branched, branches ascending to erect. Lvs ± erect, scarcely altered when dry, stem lvs slightly curved, not concave, branch lvs straight, concave, both ovate-lanceolate to lanceolate, tapering to long acuminate apex, margin usually recurved on one or both sides below, ± entire; nerve very short and double or absent; basal cells narrowly rhomboidal, not porose, angular cells quadrate to trapezoid, extending some way up margin, forming group of cells about twice as long as wide, cells above linear-rhomboidal, shorter at margin and apex, in mid-lf 5–8 μm×56–80 μm, 8–12 times as long as wide. Axillary, propaguliferous branchlets, to about 250 μm long present at tips of branches. Capsule erect, subcylindrical, straight, exothecial cell walls of ± uniform thickness; lid rostrate; spores 16–20 μm. Fr unknown in Britain. Glossy, dark green, often coppery-tinged patches or mats on bark in woodland, very rare but possibly overlooked as *Hypnum cupressiforme*. Surrey, Oxford, Berks, Huntingdon, Denbigh. 5. Europe, N. and C. Asia, Japan, Algeria, N. America.

For the occurrence of *Platygyrium repens* in Britain see Warburg & Perry, *Trans. Br. bryol. Soc.* **4**, 422–5, 1963.

167. Homomallium (Schimp.) Loeske, Hedwigia, 1907

Very close to *Hypnum*. Basal cells of lvs not porose. Perichaetial lvs not striate. A small genus of 14 species mostly occurring in eastern Asia but also found in Europe, other parts of Asia and N. and C. America.

I can see no sound reason for segregating this genus from *Hypnum* other than for dividing *Hypnum* into smaller units.

1. H. incurvatum (Brid.) Loeske, Hedwigia, 1907
Hypnum incurvatum Brid.

Autoecious. Plants very slender, stems creeping, irregularly pinnately branched, branches short, ascending or erect. Lvs erect when dry, patent to spreading, secund ± upturned when moist, concave, lanceolate to narrowly lanceolate, straight or curved, ± rapidly tapering to long acumen, margin plane, entire; nerve short and double or absent; angular cells quadrate, reaching nearly to nerve, grading into rhomboidal basal cells, forming group of cells about 1½ times as long as wide, cells above linear-rhomboidal to linear, shorter at margin and apex, in mid-lf 5.5–8.0 μm×48–68 μm, 9–12 times as long as wide. Capsule inclined to horizontal, oblong-ellipsoid, curved; exothecial cell walls ± uniformly thickened; lid conical, rostrate; spores 8–14 μm. Fr common, summer. $n=12$. Small, glossy, green or yellowish-green patches or mats on shaded, calcareous rocks, rare. Derby, Yorks, Westmorland, Cumberland, I. of Man, Ayr, W. Lothian, Perth. 12. N. and C. Europe, Caucasus, Turketan, Yenisei, Altai, Kashmir, Japan.

168. Hypnum Hedw., Sp. Musc., 1801

Dioecious or autoecious. Stems creeping to erect, ± unbranched to closely pinnately branched, branches long or short; pseudoparaphyllia usually present. Branch lvs of similar shape to but smaller than stem lvs, straight, falcate, circinate, secund, narrowly lanceolate to broadly ovate, abruptly to longly tapered to acute to filiform

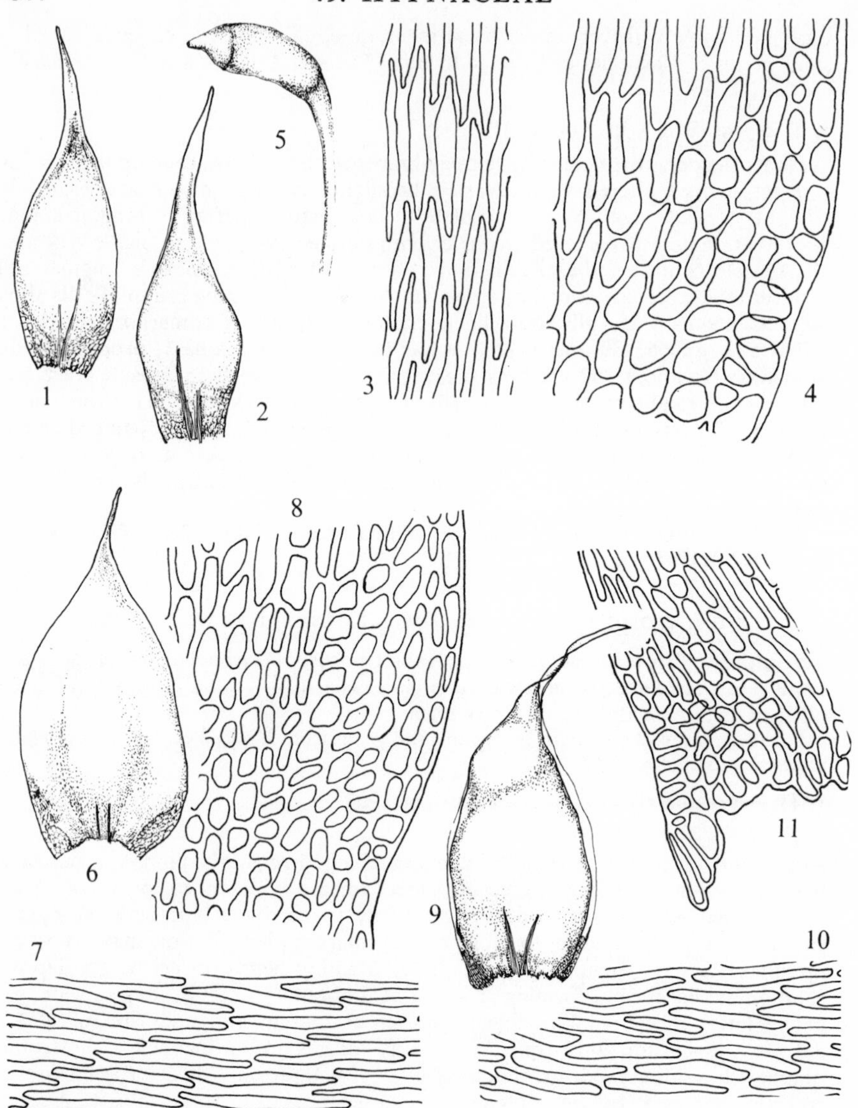

Fig. **312**. 1–5, *Homomallium incurvatum*: 1, stem leaf; 2, branch leaf; 3, mid-leaf cells; 4, angular cells; 5, capsule. 6–8, *Hypnum vaucheri*: 6, leaf; 7, mid-leaf cells; 8, angular cells. 9–11, *Hypnum revolutum*: 9, leaf; 10, mid-leaf cells; 11, angular cells. Leaves ×40, cells ×415, capsule ×10.

apex, margin plane or recurved, entire or denticulate; nerve short and double or wanting; basal cells usually narrowly rhomboidal, incrassate, porose or not, angular cells usually distinct, quadrate to rectangular, thin-walled to incrassate, hyaline or coloured, cells above elliptical to linear-vermicular. Seta long, reddish; capsule erect to horizontal, obloid to cylindrical, straight or curved; lid mamillate to subulate;

peristome double, outer entire or perforated, transversely striate below, longitudinally point-striate, papillose or smooth above, inner peristome with basal membrane and cilia. A cosmopolitan genus of about 200 species.

1 Angular cells inflated, forming distinct auricles 2
Angular cells not inflated but differentiated and sometimes forming auricles 3

2 Stems reddish-brown, irregularly branched, lvs sub-falcate, shortly tapering to apex **8. H. lindbergii**
Stems greenish, complanately pinnately branched, lvs circinate, longly tapering to filiform acumen **9. H. callichroum**

3 Lf margin recurved from base to about ¾ way up lf on one or both sides **1. H. revolutum**
Lf margin plane or recurved near base only 4

4 Differentiated angular cells very few, not distinct, plants autoecious **10. H. hamulosum**
Differentiated angular cells ± numerous, or if few then forming distinct group, plants dioecious 5

5 Angular cells few but forming distinct group, stem reddish-brown and lvs circinate-secund **7. H. bambergeri**
Angular cells ± numerous, stems greenish or if reddish-brown then lvs falcato-secund 6

6 Largest angular cells hardly more than 10 μm wide, stems and branches sub-julaceous with imbricate lvs **2. H. vaucheri**
Largest angular cells 12–20 μm wide, stems sub-julaceous or not, lvs imbricate or not 7

7 Stems regularly pinnate, branches complanate, plants very pale green or stems reddish-brown 8
Stems variously branched but not regularly pinnate, plant colour various, stems greenish 9

8 Stems greenish, pseudoparaphyllia not lobed or toothed **5. H. jutlandicum**
Stems reddish-brown, pseudoparaphyllia irregularly lobed or toothed **6. H. imponens**

9 Lid of capsule with rostellate to subulate beak, stem lvs sub-entire to denticulate in upper part, cells 49–96 μm long or if shorter then lvs straight or plants very robust **3. H. cupressiforme**
Lid mamillate, stem lvs often denticulate to midway or below, cells 32–56 μm, lvs falcate, plants filiform to medium-sized **4. H. mammillatum**

Section I. *Euhypnum*

Epidermal cells of stem narrow, incrassate; angular cells of lf usually numerous, quadrate to rectangular or irregular, forming auricles or not, not inflated.

1. H. revolutum (Mitt.) Lindb., Öfv. K. Vet. Ak. Förh., 1867

Dioecious. Plants slender; stems erect or ascending, pinnately branched; pseudoparaphyllia lanceolate, toothed. Lvs falcato-secund, concave, plicate or not, ovate to lanceolate, gradually tapering or abruptly narrowed to acuminate apex, margin recurved on one or both sides from base to ¾ way up lf, denticulate towards apex;

nerve short and double; basal cells rhomboidal, angular cells few, rectangular, poorly defined, cells above narrowly elliptical with rounded ends, in mid-lf 5.0–6.5 μm × 24–40 μm, 4–7 times as long as wide. Fr unknown in Britain. $n = 14$. Yellowish or brownish tufts on basic montane rocks, very rare. Mid Perth. 1. Northern and montane Europe, Iceland, Caucasus, N. and C. Asia, Tibet, Punjab, Kashmir, China, N. America, Greenland, Antarctica.

Although only known from Ben Lawers in Britain, there are considerable differences between the 3 gatherings made. The most recent, by H. N. Dixon and E. C. Wallace, was determined by Mönkemeyer as var. *dolomiticum* (Milde) Mönk. The first gathering made in 1890 differs in appearance but the second collection (1899) is intermediate and it seems that var. *dolomiticum* cannot be regarded as more than an extreme form.

2. **H. vaucheri** Lesq., Mem. Soc. Sc. Nat. Neuchatel, 1846

Dioecious. Primary stems stoloniform, secondary stems erect or ascending, irregularly pinnately branched, stems and branches sub-julaceous. Lvs imbricate when dry, erect or erecto-patent when moist, straight or curved above, concave, ovate or ovate-oblong, ± abruptly narrowed to long, ± channelled acuminate acumen, margin plane or recurved below, entire; nerve short and double; basal cells irregularly rectangular, angular cells numerous, ± quadrate, *ca* 10 μm wide, forming well defined group about twice as long as wide, cells above narrowly rhomboidal, (6–)8(–10) μm × 24–44 μm, 3.5–7.0 times as long as wide. Capsule erect or inclined, shortly cylindrical, curved; lid conical, acute. Fr unknown in Britain. Glossy, yellowish to brownish-green patches on basic montane rocks, very rare. Mid Perth. 1. Northern and montane Europe, Asia, N. America.

For the occurrence of this plant in Scotland see Perry & Fitzgerald, *Trans. Br. bryol. Soc.* **4**, 418–21, 1963.

3. **H. cupressiforme** Hedw., Sp. Musc., 1801

Dioecious. Plants slender to robust, stems procumbent, irregularly pinnately branched, branches procumbent to ascending; pseudoparaphyllia lanceolate-subulate. Lvs imbricate and straight to falcato-secund, concave, narrowly lanceolate and tapering to broadly ovate and abruptly narrowed to apex, apex acuminate to filiform, margin plane or recurved below, sub-entire or denticulate towards apex; nerve very short and double or wanting; basal cells narrowly rhomboidal with rounded ends, incrassate, often porose, angular cells thick-walled to strongly incrassate, quadrate to trapezoid or irregular in shape, hyaline to yellowish, forming group about twice as long as wide, middle angular cells mostly 12–20 μm wide, cells above linear, in mid-lf 5–9 μm × (32–)48–96 μm. Capsule erect or inclined, obloid to sub-cylindrical, straight or curved; lid with rostellate to subulate beak; peristome teeth longitudinally striate below, papillose or smooth above, bordered or not.

Key to intraspecific taxa of *H. cupressiforme*

1 Plants slender, lvs straight, often upturned at stem and branch tips, lanceolate, entire, lid with subulate beak — var. **resupinatum**
Plants slender to robust, lvs falcate, curving downwards, lanceolate to broadly ovate, sub-entire to denticulate towards apex, lid rostellate to rostrate — 2

2 Plants slender or medium-sized, greenish, lvs gradually tapering to apex — var. **cupressiforme**
Plants robust, yellowish-green to golden-brown or if dull green then lvs abruptly narrowed to apex — var. **lacunosum**

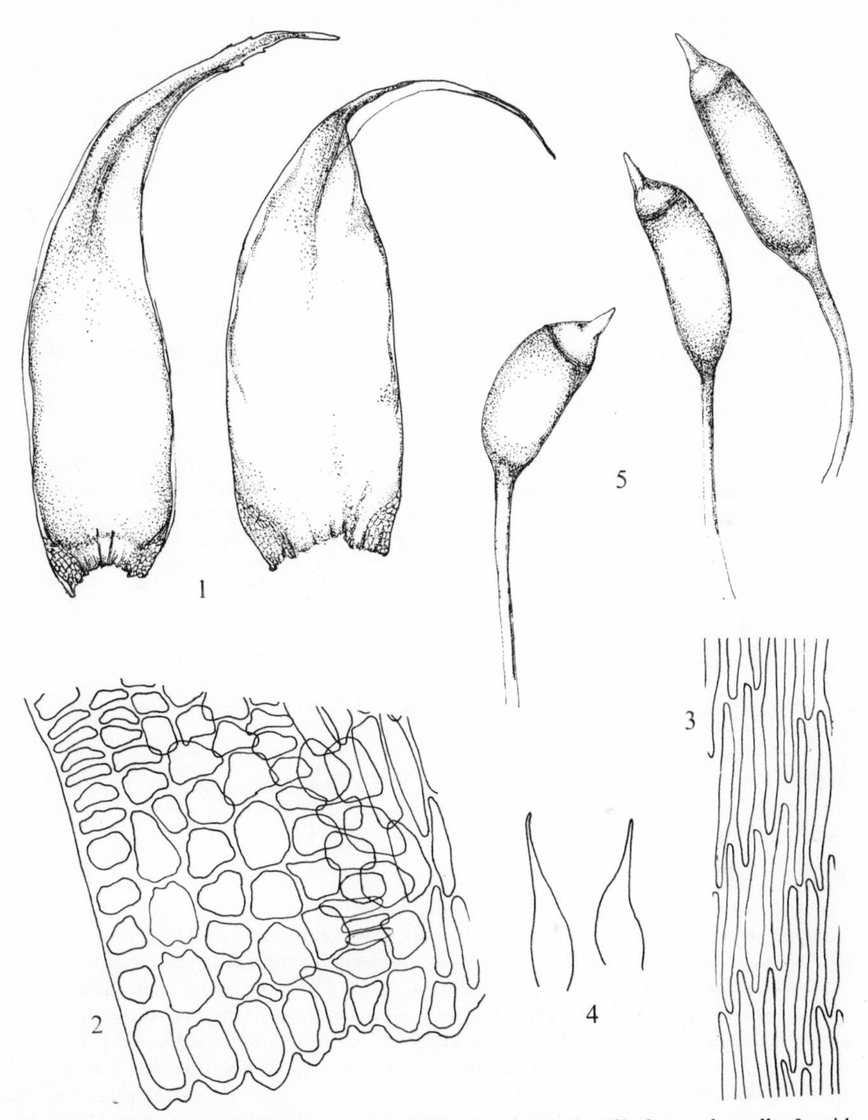

Fig. **313**. *Hypnum cupressiforme* var. *cupressiforme*: 1, leaves (× 40); 2, angular cells; 3, mid-leaf cells; 4, pseudoparaphyllia (× 100); 5, capsules (× 10). Cells × 415.

Var. **cupressiforme**

Plants slender to medium-sized. Lvs weakly falcato-secund to falcato-secund, 1.3–2.4 mm long, ovate to lanceolate, gradually narrowed to long slender acumen, margin usually plane, sub-entire or denticulate near apex; angular cells usually very incrassate, irregularly quadrate, cells in mid-lf (56–)64–90 μm long. Capsule inclined, rarely erect, obloid to sub-cylindrical, curved or straight, 1.6–2.2(–2.7) mm long (not

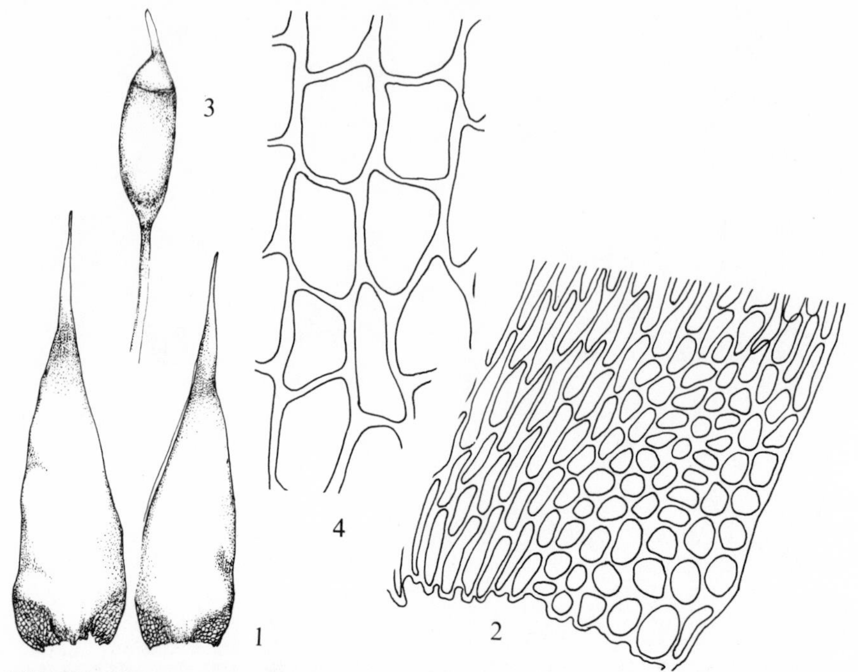

Fig. **314**. *Hypnum cupressiforme* var. *resupinatum*: 1, leaves (×40); 2, angular cells; 3, capsule (×10); 4, exothecial cells. Cells ×415.

including lid); lid rostellate to rostrate, 0.5–0.8 mm long; spores 13–19 μm. Fr occasional to frequent, autumn. $n=10$*. Green or pale green patches on rocks, walls, bark, logs, soil, etc., in sheltered or exposed habitats, especially where acidic, very common. 112, H40, C. Cosmopolitan.

Var. **resupinatum** (Tayl.) Schimp., Coroll., 1856

Plants slender. Lvs ± straight, usually somewhat sub-secund and directed upwards especially at stem and branch tips, 0.7–1.4 mm long, margin entire; angular cells ± quadrate, thick-walled, hyaline, mid-lf cells 56–80 μm long, 9–16 times as long as wide in some plants, 32–40 μm long, 6–7 times as long in others. Capsule erect, sub-cylindrical, straight or slightly curved, 1.6–2.2 mm long; exothecial cells with longitudinal walls more heavily thickened than transverse walls; lid 0.8–1.2 mm long with subulate beak; spores 18–24 μm. Fr frequent, autumn. Pale green patches on bark, logs, rocks and walls in lightly shaded habitats, frequent or common. 112, H40, C. N. and W. Europe, Azores, Madeira, Newfoundland.

Var. **lacunosum** Brid., Musc. Rec., 1801

H. cupressiforme var. *elatum* Brid., *H. cupressiforme* var. *tectorum* Brid.

Plants robust. Lvs ± straight and imbricate to falcato-secund, 1.6–3.0 mm long, concave and giving stems and branches julaceous appearance when moist, varying from ovate-oblong to oblong-lanceolate and tapering to channelled apex, to ovate-oblong or broadly ovate and abruptly narrowed to scarcely channelled apex, margin recurved below, plane or inflexed above, sub-entire to denticulate towards apex;

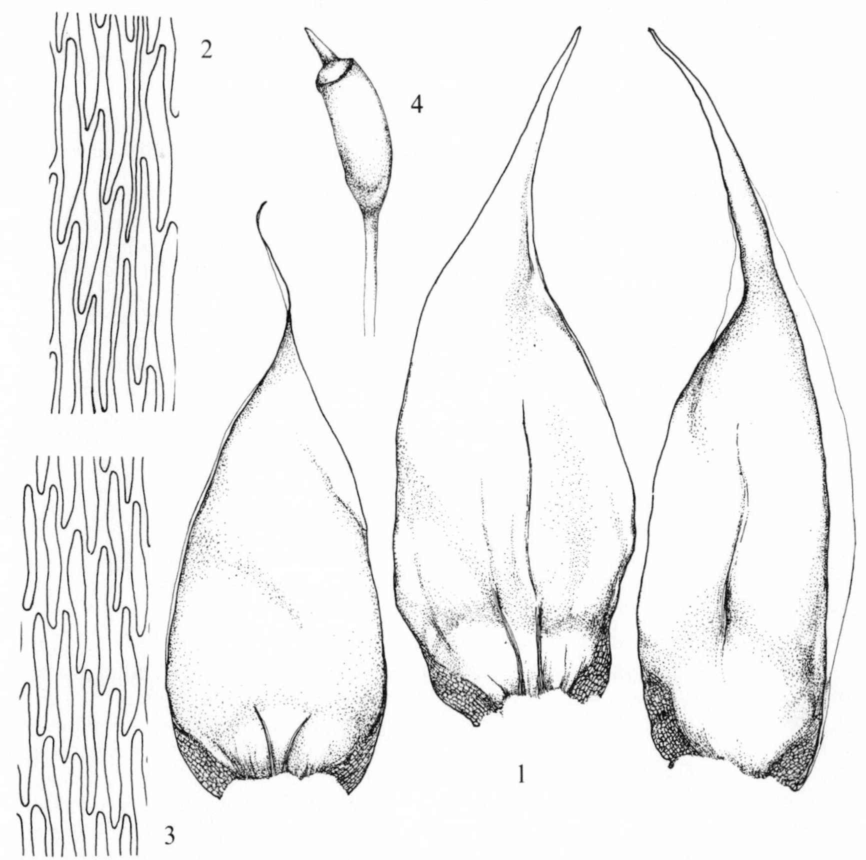

Fig. **315**. *Hypnum cupressiforme* var.*l acunosum*: 1, leaves (×40); 2 and 3, mid-leaf cells from different plants (×415); 4, capsule (×10).

angular cells quadrate to rounded-quadrate or trapezoid, incrassate but not strongly so, cells in mid-lf 32–96 μm long. Capsule ± erect, sub-cylindrical, straight or slightly curved; exothecial cells with longitudinal walls scarcely more thickened than transverse walls; lid rostrate. Fr rare, autumn, winter. Dull green to glossy, yellowish-green to golden-brown patches, sometimes extensive, on usually basic soil, rocks, wall tops, roofs, etc., common and sometimes abundant in suitable habitats. 105, H30, C. N., W., C. and S. Europe, C. America.

A very polymorphic species the nature of the variation within which is unknown and requires experimental study. It would appear that *H. cupressiforme*, *H. mammillatum* and *H. jutlandicum* form a complex within which rapid evolution and divergence is taking place. Divergence seems to have proceeded furthest along 2 lines represented by *H. mammillatum* and *H. jutlandicum*. There is a tendency for the respective characters defining var. *resupinatum* and var. *lacunosum* to correlate although they also occur at random in var. *cupressiforme*, hence the varietal status of the taxa concerned.

In var. *resupinatum* there are 2 classes of lf cell length and plants with short cells might be mistaken for slender forms of *H. mammillatum* when sterile. The lvs, however, are straight and entire. In var. *lacunosum* some plants have short lf cells, some have long cells and others have a mixture. Whether there are additional, as yet unrecognised taxa in these 2 varieties is obscure. I have seen no plants of var. *cupressiforme* with lf areolation approaching that of *H. mammillatum*.

Var. *lacunosum* and var. *tectorum* have previously been treated as separate varieties in Britain, var. *lacunosum* being said to have imbricate, ± straight lvs abruptly narrowed to the apex and var. *tectorum* to have falcato-secund lvs gradually tapering to the apex. In its extreme form var. *lacunosum* is very distinctive, resembling somewhat *Entodon concinnus* and quite different from var. *tectorum*. However forms of the latter variety occur in which some of the lvs are abruptly narrowed and there are plants that cannot be named, and hence the 2 varieties are not recognised as separate taxa. Intermediates between var. *lacunosum* and var. *cupressiforme* are also frequently encountered and it is possible that var. *lacunosum* is merely a form of calcareous habitats. It is possible, however, that it is a distinct species the limits between which and *H. cupressiforme* are obscured by phenotypic variation.

For the differences between *H. cupressiforme* and *H. mammillatum* and *H. jutlandicum* and between var. *resupinatun* and *Pylaisia polyantha* see under those species. *H. vaucheri* differs in the shorter lf cells and more numerous, smaller angular cells; *H. callichroum* and *H. lindbergii* in the large, hyaline epidermal cells of the stem and inflated angular cells; *Homomallium incurvatum* by the less well defined angular cells of the lanceolate, upturned lvs and the ± horizontal capsule, (it is also autoecious); and in *Platygyrium repens* gemmae are usually produced at the branch tips, the branch lvs are straight and the angular cells extend some distance up the lf margin.

4. H. mammillatum (Brid.) Loeske, Verh. Bot. Ver. Prov. Brandenb., 1905
H. cupressiforme var. *mammillatum* Brid., *H. cupressiforme* var. *mamillatum* auct., *H. cupressiforme* var. *filiforme* Brid.

Dioecious. Plants filiform or very slender to medium-sized, stems procumbent or creeping, irregularly pinnately branched, branches procumbent to ascending, spreading to parallel with stems. Lvs falcato-secund, ovate to lanceolate, gradually tapering to channelled acumen, margin plane or slightly recurved below, sub-entire to denticulate down to middle or in very slender forms even lower; nerve very short and double or wanting; basal cells narrowly rhomboidal with rounded ends, incrassate, porose or not, angular cells in robust forms ± irregularly quadrate, very incrassate, yellowish, in slender forms quadrate, not heavily thickened, hyaline, forming group of cells about twice as long as wide, cells above linear, in mid-lf (4–)5–7 μm × 32–56(–64) μm, 6–10(–12) times as long as wide. Capsule erect or inclined, oblong-ellipsoid to sub-cylindrical, straight or slightly curved, 1.6–2.2 mm long; exothecial cell walls ± uniformly thickened; lid mamillate, 0.4–0.5 mm long; outer peristome teeth longitudinally striate or finely papillose above, unbordered; spores often variable in size or aborted, (14–)18–24(–28) μm. Fr apparently common in larger forms, rare in slender ones, autumn. $n = 10$*. Yellowish-green to green patches on bark, fallen logs, rocks and wall tops in humid, shaded situations, rarely on soil, slender forms forming thin mats on vertical tree trunks. Probably common and generally distributed. 57, H12. N. and C. Europe.

Until recently this plant has usually been overlooked as *H. cupressiforme* and in recent years only determined when ripe undehisced capsules are present (hence the apparent common occurrence of capsules). Examination of herbarium material has shown that about 25 % of gatherings named *H. cupressiforme* belong here. All gatherings of *H. cupressiforme* var. *filiforme* studied have the areolation of *H. mammillatum* and intergrade with typical forms of that species, hence the reduction to synonymy of the 2 taxa. Usually distinct in the shape of the lid, although occasional plants with lids of an intermediate nature occur. These and sterile plants may be distinguished from *H. cupressiforme* var. *cupressiforme* by the shorter cells and more strongly toothed lvs. Although the lvs are often more strongly curved than in *H. cupressiforme* sterile plants cannot be determined except microscopically. *H.*

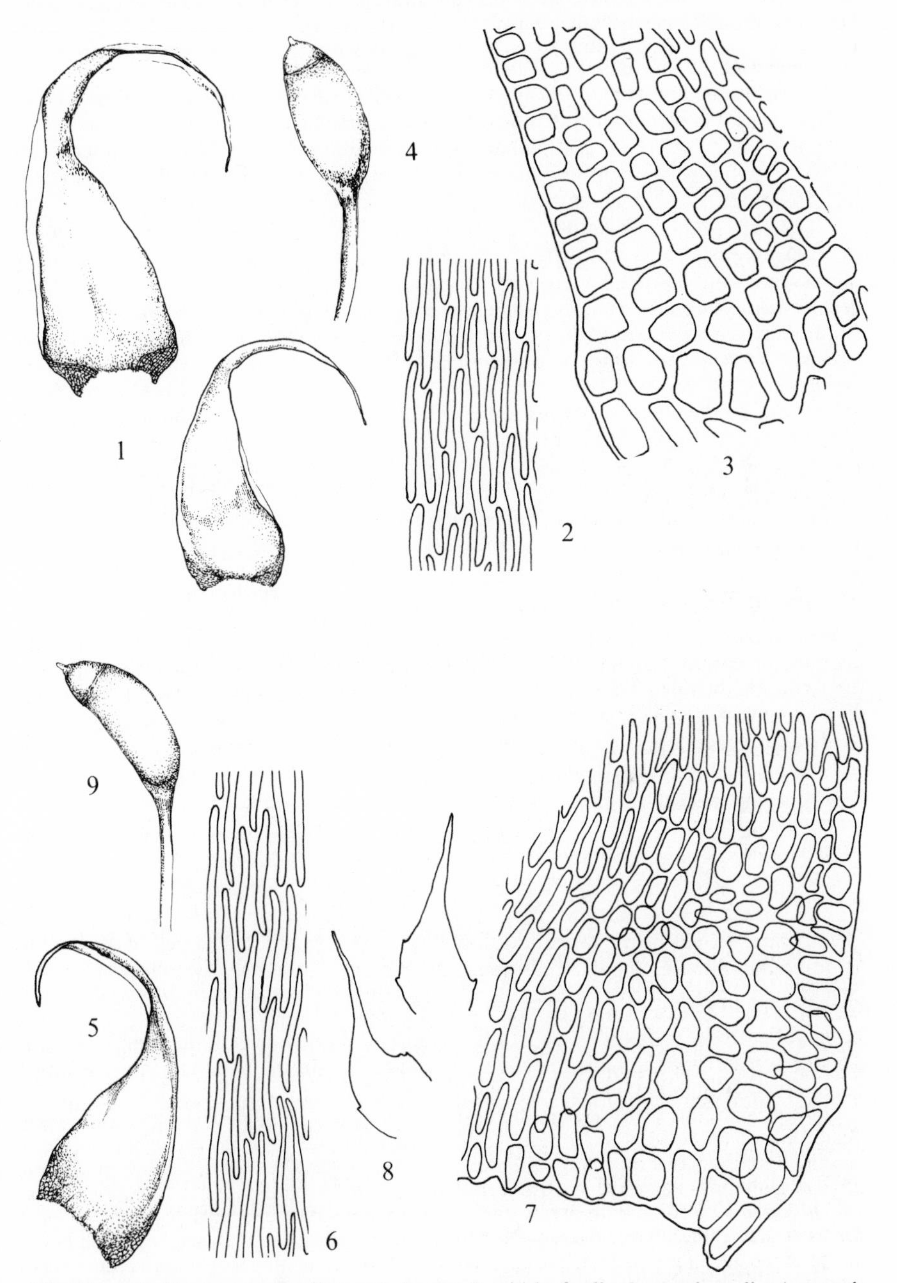

Fig. **316**. 1–4, *Hypnum mammillatum*: 1, leaves; 2, mid-leaf cells; 3, angular cells; 4, capsule. 5–9, *H. jutlandicum*: 5, leaf; 6, mid-leaf cells; 7, angular cells; 8, pseudoparaphyllia; 9, capsule. Leaves ×40, cells ×415, capsules ×10.

cupressiforme var. *resupinatum* has ± straight, entire lvs and a subulate beak to the capsule. The name *H. mammillatum* is illegitimate.

5. H. jutlandicum Holmen & Warncke, Bot. Tidsskr., 1969
H. cupressiforme var. *ericetorum* Br. Eur., *H. ericetorum* (Br. Eur.) Loeske

Dioecious. Plants usually medium-sized, stems greenish, procumbent to ascending, regularly pinnately and complanately branched; pseudoparaphyllia lanceolate-subulate to triangular with subulate apex. Lvs not crowded, only slightly overlapping, falcato-secund, sometimes sub-complanate, ovate, tapering to acuminate to filiform apex, margin plane, sub-entire to denticulate towards apex; nerve short and double or wanting; basal cells narrowly rhomboidal, incrassate, often porose, angular cells incrassate, rounded-quadrate to very irregular, forming hyaline to brownish group of cells about twice as long as wide, cells above linear, in mid-lf 5.0–6.5 μm × 44–72 μm, 9–13 times as long as wide, not porose. Capsule inclined, obloid to shortly cylindrical, curved, 1.4–1.6 mm long; exothecial cell walls ± uniformly thickened; outer peristome teeth vertically striate or smooth above, unbordered; lid rostrate; spores 13–15 μm. Fr rare, winter, spring. Lax, pale green tufts or patches on heaths, moorland and montane habitats, often associated with ericaceous plants and on litter in open places in conifer plantations, calcifuge, common in suitable habitats, generally distributed. 111, H40, C. Europe, Faroes, Algeria, Madeira, Tenerife, N. America.

It is with some reluctance that I have recognised this plant as a species. Although usually distinct in its general appearance from *H. cupressiforme*, the pale colour and the pinnately branched complanate stems being characteristic, intermediates do occur and there are no other gametophyte characters to separate the two. It is possible that the intergradation is due to phenotypic variation. It is possible that *H. jutlandicum* differs from *H. cupressiforme* in the outer peristome teeth longitudinally striate or smooth above and unbordered, but this requires confirmation. *H. mammillatum* differs in the shorter lf cells; *H. imponens* has brownish stems and broader, branched pseudoparaphyllia, also the plants are usually somewhat brownish instead of pale green when dry.

6. H. imponens Hedw., Sp. Musc., 1801
H. cupressiforme ssp. *imponens* (Hedw.) Boul.

Dioecious. Plants medium-sized; stems reddish-brown, erect or ascending, pinnately branched, branches complanate; pseudoparaphyllia lanceolate to triangular, toothed to irregularly lobed. Lvs falcato-secund, often sub-complanate, not or hardly overlapping, stem lvs strongly falcate to circinate, ovate to lanceolate, longly tapering to acuminate apex, margin plane or recurved below, remotely and obscurely denticulate above; nerve short and double or lacking; basal cells yellowish-brown, rhomboidal, porose, angular cells quadrate to rectangular, usually yellow to brown, at extreme base somewhat inflated, hyaline, forming small distinct auricles, cells above linear, frequently vermicular, 5–8 μm × 64–80(–90) μm, 10–12(–15) times as long as wide. Branch lvs smaller, less strongly curved, otherwise similar. Capsule inclined, oblong-ellipsoid, slightly curved; longitudinal walls of exothecial cells more heavily thickened than transverse walls; lid conically rostrate; spores *ca* 20 μm. Fr very rare. $n=7$, 11. Lax patches on wet heath and moorland, rare but sometimes locally frequent. Scattered localities in southern and northern England, a few in Scotland, extending north to Argyll and W. Ross, not known in Ireland. 35. N. and W. Europe, Pyrenees, Czechoslovakia, Switzerland, Iceland, Carpathians, C. Asia, Sikkim, Kashmir, Japan, Azores, N. America.

7. H. bambergeri Schimp., Syn., 1860

Dioecious. Plants small to medium-sized, stems creeping, irregularly pinnately branched, branches short or long, erect or ascending. Lvs circinate-secund, lanceolate

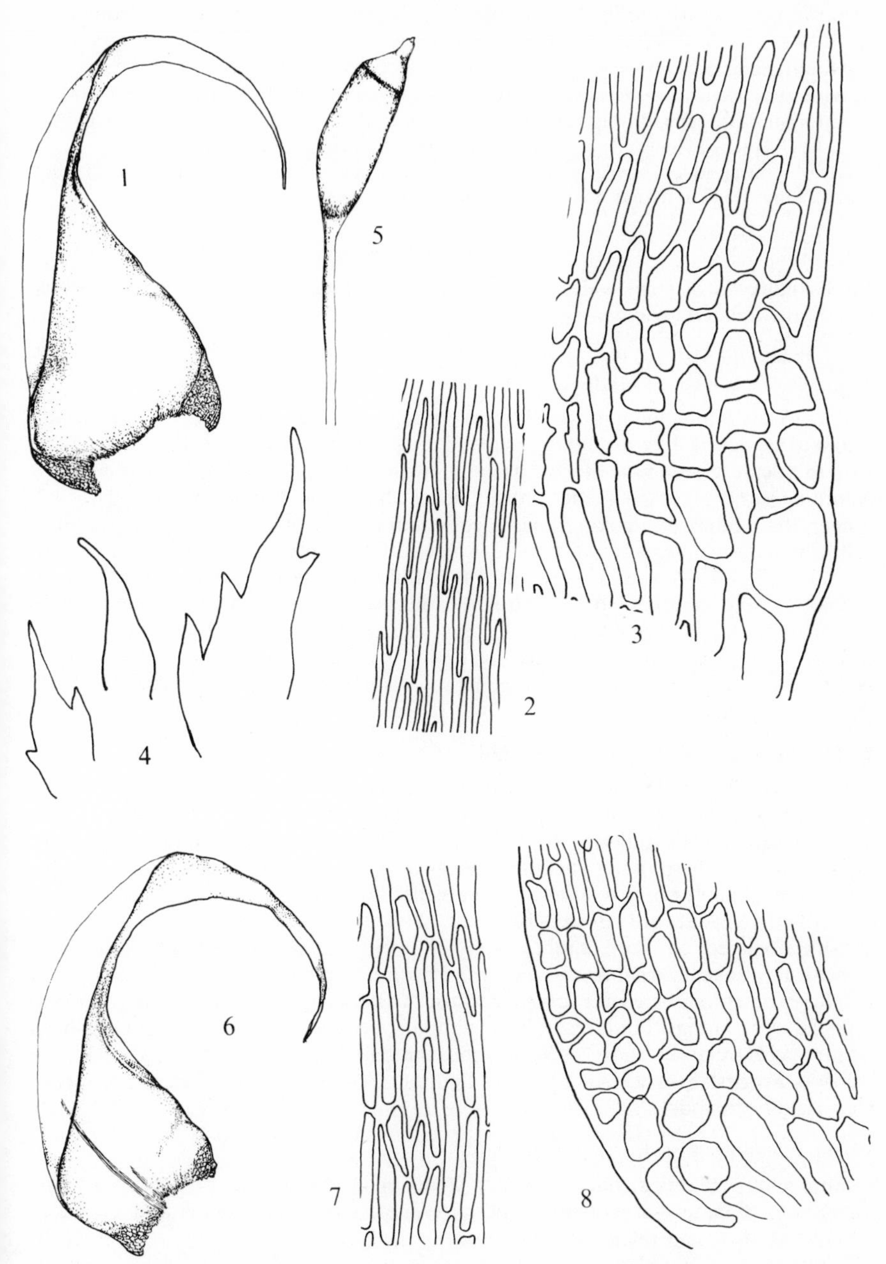

Fig. **317**. 1–5, *Hypnum imponens*: 1, leaf; 2, mid-leaf cells; 3, angular cells; 4, pseudoparaphyllia; 5, capsule (× 10). 6–8, *H. bambergeri*: 6, leaf; 7, mid-leaf cells; 8, angular cells. Leaves × 40, cells × 415.

to ovate-lanceolate, gradually tapering to long fine acumen, margin plane, entire or sinuose; nerve short and double or absent; cells porose, basal narrowly rectangular, angular quadrate or rectangular, very incrassate, orange-brown, forming small but distinct group, cells above elliptical to linear, 5–8 μm × 32–64 μm, 5–11 times as long as wide. Fr unknown in Britain. *n* = 10, 11. Yellowish-green to orange-brown patches or tufts on basic rocks at high altitudes, very rare. Perth, Angus, Inverness, Argyll. 6. N. and C. Europe, Iceland, N. Italy, Siberia, N. America, Greenland.

More robust and less regularly pinnate than *H. hamulosum* and the angular cells are much more distinctive than in that species and the cells above are more porose.

Section II. *Breidleria*

Epidermal cells of stem wide, thin-walled; angular cells of lvs few or numerous, sometimes inflated, hyaline.

8. H. lindbergii Mitt., J. Bot., Lond., 1864
H. patientiae Lindb. ex Milde

Dioecious. Plants medium-sized; stems procumbent to ascending, irregularly and sparsely branched. Lvs sub-falcate, secund, ovate or broadly ovate, shortly tapering to sub-falcate, secund, ovate or broadly ovate, shortly tapering to sub-acute to acuminate apex, margin plane, entire; nerve short and double, one branch often longer than other, or lacking; basal cells narrowly rhomboidal, porose, angular cells inflated, hyaline, forming small distinct auricles, cells above linear, ± vermicular, in mid-lf 5.0–6.5 μm × 52–112 μm, 10–20 times as long as wide. Capsule inclined, obloid, curved, mouth slightly oblique, exothecial cells uniformly thickened; spores 14–16 μm. Fr not known in Britain. Glossy, pale green to yellowish-green patches or tufts on damp, gravelly or sandy soil or paths, in quarries, by streams, in woods, on heaths, etc., occasional to frequent, generally distributed. 91, H31. Europe, Iceland, Asia, N. America, Greenland.

9. H. callichroum Brid., Br. Univ., 1827

Dioecious. Plants medium-sized, stems procumbent to ascending, complanately and pinnately branched; pseudoparaphyllia lanceolate-triangular. Lvs circinate-secund, lanceolate, gradually tapering to filiform acumen, margin plane, entire or bluntly denticulate near base; nerve lacking; basal cells narrowly rhomboidal, angular cells inflated, hyaline, forming distinct auricles, cells above linear, ± vermicular, in mid-lf 5–7(–8) μm × 48–88 μm, (6–)10–16 times as long as wide. Seta yellowish-green; capsule inclined to horizontal, oblong-ellipsoid, curved; exothecial cell walls uniformly thickened; spores 12–16 μm. Fr rare, summer. *n* = 10*. Glossy, green patches, sometimes extensive, on shaded rocks and ledges in humid woods, ravines, on cliffs and amongst boulders in basic habitats, occasional in western and northern Britain and Ireland, from N. Wales and the Lake District northwards. 33, H8. Europe, Faroes, Iceland, Asia Minor, Caucasus, N. and C. Asia, Yunnan, Japan, Canada, Greenland.

10. H. hamulosum Br. Eur., 1854

Autoecious. Plants slender; primary stems creeping, secondary procumbent to ascending, pinnately branched, branches complanate, short, stems and branches hooked at tips. Lvs falcato-secund to circinate-secund, ovate, longly tapering to filiform apex, margin plane or recurved near base, entire or sparsely denticulate from base to apex; nerve short and double or absent; basal cells narrowly rectangular to narrowly rhomboidal, angular cells very few, quadrate to trapezoid, intergrading

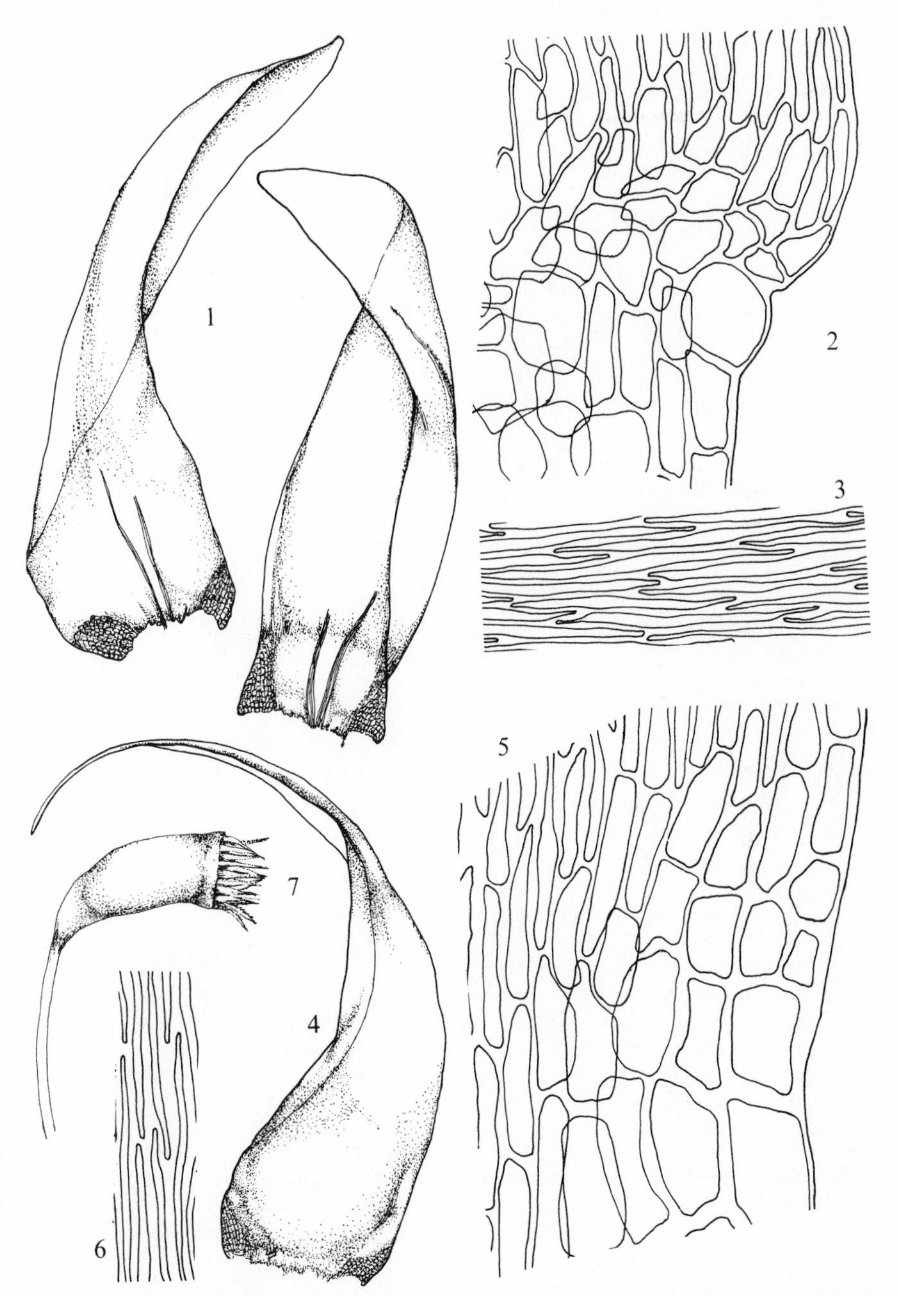

Fig. **318**. 1–3, *Hypnum lindbergii*: 1, leaves; 2, angular cells; 3, mid-leaf cells. 4–7, *H. callichroum*; 4, leaf; 5, angular cells; 6, mid-leaf cells; 7, capsule (×10). Leaves ×40, cells ×415.

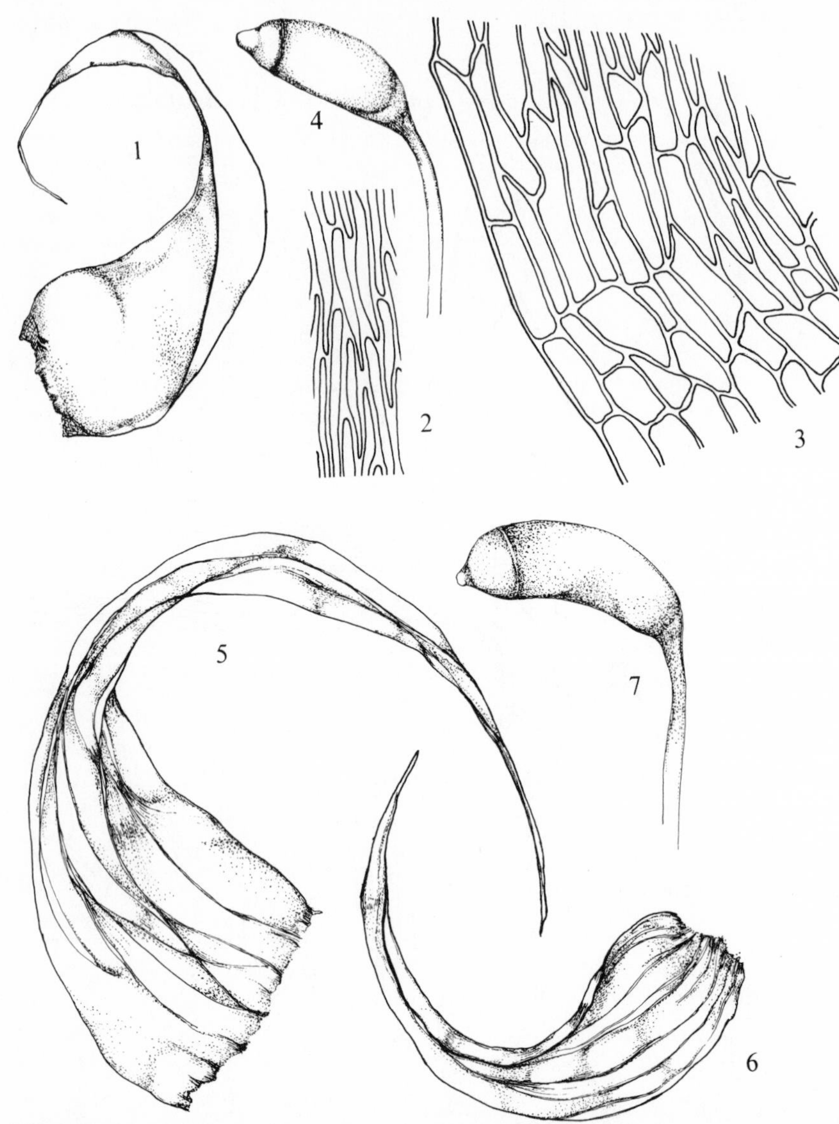

Fig. **319**. 1–4, *Hypnum hamulosum*: 1, leaf; 2, mid-leaf cells; 3, angular cells; 4, capsule. 5–7, *Ptilium cristacastrensis*: 5, stem leaf; 6, branch leaf; 7, capsule. Leaves ×40, cells ×415, capsules ×10.

with other cells and not forming distinct group, cells above linear, vermicular or not, in mid-lf 4–8 μm×(24–)40–64(–80) μm, 6–11 times as long as wide. Capsule inclined, narrowly ellipsoid, slightly curved; lid conical, obtuse; spores 12–14 μm. Fr occasional, summer. Yellowish-green to golden-brown patches on basic montane rocks, occasional. Merioneth, Caerns, Mid W. Yorks, Scottish Highlands. 21. N. and C. Europe, Pyrenees, Iceland, Faroes, Caucasus, N. Asia, Szechwan, Japan, N. America, Greenland.

169. Ptilium De Not., Cronac. Briol. Ital., 1867

Dioecious. Stems closely pinnately branched, branches complanate; pseudoparaphyllia abundant. Lvs circinate-secund, branch lvs smaller and narrower than stem lvs, both plicate, longly tapering to filiform acumen, base decurrent; nerveless; angular cells inflated, hyaline; cells above very narrow, linear-vermicular. Seta thin, reddish; capsule ± horizontal, curved. Only 2 species, one widespread in the northern hemisphere, the other occurring in the Caribbean.

1. P. crista-castrensis (Hedw.) De Not., Cronac. Briol. Ital., 1867
Hypnum crista-castrensis Hedw.

Dioecious. Plants medium sized, very elegant, shoots 5–8(–20) cm long, stems procumbent to erect, branching close, pinnate, complanate, branches of ± equal length except near stem apex. Lvs circinate-secund, longitudinally plicate, stem lvs ovate or ovate-lanceolate, longly tapering to filiform acumen, base longly decurrent; margin plane, entire or sinuose; nerve short and double or absent; basal cells linear-rhomboidal, cells at extreme angles inflated, hyaline, forming small, very distinct decurrent auricles, cells above linear-vermicular, in mid-lf 4–5 μm wide. Capsule horizontal, ovoid to obloid, gibbous to curved; lid mamillate; spores 11–13 μm. Fr rare, autumn. Lax green or yellowish-green patches or coarse wefts on litter in woodland, especially where coniferous, more rarely under *Calluna* and *Erica* or on soil amongst boulders in humid habitats, very rare in Merioneth and Caerns, occasional in the Scottish Highlands, W. Suffolk (probably introduced). 38. Europe, Caucasus, N. and C. Asia, Sikkim, Yunnan, N. America, Greenland.

170. Ctenidium (Schimp.) Mitt., J. Linn. Soc. Bot., 1869

Dioecious. Slender to robust plants; primary stems creeping, secondary stems prostrate to erect, pinnately and frequently plumosely branched; paraphyllia sparse. Stem and branch lvs differentiated, straight to circinate-secund; stem lvs with cordate-triangular base ± rapidly narrowed to acuminate apex, base broadly decurrent; nerve short and double or wanting; angular cells differentiated. Branch lvs narrowed, more gradually tapering to acuminate apex, base hardly decurrent. Seta smooth or papillose; capsule inclined to horizontal; cilia nodulose; calyptra ± hairy. A small genus of about 40 species.

Lf margin denticulate	**1. C. molluscum**
Lf margin entire or faintly denticulate only near apex	**2. C. procerrimum**

1. C. molluscum (Hedw.) Mitt., J. Linn. Soc. Bot., 1869
Hypnum molluscum Hedw.

Plants slender to robust; shoots procumbent to erect; primary stem creeping, pinnately branched or variously divided with divisions procumbent to erect and plumosely pinnately to fastigiately branched, branches crowded. Lvs smooth or more usually plicate, undulate or flexuose when dry, falcato-secund to circinate-secund, from broad base rapidly narrowed to long acumen, branch lvs smaller and narrower than stem lvs; nerve double, extending up to ¼ way up lf or absent; basal cells rhomboidal, porose, angular cells enlarged, ± incrassate, forming poorly defined

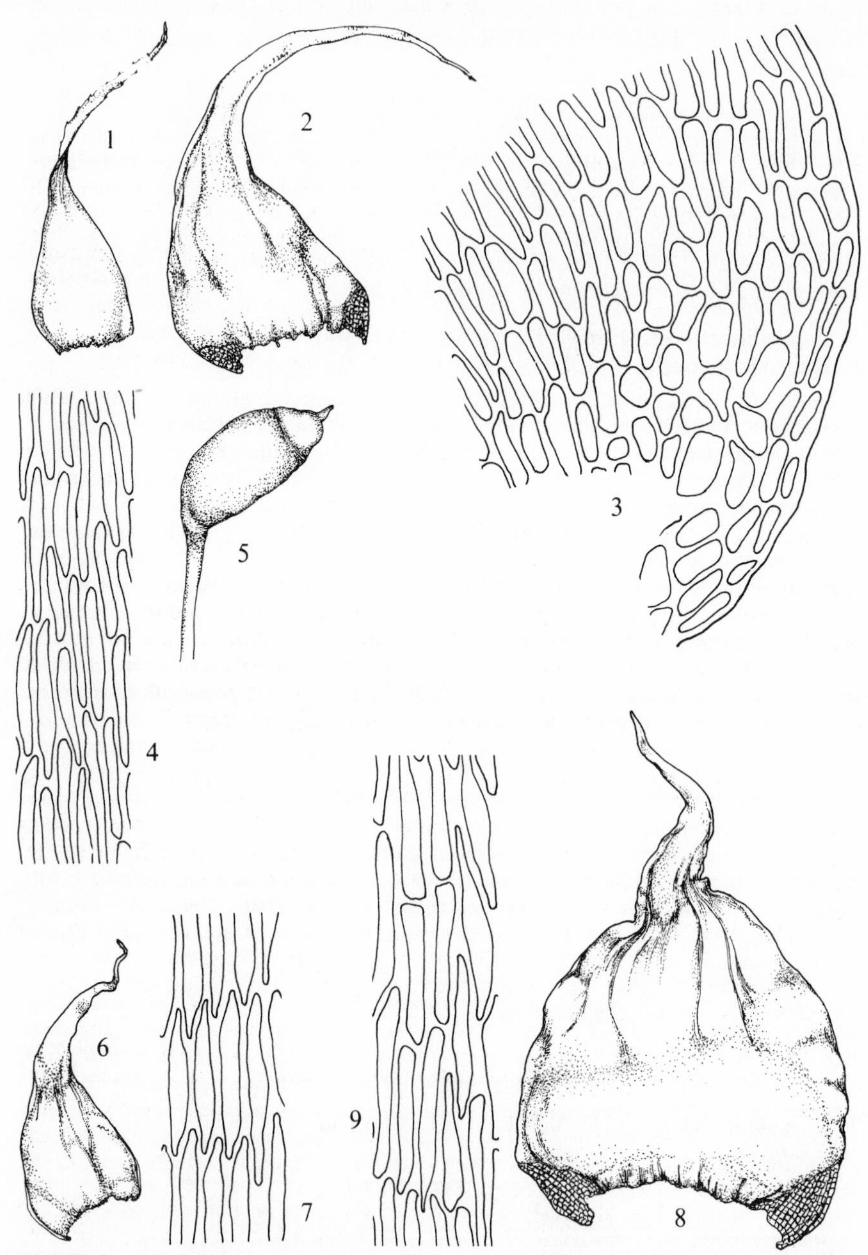

Fig. **320**. 1–5, *Ctenidium molluscum* var. *molluscum*: 1, branch leaf; 2, stem leaf; 3, angular cells; 4, mid-leaf cells; 5, capsule (×10). 6–7, *C. molluscum* var. *fastigiatum*: 6, stem leaf; 7, mid-leaf cells. 8–9, *C. molluscum* var. *condensatum*: 8, stem leaf; 9, mid-leaf cells. Leaves ×40, cells ×415.

decurrent auricles. Seta purple; capsule inclined, ellipsoid, gibbous; lid conical, acute or mamillate; spores 13–15 μm. Fr rare, winter.

Key to intraspecific taxa of *C. molluscum*

1 Plants very slender, stems erect, fastigiately branched, very crowded, stem lvs 0.8–1.3 mm long with mid-lf cells 28–48 μm long — var. **fastigiatum**
Plants slender to robust, branching various, stem lvs 1–3 mm long with mid-lf cells 36–150 μm long — 2

2 Plants robust, erect stems to 10 cm tall, irregularly branched, stem lvs 1.8–3.0 mm long with mid-lf cells 72–150 μm long — var. **robustum**
Plants slender to moderately robust, erect stems to 2 cm tall, stems ± regularly pinnately branched, stem lvs 1–2 mm long, mid-lf cells 40–76 μm long — 3

3 Stem and branch tips not conspicuously hooked, branches close, slender — var. **molluscum**
Stem and branch tips strongly hooked, branches not close, short and stout — 4

4 Branch and stem lvs of ± similar length, calcicole plants of basic rocks — var. **condensatum**
Branch lvs shorter than stem lvs, calcifuge plant of soil in woodland — var. **sylvaticum**

Var. **molluscum**

Plants small or slender; shoots procumbent to erect, when erect rarely more than 2 cm high; primary stems creeping, forked, divisions procumbent to ascending, usually closely, regularly pinnately branched, plumose, but sometimes ± fastigiately branched, branches crowded, stems and branches not markedly hooked at tips. Lvs crowded in dry habitats, distant in moist ones, stem lvs usually plicate, flexuose when dry, 1–2 mm long, strongly falcato-secund to circinate-secund, rarely almost straight, cordate-triangular to broadly cordate-triangular, rapidly narrowed to long acumen, margin often strongly denticulate; mid-lf cells 5–7 μm × 36–76 μm, 6–12 times as long as wide. Branch lvs 0.7–1.4 mm long. $n=7$, 8*. Dense, soft, glossy, yellowish-green to golden, sometimes bronze-tinged, rarely green mats or patches, sometimes extensive on compacted soil in grassland, rocks, inland and coastal cliffs, flushes or more straggling shoots or patches in fens and marshes calcicole. Common and sometimes abundant in basic habitats, rare elsewhere. 112, H40, C. Europe, Iceland, Faroes, Caucasus, N. Asia, Kamchatka, Algeria, Canaries, Azores, N. America.

Var. **fastigiatum** (Bosw. ex Hobk.) Braithw., Br. Moss Fl., 1902

Plants very slender; shoots densely crowded, to 3 cm high; primary stems creeping, secondary erect, forked, fastigiately branched. Lvs not plicate but sometimes flexuose when dry, stem lvs 0.8–1.3 mm long, ovate-cordate to cordate-triangular, shortly tapering to filiform acumen; mid-lf cells 4.0–6.5 μm × 28–48 μm, (4–)5–7 times as long as wide. Branch lvs 0.5–0.9 mm long, ovate-lanceolate, more gradually tapering. Very dense, golden-green to golden-brown tufts or deep patches on basic rocks, rare. Scattered localities from N. Devon, Glamorgan and Staffs north to W. Ross and W. Sutherland, W. Ireland. 15, H5. Endemic.

Var. **condensatum** (Schimp.) Britt., Mem. Torrey Bot. Cl., 1894

Plants moderately robust; shoots to 8 cm long but with branches hardly more than 1 cm high; stems creeping, forked, pinnately branched, branches short, spreading to erect, stem and branch tips strongly hooked. Lvs strongly plicate, frequently flexuose or undulate above, stem lvs (1.1–)1.4–2.0 mm long, falcato-secund, cordate-triangular, shortly tapering to filiform acumen, margin denticulate; mid-lf cells 5.5–8.0 μm × 40–56(–72) μm, 6–10(–13) times as long as wide. Branch lvs of similar length but narrower and more gradually tapering. Glossy, golden-green to golden-brown patches on damp, shaded, basic montane cliffs, in scree and flushes, occasional. From S. Somerset, Hereford and Staffs northwards. 50, H12. Luxembourg.

Var. **robustum** Boul. in Braithw., Br. Moss Fl., 1902

Plants robust; shoots to 10 cm tall; primary stems creeping, secondary erect, distantly and irregularly branched, branches erect, long. Lvs strongly plicate, falcato-secund, stem lvs mostly 1.8–3.0 mm long, ovate-cordate, tapering to filiform acumen, margin denticulate; mid-lf cells 5–10 μm × (40–)72–150 μm, (7–)9–20 times as long as wide. Branch lvs smaller 1.2–2.0 mm long, more strongly falcate, ovate-lanceolate, more gradually tapering. Golden-brown, often reddish-tinged, deep tufts on damp, basic montane rocks, rare. From Caerns and Mid W. Yorks north to Argyll and the Hebrides, Kerry, Waterford, Down and Antrim. 15, H5. Europe.

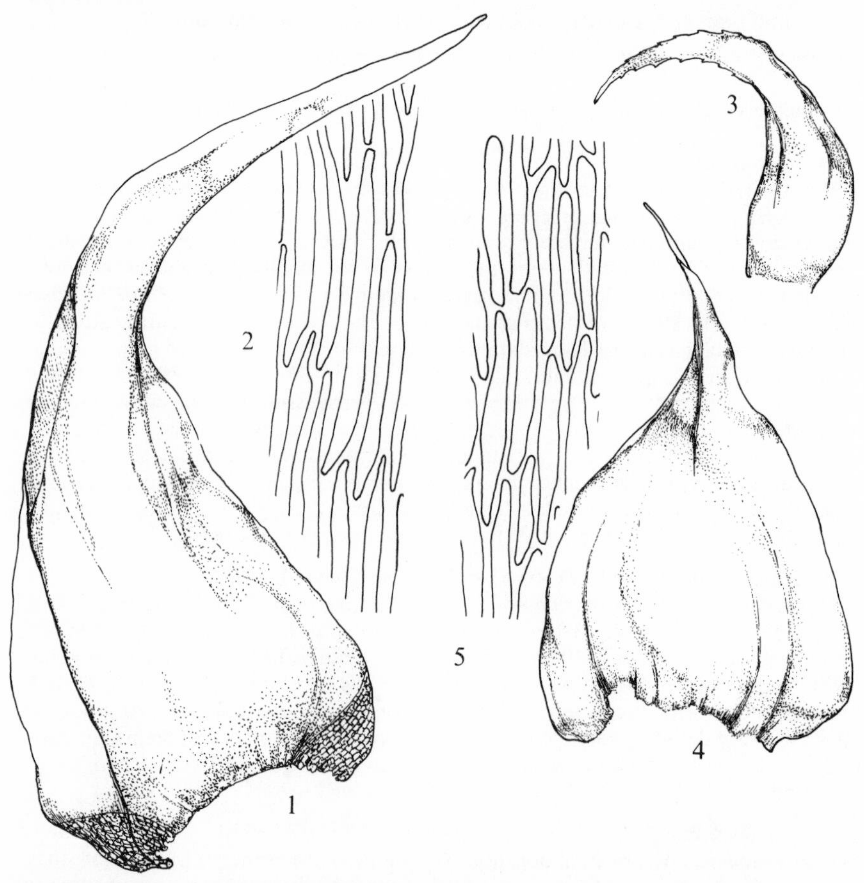

Fig. **321**. 1–2, *Ctenidium molluscum* var. *robustum*: 1, stem leaf; 2, mid-leaf cells. 3–5, *C. molluscum*, woodland taxon: 3, stem leaf; 4, branch leaf; 5, mid-leaf cells. Leaves × 40, cells × 415.

Var. **sylvaticum** F. Rose, J. Bryol., 1980

Plants medium-sized; patches rarely more than 1 cm deep; primary stems creeping, to 4 cm long, secondary stems prostrate, forked, ± pinnately branched with short, erect, ± fastigiate branches with strongly hooked, pale tips. Lvs plicate, often undulate above, stem lvs 1.3–2.1 mm long, falcato-secund, cordate-triangular, shortly tapering to filiform acumen, margin denticulate; mid-lf cells 5.5–7.0 μm× (32–)40–64 μm, 5–10 times as long as wide. Branch lvs 1.0–1.6 mm long, strongly falcate to falcate-circinate, ovate, gradually tapering to filiform apex. Dense, yellowish-green to golden-green patches with glossy, pale-tipped stems and branches on acidic soil and humus in woodland, frequent in S. England, Pembroke, Caerns, distribution elsewhere not known. 14. Brittany.

Var. *sylvaticum* is distinctive in its habit and preference for a substrate with a pH of about 5.5. It is frequent in the Weald and Cornwall and has been recorded from the following vice-counties: E. and W. Cornwall, S. Hants, Dorset, E. and W. Sussex, E. and W. Kent, Surrey, Herts, Northants, Worcs, Pembroke, Caerns. It probably occurs elsewhere particularly in southern England. Information on habitat and most of the distributional data were provided by Dr F. Rose who has also pointed out that the plant remains distinct when cultivated with var. *molluscum*. It would appear the woodland plant is a distinct taxon, probably of subspecific or even specific status, but before any firm conclusion can be reached further study is necessary. It is possible, however, that var. *condensatum* is close to the woodland taxon and that, as with some higher plants, montane cliffs and scree provide suitable ecological niches for otherwise woodland plants.

Var. *fastigiatum* is very poorly marked as intermediates between it and var. *molluscum* are frequent; var. *condensatum* is more distinctive but may merely be a more robust form of damper habitats. It somewhat resembles the woodland taxon but differs in ecology, the longer procumbent stems and the larger branch lvs; var. *robustum* is very distinctive in its size, habit, mode of branching and longer lf cells and might perhaps be regarded as a subspecies. *C. molluscum* is exceedingly variable and before the status of the varieties can be settled experimental study is required.

2. C. procerrimum (Mol.) Lindb., Bot. Not., 1882
Hypnum procerrimum Mol., *Pseudostereodon procerrimum* (Mol.) Fleisch.

Plants medium-sized, stems prostrate, ± interruptedly very closely pinnately or sometimes sub-pinnately branched, branches strongly complanate. Lvs similar moist and dry, not flexuose at tips or rugose, concave, stem lvs falcato-secund, from ovate-lanceolate basal part gradually tapering to acuminate apex, margin plane, entire or slightly denticulate above; nerve double, extending up to $\frac{1}{3}$ way up lf; extreme basal cells narrowly rectangular, angular cells quadrate to rectangular, incrassate, ± orange, forming conspicuous but ill-defined group of cells, cells elsewhere ± linear, in mid-lf 4–6 μm×32–64 μm, 7–13 times as long as wide. Branch lvs smaller, narrower, strongly falcato-secund to circinate-secund, from lanceolate or narrowly lanceolate basal part tapering to acuminate apex, margin entire. Sporophyte unknown. Dense, flat, dull green to yellowish-brown patches on basic montane rocks, very rare. Mid Perth, E. Inverness. 2. N. Europe, Alps, Tatra, Siberia, C. Asia, Yunnan, Sinkiang, arctic N. America.

171. Hyocomium Br. Eur., 1853

A monotypic genus with the characters of the species.

1. H. armoricum (Brid.) Wijk & Marg., Taxon, 1961
H. flagellare Br. Eur.

Dioecious. Plants small to robust; shoots to 15 cm long; primary stems creeping, secondary procumbent to ascending or pendulous, forked, irregularly or interruptedly

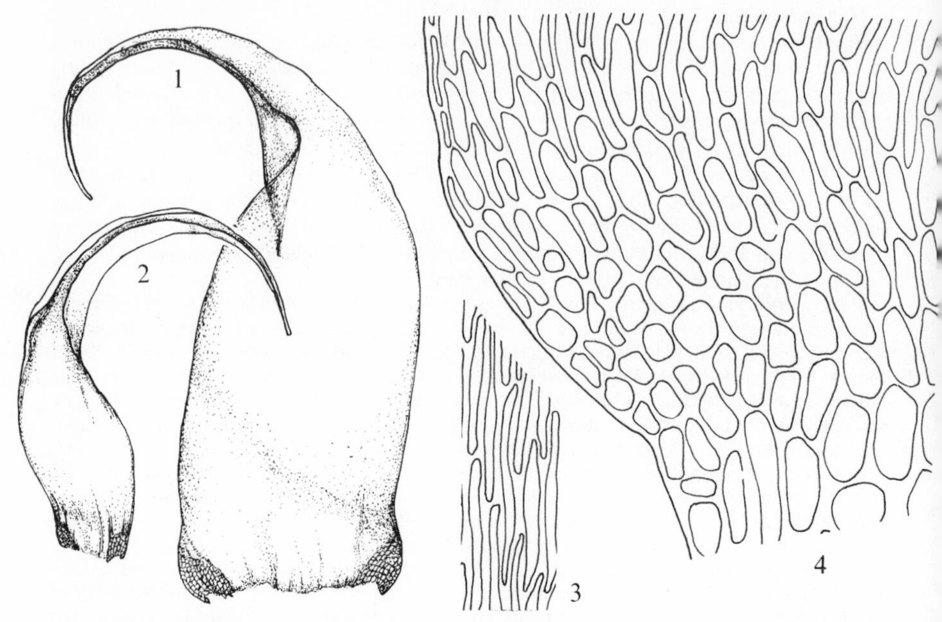

Fig. **322**. *Ctenidium procerrimum*: 1, stem leaf; 2, branch leaf; 3, mid-leaf cells; 4, angular cells. Leaves ×40, cells ×415.

pinnately branched, branches short to attenuated; paraphyllia and pseudoparaphyllia lacking. Lvs mostly patent to spreading, plicate, sometimes rugose when dry, patent to erecto-patent, sometimes slightly secund when moist, stem lvs broadly cordate-triangular, shortly tapering to acute to filiform apex, base decurrent, margin plane, irregularly toothed from base to apex; nerve short and double or wanting; cells thin-walled to incrassate, porose, basal rhomboidal to narrowly rhomboidal, angular cells enlarged, irregularly hexagonal, cells above variable in shape, narrowly rhomboidal or trapezoid to linear, in mid-lf 6–9 μm × 40–84 μm, 7–10 times as long as wide. Branch lvs smaller, less triangular, on ultimate branches ovate, acute. Seta purplish-red, papillose; capsule ± horizontal, obloid, very slightly curved; spores *ca* 16 μm. Fr rare, winter. $n=8$. Glossy, golden-green or golden-brown patches, sometimes extensive, on rocks, tree roots and boles in flood zone of streams and rivers, especially where fast-flowing, by waterfalls and on wet cliffs, rare in lowland areas, frequent or common elsewhere, especially in the west. 81, H32. W. and C. Europe, Corsica, Transcaucasia, Asia Minor, Japan, Azores.

Thamnobryum alopecurum and *Isothecium holtii* frequently grow in similar habitats and sometimes mixed with *H. armoricum* and may be confused with it. Both are readily distinguished by the narrower, nerved lvs.

172. RHYTIDIUM (SULL.) KINDB., BIH. K. SVENSK. VET. AK. HANDL., 1882

A monotypic genus with the characters of the species.

1. R. rugosum (Hedw.) Kindb., Bih. K. Svensk. Vet. Ak. Handl., 1883
Hylocomium rugosum (Hedw.) De Not.

Dioecious. Robust plants; shoots to 10 cm long; stems procumbent to ascending,

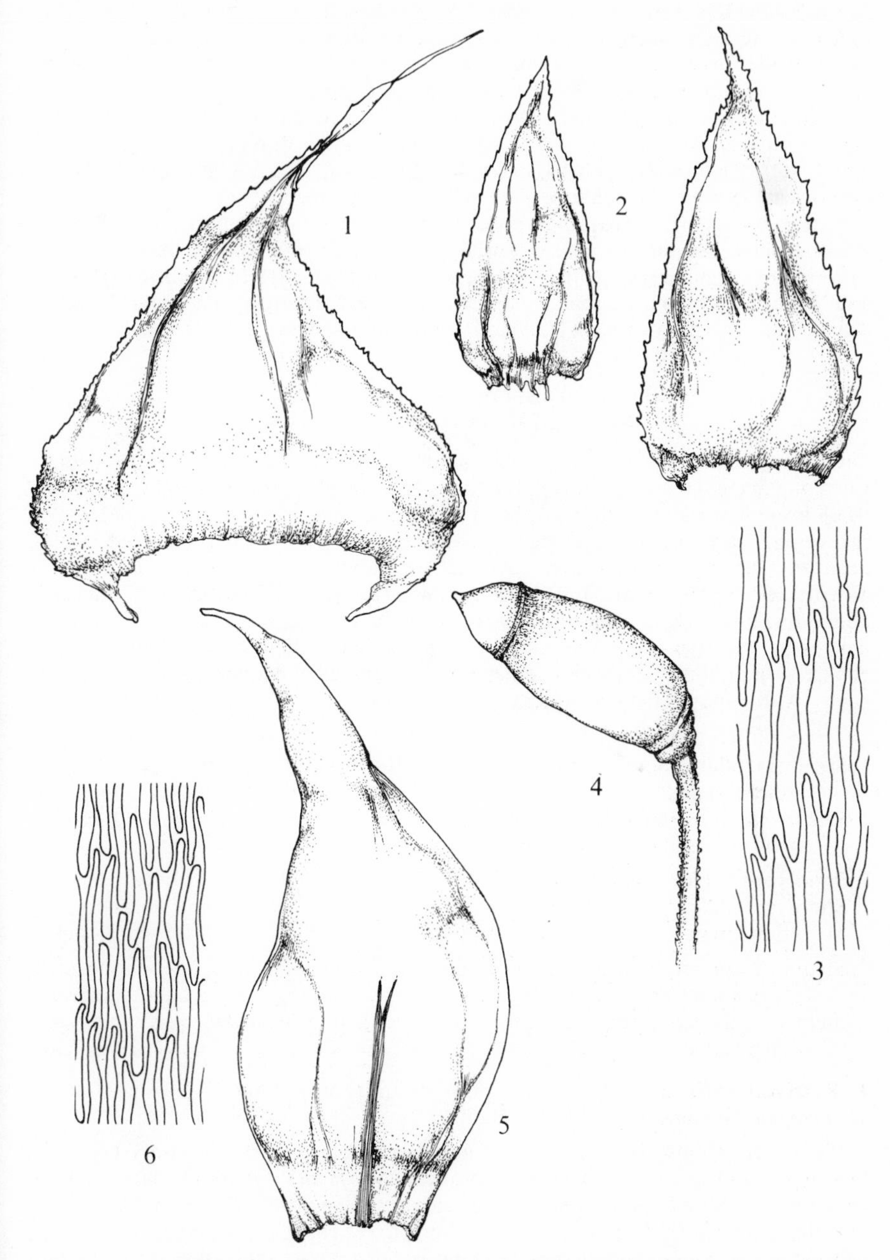

Fig. **323**. 1–4, *Hyocomium armoricum*: 1, stem leaf (×40); 2, branch leaves (×40); 3, mid-leaf cells; 4, capsule (×10). 5–6, *Rhytidium rugosum*: 5, leaf (×25); 6, mid-leaf cells. Cells ×415.

forked, irregularly pinnate, branches erect to spreading, pseudoparaphyllia lanceolate-subulate. Lvs concave, imbricate, rugose, falcato-secund, stem lvs ovate, tapering to acuminate apex, margin recurved $\pm$ from base to near apex, denticulate above; nerve single, forked above, reaching $\frac{1}{2}$–$\frac{3}{4}$ way up lf; cells porose, basal rhomboidal, angular cells numerous, quadrate to trapezoid, ascending up margin and intergrading with other cells, cells above narrowly elliptical with papillae at back above, in mid-lf 5.5–8.0 μm $\times$ 28–56 μm long, 4–7 times as long as wide. Branch lvs smaller, ovate-lanceolate to lanceolate, otherwise similar. Seta long; capsule horizontal, shortly cylindrical, curved; lid obliquely rostellate; peristome double, well developed; calyptra almost covering capsule; spores *ca* 16 μm. Fr unknown in Britain or Ireland. $n=10$. Coarse, lax patches, pale yellowish-green to golden-brown, in basic grassland or very rarely in dune grassland. N. Somerset, E. Anglia and Hereford north to E. Ross and W. Sutherland, occasional, Londonderry. 30, H1. Europe, Caucasus, N. and C. Asia, China, Japan, Morocco, N. America, Greenland, Mexico, Guatemala.

173. RHYTIDIADELPHUS (WARNST.) WARNST., KRYPT. FL. BRANDENBURG, 1906

Dioecious. Plants robust or very robust, shoots erect to procumbent, stems usually very long, irregularly to pinnately branched; paraphyllia absent, pseudoparaphyllia when present small, broad-based. Stem and branch lvs $\pm$ similar, lvs squarrose to spreading, from broad basal part gradually or abruptly narrowed to acute to acuminate acumen, margin denticulate at least above; nerve short and double or absent; cells $\pm$ incrassate, porose or not, angular cells differentiated or not, cells above linear-elliptical, smooth or papillose at back. Seta deep red, smooth, long, flexuose, sometimes convolute; capsule $\pm$ horizontal, obloid or ovoid, gibbous; lid conical or mamillate; inner peristome with nodulose cilia. A mainly north temperate genus (with 1 species from S. America) of 8 species.

1 Stem lvs plicate, not squarrose or strongly reflexed, angular cells not much differentiated — 2
  Stem lvs squarrose or strongly reflexed, not plicate, angular cells forming distinct group of cells — 3
2 Shoots erect, lvs not falcato-secund, branches not complanate — **1. R. triquetrus**
  Shoots arcuate, procumbent or stoloniform, lvs falcato-secund, branches $\pm$ complanate — **4. R. loreus**
3 Stem lvs squarrose, with erect sub-sheathing basal part completely concealing stem, stem tips stellate with crowded lvs — **2. R. squarrosus**
  Stem lvs reflexed from erecto-patent or patent, non-sheathing basal part, stem often visible between lvs, stem tips hardly stellate — **3. R. subpinnatus**

1. R. triquetrus (Hedw.) Warnst., Krypt. Fl. Brandenburg, 1906
Hylocomium triquetrum (Hedw.) Br. Eur.

Plants very robust, shoots to 20 cm long, secund at tips, stems orange-brown to brown, erect or ascending, irregularly branched, branches not complanate, short or sometimes attenuated. Lvs $\pm$ spreading, not squarrose or secund, when dry scarious but otherwise hardly altered, plicate and sometimes rugose, stem lvs from $\pm$ erect, sub-sheathing cordate basal part tapering to acute apex, base decurrent, margin plane, denticulate to dentate above; nerve double, extending to $\frac{3}{4}$ way up lf; basal cells elliptical, angular cells shorter, wider, more thin-walled but hardly distinctive,

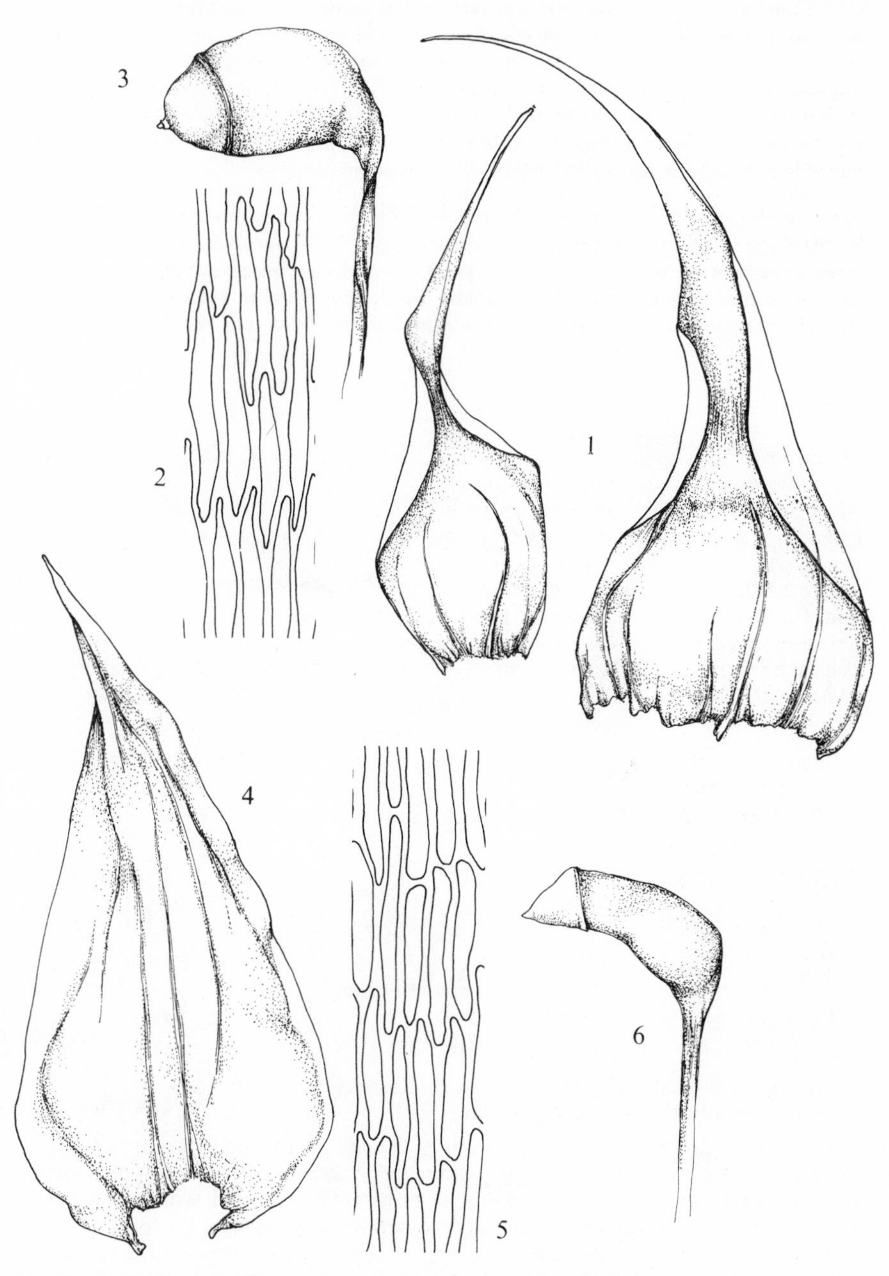

Fig. **324**. 1–3, *Rhytidiadelphus loreus*: 1, leaves (× 25); 2, mid-leaf cells; 3, capsule. 4–6, *R. triquetrus*: 4, leaf (× 15); 5, mid-leaf cells; 6, capsule. Cells × 415, capsules × 10.

cells above linear-elliptical, upper papillose at back, in mid-lf 6–9 μm × 48–72 μm, 7–11 times as long as wide. Branch lvs smaller, tapering from cordate-triangular basal part. Capsule ± horizontal, obloid, gibbous; lid conical, acute; spores 14–21 μm. Fr rare, winter. $n = 5, 6^*$. Green or yellowish-green, coarse tufts or patches, pale reddish-brown below, sometimes extensive, on grassy banks, in damp turf, open places in woodland, moorland, montane slopes and ledges, sand-dunes, etc., common and sometimes locally abundant. 112, H40, C. Europe, Faroes, Iceland, Caucasus, N. and C. Asia, Yunnan, Japan, Madeira, N. America.

2. R. squarrosus (Hedw.) Warnst., Krypt. Fl. Brandenburg, 1906
Hylocomium squarrosum (Hedw.) Br. Eur.

Plants robust, shoots to 15 cm long, erect or ascending at least at tips, irregularly and sometimes sparsely pinnately branched, branches short or sometimes long and attenuated, stems towards tips of shoots and branches green, elsewhere reddish-

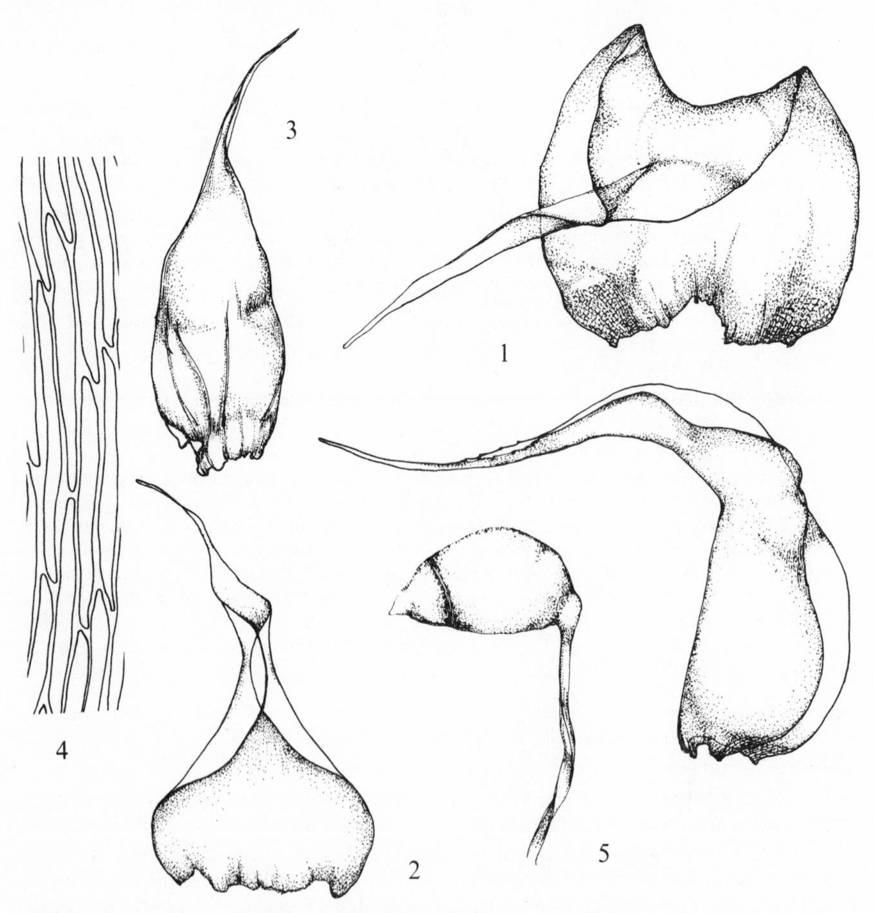

Fig. **325**. *Rhytidiadelphus squarrosus*: 1, stem leaves; 2, leaf from base of branch; 3, leaf from upper part of branch; 4, mid-leaf cells (× 415); 5, capsule (× 10). Leaves × 25.

brown. Stem lvs not plicate, strongly squarrose and at shoot tips crowded, producing stellate appearance, from erect, ± sheathing broadly ovate basal part squarrose or strongly squarrose, narrowed but not abruptly so to long acuminate apex, margin plane, denticulate above; nerve double, extending $\frac{1}{4}$–$\frac{1}{3}$ way up lf; basal cells narrowly rhomboidal, angular cells enlarged, hyaline or coloured, forming distinctive group, cells above linear-elliptical, smooth, in mid-lf 6–9 μm × 40–80(–96) μm, 7–10(–12) times as long as wide. Lower lvs of branches of ± similar shape to but smaller than stem lvs, lvs towards ends of branches ovate-lanceolate, tapering to long, fine, denticulate acumen. Seta flexuose, sometimes bent or twisted; capsule horizontal, ovoid, gibbous; lid conical, acute; spores 18–20 μm. Fr very rare, winter. $n = 10$. Pale green to yellowish, dense mats or tufts, often brownish below, sometimes extensive, in turf, on banks, streamsides, roadsides, in open woodland, heaths, marshy ground, etc., usually where damp, common and sometimes locally abundant. 112, H40, C. Europe, Faroes, Iceland, E. and C. Asia, Madeira, Azores, N. America.

3. R. subpinnatus (Lindb.) Kop., Hikobia, 1971
R. calvescens (Lindb.) Roth, *R. squarrosus* ssp. *calvescens* (Lindb.) Giac., *R. squarrosus* var. *calvescens* (Lindb.) Warnst.

Plants of similar dimensions to *R. squarrosus* but less robust, shoots procumbent, ± pinnately branched, branches complanate, short or long, stems reddish-brown almost to extreme tips. Stem lvs not crowded in stellate fashion at stem tips, not plicate, reflexed from ± erecto-patent but not erect nor sheathing basal part, basal part broadly ovate to broadly ovate-cordate, usually ± abruptly narrowed into reflexed acuminate upper part, margin plane, denticulate in upper part; nerve double, extending up to $\frac{1}{3}$ way up lf; areolation as in *R. squarrosus*. Lvs on lower parts of branches differing in shape from stem lvs, ovate-lanceolate, tapering to acuminate apex, lvs towards ends of branches ovate, abruptly narrowed or shortly tapering to denticulate acumen. Sporophyte as in *R. squarrosus*. Fr very rare. Green patches on damp soil, in grassy places and stream banks in deciduous woodland, very rare and probably over-recorded in Britain. A few localities from Hereford, Worcs and Derby north to Mid Perth and N. Ebudes. 13. Northern and C. Europe, N. Asia, Japan, N. America.

The distribution of this plant in Britain is uncertain as about half the gatherings attributed to this taxon belong to *R. squarrosus*. Correct records are from Merioneth, Cheshire, S. Lancs and Westmorland. Although confused with slender forms of *R. squarrosus*, *R. subpinnatus* may well be overlooked as that plant. It is most easily recognised by its softer appearance; the lvs are reflexed from a patent, non-sheathing base rather than squarrose from an erect sheathing base, this giving the lvs a less rigid appearance and also revealing the stem in places. The stem lvs of *R. subpinnatus* may have a much wider base more abruptly narrowed into the acumen but not always so. Depauperate forms of *R. squarrosus* growing on wet ground may be very difficult or impossible to distinguish from small plants of *R. subpinnatus*.

4. R. loreus (Hedw.) Warnst., Krypt. Fl. Brandenburg, 1906
Hylocomium loreum (Hedw.) Br. Eur.

Plants very robust, shoots ascending, arcuate or stoloniform, to 30(–50) cm long, stems reddish-brown or orange-brown, ± pinnately branched, branches short, spreading. Lvs patent, falcato-secund, hardly altered when dry, stem lvs from concave, ovate, plicate basal part gradually tapering to long channelled acumen, margin plane or narrowly recurved near base, denticulate from middle or below; nerve short, double and faint or absent; basal cells linear-rhomboidal, angular cells shorter, wider, more incrassate, sometimes coloured but not distinctive, cells above

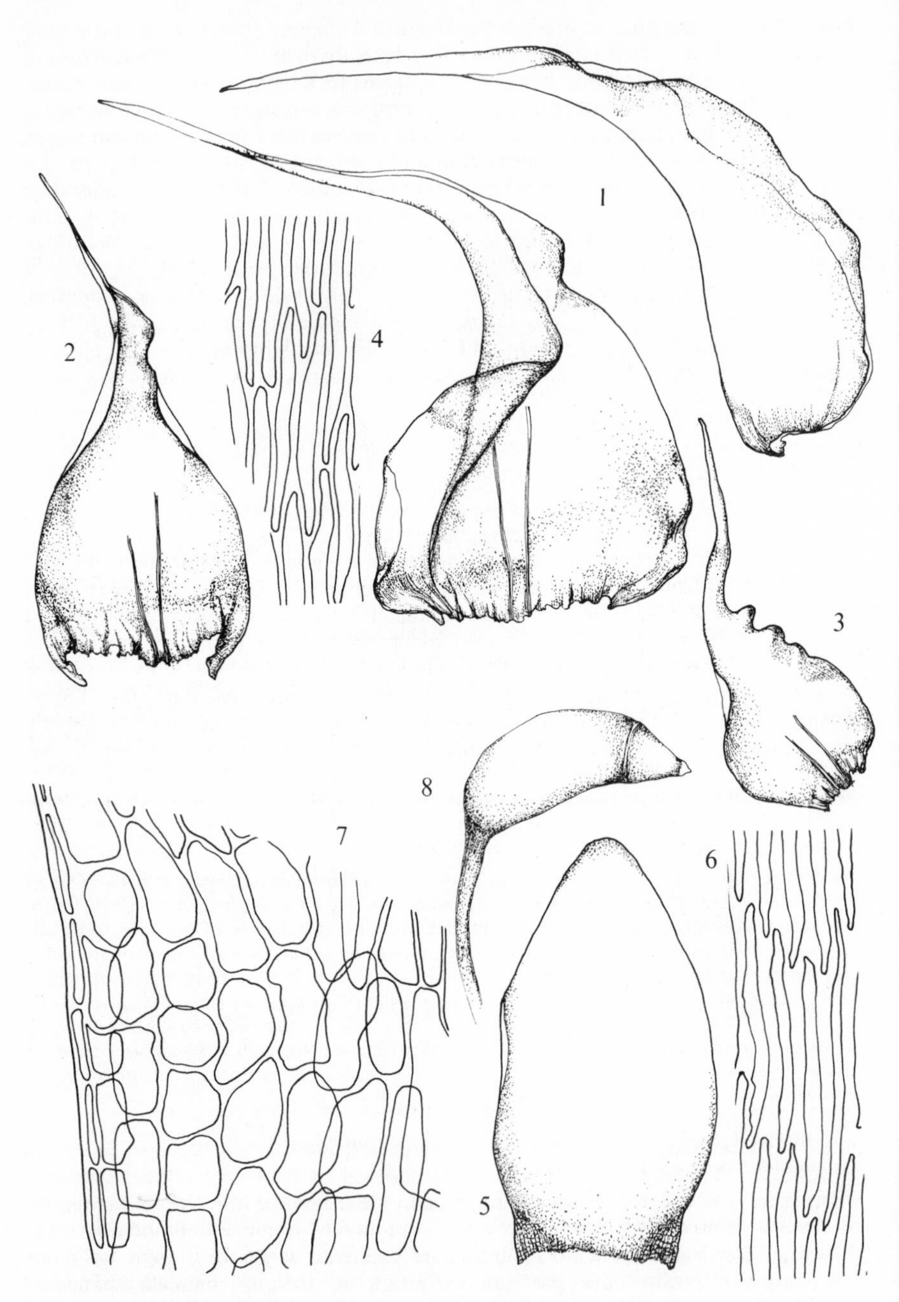

Fig. **326**. 1–4, *Rhytidiadelphus subpinnatus*: 1, stem leaves; 2, leaf from base of branch; 3, leaf from upper part of branch; 4, mid-leaf cells. 5–8, *Pleurozium schreberi*: 5, leaf; 6, mid-leaf cells; 7, angular cells; 8, capsule (× 10). Leaves × 25, cells × 415.

linear-elliptical, smooth, in mid-lf 6–8(–10) μm×40–80 μm, (5–)8–10 times as long as wide. Branch lvs smaller and narrower but otherwise similar. Capsule horizontal, ovoid, gibbous; lid mamillate or conical; spores 14–18 μm. Fr rare to occasional, winter. $n=5$*. Coarse wefts or straggling patches, yellowish-green, on soil, boulders, fallen logs in humid habitats in woodland, on banks, in boulder scree, on cliff ledges and on marshy ground, frequent or common in western and northern Britain and Ireland, rare to occasional elsewhere. 105, H39, C. Europe, Faroes, Madeira, Azores, N. America.

174. Pleurozium Mitt., J. Linn. Soc. Bot., 1869

A monotypic genus with the characters of the species.

Pleurozium is frequently placed in the Entodontaceae because of the similarity of the gametophyte to that of *Pseudoscleropodium purum* which is often similarly treated. For reasons given under *Pseudoscleropodium* that genus is placed in the Brachytheciaceae. Cytologically *Pleurozium* is unlike either *Entodon* or *Pseudoscleropodium* but very similar to *Rhytidiadelphus*, likewise with the peristome. *Pleurozium* should therefore be placed near *Rhytidiadelphus*.

1. P. schreberi (Brid.) Mitt., J. Linn. Soc. Bot., 1869
Hypnum schreberi Brid.

Dioecious. Plants medium-sized; shoots to 12 cm long, stems deep red, procumbent to ascending, pinnate, branches spreading, complanate, of ± uniform length, often slightly curved; paraphyllia and pseudoparaphyllia lacking. Lvs loosely imbricate, rendering stems and branches somewhat julaceous, stem lvs concave, ovate or broadly ovate, apex obtuse or rounded, margin incurved above, entire; nerve very short and double; cells incrassate, porose, basal trapezoid to elliptical, angular rectangular, brown, forming distinct auricles, cells above linear, 6.5–10.0 μm×60–132 μm, 8–16 times as long as wide. Branch lvs smaller, apex obtuse or truncate. Seta long, thin, red; capsule inclined, obloid, curved; lid conical, obtuse or apiculate; outer peristome teeth finely papillose throughout, inner peristome with tall basal membrane, processes keeled with widely gaping perforations, cilia well developed; spores 12–18 μm. Fr occasional in Scotland and N. Wales, very rare elsewhere, autumn, winter. $n=5$*, 7. Glossy, pale green to yellowish patches or coarse wefts in dry places on heaths, moorland, sand-dunes, banks, in scree, etc., rarely in damp habitats, calcifuge, common and sometimes locally abundant in suitable habitats. 112, H40, C. Europe, Iceland, Faroes, Asia, America, Greenland.

SUBFAMILY 2. HYLOCOMIOIDEAE

Stems with abundant branched or toothed paraphyllia, pseudoparaphyllia present.

175. Hylocomium Br. Eur., 1852

Dioecious. Plants slender to robust; stems irregularly pinnately to regularly pinnately branched; paraphyllia abundant, spinosely toothed or branched. Stem lvs usually differing in shape from branch lvs, plicate, ovate-oblong to cordate-triangular, abruptly narrowed or shortly tapering to acute apex; nerve usually double; cells narrowly rhomboidal to linear, smooth or papillose at back, angular cells not differentiated. Branch lvs ± ovate, acute. Seta long, deep red; capsule inclined to horizontal, ovoid to ellipsoid, gibbous or curved; lid conical to rostrate; calyptra cucullate. A small genus of 8 species occurring in Europe, Asia, N. and C. Africa, N. America, the Caribbean and New Zealand.

Some authors place *H. brevirostre* in *Rhytidiadelphus* or *Loeskeobryum* but I see no grounds for placing the species other than in *Hylocomium*. Similarly *Hylocomiastrum* is a very poorly marked taxon and is not recognised here.

1 Stems bipinnate, stem lvs imbricate, lvs with scattered papillae at back above **4. H. splendens**
Stems irregularly or regularly pinnate or if partially bipinnate then stem lvs distant, ± erect, lf cells smooth at back above 2
2 Stems regularly pinnate, sometimes ± bipinnate, stem lvs ± tapering to acuminate apex **3. H. umbratum**
Stems sparsely or irregularly pinnate, stem lvs abruptly narrowed to acuminate apex 3
3 Stem lvs patent to spreading or squarrose with cordate bases, paraphyllia mostly 70–130 μm long **1. H. brevirostre**
Stem lvs loosely imbricate, bases not cordate, paraphyllia mostly 130–600 μm long **2. H. pyrenaicum**

1. H. brevirostre (Brid.) Br. Eur., 1852
Loeskeobryum brevirostre (Schwaegr.) Fleisch., *Rhytidiadelphus brevirostris* (Schwaegr.) Nyholm

Plants medium-sized or robust, shoots to 12 cm; stems reddish-brown, ascending, arcuate or stoloniform, forked and irregularly pinnately branched, branches spreading, not complanate; paraphyllia abundant, small, 70–130 μm long, irregularly, often sub-stellately branched. Lvs plicate, sometimes rugose at tips especially when dry, stem lvs patent to spreading or squarrose, from cordate clasping base, ± orbicular to broadly ovate or ovate-triangular, abruptly narrowed to frequently falcate, narrowly lanceolate acumen, margin plane, bluntly to coarsely toothed; nerve double, extending up to $\frac{1}{3}$ way up lf; basal cells narrowly rhomboidal, brownish, incrassate, strongly porose, angular cells thinner, walled, rhomboidal, cells above narrowly rhomboidal to linear-rhomboidal, smooth, in mid-lf 5–8 μm × 28–64 μm, 4–9(–11) times as long as wide. Branch lvs ovate to ovate-lanceolate, ± abruptly narrowed to usually straight acumen, margin coarsely toothed. Capsule inclined, ellipsoid, slightly curved; lid with stout, obtuse beak; spores 22–26 μm. Fr rare, winter. Lax, dull or yellowish-green patches or coarse wefts on banks, rocks, wall-tops, logs and branches in shaded, humid situations especially in basic localities, rare in S.E. England and the Midlands, occasional elsewhere. 74, H37, C. Europe, Faroes, N. Asia, Kamchatka, Korea, Japan, Tunisia, Algeria, N. America.

2. H. pyrenaicum (Spruce) Lindb., Musc. Scand., 1879
Hylocomiastrum pyrenaicum (Spruce) Fleisch.

Plants medium-sized; stems reddish-brown, procumbent to ascending, occasionally forked, irregularly and sparsely pinnately branched, branches spreading, paraphyllia abundant, 130–600 μm long, much branched. Lvs loosely imbricate, concave, strongly plicate, stem lvs ovate, abruptly narrowed to apiculus to ± gradually tapering to acute, twisted apex, margin recurved below, coarsely toothed; nerve single or rarely double, faint, extending to $\frac{1}{2}$ way up lf; basal cells rhomboidal, incrassate, porose, angular cells not differentiated, cells above narrowly rhomboidal to linear-rhomboidal, smooth, in mid-lf 5–7 μm × 32–64 μm, (4–)5–11 times as long as wide. Branch lvs ovate, shortly pointed or obtuse and apiculate, coarsely toothed. Capsule inclined, ovate-ellipsoid; lid conical, acute; spores *ca* 20 μm. Fr unknown in Britain. $n=12$. Yellowish-green patches or scattered plants in turf below cliffs, on

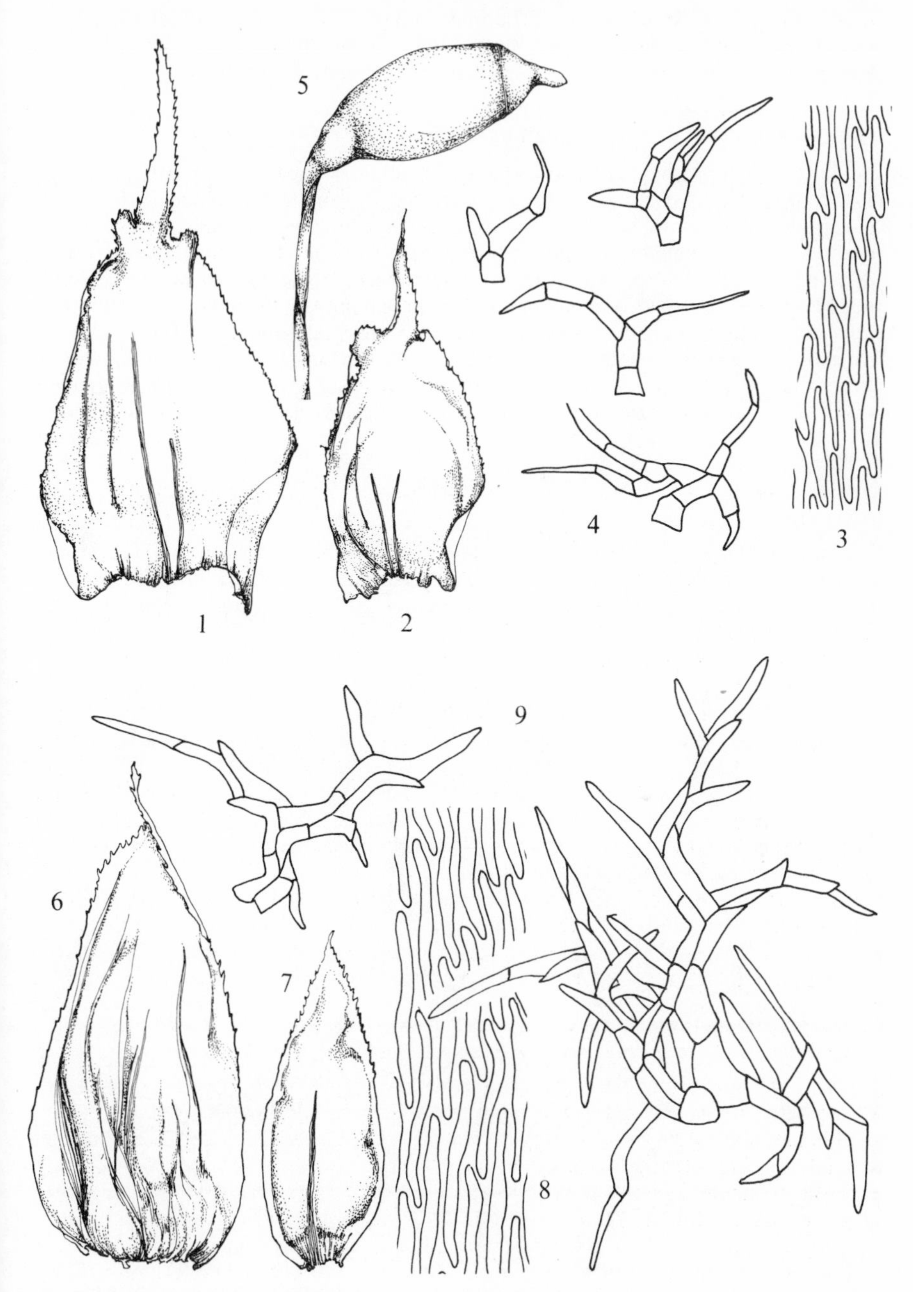

Fig. **327**. 1–5, *Hylocomium brevirostre*: 1, stem leaf; 2, branch leaf; 3, mid-leaf cells; 4, paraphyllia; 5, capsule (×10). 6–9, *H. pyrenaicum*: 6, stem leaf; 7, branch leaf; 8, mid-leaf cells; 9, paraphyllia. Leaves ×25, cells ×415, paraphylila ×250.

ledges and in boulder scree in basic habitats at high altitudes in Scotland, rare. Perth and Angus north to Ross and E. Sutherland. 10. Northern and montane Europe, Faroes, Iceland, Urals, Caucasus, N. and C. Asia, Japan, N. America.

3. H. umbratum (Hedw.) Br. Eur., 1852
Hylocomiastrum umbratum (Hedw.) Fleisch.

Plants slender, stems reddish-brown, ascending, forked, 1–2-pinnate, branches sub-complanate, sometimes crowded, variable in length, sometimes attenuated; paraphyllia abundant, variable in size, 240–1000 μm long, linear and irregularly branched to lanceolate-subulate and spinosely toothed. Lvs plicate, stem lvs ± distant, erect to erecto-patent, ovate-cordate, tapering to acute apex, base decurrent, margin plane, coarsely toothed; nerve double, extending up to $\frac{1}{3}$ way up lf; basal cells rhomboidal, angular cells scarcely differentiated, cells above narrowly to linear-rhomboidal, smooth, in mid-lf 5–7 μm × 40–64 μm, 6–11 times as long as wide. Branch lvs ± imbricate, concave, ovate to broadly ovate or ovate-triangular, coarsely toothed. Capsule horizontal, ovate-ellipsoid, gibbous; lid conical, obtuse; spores

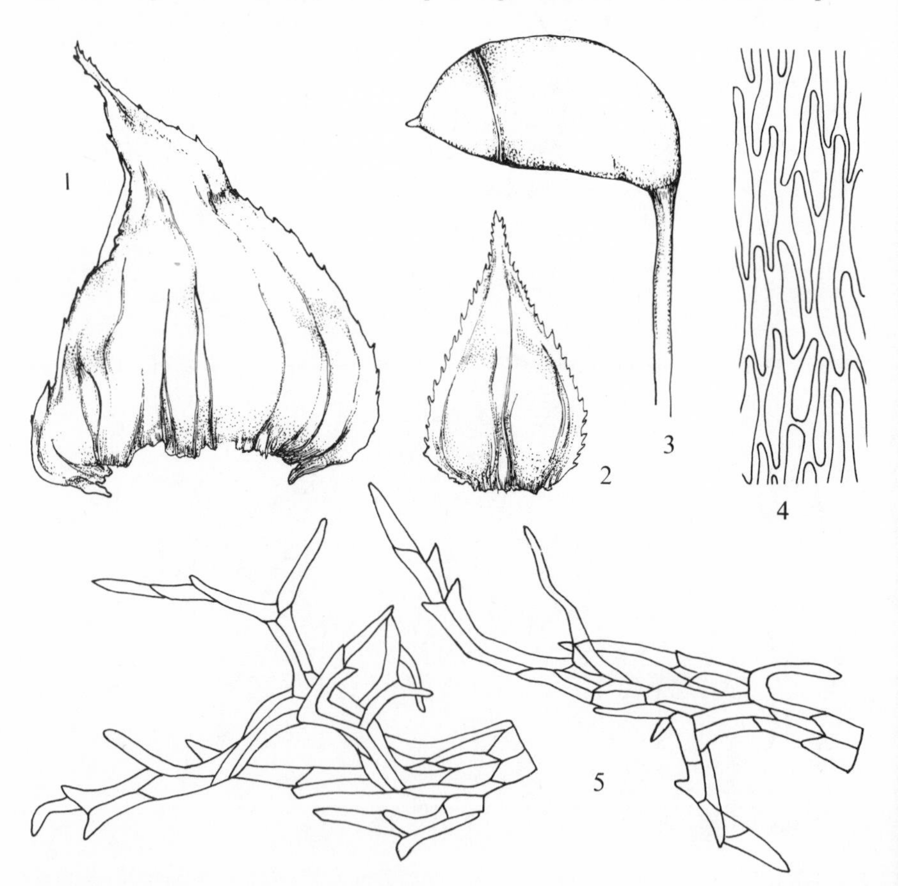

Fig. **328**. *Hylocomium umbratum*: 1, stem leaf; 2, branch leaf; 3, capsule (× 10); 4, mid-leaf cells (× 415); 5, paraphyllia (× 250). Leaves × 25.

14–18 μm. Fr very rare, autumn. $n=7$. Pale green to yellowish-green, sometimes brownish-tinged, coarse wefts or patches on soil, humus or rocks in open, humid woodland, rocky valleys, turf and rock ledges. From Mid Wales and the Lake District north to Sutherland, W. Ireland, from sea level to 1000 m, occasional. 31, H9. Europe, Faroes, Caucasus, Japan, Korea, Kamchatka, Algeria, Tunisia, N. America.

4. H. splendens (Hedw.) Br. Eur., 1852

Plants slender; shoots except in depauperate specimens with annual growth zones giving plants layered appearance; stems bright red, procumbent, forked, bipinnate or very rarely pinnate, branches complanate, not attenuated; paraphyllia abundant, 240–1000 μm long, linear-lanceolate with numerous long filiform branches. Lvs

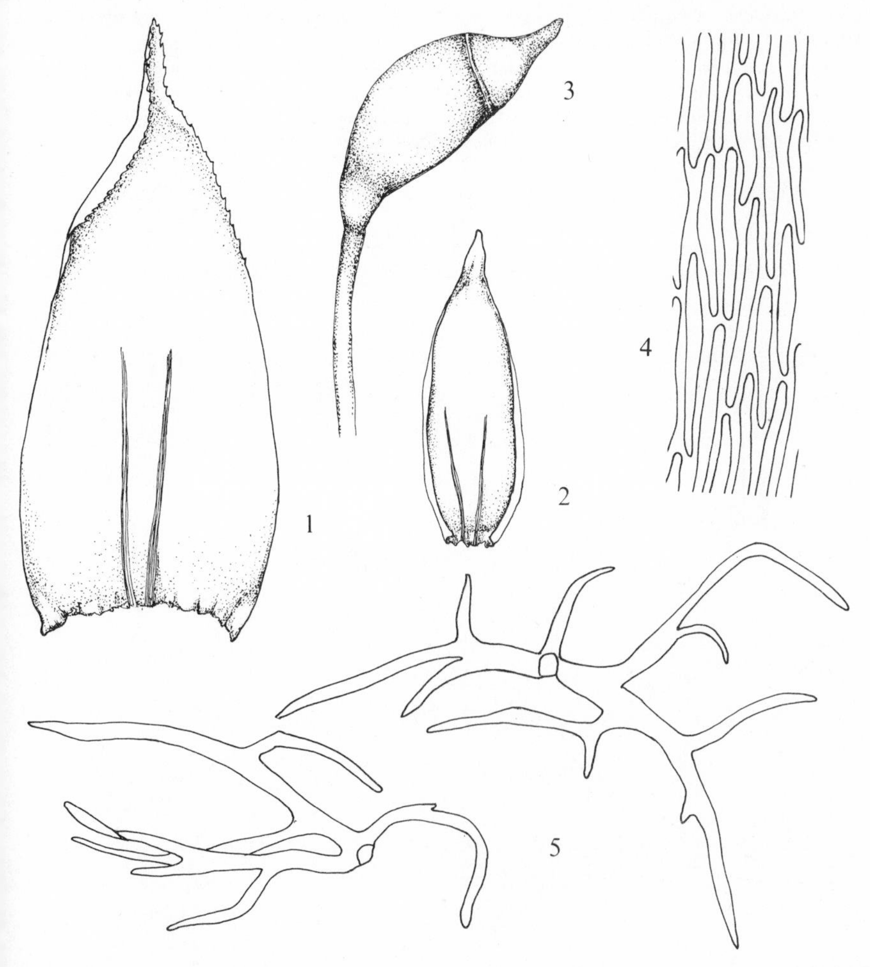

Fig. **329**. *Hylocomium splendens*: 1, stem leaf; 2, branch leaf; 3, capsule (× 10); 4, mid-leaf cells (× 415); 5, paraphyllia (× 100). Leaves × 25.

loosely imbricate, plicate below, concave, stem lvs ovate-oblong, abruptly narrowed to obtuse to acute, channelled acumen, not falcate, base decurrent, margin plane or recurved near base, denticulate or dentate; nerve double, extending up to $\frac{1}{2}(-\frac{2}{3})$ way up lf; basal cells narrowly rhomboidal, orange-brown, angular cells not differentiated, cells above narrowly rhomboidal to linear with scattered large papillae at back in upper part of lf, cells in mid-lf 5–7 μm × (28–)40–80 μm, (5–)7–14 times as long as wide. Lvs of primary branches similar to but smaller than stem lvs, those of ultimate branches much smaller, ovate-lanceolate, acute. Capsule inclined to horizontal, ovoid to ellipsoid, gibbous; lid with rostrate beak; spores 14–17 μm. Fr very rare, spring. $n=10$*, 12. Glossy, yellowish-green to yellowish-brown patches or coarse wefts on soil or in turf on banks, roadsides, heaths, moorland, sand-dunes, etc., calcifuge or if occurring in basic habitats then on leached soil, common and sometimes locally abundant. 112, H40, C. Europe, Iceland, Faroes, N. and C. Asia, Caucasus, Tibet, China, Japan, Sakhalin, Kuriles, Morocco, Canaries, Azores, N. America, Greenland, New Zealand.

Differs from bipinnate *Thuidium* species in lf shape and areolation. Simply pinnate forms may be confused with *Pleurozium schreberi* or *H. umbratum*, the former differs in the obtuse lvs and lack of paraphyllia, in the latter the branching is less regularly bipinnate and complanate and the stem lvs are more plicate and triangular in shape.

GLOSSARY

acostate Nerveless.
acrocarpous Main stem usually erect, of limited growth, terminating in an inflorescence, further growth continued by branches or innovations.
acumen Longly tapering narrow point; *adj.*, **acuminate** (fig. **331**, 21).
acute Pointed, with angle at leaf tip less than 90° (fig. **331**, 20).
alar cells Cells at basal angles of leaf.
amphithecium Outer tissue of embryo giving rise to the outer layers and spore sac of the Sphagnopsida, to the outer layers only in the Andreaeopsida and Bryopsida.
angular cells Cells at basal angles of leaf.
annulus Ring or rings of cells at mouth of capsule associated with dehiscence of lid; *adj.*, annular.
antheridium Globose to fusiform or shortly cylindrical structure containing the antherozoids.
antherozoid The biflagellate male gamete of bryophytes.
anthocyanin Reddish pigment.
apiculus Small abrupt projection at apex; *adj.*, **apiculate** (fig. **331**, 19).
apophysis Enlarged sterile basal portion of capsule (e.g. fig. **169**, 6).
appendiculate Of cilia, with transverse bars (fig. **193**, 5).
appressed Pressed against stem.
archegonium Flask-shaped structure containing the female gamete, consisting of a swollen basal part or venter surrounding the egg cell, and tubular upper part or neck.
arcuate Of capsule, curved like a bow (fig. **332**, 6).
areolation The form and arrangement of the cells (of a leaf).
arista A fine bristle-like point; *adj.*, **aristate**.
articulate Jointed, usually with reference to the transverse lines on peristome teeth (fig. **187**, 9).
ascending Growing upwards, often from an older procumbent part.
auricle Enlarged or differentiated cells at basal angles of leaf forming lobe or distinctive group of cells; *adj.*, **auriculate**.
autoecious Having antheridia and archegonia born in separate inflorescences on the same plant (fig. **330**, 3 and 4).
axil Angle between leaf base and stem; *adj.*, **axillary**.
beak Projection from lid of capsule.
bifid Divided into two segments or parts.
biflagellate With two flagella.
bipinnate Twice pinnately branched.
bistratose Two cells thick.
bulbiform Bulb-like; with imbricate radical leaves forming bulb-like plant.
bulbil Vegetative reproductive body, usually axillary, with apical projections or leaf primordia (figs. **176**, **177**).
caducous Readily falling off.
caespitose Tufted, forming tufts.
calyptra Structure developed from the venter of the archegonium covering the young capsule and at least the lid of older capsules, falling or not as capsule matures.

campanulate Bell-shaped.

capitulum Head composed of crowded branches at stem tip of *Sphagnum*.

capsule The spore-bearing structure of mosses usually differentiated into the theca or spore case and a basal sterile portion, the neck.

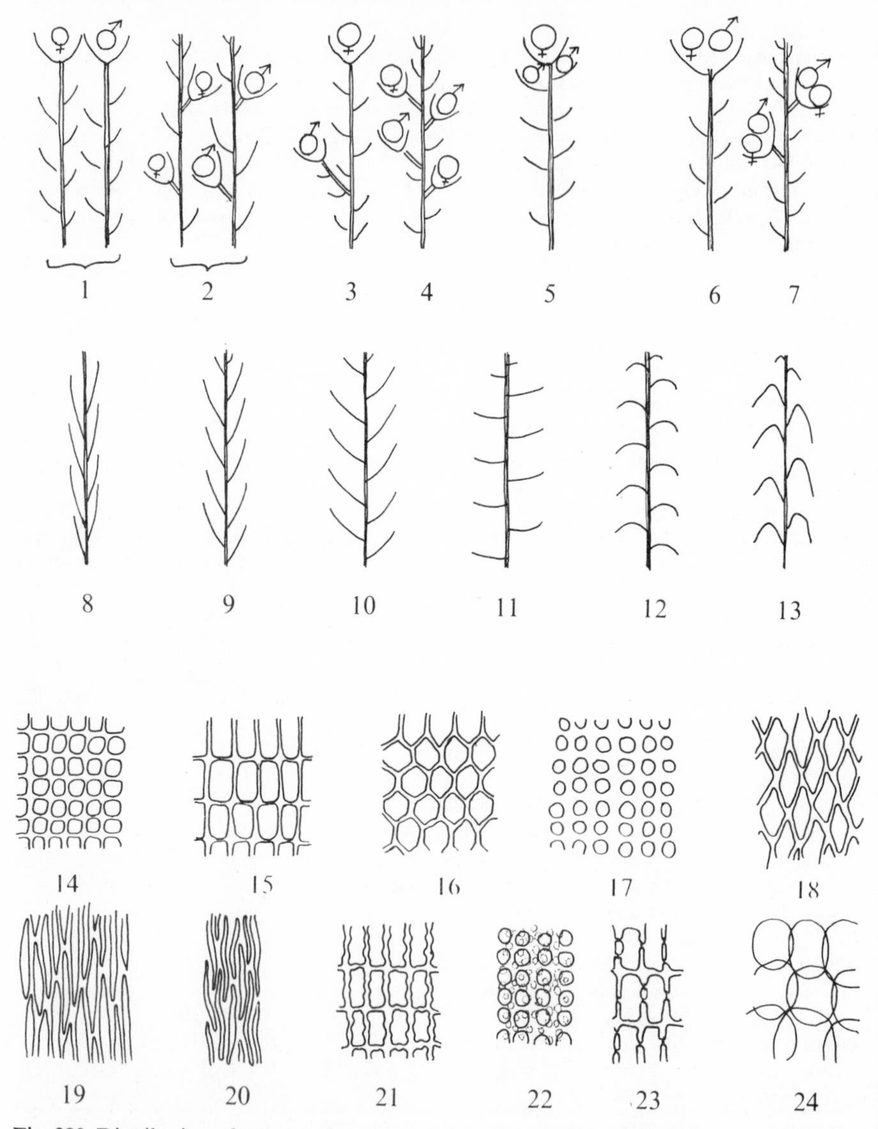

Fig. **330**. Distribution of gametangia: 1 and 2, dioecious; 3 and 4, autoecious; 5, paroecious; 6 and 7, synoecious. 8–13, leaf postures: 8, erect; 9, erecto-patent; 10, patent; 11, spreading; 12, recurved; 13, squarrose. 14–24, cell shape: 14, quadrate; 15, rectangular; 16, hexagonal; 17, rounded; 18, rhomboidal; 19, linear; 20, linear, vermicular; 21, rectangular, sinuose; 22, rounded, papillose; 23, rectangular, porose; 24, rounded, lax.

cartilaginous Thick and tough, of border of incrassate elongated cells.
catenulate Like a miniature chain in appearance.
central strand Column of small elongated cells, sometimes with thicker walls, running through centre of stem.
cernuous Drooping at an angle intermediate between horizontal and vertically downwards (fig. **332**, 7).
channelled With a groove formed by upturned margins.
chlorophyllose With chloroplasts.

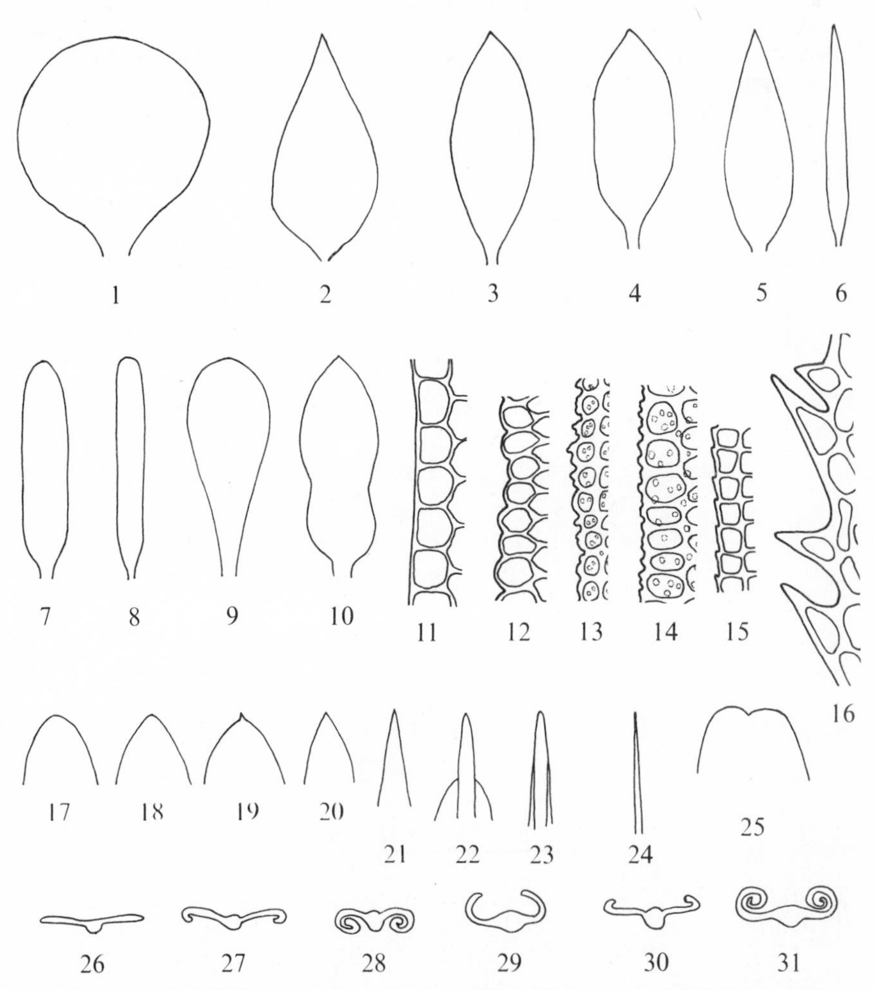

Fig. **331**. 1–10, leaf shapes: 1, orbicular; 2, ovate; 3, elliptical; 4, oblong; 5, lanceolate; 6, linear; 7, lingulate; 8, ligulate; 9, spathulate; 10, panduriform. 11–16, leaf margins: 11, entire; 12, crenulate; 13, papillose-crenulate; 14, papillose; 15, denticulate; 16, dentate. 17–25, leaf apices: 17, rounded; 18, obtuse; 19, obtuse and apiculate; 20, acute; 21, acuminate; 22, cuspidate; 23, subulate; 24, setaceous; 25, emarginate. 26–30, leaf margins as seen in section: 26, plane; 27, recurved; 28, revolute; 29, inflexed; 30, incurved; 31, involute.

cilia Hair-like structures on leaf margins or alternating with the processes of the inner peristome; *adj.*, **ciliate**.

circinate Very strongly curved and ± forming a circle (fig. **278**, 1).

cladocarpous Archegonia borne terminally on short lateral branches.

clavate Club-shaped.

cleistocarpous Not dehiscing by means of lid or valves, spores released by rupture or decay of capsule wall.

cline Continuous range or series.

cochleariform Very concave like a spoon.

collenchymatous With cell walls more heavily thickened at corners than elsewhere.

columella Central mass or column of sterile tissue surrounded by the spore sac in the capsule.

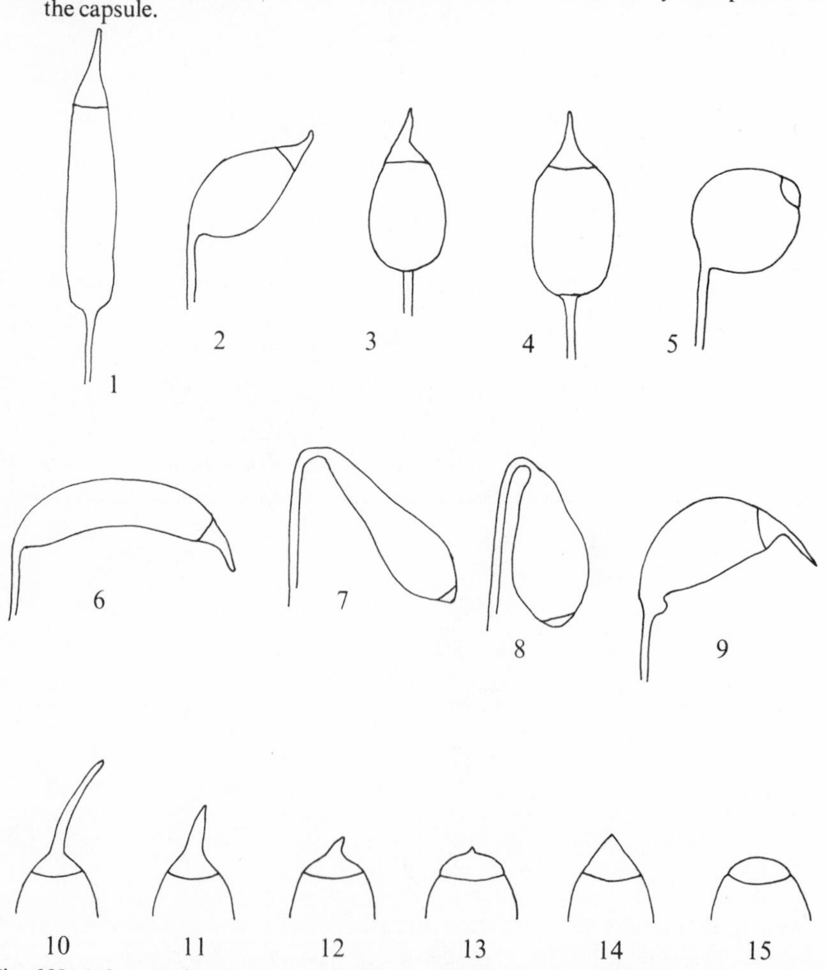

Fig. **332**. 1–9, capsules: 1, erect, cylindrical; 2, inclined, ellipsoid; 3, ovoid; 4, obloid; 5, globose; 6, horizontal, cylindrical, curved; 7, cernuous, narrowly pyriform; 8, pendulous, pyriform; 9, inclined, gibbous, strumose. 10–15, capsule lids: 10, subulate; 11, rostrate; 12, rostellate; 13, mamillate; 14, conical, obtuse; 15, convex.

coma Tuft of more crowded, usually larger leaves at stem apex; *adj.*, **comal**.
commissure In *Sphagnum*, the junction between hyaline and green cells.
complanate More or less flattened in one plane.
concolorous Of a single colour.
conduplicate Folded together face to face.
conical Cone-shaped (fig. **332**, 14).
constricted Abruptly or sharply narrowed.
convolute Clasping, enfolding, sheathing.
cordate Heart-shaped.
coriaceous Leathery.
cortex Outer layer or layers of cells of stem as opposed to central strand or cylinder.
corticolous Growing on bark.
costa Nerve; *adj.*, costate.
crenulate With minute rounded teeth, usually formed from bulging cell walls (fig. **331**, 12).
cribrose Of peristome teeth, with numerous perforations (e.g. fig. **145**, 5).
crisped Strongly curled and twisted (fig. **236**, 9).
cucullate Of leaves, with margin at apex incurved to form hood-shaped tip (fig. **137**, 2), of calyptra, hood-shaped and split down one side (fig. **162**, 10).
cushion ± hemispherical plant formed by shoots radiating from a central point.
cuspidate With a stout point, in leaves formed by thick excurrent nerve and in stem by imbricate, usually concave leaves (fig. **331**, 22).
cygneous Strongly curved in upper part like swan's neck.
cylinder See *central strand*.
cylindrical Cylinder-shaped (fig. **332**, 1, 6).
deciduous Falling, being lost at maturity.
decumbent Prostrate with ascending tips.
decurrent Of leaf base, running down stem forming wings (fig. **214**, 4).
dehisce Open by loss of lid or opening of valves; *adj.*, dehiscent.
dendroid Tree-like in habit or mode of branching.
dentate Strongly toothed, teeth usually composed of one or more cells (figs. **331**, 16).
denticulate Finely toothed, teeth usually composed of only part of a cell projecting from margin (fig. **331**, 15).
denuded Eroded away.
depauperate Small, stunted, poorly developed.
diaphanous Colourless and transparent.
dichotomous Regularly forking, with divisions of ± equal size.
dimorphism Having two states; *adj.*, **dimorphic**.
dioecious Antheridia and archegonia borne on separate plants (fig. **330**, 1 and 2).
discoid Of gemmae, disc-shaped and thickest at centre.
distal End or portion distant from base or point of origin.
distichous In two ranks, one on either side of stem (fig. **84**, 7).
divergent Spreading away from.
dorsal Of leaves, back, under surface, abaxial side.
ellipsoid Three-dimensional equivalent of elliptical (fig. **332**, 2).
elliptical With convex sides (fig. **331**, 3).
emarginate With small indentation at leaf apex (fig. **331**, 25).
emergent Capsule partly emerging above perichaetial leaves.
entire Smooth, not toothed or crenulate (fig. **331**, 11).

epiphragm Diaphragm or plate of tissue, formed from top of the columella, at mouth of capsule, attached to ends of peristome teeth.

erect Of leaves, parallel to but not pressed against stem (fig. **330**, 8); of branches or stems, in a ± vertical position relative to stem or substrate; of capsules, upright (fig. **332**, 1).

erecto-patent Of leaves, between erect and patent, at an angle of about 20–25° to stem (fig. **330**, 9).

eroded Worn away.

erose Irregularly notched.

excurrent Projecting beyond the end of the lamina.

exothecium Superficial layer of cells of capsule; *adj.*, **exothecial**.

exserted Carried well above perichaetial leaves.

falcate Curved like a sickle (fig. **278**, 5).

falcato-secund Strongly curved and pointing in one direction.

fascicle Group, bunch or tuft of branches; *adj.*, **fasciculate**.

fastigiate With branches reaching ± same height.

fibril Fibre-like thickenings of walls of hyaline cells of *Sphagnum*; *adj.*, **fibrillose**.

filamentous Thread-like.

filiform Fine or thread-like.

flagelliform Of shoots, long and thin with small leaves.

flexuose Wavy.

foot Basal part of sporophyte embedded in gametophyte tissue.

fringed With a shortly ciliate margin or edge.

fruit Loose term for sporophyte.

fugacious Readily or quickly falling.

fusiform Tapering at either end, spindle-shaped.

gametangia Antheridia and archegonia.

gametophore Structure or stem bearing gametangia.

gametophyte The dominant, leafy, haploid stage of the moss life-cycle.

gemma(e) Vegetative propagule(s) borne on rhizoids, in leaf axils, on stem or leaf surfaces or on special structures.

gibbous With one side larger than the other (fig. **332**, 9).

glaucous With whitish, greyish or bluish bloom.

globose ± spherical in shape (fig. **332**, 5).

gregarious Plants growing close together but not in mats or tufts.

guard cells The cells surrounding the opening of a stoma.

gymnostomous Without a peristome.

habit Aspect or appearance of a plant.

hair-point Hair-like leaf tip formed by attenuated apex or longly excurrent nerve, sometimes white when dry and giving leaf tips a hoary appearance.

hexagonal Six-sided (fig. **330**, 16).

hispid Rough.

homomallous Pointing in one direction.

hyaline Colourless and transparent.

imbricate Appressed and overlapping.

immersed Of capsule, surrounded by and overtopped by perichaetial leaves; of stomata, sunken below surface of surrounding epidermal cells (fig. **231**, 8).

inclined At an angle of 20°–60° to the vertical (fig. **332**, 2, 9).

incrassate Thick-walled.

incurved Leaf margin curved upwards and inwards (fig. **331**, 30).

inflated swollen.

inflorescence Structure composed of gametangia and surrounding leaves.
innovating Producing new branches, usually from below inflorescence.
insertion Line of attachment of leaf.
involute Strongly rolled upwards and inwards (fig. **331**, 31).
isodiametric Radially symmetrical, ± as wide as long.
isophyllous Stem and branch leaves similar.
julaceous Cylindrical and worm-like in appearance because of the imbricate arrangement of leaves.
keeled Sharply folded like the keel of a boat.
lamellae Flat plates of cells.
lamina Leaf blade as distinct from leaf nerve.
lanceolate Lance-shaped, about three times as long as wide (fig. **331**, 5).
lax Of cells, large with thin often bulging walls (fig. **330**, 24).
lid Operculum, dehiscent top of capsule.
ligulate Strap-shaped, narrow and parallel-sided (fig. **331**, 8).
limb Upper part of leaf when leaf is differentiated into basal and upper part.
linear Long, narrow and ± parallel-sided (fig. **330**, 19; **331**, 6).
lingulate Tongue-shaped, broad and ± parallel-sided (fig. **331**, 7).
lumen Cell cavity.
macronemata Large rhizoids produced round branch primordia and leaf axils (fig. **333**, 1, 2)
mamilla(e) Protuberance(s) from cell surface into which cell lumen projects (fig. **54**, 15), in surface view indistinguishable from papillae; *adj.* **mamillose**.
mamillate Of cells with mamillae; of lid, convex with short point (fig. **332**, 13).
m-chromosome Minute chromosome usually $\frac{1}{10}$ or less the size of other chromosomes of the complement, sometimes but not necessarily homologous with accessory chromosome, abbreviated to $+1, +2$, etc., where chromosome numbers are given.
membrane gap Irregular openings in walls of hyaline cells of stem leaves of *Sphagnum*.
micrometre (**μm**) One-millionth of a metre, one-thousandth of a millimetre; referred to colloquially as micron or mu.
micronemata Small, sparsely branched rhizoids produced on stem between leaves (fig. **333**, 1)
micronemata Small, sparsely branched rhizoids produced on stem between leaves.
mid-leaf Position approximately halfway up leaf and midway between margin and nerve or equivalent position if nerve is lacking.
mitriform Conical and symmetrical with two or more slits (fig. **163**, 4).
monoecious Having antheridia and archegonia on same plant.
monopodial Main stem of unlimited growth with inflorescences lateral.
morphology General appearance or structure; *adj.*, morphological.
mucro Short broad projection from leaf apex; *adj.*, mucronate.
muticous Lacking a point or hair-point.
n Functionally haploid chromosome number, as opposed to x, the basic chromosome number; where a number such as $n=10+1$ is given, $+1$ indicates presence of an m-chromosome.
neck Of capsule, the sterile portion between base of theca and top of seta; of retort cell, projection with pore from surface of cell.
nerve Midrib or costa of leaf.
nodulose Of cilia, with knob-like thickenings (fig. **189**, 4).
obconical Inverted cone-shaped.
obloid Three-dimensional equivalent of oblong, but with reference to capsules, with rounded edges and corners (fig. **332**, 4).

oblong Rectangular, but with reference to leaves, with rounded corners (fig. **331**, 4).
obovate Egg-shaped in outline and widest above middle.
obscure Indistinct, difficult to see.
obsolete Almost lacking.
obtuse Bluntly pointed, angle at apex more than 90°; more loosely, blunt (fig. **331**, 18).
ochrea Cylinder of thin tissue surmounting vaginula.
opaque Not transparent or translucent.
operculum Lid of capsule; *adj.*, **operculate**.
orbicular ± circular in outline (fig. **331**, 1).
ovate Egg-shaped in outline and widest below middle, about twice as long as wide (fig. **331**, 2).
ovoid Three-dimensional shape about twice as long as wide, egg-shaped (fig. **332**, 3).
panduriform Violin-shaped (fig. **331**, 10).
papilla Solid projection from cell surface; *adj.* **papillose** (fig. **330**, 22; **331**, 13, 14).
paraphyllia Small, leaf-like or variously divided structures on stem between the leaves (fig. **327**, 4 and 9).
paraphyses Unisteriate, hair-like structures mixed with the antheridia and archegonia.
parenchymatous ± hexagonal cells with upper and lower ends at right-angles to main axis and with walls of ± uniform thickness.
paroecious Having the antheridia naked in the axils of leaves immediately beneath the perichaetium (fig. **330**, 5).
patent At angle to the stem of roughly 45° (fig. **330**, 10).
pellucid Translucent, clear.
pendent Hanging.
perichaetium Structure usually composed of modified or perichaetial leaves, containing the archegonia; female inflorescence.
perigonium Structure usually composed of modified or perigonial leaves containing the antheridia; male inflorescence.
peristome The single or double ring of teeth at mouth of capsule revealed after dehiscence. In genera with a double peristome (Diplolepideae) the inner peristome is homologous with the single peristome of the genera in the Haplolepideae.
persistent Not falling or deciduous.
photosynthetic Containing chlorophyll and carrying out carbon assimilation.
phyllotaxy Spiral arrangement of leaves on stem, e.g. $\frac{2}{5}$ phyllotaxy = 5 leaves occupy 2 complete turns of the spiral.
piliferous Hair-like.
pinnate With spreading branches produced ± regularly on either side of stem.
plane Flat (fig. **331**, 26).
pleurocarpous Mosses with monopodial main stem and inflorescences produced on short lateral branches or branchlets.
plicae Longitudinal folds or ridges; *adj.*, **plicate**.
plumose Regularly pinnate and feathery in appearance.
pore Opening or pit in cell wall; in hyaline cells of *Sphagnum*, openings to the exterior; *adj.*, **porose** (fig. **330**, 23).
processes The thin teeth of the inner peristome (endostome) alternating with the outer peristome (exostome) in mosses with a double peristome (fig. **195**, 4).
procumbent Prostrate.
prosenchymatous Cells with pointed ends running in same direction as main axis.
protonema Filamentous or rarely thalloid structure produced after spore germination and from which the gametophyte shoots develop.
proximal End or part nearest base or point of origin.

pseudoparaphyllia Small, ± leaf-like structures surrounding branch primordia and found at the bases of young branches (fig. **313**, 4).

pseudopodium Leafless prolongation of stem bearing gemmae; elongated stalk of archegonium pushing capsule beyond perichaetial leaves in *Andreaea* and *Sphagnum*.

pyriform Pear-shaped (fig. **332**, 7, 8).

quadrate With four sides of ± equal length; square (fig. **330**, 14).

quinquefarious In five ranks.

radiculose With numerous, conspicuous rhizoids.

rectangular About twice as long as wide (fig. **320**, 15, 21, 24).

recurved Curved down and backwards (figs. **330**, 12 and **331**, 27).

reflexed Bent back.

resorbed Eroded parts of cell walls of *Sphagnum*; **resorption**, erosion of cell walls of *Sphagnum*; **resorption furrow**, furrow formed by erosion of cell walls at margin of leaves of some species of *Sphagnum* (fig. **3**, 6).

retort cell Cortical cells in some *Sphagnum* species ending in a projection with a pore (fig. **23**, 6).

retuse Broad apex with a slight indentation.

revolute Strongly rolled backwards or under (fig. **331**, 2, 8).

rhizoids Uniseriate branched structures occurring on stem and sometimes old leaves, sometimes anchoring plant to substrate.

rhizomatous Having horizontal underground stems analogous with rhizome of higher plants.

rhomboidal ± narrowly diamond-shaped (fig. **330**, 18).

ringed Of *Sphagnum* pores, pore surrounded by fibril ring.

rostellate With short beak (fig. **332**, 12).

rostrate With long beak (fig. **332**, 11).

rugose With irregular ± transverse undulations.

saxicolous Growing on rock.

scabrous Rough.

scarious Thin and papery in texture.

secund Pointing in one direction.

septate Of *Sphagnum* hyaline cells, having partitions or septa within cell. Estimates of the percentage of septate cells are based on regarding two hyaline cells not separated by a green cell as one cell.

seriate In rows.

serrate Regularly toothed (like a saw) with teeth composed of one or more cells.

serrulate Minutely regularly toothed (like a saw), with teeth composed of part of a cell.

sessile Without a stalk or seta.

seta Stalk of capsule.

setaceous Bristle-like (fig. **331**, 24).

sheathing Surrounding or partly surrounding.

shoulder Distal portion of basal part of leaf where leaf narrows abruptly.

sinistrose Sinistrally spiralling, i.e. turning or spiralling to the left.

sinuose Of cells, with wavy walls (fig. **330**, 21); of leaf margin, wavy but not toothed with projecting cell walls.

spathulate Spatula-shaped, narrow below, broad above (figs. **331**, 9).

spinose With sharply pointed teeth.

spinulose With minute sharply pointed teeth.

sporophyte The diploid generation of a moss, consisting of foot, seta and capsule, epiphytic upon the haploid generation.

spreading More or less at right angles to stem (fig. **330**, 11).
squarrose Upper part of leaf strongly curved back at an angle of 90° or more from the lower part of leaf (fig. **330**, 13).
stegocarpous Having a lid, dehiscent.
stellate Star-shaped.
stereids, stereid cells Long, slender, very thick-walled, fibre-like cells occurring in the nerve of the leaves of some mosses.
sterile Technically, without gametangia, but used more loosely to indicate absence of capsules.
stoloniform Stems arching and rooting at points touching substrate, analogous to the stolons of higher plants.
stoma(ta) Pore(s) surrounded by two guard cells; stomata usually occur at neck and sometimes elsewhere on capsule.
striae Small lines or ridges; *adj.*, **striate**.
struma Swelling at base of neck (fig. **332**, 9); *adj.*, **strumose**.
subula Needle-like point (fig. **331**, 23); *adj.*, **subulate** (fig. **331**, 23; **332**, 10).
sulcate With longitudinal folds.
superficial Of stomata, in the same plane as the surface of the epidermis (fig. **230**, 6).
sympodial Main stem of limited growth, further development by branches.
synoecious With antheridia and archegonia mixed in same inflorescence (fig. **330**, 6 and 7).
teeth Divisions of the peristome; serrations of leaf margin.
terete Rounded in transverse section.
terrestrial Growing on soil, or more loosely in a non-aquatic habitat.
thalloid Composed of a flat plate of tissue.
tetrad Group of four spores derived from a single spore mother cell.
theca Urn; spore-producing part of capsule.
tomentose With a felt of abundant long rhizoids.
trapezoid, trapeziform Shaped like a trapezium.
trifarious In three ranks.
triradiate With a three-pronged scar, characteristic of spores of land plants.
truncate Cut-off; ending abruptly.
tubular Of leaves, with margins inrolled and overlapping to form a tube.
tuft Clump of ± erect shoots.
tumid Swollen.
turbinate Top-shaped (fig. **186**, 4).
turf Growth form of erect but not crowded shoots.
undulate Wavy.
unilateral One-sided.
uniseriate With cells in a single row.
unistratose Cells in one layer only.
urceolate Urn-shaped; narrowed below mouth then enlarged (fig. **236**, 14).
vaginula Sheath at base of seta derived from venter of archegonium.
venter Layer of cells surrounding egg cell in swollen part of archegonium.
ventral Upper or adaxial surface of leaf.
vermicular Worm-like; long, narrow and wavy (fig. **330**, 20).
verrucose Roughened.
weft Loosely interwoven shoots.
whorled Arranged in ring.
x Basic haploid chromosome number.

BIBLIOGRAPHY

A list of the works referred to in the text or freely consulted in the compilation of the flora. References dealing with specific taxa are given in the text where appropriate.

Braithwaite, R. *The British Moss Flora*. London: L. Reeve & Co., 1887–1905.

Brotherus, V. F. Musci (Laubmoose). In Engler & Prantl, *Die Natürlichen Pflanzenfamilien*, 2nd edn, vols. 10 and 11. Leipzig: W. Engelmann, 1924–5.

British Bryological Society. Reports. Harlech, Berwick-on-Tweed, 1923–46.

The Bryologist. Journal of the American Bryological and Lichenological Society. 1898– .

Crum, H. Mosses of the Great Lakes Forest. *Contributions from the University of Michigan Herbarium* **10**, pp. 1–404. Ann Arbor, Michigan, 1973.

Demaret, F. & Castagne, E. *Flore Genérale de Belgique. Bryophytes*. Vol. II, parts i–iii, Ministère de l'Agriculture, Brussels, 1959–64.

De Sloover, J.-L. & Demaret, F. *Flore Genérale de Belgique. Bryophytes*. Vol. III, part i. Ministère de l'Agriculture, Brussels, 1968.

Dixon, H. N. & Jameson, H. G. *The Student's Handbook of British Mosses*, 3rd edn. Eastbourne: V. Sumfield, 1924.

Fritsch, B. Chromosomenzahlen der Bryophyten. *Wissenchaftliche Zeitschrift* **21**, pp. 839–944, 1972.

Gams, H. *Kleine Kryptogamenflora von Mitteleuropa*, vol. IV *Die Moose und Farnpflanzen*, 5th edn. Stuttgart: Gustav Fischer, 1973.

Grout, A. J. *Moss Flora of North America, north of Mexico*. Newfane, Vermont: A. J. Grout, 1928–40.

Husnot, T. *Muscologia Gallica*. Paris, 1884–90, 1892–4.

Ireland, R. A taxonomic revision of the genus *Plagiothecium* for North America, north of Mexico. *National Museum of Natural Sciences, Publications in Botany*, no. 1. Ottawa, 1969.

Isoviita, P. Studies on *Sphagnum* L. 1. *Annales botanici fennici*, **3**, pp. 199–264, 1966.

Jensen, C. *Skandinaviens Bladmossflora*. Copenhagen; E. Munksgaard, 1939.

Journal of Bryology. Oxford: Blackwell Scientific Publications, 1972– .

Limpricht, K. G. *Die Laubmoose Deutschlands, Österreichs und der Schweiz*. Leipzig: E. Kummer, 1890–1904.

Lindbergia. Århus & Leiden: Nordic and Dutch Bryological Societies, 1971– .

Mårtensson, O. Bryophytes of the Torneträsk Area, Northern Swedish Lappland. II. Musci. *K. svenska Vetensk-Akad. Avh.* **14**, 321 pp., 1956.

Mönkemeyer, W. Die Laubmoose Europas. In Rabenhorst, *Kryptogamen-Flora von Deutschland, Österreich und der Schweiz*. Leipzig: A. V. G., 1927.

Nyholm, E. *Illustrated Moss Flora of Fennoscandia*, II *Musci*. Lund: Gleerup, 1954–69.

Podpěra, J. *Conspectus Muscorum Europaeorum*. Prague: Československé Akademie Věd, 1954.

Revue Bryologique et lichénologique. Paris: Museum National d'Histoire Naturelle, 1928– .

Transactions of the British Bryological Society. London: Cambridge University Press, 1947–71.

Warburg, E. F. *Census Catalogue of British Mosses*, 3rd edn. Ipswich: British Bryological Society, 1963.

Watson, E. V. *British Mosses and Liverworts*, 2nd edn. London: Cambridge University Press, 1969.

Welch, W. *A Monograph of the Fontinalaceae*. The Hague: M. Nijhoff, 1960.

Wijk, R. van der, Margadant, W. D. & Florschütz, P. A. *Index Muscorum*. Utrecht: International Bureau for Plant Taxonomy, 1959–69.

Wilson, W. *Bryologia Britannica*. London: Longman, Brown, Green and Longmans, 1855.

ADDENDUM

Rhizomnium magnifolium (Horik.) Kop., Ann. bot. Fenn., 1973
R. perssonii Kop.

Dioecious. Plants *ca* 5 cm (in Britain); older parts of plant blackish; stems with micronemata and macronemata and tomentose with latter below. Lvs shrunken when dry, patent to spreading, concave when moist, orbicular to broadly elliptical, ovate or obovate, base very narrowly decurrent, apex rounded, emarginate or more often with small apiculus, margin sinuose, bordered; nerve ending below or occasionally reaching apex; cells in rows radiating from nerve, hexagonal to elongate-hexagonal, mostly 40–60 μm × 64–100 μm in mid-lf, border 2–3-stratose at base, unistratose by about middle of lf and above, border cells 15–40 μm wide at middle of lf, border cells at apex ± narrowly rectangular. Capsule similar to that of *R. punctatum*, unknown in Britain. $n=7$. Patches on damp ground by streams and near areas of late snow-lie in montane habitats, very rare. Aberdeen, Inverness. 2. Circumboreal.

The three British species of *Rhizomnium* all have macronemata, large dark brown rhizoids, *ca* 30μm diameter at base; these are confined to lf axils and appear as tufts on the stem except on youngest parts. *R. magnifolium* and *R. pseudopunctatum* also have micronemata, much smaller pale brown rhizoids, to *ca* 16 μm diameter. Micronemata arise all along the stem (although may be absent from youngest parts) and appear as a fuzzy tomentum clothing the stem, readily visible with a lens. The absence of micronemata, the border 3–4-stratose at the middle of the lf and the shape of the border cells at the lf apex distinguish *R. punctatum* from the other two species. *R. pseudopunctatum* differs from *R. magnifolium* in the synoecious inflorescence, the lvs never apiculate, the nerve never reaching the lf apex and the sub-globose capsule. In *R. pseudopunctatum* the border is unistratose almost to the base whilst in *R. magnifolium* it is unistratose only from apex to about the middle of the lf. In Britain *R. pseudopunctatum* is a calcicole whereas *R. magnifolium* occurs in acidic habitats. For the occurrence of *R. magnifolium* in Britain see Crundwell, *J. Bryol.* **10** (1), in press.

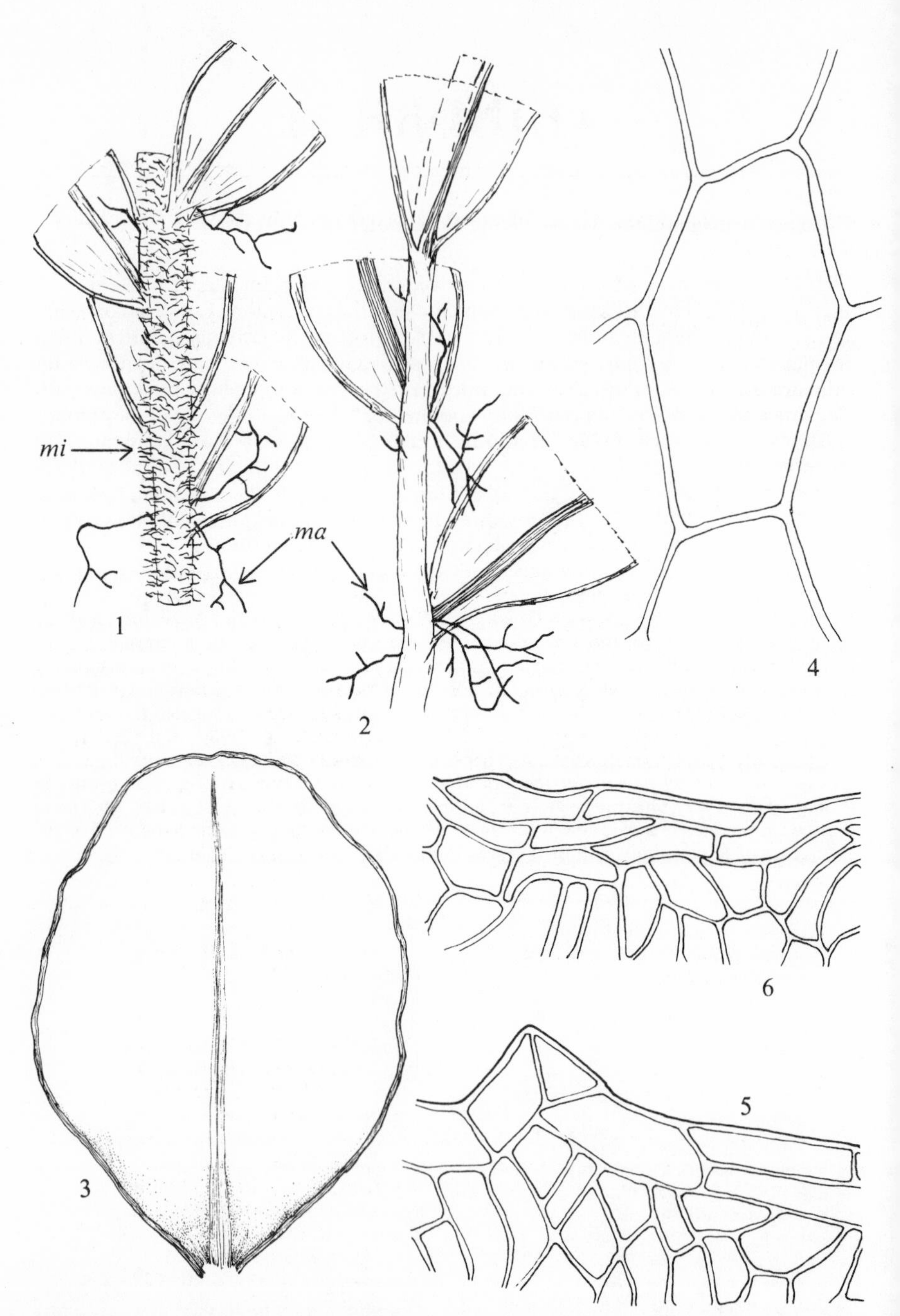

Fig. 333. 1 and 2, diagrammatic representations of stems of *Rhizomnium magnifolium* and *R. punctatum* showing macronemata (*ma*) and micronemata (*mi*), both ×10. 3–5, *R. magnifolium*: 3, leaf (×10); 4, mid-leaf cells (×415); 5, leaf apex (×165). 6, *R. pseudopunctatum*: leaf apex (×165).

INDEX

Synonyms and excluded taxa are in italics; page numbers of illustrations are in italics.